D0782153

VIROLOGY
MOLECULAR BIOLOGY AND PATHOGENESIS

LEONARD C. NORKIN

Department of Microbiology
University of Massachusetts
Amherst, Massachusetts

ASM
PRESS

Washington, DC

Address editorial correspondence to ASM Press, 1752 N St. NW, Washington, DC
20036-2904, USA

Send orders to: ASM Press, P.O. Box 605, Herndon, VA 20172, USA
Phone: (800) 546-2416 or (703) 661-1593
Fax: (703) 661-1501
E-mail: books@asmusa.org
Online: http://estore.asm.org

Library of Congress Cataloging-in-Publication Data

Norkin, Leonard C.
Virology : molecular biology and pathogenesis / Leonard C. Norkin.
p. ; cm.
Includes bibliographical references and index.
ISBN 978-1-55581-453-3 (hardcover : alk. paper) 1. Virology—Textbooks.
2. Molecular virology—Textbooks. 3. Virus diseases—
Pathogenesis—Textbooks. I. Title.
[DNLM: 1. Viruses—pathogenicity. 2. Genome, Viral. 3. Virus Diseases—etiology.
4. Viruses—genetics. QW 160 N841v 2010]
QR360.N67 2010
616.9′101—dc22
2009036895

10 9 8 7 6 5 4 3 2 1

Cover and interior design: Susan Brown Schmidler
Illustrations: Lineworks, Inc.

Cover illustration: Structure and molecular organization of a Sindbis virus particle.
Sindbis virus is a member of the togavirus family of enveloped plus-strand RNA viruses.
The surface features of the particle were determined by cryo-electron microscopy,
which yielded hundreds of highly detailed, two-dimensional images, from which
a three-dimensional image was generated using powerful computer programs.
A cross-section through the particle, showing the envelope glycoproteins
(blue), the lipid bilayer (green), the nucleocapsid (red), the mixed RNA-protein
region (orange), and the genomic plus-strand RNA (magenta), is superimposed on the
three-dimensional image. Protein structures were solved by X-ray crystallography,
and then fitted into the cryo-EM structure. See Figure 8.1 in the book for the complete
image. Adapted from W. Zhang et al., *J. Virol.* **76:**11645–11648, 2002, with permission.

I dedicate this book to my wife, Arline; my sons, Dave and Mike, and their wives, Mina and Debbie; and my grandchildren, Luke, Maya, and Theo.

"Human subtlety will never devise an invention more beautiful, more simple or more direct than does Nature, because in her inventions, nothing is lacking and nothing is superfluous."
LEONARDO DA VINCI (1452–1519)

Contents

Preface

Virology: Molecular Biology and Pathogenesis is meant to be used as a textbook for a comprehensive virology course aimed at advanced undergraduate and graduate students. With a handful of worthy virology textbooks already available, what justification might there be for yet another one, moreover one written by a single author? There are three reasons. First, and most importantly, *Virology: Molecular Biology and Pathogenesis* was conceived and organized to express my avid belief that the best way to teach virology is to discuss viruses in the context of virus families. Second, the individual virus families are viewed here within the context of the Baltimore classification system, a key unifying theme of the book and an approach that enables students to assume basic facts about the replication strategy of any virus on the basis of the nature of its genome (e.g., double-stranded DNA, plus-sense versus minus-sense single-stranded RNA, etc.). I know from more than three decades of teaching virology in this way that this is a sure-fire strategy for preparing students to approach the journal literature on any virus intelligently, key relevant knowledge already having been mastered. For the same reason, this book should continue to serve as a valuable reference for those who have studied from it. Third, I believe that this volume is unique in the uniform organization of its individual chapters and in its consistent writing style. Thus, *Virology: Molecular Biology and Pathogenesis* is intended to be a principles approach to virology, in which the same fundamental principles serve as the focal points of each chapter. In addition, the individual chapters constitute a continuous narrative in which key principles recur and are reinforced in different contexts. Finally, the writing style is often deliberately conversational so as to be more accessible and, I hope, engaging while not sacrificing rigor.

I began writing *Virology: Molecular Biology and Pathogenesis* in September 2003, and it has been the major focus of my professional life ever since. Yet its seeds may have been planted more than 35 years ago. My doctoral training in the mid-1960s was in the area of bacterial genetics. After earning my Ph.D. in 1969, I spent two years doing postdoctoral research on the mouse polyomavirus, a small DNA tumor virus. Suddenly I found myself an assistant professor at the University of Massachusetts, where I was expected to teach a virology course to senior microbiology majors and graduate students. I was an expert bacterial geneticist and could have readily taught a graduate course in microbial genetics. Although I had become familiar with the tumor viruses during my postdoctoral stint, I was far from being an expert animal virologist. Adding to my dilemma, the virology textbooks of the day were for the most part descriptive. I was thus at a loss as to how to deliver three 50-minute virology lectures per week to advanced undergraduate and graduate students. Fortuitously, I came across a review article by David Baltimore in which he put forward what is referred to as the Baltimore classification system (Baltimore, 1971). In brief, the Baltimore classification system is based on the different strategies used by viruses to express their genomes, and importantly, it recognizes that a given viral genome, by its nature, determines its expression strategy. Moreover, there were only six classes, or basic strategies of viral genome expression, in the original Baltimore scheme, and each of the numerous classic animal virus families fits into one of those classes. I had my answer regarding how to teach virology. I essentially followed Baltimore's review article and lectured from the original journal articles that were referenced therein. I continued in this way for the next several years, updating my lectures with more recent journal articles. Knowledge in the field was increasing much too rapidly, though, for me to go on in this way. I needed to turn to a textbook. I used several different ones over the years, but for one reason or another, none was the ideal teaching aid I was seeking. During this time I fantasized about writing a book that might present the field of virology as I envisioned it, and the current text is a fulfillment of that vision.

When I first taught virology in 1972, armed with the Baltimore classification system as my guide, animal virology was in its infancy. Yet it still was a challenge to cover all the pertinent material that was then available in the forty or so lectures that were allotted. Developments in the molecular biology of the animal viruses since the early 1970s have been explosive. Furthermore, viral structural biology and virus-cell and virus-host interactions, which were barely discussed in earlier decades, are now major areas of modern virology. In 1972 there was no HIV/AIDS, Ebola virus, West Nile virus, SARS, or bird flu to discuss, nor was there a vaccine to protect women from papovavirus-induced cervical cancer, and what was known of viral pathogenesis was, for the most part, descriptive. Yet, despite all the developments of the last few decades, I still have the same forty or so lectures in which to cover virology. How does one include it all in a single semester course? How does a textbook writer encompass it in one book? The answer is that one cannot include all of virology between the covers of even a large tome. One can ask, however, what readers need to know right now that will enable them to approach the journal literature successfully on any particular virus. Bearing that standard in mind, the crucial portion of each chapter describes in detail the organization of the viral genome and its pattern of expression. Bracketing that key topic are viral structure and entry, followed by assembly, release, and medical issues. Considering the vast amount of material covered in *Virology: Molecular Biology and Pathogenesis*, I suspect that most instructors will not ask their students to master it all. The portions of each chapter that concern the organization of the viral genome and its strategy of expression, however, are crucial.

Virology: Molecular Biology and Pathogenesis begins with two introductory chapters that recount the discovery of viruses and the recognition of their unique nature. Since the T-even bacteriophages and bacteriophage λ played major roles in the development of virology, these prokaryotic viruses are considered in some detail. While this book is primarily about the animal viruses, these and other bacteriophages are also discussed in later chapters, at points where it is informative to compare their unique properties to those of particular animal viruses. For example, in Chapter 6 the pattern of gene expression of the RNA phages is compared to that of the picornaviruses for the purpose of better appreciating the reasons underlying the unique aspects of each.

In the second of the two introductory chapters, we are introduced to the principles of animal virus classification, which define the classic animal virus families. We then consider the Baltimore classification scheme, which serves as a unifying principle for much of the remainder of the book. Before coming to the chapters dedicated to each of the major virus families, the student encounters chapter 3, which covers the various modes of virus infection and disease. This chapter provides background and perspective on the specific examples discussed in the chapters dedicated to the individual virus families. Chapter 4 then considers host defenses against microorganisms, on one hand, and viral countermeasures to subvert those host defenses on the other. I hope that students will read this chapter in its entirety, without necessarily studying it. Instead, I wish for readers to come away with a sense of just how potent our immune defenses are and yet how astonishingly efficiently evolution has acted to give rise to viral countermeasures against those host defenses. Without this continuing evolutionary dance between pathogen and host, there would be neither host nor virus. Readers should refer back to this chapter when viral immune evasion strategies are recounted in the contexts of the individual virus families, which we come to next, beginning with Chapter 5.

An important theme throughout the book is the manner in which some of the most dramatic breakthroughs in molecular and cellular biology sprang directly from the study of viruses. In fact, the field of molecular biology itself owes much to the earlier discovery of viruses. These early developments are woven into the history of virology, as recounted in Chapters 1 and 2; subsequent key virus-related breakthroughs are acknowledged in later chapters. Moreover, studies of viruses have led to crucial insights into the molecular basis of cancer, detailed here mainly in Chapters 15 and 20 but in other chapters as well.

Although *Virology: Molecular Biology and Pathogenesis* considers medical aspects of virology in somewhat more depth than most general texts, it is not meant to be a medical virology text per se. It does not consider diagnosis, nor is it meant to serve as a guide to therapies for the practicing physician. Rather, it is meant to give the student, the medical practitioner, and the basic scientist an understanding of fundamental virology as it relates to viral infection and disease.

Throughout the text there are boxes, which serve several purposes. Some of them review fundamental molecular and cellular biology that is relevant to the virology being discussed. Virology is also a human story, however, and I believe that as we learn the details of virology, it is important that we remain aware of the fact that we owe our knowledge to the efforts of individual scientists and physicians who at times showed exceptional brilliance and insight. Some of these individuals ran up against entrenched prevailing dogma and endured continuing humiliation. But they nevertheless persevered, until they were vindicated in the end. Still others displayed extraordinary physical courage, putting their own lives at risk for the greater public good. The stories of some of these researchers and clinicians are recounted in the boxes, and in some cases, in the main text. In addition, I have tried to convey the scientific climate of the times when some of the earlier developments took place. Perhaps the most important point of these asides is to help the student develop an appreciation of how the science of virology got to where it is today—a fascinating story in its own right. Also, since viruses are agents of life-threatening infectious diseases, virology, more than most biological sciences, impacts society in important ways. Additional boxes, therefore, address the serious public policy issues raised by viral outbreaks and epidemics, vaccines, limited resources, and so forth. Finally, throughout the text I have tried to highlight key questions that remain to be answered as well as challenge readers to suggest experimental approaches that might provide these answers. Other thought-provoking questions are scattered throughout the text simply to better engage students in the topic at hand.

The choice of specifics to include in *Virology: Molecular Biology and Pathogenesis* mainly reflected my own particular interests. But the ASM Press secured for me expert reviewers, who often steered me towards what is actually of greatest interest to those in each particular field. For one impulsive (or foolhardy) enough to single-handedly write a comprehensive textbook on virology, that expert advice has been invaluable; the efforts of these reviewers are more properly acknowledged below. I add the usual proviso that any errors of interpretation or emphasis are mine alone. I hope that users of this book will bring to my attention any errors that need correction, as well as any other advice that might make for an improved second edition.

I have not laced this book with many journal references, and those relatively few (for a comprehensive text) that are included may seem idiosyncratic to expert readers. Some articles are referenced for their technical innovations, while others were chosen for their historical interest.

Yet others were chosen for having been current at the time a particular passage was being written and for having suggested an exciting new paradigm. In these instances, only time will tell whether the new ideas they put forward will become widely accepted.

I justify this rather sparse approach to citations largely because we live today in the age of the Internet, where up-to-date information is instantly available, and because the rate of scientific progress is so rapid that any fixed set of references will be out of date virtually immediately. Thus, I urge that you, the reader, study this book with your computer at hand. When you are seeking clarification or want to know the latest information on any topic, my recommendations are as follows. While there are many Web sites that might turn up the information sought, the best, most reliable resource is the PubMed site at the U.S. National Library of Medicine. Go to the site, bookmark it, and use it often. You can search by topic or author, and the pertinent original journal articles will be listed for you, from most to least current. You will be able to view the abstract of any paper that appears on the list turned up by your search. In some cases, you will be able to access complete journal articles either directly on the site or through arrangements provided by your institution.

You will add immeasurably to your learning experience if you consult the journal literature as you read this book. Those readers who are not yet experienced at consulting the journal literature should be forewarned that it is a skill that comes with practice. Do not be discouraged if the going is rough at first. Here are some practical hints that may help. Begin by *scanning* the article's Introduction to determine whether it in fact contains, or might direct you to, the information you are seeking. Scan the Discussion as well, since it should tell you how the current paper fits into the bigger picture while it also directs you to additional references that might contain what you have been seeking. (Scanning is a valuable skill worth acquiring. Importantly, when scanning, do not stop for details that you may not understand. Keep scanning.) After scanning the Introduction and Discussion, if you have located a paper that has the answer to your specific question, or if the paper at hand is interesting to you, go back and try to read it for understanding. Most readers may not yet need to understand experimental details, but I urge you to try to understand the underlying logic of the experiments and how the data lead to the authors' conclusions. My greatest wish is that this book will serve as a foundation that will ease you into reading the virology literature, from which you will build a continuing understanding of viruses and their diseases.

As noted above, each of the chapters of *Virology: Molecular Biology and Pathogenesis* was read by one or more reviewers, each of whom provided expert commentary on an individual chapter. Other scientists, who have had much experience teaching the subject of virology, reviewed multiple chapters and, in some instances, the entire book in order to offer advice regarding its merits as a teaching tool. I am grateful to these colleagues for their time and expertise and for their special insights. As I noted above, a text of this scope by one author would not have been possible were it not for their advice. In addition, I am most grateful to those reviewers who also offered kind words of encouragement, which invariably sent me happily back to the millstone with renewed vigor. So, to each of the reviewers enumerated below I add my most sincere thanks.

In addition, I am grateful to Eva Szomolanyi-Tsuda (University of Massachusetts Medical School). A presentation that she gave at a scientific conference helped me to rethink how to present some of the material in Chapter 4. I am also grateful to Greg Payne, Senior Acquisitions Editor at ASM Press. Greg took an early interest in my book proposal and guided me through the process of securing a book contract with ASM Press. I have greatly enjoyed my interactions with him. More recently, I have enjoyed working with John Bell, Senior Production Editor at ASM Press. John gently kept me on track towards bringing this project to a conclusion while remaining sensitive to other demands on my time. Also, I am grateful to Michael Norkin, who helped me a few times when I was not confident that my own writing skills were up to the task, and to my student and friend, Dmitry Kuksin, who helped compile the figures and who provided general computer expertise.

Reviewers

Lee Abrahamsen, Bates College

Larry Anderson, Centers for Disease Control and Prevention

Walter Atwood, Brown University

Mauro Bendinelli, University of Pisa

Thomas Bleck, Northwestern University

Allan Brasier, University of Texas Medical Branch

Thomas Chambers, Merck & Co.

James Champoux, University of Washington

Richard Condit, University of Florida

Kevin Coombs, University of Manitoba

James Crowe, Vanderbilt University

Daniel DiMaio, Yale University

Ellie Ehrenfeld, National Institutes of Health

Malcolm Fraser, University of Notre Dame

Adolfo García-Sastre, Mount Sinai School of Medicine

Rebecca Horvat, University of Kansas at Kansas City

Lou Laimins, Northwestern University

Duncan McGeoch, University of Glasgow

A. Dusty Miller, University of Washington

Edward Niles, University at Buffalo

David Ornelles, Wake Forest University

Stanley Perlman, University of Iowa

Douglas Richman, University of California at San Diego

Ann Roman, Indiana University School of Medicine

Robert Rose, University of Rochester

Kathy Rundell, Northwestern University

Christian Sauder, United States Food and Drug Administration

Brad Sefton, University of California at San Diego

Peter Shank, Brown University

Guido Silvestri, University of Pennsylvania

Steven Specter, University of South Florida

Mario Stevenson, University of Massachusetts Medical School

Sankar Swaminathan, University of Florida

Peter Tattersall, Yale University

Gary Tegtmeier, University of Kansas at Kansas City

Paul Wanda, Southern Illinois University at Edwardsville

Scott Weaver, University of Texas Medical Branch

Raymond Welsh, University of Massachusetts Medical School

Reference

Baltimore, D. 1971. Expression of animal virus genomes. *Bacteriol. Rev.* **35:**235–241.

1

Introduction

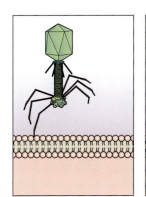 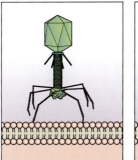  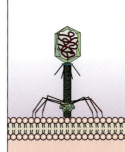 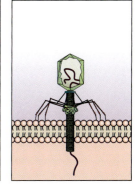

A Selective History on the Nature of Viruses

INTRODUCTION

More than 90% of all human illnesses may be caused by virus infections. To realize the validity of that estimate, we need only think of our common colds, influenza, various infirmities due to herpesviruses (e.g., genital herpes, infectious mononucleosis, chicken pox, and shingles), virus pneumonias, hepatitis, acquired immunodeficiency syndrome (AIDS), and some cancers as well. Indeed, virtually every part of our bodies is susceptible to virus infection, from our heads (viral meningitis and encephalitis) to the soles of our feet (plantar warts). Until relatively recently, we would also have had to number among agents of our common infections the viruses responsible for measles, mumps, rubella, and poliomyelitis, and earlier still, smallpox. In those instances, we can be grateful that we now have effective vaccines.

Even if the above estimate of 90% still seems high, the frequency of viral infection is actually much greater still, since most infections by viruses are either very mild or not even symptomatic and therefore go unnoticed. This is an important fact that is discussed in later chapters. Nevertheless, it was the frequency and severity of viral diseases in humans, as well as in domestic animals and plants, which led to early interest in virology. Indeed, were it not for the diseases caused by viruses, the discovery of these agents (which lie beyond the range of the best light microscopes) at the end of the 19th century may well not have occurred until the age of molecular biology, more than half a century later. As it happened, however, the development of the field of molecular biology itself owes much to the earlier discovery of viruses.

The development of the science of virology follows two somewhat separate tracks. Early studies in virology were driven by medical concerns. Later, many of the major advances in virology developed in the context of the new field of molecular biology. Today, these two tracks often intersect, with advances in the molecular and cellular aspects of virus infection leading to new insights into viral disease processes and rational approaches to the prevention and treatment of virus diseases.

At this point the reader may be looking for a definition for viruses. However, many virologists believe that although one can easily acquire a good understanding of the unique nature of viruses, one cannot readily pin down a concise definition for these entities. For that reason, rather than by beginning with a definition for viruses, we start by establishing their unique properties. Those properties are revealed in a logical sequence from a consideration of some major findings in the history of the field. And because that history, which began with the medical concerns noted above, is interesting in its own right, we start there.

THE EARLY YEARS: DISCOVERERS AND PIONEERS

Choosing a precise beginning for the history of the science of virology is somewhat arbitrary, in part because several illnesses that now are known to result from virus infections had been recognized for thousands of years without any knowledge of viruses. Regardless, there is some justification for beginning 1,000 years ago with smallpox, because that was when an empirically based measure to control this dreaded disease first became known. That measure was **variolation**, whereby uninfected individuals were inoculated with material from the scabs of individuals who survived smallpox infection. It arose in the Far East and has been carried out for the last millennium in China and India. What might have inspired the emergence of this effective prophylactic countermeasure against smallpox in 11th century China and India? Remarkably, it was the practical application of the even earlier realization by the Chinese that individuals may experience some illnesses only once in a lifetime. Variolation began to be practiced in Europe in the early 18th century, after Lady Mary Wortley Montague, the wife of the British ambassador to Turkey, had her children undergo variolation.

Despite its risks, variolation generally resulted in an infection that was milder than a naturally acquired one, and the variolated individuals were protected for life against the more severe naturally acquired smallpox infection. Why might exposing an individual, usually a child, to the dried-out scabs on the skin of a patient recovering from smallpox protect that child from a natural infection? Today we understand that the virus in a dried-out lesion of a recovering person has been partially inactivated by that person's immune response, as well as by the drying itself. Yet that partially inactivated virus can still induce immunity to natural infection in an individual undergoing variolation.

Some readers might find it remarkable that in prescientific times, and in a non-Western society as well, it was recognized that survivors of smallpox, whatever its cause,

were resistant to subsequent episodes of this disease and that an effective control measure (i.e., variolation) was developed based on that observation. Variolation encompassed risks that would be unacceptable today, in particular a fatality rate of 1 to 2%. However, those risks were acceptable in 18th century Europe when, for example, as many as 1 person in 10 died of smallpox.

A major leap forward in preventing smallpox came with the development of the smallpox vaccine in 1798 by an English country doctor, Edward Jenner. Jenner's breakthrough, though still a half-century before the proposition of the germ theory of disease and 100 years before the actual discovery of viruses, might be regarded as the most astonishing achievement in the history of medical virology. Jenner learned that some of his patients, who happened to be milkmaids, were "resistant" to smallpox. His great insight was to realize that the milkmaids' resistance to smallpox somehow came about because they had earlier acquired cowpox, which is caused by what is now recognized as the cowpox virus that infects cattle. (Cowpox virus is now known to be a relative of the smallpox virus, commonly referred to as variola virus.) This led Jenner to carry out an experiment in which he inoculated a child, James Phipps, with extract from a cowpox lesion and then demonstrated that young Phipps was resistant to a subsequent challenge with smallpox (Box 1.1).

More than a century would have to pass before it could be appreciated that this protection depended on the facts that cowpox virus is immunologically cross-reactive with smallpox virus and that it produces a relatively benign infection in humans.

The term "vaccination" is based on the Latin word for cow (*vacca*) and was coined by Pasteur in recognition of Jenner's contribution. The term "vaccine" now refers to specific and actively immunizing agents that protect against a variety of pathogenic microbes. Vaccines are one of the most effective and widely used prophylactic measures ever developed.

Interestingly, **vaccinia virus**, which is the name for the virus used in the current smallpox vaccine, is clearly different from the cowpox virus in Jenner's original vaccine, and its precise origin is not known. At one time, it was thought that the cowpox virus was inadvertently replaced by an attenuated (i.e., weakened; see below) smallpox virus. This might easily have happened since the vaccine was originally propagated by serially passing it from person to person, a practice since terminated because it was associated with the transmission of other diseases, including syphilis and hepatitis. However, more recent studies show that vaccinia virus, although related to smallpox virus, is different from the smallpox virus, perhaps reflecting the now traditional practice of propagating the vaccine

BOX 1.1

Although history usually credits young James Phipps as the first person to be vaccinated, Phipps in fact was not the subject of Jenner's first experiment. Instead, it was Jenner's first son, Edward, Jr., born in 1789, whom Jenner inoculated with swinepox virus when the infant was only 10 months old. Jenner of course did not know about microbes, and he left no records that might have revealed his purpose in inoculating Edward Jr. with swinepox. It may be relevant that cowpox was relatively rare, while a similar pox disease was more common in pigs. Regardless, the baby became sick on the eighth day with a pox disease, from which he recovered. At several unspecified times afterwards, Jenner in point of fact tried to infect Edward Jr. with actual smallpox. He failed each time, most likely because the swinepox virus may indeed have immunized Edward Jr. against smallpox.

Admittedly, this episode involving Edward Jr. somewhat muddies the waters regarding the otherwise straightforward history of the discovery of smallpox vaccination that is usually recounted. Since Jenner left no records of his experiments using his infant son, few details are actually known. We can only speculate on how Mrs. Jenner might have regarded these activities. Also, little is known about James Phipps, who was only 8 years old when he was the subject of Jenner's later experiments. Additionally, nothing is known about James's parents and whether they may have consented to Jenner's use of James. But Jenner looked after James in later years and may have felt some remorse, since he eventually built a cottage for James and even planted flowers in front of it himself. Also not often mentioned, Jenner used several other young children in his experiments, including his 11-month-old second son, Robert, in addition to Edward Jr. and James. One of these children died from a fever, possibly resulting from a contaminant (streptococcus?) in the vaccine. Thankfully, such experiments cannot be done today. Yet because of Jenner's efforts, this once dread human scourge now exists only in the laboratory.

This sidebar gives only a glimpse into the fascinating story behind Jenner's work. For more details, I recommend *Virus Hunters* by Greer Williams, Alfred A. Knopf, New York, 1960.

---

strain on the skin of horses. That is, the modern smallpox vaccine strain may be a horse virus!

In 1885, still prior to the recognition of viruses as distinct entities, Louis Pasteur developed a vaccine against rabies. Pasteur's vaccine was the second human vaccine and the first deliberately attenuated one. **Attenuation** is the conversion of a killer microbe into something that is less able to cause disease, yet still able to induce protection. In this instance, attenuation was achieved by serial passage of the rabies agent in rabbits, followed by "aging" it in vitro (see Chapter 10). Pasteur had no way of knowing what the underlying basis for attenuation actually was, nor did he even know the viral nature of the rabies agent, since the existence of viruses as a distinct class of microorganisms remained to be discovered. Knowledge of genes and mutations, like viruses, also was still in the future. Surprisingly, even today virologists know relatively little about the determinants of virus virulence (Boxes 1.2 and 1.3).

At this point in our history we go back to the mid-19th century and consider some important developments leading to the acceptance and success of the **germ theory of disease**. By this time, the existence of a variety of microorganisms, including bacteria, fungi, and protozoa, was well established. In 1867 Pasteur proposed that microorganisms might produce different kinds of diseases. This premise was based on his earlier experimental findings

BOX 1.2

When contemporary vaccinologists develop vaccines to protect against viral diseases, they are essentially tapping into biological mechanisms that already have been perfected through eons of natural selection. With many different viruses, natural infection regularly results in lifelong immunity against the same virus. This is the principal fact exploited by vaccinologists. With that in mind, note again that vaccines against smallpox and rabies were developed years before viruses were even discovered. Also, in Pasteur's day, nothing was known about how the human body generates immunity. It may be somewhat disquieting to know that even today, when so much more is known about viruses and our immune response, the development of vaccines remains largely empirical.

that different microorganisms are associated with different kinds of fermentations (Box 1.3). At about the same time, Joseph Lister, an English surgeon who admired Pasteur's work on fermentation and spontaneous generation, reasoned from Pasteur's demonstration of the ubiquity of airborne microorganisms that infections of open wounds

BOX 1.3

It is interesting that Pasteur, who was one of the greatest of all experimentalists, did not advance our understanding of the nature of the rabies agent. That is, he never came to realize that he was dealing with a new category of microbes, ones that are submicroscopic and cannot be cultivated on nutrient media. This is all the more surprising since at the time the Pasteur Institute was established, rabies actually was its main interest. Thus, Pasteur indeed attempted to find the rabies microbe, presuming that once found it could then be cultivated in vitro and serially passaged from one experimental animal to another. He was wrong on the first count, but right on the second. Unfortunately, he got sidetracked along the way. In 1880, he injected a rabbit with the saliva of a child who had died of rabies, and then, he examined the blood of the rabbit after it too died. Under his microscope, Pasteur in fact observed a microbe in the rabbit's blood. It had a mucous capsule, and Pasteur thought it might be the rabies agent. But he later found that he could isolate the same microbe from the saliva of healthy children. What was this microorganism seen by Pasteur? Ironically, it was *Pneumococcus pneumoniae*, a major bacterial pathogen that was correctly identified several years later by Albert Frankel. Thus, Pasteur missed the opportunity to identify a human pathogen much more important than rabies virus, as well as to identify a whole new class of pathogens: viruses. Pasteur in fact was frustrated by his rabies studies, largely because the rabies agent passed through filters that previously had retained bacteria (see the text). Yet he did correctly surmise that rabies is caused by an unusually small microorganism, beyond the range of his microscopes. Still, Pasteur did not suspect that it was in any basic way different from the pathogenic bacteria isolated previously. Rabies virus was identified in 1903 by Paul Remlinger, who correctly demonstrated its filterability at the Constantinople Imperial Bacteriology Institute.

---

are due to microorganisms in the environment. The aseptic techniques that Lister then introduced dramatically reduced infections during surgery. Lister also contributed the technique of **limiting dilution** to obtain pure cultures of bacteria, while Robert Koch developed the isolation of bacterial colonies on solid culture media. In addition, **Koch's postulates** provided the experimental basis for establishing that a specific microorganism is responsible for a specific disease. Koch had been studying anthrax, a disease of cattle, caused by *Bacillus anthracis*, which can be transmitted to people. He used his isolation technique to establish pure cultures of a single species of bacteria from the infected cattle. Then, he injected a sample of the pure culture into healthy animals. The healthy animals then developed anthrax. Finally, he reisolated the infectious agent from the inoculated animals. This sequence of isolation, infection, and reisolation constitutes Koch's postulates.

We now come to those developments that led to the realization that there exists a class of microbes that are fundamentally distinct from the previously recognized bacteria, fungi, and protozoa. Beginning in 1887, a Russian scientist, Dmitry Ivanovsky, was looking into the cause of tobacco mosaic disease, so named because of the dark and light spots on diseased tobacco leaves. Repeating the work of an earlier scientist, the German Adolf Mayer, Ivanovsky showed that the sap of diseased plants contained an agent that could transmit the disease to healthy plants. But Ivanovsky went an important step further. He found that the infectious agent could actually pass through so-called **Chamberland filters**. These filters, made of unglazed porcelain, contain pores that are too small to permit the passage of most bacteria. (The use of porcelain filters to retain and concentrate microbes became standard practice after 1884. Chamberland is said to have developed these bacterium-proof filters while experimenting with a broken clay pipe purchased from his tobacconist.)

It is important to appreciate that Ivanovsky's research actually might have provided an experimental precedent for demonstrating the presence and activity of a new type of infectious agent, namely, viruses. However, as the story goes, Ivanovsky, like Mayer before him, was unable to grow the organism from the infected sap. Thus, he was unable to satisfy an important component of Koch's postulates (that is, the cultivation of a single species of microorganism in pure culture). The influence of the Koch paradigm was so strong that Ivanovsky did not want to consider that he might actually have seen evidence for a previously unknown kind of microorganism. Instead, he questioned the reliability of his experimental procedures. Perhaps the causative agent was a bacterium and the filters were defective, or perhaps the causative agent was a **toxin**, a nonreproducing poisonous substance produced by an organism. Either of these possibilities could readily have been tested. Do you see how?

An important leap forward came in 1898 from the Dutch microbiologist Martinus Beijerinck, who was working with Mayer but was unaware of Ivanovsky's findings. Beijerinck, like Ivanovsky, found that the sap from diseased tobacco plants retained its ability to transmit the tobacco mosaic disease after filtration through Chamberland filters. But Beijerinck went yet another major step further. He demonstrated that the sap did not lose its ability to cause disease upon being diluted from repeated inoculations or passages through new, healthy plants. This experimental result implied that the disease-causing agent was in fact replicating in the plant tissue, thus accounting for its ability to replenish its pathogenic activity. This finding provided the means for distinguishing this new type of pathogenic agent from nonreproducing toxins, which already were known to be produced by certain bacteria.

So now we have established two fundamental properties that are characteristic of this new class of pathogens. First, they are smaller than bacteria, since they pass through filters that block bacteria. Second, they require living cells or tissue to support their propagation. The first of these properties, their small size, accounts for the fact that these agents were not seen by the microscopy procedures of the day, which readily revealed bacteria. The second property accounts for the inability of Ivanovsky to propagate the tobacco mosaic virus (TMV) in pure culture, that is, in the absence of susceptible living host tissue. Still, this part of the story is not yet complete, since it was not yet clear whether these newly discovered agents might be liquid or particulate. Indeed, during the first decades of the 20th century they were referred to simply as filterable agents. The issue was not settled until 1938, when the first electron micrographs of TMV were taken and the active agents were revealed to be tiny particles. The term "virus," from the Latin word for poison, came to be used to refer to the agents having the properties described by Mayer, Ivanovsky, and Beijerinck.

Returning briefly to the earlier pioneering days, we note some other significant achievements. First, in 1898 Loeffler and Frosch isolated the first virus obtained from animals, the foot-and-mouth disease virus (Box 1.4). Second, in 1901, in Cuba, the U.S. Army doctor Walter Reed isolated the first virus pathogenic in humans, yellow fever virus. That virus provided a new surprise. Rather than being transmitted directly from person to person, it was transmitted by mosquitoes, a hypothesis advanced earlier by Cuban physician Carlos Juan Finlay and subsequently confirmed by Reed. (The affirmation that yellow fever is transmitted by mosquitoes provides one of the more fascinating stories in the history of infectious diseases. It is told more completely in Chapter 7.) Third, and yet another surprise, was the discovery by Peyton Rous in 1911, at the Rockefeller Institute (now University) in New York, that viruses can cause cancer. Rous found that sarcomas (cancers of connective tissue) in chickens could be transmitted by a virus that is now known as the Rous sarcoma

BOX 1.4

I find the "genealogy" of leading scientists to be interesting, since influential scientists frequently were trained by mentors who likewise made extraordinary contributions. For a compelling example of this phenomenon, note the background of Max Delbrück, as well as the accomplishments of some who came under his influence, as recounted in the text. However, in this sidebar, I want to highlight that Friedrich Loeffler was trained by Robert Koch. In 1884, 14 years before Loeffler isolated the foot-and-mouth disease virus, Loeffler used Koch's postulates to identify the bacterium that causes diphtheria, *Corynebacterium diphtheriae*. In addition, Loeffler made the important observation that when he injected *C. diphtheriae* into animals, the microbe did not need to spread to the tissues that it damaged. This observation led Loeffler to propose that the bacteria were producing a poison or toxin that spread to the remote sites. This was a new concept and a prediction that was later confirmed. Interestingly, and as explained in Chapter 2 and in Box 1.9, a virus indeed plays a role in diphtheria, in a rather unexpected way.

virus, which was the first known tumor virus, as well as the first retrovirus to be discovered. More is said about that in Chapter 20. For now, see Box 1.5. Fourth was the discovery of bacteriophages in 1915 by Frederick W. Twort in London, and independently in 1917 by Felix d'Herelle in Paris. As has happened in other instances, there were quarrels over who made the discovery of bacteriophages first. Regardless, it was d'Herelle who gave the bacteriophages (phages for short) their name, which actually means bacteria eaters, which, of course, is not what actually happens. Nevertheless, since bacteriophages indeed may kill their bacterial host cells, d'Herelle spent a number of years trying to develop them for use as therapeutic agents in the treatment of bacterial disease. Those efforts never bore fruit (Box 1.6).

THE FIRST STIRRINGS OF THE MOLECULAR ERA

An experimental finding that provided a major impetus towards the eventual emergence of the modern science of virology, and indeed of molecular biology as well, involved bacteria rather than viruses. The experiment was carried out by Frederick Griffith in 1928 and involved the pneumococcus bacterium *Diplococcus pneumoniae*. Wild-type or "smooth" *D. pneumoniae* organisms contain

BOX 1.5

It is noteworthy that much of what is known today about the cellular and molecular nature of cancer, as well as of the mechanisms that control normal cell growth and proliferation, came from studies involving tumor viruses. These issues are described more extensively in Chapters 15 and 20. For now, it is noteworthy that Rous was awarded the Nobel Prize for his 1911 discovery, but not until 1966! The 55-year lag between Rous's achievement and his recognition by the Nobel Committee is the longest on record. Nobel prizes are not awarded posthumously. Fortunately, Rous had longevity on his side. He died 4 years after receiving the Prize, at age 87.

Why do you suppose Rous had to wait so long for his contribution to be recognized by the Nobel Committee? The following might contain the seed of an answer. An oncogenic virus closely related to Rous sarcoma virus actually was discovered 3 years earlier in 1908 by Vilhelm Ellereman and Olaf Bang. They found that leukemia in birds could be transmitted by a filterable agent from leukemic cells or by serum from leukemic birds. However, leukemia was not then recognized as cancer, so the significance of this discovery went unrecognized. Sarcomas certainly were recognized as cancers, but, bearing in mind the level of knowledge at the time, Rous's discovery was seen as little more than an interesting oddity. Thus, given the level of understanding during the first half of the 20th century, Rous's discovery could not have led to any breakthroughs; certainly not in 1911. Rous abandoned work on the Rous sarcoma virus around 1915 (see Chapter 20), but 20 years later he contributed significantly to the realization that the Shope papillomavirus (the first known DNA tumor virus) causes cancer in rabbits (see Chapter 16).

BOX 1.6

The novel *Arrowsmith*, by Sinclair Lewis, describes fictional efforts to develop bacteriophages for use as antibacterial therapeutic agents. I highly recommend it, if for no other reason than to get a sense of what it was like to do medical research in that earlier time. Interestingly, the idea for *Arrowsmith* was suggested to Sinclair Lewis by his friend Paul de Kruif, then a scientist at the Rockefeller Institute, and later a well-known science writer. Also, as noted in Chapter 2, there is currently renewed interest in developing bacteriophage-based approaches for antibacterial therapy.

polysaccharide capsules that enable them to avoid phagocytosis by host macrophages. This capsule is thus a virulence factor that enables the bacteria to infect the host and cause disease. Nonencapsulated or "rough" mutants are nonpathogenic. In Griffith's rather bizarre experiment, he inoculated mice with a mixture of live nonpathogenic mutant bacteria and heat-killed wild-type bacteria. Neither the nonpathogenic mutant bacteria nor the killed wild-type bacteria alone induced disease. Remarkably, mice inoculated with the mixture not only developed severe septicemia but also released live virulent encapsulated pathogens. Thus, some interaction occurred in the mouse between the live avirulent bacteria and the dead virulent organisms to generate live virulent pathogens. The following two possible explanations were considered. (Are there any others?) First, some substance may have passed from the dead virulent organisms to the live avirulent ones, enabling the latter to produce the polysaccharide capsule. Alternatively, some substance may have passed from the live avirulent organisms to the dead virulent ones, restoring the viability of the latter. The first explanation was correct, as later demonstrated in an experiment in which a cell-free extract prepared from the dead virulent cells was able to transform some live avirulent cells to virulence. The reverse transfer did not yield any viable organisms. (What might Griffith's purpose have been in carrying out so bizarre an experiment? It was, in fact, part of Griffith's efforts to make a vaccine to prevent the pneumonia epidemics that were occurring at the time. Toward that end, he inoculated mice with live avirulent and with heat-killed virulent samples of *D. pneumoniae*, as well as with both concurrently. Although Griffith's experiment is sometimes recognized as one of the keystones of modern molecular biology, he died in 1941, never knowing the impact it eventually would have.)

The next part of the story concerns the identification of the crucial transforming factor. One could argue that it is genetic in nature, since the transformed bacteria yielded progeny cells with the transformed phenotype. Moreover, the transformed bacteria and their progeny behaved like the original wild-type bacteria in subsequent transforming experiments. In 1944, Oswald Avery, Colin MacLeod, and Maclyn McCarty at the Rockefeller Institute carried out an experiment based on the earlier finding that transformation is possible using cell-free extracts. They prepared a whole-cell extract from the encapsulated cells and fractionated it into its various macromolecular species, generating protein, lipid, polysaccharide, and deoxyribonucleic acid (DNA) fractions. Next, they asked which of these fractions might have transforming activity. To the surprise of almost everyone, only the DNA fraction was capable of transforming nonencapsulated cells into capsulated ones.

It is important to our story to note that these remarkable findings, and the conclusion they implied, were met with widespread skepticism. To understand why, one must adopt the mind-set of the time. First, it was difficult for the classically trained Mendelian geneticists of the day to accept the validity of these strange new experiments. The experiments of the classical geneticists, which involved the breeding of generations of plants or animals, established that hereditary information is carried in units called "genes," and these scientists thought in terms of genes, not in terms of molecules of nucleic acid. More importantly, in those days, DNA was looked upon as a structurally uninteresting, monotonous molecule, rather like a starch. In contrast, the wide variety of proteins seemed to provide a virtually unlimited number of possible genes. Thus, it was widely assumed that the genetic composition of chromosomes depended on their proteins, not their nucleic acids. Consequently, the classical geneticists did not see a connection between the biological unit, the gene, and the chemical unit, the DNA. Thus, despite these experiments, the notion that protein constituted the genetic material generally persisted.

One objection raised against the conclusion that the transforming activity resided in DNA was that a minute amount of protein, undetectable by the methods employed, might have remained associated with the DNA during fractionation. This protein might then have been responsible for transforming activity. However, Avery, MacLeod, and McCarty showed that extensive protease treatment of the extract had no effect on its transforming activity, whereas even very brief exposure to nuclease completely destroyed biological activity. Another objection was that even if DNA was responsible for transformation in this instance, perhaps it acted in some nongenetic chemical manner to affect capsule formation. This objection was addressed by the experiments of Rollin Hotchkiss, who demonstrated that transformation also works for bacterial characteristics (e.g., antibiotic resistance) that have nothing to do with capsule formation.

At this point, we take a short step back in time to 1935 to consider a major achievement in virology that would dramatically influence later developments. First, we remind ourselves that up to this time, interest in virology was for

BOX 1.7

The heroic effort that went into purifying and crystallizing TMV cannot be appreciated from the brief account in the text. However, the Nobel Committee did appreciate it, making Stanley, in 1946, the first virologist to be awarded the Nobel Prize (he shared the Prize in Chemistry). However, even in this instance, it was impossible for many scientists of the day to accept that a crystal could have the capacity for life. In the minds of some, the specter of possible contamination hung over this work. Also, many physician scientists of the day did not care about tobacco plants with mottled skin, nor did they appreciate its possible relevancy to disease in humans.

Stanley continued to persevere, but on one important account he was wrong. Rather than being comprised entirely of protein, as Stanley had believed, TMV was shown by Frederick Bawden and Norman Pirie to also contain RNA. The composition actually was 94% protein and 6% RNA, moving TMV somewhat away from being merely a chemical to somewhat closer to being an organism. Interestingly, Max Schlesinger in 1933 was actually the first person to find a nucleic acid in a virus. Making use of high-speed centrifuges, he purified a bacteriophage to high purity and demonstrated that it is comprised of 50% protein and 50% DNA. However, Schlesinger did not crystallize his virus, as Stanley had done, and his discovery attracted much less attention than that of Bawden and Pirie.

the most part medical and agricultural. Essentially all that was known about the biology of viruses was that viruses are smaller than bacteria and that they can propagate only within suitable host cells. Moreover, the study of viruses had not yet advanced biological knowledge in general. However, biochemists had been making great strides in purifying and crystallizing proteins. (Note that solving the structure of proteins by crystallography was still well beyond the technology of the day.) Inspired by the success of the protein crystallographers and armed with indirect evidence that TMV is at least partly a protein, Wendell Stanley at the Rockefeller Institute succeeded in crystallizing TMV. Two important points now need to be noted. First, the "protein" crystals of TMV indeed were endowed with the infectious activity of TMV. Second, crystals are exquisitely pure. Thus, it could now be unequivocally demonstrated by others (Box 1.7) that TMV is not a pure protein but instead contains about 6% ribonucleic acid

(RNA) (see below). Consequently, whatever it was about TMV that enabled it to produce copies of itself, that ability resided in its protein, or in its nucleic acid, or in a combination of its two constituents. (Also note that the ability of TMV to form crystals implied that it had a regular structure. This is discussed more fully later.)

Arguably, Stanley's achievement was a major milestone in the history of not only virology but also biology. If viruses are so simple that they could be crystallized like table salt and yet express that most fundamental property of living systems, the ability to replicate, then there was reason to believe that the nature of biological replication might eventually be understood in chemical terms. To appreciate how this concept stirred the souls of some scientists of the day, we once again adopt the mind-set of the times. Whatever the chemical nature of the genetic material might be, it would have to contain information and also have the ability to be precisely copied. Because of the general ignorance regarding the structures of large molecules in the cell and the then-prevailing belief that the genetic material was protein, it was felt by many that it would be essentially impossible to understand heredity in chemical terms. Attempts were made to account for how a protein might replicate, but as you might presume, these attempts did not lead to especially satisfying models. Thus, many serious biologists and chemists expressed the belief that some "vital" force outside the known laws of chemistry would be needed to explain the phenomenon of life. This doctrine was referred to as "vitalism," and it still had its serious adherents up to the time of Stanley's work. However, crystallography is a very precise science, and the fact that TMV could be crystallized and yet retain biological activity strongly implied to at least some scientists that physical and chemical explanations indeed would suffice to explain life. Thus, Stanley's achievement would eventually mark the death knell of vitalism and spur the beginning of the field of molecular biology (Box 1.7).

THE PHAGE GROUP

Spurred on by this new line of thought, a somewhat atypical group of investigators sought to understand the nature of genes. They were atypical in that many had little or no knowledge of traditional genetics, or biochemistry, or even biology in general. Many were physicists by background, but they had a single goal in mind: to understand the physical basis of the gene. Important to our story, several recognized the advantages of focusing their research efforts on viruses. This odd group's interest in genes and its focus on viruses would lead to discoveries of singular overwhelming importance. Indeed, their research approaches and the results they generated gave rise to molecular biology.

Max Delbrück was a key player in this new group of scientists, and he is recognized as one of the principal founders of the new science of molecular biology. Delbrück originally trained as a physicist in Germany during the 1920s, studying quantum mechanics under the guidance of Max Born. Moreover, he interacted with many others of the great physicists of the day, including Wolfgang Pauli, Albert Einstein, and Erwin Schrödinger. In 1931 Delbrück went to Copenhagen for postdoctoral studies with the great Niels Bohr, and it was actually Bohr who aroused Delbrück's interest in biology. After a short subsequent stay in England, Delbrück returned to Germany in 1932 to work as an assistant to Lise Meitner, who eventually discovered nuclear fission (Box 1.8). But while he was with Meitner, Delbrück's focus was actually on developing quantum mechanical models of genes, his thinking about the nature of genes actually having been influenced by Schrödinger. By 1937, the situation in Nazi Germany became intolerable for Delbrück, and it was then that he began the journey that would lead to his move to the California Institute of Technology (Caltech), where he initiated developments of extraordinary importance for virology, which as noted above, led to the emergence of the new science of molecular biology.

Delbrück began to work with bacteriophages in 1938, recognizing that these viruses would be an ideal system for probing the nature and action of genes. Indeed, modern phage research began when Delbrück entered the field. This was the result of Delbrück's own experiments, and also the work of other members of the phage group that Delbrück and the Italian scientist Salvador Luria brought together at Caltech.

Delbrück actually began his foray into genetics by following up on H. J. Muller's discovery in the 1930s that X rays and ultraviolet (UV) irradiation could cause mutations in fruit flies. To account for Muller's findings, Delbrück spent several years working towards a quantum physics interpretation of gene mutation, which ultimately led to his thinking of genes as molecules, rather than as abstract entities, a first step on the path to molecular biology. Luria, in turn, was excited by the new idea of genes as molecules and sought a system in which to pursue their analysis. Independently of Delbrück, he too thought that bacteriophages might be an ideal system to probe the nature of genes, since bacteriophages grew rapidly and the effects of radiation on them could be measured with precision. Luria had hoped to be able to work with Delbrück. In the late 1930s, Luria, who was Jewish, faced persecution in then-fascist Italy. Independent of each other, Luria and Delbrück each left Europe and finally met in the United States in 1940. That summer, at the Cold Spring Harbor Laboratory on Long Island, they began their first experiments on the life cycle of bacteriophages.

BOX 1.8

Shortly after Delbrück left Nazi Germany, Meitner went on to discover nuclear fission; this story has fascinating ramifications because Meitner, who was Jewish, was by then a nonperson in Germany. Meitner was able to find safe haven in Holland, thanks to the efforts of Dutch physicists who persuaded their government to admit her without a visa, on an Austrian passport that no longer was valid for her. Niels Bohr subsequently found a laboratory for Meitner in Sweden where she might continue her work, which was financed by a grant from the Nobel Foundation.

During the Nazi occupation of Denmark in World War II, Bohr himself was endangered, and so he too escaped to Sweden. From there, Bohr played a crucial part in the rescue of nearly all of the Danish Jews by pressuring the Swedish government to accept responsibility for them. Rumors that German agents in Stockholm were out to assassinate Bohr led to his harrowing escape to England. He was spirited away in the unpressurized empty bomb rack of a Royal Air Force Mosquito Bomber, in which he lost consciousness and nearly died from a lack of oxygen. Bohr spent the last 2 years of the war in England and America, where he was associated with the Atomic Energy Project, and then became one of the first and most prescient arms control advocates. These incidents are chronicled in *The Making of the Atomic Bomb* by Richard Rhodes (Simon & Schuster, New York, NY, 1986).

Delbrück (jokingly?) took indirect credit for Meitner's discovery of nuclear fission, saying that his waning interest in physics was holding back Meitner's group, and thus, his leaving enabled the discovery to happen.

The original phage group founded by Delbrück and Luria trained a brilliant second generation of phage workers and virologists. What distinguished this group from others was their single-minded purpose to understand the physical basis of heredity by analyzing phage replication. (The influence of Delbrück, Luria, and the phage group was enormous. To get a feeling for the extraordinary vibrancy and accomplishments of that group and their era, I strongly urge you to spend some time with *Phage and the Origins of Molecular Biology: the Centennial Edition*, edited by J. Cairns, G. S. Stent, and J. D. Watson [Cold Spring Harbor Laboratory Press, Cold Spring Harbor, NY, 2007].)

Phage Growth: Eclipse and Replication

Delbrück's first major experiments with phage were carried out with E. L. Ellis. One of these experiments, the so-called "one-step growth experiment," characterized the parameters of bacteriophage replication. The experiment was especially important for what it revealed about phage replication. However, it also was noteworthy for having marked the beginning of quantitative virology.

We begin with a few words about experimental procedures. In the one-step growth experiment, Delbrück and Ellis needed to titrate or measure the number of infectious phages in their samples. To accomplish this, they made use of the **plaque assay**, a method that was developed earlier by d'Herelle, one of the discoverers of bacteriophages (see above). In a plaque assay, appropriate dilutions of a virus sample are added to a dense suspension of susceptible host bacteria, which is then spread over a surface of nutrient agar in a petri dish. Each infectious virus particle that is present in the sample will infect a cell in the dense bacterial lawn that develops on the agar surface. Each infected cell eventually bursts or lyses, releasing progeny viruses that infect the immediately adjacent cells in the bacterial lawn. After a few such cycles, a plaque that is comprised of a focus of lysed cells will be visible to the unaided eye (Fig. 1.1). A virus particle that registers as a plaque is referred to as a **plaque-forming unit** (PFU).

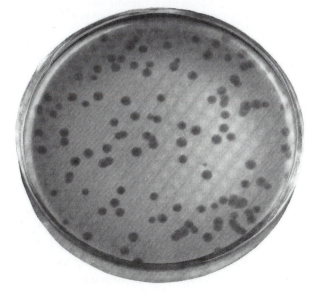

Figure 1.1 A petri plate showing growth of a lawn of *Escherichia coli* bacteria on which T2 phages have formed plaques. Reprinted with permission from G. S. Stent, *Molecular Biology of Bacterial Viruses* (W. H. Freeman and Company, San Francisco, CA, 1963).

The one-step growth experiment then went as follows. A virus inoculum was added to a dense suspension of host bacteria, which then was incubated for several minutes to allow the virus to attach to the cells. Next, unadsorbed virus in the suspension was removed, thereby synchronizing the infection among the individual infected cells. Moreover, this step ensured that any PFU that were measured during the experiment originated from the initially infected cells. Unadsorbed virus can be removed by several means. One way is by low-speed centrifugation, since the bacteria are pelleted at low speeds while the unadsorbed phages remain in the supernatant fraction. The pellet is then washed and resuspended. The resuspended bacteria are also diluted so that essentially all new viruses that are produced will result from a single cycle of infection. The infected bacterial suspension is then allowed to incubate, and at various times samples are removed from it and plated with sensitive bacteria, as in a standard plaque assay. But note that in the one-step growth experiment each intact infected cell, as well as each free virus particle in the suspension, produces a plaque on the lawn of sensitive bacteria.

Delbrück's major finding was that for the first 24 min after infection the number of PFU in the suspension remained constant (Fig. 1.2). For the next 10 min, the number of PFU rapidly increased several hundredfold, eventually reaching a plateau. How might one explain these

experimental findings, in particular, the initial 24-min lag? The answer is as follows. During the 24 min following infection, phage replication was occurring within infected cells, but none of the new phage were yet being released from the host cells. Regardless of how many new phage may have accumulated within each of the infected cells, each of those cells, if sampled while still intact, could produce only one plaque on the lawn of indicator cells. Thus, the plaques produced by the samples taken over the first 24 min were not produced by individual phage particles in the suspension. Instead, those plaques were generated from the initially infected bacterial cells. Then, at 24 min the infected cells began to burst open or lyse, thereby releasing the several hundred progeny viruses that they each now contained. Each of the released virus particles could then register as a PFU in the plaque assay.

The time elapsing between the moment of infection and the release of progeny virus is referred to as the **latent period**. This is certainly a misnomer, since much is going on during this time. The time during which individual progeny viruses are being released from the lysing cells, thereby becoming available for detection by plaque assay, is referred to as the **rise period**. The average number of PFU produced by each infected cell is referred to as the **burst size**, about 100 PFU per infected cell in the example shown.

The one-step growth experiment set the stage for investigating the events occurring during the crucial latent period, that is, the time during which the parental phage particles are replicating several hundredfold within the host cells. In 1948, A. H. Doermann began to examine events during the latent period by artificially breaking open the infected cells. This was accomplished by incubating samples from the inoculum in chloroform. Importantly, this approach enabled Doermann to follow the intracellular accumulation of phage, each particle of which could now produce a plaque. Doermann's experiment gave a most surprising and important result. He found that during the first 10 min after bacteriophage infection, the input virus disappeared (Fig. 1.3). That is, there were no PFU present in either the cells or the culture fluid during this time, despite the fact that several hundred would eventually emerge from each infected cell that was allowed to go the full course of infection. The time that elapses between infection and the intracellular appearance of infectious virus particles is referred to as the **eclipse period**. (To be more precise, it is the time that elapses between infection and the appearance of an average of one intracellular PFU per infected cell.) So now we have a new intriguing question to answer, namely, what happens to the parental virus during the eclipse period?

The experiment that would eventually answer this compelling question was inspired by information obtained by

Figure 1.2 The one-step growth experiment of phage T4. In this particular experiment, exponentially growing *E. coli* cells were infected with an average of one T4 phage per cell. The phage was allowed to adsorb to the cells for 2 min, after which the bacteria were diluted 10,000-fold into fresh media. The cells were then incubated, and samples were removed at various times and analyzed by plaque assay on sensitive bacteria. From A. H. Doermann, *J. Gen. Physiol.* **35**:645–652, 1952, as adapted by G. S. Stent and R. Calendar, *Molecular Genetics*, 2nd ed. (W. H. Freeman and Company, San Francisco, CA, 1958), with permission.

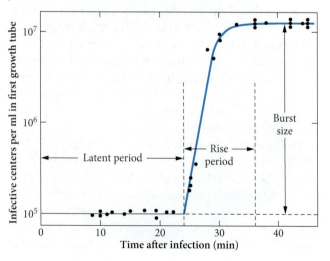

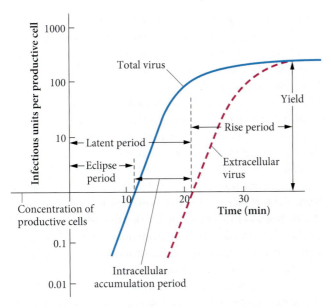

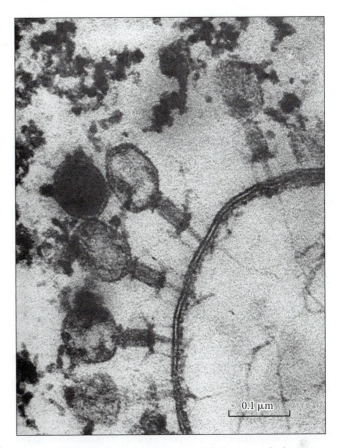

Figure 1.3 Diagram of relationships between the eclipse period, the latent period, and the rise period during one-step multiplication of bacteriophage T2. The conditions are essentially as for Fig. 1.2. However, the infection was synchronized and unadsorbed virus was eliminated by the addition of antivirus antiserum, followed by centrifugation and resuspension of the bacteria. A sample was then immediately plated for plaque assay to determine the concentration of infected cells. At subsequent times other samples were taken. Each of these samples was then divided into two aliquots: one was shaken with chloroform to break open the bacteria so that total virus might be measured by plaque assay; the other was freed of bacteria by centrifugation, and the supernatant was then assayed to measure extracellular virus. The viral titers are compared with the concentration of infected cells as 1.0. Thus, the curves indicate the average number of PFU per infected cell over time. Reprinted from B. D. Davis, R. Dulbecco, H. N. Eisen, and H. S. Ginsberg, *Microbiology*, 3rd ed. (Harper & Row, Hagerstown, MD, 1980), with permission.

Figure 1.4 Electron micrograph showing T4 phages attached to *E. coli* via their slender tails. The sheath portions of the phage tails are contracted, as occurs when the phage DNA is injected into the bacterium (see Chapter 2). Looking carefully, notice that the phages are attached to the bacterial cell wall by so-called tail pins that extend from base plates at the ends of the sheaths. Also, notice that tubes extend from the phage tails into the bacterial cell. Moreover, strands of phage DNA can be seen extending from the tips of the tubes into the cell, and the phage heads are now empty. From L. D. Simon and T. F. Anderson, *Virology* **32**:279–297, 1967, with permission.

directly visualizing phage particles. Recall that no one had yet actually seen a virus. However, viewing viruses became possible with the development of the electron microscope in the late 1930s. In 1942 Salvatore Luria and T. F. Anderson took the first electron micrographs of the T-even bacteriophages then being studied at Caltech. These micrographs showed that the phages are tadpole-shaped entities, with distinct heads and tails. In addition, phages appeared to attach to their host cells by their slender tails (Fig. 1.4). Moreover, the parental phages appeared to remain attached to the *outside* of the cell for the duration of an infection. Still, this was only one of the clues as to what might be the basis of the eclipse period. Chemical analysis of the T-even bacteriophages, carried out earlier by Schlesinger, demonstrated that these viruses are composed of roughly equal amounts of DNA and protein. A key question

is the topological relationship between the DNA and the protein in the phage particles. Anderson provided the answer in 1949, when he found that he could rupture the phage heads by osmotic shock (i.e., by sudden dilution of a concentrated phage suspension in high salt into distilled water), causing release of the DNA (Fig. 1.5). The electron micrographs of the ruptured phages clearly demonstrated that the phage heads are comprised of an outer protein shell that encloses the phage DNA within it.

So now, here is the conundrum. If a phage is comprised of this DNA-containing protein shell that appears to remain outside the cell during infection, and if the nucleic acid is the genetic material as implied by the transformation experiments, how might that DNA participate in the replication of the phage? Or if the protein were the genetic material, as still believed by many, how might it do the

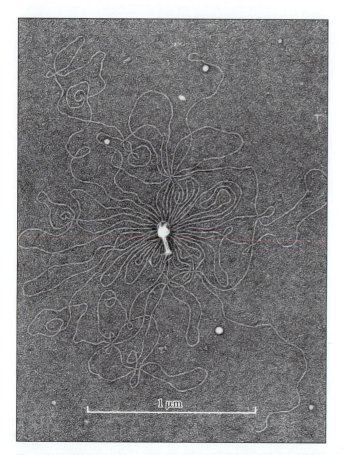

Figure 1.5 Electron micrograph showing the DNA molecule released from a T-even phage particle by osmotic shock. Contour measurements show that the DNA is a single continuous molecule that is about 50 µm long, about 550 times longer than the width of the phage head that contains it. From A. K. Kleinschmidt, D. Lang, D. Jacherts, and R. K. Zahn, *Biochim. Biophys. Acta* **61**:857–864, 1962, with permission.

same? Or put another way, what constitutes the infection, and what does this have to do with the eclipse period? Anderson provided some additional valuable information that would be important in getting to the answers. He suspected that the attachment of the phage to the bacteria via their slender tails should be unstable. He demonstrated that this was the case by shearing the phage off the bacteria in a Waring kitchen blender. Now the stage was set for another one of the most famous and significant experiments in the history of virology and molecular biology as well, the Hershey-Chase experiment.

The above experimental findings provided the impetus for Alfred D. Hershey and Martha Chase to carry out their famous experiment in 1952 at Cold Spring Harbor, Long Island. In this experiment, Hershey and Chase were able to follow the separate fates of the protein and DNA components of the T2 phage during infection. To do this, they first prepared phage stocks in which either the protein or

the DNA of the phage was radioactively labeled. This was accomplished by infecting bacteria that were being cultivated in the presence of either ^{35}S or ^{32}P. During T2 replication in the presence of the isotopes, the bacteriophage incorporated the radioactive ^{35}S or ^{32}P into their protein or DNA, respectively. Each preparation of labeled phage was then allowed to attack fresh bacteria in the absence of any isotopes. After a short period to allow phage to attach to the cells, unattached phage was removed by low-speed centrifugation of the cells and washing of the pellet. The cells then were resuspended and subjected to the shearing forces of the Waring blender, which previously had been shown to strip the phage heads from the cells. The experiment yielded two results of immense significance. First, 80% of the ^{35}S-labeled phage protein could be stripped off the bacteria, whereas essentially all the ^{32}P-labeled phage DNA remained with the cells. Moreover, the phage DNA apparently entered into the cells, since it was resistant to added DNase. Second, despite being subjected to the Waring blender, the bacteria produced progeny phage just as though they had not been blended. Thus, stripping the phage protein off the infected bacteria did not prevent the production of a normal crop of progeny phage. That is, once the phage DNA has entered the cell, the parental phage protein has no further function. Hershey and Chase provided additional evidence that the phage "injects" its DNA into the cells. In particular, they demonstrated that the phage could attach to bacterial cell wall fragments in suspension and then release its DNA into the medium on the other side of the wall fragments. As expected, once released in this way, the phage DNA was sensitive to digestion by DNase.

The singularly important conclusion that was strongly implied by the Hershey-Chase experiment is that the genetic information of the phage is embodied in its DNA. This conclusion is completely consistent with the implication of the transformation experiments conducted a decade earlier by Avery, MacLeod, and McCarty. The results of the Hershey-Chase experiment also explain the eclipse period, since the "naked" phage DNA that initially enters the host cell is not capable of initiating a plaque. That is, no PFU are present in the infected cells until the appearance of the first progeny infectious phage particles, in which the DNA is packaged into a protein coat capable of initiating infection, thereby marking the end of the eclipse period.

Importantly, for a stage in the life cycle of T2 phage, all that exists to connect one generation of phage with another is *a molecule of DNA*. This is certainly in marked contrast to the life cycle of cells, which grow larger and then divide. Thus, the eclipse period truly sets phages apart from cellular life forms in a fundamental way.

We still have to describe the events that occur during the eclipse period, as well as the mechanism of entry, but that will come in the next chapter. At this point it suffices to say that the phage DNA will somehow oversee the production of the proteins from which new phage particles can be assembled.

Shortly before Hershey and Chase carried out their classic experiment, Lwoff and coworkers showed that some phages do not necessarily replicate in, and lyse, their host cells. Instead, infection by these phages may lead to an outcome in which the phage genome is maintained in a stable state in the host cell, from one cell generation to the next. Later, it was found that in some instances the phage genome is maintained by being integrated into the cellular genome. In other instances, it is maintained as a **plasmid** (a DNA molecule that can replicate independently of the cell's chromosomal DNA). Phages capable of this "lifestyle" are said to be **temperate**, and their stable relationship with the host cell is referred to as **lysogeny**. Temperate phages are able to assume the lysogenic state because they have evolved mechanisms that enable them to control their replicative functions. That is, they are able to express only a subset of their genes when in the lysogenic state: specifically, those genes that are necessary to initiate and maintain the lysogenic state. However, as demonstrated by Lwoff, temperate phages may be induced to come out of the lysogenic state and lyse their host cells (Box 1.9).

Some animal viruses too are able to assume a relationship with the host cell that superficially resembles lysogeny. Chapter 2 contains a more detailed account of the molecular mechanisms used by bacteriophages to establish and maintain lysogeny. The mechanisms used by animal viruses to initiate and maintain latency are considered in later chapters.

We have essentially completed this part of our history of virology, but several items of related interest still need to be noted. First, the results of the Hershey-Chase experiment convinced all but the greatest skeptics that DNA indeed is the genetic material. Moreover, this experiment was a key impetus that drove Watson and Crick to solve the structure of DNA, which they indeed achieved in the next several months. Second, we must note the important fact that TMV, as well as a large number of other viruses (including such favorites as human immunodeficiency virus, influenza virus, and Ebola virus), contain RNA rather than the chemically related molecule DNA. That is, for these viruses the RNA portion alone serves as the genetic material. This was convincingly demonstrated for the first time by an experiment of Heinz Fraenkel-Conrat and Gerhard Schramm, in which they separated the RNA and protein of TMV (by shaking the particles in phenol and water; protein partitions into the phenol, and the nucleic acid

remains in the aqueous phase). When applied to wounded tobacco leaves, the RNA, but not the protein, produced a crop of complete virus particles. Thus, whereas DNA serves as the genetic material for all cellular organisms, RNA serves as the genetic material for many viruses.

BOX 1.9

The work of Victor Freeman in 1951 at the University of Washington, as well as the work of others, led to the realization that the deadly toxins produced by *Corynebacterium diphtheriae* and *Clostridium botulinum* indeed are the products of lysogenic phages carried by these bacteria. A particularly relevant experimental finding was that avirulent strains of these bacteria became virulent when infected with phages that were induced from virulent strains. In addition to carrying genes for toxins, temperate phages also may carry genes for antibiotic resistance (e.g., penicillinase). So then, are diphtheria and botulism due to phages or to bacteria?

It also is interesting to consider the origin of the term "lysogeny." It was as follows. Some bacterial strains, which were not knowingly infected with any virus, would occasionally undergo spontaneous lysis. When doing so, these bacteria would release bacteriophages. Thus, these often cryptic bacteriophages were said to be lysogenic, reflecting their ability to lyse their host cells. Consequently, it is a bit odd that their temperate relationship with their host cells is referred to as "lysogeny."

DEFINING VIRUSES

Bearing in mind the characteristics of viruses revealed by our historical survey, we return to a question posed earlier: how might we define viruses? The essential properties of viruses that were revealed by our historical account include the following. (i) Viruses are smaller than bacteria and pass through filters that block the passage of bacteria; (ii) viruses can replicate only within suitable host cells; (iii) viruses may be comprised only of protein and nucleic acid; (iv) the nucleic acid component of many viruses is RNA; (v) the nucleic acid of the virus contains its genetic information; (vi) during a stage of its replicative cycle the virus may exist only in its nucleic acid; and (vii) given the small size of viruses and their simple structure (i.e., they can be crystallized) relative to cells, we rightly would expect that they are dependent on their host cells' metabolism to provide energy and raw materials needed for their replication.

A compendium of virus characteristics is not a definition per se. Still, various attempts to define viruses have emphasized different combinations of these fundamental characteristics. Although I believe it is more important to know the distinctive features of viruses than to define them, a definition that works fairly well is the following, quoted from an earlier, superb text by S. E. Luria, J. E. Darnell, D. Baltimore, and A. Campbell (*General Virology*, 3rd ed., John Wiley & Sons, New York, 1978):

> Viruses are entities whose genomes are elements of nucleic acid that replicate inside living cells using the cellular synthetic machinery and causing the synthesis of specialized elements that can transfer the viral genome to other cells.

This definition emphasizes the facts that viruses possess genetic elements that make use of resources inside their host cells to generate their progeny and that the virus particle exists to transfer the virus genome into other cells. However, even this definition omits some of the key points enumerated above, such as the small size of viruses, the fact that viruses may have RNA genomes, and the unique eclipse period (Box 1.10).

ARE VIRUSES ALIVE?

A favorite question of students is whether viruses are alive. I feel the same about this question as I do about attempts to define viruses. That is, it is much more important to understand the nature of viruses than it is to debate whether or not they are alive. In fact, the question of whether viruses are alive is neither profound nor a matter of wonderment. Instead, it really comes down to a matter of definitions. For example, although we readily appreciate that cats and dogs and frogs and even bacteria are organisms and alive, it is not easy to state in any concise way what it is about them that makes them so. Thus, depending on our rather arbitrary definition of "alive" or "organism," we might, or might not, include viruses. For example, by some accounts, a virus is no more an organism or alive than a chromosome. The reason given is that viruses, like chromosomes, are dependent on a host cell for their replication. Thus, by the criterion of independence, viruses do not qualify as organisms. Yet by that criterion, the chlamydiae, which are obligate intracellular bacterial parasites, would likewise also not be considered to be alive. So, given the somewhat arbitrary criteria (and their interpretations) which underlie our definitions for organisms and viruses, it is much more important to appreciate what viruses are than to try to definitively answer the question of whether they are alive.

For me, the wonderment of viruses comes from considering the many different adaptations that these entities have evolved, despite having only a handful of genes in many cases, to solve the numerous problems they face in successfully exploiting their hosts. Indeed, this point helps me to personally answer the question we started with.

BOX 1.10

Didier Raoult and colleagues recently identified a previously unknown virus that challenges some of our notions of what a virus ought to be. Their discovery resulted from efforts to identify intracellular *bacteria* that might be associated with amoebae. In this regard, some important human bacterial pathogens, including *Legionella*, *Chlamydia*, *Mycobacterium*, *Shigella*, *Rickettsia*, and *Listeria*, have all adapted an intracellular lifestyle. Strangely, one amoeba-associated putative bacterium, initially classified as belonging to the *Legionella*-like amoebic pathogens, resisted all efforts to amplify its 16S ribosomal RNA (rRNA). Believing that this failure was due to an inability to digest the "organism's" cell wall, Raoult and coworkers examined it by electron microscopy, doing so before and after carrying out procedures that would have extracted the putative cell wall. To their surprise, what they saw was a giant spiked icosahedral virus-like particle (see Chapter 2). Next, they found that it contained a double-stranded DNA genome of 1,184 megabases, which is huge in comparison to that of any other known virus.

Indeed, it is larger than the genomes of at least 25 known bacteria! Consistent with its huge genome, particles of this virus contain some 438 different proteins. This is equivalent to the number of proteins associated with some smaller bacteria, such as *Rickettsia*. Moreover, this virus contains genes that were not previously seen in viruses. These genes include four that encode transfer RNA (tRNA) synthetases, five involved in DNA repair, and three chaperones. Yet the organism indeed is a virus. It lacks ribosomal protein-encoding genes and does not appear to contain any enzymes for energy metabolism. Most importantly, it does not undergo binary fission but instead demonstrates typical eclipse phase viral replication. It was named "mimivirus," which is short for "mimicking microbe." See Box 19.4 for an interesting update on mimivirus.

Reference
Raoult, D., S. Audic, C. Robert, C. Abergel, P. Renesto, H. Ogata, B. La Scola, M. Suzan, and J. M. Claverie. 2004. The 1.2-megabase genome sequence of Mimivirus. *Science* **306**:1344–1350.

As Luria, Darnell, Baltimore, and Campbell (1978) note in their book referenced above, a definition for an organism might emphasize its historical continuity, rather than its functional independence. For example, *"An organism is the unit of a continuous lineage with a continuous evolutionary history."* In that sense, viruses are most certainly organisms and alive. A chromosome, by contrast, is a constant component of a cell. It evolves in the context of the cell, which in turn may be part of a multicellular organism and the actual unit of evolution in that instance. (But as we soon will see, viruses and their hosts evolve together. For instance, in the case of animal viruses, there is a continuing "evolutionary dance" between the immune defenses of the host and the ability of the virus to evolve countermeasures to those defenses [Chapter 4]. Does that notion support or detract from the view that viruses are alive?)

ORIGIN OF VIRUSES

Unfortunately, we cannot know the evolutionary history of viruses, since viruses leave no fossils. Yet viruses may have existed since the origin of cells. That assertion is consistent with the fact that viruses seem to infect all classes of cellular organisms, from the simplest bacteria to the most complex metazoans.

We may hypothesize about the origin of viruses, provided we remain aware that these are just guesses. With that proviso in mind, a likely possibility is that at least some viruses arose from the jumping genes and replicating extrachromosomal nucleic acids that exist in cells. Those entities may have become viruses upon acquiring the ability to generate coats that would facilitate their cell-to-cell transfer and enable their extracellular survival. At that point they would embark on their independent evolutionary courses. Such a sequence of events might have occurred independently on multiple occasions.

Another possibility is that some viruses may have arisen from intracellular cellular parasites, which progressively acquired the characteristics of viruses as they became more dependent on their host cells. Regarding this possibility, see the discussion of the mimivirus in Box 1.10.

The multiple plausible possibilities suggest that some virus families have no ancestors that they share with other virus families. That is, entities that we now recognize as viruses may have arisen independently over the years. Apropos that, when we compare the patterns of gene arrangement and expression of the different families of viruses, we will appreciate much better the great differences among viruses. On the other hand, we should note that even seemingly unrelated viruses might show signs of a common ancestor. For example, a variety of so-called plus-strand RNA viruses, of dissimilar genetic sequences, contain an RNA polymerase gene that is highly conserved among them. Since varieties of these viruses are found in plants, as well as across the animal kingdom, they may have evolved in their respective hosts from an ancestral virus that was present before the separation of kingdoms. Alternatively, the ancestral virus might have evolved to spread to new hosts. We do not know.

THE MODERN ERA OF ANIMAL VIROLOGY

Recapping some of the history discussed above, the discoveries of filterable infectious agents beginning at the end of the 19th century and continuing for the next several decades were motivated by the desire to identify the etiologic agents of particular infectious diseases in humans, domestic animals, and plants. In contrast, progress towards actually understanding the nature of viruses involved mainly studies of bacteriophages and was driven primarily by the desire of the early phage workers to understand the physical basis for heredity.

Why then was there little concurrent advancement towards understanding the biology of viruses that were of medical, veterinary, and agricultural importance? Progress towards understanding phages was possible largely because of the development of the plaque assay, which enabled phage researchers to carry out quantitative studies in a simple and easy-to-manipulate system. In contrast, progress towards understanding the biology of animal viruses during those early years was hampered by the fact that there was no comparable simple experimental system in which to study or quantify them. Although some animal viruses could be grown in embryonated eggs (e.g., influenza virus), the major experimental system of the time was the whole animal.

Several major breakthroughs, beginning in the late 1940s, dramatically changed this state of affairs. First, John Enders demonstrated that poliovirus could be grown to a high titer in explanted animal cells. Then, in 1952, Renato Dulbecco went a major step further when he developed a plaque assay using cultured cells, which he used to accurately quantify Western equine encephalitis virus and poliovirus (Fig. 1.6; Box 1.11).

The plaque assay technique for animal viruses is similar in principle to the procedure for bacteriophages. However, whereas bacterial cells grow on top of, or within, agar-containing media, animal cells in culture generally must attach to a solid surface in order to grow. When adhering to the bottom of the petri dish, under appropriate nutrient media, animal cells can form a continuous monolayer of cells. With those points in mind, the plaque assay

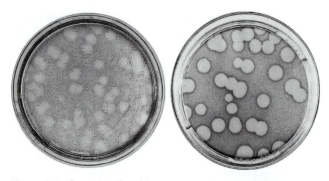

Figure 1.6 Plaques produced by Western equine encephalitis virus on chick embryo fibroblasts (left) and by poliovirus on HeLa cells, a line of cells derived from a human cervical carcinoma (right). Photo by R. Dulbecco; reprinted from S. E. Luria, J. J. Darnell, D. Baltimore, and A. Campbell, *General Virology*, 3rd ed. (John Wiley & Sons, New York, NY, 1978), with permission.

procedure for animal viruses begins by removing the medium from the cell monolayer and then adding a small aliquot of the virus sample to the cells. The virus is then allowed to attach to the cells for as long as 1 h, and the monolayer is then covered with a nutrient medium-containing agar overlay. The purpose of the agar is to prevent the diffusion of progeny virus beyond immediately adjacent cells. Plaques that develop can generally be visualized by the addition of a vital dye that is concentrated only by viable cells. These developments brought about the emergence of quantitative animal virology.

Importantly, and not mentioned before, all the viruses in a plaque are descended from a single virus. This fact enabled the isolation of virus mutants, thereby facilitating the genetic analysis of animal viruses, as well as phages. Moreover, it enabled the isolation of avirulent virus mutants for use as vaccines.

Important related developments included the establishment of continuous cell cultures by Earle in 1948 and the development by Eagle in 1960 of simple defined media that would support animal cell growth (i.e., not merely survival) in culture. Together, these developments made possible the quantitative study of animal viruses under carefully controlled conditions. The development of cell culture also made possible the large-scale industrial production of vaccines, including those for poliovirus, rubella, and measles virus.

Thus ends one chapter in our selective history. However, note that the elucidation of virus replication, as recounted thus far, has not progressed beyond establishing the parameters of one-step growth. The story continues in the next chapter, where we consider the biochemical and molecular events that occur during the eclipse period and the steps that go into generating progeny virus particles. We end the next chapter with a brief consideration of the variety of viruses and how we might sort them into families based upon the organization and expression of their genomes.

BOX 1.11

Phage and the Origins of Molecular Biology (J. Cairns, G. S. Stent, and J. D. Watson [ed.], 1966) contains a collection of essays written in tribute to Max Delbrück by some of the pioneers of molecular biology and virology who came under his influence. Renato Dulbecco was one of those scientists, and his essay relates how he came to work on animal viruses. It happened as follows.

In the late 1940s, a wealthy Californian became ill with shingles (a late complication of chicken pox, actually caused by a herpesvirus, varicella-zoster virus; see Chapter 19). The man's physician explained that nothing could be done for his shingles, and moreover, that virtually nothing was known about the viruses that infect humans. Auspiciously, the physician knew of the studies being done on bacteriophages at Caltech, and he also was aware that Caltech was *the* great center for such work. He explained to his patient that bacteriophages were only of theoretical interest regarding human disease, but he also suggested that the patient might help to develop a center at Caltech that might begin to study medically important viruses. The patient agreed, and since virology at Caltech was headed by Delbrück, the former physicist found himself with an endowment to study human viruses, with no knowledge for how to use it. Well, Delbrück summoned Dulbecco, an Italian physician who joined the Caltech phage group to learn virology, to his office, and he proposed that Dulbecco give animal viruses a try. Dulbecco was delighted by the idea, and the first of his significant accomplishments (for which he later was awarded a Nobel Prize) was to develop a plaque assay procedure for animal viruses. Thus, quantitative animal virology came to be.

Dulbecco remarks on the importance of the development of the plaque assay for animal viruses, noting that subsequent biochemical and molecular studies would have been much less meaningful without reference to the multiplication phase revealed by the plaque assay. Interestingly, although this was apparent to the phage workers, it was not equally obvious to most animal virologists of the day.

Finally, before leaving the present chapter, we note that the study of animal viruses will come to incorporate, as well as lead to the development of, cutting-edge molecular approaches. Moreover, these molecular approaches, as well as studies of animal viruses per se, will come to have a dramatic impact not only on our understanding of viruses but also on our understanding of eukaryotic cell and molecular biology as well. Research areas that will be impacted include eukaryotic gene organization, expression, and replication; intracellular signaling; endocytosis; and intracellular transport. Virus-centered studies also will come to have a major impact on our understanding of the regulation of cell growth and the molecular events that can lead to unregulated cell growth and cancer.

Suggested Readings

Cairns, J., G. S. Stent, and J. D. Watson (ed.). 2007. *Phage and the Origins of Molecular Biology: the Centennial Edition.* Cold Spring Harbor Laboratory Press, Cold Spring Harbor, NY. (This is the latest edition of the original 1966 publication.)

Girard, M. 1988. The Pasteur Institute's contributions to the fields of virology. *Annu. Rev. Microbiol.* **42:**745–763.

Luria, S. E., J. E. Darnell, D. Baltimore, and A. Campbell. 1978. *General Virology,* 3rd ed. John Wiley & Sons, New York, NY.

2

Biosynthesis of Viruses: an Introduction to Virus Classification

T-EVEN BACTERIOPHAGES AS A MODEL SYSTEM

One purpose of this chapter is to examine the biosynthetic events that occur during the eclipse period of a virus's replicative cycle. For historical reasons, and for the sake of continuity, we continue using the T-even phages as our model system (Box 2.1). In doing so, we come upon additional instances of virus-based research that led to breakthroughs that were of fundamental importance to biology in general. Since a major purpose here is to learn about features of viral replication that are rather general, we postpone until later in the chapter examining some aspects of T-even phages that are unique to these viruses.

Also, we consider in some detail the means by which the temperate bacteriophage λ (lambda) is able to follow either of two courses upon injecting its DNA into a host cell; one course leads to virus replication and cell lysis, and the other leads to lysogeny, which is the kind of latent infection manifested by temperate bacteriophages. Latent infection also occurs among the animal viruses. Indeed, it is a defining characteristic of infections involving the herpesviruses. However, the establishment and maintenance of viral latency are best understood in the case of λ phage.

The chapter concludes with an introduction to the animal viruses and a rational basis for the classification of animal viruses into families. But first, we get on with our consideration of the T-even phages.

The T-even phages were the best-studied model system of the day (the 1940s). Yet their choice as the model virus was largely a matter of chance. Although they were, at the time, thought to be the simplest possible organisms, the T-even phages are, in fact, among the largest, most complex viruses that are known, containing about 160 genes. In contrast, there are viruses that contain as few as five genes. Ideally, one would rather have simple model systems. Yet, coincident with their complexity, the T-even phages were found to have an unexpected feature that would lead to important new insights. That feature was the presence in their DNA of the unusual base hydroxymethylcytosine (HMC) in place of cytosine (see the chemical structure of HMC, below).

BOX 2.1

Where did the T-even phages come from, and what is the significance of their names? We may never know the exact origin of these viruses, but the following is known. In the late 1930s, T. L. Rakieten presented M. Demerec and U. Fano with either a mixture of raw sewerage or perhaps a lysate from *E. coli* infected with raw sewerage. These two researchers then isolated T3, T4, T5, and T6, all of which are active in *E. coli*. The precise origin of T2 phage, the most widely studied of the T phages, is less clear, although there is evidence that a researcher named J. Bronfenbrenner may have been working with T2 phage as early as 1932, and indeed Bronfenbrenner may have initially isolated the virus. If that is the case, then it is likely that the isolation was made from fecal material rather than from sewerage, as that appears to have been Bronfenbrenner's practice. At any rate, Delbruck named these *E. coli* phages Type 1 (T1), Type 2 (T2), Type 3 (T3), etc., for easy reference. Thus, the T-even phages are T2, T4, and T6, as opposed to T1, T3, T5, and T7.

Members of the T series of phages differ in size and shape. But, fortuitously (regarding their grouping as the T-even phages), phages T2, T4, and T6 are structurally similar, containing an extra row of subunits, which distorts the simple icosahedral structure. In addition, they contain similar tail structures and fibers, and they are unique in that they contain the modified base 5-HMC instead of cytosine (see the text). Moreover, they are related serologically and all have large genomes. Thus, it is *biologically* meaningful to refer to this group of viruses as the T-even phages.

What then of the T-odd phages? These phages fall into three serological groups, in which phages T3 and T7 are related, but phages T1 and T5 are each distinct. Phage T7, like the T-even phages, takes control of transcription in the infected cell, thereby converting the cell into a factory dedicated to producing progeny phages. But whereas the T-even phages modify the host RNA polymerase to recognize phage-specific promoter sequences (see the text), T7 encodes its own T7 promoter-specific RNA polymerase, while also inactivating the host RNA polymerase. Because of the specificity of the T7 RNA polymerase for T7 promoters, the T7 polymerase and promoters have been incorporated into a number of cloning/expression vectors used in recombinant DNA technology. T7 also encodes its own DNA polymerase, a modified form of which is marketed as the popular sequencing enzyme Sequenase. Phage T1, which is not related to any of the other T phages, remains the least studied of the group.

The T-even phages, together with bacteriophage λ (see the text), have been the most intensely studied of the bacteriophages. While the T-even phages played a key role in the development of virology and molecular biology (as recounted in Chapter 1 and the present chapter), bacteriophage λ served as the prototype of the lysogenic (or temperate) phages. Intense research on the T-even phages was largely due to the influence of Delbruck, who in 1944 urged the phage group to focus its research on the T series of phages. Delbruck's rationale was to get all the phage researchers to work on the same small group of phages, so that they might more easily compare their experimental findings, in that way facilitating progress. This recommendation of Delbruck was considered correct at the time but has since been criticized because it led to the early neglect of some other key issues, such as lysogeny. In that regard, none of the T series of bacteriophages are temperate (see the text).

In addition, the HMC residues on the T-even phage DNA are glucosylated in a specific pattern.

Since this form of modified DNA does not exist in uninfected cells, it was reasonable to infer that the phages somehow are responsible for generating it. Taking this argument a step further, one might hypothesize that these phages cause the formation of new enzymes that previously did not exist in the cell. Indeed, Seymour Cohen in the 1950s demonstrated the presence of deoxycytidylate hydroxymethylase, and other new enzymes, in T-even phage-infected cells.

These new enzymes in fact are encoded by phage genes, as was demonstrated at the time by the isolation of phage mutants that are defective for their expression. While the finding that phages encode enzymes might not appear to be particularly remarkable today, at the time of this discovery it was quite striking. To appreciate why this was so, we need to appreciate how little was known at that time about viruses, and of molecular biology in general. In this regard, recall where we are in our historical account of virology and molecular biology. Indeed, in our historical account of virology and molecular biology, we are only slightly past the point at which DNA was finally acknowledged to be the genetic material and Watson and Crick had solved its structure. There was little else.

The modified T-even phage DNA helped to advance the field in yet another important way. Its hydroxymethylated glucosylated DNA could be distinguished from host cell DNA, thereby making it possible to follow phage DNA replication during infection of the host cell.

T-EVEN PHAGE STRUCTURE AND ENTRY

Next, we briefly revisit the structure of T-even phage particles to describe in somewhat more detail how phage DNA enters into the cell. We recall from Chapter 1 that electron micrographs of the T-even phages showed that they are tadpole shaped, with distinct heads and tails. In addition, they appeared to attach to the outside of their host cells by their slender tails.

A more careful examination of T-even phage structure shows the presence of tail fibers, which actually mediate attachment of the phage to the host cell (Fig. 1.4, 2.1, and 2.2). Experimental proof that the tail fibers mediate phage attachment was obtained by isolating phage mutants unable to generate tail fibers and then demonstrating that these mutants are not able to attach. (A puzzle: if the phage mutant is defective for attachment and infection, how

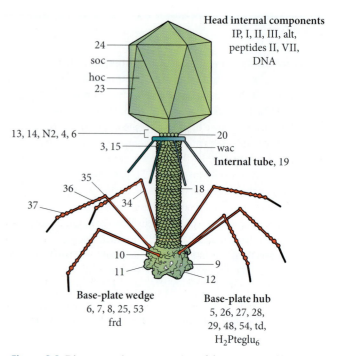

Figure 2.2 Diagrammatic representation of the structure of bacteriophage T4, with the locations of the known viral structural proteins. Compare this to Fig. 2.1. Contributed by F. Eiserling; modified from J. A. Levy, H. Fraenkel-Conrat, and R. A. Owens, *Virology*, 3rd ed. (Prentice Hall, Englewood Cliffs, NJ, 1994).

Figure 2.1 Electron micrograph of a T2 phage. Note the polyhedral capsid, the neck, the contractile sheath, the base plate, and the tail fibers. Compare this to Fig. 2.2. Photo by R. C. Williams; reprinted from J. A. Levy, H. Fraenkel-Conrat, and R. A. Owens, *Virology*, 3rd ed. (Prentice Hall, Englewood Cliffs, NJ, 1994).

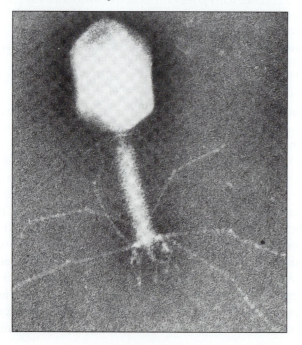

can it be isolated and propagated for further study? The answer is that the mutant is conditionally defective. Most such conditionally defective mutants contain temperature-sensitive lesions. Although these mutants express their phenotype at their restrictive temperature, they can be propagated at their permissive temperature.) In addition, attachment was impaired by antibodies reactive against the tips of the tail fibers, but not by antibodies reactive against other sites on the virus particle.

The phage tail fibers attach specifically to lipopolysaccharide receptor molecules on the outer envelope of the host cell. This binding event triggers a conformational change or bending of the tail fibers that brings the base plate at the end of the tail into contact with the cell surface (Fig. 2.3). Ensuing conformational changes then occur in the base plate and in the tail. Note that the tail is comprised of an outer sheath that surrounds a central tube (Fig. 2.2). The outer sheath of the tail contracts, and the hole in the center of the base plate enlarges. The contraction of the sheath drives the tube through the enlarged hole in the base plate, and ultimately through the cell wall. The process is likely facilitated by a viral enzyme (a lysozyme), built into the phage tail, which degrades the cell wall from the outside. The phage DNA is now able to pass from the phage head through the tube and into the cell (Box 2.2).

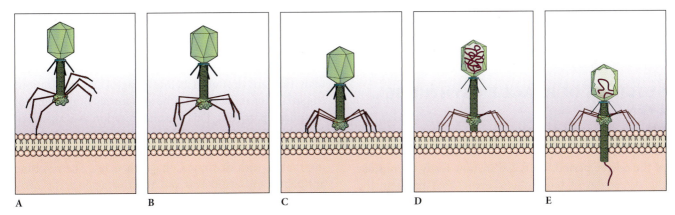

A   B   C   D   E

Figure 2.3 Diagrammatic representation of bacteriophage T4 attaching to a bacterium and "injecting" its DNA into the cell. (A and B) The phage tail fibers specifically bind to receptor molecules on the outer envelope of the bacterium. (C) This triggers a conformational change or bending of the tail fibers, which brings the base plate at the end of the phage tail into contact with the bacterial surface. (D and E) Here the tail is shown comprised of an outer sheath that surrounds a central tube. Conformational changes occur in the base plate and in the tail. The outer sheath of the tail contracts, and the hole in the center of the base plate enlarges. The contraction of the sheath drives the tube through the enlarged hole in the base plate and ultimately through the cell wall. A phage lysozyme digests away the cell below the phage, enabling the phage DNA to pass from the phage head through the tube and into the cell.

BOX 2.2

What force might drive the phage DNA from the phage head into the cell? Ever since Hershey and Chase showed that T2 phage inserts its DNA into the cell, it has been assumed that the pressure exerted by the DNA within the capsid (now known to be as much as 60 atmospheres) provides the driving force for injection and release of the DNA. Comparable pressures should exist within the capsids of other phages as well. However, consider that during bacteriophage assembly, phage genomes generally are inserted into preassembled empty capsids (see the text). Next, consider the following experimental finding from studies involving the encapsidation of phage φ29. Half of the φ29 genome can be encapsidated before substantial forces that resist further packaging become evident. Consequently, during the reverse process of DNA release, internal forces within the capsid should have largely dissipated after only about one-half of the genome has been released. Also, bear in mind that release of the phage genome involves passage through a head-tail connector and, in many instances, through a long tail as well. What then drives the complete release of the viral genome? The answer is not known in most cases. But studies of phage T7 show that the majority of its genome is actually drawn from the capsid by transcription. The first portion of the T7 genome, about 850 bp, enters the cell by a transcription-independent process, perhaps driven by pressure exerted by DNA within the capsid. Then, about 7 kilobases (kb) of DNA is "drawn out" by transcription catalyzed by the *E. coli* RNA polymerase, and the following 32 kb is drawn out by transcription catalyzed by the T7 RNA polymerase. The initial transcription-independent translocation phase exposes the gene encoding T7 RNA polymerase, enabling the phage enzyme to be produced. Phage N4 also is known to harness transcription to facilitate its entry. However, we can be rather certain that there are a variety of strategies used by different bacteriophages to release their genomes into the cell.

The mechanism by which T-even phages deliver their DNA into host cells highlights the fact that virus particles indeed may be complex molecular machines, rather than mere inert chemical configurations (Box 2.2). Examining the structure of T-even phage particles in somewhat more detail and noting the number of their distinct protein components underscore this point (Fig. 2.2). For example, consider the particle's distinct head, collar, sheath (which encloses an internal tube), base plate, and tail fibers. The head, collar, and base plate each are assembled from multiple copies of several individual proteins. Each of the tail fibers is also composed of several different proteins. The remarkable outer sheath is assembled from 144 molecules of a protein that is arranged in 24 rings. When the tail fibers attach to the cell, a change in angle between their components stimulates subsequent molecular reorganization, including contraction of the sheath, a process that involves a complex rearrangement of the sheath proteins from the initial 24 rings into 12 rings.

SEQUENCE OF PHAGE BIOSYNTHETIC EVENTS

Phage Protein Synthesis

Transport of the phage DNA into the cell marks the start of the eclipse period. Thus, we are now ready to consider the succession of biosynthetic events that result in the production of progeny phage (Fig. 2.4). We begin by noting the marked increase in protein synthesis that begins almost immediately upon entry of the phage DNA. Fortuitously (for the experimenter), the phage shuts down host protein synthesis soon after DNA entry. Thus, the protein that accumulates in the infected cells is virus specific. Not all viruses shut down host protein synthesis. Nevertheless, the redirection of protein synthesis from cellular products to phage products in this instance points to a fundamental feature of viral parasitism: the subversion by the virus of the host's biosynthetic machinery. Also, notice that a

Figure 2.4 Schematic representation of the pattern of biosynthesis in *E. coli* infected with bacteriophage T4. The broken lines in the bottom part of the figure depict the pattern of phage early protein synthesis in bacteria infected with conditional phage mutants that are unable to initiate phage DNA synthesis. "Virion" refers to a virus particle. Adapted from S. E. Luria, J. J. Darnell, D. Baltimore, and A. Campbell, *General Virology*, 3rd ed. (John Wiley & Sons, New York, NY, 1978), with permission.

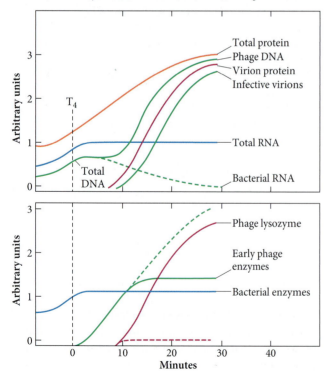

considerable amount of this phage-specific protein synthesis occurs before phage DNA replication commences. These points will be considered in more detail shortly.

What might be the functions of the early phage proteins that are produced before phage DNA replication? Might any of these proteins be structural components of the phage particles? This specific question was answered by first preparing antiserum against the proteins that constitute phage particles, which was done by simply injecting rabbits with disrupted phage particles. The resulting antiphage antiserum was not reactive against the early phage proteins, thus showing that those phage-encoded proteins that are made prior to phage DNA synthesis are not components of phage particles. In contrast, the "late" phage proteins, made after the phage DNA replicates, are reactive with the antiphage antiserum, indicating that they include structural components of phage particles. These experimental findings demonstrate that there are two functionally distinct classes of T-even phage proteins, whose expression is separated in time, and which are referred to as early and late proteins. Stated somewhat differently, there is a program of regulated phage gene expression. These points will be considered in more detail shortly. First, we need to answer the question we began with: what is the function of the early proteins?

The HMC in the phage DNA provides a clue as to the function of the early phage proteins. As argued above, since HMC does not exist in uninfected host cells, its presence in infected cells implies the existence of phage-encoded enzymes that might function in phage nucleic acid metabolism. Seymour Cohen and Arthur Kornberg demonstrated that indeed this is the case, showing that a group of different phage-encoded enzymes need to be made before phage DNA synthesis can begin. These 1950s experiments were based on genetic analysis. (Later experiments, done by others, involved in vitro translation systems prepared from the *Escherichia coli* host cells, in which protein synthesis was directed by added phage-specific messenger RNA [mRNA] that had been transcribed in vitro. However, from our historical perspective, mRNA still remains to be discovered, so these experiments cannot yet be a part of our story.)

The early phage proteins include (i) enzymes that catalyze the synthesis and glucosylation of HMC, (ii) enzymes that destroy precursors of cytosine deoxynucleotides, (iii) enzymes that synthesize thymine by a new pathway, (iv) enzymes that degrade host cell DNA to make its nucleotides available for phage DNA synthesis, and (v) a phage-specific DNA polymerase. The roles of some of these enzymes in phage DNA metabolism will be considered in more detail shortly. We briefly note here that they enable the phage to block any further cellular DNA synthesis and, moreover, they enable the phage to degrade cellular DNA. The nucleotides that are released from the degraded cellular DNA are then available to the phage to use as building blocks for phage DNA replication (Box 2.3). Yet the two essential points for the moment are, first, that the phage DNA directs protein synthesis before the phage DNA replicates and, second, that these early phage-specific proteins are not structural components of phage particles (virions) but rather are enzymes.

Once phage DNA has been replicated, phage structural component proteins then begin to appear (Fig. 2.4). Another important feature of the phage replication cycle is that these phage structural components appear before the emergence of progeny phage particles. The significance of

this fact is considered later in the chapter. For now, the appearance of intracellular progeny phage marks the end of the eclipse period and the onset of the intracellular rise period.

RNA Metabolism in Infected Cells

We now turn our attention to the pattern of RNA metabolism in T-even-phage-infected cells. To appreciate what will next unfold, note that scientists of the late 1950s still knew virtually nothing about the role of RNA. If we can imagine for the moment that we too know virtually nothing about the role of RNA, then we can better appreciate how the study of phage-infected cells led to breakthroughs of profound significance to virology and to biology in general.

BOX 2.3

What follows is an example of the T-even phages serving to test a hypothesis of paramount importance. Watson and Crick in 1953, in their very short paper reporting the structure of DNA, succinctly stated that the DNA structure in fact suggested a mechanism for its replication. Interestingly, they did not elaborate on what that mechanism might be. But what they certainly had in mind was a mechanism in which the two strands of the parental double helix might separate, with each strand then serving as a template for the formation of a new complementary strand, assembled from a pool of purine and pyrimidine bases. This process is referred to as semiconservative replication, since each daughter DNA molecule would contain one old and one new polynucleotide chain.

Well, regardless of how attractive this hypothesis might have been, it still needed to be verified experimentally before it could be accepted. In a classic and elegant experiment that was inspired in part by the finding that T-even phages shut down cellular macromolecular synthesis, Matthew Meselson and Franklin Stahl in 1958 used T-even phages to demonstrate that DNA replication in fact is semiconservative. Their experiment made use of the fact that the rare isotopes ^{15}N and ^{13}C are heavier than the common isotopes ^{14}N and ^{12}C. Thus, DNA that incorporated the heavier isotopes would be denser than DNA containing the lighter isotopes. Moreover, "heavy" DNA could be separated from "light" DNA by equilibrium density-gradient sedimentation in an ultracentrifuge.

In the centrifugation procedure, samples are suspended in a salt solution of a heavy metal, such as cesium. Under the influence of the centrifugal force generated in the

ultracentrifuge, the heavy metal ions form a concentration gradient and consequently a density gradient. At equilibrium, experimental samples applied to the heavy metal salt solution would be found in the density gradient at the positions of their buoyant densities. In this way, heavy DNA might be separated from light DNA in the density gradient.

In one version of their experiment, Meselson and Stahl proceeded as follows. They grew a batch of phage in the presence of radioactive ^{32}P, so that the input parental phage DNA strands would contain a radioactive label. They then used these phage, which contained ^{32}P-labeled DNA, to infect cells that were growing in the presence of ^{15}N and ^{13}C as the sole nitrogen and carbon sources, respectively. The DNA from the progeny phage issuing from this infection was then analyzed by equilibrium density-gradient sedimentation.

So, if Watson and Crick are correct, and DNA replication is semiconservative, where would you expect to find the ^{32}P label in the density gradient? Would it be with the heavy, dense DNA or with the light DNA? Answer: It was found in a band of hybrid density, between the heavy and the light bands. Can you explain this finding? Does it prove that DNA replication is semiconservative?

Note that Meselson invented equilibrium density-gradient centrifugation using cesium chloride, which made the experiment possible. Meselson was a graduate student at the time, and Stahl was a postdoctoral student, each at Caltech. Interestingly, Meselson first met Stahl at the Marine Biological Laboratory at Woods Hole in the summer of 1954, when Meselson was a teaching assistant in James Watson's physiology course and Stahl was a student in the course.

We begin by observing that there is a slight increase in total RNA after the start of infection and that the level of RNA then appears to remain constant over time (Fig. 2.4). The apparently constant levels of RNA seen in infected cells would seem to suggest that there is not much RNA metabolism going on. However, this conclusion would be quite incorrect. A fraction of that RNA is actually continually being synthesized and degraded. This fact was demonstrated by a "pulse-chase" experiment, done as follows. ^{32}P was added to the infected cells for a brief period of time (the pulse), to label the RNA. The cells then were washed and incubated in an excess of unlabeled precursor (the chase). At various times the RNA was extracted, and the amount of radioactivity it contained was determined. This experiment showed that the amount of label that was incorporated into the RNA during the pulse continually decreased during the chase, with a half-life on the order of minutes, thus demonstrating that this RNA rapidly turns over. That is, this RNA fraction is unstable and is continually replenished.

A major question was whether this rapidly turning over RNA is cellular or viral in origin. An early experimental approach to answer this question was to ascertain whether the base composition of that RNA might reflect the base composition of either the phage DNA or the cellular DNA. This was done in 1956 by Volkin and Astrachan, who demonstrated that the unstable RNA in T2 phage-infected cells has the same base ratio (A+U/G+C) as the phage DNA (with T in place of U), rather than that of the cell DNA. This experimental finding strongly implied that this rapidly "turning-over" RNA is phage specific, rather than cell specific, and, importantly, that it might be encoded by the phage DNA. Proof of this fact came from Hall and Spiegelman in 1961, when they found that this RNA could be hybridized with phage DNA, but not with cell DNA. (The principle behind the approach developed by Spiegelman is explained here for the benefit of those to whom it might be new. Heating double-stranded DNA molecules to temperatures above their melting point [just below 100°C] causes the hydrogen bonds between the purines and pyrimidines of the two complementary strands to break, thereby enabling the strands to separate. If this **denatured** DNA then is slowly cooled [**annealed**] to room temperature, the complementary strands reestablish their interchain hydrogen bonds, and DNA double helices re-form. If single-stranded RNA molecules are added to the denatured DNA before the slow-cooling step, and if that single-stranded RNA is complementary to some of the DNA strands, then RNA-DNA hybrid helices will form, in addition to the DNA-DNA helices, when the cooling occurs. Importantly, the hybrid molecules will form only between RNA and DNA polynucleotide chains that are complementary to each other in their base sequences. The hybridized RNA could be detected by several means. For example, it is resistant to RNase. Also, filters that allow single-stranded RNA to pass retain the hybridized RNA. The same principle of hybridization between complementary RNA and DNA strands underlies more contemporary experimental approaches, such as gene arrays.)

Our next question, and a singularly important one at that, concerns the function of the unstable phage-encoded RNA. Bearing in mind the level of knowledge at the time, now the very early 1960s, it was reasonable to suggest that this RNA has some function in protein synthesis, since a general correlation between RNA levels and protein synthesis had been apparent for several years. Moreover, it was clear that DNA could not directly serve as the template for protein synthesis. This point was obvious from the example of eukaryotic cells, in which the DNA is contained within the membrane-bound nucleus, whereas protein synthesis occurs in the cytoplasm. Moreover, cytoplasmic ribosomes were known to be the sites of protein synthesis. (This was shown by a pulse-chase experiment that was done as follows. First, radioactively labeled amino acids were injected into rats. Then, at various times rat liver tissue was removed and homogenized, and cell fractions were prepared and analyzed for their content of the radioactive label. Within 2 min, radioactively labeled polypeptides were observed in the ribosome fraction. Radioactively labeled proteins then were found loose in the cell sap [i.e., the liquid phase of the cytoplasm]. Thus, these experimental results showed that amino acids are assembled into proteins on ribosomes, and the completed proteins then are released into the cytoplasm.)

Since ribosomes were known to contain RNA, it was reasonable to suggest that the intermediary in protein synthesis might be the RNA component of the ribosomes. This hypothesis, although attractive, would prove to be wrong in a subtle but important way. It implied that each gene is transcribed to a unique RNA, which in turn is incorporated into a specialized ribosome. That specialized ribosome could then direct the synthesis of only one particular protein. If that idea were correct, then we might expect the rapidly turning over phage-encoded RNA to be associated only with the new ribosomes that are made after the start of infection. So then, what is the relationship between the phage-encoded RNA and the ribosomes? Specifically, is that RNA associated only with new ribosomes?

In a 1961 experiment notable for its logic and elegance, in addition to its exceptional importance, Sydney Brenner, Francois Jacob, and Matthew Meselson established the relationship between the rapidly turning over, phage-encoded

RNA and ribosomes. If the "one gene–one ribosome–one protein" hypothesis was correct, and if the unstable phage-specific RNA played an intermediary role in protein synthesis, then the phage-specific RNA should be found associated only with ribosomes that are made after infection is under way. Thus, Brenner, Jacob, and Meselson needed to distinguish the ribosomes made after infection had begun from the ribosomes that were present before infection. They did this by growing cells for several generations in medium containing the heavy isotopes ^{15}N and ^{13}C as the sole respective nitrogen and carbon sources. In this way, essentially all ribosomes present in the cells before infection would be "heavy." Next, the cells were washed and placed in medium containing only the common isotopes ^{14}N and ^{12}C. Moreover, the cells were immediately infected with the phage upon transfer to the fresh medium. In that way, any new ribosomes made after infection was under way would be "light."

Importantly, recall that the phage-encoded RNA could be specifically labeled by adding ^{32}P to the infected cultures. (You may have been wondering why the ^{32}P is incorporated only into phage-encoded RNA. The reason will become clear shortly.) Thus, ^{32}P was added to the cultures after infection to label any de novo phage-encoded RNA. Now, Brenner and coworkers needed only to determine whether the ^{32}P-labeled phage-encoded RNA was associated only with new light ribosomes. Heavy and light ribosomes could be separated by equilibrium density gradient centrifugation. (In 1958, Meselson and Frank Stahl used T-even phages and equilibrium density gradient centrifugation to demonstrate that DNA replication in fact is semiconservative, as recounted in Box 2.3.) When all of this was done, the ^{32}P-labeled phage-encoded RNA was found in association with the old heavy ribosomes that were made before infection. Indeed, there were no new light ribosomes in the infected cells, since the phage effectively shuts down cellular macromolecular synthesis.

The result of this experiment is of crucial importance to biology in general, since it established the existence of mRNA, which carries the information encoded in the gene to unspecialized ribosomes, which already contain their own general ribosomal RNA (rRNA). The unstable mRNAs enter into transient associations with the ribosomes, where they are translated into the proteins that they encode. Thus, rather than "one gene–one ribosome–one protein," we now have "one gene–one mRNA–one protein." These experiments soon led to the successful search for an analogous cell-specific mRNA fraction.

The experimental results of Brenner, Jacob, and Meselson are also of fundamental importance to virology, since they establish that viruses subvert the host cell protein synthesis machinery to translate their own proteins. A corollary principle is that once a virus has managed to generate mRNAs, the cellular translation machinery is available to the virus to translate that viral mRNA into protein. As we will see shortly, this concept provides the basis for a rational scheme for classifying all of the virus families into a few groups, based on the relatively few strategies by which the many different virus families transcribe their genomes.

Assembly of Progeny Phages

As noted above, the structural protein components of the phage particles are not made until after the onset of phage DNA synthesis (Fig. 2.4). This pattern of regulated late gene expression is a characteristic feature of the replication of DNA viruses in general. Soon, we will consider the advantage that this pattern of regulated gene expression might confer upon these viruses. We also will consider the mechanisms that underlie it, both in the case of the T-even phages and, in later chapters, in the cases of other viruses as well. But first we need to consider two other general characteristics of virus replication that are revealed by the analysis of the T-even phage replication cycle. The first of these is the accumulation of phage structural components in the cells, before the emergence of progeny phage particles, the significance of which is demonstrated by the phenomenon referred to as **phenotypic mixing**, as in the following example.

Before we can do our actual experiment that demonstrates phenotypic mixing, we first need to isolate particular sorts of cell and phage mutants. First, the cell mutants. Recall that T2 phage initiates infection by specifically binding via its tail fibers to receptor molecules on host bacterial cells. By exposing a culture of susceptible *E. coli* B cells to the phage, we can select for bacterial mutants that survive exposure to the phage. These T2-resistant bacteria, which are referred to as *E. coli* B/2, have mutations that alter the receptor for the phage, so that the phage can no longer attach to them.

Next, we select for phage mutants that have the ability to attach to, and infect, the T2-resistant *E. coli* B/2 cells. This is done simply by exposing a culture of *E. coli* B/2 cells to a high-titer inoculum of the T2 phages. The phage mutants that are selected in this way are referred to as T2h, where the "h" refers to the altered **host range** of the mutants. As you might predict, the mutation in the T2h mutants affects their tail fibers, enabling them to bind to the altered receptors on the *E. coli* B/2 cells. Some of these T2h mutants also retain the ability to infect the original *E. coli* B cells.

We can now proceed with our actual experiment that demonstrates phenotypic mixing. First, we expose *E. coli* B cells to an inoculum that contains a high titer of both T2 and T2h phages. In this way, every cell is infected with both **wild-type** (normal) and mutant phages. Second, we harvest the **lysate** from this mixed infection and use it as the inoculum to infect a fresh culture of *E. coli* B/2 cells. Third, we analyze the virus in the lysate from the infected *E. coli* B/2 cells. We might have expected that this lysate would contain only T2h mutants, since wild-type T2 phage would not have been able to infect the *E. coli* B/2 cells. (We can easily distinguish T2h mutants from the wild-type T2 phage by differences in the plaques [see Chapter 1] that they each produce on a mixed culture of *E. coli* B and *E. coli* B/2 cells. T2h mutants lyse both *E. coli* B and *E. coli* B/2 cells to produce a clear plaque. In contrast, wild-type T2 phage lyses only the *E. coli* B cells. Consequently, the wild-type T2 phage produces a turgid plaque on the mixed culture of *E. coli* B and *E. coli* B/2 cells.) We find that, in fact, the lysate from the infection of the *E. coli* B/2 cells contains wild-type T2 phage as well as the T2h mutants. So then, what is the explanation? Before reading on, try to solve this for yourself. Congratulations if you deduced the answer, which is as follows.

Virus particles are assembled from the pool of component protein molecules that accumulate in the infected cells. When the mutant and wild-type phages are grown together in the same cell, as in the first step of our experiment, they each contribute to the pool of phage component proteins. When progeny phage particles are assembled from the pool of phage proteins, these proteins are withdrawn from the pool at random, without regard to the genotypes of the phages that produced them. Thus, in the first step of the experiment, in which *E. coli* B cells were infected with both T2 and T2h phages, progeny phages were assembled that contained either a T2 or a T2h genome and a mixture of protein components encoded by both the wild-type T2 and mutant T2h genomes. These phages are referred to as "phenotypically mixed." With that in mind, and recalling that the tail fibers confer host range specificity, many of the phages generated in the first step of the experiment thus contain wild-type T2 genomes in phage particles with some T2h tail fibers (there are six tail fibers in all). Some of these phages have enough T2h tail fibers to enable them to infect *E. coli* B/2 cells in the second step of the experiment. However, any phage particle that has a wild-type genome will still produce a turgid plaque on the mixed culture of *E. coli* B and *E. coli* B/2 cells. This example of phenotypic mixing demonstrates the following characteristic aspect of the viral life cycle. Virus particles are assembled from a pool of precursor components. This is fundamentally different from the cellular lifestyle, in which cells grow larger and then divide.

The remaining general characteristic of viral replication that is revealed by the T-even phages concerns the actual sequence of steps by which structures as complex as these phages are assembled. The facts that phage particles are assembled from a pool of precursor components and that those individual components are taken at random from the pool do not mean that assembly is a random process. Instead, assembly occurs via a specific sequence of steps. R. S. Edgar and W. B. Wood analyzed the T4 phage assembly pathway in 1965. They made extensive use of a variety of phage mutants, each of which is unable to produce infective phage particles. These mutants were used in two different ways. In their first experimental approach, Edgar and Wood infected *E. coli* cells with the individual mutants and then determined by electron microscopy the assembly intermediates that accumulated in each instance. Their second and especially innovative approach put to use their observation that certain combinations of extracts from mutant-infected cells actually *complemented* each other in vitro, thereby generating infective particles. From their extensive electron microscopy analysis of a large number of individual viral mutants together with their in vitro complementation experiments, Edgar and Wood deduced a rather complete picture of the T4 assembly pathway (Fig. 2.5). In particular, note the three independent branches leading to the formation of heads, tails, tail fibers, and then complete phage particles.

As discussed shortly, bacteriophages and animal viruses have evolved different structural features to solve the particular obstacles that they face in gaining entry into their respective host cells. Despite these differences, important general principles of virus assembly are revealed by the T-even phages. Foremost of these is that virus particles are assembled from multiple copies of identical protein subunits. Second, viruses use subassembly pathways rather than a single linear pathway. Significantly, these two principles provide for the limited number of viral genes to be efficiently utilized and for the possibility of rejecting mistakes before the entire particle might be ruined.

Packaging DNA within the Phage Particle

The **encapsidation** of the viral genome within the protein head, or **capsid**, is a crucial step in the assembly of DNA viruses that still is poorly understood. One might have expected that the simplest way to package the phage DNA would be to assemble phage capsids around a collapsed form of the DNA, since that would avoid the energetically unfavorable process of inserting the DNA molecules, with their negatively charged phosphate groups, into preassembled empty phage heads. Regardless, pulse-chase experiments demonstrated that empty phage **proheads** indeed are precursors in the assembly pathway. That is, preassembled empty proheads are filled with phage DNA. To appreciate how remarkable this is, consider the following.

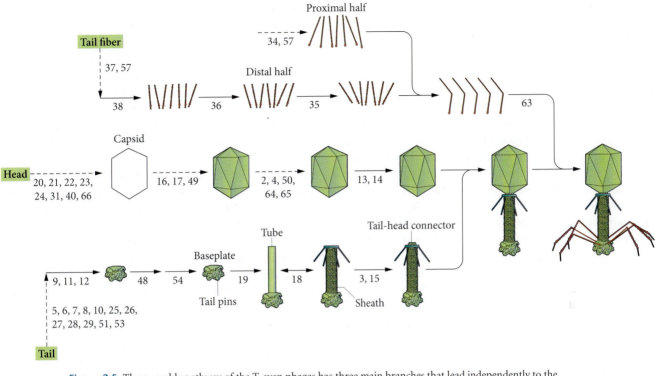

Figure 2.5 The assembly pathway of the T-even phages has three main branches that lead independently to the generation of phage heads, tails, and tail fibers. These branches then converge to yield mature phage particles. The numbers refer to the particular phage genes whose products are involved at each particular step. A mutation in any of the indicated genes leads to the accumulation of the intermediate structure seen immediately before the step in which the mutant gene is involved, as well as in the accumulation of the terminal structure of each of the other two pathways (i.e., the structures just before the convergence of the three branches). The solid arrows indicate steps that have been demonstrated in vitro. Modified from W. B. Wood, *in* F. H. Ruddle (ed.), *Genetic Mechanisms of Development* (Academic Press, New York, NY, 1973), as adapted by G. S. Stent and R. Calendar, *Molecular Genetics*, 2nd ed. (W. H. Freeman and Company, San Francisco, CA, 1958), with permission.

The diameter of the phage head is 0.085 μm. The length of the phage DNA is nearly 6,000 times greater at 50 μm (Fig. 1.5). Moreover, phosphate groups on the outside of the DNA double helices result in charge repulsion as the DNA is threaded into the proheads. Notwithstanding these facts, the DNA within the head is yet packaged to **liquid crystalline** density.

Despite the general lack of understanding of the T-even phages' encapsidation process, several aspects of the process that are known are relevant to virus assembly in general and also to the notion of viruses behaving as dynamic molecular machines. One such aspect is that proheads contain, in addition to their permanent structural components, several different proteins that transiently associate with the proheads. These transient proteins exert scaffolding and chaperone functions that facilitate the assembly of the proheads. They are eliminated from the prohead as DNA encapsidation proceeds. Importantly, exclusion of the scaffolding proteins as the prohead fills with DNA is associated with rearrangements of the permanent

structural proteins that make up the prohead. The conformational rearrangements of those proteins result in a more stable conformation of the prohead, as it matures into the head. By combining the energetically favorable conformational rearrangement of the protein lattice of the head with the energy-requiring DNA packaging, the overall process can be more energetically favorable.

The prohead is able to undergo its conformational rearrangement because it is in a **metastable** conformation. This is an extremely important point, in this instance and in general, since the release of a structure from its metastable conformation and its subsequent assumption of its most stable conformation can provide the driving force for subsequent events.

From thermodynamic considerations, we might have expected that the prohead should assemble directly into its most stable conformation. How, then, is the metastable state achieved? The general principle is to make use of specific proteolytic cleavages. In this instance, the multicomponent prohead structures are assembled from protein subunits, some of which are later cleaved. The initial structure indeed is in its most stable conformational state. However, the structure becomes metastable upon cleavage of particular subunits, in this case including the scaffolding proteins that also are extruded during encapsidation of the DNA.

Another detail of the T-even phage encapsidation process, which also is seen in the cases of some animal viruses, is the presence of a **portal complex** located at one of the 12 vertices of the phage's icosahedral head structure (see below). That portal complex (also known as a **connector complex**) is a dodecameric structure that contains a central channel that is about 30 angstroms (Å) in diameter. It functions as the entranceway for the phage DNA into the prohead. Also, it is the channel through which the DNA exits from the head during infection. ATP hydrolysis provides energy for the not-yet-understood driving force that moves the phage DNA through the channel into the prohead.

Just as the nature of the driving force for encapsidation is not clear, the driving force that causes the DNA to exit the particle at infection is not yet known either (Box 2.2). One possibility is that the pressure exerted by the DNA within the head drives it out. In the case of phage φ29 (phi 29), that pressure has been estimated to be 6 megapascals (MPa). The tensile strength needed to contain such a pressure is equivalent to that exerted by contemporary aluminum alloys. Bear in mind that the molecular motor that packages the DNA into the head must work against that pressure.

The exceptional tensile strength of the phage head, and of virus capsids or shells in general, results from the strong interactions between the virus's structural proteins. Features of virus structure that underlie these strong interactions are discussed shortly. Here we note that in addition to being able to sequester the phage DNA in the face of the enormous pressure exerted by the DNA, the capsid also protects the DNA from extremes of temperature, pH, and potentially lethal chemicals and radiation. Yet under appropriate specific circumstances, which are recognized by the very same exceptionally stable capsid, it releases its DNA so that the next replication cycle can begin. The explanation for this seeming paradox lies partly in the fact that even after the virus particles mature they still are in a metastable state. These facts further highlight a level of functional elegance that virus particles have evolved, which we probably did not appreciate earlier.

UNIQUE FEATURES OF T-EVEN PHAGES
Modified Bases

We now reconsider in somewhat more detail some special features of the T-even phage replication cycle that are unique to these particular viruses. The first of these features is the presence in their DNA of the modified base HMC. An examination of how this strange base is generated and then incorporated into the phage DNA will enable us to appreciate how it serves the phage's purpose.

First, a phage-encoded enzyme (deoxycytidine tri- and diphosphatase) dephosphorylates the intracellular deoxycytidine triphosphate (dCTP) and deoxycytidine diphosphate (dCDP) pools to deoxycytidine monophosphate (dCMP). At this point, DNA synthesis cannot occur for want of dCTP as a precursor. Next, the phage-encoded enzyme deoxycytidylate hydroxymethylase adds a hydroxymethyl (HM) group to dCMP to yield HMdCMP. Another phage-encoded enzyme, deoxynucleotide kinase, phosphorylates HMdCMP, yielding HMdCDP, which it in turn phosphorylates to yield HMdCTP. The virus-encoded DNA polymerase then uses HMdCTP as a precursor for the synthesis of progeny phage genomes, which thereby come to contain HMC.

The clever phage is here exploiting the important fact that the phage DNA polymerase, but not the cellular DNA polymerase, can use HMdCTP as a precursor for DNA synthesis. Since the phage has converted the intracellular pool of dCTP to a pool of HMdCTP, the phage has essentially commandeered the precursor pool of nucleotide bases to its own purpose. In addition, the presence of the strange base in the phage DNA protects it from a phage-encoded **endonuclease** that is specific for unmodified cytosine-containing host DNA. The nucleotides that are released from the degraded cell DNA are used by the phage as building blocks for its own DNA replication. Moreover, glucosylation of the HMCs in the phage DNA by yet other phage-encoded enzymes (α- and β-glucosyltransferase) protects the phage DNA from the defensive cellular restriction endonucleases (Box 2.4). This is all rather diabolical!

Modified bases in biological nucleic acids are of course found in cellular transfer RNA (tRNA). However, large-scale replacement of a standard base with a modified one is found only in the DNA of certain bacteriophages. Other examples include the *Bacillus subtilis* phages SPO1 and PBS1, which contain hydroxymethyluracil instead of thymine and uracil instead of thymine, respectively. These strange bases in the DNA of the *B. subtilis* phages likely serve the same purpose as the HMC in the T-even phages.

Regulated Gene Expression

The temporal regulation of T-even phage gene expression is an important feature of the replication cycle of these viruses, which was too briefly discussed above. We continue by first considering the possible advantages to these

BOX 2.4

Glucosylation, as the means by which T-even phages undermine host cell restriction, was demonstrated as follows. Phages with glucosylated DNA were used to infect mutant *E. coli* B cells that were unable to make uridine diphosphate glucose (UDPG). Since UDPG is a necessary substrate for glucosylation of the phage DNA, its absence meant that progeny phage DNA could not be glucosylated. However, phage DNA still could be replicated in the UDPG-negative cells, probably because host restriction enzymes are membrane bound and act on the incoming phage DNA at entry. Also, at an early point in infection, the phage makes a protein that inhibits any cytoplasmic restriction enzymes. The phage encodes such a protein because new phage DNA is not immediately glucosylated. When T2 phage with nonglucosylated DNA (referred to as T2*) was allowed to infect fresh *E. coli* B host cells, its DNA was rapidly degraded by restriction endonucleases present in those cells. However, T2* phage was able to replicate in *Shigella*, which lacks the restriction enzymes. Interestingly, its DNA was glucosylated in the *Shigella* cells, so that it could then infect *E. coli* B.

---

phages of their temporally regulated pattern of gene expression, and then we consider the mechanisms by which it is accomplished.

It is logical that phage genes, which encode enzymes required for phage DNA replication, are expressed prior to DNA replication. However, what might be the benefit to the virus of postponing the expression of genes that encode structural and other proteins that are required for generating phage particles, until after the phage DNA has been replicated? The relevance of this issue extends beyond the DNA bacteriophages, since the regulated pattern of T-even phage gene expression is characteristic of the replication of DNA viruses in general.

A likely advantage to the virus of this pattern of regulated gene expression is that it maximizes the efficiency of viral protein synthesis. Consider that at the start of the infection, when there may be only one viral genome in the cell, the virus has relatively little capacity to generate large amounts of any particular protein. However, the virus must make those virus-encoded proteins that are necessary for it to replicate its genome. Fortunately, these early proteins have an enzymatic role and thus are not needed in great amounts since they are used over and over again. In contrast, the virus will need many copies of the different proteins that make up the virus particle. Also, each of

those structural component molecules is incorporated into a progeny virus particle only once. Thus, it is logical for the virus to begin by synthesizing the small amount of the catalytic proteins necessary for it to amplify the number of viral genomes. Likewise, it is logical for the virus to wait to make its structural components until a much larger number of progeny DNA genomes have become available to direct that process.

It also is advantageous to the virus to make very large amounts of its structural components because these components must find, and interact with, each other in an intracellular milieu in which the concentration of cellular proteins may be as high as that attained in protein crystals (i.e., 20 to 40 mg/ml). Since many viral structural subunits may inadvertently interact with irrelevant cellular proteins, and since viral assembly is driven in the forward direction by the high concentration of viral structural subunits, it is beneficial to the virus to synthesize a great excess of structural proteins; this can be done most efficiently by first amplifying the genomic templates as described above. (The fact that virus assembly takes place in an intracellular environment in which the concentration of irrelevant cellular proteins is strikingly high also may account in part for the strategy of many viruses, animal viruses as well as bacteriophages, to shut down cellular protein synthesis.) A further and obvious advantage to the virus of maximizing its capacity for directing protein synthesis is that it enables the virus to generate the maximal number of progeny virus particles.

The production of large amounts of foreign viral proteins surely must be disadvantageous to the cell, particularly since at least some of these proteins may promote virus release by compromising the integrity of cellular membranes. Other virus proteins may compromise other cellular processes that support virus replication. Thus, by synthesizing late proteins only after the viral genome has replicated, late protein synthesis does not compromise genome replication. And it is less likely to compromise final virus assembly.

We next consider the mechanism by which T-even gene expression is regulated. We begin by noting that the pattern of T-even gene expression is characterized by the following three features. First, phage DNA synthesis is dependent on previous phage-specific protein synthesis. Second, early protein synthesis ceases at the time that DNA synthesis begins. Third, phage DNA synthesis is necessary for the switch from early to late gene expression to occur. Thus, if phage DNA synthesis is blocked (e.g., by ultraviolet [UV] irradiation of the phages or by mutation), then early genes continue to be expressed and late proteins are not made (see Fig. 2.4, in which the broken lines in the bottom part of the figure depict the pattern of phage early protein synthesis in bacteria infected with conditional phage mutants that are unable to initiate phage DNA synthesis).

The experimental findings described above demonstrate that phage DNA synthesis is the event that triggers the switch from early to late phage protein synthesis. How might phage DNA synthesis bring about this change in the pattern of phage gene expression? We might hypothesize that the switch depends on phage DNA replication because DNA replication brings about an actual change in the phage DNA. An alternative hypothesis is that phage DNA synthesis might cause a change in cellular metabolism that might then affect viral gene expression.

Bruner and Cape distinguished between these alternative hypotheses in 1970. Since they used an experimental approach that exemplifies the clever molecular genetics of the day, we recount their experiment here. First, some necessary background that is probably already familiar to most readers is provided. A single-base change in the DNA can result in a mutant that is **conditionally defective**. These **conditional mutants** are usually **temperature sensitive** (ts). That is, they produce a protein that is not functional over as wide a range of temperatures as the wild-type protein. Generally, they are inactive at 39°C and higher, conditions under which wild-type proteins are still active. Other single-base changes can result in the conversion of a triplet that encodes an amino acid into one of the three triplets that cause chain termination. When such mutations occur in the coding region of a gene, they result in premature chain termination, and the mutants are usually nonconditionally defective. (They are not strictly nonconditional, since there are other mutations that can suppress their effects.) They are often collectively referred to as **amber mutations**, since the amber triplet UAG is one of the three chain-terminating triplets. Here is the actual experiment. First, cells are infected at 30°C with a phage that carries two mutations. One is a ts mutation (DNA-ts) in a gene that is required for DNA synthesis. The second is an amber mutation in a late gene (am-late). At 30°C, the DNA-ts mutant protein is functional, and the double mutant replicates its DNA. However, it cannot express the late function that contains the nonconditional amber lesion. Consequently, no progeny phage can be produced. The culture is then shifted to 42°C and **superinfected** (i.e., infected with a second virus) with a virus that carries only the DNA-ts lesion. This second virus cannot replicate its DNA at the restrictive high temperature. However, it contains a normal complement of late genes. Will it be able to express its late genes, since they are normal, and since viral DNA replication has already occurred in the cells? Or will the superinfecting virus be unable to express its late genes, since it has not replicated its own DNA? The answer is

that the superinfecting virus cannot express its late genes unless it is permitted to replicate its own DNA. The actual explanation is that replication introduces nicks into the DNA, which are necessary to activate the late transcriptional promoter. As then might be expected, the late promoter can be activated in vitro by introducing nearby nicks.

The role of DNA replication in the switch from early to late protein synthesis is only a part of the story of the regulation of T-even phage gene expression. The story also involves sigma (σ) factors, which confer specificity on the phage RNA polymerase. In the absence of an associated σ factor, an RNA polymerase initiates transcription at random points along a DNA template. But when an RNA polymerase is associated with a σ factor, the polymerase initiates transcription at those promoters that are specifically recognized by that particular σ factor. The role of σ factors in transcription in fact was revealed by the analysis of the role they play in the regulation of T4 transcription.

Those phage genes that encode the DNA polymerase, the nuclease (which degrades host cell DNA), and the activities responsible for nucleotide metabolism are all expressed soon after infection. This is possible because the cellular RNA polymerase, in association with a cellular σ factor, σ70, recognizes the phage promoters for these genes. Somewhat later, but still before phage DNA synthesis, host RNA polymerase, still in association with σ70, but with the addition of the T4-specified positive regulatory factor, *motA*, transcribes yet other phage genes. Transcription of the late genes requires, in addition to phage DNA replication, the replacement of σ70 with the phage-specified σ factor, gp55. That switch of σ factors is mediated by yet another phage protein. Phage-specified regulators of transcription and translation mediate the shutoff of the phages' early genes. These regulators are among the late proteins that the virus encodes.

The shutoff of host transcription is mediated by a phage-induced modification of the host RNA polymerase core enzyme (i.e., the cellular core RNA polymerase is adenosine diphosphate [ADP]-ribosylated by the phage) a few minutes after infection. One of the activities responsible for this modification is the protein gpalt, which the phage injects into the cell along with its DNA. The other is gpmod, one of the phage proteins synthesized immediately after infection.

Phage Release: Lysozyme and the *rII* Region

Next we consider how progeny phages manage to ensure their release from the host cell at the end of their replication cycle, so that they are free to initiate another cycle of infection. Since bacterial cells are encased within rigid cell walls, it might be expected that phages require phage-encoded functions to facilitate their release. This assumption is correct. At late times in infection, T-even phages express a **lysozyme**, which is a lytic enzyme that attacks the peptidoglycan layer in the bacterial cell wall, leading to cell lysis and liberation of the progeny phages. The phage expresses another protein, encoded by its *t gene*, which may disrupt the cytoplasmic membrane, in that way enabling the lysozyme to reach the peptidoglycan layer.

It is important that the timing of lysozyme expression is regulated, since if lysozyme were expressed too soon, then cell lysis might occur before infectious progeny phages were completely assembled (Fig. 2.4). The *rII* genes of bacteriophage T4 are responsible for preventing premature expression of T4 lysozyme and consequent early cell lysis. The *r* in *rII* stands for "rapid lysis," since that is the phenotype of *rII* mutants (Box 2.5).

In contrast to the T-even phages, not all phages lyse their bacterial hosts. The **filamentous** phages provide a particularly noteworthy example of such nonlytic phages. (These phages have circular, single-stranded DNA genomes that are enclosed within a 1-μm-long tube composed of about 2,000 copies of a single hydrophobic protein.) Rather than lysing their bacterial hosts, progeny filamentous phages are continually extruded through the cell envelope, while cell growth is affected only moderately by the infection. (Cell growth is probably affected because a portion of the cellular biosynthetic capacity is supporting replication of the phage.) Filamentous phage growth can continue for long periods, making for a situation intermediate between that of the lytic phages, such as the T-even phages, and temperate phages, such as λ phage, described below. Importantly, animal viruses too have their counterparts to this range of interactions between bacteriophages and their bacterial hosts. In the case of animal viruses, these types of infections are referred to as **acute**, **chronic**, and **latent**.

BACTERIOPHAGE λ: LYSOGENY AND TRANSDUCTION

In Chapter 1, we briefly introduced the temperate bacteriophages, noting that they do not necessarily replicate in, and kill, their usual host cells. Instead, infection by these viruses may follow a course in which a mostly quiescent phage genome is stably maintained in the host cell; this virus-cell relationship is referred to as lysogeny (Box 1.9). In some instances (e.g., bacteriophage λ), the lysogenic phage genome is integrated into the cellular genome, whereas in other instances (e.g., phage P1), it is maintained as an extrachromosomal plasmid. In either case, the lysogenic phage genome is typically replicated in

conjunction with the replication of the cellular genome, and a copy of the phage genome is passed on to each daughter cell at division.

In order for temperate bacteriophages to follow the lysogenic course of infection, they must be able to restrict the expression of their replicative functions. Thus, temperate phages make use of mechanisms that enable them to express only a subset of their genes when in the lysogenic state, in particular those genes which enable them to establish and maintain the lysogenic state.

Since lysogenic viruses express a small portion of their genomes when in the lysogenic state, Lwoff was led to state that lysogens "are the first example of a specific hereditary property being conferred on an organism by a specific extrinsic particle." In that regard, lysogens can be medically relevant, since the lysogenic virus may express deadly toxins, as in the case of *Corynebacterium diphtheriae* and *Clostridium botulinum*, in which the toxins produced by these potentially deadly bacteria are encoded by the lysogenic viral genomes that they carry.

During the time that a bacteriophage is in the lysogenic state, the infection is said to be latent. Importantly, when it might serve its purpose, a phage in the lysogenic state can activate and take up the course of lytic growth. In so doing, the phage has the opportunity to infect new hosts. The λ phage of *E. coli* is the paradigm of a temperate phage. As we soon will see, the ability of λ phage to "choose" between the alternate existences of lysogeny versus lytic replication depends upon a complex regulatory mechanism of extraordinary elegance (Box 2.6).

Before we consider the mechanism that determines whether λ phage will initiate a lytic versus a lysogenic course of infection, there are several other matters to attend to. The first of these is the question of what advantages λ phage might derive from having these alternative choices. The answer appears satisfyingly straightforward. When λ phage infects healthy, metabolically active cells, it chooses the lytic course. However, when λ phage infects metabolically listless cells, it chooses the lysogenic course. Actually, the metabolic state of the host cell directly affects the viral switch that oversees the lytic versus lysogenic decision. Consequently, when conditions within a lysogen improve, the λ genome can then activate and take up the course of lytic growth. Moreover, if the cellular genome should be damaged by UV irradiation or by some other means, the lysogenic virus can be activated to initiate lytic replication in order to find more fertile ground in a new cell. However, while in the lysogenic state, λ is able to remain within the protective environment of the host cell and replicate as a lysogen, concomitant with the replication of the host cell.

BOX 2.5

"Adventures in the *rII* region"

In the late 1950s, Seymour Benzer carried out genetic experiments involving the *rII* genes of T4 phage, which produced results of major importance to the newly emerging science of molecular biology. First, Benzer's analysis of the T4 *rII* region provided the most detailed fine-structure genetic map in existence. Second, he was the first to document genetic recombination between adjacent nucleotides. Third, he demonstrated that single nucleotides are the minimal unit of mutation (muton). Fourth, and perhaps most important, he devised the *cis-trans* test to define the unit of gene expression, the **cistron**.

The background for the last contribution is as follows. When Benzer infected *E. coli* K cells with pairs of independently isolated T4 *rII* mutants, he found that wild-type T4 recombinants are generated at a very low frequency, indicating that all of the *rII* mutations are very close together on the phage chromosome. In addition to finding rare genetic recombinants between *rII* mutations, he also found that certain pairs of *rII* mutants actually replicated together in *E. coli* K. That is, they complemented each other.

Next, Benzer found that all of the *rII* mutants could be placed in either of two groups, designated *A* and *B*. All *A* mutants complemented all *B* mutants and vice versa. However, mutants within the same group could not complement each other. Moreover, for complementation to occur, the mutations also had to be on separate phage chromosomes; that is, they had to be in *trans*. Complementation did not occur if the mutations were on the same phage chromosome, that is, in *cis*.

Together, these experimental results showed that whereas all of the *rII* mutations are very close together on the T4 chromosome, they nevertheless define two distinct functions and that a mutation in either function results in the mutant phenotype. But what might be the mechanistic explanation for complementation between *rIIA* and *rIIB* mutants when expressed in *cis*, but not when expressed in *trans*? The answer is that the *rIIA* and *rIIB* regions of the phage chromosome define separate units of transcription. These units of transcription, or cistrons, are operationally defined by the *cis-trans* test. That is, mutations in separate cistrons complement each other when expressed in *cis*, but not when expressed in *trans*. Cistrons are thus the genetic units of function. Today, the term "cistron" is rarely used. Instead, the word "gene" has taken on the same meaning as "cistron."

To appreciate the immense significance of Benzer's contributions, we need to be reminded that classical genetics made no distinction between genes as units that specified a particular phenotypic trait, as units of mutation, and as units of recombination. Indeed, classical genetics envisioned a gene as a single indivisible unit that embodied all three of these properties. Benzer's experiments provided the distinctions between genetic units of function, of recombination, and of mutation, showing that a single pair of nucleotide bases is the unit of recombination and that a single nucleotide is the minimal unit of mutation; and for the first time, he gave the gene a clear operational definition. Moreover, the *cis-trans* test could be used and indeed was used to determine whether any two mutations are in the same or different functional genetic units.

As a graduate student in physics during World War II, Benzer participated in a project to develop better crystal rectifiers, a crucial component for radar. His research findings on semiconductors later contributed to the development of the first transistors. When his interested shifted to biology, he enrolled in Delbruck's summer phage course at the Cold Spring Harbor Laboratory, eventually joining Delbruck's Phage Group at Caltech.

From a human-interest standpoint, when Max Delbruck summoned Renato Dulbecco to his office to propose that he, Dulbecco, work on animal viruses (Box 1.11), Benzer was called in along with him. Dulbecco relates, "I immediately expressed my interest, before Benzer could say anything. Benzer, on the other hand, was not interested, so everything was settled without delay. At that moment I became an animal virologist." (See Dulbecco's chapter in *Phage and the Origins of Molecular Biology*). Benzer's passion seems to have been the nature of the gene, and his contributions there were huge.

In the 1960s, Benzer's scientific interest shifted again, this time to the goal of developing a model system that might lead to insights into the genetic basis for behavior. He eventually settled on *Drosophila melanogaster* and founded the field of neurogenetics. Benzer passed away in November 2007.

Francis Crick in 1961 used the T4 *rII* region to make yet another fundamental contribution to molecular biology, in this instance demonstrating that the genetic code is read three nucleotides at a time. In brief, he found that certain T4 mutant phages, which contained two *rII* mutations in *cis*, displayed a wild-type-like phenotype. Specifically, some of these mutations, referred to as "+" (plus), gave a wild-type

Box continues

BOX 2.5 (continued)

phenotype when in *cis* with other mutations, referred to as "−" (minus). In contrast, + + and − − double mutants never displayed a wild-type phenotype. However, + + + and − − − triple mutants could display a wild-type phenotype. The explanation is that the + mutations are single-nucleotide insertions that change the reading frame by +1. Likewise, − mutations are single-nucleotide deletions that change the reading frame by −1. When a + mutation is paired with a − mutation in *cis*, the reading frame downstream of the second mutation is restored, leaving the possibility that a functional protein might yet be translated. Similarly, if the

code was read as triplets, with each triplet specifying an amino acid, then a +++ or a − − − triple mutant would also not change the reading frame downstream of the last mutation.

The title at the head of this box is that of Benzer's chapter in *Phage and the Origins of Molecular Biology*.

Reference
Cairns, J., G. S. Stent, and J. D. Watson (ed.). 2007. *Phage and the Origins of Molecular Biology: the Centennial Edition*. Cold Spring Harbor Laboratory Press, Cold Spring Harbor, NY.

BOX 2.6

Although λ phage resembles the T-even phages in key ways (e.g., the λ head is filled with DNA and it is attached to a slender tail, through which it "injects" its DNA into susceptible host cells), Delbruck and his group of phage researchers were initially not interested in lysogeny, primarily because the T-even phages they had been studying did not display this phenomenon. They also may have been influenced by the view of d'Herelle, one of the discoverers of bacteriophages (Chapter 1), who believed that infection by a bacteriophage inexorably leads to cell death. d'Herelle thought that so-called lysogens were merely contaminated with a bacteriophage.

It also is historically interesting that the original laboratory strain of *E. coli* (i.e., *E. coli* K-12), which was isolated in 1922 from a patient with an intestinal disorder, was a λ

lysogen, as later discovered by Esther Lederberg in 1951. Esther Lederberg was the wife of Joshua Lederberg, who was awarded the Nobel Prize in 1958 for discovering genetic recombination associated with sexual conjugation in bacteria. The couple worked together to develop the technique of **replica plating** in 1952, a technique that enabled the selection of bacterial mutants from among hundreds of bacterial colonies on a plate and, more importantly perhaps, proof of the spontaneous origin of mutants with adaptive advantages.

Esther Lederberg "accidentally" isolated nonlysogenic or "cured" derivatives of *E. coli* K-12 that could be infected by samples of culture fluid from the parental K-12 strain, which was sporadically producing low levels of λ phage. As explained in the text, bacterial strains that carry a lysogenic virus are "immune" to superinfection by the same virus.

Next, note that the λ genome is linear in the virus particle but that it circularizes upon injection into the cell, doing so via cohesive sites at its ends. The ability of the λ DNA to circularize is a key factor that enables it to integrate into the cellular genome and thus be maintained in its lysogenic state. During lytic growth, the circular λ genome is replicated by means of a "rolling-circle" mechanism, which is discussed in detail in Chapter 14. For the sake of comparison, the λ genome is considerably smaller than the genomes of the T-even phages, encoding about 60 proteins, versus the 160 or so proteins encoded by the larger viruses.

Before we can understand how λ phage "decides" whether infection will be lytic versus lysogenic, we also

need to consider the elements of the critical λ control region, which is located in the right-most third of the linear λ genome (Fig. 2.6). The λ control region is referred to as the **immunity region**. It encodes the diffusible λ CI protein, which is the key λ repressor that suppresses lytic phage growth. The immunity region derives its name from the fact that CI also suppresses the growth of any superinfecting λ phages, as well as of the resident integrated λ **prophage**. The suppressing effect of CI is specific to λ phage. CI has no effect on infection by other bacteriophages. The immunity region also contains the binding sites for the CI protein, which are known as **operators**, since they are analogous to the operators that control transcription of the *lac* operon of *E. coli*.

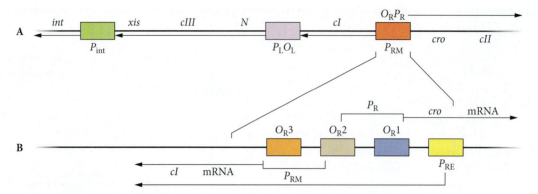

Figure 2.6 (A) The λ control region, referred to as the immunity region, is responsible for the establishment and maintenance of lysogeny. It encodes the diffusible λ CI protein, which is the key λ repressor that suppresses lytic phage growth. The immunity region also contains the binding sites for the CI protein, designated O_L and O_R, which are located on each side of the *cI* gene. The λ genome also contains two main promoters where the *E. coli* RNA polymerase can initiate transcription. Each of these promoters is located in the immunity region, one on each side of the *cI* gene. They are designated P_L and P_R and overlap O_L and O_R, respectively. O_L initiates transcription leftward, and O_R initiates transcription rightward. Another promoter, designated P_{RM} (RM stands for repressor mainte-nance), is located near P_R but initiates transcription in the opposite direction, to generate the mRNA that encodes the CI protein. P_{int} is the promoter for the *int* gene, which encodes the integrase that catalyzes the site-specific re-combination between the phage DNA and the cell DNA. (B) O_L and O_R are each comprised of three segments, O_R1, O_R2, and O_R3 in the case of O_R, and each of these segments binds a dimer of CI. The figure shows the topological relationship between the elements of O_R and P_{RM} and P_R. When CI binds to O_L and O_R, it inactivates P_L and P_R, re-spectively, while activating P_{RM}, thereby ensuring continued synthesis of CI while the phage is in the lysogenic state. When λ phage DNA first enters a cell, early viral transcription can begin from P_L and P_R, resulting in the produc-tion of the N and Cro proteins. N is a transcriptional antiterminator that enables the cellular RNA polymerase to continue transcription beyond the ends of the *N* and *cro* genes, so that the *cII* gene, among other early genes, might be transcribed. Cro is a repressor, which, like the CI protein, binds to O_L and O_R. But whereas CI has a preference for O_R1, Cro has a preference for O_R3, but not when CI is bound to O_R1. P_{RM} is blocked when Cro is bound to O_R3, but not when CI is bound to O_R1 and O_R2. CI dimers bind cooperatively, such that when a CI dimer binds to O_R1, it facilitates the binding of a sec-ond dimer to O_R2. In contrast to CI, Cro does not bind cooperatively. Thus, CI can repress expression of P_L and P_R, thereby blocking lytic infection, whereas Cro can repress expression of P_{RM}, and thus of *cI*, thereby enabling lytic in-fection. P_{RE} is a second promoter from which transcription of *cI* can be initiated. The decision as to whether infec-tion will be lytic or lysogenic actually depends on the amount of CI that might be present as a result of transcrip-tion initiated at P_{RE}. If there is not enough CI present to prevent transcription from P_L and P_R, then the outcome will be lytic infection. But if sufficient CI is produced from P_{RE} to repress P_L and P_R, then production of early pro-teins will be blocked, including that of Cro. Moreover, CI produced as a result of transcription from P_{RE} can activate further expression of *cI* from P_{RM}, thereby facilitating the establishment and maintenance of lysogeny. The level of CI generated from P_{RE} is determined by the level of CII, which positively regulates initial transcription of *cI* by binding to P_{RE}, thereby activating the otherwise weak P_{RE}. The level of CII in turn depends on the metabolic state of the cell. CII is inactivated by cellular proteases that are more abundant in cells that are actively replicating.

The λ genome contains two main promoters where the *E. coli* RNA polymerase can initiate transcription. Each of these promoters is located in the immunity region, one on each side of the *cI* gene. They are designated P_{left} (P_L) and P_{right} (P_R), and they initiate transcription leftward and rightward, respectively, from *cI*. Another promoter, desig-nated P_{RM} (RM stands for repressor maintenance), is lo-cated near P_R but initiates transcription in the opposite direction, to generate the mRNA that encodes the CI protein.

The binding sites in the immunity region for the CI protein are designated O_L and O_R, since they are analo-gous to operators on bacterial genomes. O_L overlaps P_L, and O_R overlaps P_R (Fig. 2.6). Each operator actually is comprised of three distinct segments (e.g., O_L1, O_L2, and O_L3), and each of these segments binds a dimer of CI.

When λ phage initially injects its DNA into a cell, and if there is no prophage present that might be synthesizing the CI immunity repressor, then immediate early viral transcription begins from P_L and P_R, resulting in the pro-duction of the N and Cro proteins (Fig. 2.6). The N pro-tein is a transcriptional antiterminator that enables the cellular RNA polymerase to continue transcription be-yond the ends of the *N* and *cro* genes, into the adjacent viral early genes. N acts by a mechanism analogous to that of the human immunodeficiency virus (HIV) Tat protein (see Chapter 21). In brief, N binds to a sequence called NUT (which stands for *N* utilization), which is present on nascent λ transcripts that are initiated at P_L and P_R. NUT positions the bound N protein so that it is able to interact with cellular factors that enhance the processivity of the cellular RNA polymerase. Thus, when N is present, it enables the polymerase to transcribe the cII gene, immediately adjacent to cro. As soon will be seen, cII plays a crucial role in determining whether the infection will be lytic or lysogenic. Moreover, N enables the polymerase to continue transcribing leftward and rightward (past cro), to express the remaining early genes, including the late gene activator.

The Cro protein is a repressor, which, like the CI protein, binds to O_L and O_R. However, recall that the operators are comprised of three distinct segments (Fig. 2.6; look carefully at the topological relationships of P_{RM} and P_R with O_R3, O_R2, and O_R1.). Whereas CI has a preference for O_R1, Cro has a preference for O_R3. Importantly, P_{RM} is blocked when Cro is bound to O_R3, but not when CI is bound to O_R1 and O_R2. Apropos that, CI dimers bind cooperatively, such that when a CI dimer binds to O_R1, it facilitates the binding of a second dimer to O_R2. In contrast to CI, Cro does not bind cooperatively. Thus, CI can repress expression of P_L and P_R, thereby blocking lytic infection, whereas Cro can repress expression of P_{RM}, and thus of cI, thereby enabling lytic infection.

The plot thickens somewhat because the cI promoter, P_{RM}, is a weak promoter that actually is activated when CI binds to O_R2 (Fig. 2.6). CI has this effect on P_{RM} by directly facilitating binding of the RNA polymerase to P_{RM}. Thus, CI activates its own gene while suppressing transcription from P_R. This raises the following question. If CI is necessary to activate its own gene, then where does the CI come from to do this? The answer is crucial, since CI is necessary to establish and maintain the lysogenic state. Here is the explanation.

The λ control region has yet another promoter, designated P_{RE} (RE stands for repression establishment), from which transcription of cI can be initiated (Fig. 2.6). P_{RE} is located just downstream of cro, so that transcription from P_{RE} is in the opposite direction (and from a different DNA strand) from transcription from P_R. Now, the decision as to whether infection will be lytic or lysogenic depends upon the amount of CI that might be present as a result of transcription initiated at P_{RE}. If there is not enough CI to prevent transcription from P_L and P_R, then the outcome will be lytic infection. But if sufficient CI is produced from P_{RE} to repress P_L and P_R, then production of early proteins will be blocked, including that of Cro. Moreover, CI produced as a result of transcription from P_{RE} can activate further expression of cI from P_{RM}, thereby facilitating the establishment and maintenance of lysogeny. So, the crucial question is, what determines the critical level of CI?

The level of CI generated from P_{RE} is determined by the level of CII. (Recall that cII is located just downstream of cro and is transcribed from P_R [Fig. 2.6].) CII positively regulates initial transcription of cI by binding to P_{RE}, thereby activating the otherwise weak P_{RE}. So, the lytic replication versus lysogeny decision depends on the level of CII. Now here is a really neat part of the story. The level of CII in turn depends on the metabolic state of the cell. CII is inactivated by cellular proteases that are more abundant in cells that are actively replicating. Thus, infection tends to be lytic in exponentially replicating cells and lysogenic in stationary-phase cell cultures. The regulatory machinery that maintains the lysogenic state is indeed very effective, since lysogeny is as stable a characteristic of a bacterial strain as any other inherited characteristic.

The above account of the λ lysis-versus-lysogeny decision is somewhat watered down, since it does not consider items such as the interaction between CI dimers in the regulatory region and the role of CIII in inhibiting the cellular proteases that inactivate CII. Nevertheless, I hope that it succeeds in conveying the elegance of the λ genetic switch. For those whose interest may have been piqued by the above, I strongly suggest a marvelous text by Mark Ptashne (2004), who contributed much to the understanding of the lysis-versus-lysogeny decision.

Next, we briefly consider the means by which the circularized λ genome is integrated into the cellular chromosome, a process that perpetuates the phage genome in the cell and its descendants, until an opportune moment might arise when the viral genome might be activated to undergo lytic replication. In addition, we consider the means by which the lysogenic phage can resume productive infection, when it becomes advantageous for it to do so.

Concerning integration, when λ phage makes the "decision" to lysogenize, it expresses a site-specific recombination system that recombines the $attP$ site on the circularized phage genome with the $attB$ site on the E. coli chromosome (located between the galactose [gal] and biotin [bio] operons). By recombining a circularized phage genome with the bacterial chromosome, a linear viral genome is integrated into the bacterial chromosome.

The $attP$ site is about 250 base pairs (bp) in length, whereas $attB$ is about 25 bp long. Each of these sequences contains a common 15-bp-long sequence, at which the integration of the λ chromosome takes place. Nevertheless, the site specificity of this crossover reaction is largely due to the phage Int protein, which recognizes the $attP$ and $attB$ sites on the phage and bacterial chromosomes, respectively, and catalyzes the integration event.

What might control the expression of int? Congratulations if you correctly surmised that in addition to positively controlling P_{RE}, CII also positively regulates P_{int}, the weak promoter for *int* (Fig. 2.6). A very neat picture!

Integration and lysogeny have a flip side: induction of the lysogenic viral genome and the excision reaction that frees it from its integrated state. As noted above, the relative levels of CI and Cro, which determine whether the infection will be lysogenic or lytic, are affected by the metabolic conditions of the host cell. Thus, under conditions of robust cell growth, the regulatory circuit may favor induction of the lysogenic phage and its vegetative replication. The regulatory circuits may also occasionally fail "spontaneously," so that there usually are a few cells in a lysogenic culture that are producing viruses at any moment. Apropos the excision process, another early gene, *xis* (Fig. 2.6), encodes the excisionase, Xis, which together with Int carries out the reverse process of excision, thereby freeing a circularized λ genome from the bacterial chromosome. The mRNA that encodes both Int and Xis is transcribed from the reactivated P_L promoter.

Before leaving the topic of λ lysogeny, note the following two points regarding the induction of a λ lysogen and its excision from the cellular genome. First, induction of the lysogen also occurs under conditions in which the host cell has been damaged, perhaps irreversibly. Induction under these circumstances may enable the phage to find a new healthy host cell in which it might reestablish itself. But how might the phage be so clever as to "know" to reactivate in the traumatized cell? The answer is that when the cell undergoes extensive damage, such as from UV irradiation, it activates its so-called **SOS repair system**. To do so, the cell produces a particular protease, which attacks and destroys the repressor of the host SOS genes. So, our clever λ phage has evolved its CI protein so that it too is inactivated by that host protease, and in that way the λ lysogen is activated in the traumatized cell.

The second point is that whereas excision is generally a precise reversal of the site-specific integration reaction, nevertheless, it occasionally goes awry, so that the released phage genome may contain a portion of the bacterial chromosome. Since the λ genome specifically inserts between host genes involved in galactose metabolism and biotin synthesis, an incorrectly excised λ genome may contain a portion of one of these host genome sequences. Some of these incorrectly excised phage genomes may yet be able to replicate and even integrate into the genome of a new host cell. In that case, when in the lysogenic state in the new host cell, they may be able to express the captured host cell gene sequence. This phenomenon is referred to as **transduction**.

Since the λ genome inserts at only one specific site in the bacterial chromosome, it only transduces cellular gene sequences adjacent to its integration site. That is, λ phage is a **specialized transducing phage**. However, other bacteriophages, such as phage P1 of *E. coli* and P22 phage of *Salmonella enterica* serovar Typhimurium, can pick up any gene segment from the host chromosome, a phenomenon referred to as **generalized transduction**. (Experiments with phage P22, carried out by Norton Zinder and Joshua Lederberg in 1952, led to the discovery of transduction.) In these instances, transduction results from the "headfull packaging" mechanism by which the phage heads package DNA. Each phage head is normally filled with a little more than one phage length of DNA from a concatemer of progeny phage genomes, which have been generated by rolling-circle replication. Although these phages typically recognize their own DNA as appropriate for packaging, they occasionally incorrectly package host DNA, presumably because they mistake a sequence in the cellular DNA for the special phage packaging signal. Thus, these transducing phage particles carry cellular DNA sequences only. Also in contrast to λ phage, the P1 and P22 genomes are maintained as extrachromosomal plasmids.

The cellular DNA carried by a generalized transducing phage is likely to be linear due to an absence of cohesive ends that might facilitate its circularization. Thus, for the transduced DNA to integrate so that it might be stably inherited, it must undergo two homologous recombination events with the recipient host chromosome, catalyzed in this instance by the host recombination enzymes. One consequence is that a portion of the recipient chromosome is replaced by a portion of the incoming DNA. Despite the requirement for two homologous recombination events, there still is about a 10% probability that the incoming chromosomal DNA will be recombined into the recipient's chromosome.

Generalized transducing phages played a major role in the analysis of several bacterial genomes. What sorts of information might they have provided? Importantly, the distance between two genes on the bacterial chromosome can be inferred by the frequency at which they are cotransduced. Thus, it was possible to use generalized transducing phages to generate fine-structure maps of small regions of bacterial chromosomes.

Interestingly, whereas λ phage inserts its DNA into the host cell chromosome at a specific site, Mu phage (not previously mentioned) inserts its DNA anywhere in the host chromosome. (Although Mu may incorporate host sequences into its DNA when it is excised, it is not a transducing phage, because when Mu infects a new host, the Mu DNA, but not the variable sequences of DNA acquired from the previous host, is incorporated into the recipient's chromosome by the Mu integration process.)

As we will see in subsequent chapters, many animal viruses too establish long-term associations with their eukaryotic hosts, which may have serious clinical consequences. Examples include HIV, hepatitis B virus, the papillomaviruses, and herpesviruses. Each of these different animal viruses has evolved a different means for remaining associated with the individuals it has infected, and each of these means is different from that evolved by bacteriophage λ; this is yet another example of different viruses evolving different strategies to achieve the same ends. Nevertheless, bacteriophage λ is able to select between two alternative existences, lytic growth versus latency, choosing the more suitable of these outcomes in light of the health and growth state of the host, a feat that is beyond the capabilities of any known animal virus.

SOME FINAL COMMENTS ON BACTERIOPHAGES

This chapter began with the continuing account of how analysis of the T-even phages revealed key general features of virus replication. In addition, we also considered some aspects of T-even phage replication that are specific for these viruses, as well as features of bacteriophage λ that are relevant to the phenomenon of lysogeny. Still, despite the fact that many years have passed since bacteriophages served as models for animal viruses, much more could be said about bacteriophages. Other bacteriophages, such as the small phages with single-stranded DNA genomes (in particular, φX174) and the filamentous DNA phages (the prototype is fd), also played important roles in the development of virology.

The single-stranded RNA phages (e.g., Qβ and MS2) also were not discussed above. However, these phages are considered in detail in the chapter on the *Picornaviridae*, a family of small RNA viruses that includes the rhinoviruses, which are responsible for many of our common colds, and poliovirus. The reason for including the RNA phages in that context is to compare their genome organization and patterns of gene expression with those of the eukaryotic picornaviruses.

Before leaving the bacteriophages, we should be aware that there has been a recent resurgence of interest in them for their own sake. One reason is that new genomics-based studies revealed a more pervasive relationship between these parasites and their prokaryotic host cells than was previously imagined. Consequently, since bacteria are estimated to constitute more than one-half of the Earth's biomass, and since phages are widespread among the bacteria, virulent phages may profoundly affect more than one-half of the Earth's biomass. Indeed, it is estimated that every 2 days, one-half of the bacteria on the planet are killed by bacteriophages!

Also, since (i) bacteria are haploid and (ii) phages are capable of transducing host cell genes, phages make an important contribution to the genetic diversity of bacteria. Moreover, since temperate phages often encode bacterial virulence factors (e.g., toxins) and antibiotic resistance genes, phages play important roles in particular instances of bacterial pathogenesis and contribute to the general problem of antibiotic resistance.

Finally, despite the early unsuccessful attempts by d'Herelle to use bacteriophages as therapeutic agents in the treatment of bacterial diseases (see Chapter 1), this approach has reemerged in new guises, largely because of the growing problem of antibiotic resistance among bacterial pathogens. For example, the recently isolated lytic φKMV phage of *Pseudomonas aeruginosa* is highly virulent against many clinical isolates of its host cell. That fact and the short replication time of φKMV make the phage or a phage product (e.g., the phage lysozyme) potentially attractive as therapeutic agents against *P. aeruginosa*.

INTRODUCTION TO THE ANIMAL VIRUSES

We begin our consideration of the animal viruses with a brief introduction to virus structure and follow with a short account of how animal viruses invade their host cells. These topics are covered in more detail in subsequent chapters, in the context of the individual virus families.

Animal Virus Structure

As noted earlier, viruses are assembled from multiple copies of one or more kinds of identical protein subunits. This is partly, but not entirely, because viruses are limited in the number of genes that they might be able to package within their capsids. Indeed, some viruses must make do with as few as five genes.

The protein building blocks or subunits from which viruses are assembled are arranged to generate either of two general types of capsid structures. One type displays **helical symmetry**. The other type displays **icosahedral symmetry**. In order to generate symmetry of either type, each of the identical subunits makes the same sorts of contacts with its neighboring subunits. This is an important point that helps to account for the extraordinary tensile strength that is characteristic of virus capsids. If the contacts between subunits are the same, then they can be optimized to be most favorable in all instances. But if they are not the same, then some contacts will be weaker than others. Thus, another reason for viruses to build their capsids from identical subunits is that it enables them to assemble capsids of extraordinary tensile strength. Remarkably, in

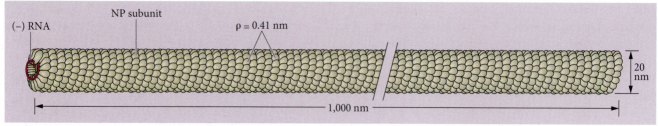

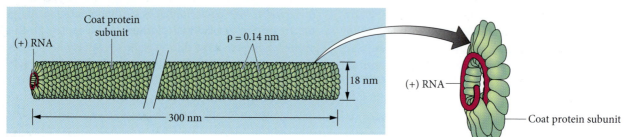

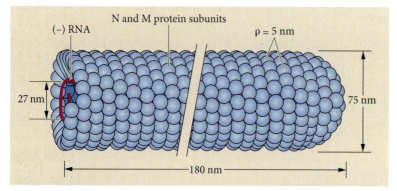

Figure 2.7 Examples of virus structures with helical symmetry. Tobacco mosaic virus (TMV) is a prototypical single-stranded RNA virus. Its RNA genome is of the plus (+) sense (see the text). A single copy of the TMV RNA genome is enclosed within a helical structure comprised of a single viral protein. There are $16^{1}/_{3}$ copies of that protein per turn of the helix; each of these protein subunits makes the same interactions with its neighbors, and each interacts with three nucleotides in the RNA. Sendai virus (a paramyxovirus) has a single-stranded RNA genome of the minus (−) sense (see the text). Its structure resembles that of TMV, but whereas the TMV helix is right-handed, the Sendai virus helix is left-handed. (Viruses having minus-sense RNA genomes generally contain nucleocapsids that possess helical rather than icosahedral symmetry. Moreover, the helical nucleocapsids of minus-sense RNA viruses are enclosed within lipid envelopes [not shown here, but see Fig. 2.11]. Also, the genomes of these viruses are associated with multiple copies of the viral RNA polymerase molecules, in addition to the NP protein, which serves to encapsidate the RNA.) Vesicular stomatitis virus (a rhabdovirus), like Sendai virus, contains a minus-sense single-stranded RNA genome. Notice that the Sendai virus helical capsid contains about 30 turns of uniform diameter, followed by 5 or 6 turns of decreasing diameter, thereby conferring on these viruses their characteristic bullet shape. Adapted from S. J. Flint, L. W. Enquist, R. M. Krug, V. R. Racaniello, and A. M. Skalka, *Principles of Virology: Molecular Biology, Pathogenesis, and Control*, 2nd ed. (ASM Press, Washington, DC, 2003), with permission.

most, but not all instances, the protein subunits interact via noncovalent contacts.

Viruses displaying helical symmetry enclose their genomes within long tubes that are comprised of helically arranged protein subunits (Fig. 2.7). The length of the tube reflects the length of the particular viral genome.

An icosahedron is a regular 20-sided polygon, where each face is an equilateral triangle (Fig. 2.8). Note that

icosahedra have 12 vertices, each of which is a point at which five of the triangular faces come together. The simplest icosahedral viruses contain 3 protein subunits per face and thus 60 subunits in the entire capsid. In point of fact, an icosahedron containing 60 subunits is the simplest structure, containing the least number of subunits, which can form a shell that completely encloses a space of maximal volume. Interestingly, the icosahedral structure is also

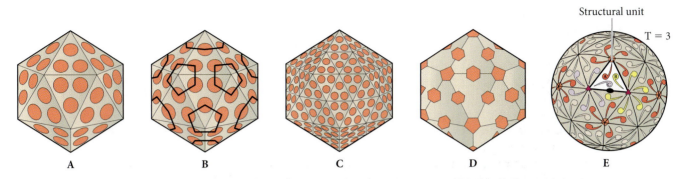

Figure 2.8 The building plan of viruses that have icosahedral symmetry. An icosahedron is a regular 20-sided polygon, where each face is an equilateral triangle. Note that icosahedra have 12 vertices, located at points where five of the triangular faces come together. (A) The simplest icosahedral viruses contain 3 protein subunits per face and thus 60 subunits in the entire capsid. (B) Icosahedral capsids are assembled from pentamers of the individual subunits, which become the vertices of the icosahedron. The five subunits that form the vertices are connected by lines in the diagram. (C) Making a larger icosahedral capsid requires adding more protein subunits, in accord with strict rules for how this may be done. Essentially, each of the 20 triangular faces of the simplest icosahedron may be divided into smaller triangles, each of which contains three subunits. Thus, icosahedral capsids always contain a number of subunits that is a multiple of 60. The resulting structure retains the fivefold, threefold, and twofold axes of symmetry that are characteristic of icosahedra. Looking carefully at the arrangements of subunits in the larger icosahedral capsids, notice that whereas the subunits at the 12 icosahedral vertices are arranged as pentamers, all other subunits are arranged in groups of 6 (i.e., as hexamers). In some instances, the hexamers or hexons may be comprised of the same protein subunits that constitute the pentons at the vertices. In other instances they may be comprised of a different protein. (D) In the larger icosahedral virus, the pentons at the vertices are surrounded by five hexons. Each of the hexons is surrounded here by four hexons and two pentons. Thus, notice that the pentons are foci of fivefold rotational symmetry, and the hexons are foci of twofold rotational symmetry. (E) Interactions between subunits in adjacent pentamers and hexamers cannot all be the same. Yet they can be very similar, according to the principle of quasiequivalent bonding, in which each of the subunits can be bound to its neighbors in similar, but not exactly identical, ways (e.g., the head-to-head interactions are very similar throughout the particle, regardless of whether they are between pentamers and hexamers or between hexamers and hexamers). Panels A to D were adapted from J. A. Levy, H. Fraenkel-Conrat, and R. A. Owens, *Virology*, 3rd ed. (Prentice Hall, Upper Saddle River, NJ, 1994), and panel E was adapted from B. N. Fields, D. M. Knipe, and P. M. Howley (ed.), *Fundamental Virology*, 3rd ed. (Lippincott-Raven, Philadelphia, PA, 1996), with permission.

the basis for the geodesic dome, which is exceptionally rigid, and which is said to require the least energy to assemble (Box 2.7).

Icosahedral capsids are actually assembled from pentamers of the individual subunits, which become the vertices of the icosahedron. That is, the capsids are assembled from morphological units that correspond to their vertices, rather than their faces. In the simplest cases, each of the identical triangular faces contains three identical subunits, as noted above.

As in the case of helical viruses, making a larger icosahedral capsid requires adding more protein subunits. However, whereas helical viruses can be made larger by simply extending the length of the helix, making a larger icosahedral virus is somewhat more complicated. Casper and Klug discovered the strict rules that determine how this can be done. Essentially, each of the 20 triangular faces of the icosahedron may be divided into smaller triangles, each of which contains three subunits. Thus, icosahedral capsids always contain a number of subunits that is a multiple of 60 (Fig. 2.8). The structure may no longer

look like an icosahedron per se; however, it always retains the fivefold, threefold, and twofold axes of symmetry that are characteristic of icosahedra (see if you can identify these axes of symmetry in Fig. 2.8).

Looking carefully at the arrangements of subunits in the larger icosahedral capsids, notice that the subunits at each of the 12 icosahedral vertices are arranged in groups of 5. However, all of the remaining subunits are arranged in groups of six, that is, as hexamers. In some instances (e.g., many bacteriophages), these **hexamers** or **hexons** may be comprised of the same protein subunits that constitute the pentons at the vertices. In other instances (e.g., the adenoviruses), they may be comprised of a different protein. In cases where pentons and hexons are comprised of the same protein subunit, the interactions between subunits in the pentons cannot be exactly identical to their interactions in the hexons. Yet they can be very similar. Moreover, the interactions between subunits in adjacent pentamers and hexamers cannot all be the same. Yet they too can be very similar. For these reasons, Casper and Klug introduced the concept of **quasiequivalent bonding**,

Buckminster Fuller (1895–1983) was an American architect and an early environmentalist, best known for designing the geodesic dome and for popularizing the term "Spaceship Earth." More recently, carbon molecules, known as buckyballs, or fullerenes, were named for their resemblance to Fuller's domes.

The geodesic dome is essentially an icosahedron, with its faces subdivided into equilateral triangles. However, the vertices of the geodesic dome are projected onto the surface of a circumscribing sphere. Regardless, the icosahedral building plan provides the same advantages to the architect as it does to viruses. First, there is its inherent stability, which results from the rigidity of triangles. Second, a sphere, which is more closely approached by an icosahedron than by any other polygon, has the least surface area for a given volume. This is important to the architect, since it minimizes construction costs and heat loss as well. Thus, the geodesic dome and the icosahedral viral capsid each benefit from the advantageous shape of the icosahedron. And, in each instance, by subdividing the triangles, according to certain rules, the structure comes closer to an actual sphere, while maintaining its stability.

Geodesic domes were being constructed for at least a decade before electron microscopy revealed the icosahedral nature of virus shells. But bear in mind that viruses came up with the idea well before Buckminster Fuller did.

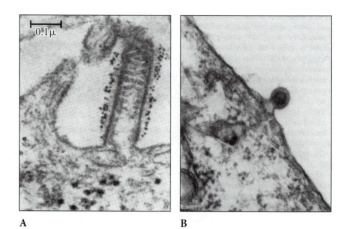

A   **B**

Figure 2.9 (A) Animal viruses with helical symmetry are generally enclosed within lipid envelopes. The electron micrograph shows a filamentous form of simian virus 5 (a paramyxovirus) budding from the surface of an infected cell. (B) Some viruses that have icosahedrally symmetric capsids also "bud" through a particular cellular membrane, thereby acquiring envelopes. The electron micrograph shows a budding type C retrovirus. Note the continuity of the viral envelope with the cellular plasma membrane. Electron micrographs by R. W. Compans and P. W. Chopin (A) and C. Moore (B); reprinted from S. E. Luria, J. J. Darnell, D. Baltimore, and A. Campbell, *General Virology*, 3rd ed. (John Wiley & Sons, New York, NY, 1978), with permission.

Depending on the particular virus, that membrane might be the plasma membrane, the nuclear membrane, the endoplasmic reticulum membrane, or a Golgi membrane.

Entry of Animal Viruses

All virus infections begin when the virus attaches to a specific receptor on the surface of a **susceptible** or **permissive** host cell. (A susceptible cell is sometimes taken to mean one that expresses a receptor that the particular virus can bind to. A permissive cell additionally expresses whatever other factors might be required to support productive replication of the virus.) The particular receptor used by an animal virus largely determines its tissue **tropism** (i.e., choice of cell type) within the animal host and, consequently, the pattern of pathology that might result from the infection. HIV, which causes acquired immunodeficiency syndrome (AIDS), provides an excellent example of this point (Chapter 21). HIV uses the helper T-cell protein CD4 (Chapter 4) for its receptor. This explains why HIV targets helper T cells and thus causes AIDS, which is the severest form of immunodeficiency known.

After binding to their receptors, animal viruses, like bacteriophages, must transport their genomes into the host cell for replication to begin. However, their means for doing so is in marked contrast to the corresponding step in a bacteriophage infection. Phages "inject" their genomes into their bacterial hosts. They must do so in order to

whereby each of the subunits can be bound to its neighbors in similar, but not exactly identical, ways (e.g., the head-to-head interactions are very similar throughout the particle, regardless of whether they are between pentamers and hexamers or between hexamers and hexamers [Fig. 2.8]). Importantly, in most viruses, all subunits interact via noncovalent bonds and, therefore, are arranged to make maximal contact.

Note that even in the cases of the smallest icosahedral viruses, which contain 60 subunits in the capsid, the subunits need not be identical. For example, the poliovirus capsid shell is comprised of three different kinds of protein subunits. Other variations of the above general themes are discussed in the context of the individual virus families.

Animal viruses with helical symmetry are generally enclosed within lipid **envelopes**. In addition, some icosahedral viruses also are enclosed within lipid envelopes. Enveloped viruses generally acquire their envelopes when they bud through a particular cellular membrane (Fig. 2.9).

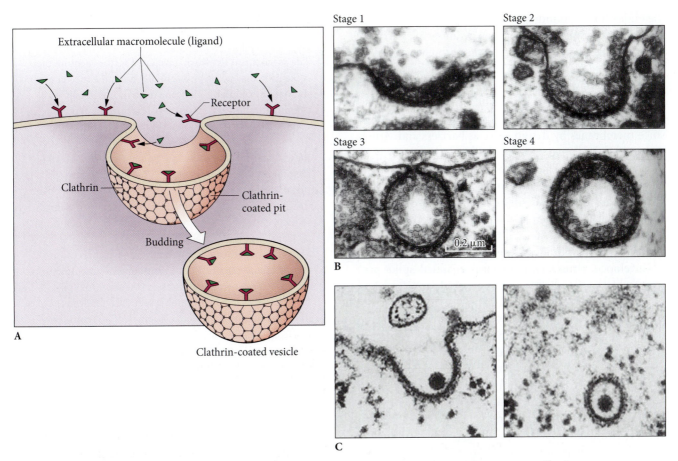

Figure 2.10 Approximately two-thirds of animal viruses are estimated to hijack the cellular receptor-mediated endocytosis pathway, which the cell normally uses for the uptake of specific macromolecules. (A) Receptor-mediated endocytosis has the following characteristics. Ligands bind to their specific receptors, which diffuse in the plane of the membrane into clathrin-coated pits. These coated pits continuously and constitutively mature into coated vesicles that "pinch off" from the plasma membrane, thereby delivering their cargo into the cell. The endocytic vesicle then traffics to, and fuses with, an early endosome, into which it releases its cargo. Note that early endosomes are somewhat acidic because of the activity of a vacuolar proton pump. Some receptors traffic back to the plasma membrane via recycling endosomes that bud from the early endosome, whereas others traffic to late endosomes, which then fuse with, or mature into, lysosomes. (B) The electron micrographs show four sequential stages from the development of a clathrin-coated pit to the release of a clathrin-coated vesicle. Adapted from M. M. Perry and A. B. Gilbert, *J. Cell Sci.* **39:** 257–272, 1979, with permission. (C) Semliki Forest virus (a togavirus) is seen associated with a coated membrane invagination (left) and enclosed within a coated vesicle (right). Electron micrographs by A. Helenius, reprinted from K. Powell, *J. Cell Biol.* **170:**1018, 2005.

breach the cell walls that are characteristic of prokaryotic cells. Since animal cells do not have cell walls, many animal viruses are able to initiate entry by subverting the very mechanisms that their eukaryotic host cells use for their normal ingestive processes.

Most animal viruses hijack the cellular **receptor-mediated endocytosis** pathway to mediate their entry. The cell normally uses this process to specifically internalize particular macromolecules (Fig. 2.10). Receptor-mediated endocytosis has the following characteristics. Ligands bind to their specific receptors, which associate with clathrin-coated pits in the plasma membrane. These coated pits continuously and constitutively mature into coated vesicles that pinch off from the plasma membrane, thereby delivering their cargo into the cell. However, bear in mind that the endocytosed virus, as well as any other endocytosed cargo, is still contained within the endocytic vesicle. Thus, the virus still must breach at least one cellular membrane in order to initiate its replication.

The endocytic vesicle generally traffics to, and fuses with, an endosomal compartment, thereby releasing its cargo into the endosomal lumen. The mildly acidic pH of the endosome then induces the virus particle to undergo structural rearrangements that can promote release of the virus or its genome into the cytosol. Both enveloped and nonenveloped viruses may access the cytosol in this way. However, we have not actually explained the processes by which animal viruses breach a cellular membrane. Breaching a biological membrane is by no means a trivial event, since membranes by design serve as barriers and viruses are very large relative to other entities that necessarily do cross membranes. The means by which animal viruses cross cellular membranes are better understood for some viruses than for others. One well-studied case is that of influenza virus (Chapter 12), and another is that of HIV (Chapter 21). Some general points regarding how viruses cross cellular membranes are as follows.

Enveloped viruses contain virus-encoded **spike proteins**, which are embedded in their envelopes (Box 2.8). The low pH of the endosomal compartment may trigger conformational changes in these spike proteins, which has the effect of exposing hydrophobic regions on them. These regions, known as fusion peptides, were not exposed when the proteins were in their native conformation. Importantly, the now-exposed fusion peptides are able to insert into the endosomal membrane. The spike proteins then undergo further conformational rearrangements, which stress the viral envelope and endosomal membrane, while also bringing them into apposition, thereby promoting fusion between them. This process releases the viral core into the cytosol.

The low pH of the endosomal compartment is able to trigger the conformational rearrangements of the spike proteins because the original native conformations of these proteins are metastable. The concept that a virus component, or indeed the entire capsid, may initially be metastable or in a metastable arrangement is extremely important apropos the seeming paradox that viruses assemble into quite stable structures at the end of one replication cycle but readily disassemble to initiate the next replicative cycle. Do you see how the notion of initially metastable structures might explain this paradox? This issue, as well as the actual conformational rearrangements of virus spike proteins and how they facilitate membrane fusion, is discussed in detail in the chapter on influenza viruses (Chapter 12).

Other enveloped viruses carry out fusion at the plasma membrane, at the time of initial attachment. In those instances, the fusion-promoting conformational rearrangements of the spike proteins are triggered by the interaction of the spike proteins with their receptors at the time of attachment.

The mechanisms by which nonenveloped viruses breach cellular membranes must be fundamentally different from

BOX 2.8

The mechanism by which enveloped viruses acquire their spike proteins begins by essentially the same mechanism that targets cellular membrane proteins to specific host cell membranes. This process is described in detail in Chapter 8. In brief, it occurs as follows. Integral membrane proteins contain so-called **signal sequences**. As these signal sequences emerge on the ribosome-bound nascent polypeptide, they target the nascent polypeptide and ribosome to translocation complexes located at the endoplasmic reticulum (ER) membrane. The signal sequence then inserts into channels in the ER membrane. As translation of the nascent protein continues, the protein moves through the channel. What then causes the protein to remain as an integral membrane protein? Like cellular integral membrane proteins, virus spike proteins contain a hydrophobic sequence that causes them to remain embedded as transmembrane proteins in the ER membrane. How then does the virus acquire its envelope, and how does the envelope acquire its integral spike proteins? Membrane vesicles containing the spike proteins pinch off from the ER and then traffic along the normal vesicle-mediated secretory pathway, fusing first to the Golgi body. Depending on the particular targeting signal on the spike protein, it may remain in the Golgi membrane, or it may undergo vesicle-mediated transport to the plasma membrane. By processes that are described in subsequent chapters, the viral core structure may assemble at the spike protein-containing membrane, thereby becoming enclosed within the envelope. If this occurs at the plasma membrane, the virus is released in the process. If it occurs at the Golgi body, the virus buds into the Golgi lumen. It then may be released by exocytosis.

that of enveloped viruses. The cellular factors that trigger the necessary conformational changes in these viruses are slowly being identified, but neither the molecular nature of the intermediate viral structures that actually breach the membrane nor the mechanisms by which they do so are well known.

Viruses that breach an endosomal membrane have adapted to undergo their conformational transitions at a pH above that which exists in lysosomes. The reason is as follows. Many cellular receptors and their ligands are destined to be degraded in lysosomes. Toward that end, endocytic carrier vesicles transport those receptors and ligands from early to late endosomes, which fuse with, or mature into, lysosomes. These organelles are more acidic than endosomes, and they also contain powerful degradative enzymes that have low-pH optima. Consequently, lysosomes would be a lethal environment for many viruses. For this reason, many viruses are adapted to undergo the low-pH-induced entry step under the milder conditions that exist in the endosomal compartment, thereby avoiding exposure to the harsher conditions within lysosomes.

The endocytic virus entry pathway confers two obvious advantages. (Are there others?) First, it is an efficient means for the virus to cross the plasma membrane and the underlying cytoskeleton, thereby gaining access to the deeper regions of the cell, where it might then breach a cellular membrane and replicate. Second, were an animal virus to leave its capsid at the cell surface (like a phage does), it would quickly alert the host's immune system to its presence there.

Specific examples of animal virus entry are discussed in detail in later chapters. For now, we can appreciate why the animal viruses we are about to encounter look so different from the T-even phages discussed above.

THE FAMILIES OF ANIMAL VIRUSES: PRINCIPLES OF CLASSIFICATION

Figure 2.11 depicts a representative sampling of some, but by no means all, of the virus families that infect vertebrate hosts. This first encounter with the animal viruses may give the impression of a somewhat daunting number of individual viruses and virus families. Moreover, animal viruses seem to come in an intimidating variety of sizes and shapes. Thus, our purpose here (believe it or not) is to simplify matters. We do this by first asserting that the most fundamental problem faced by any virus is to express its genome. Then, we see that there are only a few more than a handful of strategies that viruses use to solve this most fundamental problem. Consequently, each of the numerous families of viruses that infect humans and animals uses one of these strategies. But before we consider how these points will simplify our study of the animal viruses, we begin with a few general comments concerning the principles that have been the basis for the classification of animal viruses in the past, as well as at present.

In the early days of virology, when little was known about the biological properties of viruses, one criterion for classifying animal viruses was based on the pathologies they caused. This was not a particularly useful approach to virus classification, since the pathology of an infection had little, if any, relationship with more fundamental biological features of viruses, in particular, the strategies by which they express their genomes and replicate. Also, many viruses do not cause disease.

Another early approach to classifying viruses, which was favored at the time by epidemiologists, was based on their modes of transmission. According to this approach, some viruses were classified as "respiratory viruses," others as "enteric viruses," and yet others as "arthropod-borne viruses" (arboviruses). These terms are still used, but as in the case of a pathology-based approach to classification, they do not tell us anything about the more fundamental biological aspects of the virus.

Many biologists might rightly argue that a system of biological classification is useful only to the extent that it reflects evolutionary relatedness. But since viruses do not leave fossils, how might we infer evolutionary relationships among viruses? The answer is that we do so from morphological, from serological (i.e., immunological cross-reactivity), and, most critically, from molecular features of current viruses. Indeed, the emergence of molecular biology made it possible to classify viruses as DNA or RNA viruses, as single stranded versus double stranded, and with respect to the sizes of their genomes. Later, virus genomes were sequenced, and the arrangements of the genes within their genomes were determined. Based on this sort of information, viruses could unambiguously be placed into distinct families, the members of which clearly are related (see Box 18.11).

As seen in Fig. 2.11, viruses within each of some of the major currently recognized virus families not only have a similar size and morphology but also have genomes of common size and type (i.e., DNA versus RNA and single stranded versus double stranded). Most important, and not represented in the figure, is the fact that viruses within a family have a similar set of genes that are similarly arranged and expressed. Note that the viruses constituting some virus families have as few as 5 genes, whereas members of other virus families have 200 or more genes. Note that the arrangement of virus families in the figure is not meant to infer evolutionary relatedness between families, since these relationships are largely unknown.

Viral Genetic Systems: the Baltimore Classification Scheme

Although classifying viruses into families by the molecular criteria recounted above makes good biological sense, we are still left with a large number of diverse virus families, which might, at first sight, appear overwhelming. Therefore, we ask whether there might be some rational conceptual framework that might simplify our first encounter with the large number of animal virus families and that might also help us to keep abreast of future developments in the journal literature. David Baltimore provided just such a framework, based on the notion of grouping virus families into classes whose members have a common basic strategy for dealing with the critical issues of viral genome expression and replication (Baltimore,

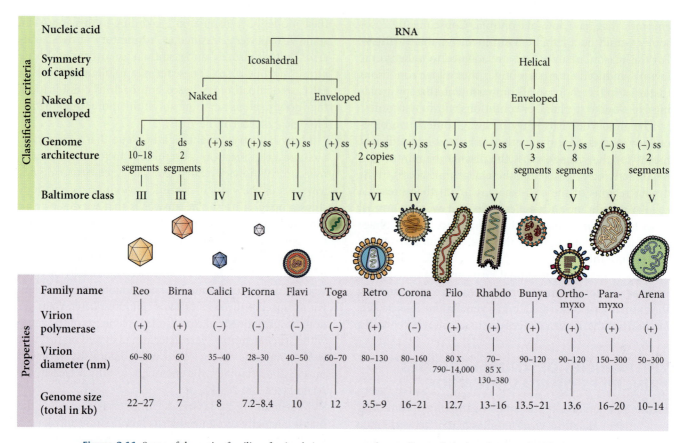

Figure 2.11 Some of the major families of animal viruses, grouped according to their shared properties. The most fundamental shared distinguishing characteristics of members of a family are the organization of their genomes and their patterns of gene expression. Other shared characteristics are shown as well. The Baltimore classification system is superimposed on the figure (see the text). The figure is not meant to indicate evolutionary relationships between virus families, since these are largely unknown. Adapted from M. H. V. van Regenmortel et al. (ed.), *Virus Taxonomy Classification and Taxonomy of Viruses: Sixth Report of the International Committee on Taxonomy of Viruses* (Springer-Verlag, Vienna, Austria, 1995), as appears in S. J. Flint, L. W. Enquist, R. M. Krug, V. R. Racaniello, and A. M. Skalka, *Principles of Virology: Molecular Biology, Pathogenesis, and Control,* 2nd ed. (ASM Press, Washington, DC, 2003).

1971). Baltimore's original classification scheme defined six such fundamental strategies and six corresponding virus classes (Fig. 2.12). We might now add a seventh class, to include the hepadnaviruses, which do not easily fit into any of the original six classes. Then, each of the recognized virus families can be grouped into one of the seven Baltimore classes. But see also Box 22.1.

To appreciate the rationale for Baltimore's approach, consider as an example the particular strategy adapted by the reoviruses, which have double-stranded RNA genomes (Fig. 2.11). We begin by noting that there is no enzymatic machinery in the cell that can replicate double-stranded RNA. Thus, if the reovirus is to replicate, it will have to produce its own replicase. However, to do this, the virus must first transcribe the viral gene that might encode its replicase. But the dilemma for the virus continues because there also is no cellular enzyme that can use double-stranded

RNA as a template for transcription. Moreover, reovirus double-stranded RNA genomes do not have mRNA activity of their own. Yet if the virus could somehow transcribe the gene that encodes its own RNA-dependent RNA polymerase (or **transcriptase**), then its predicament would be solved. The reason for this is based on a point that we developed at length earlier in the chapter, i.e., the fact that virus mRNA is translated by the cell's own translation machinery. Thus, a virus can generate any special function that it might require, provided it can transcribe the viral gene that might encode the needed function. In the case of the reoviruses, they cannot use a host RNA polymerase to make the necessary mRNA. So how do reoviruses escape from the circular dilemma we have posed? The answer is that these viruses bring the necessary transcriptase activity into the cell, contained within the virus particle. Those virus particle-associated transcriptase

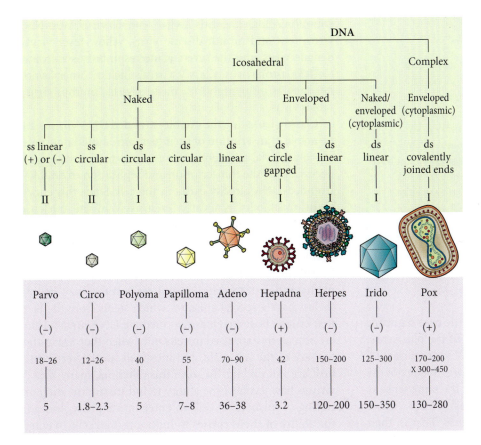

Figure 2.11 Continued

molecules were produced during the preceding cycle of infection and were incorporated into the progeny reovirus particles during their maturation. We can now appreciate what is meant by patterns of strategies of virus gene expression and how these might vary from one virus family to another. Reoviruses are the only known double-stranded RNA virus family that infects vertebrates. Bacteriophage φ6, a double-stranded RNA phage that naturally infects *Pseudomonas phaseolicola*, is discussed in Box 5.8. Double-stranded RNA viruses are class III viruses in the Baltimore classification scheme (Fig. 2.12).

The class IV viruses in the Baltimore scheme provide an interesting contrast to the reoviruses, as well as to the class V viruses. Class IV and class V viruses each have single-stranded RNA genomes, but with one fundamental difference. To appreciate that key difference, first note that at any site along a double-stranded DNA molecule, only one of the DNA strands usually serves as a template for transcription. That strand, which is complementary in base sequence to the mRNA, is referred to as the minus strand, since by convention, mRNA is always of the plus

Figure 2.12 The Baltimore classification scheme, depicting the relationship of viral genomes to the mRNAs they produce. Viruses are classified based on the nature and polarity of their genomes. Each class uses a different strategy to generate its mRNA. This provides a fundamental distinction between classes, since all virus mRNAs are translated by the cellular translation machinery. An example of each class is indicated. From D. Baltimore, 1971, as adapted in S. E. Luria, J. J. Darnell, D. Baltimore, and A. Campbell, *General Virology*, 3rd ed. (John Wiley & Sons, New York, NY, 1978), with permission.

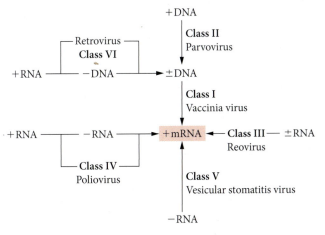

BOX 2.9

When single-stranded RNA viruses generate progeny genomes, as well as mRNAs, there is no true double-stranded RNA intermediate per se. Instead, each single-stranded RNA template simultaneously transcribes several staggered nascent single-stranded complementary RNA chains. These replicative intermediates do not collapse into double-stranded RNA forms. Perhaps single-stranded RNA viruses evolved this mode of replication to avoid inducing interferon (see Chapter 4). How then might the reoviruses avoid inducing interferon when replicating their double-stranded RNA genomes? The answer is quite astonishing, at least to me (see Chapter 5).

sense. The other DNA strand has the same base sequence as mRNA. Hence, like the mRNA, it is of the plus sense. Returning now to the class IV and class V viruses, class IV viruses have single-stranded RNA genomes that are of the plus sense. That is, their genomes are of the same sense as the mRNA that they generate. In contrast, class V viruses have genomes that are of the minus sense. That is, their genomes are complementary in base sequence to the mRNA that they transcribe.

Both class IV and class V viruses must generate complementary RNAs in order to produce mRNAs and progeny viral genomes (Box 2.9). And as in the case of the class III reoviruses, there is no enzymatic machinery made by the cell that can either transcribe or replicate RNA. How then do class IV and class V viruses manage to carry out their transcription and replication?

For class IV viruses the solution is relatively straightforward. Since their single-stranded RNA genomes are of the plus sense, their genomes may be directly translated on host cell ribosomes to yield the required virus-encoded transcriptase and replicase activities that they encode. Indeed, the first step in the replication cycle of class IV viruses, after uncoating, is the translation of their genomes on cellular ribosomes.

What then of the class V viruses? Since their single-stranded RNA genomes are of the minus sense, their genomes cannot be translated. You may have correctly recognized that their situation is essentially the same as that of the class III reoviruses. Not surprisingly, then, their solution is the same as that of the reoviruses. That is, the required polymerase activity enters the cell as a component of the class V virus particle.

Class IV viruses include such well-known examples as poliovirus, the rhinoviruses (responsible for many of our common colds), hepatitis A virus, rubella virus, West Nile virus, foot-and-mouth disease virus, yellow fever virus, dengue virus, and the severe acute respiratory syndrome (SARS) coronavirus. Class V viruses include other favorites such as influenza virus, measles virus, mumps virus, rabies virus, and Ebola virus.

Class I viruses are the double-stranded DNA viruses. In principle, transcription of their genomes should not present any special problem, since the enzymatic machinery necessary for the transcription and replication of double-stranded DNA already exists in the cell. Since this enzymatic machinery is located exclusively in the nucleus, most class I viruses indeed replicate in the nucleus. The poxviruses (Chapter 19) are the only well-studied exception to this generalization.

Although most class I viruses use cellular enzymes to mediate their transcription and replication, they nevertheless have a special concern, which stems from the fact that crucial host enzymes that mediate DNA metabolism are efficiently expressed in cells only when they are in the S phase of the cell cycle, and most cells in the vertebrate host are not cycling. To solve this dilemma, many class I viruses have evolved the ability to induce nondividing or resting (a misnomer) cells to enter into S phase. Indeed, the ability of these viruses to induce resting cells to enter into S phase underlies the ability of some of them to induce tumors in the laboratory in some instances and in their human and animal hosts, as well, in other instances. Class I viruses include papillomaviruses (responsible for cervical carcinoma and common warts), adenoviruses, herpesviruses (including the herpesviruses responsible for genital herpes, chicken pox and shingles, and mononucleosis), and smallpox virus.

Class II viruses have single-stranded DNA genomes. The parvoviruses, which include some of the smallest known viruses, are the only single-stranded DNA viruses known to infect animals. Interestingly, a subgroup of the parvoviruses, the dependoviruses (also known as adeno-associated viruses), are able to replicate only in cells that are coinfected with an adenovirus or a herpesvirus, which serves as a helper for the dependoviruses. That is, wild-type dependoviruses require one of several different unrelated helper viruses to promote their replication. Dependoviruses may encapsidate either a plus-DNA strand or a minus-DNA strand. In contrast, **autonomous parvoviruses**, which do not require a helper virus, usually package only minus-DNA strands. Regardless, transcription of parvovirus DNA by the host cell RNA polymerase requires that the parvovirus first generate a double-stranded DNA replicative form; this step is catalyzed by host cell enzymes. And, as you might then have supposed, class II viruses, like class I viruses, replicate in the nucleus.

Class VI viruses are the remarkable retroviruses. These viruses have single-stranded RNA genomes of the plus sense. However, unlike class IV plus-sense RNA virus genomes, retrovirus plus-sense RNA genomes do not serve as mRNAs. Instead, in a process catalyzed by a virus-encoded enzyme (the **reverse transcriptase**), the retrovirus RNA genome is **reverse transcribed** to generate double-stranded DNA, which then is integrated into a cellular chromosome. The integrated viral DNA, referred to as a provirus, is transcribed by the cellular pol II RNA polymerase to generate all retroviral mRNA molecules and all retroviral progeny RNA genomes as well. As you may already have surmised, the retroviral reverse transcriptase enters the cell with the viral RNA genome. There is much to say about this extraordinary virus family that will have to wait (Chapters 20 and 21). For the moment, note that this family includes the RNA tumor viruses (including the Rous sarcoma virus), studies of which led to enormous insights into cellular growth control and cancer, and HIV, the causative agent of AIDS.

The reproductive cycle of the hepadnaviruses, which include hepatitis B virus and its relatives, was determined after the Baltimore scheme was first put forward. Although hepadnaviruses have double-stranded DNA genomes, their replication cycle differs fundamentally from that of the class I viruses, and consequently, this virus family might now be seen to merit a Baltimore class of its own (i.e., as class VII viruses). In brief, the hepadnavirus replication cycle is as follows. Cellular RNA polymerase transcribes the viral circular double-stranded DNA genome into a longer-than-genome-length RNA molecule. This RNA then serves as mRNA that is translated into capsid proteins and, also, an enzyme that has reverse transcriptase activity. The next steps are truly remarkable. The long RNA, in association with the reverse transcriptase, is then packaged into capsids. Within the capsids, the RNA is reverse transcribed to generate progeny double-stranded DNA genomes. The template RNA is degraded in the process. Note that hepatitis B virus is a major human pathogen. It is a major cause of hepatitis, and it is responsible for over a million cases of fatal hepatocellular carcinoma (a liver cancer) per annum, primarily in Asia.

Since hepatitis B virus and the retroviruses use reverse transcription in their replication, we recap the essential differences in their replication strategies. Hepatitis B virus uses an RNA intermediate to replicate its DNA genome, whereas retroviruses use a DNA intermediate to replicate their RNA genomes. For further discussion of whether the hepadnaviruses should constitute a separate Baltimore class of their own, see Box 22.1.

Before moving on, note that the Baltimore classification scheme is not a formal means for classifying viruses. Instead, it provides a rational basis for grouping virus families, based on their strategies for gene expression. Knowing where any virus fits in the Baltimore scheme enables one to approach the journal literature discussing that virus with a wealth of background information about the replication strategy of that virus already at hand.

From the experience of many years of teaching virology, I am certain that I have not overestimated the value of the Baltimore scheme as a didactic approach to the animal viruses. On the other hand, be aware that there are important features other than the general patterns of transcription that distinguish one virus family from another within a Baltimore class, as well as features that distinguish between members of the same family. Apropos that point, think of the vertebrate host and its cells and tissues as the virus's environment. Also, think of the highly effective innate and adaptive immune responses of the host (Chapter 4), as well as the defensive measures that individual cells are endowed with, as factors in that environment. While these host factors would seem to make the vertebrate host a not particularly hospitable environment for any would-be microbial invader, once immersed within their hosts, viruses are exquisitely endowed to thrive. Like any other successful "organism," viruses are adapted to their environments. Virus adaptations enable these parasites to enter permissive cells within their hosts, release and express their genomes, produce and release progeny viruses, and transmit the infection to new hosts before they might be cleared by host immune functions. The variety of adaptations that different viruses have evolved, both within and between families, to carry out each of the steps of their unique lifestyles, as well as to evade and subvert host defenses, is one of the most fascinating aspects of virology.

Suggested Readings

Baltimore, D. 1971. Expression of animal virus genomes. *Bacteriol. Rev.* 35:235–241.

Cairns, J., G. S. Stent, and J. D. Watson (ed.). 2007. *Phage and the Origins of Molecular Biology: the Centennial Edition.* Cold Spring Harbor Laboratory Press, Cold Spring Harbor, NY. (This is the latest edition of the original 1966 publication.)

Ptashne, M. 2004. *A Genetic Switch: Phage Lambda Revisited*, 3rd ed. Cold Spring Harbor Laboratory Press, Cold Spring Harbor, NY.

Modes of Virus Infection and Disease

INTRODUCTION

Our purpose in this chapter is to provide an overview of the various ways different viruses may interact with their hosts in vivo. We begin by considering how different viruses are transmitted within their host populations, after which we discuss the means by which they then disseminate within an infected host. Next, we consider the diverse patterns of infection within the host, paying particular attention to the distinctions between rapidly resolving acute infections and the several kinds of persistent infections, in which the virus may persist for the lifetime of the host.

But again, this chapter is an overview, in which several key points are illustrated by a relatively few selected examples. Thus, some medically important viruses, such as the human immunodeficiency virus (HIV), the influenza viruses, and Epstein-Barr virus (EBV), may seem to be mentioned all too briefly here, and others, such as some notable hemorrhagic fever viruses, are not mentioned at all. However, all of these viruses are considered at greater lengths in later chapters.

One point to be made before moving on is that Chapter 4, which follows, is meant to provide an immunology background for all immunity-related issues that might come up anywhere in the text. Thus, the information contained in Chapter 4 would certainly make the present chapter more accessible for many readers. Nevertheless, I chose to place Chapter 4 after the present chapter for two noncritical reasons. First, I wanted to get the reader further into the subject of virology before taking a big detour into immunology. Second, the information in the present chapter does provide an important context for the material in the following chapter. So, which chapter to place first was actually a toss-up. Regardless, the real point is this: please read Chapter 4 first if that works better for you, or perhaps you might skim or read Chapter 4 along with the present chapter.

PORTALS OF ENTRY

Viruses may cause disease only after they have gained access to susceptible cells and tissues of the host. But before disease symptoms might emerge, the virus must replicate at the initial or primary site of infection, and in the case of some viruses, the infection must disseminate to secondary sites as well before disease might emerge.

Most viral infections are transmitted via the respiratory route (i.e., via inhalation). The reasons why many viruses exploit the respiratory tract as a site of entry become clear when we consider the following points regarding the respiratory tract. The purpose of the respiratory tract is to exchange oxygen for CO_2. To accomplish this task, humans respire about 6 liters of air per min. Moreover, the surface area of the respiratory **mucosa** (mucosa refers to the epithelial layer lining the respiratory, digestive, and urogenital tracts) must be large. Indeed, it is about 140 m². For comparison, the entire surface area of the skin is a mere 2 m². Furthermore, the respiratory mucosa is delicate and richly vascularized.

Since the respiratory tract has features that make it a singularly attractive portal of entry for pathogens, Nature has seen fit to endow it with an impressive array of defensive mechanisms (Chapter 4). Nevertheless, some viruses have evolved effective countermeasures that enable them to yet exploit the respiratory tract as an entranceway into the host.

The human rhinoviruses (the major cause of the common cold) and influenza viruses are among the viruses that exploit the respiratory tract as their portal of entry. In the case of these two virus groups, all of their replication may occur in the respiratory mucosa, accounting for their

all too familiar symptoms. Other viruses, such as measles virus and varicella-zoster virus (VZV) (a herpesvirus responsible for chicken pox and shingles), also enter via the respiratory tract but then disseminate to secondary sites of infection.

The skin is an inherently more formidable barrier to viruses than the respiratory tract since (i) it is several cell layers thick; (ii) the outermost layer is comprised of tough, dead, highly keratinized cells; and (iii) the skin surface is acidic due to the fatty acids secreted by the sebaceous glands (Fig. 3.1). Still, some viruses do invade the skin by taking advantage of breaks caused by injury. When the integrity of the skin is compromised, these viruses might target the mitotically active **keratinocytes** of the germinal stratum or other cells of the basal layer, as well as cells beneath the skin. (Keratinocytes are the epidermal cells that synthesize keratin and undergo characteristic changes as they move upward from the basal layers to the cornified layers of the skin.) Viruses that replicate in the germinal stratum include the herpes simplex viruses (HSVs) (the cause of cold sores, fever blisters, and genital herpes) and some poxviruses. The papillomaviruses (the cause of common warts and cervical carcinoma) have a more complex replication cycle, in which successive stages of their replication take place in different layers of the epithelium or the genital mucosa.

Other viruses initiate infection across the skin by making use of insect vectors. Yellow fever virus provides a prominent example, as noted in Chapter 1. Other notable

Figure 3.1 A schematic representation of the skin. Note the multilevel nature of the epidermis, which contains the stratum corneum, consisting of several rows of dead keratinized cells. The stratum corneum in turn overlies the stratum lucidum, the stratum granulosum, and the basal layer (stratum germinativum). Also note that the epidermis is purely a cellular structure, without blood vessels or nerves. Below the epidermis are the basement membrane and the dermis. The latter contains blood vessels, lymphatic vessels, macrophages, and fibroblasts (which form the fibrous tissue and the matrix or ground substance). Adapted from F. Fenner et al., *The Biology of Animal Viruses*, Academic Press, New York, NY, 1974, with permission.

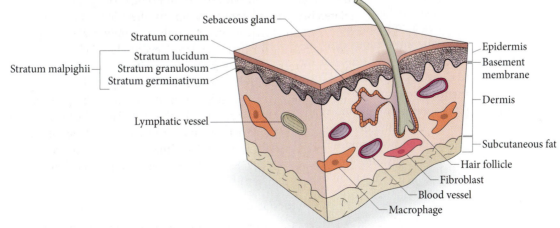

BOX 3.1

In Chapter 2, we discussed criteria that were used for classifying viruses before the molecular biology era, noting that one early classification scheme was based on mode of transmission. This approach to virus classification gave rise to the designations respiratory virus, enteric virus, and arbovirus (derived from arthropod-borne virus). The term "arbovirus" is still sometimes used today to refer to a vertebrate-infecting virus that is transmitted by an arthropod vector. But on first encountering this term, one might well conclude that it refers to a virus that actually spends its life cycle in an arthropod host. Contrary to that expectation, arboviruses are actually dependent on their vertebrate hosts, as follows. The arthropod becomes infected upon taking a blood meal from an infected vertebrate. The virus then replicates in the arthropod's gut epithelium, from which it disseminates to the salivary gland, where it mixes with the saliva. Thus, a bite can transmit the virus to the next vertebrate victim. Arboviruses generally produce only persistent infections in the arthropod host, while producing either acute or persistent infections in the vertebrate host. There are genuine insect viruses, such as the baculoviruses (not discussed further in this text), which indeed spend their entire life cycles in the arthropod host.

examples include West Nile virus, Japanese encephalitis virus, dengue virus, and St. Louis encephalitis virus, among others (Box 3.1).

Finally, we should acknowledge that the skin may be penetrated and infection may occur as a direct result of human activity. This may happen inadvertently during medical procedures, such as blood transfusions. Such medically induced events are said to be **iatrogenic** (from *iatros*, the Greek word for physician). More commonly, penetration of the skin occurs as a consequence of the sharing of hypodermic syringes by intravenous drug users. Note that this is a major route of transmission for HIV and for hepatitis B and C viruses (HBV and HCV).

The mucosa of the **alimentary tract**, which lines the entire cavity running from the mouth to the anus, is another major portal of virus entry. As you might expect, the food we eat and the fluids we drink may be contaminated with a variety of pathogens. Accordingly, the alimentary tract, like the respiratory tract, is armed with a formidable array of defensive measures. These begin in the oral cavity, where the alimentary tract is protected by an epithelium that is several cell layers thick. Moreover, the oral cavity is bathed in saliva, which contains several important immune

mediators, including phagocytes and complement, and also enzymes (e.g., amylases) that may destroy viruses and virus-infected cells. But then, as you might have expected, viruses that exploit the oropharynx as a site of entry have evolved countermeasures to subvert these host defenses. Examples of these viruses include HSV (oral herpes), EBV (a herpesvirus that causes infectious mononucleosis), and some papillomaviruses.

The oral cavity leads to the gastrointestinal tract, which includes the stomach, small intestine, large intestine, rectum, and anal canal. The gastrointestinal tract too is a harsh environment for most viruses. Wild fluctuations in pH are found there, as the stomach is acidic (epithelial cells lining the gastric glands secrete hydrochloric acid, reducing the pH of the stomach to as low as 2), whereas the intestine is alkaline. In addition, there are digestive enzymes (e.g., proteases, amylases, and lipases) and bile salts (which emulsify fats and destroy viral envelopes), and the luminal surfaces of the tract are covered with mucus that also harbors antibodies and phagocytes.

The mucosa of the small intestine also includes the **Peyer's patches**, which are focal aggregates of organized lymphoid tissue (Fig. 3.2 and 3.3). Like lymph nodes, Peyer's patches are sites at which antigens are presented to the effector cells (i.e., B and T lymphocytes) of the adaptive immune system, thereby activating an adaptive immune response against the invader (Chapter 4). Lymphocytes enter Peyer's patches, as well as lymph nodes, from the blood circulation. However, whereas antigen enters lymph nodes from the lymphatic vessels that drain the infected sites, there are no lymphatics that drain into Peyer's patches. How then does antigen enter Peyer's patches? The answer is through specialized cells of the Peyer's patches called **M cells**, which specialize in transporting antigens from the intestinal lumen to the underlying cells of Peyer's patches. In order to carry out their task, M cells have a very thin cytoplasm, and they transport antigens by **transcytosis**, delivering them virtually intact to the underlying lymphoid tissue. (Transcytosis is a process that occurs in polarized epithelial and endothelial cells [see below]. Ligands are taken up in endocytic vesicles on one surface of the polarized epithelium. Then, the cargo is transported to the opposite surface of the epithelium by a vesicle-mediated process. At the opposite surface, the cargo is released intact by fusion of the transport vesicle with the plasma membrane. As noted above, M cells are highly specialized for transcytosis.)

Unfortunately for the host, some pathogens subvert the very properties of M cells that enable those cells to transport antigen to the underlying Peyer's patch, doing so to cross the intestinal epithelium. For example, when reoviruses are taken up by M cells and transferred by

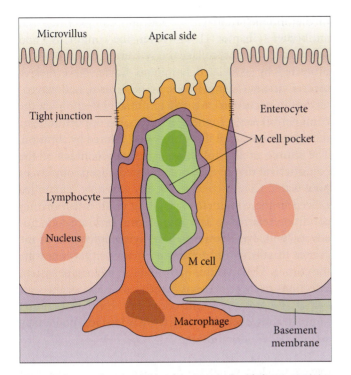

Figure 3.2 Drawing of a section of the intestinal epithelium containing an M cell connected to two **enterocytes** (i.e., simple columnar epithelial cells found in the small intestines and colon). Lymphocytes and macrophages move in and out of the invagination of the M cell at its basolateral side. Notice the thinness of the M-cell cytoplasm. Although M cells are present for defensive purposes, a variety of microbial pathogens exploit M cells to cross the intestinal epithelium. Adapted from A. Siebers and B. B. Finlay, *Trends Microbiol.* **4**:22–28, 1996, and B. Alberts et al., *Molecular Biology of the Cell*, Garland Publishing, New York, NY, 1994, with permission.

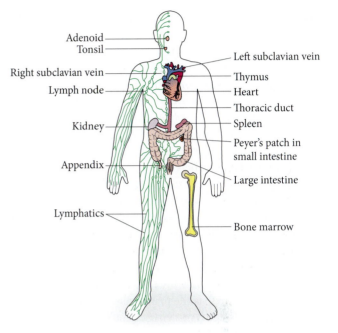

Figure 3.3 Distribution of lymphoid tissues in the body. Note the concentration of lymph nodes in the region of the nasopharynx, the lungs, and the gastrointestinal and urogenital tracts. Modified from R. Coico, G. Sunshine, and E. Benjamani, *Immunology, A Short Course*, 5th ed. (John Wiley & Sons, Hoboken, NJ, 2003), with permission.

transcytosis to underlying Peyer's patches, these viruses then use the occasion to disseminate through the host via the circulation. In contrast, the related rotaviruses replicate in, and destroy, M cells. This brings about the inflammation of the gut and diarrhea characteristic of rotavirus infections.

Despite the above two examples of viral invasion via M cells, the epithelial cells and white blood cells of the gastrointestinal mucosa are the principal cells that viruses target in that tissue. The corresponding cells of the oropharynx likewise are the principal cells targeted by viruses at that site.

Other virus types that initiate infection in the gastrointestinal tract include adenoviruses, parvoviruses, picornaviruses (a family that includes poliovirus, hepatitis A virus, and enteroviruses), coronaviruses, and the Norwalk calicivirus (renamed norovirus). It is not entirely clear how these viruses might breach the defensive barriers of the gastrointestinal tract. Yet each of these viruses is resistant to low pH, to gastrointestinal proteases, and to the emulsifying action of bile detergents. (With the exception of the coronaviruses, none of these viruses are enveloped.)

Note that the rhinoviruses, which cause common colds, are picornaviruses. As such, they are closely related to the other picornaviruses enumerated above. Yet the rhinoviruses are sensitive to low pH, and they have adapted to infecting the respiratory tract rather than the alimentary tract (Chapter 6). Thus, closely related viruses may use different portals of entry and different target cells in which to replicate.

HIV, when transmitted via anal intercourse, can infect the lower gastrointestinal tract without having to traverse the upper respiratory tract and all the inherent hazards it poses to a virus. Under these circumstances, HIV can infect **dendritic cells** in the epithelium of the anus, another instance in which a virus subverts an immune effector cell for its own purposes. Dendritic cells are called "professional antigen-presenting cells" (Chapter 4). They continually sample the environment at sites of pathogen entry. Upon being activated by a pathogen, an immature sentinel dendritic cell matures into an effector dendritic cell and migrates to a draining lymph node, where it presents antigen to a naïve T cell (i.e., a T cell that has not previously encountered its specific antigen and been activated), thereby activating an adaptive immune response (Chapter 4). In the case of HIV infection, when the dendritic cell delivers HIV to the lymph node, in point of fact it is delivering the virus to the very site at which its principal host cell, the CD4 T lymphocyte, is concentrated. Importantly, the tropism of HIV for CD4 T cells is the major determinant of the pathogenesis of **acquired immunodeficiency syndrome (AIDS)** (Chapter 21). Note that the virus may not need to be internalized by the dendritic cell during these events (a point discussed at greater length in Chapter 21). A similar course of events may occur during vaginal intercourse (see below), but the risks of infection are greater during anal intercourse because of the greater risk of tissue damage there.

The **urogenital tract** can be invaded by several viruses, despite the fact that it too manifests several barriers to infection. Regular urination is one such barrier, since it flushes out most pathogens before they might initiate infection. In addition, urine contains immunoglobulin A (IgA) antibodies (the type of antibodies secreted by mucosal lymphoid tissue [Chapter 4]), and it is also slightly acidic.

Sexual intercourse is the principal means by which several medically important viruses invade the urogenital tract. The tears and abrasions that result from normal sexual activity may facilitate infection via this route. We noted above that HIV can be transmitted via anal sex. However, the mucosal surfaces of the female genital tract are actually the primary portal by which HIV is spread. That is to say, heterosexual transmission is the major route by which HIV is spread worldwide (Chapter 21). Other viruses that invade the urogenital tract include HSVs, papillomaviruses (the cause of genital warts and cervical cancer, as noted before), and HBV.

For HIV to be transmitted via the female urogenital tract, the virus must cross the genital mucosa so that it may make its way to a lymph node, where it finds its principal target cell, the CD4 T lymphocyte. The mechanism by which HIV accomplishes these steps is not well understood. One possibility is that HIV crosses the mucosa via lesions caused by other infections. However, experiments with rhesus macaques show that mucosal lesions are not necessary for infection with the related simian immunodeficiency virus, which appears to be able to cross an intact epithelium. HIV might cross the urogenital epithelium by M-cell-mediated transcytosis. The virus might then exploit the fact that Langerhans cells, which are the dendritic cells at that site, lie directly under the mucosa. That is, HIV might use the Langerhans cells for transport to the lymph node (see above). Coincidentally, these Langerhans cells might be especially well suited to this task because they express a particular lectin, DC-SIGN, that binds the HIV envelope glycoprotein gp120. As noted above, the virus may not need to be internalized by the dendritic cell during these events.

The **cornea** and **conjunctiva** (i.e., the inner surface of the eyelid) are protected by the continual cleansing action of the blinking response. In addition, tears contain IgA antibodies (Chapter 4). Moreover, the conjunctiva and **sclera** (i.e., the fibrocollagenous covering of the eye) are protected by associated lymphoid tissue. Nevertheless, several viruses (adenoviruses, herpesviruses, and enteroviruses in particular) are able to initiate infection via the conjunctival route. In these instances, the opportunity for infection is increased if the mucosa is injured by abrasion. The source of the infection is usually contact of the conjunctival tissue with contaminated fingers, objects, or pool water. Conjunctival infections generally remain localized in the conjunctiva, resulting in an inflammatory condition referred to as **conjunctivitis**, or more colloquially as "pink eye." However, herpesvirus infections of the cornea usually spread to sensory neurons that innervate the cornea. These viruses then establish lifelong persistent infections in these neurons, which rarely are serious (see below). Enterovirus 70 infections of the eye seldom disseminate, but when they do, they may lead to a potentially serious paralytic condition.

ROUTES OF DISSEMINATION

Hematogenous and Neural Dissemination

As we have seen, viruses invade the body through breaks in the skin or by infecting the mucoepithelial membranes that line the respiratory tract, oropharynx, gastrointestinal tract, urogenital tract, and eyes. In many instances, the virus remains at the initial site of infection, and any symptoms that may arise emanate from that site. For example, the rhinoviruses, which cause common colds, invade the upper respiratory tract and do not disseminate from that initial site of infection. Other viruses may replicate at the primary site but then disseminate to secondary sites. In these cases, symptoms might result from tissue damage at the secondary site.

The lymphatic system and the blood circulation are the major conduits for virus dissemination in the body. As virus particles are released from infected cells into the extracellular space, they may then enter a local **afferent lymphatic vessel**, which drains the infected site. These afferent lymphatics normally deliver antigen and dendritic cells to the nearest draining lymph node (Fig. 3.3). Although free virus particles may drain to the lymph nodes, some viruses may be phagocytosed by sentinel dendritic cells at the infected site (Chapter 4) and then subvert those phagocytes to transport them to the nearest lymph node. In that regard, recall that during vaginal and anal intercourse, HIV interacts with dendritic cells in the epithelium. Those dendritic cells might then deliver the virus to lymph nodes, where the virus is able to infect its principal target cell, a CD4 T cell.

Other viruses, such as several enteric viruses of the picornavirus and reovirus families, bind to receptors on M cells, which then transport them to underlying Peyer's patches of the lymphatic system. But by whatever means virus particles enter into the lymphatic circulation, they are delivered to local draining lymphoid organs via afferent lymphatic vesicles. To appreciate what happens next, we have to briefly consider the lymphatic circulation. Blood-borne lymphocytes enter the lymph nodes through **postcapillary venules**. They return to the blood circulation by leaving the lymph nodes via **efferent lymphatic vesicles**, which converge in the **thoracic duct** (Fig. 3.3). The thoracic duct empties into the **vena cava**, the vessel that returns blood to the heart, and thus returns lymphocytes back to the blood circulation. This lymphatic circulation is crucial in immunity, because it enables antigen, antigen-presenting cells, and lymphocytes to come together in lymphoid organs to initiate an adaptive immune response (Chapter 4). But some viruses exploit the lymphatic circulation, such that *the lymphatic system is the major route* by which viruses invade the bloodstream.

There are still other means by which a virus may enter the blood circulation. For instance, infection often leads to a local inflammatory response, the extent of which depends on several factors (Chapter 4). Inflammation is characterized by the dilation of local blood vessels, rendering them more permeable to the cellular and humoral mediators of immunity. Although the inflammatory response is a host defensive measure, the increased permeability of capillaries at the infected site may enable some viruses to enter the blood circulation. Viruses also may enter the blood by replicating in the endothelial cells that line the blood capillaries. Arboviruses may be transmitted to the blood by the bite of an arthropod vector. Also, virus entry into the blood circulation may be iatrogenic or, more commonly, may occur as a result of needle sharing by drug users, as in the cases of HIV and hepatitis B and C viruses (see above).

The initial appearance of virus in the blood is referred to as a **primary viremia**. If the virus should then replicate in a secondary site of infection, such as the endothelial lining of blood vessels or the liver, spleen, or lungs, then progeny virus from the secondary site also may enter the blood circulation, resulting in a **secondary viremia**. Secondary sites of infection, such as the liver, may support extensive virus replication. Accordingly, secondary viremias usually lead to much greater levels of virus in the blood than primary viremias. Measles virus provides a good example of a virus that gives rise to acute primary and secondary viremias (see below and Chapter 11). The high levels of virus in the blood resulting from the secondary viremia may enable the virus to infect yet other generally inaccessible sites, such as the central nervous system and brain. When many organs become infected, the infection is said to be **systemic**.

Virus in the blood may be free, or it may be associated with lymphocytes or **macrophages** (macrophages are cells that are specialized to phagocytose microbial pathogens [Chapter 4]). Most viruses that are taken up by macrophages are likely to be destroyed by them. However, some viruses may replicate in macrophages, using them as a metaphorical Trojan horse to mediate their transport to other tissues of the body. Indeed, some viruses may enter the central nervous system by infecting migrating macrophages, while other viruses may use lymphocytes for that purpose.

While some viruses can invade the central nervous system or brain via the blood circulation, other viruses can spread from the primary site of infection to the central nervous system by infecting the endings of neurons that innervate or are near the infected site (Fig. 3.4). Indeed, neural spread is second only to the blood circulation as the means by which viruses disseminate in the host.

In order for a virus to replicate in a neuron, it must make its way to the cell body of the neuron, since protein synthesis does not occur in the extended processes of these cells. Viruses access the neuronal cell body and spread along neurons by microtubule-associated fast axonal transport. Indeed, neural spread is a defining characteristic of infections by certain viruses, including rabies virus, HSV, and VZV. For poliovirus, neural invasion occurs inadvertently and rarely, but it is a critical factor in the pathogenesis of paralytic poliomyelitis (Chapter 6).

Only olfactory neurons have nerve endings that are exposed to the environment. Accordingly, **neurotropic** viruses generally must replicate in nonneuronal cells before they are able to access neurons. Since the epithelial mucosas are usually the primary sites of virus infection, the first neurons to be infected are commonly components of the peripheral nervous system. Rabies virus, which is transmitted by an animal bite, is generally introduced into muscle tissue that is rich in nerve endings. Thus, it can then invade and ascend peripheral neurons, eventually crossing their synapses by transsynaptic spread. Virus spread can continue in this way, with the virus eventually making its way to the central nervous system or brain.

HSV and VZV (the neurotropic herpesviruses) typically become latent in the cell bodies of peripheral neurons (see below and Chapter 18). When reactivated, they usually make their way back down the neuron and reinfect the original epithelial site of infection. Only rarely will these herpesviruses also move in the **retrograde** direction to the central nervous system. Thus, the more common direction of herpesvirus movement following

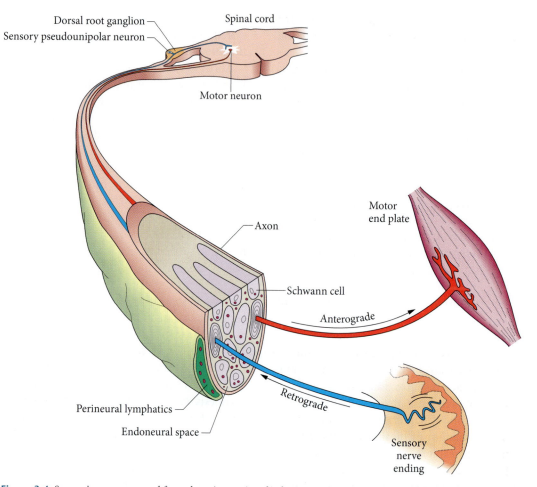

Dorsal root ganglion

Sensory pseudounipolar neuron

Spinal cord

Motor neuron

Axon

Motor end plate

Schwann cell

Anterograde

Perineural lymphatics

Endoneural space

Retrograde

Sensory nerve ending

Figure 3.4 Some viruses can spread from the primary site of infection to the central nervous system by infecting the endings of neurons that innervate or are near the infected site. Directed transport within the neuron is defined as anterograde (movement from the minus to the plus ends of microtubules) or retrograde (movement from the plus to the minus ends of microtubules). Adapted from R. T. Johnson, *Viral Infections of the Nervous System*, Raven Press, New York, NY, 1982, with permission.

reactivation results in a cold sore, whereas the rarer retrograde transport leads to lethal viral encephalitis (defined below). (Note that **anterograde** versus **retrograde** neuronal transport refers to the direction of microtubule movement. Anterograde is from the minus ends to the plus ends of microtubules. Thus, movement toward the cell body is called retrograde transport and movement toward the synapse is called anterograde transport, irrespective of the direction with regard to the central nervous system.)

Encephalitis refers to infection of the actual neurons of the brain and brain stem, as distinct from their supportive tissue. Rabies virus provides a well-known example of a virus that gives rise to encephalitis in the brain. Note that rabies encephalitis is almost invariably fatal, whereas other instances of viral encephalitis can have a more favorable outcome with appropriate care.

Infection of the brain may give rise to other distinct and important clinical conditions, in particular **aseptic meningitis** and **myelitis**. Beginning with aseptic meningitis, many viruses, including some enteroviruses (Chapter 6) and HSV type 2 (HSV-2) (sometimes referred to as genital herpes [Chapter 18]), infect the **meninges** (the lining of the brain and the brain stem), where they give rise to viral or aseptic meningitis. (Aseptic meningitis is a misnomer used to distinguish viral meningitis and meningitis from noninfectious causes [e.g., a side effect of medications] from meningitis caused by other infectious agents.)

When viruses infect the neurons of the spinal cord, they can give rise to a condition known as myelitis. For example, poliovirus, an enterovirus of the picornavirus family (Chapter 6), which typically infects the gastrointestinal tract, can give rise to poliomyelitis, a condition characterized by neuronal loss and rapidly progressive paralysis. HSV-2 too can cause myelitis, as well as meningitis, as noted above.

Before a virus can invade the central nervous system, it first must cross the so-called **blood-brain barrier**. The physical nature of the blood-brain barrier and the means by which certain viruses are able to traverse it are discussed below. But first, consider that the different clinical outcomes that result from virus infections of the central nervous system result in large part from differences in the cell and tissue **tropism** of different viruses within the central nervous system. Tropism refers to the specific cells and tissues that a virus infects within its host. For RNA viruses, tropism is largely (but not entirely) determined by the cell- and tissue-specific expression of the virus receptor. For DNA viruses, receptor distribution also is a crucial determinant of tropism, but other cellular factors come into play as well. Regardless, given the vital role of the central nervous system in the life of a vertebrate, infection of this organ system poses serious hazards.

Rabies virus is much more efficient than HSV at invading the central nervous system. That is, it is much more **neuroinvasive**. However, HSV, like rabies virus, can also cause fatal disease upon infecting the brain. That is to say, HSV and rabies virus are each highly **neurovirulent** (Box 3.2). Mumps virus is highly neuroinvasive; indeed, it is more so than even rabies virus. However, mumps virus is not particularly neurovirulent. Thus, neuroinvasiveness and neurovirulence do not necessarily go hand in hand.

Next, we consider the factors that determine whether an infection remains local or instead disseminates to become systemic. The *distribution of the receptor or receptors* used by a particular virus is one key determinant of its pattern of dissemination. For example, the rhinoviruses, coronaviruses, influenza viruses, and measles virus are all transmitted by the respiratory route, and all infect the respiratory tract. However, measles virus is the only one of these viruses that disseminates throughout the body. One reason why measles virus disseminates whereas these other viruses remain in the respiratory tract is that measles virus can bind to several different receptors, including CD46, which is present on most cells of the body. The consequence of this fact is that serious complications from measles can occur in almost every organ system of the body.

Since blood-borne dissemination is the most common means by which viruses spread to infect secondary sites, the ability of a particular virus *to cross the mucosal barrier* and then invade the underlying tissue and mucosal blood vessels is another determinant of whether a virus can spread systemically. Apropos this discussion, the cells of the mucosal epithelium and of the vascular endothelium are **polarized**. That is, **tight junctions** between these cells separate their **apical surface** from their **basolateral surface**, thereby preventing the free diffusion of membrane constituents from one surface to the other. (In the case of epithelial cells, the apical surface is the exposed free surface, and the basal surface contacts underlying cells of the epithelium or the extracellular matrix. In the case of endothelial cells, the apical surface faces the lumen of the vessel and is in direct contact with the blood circulation, whereas the basal surface faces the tissue and contacts the extracellular matrix.) The physical as well as physiological separation of the apical from the basolateral surfaces enables each of these surfaces to be specialized to carry out its particular functions. In that regard, each surface contains receptors unique to that surface.

If a virus that was replicating in an epithelium was to assemble and release its progeny from the apical surface exclusively, as the rhinoviruses do, then the progeny virus would end up where the infection began, at the mucosal surface. In that event, the infection would remain local. On the other hand, if the virus were to assemble and release its progeny from the basolateral surface, then the virus would be in the subepithelial connective tissue, positioned to invade and cross the vascular endothelium to invade the blood circulation. Actually, it is a bit more complicated than that, since viruses that disseminate in this way generally replicate in the local subepithelial connective tissue, before succeeding in crossing the endothelial barrier. Some viruses cross endothelial barriers by infecting and replicating in the vascular endothelial cells, whereas others may cross the endothelium by transcytosis.

BOX 3.2

Rabies virus may well be the most lethal virus that infects humans. The virus typically invades the peripheral nervous system following infection by a bite, and mortality is nearly 100% once clinical disease occurs. On the other hand, most of those individuals exposed to rabies virus (i.e., by the bite of an infected animal) do not get infected themselves. For example, in a classic study of this phenomenon, only 2 of 27 people who were bitten by a rabid wolf in Tehran, Iran, contracted clinical rabies. Regardless, if rabies infection should occur, it is almost invariably fatal, whereas untreated HSV infections of the brain are lethal in about 50% of cases. Moreover, while treatment against rabies is not effective once the virus has invaded the brain, treatment of neural HSV infections results in a significant drop in fatalities.

Apical release and basolateral release each provide distinct advantages to a virus. For example, copious apical release of rhinoviruses from the nasal mucosa facilitates highly efficient transmission of the virus via droplets from nasal secretions. Likewise, apical release of poliovirus from the gastrointestinal mucosa facilitates transmission of that virus via the fecal-oral route. In contrast, viruses released from the basal surface have the advantage of moving away from the mucosal defenses. This ability to disseminate from the initial site of infection also creates the opportunity for some viruses to establish a persistent infection at a secondary site. The ability to establish persistent infection is an important viral strategy, since it enables a virus to survive in small host populations. That is so because the virus has a home while awaiting new susceptible (i.e., nonimmune) individuals to be born into the population (see below).

As noted above, some viruses are able to travel via the lymphatic vasculature to nearby draining lymph nodes. Some viruses do so as free viruses, while others do so by infecting phagocytes, which then migrate to the nearest lymph node. Regardless, these viruses then may enter the blood circulation by passing via the efferent lymph to the thoracic duct (Fig. 3.3). Some viruses may undergo further replication in the lymph nodes before being disseminated via the efferent lymph.

Viruses that are carried by infected lymphocytes or phagocytes are protected from antibodies and complement in their travels to distant sites. This cell-associated form of viremia is a feature of measles virus infection and of several herpesvirus infections. Other viruses, such as HBV, rubella virus, enteroviruses, and flaviviruses (e.g., West Nile virus, St. Louis encephalitis virus, and HCV), circulate as free viruses in the blood.

Viremia usually persists until the adaptive immune system brings the infection under control (Chapter 4). However, in some cases, the immune system cannot clear the virus from the host, and the result is a persistent viremia. Persistent viremia is characteristic of HIV and HBV infections, playing an important role in the pathogenesis and transmission of each (see below).

While it is true that a disseminated infection can be very serious, it would be wrong to assume that localized infections are necessarily mild. On the contrary, some localized infections can be quite severe. For example, rotaviruses can do considerable damage to the intestinal epithelium, leading to severe diarrhea that may be life threatening in infants and young children (Chapter 5). Likewise, influenza virus infections of the respiratory tract epithelium produce a serious and life-threatening illness in many individuals (Chapter 12).

Having accounted for how viruses enter the blood circulation, we now need to account for how they exit from

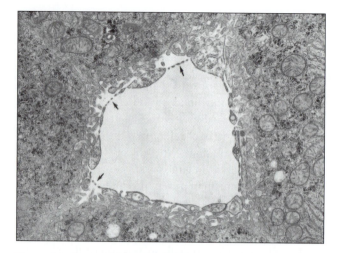

Figure 3.5 Electron micrograph of a sinusoid in fat liver. The endothelium is extremely attenuated in some areas, where there are fenestrations that give it a sieve-like structure (arrows). From W. Bloom and D. W. Fawcett, *A Textbook of Histology*, 10th ed. (W. B. Saunders Company, Philadelphia, PA, 1975), with permission.

the circulation so that they might infect secondary sites. Although almost any organ in the body may be infected by some virus, individual viruses tend to have preferred exit sites from the circulation, which are based on special features of the blood-tissue junctions at those sites.

The liver, spleen, bone marrow, and adrenal glands constitute one group of exit sites from the blood that are preferred by viruses. The common feature of the blood-tissue junctions at these sites is that in each instance they contain **sinusoids**, which serve to filter the blood and remove foreign particles. (A sinusoid is a small blood vessel, similar to a capillary, but with a discontinuous endothelium. Some gaps between the sinusoid endothelial cells are large enough for blood cells to pass through [Fig. 3.5].) To carry out their filtering function, sinusoids are lined with macrophages, which are a component of the **mononuclear phagocyte system**, previously known as the **reticuloendothelial system**. Macrophages of the mononuclear phagocyte system include the peritoneal macrophages of the peritoneal fluid, alveolar macrophages of the lung, and microglial cells of the central nervous system. While the major function of the mononuclear phagocyte system is to phagocytose microorganisms and foreign substances that are in the blood and various tissues, the anatomy and histology of the blood-tissue junctions in the liver, spleen, bone marrow, and adrenal glands result in a barrier that is more permeable to viruses than the blood-tissue barrier at other sites.

Viruses in the blood generally can infect reticuloendothelial system macrophages of the liver, which are known as **Kupffer cells**. The outcome of that infection depends

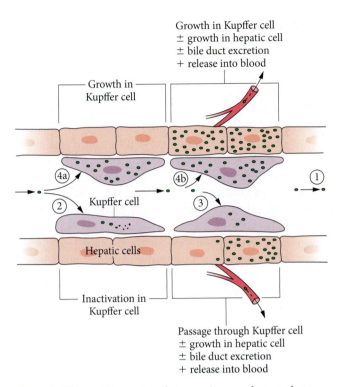

Growth in Kupffer cell
± growth in hepatic cell
± bile duct excretion
+ release into blood

Growth in Kupffer cell

Kupffer cell

Hepatic cells

Inactivation in Kupffer cell

Passage through Kupffer cell
± growth in hepatic cell
± bile duct excretion
+ release into blood

Figure 3.6 Types of interactions between viruses and macrophages, exemplified by the Kupffer cells lining a sinusoid in the liver. (1) Macrophages may fail to phagocytose virions; for example, in Venezuelan equine encephalitis virus infection, this is an important factor favoring prolonged viremia. (2) Virions may be phagocytosed and destroyed. Because the macrophage system is so efficient, viremia can be maintained only if virions enter the blood as fast as they are removed. (3) Virions may be phagocytosed and then passively transferred to adjacent cells (hepatocytes in the liver). If, like Rift Valley fever virus or HBV, the virus replicates in these cells, it can cause clinical hepatitis, and the virus produced in these cells in the liver can produce a high level of viremia. (4) Virions may be phagocytosed by macrophages and then replicate in them. With some viruses, such as lactate dehydrogenase virus in mice, only macrophages are infected (4a), and progeny virions enhance the viremia, which reaches an extremely high level. More commonly (4b), as in yellow fever, virus replicates in both macrophages and hepatic cells, producing severe hepatitis. Adapted from D. O. White and F. J. Fenner, *Medical Virology*, 4th ed. (Academic Press, Inc., San Diego, CA, 1994), with permission.

on the state of activation of the Kupffer cell, as well as on the particular virus (Fig. 3.6). Some viruses may be destroyed by Kupffer cells. Other viruses may replicate in Kupffer cells, thereby creating an opportunity to infect the liver parenchyma. Still other viruses may access the underlying hepatocytes by transcytosis across the endothelium. Regardless, viral invasion of the liver may cause severe hepatitis, as in the case of yellow fever. In the instances of HBV and HCV, infection of the liver may be chronic, with the possible development of **cirrhosis** and, less commonly, fatal **hepatocellular carcinoma** (see below).

BOX 3.3

Viral infection of the pancreas has long been thought to be a possible cause of Type I (insulin-dependent) diabetes mellitus (Chapter 4). Evidence from seroepidemiological surveys, and more recently from PCR-based procedures, implicates enteroviruses, as well as several other viruses, as etiologic agents for Type I diabetes. For example, studies show that enterovirus infections accompany or precede the onset of Type I diabetes in many children. Moreover, infection with enteroviruses appears to be linked to the induction of autoantibodies against pancreatic islet cells, as well as to the expression of IFN-α. Both of these events are connected with islet cell destruction. Experiments in the mouse model provide compelling evidence that viruses may cause diabetes in that system, but there is still no definitive evidence that viruses cause diabetes in humans.

The kidney glomeruli, the pancreas, the small intestine, and the colon constitute another preferred set of virus exit sites from the blood. The common feature of the blood-tissue junction at these sites is that the endothelium there is "fenestrated." That is, there are **fenestras**, or windows (more precisely, loose junctions), between the endothelial cells that enable viruses to pass more easily to the underlying tissue. Some viruses, such as HSV, measles virus, and yellow fever virus, cross these endothelia as passengers within macrophages or lymphocytes. These cells enter the underlying tissue by squeezing between the endothelial cells in a process referred to as **diapedesis** (Box 3.3).

Virus exit from the circulation is not restricted to those preferred sites enumerated above. For example, viruses also exploit the fact that the blood flow is slowest, and the endothelial barrier is thinnest, within capillaries and venules. Making use of that fact, a virus can initiate its passage across the vascular endothelium by binding to a specific receptor on the apical surface of the endothelium. The virus might then be transcytosed across the endothelium, or it might have to replicate in the endothelium in order to traverse it. (What factors might determine whether a virus is transported across an endothelium by transcytosis?) Since virus dissemination across the endothelium by either of these possibilities is thought to be inefficient, infection of secondary sites via these routes likely requires high concentrations of virus in the blood, a possibility that is favored by extensive replication at the primary site of infection.

As mentioned previously, some viruses may invade the central nervous system from the blood circulation, and as just noted, the chances of this happening depend on the level of the viremia. Moreover, there are two routes by which the viruses might invade the central nervous system (Fig. 3.7). In the first of these routes, the virus crosses the endothelium of the blood vessels that vascularize the brain parenchyma. But note that the endothelia of these vessels are characterized by unique features to prevent viruses from doing just that. The endothelial cells of these capillaries are connected to each other by special tight junctions, and they are backed by a dense basement membrane that is comprised of an extracellular condensation of mucopolysaccharides and proteins. These tight junctions and the underlying basement membrane constitute the blood-brain barrier. Yet despite this blood-brain barrier, most viruses that invade the central nervous system do so by crossing these endothelia. Examples include poliovirus, togaviruses, bunyaviruses, and parvoviruses. Some viruses are transported across this endothelium, courtesy of infected monocytes or lymphocytes. Examples include HIV, measles virus, mumps virus, and JC virus (a human polyomavirus).

The second viral route from the blood circulation to the brain is via the capillaries in the meninges and choroid plexus. The endothelial cells of these capillaries do not form tight junctions characteristic of the capillaries that vascularize the brain parenchyma. Instead, these endothelia are fenestrated, and their basement membranes are sparse. This anatomy of the blood-tissue junctions in the meninges and choroid plexus enables some viruses to pass from the blood circulation into the cerebrospinal fluid. Viruses that cross the blood vessels of the meninges or that grow in the epithelium of the choroid plexus may then give rise to meningitis. In these instances, virus particles are found in the cerebrospinal fluid. Additionally, some viruses in the cerebrospinal fluid may be able to infect the ependymal cells that line the ventricles and thereby invade the underlying brain tissue, where they give rise to encephalitis (Fig. 3.7).

The Placenta and the Fetus

The placenta is the organ that exchanges nutrients and waste products between the maternal and the fetal circulatory systems. It is absolutely essential to the survival of all mammalian species. Here, we consider another role for the placenta, as a protective interface between the mother and the fetus.

We begin our discussion of the placenta by first noting that it is derived entirely from the embryo. Thus, the placenta, like the fetus, is a foreign tissue within the mother's body. This fact has given rise to the notion that pregnant women are in an immunologically compromised state, so as not to reject either the placenta or the fetus. Another widely held concept is that newborns are immunologically ineffectual.

Neither of the above ideas can be entirely true. Indeed, if they were true, then it would be most difficult to account for our survival as a species. That is so because we are in effect barraged by a battery of potentially lethal infectious agents. Yet, remarkably, among viruses only a handful or so infect the human fetus with any notable frequency. These include rubella virus, cytomegalovirus (a herpesvirus), parvovirus B19, and HIV. Other viruses, including HSV-1 and HBV, also may be transmitted

Figure 3.7 Two routes by which viruses might invade the central nervous system from the blood circulation. In the first, the virus crosses the endothelium of the blood vessels that vascularize the brain parenchyma. But note that the tight junctions between the cells of the central nervous system endothelia and the underlying basement membrane constitute the so-called blood-brain barrier. Yet despite this blood-brain barrier, most viruses that invade the central nervous system do so by crossing these endothelia. Some viruses do so via infected monocytes or lymphocytes. The second route by which viruses pass from the blood circulation to the brain is via the capillaries in the meninges and choroid plexus. The endothelial cells of these capillaries do not form tight junctions characteristic of the capillaries that serve the brain parenchyma. Instead, these endothelia are fenestrated, and their basement membranes are sparse. This anatomy of the blood-tissue junctions in the meninges and choroid plexus enables some viruses to pass from the blood circulation into the cerebrospinal fluid. Viruses that cross the blood vessels of the meninges or that grow in the epithelium of the choroid plexus may then give rise to meningitis. In these instances, virus particles are found in the cerebrospinal fluid. Additionally, some viruses in the cerebrospinal fluid may be able to infect the ependymal cells that line the ventricles and thereby invade the underlying brain tissue, where they give rise to encephalitis. Adapted from C. A. Mims et al., *Mims' Pathogenesis of Infectious Disease*, Academic Press, Orlando, FL, 1995, with permission.

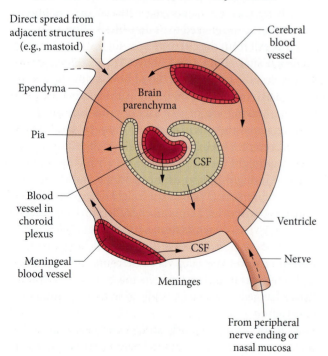

transplacentally. However, transplacental infections by these viruses are rarer still.

Another point to bear in mind is that during birth, the neonate is virtually bathed in a sea of microbes that are present in the mother's genital tract and anorectal sites. Consequently, the baby's skin, eyes, oropharyngeal mucosa, and possibly upper and lower respiratory tracts are exposed to these microbes. One could hardly expect the baby to survive if it were defenseless against this microbial onslaught. Likewise, despite anecdotal reports to the contrary, objective studies of matched pregnant and nonpregnant women have shown no significant difference between them regarding the severity of most infections.

Bearing the above points in mind, it is not yet clear how the fetus avoids being rejected by the mother's immune system, but it is not because the mother is immunologically impaired. Quite the contrary is the case, since preexisting immunity in the mother likely acts to impede transmission of pathogens to the fetus. (One factor that might act to protect the placenta from maternal immunity is that the fetal trophoblast cells, which constitute the outer layer of the placenta and contact the maternal tissue, do not express either class I or class II major histocompatibility molecules [Chapter 4].)

The mother's antibodies of course protect the fetus by protecting the mother from infection. Moreover, maternal antibodies by design (i.e., those of the IgG class [Chapter 4]) are transferred across the placenta to the fetus. Thus, maternal antibodies protect the fetus on both sides of the placenta (Box 3.4).

Considering the importance of the placenta to the survival of humans as a species, it is remarkable how little is known about how viruses might cross it. One reason for our ignorance regarding this issue is that placentas are not routinely examined in cases of transplacental infections. Moreover, even in instances where the placenta is examined, infection of the fetus might have occurred without any obvious signs of placental involvement.

Since most viruses disseminate via the circulation, the great majority of transplacental infections of the fetus occur when the mother is **viremic**. This fact leads us to two key questions. First, how is the placenta able to prevent most of the different viruses that might be in the mother's blood from passing to the fetus? Second, what are the special features of those few viruses that do cross the placenta?

Regarding the second question, we can eliminate size as the viral factor, since cytomegalovirus, a large virus, and parvovirus B19, a small virus, both efficiently cross the placenta. Moreover, EBV, which like cytomegalovirus is a herpesvirus, only rarely crosses the placenta, despite the fact that it too is also commonly in the maternal blood. Thus, there must be virus-specific factors that confer the

---

BOX 3.4

Since the baby automatically carries the mother's antibodies for as long as 2 years or more, there is a problem in determining whether babies of HIV-infected mothers who test HIV positive by antibody-based tests are actually infected, as opposed to just carrying the mother's anti-HIV antibodies. On the same point, in the developed world, mothers of newborns who test HIV positive are advised not to breast-feed their babies. What do you suppose is the reason behind this measure? The answer is that the baby might test positive only because it carries the mother's antibodies, not because it is actually infected. If that were the case, then the mother might yet transmit the infection to her baby via her breast milk. But what then is the policy in underdeveloped nations? There, mothers are advised to breast-feed their babies because of the lack of available health care and adequate nutrition.

An interesting social commentary is that in much of the underdeveloped world, stigma is attached to a mother who transmits the infection to her baby, but not to the man who transmitted the infection to the mother.

---

ability to cross the placenta, although size does not appear to be one of them.

There are several potential mechanisms by which viruses might cross the placenta, but there is little hard evidence to support any one of them. But before considering these putative mechanisms, it would be most useful to have a look at the anatomy of the placenta (Fig. 3.8). Pay particular attention to the patterns of the fetal and maternal circulation, noting that the placental villi absorb nutrients from the mother's blood in the intervillous space and excrete wastes into that space. On the maternal side of the placenta, blood from branches of the uterine arteries passes through openings in the basal plate of the placenta. From there, maternal blood passes into the intervillous space. Thus, it is reasonable to postulate that viruses can enter the intervillous space in the mother's blood. A virus might then infect the cells of the villi, thereby enabling the virus to invade the fetal blood circulation. The villi are comprised of trophoblasts, which also line the intervillous space. Consequently, trophoblasts would be the major target cell in this scenario, one that is supported by observations that these cells can be infected by some viruses.

In order for a virus to infect trophoblasts, the virus would need to express an attachment protein that could recognize a receptor on those cells. Moreover, the trophoblasts

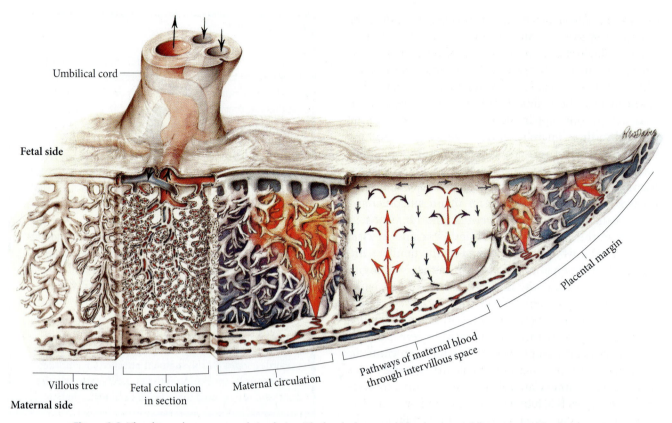

Umbilical cord —

Fetal side

Placental margin

Pathways of maternal blood through intervillous space

Villous tree — Fetal circulation in section — Maternal circulation

Maternal side

Figure 3.8 The placenta's structure and circulation. The head of maternal blood pressure drives entering blood toward the chorionic plate in fountain-like spurts. As the head of pressure is dissipated, lateral dispersion of blood occurs. Inflowing arterial blood pushes venous blood out into the endometrial veins. The intervillous space is lined by trophoblasts. The villi absorb nutrients from the maternal blood in the intervillous space and excrete wastes into it. From W. Bloom and D. W. Fawcett, *A Textbook of Histology,* 10th ed. (W. B. Saunders Company, Philadelphia, PA, 1975), with permission.

would then need to provide all the factors that the virus might need to support its intracellular replication.

Although the requirements for viral infection of the trophoblasts appear rather simple, the placenta still constitutes a formidable defensive barrier against transplacental virus infections. That might be so, at least in part, because even if a virus were to succeed in entering the placental villi, it still would encounter intervillous defenses, such as the placental macrophages known as **Hofbauer cells.** Moreover, even if a virus were to succeed in making its way to the fetus, it would still encounter the fetal defenses, which become more imposing as development proceeds.

As noted above, rubella virus, human cytomegalovirus, parvovirus B19, and HIV are responsible for the majority of transplacental viral infections. So, what might be the consequences of these infections to the fetus? B19 infections of the fetus usually cause no permanent harm. But less than 10% of the time, these infections lead to death of the fetus and spontaneous abortion. In the case of HIV, children who are infected with this virus in utero and who are not given antiretroviral therapy will eventually develop

fatal AIDS. In contrast, rubella virus and cytomegalovirus are not fatal to the fetus or to the child after birth. Instead, they are **teratogenic.** That is, they cause abnormal development, which results in birth defects. Indeed, congenital cytomegalovirus infection is the leading infectious cause of mental retardation, deafness, and visual impairment. These viruses are each discussed further in their respective chapters. The emphasis here is on their interactions with the placenta.

Placental involvement in rubella virus infection in utero has been well documented for more than 40 years. In these instances, damaged foci have been seen in the trophoblasts that line the villi and the intervillous space. Moreover, the endothelia of the placental capillaries and larger vessels also show signs of damage. In addition, rubella virus often can be isolated from the placentas of early abortion specimens. Nevertheless, by the end of pregnancy, the virus is generally not detected in the placenta, even though it persists in the fetus. What hypothesis might you formulate to explain those facts?

Cytomegalovirus reaches the fetus in about 40% of primary maternal infections, and this virus is present in as many as 2.3% of all newborns. (A primary infection is the initial introduction of an infectious agent into the body. Cytomegalovirus gives rise to latent infections that spontaneously reactivate [see below]. A mother initially infected with the virus during pregnancy will not have antibodies in place to control her infection or to pass on to the fetus. In contrast, if a mother's latent infection reactivates during pregnancy, she will have antibodies in place to control her infection and to pass on to the fetus.) Evidence for placental involvement in cytomegalovirus infections in utero comes primarily from the analysis of abortion specimens. In these specimens, virus can be detected in villus trophoblasts and in the vascular endothelium. In addition, explants of fetal trophoblasts can be infected with cytomegalovirus in vitro. As noted previously, most transplacental infections in utero are transmitted from the maternal blood circulation. But, in the case of cytomegalovirus, virus that is excreted from the cervix of infected women (after a recurrence) also can get to the fetus by ascending up the genital tract to the amnion.

About 30% of women who are infected with parvovirus B19 during pregnancy pass that virus on to the fetus. Thus, parvovirus B19, like cytomegalovirus, appears to be rather efficient at crossing the placenta. Placental involvement in these fetal infections is evident from the considerable damage to the placental vascular endothelium.

The consequence of HIV infection in utero is enormous. It is estimated that during 2002 and 2003, some 1.6 million children worldwide were born infected with HIV. Although this virus also may be transmitted to a newborn via the mother's milk, experimental evidence shows that over 90% of pediatric AIDS patients acquired the virus in utero from an HIV-infected mother or during passage through the birth canal. In most instances of HIV infection in utero, the fetus is infected sometime after the first trimester, but infection can occur as early as the eighth week of gestation. The likelihood that the fetus will be infected by an untreated infected mother is estimated to be as high as 30 to 50% (Box 3.5).

The mechanism of HIV transmission in utero is not well understood, but HIV type 1 is detected in the placenta in vivo, and it can replicate in placental explants in vitro. Indeed, placental trophoblasts are susceptible to HIV infection in vitro, although about 10-fold less so than CD4 T cells, which are the primary target cells of HIV. In addition, there are research reports that HIV may also cross the polarized trophoblast cell layer by transcytosis. This process may occur independently of, or concurrently with, virus replication.

ACUTE VERSUS PERSISTENT INFECTIONS

This section considers the key distinctions between acute and persistent virus infections and the variety of patterns

BOX 3.5

As is discussed at greater length in Chapter 21, the AIDS epidemic has compelled individuals, and society as a whole, to confront old attitudes and prejudices and to make difficult moral judgments. One such complicated moral judgment concerns making choices in the administration of AIDS therapies, particularly in underdeveloped countries, in the face of high costs. Consider that antiretroviral therapies administered to the mother during pregnancy significantly reduce the incidence of HIV transmission to the fetus. At first glance, the decision here would appear to be obvious. Tragically, however, this practice also leads to many abandoned orphans, many of whom are HIV infected despite treatment and who will eventually themselves succumb to AIDS. In the countries that are hardest hit by HIV, including Zimbabwe, Botswana, Namibia, Swaziland, and South Africa, one in three children are presently orphans of parents who succumbed to AIDS. In point of fact, 50% of new mothers in South Africa likely will die of AIDS. Bearing these facts in mind, it is not universally agreed that saving an infant (perhaps only temporarily) of an HIV-infected mother is the wisest use of the limited funds that might be available for all worthy public health measures in resource-limited, underdeveloped countries. What do you think?

that the latter may display. We begin by enumerating the defining characteristics of acute infections. Viruses are rapidly produced during the course of an acute infection, and symptoms appear suddenly if there is an associated illness. Acute infections terminate in either of two possible outcomes. In one outcome, the infection is rapidly cleared by host immune defense mechanisms, and the patient recovers and is immune to the specific virus. The alternative outcome is for the host to quickly succumb to the infection. In either case, the duration of the episode is generally less than 2 weeks.

The alternative courses that a virus infection might follow are referred to collectively as **persistent infections**, in which the virus may persist for years, perhaps even for the normal life span of the infected individual. Such persistent infections are categorized as **latent**, **chronic**, or **slow**, depending on the pattern of virus production and actual disease episodes (Fig. 3.9). The defining features of these different sorts of infections are noted below, and several examples are considered in each instance. Some specific diseases are discussed in more detail in the chapters pertaining to the particular virus families.

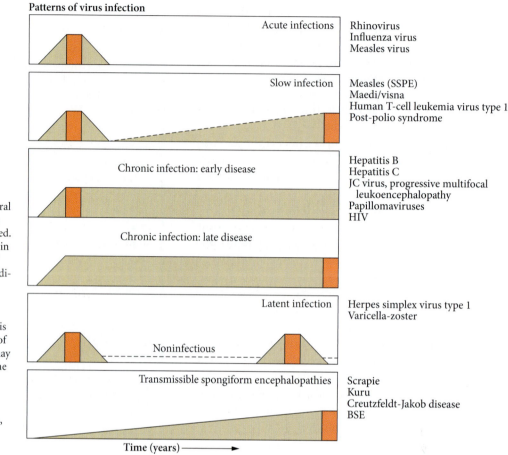

Patterns of virus infection

Acute infections — Rhinovirus / Influenza virus / Measles virus

Slow infection — Measles (SSPE) / Maedi/visna / Human T-cell leukemia virus type 1 / Post-polio syndrome

Chronic infection: early disease — Hepatitis B / Hepatitis C / JC virus, progressive multifocal leukoencephalopathy / Papillomaviruses / HIV

Chronic infection: late disease

Latent infection — Herpes simplex virus type 1 / Varicella-zoster

Noninfectious

Transmissible spongiform encephalopathies — Scrapie / Kuru / Creutzfeldt-Jakob disease / BSE

Time (years) →

Figure 3.9 Diagram of the general patterns of a viral infection, with several selected examples indicated. The time scale is arbitrary. Areas in orange indicate the presence of virus and disease. Areas in tan indicate the presence of virus but no disease. When under a solid line, virus is readily demonstrable. When under a dashed line, virus is not readily demonstrable. Some of the viruses named as examples may give rise to infections that have the characteristics of more than one category. Modified from D. O. White and F. J. Fenner, *Medical Virology*, 4th ed. (Academic Press, Inc., San Diego, CA, 1994), with permission.

Acute Infections

Viruses such as the rhinoviruses and influenza viruses, which remain localized at the primary site of infection, generally give rise to acute infections. Although some localized acute infections (for example, those involving influenza viruses) can be lethal, localized acute infections generally are less life threatening than disseminated infections, and they are commonly resolved with the elimination of the virus from the body.

Because those viruses that give rise to acute infections generally infect cells that make up the mucosal surfaces, they usually are efficiently transmitted from one individual to another. This is so despite the fact that these infections are usually quickly resolved. As a consequence of their high transmission rates, viruses that give rise to acute infections tend to be associated with widespread epidemics that might affect millions of individuals. Influenza is a good case in point. (See Box 11.14 for the other characteristics of viruses that give rise to periodic epidemics.)

The **rhinoviruses** provide a good example of a virus group that is well adapted to exploit the acute-infection lifestyle. To better appreciate how rhinoviruses do so, it

would be useful to first note that the mammalian immune system is comprised of two branches: the innate immune system and the adaptive immune system (Chapter 4). The innate immune system is ready to respond immediately and nonspecifically to a broad range of invaders. In contrast, the adaptive immune system needs to be activated to tailor a specific response to each particular invader. Whereas adaptive immunity is required to resolve infections by most viruses, some sources state that the innate immune system is often sufficient to resolve rhinovirus infections, since these infections resolve within just a few days, that is, before the adaptive immune system might be fully activated. Whether or not a rhinovirus infection might actually be resolved by the innate immune system alone is discussed below and in Chapter 4. Regardless, the early resolution of rhinovirus infections poses no particular obstacle to the success of these viruses, since they replicate very efficiently during that brief interval of only several days. Moreover, the inflammatory response, which in reality is a defensive stratagem of the innate immune system, produces the runny nose, sneezing, and cough reflex that facilitate the efficient transmission of rhinoviruses from one individual to another. Furthermore, if the adaptive immune system is not fully activated, then there may be little acquired immunity against the same rhinovirus **serotype**. (Serotypes are groups of viruses that are defined by their interactions with antibodies. Neutralizing antibodies against one serotype do not neutralize another serotype. As noted below, there are more than 100 distinct rhinovirus serotypes that circulate in humans.)

Considering the unique capability of the lymphocytes of the adaptive immune system to target and destroy virus-infected cells (Chapter 4), many virologists and immunologists would not expect that a virus infection could be completely resolved in the absence of adaptive immunity. Also arguing against the premise that rhinovirus infections might be resolved by innate immunity only is the fact that increased numbers of T lymphocytes (which mediate the adaptive immune response against virus-infected cells) are found in nasal secretions only 3 to 4 days following experimental inoculation with rhinoviruses.

But if adaptive immune responses do play a role in the resolution of rhinovirus infections, then infection should also result in specific immunity against reinfection by the same virus (Chapter 4). Why then do we have rhinovirus common colds throughout our lifetimes, sometimes as many as several per year? First, there are more than 100 antigenically distinct rhinovirus serotypes circulating in humans. Consequently, there are always some serotypes against which a person may have little, if any, acquired immunity. Second, human rhinoviruses use several different strategies to evade neutralization by antibodies, as discussed in detail in Chapters 4 and 6.

Influenza viruses, like rhinoviruses, cause only acute infections, and in the case of influenza, the adaptive immune system is clearly needed to resolve the infections. Thus, influenza virus may have several days more than the rhinoviruses to transmit the infection from one individual to another. The first new influenza viruses appear in the respiratory mucosa, the first day after infection. Virus levels peak at about the fourth day, when the adaptive immune system springs into action. Virus levels then begin to decline by day 6, but the patient may continue to be infectious for several days more. One distinction between a rhinovirus common cold and influenza is that the patient may feel well soon after the rhinovirus infection has cleared, while the influenza patient may continue to feel ill for days, even weeks afterwards, due primarily to lingering effects of the immune responses (Chapter 4).

Recalling that rhinoviruses and influenza viruses remain at the initial site of infection, note that some disseminating viruses also give rise to acute infections. In those instances, the infections at both the primary and the secondary sites are acute. **Measles virus** is an example of a disseminating virus that gives rise to acute infections. Like influenza viruses and the rhinoviruses, measles virus spreads by the respiratory route. However, the course of a measles virus infection is quite different from that of the two classic respiratory viruses. The initial measles virus infection of the respiratory mucosa produces little transmissible virus. The reason is that innate immunity is extremely effective at controlling the initial measles virus infection at that site. Thus, measles virus would likely not do very well as a respiratory virus unless it had some additional stratagem up its viral sleeve. Indeed, measles virus has adapted to infect dendritic cells at the respiratory mucosa. (Dendritic cells, mentioned above, are discussed at length in Chapter 4. We note here that they serve as sentry cells in the tissues, where they phagocytose pathogens and their antigens, which they then deliver to lymph nodes. In the lymph nodes, they present these antigens to the effector cells of adaptive immunity, thereby activating an adaptive immune response.) Measles virus not only survives in dendritic cells but also replicates in them. Thus, when a measles virus-infected dendritic cell arrives at a lymph node, it turns the lymph node into a secondary site of infection. But measles virus is yet more diabolical; infection of lymph nodes temporarily impairs the function of the T cells and B cells that mediate adaptive immunity, thereby increasing the efficiency of the secondary infection.

The measles virus subterfuge described above results in a massive secondary viremia that enables the virus to infect other tissues, as well as to reinfect the respiratory mucosa. The secondary attack on the respiratory mucosa is initiated by many more viruses than initiated the primary infection. Consequently, sufficiently large numbers of virus are produced during the reinfection of that mucosa to enable very efficient transmission of the virus to new individuals. Indeed, measles virus is one of the most contagious viruses in humans, infecting about 40 million individuals worldwide per annum. One to two million of those infections are fatal. This complex course of events during measles virus infections explains why there usually are no symptoms until 10 days or so. (A recent new study implies a different scenario of measles virus infection, in which lung lymphocytes are the first cells to be infected. These primarily infected lymphocytes then transport the virus to the basolateral surface of the airway epithelium, which then sustains measles virus shedding into the airway lumen, a step that is crucial for transmission of the virus. Importantly, primary infection of the lymphocytes was found to be sufficient to give rise to viremia and clinical measles. Experimental details are recounted in Box 11.7.)

In approximately 1 of every 300,000 measles infections, the virus persists in the brain of the infected individual for years, eventually producing an invariably fatal neurodegenerative disease called subacute sclerosing panencephalitis (SSPE). Persistent virus infections are discussed at some length below. We mention SSPE here because this late complication of measles virus infections demonstrates that some viruses that are generally associated with acute infections also may give rise to persistent infections. See also "Slow Infections," below, for more information on SSPE.

Persistent Infections

Persistent virus infections usually begin as typical acute infections but then develop into one of the recognized categories of long-term persistent infections (i.e., latent, chronic, and slow). Some persistent infections are associated with very serious human illness, as evidenced by such diseases as AIDS, resulting from infection with HIV; cirrhosis and hepatocellular carcinoma, resulting from chronic infections with HBV and HCV; and cervical carcinoma, associated with the human papillomaviruses.

From the point of view of the virus, persistent infection is an important strategy that enables it to maintain itself in the host population, particularly in small populations. This is a very important point. Virus persistence provides time for new susceptible individuals to be born or migrate into the host population. These new members of the host population then might themselves become carriers of the virus. (Measles virus, which gives rise only to acute infections, requires a host population of at least 500,000 individuals in order to ensure a sufficient quantity of new susceptible individuals to sustain the virus in the population [Chapter 11]. Because populations of such size did not exist in preagricultural times, measles virus could not have been sustained in the human population until relatively recently in human history. But humans are the only host for measles virus. This seeming paradox is explained in Chapter 11. In contrast to the acute measles virus, persistent viruses, such as the herpesviruses, have been coevolving with their hosts for millions of years [Box 18.11].)

For a virus to establish a long-term persistent infection, it must avoid being cleared by the immune system of the host. Moreover, in instances where the virus is cytocidal, the virus must also be able to modulate expression of those viral genes that sustain lytic infection, so as not to kill the host cell. The first need might be met in part by the persistence of the virus in immunologically privileged sites, such as the central nervous system. Perhaps this explains why so many persistent virus infections affect the central nervous system. Additionally, persistent viruses generally have evolved other rather clever means for evading host immunity. These strategies are reviewed in Chapter 4 and are discussed further in chapters dealing with the particular virus families. The second need also has immune implications, since the death of infected cells is one of the main triggers of innate immunity, which in turn initiates adaptive immune responses (Chapter 4).

It is convenient to categorize persistent virus infections as slow, chronic, or latent, based on the pattern of virus production and the emergence of symptoms (Fig. 3.9). Yet some persistent infections show features of more than one of these categories and cannot be pigeonholed neatly into any one of them. Thus, note that other authors may use examples that here illustrate one category of persistent infection as an example of another.

Slow Infections

Slow virus infections may begin as **fulminating** acute infections (i.e., ones where illness comes on suddenly, with severe symptoms of short duration). However, a defining characteristic of slow infections is that in virtually all instances disease progresses slowly and relentlessly towards death. Note that "slow" more appropriately refers to the time course of the disease per se, not the rate at which the virus replicates. Virus replication may very well be rapid, during the acute phase at least. Afterwards, however, replication may occur at a low level. For this reason, some authors prefer the term "smoldering infection" when referring to this class of infections. Another characteristic of slow infections is that virus generally can be recovered from the experimental animal or patient during the long preclinical phase, as well as after symptoms have emerged. These infections also have been called "acute infections with rare late complications." The continuous low levels of virus replication characteristic of slow virus infections tend to continually stimulate the T-cell population, leaving little time for its recovery. This may lead to impaired T-cell function, thereby facilitating viral persistence.

SSPE exemplifies a slow virus infection (Fig. 3.9). As noted above, measles virus persists in the brain of approximately 1 of every 300,000 individuals who are naturally infected with measles virus. After an incubation period of 1 to 7 years, the virus may give rise to SSPE, which is invariably fatal. This disease is discussed in greater detail in Chapter 11. For now, note that SSPE patients display little, if any, measles virus in the brain. However, they have exceptionally high titers of neutralizing anti-measles virus antibodies in the cerebrospinal fluid. Together, these facts imply an ongoing low level of virus production in the brain (i.e., a smoldering infection). This slow replication and spread of measles virus in the brain and the high antibody titers in the cerebrospinal fluid are defining characteristics of SSPE. The virus enters the brain via lymphocytes, which it infects during the primary or secondary viremia (see above). Much remains to be learned concerning the viral and host factors that underlie virus persistence and pathogenesis in SSPE.

Maedi/visna (initially recognized in the 1930s, when it appeared in Icelandic sheep) was in actuality the first slow virus disease to be characterized. In fact, the term "slow infection" was originally coined to describe this condition. Maedi/visna is caused by a lentivirus, a subgroup of the retroviruses that also includes HIV. It is a demyelinating disease of sheep, with an incubation period of about 2 to 3 years, which progresses slowly but inevitably to death. Virus is found intermittently in the cerebrospinal fluid and blood, and neutralizing antibodies are present, although they obviously do not eliminate the virus. The virus is not **cytopathic** (i.e., virus replication and release per se do not lead to cell death), and disease may be due to an antigen-antibody response on the surface of infected glial cells.

Another retrovirus, the **human T-cell leukemia virus type 1 (HTLV-1)**, is asymptomatic in more than 95% of infected individuals. Still, it is responsible for several rare, late complications. One is **adult T-cell leukemia**, which has an incidence of about 2% in infected individuals. Another is **HTLV-1-mediated tropical spastic paraparesis**, again with an incidence of about 2%. This is a paralytic disease, characterized by demyelination of long neurons in the spinal cord. This pathology is believed to result from an autoimmune attack on the myelin, brought on by the persistent HTLV-1 infection. Each of these HTLV-1-mediated diseases emerges 30 to 50 years postinfection!

HTLV-1 is found almost exclusively in association with CD4 T cells in infected individuals. That is to say, there is little, if any, free virus that can be detected in these patients. Consequently, HTLV-1 is believed to be transmitted by means of these cells, which are present in semen, blood, and breast milk. Accordingly, routes of transmission are believed to be via sexual contact, through exposure to contaminated blood (either through blood transfusion or sharing of contaminated needles), and from mother to child via breast-feeding. Surprisingly, perhaps, free HTLV-1 is actually very inefficient at initiating infections.

Here now is an interesting point. Since in each of the above modes of HTLV-1 transmission infected cells, rather than free virus, must be passed from an infected individual, how does the virus actually infect the recipient's cells? Remarkably, when an HTLV-1-infected lymphocyte contacts an uninfected lymphocyte, a "virological synapse" forms between the two cells. Both cell-to-cell contact (involving the cell surface proteins ICAM-1 and LFA-1 [Chapter 4]) and expression of the HTLV-1 Tax protein are necessary to induce the synapse. One study (Igakura et al., 2003) reported that this contact between an HTLV-1-infected lymphocyte and an uninfected lymphocyte induces the formation of macromolecular complexes in the infected

cell, localized at the synapse, which contain the viral RNA genome and the precursor polyprotein (Gag) for the viral reverse transcriptase (Chapter 20). Moreover, cell-to-cell contact induces the rearrangement of the cytoskeleton beneath the complexes, which implies that microtubules are involved in transporting RNA/Gag complexes to the synapse and into the recipient cell. These experimental findings are consistent with the apparent fact that lymphocytes that are naturally infected with HTLV-1 do not readily produce free enveloped extracellular virus particles. Note that HIV too can be transmitted by direct cell-to-cell contact via a virological synapse (Chapter 21).

The ensuing replication cycle of HTLV-1 is most unusual, even for a retrovirus. And the atypical HTLV-1 replication cycle is important in the current context because it explains why, in contrast to other infections that lead to slow virus diseases, HTLV-1 infections do not begin as fulminating acute infections. First note that a key step in the replication of all retroviruses, including HTLV-1, is for the plus-strand RNA genome to be reverse transcribed by the viral reverse transcriptase into a double-stranded DNA genome, which is then integrated by covalent bonds into the cellular genome (Chapter 20). Importantly, reverse transcription per se does not replicate the viral genome. Instead, it merely exchanges the single-stranded RNA genome for one comprised of double-stranded DNA. Most retroviruses then amplify their genomes by transcribing the integrated **proviral** DNA genome, as mediated by the cellular RNA polymerase II. (The term "provirus" usually refers to the integrated DNA form of a retrovirus genome.) In contrast, HTLV-1 replication and persistence occur mainly as follows. Like other retroviruses, HTLV-1 generates a DNA copy of its genome, which is then integrated into the host cell genome. When the cell divides, the integrated provirus genome is replicated along with the cellular DNA and is passed in the integrated state to each of the daughter cells. Since free virus is very inefficient at infecting new cells, this is the principal means by which the HTLV-1 infection is amplified. And, as noted above, the infection is transmitted to new individuals via these infected cells. (This means of viral replication might be referred to as "clonal proliferation." What experimental evidence might support this mechanism? Most importantly, treatment of HTLV-1-mediated tropical spastic paraparesis patients with reverse transcriptase inhibitors, which are routinely used in antiretroviral therapy [Chapter 4], does not change the proviral loads in these patients. This would be expected if free virus does not contribute to the proviral burden.)

HTLV-1-infected cells are not killed by these events. This is an important point, since the nonlytic nature of the virus is a factor enabling it to persist. Another point to note is that the proviral DNA can be integrated into the host cell DNA only when the host cell is replicating. Thus, the virus remains in an as yet poorly defined kind of latent state in an infected cell until such time as the cell might replicate, which, in the case of a CD4 T cell, occurs when the cell is presented with its cognate antigen (Chapter 4). As we will see when we consider the features of chronic and latent infections, HTLV-1-associated infections display features of these other categories of persistent infections, as well as of slow infections.

One last thought regarding HTLV-1 is that the mechanism by which this leukemia virus induces neoplasia is different from that of other oncogenic retroviruses. Unlike many of these other retroviruses, HTLV-1 does not harbor a viral homolog of a cellular **proto-oncogene**. (An oncogene is a gene that can cause a cell to become malignant. Retroviral oncogenes are generally derived from normal cellular counterparts that have been incorporated into viral genomes and altered so they malfunction when expressed in an infected cell [Chapter 20].) They do not play a role in virus replication. In contrast, the primary neoplastic activity of HTLV-1 maps to a viral gene that encodes its Tax protein, which plays a key role in virus replication and which has multiple effects on intracellular signaling pathways, cell cycle progression, and chromosome stability.

Postpolio syndrome, a late **sequela** (i.e., a disease that is caused by a preceding disease or injury in the same individual) of poliomyelitis (Chapter 6), illustrates another variation of a slow virus infection. This condition may occur 30 to 40 years after the original poliovirus infection, in as many as 80% of the original victims. Affected individuals suffer a further deterioration of the originally affected muscles. In this condition, the virus cannot be detected and may very well not even be present. It is believed that the syndrome results from a later loss of neurons in the initially infected nerves, due to the greater burden borne by them over the years.

Chronic Infections

Chronic virus infections are generally characterized by continuing high levels of virus replication and viremia. However, disease may be absent, or it may be chronic, or it may develop late in some instances (Fig. 3.9). In any case, and in contrast to slow virus infections, chronic infections do not lead inexorably to death. (Chronic infections with some viruses may, in rare instances, lead to fatal neoplasia. For example, chronic HBV infections may progress to fatal hepatocellular carcinoma [see below]. Nevertheless, hepatocellular carcinoma is rare, even among HBV carriers. Still, it is about 100-fold more prevalent in these individuals than in controls.)

HBV and HCV are two unrelated viruses, each of which gives rise to medically important chronic infections of the liver. However, despite infecting the same target organ and causing similar clinical outcomes, there nevertheless are marked differences in the patterns of chronic infection of these two distinct viruses. JC virus, a human polyomavirus that causes a demyelinating disease in the brain, and human papillomaviruses, which cause benign warts and cervical carcinoma, demonstrate still other patterns of chronic virus infection. Finally, HIV, which causes AIDS (a disease that is particularly difficult to place in any one category), is briefly considered here among the chronic infections.

HBV causes the most important chronic virus infection in humans. Actually, HBV is one of the most important human pathogens of any kind, as there are about 500 million chronic human carriers worldwide and over 1 million fatalities each year due to HBV-associated liver disease. In some areas of the world, including Southeast Asia and southern Africa, as many as 50% of all individuals are infected. (HBV, a member of the hepadnavirus family [Chapter 22], is ostensibly a double-stranded DNA virus. But in actuality, its genome is partially double stranded and partially single stranded. While this may seem a bit bizarre, it is not nearly as bizarre as the overall replication strategy of this virus. In brief, the HBV replication strategy involves an RNA intermediate transcribed from the viral genome, which is packaged into empty virus capsids, where it is then reverse transcribed by a virus-encoded reverse transcriptase activity, thereby generating the DNA genome.)

Several features of HBV are thought to facilitate establishment of a chronic infection, two of which are noted here. First, HBV is not cytocidal, and second, it does not induce the early production of **interferons (IFNs)**. (The IFNs are briefly introduced in Box 3.6 and are discussed at length in Chapter 4 and elsewhere in the text as well.) Since tissue damage and IFN production together generally induce an inflammatory response in the host, which usually is necessary to activate a specific adaptive immune response against a virus (Chapter 4), these features of HBV enable it to steer clear of alarming the immune system, while it establishes persistence.

Eventually, HBV does activate the adaptive immune system, although it is not entirely clear how it does so. Maybe it is because in due course a sufficient number of infected cells are killed. Regardless, B lymphocytes begin to produce antibodies against the virus, and virus-specific cytotoxic T cells are activated as well. The adaptive immune response then resolves the infection in more than 90% of adult cases. Moreover, these individuals experience few, if any, symptoms, and upon recovering they have lifelong protective immunity. In the remaining infected adults, there is sufficient liver damage to cause the characteristic nausea, jaundice, and liver pain associated with hepatitis.

BOX 3.6

IFNs are a group of antiviral proteins that are expressed by cells in response to a viral infection (Chapter 4). There are three classes of IFNs, IFN-α, IFN-β, and IFN-γ. IFN-α and IFN-β are referred to collectively as IFN-α/β or type I IFNs. They are synthesized by, and secreted from, almost any type of virus-infected cell. They bring about an antiviral state in neighboring cells by transmitting signals mediated by IFN receptors on those cells which upregulate the expression of IFN-responsive genes. IFN-γ is produced by certain lymphocytes after they have been exposed to a foreign antigen, and it is capable of enhancing the efficacy of the immune system.

Since the genes for IFN-α/β are expressed in response to a virus infection, note that HBV does not seem to affect the expression of any liver cell genes, including the IFN-responsive genes, at least as indicated within the sensitivity of microarray analysis. On the other hand, experimental results from a chimpanzee model suggest that IFN-γ may have a role in regulating HBV infections. In that model, most of the virus is cleared before the development of an adaptive immune response, and IFN-γ may be a key factor in that clearance (see the text). Much remains to be learned about this intriguing infection.

Interestingly, although recovered individuals are disease free and immune to reinfection with HBV, they yet may produce trace amounts of the virus, with the low-level infection regulated by the adaptive immune response. Actually, it is not clear whether or not HBV is ever completely eliminated from infected individuals. For example, clinically recovered patients who have later been immunosuppressed by cancer chemotherapy are known to have experienced reactivation of their hepatitis B infections.

T lymphocytes were shown recently to have a crucial role in the control of HBV infections. This was demonstrated in chimpanzees, in which experimental depletion of either helper T cells or cytotoxic T cells prevented virus clearance and clinical recovery. In addition, a still-mysterious noncytolytic antiviral mechanism also is at work, as implied by the observation that most of the virus can be cleared from the liver and blood of infected chimpanzees before there is any detectable T-cell infiltration of the liver or liver injury. Studies using a transgenic mouse model suggest that this mysterious factor might be gamma interferon (IFN-γ) released by activated cytotoxic T cells (Box 3.6).

Now here is a bit of a puzzler. Since HBV is itself noncytocidal, what processes might underlie the liver damage and hepatitis symptoms associated with HBV infections? In point of fact, the inflammatory response of the innate immune system and cytotoxic T cells of the adaptive immune system are together responsible for HBV pathogenesis. Indeed, the immunopathological responses that are induced by some infections are the inadvertent price we sometimes must pay to enjoy the protection afforded by our otherwise quite remarkable immune systems (Chapter 4). Regardless, the conclusion that immunopathology underlies the liver damage in HBV-infected individuals is consistent with the observation that infections in patients may change over from an **immunotolerant** phase to an **immunoactive** one. (An individual is said to be immunotolerant when he or she is nonreactive to a particular antigen. Generally, we are immunotolerant of self antigens.) The immunotolerant phase is characterized by higher virus titers but less-severe liver disease, whereas the immunoactive phase is characterized by low virus titers and more-severe hepatitis. Note that the immunoactive, low-replication phase can last for the lifetime of the patient (Box 3.7).

It is important to distinguish HBV infections in adults from infections by that virus in the very young. This distinction is clinically important and also important with respect to the natural history of the virus. We begin by comparing routes of transmission in these two instances. Important routes of HBV transmission between adults include the sharing of syringes among intravenous drug users, transfusions of contaminated blood, acupuncture, body piercing, tattooing, and sexually via blood and saliva. Of these routes of transmission between adults, only the last might be considered a natural route. With that in mind, the most important *natural* route of HBV transmission is

BOX 3.7

Other evidence that immunopathology underlies HBV pathogenesis includes experimental results obtained in a mouse model. Transgenic mice, which express the hepatitis B surface antigens, display no symptoms whatsoever. Moreover, these mice are immunotolerant of those virus antigens. That is, they do not recognize those virus antigens as foreign. (Why do you think that might be so? See Chapter 4.) However, when these mice are injected with hepatitis B-specific cytotoxic T cells, they develop pathologies that strongly resemble pathogenic human HBV infections.

actually from a chronically infected mother to her child at birth. When virus levels are sufficiently high in a chronically infected mother, then there may be a high rate of **vertical transmission** (i.e., from mother to offspring). To reiterate, the most important natural route of HBV transmission is vertical transmission from mother to offspring.

HBV infection during adulthood typically does not result in chronic hepatitis but instead leads to protective immunity. However, in about 3% of individuals infected as adults, the virus prevails against the immune system to establish a long-term, chronic infection that can persist for life. Those individuals infected as adults who become chronically infected might have mounted an inefficient immune response. In addition, the clever virus itself contributes to the inefficiency of the immune response by generating noninfectious empty particles, which are lacking the viral genome. These particles may exceed the number of infectious particles by a factor of 10^4 to 10^6 and are believed to serve as decoys by absorbing most or all of the neutralizing antiviral antibodies.

In contrast to the majority of individuals infected as adults, virtually all infected newborns go on to become chronic carriers themselves, perhaps because of their immature (but hardly inert; see above) immune systems, although other explanations are plausible (e.g., route of infection or dosage). Moreover, individuals infected as newborns suffer less tissue damage and have milder symptoms than individuals contracting the infection as adults, again possibly because of a less effective immune response. The infected newborns, as chronically infected carriers, will one day be able to efficiently pass the infection on to their offspring, thereby accounting for the fact that vertical transmission is the most important natural route of hepatitis B transmission. Indeed, HBV has evolved a very efficient means to ensure its maintenance in the host population.

The chronic HBV infection may be asymptomatic, or it may be associated with chronic active **hepatitis**. The latter condition may progress to potentially fatal **cirrhosis** (a chronic progressive disease of the liver characterized by the replacement of healthy liver tissue with scar tissue) and, more rarely, to hepatocellular carcinoma, a usually fatal malignancy with a latency period of between 9 and 35 years.

HBV does not express an oncogene (see above). Instead, the development of hepatocellular carcinoma is believed to result from the extensive cell proliferation that takes place in the liver, which occurs for the purpose of replacing the cells that are continually destroyed by the immunopathology associated with the infection. This extensive cell proliferation may increase the risk that a cell will acquire cancer-causing mutations. Yet this cannot be the whole story, since all hepatitis B-associated liver cancers contain the viral genome, strongly implying that expression of a viral function in the neoplastic cell in some way promotes cancer development.

As stated above, HBV is a member of the hepadnavirus family of partially double-stranded DNA viruses, which replicate via the rather astonishing reverse transcription of an RNA intermediate (Chapter 22). In contrast, **HCV** is a member of the flavivirus family of conventional, positive-strand RNA viruses (Chapter 7). But despite the fact that HBV and HCV are unrelated, each gives rise to medically important chronic infections of the liver, which in each instance may lead to chronic hepatitis, cirrhosis, and hepatocellular carcinoma. These facts illustrate the point that biologically unrelated viruses may infect the same target organ and cause similar clinical outcomes.

HCV infects about 170 million people worldwide. Still, despite the fact that HCV affects so many individuals, aspects of the virus-host interaction that might underlie chronic HCV infection in vivo are not yet understood. This is so largely because a system for the study of HCV in cell culture has not yet been developed. Also, little is known regarding the natural routes of HCV transmission. Like in the case of HBV, vertical transmission from mother to offspring at birth is the only well-documented natural route of HCV transmission. There is evidence that the virus also may be transmitted sexually, but that appears to be a rather inefficient route. As expected, HCV is very efficiently transmitted via infected blood.

In contrast to HBV, which establishes a chronic infection in only about 3% of individuals infected as adults, HCV establishes chronic infections in most (i.e., 60 to 80%) of those infected as adults. Moreover, most individuals infected with HCV as adults eventually develop chronic hepatitis. Also, the development of chronic hepatitis actually correlates with a 100- to 1,000-fold drop in the virus titer. (How might you explain the correlation between the drop in the HCV titer and the emergence of chronic hepatitis? See above with respect to hepatitis B.)

HCV must evade innate immunity at early times in the infection, in order to establish persistent infection. In that regard, the virus evades the effects of the IFNs, which are among the effectors of innate immunity (Box 3.6). Recall that HBV avoids the effects of the antiviral type I IFNs by not inducing their expression in the first place. In contrast, HCV does upregulate the expression of many liver cell genes, including those involved in the production of type I IFNs. Nevertheless, HCV avoids the effects of these IFNs by producing two proteins, E2 and NS5A, each of which inhibits the activity of RNA-activated protein kinase (PKR), a mediator of the IFN antiviral effect (Chapter 4). E2 acts as a decoy substrate for PKR, whereas NS5A forms heterodimers with PKR to inhibit its activity.

Since the most important natural route of HCV transmission may be from an infected mother to her child at birth, an infected female child must carry the virus until she reaches maturity if she is to be able to transmit the infection to her own offspring. Thus, the virus needs some means for persisting in the face of virus-specific antibodies and cytotoxic T cells of adaptive immunity. A means by which HCV is thought to evade adaptive immune defenses is revealed by the following observations. First, compared to the high fidelity of cellular DNA replication, all viruses replicate their genomes with relatively low fidelity. Moreover, the error rate is vastly greater in the case of RNA viruses (i.e., one error in every 10^4 nucleotides in the case of HIV, and about one error in every 10^5 nucleotides in the case of most other RNA viruses). Thus, if one were to sample the HCV population in any infected individual, one would find the simultaneous presence of a large number of genetic variants with different antigenicities. Second, if one were to monitor an infected individual's liver enzyme levels in the blood, which are indicative of virus-mediated liver damage, one would notice cycles of approximately 6 weeks in length. Together, these observations suggest that every 6 weeks or so, the error-prone virus polymerase manages to generate "immune escape mutants" that are able to evade the antigen-specific adaptive defenses that are presently expressed by the host. These mutants prevail until the adaptive immune system catches up with new antibody-producing B cells and cytotoxic killer T cells, at which time the cycle repeats.

Importantly, patients infected with HCV as adults, who have become chronically infected, rarely recover. Yet when they do recover, it is not clear whether or not the infection is ever eliminated. (Recall that it also is not clear whether HBV-infected individuals ever clear that infection.) On the one hand, HCV reactivation has not been reported in those rare recovered individuals who subsequently undergo immunosuppression (compare to HBV infections, above). But on the other hand, by means of very sensitive polymerase chain reaction (PCR) techniques, viral sequences have been detected in peripheral-blood lymphocytes of recovered patients. It is not known whether or not such individuals might be infectious. (There is evidence that some patients who underwent antiviral drug therapy may have been cured of their HCV infections [Chapter 7].) As in the cases of patients who recover from chronic HBV infections, patients who recover from HCV infections likewise mount strong, adaptive antiviral immune responses.

Chronically infected HCV carriers usually live symptom free for many decades, implying that the viral immune evasion strategies also modulate immunopathology. However, in about 20% of chronic carriers, sufficient accumulated liver damage transpires over the course of 20 to 30 years to lead to cirrhosis. Within 10 more years, about 10% of these affected individuals eventually develop hepatocellular carcinoma.

Before moving on, it would be worthwhile to underscore several key differences between infections involving HCV and those infections involving HBV. In each instance, the immune response plays a major role in (i) controlling the infection, (ii) clinical recovery and protective immunity, and also (iii) pathogenesis. However, when adults are infected by HCV, the outcome is generally chronic infection, whereas infection of adults with HBV generally results in protective immunity. Thus, these two viruses must interact differently with the immune system. Although much remains to be learned about the interactions of these viruses with host immunity, we might note that HBV, which has a double-stranded DNA genome, evades immune defenses by generating an excess of empty decoy particles that soak up neutralizing antibodies. HCV, which has a single-stranded RNA genome, uses its inherently error-prone RNA polymerase to generate large numbers of antigenic variants (immune escape mutants). Yet these are surely not the only viral immune escape and survival strategies at play in each instance (Rehermann and Nascimbeni, 2005). Another difference to note is that hepatocellular carcinoma in hepatitis C patients almost always develops against a background of cirrhosis, whereas in hepatitis B patients it might develop in the absence of cirrhosis. Thus, there are different virus-specific factors that contribute to hepatocarcinogenesis in each instance.

The differences noted above regarding the known means by which HBV and HCV evade host adaptive immunity are relevant to the development of practical countermeasures against these viruses. For example, since HBV produces essentially one serotype, it was possible to develop an effective vaccine against that virus. In contrast, the vast number of HCV serotypes that are generated may preclude the development of a vaccine against that agent.

JC virus is a member of the polyomavirus family of small, double-stranded DNA viruses (Chapter 15). (JC virus is closely related to simian virus 40 [SV40], a virus that has been extensively studied for its ability to induce neoplastic transformation of cells in culture and tumors in laboratory animals. See Chapter 15 for a detailed account of JC virus and SV40.) JC virus is ubiquitous in humans, infecting more than 85% of adults, with most infections occurring by the age of 15 years. The virus is believed to initially infect the respiratory tract and to then disseminate to the kidneys and lymphoid organs. Moreover, JC virus establishes a lifelong chronic infection, during which it may replicate in tubular kidney cells. The primary and secondary infections, as well as the chronic infection, are usually not associated with any illness. However, in individuals who are severely immunosuppressed, JC virus can spread to the central nervous system, where it causes the usually fatal neurodegenerative disease **progressive multifocal leukoencephalopathy** (**PML**). The term "leukoencephalopathy" refers to the fact that lesions are seen only in the white matter.

Historically, PML is the first demyelinating disease in humans that was found to be caused by a conventional virus (i.e., in contrast to the transmissible spongiform encephalopathies, which are caused by proteinaceous infectious particles, or prions; see below). PML was an extremely rare condition when it was first identified as a distinct disease entity in the 1950s. At that time, it was essentially seen only in organ transplant recipients on immunosuppressive regimens and in cancer patients whose neoplasms compromised immunity (e.g., Hodgkin's disease and chronic lymphocytic leukemia). Immunocompetent individuals who are chronically infected with JC virus generally do not develop PML, presumably because their JC virus-specific CD4 and CD8 T cells control the infection. These virus-specific T cells likewise play a role in determining the clinical course of PML, since disease progression is associated with declining T-cell levels.

At present, the majority of PML cases occur in patients with AIDS, which is the most severe condition of immunosuppression known. Indeed, because of the growing size of the AIDS population, PML has become a much more common condition in recent years, developing in as many as 5% of all AIDS patients and presently affecting nearly 1 in every 200,000 persons overall. Once underway, it is relentlessly progressive, with death generally occurring 3 to 6 months after the onset of symptoms.

It is not known with certainty how JC virus invades the central nervous system in order to give rise to PML, although some researchers suggest that it may do so via infected B cells. Once in the central nervous system, JC virus has a tropism for **oligodendrocytes** (the myelin-generating cells in the brain) and **astrocytes** (star-shaped cells that support and nourish nerve cells in the brain) (Fig. 3.10). Lytic JC virus infection of oligodendrocytes causes the demyelination characteristic of PML lesions.

The receptor for JC virus was recently identified as 5-HT2A, a receptor of the family of serotonin receptors. The serotonins are neurotransmitters, and consequently, this finding neatly accounts for the neurotropism of JC virus. (Incidentally, 5-HT2A also is the receptor for the hallucinogen LSD.)

How might we classify JC virus infections? On the one hand, the generally benign persistent JC virus infections in immunocompetent individuals might be classified as chronic in nature. On the other hand, the emergence

Figure 3.10 The cellular organization of the central nervous system. Modified from M. McKinley and V. O'Loughlin, *Human Anatomy* (McGraw-Hill, New York, NY, 2006), with permission.

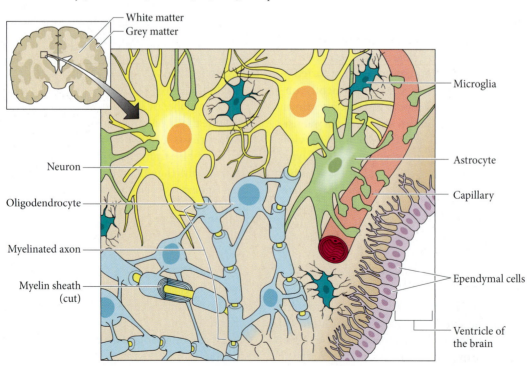

of PML in immunocompromised individuals might be thought of as a slow virus disease, as well as a chronic infection. In any case, JC virus infections cannot be neatly categorized as either chronic or slow.

Papillomaviruses are small double-stranded DNA viruses that cause benign skin **warts**, and more importantly, they are the primary cause of **carcinoma of the cervix**, one of the most common cancers in women (Chapter 16). These pathologies may start to develop several months to years after the initial infection. Indeed, invasive cervical carcinomas may emerge as late as 20 to 50 years after the initial papillomavirus infection.

The establishment and maintenance of chronic papillomavirus infections are part and parcel of the papillomavirus replication strategy. In brief, these viruses have a rather strict tropism for the basal cells that constitute the lowest of the several layers of the epidermis (Fig. 3.1). Papillomaviruses access these cells through tears in the skin and mucosal layers. Basal cells continually proliferate in order to replenish the nonproliferating cells of the overlying layers, which continually die and slough off. As basal cells divide, one daughter cell migrates into the upper layers and begins a course of differentiation to a keratinocyte while the other daughter cell remains behind as a basal cell. Importantly, papillomaviruses do not carry out a complete replication cycle in the basal cells. Instead, they only replicate their genomes, producing only a few copies in each basal cell. Then, these viral genomes are distributed at random between the daughter cells when the basal cell divides. Again, one of the daughter cells remains behind in the basal layer, while the other migrates to the upper layers of the epithelium.

As the basal cells migrate to the surface, they eventually differentiate into keratinocytes, which acquire the ability to support the complete papillomavirus replication cycle. For its part, the virus enhances the efficiency of its replication by producing two proteins, E6 and E7, which keep the keratinocytes in a vegetative state; this is a necessary condition for virus replication, since the virus is dependent on the cellular DNA replication machinery, which otherwise is active only in replicating cells. The mechanisms of action of the papillomavirus E6 and E7 proteins are described in Chapter 16. The mechanism by which certain papillomaviruses induce cervical carcinoma is also described in Chapter 16. As you may have surmised, expression of E6 and E7 is required for the development of this cancer, but as will be seen, there is more to this part of the story (Chapter 16).

We can now appreciate that papillomavirus persistence results from a strategy in which these viruses infect the basal cells beneath the skin and mucosa, in which limited viral genomic replication occurs and in which the infection is otherwise abortive. But as these same cells migrate to the surface of the epithelium and differentiate, productive virus replication occurs, thereby facilitating the efficient transmission of the infection to new individuals. All the while, viral genomes are maintained in the basal layer, thereby sustaining the long-term persistent infection. Interestingly, replication of papillomavirus genomes in the basal cells is regulated by the fact that the viral origin of DNA replication (ori) sequence interacts like a cellular ori with the cellular DNA replication machinery. In this way, the average number of papillomavirus genomes per basal cell remains constant over time.

Like other viruses that establish persistent infections, papillomaviruses likewise have evolved strategies that enable them to evade immune clearance. For example, these double-stranded DNA viruses transcribe only one of their DNA strands. Thus, they do not inadvertently generate the double-stranded RNA that might induce IFN (Chapter 4). Moreover, papillomaviruses do not harm the abortively infected basal cells, which sustain the persistent infection. For these two reasons, papillomaviruses do not initially activate the innate immune system. And if the innate immune system is not activated, there is no adaptive response. (This key feature of adaptive immunity is explained at length in Chapter 4.) Eventually, the adaptive immune system is activated, presumably because some virus-producing cells are killed. Nevertheless, the papillomavirus stratagem of restricting late protein synthesis and virus production to the terminally differentiated keratinocytes at the outermost layers of the epithelium places these virus-producing cells beyond the reach of immune attack by cytotoxic T lymphocytes (CTLs), while also preventing viruses from being exposed to neutralizing antibodies. (CTLs and neutralizing antibodies are the key effectors of adaptive immunity [Chapter 4].) Overall, this strategy is so successful that at least 25 million Americans currently have active genital papillomavirus infections.

We conclude this section with a brief consideration of HIV, mainly to note that infections with this virus display defining features of each of our three categories of persistent virus infections (i.e., slow, chronic, and latent). We begin by noting that the average time that elapses between HIV infection and the onset of full-blown AIDS is about 10 years (Chapter 21), thereby qualifying AIDS as a slow virus disease. Next, infectious virus always is present and being shed, a characteristic of a chronic infection. Finally, at any given instant, there are many more latently infected cells than productively infected cells, a characteristic of a latent infection (see below). Still, for me at least, HIV infections are most appropriately categorized as chronic. The reasons are as follows.

Latency in the case of HIV is distinctly different from the quintessential latency that characterizes infections with the human herpesviruses, such as HSV-1, a virus that establishes lifelong latent infections, which periodically may reactivate to give rise to recurrent acute episodes. In contrast, while HIV is latent in most infected cells of an HIV-infected individual, a subset of infected cells is producing virus at any given time, such that untreated individuals are always viremic. Moreover, herpesviruses spend most of their time hiding from the host immune system, and at no time are they actually attacking it. In contrast, although HIV is hiding from the immune system in latently infected cells, a striking feature of an HIV infection is the unrelenting direct attack by the virus on the immune system. Additionally, the continued production and shedding of HIV and the delayed onset of AIDS, about 10 years after the initial infection on average, are more typical of chronic infections, hence the inclusion here of HIV under chronic infections.

Some of the numerous mechanisms used by HIV to evade innate and adaptive immune responses are recounted in Chapters 4 and 21. HIV persistence and the pathogenesis of AIDS are intimately related to the molecular biology of the virus and its interaction with its host cells. These issues are considered in detail in Chapter 21.

Latent Infections

As mentioned above, herpesviruses give rise to quintessential latent infections. The known human herpesviruses are HSV-1 and -2, human cytomegalovirus, EBV, VZV, human herpesvirus 6 (HHV-6), HHV-7, and the more recently discovered HHV-8, which is associated with Kaposi's sarcoma in AIDS patients (Chapters 18 and 21).

Each of the human herpesviruses has evolved a somewhat different strategy for establishing latency, in part because each targets a particular host cell type in which to do so. HSV-1, HSV-2, and VZV establish latency in neurons, EBV in B lymphocytes, cytomegalovirus in monocytes and lymphocytes, and HHV-6 and HHV-7 in T lymphocytes. The cell type in which HHV-8 becomes latent is not yet known. Regardless, the herpesvirus strategy for establishing and maintaining latency depends on cellular factors, such as whether the host cell might be dividing (which it is not, in the case of neurons) or long-lived (which it is not, in the case of lymphocytes). Infections by each of the herpesviruses are considered in some detail in Chapter 21. Our focus here is on HSV-1, as an example of a classic latent infection.

HSV-1, which gives rise to both oral and genital herpes, infects more than 60% of all Americans by late adolescence. The virus replicates in epithelia and is usually transmitted by direct intimate contact between individuals, reflecting the facts that it is present in the saliva of those with oral HSV-1 infections and is shed from the genitalia of those with genital infections.

The acute phase of an HSV-1 infection is generally localized to the initial site of infection. That is so because the virus receptor, heparin sulfate, is also a major constituent of the extracellular matrix. Thus, despite the fact that heparin sulfate is present on nearly all cells, extracellular virus cannot diffuse very far before being bound by heparin sulfate molecules of the extracellular matrix.

HSV-1 has evolved numerous strategies for evading innate and adaptive immune responses during the acute phase of an infection (recounted in detail in Chapters 4 and 18). Although these viral countermeasures against host immune defenses are clever, perhaps even diabolical, within about 2 weeks' time the immune system generally does succeed in resolving the acute phase of the infection. Nevertheless, and importantly, the viral immune evasion strategies provide a window of opportunity for the virus to establish latent infection. This occurs as follows. During the acute phase of the HSV-1 infection, the virus reproduces quickly in the epithelial cells at the initial site of infection. Some of the progeny virus that is produced there then invades the endings of the sensory neurons that innervate the epithelium. These virus particles are conveyed by retrograde axonal transport to cranial or spinal sensory ganglia, where the viral genomes persist essentially indefinitely in the nuclei of a minority of neurons (Fig. 3.4 and 3.11).

The establishment of HSV-1 latency would be noteworthy if it merely provided some advantage to the virus or if it were merely clinically relevant. But it is both advantageous to the virus and clinically relevant. We begin by considering the advantage to the virus of its latency lifestyle. Bear in mind the following points: (i) HSV-1 is transmitted by intimate contact between individuals, (ii) the acute phase of the infection is resolved within a time interval of 2 weeks, (iii) humans are generally selective regarding their intimate contacts, and (iv) humans are the only host for HSV-1. Considering these points, HSV-1 would be in danger of not surviving in Nature unless it had some special strategy for perpetuating itself in its host population. This special strategy is the ability of the virus to establish latency in the ganglia of neurons that innervate the initial site of infection. These neurons provide an immunologically privileged site, in which the virus might find a long-term safe haven. The immunologically privileged status of neurons is due in part to the fact that they express only low levels of major histocompatibility complex class I molecules (which display virus antigens to cytotoxic T cells [Chapter 4]). Moreover, the virus contributes to its stealthiness in neuronal cells by generating only negligible amounts of proteins that might be recognized by immune effectors (Box 3.8).

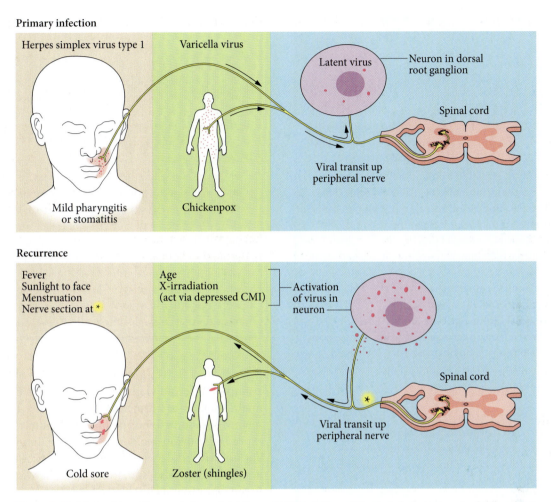

Figure 3.11 Mechanisms of latency in herpes simplex and VZV infections. Primary infection occurs in childhood or adolescence; latent virus is located in the cerebral or spinal ganglia. Reactivation of HSV causes recurrent herpes simplex; reactivation of VZV causes zoster (shingles). CMI, cell-mediated immunity. Adapted from D. O. White and F. J. Fenner, *Medical Virology*, 4th ed. (Academic Press, Inc., San Diego, CA, 1994), with permission.

Now here is an important point that is relevant both to the natural history of HSV-1 and to herpesvirus disease. It would do HSV-1 little good if the latent phase of the infection were permanent, since latent virus cannot be transmitted to susceptible individuals. Consequently, latent HSV-1 genomes are able to periodically reactivate, thereby reinitiating the productive infection (Fig. 3.11). It is not yet entirely clear how reactivation occurs at the molecular level (Box 3.9). However, reactivation is known to be triggered by hormonal factors (e.g., the menstrual cycle), as well as by external factors such as psychological stress, physical trauma, and ultraviolet (UV) irradiation (i.e., time spent in bright sunlight). In addition, reactivation is likely to occur when an individual is immunosuppressed. At any rate, when HSV-1 reactivation occurs, the virus replicates in some of the neurons in which it has been latent. It then is transported back down the axon to

the original site of infection, where it can undergo a new cycle of lytic replication in the epithelium and thus be transmitted to new individuals (Box 3.10).

A key feature of the above HSV-1 strategy is that it makes use of two cell types: epithelial cells, in which the virus undergoes productive infection, and neuronal cells, in which it establishes latency. For years it was believed that during the latent periods no infectious virus was produced but viral genomes instead remained cryptic in certain protected cells. However, newer, more sensitive molecular procedures show that infectious HSV-1 may be shed continuously by asymptomatic individuals. Thus, latency in the case of HSV-1 provides a means for an infected individual to be a continuing source of infection for a lifetime, perhaps accounting for the ability of this virus to be so ubiquitous in the human population. Two other points to note are as follows. First, transmission by

BOX 3.8

Whether the latent viral genome might persist as an **episome** or is integrated by covalent bonds into the cellular chromatin is one of the defining characteristics of a particular latent infection. In the case of HSV-1, the viral genomes persist as free episomes in neuronal cell nuclei, approximately 10 to 100 genomes per nucleus. But strictly speaking, by definition an episome can exist either autonomously or as part of a chromosome. In contrast, a **plasmid** is an extrachromosomal DNA molecule that is capable of replicating independently of the chromosomal DNA. Although it may be more appropriate to refer to the latent extrachromosomal HSV-1 genome as a plasmid, the journal literature often refers to it as an episome.

It is enlightening to compare the state of latent HSV-1 genomes in neuronal cells with the state of latent HIV genomes in CD4 helper T cells. Whereas latent HSV-1 genomes persist in neuronal cells as extrachromosomal episomes, latent HIV genomes persist in CD4 helper T cells in an integrated state. Why the difference? Perhaps the reason is as follows. Neurons do not replicate, whereas CD4 helper T cells do. By integrating its genome into the cellular chromatin, HIV ensures that it will be passed on to each daughter cell at division. HSV-1 has no such concern in nondividing neurons. Alternatively, or in addition, the integration of the HIV genome is a crucial step that enables it to subsequently be reactivated and replicated (Chapter 21).

BOX 3.9

The expression of HSV-1 genomes is not completely shut down in latently infected neurons. Indeed, the establishment and maintenance of latency require the expression of specific HSV-1 genes in those cells. The family of HSV-1 messenger RNAs (mRNAs), referred to as the **latency-associated transcripts**, is specifically expressed in latently infected neurons. Moreover, latency depends on cellular factors as well. Actually, it may depend on the *absence* of cellular factors, such as particular cellular transcription factors that might be needed for the expression of crucial viral early genes. With appropriate stress to these cells, the necessary cellular genes might be activated so that viral reactivation might occur in at least a few neurons. Also, reactivation from latency may require the expression of specific reactivation-associated viral genes as well (Fig. 3.11). Current understanding of the workings of these viral factors is recounted in Chapter 18.

BOX 3.10

It is worth reiterating that any virus that gives rise only to acute infections faces the dilemma that eventually all the individuals in its host population may have been infected and thus become immune. For this reason, acute viruses require some minimal host population size to allow for a sufficient number of new susceptible individuals to be born into it to sustain the virus. In the case of measles virus, that minimal population size is about 500,000 individuals. Thus, measles virus, which has no animal reservoir, can only recently (i.e., within the last 10,000 years or so) have become a human virus, since such large human populations did not exist before then (Chapter 11). On the other hand, latency provides a means for a virus to sustain itself in populations of much smaller size. Do you see why that is so? HSV-1 has long been a human virus, coevolving with its host before and ever since humans diverged to become a distinct species.

asymptomatic individuals is more important than you might think at first, since many candidates for infection might resist some kinds of intimate contacts with those expressing obvious cold sores or other herpetic lesions, but not with those who are asymptomatic. Second, these newer findings also somewhat blur the distinction between chronic and latent infections.

Finally, coming full circle back to portals of entry and transmission, with which this chapter began, the ability of HSV-1 to thrive in the human population demonstrates how strongly we crave intimate contact. We could not avoid becoming infected by acute respiratory viruses, such as the rhinoviruses and influenza virus, short of becoming hermits. But we could avoid becoming infected by HSV-1 simply by avoiding intimate physical contacts. It is not a stretch to presume that HSV-1 evolved to exploit our human urges for intimate contacts.

Transmissible Spongiform Encephalopathies: Prions

The **transmissible spongiform encephalopathies**, also referred to as subacute spongiform encephalopathies, are an especially intriguing group of slow, infectious, neurodegenerative diseases. **Scrapie** in sheep is the prototype for this group of diseases, which also includes **kuru, Creutzfeldt-Jakob disease (CJD), fatal familial insomnia**, and the **Gerstmann-Sträussler-Scheinker syndrome** in humans and **bovine spongiform encephalopathy (BSE)** in cattle, known colloquially as "**mad cow disease**." (Actually, fatal familial insomnia and the Gerstmann-Sträussler-Scheinker syndrome are inherited, rather than infectious, diseases; see below.)

Importantly, although the agents responsible for these diseases easily pass through filters, they are not viruses! What then are they? First, let us briefly consider what we mean by the term "virus." Admittedly, we do not have any consensus for an all-inclusive definition for a virus (Chapter 1). Yet we likely would agree that viruses contain genes, have an apparent structure, and elicit an immune response. Unlike the agents that we agree are viruses, the agents responsible for the transmissible spongiform encephalopathies have no apparent genome or virus structure, nor do they elicit an immune response. Moreover, they are resistant to physical and chemical treatments (e.g., heat, radiation, and formaldehyde) that would inactivate a conventional virus. These facts, while intriguing, do not tell us the nature of the agents responsible for the transmissible spongiform encephalopathies. So then, what are they?

The agents responsible for the transmissible spongiform encephalopathies actually appear to be variant forms of normal cellular proteins. Both the normal and infectious forms of these proteins are referred to as **prions**, an acronym derived from "proteinaceous infectious particle." This name was coined by Stanley Prusiner, who won a Nobel Prize in 1997 for his breakthrough studies into the proteinaceous nature of prions. The aberrant prion conformation responsible for scrapie is termed PrP^{Sc}. (Although PrP^{Sc} now designates "scrapie prion protein," the origin of the PrP abbreviation is "protease resistant protein.") The corresponding normal host cell prion protein is termed PrP^{C}. Both PrP^{Sc} and PrP^{C} are about 27,000 to 30,000 daltons (Da). However, differences in folding between PrP^{Sc} and PrP^{C} account for the differences in their biological effects. PrP^{C} is predominantly α-helical, whereas PrP^{Sc} is a β-sheet-rich structure.

Prusiner's studies demonstrated to the satisfaction of many researchers (not all are convinced) that PrP^{Sc} replicates and causes scrapie by somehow catalyzing the conversion of the benign cellular PrP^{C} conformation to the PrP^{Sc} conformation. Once underway, this conversion process might escalate exponentially. The accumulation of the PrP^{Sc} isoform in the brain then leads to the disease. The fact that the aggregates of PrP^{Sc} that accumulate are comprised of host proteins may account for why there is no immune response to these agents in infected animals and patients.

In the transmissible spongiform encephalopathies, such as kuru and CJD, the transformation of PrP^{C} to PrP^{Sc} is initiated by an exogenous source of PrP^{Sc}. In the inherited forms, such as fatal familial insomnia and the Gerstmann-Sträussler-Scheinker syndrome, the transformation is triggered by mutation(s) in the prion protein gene. For example, in the Gerstmann-Sträussler-Scheinker syndrome, a change in PrP^{C} amino acid 102 from proline to leucine is found in most affected individuals. Sporadic cases of these diseases are then caused by random, spontaneous events.

What sort of experimental evidence might support the above prion model? In one early study, transgenic mice were generated which lacked both copies of the mouse PrP^{C} gene. These mice were completely resistant to mouse scrapie prions. However, when hamster PrP^{C} genes were introduced into these mice so that they expressed hamster PrP^{C}, these mice then were susceptible to hamster scrapie prions but remained resistant to mouse scrapie prions. Presumably, hamster scrapie prions, but not mouse scrapie prions, could interact with the hamster PrP^{C} to catalyze their conversion to the scrapie isoform.

More recently, Prusiner's research group produced neurologic dysfunction in mice by inoculating them intracerebrally with a recombinant mouse prion protein generated in *Escherichia coli* (Legname et al., 2004). The recombinant mouse prion protein contained a mutation analogous to the human prion protein mutation that causes Gerstmann-Sträussler-Scheinker syndrome (see above).

The preclinical phase (incubation period) for scrapie in sheep is about 3 years. For CJD and kuru in humans, it may be as long as 30 years. But once symptoms appear, patients generally succumb within a year.

Clinically, the subacute spongiform encephalopathies are characterized by a loss of muscle control, shivering, tremors, and loss of coordination, and in humans dementia is also one of the characteristic symptoms. The term "spongiform" refers to the characteristic vacuolated neurons. Histological changes also include the proliferation and hypertrophy of astrocytes and the fusion of neurons with adjacent glial cells. PrP^{Sc} is taken up by neurons and probably contributes to the vacuolization of the brain tissue. Interestingly, prions are found in other tissues as well, but only the brain shows any signs of disease. It is not known how prions cause the fatal pathogenesis of the spongiform encephalopathies.

Although kuru is a disease that readers of this book will probably never actually see, it makes for a most compelling story. This transmissible spongiform encephalopathy is limited to a tribe of Fore-speaking people, who live in the remote mountains of eastern New Guinea. The term "kuru" means "shivering" or "trembling" in the Fore language.

In the late 1950s, as the Fore people were just emerging from the Stone Age, stories about their strange disease began to leak out to the modern world. These stories intrigued virologist Carlton Gajdusek at the U.S. National Institutes of Health, who came to the Fore people to see the disease for himself. What he found was as follows. The

Fore people numbered about 35,000 individuals, and kuru affected about 1% of their population. Importantly, the disease occurred primarily in women, with many fewer cases seen in children (bear in mind the generally long incubation periods for these diseases, which explains the rarity of disease in children; see below). Interestingly, children of both sexes were affected in equal numbers. However, the disease was most prevalent in women, and for a time the majority of all deaths in women of the tribe were due to kuru. Consequently, men outnumbered women by 3 to 1 in some villages.

Gajdusek was puzzled by (i) the strange demographics of kuru in the Fore population, (ii) its slow progression, and (iii) the fact that there was no inflammation or fever associated with the condition. For these reasons, he thought that the disease might be caused by genetic factors, or perhaps by malnutrition. Then, while Gajdusek was still searching for the cause of kuru, William Hadlow, an American who had been studying scrapie, entered the scene. Fortuitously, while Hadlow was visiting the Wellcome Medical Museum in London, he happened to look at a display on kuru that Gajdusek had prepared. It was there that Hadlow noticed that the pathological changes occurring in human kuru were virtually identical to those seen in ovine scrapie. Importantly, it was already well known that scrapie is due to a transmissible agent.

Following up on Hadlow's insight, Gajdusek attempted to transmit kuru to chimpanzees. His success in that effort demonstrated that kuru too is caused by a transmissible agent. Moreover, it was the first demonstration of a human slow infection, conventional or otherwise. The incubation period for kuru in the chimpanzees ranged from 14 to 82 months, and the disease could be serially propagated from one chimpanzee to another. (Do you think that the study of kuru transmission might have been a desirable topic for a Ph.D. dissertation?)

Considering that kuru is caused by a transmissible agent, rather than an inherited mutation, what might explain its strange demographics in the population? The answer in a word is cannibalism. The Fore people were cannibals, and kuru was transmitted from person to person by cannibalism. Actually, the Fore people ate their dead relatives as an act of homage during funeral rites. The bodies of the deceased would be cut up into parts. As might be expected, the men took the meatiest parts for themselves, leaving the pancreas, liver, kidneys, and, importantly, the brain for the women and the children. Many of the women would develop kuru after eating the infected brains. When they died, they too were ritually eaten, thereby escalating the epidemic.

Gajdusek came to realize the connection between kuru and this ritualistic cannibalism after living among the Fore people for several years. This cycle finally was broken when cannibalism among the Fore people ended, in the 1960s.

Gajdusek's discovery that kuru is a transmissible disease took on new importance when the more widespread human presenile dementia, CJD, was found to be caused by an agent strikingly similar to the agents responsible for scrapie and kuru. CJD has a frequency of about one per million individuals worldwide. Interestingly, CJD cases have been traced to medical interventions such as neurosurgery, corneal transplants (a donor was diagnosed postmortem with CJD), and administration of growth hormone derived from pituitary extracts.

In addition to the spread of CJD by a transmissible agent, the disease also may appear suddenly, without any obvious epidemiological signs. Those random instances are referred to as sporadic CJD. Presumably, sporadic CJD occurs when a PrP^C isoform spontaneously assumes the PrP^{Sc} conformation. Perhaps kuru originally arose among the Fore people from the consumption of an individual with sporadic CJD.

Fatal familial insomnia, which is a very rare inherited prion disease due to a specific mutation in the PrP^C gene, differs from other prion diseases in that it predominantly affects one area of the brain, the thalamus, which influences sleep. Affected individuals eventually become completely sleepless. Like other prion diseases, this condition is invariably fatal. Note that the dominant gene responsible for this disease has been found in only 28 families worldwide.

BSE (referred to informally as mad cow disease) made its first appearance in 1987, in Great Britain. Interestingly, a cannibalism of sorts may underlie Britain's mad cow outbreaks; cannibalism, that is, between cows and sheep. Scrapie in sheep, which is very similar to BSE in cattle, has existed in England for at least 200 years, where it affects 0.5 to 1% of the sheep population per annum. Here now is the connection between "cannibalism" and BSE. For several decades, cattle feed had included protein supplements made from the carcasses of other animals, including sheep (this material is referred to as "offal"). For most of that time, scrapie never crossed into cows. However, this changed in the 1980s because of a new way of rendering the carcasses, which led to feed supplements that were contaminated with the scrapie agent. In earlier times, carcasses were exposed to prolonged heat treatment, which apparently was sufficient to inactivate PrP^{Sc} to safe levels. In the 1980s, there was a changeover to rendering the carcasses by a chemical procedure. This was done in order to cut costs; the heat treatment required the burning of increasingly expensive fossil fuels. Importantly, the new procedures apparently were not sufficient to inactivate PrP^{Sc}.

BSE has been in the news in recent years, not only for its effect on cattle but also because of evidence that eating meat from infected cows can give rise to CJD in humans. Because CJD is so rare (even among those exposed to BSE-contaminated beef), it is difficult to demonstrate a statistically significant increase in the number of CJD cases that might be due to eating contaminated meat. However, the change in the demographics of CJD in Great Britain is striking, and it strongly implies that something new is amiss. Previously, CJD was a disease of the elderly. But with the emergence of the BSE epidemic in British cattle, CJD in Great Britain emerged in young adults. Moreover, the PrP^{Sc} isoform found in the younger CJD patients has a glycosylation pattern similar to that found in cattle with BSE, and it is significantly different from the isoform seen in older CJD patients.

Because (i) BSE has a long incubation period in cattle and (ii) CJD has a long incubation period in humans, the problem of BSE in British cattle and its danger to humans became apparent only recently. Alarmingly, however, for 2 years after BSE became known, infected cattle were still allowed into England's food supply.

Some of the early complacency surrounding BSE may have been due to the fact that whereas scrapie had been around in sheep for several centuries, it was not known to cause a problem in humans. However, there have been accusations that the interests of the British meat industry were placed above those of the public. (Indeed, one investigating British scientist is quoted as saying, "And if we had been too alarmist, we were in danger of upsetting the whole of the meat industry in Britain and elsewhere in Europe.") Yet outside England, restrictive steps were taken. European countries and the United States imposed a ban on the import of British cattle. Finally, in response to public pressure, the British government banned the beef-processing industry from the practice of feeding cows and sheep back to each other. In addition, cattle already showing signs of BSE were excluded from use in human food and destroyed. (What is your reaction to the fact that belated governmental intervention was necessary to impose these changes?)

Selected Readings

Igakura, T., J. C. Stinchcombe, P. K. C. Goon, G. P. Taylor, J. N. Weber, G. M. Griffiths, Y. Tanaka, M. Osame, and C. R. M. Bangham. 2003. Spread of HTLV-I between lymphocytes by virus-induced polarization of the cytoskeleton. *Science* **299:** 1713–1716.

Legname, G., I. V. Baskakov, H. O. Nguyen, D. Riesner, F. E. Cohen, S. J. DeArmond, and S. B. Prusiner. 2004. Synthetic mammalian prions. *Science* **305:**673–676.

Rehermann, B., and M. Nascimbeni. 2005. Immunology of hepatitis B virus and hepatitis C virus infection. *Nat. Rev. Immunol.* **5:**215–229.

4

Host Defenses and Viral Countermeasures

INTRODUCTION

Much of modern virology concerns the molecular biology of viral gene expression and virus replication. Indeed, that is the emphasis in this book. However, it is important to bear in mind that the environment for a mammalian virus consists of more than just the susceptible cells of the host in which the particular virus replicates. Indeed, other specialized cells and tissues of the host are responding to the infection to contain and eliminate it. Moreover, even the susceptible cells themselves respond to control the infection. Therefore, we need to take account of the entirety of the defensive stratagems of the host and of the countermeasures that viruses have evolved in order to evade and even subvert them.

We begin by reminding ourselves that from the point of view of their vertebrate hosts, viruses, as well as other pathogenic microbes, are potentially lethal invaders. Consequently, over the course of hundreds of millions of years, vertebrates have had to evolve defenses not only against viruses but also against bacteria, fungi, protozoa, and metazoan parasites (e.g., worms).

Each of these different classes of invasive pathogens uses a different strategy to establish itself in the host. Moreover, there are a variety of strategies used by different pathogens within each class, as well as between classes. In response, vertebrates have evolved a network of multiple defensive measures, and they are able to bring to bear the most appropriate combinations of these defensive measures to counter each particular class of invader. Although we are primarily concerned with host defenses against viruses, note that some but not all antiviral defenses overlap defenses used against other classes of invaders as well.

Our antivirus defenses indeed are formidable. This may reflect the fact that viruses can evolve much more quickly than their more slowly evolving vertebrate hosts. Thus, the host needs to have potent and comprehensive defenses in place that might cope with whatever countermeasures the rapidly evolving viruses might adapt.

One of my hopes in writing this chapter has been to convey just how formidable, and one might say refined and elegant, our immune defenses are. Yet, as we just have noted, rapidly evolving viruses continue to devise countermeasures that challenge those host defenses. Thus, much as one may be struck by the potency and complexities of the host defenses, many of the most intriguing viral adaptations are those that enable these microbes to evade and even subvert those host defenses. Indeed, only those viruses which have succeeded in doing so can be sustained within any host population. Thus, I hope that after reading this chapter you will have a high regard for just how formidable our antiviral defenses actually are, and in turn, I hope you also will appreciate the variety of countermeasures that viruses have evolved to evade those defenses. The variety of these countermeasures is vast. For example, among the herpesviruses (a family of large DNA viruses), perhaps one-half of the entire complement of the 80 or more viral genes are concerned primarily with immune evasion. (If a virus were able to completely countermand our immune defenses, then neither we nor the virus would still be here. Importantly, the virus needs to evade immune clearance only long enough to pass the infection on to a new susceptible individual.)

It is impossible to cover all of the important details of the mammalian immune responses and all of the viral immune evasion strategies in just one chapter, albeit a long one. Instead, we consider a few examples of some of the best-understood or especially intriguing (at least to the author) viral evasion strategies. Some of these, as well as ones not covered here, are discussed in more detail in the chapters that follow. (Many readers will likely be more than satisfied with the amount of detail crammed into this chapter. However, rest assured that it is but a summary of how the immune system works, hopefully containing enough details to impress you with its elegance and potency, and with the sophistication of the many viral countermeasures against it.)

OVERVIEW OF DEFENSES

The first aim of our antiviral defenses is to prevent infection. If infection should occur, the purposes are then to slow the replication and spread of the virus, followed by its elimination. Towards those ends, our defenses may be thought of as comprised of a three-tiered system. The first of these tiers consists of the physical barriers presented by the skin and by the mucosae that line the respiratory, the gastrointestinal, and the urogenital tracts. The skin and the aforementioned mucosae provide an effective shield that prevents most viruses from gaining access to the cells and tissues of the body. Should a virus invade or penetrate one of these physical barriers, it then confronts the second tier of defense, the **innate immune system**, and if necessary, the third tier, **the adaptive immune system.**

Key differences between the innate and adaptive immune systems are as follows. The innate immune system recognizes general features present on a variety of pathogens, and it does so by means of a fixed collection of unchanging receptors which are encoded in the germ line of the host. These receptors recognize specific ligands that are present on a variety of different pathogens (e.g., bacterial lipopolysaccharides). In contrast, the adaptive immune system uses receptors that are specific for each particular pathogen that might be encountered. The receptors of the adaptive immune system are not encoded in the germ line and must be generated by a special process, described below. Because of the difference in the origins of the receptors of the innate and the adaptive immune systems, the innate immune system employs a repertoire of relatively few receptors, each of which is always present in plentiful amounts, whereas the adaptive system has a very large repertoire of receptors, but the number of receptors with any particular specificity is initially low.

Each cell of the adaptive immune system expresses immune receptors of only a single specificity. However, cells of that specificity will undergo a dramatic expansion if they should encounter an invader that is recognized by their particular receptor. Indeed, the adaptive immune system is so named because its responses are tailored to each particular pathogen. In fact, the specificity and effectiveness of the adaptive immune response are continuously refined throughout the course of the infection.

The differences between the innate and adaptive immune systems outlined above have the following consequences. First, the innate immune system can begin to effectively control an infection within the first several hours. In contrast, the adaptive immune system has to be activated by a special process to expand the number of effectors that can respond to the particular pathogen being encountered. For this reason, the adaptive immune system is not effective until the fourth day after infection or later. Another important feature of the adaptive response is that it has a memory of past encounters, which leaves it primed to react quickly should it reencounter a pathogen that it responded to earlier. This memory is the basis for immunity. Indeed, one objective of vaccination is to generate memory. The innate immune system has no memory of past infections.

Despite the distinctions between the innate and adaptive systems, separating them is somewhat artificial, since they frequently work together. Indeed, a major difficulty in describing the immune system is that it is a network that involves interactions between many different players. Thus, in describing the individual players, it is important not to lose sight of the forest for the trees. To merely catalog the actions of each individual player would miss the important interactions that together constitute the immune system's response to the infection. Because of these interactions, we will see individual players reemerge multiple times, as we introduce new players with which they interact.

PHYSICAL BARRIERS AGAINST INFECTION

The skin is an effective barrier against viruses in general. That is so because viruses require living cells to support their replication, and the skin is covered with multiple layers of dead cells that are filled with keratin. Yet some viruses, for example, the papillomaviruses (Chapter 16), can exploit cuts or abrasions in the skin to target the living basal layer of the epidermis. Other viruses (e.g., yellow fever virus [Chapter 7]) take advantage of insects or other vectors, which penetrate the skin via a sting or bite.

Since viruses cannot replicate in the outer dead layers of the skin, most use the living epithelia of the respiratory, gastrointestinal, and urogenital tracts for their portals of entry into the host. In contrast to the multilayered skin, these epithelia are only one or a few cell layers thick. Moreover, they are actively transporting oxygen and nutrients to underlying tissue. In addition, the area covered by our skin is only about 2 m^2, in contrast to the ~400 m^2 that are covered by the mucous membranes lining our respiratory, gastrointestinal, and urogenital tracts. Thus, these epithelia constitute a vulnerable portal to potential invaders.

Yet because of their vulnerability, these epithelia have evolved formidable barriers of their own against infection. For example, in the case of the respiratory mucosa, the mucous coating and the beating cilia combine to impair virus binding to the cell surfaces and to sweep the viruses up to the throat, where they may be coughed away or swallowed.

As for the gastrointestinal tract, the low pH within the gut provides a major block to infection by that route. In the stomach, the pH may be as low as 2.5, a level of acidity that is lethal to most viruses. Moreover, powerful proteolytic enzymes such as pepsin provide yet another barrier to infection via the gut. Should a virus somehow survive passage through the stomach, it then encounters the digestive enzymes of the small intestine, as well as the bile salts contributed by the liver. Together, the degradative activities of the enzymes and the detergent activity of the bile salts make for a most formidable impediment against any viral pathogen.

The epithelium of the vagina, like the skin, is several cell layers thick. Moreover, only the deepest, most inaccessible basal cells of the vaginal epithelium are actively dividing. Since viruses replicate best in cells that are actively replicating, these properties of the vaginal epithelium combine to create an effective barrier against most viruses. Moreover, the vagina is well colonized in healthy subjects by a bacterial flora that produces lactic acid. The resulting acidity, as in the case of the gut, protects the vagina from infection by most viruses. In addition, the epithelial cells that line the reproductive tract, like those that line the respiratory and gastrointestinal tracts, are covered by a layer of protective mucus. Nevertheless, the vaginal epithelium can be torn, for example, during sexual intercourse, providing a site of infection for viruses such as genital papillomaviruses, herpesviruses, and human immunodeficiency virus (HIV).

THE INNATE IMMUNE SYSTEM

Should an invader succeed in breaching a physical barrier, it then encounters the components of the innate immune system. This arm of immunity is so named because it is in place to function nonspecifically against a broad range of invaders. Like the adaptive immune system, its components need to be activated by the invader. But since the innate immune system does not need to tailor its response to each particular invader, it can begin acting against the invader days before the adaptive immune system is able to do so. Thus, the principal role of the innate immune response is to provide an early, nonspecific, effective line of defense against a broad range of potential invaders.

As noted above, the innate immune system recognizes molecular patterns that are characteristic of broad classes of microorganisms, doing so via receptors that are encoded in the germ line. In contrast, the adaptive immune system recognizes **antigenic determinants** that are unique to each particular invader, doing so via receptors that are *not* encoded in the germ line and that are generated by a specific process, described below. When activated by a particular invader, the adaptive immune system tailors its response to that invader. However, the price for that specificity is that activation can take 1 week or longer.

Since virus infectious cycles can be as short as 4 to 5 h, the host does not have the luxury of doing nothing while waiting for the adaptive immune system to respond appropriately. Consequently, the innate immune system is nearly entirely responsible for controlling virus infections for the first 3 days or so after a primary (first time) infection with a virus. (Recall that the adaptive immune system has a memory of previously encountered pathogens. This memory enables adaptive immunity to already be in place against previously encountered pathogens, even years after the primary infection. Importantly, memory enables adaptive immunity to spring into full action much more quickly during reinfections than during primary infections.)

Indeed, the host's need to mount an immediate response to a primary invasion almost certainly was a major selective pressure behind the evolution of the innate immune system. Yet the role of innate immunity does not end there, since the innate immune response plays key roles in orchestrating a specific adaptive immune response when the latter is needed.

The crucial role of the innate response as "quarterback" of the adaptive response is illustrated by the substantial impairment of immune protection in individuals lacking innate immune function. Importantly, such individuals have vastly more severe infections than individuals who are deficient for adaptive immunity only (Fig. 4.1). Infections go largely uncontrolled in individuals with inherited deficiencies of innate immunity, because the adaptive immune response cannot be deployed in the absence of the quarterback role of the innate response. In contrast, in individuals who are deficient in adaptive immunity only, the innate response may be able to initially control the infection, although it likely will not be able to clear it. These findings may account for the rarity of individuals with inherited deficiencies in innate immune function.

Experiments using knockout mice that were made deficient for effectors of either innate immunity or adaptive immunity lead to the same conclusion. Mice infected with vaccinia virus (a poxvirus) or herpes simplex virus (HSV) undergo much more severe infections in the absence of

Figure 4.1 The crucial quarterback role of the innate response is illustrated by the relative effects of deficiencies in innate immunity versus deficiencies in adaptive immunity. Infections go largely uncontrolled in individuals with an inherited innate immunity dysfunction because an adaptive immune response cannot be effectively established in the absence of the quarterback role of the innate immune response. In contrast, in individuals who are deficient in adaptive immunity functions only, innate immunity may be able to initially contain the infection, even if it cannot clear it. Adapted from P. Parham, *The Immune System*, 2nd ed. (Garland Science, New York, NY, 2005), with permission.

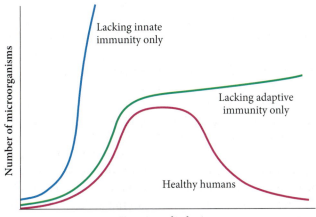

innate immunity effectors than in the absence of adaptive immunity effectors only. The effectiveness and crucial role of innate immunity are likely the reasons why vaccinia virus and the herpesviruses express a variety of proteins that are not essential for virus replication in vitro but help these viruses to evade innate immune responses in vivo. Such viral proteins impede interferon (IFN), complement, and inflammatory responses and oppose apoptosis (see below). The quarterback role of the innate response in activating adaptive immunity is considered in more detail later.

As noted above, the innate immune system recognizes molecular patterns that are common to a variety of pathogens but are not present on uninfected host tissues. The lipopolysaccharide components of the coats of gram-negative bacteria provide a well-known example of a common molecular pattern that is recognized by the innate immune system. In the case of viruses, the structures of both enveloped and nonenveloped viruses contain regularly repeating surface features that may be recognized by the innate immune system. Moreover, as described below, the innate immune system can also react to virus-specific nucleic acids.

When the innate system senses the presence of these foreign molecular patterns, it attempts to wall off, contain, and then eliminate the invading pathogen. To do so, the cells of the innate system respond by secreting a set of chemical effectors called **cytokines**. (Cytokines are proteins that are secreted by cells in order to mediate specific effects on other cells that express specific cytokine receptors.) These secreted cytokines trigger the **inflammatory response**, which is exemplified by the swelling, redness, and flu-like symptoms that are characteristic of many virus infections. How the innate system knows what cytokines to secrete and when to do so is as follows.

Cells of the innate immune system, in particular **macrophages**, are present behind the physical barriers, where they function as sentries, on the alert for invading pathogens. They express a family of receptors known as **Toll-like receptors (TLRs)**, each of which recognizes a particular kind of molecular feature that is common to a broad class of pathogens. For example, TLR4 at the cell surface binds to the lipopolysaccharides of gram-negative bacteria. Together, the 10 known TLRs can recognize virtually every pathogen that can cause infection. (Mammalian TLRs were originally identified and studied in macrophages. However, TLRs have also been found in nonmyeloid cells, in which their role is not yet well clarified. Recently, functional TLRs have been found in nonmyeloid human cells in vivo [e.g., gingival fibroblasts] and in a variety of nonmyeloid human and mouse cell lines in culture.)

A number of viruses now are known to trigger innate responses via TLRs. For example, respiratory syncytial virus and mouse mammary tumor virus have been shown to trigger an innate response through TLR4, whereas measles virus and human cytomegalovirus (HCMV) appear to trigger an innate response via TLR2. Also, TLR3 and TLR8, which are expressed on intracellular membranes, bind to double-stranded and single-stranded RNA molecules, respectively, which are generated during a variety of virus infections (see below). Thus, at the very earliest stages of the virus-cell interaction, TLRs seem to sense the virus and activate an inflammatory response against it.

Once TLRs of a macrophage are engaged, they transmit a signal that induces it to release a particular set of cytokines appropriate for dealing with the particular class of invader. For example, TLR3 signals induce the expression of IFNs, which are the major antiviral cytokines (see below). Cytokines also recruit additional innate immune effector cells to the infected site and induce inflammation (Box 4.1).

The components of the innate immune response include the **professional phagocytes** (i.e., macrophages and **neutrophils**), **natural killer** (NK) cells, and a variety of cytokines that these and other cells produce. The **complement system** of proteins too constitutes an effector mechanism of innate immunity. Importantly, bear in mind that these

different players interact and synergize with each other to provide an impressive degree of effectiveness. Moreover, they likewise orchestrate and synergize with the mediators of adaptive immunity (see below). We now consider the individual players of innate immunity.

Cytokines: the IFNs

In the late 1950s, A. Isaacs and J. Lindenmann observed that culture media from chicken cells exposed to *inactivated* influenza virus contained an activity that would render fresh uninfected cells more resistant to infection with certain viruses. Moreover, this antiviral activity could be detected within hours of exposure to the virus. Subsequent protein fractionation showed that the interfering activity actually is comprised of a set of related proteins, now known as alpha-interferon (IFN-α) and IFN-β (also referred to as type I IFNs). Thirteen related IFN-α proteins are presently known.

Since type I IFNs could be induced by inactivated influenza virus, virus replication per se is not required for their induction. What then induces these IFNs?

Double-stranded RNA produced inadvertently during virus replication is thought to trigger IFN induction, as implied by the experimental finding that type I IFNs can be induced with *synthetic* double-stranded RNA (e.g., polyinosine/polycytosine; see below). Consequently, one might expect that IFN is not specific for any particular virus and that the IFN activity induced by one kind of virus might have an antiviral effect against unrelated viruses. Indeed, that is so. The antiviral state induced by type I IFN is not virus specific. In this regard, recall that nonspecificity is a characteristic feature of innate immune responses in general. However, IFN is host specific. For example, human IFN induces an antiviral effect in human cells but not in mouse cells.

There is a second pathway of IFN-α and IFN-β induction, which like the first is independent of virus replication but, in this instance, is even independent of virus gene expression. It is observed in so-called "natural" IFN-producing cells (IPCs) of hematopoietic origin, the principal one of which is a particular type of immature dendritic cell (see below). Induction occurs when these IPCs are exposed to any of a variety of enveloped viruses, even when those viruses have been physically or chemically inactivated. Induction is thought to result from the direct interaction of viral envelope glycoproteins with the surface of the IPC. (Proviso: Much of what is known about the activities of the IFNs [and of other cytokines as well] is known from studies of model systems in cell culture. Yet cell cultures cannot be expected to reflect the complexities of the mixed cytokine responses that likely take place in vivo [see below]. Consequently, the actual workings and

BOX 4.1

The discovery of TLRs in the late 1990s helped to clarify the workings of the innate immune system. Moreover, their discovery resulted in a better appreciation of the importance of innate immunity in host defenses. Although much was known earlier about the IFNs and NK cells (see the text), innate immunity still received relatively little attention compared to the plentiful attention lavished on the adaptive immune system, which seemed intrinsically more glamorous with its specific antibody-producing and cell-killing functions.

The TLRs derive their name from their resemblance to a fruit fly protein called Toll. Flies that lack Toll are defective for their dorsal-ventral differentiation. The gene is called *Toll*, since that is the German word for "fantastic," although in this instance it may have the connotation "weird." Interestingly, Toll in flies also acts to protect these insects from fungal infection. In addition, plants too contain TLRs, one of which (called N protein) protects tobacco against tobacco mosaic virus. Thus, Toll-like proteins appear to have arisen early in evolution, and they likely played a crucial role in making multicellular organisms possible.

effects of these cytokines, as they occur in vivo, are only partly understood.)

Before continuing our discussion of the two type I IFNs, we should note that they, together with gamma IFN (IFN-γ) (a type II interferon), are among the most important cytokines that are induced at early times in a variety of viral infections. The type I IFNs are best known for their antiviral activities, whereas the type II IFN, IFN-γ, is best known as an important modulator of immune system function. Yet the type I IFNs are also important modulators of immune system function, and IFN-γ also expresses an antiviral activity, but one that is different from that of the type I IFNs.

For convenience, IFN-α and IFN-β are often referred to collectively as IFN-α/β. Yet there are differences between them. IFN-α is produced in leukocytes, and IFN-β is broadly produced in fibroblasts and epithelial cells. Almost any cell will produce either IFN-α or IFN-β in response to being infected by a virus. In contrast, IFN-γ is produced by T cells of the adaptive immune system and by NK cells of the innate immune system, in each case doing so in response to extracellular signals. The crucial importance of the IFN responses is demonstrated by animals that cannot express IFN defenses. These animals are more quickly killed by viruses, and at lower infectious doses than are required to kill animals that express normal IFN responses.

Viruses are known to induce IFN-α/β early after the onset of infection, as shown in chickens, mice, nonhuman primates, and humans, and they presumably do so in other avian and mammalian species as well. The mechanisms for inducing IFN-α/β are known only in part. The most potent known inducer of these IFNs is double-stranded RNA, which many different viruses generate during their replication. In the case of single-stranded RNA viruses, IFN induction happens as follows. Single-stranded RNA viruses replicate by first generating single-stranded RNA molecules that are complementary to the genomic RNA. These complementary RNA molecules then serve as templates for the production of progeny viral genomes. Usually, there is no double-stranded RNA replicative intermediate as such (might this be an IFN evasion strategy?), but the complementary RNA molecules produced during replication may inadvertently anneal with genomic RNA, forming double-stranded RNA molecules that can induce IFN-α/β. In the case of DNA viruses, many of these viruses transcribe both strands of certain regions of their genomes. Thus, these DNA viruses also produce RNA molecules with complementary sequences that inadvertently may anneal, thereby generating double-stranded RNA that can trigger IFN production.

TLR3 recognizes and responds to double-stranded RNA, as noted above, and it then triggers expression of IFN-α/β.

For example, TLR3 of human corneal epithelial cells experimentally triggers production of IFN-β in response to the synthetic double-stranded RNA polyinosine/polycytosine. (Polyinosine/polycytosine also triggers the activation of proinflammatory cytokines interleukin 6 [IL-6] and IL-8 by these cells.) Importantly, corneal epithelial cells are a first line of defense against ocular pathogens. (As noted above, functional TLRs have been found in mammalian cells other than macrophages and other cells of the myeloid lineage. Human corneal epithelial cells are an example of a cell type that is not of the myeloid lineage but nevertheless expresses TLRs.)

Other TLRs can trigger expression of IFN-α/β. Furthermore, recent experimental results show that the retinoic acid-induced gene I product (**RIG-I**) also may be an important intracellular sensor of double-stranded RNA in virus-infected cells. Indeed, RIG-I appears to be essential for activating transcription of IFN-α/β during infection by several RNA viruses, including paramyxoviruses, flaviviruses, and influenza viruses. These experimental findings were obtained using mouse knockout cells and animals (see Box 4.4).

Next, we consider the means by which type I IFNs bring about an antiviral state and the means by which that antiviral state protects the host. We begin by noting that IFN-α/β do not necessarily protect the virus-infected cells in which they are first induced. Instead, the IFN molecules secreted by virus-infected cells have a more important effect on as yet uninfected neighboring cells, in which they actually induce an antiviral state. Upon binding to specific receptors on these neighboring cells, IFN-α/β transmit a signal that results in the transcriptional up-regulation of about 300 genes (as demonstrated by gene array technology). The expression of these genes establishes an antiviral state that prepares IFN-activated cells to defend themselves against viral invaders. The up-regulation of IFN-responsive genes occurs only when extracellular IFN binds to IFN-specific receptors on the plasma membrane.

IFN-α/β, as well as IFN-γ and other cytokines (e.g., interleukins; see below), transmit signals through receptors that activate the **JAK/STAT signaling pathway** (Fig. 4.2). The significance of JAK/STAT signaling is as follows. Recall that mitogenic transmembrane signals, which are induced by growth factors (e.g., the epidermal growth factor), generally are propagated via a series of protein kinases that activate each other in a particular sequence, usually beginning with a tyrosine kinase and followed by a succession of serine/threonine kinases. In contrast, the JAK/STAT signaling pathway is a short, direct pathway that enables cells to respond quickly to cytokine stimulation. Its key features are as follows.

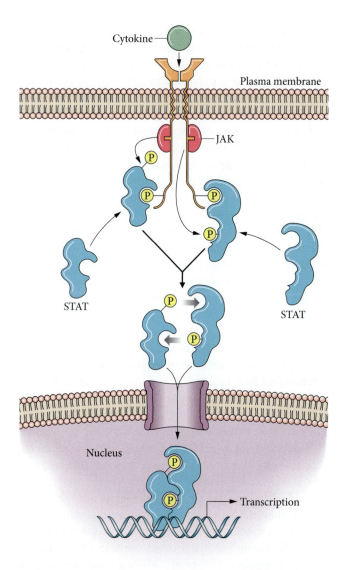

Cytokine

Plasma membrane

JAK

P

P
P

P

STAT

STAT

P

P

P

Nucleus

P

P
Transcription

Figure 4.2 Simplified depiction of the JAK/STAT signaling pathway, by which IFN-α/β, as well as IFN-γ and other cytokines (e.g., interleukins [see the text]), transmit signals. IFN binding causes dimerization of the individual IFN receptor chains, enabling their associated JAKs to phosphorylate tyrosine residues in the cytoplasmic domains of the receptor polypeptides. Transcription proteins named STATs bind to the phosphorylated receptors and in turn are phosphorylated by the JAKs. The phosphorylated STATs dimerize and move into the nucleus, where they bind to specific transcriptional control sequences called IFN-stimulated response elements, thereby activating specific target genes. Adapted from G. M. Cooper and R. E. Hausman, *The Cell: a Molecular Approach*, 4th ed. (ASM Press, Washington, DC, 2007), with permission.

There is one cell surface receptor for IFN-α/β and another for IFN-γ. The cytoplasmic domains of these receptors are associated with protein kinases of the Janus kinase (JAK) family, named for Janus, a god of Roman mythology, who stood watch over gates and doorways. (Janus saw in two directions at once and thus is depicted with two faces.)

IFN binding causes dimerization of the individual IFN receptor chains, so that their associated JAKs can phosphorylate tyrosine residues in the cytoplasmic domains of the receptor polypeptides. Transcription proteins named STATs (acronym for signal transduction and transcription factors) bind to the phosphorylated receptors and in turn are phosphorylated by the associated JAKs. The phosphorylated STATs dimerize and move into the nucleus, where they bind to specific transcriptional control sequences called IFN-stimulated response elements, and in so doing they activate specific IFN-regulated target genes.

There are seven distinct STAT genes in mice. Inactivating *stat-1* causes the mouse to be unable to mount an innate immune response against bacterial and viral infections. Inactivating *stat-4* or *stat-6* severely compromises the ability of the mouse to mount an adaptive immune response. These experimental findings underscore the crucial roles played by the JAK/STAT signaling pathway in both innate and adaptive immunity systems. As seen below, a variety of cytokines signal via the JAK/STAT signaling pathway.

Next, we see that IFNs induce a variety of innate antiviral processes.

Of the several hundred cellular genes that are upregulated by IFN-α/β, three in particular are known to play important roles in the antiviral state. (It is a safe bet that more than a few others will also be found to play important roles.) These three genes encode the RNA-activated protein kinase **PKR**, RNA-activated **RNase L**, and the double-stranded RNA-specific adenosine deaminase **ADAR**. We consider each of these host gene products in turn.

PKR levels are upregulated 5- to 10-fold by IFN-α/β. Yet, although PKR levels are up-regulated by IFN-α/β, this serine/threonine protein kinase remains inactive in uninfected cells. However, it is ready to be activated if the cell should become infected. This occurs as follows. PKR has a double-stranded RNA-binding motif. When two molecules of PKR bind to the same molecule of double-stranded RNA, these two PKR molecules cross-phosphorylate each other, thereby activating each other's serine/threonine kinase activity. Thus, if a cell in the antiviral state is infected, the double-stranded RNA that is generated by the virus binds and activates PKR.

The activated PKR acts by phosphorylating numerous target proteins, but the best-understood antivirus effect of PKR involves **eIF2a**, an initiation factor for protein synthesis. PKR phosphorylates eIF2a, thereby inactivating its ability to initiate translation, which brings about an abrupt shutdown of both viral and cellular protein synthesis in the infected cell. Since cellular protein synthesis is inhibited by the IFN-activated PKR, the cell may be sacrificed in the process. However, this is usually preferable to having it live to support the replication and dissemination of the virus.

RNase L is a nuclease that degrades most viral and cellular RNAs. Like PKR, it is activated by a multistep process. First, IFN-α/β induces a 10- to 1,000-fold up-regulation of RNase L production, as well as that of another cellular enzyme, **2′-5′-oligo(A) synthetase.** But as in the case of PKR, these enzymes remain inactive until the cell might be infected. Should the cell be infected, the double-stranded RNA that is generated by the virus activates the 2′-5′-oligo(A) synthetase, which then produces 2′,5′ oligomers of adenylic acid. This unusual polynucleotide then binds to, and activates, RNase L, which goes on to degrade most viral and cellular RNAs, thereby shutting down viral replication. Since cellular RNAs too are degraded by activated RNase L, this enzyme, like PKR, may kill the cell while blocking viral replication.

ADAR, a double-stranded RNA-specific adenosine deaminase, catalyzes the deamination of adenosine to inosine on viral and cellular RNAs. This site-specific RNA editing results in the translation of defective proteins.

Still another IFN-induced antiviral effect, in this instance resulting from long-term exposure to IFN-α/β, involves the induction of a number of cellular genes that lead to **programmed cell death (apoptosis).** Whereas this leads to the early suicidal death of infected cells, thereby blocking virus production within them, it also leads to the suicide of nearby uninfected cells. This seemingly extreme response benefits the host, since (as argued above) blocking replication of a viral pathogen is well worth the loss of a relatively few cells. Moreover, these cells might have been killed anyway by the virus, but only after producing many progeny viruses. (Note that apoptosis is a normal host process, which serves several important functions not related to immune defenses. For example, apoptosis eliminates cells in which the DNA might be damaged, in that way eliminating cells that might go on to become cancerous. It also plays an important role in development. A well-known example is in the elimination of tissue to generate the spaces between the fingers and toes.)

IFN-α/β also may cause an inflammatory response. Inflammation is discussed in more detail below in the context of tumor necrosis factor alpha (TNF-α). For now, note that this effect of IFN-α/β can lead to the appearance of common **flu-like symptoms,** such as fever, chills, muscle aches, nausea, and malaise. Indeed, these symptoms, which are typically experienced during an influenza virus infection, result from the inflammatory response rather than from the direct action of the virus per se. This explains why a variety of viruses, despite being unrelated to influenza virus, cause a similar set of flu-like symptoms.

Yet another role for IFN-α/β is to have an effect on NK cells. As described below, NK cells are ready to kill virus-infected cells without having to first be activated. However, NK cells become more efficient killers of virus-infected cells when they are activated by signals alerting them that an infection is under way. IFN-α/β are among the most potent of these NK cell activators, enhancing the base level of NK cell cytotoxicity 20- to 100-fold. This example is one of several that illustrate how two or more players of the innate immune response synergize to strengthen innate defenses.

Next, we consider the induction and activities of IFN-γ. Recall that IFN-α/β may be induced in almost any cell that is infected by a virus. In contrast, IFN-γ is produced and secreted *nearly exclusively* by T cells and NK cells. T cells are key effectors of adaptive immunity. They generate IFN-γ only after their specific immune receptors recognize their **cognate antigenic determinants** (i.e., the specific antigenic determinants that a particular T cell's or B cell's receptor recognizes and binds to), which are presented to them via a special process, described later. In contrast, NK cells, which are effectors of innate immunity, are induced to produce IFN-γ by the IFN-α/β given off by virus-infected cells, as well as by the cytokines TNF-α and IL-12. The last two cytokines are secreted by macrophages in a positive-feedback loop with NK cells, as described in the next paragraph. Like IFN-α/β, IFN-γ transmits signals to target cells by binding to a heterodimeric receptor, which activates the JAK/STAT signaling pathway (Fig. 4.2). IFN-γ receptors are present on the plasma membranes of most cell types.

IFN-γ mediates a variety of antiviral and immune regulatory effects, involving both innate and adaptive immune functions. The following example involves NK cells of the innate immune system. IFN-γ secreted by NK cells can activate macrophages. These activated macrophages then secrete TNF-α and IL-12. (The designation "IL" is for interleukin. **Interleukins** are cytokines that mediate communication between leukocytes.) TNF-α and IL-12, which are thus produced by the activated macrophages, in turn stimulate the NK cells to produce more IFN-γ, in that way establishing a positive-feedback loop between the NK cells and macrophages (Fig. 4.3). The net result is that these crucial immune effector cells are primed to efficiently carry out their immunity functions.

One of the best-characterized effects of IFN-γ on macrophages is the induction of the enzyme **nitric oxide synthetase 2 (NOS2).** NOS2 generates **nitric oxide (NO),** which in turn reacts with oxygen to yield other reactive molecules. Together, these molecules can inactivate essential virus proteins. A variety of important human pathogenic viruses (e.g., HSV, Epstein-Barr virus [EBV], hepatitis B virus [HBV], and vaccinia virus) were shown to be sensitive to the action of NOS2. Indeed, experiments

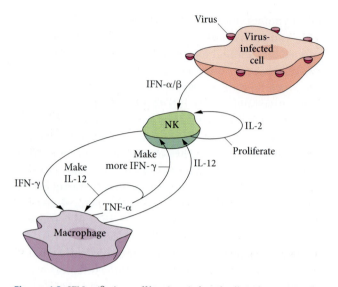

Figure 4.3 IFN-α/β given off by virus-infected cells induces NK cells to produce IFN-γ. The IFN-γ secreted by the NK cells in turn activates macrophages. The activated macrophages then secrete TNF-α and IL-12. TNF-α and IL-12 thus produced by the macrophages in turn stimulate NK cells to produce more IFN-γ, in that way establishing a positive-feedback loop between the NK cells and the macrophages. The net result is that these crucial immune effector cells are each primed to efficiently carry out their functions.

with mice showed that impairing NO production results in uncontrolled infection by a variety of viruses.

IFN-γ also influences B cells and T cells of the adaptive immune system. How it does so is explained in "The Adaptive Immune System," below. Thus, do not worry if the remainder of this paragraph is not yet comprehensible. The material discussed therein will come up again later. IFN-γ helps induce B cells to undergo antibody class switching to produce immunoglobulin G (IgG) antibodies, which are good at **opsonizing** virus and binding **complement**. IFN-γ also affects T cells, doing so indirectly by inducing macrophages to secrete IL-12. This cytokine in turn induces T cells to secrete a **cytokine profile** that is especially effective against bacterial and viral infections, as opposed to defending against a parasite.

Cytokines: TNF-α, Some Other Cytokines, and Inflammation

TNF-α regulates immune defenses against viruses as well as against other pathogens. Although it affects immune defenses in a variety of different ways, the major role of TNF-α is to *heighten inflammation*. Inflammation is a rather general term, referring to the local accumulation of fluid, plasma proteins, and leukocytes (mainly neutrophils and NK cells) at the infected site. This accumulation causes the familiar swelling, reddening, heat, and pain associated with inflammation. TNF-α induces these changes

by affecting the local blood capillaries, thereby enabling fluid, plasma proteins, and cells to pass from the circulation to the infected tissue. Inflammation is a *critically important* antiviral response. It is essential to initiate antiviral defenses, while also contributing significantly to the overt symptoms that result from infection.

TNF-α is produced initially by macrophages at the infected site. These cells, which are described in more detail below, are stationed under the physical barriers and in the tissues as well. They might be thought of as both immune sentries and first responders. Tissue macrophages may be able to deal with the few viruses that penetrate the physical barriers. However, if the tissue macrophages cannot handle the situation, they call for reinforcements. They do so by secreting TNF-α. Importantly, macrophages produce inflammatory levels of TNF-α only when there is a sustained engagement of a number of macrophages, an event which occurs only when the infection is outpacing the macrophages at the infected site. Thus, inflammation occurs only when local innate defenses are seriously engaged and in danger of being overrun.

Neutrophils are the major leukocytes of the inflammatory response. They leave the blood and hone in on the infected site in response to TNF-α (and also cytokines IL-1 and IL-8), which is released by the embattled macrophages. Neutrophils and macrophages are described in more detail below. However, it is important to note here the following distinction between these cell types. Although both macrophages and neutrophils are *professional phagocytes*, macrophages are stationed in the *tissues*, whereas neutrophils remain in the *blood circulation*, unless recruited to an infected site. The reason neutrophils are normally retained in the circulation is that they are "professional killers." When they move from the blood to the surrounding tissues, they are immediately ready to kill target cells. Consequently, they have the potential to do much unintended damage to uninfected host tissue, and it is therefore important that they do not leave the circulation, unless they are needed to contain an infection. Neutrophils make up about 70% of the white blood cells in the circulation.

For the reason just given, the exit of neutrophils from the blood circulation is carefully regulated. Indeed, exit of neutrophils from the circulation requires multiple signals from the infected site. Moreover, these signals induce local changes both in circulating neutrophils and in the local capillary endothelial cells. Each of these changes is necessary to trigger **extravasation**, the movement of cells from the blood circulation to the surrounding tissue. We consider the steps of neutrophil extravasation in the next paragraph. Note the multiple roles that TNF-α plays in the process.

Both TNF-α and IL-1 are secreted by activated macrophages at the infected site. Together, they induce the expression of adhesion molecules on the luminal plasma membranes of local capillary endothelial cells. As neutrophils (and other leukocytes as well) circulate through those local capillaries, TNF-α and IL-1 also induce the expression of adhesion molecules on these neutrophils. The upregulated adhesion molecules on the neutrophils and on the luminal surfaces of the capillary endothelial cells act together to cause the neutrophils to slow down near the infected site and "roll along" the luminal surface of the capillaries. IL-8, which also is secreted by activated local macrophages, induces conformational changes in these upregulated adhesion molecules, which has the effect of strengthening the adhesions between the neutrophils and capillary endothelial cells. The net result is that the neutrophils stop rolling, thereby enabling them to respond to yet other alarm signals coming from the infected site. These additional signals prompt the neutrophils to cross the capillary endothelia to the infected region, by a process known as **diapedesis**. TNF-α, produced at the infected site, also affects the activities of neutrophils and other leukocytes appearing there. Moreover, TNF-α induces vascular endothelial cells to make **platelet-activating factor**, which causes blood clotting and blockage of local blood vessels. This prevents pathogens at the infected site from entering into the circulation and using it as a conduit to disseminate the infection.

In addition to the local effects caused by cytokines at an infected site, cytokines also induce distant effects that also heighten host defenses. For example, TNF-α, IL-1, and IL-6 elevate body temperature, giving rise to fever, by acting on temperature control sites in the hypothalamus. They also act on muscle and fat cells, altering their metabolism, thereby causing them to generate heat. These cytokines are referred to as **endogenous pyrogens**, to distinguish them from the **exogenous pyrogens** produced by certain bacterial pathogens, which also cause fever.

Although we can attest to the fact that fever can cause considerable discomfort, its purpose is to enhance the ability of the immune system to fight the infection. Fever enhances host defense because most microbial pathogens (including viruses) replicate less efficiently at elevated body temperatures. Moreover, adaptive immune responses are improved at elevated temperatures.

We noted above that IFN-α/β may trigger apoptosis. TNF-α likewise may trigger apoptosis when it binds to its receptors on virus-infected cells.

Once again, note that these cytokines generally work in various combinations to generate their effects. For example, activated macrophages produce IL-1 as well as TNF-α. These cytokines act together to induce expression of adhesion molecules on leukocytes and capillary endothelial cells. Similarly, IFN-γ and IL-12 cooperate with TNF-α to affect the activities of leukocytes at the infected sites.

Macrophages, Neutrophils, NK Cells, and Antibody-Dependent Cellular Cytotoxicity

Macrophages are large professional phagocytes that are stationed in the tissues beneath the mucosa, positioned so as to engulf viruses or bacteria that may have succeeded in penetrating the mucosa. Note that other macrophages are stationed in the lymph nodes, where they are positioned to trap and degrade pathogens that arrive in the lymph from the sites of infection. In the lymph nodes, these macrophages may play an important role in the activation of adaptive immunity. That is, they "present" antigens derived from the trapped pathogens to T cells, as explained in "The Adaptive Immune System," below.

Above, we discussed the positive-feedback loop between macrophages and NK cells, in which TNF-α and IL-12 secreted by macrophages activate NK cells, while IFN-γ from NK cells activates macrophages (Fig. 4.3). Indeed, macrophage activation results primarily from the influence of IFN-γ. IFN-γ also plays a role in the activation of the helper T lymphocytes and cytotoxic T lymphocytes (CTLs) of the adaptive immune response (an example of the interaction between innate and adaptive immunity; see below). Nevertheless, macrophages are the main targets of IFN-γ activity. (**Lymphocytes** are collectively a class of white blood cells that includes the B lymphocytes [B cells] and T cells of the adaptive immune system and the NK cells of the innate immune system. Importantly, B cells and T cells have a large number of variable receptors, each of which is specific for a particular antigen, while NK cells have a small fixed number of receptors against general molecular features present on a variety of pathogens.)

Many studies have shown that IFN-γ enhances the ability of macrophages to destroy ingested bacterial and protozoan pathogens. Yet the importance of IFN-γ in the innate immune response against viruses appears to depend on the particular virus. For example, *ifn-γ*-deficient mice show enhanced susceptibility to vaccinia virus and lymphocytic choriomeningitis virus (LCMV), while their susceptibility to vesicular stomatitis virus and Semliki Forest virus is unimpaired. Note that one can compare *ifn-γ* knockout mice with *ifn-α/β* knockouts, as well as with double knockouts, with respect to their susceptibility to vaccinia virus and LCMV. The results show that mice in which the two IFN types are knocked out are more susceptible to these viruses than are single knockouts, showing that the two IFN types are nonredundant and complementary in their effects.

Note that activated T cells (see below) as well as NK cells produce IFN-γ, and either of these cell types can activate macrophages, but only NK cells can do so at early times. Indeed, the early production of IFN-γ by NK cells makes these cells a central player in the regulation of the early stages of the immune response, especially regarding intracellular pathogens.

Usually, local tissue macrophages can cope with the few viruses that might penetrate the mucosa. In those instances where the macrophages might be overwhelmed by the invader, help is provided by another professional phagocyte, the neutrophil. As noted above, neutrophils comprise about 70% of all circulating white blood cells. They leave the blood and hone in on the infected site in response to the cytokines IL-1 and TNF-α, which are released by the embattled macrophages. In point of fact, neutrophils are the major leukocyte in the inflammatory response (see above). They too release cytokines that can influence subsequent events, in particular, Mip-1α, which is a chemoattractant for T cells and B cells of the adaptive immune response.

Neutrophils have been well studied as key components of the innate immune response against bacteria and fungi. Although they are less studied in the context of virus infections, there is evidence that neutrophils do help to control infections by some important viral pathogens, including HSV and influenza viruses. Still, while neutrophils are known to be a major component of inflammatory infiltrates and to secrete cytokines that promote inflammation (and vascular damage as well) and activate adaptive immunity, there is relatively little information available concerning how they might directly defend against viruses. Neutrophils are known to phagocytose bacteria, which they then degrade via an assortment of degradative enzymes. However, during virus infections, neutrophils generally do not persist at viral lesions but are replaced there by macrophages and lymphocytes.

One conceivable mechanism by which neutrophils might help contain viral infections is referred to as **antibody-dependent cellular cytotoxicity (ADCC)**, a mechanism that might be particularly effective against enveloped viruses. ADCC works as follows. Recall that the envelope glycoproteins of many enveloped viruses are exposed at the host cell plasma membrane prior to virus maturation and budding (Chapter 2). Consequently, these membrane-associated viral glycoproteins can be recognized by virus-specific antibody molecules, which are generated by the adaptive immune response. Now, note the following feature of antibody molecules (discussed in greater detail later in the chapter). The two antigen-binding domains of an antibody molecule, which constitute the so-called Fab region, are at the two amino termini of the molecule, while the single invariant portion, referred to as the Fc region, is at its single carboxy terminus (Fig. 4.4). When an antibody binds to its cognate antigen via the Fab region, the Fc region remains free. Neutrophils, macrophages, and NK cells all have receptors on their surfaces that can bind to the free Fc regions of antigen-bound antibody molecules. Thus, antibodies against viral glycoproteins can act as a bridge, linking the above immune effector cells to virus-infected target cells that display viral glycoproteins at their surfaces. Neutrophils contain granules (actually modified lysosomes), which sequester hydrolytic degradative enzymes, nitric oxide, and toxic oxygen radicals. The contents of these granules are released by contact of the neutrophil with the target cell, leading to the target cell's demise. The purpose of this lethal assault on the infected cell is to terminate virus replication within it.

Bear in mind that in ADCC, killing is carried out by nonspecific neutrophils of the innate immune system. However, antibodies that are produced by the adaptive immune response make ADCC specific against virus-infected cells. Thus, ADCC offers an example of the innate and adaptive immune systems working in concert to contain an infection.

Since ADCC requires the participation of specific antibodies, it cannot occur early during a *primary* infection by the invader (i.e., the adaptive immune system needs to be activated to produce antibodies, a process that takes several days or more). Yet ADCC might play a role in controlling human infections involving HIV and herpesviruses. (Why might ADCC be effective in these instances? Have a look back at Chapter 3.) Also, we might imagine that ADCC can provide an early defense during a second encounter with a virus, in which case preformed antibodies already are present (see below).

Another means by which neutrophils, and macrophages too, are thought to contribute to antiviral defenses is by phagocytosing extracellular viruses. This occurs as follows. First, antibodies also bind directly to free virus particles, as well as to viral glycoproteins in the plasma membranes of infected cells. When an extracellular virus is "decorated" with bound antibodies, it is readily recognized by the Fc receptors on professional phagocytes, thereby enhancing phagocytosis of the invader. A pathogen so decorated by antibodies is said to be **opsonized**. When the pathogen has been phagocytosed, the endocytic vesicle that contains the pathogen fuses with a granule, and the invader is destroyed. Fc receptors on macrophages enable these cells to phagocytose and digest whole virus-infected cells that display viral antigens on their surfaces (Box 4.2).

The granules of neutrophils also contain antimicrobial peptides called **defensins**, which are 29 to 34 amino acids long. Defensins are active in vitro against bacteria, fungi, and various viruses, and they presumably act in vivo as well. Viruses that are sensitive to defensins in vitro include

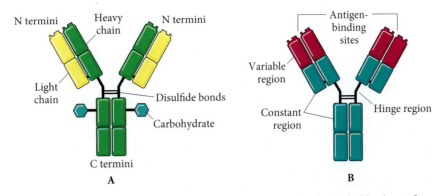

Figure 4.4 IgG antibodies, which are the most common antibody class in the blood, are often used to illustrate common structural features of antibodies. (A) Each IgG molecule is comprised of two identical heavy chains (green) and two identical light chains (yellow). The two heavy chains are linked by disulfide bonds, and each contains a carbohydrate moiety (blue). (B) The antigen-binding sites are located at the N termini of the variable (V) region (red), which is comprised of sequences on both the heavy and light chains. These sequences vary from one IgG molecule to another and confer the specificity for a particular antigen. The constant region (blue), which is the same for all antibodies in the class, is likewise comprised of sequences on the heavy and light chains. IgG antibodies contain a flexible hinge region at about the middle of the two heavy chains. The protease papain experimentally cleaves IgG molecules at the hinge region to generate three fragments: two Fab fragments and one Fc fragment. The Fab fragments contain the antigen-binding sites on the antibody, whereas the Fc fragment provides the link to other immune effectors. For simplicity, the carbohydrate is not shown here. Adapted from P. Parham, *The Immune System*, 2nd ed. (Garland Science, New York, NY, 2005), with permission.

HSV types I and II, CMV, type A influenza viruses, and HIV. In contrast, echovirus and rhinovirus are defensin resistant. Each of the defensin-sensitive viruses noted above is enveloped, whereas the defensin-resistant examples are nonenveloped. This raises the possibility that defensins attack viral lipid envelopes. Much remains to be learned about the defensins, as these versatile peptides appear to have other important immune activities, including serving as chemoattractants for phagocytes and T lymphocytes.

The primary functions of NK cells are (i) to kill virus-infected cells and (ii) to produce cytokines. In point of fact, NK cells are the *major producers* of IFN-γ early in an infection. Like B cells and T cells of the adaptive immune system, NK cells are white blood cells or **leukocytes**. However, NK cells are key players in the innate immune system response to virus infections, whereas T cells and B cells are the principal cells of adaptive immunity. To fulfill their functions in adaptive immunity, T cells and B cells rearrange and express particular gene clusters to generate antigen-specific cell surface receptors and soluble antibody molecules. These developments are explained in detail later. In contrast, as effectors of innate immunity, NK cells do not rearrange or express genes that might enable antigen-specific responses.

Recalling that a primary function of NK cells is to secrete cytokines, in particular IFN-γ, NK cells are induced to do so by macrophages as follows. As discussed above,

macrophages have TLRs that recognize a variety of molecular features common to broad classes of pathogens, and these receptors transmit signals that induce macrophages to secrete cytokines. IL-12, which is one of the important proinflammatory cytokines produced by TLR-stimulated macrophages, activates NK cells to secrete IFN-γ. The IFN-γ produced by the NK cells in turn induces the macrophages to secrete TNF-α. Macrophages have receptors for TNF-α as well as IFN-γ, and the combination of these two cytokines hyperactivates the macrophages to secrete even more IL-12. NK cells have specific receptors for both IL-12 and TNF-α, and the combination of these cytokines produced by the hyperactivated macrophages causes the NK cells in turn to produce even more IFN-γ. Thus, the positive-feedback loop noted above is established between macrophages and NK cells at the infected site (Fig. 4.3).

Importantly, synthesis of IFN-γ early during infection depends upon the production of IL-12 by macrophages. Now, also recall that the ability of NK cells to kill virus-infected cells is stimulated 20- to 100-fold by IFN-α/β produced by virus-infected cells. Thus, stimulation of NK cells with IFN-α/β favors induction of NK cytotoxicity, whereas stimulation of NK cells with IL-12 favors production of IFN-γ.

NK cells first were recognized for their ability to kill virus-infected target cells, with the important qualification that they did not need to be activated to do so. This may

As you are likely aware, antibodies also may neutralize viruses. Neutralizing antibodies are discussed in "The Adaptive Immune System," below. Yet it is worth noting here that antibodies alone, in actuality, do not kill anything. Instead, the role of antibodies is to bind to a specific target. The role played by antibodies in ADCC and in opsonizing viruses underscores the fact that antibody binding marks a target for destruction by other immune effectors. In this regard, note that antibodies also tag viruses for destruction by the classical pathway of complement activation, as described in the text.

What then of neutralizing antibodies? First, only some of the antibodies that bind to a virus actually neutralize it, that is, actually prevent it from initiating an infection. Second, the mechanisms by which neutralizing antibodies block infection are not well understood in all instances. When some neutralizing antibodies bind to a virus, they impair the ability of the virus to attach to the surface of a cell. However, other neutralizing antibodies do not prevent entry of the virus into the cell.

seem to contradict the preceding paragraph. But note that NK cells in fact exist in several stages of readiness to kill. "Resting" NK cells are competent to kill. However, they become much more efficient killers when activated by IFN-α/β. At any rate, the ability of "resting" NK cells to kill virus-infected cells is in marked contrast to CTLs of the adaptive immune system, which need to be activated by a rather complex sequence of events (see below) before they can kill virus-infected cells.

Since NK cells, as components of innate immunity, do not have receptors for specific viruses, how do they recognize virus-infected cells? Current understanding is that NK cells use two sets of receptors for this purpose. One set comprises the **activation receptors**, about which little is known. The other set, the **inhibitory receptors**, recognize **major histocompatibility complex (MHC)** class I molecules. (The role of MHC class I molecules is described in detail below. Briefly, these proteins are present on the surfaces of virtually all host cells. They are used by infected cells to display antigenic peptides that are produced during infection by intracellular pathogens. CTLs have antigen-specific receptors that recognize the combination of antigenic peptide and MHC molecule at the surface of the infected cell.) NK cells can attack and destroy only those cells that do not activate the inhibitory receptors. Thus, NK cells preferentially kill cells displaying low levels of MHC class I molecules.

The rationale for NK cells preferentially killing cells that display low levels of MHC class I molecules is that many viruses attempt to evade CTL-mediated attack by down-regulating host cell MHC class I expression. NK cells counter this viral immune evasion strategy by attacking and destroying cells that do not display sufficient levels of MHC class I molecules to deliver the inhibitory signal to NK cells (Fig. 4.5). Many tumor cells likewise bear low levels of class I MHC molecules and are destroyed by NK cells. (Proviso: The concept that NK cells kill virus-infected cells because of MHC down-regulation is satisfying and indeed is a widely accepted dogma. Nevertheless, there is not a lot of experimental evidence to substantiate it.)

Once NK cells have identified a virus-infected target, how then do they administer the lethal blow? The answer is that NK cells induce infected target cells to commit suicide; that is, NK cells induce target cells to undergo apoptosis. Their modes of inducing apoptosis are similar to those used by CTLs (see below). In one such mode, the adhering killer cell releases a mixture of **perforin** (a membrane active protein similar to the complement protein C9 [see below]) and the enzyme **granzyme B** onto the surface of the target cell. The target cell then takes up these ligands by endocytosis. Within the target cell, the perforin molecules then create pores in the membranes of the endocytic vesicles. This allows the granzyme B to enter the cytoplasm, where it induces a chemical chain reaction leading to apoptosis. (Note that other descriptions of this process have the released perforin acting at the plasma membrane of the target cell, thereby creating pores that enable the granzyme to enter the target cell, where it induces apoptosis.) In another pathway, NK cells use the **Fas ligand (FasL)** on their surfaces to transmit an apoptotic signal into the target cell via the Fas protein on the target cell surface. These apoptotic pathways are described in greater detail later.

NK cells, like neutrophils, have Fc receptors. Thus, NK cells likewise may also kill virus-infected cells by ADCC. As in the case of neutrophil-mediated ADCC, antiviral antibodies bound to the infected cell trigger the NK cell attack. That is, antibodies against the viral glycoproteins in the host cell plasma membrane can act as a bridge between the NK cell and the target cell. Moreover, when the NK cell's Fc receptors bind to the antibody molecules on the target cell, the NK cell actually becomes a more effective killer.

Also like neutrophils, NK cells may have the potential to do much damage in the tissues. Thus, again like neutrophils, NK cells migrate from the circulation to infected sites only in response to multiple signals mediated by inflammatory cytokines released from the infected site. And they exit from the blood circulation by the cytokine-triggered "roll, stop" process, by which neutrophils likewise leave the blood circulation (Box 4.3).

Interaction of NK cell with
uninfected healthy cell

Interaction of NK cell with target cell
in which MHC class I expression is lost

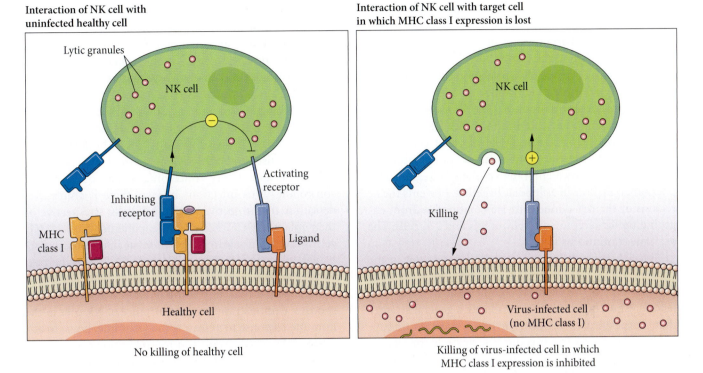

No killing of healthy cell

Killing of virus-infected cell in which
MHC class I expression is inhibited

Figure 4.5 A possible mechanism by which NK cells recognize virus-infected cells. NK cells use two sets of receptors to determine whether to attack a target cell. One set comprises the activation receptors, about which little is known. The other set, the inhibitory receptors, recognizes MHC class I molecules. NK cells attack and destroy only those cells that do not activate the inhibitory receptors. Thus, NK cells preferentially kill cells displaying low levels of MHC class I molecules. The rationale for this is that many viruses attempt to evade CTL-mediated attack by down-regulating host cell MHC class I expression. NK cells provide a host countermeasure to this viral evasion strategy by destroying cells infected by these viruses, since these cells do not deliver the inhibitory signal to NK cells. Adapted from P. Parham, *The Immune System*, 2nd ed. (Garland Science, New York, NY, 2005), with permission.

Since NK cells and neutrophils receive activating signals when their Fc receptors bind antibodies, and since activated NK cells and neutrophils can cause serious damage to tissues, it is important that Fc receptors on these cells be able to distinguish between antibody molecules that are bound to a target cell and the majority of free antibody molecules. How might this be accomplished? A probable answer is that activation of the killing response of these effector cells requires a clustering of their Fc receptors, which occurs only when they bind to an antibody-decorated target cell. Alternatively, or in addition, the binding of Fc receptors to antibodies is a low-affinity interaction. However, the effector cell can bind to the antibody-coated infected cell with high affinity. That is so because many viral envelope proteins accumulate in the target cell plasma membrane, to which the Fab arms of many antiviral antibodies might bind. And many Fc receptors are present on the NK cells and neutrophils, which can bind to the Fc regions of those target-cell-bound antiviral antibodies and, in that way, be cross-linked.

BOX 4.3

At several points in this chapter, we encounter the principle that the launching of certain potent weapons of the immune system requires multiple signals. This is reminiscent of dual-key activation, in which two individuals must act in concert to launch a ballistic missile. Such a fail-safe system is installed to safeguard against the inappropriate launching of a highly destructive weapon. Leave it to Mother Nature to have invented fail-safe systems well before Dr. Strangelove.

While there are numerous studies demonstrating the efficacy of ADCC in cell culture, there is not a lot of experimental evidence that might support a role for this process in vivo. However, the following example, involving HSV, provides a tantalizing suggestion that ADCC indeed is relevant in vivo, while also showing why it is frustratingly difficult to prove it is so.

HSV and other herpesviruses produce envelope glycoproteins that can bind the Fc portion of human antibody molecules. This envelope glycoprotein in the case of HSV is called gE-gI. It protects HSV-infected cells from ADCC in vitro, presumably as follows. The Fab arms of anti-HSV-specific antibodies bind to the other HSV envelope glycoproteins (i.e., gC and gD) in the host cell plasma membrane. But these antibodies simultaneously bind to gE-gI in the plasma membrane via their Fc region. This phenomenon, known as bipolar bridging, prevents the Fc regions of the cell-bound anti-HSV antibodies from engaging the Fc receptors of neutrophils and NK cells. The example continues as follows.

Since the herpesvirus gE-gI protein is present in the viral envelope, this protein also protects the virus from destruction via the classical complement pathway (see below). Thus, bipolar bridging is likely a general strategy that herpesviruses evolved to evade host antibody-mediated defenses. Yet it is difficult to study the significance of bipolar bridging in vivo. For example, gE-gI also mediates the spread of the virus in both epithelia and

nervous tissue in vivo. Moreover, bipolar bridging cannot be evaluated in the mouse model, since gE-gI does not bind well to the Fc region of mouse antibodies. This is an important obstacle, since the mouse model is particularly valuable because it is an in vivo system in which it is possible to generate knockouts (Box 4.4). (If gE-gI did react with the Fc regions of mouse antibodies, then what sort of experiments might you carry out in the mouse model to demonstrate the in vivo efficacy of bipolar bridging on the part of the virus and of ADCC on the part of the host?)

Despite uncertainty concerning a role for ADCC in controlling virus infections in vivo, there is ample evidence for the protective role of NK cells, irrespective of their role in ADCC. For example, patients who lack NK cells suffer from severe persistent virus infections, in particular, infections involving HSV and CMV (also a herpesvirus). Yet those rare NK cell-deficient individuals indeed remain able to mount normal adaptive immune responses, demonstrating the importance of NK cells in the immune response to at least some virus infections. In addition, NK cell depletion and reconstitution experiments with mice also demonstrate the importance of NK cells in controlling at least some virus infections in vivo (Box 4.4).

BOX 4.4

Much of our thinking about the human immune response to microbial pathogens in vivo is based on experiments using animal models. These model systems indeed have provided important insights into the multifaceted mammalian immune response to virus infections. Nevertheless, bear in mind that although a human virus might infect an animal model, the features of the infection in the animal (e.g., portal of entry, dissemination, sites of replication, and pathogenesis) might not entirely reflect the pattern of events in the human patient. Furthermore, there are differences between the immune systems of different mammalian species, and different host species may use different components of the immune system to combat the same virus. For these reasons, a particular animal model is not likely to reflect all the important features of the corresponding infection in humans. Nevertheless, despite these stipulations, animal models have provided singularly important insights into the workings of mammalian immunity and of viral countermeasures.

The most useful experimental animal models have involved mice. Among these are models involving highly inbred strains, in which all mice are of the same MHC haplotype, thereby allowing experiments involving the **adaptive transfer** of T cells, B cells, and antibodies.

Also useful are mice with **severe combined immunodeficiency disease (SCID)**. **SCID mice** are homozygous for a rare recessive allele that prevents them from making an enzyme necessary for assembling the DNA sequences encoding the variable regions of antibody molecules and T-cell receptors [i.e., they are defective for V(D)J recombination; see the text]. Consequently, these mice fail to develop both humoral and cellular adaptive immune responses, and thus, they are largely helpless against viruses. Since B cells and T cells fail to develop in these mice, they accept cell and tissue grafts from other strains of mice and even from humans, making them a useful model for modeling the human immune system in vivo. SCID mice are used experimentally to assess the importance in vivo of particular effectors of immunity (which may be added by passive transfer) against particular pathogens.

Another mouse model involves thymectomized and congenitally athymic or **nude mice**, in which T-cell development and adaptive immunity can be restored with thymic tissue grafts. The ability to generate **transgenic mice** and **knockout mice** also provides important opportunities to evaluate specific immune effectors. See also Box 4.6.

Moreover, increased NK cell-mediated cytotoxicity always follows induction of IFN-α/β in vivo, implying that the response plays a physiologically important role. Furthermore, NK cells also produce an important group of cytokines, including IFN-γ and TNF-α, that restrict virus infections in multiple ways. Finally, the relevance of NK cells in the control of at least some virus infections is underscored by the various strategies that viruses have evolved to evade NK cell-mediated defenses (see below).

The Complement System

The complement system is a key component of innate immunity. It is comprised of about 30 different proteins that are synthesized in the liver and circulate in the plasma. Several of these proteins have protease activity, but they are released in an inactive form, known as **zymogens**. Thus, like other immune system players, the complement system needs to be activated. As described below, activation involves a cascade of proteases, in which each enzyme cleaves and activates the next enzyme in the pathway.

When activated, the complement components work together to carry out several distinct functions. First, these proteins cooperate to destroy some viruses, as well as virus-infected cells. They do this by forming a **membrane attack complex (MAC)** that can destroy viral envelopes and punch holes in the membranes of infected cells. Second, the deposition of complement on viruses may neutralize them, consistent with the concept that neutralization of viruses can be explained by the progressive accumulation of exogenous protein on the virus surface. (As described below, viruses can bind complement via both an antibody-dependent and an antibody-independent pathway, and either of these pathways can result in virus neutralization. Neutralization via complement alone *may* be important in the early phase of a primary infection, before antibody is present.) Third, phagocytes have receptors for complement. Consequently, *complement can opsonize pathogens*, thereby promoting their destruction by phagocytosis. Finally, fragments of the complement proteins that are generated during its other functions also serve as chemoattractants, which recruit other immune system players to the infected site.

The crucial importance of the complement system is demonstrated by the fact that the rare humans who are born with a defect in one of the complement proteins usually do not survive infancy. Yet despite the rarity of humans with complement deficiencies, defects in a variety of different complement components have been observed. Among these are defects in the C3 complement component, which occupies an important place in both the classical (antibody-dependent) and alternative (antibody-independent) complement pathways (see below).

Despite the above, the actual significance of the complement system in the human response to virus infections is not known with certainty. This is so in part because complement deficiency conditions in humans have thus far been associated only with recurrent or chronic bacterial infections. However, mice depleted of complement components were shown to have more severe infections with Sindbis and influenza viruses. Moreover, early in infection, before the appearance of virus-specific antibody, complement in some way controls the viremia in complement-expressing animals. Another good reason for believing that complement plays a significant role in the control of virus infections is that pathogenic viruses have evolved a variety of strategies to evade complement activation, as described later in the chapter.

Activation of the complement system can occur via several pathways. The first of these is the **classical pathway**, so named because it was the first complement activation pathway to be discovered. Activation via the classical complement pathway is dependent on antibodies. This is in contrast to the alternative complement pathway, which is antibody independent.

Activation of the classical complement pathway occurs as follows. Complement protein C1 binds to the Fc region of either IgG or IgM antibodies that are bound to the target (Fig. 4.6A). (There are four main classes of antibodies: IgM, IgG, IgA, and IgE. They differ in their Fc regions, and those differences make each class particularly well suited to carry out its particular tasks. IgM and IgG antibodies are superb and good, respectively, at fixing complement. The roles of the different antibody classes are discussed below, in "The Adaptive Immunity System.") C1 molecules contain a serine protease activity that is activated when C1 binds to the antibody. The activated C1 then sequentially cleaves complement proteins C4 and C2, to generate C4b and C2a, respectively (Fig. 4.6B). Some of the C4b molecules that are generated bind covalently to the target membrane, where they interact with soluble C2b to form membrane-bound C4b2a. Importantly, C4b2a serves as a C3 convertase, cleaving C3 to generate C3a and C3b (Fig. 4.6C).

The cleavage of C3 into C3a and C3b is the key event in the activation of the complement system by both the classical and alternative pathways. The reasons are as follows. C3b binds covalently to proteins and carbohydrate groups on the pathogen surface, a step referred to as **complement fixation**. C3b is then able to bring together the proteins that constitute the MAC. (C3b is also the complement product that opsonizes the pathogen for destruction by phagocytosis.) First, C3b forms a complex with C4b2a, designated C4b2a3b (Fig. 4.6C), which cleaves C5 to generate membrane-associated C5b and the free fragment C5a (Fig. 4.6D). The last four components of

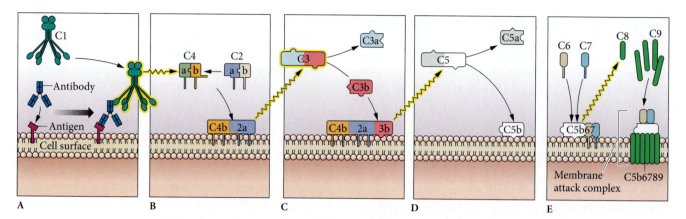

Figure 4.6 The classical pathway of complement activation. (A) Complement protein C1 binds to the Fc region of either IgG or IgM antibodies that are bound to the target. (B) C1 molecules contain a serine protease activity that is activated when C1 binds to the antibody. The activated C1 then sequentially cleaves complement proteins C4 and C2, to generate a complex designated C4b2a, which binds to the cell surface. (C) C4b2a then activates C3. That is, C4b2a serves as a C3 "convertase," cleaving C3 to generate C3a and C3b, the latter of which binds to proteins and carbohydrate groups on the cell surface, forming a complex with C4b2a, designated C4b2a3b. (D) C4b2a3b then cleaves C5 to generate membrane-associated C5b and the free fragment C5a. (E) The last four components of the complement cascade, C6, C7, C8, and C9, then bind to the cell surface to form the MAC that lyses the pathogen.

the complement cascade, C6, C7, C8, and C9, then bind to the target surface to form the MAC (Fig. 4.6E), which lyses the pathogen by forming pores in its membrane. Thus, the lesions generated by these MACs have the same effects as those generated by perforin released by CTLs and NK cells. Actually, the C9 complement protein is homologous to perforin.

Several other points about the classical pathway should be noted. First, the C4b2a complex is active only briefly. But in that time it generates 1,000 or so C3b molecules, many of which are membrane bound. These membrane-bound C3b molecules can bind complement factor B, which is then cleaved by complement factor D to yield the Bb fragment, which forms a complex with C3b known as C3bBb. Importantly, C3bBb, like C4b2a, has C3 convertase activity. Thus, the number of C3b molecules deposited at the site of complement fixation, and hence the number of MACs that assemble there, is much greater than the number of C4b2a complexes.

Second, since the classical pathway for complement activation is dependent on antibodies, it provides an excellent example of the cooperation between effectors of the innate and adaptive immune systems. Moreover, it underscores the fact that antibodies by themselves do not kill pathogens. Rather, they mark their targets for destruction by other immune effectors. In fact, the complement proteins are so named because they "complement" the antigen-binding function of antibodies.

The alternative pathway for complement activation differs from the classical pathway in that it does not require

antibody for its activation. Instead, it depends on the slow spontaneous cleavage of C3, the most abundant complement protein. This spontaneous cleavage of C3 generates a C3b-like molecule that binds to membranes in a fashion similar to the C3b molecules generated by the classical pathway (Fig. 4.6). As in the classical pathway, the spontaneously generated C3b binds B, which is next activated by factor D to form Bb, which is then bound in the C3bBb complex. Importantly, C3bBb is the C3 convertase of the alternative complement pathway. A positive-feedback loop is thereby established, generating more C3b and C3bBb, which greatly accelerates the generation of MACs (Box 4.5).

If the C3b-like protein of the alternative complement pathway does not react with a target membrane in about 60 µs (microseconds), it is inactivated by two plasma proteins: complement factor I (CFI) and complement factor H (CFH). This inactivation of C3b is important to maintain functionality of the complement system. For example, in patients with factor I deficiency, C3b accumulates unchecked, leading to the accumulation of the C3 convertase, C3bBb. The end result is that the reservoir of C3 in the blood is depleted, and factor-I-deficient individuals are thus more susceptible to a variety of infections.

Has the following occurred to you? Since the alternative pathway for complement activation occurs spontaneously, and since C3b binds nonspecifically to proteins and carbohydrate groups on membranes, what is there to prevent that nasty molecule from initiating attack on our own healthy cells? In point of fact, since these complement proteins can be devastating if inappropriately activated on

BOX 4.5

Here are a couple of unexpected developments. First, apropos the classical complement pathway, several viruses, including retroviruses, Sindbis virus, and Newcastle disease virus, activate the classical complement pathway in the complete absence of antibody. Second, apropos the alternative complement pathway, although that pathway is activated in the absence of antibody, in some instances the biological activity of the alternative pathway is augmented by specific antibody. For example, in the case of measles virus-infected HeLa cells, although the alternative complement pathway is activated in the absence of antibody, infected cells are not lysed unless IgG antibodies also are present. In an important control experiment, lysis of these measles virus-infected cells was not diminished by the absence of complement components specific to the classical (i.e., the antibody-dependent) complement pathway. Thus, the effect of antibody in this example does not involve the classical pathway, and the role of antibody in this lytic process is unknown. Similar experimental results were obtained in studies of cells infected by HIV, HSV, influenza

virus, and mumps virus. The explanation for these seemingly paradoxical observations is not clear, despite the fact that this phenomenon was first observed more than 25 years ago. Moreover, it is not even clear how virus-infected cells are able to activate the alternative complement (i.e., the antibody-independent) pathway, although it presumably involves some effect of the virus on the plasma membrane. As you may have supposed, there also are reports of virus-infected cells being lysed by the classical pathway.

While noting the gaps in our understanding of complement-mediated lysis of virus-infected cells via the alternative pathway, we should add that its significance as an antiviral measure in vivo also is not known with certainty. However, it well may be an important antiviral defense in vivo, since it occurs early to an infected cell, before the release of progeny virus. Moreover, it might occur early in the course of infection, before the adaptive immune system can be activated to clear the infection.

---

healthy cells, there are about as many proteins that regulate complement activation as there are proteins facilitating it. One such regulatory protein is the **decay-accelerating factor** (DAF; also called CD55). DAF is present on the surfaces of human cells, where it accelerates the breakdown of C3b, as well as C4b of the classical pathway. This action by DAF prevents formation of both the alternative and classical C3 convertases. The host membrane proteins protectin (CD59) and membrane cofactor protein (CD46) have similar activities.

Before moving on, we should note that there is yet a third pathway for complement activation: the **lectin activation pathway**, which is known to be effective in the defense against bacterial and yeast pathogens that have mannose-containing glycoproteins exposed on their surfaces. Importantly, there are no mannose-containing carbohydrates on human cells. The key player in this pathway is the protein **mannose-binding lectin** (MBL; a lectin is a protein that can bind to a carbohydrate). MBL circulates in a complex with an inactive enzyme called MBL-associated serine protease, which is activated when MBL binds to a target. Activated MBL-associated serine protease cleaves C4 and C2 (much as C1 cleaves C4 and C2 in the classical pathway) to initiate complement activation. Note that MBL also was shown to bind to surface carbohydrates of viruses such as HIV, HBV, and influenza virus, implying that complement activation via the lectin pathway may help to regulate infections by these viruses.

Since MBL binds to carbohydrate moieties found on common pathogens, the lectin activation pathway is more specific in its targeting of an invader than the alternative pathway, which is indiscriminate. Yet the lectin activation pathway is less specific than the classical pathway, in which antibodies direct complement activation against specific pathogens. In any event, in the lectin activation pathway, MBL has a function similar to, but less specific than, that of antibodies in the classical pathway. Also, note that MBL is one of a family of proteins known as **collectins**, all of which are soluble pattern recognition receptors. All interact with glycoproteins and carbohydrates on microbial invaders. Some cause microorganisms to aggregate, or they may opsonize microorganisms to increase their phagocytosis. They are considered a first line of defense against pathogens that display appropriate residues. MBL is present only at low levels in the absence of infection. Together with the complement proteins, it is one of the so-called **acute-phase proteins** that are produced by the liver after an inflammatory infection. Their generation is stimulated by the cytokines produced at the site of infection.

As we have noted, neutrophils and macrophages bear surface opsonic receptors for complement. Consequently, complement, like antibodies, may facilitate the destruction of pathogens by phagocytosis. However, while opsonization by antibodies must await the induction of adaptive immunity, complement deposition via the alternative pathway occurs spontaneously, and opsonization by complement can thus occur early during a primary infection. Opsonization by antibodies can yet play an early role during a second encounter with a pathogen.

Have you been wondering what role, if any, might be played by the free C3a, C4a, and C5a fragments that are generated during activation of the complement pathways? These fragments do not bind to the invader. Instead, each functions as a chemoattractant that recruits inflammatory cells to infected sites. For example, C5a is a chemoattractant that targets the transport of neutrophils from the blood circulation to an infected site, as discussed above. Moreover, at the infected site these complement protein fragments activate macrophages and neutrophils to be more effective killers. Thus, the complement system makes efficient use of its various products.

So, we have the complement proteins being able to destroy viruses by (i) building MACs, (ii) facilitating the phagocytic destruction of the pathogen, and (iii) enhancing the inflammatory response to control the infection. Moreover, the complement proteins may also (iv) neutralize some viruses and (v) destroy virus-infected cells. Importantly, the complement system can do these things very fast.

There is yet one other function of the versatile complement system that merits noting. Antibodies generated by adaptive immunity interact with the invader to produce a potentially pathologic accumulation of antigen/antibody complexes in lymphoid organs and kidneys. The complement system acts on these immune complexes in several ways. First, it opsonizes the larger complexes, thereby facilitating their elimination by phagocytosis. Second, the fixing of complement also leads to the breakdown of the immune complexes by a mechanism referred to as the **detergent-like effect**.

Viral Evasion of Innate Immunity

Evasion of IFNs

Since IFN-α/β and IFN-γ mediate a variety of crucial early host antiviral defenses, we might expect that viruses have evolved a variety of countermeasures to elude these IFN-based defenses. Indeed, viral countermeasures are known that act against every step in the sequence of IFN-related events, from IFN induction to activation and expression of effector function (Table 4.1).

The reoviruses have evolved an especially remarkable stratagem to avoid inducing IFN-α/β. Despite the facts that reoviruses have double-stranded RNA genomes and that double-stranded RNA is the major inducer of IFNs, reoviruses nevertheless manage to go through their entire replicative cycle without ever generating nonencapsulated free double-stranded RNA that might otherwise induce IFN expression. And, for good measure, the reovirus σ3

protein prevents PKR activation by competing for binding to double-stranded RNA. (The reovirus replication cycle is covered in detail in Chapter 5. For now, try to conceive of stratagems by which reoviruses might conceal the replication of their double-stranded RNA genomes from the host cell. Later, you can ask whether reoviruses actually use one of your stratagems.)

Influenza viruses have single-stranded RNA genomes. Yet, like other viruses that have single-stranded RNA genomes, influenza virus may inadvertently form double-stranded RNA during infection. (As noted above, single-stranded RNA viruses generally do not replicate via double-stranded RNA replicative intermediates. Still, double-stranded RNA may inadvertently form if a viral RNA were to hybridize with its RNA template postsynthesis.) Yet influenza virus is somewhat effective at countering the action of IFN-α/β by producing a protein, NS1, which impairs several pathways of interferon induction and blocks several IFN effectors as well (Chapter 12). For example, NS1 binds to double-stranded RNA to prevent it from activating PKR (Table 4.1).

Here is an immune evasion strategy that I find particularly intriguing. Some viruses generate soluble homologs of IFN receptors, which are secreted from infected cells, so that they might inhibit extracellular IFN from signaling neighboring cells. That is, these virus-encoded IFN receptor homologs act as decoy receptors that intercept extracellular IFN molecules, not unlike sending out decoys as a countermeasure to an antiballistic missile system. Examples of this stratagem are provided by vaccinia virus and myxoma virus, a poxvirus of rabbits (Table 4.1).

Another viral anti-IFN strategy is to block the JAK/STAT signaling pathway, which mediates signaling via the IFN receptor. JAK/STAT signaling can be impaired by increasing the degradation rate of JAK/STAT or by blocking STAT phosphorylation. Some viruses that are known to employ these stratagems are listed in Table 4.1.

Another strategy is to block the activities of IFN-induced effectors, such as RNase L and PKR. For example, hepatitis C virus (a flavivirus) blocks the activity of PKR by producing two proteins, E2 and NS5A. E2 acts as a decoy substrate for PKR, whereas NS5A forms heterodimers with PKR to inhibit its activity.

The poxviruses, which are the largest of the animal viruses, display the most extensive array of countermeasures against IFN. For example, vaccinia virus secretes homologs of IFN receptors, as noted above. Remarkably, poxvirus-encoded IFN receptor homologs include proteins that bind to each of the three IFNs (i.e., IFN-α, -β, and -γ), effectively neutralizing their activity at the infected site. Moreover, vaccinia virus produces a protein, A18R, which acts to prevent aberrant transcription of opposite DNA

Table 4.1 Viral anti-IFN strategies

Strategy	Virus examples[a]	Mechanism
Secreted homologs of vIFN-α/β receptors	B18R (VV)	Inhibits extracellular IFN-α/β
Secreted homologs of vIFN-γ receptors	B8R (VV) M-T7 (MYX)	Inhibits extracellular IFN-γ
Intracellular homologs of vIRF	vIRF-1/K9 (HHV-8) vIRF-2 (HHV-8) vIRF-3/K10.5 (HHV-8)	Represses IFN-inducible genes
Inhibition of PKR dsRNA binding	E3L (VV) σ3 (Reo) NSP3 (Rota) NSI (IV)	Competes with PKR for binding to dsRNA
dsRNA-like analogs	VA1-RNA (Ad) EBER-1 RNA (EBV) TAR-RNA (HIV)	Binds and inactivates PKR
Substrate modifications or competition	K3L (VV) ICP 34.5 (HSV)	Homolog of eIF2α Dephosphorylation of eIF2α
Enzyme inhibition	US11 (HSV) PK2 (Bac) Tat (HIV) E2 (HCV)	Blocks PKR activity
Inhibition of RNase L	? (HSV) ? (HIV, EMCV)	Synthesis of 2′-5′ oligoadenylate analogs Induction of RNAse L inhibitors
Translation protection	γ34.5 (HSV-l)	Interferes with translational shutoff
Inhibition of IFN-induced gene expression	ElA (Ad) TP (HBV) EBNA-2 (EBV) NS1 (IV)	Represses IFN-specific transcriptional responses
Interference with JAK/STAT pathway	ElA (Ad) T-Ag (Py) V (SV5) TP (HBV) Tax (HTLV) gp55 (SFFV) E7 (HPV-16) HCMV ?	Inhibits signaling downstream of IFN receptors (e.g., JAK/STAT)

[a]Reo, reovirus; Rota, rotavirus; IV, influenza virus; Bac, baculovirus; EMCV, encephalomyocarditis virus; Py, polyomavirus; SV5, simian virus 5; HBV, hepatitis B virus; HTLV, human T-cell leukemia virus; SFFV, spleen focus-forming virus; VV, vaccinia virus; myx, myxoma virus; HHV-8, human herpesvirus 8; Ad, adenovirus; EBV, Epstein-Barr virus; HIV, human immunodeficiency virus; HSV, herpes simplex virus; HCV, hepatitis C virus; HCMV, human cytomegalovirus. From D. C. Johnson and G. McFadden, Viral immune evasion. *In* S. H. E. Kaufmann, A. Sher, and R. Ahmed (ed.), *Immunology of Infectious Diseases* (ASM Press, Washington, DC, 2002), with permission.

strands that might then give rise to double-stranded RNA. For good measure, this virus also expresses proteins that directly block activation of PKR and of RNase L (Box 4.6).

Together, the multifaceted ways in which IFNs can impair virus replication and the variety of anti-IFN countermeasures that viruses have evolved underscore the importance of IFNs in host defense. Yet there are viruses that do not display any known anti-IFN countermeasures. Moreover, replication of these viruses (e.g., vesicular stomatitis virus) in animals may provoke IFN responses, and these viruses indeed may be severely sensitive to IFN treatment in cell culture. Nevertheless, they remain successful viruses in vivo. The explanation for this seeming paradox is not yet known.

Recalling that the IFNs and NK cells can activate cellular apoptotic pathways, we might expect that viruses have evolved countermeasures to deal with the premature cell death induced by those effectors of innate immunity. That indeed is the case. But since the destruction of infected cells by CTLs, which are key effectors of adaptive immunity, is of paramount importance for the clearance of a virus infection, and since the induction of apoptosis is the sole means by which CTLs destroy virus-infected cells, we will take up viral countermeasures against apoptosis

BOX 4.6

The variety of countermeasures that viruses have evolved to counter IFNs, and indeed to counter other host defenses as well, is rather remarkable. However, these viral countermeasures are known for the most part from studies in cell culture. Thus, it is important to consider what sort of experimental approaches might verify that a particular putative viral countermeasure indeed plays a role in vivo.

The following example, involving the HSV ICP34.5 gene, which blocks PKR (Table 4.1), demonstrates that this activity of ICP34.5 indeed plays a role during infection in vivo. Moreover, it provides an example of a valuable experimental approach for evaluating the importance of putative virus virulence factors. The first experimental step was to delete the ICP34.5 gene from the viral genome. Second, normal mice were infected with either the engineered mutant virus or the wild-type virus. When normal mice were infected

with the mutant virus, infection was found to be attenuated relative to infection with wild-type virus. Since attenuation might very well have resulted from some reason unrelated to the mutant's inability to countermand PKR, such as impaired replication of the mutant virus, the experiment was repeated in mice in which the PKR gene was knocked out. In these *PKR* knockout mice, the mutant virus displayed wild-type levels of virulence (Leib et al., 2000). The combined use of virus genetic manipulation and knockout mice, as illustrated in this example, is a powerful general approach to evaluating possible virus virulence factors.

Reference
Leib, D. A., M. A. Machalek, B. R. G. Williams, R. H. Silverman, and H. W. Virgin. 2000. Specific phenotypic restoration of an attenuated virus by knockout of a host resistance gene. *Proc. Natl. Acad. Sci. USA* **97:**6097–6101.

---

after we have considered the effector mechanisms of CTLs (see below).

Evasion of Cytokines

Cytokines other than IFNs, such as the **chemokines**, which guide lymphocytes to sites of infection, and the **interleukins**, which are produced by leukocytes, also play key roles in activating and regulating antiviral immune defenses. Thus, it is not surprising that viruses have evolved multiple strategies to impair and subvert cytokines in general. Moreover, since the IFNs are a subset of the cytokines, some of the strategies used by viruses against IFNs are employed against other cytokines as well. For instance, herpesviruses and poxviruses generate homologs of cytokines and cytokine receptors as a cytokine evasion strategy. A sampling of some of the proteins encoded by poxviruses to evade host immunity, including molecular homologs of cytokines and cytokine receptors, are depicted in Fig. 4.7. Some specific examples of viral anticytokine factors and stratagems are as follows.

Members of the herpesvirus and poxvirus families encode cytokine receptor homologs (**viroreceptors**) that are expressed at the cell surface. For example, the K2R protein of swinepox virus is a decoy receptor for IL-8, thereby impeding binding of the cytokine to its cellular receptors. (Interestingly, the US28 protein of HCMV [a herpesvirus] and the ORF74 protein of human herpesvirus 8 [HHV-8], also known as the Kaposi's sarcoma herpesvirus, are viroreceptors that actually transmit signals in response to host chemokines. The functional significance of these viroreceptor-mediated signals remains controversial, but

it is believed by some investigators that they might promote the development of malignancies in the HCMV and HHV-8 examples.)

In contrast to the viroreceptors that are expressed at the cell surface, the secreted cytokine-binding proteins constitute a second major class of cytokine modulators. The myxoma virus (a poxvirus) T2 protein, which mimics the binding domain of the TNF receptor, was the first of these soluble decoys to be discovered. In this instance, myxoma virus apparently appropriated the segment of the TNF receptor gene that encodes that protein's ligand-binding domain, while leaving behind the gene segment that encodes the transmembrane anchoring sequence. Thus, myxoma virus is able to produce a soluble protein that impedes TNF from interacting with the infected cell, in that way protecting the cell from TNF-induced lysis. The herpesvirus saimiri ECRF3 protein, which binds IL-8, provides an example of a soluble cytokine-binding protein encoded by a herpesvirus.

Another viral anticytokine strategy that employs molecular mimicry is for viruses to generate cytokine homologs, which act as **antagonists** (i.e., nonsignaling competitors) of the cellular ligand. The MC148R protein of molluscum contagiosum virus (MCV) (a poxvirus) is one of the best-studied examples of a viral cytokine antagonist. This chemokine homolog inhibits the binding of both CXC and CC chemokines to their respective receptors (see Chapter 21 regarding CXC and CC chemokine receptors and the roles they play during HIV infection). MC148R is able to act as an antagonist, rather than as an agonist, because it lacks the N-terminal region of chemokines, which

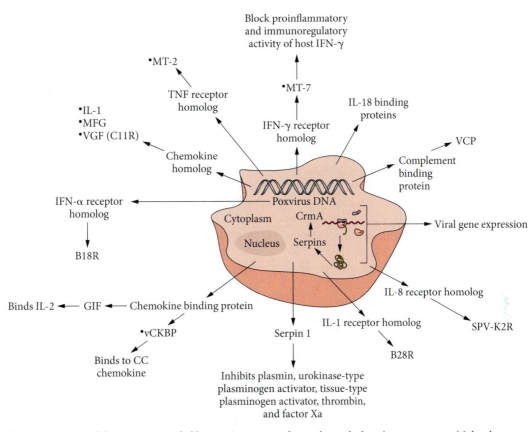

Figure 4.7 Some of the proteins encoded by poxviruses to evade or subvert the host immune system. Molecular mimicry of cytokines and cytokine receptors is a common immune evasion strategy adopted by members of the herpesvirus and poxvirus families of large DNA viruses. Note the soluble proteins that are secreted from infected cells, which function as decoy receptors for IFN-γ, TNF, and various cytokines (including chemokines and interleukins). Poxviruses also secrete homologs of humoral immune regulators (virokines), such as the viral IL-10, vascular endothelial growth factor (VGF), and myxoma viral growth factor (MGF), which is a homolog of epidermal growth factor (EGF). The last two factors may stimulate neoplastic cell proliferation. Pathogenic poxviruses also encode a complement control protein. In fact, the vaccinia virus complement control protein (VCP) was one of the first soluble microbial proteins thought to have a role in immune evasion. The figure does not depict the activities of a particular poxvirus but instead is a composite of activities detected during infections of several different poxviruses, including vaccinia virus, cowpox virus, molluscum contagiosum virus, and myxoma viruses. Adapted from P. Jha and G. J. Kotwal, *J. Biosci.* **28:**265–271, 2003, with permission.

is involved in receptor activation. The observation that inflammatory cells are generally absent from MCV lesions in vivo, despite extensive virus replication at infected sites, supports the premise that the antagonistic effect of MC148R provides an effective MCV chemokine evasion stratagem.

On the other hand, some secreted viral cytokine homologs, such as the UL146 protein of HCMV, are **agonists** (i.e., proteins or peptides that activate a receptor), rather than antagonists, of chemokine receptors. The result in the HCMV example is that more leukocytes are recruited to the infected site, where they may facilitate dissemination of the infection. The HIV Tat protein likewise acts as a chemoattractant, at least in vitro.

A fourth anticytokine strategy used by a variety of viruses is to impair cytokine-mediated signal transduction downstream of the cytokine receptors by interfering with the JAK/STAT pathway (Table 4.1). Finally, some viruses impair the secretion of proinflammatory cytokines by blocking the activating enzyme caspase-1. Note that caspase-1 also plays a key role in activating apoptosis, and the viral anti-caspase-1 activity may work to block apoptosis as well (see below).

Evasion of NK Cells and ADCC

The human and murine cytomegaloviruses are known to employ an ingeniously diabolical strategy to defeat NK cell-mediated defenses. These members of the herpesvirus family actually produce an MHC class I molecule look-alike, which engages the NK cell inhibitory receptor to generate a "don't kill" signal.

What evidence might there be that the abilities of these herpesviruses to produce an MHC class I homolog, and thereby evade NK cells, might be functionally relevant in vivo? Consider the following. All of the known herpesviruses have the ability to give rise to lifelong persistent infections. Indeed, this is a hallmark of the family. In contrast, among the poxviruses, MCV is thus far unique for its ability to establish persistent infections. Moreover, MCV also is unique among poxviruses in its ability to express an MHC class I homolog. Thus, the persistent virus lifestyle in some cases may depend on the ability of the virus to evade NK-cell-mediated lysis. (Also, recall that HSV and other herpesviruses produce membrane glycoproteins that bind antibody Fc regions, thereby thwarting ADCC.)

HIV is a champion among viruses at evading host immunity. As such, HIV has several mechanisms for down-regulating expression of MHC class I molecules on infected cells, in that way enabling the virus to evade attack by CTLs of the adaptive immune system. That fact and the ability of HIV to persist throughout the 10 years or more of an HIV infection (Chapter 21) might well lead you to expect that HIV also is able to impede NK-cell-mediated cytolysis of HIV-infected cells. Indeed it is, and its overall strategy in that regard is rather remarkable. The HIV Nef protein selectively down-regulates the human MHC class I genes *HLA-A* and *HLA-B*, while not affecting *HLA-C*. (The group of genes encoding the human MHC proteins is referred to as the **human leukocyte antigen [HLA] complex** because MHC proteins are found on leukocytes [white blood cells]). Although HLA-C proteins can present viral peptides to CTLs, they are the predominant molecules that transmit inhibitory signals to NK cells, thereby preventing killing of a target cell by an NK cell. The net result of the HIV strategy is that the MHC class I peptide-presentation pathway is sufficiently compromised in HIV-infected cells to impair attack by CTLs, while the unimpaired expression of *HLA-C* enables HIV-infected cells to also resist attack by NK cells.

There also is a report that HIV up-regulates expression of the nonclassical MHC class I molecule HLA-E on the surface of lymphocytes. HLA-E molecules have limited sequence variability and do not play a role in antigen presentation, as do HLA-A, HLA-B, and HLA-C molecules. Instead, HLA-E is believed to negatively regulate NK cell activity by delivering an inhibitory signal to NK cells. Thus, the HIV-induced up-regulation of HLA-E expression may further enable HIV-infected cells to evade attack by NK cells, despite the fact that the virus down-regulates expression of HLA-A and HLA-B molecules. Together, these effects may help HIV-infected cells to evade both NK cells and CTLs and to establish chronic infection.

Evasion of Complement

Viruses have evolved a variety of strategies to evade complement-mediated defenses. For example, both HIV (Box 4.7) and vaccinia virus incorporate host cell CD46, CD55, and CD59 into their envelopes when they bud from the host cell plasma membrane. As noted above, these host proteins are present in cellular plasma membranes to prevent the complement system from initiating attack on healthy host cells.

HCMV not only incorporates CD46 and CD55 into progeny virus particles but also up-regulates their expression at the plasma membrane. Thus, free virus is protected against complement and, moreover, virus-infected cells are protected against lysis by antibody plus complement.

Vaccinia virus and the related variola and cowpox viruses display yet another countermeasure against complement.

BOX 4.7

HIV may very well be the most diabolical of all viruses, as it is able to resist virtually all of the immune defenses that humans are able to mount against it. Regarding HIV and complement, activation of the complement system indeed leads to the deposition of complement on free HIV particles and on HIV-infected cells as well. Yet HIV infection is nearly entirely unaffected by complement, even when anti-HIV antibodies are present. As noted in the main text, HIV particles incorporate CD55, the complement decay-accelerating factor, at budding. Yet CD55 is only partly responsible for the resistance of HIV to complement. Another HIV countermeasure against complement results from the ability of the HIV envelope glycoproteins gp120 and gp41 to interact with complement factor H (CFH), a humoral negative regulator of complement activation (see the text). This interaction is very likely relevant in vivo, as shown by the following. Although HIV-infected cells are ordinarily resistant to complement, when infected cells are cultured in medium that is depleted of CFH, the addition of anti-HIV antibodies plus complement results in cell destruction. In contrast, uninfected control cells are not affected by the addition of anti-HIV antibodies plus complement. These experimental findings demonstrate that the ability of HIV-infected cells to resist the effects of complement depends primarily on the ability of the HIV envelope glycoproteins to associate with CFH. Interestingly, the streptococci have evolved a similar strategy for resisting complement; the streptococcal M protein also binds CFH.

These viruses encode a protein that is similar to the human serum complement control protein C4, which accelerates the breakdown of C3b and C4b and consequently the decay of both the alternative and classical C3 convertases.

HSV likewise encodes a protein, gC, which accelerates decay of the alternate C3 convertase. In addition, the HSV gC protein also blocks the interaction of C5 with C3 (Fig. 4.6). Together, these effects of gC protect both free virus and HSV-infected cells from complement. (In mice, HSV gC⁻ mutants produce less severe skin lesions than does wild-type virus, consistent with the premise that complement would act to contain HSV infection in vivo, were it not for the action of gC. However, since the HSV gC protein also functions as a virus attachment protein, how might one ascertain that the less severe pathology of the *gC* mutants results from an inability to impair complement action rather than from less efficient virus attachment? Answer: C3 knockout mice were used. In these knockout mice, HSV *gC* mutant virus and wild-type virus were equally pathogenic. Therefore, gC can mediate resistance to complement in vivo [Box 4.6].)

Antibodies are properly thought of as crucial effectors of adaptive immunity. Yet antibodies also interact with key players of innate immunity in their role as opsonizers that enhance phagocytosis and as effectors of ADCC and of the classical pathway of complement activation. Thus, it is expected that viruses would have evolved countermeasures that impair these antibody-dependent innate immunity defenses by targeting the antibody. HSV and other herpesviruses provide an example of a viral countermeasure that targets the antibody. As noted above, these viruses produce envelope glycoproteins that bind the Fc portion of antibody molecules, by that means protecting HSV from opsonin-enhanced phagocytosis. Moreover, those viral glycoproteins protect HSV-infected cells from ADCC and also protect both free virus particles and infected cells from complement-mediated lysis via the classical complement pathway.

Viruses have other strategies for evading the effects of antibodies that are more properly considered in the context of evading adaptive immunity. Some of these are simply mentioned here and discussed in more detail later. They include (i) antigenic variation via several stratagems; (ii) being inherently poor antigens; (iii) replicating in **immunoprivileged sites** (e.g., the central nervous system); and (iv) establishing latent infection, in which virus proteins are not expressed.

APOBEC3G and the HIV Vif Protein

The HIV Vif (virion infectivity factor) protein is necessary for HIV to replicate in human T lymphocytes. Recent studies of how Vif promotes HIV infection of T cells revealed a new intracellular innate defense mechanism, as well as a new viral evasion strategy, as manifested by Vif. T cells express a protein known as APOBEC3G, which is a cytidine deaminase that converts cytidines to uridines. The unabated activity of APOBEC3G during HIV infection would hypermutate the HIV genome, thereby rendering it noninfectious. Vif binding to APOBEC3G targets that host protein for degradation by sending it to the proteasome. (The actual state of affairs may be more complicated, since APOBEC3G can restrict HIV, even when missing its cytidine deaminase activity, and Vif-induced degradation of APOBEC3G may not be the only means by which Vif counteracts that protein [Chapter 21].)

It is not yet known whether APOBEC3G has any other role in the cell. Also, it is not known how many other viruses APOBEC3G might act against. However, APOBEC3G appears to be a major factor that impedes HBV from establishing a chronic infection. In this regard, HBV, like HIV, makes use of reverse transcription during its replication. (Interestingly, an enzymatic activity similar to that of APOBEC3G is responsible for initiating the process of **somatic hypermutation**, which introduces mutations into specific regions of the Ig heavy- and light-chain genes, serving to greatly increase the affinity of antibody molecules for their cognate antigens [see below].)

THE ADAPTIVE IMMUNE SYSTEM

Macrophages, neutrophils, NK cells, IFN, and complement of the innate immune system slow down most virus infections and *may* even be sufficient to contain some of them. However, in most cases, virus infections progress to a point where the innate immune system needs reinforcements. Under those circumstances, the adaptive immune response is activated.

In point of fact, because all viruses exploit an intracellular lifestyle, the innate immune system is generally much less effective at overcoming viruses than it is at overcoming pathogens that are strictly extracellular. Fortunately for us, effectors of the adaptive immune system are especially well suited to the task of clearing virus infections. The killer cells or CTLs of adaptive immunity are specialized to attack and destroy host cells that harbor intracellular virus, while B cells produce soluble factors (antibodies) that specifically neutralize extracellular virus. Together, the T cells and B cells of adaptive immunity work to incapacitate both intracellular and extracellular virus. Moreover, the adaptive immune system has a memory of its encounter with each pathogen. This **immunological memory** allows the adaptive immune system to respond faster and stronger to a subsequent encounter with the same pathogen, a phenomenon referred to as **acquired immunity**, which may be lifelong against some viruses.

Adaptive immunity also can be important in overcoming bacterial infections. Antibodies of adaptive immunity act against all bacteria, and CTL-mediated defenses *may* be important against the relatively few bacteria that have an intracellular lifestyle (e.g., the chlamydiae). And, as might be expected, intracellular bacteria, like viruses, have evolved countermeasures against CTL-mediated defenses. One such countermeasure used by intracellular bacteria is to reside in an intracellular compartment that they modify so that it does not intersect the pathways of antigen presentation by MHC molecules (see below). Yet bear in mind that *all* viruses are obligate intracellular parasites, and consequently, the objective of the adaptive immune response to a virus infection is to *eradicate* virus-infected cells as well as free extracellular virus particles. Because it is crucial to eliminate intracellular as well as extracellular virus, and because the adaptive immune system is uniquely suited to destroying intracellular virus, it is plausible that the adaptive immune response may have evolved *primarily* to respond to the threats posed by viruses.

The adaptive immune response is so named because it *adapts* to each specific pathogen that it encounters, tailoring a specific response to each. As noted above, adaptive immunity is mediated by B cells and T cells, which contain cell surface receptors that are specific for particular antigens. Moreover, these receptors vary from one T cell to another. This is in contrast to NK cells of innate immunity, which contain a fixed set of receptors that are the same on all NK cells.

Adaptive immunity comprises two distinct but complementary branches for containing and eliminating an infection. One branch of adaptive immunity, referred to as **humoral immunity**, is based on antibodies, which are synthesized by B cells (i.e., lymphocytes that mature in the bone marrow). The other branch of adaptive immunity, the **cell-mediated immune response**, is carried out by T cells (i.e., lymphocytes that mature in the thymus).

The mechanisms by which antibodies control virus infections are described below. For now, a major role for antibodies is to directly bind to viruses, thereby preventing them from infecting susceptible cells. An antibody that does that is said to **neutralize** the virus.

A shortcoming of neutralizing antibodies is that the infection inside a cell is generally inaccessible to antibodies. Thus, intracellular virus replicates in an antibody-free environment, generating thousands of progeny virus particles that are released into the extracellular environment. A sufficient number of these released viruses may escape being neutralized by extracellular antibodies and succeed in beginning new replication cycles in new cells. Thus,

antiviral antibodies alone are generally not sufficient to eliminate viral pathogens.

Fortunately, intracellular virus can be eliminated through the action of the second branch of adaptive immunity, the cell-mediated immune response. CTLs, which mature in the thymus, are the principal effectors of cell-mediated immunity against viruses. These cells specifically recognize and destroy virus-infected cells, destroying nascent intracellular virus along with the cell. The sacrifice of an infected cell is more than compensated for by preventing the spread of the infection to healthy cells.

Generally speaking, while the cellular immune response is generally necessary to resolve a viral infection, the humoral and cellular arms of adaptive immunity in fact cooperate in that effort. In addition, having the two arms provides a safety net, should one of the arms be impaired. Nevertheless, the relative importance of humoral versus cellular immunity actually varies from one virus to another. This is readily demonstrated using knockout mice, in which different components of the immune system are specifically depleted. In addition, humans with agammaglobulinemia (i.e., impaired synthesis of Igs) provide a real-life model for assessing the importance of humoral immune responses against particular virus infections. For instance, these individuals are more at risk for developing paralytic poliomyelitis after receiving the live oral polio vaccine (Chapter 6). Moreover, patients with agammaglobulinemia may continually excrete **enteroviruses**, including vaccine-derived poliovirus, for extended times. (There is some concern that long-term poliovirus shedders, although rare, might in the future create a risk to the general population at a time after poliovirus might be considered to have been eradicated and routine vaccination is no longer practiced.)

A most important feature of the adaptive immune response noted above is that it has a **memory**. That is, it remembers its first or primary encounter with a pathogen. Because of this memory feature, the adaptive immune system can quickly mount a vigorous immune response to a previously encountered pathogen, in just a matter of hours after reexposure. This aspect of the memory response is in contrast to the primary response, which can take many days to reach effective levels. The rapid-recall memory response can prevent or greatly diminish clinical symptoms. It is the basis for acquired immunity and also the rationale underlying vaccination. Specificity and memory are the two hallmarks of adaptive immunity.

Antibodies and B Cells

Antibody responses against viruses are generally detectable by 3 to 5 days after infection, and antivirus antibodies have been shown to help clear primary infections involving a variety of different viruses. Moreover, they are a key defense against previously encountered viruses, and in the cases of viruses that establish latent or persistent infections, antibodies can control new outbreaks in an individual that may occur after a period of viral inactivity. The ability of antibodies to help contain persistent virus infections is largely due to immunological memory.

Antibodies can impede a virus infection by several different mechanisms (Fig. 4.8). First, some antibodies may interfere with the ability of a virus to attach to, and infect, a host cell. This can happen if the antibody binds directly to the viral attachment sites. For some time it was believed that other antibodies, which bind elsewhere on the virus's surface, may yet block infection by causing conformational changes on the virus particle that affect its attachment

sites. However, this premise is now controversial because it could not be confirmed by structural studies.

Second, antibody molecules have multiple antigen-binding sites (Fig. 4.4). Consequently, antibodies can agglutinate virus particles. This reduces the number of infectious units and facilitates clearance of the virus by phagocytes.

Each of the antibody mechanisms enumerated above is said to neutralize the virus, that is, impair its infectivity. Recall that antibodies also may interact with effectors of innate immunity to help contain the infection. For example, antibodies may opsonize the virus and participate in ADCC and complement fixation.

Despite the notion that antibodies act against extracellular viruses only, antibodies can disrupt the intracellular replication cycles of some viruses. For example, antibodies against measles virus can suppress measles virus transcription and translation, at least in cell culture. The underlying mechanism is not clear, but it may be the result of an effect mediated by the antibodies from the cell surface. In another example, passive transfer of antibodies against Sindbis virus can clear that virus from the brains and spinal cords of persistently infected SCID mice (see Box 4.4 for a description of SCID mice). The mechanism by which antibodies clear the Sindbis virus infection is not entirely known, but studies in vitro show that it requires antibody-mediated cross-linking of viral glycoproteins on the cell surface and involves inhibition of virus budding (Burdeinick-Kerr et al., 2007). Thus, these antibodies against Sindbis virus impair the maturation and release of intracellular Sindbis virus.

All antibody molecules are comprised of four polypeptide chains: two identical heavy chains and two identical smaller light chains, which are linked together by disulfide bonds (Fig. 4.4). This combination of heavy and light chains appears as a Y-shaped structure. Each branch of the Y is comprised of a complete light chain and the N-terminal portion of a heavy chain. The stem of the Y contains the paired C termini of the heavy chains.

Before going on with our description of antibody molecules, it would be useful to specify more precisely what antibodies actually recognize. Up to now, we have been using the term "antigen" without rigorously defining it. Actually, **antigen** is a rather loose term referring to an antibody target. At one extreme, an antigen may be the whole pathogen, or it may be a particular protein. Importantly, each antibody (and T-cell receptor as well; see below) recognizes and binds to a specific site or determinant on the antigen, referred to as an **antigenic determinant** or **epitope**. The epitope recognized by an antibody is comprised of a cluster of only 3 to 5 amino acids. (In contrast, T-cell receptors recognize between 8 and 20 amino acids.) Thus,

Figure 4.8 Antibodies can impede a virus infection by several different mechanisms. (A) Some antibodies may interfere with the attachment of a virus to the host cell. This can occur if the antibody binds directly to the viral attachment sites. (B) It once was believed that some antibodies may cause conformational changes on the virus particle that affect but do not directly block its attachment sites. This premise is now controversial because it could not be confirmed by structural studies. (C) Aggregation of viruses by antibodies reduces the number of infectious units and facilitates clearance of the virus by phagocytes. (D) Contrary to the notion that antibodies act only against extracellular viruses, in some instances antibodies appear to be able to disrupt the intracellular replication of viruses, by mechanisms that are not entirely clear. (E) Opsonization. (F) Lysis of virus/antibody complexes by complement. (G) Lysis of virus-infected cells by antibody plus complement. (H) Antibody-dependent cellular cytotoxicity (ADCC). (I) Inhibition of virus budding and release. Abbreviations: VR, virus receptor; V-ag, viral antigen; mφ, macrophage; CR2, complement receptor. Adapted from H. E. Kaufmann, A. Sher, and R. Ahmed (ed.), *Immunology of Infectious Diseases* (ASM Press, Washington, DC, 2002), with permission.

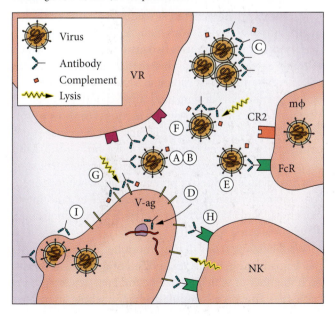

even a small protein may contain numerous epitopes, and as a result, a protein can be recognized by several different antibodies, each of which is specific for a particular epitope on the protein.

The **antigen-binding sites** on an antibody molecule are located at the ends of the branches of the Y-shaped structure. Thus, each antibody molecule has two antigen-binding sites. Also, amino acids on both a heavy chain and a light chain contribute to each antigen-binding site. As we soon will see, each of the antigen-binding sites on an antibody is specific for the same epitope. Consequently, antibodies are able to aggregate viruses into clusters, thereby reducing the number of infectious units and facilitating clearance of the virus by phagocytes. Moreover, when antibodies bind to their cognate antigens, the single invariant portion, referred to as the Fc region (see below), is available to interact with other immune effectors. Recall that neutrophils, macrophages, and NK cells all have receptors on their surfaces that can bind to the free Fc regions of antigen-bound antibody molecules. Furthermore, bound antibodies can trigger the classical complement pathway.

There are five functionally distinct classes of antibodies. They are IgA, IgD, IgE, IgG, and IgM. These antibody classes are distinguished at the molecular level by differences in their stem structures. Consequently, the particular specialized functions of these different antibody classes are determined by their stem structures, as discussed below.

IgG antibodies, which are the most common antibody class in the blood, are generally used to illustrate common structural features of antibodies (Fig. 4.4). IgG antibodies contain a relatively unstructured tract in the middle of the heavy chain, which forms a flexible "hinge" region. This is useful for our description of antibodies because the protease papain cleaves IgG molecules at the hinge region to generate three fragments: two **Fab** fragments and one **Fc** fragment. The Fab (acronym for fragment antigen-binding) fragments contain the antigen-binding sites on the antibody, whereas the Fc (acronym for fragment-crystallizable, because it readily crystallized in early experiments) fragment provides the link to other immune effectors. (It may be useful to think of the Fc fragment as the constant fragment, since it varies little, if at all, within an antibody class [see below].)

The heavy and light chains each contain reiterations of a particular sequence motif referred to as an **Ig domain** (Fig. 4.9). Each such domain is about 100 amino acids in length. Notice how Ig domains consist of two β-sheets that are connected by loops. An Ig domain that participates in antigen binding is referred to as a **variable (V) domain**. V domains are so named because they contain the sequence differences between antibody molecules that enable them to specifically recognize their cognate epitopes.

Like the variable domains, the **constant (C) domains**, which make up the stem of the antibody molecule, also are comprised of Ig domains. However, in contrast to the variable domains, there is little or no sequence diversity in the constant domains within an antibody class.

The sequences within a V domain that actually determine epitope-binding specificity are comprised of three loops at the end of the V domain referred to as **hypervariable regions**, and also as **complementarity-determining regions (CDRs)**, because they have a surface that is complementary to their cognate epitopes (Fig. 4.10). The CDRs are flanked by so-called **framework regions**.

Antibody Diversity

Here is an interesting question. Since each antibody molecule is specific for a particular epitope, how many different antibody specificities are available within an individual? Well, current estimates range from as high as 10^8 to an astonishing 10^{16} different specificities! Presumably, this variety of antibody specificities available within an individual, known as the **antibody repertoire**, suffices to recognize any pathogen that might be encountered.

This brings up what may be a more compelling question. Considering that every antibody is a protein, how many genes might be required to generate 10^8 different antibodies (the conservative estimate of our antibody repertoire)? Well, it is much less than 10^8, which is fortunate for us since we only have about 3×10^5 genes. But recall that each antibody variable region is comprised of a variable domain from a heavy chain (V_H) and a variable region from a light chain (V_L). Thus, mixing and matching 10^4 heavy chains with 10^4 light chains could do the trick. But that would still require a lot of genes (i.e., 2×10^4, or about 10% of our entire genome). Fortunately, elegant mechanisms have evolved which generate a seemingly infinite number of antibody specificities from many fewer genes than that!

The immune system uses two general principles, which it builds upon, to generate this virtually limitless number of different antibody specificities. The first of these principles is **modular design**, and the second principle is **clonal selection**. Modular design accounts for how a population of **naïve B cells** (i.e., B cells that have never encountered an antigen) preexists in the host, ready with receptors that can recognize and respond to virtually any new invader. Clonal selection explains how when naïve B cells encounter their cognate antigen, they expand into a clone of identical B cells that secrete antibody molecules, all with the same specificity.

We begin by considering modular design, i.e., the mechanism for generating a population of naïve B cells, each member of which is specific for a particular epitope, such that the B-cell population as a whole may have

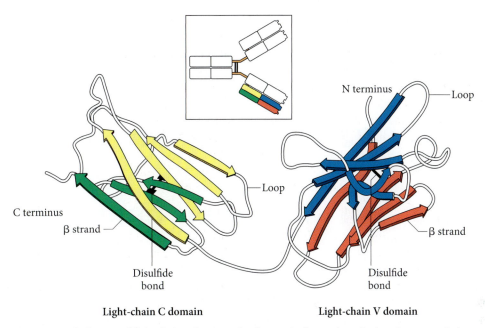

N terminus — Loop

Loop

C terminus

β strand

β strand

Disulfide bond

Disulfide bond

Light-chain C domain   **Light-chain V domain**

Figure 4.9 The heavy and light chains of an Ig molecule contain four and two Ig domains, respectively. Shown is an individual light chain of an IgG molecule. Each Ig domain contains about 100 amino acids, which are folded into two β sheets that are connected by loops. An Ig domain that participates in antigen binding is referred to as a variable (V) domain (right panel). V domains are so named because they contain the sequence differences between antibody molecules that enable them to specifically recognize their cognate epitopes. Like the variable domains, the constant (C) domains, which make up the stem of the antibody molecule, also are comprised of Ig domains. However, in contrast to the variable domains, there is little or no sequence diversity in the constant domains within an antibody class. Arrows point from the amino terminus to the carboxy terminus. The inset depicts the location of the light chain in an IgG molecule. Adapted from P. Parham, *The Immune System*, 2nd ed. (Garland Science, New York, NY, 2005), with permission.

members ready to respond to virtually any pathogen that might be encountered. The underlying basis for modular design is that the Ig genes are arranged on the chromosome in a special kind of fragmented form, from which *mature* Ig genes may be assembled by a special genetic recombination process that occurs only in B cells (Fig. 4.11 and 4.12). More precisely, a B cell assembles a mature

heavy-chain gene by splicing together a V, a D, a J, and a C segment (Fig. 4.12). The combined V, D, and J segments encode the variable region, and the C segment encodes the constant region. And, a mature light-chain gene is assembled from a V segment and a J segment.

Now here is a really important point. Many alternative forms of each of the V, D, and J segments are present

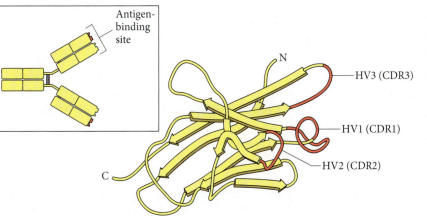

Antigen-binding site

N

HV3 (CDR3)

HV1 (CDR1)

C

HV2 (CDR2)

Figure 4.10 The hypervariable regions (HV1, HV2, and HV3) of an antibody molecule, which determine epitope-binding specificity, are comprised of three loops at the end of the V domain that also are referred to as complementarity-determining regions (CDRs), because they have a surface that is complementary to their cognate epitopes. These regions are flanked by so-called framework regions. Adapted from P. Parham, *The Immune System*, 2nd ed. (Garland Science, New York, NY, 2005), with permission.

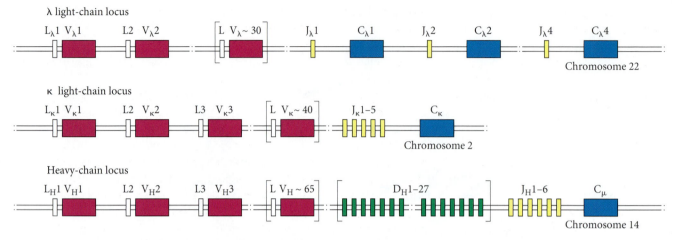

Figure 4.11 The basis for modular design is that the Ig genes are arranged in the genome in a special kind of fragmented form. In the human genome there are two loci, λ and κ (kappa), that encode light-chain segments, and one locus that encodes heavy-chain segments. Different Ig gene segments encode the leader peptide (L), the V region (V), and the constant region (C) of both the heavy and light chains. Note the multiple number of different V gene segments for the light and heavy chains. Also note the multiple joining (J) gene segments and the additional set of diversity (D) gene segments in the heavy-chain locus. The heavy-chain locus also contains C gene segments for each of the antibody classes. Introns separate each of these gene segments. A mature light-chain gene contains one V segment and one J segment, which together constitute the V region, and one C segment. A mature heavy-chain gene is comprised of a V, a D, a J, and a C segment. These are joined together by DNA recombination (Fig. 4.12). Humans have two functionally equivalent light chain isotypes, kappa (κ) and lambda (λ), and five functionally distinct heavy chain classes (see text). Adapted from P. Parham, *The Immune System*, 2nd ed. (Garland Science, New York, NY, 2005), with permission.

within the unrecombined DNA (Fig. 4.11). Moreover, these alternative forms can be spliced together by DNA recombination such that each of the alternative V, D, and J segments is selected at random. The course of events for generating the mature heavy-chain gene is as follows.

One of the 27 D segments is recombined with one of the 6 J segments. One of the 65 V segments is then joined to the DJ sequence, and the resulting VDJ sequence is then joined to a C segment (Fig. 4.12). The mature light gene is likewise assembled using this mix-and-match strategy.

Figure 4.12 Many alternative forms of each of the V, D, and J segments are present within the germ line DNA (Fig. 4.11). Importantly, these alternative forms can be spliced together by DNA recombination such that one of each of the alternative V, D, and J segments is selected at random. The course of events for assembling a mature light-chain gene is as follows. One of 30 to 40 V segments is joined to one of the J segments to form a complete V region, which is then joined to a C segment. The course of events for generating the mature heavy-chain gene is as follows. One of the 27 D segments is recombined to one of the 6 J segments. One of the 65 V segments is then joined to the DJ sequence, and the resulting VDJ sequence is then joined to a C segment (Fig. 4.10). Adapted from P. Parham, *The Immune System*, 2nd ed. (Garland Science, New York, NY, 2005), with permission.

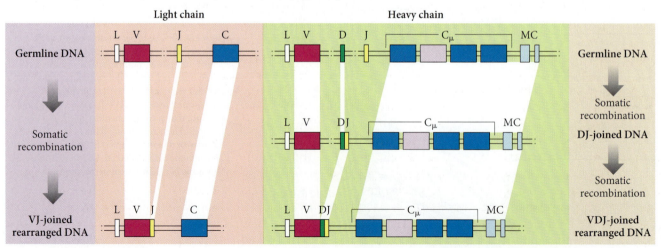

These rearrangements of heavy- and light-chain gene loci occur only in B cells, and the process is referred to as **somatic recombination**, since it does not occur in the germ line.

Because of the number of different V, D, and J DNA segments that are recombined together at random to make up a variable region, this strategy generates about 10^7 different antibody specificities, which is an impressive number, but not quite enough to account for the estimated number of different antibody specificities. Well, additional diversity is created as a consequence of the imprecision of the recombination process. Because of this imprecision, extra nucleotides are inserted at the recombinational joints. Thus, evolution has given rise to a means for generating a seemingly limitless number of antibody specificities from only about 300 genes! However, that is still not the end of the story, since Mother Nature has invented a way to improve the affinities of those antibodies for their cognate antigens over the course of an infection, as will be seen shortly. Moreover, recalling that there are five functionally distinct classes of antibodies, each of which is specialized for a particular purpose, the immune system selects the antibody class that is most appropriate for the particular circumstance.

An important consequence of modular design is that each B cell makes identical antibodies that are specific for one particular epitope. That is, each B cell makes heavy chains with only one V_H region and light chains with only one V_L region. You may have noticed a bit of a conundrum here because of the fact that B cells are diploid, with modular rearrangements likely occurring independently at each of the heavy- and light-chain loci. Modular rearrangements indeed do occur independently at each heavy- and light-chain locus. However, a successful rearrangement occurs first at one of the loci. When that happens, the other locus is silenced.

Clonal selection, the second general principle underlying the generation of antibody-producing B cells, occurs as follows. When a B cell matures and begins to make its particular Ig, the heavy chains contain a hydrophobic sequence near their carboxy termini, causing the antibody molecules to become integral membrane proteins that are displayed on the surface of the B cell. No soluble antibodies are yet released by these cells. However, the surface antibodies, called B-cell receptors, are able to recognize their cognate antigen. When the B-cell receptor of a particular *naïve* B cell (i.e., a mature B cell that has left the bone marrow but has not yet encountered its cognate antigen) encounters and binds its cognate antigen, a signal is transmitted into the B cell, causing it to proliferate and generate a clone of about 20,000 identical B cells. Most of the B cells in the clone mature into **plasma B cells**, each of which produces and secretes huge quantities (about 2,000 molecules per s) of the soluble form of the antibody into the blood and tissues. Activation of the B cell and its **clonal expansion** may take a week or longer. During this period, the innate immune system is responsible for containing the infection.

We noted above that Mother Nature has evolved a way to improve upon modular design, to actually perfect antibodies over the course of an infection. This happens as follows. When a naïve B cell encounters its cognate antigen and is activated to undergo clonal expansion, its V-domain coding sequences go through a process called **somatic hypermutation**, which randomly introduces point mutations at a high rate (as high as one mutated base per 1,000 bases per generation) throughout the rearranged heavy- and light-chain genes. Importantly, somatic hypermutation is restricted to the variable regions of the rearranged Ig genes. The process involves an enzyme, activation-induced cytidine deaminase, which is induced when the naïve B cell is activated. Activation-induced cytidine deaminase converts cytidine residues in the DNA to uracil. The uracil residues are excised, and replacement nucleotides are randomly incorporated when the DNA is replicated. (Recall that T cells express a cytidine deaminase, $APOBEC_3G$, which can hypermutate the HIV genome, rendering it noninfectious [see above].)

Somatic hypermutation generally leads to antibody V regions that have higher affinities for their cognate epitopes than those encoded by the original rearranged Ig genes. This happens as follows. Recall that a transmembrane form of the antibody is present as the B-cell receptor. Those B cells that express higher-affinity receptors are selectively activated for further clonal expansion. This occurs because maturing B cells must continually be restimulated by binding to their cognate antigen in order to continually proliferate. Consequently, those B cells that express the highest-affinity B-cell receptors are preferentially selected to continue dividing and become antibody-producing plasma B cells. The end result is that the antibody response is fine-tuned to fight the invader. The process is called **affinity maturation**. This is still not the end of the story.

Recall that there are five different classes of antibodies: IgA, IgD, IgE, IgG, and IgM. These classes are distinguished by their different constant regions. Importantly, whereas the variable regions determine an antibody's specificity, the constant regions specify how and where the antibody will function. This plays out as follows.

When a B cell is born in the bone marrow, it rearranges its V, D, and J Ig gene segments and initially joins them to an M-type constant region. Consequently, the B-cell receptors are initially of the IgM class, and when the B cell is activated it will initially produce only **IgM antibodies**.

This is advantageous to the host, because IgM antibodies are exceptionally good at fixing complement, which may be important at a time when antibodies are just beginning to be made. In addition, IgM antibodies are generally good at neutralizing viruses, although affinity maturation is sometimes necessary. Importantly, however, the B cell has the opportunity to change the class of antibody it makes from IgM to any of the other classes, each of which provides special advantages.

IgG antibodies also are very good at neutralizing viruses and offer the additional unique asset of being able to pass from the mother's blood across the placenta into the fetal circulation. These maternal antibodies provide the newborn with a measure of protection against currently circulating pathogens, at a time when the newborn's own immune system needs to become operational. The fact that IgG antibodies are much longer-lived than IgM antibodies also suits them to this purpose. Moreover, NK cell receptors specifically interact with the Fc segment of IgG antibodies. Consequently, this is the antibody class that mediates ADCC.

Whereas IgG antibodies are the most common antibodies in the blood, **IgA antibodies** are the most abundant antibody class in the body. The reason for the abundance of IgA antibodies is that these antibodies guard the mucosal surfaces of the body and the human body has about 400 m^2 of these mucosal surfaces to defend. There are more IgA-producing B cells located under the mucosal epithelia than there are B cells in any other organ of the body, including the spleen, thymus, and lymph nodes, which are the major sites in the body where cells of the immune system interact. Bear in mind that these delicate mucosal surfaces, which include those of the respiratory, digestive, and urogenital tracts, are all that separate those tracts from the outside world. Since the rest of the body surface is a *mere* 2 m^2 and is covered by dead keratinized skin cells, the mucosal surfaces are where the vast majority of viruses invade the body. Thus, one can appreciate why they are so immunologically active. IgA antibodies also are secreted into the milk of nursing mothers, thereby protecting the baby's digestive tract against currently circulating pathogens that might be ingested.

IgA molecules reach their site of action on the mucosal surface by a process known as **transcytosis**, whereby they first bind to an IgA receptor on the basolateral surface of an epithelial cell and then undergo receptor-mediated transport to the apical side of the cell, where they are released. Importantly, IgA antibodies do not fix complement. This is good, considering all the foreign microorganisms coming into contact with mucosal surfaces. Thus, if IgA antibodies fixed complement, then mucosae would be in a constant state of inflammation.

The same rearranged Ig gene encodes IgM and **IgD antibodies**, but the messenger RNA (mRNA) is spliced one way to encode an IgM antibody and another way to encode an IgD antibody. IgD antibodies make up only a small percentage of the circulating antibodies in humans and are found in small amounts in the tissues that line the belly and chest. How they might work is not clear. We have little to say here about **IgE antibodies**, since they confer protection against parasites (e.g., worms).

As noted above, B cells initially produce IgM antibodies. Class or **isotype switching**, to produce other antibody classes, is accomplished at the genome level by splicing an IgA, IgG, or IgE C gene segment to the VDJ sequence that was assembled for the original IgM antibody. The original IgM C segment is spliced out in the process. Notice that by means of affinity maturation and class switching, the B cell is able not only to fine-tune the specificity of the antibody but also to fine-tune it with respect to its sites of action and function. But that is still not the end of the story.

As the clone of activated B cells proliferates and matures, not all of its members become antibody-producing plasma cells. Instead, some of those B cells mature into **memory B cells**. Importantly, upon subsequently reencountering the same pathogen, these memory cells elicit a much faster and stronger immune response. There is actually more to the story of humoral immunity than these memory B cells. In addition, there is a population of plasma B cells that persist, constitutively producing antibodies (see below). The preexisting antibodies produced by these cells provide the first line of defense against reinfection with a previously encountered microbial pathogen. (T cells likewise generate a population of memory cells with attributes similar to those of memory B cells [see below].)

Viral Evasion of Antibodies

So, we have a preexisting population of naïve B cells that is diverse enough to recognize and respond to virtually any microbial invader. Moreover, we have a population of memory B cells ready to spring into action more quickly and strongly should we reencounter an earlier invader. In addition, after a primary infection by a pathogen, a population of plasma cells remains that constitutively produces antibodies against that pathogen. How then do some viruses still manage to evade humoral immunity and cause illness, and why, in some instances, do we have multiple episodes with the same virus?

Regarding the illness issue, in some instances the immune response itself contributes significantly to the symptoms of the infection (see below). Yet there also are a variety of different strategies that viruses have evolved to evade humoral immunity. For example, we already noted how herpesviruses express receptors for the Fc domains of antibodies, enabling them to thwart ADCC and antibody-mediated interactions with complement. Some additional examples of viral antibody evasion strategies are as follows.

A major means by which viruses evade antibody-mediated host defenses is by generating so-called **immune escape mutants**, which display altered antigenicity. In actuality, a high mutation rate is virtually an inevitable feature of the viral lifestyle. In the case of RNA viruses, their RNA-dependent RNA polymerases do not have a proofreading function (as DNA polymerases do) in the form of an exonuclease that can excise misincorporated nucleotides. Consequently, their error frequency can be as high as one misincorporated nucleotide for every 10^3 to 10^4 nucleotides incorporated into RNA. Thus, essentially every individual virus in an RNA virus population has a unique RNA genome sequence. DNA viruses are more genetically stable than RNA viruses, but even they display mutation rates several orders of magnitude greater than that of the host. Making matters worse yet for the host, all viruses generate huge numbers of progeny in a relatively short period of time. Bearing in mind that a change in only one amino acid at a neutralizing antibody-binding site can drastically reduce the affinity of an antibody against that site, we can appreciate how immune escape mutants can readily emerge. Thus, the humoral immune response is often playing a game of catch-up, continually having to amplify its response against each new serological variant of the virus. Generally, immune escape mutants may be more of a problem in the cases of RNA viruses than for DNA viruses, because of the higher mutation rates of the former. But immune escape mutants are an issue in the cases of all viruses, DNA as well as RNA, that establish persistent infections.

HIV is especially adept at evading nearly all measures of host immunity, employing a number of different strategies toward that end. The generation of immune escape mutants is one of the key means by which that virus evades antibody-mediated defenses. HIV infections begin like many other virus infections, with an early acute phase characterized by extensive virus replication. This acute phase is followed by a marked decrease in virus levels, as the adaptive immune response comes into play to bring the infection under control. However, HIV persists in infected individuals for years and is rarely and perhaps never eliminated. That is so, despite the fact that infected individuals mount vigorous adaptive immune responses against the virus until nearly the end. During much of this time, the patient does not display symptoms. This period of *clinical* latency can last for 10 years or more. However, again note that this period is actually characterized by immense levels of immune activity, as well as virus replication.

Humoral immunity likely plays an important part in maintaining the long period of clinical latency, since huge numbers of virus particles are neutralized, most probably by antibody-mediated opsonization and phagocytosis. Yet the high mutation rate of HIV ensures that an untreated patient will have viruses circulating that contain virtually every possible base substitution. Among these will be a large variety of immune escape mutants.

The high HIV mutation rate is but one of several means this virus has for evading the humoral immune response and ultimately destroying adaptive immunity. The HIV envelope glycoprotein gp120, which is the predominant antigen on the viral surface, has features that make it a difficult target for neutralizing antibodies from the very start of the infection. This protein is one of the most extensively glycosylated proteins known, making it a weak antigen. That is, relatively little neutralizing antibody is generated during an HIV infection. HIV has several other immune escape mechanisms that target cell-mediated immunity (see below).

Human rhinoviruses, which cause the common cold (Chapter 6), use several different strategies to evade neutralization by antibodies. First, the rhinovirus receptor-binding sites, on the surface of their icosahedral capsids, are recessed in what are referred to as "canyons." Importantly, these canyons are narrower than the "footprint" made by the antigen-binding site of an antibody. Thus, the receptor-binding sites, while accessible to the receptor, are inaccessible to antibodies. Consequently, these viruses are somewhat resistant to potentially neutralizing antibodies that might recognize those sites. Moreover, having receptor binding sites that are inaccessible to antibodies implies that these viruses are under little, if any, selective pressure from the host immune system to modify their receptor binding sites, something they could not likely do and remain viable.

Second, there are more than 100 antigenically distinct rhinovirus serotypes circulating in humans, so that there are always some serotypes against which a person may have little, if any, acquired immunity. Third, some sources state that the innate immune system is often able to resolve rhinovirus infections within just a few days, so that the adaptive immune system may not get fully activated. If that is so, then these viruses yet remain quite successful since they are so efficiently transmitted that the infection can be passed on, even in the short time before it is cleared by innate immunity. In fact, the viruses may benefit, since if the adaptive immune system is not activated, then there is no acquired immunity, and one can be reinfected with the same rhinovirus serotype. (Yet I believe it is unlikely that any virus infection might be resolved completely in the absence of adaptive immunity.)

In contrast to the rhinovirus case, in which more than 100 stable serotypes may circulate simultaneously in the human population, there are usually only one or two predominant strains of influenza virus that circulate in humans at any time (Chapter 12). Yet influenza virus undergoes continuous antigenic change from one year to the next, such that acquired immunity does not carry over well from year to year, thereby ensuring the survival of the virus in the population. There are two basic means by which influenza virus undergoes antigenic change to keep ahead of acquired immunity. The first is the simple accumulation of mutations that results from the high rate of misincorporation of nucleotides while replicating the influenza virus RNA genome. This type of change is referred to as **antigenic drift**. The second type of change, referred to as **antigenic shift**, depends on the fact that the influenza virus genome is comprised of eight distinct RNA segments. For this reason, if a host were simultaneously infected with two different influenza virus strains, their gene segments might reassort with very high frequency to yield a completely new strain. Serious human **pandemics** (worldwide epidemics) are thought to result in part from the genetic reassortment of human influenza virus strains with avian influenza virus strains. Humans may have little to no immunity to these reassortant viruses. (This issue is discussed at length in Chapter 12.)

Yet another strategy for evading humoral immune responses is to replicate within cells or tissues that are said to be in **immunologically privileged sites**, that is, sites that are subject to *reduced* immune surveillance. The kidney tubules are such a site. They are continually exposed to foreign matter, and they employ measures other than immune surveillance to avoid a continuous state of inflammation. Possibly, it is for this reason that the kidneys are a favored site for the replication of several polyomaviruses, including simian virus 40, JC virus, and BK virus. Reduced immune surveillance of the kidney tubules also may be a factor that favors the establishment of persistent infections by these polyomaviruses.

As explained below, activation of the adaptive immune response generally requires that antigen or antigen-presenting cells enter lymph nodes, where antigen can be presented to helper T cells. Importantly, the brain does not have classical antigen-presenting cells such as **dendritic cells**, and it does not drain through conventional lymphatics (see the note at the end of this paragraph). Thus, infection of brain sites does not readily lead to activation of adaptive immunity (Box 4.8). Moreover, it is reasonable to suggest that the brain is devoid of mediators of inflammation, since inflammation within the brain could be disastrous for the host. Another important factor making the brain an immunoprivileged site is that neurons do

---

BOX 4.8

The term "immune privilege" is meant to convey the idea that there is an absence of immunosurveillance within certain organs and tissues, such as in the central nervous system. This notion is now known to be too extreme. For example, tissue grafts within the brain are eventually rejected. Also, there is the important example of **experimental autoimmune encephalitis**, in which immunizing mice with myelin basic protein (a principal protein of the myelin sheaths in the brain) induces brain invasion by lymphocytes, destruction of myelin, and paralysis. Experimental autoimmune encephalitis is considered to be an important model of multiple sclerosis (see the text and Box 4.19). Yet autoimmune attack on our myelin is fortunately rare. What factors might account for that? The answer appears in part in the text.

---

not express MHC molecules, which are needed to activate and mediate adaptive immunity (see below). (Note that dendritic cells, like macrophages, are "sentry cells" stationed beneath the mucosa and other epithelia. Although they have many properties in common with macrophages, an important difference is that while activated macrophages remain in place at an infected site, activated dendritic cells migrate from the infected site to the nearest draining lymph node, where they activate B cells and T cells of adaptive immunity.)

Fortunately for us, tight junctions between the cells that line the brain capillaries and ventricles constitute a so-called **blood-brain barrier**, which limits the entry of many pathogens from the blood circulation into the brain. Yet the blood-brain barrier does allow the limited entry of immune cells circulating in the periphery (Box 4.8). Also, resident glial cells in the brain can function as antigen-presenting cells, the significance of which is explained later. Nevertheless, and predictably perhaps, the factors that make the brain a *somewhat* immunologically privileged site together make it a favored site for some viruses that establish persistent infections there. The herpesviruses, polyomaviruses, complex retroviruses (Chapter 21), and papillomaviruses are examples of virus families that thrive by establishing persistent infections, often in immunologically privileged sites, such as the central nervous system. HIV and HSV are important examples of viruses that establish persistent infections in the central nervous system. Moreover, the ability of some of these viruses to establish a latent state in some cells, in which viral genes are temporarily not expressed, contributes to their ability to evade immune detection and their success in vivo.

The *neurotropic* viruses enumerated above are able to gain access to the central nervous system despite the blood-brain barrier, doing so by several different routes. For example, viruses in the blood circulation might infect and disrupt the capillary endothelial cells or the meninges that constitute the blood-brain barrier. HSV, varicella-zoster virus (a herpesvirus responsible for chicken pox and shingles), and rabies virus employ another stratagem to access the central nervous system. They initially infect mucoepithelium, skin, and muscle, respectively. From their primary sites of infection, these viruses can infect peripheral neurons that innervate those sites. Those neurons then provide a conduit to the central nervous system.

Cell-Mediated Immunity

Whereas the humoral immune response generates antibodies that interact with and neutralize extracellular viruses in the blood and tissues, CTLs of the cellular arm of the adaptive immune response recognize and kill virus-infected cells. To understand how CTLs might carry out these crucial functions, we need to answer at least four basic questions. First, by what means does an infected cell alert the effectors of adaptive immunity that it is, in fact, infected? Second, by what means are the effector cells of the adaptive response able to recognize that a cell is infected? Third, how do the effector cells of adaptive immunity destroy an infected target cell? Pertinent to all three questions is the fact that all the effectors of adaptive immunity, humoral as well as cellular, are specific for their particular cognate antigens. So, fourth, by what means are the mediators of the cellular response able to *specifically* recognize any of the particular innumerable antigenic determinants that might be generated by any of the many different pathogens that might infect a cell?

The following points are relevant to the first two of the above questions. Whereas B-cell receptors and soluble antibodies recognize specific epitopes on the whole pathogen or on whole proteins of the pathogen, T-cell receptors recognize particular short peptides derived from the pathogen's proteins. Most importantly, T-cell receptors recognize those pathogen-derived peptides only when they have been assembled into a complex with MHC molecules, which are then expressed at the surface of the infected cell. That is, MHC molecules are integral membrane glycoproteins that present antigenic peptides at the cell surface, where they might then be seen by effector cells of cell-mediated immunity.

The interactions between a T-cell receptor and an MHC molecule/peptide complex are shown schematically in Fig. 4.13. Notice how the components of the T-cell receptor make contact with *both* the antigenic peptide and the MHC molecule. **Antigen presentation** by MHC molecules is one of the central and perhaps most elegant aspects of the adaptive immune response. (Importantly, in contrast to macrophages and neutrophils of the innate immune system, which recognize regularly repeating arrays of proteins and carbohydrates that are shared among broad classes of microbes, lymphocytes of the adaptive immune system specifically recognize small segments of proteins, either directly as in the case of B cells or following processing and presentation as in the case of T cells.)

Note that humans are highly **polymorphic** for the genes that encode MHC molecules. That is, there are numerous alternative forms, or **alleles**, of the individual MHC genes in the human population. The advantages of that polymorphism are discussed later. For the moment, note the following. A T-cell receptor recognizes a particular antigenic peptide only when it is associated with an MHC molecule that is encoded by a particular MHC allele. That is, a T-cell receptor recognizes the complex of a particular antigenic peptide and a particular MHC molecule. Consequently, a T cell that recognizes a peptide presented by a particular MHC **allotype** will not recognize that same peptide when it is presented by a different MHC allotype. This phenomenon is known as **MHC restriction** (Boxes 4.9 and 4.10).

A typical T-cell receptor is comprised of an α chain and a β chain, both of which are anchored to the plasma membrane by a transmembrane domain (Fig. 4.13). Like the heavy and light chains that make up an antibody molecule, the α and β chains of the T-cell receptor contain a variable region and a constant region. As you might have surmised, the variable regions of T-cell receptors contain the binding sites for the complexes of MHC molecules and antigenic peptides.

The varied T-cell receptors are generated by modular design, in much the way that the diverse B-cell receptors and antibodies are generated. That is, the variable regions of T-cell receptor chains are encoded in a series of sets of separate gene segments. Alternative forms of each segment are present on the DNA, and they can be recombined in many different combinations. During recombination, additional nontemplated nucleotides are inserted to generate greater T-cell receptor diversity. As in the case of the gene segments making up B-cell receptors and antibodies, the segments forming T-cell receptors are referred to as V, D, J, and C segments. As then might be expected, each mature T cell produces only one functional α chain and one functional β chain, which together make up its T-cell receptor. (Note that rearrangements of heavy- and light-chain gene loci, which result in B-cell receptors and antibodies, occur only in B cells and that rearrangements at α and β chain loci, which give rise to T-cell receptors, occur only in T cells.) Importantly, T cells, like

Figure 4.13 The left panel depicts a T-cell receptor on a CTL interacting with an MHC class I molecule/antigenic peptide complex on a target cell. The right panel depicts a T-cell receptor on a helper T cell interacting with an MHC class II molecule/antigenic peptide complex on an antigen-presenting cell. Notice how the components of the T-cell receptor contact the antigenic peptide, as well as the MHC molecule. The CD8 molecule on the CTL interacts with the α_3 domain of the MHC class I molecule, ensuring that MHC class I molecules present peptides to CD8 cells only (left panel). Similarly, the CD4 molecule on the helper T cell interacts with the β_2 domain on the MHC class II molecule, ensuring that MHC class II molecules present antigenic peptides to CD4 cells only (right panel). Adapted from P. Parham, *The Immune System*, 2nd ed. (Garland Science, New York, NY, 2005), with permission.

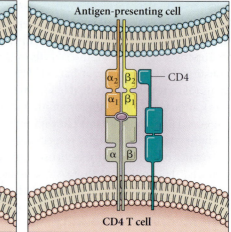

BOX 4.9

MHC restriction was discovered by Rolf Zinkernagel and Peter Doherty in 1974 at the John Curtin School of Medical Research in Canberra, Australia. The discovery was made while Zinkernagel and Doherty were trying to explain how LCMV, an arenavirus, although noncytopathic, nevertheless gives rise to lethal infections in the brains of mice. They hypothesized that death in vivo was due to immunopathology resulting from CTLs attacking infected brain. In one experiment to test this hypothesis, they isolated CTLs from infected mice and examined whether they might attack LCMV-infected cells in culture. Their key observation was that these CTLs indeed destroyed LCMV-infected cells in vitro but, unexpectedly, only if the target cells expressed the same MHC isoform as the CTLs. The CTLs did not attack uninfected cells, regardless of their MHC isoform, nor did they attack cells infected with an unrelated virus. Thus, a T-cell receptor recognizes a specific antigenic peptide in the context of a particular MHC isoform. This study provides yet another example of virology-based studies leading to major breakthroughs in other disciplines, in this instance in immunology. "It was a wonderful example of how certain things cannot be planned," said Zinkernagel. "Absolutely, this was a miracle of chance."

Reference
Zinkernagel, R. M., and P. C. Doherty. 1974. Restriction of in vitro T-cell-mediated cytotoxicity in lymphocytic choriomeningitis within a syngeneic or semiallogeneic system. *Nature* (London) **248**:701–702.

BOX 4.10

MHC molecules play an essential role in adaptive immunity. What then is the meaning of "major histocompatibility complex?" The answer is as follows. Because of the high level of polymorphism at the MHC locus in humans, there are an exceptionally large number of combinations of antigenically distinct MHC molecules expressed by different individuals in the human population. The differences between the MHC molecules expressed by different individuals are *the* major cause of graft rejection of transplants from unrelated donors. Historically, the MHC gene complex was recognized for this graft rejection phenomenon before its actual physiological function in immunity was understood.

B cells, are amplified by clonal selection, with different T-cell clones recognizing different antigens.

As might be expected from the above, there are notable similarities between T-cell receptors and B-cell receptors. For example, the α and β chains of T-cell receptors are each folded into conformations that resemble Ig domains, and the three-dimensional structures of their antigen-binding regions resemble those of antibodies.

Despite the similarities between T-cell receptors and B-cell receptors, they differ in important ways. For example, T-cell receptors do not undergo mutation or class switching after gene rearrangement. The latter is so because there is only one C_α segment and two functionally equivalent C_β segments.

Another important difference between T-cell receptors and B-cell receptors is that there is no soluble form of the T-cell receptor. T-cell receptors function solely as cell surface receptors for MHC molecule/antigenic peptide complexes.

Despite the facts that T-cell receptor genes are assembled by modular design and that T cells undergo clonal selection, T-cell receptors also differ from B-cell receptors and antibodies with respect to their degree of degeneracy. That is, T-cell receptors are less specific for a particular antigen than are B-cell receptors, and thus are more likely than B-cell receptors to **cross-react** with unrelated antigens. The reason for the degeneracy of the T-cell receptor is that *all* T-cell receptors recognize the same common structure of a peptide situated in the groove of an MHC molecule (Fig. 4.13). The structure of the groove is described in detail below. For now it is sufficient to note that the MHC/peptide complex appears as a relatively flat surface to the T-cell receptor. Moreover, the bound peptide represents less than one-third of that surface. In contrast, B-cell receptors and antibodies recognize antigens that are not submerged within the context of an antigen-presenting MHC molecule.

T-cell cross-reactivity is not necessarily disadvantageous to the host. For example, it can be beneficial in the case of a T-cell receptor that cross-reacts with serological variants of a particular virus, such as influenza virus (Chapter 12). Cross-reactive T cells also account, at least in part, for the finding that earlier virus infections can have a profound effect on a host's response to later infections with unrelated viruses. For example, mice that are immune to LCMV clear Pichinde virus or vaccinia virus faster than mice not immune to LCMV. This enhanced clearance of the unrelated viruses is due to cross-reactive LCMV-specific memory CTLs. On the other hand, cross-reactive T cells can be harmful if they lead to immunopathology, resulting from T cells cross-reacting with tissues of the host (see below).

Now here is a very important proviso to our story. There are two functionally distinct classes of T cells, each of which expresses T-cell receptors. Thus far, we have mentioned only the CTLs that directly attack and destroy virus-infected cells. In addition, there are **helper T cells (Th cells)**, which orchestrate the adaptive immune response. They do so by secreting the particular cytokines that the particular situation calls for, in that way activating either CTLs, or B cells, or both. How Th cells might know what is needed is described below. But first, there are several other points to note.

Some Th cells circulate in the blood, while others congregate in lymph nodes. The reason why some Th cells are found in lymph nodes is because that is where Th cells and B cells often first encounter their cognate antigen and where they can interact with each other when they do so. Interactions between these lymphocytes are important because B cells, as well as CTLs, generally must interact with Th cells to be activated and undergo clonal expansion. Also note that lymph nodes are anatomically organized to facilitate the interactions between Th cells and B cells and CTLs. Other Th cells leave the lymph nodes and travel via the efferent lymph and blood to the infected tissue, where they act together with macrophages to enhance the inflammatory response.

Here is another important particular. As there are two classes of T cells, likewise, there are two classes of MHC molecules. MHC class I molecules present or display *endogenous* antigenic peptide fragments. That is, they present antigenic peptides that have been generated from proteins that are produced by the pathogen during its replication cycle in the cell displaying those peptides. In contrast, **MHC class II molecules** present *exogenous* antigenic peptides. That is, they present peptides that are generated by special phagocytic cells, which phagocytose the pathogen and process it for presentation by MHC class II molecules. These special phagocytic cells are the **professional antigen-presenting cells**, which include dendritic cells, macrophages, and B cells. Their job as professional antigen-presenting cells is to alert Th cells that an infection is under way.

The above seemingly disjointed facts come together as follows. Virtually all cells in the body express MHC class I molecules, which present endogenous antigenic peptides that inform CTLs when a cell is infected and should be destroyed. In contrast, only professional antigen-presenting cells express MHC class II molecules, which present exogenous antigenic peptides to inform Th cells that an infection is under way. When so informed, Th cells can take up their role as organizers of the adaptive immune response. (Neurons do not express MHC class I molecules, as noted above. Erythrocytes, which do not contain nuclei, also do not express MHC class I molecules. Perhaps this is the factor that predisposes them to persistent infection with malarial parasites.)

Since T-cell receptors recognize an antigen in the form of a peptide fragment in association with an MHC molecule, how are these antigenic peptides generated, and how are they coupled with MHC molecules so that they might be presented to T cells? These processes are described in some detail below. However, before considering them, we first need to consider some other issues. We begin by having a more detailed look at the structures of MHC molecules.

MHC class I molecules are comprised of a membrane-bound heavy or α chain and a noncovalently bound β chain, also called β_2-microglobulin. The α chain contains three extracellular domains: $\alpha1$, $\alpha2$, and $\alpha3$. The two most membrane-distal domains, $\alpha1$ and $\alpha2$, fold together to

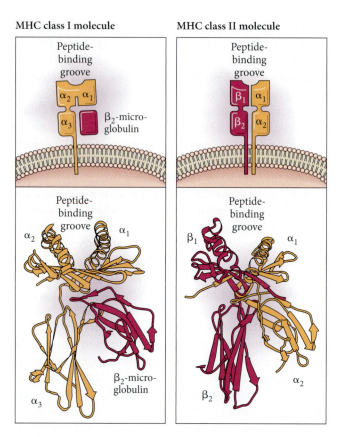

Figure 4.14 MHC class I molecules (left panels) are comprised of a membrane-bound heavy or α chain and a noncovalently bound β chain, also called β_2-microglobulin. The α chain contains three extracellular domains: α1, α2, and α3. The two most membrane-distal domains, α1 and α2, fold together to constitute the deep peptide-binding groove. MHC class II molecules (right panels) are comprised of two membrane-bound polypeptide chains, referred to as the α and β chains. Each of the MHC class II chains contains two extracellular domains: α_1 and α_2, and β_1 and β_2, respectively. The most membrane-distal (α_1 and β1) domains come together to form the peptide-binding groove. The α_3 domain and β_2-microglobulin of MHC class I molecules, as well as the α_2 and β_2 domains of MHC class II molecules, are Ig-like domains. In each instance, they form the stem structures that support the peptide-binding domains of the respective MHC class molecules. Thus, MHC class I and MHC class II molecules each contain two similar domains that come together to form a peptide-binding groove, which sits on two Ig-like domains. Adapted from P. Parham, *The Immune System*, 2nd ed. (Garland Science, New York, NY, 2005), with permission.

comprise the deep peptide-binding groove (Fig. 4.14 and 4.15).

MHC class II molecules are comprised of two membrane-bound polypeptide chains, referred to as the α and β chains (Fig. 4.14 and 4.15). Each of the MHC class II chains contains two extracellular domains: α_1 and α_2, and β_1 and β_2, respectively. The most membrane-distal (α_1 and

β1) domains come together to form the peptide-binding groove on MHC class II molecules.

The α_3 domain and β_2-microglobulin of MHC class I molecules, as well as the α_2 and β_2 domains of MHC class II molecules, are Ig-like domains. In each instance, they form the stem structures that support the peptide-binding domains of the respective MHC molecules. Thus, MHC

Figure 4.15 Looking down at the top of an MHC class I peptide-binding groove (left panel) and an MHC class II peptide-binding groove (right panel), notice how the structure of the MHC class II peptide-binding groove resembles that of MHC class I molecules. In each instance, the sides of the groove are comprised of α helices and the floors are formed by a sheet of eight antiparallel β strands. A bound peptide is shown in each instance. Peptides are bound to MHC class I molecules by interactions at their ends and to MHC class II molecules by interactions along their length. Adapted from P. Parham, *The Immune System*, 2nd ed. (Garland Science, New York, NY, 2005), with permission.

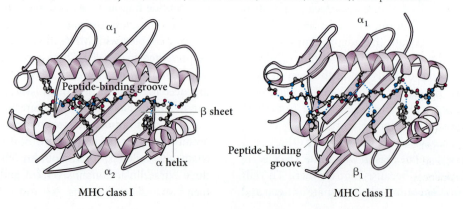

class I and MHC class II molecules each contain two similar domains that come together to form a peptide-binding groove, which sits on two Ig-like domains. Note how the structure of the MHC II peptide-binding groove resembles that of MHC I molecules. In each instance, the sides of the groove are comprised of α helices, and the floors are formed by a sheet of eight antiparallel β strands.

In contrast to the genes that encode T-cell and B-cell receptors and antibodies, the genes of the MHC complex are conventional and do not undergo rearrangement. Also, in contrast to the specificity displayed by T-cell and B-cell receptors and antibodies, each MHC molecule can present a variety of antigenic peptide fragments. Yet no MHC molecule can present all antigenic peptide fragments. How then might an individual be able to display antigenic peptide fragments from whatever pathogen might be encountered? The answer is that each individual has six different genes for the MHC class I α chain, which forms the peptide-binding groove on MHC class I molecules. Importantly, with respect to the population per se, three of these genes, HLA-A, HLA-B, and HLA-C, are highly polymorphic (Box 4.11). That is, there are 218, 439, and 96 known variants of HLA-A, HLA-B, and HLA-C, respectively, in the human population. Moreover, an individual is likely to be heterozygous at each of these loci. (The β_2-microglobulin is monomorphic. Regardless, it does not contribute to the peptide-binding site.) Likewise, there is a high level of polymorphism for several of the genes that encode the MHC class II α and β chains, which together form the peptide-binding groove on MHC class II molecules.

Each variant or **isoform** of an MHC molecule is different with respect to the antigenic peptides it can present to T-cell receptors. Consequently, the relatively large number of MHC alleles within each individual and the high level of polymorphism within the population ensure that many individuals, and the population in general, can respond to virtually any pathogen that might be encountered. (What might be the implications of these points with regard to highly inbred human and animal populations?)

BOX 4.11

The human MHC is also referred to as the human leukocyte antigen (HLA) complex because the proteins it encodes are found on leukocytes (white blood cells). MHC (HLA) proteins should not be confused with the perhaps better known ABO cell surface antigens, which are found on red blood cells. An individual's ABO type is determined by the inheritance of 1 of 3 alleles from each parent (A, B, or O), and these do not change as a result of environmental influences.

Although the peptide-binding grooves of MHC class I and MHC class II molecules resemble each other in form, there are differences between them. One of the most important differences lies at the ends of the grooves. These ends are more open in the case of MHC class II molecules. As a result, peptides that bind to MHC class I molecules are usually 8 to 10 amino acids long (restricted in length by the ends of the groove), whereas peptides that bind to MHC class II molecules are at least 13 amino acids long and may be much longer. In each instance, the peptide lies in an elongated conformation within the groove (Fig. 4.15).

Now, consider the following puzzler. On the one hand, a particular MHC isoform must be able to present a variety of antigenic peptide fragments. Moreover, it must be able to stably bind each of those antigenic peptides. If that were not the case, then peptide exchanges could occur at the cell surface, in that way compromising the ability of MHC molecules to accurately reflect the events taking place within a cell. It would be especially troubling if MHC class I molecules at the surface of an uninfected cell were to acquire antigenic peptides released from an infected cell. If that were to happen, then that uninfected cell might be attacked by CTLs. That is, the uninfected cell might become an **innocent bystander**. On the other hand, this ability of individual MHC isoforms to stably bind a variety of peptides is very different from the behavior of peptide-binding receptors (e.g., hormone receptors), which usually are specific for just one peptide. So, how then do MHC isoforms manage to stably bind a variety of different peptides?

The rather elegant solution conceived by Mother Nature is as follows. The peptides that bind to any particular MHC isoform have the same or very similar amino acids at two or three specific positions within the chain. There are pockets within the groove of that MHC molecule that can accommodate those specific amino acids on the peptide. Importantly, the amino acids that line those particular pockets in the groove are the very ones that are highly polymorphic within the population (Fig. 4.16). Thus, the groove is specific for only a few sites on the antigenic peptide, and several variants of the groove are present in every individual. Moreover, as is discussed below, the antigenic peptide actually stabilizes the dimeric MHC molecule. Were it to be lost at the cell surface, the MHC molecule would disassemble and consequently not be able to acquire an exogenous peptide. Furthermore, disassembled MHC class I molecules are in fact released from the cell surface by proteolytic cleavage, further ensuring against "innocent bystander" events.

Here is an important point that has not yet been explicitly stated. The T-cell receptors of CTLs and Th cells are exactly the same. That is, they are generated from the very same genomic sequences. Indeed, T-cell receptors are generated before T-cell progenitors decide to become CTLs

MHC class I variability

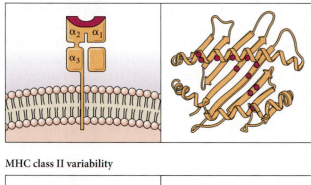

MHC class II variability

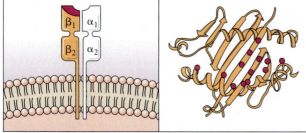

Figure 4.16 MHC molecules bind peptides at specific sites within the peptide-binding groove, as highlighted by the red dots (top, MHC class I; bottom, MHC class II). The amino acids that constitute those particular sites are the very ones that are highly polymorphic within the population. Thus, the groove is specific for only a few sites on the antigenic peptide, and several variants of the groove are present in every individual. Adapted from P. Parham, *The Immune System*, 2nd ed. (Garland Science, New York, NY, 2005), with permission.

or Th cells. Next, let us recall that MHC class I molecules present endogenous foreign peptides for recognition by CTLs, whereas MHC class II molecules present exogenous foreign peptides for recognition by Th cells. How then are T-cell receptors of CTLs able to be specific for MHC class I molecules, and how are T-cell receptors of Th cells able to be specific for MHC class II molecules? The answer is that T cells express coreceptor molecules, referred to as **CD4** and **CD8**. Th cells generally express CD4 coreceptors, and CTLs generally express CD8 coreceptors. Indeed, Th cells are often called "CD4 T cells," and CTLs are often called "CD8 T cells." The CD4 and CD8 coreceptors are responsible for the specificity of Th cells for MHC class II molecules and of CTLs for MHC class I molecules, respectively. The coreceptors confer this specificity by coupling to their respective cognate MHC molecule, thereby strengthening the adhesion between the T-cell receptor and the appropriate class of MHC molecule (Figure 4.13). Consequently, CTLs can focus their attention on virus-infected cells, and Th cells can get their cues from professional antigen-presenting cells (Boxes 4.12 and 4.13).

When T-cell receptors of a CTL bind to MHC class I molecules on an infected cell, a signal is transmitted into the CTL informing it to kill. When T-cell receptors of a Th cell bind to MHC class II molecules on a professional antigen-presenting cell, a signal is transmitted into the Th cell, activating it to provide help (i.e., to help activate a B cell or CTL). Yet as we noted above, the T-cell receptors

BOX 4.12

Biological systems often turn out to be considerably more complex than our sometimes simplistic working models would have them be. In this particular instance, some CD4 T cells are known to express CTL activity, in addition to their better-known helper activity. Indeed, CD4 and CD8 CTLs against chicken pox and parainfluenza viruses exist in equal frequencies, and up to 50% of the CD4 T cells in some HIV-infected individuals display clear cytotoxic potential.

Although the contribution of CD4 CTLs to antivirus defenses in vivo is not yet clear, there is some indication that these cells might play a role in the clearance of HSV from recurring herpetic lesions. For example, CD4 T cells are detected in herpetic lesions at a time when viral titers are declining, whereas CD8 T cells are found in these lesions only several days later, after virus titers have already declined.

In addition to evidence suggesting that CD4 CTLs might act to promote the clearance of virus infections, there also is evidence that these cells might account in part for the immunopathology associated with many, if not most, virus infections (see "Immunopathology" in this chapter). LCMV provides a well-known example of this phenomenon. In LCMV-infected mice that are depleted of CD8 T cells, passive transfer of MHC class II-restricted CD4 CTLs results in LCMV disease.

Also note that whereas CD4 T cells are better known as cytokine producers than are CD8 T cells, much CD8 T-cell activity is mediated by IFN-γ, in addition to CTL-specific killing mechanisms (see the text). Perhaps we should be thankful that CD8 T cells are not known to express helper activity, since that would complicate matters even further.

are the same on each of these cell types. How then are these identical T-cell receptors able to transmit distinctly different signals? The answer is that CD4 and CD8 are also signaling molecules and the signals they transmit are likely quite different. Thus, the respective signals transmitted by CD4 and CD8 are probably in part responsible for the functional differences between CTLs and Th cells. Other functional differences between CTLs and Th cells may also affect the signals transmitted by their respective T-cell receptors.

Antigen Presentation by MHC Class I Molecules

Recall that virtually all cells express MHC class I molecules and that these molecules serve to present *endogenously* synthesized peptides to CTLs. The process of antigen presentation by MHC class I molecules thus begins with the degradation of endogenously synthesized viral proteins. In this regard, eukaryotic cells have two major protein degradation pathways. The ubiquitin-proteasome pathway mediates degradation of endogenously synthesized proteins, whereas exogenous proteins are degraded in lysosomes. Thus, MHC class I molecules present peptides generated by proteasomal degradation of endogenously synthesized viral proteins.

In actuality, a fraction of *all* the proteins made in the cell are degraded by the ubiquitin-proteasome pathway, regardless of whether the cell is infected. Moreover, some of the peptides that are thus generated, host as well as viral, are presented by MHC class I molecules at the cell surface. Importantly, if the cell is infected, then some of the peptides presented at the cell surface will be derived from virus-encoded proteins that were in the cytosol before the emergence of progeny virus particles. That is a key point, since it means that a cell can be recognized and killed by CTLs, in the early stages of infection of that cell, before any progeny infectious virus particles might be assembled within it.

The ubiquitin-proteasome pathway begins with proteins (both self proteins and ones encoded by an intracellular pathogen) being marked for degradation by **ubiquitin**, a 76-amino-acid polypeptide that the cell attaches to the amino groups on lysine residues of proteins destined for destruction (Fig. 4.17). Ubiquitination is the principal means by which a cell marks any protein for degradation. Additional ubiquitins are added to form a long polyubiquitin chain. The polyubiquitinated protein then is recognized by a **proteasome**, a large barrel-shaped protein complex comprised of some 28 subunits, which expresses several different protease activities and degrades the protein into polypeptides.

The peptides that are generated by the proteasome are released into the cytosol. Those peptides that are 9 amino acids long or longer are then transported into the lumen of the endoplasmic reticulum (ER) by the protein complex referred to as the **transporter associated with antigen processing (TAP)**, which is embedded in the ER membrane. TAP is actually a heterodimer, comprised of TAP-1 and the related TAP-2 (Box 4.14).

Since MHC class I α chains are transmembrane proteins, they contain signal sequences that target them to the ER membrane while they are still ribosome-bound nascent polypeptides (see Chapter 2 for the generation of transmembrane proteins). Likewise, β_2-microglobulin is targeted to the ER. Within the ER, the newly synthesized MHC class I α chains and β_2-microglobulins associate with each other and with peptide fragments generated in the cytosol by the proteasome. Then, the complex of MHC class I molecule and peptide is transported by the usual vesicle-mediated secretory pathway to the plasma membrane.

Before moving on, there are several points that bear reiterating. First, the MHC class I peptide presentation pathway occurs in noninfected as well as infected cells. Moreover, it presents host- as well as pathogen-derived peptides. Because or in spite of this fact, it informs CTLs as to whether or not the cell might be infected. Second, nearly all cell types are able to express MHC class I molecules. Here then is a new point. IFN-α/β greatly increase transcription of MHC class I genes and of the genes encoding the TAP transporter proteins and the LMP2, LMP7, and MECL-1 proteasome subunits (Box 4.14). Thus, IFN-α/β enhance the efficiency of the MHC class I antigen presentation pathway, in that way causing a virus-infected cell to be more discernible to CTLs, another example of the interplay between the innate and adaptive immune systems.

Antigen Presentation by MHC Class II Molecules

We begin our consideration of antigen presentation by MHC class II molecules by recalling that virtually all cells

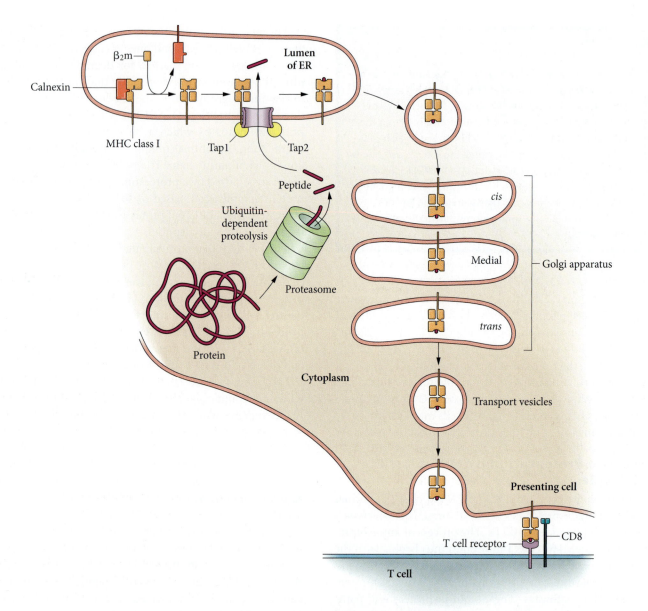

Figure 4.17 The pathway for the processing and presentation of endogenous proteins by MHC class I proteins begins with both self proteins and ones encoded by an intracellular pathogen being marked for degradation by ubiquitin, a 76-amino-acid polypeptide that the cell attaches to the amino groups on lysine residues of proteins destined for destruction. Additional ubiquitins are added to form a long polyubiquitin chain (not shown). The polyubiquitinated protein then is recognized by a proteasome, a large barrel-shaped protein complex comprised of some 28 subunits, which expresses several different protease activities and degrades the protein into polypeptides. The peptides that are generated by the proteasome are released into the cytosol. Those peptides that are nine or more amino acids long are transported into the lumen of the ER by the protein complex referred to as the transporter associated with antigen processing (TAP), which is embedded in the ER membrane. Since MHC class I α chains are transmembrane proteins, they contain signal sequences that target them to the ER membrane while they are still ribosome-bound nascent polypeptides. Likewise, β₂-microglobulin is targeted to the ER. Within the ER, the newly synthesized MHC class I α chains are stabilized by the membrane chaperone protein calnexin, until the MHC class I α chains associate with β₂-microglobulin. Next, the peptide fragments generated in the cytosol by the proteasome bind to the MHC class I molecule, and the MHC class I/peptide complex is transported by the usual vesicle-mediated secretory pathway to the plasma membrane. At the cell surface, the MHC class I/peptide complexes may then interact with the T-cell receptors of a cytotoxic CD8 cell. Adapted from S. J. Flint, L. W. Enquist, R. M. Krug, V. R. Racaniello, and A. M. Skalka, *Principles of Virology: Molecular Biology, Pathogenesis, and Control*, 2nd ed. (ASM Press, Washington, DC, 2003), with permission.

Here is yet another example of the elegance of the mammalian immune response. First, bear in mind that a major purpose of the ubiquitin-proteasome pathway is to degrade faulty or misfolded *cellular* proteins. Second, when an infection is under way, IFN-γ induces the expression of three proteins, LMP2, LMP7, and MECL-1, which replace three constitutive components of the proteasome. The incorporation of these IFN-γ-inducible components into the proteasome modifies its specificity, such that it generates peptides that contain residues that are preferred by the anchoring sites of the peptide-binding grooves of MHC class I molecules. Moreover, the incorporation of these IFN-γ-inducible components into the proteasome also causes it to generate peptides that are the preferred size for transport into the ER by TAP (see the text).

express MHC class I molecules and that MHC class I molecules present endogenously synthesized foreign peptides to CTLs, so that infected cells might be recognized and killed by CTLs. In contrast, MHC class II molecules are expressed only by professional antigen-presenting cells, that is, macrophages, dendritic cells, and B cells. Moreover, the function of MHC class II molecules is to present peptides generated from exogenous (i.e., endocytosed) proteins, in order to alert Th cells that an infection is under way in the host. As we will see, these Th cells then activate other immune effectors. For example, they stimulate B cells to produce antibodies and may help activate CTLs. Importantly, Th cells also secrete cytokines (e.g., IFN-γ), which enhance the expression of MHC molecules and other components of the antigen presentation machinery.

As noted above, the ubiquitin-proteasome pathway, which mediates the degradation of endogenously synthesized proteins for MHC class I peptide presentation, is one of the two major protein degradation pathways in eukaryotic cells. The other is lysosomal proteolysis, by which exogenous proteins that have been taken up by endocytosis are degraded. Lysosomes contain several proteases for this purpose. With these points in mind, professional antigen-presenting cells acquire *exogenous* antigenic proteins by endocytosing whole pathogens and generate antigenic peptides from them by lysosomal proteolysis.

Before we consider the pathway for MHC class II peptide presentation, we first need to take into account the following. Since MHC class II molecules, like MHC class I molecules, are transmembrane glycoproteins that are expressed at the plasma membrane, they likewise are synthesized at the ER and are transported to the plasma membrane via the secretory pathway (Fig. 4.18). Thus, evolution had to devise a means to prevent MHC class II molecules from acquiring *endogenous* peptides that are generated in the cytosol by the ubiquitin-proteasome pathway and transported into the ER by the TAP transporter. To avoid this inappropriate loading of *endogenous* peptides onto MHC class II proteins, professional antigen-presenting cells express a molecule referred to as the MHC class II-associated **invariant chain** (Ii), which binds to partially folded MHC class II molecules in the ER. Ii binds to MHC class II molecules in such a way that part of Ii is in the peptide-binding groove, thereby blocking any interaction of MHC class II molecules with endogenous peptides in the ER.

Evolution is not only clever but also efficient, for Ii also facilitates the export of MHC class II molecules from the ER. Moreover, Ii also targets MHC class II molecules to the endosomal/lysosomal compartment, where *exogenous* peptides are generated and loaded onto MHC class II molecules. This Ii-mediated targeting of MHC class II molecules is noteworthy because the secretory pathway does not usually intersect the endocytic pathway. Before loading of exogenous peptides can happen, MHC class II-associated Ii is first cleaved and released by the action of lysosomal proteases.

One additional point about antigen presentation by MHC class II molecules is that MHC class II molecules are generated and intersect the endosomal/lysosomal compartment in noninfected as well as infected cells. Consequently, like MHC class I molecules, MHC class II molecules also present self peptides. This brings up what is perhaps the most intriguing aspect of the mammalian immune response: how does it distinguish nonself from self (see below)? However, there are several other issues that need to be considered first.

The Rationale for MHC Restriction

Presentation of antigen by MHC molecules is certainly elegant, but is it a needlessly complex process? That is, why not have CTLs simply recognize antigen that is presented in a non-MHC-restricted manner, the way B-cell receptors do? We cannot know precisely what evolution had in mind when concocting MHC restriction, but a reasonable supposition is that MHC restriction enables CTLs to focus their attention on virus-infected cells, without having their potent and energetically costly effector functions sidetracked by free extracellular virus particles or proteins. Antibodies can deal well enough with free virus, and they are much more plentiful and cost-effective for that purpose. Recall that each plasma B cell can pump out thousands of antibody molecules every second! Yet

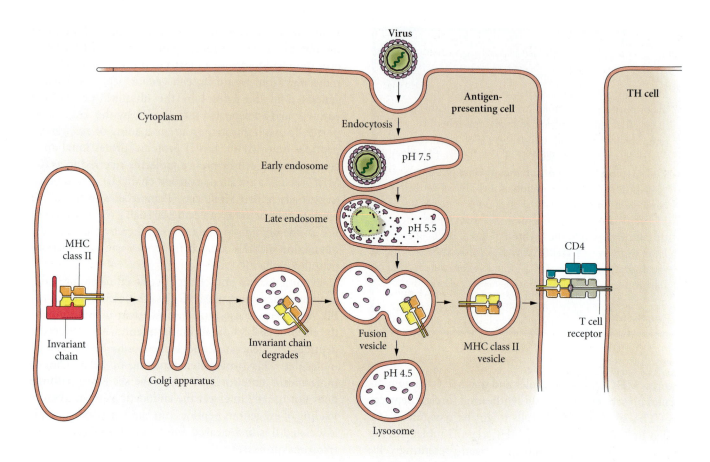

Figure 4.18 The MHC class II peptide presentation pathway presents peptides generated by lysosomal proteolysis of endocytosed pathogens and proteins. MHC class II molecules are synthesized and assembled in the ER. During their transport to the plasma membrane via the secretory pathway, they are diverted to the internalized endosomes that contain the pathogen-derived peptides. The secretory vesicles containing the MHC class II molecules then fuse with these endosomes, and the MHC class II molecules acquire the peptides that they present at the cell surface. MHC class II molecules are prevented from acquiring endogenous peptides, which are generated in the cytosol by the ubiquitin-proteasome pathway and transported into the ER by the TAP transporter, as follows. Professional antigen-presenting cells express a molecule referred to as the MHC class II-associated invariant chain (Ii), which binds to partially folded MHC class II molecules in the ER in such a way that part of Ii is in the peptide-binding groove, thereby blocking any interaction of MHC class II molecules with endogenous peptides in the ER. Ii also facilitates the export of MHC class II molecules from the ER. Moreover, Ii also targets MHC class II molecules to the endosomal/lysosomal compartment, where loading of exogenous peptides occurs. This Ii-mediated targeting is noteworthy because the secretory pathway does not usually intersect the endocytic pathway. Before loading of exogenous peptides can happen, MHC class II-associated Ii is first cleaved and released by the action of lysosomal proteases. At the cell surface, the MHC class II/peptide complex is available to the T-cell receptor of CD4 Th cells. Adapted from S. J. Flint, L. W. Enquist, R. M. Krug, V. R. Racaniello, and A. M. Skalka, *Principles of Virology: Molecular Biology, Pathogenesis, and Control*, 2nd ed. (ASM Press, Washington, DC, 2003), with permission.

those antibodies cannot reach intracellular virus, whereas the vastly more elaborate CTLs can. Why then have those very special CTLs distracted by extracellular virus?

Here is another important reason for MHC restriction. Suppose that CTLs did recognize non-MHC-associated antigens at the cell surface. Then, consider the consequences if an uninfected cell were to have viral antigens bound to it. Such antigens might be contained in the debris from neighboring infected cells. If CTLs were to recognize

antigens bound to uninfected cells, the result would be the innocent-bystander effect mentioned earlier.

The above two arguments pertain to antigen presentation by MHC class I molecules, which results in CTLs recognizing and attacking infected cells. But why have MHC restriction in the case of MHC class II molecules, which present antigen to Th cells? A plausible answer is based on three points. First, Th cells must be activated to contribute to the immune response, and second, professional antigen-presenting cells must deliver *two* signals to Th cells in order to activate them. Before considering the third point, note that the above two points play out as follows. The first signal from the professional antigen-presenting cell is triggered by protein B7, on its surface, which interacts with protein CD28 on the surface of the Th cell (Fig. 4.19). Note that the signal triggered by B7 is nonspecific, in the sense that CD28 on any Th cell might respond to it. However, the second signal is specific, since it is triggered by the MHC class II/peptide complex, which is recognized specifically by T-cell receptors on particular T cells. Actually, these two signals may be delivered at the same time. (Note that the capability of professional antigen-presenting cells to transmit a stimulatory signal via MHC

Figure 4.19 Professional antigen-presenting cells must deliver two signals to Th cells in order to activate them. One signal (diagram 1) is triggered by the MHC class II/peptide complex on the surface of the professional antigen-presenting cell, which interacts specifically with T-cell receptors on particular T cells. The second signal (diagram 2) is triggered by protein B7 on the surface of the professional antigen-presenting cell, which interacts with protein CD28 on the surface of the Th cell. The capability of professional antigen-presenting cells to transmit a signal via MHC presentation, as well as a costimulatory signal via B7, is a defining property of these cells. The T-cell receptor per se comprises two proteins, an α chain and a β chain. The cytoplasmic tails of the α and β chains consist of only about three amino acids. Thus, for the T-cell receptor to signal, it is associated with the CD3 complex, which comprises four different proteins: one γ chain, one δ chain, two ε chains, and two ζ chains. The CD4 molecule on the helper T cell is shown interacting with the β₂ domain on the MHC class II molecule. Adapted from P. Parham, *The Immune System*, 2nd ed. (Garland Science, New York, NY, 2005), with permission.

The co-stimulatory molecule B7 on the dendritic cell binds CD28 on the naive T-cell

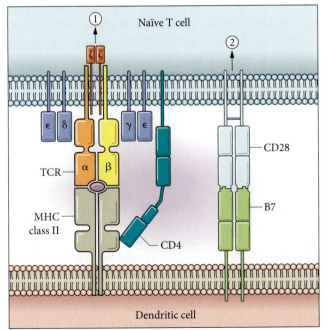

presentation, as well as a costimulatory signal via B7, is a defining property of these cells. Indeed, that is why they are uniquely endowed to activate Th cells.)

Now here is the third point. In noninfected tissues, professional antigen-presenting cells are in a "resting state," in which they express very low levels of MHC class II molecules and of B7. However, when an infection is under way, inflammatory cytokines such as IFN-γ are produced, which induce the expression of elevated levels of MHC class II molecules and B7. The professional antigen-presenting cell thus activated can then deliver its two signals: a general one in response to the infection and a specific one in response to the particular pathogen. Herein may lie the rationale for the antigen presentation by MHC class II molecules expressed by professional antigen-presenting cells. Since the effectors of adaptive immunity are powerful, energy-costly, and highly specialized weapons, they must not be activated at inappropriate times. That is, they must be activated only in the event of an infection. By making activation of adaptive immunity dependent on a nonspecific signal in response to a general infected state as well as on a specific signal in response to a particular pathogen, inappropriate activation is not likely to happen. Note that a Th cell that receives the two signals then undergoes clonal expansion. (Note the analogy between the dual-key activation system to provide for the failsafe launching of ballistic missiles and the requirement for two signals to activate affectors of immunity [Box 4.3].)

Another advantage of peptide presentation via MHC molecules, apropos both the MHC class I and MHC class II peptide presentation pathways, stems from the fact that peptide fragments rather than whole proteins are displayed. Since many potential antigenic determinants might be buried within the folded structure of an intact protein, displaying peptide fragments significantly increases the number of different epitopes that might be recognized by T-cell receptors. This creates a stronger, more diverse adaptive immune response. Furthermore, most intact virus proteins are never present at the cell surface during the virus's replication cycle. Yet MHC presentation can display epitopes from essentially every virus protein that is present in the cell.

Relevant to the preceding argument, recall that different MHC allotypes contribute different peptide-binding specificities. Therefore, since each individual is polymorphic for MHC class I and MHC class II genes, the number of foreign peptides that can be presented during any infection is appreciably amplified. This strengthens the immune response by increasing the number of activated pathogen-specific T cells. This argument applies to each individual within a population. It also applies in an important way to the population as a whole. Variability of MHC alleles within a population means that individuals vary in their abilities to respond to particular pathogens, thereby helping to ensure that at least some individuals survive any particular epidemic. Small or highly inbred populations suffer in this regard.

Finally, why have antigenic peptides presented by both MHC class I and MHC class II molecules? Might only one form of MHC molecule be sufficient? Perhaps, but having only one form of MHC presentation almost certainly would not be as effective as the system we have. First, if there were only one form of MHC molecule recognized by the T-cell receptors of both CTLs and Th cells, then a professional antigen-presenting cell in the process of activating a Th cell might be the object of attack by a CTL. Second, keeping the activation and targeting of Th cells distinct from those of CTLs provides a measure of control that almost certainly benefits the host.

Activation of Th Cells: Dendritic Cells and B Cells

Before detailing the process of Th cell activation, it would be good to recap the following points.

As noted before, Th cells must be activated before they can carry out their crucial functions. Actually, Th cell activation may be considered to be the first step of a primary adaptive immune response. We also emphasized that professional antigen-presenting cells must deliver both a stimulatory signal and a costimulatory signal to activate resting or naïve Th cells. As also noted, the defining property of the professional antigen-presenting cells (i.e., macrophages, dendritic cells, and B cells) is their capacity to transmit a stimulatory signal via the interaction between an MHC class II/peptide complex and a T-cell receptor, as well as a costimulatory signal via the interaction between B7 and CD28 (Fig. 4.19).

Also recall that both macrophages and dendritic cells of the innate immune system are positioned throughout the body, on guard against invading pathogens. However, at uninfected sites, these cells are in a resting state, in which they express low levels of MHC and costimulatory molecules. But if the site should become infected, then cytokines are produced by the innate immune response and activate local macrophages and dendritic cells, causing them to express elevated levels of MHC molecules and B7. In addition, these cells begin to vigorously internalize antigen from the extracellular fluid or from infected cells, which they then can present via MHC molecules. The inflammatory cytokines produced by the innate immune response, together with the encountered antigen, stimulate macrophages and dendritic cells to mature into fully functional antigen presenters.

Dendritic cells are more important than macrophages for initially presenting antigenic peptides to T cells. That is so because macrophages remain at the local site of infection, even after being activated. In contrast, activated dendritic cells leave the infected tissue and migrate through the lymphatic system to the nearest draining lymph node. This is a key point because naïve CTLs and Th cells continually circulate from lymph node to lymph node, searching for an antigen-presenting cell that might be displaying their cognate antigen. In fact, T cells do not leave the circulation unless they are activated by an antigen-presenting cell which itself has been activated. (Activated macrophages may provide activating signals to T cells at the site of infection, to keep them active there until the infection might be cleared.)

Why might the initial activation of Th cells occur in the lymph nodes, rather than at the site of infection? First, consider that for any given infection, the fraction of circulating naïve T cells that might be specific for the invading pathogen might be as few as 1 in 10^6. Thus, there needs to be some mechanism for rather quickly and efficiently bringing the minute fraction of naïve Th cells which are specific for a particular pathogen into contact with professional antigen-presenting cells that display that pathogen's antigens and that can then activate the Th cell. The migration of dendritic cells from the site of infection to the nearest draining lymph node, together with the circulation of naïve Th cells from lymph node to lymph node, is the means by which the immune system solves this problem. This is a rather important concept. Remember that the adaptive immune system does not attempt to initiate an adaptive immune response at the site of infection. Instead, it uses the sentinel dendritic cells to capture antigen from the infected site and transport that antigen to the nearest lymphoid organs, where the effectors of adaptive immunity that might recognize that antigen are concentrated. Moreover, the fact that T cells do not leave the circulation unless activated is important with respect to the maintenance of **peripheral tolerance** (see below).

Once the naïve Th cell encounters its cognate antigen-presenting cell, it matures and undergoes clonal expansion, a process that takes several days. The activated Th cell then must play its roles helping to activate CTLs and B cells. During this entire period, in which first dendritic cells are activated at the infected site, followed by activation of Th cells in the lymphoid organs and finally activation of B cells and CTLs, the innate system is doing its best to contain the infection.

Here is another detail to note. Th cells tailor an adaptive immune response to the particular situation by secreting particular combinations of cytokines. For this purpose, naïve Th cells can mature into either of two classes of Th cells, referred to as Th1 and Th2. Th1 cells secrete cytokines IFN-γ, IL-2, and TNF-α, which favor a CTL response, whereas Th2 cells secrete cytokines IL-4, IL-5, and IL-10, which favor antibody production.

There is much research and discussion regarding which of these responses is more important in resolving virus infections. The answer may vary from one virus to another. Yet the optimal outcome will likely occur when both arms of adaptive immunity are engaged. Apropos that point, although the Th1/Th2 paradigm is useful for the purpose of teaching, it is a simplification of the actual state of affairs in vivo, since the diversity of T-cell responses is more varied than merely the strict Th1 versus Th2 distinction. That is, T cells commonly secrete particular combinations of Th1 and Th2 cytokines. And, in point of fact, there also are other types of T cells in addition to Th cells and CTLs (Box 4.15).

Whether a naïve T cell matures into primarily a Th1 or Th2 cell is partially determined by the costimulation it receives during activation. This occurs as follows. First, at the infected site, dendritic cells receive inputs through their various cell surface receptors, particularly their Toll-like and cytokine receptors, which inform them as to the nature of the invading pathogen. For example, recall how TLR4 senses bacterial lipopolysaccharides, whereas TLR3 recognizes the double-stranded RNA produced during infections by a variety of viruses. Thus, dendritic cells are able to recognize a diversity of signals, each characteristic of a particular class of pathogen. A second key point is that dendritic cells can pass this information on to naïve Th cells when activating them. The dendritic cells do so as follows. Whereas the costimulatory signal transmitted by B7 is necessary to activate a naïve Th cell, B7 is not the only costimulatory molecule expressed by the dendritic cell. In particular, dendritic cells also secrete cytokines which help to activate the Th cell, and the cytokine profile secreted by the dendritic cell reflects the inputs that the dendritic cell received at the infected site. The combination of signals transmitted by the dendritic cell when it is activating the Th cell determines the set of cytokines that the Th cell will then generate. Consequently, the activated Th cell can, in turn, orchestrate the most effective adaptive immune response against the invader.

B cells also are important antigen presenters during virus infections. They carry out this function by specifically

BOX 4.15

Regulatory T cells are also part of the T-cell repertoire, serving to shut down T-cell-mediated immunity towards the end of an immune response. Regulatory T cells also help to prevent autoimmune reactions, that is, reactions against self. They do this by suppressing autoreactive T cells that may have escaped the negative selection process in the thymus (see the text).

You may be thinking (mockingly perhaps) that it would be too simple if there were only one type of regulatory T cell. If so, you indeed are correct. In actuality, there are two known classes of regulatory T cells: the "naturally occurring" regulatory T cells (T$_{reg}$ cells) and the adaptive regulatory T cells (also known as Tr1 cells and Th3 cells). Each of these regulatory T-cell subsets provides another example of the phenomenon whereby one class of lymphocytes can regulate the actions of another.

T$_{reg}$ cells appear to originate from a unique lineage of CD4 T cells in the thymus, and they are dedicated to expressing only a regulatory role. In order to carry out their regulatory role, T$_{reg}$ cells must make direct contact with responder T cells. In contrast, the adaptive regulatory T cells usually start out as effector T cells that express immunity-promoting cytokines. Afterwards, they switch to producing inhibitory cytokines. They convey their regulatory effect to responder T cells via these cytokines.

T$_{reg}$ cells may be identified by their expression of a transcription factor known as FoxP3, which is not known to be expressed by any other type of cell. Mutations of *FOXP3* can prevent development of T$_{reg}$ cells, resulting in a fatal autoimmune disease known as IPEX. Since FoxP3 is encoded by a gene on the X chromosome, IPEX principally affects boys. The only treatment for IPEX is a bone marrow transplant from a healthy, HLA-identical sibling.

The two known types of adaptive regulatory T cells, Tr1 and Th3, are distinguished by their cytokine profiles. Tr1 cells produce high levels of IL-10 and transforming growth factor beta (TGF-β), whereas Th3 cells primarily produce TGF-β. These cells mediate immunosuppression via the secretion of these cytokines. Both IL-10 and TGF-β inhibit helper T cells, while TGF-β may suppress immunity in general. The possible relationship between Tr1 and Th3 cells is not yet clear.

Memory T cells are discussed in the text. Antigen-specific memory T cells, which persist long after an infection has resolved, exist for both CD4 Th cells and CD8 CTLs. If at some later time these cells should reencounter their cognate antigen, then they are able to quickly activate and multiply. They are key components of immunological memory. Three distinct subpopulations of both CD4 and CD8 memory T cells are presently known but are not discussed here.

binding antigens to their B cell receptors (i.e., cell-bound antibodies), internalizing them by receptor-mediated endocytosis and then presenting those antigens via the regular MHC class II peptide presentation pathway. However, B cells cannot initiate a primary adaptive response. The reason is that naïve B cells express only low levels of MHC class II molecules and B7. However, once a B cell has been activated by an activated Th cell (see below), it expresses greatly increased levels of these molecules. Consequently, B cells may not be important antigen-presenting cells during the early stages of infection, because their activation is dependent on the prior activation of naïve Th cells, which in turn depends on the prior activation of dendritic cells. Yet B cells play important roles as antigen-presenting cells during the later stages of a primary infection, and memory B cells are important antigen presenters during subsequent encounters with the same pathogen. (T-cell-independent activation of B cells is possible in some instances [see below].)

A reason that B cells are especially efficient antigen presenters is that B-cell receptors *specifically* bind antigen, which is then taken up by endocytosis for processing and presentation by the MHC class II peptide presentation pathway. In contrast, dendritic cells are indiscriminate phagocytosers. Consequently, B cells can specifically concentrate antigen for presentation via the MHC class II peptide presentation pathway, whereas much of what dendritic cells present is from internalized cellular debris. This distinction is important because activation of Th cells requires cross-linking of their T-cell receptors. Indeed, there is a threshold number of T cell receptors that must be cross-linked for the Th cell to be activated. Since B cells specifically concentrate and present specific antigen rather than other debris, they achieve a higher level of T-cell receptor cross-linking than other professional antigen-presenting cells. This is especially consequential when low levels of antigen might be present.

Th1 and Th2 cells have different fates following their activation. Once CTLs and Th1 cells have been activated (and undergone clonal expansion), they migrate from the lymph nodes, where activation occurs, to the infected sites. In contrast, Th2 cells stay in the lymphoid tissues, where they transmit signals that help to activate B cells. When Th1 cells reach the infected site, their commitment to their Th1 cytokine profile may be strengthened or altered by the particular cytokines they encounter there. Next, we consider the actual functions of activated Th cells.

Activation of B Cells

One of the major roles for Th cells is to help B cells mature into antibody-producing plasma cells. To appreciate why B cells might require help from T cells, recall how naïve T cells require two separate signals in order to be activated

(Fig. 4.19). One of these signals is the pathogen-specific one, which is triggered by the interaction of the T-cell receptor on the naïve T cell with an MHC class II/peptide complex on the professional antigen-presenting cell. The other signal is a general one, triggered by the interaction of CD28 with B7. Since professional antigen-presenting cells express B7 only in response to inflammatory cytokines that are produced when an infection is under way, the requirement for the two signals ensures against the inappropriate activation of helper T cells. Consequently, the B-cell requirement for T-cell help serves as a means to ensure that B cells likewise will be activated only in response to an infection.

Yet the B cell itself also requires two signals in order to become a plasma cell. In addition to the signal from the Th cell, the B cell also must recognize its cognate antigen, further ensuring against its inappropriate activation. In this regard, recall that virus particles and bacteria are characterized by regularly repeating surface features (more importantly, epitopes). Indeed, these epitopes are spaced such that they very effectively cross-link B-cell receptors, an important point since cross-linking of the B-cell receptors is the key to triggering the signal (Fig. 4.20). The details are as follows. The B-cell receptor has only a short cytoplasmic domain and therefore requires associated signaling molecules to transmit a signal into the cell. These B-cell-receptor-associated signaling molecules are referred to as Igα and Igβ (Fig. 4.20). Together with the plasma membrane-associated IgM molecules (i.e., the antibody class made by naïve B cells), they form the functional B-cell receptor. On the cytoplasmic side of the plasma membrane, Igα and Igβ associate with the tyrosine kinase Blk, Fyn, or Lyn. When the receptor complexes are cross-linked by the pathogen, the receptor-associated tyrosine kinases cross-phosphorylate the cytoplasmic tails of Igα and Igβ on neighboring cross-linked receptor complexes, thereby initiating the signal cascade. It is likely that this B-cell receptor signaling process evolved to exploit the regularly repeating surface features on viral and bacterial pathogens. (As in the case of the heavy chains of the B-cell receptor, the α and β proteins of the T-cell receptor have short cytoplasmic domains, consisting of only several amino acids. Consequently, they too are not able to signal by themselves. However, the cytoplasmic domains of T-cell receptors are associated with the **CD3** complex, comprised of four different proteins, γ, δ, ε, and ζ [zeta], which transmit the signal when the T-cell receptor binds to an MHC/peptide complex [Figure 4.19].)

The signal provided by the activated Th cell to the B cell is referred to as the **costimulatory signal** (Fig. 4.21). Yet the B cell itself must first play a key role before it can receive this signal. Recall that B-cell receptors bind their

Clustering of antigen receptors allows receptor-associated kinases to phosphorylate the ITAMs

Syk binds to doubly phosphorylated ITAMs and is activated on binding

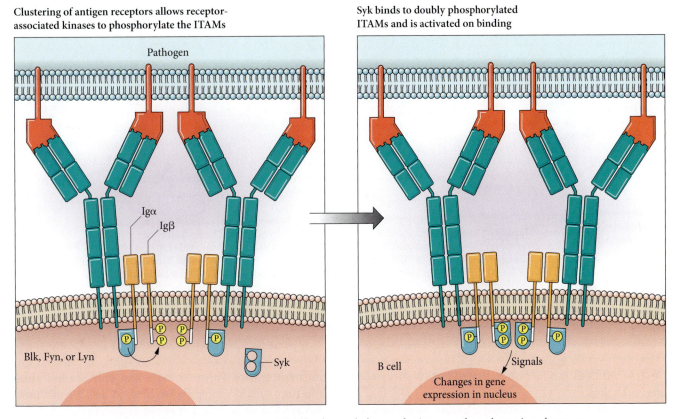

Figure 4.20 B-cell receptors must be cross-linked by the regularly spaced epitopes on the pathogen in order to transmit an activating signal. Since the B-cell receptor has only a short cytoplasmic domain, it requires associated signaling molecules, referred to as Igα and Igβ, to transmit a signal into the cell. On the cytoplasmic side of the plasma membrane, Igα and Igβ associate with the tyrosine kinase Blk, Fyn, or Lyn. When the receptor complexes are cross-linked by the pathogen, the receptor-associated tyrosine kinases cross-phosphorylate the cytoplasmic tails of Igα and Igβ on neighboring cross-linked receptor complexes, thereby enabling the Syk tyrosine kinase to bind. Syk molecules, bound to neighboring B-cell receptors, cross-phosphorylate each other, thereby triggering the signal cascade. The signal reaches the nucleus, where it induces changes in gene expression, resulting in B-cell activation. Adapted from P. Parham, *The Immune System*, 2nd ed. (Garland Science, New York, NY, 2005), with permission.

cognate antigens, which they then internalize by endocytosis and process for presentation by the MHC class II peptide presentation pathway. Only upon presenting its cognate antigen by the MHC class II peptide presentation pathway can the B cell be recognized by the particular clone of activated Th cells. Importantly, while these B cells recognize the same pathogen as the activated Th cells, recall that B-cell receptors recognize epitopes on whole antigens (pathogens), whereas T-cell receptors recognize peptides presented in the context of MHC molecules. Therefore, although the interacting Th cell and B cell are specific for the same pathogen, they may well recognize different features on the pathogen.

When the T-cell receptors on the Th cell bind to the MHC class II/peptide complexes on the B cell, the Th cell responds by synthesizing the **CD40 ligand**, which interacts with **CD40** on the surface of the B cell. This interaction

transmits a signal back into the Th cell, causing it to release cytokines that, together with CD40 ligand and the signal transmitted by the B-cell receptors, cause B cells to proliferate and mature into antibody-secreting plasma cells. Important points to bear in mind are that Th cells must be activated to deliver costimulatory signals to B cells and Th cells can be activated only in response to an ongoing infection.

These interactions of naïve B cells with the pathogen and with activated Th cells occur in the lymphoid organs. The pathogen and antigen-presenting dendritic cells are carried to a lymph node primarily by the lymph that drains the infected tissue. In contrast, Th cells enter lymphoid organs from the blood circulation, searching from one to another for their cognate antigen. Since activation of B cells requires prior activation of naïve Th cells, which in turn was preceded by activation of dendritic cells, the

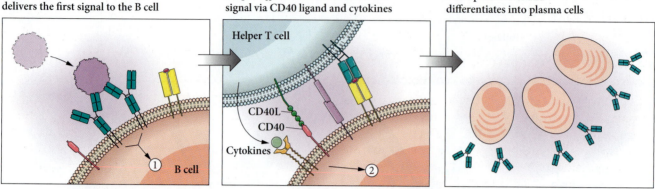

Antigen binding to B cell receptor delivers the first signal to the B cell

Helper T$_H$2 cell delivers the second signal via CD40 ligand and cytokines

B cell proliferates and differentiates into plasma cells

Helper T cell

CD40L
CD40
Cytokines

① B cell

②

Figure 4.21 B-cell activation requires two signals. The first is provided when the B-cell receptors are cross-linked by the pathogen (left panel). The second or costimulatory signal is provided by activated Th cells (center panel). In order for the B cell to receive the costimulatory signal, the B-cell receptors must first bind and internalize their cognate antigen for processing and presentation by the MHC class II peptide presentation pathway. When the T-cell receptors on the Th cell bind to the MHC class II/peptide complexes on the B cell, the Th cell responds by synthesizing the CD40 ligand, which interacts with CD40 on the surface of the B cell. This interaction transmits a signal back into the Th cell, causing it to release cytokines that, together with CD40 ligand and the signal from the B-cell receptor, cause B cells to proliferate and mature into antibody-secreting plasma cells (right panel). Adapted from P. Parham, *The Immune System*, 2nd ed. (Garland Science, New York, NY, 2005), with permission.

onset of antibody production is delayed until a week or so after infection.

Here now is another of those irksome exceptions to our neat generalizations. Experiments with T-cell-deficient mice show that some pathogens can activate B cells to produce IgM and IgG antibodies without T-cell help. Antigens on these T-cell-independent pathogens are generally arranged in a very organized, highly repetitive fashion, which enables them to very extensively cross-link the B cell receptor and thereby to deliver a strong activating signal to B cells, which can suffice to activate them. For example, some bacteria have highly organized repetitive polysaccharides on their surfaces, which extensively cross-link B-cell receptors, thereby activating B cells without T-cell help.

Experiments involving the mouse polyomavirus provide an example of virus-induced T-cell-independent activation of B cells. T-cell-deficient mice were infected with live polyomavirus, or they were immunized with empty virus-like particles (VLPs) that were assembled from pentamers of the major polyomavirus capsid protein, VP1 (which was generated in recombinant insect cells). These VLPs have the same antigenic structure as live virus. Other T-cell-deficient mice were immunized with an equal amount of soluble VP1. In each instance, IgM production was elicited in the T-cell-deficient mice. However, live virus and the VLPs elicited a 10-fold-stronger response than soluble VP1 molecules. These experimental results show that when the viral antigens are present in a highly organized, repetitive capsid structure, they more efficiently induce B cells to produce IgM, such that T-cell help may

not be necessary. However, note that only the live virus infection caused isotype switching to IgG production. Perhaps live virus infection causes signals to be generated by the innate immune system that might trigger the class switch. However, T-cell costimulatory signals may yet be required for further antibody class switching and for somatic hypermutation.

Does T-cell-independent activation of B cells play a role in natural virus infections in an immunocompetent host? The answer is not known with certainty. However, one might suspect that the answer is yes, in at least some instances, since T-cell-independent activation of B cells could occur before T cells might be activated. Thus, T-cell-independent activation of B cells could jump-start antibody production, thereby helping to bring about earlier control of the infection.

Activation of CTLs

The activation of naïve CD8 T cells to effector CTLs, like the activation of Th cells and B cells, requires two signals. Reviewing the activation of Th cells, we saw that Th cells are activated when a dendritic cell delivers (i) a general signal via B7, which is induced on the dendritic cell by cytokines and antigens at the infected site; and (ii) a specific signal via MHC class II molecules, which present antigen from the invading pathogen. In the case of B-cell activation, one of the activating signals is triggered when the B-cell receptors are cross-linked by their cognate antigen. The second or costimulatory signal is triggered by a Th cell that recognizes its cognate antigen, as presented by the B cell in the context of an MHC class II molecule. The activation of naïve CD8 T cells may result from several variations on this two-signal scheme. However, in each case, activation requires the participation of an activated dendritic cell.

Why the particular requirement for dendritic cells to activate a naïve CD8 T cell? The answer is in keeping with an earlier maxim stating that the large energy outlay necessary to generate CTLs and ensure their potency necessitates ensuring that CTLs be activated only in response to an actual infection. Apropos that premise, recall that dendritic cells can become antigen presenters only in response to inflammatory cytokines and antigens at the infected site. Moreover, only dendritic cells, which are the most potent of the professional antigen-presenting cells, can provide sufficient stimulation, via the several different pathways described below, to activate a naïve CD8 T cell to become a CTL.

Have you thought about how a dendritic cell might deliver an antigen-specific signal to a naïve CD8 T cell? This is not a trivial matter. A dendritic cell, as a professional antigen-presenting cell, might be expected to present *exogenous* peptides to the naïve CD8 T cell via its MHC class II peptide presentation pathway. But do you see a problem here? Recall that CD8 T cells specifically recognize peptides presented on MHC class I molecules. So, what might be the way out from this conundrum? The answer is that dendritic cells have the special ability to present *exogenous* peptides on both MHC class I and MHC class II molecules, another feature of dendritic cells making them the most important of the professional antigen-presenting cells. Thus, naïve CD8 T cells can recognize their cognate antigen when presented by the dendritic cells' MHC class I peptide presentation pathway.

The pathway by which dendritic cells present *exogenous* antigenic peptides to naïve CD8 T cells via MHC class I molecules is known as **cross-presentation**. It is not yet well understood, but there is experimental evidence which suggests that some internalized proteins (in particular, those proteins that have formed immune complexes with antibodies) can somehow escape from the endosome/lysosome compartment into the cytosol, where they are exposed to the ubiquitin/proteasome pathway, followed by transport of resulting peptides to the ER and presentation by the MHC class I peptide presentation pathway. (What might be the consequences if other types of cells were able to present *exogenous* antigen via MHC class I molecules?)

As noted above, the activation of CTLs, like that of other immune effectors, requires two signals. Consequently, the most straightforward pathway by which a dendritic cell can activate a naïve CD8 T cell is for the dendritic cell to transmit both of these signals to the CD8 T cell.

One signal is the antigen-specific activating signal, transmitted via the interaction between the MHC class I/peptide complex on the dendritic cell and the T-cell receptor on the naïve CD8 T cell. The other is a nonspecific costimulatory signal, transmitted by the interaction between B7 on the dendritic cell and CD28 on the naïve CD8 T cell (Fig. 4.22, left panel). However, the dendritic cell will be ready to transmit a sufficiently strong costimulatory signal only if it has been activated sufficiently at the infected site to express adequate levels of B7 to do so. The now-activated CTL then produces IL-2, driving its own proliferation. Notice that this means by which a dendritic cell activates a CTL is much the same as the way by which it activates a Th cell (see above).

Even if the dendritic cell can offer only suboptimal costimulation via B7, it still can activate a naïve CD8 T cell, but with the help of a Th cell. This can happen whether or not the CD4 T cell is already an activated Th cell. If the CD4 T cell is already an activated Th cell, then, upon binding to the dendritic cell, it induces the dendritic cell to increase its level of costimulators to levels that are sufficient to activate the CD8 T cell (Fig. 4.22, center panel). Once again, the activated CTL produces IL-2, driving its own proliferation. Note that the Th cell recognizes MHC class II/peptide complexes on the dendritic cell, while the CTL and its naïve CD8 T cell precursor recognize MHC class I/peptide complexes on the dendritic cell.

Even if the CD4 T cell is still in its naïve state, it still might be able to help activate the CTL. That can happen if the naïve CD8 T cell and the naïve CD4 T cell should simultaneously recognize the same antigen-presenting dendritic cell (Fig. 4.22, right panel). The naïve CD4 T cell can then be activated by the dendritic cell to produce IL-2, which in turn promotes the activation and proliferation of the naïve CD8 T cell to generate a clone of mature CTLs.

Mechanism of Action of CTLs

Once a naïve CD8 T cell has been activated and undergone clonal expansion, the resulting mature CTLs leave the lymph node and enter the circulation in search of infected target cells to attack and destroy. Circulating T cells are directed to infected sites by particular chemokines, i.e., small proteins that guide lymphocytes to sites of infection. These include macrophage inflammatory protein 1α (MIP-1α; HIV uses the MIP-1α receptor, CCR5, for a coreceptor [Chapter 21]) and IFN-γ-inducible protein-10, which are induced by IFN-γ in a variety of cell types at the infected site. When a CTL is thus directed to an infected site, it scans the tissue for cells displaying MHC class I/peptide complexes that its T-cell receptor might specifically recognize. Upon encountering such a target cell, the CTL then initiates the steps leading to its destruction.

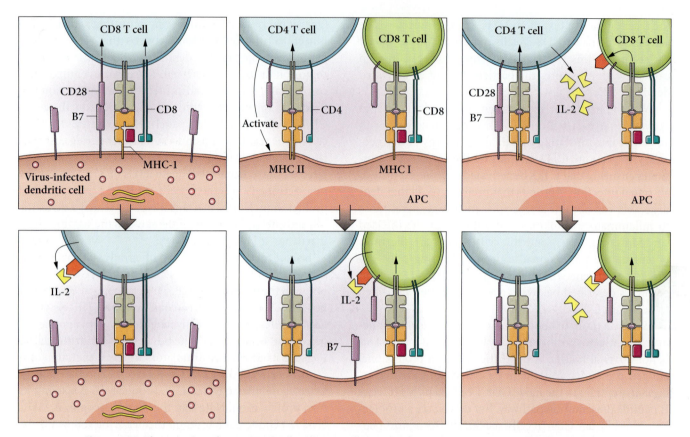

Figure 4.22 The activation of naïve CD8 T cells to become effector CTLs requires two signals. (Left panels) A fully activated dendritic cell can transmit both of these signals: (i) an antigen-specific activating signal transmitted by the interaction between the MHC class I/peptide complex on the dendritic cell and the T-cell receptor on the naïve CD8 T cell and (ii) a nonspecific costimulatory signal transmitted by the interaction between B7 on the dendritic cell and CD28 on the naïve CD8 T cell. However, this can occur only if the dendritic cell expresses levels of B7 that are sufficient to transmit a strong enough costimulatory signal. (Center panels) Even if the dendritic cell can offer only suboptimal costimulation via B7, it still can activate a naïve CD8 T cell, but with the help of a CD4 T cell. If the CD4 T cell already is an activated Th cell, then, upon binding to the dendritic cell, it induces the dendritic cell to increase its level of costimulators to levels that are sufficient to activate the naïve CD8 T cell. For example, note the upregulation of the B7 costimulatory molecule. (Right panels) Even if the CD4 T cell is still in its naïve state, it still might be able to help activate the naïve CD8 T cell, if the naïve CD8 T cell and the naïve CD4 T cell should simultaneously recognize the same antigen-presenting dendritic cell. The naïve CD4 T cell can then be activated by the dendritic cell to produce IL-2, which in turn promotes the activation and proliferation of the naïve CD8 T cell to generate a clone of mature CTLs. Adapted from P. Parham, *The Immune System*, 2nd ed. (Garland Science, New York, NY, 2005), with permission.

The CTL kills the target cell by inducing it to undergo programmed cell death, also referred to as apoptosis (see above). The apoptotic process is characterized by the fragmentation of cellular DNA, breakup of the nucleus into smaller bodies, and finally, the shrinkage and breakup of the cell into membrane-enclosed vesicles called apoptotic bodies (Fig. 4.23).

As noted earlier, apoptosis is a normal process in the host, in which it serves several important functions not necessarily related to immune defenses. Regardless, the CTL-mediated destruction of virus-infected target cells by apoptosis rather than by **necrosis** (the type of cell death resulting from acute cellular injury or trauma) has several

important advantages for the host. Necrotic cells eventually suffer loss of membrane integrity and thus release potentially damaging cellular contents into the tissues. Were a necrotic cell infected with a virus, infectious virus particles would very well be released as well. In contrast, the apoptotic process rapidly degrades viral nucleic acids, resulting in the rapid cessation of virus replication in the cell. Moreover, since the apoptotic cell does not lyse, no intracellular virus that might be present is released. In addition, no potentially damaging cellular constituents are released into the tissues. Also, the apoptotic bodies are readily phagocytosed by tissue macrophages and cleared from the infected site. Finally, and importantly, apoptotic

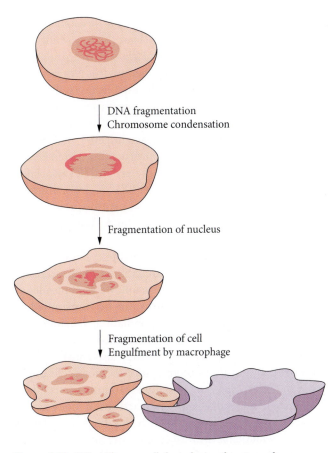

↓ DNA fragmentation
Chromosome condensation

↓ Fragmentation of nucleus

↓ Fragmentation of cell
Engulfment by macrophage

Figure 4.23 CTLs kill target cells by inducing them to undergo programmed cell death, also referred to as apoptosis. The apoptotic process is characterized by several distinct stages, beginning with the fragmentation and compressing of cellular DNA, which is seen in distinct masses at the nuclear envelope. The cytoplasm then condenses, and the nucleus breaks into smaller bodies. In the end, the cell breaks up into membrane-enclosed vesicles called apoptotic bodies, which may be engulfed and destroyed by macrophages. Adapted from G. M. Cooper and R. E. Hausman, *The Cell: a Molecular Approach*, 4th ed. (ASM Press, Washington, DC, 2007), with permission.

bodies also are taken up by dendritic cells, which then process and present the antigen within them to T cells. If apoptosis is blocked in infected cells, then antigen acquisition by dendritic cells is impaired, and consequently, antigen presentation to T cells is impaired as well.

CTLs can induce apoptosis by two distinct mechanisms. Actually, these mechanisms are the same as those used by NK cells (see above). The first of these mechanisms, referred to as the **granule exocytosis pathway**, is dependent on the protein perforin and an enzyme called **granzyme B**. Perforin is a close relative of complement protein C9, with which it shares the ability to poke holes in cell membranes. These proteins are stored in granules within the activated CTL. When the CTL recognizes and binds to a target, the granules migrate to the CTL's plasma

membrane, and the granule's contents are released onto the target cell. The free perforin then may assemble membrane attack complexes on the target cell membrane, thereby creating pores that enable the granzyme to enter the target cell, where it induces apoptosis. (Other descriptions of this process have the target cell endocytose the perforin and granzyme that were released by the CTL. Within the target cell, the perforin molecules then create pores in the membranes of the endocytic vesicles. This allows the granzyme B to enter the cytoplasm, where it induces the sequence of reactions leading to apoptosis.)

The second mechanism for inducing apoptosis is the well-characterized pathway involving the cell surface proteins Fas and FasL. FasL, on the surface of the CTL, binds to Fas on the surface of the target cell, thereby transmitting a signal into the target cell that triggers apoptosis.

Why have two mechanisms by which CTLs can induce apoptosis? Here are several possible reasons. First, not all cells express Fas, so the Fas-mediated process cannot suffice to kill all virus-infected target cells. Second, as demonstrated by experiments in which perforin and Fas knockout mice were infected with influenza virus, optimal clearance of the virus occurs when both perforin and Fas can act. Thus, the two mechanisms have an additive effect, at least in some cases. Third, the perforin and Fas pathways generally serve different primary purposes. The perforin pathway is a lot quicker than the Fas pathway. Perhaps that is why it is the major pathway for killing virus-infected cells. In contrast, the Fas pathway is the primary means for disposing of self-reactive T cells during negative selection in the thymus (see "Self Tolerance," below). This distinction raises the following thought. Once an infection has been cleared, it is important to eliminate the expanding population of activated, cytokine-secreting T cells. Thus, just as apoptosis, as mediated by the Fas pathway, is the primary means for disposing of self-reactive T cells during negative selection in the thymus, apoptosis induced by the Fas-FasL interaction between lymphocytes is the main means for disposing of unwanted lymphocytes after an infection has been resolved. Individuals with a mutant *FAS* gene have a condition known as **autoimmune lymphoproliferative syndrome**. They cannot remove autoreactive T cells, nor can they control the size of their lymphocyte population.

T Cells and Antiviral Cytokines

Despite the fact that Th cells and CTLs have crucially important roles in helping to activate B cells and CTLs and in killing virus-infected cells, these cells have yet other vital functions to perform. In particular, when both CD4 and CD8 T cells are activated, they produce large quantities of antiviral cytokines, including IFN-γ and TNF-α. Production of these cytokines by activated T cells very likely contributes to the resolution of at least some virus infections. It is probably important in this regard that IFN-γ up-regulates expression of MHC molecules and other antiviral proteins, including PKR, 2′-5′-oligo(A) synthetase, ADAR, and NO synthetase (see above). Also, since T-cell expression of IFN-γ and TNF-α is dependent on contact with antigen, production of these cytokines and their potentially damaging impact are thereby restricted to the infected site.

Another important feature of cytokines such as IFN-γ and TNF-α is that their antiviral effects can occur in the absence of target cell destruction. This may be particularly important in instances where vital organs might be infected and excess tissue damage must be avoided. For example, IFN-γ and TNF-α, as expressed by CTLs, were shown to inactivate HBV within hepatocytes by a noncytolytic process.

So, a message to take away is that T cells act to resolve virus infections by means of the cytokines they secrete as well as by their cytolytic mechanisms. The effectiveness of these respective defensive stratagems depends, as you might have supposed, on the particular virus and the particular tissues that are infected. Also, as you might expect, the optimum immune response in most instances requires the combined action of both cytokines and cytolytic mechanisms.

Viral Evasion of Cell-Mediated Immunity

I hope that I have succeeded in conveying a sense that we possess cell-mediated immune defenses of astonishing sophistication and potency. Yet the same evolutionary forces responsible for these immune defenses also have been at work in the service of viruses. Moreover, as we noted above, viruses are able to evolve much faster than their hosts. The results of virus evolution are perhaps nowhere more impressive than in the range of countermeasures, discussed below, which viruses have adapted to cope with cell-mediated immune defenses. Indeed, virtually every stage of the MHC class I and MHC class II peptide presentation pathways is impeded by one or another virus. (It is historically interesting that many of the steps of antigen presentation via MHC molecules actually were discovered in the course of investigating virus mechanisms for evading adaptive immunity.) In addition, viruses also have evolved measures that obstruct apoptosis as a host antiviral defense. Below, we consider but a few examples to acquire a further appreciation of the seemingly unlimited resourcefulness of viruses to find ways to evade host defenses.

Inhibition of Antigen Presentation to CTLs

A rather direct strategy for avoiding antigen presentation via the MHC class I peptide presentation pathway is for the virus to generate proteins that are resistant to the ubiquitin/proteasome pathway. (Had this possibility occurred to you?) This is a particularly effective strategy in the case of EBV, a member of the herpesvirus family which, like other herpesviruses, generally establishes latent infections. While in the latent state, herpesviruses express very few proteins. (Note that this is in itself a CTL evasion strategy.) In fact, the EBV nuclear antigen-1 may be the only viral protein expressed in some latently infected B cells, and it is the only viral protein expressed in EBV-associated malignancies (see Chapter 18). Yet EBV nuclear antigen-1 is not presented by the MHC class I peptide presentation pathway because it contains an amino acid sequence that inhibits the host proteasome.

Latency is a central feature of the life cycles of all herpesviruses. Yet for latency to contribute to the success of these viruses in Nature, it must be followed by episodes of reactivation so that these viruses can be transmitted to susceptible individuals in the population. During these reactivation episodes, herpesviruses need to express the full range of their proteins, and they also must contend with primed immune systems, since memory cells are in place in latently infected individuals to jump-start adaptive immune responses when reactivation does occur. Therefore, as we see immediately below, the herpesviruses display a remarkable array of countermeasures against cell-mediated immunity.

Several herpesviruses block TAP-mediated peptide transport into the ER. One such example is provided by an HSV protein, ICP47, which binds to the cytoplasmic side of the Tap transporter (Fig. 4.24). Another herpesvirus, HCMV, encodes a protein, U_S6, which blocks the Tap transporter from the luminal side of the ER membrane.

The herpesviruses, as well as members of other virus families, generate proteins that either directly or indirectly inhibit the ability of MHC class I molecules to present antigenic peptides. For example, the HCMV U_S3 protein and the adenovirus E3 19K protein each physically associate with peptide-loaded MHC class I molecules in the ER, thereby preventing their transport to the cell surface. Another stratagem is displayed by the HCMV U_S11 and U_S2 proteins, which cause MHC class I proteins to be shed from the ER into the cytoplasm, where they are then degraded. (The interactions of U_S11 and U_S2 with MHC class I molecules in the ER apparently target the MHC class I molecules to the normal cellular pathway for the degradation of misfolded proteins. That is, U_S11- and U_S2-bound MHC class I molecules are removed from the ER via Sec61 pores and degraded by cytosolic proteasomes.)

If you have been counting, you may have noticed that HCMV encodes at least four inhibitors of the MHC class I peptide presentation pathway: U_S2, U_S3, U_S6, and U_S11.

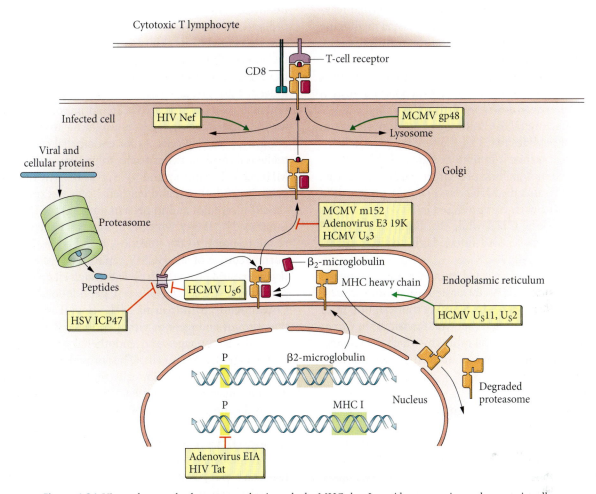

Figure 4.24 Viruses have evolved stratagems that impede the MHC class I peptide presentation pathway at virtually every stage, including the processing of viral proteins in the proteasome and the transport of the resulting peptides to the ER, the transcription of MHC class I proteins, and their transport to, and retention at, the plasma membrane. Specific viral gene products are highlighted in the yellow boxes. Steps in the antigen presentation pathway that are blocked by particular viral gene products are indicated by blunt-ended red bars, whereas steps that they stimulate are indicated by green arrows. Adapted from S. J. Flint, L. W. Enquist, R. M. Krug, V. R. Racaniello, and A. M. Skalka, *Principles of Virology: Molecular Biology, Pathogenesis, and Control,* 2nd ed. (ASM Press, Washington, DC, 2003), with permission.

One might ask why such apparent redundancy of function is necessary. A straightforward answer is that the multiple gene products likely act additively to impair MHC class I function. Also, individual viral proteins might work better against different MHC class I isoforms. Regardless, the evolution of these multiple virus countermeasures against MHC class I-mediated defenses strongly implies that these defenses would otherwise be a singularly important obstacle to these viruses.

The mouse cytomegalovirus (MCMV) likewise encodes multiple proteins that block the MHC class I peptide presentation pathway. One of these, gp48, binds to MHC class I complexes and redirects them to the endocytic route for rapid proteolytic destruction in lysosomes. Another

MCMV protein, m152, causes MHC class I/peptide complexes to be retained in the ER-Golgi intermediate compartment. In addition, m152 is thus far unique in that it also interferes with the expression of the cellular ligand for the NK cell receptor, whose engagement provides activating signals to NK cells. (We noted above that HCMV, the human cytomegalovirus, also resists attack by NK cells, in that instance by expressing an MHC class I homolog that acts as a decoy for NK cell inhibitory receptors.)

HIV expresses two proteins that ultimately cause the down-regulation of MHC class I molecules at the cell surface. One of these proteins, Vpu, accomplishes this by destabilizing newly synthesized MHC class I molecules in the ER, thereby triggering their destruction by proteasomes.

The other HIV protein, Nef, interrupts the normal recycling of MHC class I α chains from the plasma membrane back to the ER and instead diverts them to lysosomes, where they are degraded.

The HIV Nef protein has another effect, which provides one of many examples of how diabolical a virus is HIV. Nef selectively down-regulates *HLA-A* and *HLA-B*, while not affecting *HLA-C*. Although HLA-C proteins can present viral peptides to CTLs, they are the predominant molecules that transmit inhibitory signals to NK cells, in that way preventing killing of a target cell by an NK cell. The net result is that the MHC class I peptide-presentation pathway is sufficiently compromised in HIV-infected cells to impair attack by CTLs, while the unimpaired expression of *HLA-C* enables HIV-infected cells to also resist attack by NK cells.

In the last of these selected examples, the adenovirus E1A protein and the HIV Tat protein each block expression of MHC class I genes by blocking transcription from the MHC class I promoter. (See "Tat Revisited" in Chapter 21 for more of the several means by which the remarkable HIV Tat protein impairs antigen presentation by HIV-infected cells.)

Inhibition of Antigen Presentation to Helper T Cells

Considering that the recognition of MHC class II/peptide complexes by Th cells is essential for the activation of both antibody and cell-mediated antiviral immune responses, you might expect that some viruses have evolved ways to impair the MHC class II peptide presentation pathway. Also, since as many as one-half of the 80 or more herpesvirus genes play a role in evading host immune defenses (Chapter 18), you might expect that some of the herpesvirus immune evasion genes might serve to impede the MHC class II peptide presentation pathway. And indeed, that is the case. For example, the HCMV U_S2 protein, which causes MHC class I proteins to be shed from the ER into the cytoplasm to be degraded, likewise targets MHC class II α chains for degradation. The HCMV U_S3 protein, which retains MHC class I proteins in the ER, also destabilizes MHC class II molecules, perhaps via an effect on Ii.

On the other hand, the MHC class II antigen presentation pathway occurs primarily in professional antigen-presenting cells. Thus, considering that the HCMV U_S2 and U_S3 proteins are expressed only within cells that are productively infected, and since HCMV would not be able to infect more than a very small minority of MHC class II-expressing cells, what actual advantage might HCMV derive from its U_S2- and U_S3-based mechanisms for impairing the MHC class II peptide presentation pathway? Indeed, the vast majority of these cells would remain unaffected and thus be able to present exogenous HCMV proteins via the MHC class II pathway. How then is the virus benefited by its ability to shut down antigen presentation via MHC class II molecules during productive infection? The answer to this question is as follows. HCMV indeed does productively infect macrophages and endothelial cells, and these cells are an important reservoir for the virus. Macrophages, as we know, are professional antigen-presenting cells. As for endothelial cells, they actually acquire a limited capacity to present antigen via the MHC class II pathway and to express B7. The limited capacity of endothelial and epithelial cells to express MHC class II molecules and to present antigen to CD4 Th cells is induced by IFN-γ, which is produced by the innate immune response to the infection. In addition, during the normal course of the HCMV replication cycle, some *endogenous* viral proteins do cycle through endosomal compartments where MHC class II peptide loading occurs. Consequently, by impairing MHC class II-mediated presentation of endogenous viral peptides, HCMV prevents infected macrophages and endothelial cells from being detected by CD4 Th cells. This provides an important advantage, since CD4 Th cells that recognize HCMV-infected macrophages and endothelial cells can respond by producing antiviral cytokines. Moreover, a subset of CD4 T cells actually have CTL function (Box 4.12).

Since (i) the ability of endothelial cells to express MHC class II molecules and present antigen to CD4 Th cells is induced by IFN-γ, and (ii) IFN-γ transmits signals via the JAK/STAT signaling pathway, it would be advantageous to a virus to be able to short-circuit the JAK/STAT pathway. Members of the herpesvirus and adenovirus families and other viruses indeed inhibit the JAK/STAT pathway, thereby blocking the effects of IFN-γ, as well as of IFN-α, on infected cells. These viral countermeasures can be particularly important when endothelial and epithelial cells are infected, since MHC class II expression can be up-regulated in these cells from low basal levels by cytokines produced at the infected site.

Inhibition of Apoptosis

Apoptosis is triggered by several effectors of both the innate and adaptive immune responses, including cytotoxic cytokines such as TNF-γ, IFN-α, and IFN-β, as well as by NK cells and CTLs. Viruses have countered the threat of premature apoptotic host cell death by discovering how to obstruct virtually every step in the pathways by which the immune system triggers apoptosis (Fig. 4.25).

Several viruses (e.g., vaccinia virus, a poxvirus) generate cytokine homologs, which interact with cytokine receptors. In so doing, they compete with the normal cytokines for receptor binding, while not triggering an apoptotic

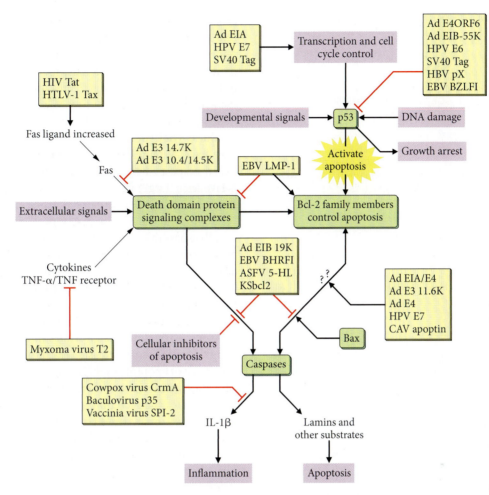

Figure 4.25 Viruses have countered the threat of premature apoptotic host cell death by discovering how to obstruct virtually every step in the pathways by which apoptosis might be triggered either by the immune system or by the cellular response to the infection. There are two cellular pathways to apoptosis. First, there is the extrinsic pathway that is triggered when a proapoptotic ligand (e.g., TNF-α or FasL) binds to its receptor (e.g., TNF-α receptor or Fas) at the cell surface. The extrinsic pathway is also activated when CTLs or NK cells release perforins and granzyme proteins. Second, there is the intrinsic pathway that is triggered in response to internal cues, such as DNA damage or inappropriate cell divisions. Each pathway is mediated by a convergent caspase cascade. In the intrinsic pathway, p53 is activated and goes to the nucleus, and Bax goes to the mitochondria to trigger activation of the caspase cascade. Bcl-2 family members negatively regulate apoptosis. Examples of viral proteins that affect particular stages in the activation and regulation of apoptosis are listed in the yellow boxes. Steps in the apoptosis pathways that are blocked by particular viral gene products are indicated by blunt-ended red bars, whereas steps that they stimulate are indicated by arrows. Ad, adenovirus; HPV, human papillomavirus; SV40, simian virus 40; CAV, chicken anemia virus; ASFV, African swine fever virus; KS, Kaposi's sarcoma herpesvirus (human herpesvirus 8); HTLV-1, human T-cell leukemia virus type 1. Adapted from S. J. Flint, L. W. Enquist, R. M. Krug, V. R. Racaniello, and A. M. Skalka, *Principles of Virology: Molecular Biology, Pathogenesis, and Control*, 2nd ed. (ASM Press, Washington, DC, 2003), with permission.

signal. Another strategy used by some viruses is to produce secreted cytokine-binding proteins (i.e., soluble decoy receptors), which bind to the cytokines in the extracellular fluid, thereby preventing the cytokines from binding to their cellular receptors. Yet another strategy used by some poxviruses and herpesviruses (e.g., HCMV)

is to produce integral membrane cytokine receptor homologs (viroreceptors), which compete with the cellular receptor for cytokine binding at the plasma membrane but are incapable of transmitting a signal. The retroviruses HIV and human T-cell leukemia virus type 1 employ yet another rather remarkable variation of this theme.

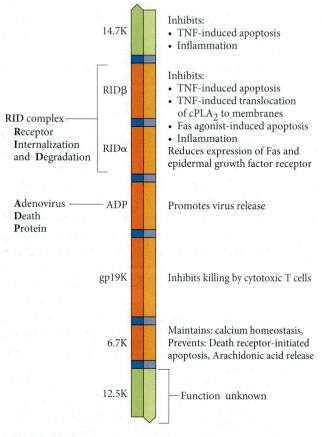

	Inhibits:
14.7K	• TNF-induced apoptosis • Inflammation

RID complex — Receptor Internalization and Degradation

RIDβ — Inhibits:
• TNF-induced apoptosis
• TNF-induced translocation of cPLA$_2$ to membranes
• Fas agonist-induced apoptosis
• Inflammation

RIDα — Reduces expression of Fas and epidermal growth factor receptor

Adenovirus Death Protein — **ADP** — Promotes virus release

gp19K — Inhibits killing by cytotoxic T cells

6.7K — Maintains: calcium homeostasis, Prevents: Death receptor-initiated apoptosis, Arachidonic acid release

12.5K — Function unknown

Figure 4.26 The adenovirus E3 region contains seven genes that have functions related to immune evasion. Glycoprotein 19K prevents MHC class I proteins from presenting viral antigens on the surface of the infected cell (Fig. 4.24). RID is an acronym for "receptor internalization and degradation." The RIDα/β complex down-regulates expression of the TNF receptor and Fas at the plasma membrane. RID is thought to associate with the TNF receptor at the plasma membrane and with FAS in a cytoplasmic compartment. Protein 14.7K has been found to interact with several host proteins, and the mechanism by which it prevents apoptosis is not yet clear. In Fig. 4.25, 14.7K is shown blocking the Fas-mediated death signal. Recent studies show that 14.7K also inhibits ligand-induced internalization of the TNF receptor. Adapted from W. S. M. Wold and A. E. Tollefson, *Semin. Virol.* **8:**515–523, 1998, with permission.

They induce the up-regulation of FasL in the infected cell, so that the attacking CTL rather than the infected target cell might be on the receiving end of the apoptotic signal.

The adenoviruses provide an example of still another strategy to block apoptosis. Proteins encoded by their E3 region induce the endocytosis and lysosomal degradation of the TNF-α receptor and Fas, in that way inhibiting TNF-α- and Fas-mediated apoptosis. Indeed, the adenovirus E3 region contains seven genes that have functions related to immune evasion (Fig. 4.26).

Still other viruses impair signal transduction downstream of the cytokine receptors. Major cellular targets of

this virus stratagem include the **tumor suppressor protein** p53, the caspases, and the apoptosis regulatory proteins Bcl-2 and Bax (Fig. 4.25). p53 normally is activated by a variety of cellular signals, including DNA damage. Activation of p53 in turn activates the **caspases**, which are the actual effectors of apoptosis. The caspases are a family of **cysteine proteases** that cleave proteins after asparagine residues. They are known as cysteine proteases because they have cysteine residues at their active sites. They are called caspases because they are cysteine proteases that cleave their target proteins after an aspartic acid residue (the term "caspase" is derived from the first letter of "cysteine" and the first three letters of "aspartic").

All caspases are synthesized as inactive precursors, which are converted to their active form by proteolytic cleavage, as catalyzed by other caspases. Caspase-1 and caspase 8 are **initiator caspases.** That is, they activate downstream **effector caspases** (e.g., caspases 3, 6, and 7) by cleaving their inactive proforms. The activated effector caspases then cleave as many as 100 different substrates in the cell, resulting in apoptosis. Thus, some viruses (e.g., poxviruses) may block activation of apoptosis by blocking the caspase-1-mediated activation pathway. Moreover, since expression of caspase-1 also activates the secretion of proinflammatory cytokines (see above), some viruses (e.g., again, the poxviruses) also evade cytokines by blocking caspase-1 activity.

Bcl-2 and Bax serve to regulate apoptosis; Bcl-2 is a negative regulator, and Bax is a positive regulator. The life or death fate of the cell is determined by the balance of these proapoptotic and antiapoptotic factors. There are examples of viruses that can act against each of these host factors to obstruct apoptosis. For example, several members of the herpesvirus family express Bcl-2 homologs to negatively regulate apoptosis (Fig. 4.25). The adenovirus E1B-19K protein is perhaps more remarkable still. It not only up-regulates transcription of host Bcl-2 to negatively regulate apoptosis but also binds to Bax. The latter binding activity prevents Bax from activating the caspase cascade, a rather diabolical dual activity.

Before moving on from the general topic of viral evasion of host immune defenses, there is a rather extraordinary instance of immune evasion, in which a parasite (a lepidopteran insect) uses virus-derived genes to transfer genes to its host that undermine its host's immune defenses. We also note that mammals, too, use genes derived from viruses, and when inappropriately expressed, those genes may lead to serious human disease. See Box 4.16 for details.

IMMUNOLOGICAL MEMORY

Specificity and memory are the two hallmarks of the adaptive immune system. Memory enables the adaptive immune system to respond to a reencounter with a pathogen more quickly and potently than it responded to that pathogen during the primary infection. This rapid and potent response may greatly lessen the severity of the secondary infection and perhaps even completely prevent disease in that instance. Consequently, immunological memory is the basis for adaptive immunity.

Recall that during a primary infection, antibodies defend against free extracellular viruses, whereas T cells are on the lookout for infected cells. The humoral and cellular branches of the adaptive response likewise play the same roles during a second exposure to a pathogen; but, as just noted, their respective responses are much faster and more potent. Indeed, these memory responses are so potent that in some instances either branch of adaptive immunity alone can completely control an infection. This can be demonstrated for the humoral response by passive antibody transfer (i.e., inoculation of a naïve subject with serum from an immune individual). It can be demonstrated for the cell-mediated response by the transfer of CD8 T cells from an individual vaccinated with an antigen that contains only T-cell epitopes. However, the highest levels of protective immunity occur when the humoral and cell-mediated responses act together.

When considering immunity, it is very important to make the distinction between infection and disease. Disease symptoms usually do not emerge until an infection is well under way. Often, the virus must spread from primary to secondary sites of infection before symptoms might surface. With the distinction between infection and disease in mind, we might ask whether adaptive immunity can actually prevent infection or instead merely slow the infection so that it might be contained or eliminated before the onset of disease. To the best of my knowledge, no vaccine or natural immunity has ever been shown to prevent infection per se. More likely, protective immunity provides a quick and powerful response that brings the infection under control before symptoms might arise. (As we will see later [Chapter 21], the likely inability of protective immunity to block infection per se is one reason why some acquired immunodeficiency syndrome [AIDS] researchers are not optimistic that an effective vaccine against HIV might be developed. In brief, HIV establishes a chronic infection whereby 10 years pass by, on average, before the onset of AIDS. Yet during this entire time, HIV-infected individuals mount a strong adaptive immune response against the virus but cannot clear it. Thus, unless an AIDS vaccine is able to prevent the initial HIV infection, it may not be expected to prevent the onset of disease.)

Next, we consider the nature of B-cell and T-cell memory, beginning with the former. When naïve B cells are activated, a population of memory B cells, as well as plasma B cells, is generated. Since these memory B cells have undergone clonal expansion, there are many more of them than of those which constituted the original set of naïve B cells with the same specificity. Moreover, these memory B cells have already undergone affinity maturation and class switching. Thus, memory B cells can respond to a second exposure to the same pathogen quickly and potently.

But there is more to humoral immunity than memory B cells alone. Importantly, a subset of the plasma B cells are long-lived and constitutively produce antibody. These plasma B cells are terminally differentiated. That is, they undergo no further affinity maturation or class switching. The antibodies that they continually produce constitute a first line of defense when reexposed to a pathogen.

The long-lived plasma B cells are activated in the lymphoid organs, but they eventually migrate to the bone marrow and then reside there. In contrast, the memory B cells continue to reside in the secondary lymphoid organs, a site where they might quickly be mobilized when needed. Unlike plasma B cells, memory B cells do not constitutively secrete antibodies. But they respond to reinfection by quickly dividing to generate new antibody-secreting plasma cells and a new population of memory B cells. Since the initial memory B cells had already undergone affinity maturation and class switching, the new plasma cells that they generate immediately secrete effective high-affinity antibodies.

In contrast to B-cell memory, there is no subpopulation of T cells that remain active after the primary infection has been resolved. Instead, a population of immunologically quiet **memory T cells** persists in greater numbers than the original population of naïve T cells of the same specificity. Moreover, memory T cells respond to a second encounter with a pathogen more quickly than naïve T cells. The combination of increased numbers and faster response times is the basis for T-cell protective immunity. A possible reason why there is no T-cell equivalent to the long-lived plasma B cells is that T cells are efficient cytokine producers and continued production of cytokines in the absence of infection can do serious harm. Yet, some memory T cells, although not vigorously secreting cytokines, are sufficiently ready to immediately kill targets.

SELF TOLERANCE

We have repeatedly emphasized that activating the individual effectors of adaptive immunity generally requires two or more signals that are triggered by the infection. A reason why multiple signals are required to activate these

BOX 4.16

Here is a recently described and rather bizarre example in which a parasitic eukaryote actually makes use of virus-derived genes to undermine the immune defenses of its host. The parasites in this instance are species of wasps that inject their eggs into caterpillars, where the wasp larvae then develop. But, along with their eggs, the female wasps also inject virus-like particles, called polydnaviruses, which carry wasp genes that undermine the caterpillar's immune defenses, thereby enabling the wasp larvae to grow. Production of these virus-like particles occurs exclusively in the female wasp's ovaries.

New research shows that the genes encoding the polydnavirus coats were acquired from insect viruses known as nudiviruses, and were incorporated into the wasp DNA 100 million years ago. But the polydnaviruses do not actually carry the genes that encode their coats, and consequently, they do not replicate per se in either the wasp or the caterpillar. Rather, the polydnavirus particles carry only bona fide wasp genes that undermine the caterpillars' immune defenses.

Oddly, the genomes packaged within the polydnaviruses are comprised of multiple, circular, double-stranded DNA molecules. Because that DNA does not encode viral structural proteins, the relationship between the polydnaviruses and authentic viruses was not clear until now. The vertically transmitted nudivirus-derived genes are thought to have been maintained in the wasp genome because of their contribution to successful wasp parasitism.

We return now to our more familiar animal viruses, first noting that mammals were earlier known to make use of genes derived from viruses. For example, the syncytins, which are highly fusogenic membrane glycoproteins that act in the differentiation of the human placenta, are actually encoded by the envelope gene of an endogenous human retrovirus (i.e., a remnant of an ancestral germ line infection by a then-active retrovirus, which is now transmitted in the human germ line [Chapter 20]). The syncytins are specifically expressed in placental trophoblasts, in which they promote the fusion of placental trophoblasts, to generate the syncytiotrophoblast layer of the human placenta (Fig. 3.7).

Most interestingly, the endogenous retrovirus-encoded syncytins may be relevant to human neurodegenerative disease. In particular, several studies have provided evidence that syncytin mRNA levels may be higher in brain tissue from multiple sclerosis patients than in brain tissue from controls. Moreover, expressing synctin in a mouse model caused demyelination in vivo. These effects of synctin may be due to its apparent proinflammatory properties in vivo.

References

Bézier, A., M. Annaheim, J. Herbinière, et al. 2009. Polydnaviruses of braconid wasps derive from an ancestral nudivirus. *Science* 323: 926–930.

Mameli, G., V. Astone, G. Arru, S. Marconi, L. Lovato, C. Serra, S. Sotgiu, B. Bonetti, and A. Dolei. 2007. Brains and peripheral blood mononuclear cells of multiple sclerosis (MS) patients hyperexpress MS-associated retrovirus/HERV-W endogenous retrovirus, but not *Human herpesvirus 6*. *J. Gen. Virol.* 88:264–274.

---

immune effectors is that they are powerful weapons that must not be activated in the absence of infection, lest they have deleterious effects for the host.

Now recall that while MHC class I molecules present foreign antigenic peptides when an infection is under way, they also continually present self peptides, even in the absence of infection. Also, when professional antigen-presenting cells phagocytose foreign antigens and process them for presentation via the MHC class II peptide presentation pathway, they present self peptides, as well as the foreign antigens. Moreover, the mechanisms that generate the B-cell and T-cell receptors produce an assortment of receptor specificities that are capable of recognizing host peptides as well as pathogen-derived peptides. These facts raise one of the most fundamental and fascinating questions about the immune system: how do lymphocytes distinguish self from nonself? If they were not able to do so, we would all die from autoimmune disease in short order.

The answer is not known in its entirety, nor is it possible here to give more than a brief outline of what is known. But before we do that, note that serious attacks by the immune system against self, which result in autoimmune disease, indeed do occur, although rarely. Also, as we will see, autoimmune disease can be triggered by infection.

The term **tolerance** refers to the failure of the immune system to respond to an antigen, and we are said to be tolerant of antigens that we do not respond to. Our concern here is with how we acquire **self tolerance** (i.e., tolerance to self). Then, we consider how infection might trigger a breakdown of self tolerance, leading to autoimmune disease.

T cells are born in the bone marrow but mature in the thymus, and part of their maturing process is learning tolerance to self. Actually, their education in the thymus involves more than learning *not to* recognize self antigens, for in addition, they also must learn *to* recognize self MHC molecules. Recall that by recognizing antigen only when presented in the context of MHC molecules, T cells ignore free antigen and instead concentrate on infected cells. So the education of a T cell involves two important components. First, in the cortical region of the thymus, **positive selection** occurs, which chooses for survival only those T cells whose receptors interact with the individual's own MHC molecules. (Actually, these MHC molecules are loaded with self peptides at the time.) This test is administered by resident cortical epithelial cells. The T cell passes the test if its T-cell receptor transmits a positive signal, which prevents apoptosis. If a T cell does not recognize self MHC molecules at this time, it fails the test and dies by apoptosis. Killed T cells are quickly phagocytosed by thymic macrophages.

If a T cell passes the first test, it then goes to the central region of the thymus, called the medulla, where it undergoes its second test, referred to as **negative selection**, which selects against T cells that recognize the combination of self MHC molecules and self peptides. This test is administered by thymic dendritic cells. These cells are different from the dendritic cells discussed above, since they come to the thymic medulla exclusively from the bone marrow and, importantly, they generally present only self antigen. That is so (i) because they come to the thymus directly from the bone marrow and (ii) because they survive in the thymus for only a few days. If a T cell binds too well to these thymic dendritic cells, it fails the test and its T-cell receptor transmits a signal that causes it to die by apoptosis.

Before moving on, here is a brief recap of the two tests the T cell must pass in the thymus. First, to pass positive selection, the T cell must bind at least *weakly* to MHC/peptide complexes on thymic epithelial cells. Second, to pass negative selection, the T cell must not bind strongly to MHC/peptide complexes on thymic dendritic cells.

Thymic selection is so stringent that less than 1% of all immature T cells survive it to become circulating T cells. These survivors constitute a population of T cells with receptors that for the most part can recognize only foreign antigen, and only when that antigen is presented by self MHC molecules.

Although tolerance selection in the thymus is extremely stringent, it is not perfect. What then might be the consequence if a T cell expressing a TCR that recognizes a self protein were to survive thymic selection? Would self tolerance be compromised? Not necessarily. One mechanism for maintaining tolerance in the periphery results from the fact that naïve CTLs and Th cells continually circulate from lymph node to lymph node, searching for an antigen-presenting cell that might be displaying their cognate antigen. Importantly, these T cells do not leave the circulation unless they are stimulated to do so, in a lymphoid organ, by an activated professional antigen-presenting cell. Now, recall that activation of a naïve T cell requires two signals from a professional antigen-presenting cell; one signal is generated by the clustering of the T-cell receptors by MHC/peptide complexes, and the other is the costimulatory signal transmitted by B7. Professional antigen-presenting cells generally cannot present self peptide and also transmit the costimulatory signal. To do the latter, they need to be activated at an infected site by inflammatory cytokines such as IFN-γ and TNF-α, which are secreted by cells of the innate system only during an infection. Now here is an important additional part of the story. A naïve T cell does not need the costimulatory signal merely to be activated. In point of fact, if its T-cell receptors are clustered by MHC/peptide complexes but it does not receive a costimulatory signal, it is permanently inactivated or **anergized**, further ensuring that tolerance will not be broken. Note that other mechanisms also act to ensure tolerance in the periphery (Box 4.15).

B cells, like T cells, also must be educated, in their case to prevent the emergence of plasma B cells that might produce antibodies reactive against the cells and tissues of the body. B cells, like T cells, are also produced in the bone marrow. But unlike T cells, which mature in the thymus, B cells mature in the bone marrow. B cells whose B-cell receptors bind strongly to self proteins on bone marrow cells undergo apoptosis. Notice the general role played by apoptosis in the selection of self-tolerant effector cells of adaptive immunity.

There is much more to this fascinating story than can be presented in this general chapter, but here is an interesting puzzler that may have occurred to you. How is it possible for a T-cell receptor to mediate both positive and negative selection in the thymus? That is, during positive selection for MHC restriction, the T-cell receptor transmits a signal to prevent apoptosis. Yet during negative selection for tolerance, it transmits a signal that triggers apoptosis. Moreover, provided that a T cell passes both tests in the thymus, its T-cell receptor is able to transmit yet a third kind of signal, an activation signal in response to foreign antigen presented by self MHC molecules. Thus, how does the T-cell receptor transmit signals that result in three entirely different outcomes (i.e., survival, death, and activation)? The answer almost certainly involves the fact that the T cell interacts with a different type of host cell in each instance: a thymic epithelial cell during positive selection, a thymic dendritic cell during negative selection, and a professional antigen-presenting cell during activation. Chances are that each of these different cell types expresses a different set of costimulatory molecules (including secreted cytokines) and adhesion molecules that tailor the particular T-cell response. Moreover, the T cell itself may well change following each of its tests, resulting in different responses to similar signals.

THE IMMUNE SYSTEM IN DISEASE

Since the immune system is able to mount a response that is deadly to most microbial invaders, its effectors may inadvertently harm the host as well. In truth, unintended effects of host immune activities directed against the virus, rather than virus-induced cytopathic effects per se, are responsible in most instances for virus disease syndromes. This explains why many viruses that replicate in cell culture without causing any noticeable cytopathic effects are yet pathogenic in vivo.

Immunopathology

Damage to the host that results from the immune response to the infecting organism is known as **immunopathology**. We might think of it as being analogous to what is euphemistically referred to in military terminology as "collateral damage." Or, perhaps as in Chinese philosophy, in which the dual, opposite, and complementary principles of yin and yang are thought to exist in varying proportions in all things, it is a price we pay for possessing otherwise remarkably effective immune defenses that are able to resolve otherwise life-threatening virus infections. Yet it seems paradoxical and perhaps ironic that immunopathology is most often associated with those viruses that are noncytolytic rather than with those viruses that are cytolytic.

How might we explain this apparent paradox? Here is one possibility. If antibodies and CTLs were our only defenses against viruses, then cytolytic viruses might commonly cause life-threatening tissue damage before our adaptive immune responses might be set in motion against them. Thus, our innate immune system and lymphocytes have evolved to quickly respond to cytolytic viruses by immediately releasing cytokines that can be brought to bear against the infection. Importantly, cells of the innate immune system release these proinflammatory cytokines in response to the tissue damage caused by the cytolytic virus. Adaptive immune responses are almost certainly needed to completely clear the pathogen, but the innate immune system is acting to contain the infection while the adaptive immune response is working up the appropriate response. In contrast to cytolytic viruses, noncytolytic viruses, because they directly cause little, if any, tissue damage, may induce a weaker or delayed innate immune response, thereby enabling the infection to spread. Also, cytokines released at the infected site are required to activate professional antigen-presenting cells, thereby enabling them to deliver a costimulatory signal to T cells. Consequently, the weaker or delayed innate immune response against the noncytolytic virus results in a delayed adaptive immune response as well. Thus, by the time adaptive responses are brought to bear, the infection may have become more widely disseminated, resulting in potentially greater immunopathology.

Note that both the innate and adaptive immune systems can contribute to immunopathology, either separately or together. Beginning with the innate immune system, the flu-like symptoms (e.g., fever, congestion, headache, malaise, and myalgia) associated with many viral infections are caused by the inflammatory response to the infection, as mediated by the effectors of the innate immune system. In the case of actual influenza virus infections, the virus initially targets the ciliated epithelial cells that line the upper airways. As a consequence, inflammatory cells (initially neutrophils and later macrophages, lymphocytes, and plasma cells) accumulate in the airways, thereby impeding respiration and, for reasons that are not entirely clear, actually predisposing the host to potentially dangerous secondary bacterial infections. (Perhaps the Th1 cytokine profile at the infected site prevents a Th2 response that might control a bacterial infection.) In more severe influenza cases, the infection may spread to the epithelial cells lining the terminal bronchioles and alveoli. In that event, the accumulation of fluid (edema) and inflammatory cells in the alveoli further impair breathing. The characteristic symptoms of an influenza infection noted above (i.e., fever, headache, malaise, and myalgia) reflect the **systemic** effects (i.e., effects throughout the whole body) of the cytokines that are released by the innate immune system from the extensively vascularized lung tissue (Box 4.17).

While common flu-like symptoms are caused by effectors of innate immunity, most instances of virus-induced immunopathology are due to the actions of the adaptive immune system, primarily T cells, but antibodies too can cause disease. A well-studied example of virus-induced, cell-mediated immunopathology is that associated with LCMV, a noncytocidal virus that causes disease in mice, but only if they are immunocompetent. Tissue damage resulting from LCMV infection is the direct result of virus-specific CD8 CTLs, as demonstrated in numerous experiments using knockout mice and transgenic mice, or mice in which specific lymphocyte types have been deleted and then transferred back. But CD4 CTLs too may contribute to LCMV-associated immunopathology (Box 4.12).

HBV is a major human pathogen, infecting about 5% of the world's population, in which it can cause a potentially long-lasting and severe inflammatory liver disease that can progress to cirrhosis and fatal hepatocellular carcinoma (Chapter 22). Liver pathology caused by HBV results primarily from the action of the adaptive immune system, specifically, CD8 CTLs. This was demonstrated experimentally in a mouse model in which transgenic mice were engineered to express HBV transgenes. Generally, these mice suffer no ill effects from expression of the

The 1918 influenza **pandemic**, which killed as many as 50 million people worldwide, was the severest epidemic in human history. The basis for the exceptional virulence of the 1918 pandemic influenza virus strain is currently a matter of great concern (see Chapter 12). During the last years of the 20th century, it became technically feasible to reconstruct the 1918 pandemic virus and to then demonstrate that it is extraordinarily lethal in mice. It also became technologically feasible to identify the particular genes of the 1918 virus that might have been the cause of its extreme virulence. Several of the 1918 virus's genes, including those encoding the two major envelope glycoproteins (HA and NA) and the components of the viral RNA polymerase (PB1, PB2, and PA), contributed to the extreme virulence of this virus, but in ways that are not yet clear. However, note that the "reconstructed" 1918 virus also caused an extraordinarily intense inflammatory response in mice, which was mediated by elevated levels of alveolar macrophages, cytokines, and neutrophils. This intense inflammatory response resulted in extreme hemorrhaging, reminiscent of the influenza illness in humans during the 1918 pandemic.

HBV transgenes. However, the transfer of HBV-specific CTLs into these mice resulted in severe inflammatory liver disease, in which the lesions resembled those seen in human viral hepatitis. (Why might the transgenic mice be tolerant of the HBV transgenes?)

How might immunopathology lead to hepatocellular carcinoma? It is believed that the extensive immune-mediated tissue damage in the liver causes compensatory regenerative cell proliferation that inadvertently results in DNA damage and mutations, which can lead to the eventual generation of malignant cells (Chapter 22).

Interestingly, although the cell-mediated adaptive response against HBV infection is the cause of the liver disease, only those patients who mount a vigorous CTL response against the virus may successfully clear it. Those patients who mount a relatively weak CTL response will become chronically infected. It is not clear why some adults respond poorly to the infection. Also, while only 3% of individuals infected as adults become chronically infected, nearly all patients infected at birth become chronically infected. (Since the classic experiments of Peter Medawar and colleagues in the 1950s, it has been thought that newborns exposed to foreign antigens are at risk of developing tolerance rather than immunity, a phenomenon known as **neonatal tolerance**. A variety of distinct factors appear to determine whether exposure of the

neonate to an antigen will lead to tolerance or immunity. These factors include the nature of the antigen-presenting cell, the particular antigen, and its dosage.)

Effectors of innate immunity also may contribute significantly to the immunopathology associated with HBV infection, doing so in conjunction with effectors of adaptive immunity. For example, cytokines, which are released locally by Th cells at the infected sites, recruit nearby macrophages to those sites. IFN-γ secreted by the Th cells, together with the signal triggered by the interaction between the CD40 ligand on the Th cell and CD40 on the macrophage, activates the macrophage to synthesize TNF-α, which synergizes with IFN-γ to raise the level of macrophage activation (Fig. 4.3 and 4.21). These events contribute to the HBV immunopathology, as shown in the mouse model by experimentally administering antibodies against IFN-γ or by depleting macrophages, either of which may prevent death from hepatitis.

That CD4 Th cells should be an effector of immunopathology, as in the above example of HBV, should not come as a surprise, since CD4 Th cells may secrete far greater amounts of cytokines than CTLs and thus may recruit many nonspecific effector cells of the innate immune system to the infected site.

Theiler's murine encephalomyelitis virus (a picornavirus [Chapter 6]) persistently infects the mouse central nervous system, in which it induces inflammatory reactions that lead to demyelination of neurons. Myelin enables rapid neuronal transmission in the central nervous system, and its breakdown leads to weakness and possibly paralysis (see below with respect to multiple sclerosis). In this instance, Th cells trigger an inflammatory response by releasing cytokines that activate macrophages, and perhaps NK cells, which then mediate the demyelination of neurons. The Theiler's virus/mouse model thus provides another example in which Th cells trigger inflammatory immunopathology by cytokine release. It also serves as a model of virus-induced autoimmune disease, as described below.

Coxsackievirus B3 (a picornavirus [Chapter 6]) provides another example in which CD8 CTLs play a role in the immunopathology associated with the infection. B3 virus can cause myocarditis (inflammation of heart muscle) that can be life threatening, particularly in newborns. Older children and adults often recover, but permanent heart damage may remain. In the mouse model, development of coxsackievirus-induced myocarditis depends on the presence of CD8 CTLs. But cytokines such as TNF-α and IL-β also contribute to the pathogenesis.

Autoimmune Disease

Immunopathology, as discussed above, refers to damage to the host that inadvertently results from host immune responses directed against the invader. Yet there is another kind of immune system pathology, referred to as **autoimmune disease**, in which adaptive immunity becomes misdirected towards healthy cells and tissues of the body. Generally speaking, autoimmune disease occurs when the mechanisms that maintain self tolerance break down.

Autoimmune disease may result from genetic defects that compromise one of the several processes that underlie tolerance induction and maintenance. Yet apropos to the subject of virology, there is evidence that most autoimmune disease is actually triggered by an infection. Anecdotally, physicians have noticed for years that autoimmune diseases generally follow bacterial or viral infections. For some autoimmune diseases, the evidence for an infectious etiology is compelling. In some of those cases, a particular infectious agent is consistently isolated from affected patients. In others, an immune response to the particular pathogen always can be demonstrated. The consistent association of the bacterium *Streptococcus pyogenes* with rheumatic fever provides a well-known example, in this instance involving a bacterium.

But what then of viruses? Can they too cause autoimmune disease? The answer is yes, as shown by the several examples noted below. Nevertheless, it is usually difficult to prove that a particular virus actually causes any particular autoimmune syndrome in humans (Box 4.18). On the other hand, animal models demonstrate that infections with some viral pathogens can readily trigger autoimmune disease. Moreover, animal models have been instructive at revealing potential underlying mechanisms.

Autoimmune tissue damage that is triggered by infection can be mediated either by self-reactive antibodies or by self-reactive T cells. In the above example of rheumatic fever, some of the antibodies generated against cell wall components of *S. pyogenes* inadvertently happen to **cross-react** with host epitopes on heart tissue. When these antibodies bind to the heart, they fix complement and activate an inflammation (rheumatic fever) that can lead to heart failure. This chance resemblance between pathogen and host antigens is referred to as **molecular mimicry**. (Empirical evidence for the association between *S. pyogenes* and rheumatic fever includes the observations that [i] outbreaks of rheumatic fever closely follow epidemics of streptococcal sore throats and [ii] treatment of strep throats with antibiotics markedly decreases the incidence of rheumatic fever.)

Generally, there should be no autoimmunity in the absence of infection. That is so because there should not be any autoreactive Th cells available to provide help against self antigens, since such cells usually do not survive tolerance selection in the thymus, or they are anergized in the periphery, as described above. However, infection can cause autoimmunity if the pathogen expresses an epitope with the following properties. (i) It resembles a host epitope, yet is sufficiently different from the host epitope that it can activate a clone of Th cells that has not been tolerized. (ii) The foreign epitope is similar enough to the host epitope that the activated Th cells are still able to recognize the host epitope. If these conditions apply, then the pathogen can elicit autoimmunity.

Here is another factor that may feature in the induction of autoimmunity by a pathogen. Recall that T-cell receptors are less specific for a particular antigen than are B-cell receptors, since all T-cell receptors recognize the

BOX 4.18

Despite the obvious importance of verifying that an autoimmune disease might be caused by a microbial pathogen, demonstrating such a cause-and-effect relationship in humans is generally challenging. One reason is the long period that might elapse between the moment of infection and the onset of disease. By the time symptoms emerge, the putative etiologic agent may in fact be long gone. Even if the etiologic agent is still present, too much time may have passed to establish a clear causal relationship. For example, IDDM, also known as type I diabetes, is preceded for months to years by a prediabetic state during which pancreatic cells are being lost but frank diabetes symptoms have not yet emerged. By the time symptoms do emerge, it may be too late to establish a cause-and-effect relationship between a putative etiologic agent (e.g., coxsackievirus B4) and IDDM.

Yet another complicating factor is the difficulty of establishing appropriate control groups. Household and community controls are critical. Moreover, both experimental and control groups must be sufficiently large in order to ensure that statistically significant results can be obtained. Ensuring that these stringent requirements are met often requires input from professional biostatisticians. In a typical recent study involving coxsackievirus B4 and IDDM, more than one-third of patients with recent-onset IDDM had antibody to coxsackievirus B4. Yet a systematic review of evidence from all published controlled studies of the relationship between coxsackie B virus serology and IDDM, dating from 1966 to 2002, concluded that there is not yet convincing evidence for or against an association between coxsackievirus B4 infection and IDDM.

same common structure, which consists of a peptide bound to an MHC peptide-binding groove. Moreover, the surface of the MHC/peptide complex, which the T-cell receptor contacts, is very flat, and the bound peptide makes up less than one-third of that surface. Since Th cells activate all adaptive immune responses that feature in autoimmune disease, and since CTLs can be effectors of autoimmune pathology, the degeneracy of the T-cell receptor may be a factor that enables molecular mimicry.

With the above in mind, consider the following example in which a virus appears to trigger autoimmune disease via molecular mimicry. Herpes stromal keratitis (HSK) is a syndrome caused by infection with HSV-1 and a leading cause of blindness in humans. HSK results from T-cell-mediated destruction of corneal tissue, as demonstrated by adaptive transfer of the disorder into nude mice. More recently, HSK was shown to be induced by an epitope on a coat protein of HSV-1 that activates CD4 T cells, which cross-react with unidentified corneal antigens. Mutant HSV-1 strains that lack this epitope do not induce the autoimmune disease.

Viruses also may trigger autoimmune disease by a process referred to as **epitope spreading.** (The basis for this term is explained at the end of the paragraph.) In these instances, self-reactive T cells are generated when so-called **sequestered host antigens** are released by virus cytopathic effects, or by virus-induced inflammatory responses that result in release of self antigens, or by tissue destruction mediated by virus-specific T cells. (Some proteins of the central nervous system and of the heart provide examples of sequestered antigens.) Autoreactive T cells against these self antigens may exist in the host, but they usually are not activated to initiate autoimmunity because they do not come into contact with the autoantigen. Moreover, even if they were to come into contact with the host antigen, in the absence of infection they do not receive a costimulatory signal. (In a mouse model, injection of self tissue generally does not lead to autoimmune disease. However, if these same self constituents are mixed with an infectious agent or a microbial product that induces an inflammation at the injected site, then autoimmune disease is *consistently* induced.) Epitope spreading derives its name from early experiments in a mouse model in which the animals were immunized with myelin basic protein. At first, the T cell response was directed against an N-terminal peptide of myelin basic protein, but it subsequently spread to sites throughout the protein. This animal model is referred to as experimental autoimmune encephalomyelitis (Box 4.8).

Theiler's murine encephalitis virus gives rise to a chronic T-cell-mediated demyelinating disease in mice, as noted above. Additionally, the following points about this mouse infection show that it also provides an example of epitope spreading arising during a chronic virus infection. The virus targets the central nervous system, and neurological symptoms first appear at 30 to 35 days. These initial symptoms are caused by virus-specific T cells that target central nervous system cells, resulting in damage to myelin (see below). By 60 to 90 days, T-cell-mediated responses are detected against several central nervous system antigens, first against a particular peptide of proteolipid protein, followed by responses against other sites on proteolipid protein, myelin basic protein, and myelin oligodendrocyte protein. Now here is a very important point. Mice that are tolerant to Theiler's virus peptides can be generated. That is, they do not express Theiler's virus-reactive T cells. (How do you suppose these mice might be generated?) When such mice are infected with Theiler's virus, there is no disease, implying that virus-specific T cells are a key factor in triggering this model chronic neurodegenerative disease. Moreover, this model demonstrates that chronic virus infections in general and those involving the central nervous system in particular can prime autoreactive T cells. In addition, this mouse infection points out processes that might be relevant in human disease. Indeed, the Theiler's virus/mouse system has served as a model of multiple sclerosis in humans. (Multiple sclerosis is discussed below, and its possible viral etiology is discussed at length in Boxes 4.18 and 4.19.)

Insulin-dependent diabetes mellitus (IDDM), also known as **type I diabetes,** is a form of diabetes that involves the chronic inflammatory destruction of the insulin-producing β-islet cells of the pancreas. IDDM has long been associated with coxsackievirus B4 infection, an association supported by the ability of this virus to induce IDDM in mice. Coxsackievirus B4 is known to target the pancreas, but how it might cause the tissue destruction underlying IDDM is only recently being understood. Earlier, it was thought that the virus might directly destroy pancreatic β cells. However, the major effectors of β-cell destruction in IDDM are CTLs and cytokines, followed by antibodies. Recent studies suggest that proinflammatory cytokines of innate immunity, such as IL-1 and TNF-α, which are induced by the infection, might initiate the damage to the pancreas, which then leads to the breakdown of tolerance. TLR4 appears to be the main pattern recognition receptor that triggers the cytokine response to coxsackievirus B4 (Box 4.18).

For most autoimmune diseases, the likelihood that a specific virus might be the etiologic agent is only speculative. **Multiple sclerosis** is certainly a most important example of an autoimmune disease of unknown etiology. It is characterized by a T-cell-mediated attack against myelin sheaths, resulting in lack of coordination, potentially crippling muscle weakness, and impaired vision. It is both

BOX 4.19

Over the years, the possibility that multiple sclerosis might be caused by a virus has been extensively discussed and investigated. Early interest in the possibility that an environmental factor (perhaps an infectious agent) might trigger multiple sclerosis was kindled by certain key epidemiological findings. For example, consider the implications of the following, when taken together. First, multiple sclerosis rarely begins before the age of 15 years. Second, prevalence rates tend to be high (i.e., from 30 to 80 cases per 100,000 individuals) in northern Europe, Canada, and the northern United States and low (i.e., 4 to 6 cases per 100,000 individuals) in southern Europe and the southern United States. Third, individuals migrating after age 15 from an area of high prevalence to an area of low prevalence retain the risk associated with the birthplace. In contrast, migration in early childhood leads to acquisition of the risk associated with the new area. Together, these facts are consistent with the possibility that multiple sclerosis is triggered by a persistent virus infection that generally begins before puberty and in which there may be a long latent period before the onset of symptoms. Alternatively, the virus may not need to be persistent if memory cells made in response to a virus in the high-risk area might be stimulated later by an unrelated virus in the low-risk area. This is not unreasonable, since unrelated viruses, as well as the host, are known to have proteins with similar functional domains. For example, proteins that interact with DNA or RNA have similar DNA/RNA-binding regions. The recognition by adaptive immune cells of such conserved domains on a protein of the pathogen could result in cross-recognition between pathogen and human proteins, resulting in an autoimmune disease in the host. See the text and Box 4.20.

The following facts are also consistent with a possible viral etiology for multiple sclerosis. First, one form of multiple sclerosis is characterized by periods of remissions and exacerbations, and these are not unlike the course of remissions and exacerbations which occur in a herpesvirus infection. Second, JC virus, a human polyomavirus, is unequivocally responsible for a neurodegenerative disease called progressive multifocal leukoencephalopathy. Consequently, JC virus provides a precedent for a conventional virus (i.e., not a prion) that induces demyelination in humans (Chapter 15). However, note that progressive multifocal leukoencephalopathy is not an autoimmune disease. Instead, it results from lytic JC virus infection of oligodendrocytes, the myelin-generating cells in the brain.

Over the years, a number of different viruses have been associated with multiple sclerosis, either by direct visualization in micrographs, by isolation from biopsy samples, or by serology. Among the viruses that have been considered to be likely triggering agents for multiple sclerosis are measles virus, mumps virus, parainfluenza virus, canine distemper virus, varicella-zoster virus, and EBV (the last two are herpesviruses). Newer, more powerful molecular techniques, such as polymerase chain reaction (PCR) and in situ hybridization, have provided exciting new evidence implicating yet other viruses. These include retroviruses, coronaviruses, and several newly discovered human herpesviruses (e.g., HHV-6).

Recently, CD4 T cells that were isolated from the cerebrospinal fluid of a multiple sclerosis patient during disease exacerbation and clonally expanded in vivo were found to react with an epitope of the nonpathogenic and ubiquitous Torque Teno virus. (Torque Teno virus is present in more than 90% of adults worldwide and is currently classified as a circovirus [Chapter 14].) Interestingly, these T-cell clones also can be stimulated by related protein domains from other common viruses, and they recognize multiple self antigens, consistent with the possibility that cross-recognition between pathogen and human proteins might lead to the disease. Still, none of the above viruses, including Torque Teno virus, has yet unequivocally been shown to trigger or exacerbate multiple sclerosis.

It is now believed that activated Th1 cells and the IFN-γ that they secrete are the effectors of multiple sclerosis. These Th1 cells are enriched in the blood and cerebrospinal fluid. In this regard, recall that whereas the blood-brain barrier restricts the entry of many pathogens, it does allow the limited entry of immune cells circulating in the periphery. Thus, activated lymphocytes indeed circulate across the blood-brain barrier, and they are retained in the brain if they encounter their cognate antigen there.

The autoantigen in multiple sclerosis appears to be myelin basic protein, a constituent of the myelin sheaths. (What sort of experiment might have demonstrated that myelin basic protein is the autoantigen in multiple sclerosis?) An exciting related finding is that CD4 T cells specific for a peptide present in the EBV DNA polymerase were present in the cerebrospinal fluid from a multiple sclerosis patient, and, moreover, these T cells cross-recognized a myelin basic protein peptide (Holmoy et al., 2004). These cross-reactive EBV-specific CD4 T cells could be involved in the pathogenesis of multiple sclerosis, since activated T cells can gain entry into the brain compartment and, therefore, target myelin basic protein in the central nervous system (Box 4.20).

Box continues

BOX 4.19 (continued)

Still, how might we interpret the fact that numerous viruses have been implicated as a triggering agent for multiple sclerosis? Stated differently, is it possible that at least some of these many viruses provoke multiple sclerosis? Perhaps. Recall the promiscuous nature of the T-cell receptor. This explains the experimental observation that T-cell receptors that recognize self antigens often cross-react with multiple different foreign antigens. Moreover, unrelated viruses can express proteins containing similar functional domains that might cross-react, and similar domains also may be present on host proteins as well, as noted above. Thus, repeated

infections with different viruses could expand T-cell clones that are specific for these similar protein domains, thereby provoking an autoimmune attack against host targets. And since microbial pathogens can provoke autoimmune diseases, we probably should not be surprised when several or more different microbes are found to trigger the same autoimmune disease.

Reference
Holmoy, T., E. O. Kvale, and F. J. Vartdal. 2004. Cerebrospinal fluid CD4+ T cells from a multiple sclerosis patient cross-recognize Epstein-Barr virus and myelin basic protein. *J. Neurovirol.* **10:**278–283.

---

interesting and frustrating that a variety of different experimental approaches have implicated a number of different viruses as potential etiologic agents of multiple sclerosis. Moreover, plausible hypotheses have been put forward to account for how each of these diverse viruses might be able to trigger the disease. On the one hand,

involvement of any particular virus in multiple sclerosis is still largely speculative, as is the basis for disease pathogenesis. On the other hand, since multiple sclerosis is a major human disease, its possible viral etiology is discussed in some detail in Box 4.19. For more on this issue, see also Box 4.20.

BOX 4.20

An experimental approach to identify pathogen-encoded proteins that might cross-react with a host protein is to scan protein databases for pathogen proteins with properties similar to those of the particular host protein. In one such study, this approach was taken to identify pathogen-encoded proteins that might induce a T-cell response against myelin basic protein. The rationale behind this study was that T cells, which are autoreactive against myelin basic protein, might normally be present in an individual. However, they do not induce disease because the blood-brain barrier limits their access to the central nervous system, where they might recognize myelin basic protein and be activated. However, invasion of the central nervous system by these T cells might occur if they were activated in the periphery, which could occur in the absence of the self antigen if a viral peptide had sufficient sequence similarity with a T-cell immunodominant self-peptide. (**Immunodominant peptides** are those peptides that bind to an individual's MHC molecules and activate CD4 T cells.)

With the above premise in mind, the following structural criteria were applied in a search for a pathogen protein that might trigger a T-cell response against myelin basic protein. First, the pathogen protein had to contain sequences that might bind to MHC class II molecules, and second, it had

to contain those amino acid side chains required for T cell activation.

This approach led to the identification of a set of 129 candidate peptides, each of which was tested against seven myelin basic protein-specific T-cell clones from multiple sclerosis patients. Of these 129 peptides, seven viral peptides (including ones from influenza virus, human papillomavirus, HSV, and the DNA polymerase of EBV [Box 4.19]) and one bacterial peptide efficiently activated three of the myelin basic protein-specific T-cell clones. Only one of these peptides would have been identified based on sequence alignment only.

This study also demonstrated that a single T-cell receptor can recognize quite distinct peptides from different pathogens, which is significant with respect to the paradigm that autoimmune disease can arise from molecular mimicry. Moreover, these experimental findings suggest how various unrelated viruses might each be able to trigger multiple sclerosis. It also is important that the vast majority of pathogen peptides that were identified did not result in T-cell activation.

Reference
Wucherpfennig, K. W., and J. L. Strominger. 1995. Molecular mimicry in T cell-mediated autoimmunity: viral peptides activate human T cell clones specific for myelin basic protein. *Cell* **80:**695–705.

Guillain-Barré syndrome (GBS) is a serious autoimmune neurological disease that is known to be brought on by several different pathogens. It occurs throughout the world with an incidence of 1.3 cases per 100,000 individuals and is characterized by limb weakness that progresses over a period of days or weeks, resulting in partial to total paralysis, usually in more than one limb. Approximately 20% of GBS patients are left with severe disability, and there is a mortality rate of about 10%. The gram-negative bacterium *Campylobacter jejuni* is the most common cause of GBS, but EBV and HCMV also are commonly associated with GBS. Indeed, HCMV is identified in 10 to 15% of GBS patients.

GBS symptoms usually appear several weeks after the initial infection. And in contrast to multiple sclerosis, the cause of which is not yet clear, the links of EBV and CMV to GBS are well established. That is so because in the case of GBS the antiviral antibody response lasts for up to 20 weeks after infection, thereby enabling the viral infection to be diagnosed when neurological disease emerges. Infection with CMV is associated with cross-reactive antibodies that bind to the ganglioside GM2, which is present on motor neurons. In contrast, infection with the bacterium *C. jejuni* is associated with cross-reactive antibodies against ganglioside GM1.

Suggested Reading

Burdeinick-Kerr, R., J. Wind, and D. E. Griffin. 2007. Synergistic roles of antibody and interferon in noncytolytic clearance of Sindbis virus from different regions of the central nervous system. *J. Virol.* **81:**5628–5636.

II

Virus Replication and Pathogenesis

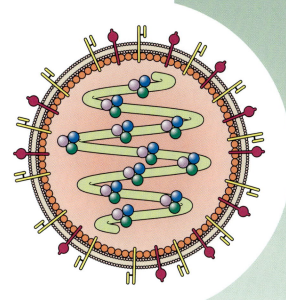

RNA VIRUSES: DOUBLE STRANDED

Reoviruses

INTRODUCTION

The *Reoviridae* are distinguished by their unique double-stranded RNA genomes, which are comprised of multiple (10, 11, or 12) individual genomic segments. Yet their highly segmented double-stranded RNA genomes are not their only exceptional feature. As we will see, reoviruses display a notable set of characteristics that challenge some of our most basic assumptions about the replication cycles of viruses. As double-stranded RNA viruses, reoviruses are class III viruses in the Baltimore classification scheme (Chapter 2).

Reoviruses are remarkably widespread among eukaryote hosts, having been found in mammals, birds, reptiles, fish, crustaceans, insects, ticks, and arachnids, as well as in plants and fungi. Some 75 reovirus species have been identified to date, and these are classified into 12 genera. About 30 more virus isolates are candidates for inclusion among the reoviruses. One recent candidate, isolated from mosquito cells, was found to contain nine genomic segments and may represent a new reovirus genus.

The best known of the 12 recognized reovirus genera are the orthoreoviruses, rotaviruses, orbiviruses, and coltiviruses. Orthoreoviruses and rotaviruses replicate only in vertebrates. They are ubiquitous in humans worldwide and, together with several other reovirus genera, are believed to be transmitted between individuals by the oral/fecal route. In contrast, the orbiviruses, coltiviruses, and seadornaviruses include members that infect both insects and animals. Thus, members of those genera are considered to be arboviruses (Chapter 2). Some of the plant reoviruses (i.e., the phytoreoviruses) also are transmitted via insect vectors and thus are considered to be plant arboviruses. Of note, the orthoreoviruses and orbiviruses contain 10 genomic segments, whereas rotaviruses contain 11 segments and coltiviruses contain 12 segments (Box 5.1). (As you read on, be alert for a reason why there is no reovirus known to contain more than 12 genomic segments.)

The orthoreoviruses were initially isolated from the stools of young children in the 1950s and were the first of the reoviruses to be discovered.

BOX 5.1

As might be expected from the text, viruses with double-stranded RNA genomes are found throughout the living world. In general, these viruses can be divided into two groups: those like the reoviruses, which contain 10, 11, or 12 genomic segments, and those that contain one to three genomic segments. The *Birnaviridae*, whose double-stranded RNA genomes are comprised of two segments (i.e., *bi*-RNA), provide an example of the latter. There are birnaviruses known to infect fish, birds, and insects, but none are yet known that infect mammals.

The **picobirnaviruses** ("pico" for small) are a more recently discovered group. Like the birnaviruses, they contain two genomic segments. However, whereas the birnaviruses have icosahedral capsids that are 70 nm in diameter, the nondescript picobirnavirus capsids are only 35 nm in diameter. Also, the genomic segments of picobirnaviruses are smaller than those of the birnaviruses. Interestingly, picobirnaviruses have been isolated from stool samples of mammals, including humans, as well as from birds. Picobirnaviruses have been implicated as a cause of gastroenteritis in humans and animals, but this issue requires further study. The genomes of **plant cryptoviruses** similarly contain two segments. *Pseudomonas* phage φ6 (phi 6) contains three genomic segments (see also Boxes 5.8, 5.9, and 5.12).

Since then, the orthoreoviruses have been the most extensively studied of all double-stranded RNA viruses. Also, double-stranded RNA was detected for the first time in nature in these reoviruses.

The orthoreoviruses are ubiquitous in humans, generally giving rise to clinically harmless infections of the respiratory and digestive tracts. In contrast, the rotaviruses, which likewise are ubiquitous in humans, often cause severe diarrhea in young children. Indeed, members of the rotavirus genus are responsible for about 500,000 deaths of young children per year in developing countries, where malnutrition and lack of adequate amounts of safe water compound the problem of diarrhea-associated dehydration. Coltiviruses cause feverish illnesses in humans, in which headache and muscle ache are accompanying symptoms (e.g., Colorado tick fever; see below). Orbiviruses cause disease mainly in animals (e.g., bluetongue disease of sheep).

How did the reoviruses acquire their name, and what might be its significance? The story begins with the fact that the reoviruses were misclassified prior to 1959 (before the dawning of the age of molecular animal virology), as echoviruses. The echoviruses are actually a genus within the *Picornaviridae*, a family of single-stranded RNA viruses that also includes poliovirus, foot-and-mouth disease virus, and the rhinoviruses (causing the common cold) (Chapter 6). A reason for the initial misclassification of the reoviruses was that orthoreoviruses, like echoviruses, are associated with clinically harmless infections of the gastrointestinal tract. The acronym "echo" is actually derived from "enteric cytopathic human orphan," in which "orphan" refers to the fact that these viruses are not associated with any human illness. Thus, reoviruses were initially classified as echoviruses because they gave rise to clinically unapparent infections of the gastrointestinal tract.

With the emergence of molecular animal virology, echoviruses were properly classified as a genus within the picornavirus family of plus-strand RNA viruses. But the reoviruses actually were reclassified earlier for a quite anomalous reason. It was because the orthoreoviruses, unlike echoviruses, also are isolated from the respiratory tract, where they also are clinically harmless. For that reason, back in 1959, Albert Sabin (of poliovirus vaccine fame [Chapter 6]) proposed that reoviruses be classified as a new family (not a very rational basis for virus classification, as discussed in Chapter 2). Also, he proposed the new name for the family, in which "reo" is an acronym derived from "respiratory enteric orphan." Only afterwards was the distinctive nature of the reovirus double-stranded RNA genome realized. Finally, as noted above, the family *Reoviridae* is not entirely harmless in humans, since the rotaviruses can cause serious disease, although usually only in the very young.

Regarding terminology, "reovirus" often is used to refer to the specific genus *Orthoreovirus* (the prototype genus of the family), as well as to the entire family *Reoviridae*. The term is used both ways in this chapter.

STRUCTURE, BINDING, ENTRY, AND UNCOATING

Reovirus Binding and Entry into the Cell

Orthoreoviruses use junctional adhesion molecule-1 (JAM1) for their cell surface receptor (Box 5.2). This molecule contains a single extracellular immunoglobulin (Ig)-like domain, and thus, it is a member of the Ig superfamily of proteins (Box 5.3). JAM1 is found in the tight junctions between epithelial cells and endothelial cells, where it plays an important role in maintaining epithelial and endothelial barrier function. The residues that form the interface between JAM1 molecules on neighboring cells are in the JAM1 Ig domain. Interestingly, these residues also appear to be crucial for reovirus binding, implying that the affinity of

BOX 5.2

What kinds of experimental evidence might be used to identify the receptor for a virus? There are a variety of approaches to this issue, as is discussed in later chapters. The following generally applicable methodology was used in the case of the reoviruses, and it is in fact a "gold standard," so to speak. First, a cDNA library was prepared from **susceptible cells** (that is, cells to which the virus can bind). Then, the cDNA library from the susceptible cells was used to transfect **nonsusceptible cells**, to which the virus cannot bind. From the population of transfected cells, four clones that conferred virus binding were identified. This was accomplished by using fluorescence-activated cell sorting to isolate transfected cells that bound fluorescent reovirus particles.

Other strategies for identifying transfected cells that have acquired the receptor include using receptor-specific monoclonal antibodies (i.e., monoclonal antibodies that previously were demonstrated to compete with the virus for binding to cells) or attenuated virus mutants that carry a drug resistance marker.

There are two general approaches that might then be used to identify the receptor-encoding DNA sequences in those transfected cells that acquired the ability to express the receptor. The first is used when the donor DNA library is derived from genomic DNA. In that instance, the donor DNA contains species-specific repetitive sequences that allow it to be distinguished from the recipient cell DNA, using a hybridization probe that contains the repetitive sequence. The second approach is to first clone the cDNA into a plasmid library. A pool of the plasmids is grown and subdivided into subpools. Each of the subpools is then tested for its ability to confer receptor activity. The positive subpools are continually subdivided and tested until a single receptor-encoding plasmid is identified.

Each of the four cell clones that bound reovirus contained a transfected sequence that expressed JAM1. To corroborate that JAM1 indeed is a reovirus receptor, monoclonal antibodies against JAM1 were preadsorbed to cells to see if they might inhibit reovirus binding, which they indeed did.

Reference
Barton, E. S., J. C. Forrest, J. L. Connolly, J. D. Chappell, Y. Liu, F. J. Schnell, A. Nusrat, C. A. Parkos, and T. S. Dermody. 2001. Junction adhesion molecule is a receptor for reovirus. *Cell* **104:**441–451.

BOX 5.3

A number of proteins that are not Igs (i.e., antibodies) nevertheless contain Ig-like domains, which are characterized by two antiparallel β-pleated structures: one containing four chains and the other containing three chains (Fig. 4.9 and 4.10). On antibody molecules, the loops between the chains and between the sheets come together to form antigen-specific binding sites.

Molecules that contain Ig-like domains, which may express a variety of functions, constitute a family of proteins referred to as the Ig superfamily. Many of these proteins, like JAM1 (and ICAM-1, which is the receptor for most human rhinoviruses [Chapter 6]), have a recognition or binding role at the cell surface.

reovirus particles is discussed below. For the moment, note that the viral genome is encased within a capsid comprised of two concentric icosahedral protein shells and that σ1 is a component of, and projects from, the outer shell (Fig. 5.1).

Although many viruses interact with only a single cell surface receptor to mediate their binding and subsequent entry, other viruses interact with different receptor and coreceptor molecules to mediate the separate processes of attachment and internalization. Human immunodeficiency virus (HIV) (Chapter 21) is but one such example of a virus that interacts sequentially with a receptor (CD4) and coreceptor (a chemokine receptor) to accomplish binding and entry.

The second interaction (i.e., with the coreceptor) facilitates conformational rearrangements in the viral envelope glycoproteins that result in fusion of the viral envelope with the cell membrane and, consequently, the release of the viral core into the cell.

Despite the fact that reoviruses are nonenveloped, some reoviruses too may need to interact with another cell surface molecule, in addition to JAM1, to achieve the separate steps of binding and entry. This notion is suggested by the experimental finding that other cell surface molecules, particularly those that contain sialic acid moieties, enhance

the virus for those residues must be strong enough for the virus to disrupt the JAM1 homodimers.

Reoviruses attach to JAM1 via their σ1 (sigma 1) proteins and do so with high affinity, consistent with the argument in the preceding paragraph. The structure of

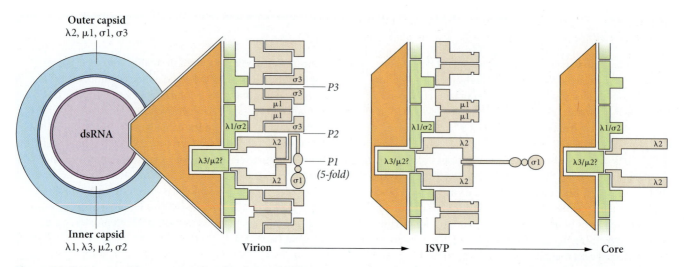

Figure 5.1 Reovirus particles are nonenveloped and 85 nm in diameter. Structurally, they are comprised of two concentric protein shells, each of which displays icosahedral symmetry. The outer shell consists of proteins μ1, σ1, and σ3. The inner shell contains proteins λ1, σ2, λ3, and μ2. Protein λ2 spans both shells. σ1 is the receptor recognition protein, and it is inserted into λ2. These structural elements are depicted schematically; the wedge represents their organization at a five-fold axis of symmetry. Positions of open or potential channels through the outer capsid at positions of fivefold (P1) or sixfold (P2 and P3) symmetry are indicated. Comparable wedges from a reovirus ISVP and core are shown to depict the loss and change in conformation of outer capsid proteins as entry progresses. ISVPs are distinguished by the loss of σ3, by endoproteolytically cleaved μ1 (indicated by the notch), and by conformationally altered (extended) σ1. The core is distinguished by the additional absence of μ1 and σ1 proteins and by conformationally altered λ2 proteins. The conformational change in λ2 and loss of σ1 open a channel through the λ2 "spike" that protrudes from the core surface. Adapted from M. L. Nibert and B. N. Fields, p. 341–364, *in* E. Wimmer (ed.), *Cellular Receptors For Animal Viruses*, Cold Spring Harbor Laboratory Press, Cold Spring Harbor, NY, 1995, with permission.

infection by some reoviruses. Also, Type 3 reoviruses bind to sialic acid-linked molecules, as well as to JAM1, using distinct regions of their σ1 proteins to bind to each of these possible receptors.

Together, the above facts led to the suggestion that reovirus entry is facilitated by a multistep binding process involving a primary receptor and a coreceptor. In the reovirus case, the virus is thought to initially bind to sialic acid. In this way, the virus first engages an abundant cell surface molecule, albeit via a low-affinity adhesion. This initial interaction with sialic acid may then facilitate the subsequent engagement of the less accessible JAM1 proteins via a high-affinity adhesion.

Yet even the latter interaction of reovirus with JAM1 is probably not sufficient to mediate its uptake into the cell. Indeed, recent experimental results suggest that after the reovirus σ1 proteins bind to JAM1, λ2 (lambda 2) proteins on the viral surface then bind to β1 (beta 1) integrin, and it is actually the latter interaction that facilitates reovirus internalization by receptor-mediated endocytosis (Fig. 5.2). Note that a major function of integrins is to mediate cellular adhesion to the extracellular matrix. In addition, integrins transmit signals that act to reorganize the cytoskeleton. In principle, integrin-transmitted signals could mediate cytoskeletal rearrangements that facilitate virus internalization.

Notice that this reovirus entry scenario is different from that of HIV, noted briefly above. That is so because HIV is an enveloped virus, whereas the reoviruses are nonenveloped or "naked" viruses. This distinction is relevant as follows. All viruses must cross at least one cellular

Figure 5.2 Reovirus initially binds to sialic acid via a low-affinity adhesion, which then enables the virus to engage the less accessible JAM1 via a high-affinity adhesion. The latter interactions position the virus on the cell surface so that λ2 proteins on the viral surface can then bind to β1 integrin, and it is actually the latter interaction that facilitates reovirus internalization by receptor-mediated endocytosis. Adapted from M. S. Maginnis et al., *J. Virol.* **80:**2760–2770, 2006, with permission.

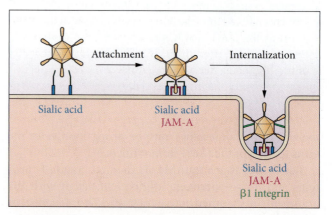

membrane during entry. Enveloped viruses are able to cross that membrane by membrane fusion, as facilitated by their fusion proteins (Chapter 2). Generally, less is known about how a nonenveloped virus or its genome might cross a cellular membrane, a necessary step for replication to occur (see below). Regardless, quite an array of evidence was brought to bear in support of the premise that the interaction between the viral λ2 protein and the cellular β1 integrin actually leads to viral entry.

First, what evidence shows that JAM1 alone does not mediate viral entry? The relevant evidence involved a mutant JAM1 molecule, which lacks a cytoplasmic tail. An aberrant receptor, which lacks a cytoplasmic tail, is unable to undergo receptor-mediated endocytosis. That is so because the cytoplasmic tail contains an internalization signal that binds to an adaptor protein on the cytosolic side of the plasma membrane, which in turn binds to clathrin. Thus, the cytoplasmic tail is necessary for the receptor to concentrate in clathrin-coated pits and then undergo endocytosis. Nevertheless, despite the inability of the tail-less JAM1 to undergo endocytosis, it still supported virus binding that led to entry. Thus, some cell surface molecule other than JAM1 must mediate viral entry.

Next, what is the evidence that β1 integrin mediates reoviral internalization? First, the reovirus outer capsid protein λ2 (see below), which serves as the insertion site on the virus for σ1, contains the conserved integrin-binding sequences Arg-Gly-Asp (RGD) and Lys-Gly-Glu (KGE). Second, exposing susceptible human cells in culture to an antibody specific for β1 integrin inhibits reovirus infection of those cells. In control experiments, exposing those cells to antibodies specific for other integrins has no effect on infection. Third, chick embryo cells, which are otherwise resistant to orthoreovirus infection, are rendered susceptible by transfection with a JAM1-expressing plasmid, together with a murine β1 integrin-expressing plasmid, but not by the JAM1-expressing plasmid alone. Fourth, β1 integrin-deficient mouse embryonic stem cells are much less susceptible to reovirus than isogenic cells which express β1. However, the virus binds as well to the β1 integrin-deficient cells as to the β1 integrin-expressing cells, showing that β1 integrin is involved in a postattachment entry step (Maginnis et al., 2006). Members of the rotavirus genus of the *Reoviridae* also engage a variety of integrins to mediate binding and entry (Box 5.4).

Before leaving this section, it is noteworthy that the reovirus attachment protein, σ1, is a major determinant of reovirus tropism and, consequently, of reovirus pathogenesis in vivo. This key point was first demonstrated in classic studies by Bernard Fields, carried out with newborn mice. The experimental approach made important use of the segmented nature of the reovirus genome

BOX 5.4

Integrins also serve as attachment and entry receptors for members of several other virus families, including picornaviruses (e.g., echovirus and foot-and-mouth disease virus), hantaviruses (e.g., hantavirus NY-1 and Sin Nombre virus), adenoviruses, and herpesviruses (e.g., Kaposi's sarcoma herpesvirus and cytomegalovirus). Integrins also act as adhesion molecules for several bacterial pathogens (e.g., *Bordetella pertussis*), enabling them to colonize their hosts. Moreover, β1 integrin is the cell surface receptor used by the bacterium *Yersinia pestis*, the etiologic agent of bubonic plague, to invade M cells, special cells in the intestinal epithelium that sample antigens in the intestine (Chapter 3). The invasion of M cells is the first step in a *Y. pestis* infection. After using M cells to cross the intestinal epithelium, the bacterium replicates extracellularly.

(recall that orthoreovirus genomes are comprised of 10 individual double-stranded RNA segments). Viruses with segmented genomes readily generate genetic reassortants in cells that are coinfected with two genetically distinct virus strains. In this instance, genetically reassorted viruses were generated in cells coinfected with sialic acid-binding and nonbinding reovirus strains. Importantly, the patterns of electrophoretic mobilities of the genomic segments of each individual strain are unique (see Fig. 5.4). Thus, by examining numerous reassorted viruses, it was possible to demonstrate that sialic acid-binding and neurotropism each segregated with the reovirus gene segment encoding the σ1 protein. In this regard, by using the same strategy of examining reassortant viruses it was possible to correlate specific genomic segments with specific proteins (see below).

These studies demonstrated that in the mouse model, the ability of the reovirus σ1 protein to bind to sialic acid is a major determinant of the virus tropism for the central nervous system and the consequent neuropathology. Also, the ability of these viruses to infect and replicate in intestinal tissue also is dependent on their σ1 protein. More importantly, perhaps, these studies provided an early demonstration of the general truth that the preference of a virus for a particular receptor is a major determinant of its tissue and organ tropism and, consequently, of its pattern of pathogenesis in the host.

The experimental approach used by Fields resulted in a sweeping generalization concerning the relationship between cell and tissue tropism and pathogenesis. Nevertheless, progress towards characterizing σ1 functions, as well as the particular roles of other reovirus proteins that facilitate entry (i.e., μ1 [mu 1] and σ3; see below), was hampered by the inability to produce infectious virus particles containing defined mutations in those proteins. For this reason, M. L. Nibert and coworkers developed procedures to generate infectious virus-like particles that contain engineered mutations in these proteins. First, as briefly noted above, reovirus particles are comprised of two concentric protein shells. By treating purified virus particles with chymotrypsin, it is possible to remove the outer shell, which consists of σ1, μ1, and σ3. Nibert was then able to generate infectious virus-like particles in vitro by "recoating" the genome-containing core particles with recombinant forms of σ1, μ1, and σ3, which were produced by a baculovirus vector in insect cells. (Baculoviruses are very large double-stranded DNA viruses that infect insects only.) Experiments with these recoated reovirus particles, using recombinant σ1 proteins from a variety of reovirus strains and a variety of cell types, confirmed that differences in cell attachment and infectivity previously observed between those strains are determined by the σ1 protein. This experimental approach is referred to as **recoating genetics**.

Structure, Uncoating, and Entry into the Cytoplasm

Reovirus structure is described in detail as we follow the rather remarkable programmed sequence of structural alterations that reovirus particles undergo during their entry. As you follow this sequence, be particularly attentive to those structural transitions that enable the virus to express its genome upon entry into the cytosol. Doing so should dispel any notions of virus particles as merely inert entities that are passively taken up into host cells. With this recommendation, the story begins as follows.

In Chapter 2, much was made of the implications of the eclipse period, as it typically occurs in the viral life cycle. In particular, the eclipse period underscored the fact that at a moment during virus replication, a molecule of naked nucleic acid may be all that connects one virus generation with another. This was demonstrated most dramatically in the case of the T-even bacteriophages, where the genome is the only component of the virus to enter into the cell. In the brief overview of animal virus entry in Chapter 2, we noted that the complete animal virus particle often enters the host cell. Nevertheless, the subsequent *intracellular* uncoating of animal virus genomes generally also leads to an eclipse period, in which the infecting virus particle ceases to exist as an organized entity.

In 1968, Silverstein and Dales provided the first indication that the classic concept of the eclipse period (i.e., that which follows the release of the viral genetic material from the particle) might not apply to the reoviruses. In brief, they found that reovirus uncoating occurs in lysosomes but, importantly, only about 25% of the ^{14}C-labeled virus coat is degraded in the process. Subsequent studies demonstrated that the 85-nm-diameter parental reovirus particles are converted in lysosomes to partially uncoated 70-nm-diameter structures, which then enter the cytoplasm. Importantly, these 70-nm particles do not release their genomes at any stage of the reovirus replication cycle. We will consider the transition of the 85-nm parental particles to the 70-nm structures in detail below. For now, note that nonenveloped native reovirus capsids are about 85 nm in diameter and are comprised of two concentric protein shells, each of which displays icosahedral symmetry (Box 5.5).

So then, how does the fact that reovirus particles never release their genomes affect our concept of the eclipse as a fundamental characteristic of virus replication? Perhaps it broadens our conception of the eclipse period, since the incompletely uncoated reovirus core particles that emerge from the lysosome are, like a naked bacteriophage genome, noninfectious (see below). So, we might choose to say that reoviruses do undergo an eclipse period, although one that differs from the original notion of the eclipse as representing the period following genome release. Regardless, the classic notion of the eclipse enabled us to make the key point that in no case does a virus follow the pattern of growth and division that characterizes cellular replication. That assertion is not affected by the atypical nature of the reovirus eclipse.

Before going on it would be good to bear in mind the following important points. All viruses must penetrate at least one cellular membrane for their genomes to gain access to a cellular compartment where they might be transcribed and replicated. Enveloped viruses generally cross the crucial membrane by membrane fusion, as facilitated by their fusion proteins (Chapter 2). For nonenveloped viruses such as the reoviruses, the processes by which membrane barriers are crossed are still not clear. How a

---

BOX 5.5

The point is often made in this text and elsewhere that the study of animal viruses may lead to important insights into eukaryotic cell biology. The detailed description of reovirus uptake by Silverstein and Dales, in which reovirus particles were seen to be enclosed within endocytic vesicles, and later in lysosomes, was a key observation leading to the characterization of the steps in receptor-mediated endocytosis in general.

Table 5.1 Reovirus proteins[a]

Encoding segment	Protein	Mass kDa	Copy number per virion	Location in virions	Presence in particle forms	Function or property
L1	λ3	142	12	Inner capsid	VIC	RNA-dependent RNA polymerase
L2	λ2	145	60	Outer capsid, core spike	VIC	Guanylyltransferase, methyltransferase?
L3	λ1	137(143)	120	Inner capsid	VIC	Binds dsRNA, zinc metalloprotein
M1	μ2	83(78)	20–24	Inner capsid	VIC	Unknown
M2	μ1	76	600	Outer capsid	VI	N-myristoylated, cleaved into fragments, role in penetration, role in transcriptase activation
M3	μNS	80	0	Nonstructural	—	Binds ssRNA, associates with cytoskeleton, role in assortment, role in secondary transcription
S1	σ1	49	36	Outer capsid	VI	Cell-attachment protein, hemagglutinin, primary serotype determinant
	σ1s	14	0	Nonstructural	—	Unknown
S2	σ2	47	150	Inner capsid	VIC	Binds dsRNA
S3	σNS	41	0	Nonstructural	—	Binds ssRNA, role in assortment
S4	σ3	41	600	Outer capsid	V	Sensitive to protease degradation, binds dsRNA, zinc metalloprotein, effects on translation

[a]V, virus particle; I, ISVP; C, core. Modified from B. N. Fields, D. M. Knipe, and P. M. Howley (ed.), *Fundamental Virology*, 3rd ed. (Lippincott-Raven, Philadelphia, PA, 1996), with permission.

particle as large as 70 nm might cross a membrane is especially enigmatic. To gain insight into these facts and into the way reoviruses deal with this dilemma, we next consider the structure of reovirus particles in more detail.

As previously noted, native (i.e., unaltered) reovirus particles are nonenveloped and 85 nm in diameter. Structurally, they are comprised of two concentric protein shells, each of which displays icosahedral symmetry. The outer shell consists of proteins μ1, σ1, and σ3 (Fig. 5.1 and Table 5.1). The inner shell contains proteins λ1, σ2, λ3, and μ2. Protein λ2 spans both shells. As noted above, σ1 is the receptor recognition protein, and it is inserted into pentamers of λ2.

Lysosomal proteases convert the input native reovirus particles into so-called "intermediate or infectious subviral particles" (**ISVPs**). In the process, σ3, which stabilizes the outer shell, is proteolytically degraded and lost, μ1 is cleaved to μ1N and μ1C (both of which remain particle associated at this time), and σ1 acquires an extended conformation. Consequently, the reovirus particle goes from a structure that is 85 nm in diameter to an ISVP that is 70 nm in diameter (Fig. 5.1). This is the phenomenon reflected by the early experimental findings of Silverstein and Dales, as mentioned above.

As their name implies, ISVPs indeed are infectious. Moreover, ISVPs are uniquely able to induce some sort of perforation in target membranes, which likely enables them to initiate membrane crossing. The ability of ISVPs to disrupt membranes was demonstrated in several ways. In one experimental approach, ISVPs but not native reovirus particles caused the release of ^{51}Cr from preloaded mouse cells. In another experimental approach, ISVPs but not virus particles caused release of hemoglobin from erythrocytes. In yet another approach, only ISVPs opened current-conducting pores in lipid bilayers. Each of these ISVP-mediated effects is consistent with the premise that ISVPs are functional intermediates in the infection process. Why then is there an 85-nm virus particle? Presumably, 85-nm particles exist because of their greater stability and because they are better able to protect the virus in the environment.

The acidic milieu within lysosomes (or within the gut lumen in some instances [Box 5.6]) is needed for the conversion of reovirus particles to ISVPs. That is so because of the low pH optima of lysosomal (and gut) proteases. The fact that conversion of native virus particles to ISVPs is sensitive to bafilomycin A1 and concanamycin A substantiates this point. These drugs inhibit the lysosomal ATPase activity necessary to maintain the low pH of that organelle. (Note that drugs such as bafilomycin A1 and concanamycin A are routinely used to demonstrate a pH-dependent step in a virus's life cycle.) Interestingly, these drugs impair infection by native reovirus particles, but not by ISVPs, consistent with the premise that ISVPs are functional intermediates in the reovirus entry pathway that have acquired the capacity to penetrate membranes.

Exactly how ISVPs disrupt and cross membranes is not entirely clear. However, it is believed that this ability is related to the fact that the outer capsid protein, μ1, is modified with a hydrophobic myristoyl (i.e., C14 fatty-acyl) group at its N terminus. Myristoyl groups on proteins generally have membrane-targeting functions. In the native virus particle, the myristoyl group and other hydrophobic sequences are buried in each of the 200 trimers of μ1 per particle. The disruption of intersubunit interactions that occurs as a result of μ1 cleavage by lysosomal proteases exposes these sequences, which then act in membrane penetration and disruption.

As is discussed in later chapters, several features of reovirus entry that are recounted here also are relevant to the entry of other virus families. For example, myristoylation is a modification that also is found on capsid proteins of nonenveloped picornaviruses (plus-strand RNA viruses) and polyomaviruses (small, double-stranded DNA viruses) and is believed to play a role in entry in each of those instances. Moreover, in the case of both picornavirus and reovirus, the myristoylated protein undergoes a cleavage that primes it to undergo the conformational change that exposes its hydrophobic sequences. In all instances, major alterations of the capsid structure are necessary to expose the hydrophobic sequences that facilitate entry. Presumably, these hydrophobic sequences insert into cell membranes, subsequently disrupting them.

Another recurring theme, alluded to in the preceding paragraph, is that viruses make use of protein cleavage to generate metastable structures, which might undergo subsequent conformational rearrangements when properly triggered. Cleavage of the rotavirus VP4 protein provides another example of this principle as it applies to the rotavirus genus of the *Reoviridae*. Like the orthoreovirus μ1 protein, the rotavirus VP4 protein is a major component of the entry machinery of the related rotaviruses. Moreover, like reovirus μ1, the rotavirus VP4 protein is primed by proteolytic cleavage to mediate entry. Cleavage of the rotavirus VP4 protein occurs in the gut lumen (as might be expected in the case of these enteric viruses), and it results in the VP4 spikes attaining a more rigid state, in which they assume the form of a **coiled-coil** trimer. This molecular structure is described in detail in Chapter 12. Importantly, it is the structure assumed by the fusion proteins of a variety of *enveloped* viruses, including influenza virus, HIV, and Ebola virus. Importantly, in the case of each of these enveloped viruses, this structural transition exposes a hydrophobic segment of the envelope proteins that can interact with a cellular membrane, thereby promoting fusion of the viral envelope with the cellular membrane and consequent entry.

Although there is no membrane fusion in the case of the nonenveloped reoviruses, the similar structural transition on the rotavirus VP4 protein exposes a hydrophobic segment that could potentially destabilize a cellular membrane, thereby promoting entry. Consistent with that premise, rotavirus VP4 undergoes a second conformational transition in which each subunit of the trimer folds back on itself, in that way bringing the viral core into closer apposition to the membrane. Importantly, this rearrangement also occurs in the cases of the fusion proteins of the several enveloped viruses noted above. Hence, the conformational rearrangements of the rotavirus VP4 protein and its ability to promote reovirus entry are reminiscent of the examples of the fusion proteins of several clinically important enveloped viruses.

Reovirus ISVPs undergo a further remarkable conformational rearrangement to yield structures known as "cores" (Fig. 5.1). To fully appreciate the nature and significance of this change in structure, first we must note a few special features of the reovirus genome and its pattern of transcription.

THE REOVIRUS GENOME: TRANSCRIPTION AND TRANSLATION

The Particle-Associated RNA Polymerase

Recall, first, that reoviruses contain double-stranded RNA genomes and, second, that uninfected host cells contain no enzymatic machinery that might use double-stranded RNA as a template for transcription or replication. Moreover, double-stranded RNA itself cannot function as messenger RNA (mRNA). How then does a reovirus initiate its replication cycle? We know the answer from our discussion of the Baltimore classification scheme in Chapter 2. Parental reovirus particles contain an RNA-dependent RNA polymerase, which enters the cell as a component of the virus, in that way being available to transcribe the parental viral genome.

The first experimental evidence that reovirus particles might contain a particle-associated RNA-dependent RNA polymerase was the following. Treating infected cells with inhibitors of translation, such as cycloheximide, did not affect initial reovirus transcription. Since uninfected cells do not contain an RNA-dependent RNA polymerase activity and that activity appeared in infected cells even in the presence of inhibitors of translation, it already had to be associated with the parental virus particle before infection.

In 1968, Shatkin and Sipe directly demonstrated that reovirus particles indeed contain an RNA polymerase activity that could be expressed in vitro. But an important proviso was that this RNA polymerase activity was demonstrable in vitro only after the virus particles were exposed to the protease chymotrypsin. Later experiments would demonstrate that treating virus particles with protease in vitro causes conformational transitions similar to those that generate viral "cores" in vivo (see below). Moreover, the reovirus-associated polymerase activity was truly an RNA polymerase, since it could use ribonucleoside triphosphates, but not deoxyribonucleoside triphosphates, as precursors. Additionally, it required RNA, rather than DNA, for its template.

In 1970, Acs and coworkers characterized the reovirus RNA polymerase activity, as expressed by the chymotrypsin-generated cores. First, the product of the core-associated RNA-dependent RNA polymerase activity is single-stranded RNA, as shown by the fact that it is sensitive to digestion by RNase. Second, the reovirus genomic RNA *within* the subviral cores is the template for the transcriptase activity, as demonstrated by the fact that 90% of the ^{3}H-labeled single-stranded RNA product anneals to denatured genomic RNA. (More will be said about this remarkable feature of reovirus transcription. For the moment, bear in mind that ISVPs and cores do not release their genomes at any stage of the reovirus replication cycle. Also, consider the implications of the fact that the double-stranded reovirus genome is packed to liquid crystalline density within the cores.) Third, only one strand of the double-stranded RNA template is transcribed. This was indicated at the time by the fact that the transcription products did not self-anneal. Later experiments demonstrated that the single-stranded RNA product is of the same polarity as reovirus mRNA (that is, the plus sense) (Box 5.7).

It makes eminent sense that the product of the reovirus core-associated RNA polymerase activity should be mRNA, since the virus must transcribe genes that code for the enzymatic machinery it needs to further advance its replication. So then, when and how are minus strands of the double-stranded progeny genomes made? For now, hold that question in mind, because we still need to note other important features of reovirus transcription and genome organization.

One can put forward two general mechanisms by which the reovirus double-stranded RNA genomes might be transcribed to generate single-stranded RNAs. The first mechanism is referred to as "semiconservative." In that scenario, nascent plus strands remain bound to the minus-strand templates, while displacing the original plus strands from the double-stranded RNA molecule (Fig. 5.3). In that way, the parental genome is semiconserved. Note that this process is somewhat reminiscent of the semiconservative replication of double-stranded DNA. The second mechanism is referred to as "conservative" because the parental genome is entirely conserved (Fig. 5.3). This process is reminiscent of the transcription of a double-stranded DNA template. (At this moment it would be good to pause and think of how you might experimentally discriminate between these two plausible potential mechanisms.)

Acs and coworkers distinguished between these alternative possibilities by taking advantage of the fact that RNA within the cores is beyond the reach of added RNase. Their experiment was as follows. First, they prepared reovirus particles that contained ^{3}H-labeled genomic RNA. This was accomplished by growing virus in the presence of ^{3}H-uridine triphosphate (UTP). Next, they prepared subviral cores from these virus particles by treating them with chymotrypsin. Then, under conditions in which the cores generated single-stranded RNA products that were sensitive to RNase, they demonstrated that the H^{3}-labeled parental RNA remained completely protected within the cores against added RNase. Thus, transcription of the

BOX 5.7

Evidence for the assertion that the single-stranded RNA made by the reovirus cores in vitro is of the same polarity as mRNA was as follows at the time the assertion was made. The RNA generated by the reovirus cores in vitro and RNA isolated from ribosomes of reovirus-infected cells each annealed with denatured reovirus genomic RNA, but these two RNAs did not anneal with each other. This was the best experiment that could be done at the time. What experiment might you do now to provide direct and critical evidence that the in vitro transcription product indeed is mRNA?

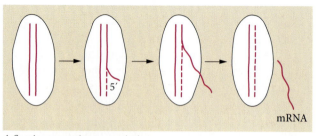

A Semiconservative transcription

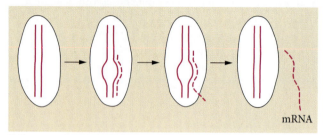

B Conservative transcription

Figure 5.3 Schematic of the alternative mechanisms by which the double-stranded RNA-containing reovirus cores might generate mRNA. The solid lines represent parental RNA strands, and the dashed lines represent transcripts. In actuality, the parental RNA was completely conserved within the reovirus cores. Redrawn from J. A. Levy, H. Fraenkel-Conrat, and R. A. Owens, *Virology*, 3rd ed. (Prentice Hall, Englewood Cliffs, NJ, 1994), with permission.

reovirus double-stranded RNA genome is conservative, analogous to transcription of a double-stranded DNA template (Box 5.8).

The Segmented Reovirus Genome

Early sedimentation analysis of the reovirus double-stranded RNA genome provided the first suggestion that

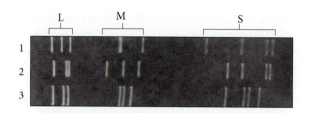

Figure 5.4 Electropherotyping gel for reovirus gene segments. The 10 double-stranded RNA segments from strains T2J (lane 1), TL1 (lane 2), and T3D (lane 3) were separated by electrophoresis on a 10% polyacrylamide gel. Notice that they cluster as three large (L), three medium (M), and four small (S) segments. Differences between analogous segments from these strains can be exploited to define the electropherotypes of reassortant viruses for use in genetic studies. From B. N. Fields, D. M. Knipe, and P. M. Howley, (ed.) *Fundamental Virology*, 3rd ed. (Lippincott-Raven, Philadelphia, PA, 1996), with permission.

it might be segmented. Later examination, by polyacrylamide gel electrophoresis, demonstrated that the orthoreovirus genome actually is comprised of 10 distinct double-stranded RNA segments that fall into three size classes: large (L), medium (M), and small (S) (Fig. 5.4). Moreover, the transcripts produced in vitro and in vivo by the reovirus cores also comprise three size classes, and as might be expected, the reovirus proteins also fall into three size classes, designated λ, μ, and σ (Fig. 5.5). Each of the RNA size classes is just sufficiently long to encode a protein of the corresponding protein size class, implying that a distinct RNA segment encodes each of the reovirus proteins (this is almost correct, but see below). Each plus strand of the double-stranded RNA segments contains an identical "cap" at its 5′ end (see below).

BOX 5.8

For the sake of comparison, consider bacteriophage φ6, which naturally infects *Pseudomonas phaseolicola*. This double-stranded RNA phage enters its bacterial host through the host pilus, in that way resembling the single-stranded RNA phages discussed in Chapter 6. However, our reason for making note of these phages here is that like the reoviruses, they have segmented double-stranded RNA genomes (three genome segments in this instance) and contain a particle-associated transcriptase activity. Moreover, the particle-associated transcriptase generates three mRNAs in vivo, from within the virus core, with one mRNA species corresponding to each genomic segment. However, in contrast to the reovirus instance, particle-associated φ6 transcription is semiconservative.

Figure 5.5 Polyacrylamide gel electrophoresis radioautogram of reoviral proteins, ranging from λ1, the largest, to σ3, the smallest. Reprinted from J. A. Levy, H. Fraenkel-Conrat, and R. A. Owens, *Virology*, 3rd ed. (Prentice Hall, Englewood Cliffs, NJ, 1994), with permission.

Protein-coding assignments for the individual reovirus genomic segments were initially established using the experimental approach described above, by which Fields (at a later time) demonstrated that reovirus neurotropism segregates with the reovirus genomic segment encoding the σ1 protein. First, cells were coinfected with two different reovirus strains in order to obtain genetically reassorted viruses. Such reassorted viruses are readily generated, since during coinfection each of the reovirus strains generates RNA segments that together constitute a pool from which progeny reovirus genomes are assembled at random. That is, when new progeny viruses are assembled, each can incorporate any of its 10 genomic segments without regard to which parental strain might have encoded it. (This is reminiscent of phenotypic mixing [Chapter 2], which results from the fact that viruses are assembled from pools of subunits. However, in the reovirus case in point, this fundamental aspect of virus replication results in what might be dubbed "genotypic mixing.") Importantly, the electrophoretic mobilities of the genomic segments, as well as of the proteins, of each individual strain are unique (for an example, see Fig. 5.4). Thus, by electropherotyping the genomic segments and proteins of numerous reassorted viruses, it was possible to correlate specific genomic segments with specific proteins. Protein-coding assignments were confirmed and expanded upon when it became possible to carry out in vitro translation of individual mRNAs.

The coding assignments for each of the 10 reovirus genome segments are shown in Fig. 5.6. Note the correspondence between the length in nucleotides of each of the RNA segments and the lengths in amino acids of the corresponding encoded proteins. Also note that segment S1 is unique in that it encodes a second protein product, σ1s, in a reading frame that overlaps that of the larger product, σ1. Thus, the orthoreovirus genome encodes a total of 11 proteins (Fig. 5.6 and Table 5.1). Eight of those proteins are the previously noted structural components of reovirus particles, whereas three proteins, μNS, σNS, and σ1S, are nonstructural proteins that mediate other functions during the reovirus replication cycle.

The relative translation efficiencies of the individual reovirus mRNAs vary as much as 100-fold. This differential translation efficiency makes it possible for very low levels of proteins that function as enzymes (e.g., transcriptase and guanylyltransferase) to be made, relative to the much higher levels of other proteins that are structural components of the virus. As discussed below, there are only 12 λ3 polymerase proteins per virus particle. In contrast, there are 60 copies of λ2 and 600 copies of σ3 (Table 5.1). Thus, translation of the reovirus mRNAs is regulated to produce appropriate relative amounts of the

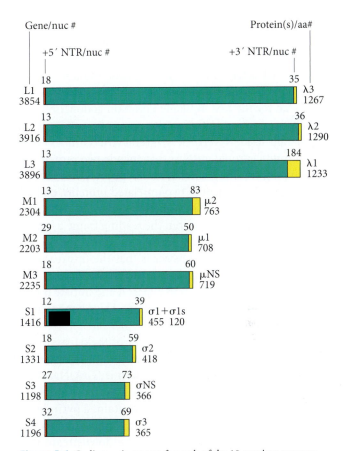

Figure 5.6 Coding assignments for each of the 10 reovirus genome segments. Segments are shown approximately to scale and are oriented so that left to right corresponds to 5′ to 3′ for the protein-coding plus strands. The designation of each genome segment and its length in nucleotides are shown on the left, and the corresponding encoded protein and its length in number of amino acids are shown on the right. The portion of each genome segment that is translated is shaded, and the nucleotide number of the first translated nucleotide is indicated. Notice that the S1 genome segment encodes two proteins, σ1 and σ1s. See the text. Modified from B. N. Fields, D. M. Knipe, and P. M. Howley (ed.), *Fundamental Virology*, 3rd ed. (Lippincott-Raven, Philadelphia, PA, 1996), with permission.

individual protein products. The mechanisms that might underlie the differences in the efficiencies of translation of the individual proteins are not entirely clear. However, there is evidence that nucleotide sequences at positions flanking the initiator AUGs, as well as other sequences, may affect the efficiencies of translation of the individual reovirus mRNAs.

What more might be said about the reovirus S1 mRNA, which, as noted above, encodes two unrelated proteins, σ1 and σ1s? First, recall that the vast majority of eukaryotic mRNAs are monocistronic. That is so because unlike the translation of prokaryotic mRNAs, in which translation can initiate at multiple internal translation initiation sites, translation of eukaryotic mRNAs generally initiates only at the most 5′ terminal translation initiation site on the mRNA. In this regard, the vast majority of eukaryotic mRNAs begin with the 5′ 7-methylguanosine cap structure (see Chapter 6). Because eukaryotic initiation complexes attach to an mRNA only at its 5′ cap structure and then scan for the first available initiation codon, only the most 5′ translation initiation site on a eukaryotic mRNA is generally used. Nevertheless, several virus families make use of multicistronic mRNAs that encode two proteins in overlapping reading frames. A case in point is the reovirus S1 mRNA, which encodes two unrelated proteins, σ1 and σ1s. On the S1 mRNA, the initiation codon for σ1S is the first out-of-phase AUG within the reading frame for σ1.

How might the internal translation initiation site for σ1S be recognized? The answer is that the upstream σ1 translation initiation site is a relatively inefficient one, thereby enabling ribosomes to bypass it and initiate translation at the downstream σS1 start site. This phenomenon is referred to as leaky scanning.

What accounts for the inefficiency of the σ1 start site? The answer is as follows. The most efficient consensus sequence for the initiation of translation is (G/A)NNAUG, where N is any nucleotide and where a purine is at the −3 position. However, the σ1 start site contains a pyrimidine at its −3 position, which enables some ribosomes to bypass it and initiate translation at the downstream σ1s start site. There also is evidence suggesting that σ1 translation is modulated by ribosomes pausing at the downstream σ1S start site. Nevertheless, more of the σ1 protein than of the σ1s protein is translated from the S1 mRNA (Box 5.9).

We already noted that the reoviral double-stranded RNA genome within the cores is the template for the subviral particle-associated transcriptase. Our next question is whether or not each of the 10 core-associated reovirus genomic RNA segments might be transcribed from within the cores. (Once again, it would be good to consider what experiment you might do to answer this question before reading on.) In 1972, in a rather straightforward experiment, Acs and coworkers answered this question. They isolated reovirus cores from infected cells and used them to produce ^{3}H-UTP-labeled transcripts in a cell-free system. The ^{3}H-UTP-labeled single-stranded transcripts next were annealed to unlabeled denatured reovirus genomic RNA isolated from virus particles. Then, polyacrylamide gel electrophoresis was used to compare the double-stranded RNA product of the annealing reaction with ^{3}H-UTP-labeled double-stranded RNA isolated from virus particles. Each of the two samples generated the same 10 bands in the polyacrylamide gels. Thus, each one of the 10 reovirus genomic RNA segments is transcribed from within the cores.

Bearing in mind that the 10 reovirus genomic segments are packed to liquid crystalline density within the cores, one indeed might be surprised that *any* of those encapsidated genome segments actually are transcribed. Our next question asks how many of the 10 reovirus genome segments can be transcribed *simultaneously* within any core. The answer is all ten! This is most dramatically demonstrated by electron microscopy, which shows multiple transcripts being simultaneously extruded from individual subviral cores (Fig. 5.7).

Considering again the tight packing of the reovirus genomes within cores, we would be correct to suppose that there is no room for nascent transcripts to accumulate within those cores. Thus, it would seem that nascent transcripts must be extruded from the cores as they are synthesized, such that at no time do they actually occupy space within the cores. Please pause again to consider how you might test that hypothesis. The way in which it actually was tested was to once again make use of the fact that the interiors of the cores are inaccessible to RNase. So, cores were used to generate transcripts in the presence of ^{3}H-UTP. Since the interiors of the cores are inaccessible to RNase, if the mRNA were being extruded as it was being synthesized, then none of the ^{3}H-labeled product should

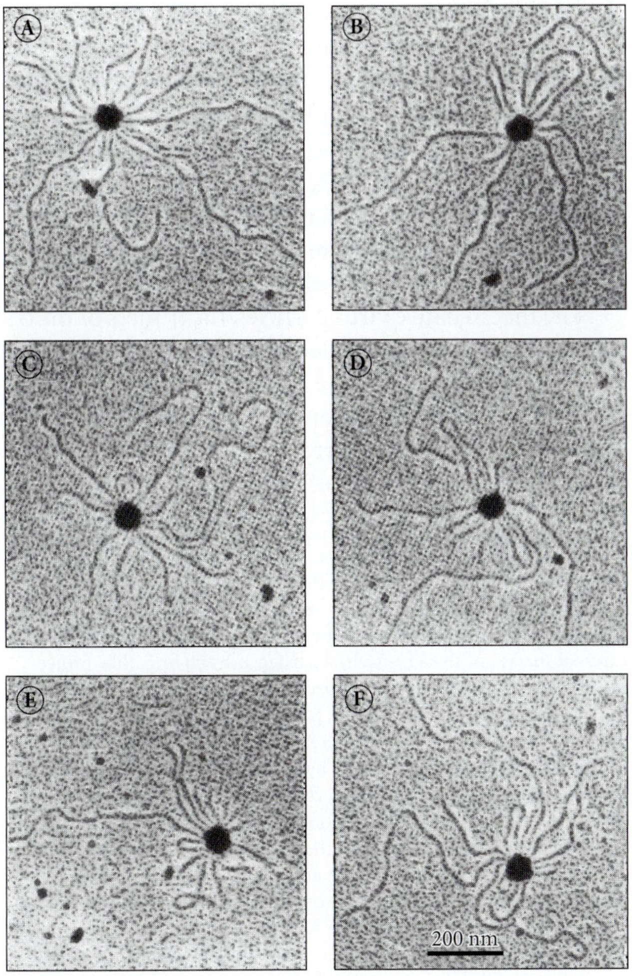

Figure 5.7 Electron micrograph of reovirus cores in the process of simultaneously transcribing 10 mRNA molecules. Bar, 200 nm. Reprinted from J. A. Levy, H. Fraenkel-Conrat, and R. A. Owens, *Virology*, 3rd ed. (Prentice Hall, Englewood Cliffs, NJ, 1994), with permission.

have been resistant to RNase at any time. That was indeed the case. Whereas several minutes were required for the synthesis and release by the cores of complete mRNA molecules, nascent mRNA molecules became sensitive to RNase digestion within seconds of the initiation of their synthesis. Thus, reovirus nascent mRNA is extruded from the cores concomitant with its synthesis, and there is virtually no accumulation of mRNA within the cores. This fact helps to explain in part how reovirus cores are able to simultaneously transcribe 10 genomic segments, since crowding of the transcripts within the cores is not an issue.

Still, we have not yet explained how the cores might generate and secrete even 1 mRNA molecule, much less 10 molecules, simultaneously yet. A deeper understanding of reovirus transcription awaits the consideration that follows of the conformational transitions that lead to the

core structure, and the functions and locations of critical core components that might enable simultaneous transcription of 10 mRNAs, which is concomitant with their extrusion.

Conversion of ISVPs to Cores

ISVPs are generated from reovirus particles by lysosomal proteases after the virus particles are taken up by endocytosis from the cell surface. The ISVPs then play a crucial role in reovirus entry since they have a special ability, which is not possessed by viruses or cores, to permeabilize membranes. This enables ISVPs to cross the lysosomal membrane into the cytoplasm, where replication can then ensue. Yet despite their remarkable ability to cross membranes, ISVPs do not demonstrate transcriptase activity. Rather, transcriptase activity is unmasked during the conversion of ISVPs to cores.

As noted above, conversion of virus particles to ISVPs involves degradation of σ3, cleavage of μ1 to μ1N and μ1C, and the acquisition by σ1 of an extended conformation. These transitions expose the hydrophobic myristoyl moiety at the N terminus of μ1N, enabling it to fulfill its putative role in membrane penetration.

ISVPs become transcriptionally active cores by shedding the 12 σ1 trimeric complexes located at the vertices of the icosahedral shell and by shedding the 200 trimers of μ1 subunits (Fig. 5.1). Indeed, shedding of σ1 may be necessary to disengage the cores from the receptors, which bind the ISVPs to the lysosomal membrane, in that way allowing the cores to be released into the cytoplasm. For that reason, and because ISVPs are intermediates in the transition of virus particles to cores, membrane penetration and the structural transitions leading from ISVPs to cores may occur concurrently.

Shedding of σ1 may also be necessary in order to open up the cores, so that transcription might occur. More is said on that point shortly. First, since the release of the membrane-penetrating μ1N fragment likewise is a key and perhaps final step in reovirus outer-capsid disassembly and entry, we consider that step in more detail.

Cores appear in the cytoplasm at the same time as free μ1N fragments do, consistent with the notion that the release of μ1N fragments occurs coincidentally with, or immediately after, membrane crossing by the cores. Also apropos that observation, the journal literature sometimes refers to "ISVP*s," which are thought to be transient intermediates in the conversion of ISVPs to cores, and which possess some of the properties of each. That is, ISVP*s have shed their σ1 fibers and their μ1C subunits, and their core-associated transcriptase is activated. However, they still retain their μ1N fragments. Thus, the transient existence of ISVP*s is consistent with the premise that μ1N is shed coincidentally with the generation of cores and their release into the cytoplasm.

Molecular chaperones are increasingly recognized as facilitators of a variety of steps in the replication cycles of different viruses, including their disassembly (Box 5.10). Nibert and coworkers recently tested the premise that molecular chaperones might facilitate release of the reovirus μ1N segment. They did so by establishing a reticulocyte cell-free system that carried out release of the μ1N fragments in vitro. To the point, they demonstrated that immunodepleting members of the HSP70 family of molecular chaperones from the reticulocyte lysate has the effect of impairing the release of the μ1 N-terminal fragments. Moreover, Hsc70 was found in association with the released μ1N segments in the cell-free system, as well as in vivo. Thus, these experimental results are completely consistent with the premise that Hsc70 plays a role in reovirus outer capsid disassembly, coincidentally with, or immediately after, membrane penetration.

A superb study by Dryden and coworkers (Dryden et al., 1993) made use of cryo-electron microscopy (cryo-EM) and computer-based three-dimensional image reconstruction to reveal the dramatic changes in structure and protein conformation (at 27- to 32-angstrom [Å] resolution) that characterize the transition of parental reovirus

particles to ISVPs and of the latter to cores. Most importantly, they revealed the striking alteration in the conformation of the λ2 pentamers. In virus particles and ISVPs, flower-shaped pentamers of the λ2 protein are located at each of the 12 icosahedral vertices (Fig. 2.8, 5.8, and 5.9). In the transition from ISVPs to cores, domains of λ2 rotate upward and outward to transform the flower-shaped λ2 pentamers into turret-like structures that now contain an 84-Å-wide central channel.

Note that in native virus particles, the trimers of the σ1 attachment protein are located at the center of each of the 12 λ2 pentamers. The release of σ1 may trigger the transition of the λ2 pentamers from the closed flower-like conformation to the open turret-like structure, while also freeing the central channel of any obstruction.

BOX 5.10

Molecular chaperones are proteins best known for facilitating the correct folding or assembly of other proteins. Since chaperones are often required for assembly of protein complexes, it might be expected that they would also play a role in the reverse process of disassembly of protein complexes, as indeed is the case.

The HSP70 family of 70-kilodalton (kDa) molecular chaperones includes Hsc70, which is constitutively expressed, and Hsp70, which is induced in response to heat shock, ischemia, and other stresses.

Members of the HSP70 family are activated by association with other molecular chaperones (cochaperones) that contain a so-called DnaJ domain. Interestingly, the polyomavirus-encoded LT protein contains a DnaJ domain that associates with, and activates, Hsc70. Moreover, Hsc70 is able to disassemble mouse polyomavirus in vitro. Hsc70 also plays roles in regulating polyomavirus gene expression, genome replication, and final virus assembly (Chapter 15).

Figure 5.8 (A) Twofold (ellipse), threefold (triangle), and fivefold (pentagon) axes of symmetry on a reovirus particle. (B) A shaded representation of the virus surface. The most prominent features on the external surface of the 850-Å virus particle are 600 fingerlike projections (subunits) and a flower-shaped structure in a depression at each of the 12 icosahedral fivefold axes. Each flower-like structure is comprised of five 90-Å-long petals. (C) An ISVP is depicted, which is distinguished from virus particles by the loss of the 600 fingerlike subunits seen in the virus particle, consistent with the loss of σ3 and its smaller diameter of 795 Å. The ISVP has the flower-shaped features at the fivefold vertices, but notice the thin, beaded, spike-like features (σ1) that extend from them for an additional 110 Å or so. (D) Representation of a core. In particular, note how the flower-like feature on virus particles and ISVPs has morphed into the 12 turret-like features at the fivefold vertices. The external radius of the core varies from a maximum of 400 Å for the prominent, turret-like spikes surrounding the fivefold vertices to a minimum of 305 Å at the threefold axes, reflecting the loss of σl, μl, and σ3. Bar, 500 Å. From K. A. Dryden et al., *J. Cell Biol.* **122:**1023–1041, 1993, with permission.

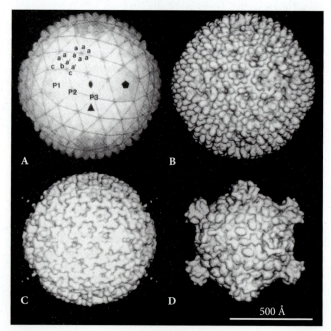

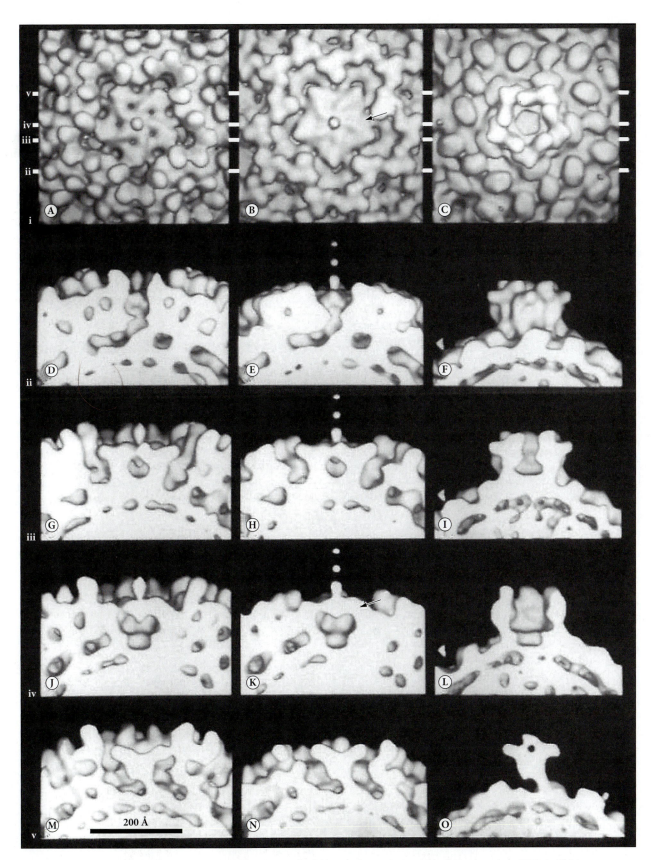

Figure 5.9 Close-up surface-shaded views down a fivefold axis of a reovirus particle (left column), an ISVP (middle column), and a core (right column). The tick marks in the top panels (A, B, and C in row i) identify the locations of cross-sectional views displayed in rows ii through iv. For instance, the topmost tick marks correspond to the side views in row v. The bottom-most tick marks correspond to the side views in row ii. Arrows in B and K represent a putative hinge region of λ2. Bar, 200 Å. From K. A. Dryden et al., *J. Cell Biol.* **122**:1023–1041, 1993, with permission.

So, then, how are these conformational transitions relevant to the activation of the core-associated transcriptase activity and the simultaneous transcription of 10 mRNAs by the tightly packed cores? For starters, the λ2 protein itself is an essential component of the reovirus transcriptional machinery, containing three of the four enzyme activities necessary for adding the 5′ cap that is generally present on eukaryotic mRNAs (Box 5.11). The λ2-associated capping activities include the 5′-guanyltransferase, the 7-N-methyltransferase, and the 2′-O-methyltransferase (Fig. 5.10). Also, μ2 and λ1 are each candidates for expressing the RNA 5′-triphosphatase activity, which mediates the first reaction in capping. Notice how the conformational change of λ2 and the losses of σl, μl, and σ3 expose these activities. Furthermore, and most importantly, the nascent viral mRNAs actually are extruded through the channels of the 12 λ2 pentamers. In this regard, the release of σ1 leaves these channels free of any obstruction.

To better understand how cores are able to carry out simultaneous transcription and concurrent capping of 10 different mRNAs, first note that the λ3 protein is the reovirus RNA polymerase (transcriptase) and that there are 12 λ3 molecules per reovirus particle (Table 5.1). Next, we need to know the location and orientation within the cores of these 12 λ3 molecules. The more recent study of Zhang and coworkers (Zhang et al., 2003) determined just that, using more-refined cryo-EM to reveal the structure of cores at 7.6 Å of resolution. They then fitted the X-ray crystal structure of the λ3 protein into the cryo-EM core structure (Fig. 5.11). Each of the 12 copies of λ3 was found to be anchored to the inner surface of the icosahedral core shell and overlapping, but not centering on, the fivefold axes, which are the sites of the λ2 pentamers. In fact, the bulk of each λ3 molecule lies to the side of a fivefold axis. However, there is enough room for only one molecule of λ3 at each of these 12 vertices.

The structure of λ3 resembles that of other RNA polymerases. The actual catalytic site is in a large cavity within

BOX 5.11

The 5′ cap, which begins with a 5′ 7-methylguanosine moiety that is joined to the second nucleotide by the atypical 5′-5′ phosphodiester linkage, targets eukaryotic pre-mRNAs for further processing and transport from the nucleus. In the cytoplasm, the cap is necessary for the efficient initiation of translation from the 5′ end of the mRNA. This fact takes on added relevance to reovirus infection regarding the takeover by the virus of protein synthesis within the cell (see the text). Also, bearing in mind that cellular capping enzymes are located in the nucleus, we can understand why a virus that carries out transcription in the cytoplasm must provide for its own capping enzymes.

What experiment might you do to demonstrate that capping activity is intrinsic to reovirus cores?

Figure 5.10 Enzymatic activities in transcription and capping by reoviruses. Polymerization of nucleoside triphosphates (shown here for the first nucleotide pair) as specified by a genomic double-stranded RNA template is mediated by the double-stranded RNA-dependent RNA polymerase. The RNA triphosphatase generates a nucleosidyl diphosphate at the 5′ end of the nascent mRNA and sets the stage for formation of the dimethylated 5′ cap by sequential actions of the guanylyltransferase and methyltransferase(s). The role of a helicase in reovirus transcription is proposed but not characterized. Individual reovirus proteins proven or suspected to mediate each activity are indicated (right). S-AdoMet, S-adenosylmethionine; S-AdoHcy, S-adenosyl homocysteine. From B. N. Fields, D. M. Knipe, and P. M. Howley (ed.), *Fundamental Virology*, 3rd ed. (Lippincott-Raven, Philadelphia, PA, 1996), with permission.

Reaction	Enzyme	Product	Protein
pppG + pppC	dsRNA-dependent RNA polymerase	pppGpC + pp$_i$	λ3
pppGpC	RNA Triphosphatase	ppGpC + p$_i$	μ2, λ1
pppG + ppGpC	Guanylyltransferase	GpppGpC + pp$_i$	λ2
S-AdoMet + GpppGpC	Methyltransferase 1	^{m7}GpppGpC + S-AdoHcy	λ2?
S-AdoMet + ^{m7}GpppGpC	Methyltransferase 2 + Helicase	^{m7}GpppG^{m2}pC + S-AdoHcy	λ2?

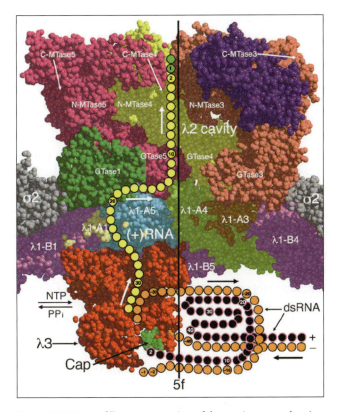

Figure 5.11 Space-filling, cutaway view of the reovirus core, showing a proposed exit pathway for newly synthesized plus-strand (+) RNA transcripts leading from λ3 through the λ1 shell to the λ2 cavity. The approximate locations of four RNA capping guanylyltransferases (GTase1 through 4), three N-terminal methyltransferases (N-MTase1 through 3), and three C-terminal methyltransferases (C-MTase1 through 3) in the λ2 pentamer subunits are shown. The minus and plus strands of the template double-stranded RNA (dsRNA) and the nascent plus-strand transcript are orange, black, and yellow, respectively. The transcript is shown exiting the channel at the top of λ3, traversing a putative channel formed at the interfaces between two λ1-A subunits (A1 and A5) and between λ2 and the λ1-A5 subunit, and finally entering into the large λ2 channel. The cap structure (m7GppppG m2′) at the 5′ end of the dsRNA plus strand is shown tethered to the cap-binding region (green) on the λ3 molecule. The five identical λ1-A subunits (those closest to the fivefold axis) are highlighted in yellow (A1), blue (A2), brown (A3), green (A4), and cyan (A5). The nascent transcript is shown as having already been modified with a 5′ cap (green) by the RNA 5′ triphosphatase (position unknown) and λ2 capping domains. From X. Zhang et al., *Nat. Struct. Biol.* **10**:1011–1018, 2003, with permission.

a central domain of the protein (Fig. 5.11). Two channels lead to that catalytic domain. One of these is the entry channel for the RNA template, and the other is the entry channel for nucleoside triphosphate precursors. Likewise, two channels lead from the central catalytic site. One of these is the exit channel for the RNA template, and the other is the exit channel for the nascent mRNA product. Importantly, the high-resolution cryo-EM studies

demonstrated that the mRNA exit channel of the λ3 protein faces directly towards the inner surface of the core shell, at a site identified as an opening through the shell. This opening allows the nascent mRNAs to pass into the large central channel of the λ2 turrets, where they are capped and extruded from the cores. In contrast, the template entry channel faces toward the center of the core, and the template exit channel lies parallel to the core's inner surface.

Together, the experimental findings recounted above demonstrate that each of the 12 vertices of the reovirus cores contains an RNA polymerase that is ideally oriented with respect to both the core interior and an exit channel. With that in mind, we can begin to appreciate how the reovirus cores are able to simultaneously transcribe and cap 10 distinct mRNAs, despite the fact that they are packed to liquid crystalline density. It is generally believed that each double-stranded genomic segment moves past the fixed λ3 transcriptase molecules and that the nascent RNA is directed through the adjacent overlying λ2 pentamers for capping and extrusion. The closely related rotaviruses and coltiviruses contain 11 and 12 genome segments, respectively. The relationship of the core's transcriptional machinery to the exit channels at the 12 vertices is presumably a reason why these double-stranded RNA viruses do not contain more than 12 genome segments.

While bearing in mind the rather striking relationship between structure and function within reovirus cores, remember that these subviral cores are generated from stable virus particles that serve to protect the virus genome from the environment and to mediate the invasion of susceptible host cells. Furthermore, the native virus particles are also programmed to undergo the structural transitions that promote entry into the cytoplasm and the generation of the transcriptionally active cores. The reovirus particles, ISVPs, and cores together indeed provide impressive examples of the dynamic relationships between molecular form and function.

REPLICATION AND ENCAPSIDATION OF THE REOVIRUS GENOME

Synthesis of Double-Stranded RNA

The experiments described above demonstrated that reovirus cores generate only single-stranded RNA that is exclusively of the plus sense. We now ask when and where the minus strands of the progeny double-stranded RNA genomes are synthesized.

A consideration of the experimental findings noted above gives us a clue from which to start our inquiry. Since minus-sense RNA strands are found only in double-stranded RNA, it is reasonable to hypothesize that minus strands are synthesized on plus-strand templates, part and parcel with their incorporation into double-stranded RNA products. That is, we predict that minus strands are synthesized on preformed plus-strand templates, in that way giving rise to double-stranded RNA genomes. Notice that this hypothesis implies that there is an asynchrony in the synthesis of plus strands and minus strands: plus strands first, and then minus strands.

In 1971, Levin and coworkers demonstrated that this predicted asynchrony in the synthesis of plus strands and minus strands is indeed the case. They reasoned that if double-stranded RNA is formed by the synthesis of minus strands on preformed plus strands, then the *double-stranded* RNA formed during a sufficiently brief exposure of infected cells to ^{3}H-UTP should be preferentially labeled in the minus strand. That is, if the plus strand needs to be synthesized first, then a very brief labeling period (followed by the immediate extraction of the double-stranded RNA) would not provide enough time for the label to be incorporated into both the plus strand and the minus strand, and the entire label incorporated into the new double-stranded RNA molecules would be exclusively in the minus strand.

To be able to test whether very short labeling periods indeed would preferentially label the minus strand of the double-stranded RNA product, the researchers first needed to obtain an excess of RNA plus strands that might be used as a probe to detect the minus strands. To obtain those RNA plus strands, they made use of the fact that in a cell-free system, reovirus cores generate large amounts of single-stranded RNA that is entirely of the plus sense. The actual experiment was then as follows. The double-stranded RNA made during a very short labeling period in cell culture was melted and annealed with an excess of plus strands that were generated by reovirus cores in the cell-free system. The researchers found that 100% of the label in that *double-stranded* RNA could be annealed to the excess plus RNA strands, demonstrating that in the short labeling period, the label incorporated into the double-stranded RNA product indeed was incorporated exclusively into minus RNA strands.

As might be expected, when ^{3}H-UTP was present continuously during reovirus replication in cells, then only one-half of the label in the new double-stranded RNA molecules could be annealed to the excess plus strands. (Do you see why this is so?) However, a half-hour pulse at late times in the infection cycle labeled only the minus strands of the double-stranded RNA. This experimental result demonstrated that at late times in the infection cycle only minus RNA strands are generated, confirming that these RNA minus strands are produced on RNA plus strands that were synthesized earlier during infection.

Thus, reovirus double-stranded RNA replication is dissimilar to DNA replication, with respect to being asynchronous, as well as with respect to being conservative.

As already described, reovirus plus RNA strands are generated exclusively by reovirus cores. Where then are the minus RNA strands generated? The answer is rather extraordinary. The minus strands too are generated exclusively within cores. That is, reovirus plus-sense RNA strands are packaged within new progeny subviral cores. Within those progeny cores, each plus RNA strand then serves as the template for the synthesis of just one minus RNA strand, thereby generating the progeny double-stranded RNA genomes. The channel of the λ3 polymerase, which serves as the exit channel for the template RNA during synthesis of plus strands, is also the exit channel for the double-stranded RNA product during replication.

One might ask how this rather atypical mode of nucleic acid replication might be advantageous to reoviruses. A plausible answer is suggested by the fact that double-stranded RNA is inherently stable in an aqueous environment, thereby impeding its replication. Thus, the partial strand separation required to synthesize the plus strands (i.e., the first stage of reovirus RNA replication) is energetically more favorable within the relatively nonaqueous environment of the subviral core. In the relative absence of polar water molecules, charge repulsion between the phosphate backbones of the RNA strands is more pronounced, thereby favoring strand separation. Another possible advantage is described below in "Reoviruses and IFN."

Assembly of Progeny Subviral Particles and Encapsidation of RNA Segments

While the structure of assembled reovirus cores has been described at the molecular level, the mechanism of their assembly at the molecular level has not yet been characterized. Specific interactions between component proteins and any subsequent conformational transitions remain to be determined.

Another unresolved issue is the means by which each of the 10 unique reovirus genomic double-stranded RNA segments is selected for encapsidation within each particle. Although the mechanism for this selective packaging of the individual genomic segments is not yet known, each of the 10 segments has been sequenced, revealing the presence of small, nontranslated regions at the 5′ and 3′ ends of each plus-sense RNA strand. These nontranslated regions may contain RNA packaging sequences (Box 5.12).

Reoviruses and IFN

The two-stage mode by which reoviruses replicate their double-stranded RNA genomes might be advantageous to reoviruses with respect to evading antiviral defenses

BOX 5.12

One possible means by which a virus with 10 genome segments might assemble a complete genome is by the *random* assembly of more than 10 segments per virus particle. Not all virus particles would acquire complete genomes by that means, but a sufficient percentage would ensure the maintenance of the virus in the host population. However, this possibility cannot account for how reovirus particles acquire complete genomes, since calculations of the space occupied by a complete reovirus genome and of the volume inside the capsid indicate that there is just enough room to accommodate only one complete genome. Moreover, essentially all reovirus particles are able to generate plaques, demonstrating that each contains a complete genome comprised of each of the 10 genomic segments. Also, purified virus particles contain equimolar amounts of the 10 genomic segments. Thus, there must be some mechanism to ensure that exactly one copy each of the 10 reovirus genome segments is encapsidated within each reovirus particle.

Phage φ6, which also contains a segmented double-stranded RNA genome, provides a precedent for a selective packaging mechanism. In that instance, each of the three φ6 double-stranded RNA segments is packaged in a specific sequence. The S segment is the first to be packaged, triggering packaging of the M segment, which in turn triggers packaging of the L segment.

eukaryotic translation initiation factor eIF2-α, thereby impairing its activity and blocking all protein synthesis in that cell. Consequently, the cell is sacrificed, but virus replication within it is aborted.

The reovirus replication cycle, in which viral double-stranded RNA is always encased within a protein coat, *may* have evolved, in part, to avoid inducing a massive IFN response that might limit viral spread. Free genomic double-stranded RNA is not available to upregulate IFN-α/β or to activate PKR.

Despite the fact that reovirus genomic double-stranded RNA is present only in its encapsidated state, reovirus still induces IFN defenses, since single-stranded RNA can form double-stranded stems. However, reovirus has yet other means by which to counteract the IFN-based host defense. The reovirus σ3 protein prevents PKR activation by competing for binding to double-stranded RNA.

PRIMARY VERSUS SECONDARY TRANSCRIPTION

The transcripts produced by the *parental* subviral cores, which are derived from input virus particles, are referred to as **primary transcripts**. As already noted, these transcripts serve as mRNA and also as intermediates in the synthesis of progeny double-stranded RNA genomes within progeny cores. Note that primary transcripts are capped by core-associated capping enzymes. The newly assembled progeny subviral cores also generate mRNA, referred to as **secondary transcripts**. Indeed, secondary transcripts constitute the major fraction of reovirus mRNA that is produced during infection. Importantly, in contrast to primary transcripts, secondary transcripts are not capped.

What might be the significance of the fact that primary transcripts are capped, whereas secondary transcripts are not? To begin answering this question, first note that progeny cores can be isolated from infected cells by density gradient centrifugation. When this was done by Millward and coworkers in 1980, they made several key observations. They began by comparing the enzyme activities expressed by the progeny cores with the activities expressed by native virus particles. In the case of the native virus, as expected, no polymerase or capping activity was detected, unless the particles were treated with chymotrypsin. In contrast, the progeny cores displayed polymerase activity *prior* to chymotrypsin treatment, but with no capping activity. However, treating the progeny cores with chymotrypsin did enable them to also express the capping activity. Thus, whereas the polymerase is active in progeny cores, capping activities are present but masked in those particles.

The above experimental findings, together with the fact that progeny cores appear when viral mRNA synthesis is

mediated by alpha interferon (IFN-α) and IFN-β, referred to collectively as IFN-α/β (the type I interferons). The following is a short, partial account of the biology of IFN-α/β, to provide background for how reoviruses might thwart those antiviral cytokines. See Chapter 4 for a more complete account of IFNs and their manner of action.

Double-stranded RNA, which is generated during infection by a variety of viruses, is a potent inducer of IFN-α/β by the infected cell. However, IFN-α/β usually does not protect the initially infected cells. Instead, it is induced in, and released from, these cells to alert uninfected neighboring cells to the presence of the virus and to prepare them to counter the infection. It does that in part by inducing the neighboring cells to upregulate synthesis of the double-stranded RNA-activated protein kinase **PKR**. However, PKR remains inactive in those bystander cells until they too might become infected. If those cells should then be infected, the double-stranded RNA, which might be produced by the infection, will activate their PKR. The activated enzyme then phosphorylates the maximal, imply that the vast majority of viral transcripts are uncapped. This indeed was found to be the case, a finding which leads to the rather remarkable expectation that the translational machinery of reovirus-infected cells is modified to preferentially translate uncapped mRNA. Indeed, cell-free extracts prepared at late times after infection efficiently translate uncapped reovirus mRNA made late in infection, but inefficiently translate capped reovirus mRNA made early in infection, and cellular mRNA as well. Thus, by modifying the host cells' translational machinery to preferentially translate uncapped mRNA, reovirus subverts the host's capacity to support protein synthesis to favor the purposes of the virus.

It is not yet clear how exactly reoviruses cause translation of uncapped messages to be favored. There is some evidence that reovirus does not actually alter the cellular translational machinery but instead produces a factor that stimulates translation of uncapped mRNAs. Moreover, protein σ3 can exert that effect in vitro.

Interestingly, the 5' ends of all reovirus genomic RNA plus strands are capped. The reason for this is that only the capped RNA produced by the parental cores is packaged in progeny cores, while the uncapped plus strands produced by the progeny cores serves only as mRNAs. Perhaps the cap constitutes part of the packaging signal. Once encapsidated, capped plus strands serve as the templates for the synthesis of minus strands, which then remain in the nascent progeny particles as the minus strands of progeny double-stranded genomes. (Can you see why the minus strands of the double-stranded genomes are not capped? Can you see more than one reason?)

FINAL VIRUS ASSEMBLY

As in the case of the assembly of subviral cores, the molecular mechanisms that underlie final reovirus assembly and maturation are not well understood. Also, the precise order in which reovirus proteins come together to form nascent particles and the role of virus RNA in the assembly process are not yet known. Still, it is known that minus strands are generated within the progeny cores before the outer capsid proteins are added to them.

Also recall that the outer shell of the capsid is comprised of μ1, σ1, and σ3. In addition, recall that σ1 is located in the centers of the λ2 pentamers. Before μ1 and σ3 can add on to the nascent cores to assemble the outer shell of the capsid, they first must form complexes in solution that contain equal numbers of μ1 and σ3. Noting that μ1 exists as a trimer, these hexameric complexes then add on to the nascent cores. The addition of σ1 then seals the particles so that no further transcription is possible, while also completing virus assembly. Some of those viral

proteins may undergo conformational transitions concomitant with assembly. The result is the maturation of progeny cores into mature progeny virus particles.

PATHOGENESIS
Orthoreoviruses

As noted above, **orthoreoviruses** are ubiquitous in humans. Yet orthoreovirus infections of humans are clinically mild or unapparent.

Rotaviruses

Like the orthoreoviruses, rotaviruses too are ubiquitous in humans, with 95% of children worldwide infected by 3 to 5 years of age. But, in contrast to the orthoreoviruses, rotaviruses are important human pathogens, accounting for one-third of all hospitalizations for serious diarrhea worldwide. Moreover, rotavirus diarrhea results in about 500,000 deaths annually in children worldwide.

Since these childhood fatalities generally result from dehydration, disease is most severe in those children who are dehydrated or malnourished prior to infection. Consequently, and despite the fact that infection occurs with equal frequency throughout the world regardless of the level of sanitation, the highest percentages of fatal infections occur in developing countries, where children are more likely to be malnourished or dehydrated before infection. In the United States, rotavirus infections are responsible for about 70,000 cases of severe diarrhea in children that require hospitalization (because of dehydration). Indeed, about 50% of all such cases of severe diarrhea in children are due to rotavirus infections.

It is thought that rotaviruses are passed from person to person by the fecal-oral route. Consistent with that belief, rotavirus is shed in high concentration in the stool (i.e., as many as 10^{10} virus particles per g of stool) both before the onset of diarrhea and up to 10 days after symptoms appear. Moreover, the virus is usually not found in any other specimens, implying that the fecal-oral route is the predominant means of transmission. Only a few infectious virus particles are needed to cause disease.

Until recently, rotavirus infections were thought to be restricted to enterocytes, the cells that constitute the villous epithelium of the small intestine. However, there are now several reports that rotavirus can escape and spread from the gastrointestinal tract into the blood circulation. For example, rotavirus antigens and RNA were identified in serum samples from approximately 65% of children suffering from rotavirus diarrhea, implying that viremia can occur during rotavirus infection. Moreover, there also are reports of rotavirus antigen and RNA in the central nervous systems, spleens, hearts, kidneys, testes, and bladders of children who succumbed to rotavirus infections. Also, there is evidence for the virus in the heart microvasculature of children and adults with either respiratory infections or cardiorespiratory failure. Together, these findings imply that rotaviruses might cause a wider range of clinical disease than just gastroenteritis. Additional studies will be necessary to determine the actual extent of rotavirus nonintestinal disease and also to account for how the virus disseminates from the intestine.

Since rotaviruses are ubiquitous and infections are acquired early in life, it is difficult to prevent virus spread. Also, there is no specific therapy for rotavirus infection, and treatment is generally supportive, involving fluid replacement.

Considering the potential severity of rotavirus infections, especially in children, vaccines have been developed against these agents. RotaShield was the first rotavirus vaccine to be licensed by the U.S. Food and Drug Administration. As such, it was expected to play a major role in global efforts to protect children, especially in developing countries, from potentially fatal rotavirus infections. However, the vaccine program suffered a major setback in 1999, when RotaShield was withdrawn from the U.S. market, less than a year after its introduction. The reason for the withdrawal was that the vaccine was associated with an uncommon but potentially life-threatening adverse occurrence, **intussusception**, in about 1 of every 10,000 vaccine recipients. (Intussusception is a form of intestinal obstruction in which a segment of the bowel prolapses, or slips into a more distal segment.)

Yet here was an interesting issue. In the United States there are "only" about 20 deaths annually from rotavirus diarrhea. Thus, in the United States, the potential risk of RotaShield clearly exceeded its benefits. On the other hand, in some developing countries, a child's risk of death from rotavirus diarrhea is 1 in 200 or greater, clearly exceeding the potential risks of the vaccine. So, should vaccine standards that apply in the United States be applied globally?

To at least some readers it may seem that the vaccine standards should be determined by the particular circumstances in a region. Nevertheless, debate over this question took place, and it soon became obvious that it would not be tenable to send a vaccine to developing countries that had been withdrawn in the United States, and thus, RotaShield is no longer produced. So, is this an example of political correctness trumping sound public health policy? If you were a public health official in a developing country, where the risk of a child dying from rotavirus diarrhea is 1 in 200, how would you feel about the nonavailability of RotaShield in your country?

During the brief time that RotaShield had been in use, about 1 million children received the vaccine, a sufficient number to establish the risk of vaccine-associated intussusception. Other rotavirus vaccines in development have (as of February 2007) been administered to far too few individuals to establish the frequency of inadvertent complications. Thus, it was not yet known at the time this book went to press whether any of these other vaccines might be an acceptable means for controlling rotavirus infection. Moreover, it is conceivable that rotavirus itself may cause intussusception in children who are prone to this problem and that vaccination with any live oral rotaviral agent may be a triggering event.

At present, China is the only country that has a rotavirus vaccine in use. There is little, if any, information available regarding its efficacy and safety.

Coltiviruses

Coltiviruses are transmitted by arthropod vectors, specifically, by the wood tick *Dermacentor andersoni*. Also, it is interesting that coltiviruses have 12 genomic segments, rather than the 10 genomic segments characteristic of the orthoreoviruses.

The **Colorado tick fever virus** is the prototype coltivirus, and it is responsible for the illness bearing its name, one of the most common tick-borne viral diseases in the United States. Apropos its name, the rate of Colorado tick fever is particularly high in Colorado, where blood tests have shown that up to 15% of campers have previously been exposed to the virus. Infection is much less common in the rest of the United States.

Risk factors for Colorado tick fever virus infection are recent outdoor activity (particularly in Colorado) and recent tick bite. The infection is acute, and symptoms, which first emerge 3 to 6 days after the tick bite, include fever, sweating, muscle aches, stiff joints, headache, sensitivity to light, nausea and vomiting, and occasional rash. (It is important to not confuse Colorado tick fever with **Rocky Mountain spotted fever**. The latter illness is caused by a tick-borne rickettsial infection and is therefore susceptible to antibiotic therapy, which may be necessary in some instances.)

Although Colorado tick fever virus infection is usually self-limiting and not dangerous, serious hemorrhagic disease can result from infection of endothelial cells and pericytes (the two cell types that predominate in the endothelium of the microvasculature), thereby weakening capillary structure. The hemorrhaging can result in hypotension and shock. Other possible severe complications that can result from Colorado tick fever virus infection include meningitis and encephalitis. Hemorrhagic fever and encephalitis are seen in about 5% of patients, most of whom are children.

Colorado tick fever virus also infects erythroid precursor cells, doing so without causing lethal cell damage. This enables the virus to persist in these cells, even after they mature into red blood cells, and explains how viremia can persist for months after recovery from symptoms, a feature that promotes transmission of the virus back to the tick vector.

There is no specific treatment for Colorado tick fever. A vaccine against the virus has been developed. However, since the infection is usually not life threatening, its general use has not been justified.

Another coltivirus, **Eyach virus**, was isolated from ticks in France and Germany, where it has been incriminated in febrile illnesses and neurologic syndromes in humans. The natural cycle of Eyach virus is not yet known, but serological surveys in mammals suggest that its reservoir host may be the European rabbit *(Oryctolagus cunniculus)*. Eyach virus may be widespread in Europeans, as suggested by one survey that detected anti-Eyach virus antibodies in sera and cerebrospinal fluid from 12% of patients in the former Czechoslovakia. Most of those patients were initially diagnosed as suffering from tick-borne encephalitis, an infection caused by an arthropod-borne virus belonging to the flavivirus family (Chapter 7). To date, it has not been possible to propagate Eyach virus in mammalian cell cultures.

Selected Readings

Dryden, K. A., G. Wang, M. Yeager, M. L. Nibert, K. M. Coombs, D. B. Furlong, B. N. Fields, and T. S. Baker. 1993. Early steps in reovirus infection are associated with dramatic changes in supramolecular structure and protein conformation: analysis of virions and subviral particles by cryoelectron microscopy and image reconstruction. *J. Cell Biol.* **122:**1023–1041.

Maginnis, M. S., J. C. Forrest, S. A. Kopecky-Bromberg, S. K. Dickeson, A. Samuel, S. A. Santoro, M. M. Zutter, G. R. Nemerow, J. M. Bergelson, and T. S. Dermody. 2006. β1 integrin mediates internalization of mammalian reovirus. *J. Virol.* **80:**2760–2770.

Zhang, X., S. B. Walker, P. R. Chipman, M. L. Nibert, and T. S. Baker. 2003. Reovirus polymerase λ3 localized by cryo-electron microscopy of virions at a resolution of 7.6 Å. *Nat. Struct. Biol.* **10:**1011–1018.

6

RNA VIRUSES: SINGLE STRANDED

Picornaviruses

INTRODUCTION

The *Picornaviridae* are small plus-strand RNA viruses. In fact, *pico* means small; consequently, picornavirus means small RNA virus. But bear in mind that the genomes of RNA viruses are small in general (Fig. 2.11). (The high error frequency of RNA polymerases places a theoretical upper limit on the size of RNA virus genomes at about 30 kb. Indeed, this is the genome size of the coronaviruses, the largest of the RNA viruses.) What then might distinguish the picornaviruses from other families of plus-strand RNA viruses? First, they, together with the caliciviruses (Chapter 13), in fact are the smallest of the plus-strand RNA viruses. Second, the family is further characterized by its nonenveloped icosahedral capsids, which are assembled from four nonidentical protein subunits. As plus-strand RNA viruses, picornaviruses are Class IV viruses in the Baltimore classification scheme.

Picornaviruses occupy a prominent place in the history of virology. A picornavirus, foot-and-mouth disease virus (FMDV), was the first animal virus to be discovered (Chapter 1). Also, quantitative animal virology had its beginnings when Enders and Dulbecco developed methods for propagating poliovirus in cell culture and for measuring poliovirus infectivity by plaque assay (Box 1.11). These studies enabled Salk and Sabin to develop their respective poliovirus vaccines, which are among the greatest achievements of 20th century medical research. Moreover, the first molecular studies of animal virus replication were carried out using picornaviruses, particularly poliovirus. The first RNA-dependent RNA polymerase was discovered by David Baltimore in cells infected with mengovirus, a member of the cardiovirus genus of the picornavirus family (see below).

The picornavirus family currently contains nine genera, and several more were recently proposed. The best-known picornavirus genera are *Rhinovirus*, *Enterovirus*, *Hepatovirus*, *Cardiovirus*, *Parechovirus*, and *Aphthovirus*. The other genera are *Erbovirus*, *Kobuvirus*, and *Teschovirus*.

The various picornavirus genera contain important pathogens of humans and animals. Some of these are briefly enumerated here and are discussed in greater detail later.

We begin our brief survey with the rhinoviruses, since they are the major cause of the common cold. Although not life threatening, the common cold is the most frequent reason why people seek medical attention. More than 100 separate rhinovirus serotypes cause the common cold. However, note that other unrelated viruses (i.e., coronaviruses, respiratory syncytial virus, parainfluenza virus, influenza virus, and adenovirus) together cause somewhat more than one-half of all common colds. These important facts account for why modern medicine has been unable to make any major inroads against this most pervasive human illness.

The enteroviruses include poliovirus (the agent of poliomyelitis and the prototype for the entire picornavirus family), coxsackievirus groups A and B (named after Coxsackie, NY, where they first were isolated), and echoviruses. The medical significance of these viruses and of members of other picornavirus genera is noted briefly here and discussed at greater length later.

During the first half of the 20th century, paralytic poliomyelitis was one of the most feared diseases in the industrialized world. However, the success of the poliovirus vaccines has led to the near disappearance of poliovirus from the Western world and, indeed, from much of the rest of the world as well. Yet poliovirus is still endemic in Nigeria, India, Pakistan, and Afghanistan, and infections occur in several other countries as well.

Coxsackieviruses cause a range of illnesses in humans, including a common-cold-like syndrome, myocarditis, and aseptic (i.e., viral as opposed to bacterial) meningitis and paralysis. The best-known coxsackie A virus disease is hand, foot and mouth disease, which is a common childhood illness that is often produced by coxsackievirus A16. Despite its name, this illness is not related to, and should not be confused with, foot-and-mouth disease, which is caused by FMDV, a member of the genus *Aphthovirus* of the *Picornaviridae*. Coxsackie B virus diseases can range from relatively minor gastrointestinal symptoms to paralysis, cardiac damage (myocarditis), and birth defects. Indeed, coxsackie B virus is the most common agent for myocarditis and dilated cardiomyopathy. Nevertheless, most coxsackie B virus infections are mild and subclinical.

The acronym *echo* (derived from "enteric cytopathic human orphan") denotes that these viruses were initially not associated with any disease and were thought to be nonpathogenic in humans. That acronym is now known to be misleading, since echoviruses cause two-thirds of the 30,000 to 50,000 cases of viral meningitis in children and adolescents that require hospitalization each year in the United States.

Hepatitis A virus was originally classified within the genus *Enterovirus*. But while hepatitis A virus shares some characteristics with the enteroviruses, it is sufficiently different from them that it is now classified as the only species within the genus *Hepatovirus*. Distinctive characteristics of hepatitis A virus include its unique stability at pH 1 and resistance to heat (56°C for 30 min). Also, in contrast to the other picornaviruses, hepatitis A virus is not cytolytic. As its name implies, hepatitis A virus is one of the viral agents responsible for clinical hepatitis.

The genus *Cardiovirus* is generally considered to be comprised of murine viruses, particularly encephalomyocarditis virus (EMCV) and Theiler's murine encephalitis virus. The latter virus persistently infects the mouse central nervous system, where it causes a chronic T-cell-mediated demyelinating disease that has been an important model of multiple sclerosis in humans (Chapter 4).

The aphthoviruses are the FMDVs. These viruses are the single most important pathogens of livestock, infecting mainly cattle, sheep, goats, and pigs. They rarely infect humans.

The picornaviruses are the first family of eukaryotic plus-strand RNA viruses to be dealt with in this book. They and several other families of eukaryotic plus-strand RNA viruses employ a pattern of gene expression that may appear bizarre to at least some readers. In brief, they generate their individual gene products by proteolytic cleavage of large precursor **polyproteins**. In contrast, the individual proteins of the RNA phages are each the product of a separate translation initiation event. To appreciate why the picornaviruses and other families of eukaryotic plus-strand RNA viruses take their particular approach to translation, later in the chapter we digress to consider the RNA phages, and then we see why the eukaryotic RNA viruses must express their genomes so differently from the RNA phages.

STRUCTURE, BINDING, AND ENTRY
Picornavirus Structure

Picornavirus capsids display icosahedral symmetry and are typically constructed from four distinct proteins, VP1, VP2, VP3, and VP4. These four nonidentical proteins comprise a subunit structure called a **protomer**, which contains one of each protein. The icosahedron is built from pentamers of those protomers (Fig. 6.1). Note that VP1, VP2, and VP3 are at the surface of the particle, whereas VP4 is located on the inner surface of the shell. As described below, VP1, VP2, and VP3 each have a wedgelike shape that enables them to pack tightly into the virus structure. The "prows" of the VP1 proteins come together to make up the fivefold symmetry axes, whereas the prows of VP2 and VP3 come together to form the threefold symmetry axes. Notice the similarity between the structure of

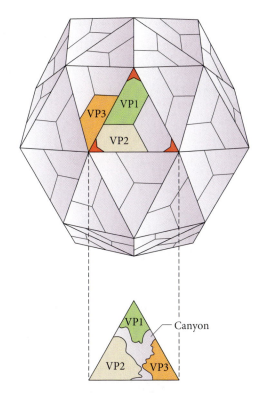

Figure 6.1 Picornavirus capsids display icosahedral symmetry and are typically constructed from four nonidentical proteins, VP1, VP2, VP3, and VP4. These four proteins constitute a subunit structure called a protomer, which contains one of each protein. VP1, VP2, and VP3 are at the surface of the particle, whereas VP4 (not seen in the figure) is located on the inner surface of the shell. The icosahedron is built from pentamers of those protomers. VP1, VP2, and VP3 each have a wedge-like shape, featuring a prow (indicated by red arrowhead) at one end of the protein. The wedge-like shape of VP1, VP2, and VP3 enables them to pack tightly into the virus structure. The prows of the VP1 proteins come together to make up the fivefold symmetry axes, whereas the prows of VP2 and VP3 come together to form the threefold symmetry axes. (Despite the fact that six equilateral triangular faces come together at the latter points, they nevertheless are points of threefold symmetry. Can you see that?) As illustrated in the drop-down insert, the orientation of the capsid proteins results in the surfaces of VP1 proteins being separated from the surfaces of VP2 and VP3 by a valley or cleft that runs around each fivefold axis. Note the overall similarity between the structure of the picornavirus capsid and that of a capsid comprised of identical subunits, as seen in Fig. 2.8B. Modified from B. N. Fields, D. M. Knipe, and P. M. Howley (ed.), *Fundamental Virology*, 3rd ed. (Lippincott-Raven, Philadelphia, PA, 1996), with permission.

the picornavirus capsid (Fig. 6.1) and that of a capsid comprised of identical subunits, as seen in Fig. 2.8B. The assembly of VP1, VP2, VP3, and VP4 into protomers will make eminent sense after considering the mode of their translation, later in the chapter.

VP1, VP2, and VP3 do not contain related amino acid sequences. Nevertheless, each of these proteins contains a central β-sheet region, referred to as a β-barrel jelly roll, Swiss jelly roll, or eight-stranded antiparallel β-barrel (Fig. 6.2). This β-barrel structure is the basis for the

wedge-like shape of the picornavirus capsid proteins, which enables them to fit together at the vertices of the icosahedral particle. The walls of the wedge are formed by two sets of β-strands (G D I B and C H E F in the schematic), which come together to form the prow. The loops that connect the β-strands vary between the three proteins, and these variations are responsible for the structural and antigenic differences among the picornaviruses. Interestingly, the β-barrel structure is found in the capsid proteins of a variety of unrelated viruses, including those of plants and invertebrates, regardless of their size and whether they contain RNA or DNA genomes. This fact implies either that some of these viruses have a common ancestor or that convergent evolution has selected for similar protein structural domains in unrelated viruses.

The discussion under the remaining subheadings in this section of the chapter concerns picornavirus receptor binding and entry. It is somewhat lengthy overall to underscore the point that different species of the *Picornaviridae*, although closely related, may interact differently with their respective receptors and have dissimilar pathways for entry. These differences highlight for us the potential hazard of making generalizations for an entire virus family, or even for a genus, based on the experimental findings for any particular member of the family or genus. Also, they illustrate the variety of receptors and modes of entry that might be used even within a family or genus of closely related viruses. Moreover, they enable us to formulate several important general principles concerning virus-receptor interactions and entry. Finally, the analysis of virus entry in general has become one of the most active areas of modern virology. We begin with the major subgroup of the human rhinoviruses.

Rhinovirus Receptor and Binding: the Canyon Hypothesis

The major subgroup of human rhinoviruses, which comprises about 90% of the more than 100 rhinovirus serotypes, is distinguished from the minor subgroup by its receptor preference. Major subgroup rhinoviruses use the intracellular adhesion molecule 1 (ICAM-1) for their cell surface receptor. ICAM-1 is present on the nasal epithelium, as well as on most other cells of the body. The normal function of ICAM-1 is to bind to the protein LFA-1, which is expressed on the surfaces of lymphocytes. The interaction between LFA-1 on lymphocytes and ICAM-1 on tissue cells strengthens the adhesion between lymphocytes and their target cells during immune interactions (e.g., during T-lymphocyte-mediated lysis of virus-infected cells) (Box 6.1).

Interestingly, cell surface expression of ICAM-1 is upregulated by several cytokines (i.e., mediators of inflammation

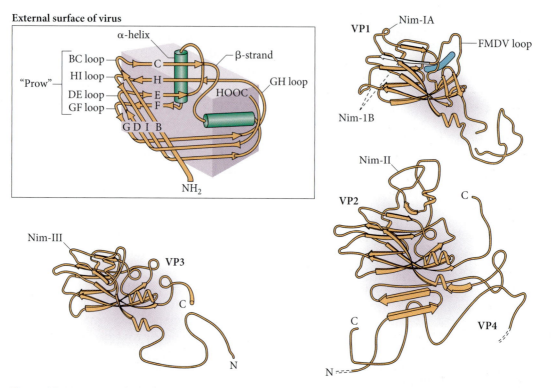

Figure 6.2 VP1, VP2, and VP3 do not contain related amino acid sequences. Nevertheless, each of these proteins contains a central β-sheet region, referred to as a β-barrel jelly roll, Swiss jelly roll, or eight-stranded antiparallel β-barrel. This β-barrel structure gives each of these proteins their wedge-like shape, which enables them to fit together at the vertices of the capsid. The walls of the wedge are formed by two sets of β-strands (GDIB and CHEF in the schematic), which come together to form the prow (Fig. 6.1). The loops that connect the β-strands vary between the three proteins, and these variations are responsible for the structural and antigenic differences among the picornaviruses. Note that VP4 is depicted as it would sit relative to VP2, on the inner surface of the shell, and that VP2 and VP4 are generated by the proteolytic cleavage of a VP0 (VP4VP2) polyprotein precursor (Fig. 6.9). The blue sausage-shaped structure in VP1 is a WIN compound, an antiviral that inhibits picornavirus attachment or uncoating (Box 6.6). Modified from B. N. Fields, D. M. Knipe, and P. M. Howley (ed.), *Fundamental Virology*, 3rd ed. (Lippincott-Raven, Philadelphia, PA, 1996), with permission.

such as gamma interferon) that are released at sites of infection. This process serves the host by localizing lymphocytes to those sites. However, it also results in enhanced expression of the rhinovirus receptor on neighboring uninfected cells, thereby facilitating spread of the rhinovirus infection. Thus, by using ICAM-1 for their receptor, rhinoviruses actually may be subverting the host's inflammatory response to further their own replication and spread. (If this is the reason rhinoviruses evolved to use ICAM-1 for their receptor, then this is indeed a rather diabolical adaptation on the part of the virus.)

ICAM-1 is a member of the immunoglobulin (Ig) superfamily of proteins. Recall that the reovirus receptor JAM1 also is an Ig superfamily protein (Chapter 5). However, whereas JAM1 has one extracellular Ig-like domain, ICAM-1 has five such extracellular domains that are arranged head to tail (Fig. 6.3). Rhinoviruses interact with the two most membrane-distal Ig-like domains of ICAM-1

(D1 and D2). This detail was demonstrated by an experiment that made use of the finding that human rhinoviruses can bind to human ICAM-1 but not to mouse ICAM-1. Recombinant ICAM-1 genes were constructed that encoded chimeric molecules which contained human and mouse Ig-like domains. Human rhinoviruses were able to bind only to those chimeric ICAM-1 molecules that derived the two most membrane-distal Ig-like domains from human ICAM-1. Studies using deletion mutants corroborated the conclusion that the two membrane-distal Ig domains are necessary for the rhinoviruses to bind and also showed that they are sufficient to support binding. Analysis of amino acid substitutions within those domains demonstrated that the most critical binding sites for the virus, and for LFA-1 as well, are in the most membrane-distal Ig-like domain (Fig. 6.3). However, the sites in that domain used by rhinoviruses for binding are distinct from the binding sites for LFA-1.

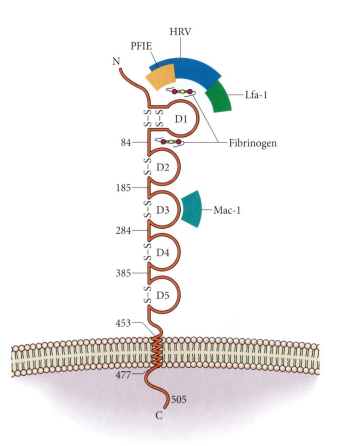

Figure 6.3 Major subgroup rhinoviruses use the intracellular adhesion molecule 1 (ICAM-1) for their cell surface receptor. ICAM-1 is a member of the Ig superfamily of proteins and has five extracellular Ig domains (Chapter 4) that are arranged head-to-tail. Rhinoviruses interact with the two most membrane-distal Ig domains (D1 and D2) of ICAM-1. The binding sites on ICAM-1 for human rhinoviruses (HRV), LFA-1 (a natural ligand of ICAM-1), malaria-infected erythrocytes (PFIE), fibrinogen, and MAC-1 (a glycoprotein related to LFA-1 that has a role in neutrophil adhesion to endothelial cells [Chapter 4]) are indicated. Modified from J. Bella et al., *Proc. Natl. Acad. Sci. USA* **95:**4140–4145, 1998, with permission.

Now let us look again at the rhinovirus structure. First, the loops that connect the β-strands and the β-sheets within the individual picornavirus capsid proteins, VP1, VP2, and VP3, comprise the surface features of these viruses (Fig. 6.1, 6.2, and 6.4). Second, note how the prows of VP1 concentrate at the fivefold axes of symmetry and how the prows of VP2 and VP3 concentrate at the threefold axes of symmetry (Fig. 6.1). This orientation of the capsid proteins results in the surfaces of VP1 proteins being separated from the surfaces of VP2 and VP3 by a valley or cleft that runs around each fivefold axis (Fig. 6.1 and 6.4). In the human rhinoviruses, these clefts are so wide and deep (approximately 1.2 to 1.5 nm wide and 1.2 nm deep) that they are referred to as **canyons**. The "north wall" of each canyon is formed by VP1, while VP2 and VP3 form the south wall. Importantly, these canyons on the surfaces of the rhinoviruses contain the binding sites for the receptor. Evidence in support of this fact is provided by genetic analysis, in which substitutions of certain amino acids that line the canyons but not of amino acids in the surface loops abolish rhinovirus binding to ICAM-1. Moreover, reconstructed images from cryo-electron microscopy of rhinovirus particles bound to the two most membrane-distal

Ig-like domains of ICAM-1 confirm that the receptor binds to the sites in the canyon predicted by the genetic analysis (Fig. 6.5). Furthermore, in those images, ICAM-1 appears to bind to each of the 60 symmetrically placed depressions on the virus, consistent with the capsid structure. In addition, the ICAM-1 fragments are oriented roughly perpendicular to the virus surface, making more extensive contacts with the southern than the northern wall and rim of the canyon.

The antigenic sites on the virus surface that are recognized by host antibodies are located in the exposed loops on VP1, VP2, and VP3 (Fig. 6.4). In the case of the human rhinoviruses, these sites appear to have undergone extensive antigenic variation in response to antibody-driven selection (note the approximately 100 human rhinovirus serotypes).

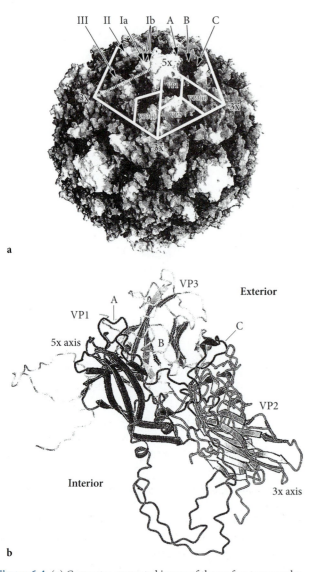

a

b

Figure 6.4 (a) Computer-generated image of the surface topography of poliovirus, which like the rhinoviruses has a canyon that runs around each fivefold axis of symmetry (shaded dark and indicated by B). A points to the north rim, and C points to the south rim. Ia, Ib, II, and III show the locations of the binding sites for neutralizing antibodies against poliovirus. (b) Drawing of a ribbon model depicting the canyon relative to A, B, and C, as shown in panel a, highlighting the eight-stranded β-barrel structure of VP1, VP2, and VP3. VP1 and VP2 build up the side walls, where the BC loop (Fig. 6.2) at position A and the GH loop at position C are shaded black. The back wall, given by VP3, is lightly shaded. In the human rhinoviruses, the canyon is approximately 1.2 to 1.5 nm wide and 1.2 nm deep. From E. Wimmer, J. J. Harber, J. A. Bibb, M. Gromeier, H. H. Li, and G. Bernhardt, Poliovirus receptors, p. 101–127, *in* E. Wimmer (ed.) *Cellular Receptors for Animal Viruses*, Cold Spring Harbor Laboratory Press, Cold Spring Harbor, NY, 1994, with permission.

Importantly, despite the impressive width of the canyons, at 1.2 to 1.5 nm they still are much too narrow to permit penetration by the Fab regions of antibody molecules (Fig. 6.6). Fab regions are the portions of antibody

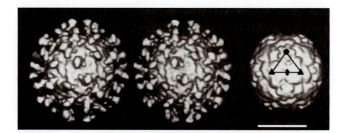

Figure 6.5 A reconstructed stereoview from cryo-electron microscopy of a rhinovirus particle binding the two most membrane-distal Ig-like domains (D1 and D2) of ICAM-1, as viewed along an icosahedral twofold axis. A molecule of ICAM-1 appears to bind to each of the 60 symmetrically placed depressions on the virus. In addition, the ICAM-1 fragments are oriented roughly perpendicular to the virus surface, making more extensive contacts with the southern than the northern wall and rim of the canyon. Try to see the stereo image. Hint: Place your first finger against your nose and your pinky between the left and right images. From E. Wimmer, J. J. Harber, J. A. Bibb, M. Gromeier, H. H. Li, and G. Bernhardt, Poliovirus receptors, p. 101–127, *in* E. Wimmer (ed.), *Cellular Receptors for Animal Viruses*, Cold Spring Harbor Laboratory Press, Cold Spring Harbor, NY, 1994, with permission.

molecules that bind to antigens (Chapter 4). They consist of a segment of a longer heavy chain and a shorter light chain. The heavy chain and light chain interact with each other such that each of the two Ig domains on the light chain is paired with an Ig domain on the heavy chain. The pair of Ig domains at the distal end of the Fab region comprises the antigen-binding site, which has a diameter of approximately 3.5 nm, which is too wide to penetrate into the 1.2- to 1.5-nm-wide canyon. In contrast to the dimeric Fab regions, ICAM-1 molecules are monomers with linearly arranged Ig-like domains, which readily penetrate into the rhinovirus canyons. Consequently, since the canyons are inaccessible to antibodies, the virus can evade antibodies by varying exposed amino acids in the surface loops, without the pressure of having to alter its crucial receptor-binding sites in the canyons. This is the so-called canyon hypothesis (Box 6.2).

The canyon hypothesis is consistent with the following observations. Among rhinovirus serotypes, the most variable amino acids indeed are found in the loops at the rim of the canyon, which, as noted, are the major attachment sites for neutralizing antibodies. In contrast, amino acids that line the canyon tend to be conserved. Also, when rhinovirus mutants were selected for their abilities to grow in the presence of neutralizing antibodies, the substituted amino acids were found in the portions of the loops at the canyon rims. Moreover, site-directed mutagenesis showed that specific amino acids on the floor of the canyon are necessary for receptor binding. Finally, as noted above, cryo-electron microscopy confirmed that ICAM-1 binds in the 1.2-nm-deep canyons.

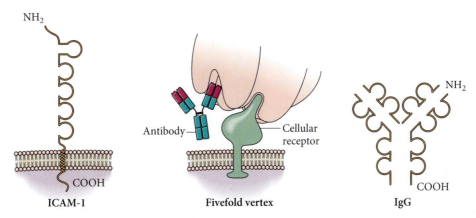

Figure 6.6 The antigen-binding sites of an antibody molecule consist of two polypeptide chains, each of which contains Ig-like domains. In contrast, ICAM-1, the receptor for the major subgroup human rhinoviruses, consists of a single polypeptide chain that contains Ig-like domains. Consequently, the binding site of an ICAM-1 molecule is about half as wide as that of an antibody molecule. Thus, despite the impressive width of the rhinovirus canyons, at 1.2 to 1.5 nm they still are much too narrow to permit penetration by antibody molecules but do allow the penetration of the receptor. Modified from B. N. Fields, D. M. Knipe, and P. M. Howley (ed.), *Fundamental Virology*, 3rd ed. (Lippincott-Raven, Philadelphia, PA, 1996), with permission.

BOX 6.2

Strictly speaking, it is not entirely clear how antibodies neutralize a virus. Operationally, neutralization of a virus by antibodies is measured as a reduction in virus infectivity. Yet not all antibodies that bind to a virus reduce its infectivity. However, an antibody that binds to a virus's receptor-binding site will reduce infectivity. Thus, having its receptor-binding site in a canyon that is inaccessible to antibodies may be a stratagem for the virus to escape one means of antibody neutralization. Considering the other side of the coin (i.e., the receptor), residues on ICAM-1 that are important for binding of LFA-1 are more highly conserved than residues involved in binding rhinovirus. Is it clear why that should be so?

Although the canyon hypothesis has indisputable appeal, it may be premature to accept it as entirely correct or to perhaps risk overstating its utility to the virus. These provisos are implied by X-ray crystallographic analysis of a complex between a rhinovirus particle and a specific antibody molecule, which showed that the antibody molecule indeed could penetrate into the canyon. The ability of the antibody to penetrate into the canyon despite the fact that the canyon is narrower than the footprint of the antibody might be explained by the tendency of the rhinovirus capsid to "breathe." That is, the rhinovirus capsid may well be a pulsating, breathing structure that spontaneously undergoes conformational transitions that might

make the canyon momentarily accessible to antibodies. This premise is supported by the experimental finding that residues of VP4 and internal residues of VP1 spontaneously become accessible to trypsin. This breathing on the part of the virus may play a role in entry (see below).

The approximately 10 serotypes of the minor subgroup of the human rhinoviruses use members of the low-density lipoprotein (LDL) receptor family for their receptor. The LDL receptor is not an Ig superfamily protein, and it is neither functionally nor structurally related to ICAM-1 (Fig. 6.7). Interestingly, comparison of the amino acid sequences within the canyons of the major and minor subgroups of the human rhinoviruses does not provide an obvious reason for why the latter viruses do not use ICAM-1 for their receptor. A recent analysis using cryo-electron microscopy showed that in contrast to the rhinovirus major subgroup, the minor subgroup receptor binds to the star-shaped dome on the fivefold axis rather than in the canyon (Fig. 6.4a; note the star-shaped structures at the fivefold axes).

The Poliovirus Receptor

The gene for the poliovirus receptor (PVR) was isolated by making use of the fact that rodent cells are not susceptible to poliovirus, primarily because they do not express the human PVR, and the virus does not bind to the murine homologue of that protein. (Indeed, poliovirus has no known nonhuman host, presumably because it is exacting in its specificity for the human PVR. This fussiness and the consequent absence of a nonhuman reservoir for poliovirus are important factors in the anticipated vaccine-based global eradication of poliovirus.) Thus, the

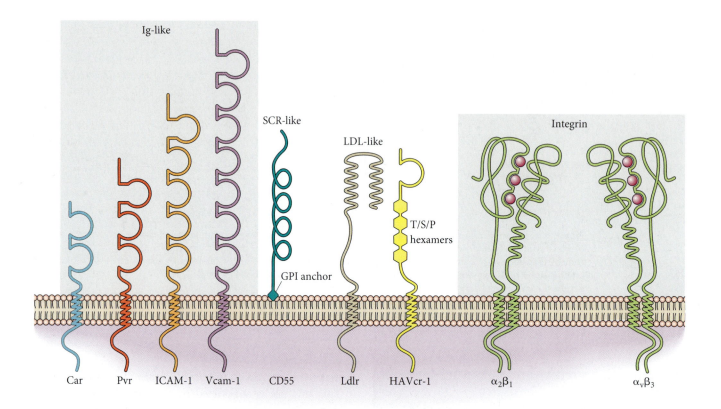

Figure 6.7 Schematic of the cell proteins that function as receptors for different picornaviruses. The different domains, Ig-like, short consensus repeat-like (SCR-like), low density lipoprotein-like (LDL-like), and threonine/serine/ proline (T/S/P), are labeled. Abbreviations: Car, coxsackievirus and adenovirus receptor; Vcam-1, vascular cell adhesion molecule 1; Ldlr, low-density lipoprotein receptor; HAVcr-1, hepatitis A virus cellular receptor 1. Adapted from J. W. Almond, *Trends Microbiol.* **6:**198–202, 1998, with permission.

strategy for identifying the poliovirus receptor was to transfect poliovirus-resistant mouse cells with a DNA library from susceptible human cells. Transfectant mouse cells that expressed the human PVR were identified using a monoclonal antibody (MAb) which had been selected for its ability to protect human cells against infection by poliovirus. It protects cells from infection by competing with virus for binding to PVR. The human PVR gene then was cloned from the PVR-expressing transfectant mouse cells, and its product was identified as protein CD155, which like ICAM-1 is a member of the Ig superfamily of proteins.

Whereas ICAM-1 contains five extracellular Ig-like domains, PVR (CD155) contains three such domains (Fig. 6.7). A known normal role for CD155 is to bind the extracellular matrix protein vitronectin, thereby mediating cell-to-matrix contacts. CD155 appears to be a member of a small family of related proteins, whose other members mediate cell-to-cell adhesions via homophilic contacts. Moreover, recent experimental results show that the cytoplasmic domain of CD155 can interact specifically

with the light chain of a dynein motor complex, thereby perhaps facilitating viral invasion of the central nervous system through the retrograde axonal pathway (see below).

The principle that the susceptibility of a cell to infection by an RNA virus is generally dependent on the expression of an appropriate receptor at the cell surface was put to use to generate transgenic mice that express the human PVR (Box 6.3). These mice indeed are susceptible to poliovirus and provide researchers with a small-animal model for human poliomyelitis.

An especially interesting finding from the mouse model was that when poliovirus was injected into a hind limb, it replicated in the muscle and then spread to the upper spinal cord and brain via retrograde neuronal transport. Transport to the brain via the nervous system was verified by the experimental finding that severing the sciatic nerve prevented the lethal brain infection. Thus, while poliovirus may spread to the central nervous system in humans via the blood, the transgenic mouse model provided convincing evidence for neuronal transport of the virus, which might be a factor in paralytic poliomyelitis. (These mice cannot be infected by the oral route. Thus, some aspects of the disease [e.g., spread from the gut to the central nervous system] cannot be studied in this model [Ren and Racaniello, 1992].)

Chimeric and mutant PVR molecules were generated to identify the regions of PVR that mediate poliovirus binding. These studies demonstrated that the most membrane-distal Ig-like domain of PVR is sufficient for virus binding but that the adjacent Ig-like domain enhances binding. These experimental results are reminiscent of those observed for the rhinovirus-ICAM-1 interaction.

The poliovirus surface, like that of the rhinoviruses, contains deep canyons surrounding each of the 12 fivefold axes of symmetry (Fig. 6.4). For this reason, and because of similarities between PVR and ICAM-1, it was expected that the receptor-binding site on poliovirus is also in the canyon. Actually, there are three poliovirus serotypes (PV1, PV2, and PV3). The structures of all three poliovirus serotypes, complexed with CD155, were determined by cryo-electron microscopy. Each poliovirus serotype bound to CD155 by using the same binding sites and orientations in the viral canyon, implying that all three serotypes use a common mechanism for cell entry. Interestingly, although the most membrane-distal Ig-like domain (D1) of CD155 binds into the poliovirus canyon, its orientation in the canyon is quite different from that of ICAM-1 in the rhinovirus canyon. CD155 binds tangentially to the poliovirus surface, whereas ICAM-1 binds perpendicular to the rhinovirus surface. Also, CD155 has a larger footprint than that of ICAM-1 because of an additional binding region on the east side of the canyon.

When poliovirus binds to the D1 domain of CD55, it interacts with amino acids that are present in at least two β-strands of that domain, as well as with amino acids in adjacent loops. In contrast, when rhinoviruses bind to ICAM-1, they interact only with amino acids in the loops of the most membrane-distal Ig-like domain. Thus, the closely related poliovirus and rhinoviruses, which share a similar structure, interact in different ways with their structurally related receptors. These differences may help to account for differences in the entry pathways of these respective viruses.

Receptors for Coxsackieviruses and Other Enteroviruses

Coxsackieviruses, like poliovirus, are members of the genus *Enterovirus* of the *Picornaviridae*. Coxsackievirus A21 (CVA21), the prototype coxsackievirus, attaches initially to a primary receptor, but its entry is subsequently triggered by its interaction with a secondary receptor, or **coreceptor**. (The concept of coreceptors came up in Chapter 5, in which it was noted that some members of the reovirus family likewise interact with a receptor and coreceptor to mediate their binding and entry.) CVA21 uses the decay-accelerating factor (DAF) protein for its primary receptor. The normal function of DAF is to prevent inappropriate complement activation on normal host cells. (Complement is a group of sequentially acting host proteins that form a destructive complex on virus-infected cells, as well as on some viruses and on other microbes [Chapter 4].) The binding of CVA21 to DAF then enables the virus to interact with its secondary receptor, ICAM-1. The interaction of CVA21 with ICAM-1 induces conformational transitions of the capsid which then lead to infection. CVA21 interacts with the most membrane-distal Ig-like domain of ICAM-1, reminiscent of the interaction of the major subgroup rhinoviruses with ICAM-1.

DAF is not an Ig-like protein. Instead, it contains four so-called "short consensus repeats," which are characteristic of complement proteins (Fig. 6.7). Also, rather than being a transmembrane protein, DAF is attached to the plasma membrane by a glycosylphosphatidylinositol linkage.

The fact that the 90 or so serotypes of the major subgroup of the human rhinoviruses, as well as poliovirus, use an Ig superfamily protein for their receptor might have led us to expect that all picornaviruses use Ig superfamily proteins for their receptors. However, the example of the minor subgroup rhinoviruses, and now of coxsackievirus A21, shows that this is not so. The enteroviruses EV7 and EV11 also use DAF as their receptor. The DAF binding site on EV7 is outside the canyon, close to the twofold axis of the icosahedron, as shown by cryo-electron microscopy. The DAF binding site on EV11 also is outside the canyon, as demonstrated by genetic analysis. However, in contrast to the EV7 DAF-binding site, the EV11 DAF-binding site is close to the fivefold icosahedral axis. These experimental observations, as well as the example of the minor subgroup of the human rhinoviruses (above), show that picornavirus receptor-binding sites are not necessarily in their surface canyons. This fact is not surprising since not all picornaviruses, including EV7 and EV11, contain prominent canyons. Moreover, bearing in mind that poliovirus, EV7, and EV11 are all members of the genus *Enterovirus* of the *Picornaviridae*, we also can make the more general statement that closely related viruses do not necessarily interact with their receptors via the same sites or surface features on the viruses.

In contrast to coxsackievirus A21, EV6 and EV11 do not require a coreceptor. That is, DAF alone mediates their binding and entry. This difference between these viruses might be related to the fact that EV6 and EV11 enter via **caveolae**, rather than via the more typical clathrin-coated pit-mediated endocytic pathway. Caveolae are small invaginations of the plasma membrane that are distinguished from clathrin-coated pits by their size (70 to 100 nm versus about 250 nm for clathrin-coated pits), distinctive flask-like shape, and lack of a visible coat in thin-section electron micrographs (for an example, see Fig. 15.2). Caveola-mediated entry is discussed in greater detail in later chapters. For the moment, note that caveola-mediated endocytic pathways deliver their cargos to unusual destinations, such as the endoplasmic reticulum or the Golgi apparatus, rather than to the more typical endosomal/lysosomal compartment. (Simian virus 40, a polyomavirus, enters by a caveola-mediated endocytic pathway, which delivers it to the endoplasmic reticulum. There, endoplasmic reticulum-specific factors [i.e., particular chaperones] appear to promote subsequent steps in simian virus 40 entry [Chapter 15].)

The notion that EV6 and EV7 do not require a coreceptor because they enter via a caveola-mediated pathway is consistent with the finding that CVA21 too can enter via DAF alone, provided that DAF is cross-linked with MAbs. Cross-linking of DAF apparently causes it to then associate with caveolae, so that caveolae might then mediate CVA21 entry. Perhaps binding of EV6 and EV11 to DAF induces sufficient cross-linking of that protein to favor the caveola-mediated entry pathway. Interestingly, EV1 also enters via caveolae, but this virus uses α2β1 **integrin** for its receptor. More details about the integrins are forthcoming below.

Receptors for FMDVs

FMDVs are members of the genus *Aphthovirus* of the *Picornaviridae*. FMDVs use integrins for their cell surface receptors. Integrins are cell surface α/β heterodimeric glycoproteins (Fig. 6.7) that have a variety of cellular functions, including cell-to-cell and cell-to-matrix adhesion and induction of signal transduction pathways.

The FMDVs, like the coxsackieviruses and EV7 and EV11, do not have prominent canyons on their surfaces that might contain receptor-binding sites. Instead, FMDVs bind to integrins via the long GH loop of their VP1 proteins (Fig. 6.2). The GH loop on FMDV VP1 contains a conserved arginine-glycine-aspartic acid (RGD) sequence. This sequence is a common motif in a variety of extracellular adhesion proteins (e.g., fibronectin, vitronectin, and type 1 collagen) and is part of the sequence recognized by most members of the integrin superfamily of receptors, which bind to those proteins (Box 6.4).

The exposed receptor-binding domains of the FMDVs and coxsackieviruses imply that these sites are not protected to any extent from immune surveillance. However, as a general principle, it may be possible to have antigenic diversity at these exposed sites without compromising receptor binding, provided that the footprint of the receptor on the viral surface is smaller than the footprint made there by an antibody molecule. This fact may well account for the variation in the amino acid sequences that flank the RGD motifs in the various coxsackievirus isolates.

BOX 6.4

Several other picornaviruses also bind to integrin receptors. These include coxsackievirus A9, the Barty strain of echovirus type 9, and EV1 (as noted above). As in the case of FMDV, coxsackievirus A9 likewise contains the RGD motif in its surface loops. Thus, the FMDVs and coxsackievirus A9 specifically bind to integrins by containing the key recognition motif of the receptor's natural ligand.

Coxsackievirus A9 was used in a straightforward experiment that investigated whether the RGD motif on that virus's VP1 protein might be the viral site that recognizes the receptor. Briefly, synthetic peptides containing the RGD motif blocked infectivity, implying that the receptor indeed is recognized by the RGD motif on the virus.

Interestingly, unrelated adenoviruses (a family of fairly large double-stranded DNA viruses) also use RGD-specific integrins to initiate infection. Thus, not only may related viruses within the same family use different receptors, but also unrelated viruses in different families may use the same or similar receptors. For example, coxsackievirus A9 and adenovirus each use integrin αvβ3.

Poliovirus and Rhinovirus Entry: Some General Points

All viruses must ensure that their genomes cross at least one cellular membrane in order to initiate an infection. In the case of enveloped viruses, transport of the virus across a cellular membrane is rather straightforward in principle. In brief, virus-encoded glycoproteins in the viral envelope undergo triggered conformational rearrangements that promote fusion of the viral envelope with the cellular membrane, thereby releasing the viral core or genome on the other side of the membrane. The conformational rearrangements of the envelope proteins are triggered either by their interaction with their receptors and coreceptors at the cell surface or by the low pH of the endosomal compartment. In contrast, the means by which nonenveloped viruses, such as picornaviruses, penetrate membranes is not nearly as well understood, but it is thought to involve the triggered exposure of a hydrophobic moiety on the particle or the release of a lytic factor.

In the preceding chapter, we discussed how the double-stranded RNA genomes of nonenveloped reovirus particles cross a host membrane into the cytosol, noting that they did so as an internal component of the so-called intermediate subviral particle. Moreover, a key feature of the reovirus replication cycle is that the viral double-stranded RNA genome remains conserved within the intermediate subviral particle throughout the entire replication cycle. But that case in point is rather exceptional. In general, nonenveloped viruses face the double problem of having to release their genomes from a capsid shell, as well as to transport their genomes across a cellular membrane. Despite extensive studies, molecular details of how poliovirus and human rhinoviruses solve this twofold problem are not yet known. Yet much is known, and the comparisons of these picornaviruses from two different genera are interesting for showing how structural differences between these viruses account for their different modes of interacting with their receptors and their different entry pathways. So, the comparison begins.

When either poliovirus or rhinoviruses bind to their respective receptors, the virus particles are induced to undergo conformational transitions that lead to partial disruption of the particle and release of the genomic RNA. But events following virus binding are quite different in each instance. In the case of poliovirus, the entire course of conformational transitions of the capsid and release of the genome into the cytosol happens entirely at the plasma membrane. In contrast, the capsids of most rhinovirus strains are only *primed* to undergo a conformational transition by their interaction with their receptor. These rhinovirus strains are then internalized by receptor-mediated endocytosis, and their capsids are then induced to undergo

further conformational transitions by the low-pH conditions within an endosomal compartment. These low-pH-induced disruptions of the capsids then enable them to release their genomes across the endosomal membrane into the cytosol.

Consistent with the above, poliovirus infection, but not rhinovirus infection, can occur in the presence of inhibitors of receptor-mediated endocytosis and of lysosomal acidification. (Note that poliovirus, as an enteric virus, should be stable under acidic conditions and should not depend on lysosomal acidification in its entry pathway.) But despite the differences in the course of events following the interactions of poliovirus and these rhinoviruses with their respective receptors [PVR, or CD155, and ICAM-1], each of these receptors triggers conformational changes in the particular picornaviruses that bind to them, which go beyond the role of viral receptors as mere recognition proteins at the cell surface. The important differences between the ways that these picornaviruses interact with their respective receptors are discussed next.

Poliovirus Entry

Poliovirus particles contain surface canyons with dimensions that are similar to those of rhinoviruses. Moreover, genetic evidence demonstrates that the poliovirus binding sites for CD155 are located in the canyons. However, as noted above, although CD155 and ICAM-1 bind into the canyons of poliovirus and the rhinoviruses, respectively, the tangential orientation of CD155 relative to the poliovirus surface is quite different from the perpendicular orientation of ICAM-1 relative to the rhinovirus surface. Because of its tangential orientation, CD155 makes a larger footprint in the poliovirus canyon than ICAM-1 makes in the rhinovirus canyon. This larger footprint may enable CD155 to induce dramatic and irreversible structural changes in poliovirus particles, sufficient to trigger the release of the genome across the capsid shell, across the cell membrane, and into the cytoplasm (Box 6.5). Indeed, the interaction of poliovirus with CD155 triggers the complete externalization and release of VP4 from its location under the capsid shell and the exposure of the hydrophobic N terminus of VP1, which had been located on the inner surface of the capsid. The exposed hydrophobic N terminus of VP1 allows the particle to attach to synthetic membranes in vitro, demonstrating that this structural transition might generate an intermediate in the infection pathway.

The conformationally altered particles that result from the interaction of poliovirus with CD155 are referred to as "A particles." These A particles can be eluted from the plasma membrane before they release their RNA genomes. Even though poliovirus A particles still contain their

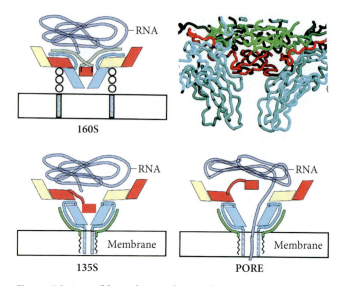

Figure 6.8 A possible mechanism for transferring poliovirus RNA across the cell membrane. VP1, VP2, VP3, and VP4 are colored cyan, yellow, red, and green, respectively. In the crystal structure of the virus particle (upper right), the beta-tube of VP3 (red) forms a plug at the fivefold axis that separates the virus interior from the outer surface. Attachment of the 160S particle (upper left) to the PVR (three gray circles) triggers conversion to the 135S form (lower left). Upon conversion, cell attachment is mediated by externalized VP4 (green tubes) and the N termini of VP1 (blue tubes). The N termini emerge from the bottom of the canyon and extend along the sides of the fivefold mesa towards the apex. Once the N-terminal helices of VP1 have inserted into the membrane, they rearrange to form a pore (lower right). In order for the RNA (purple tube) to pass through the pore into the cytoplasm, it would be necessary for the VP3 beta-tube (red rectangle) to shift on its 40-residue tether (red tube) and for the VP1 barrels to splay farther apart. From D. M. Belnap et al., *J. Virol.* 74:1342–1354, 2000, with permission.

genomes, they nevertheless have undergone a change in their sedimentation coefficient from 160S for native virus particles to 135S for A particles. This change in sedimentation coefficient is largely accounted for by the fact that A particles have somehow released their VP4 proteins, despite the fact that VP4 is located *under* the capsid shell in the native virus particle. In addition, the normally internal N terminus of VP1 is translocated to the particle surface, thereby making the capsid hydrophobic.

The interaction of CD155 with poliovirus particles is indeed sufficient to induce the conformational transition to 135S A particles, as shown by the fact that this conformational transition can be triggered by exposing the virus to soluble CD155. Importantly, the 135S A particles are not dependent on the receptor to initiate an infection (recall a similar finding in the case of reovirus intermediate subviral particles [Chapter 5]). These observations and the demonstration that A particles can form ion channels in synthetic bilayers without acidification led to the conclusion that A particles, or something like them, are functional intermediates in the poliovirus entry pathway.

The following model for poliovirus entry (Fig. 6.8) is based on the above facts. The interaction of poliovirus particles with CD155 somehow induces the loss of VP4 from its internal locations in the particles and the externalization of the normally internal N termini of the VP1 proteins. The hydrophobic N termini of VP1 are extruded from the bottom of the canyon. Since they are arranged around the fivefold axis, a group of five hydrophobic VP1 N termini are ideally positioned for membrane insertion and the formation of a transmembrane pore through which the genome might pass. The fate of VP4 is not known. However, VP4 contains an N-terminal myristate residue that might possibly become embedded in the plasma membrane to facilitate the insertion of the VP1 N termini into the membrane. Also not clear is how the RNA genome

exits from the virus. No holes are seen either in the 135S A particles or in the subsequent "empty" 80S particles.

The assembly and maturation of picornaviruses are described below. For now, note that a last step in the maturation of picornavirus particles is for the precursor polypeptide, VP0, to undergo proteolytic cleavage within nascent particles to generate VP2 and VP4. Maturation cleavage of VP0 is believed to trap the virus in a metastable state that is primed to undergo the irreversible conformational transitions that facilitate entry at the start of the next round of infection. (The notion of trapping a virus particle in a metastable state that is primed to undergo subsequent conformational transitions should now be a familiar concept.)

Analysis of poliovirus structure by X-ray crystallography revealed the presence of a fatty acid-related molecule, referred to as "pocket factor," in a cavity present in the capsid protomers. This molecule appears to stabilize the metastable poliovirus particles. A similar pocket factor is present in some rhinovirus serotypes but not others. The possible significance of the pocket factor is considered below.

Human Rhinovirus Entry

The major subgroup human rhinoviruses indeed undergo a conformational transition when they bind to their receptor, ICAM-1, at the plasma membrane. However, there is no concomitant release of the rhinovirus RNA at this point, as occurs when poliovirus binds to its receptor, CD155. Instead, rhinovirus particles are only *primed* to release their genomes by their interaction with ICAM-1 at the plasma membrane. They release their RNA after they are taken up by receptor-mediated endocytosis, in response to the low pH of the endosomal/lysosomal compartment. Low pH triggers release of the rhinoviral RNA by causing the externalization of the hydrophobic N termini of rhinovirus VP1 molecules and the release of VP4 molecules, events reminiscent of the conformational rearrangements of poliovirus that are triggered by its binding to CD155 (Fig. 6.8).

For both poliovirus and the rhinoviruses, these structural transitions enable the viral genomes to exit the capsid while also creating a membrane pore through which the viral genomes might pass into the cytoplasm. However, in the case of these rhinoviruses, their genomes pass through an endosomal/lysosomal membrane rather than the plasma membrane, and their final conformational rearrangements are set in motion by low pH rather than by the interaction with the receptor.

Thus far, the distinction between the poliovirus and rhinovirus entry pathways has been rather clear-cut. But in actuality, the binding of some serotypes of the major subgroup rhinoviruses to ICAM-1 is sufficient to induce those serotypes to release their RNA at neutral pH. Why then might other major subgroup serotypes require low pH to trigger RNA release? Recall the particle-stabilizing pocket factor noted above. Human rhinovirus serotypes 3 and 14 (HR3 and HR14) lack the stabilizing pocket factor and can release their genomes upon receptor binding at neutral pH. In contrast, HR16, which contains the pocket factor, does not undergo receptor-mediated uncoating at neutral pH. So, less stable major subgroup rhinovirus serotypes, which lack the pocket factor, might be uncoated at the plasma membrane, whereas the more stable ones, which contain the pocket factor, might require an acidic endosomal compartment to activate release of the RNA.

How might the interaction of major subgroup rhinoviruses with ICAM-1 promote the conformational transitions that facilitate entry? In the case of those serotypes that contain the stabilizing pocket factor, binding of the receptor into the canyon might sterically alter the hydrophobic pocket underneath the canyon, thereby displacing the pocket factor, and in that way destabilize the virus and initiate uncoating (Box 6.6).

What then of those major rhinovirus subgroup serotypes that lack the pocket factor? Apropos those viruses, recall that the rhinovirus capsid is actually thought of as a pulsating, breathing structure, from which residues of VP4 and internal residues of VP1 spontaneously become exposed at the virus surface. Thus, the receptor might catalyze uncoating by catching the viral capsid in a spontaneous expanded open state and locking it there. In this regard, HR3 (which lacks the stabilizing pocket factor), but not the more stable HR16 (which contains the pocket factor), is trapped by ICAM-1 in the expanded conformation. Moreover, capsid expansion of receptor-bound HR3 was found to correlate with uncoating. Presumably, in the expanded conformation, critical interprotomer connections are relaxed to facilitate subsequent extrusion of capsid proteins and RNA. These expanded rhinovirus particles might be the functional equivalents of the conformationally altered poliovirus A particles that are believed to be intermediates in poliovirus entry.

BOX 6.6

A number of compounds are known to block uncoating of rhinoviruses. Among the best studied of these are the so-called WIN compounds developed by Sterling Winthrop. Each of these hydrophobic WIN compounds binds in the rhinovirus pocket, thereby displacing the pocket factor (Fig. 6.2). The WIN compounds are able to do so because their affinity for the pocket is greater than that of the pocket factor. They then prevent the capsid from undergoing the receptor-mediated structural transitions that lead to infection.

Although these compounds are effective at impairing rhinovirus infection in cell culture, clinical trials of their usefulness as antirhinovirus therapies have been disappointing. Possible problems include the likelihood that effective concentrations cannot be attained in patients at the site of infection. Also, there is the problem of the large number of rhinovirus serotypes, each of which has a distinct pocket into which the drugs must bind. In addition, escape mutants are readily generated. Despite these difficulties, researchers in this field are optimistic that effective antiviral agents based on these principles can be developed.

The minor subgroup human rhinoviruses bind to the LDL receptor, rather than to ICAM-1. Moreover, they bind to the LDL receptor at a site on the virus particle close to its icosahedral fivefold axis, rather than in the canyon. The enteroviruses EV7 and EV11 likewise do not bind their receptor, the DAF, in their canyons. Instead, EV7 and EV11 bind to DAF near their respective twofold axes and fivefold icosahedral axes. Perhaps it is not a coincidence that in each of these instances, in which the receptor-binding site on the virus particle is not in the canyon, the receptors serve merely as recognition proteins for the virus. They neither prime nor trigger RNA release. These viruses are taken up by endocytosis and require the low pH of the endosome to both prime and trigger RNA release.

The differences in the uncoating mechanisms of the various picornaviruses likely reflect, at least in part, the differing extents to which the diverse virus-receptor interactions might affect particle stability. For instance, the binding site of the minor subgroup human rhinovirus receptor lies entirely on the dome on the fivefold axis. It does not overlap the canyon or the pocket in the canyon and thus cannot affect particle stability in the ways that ICAM-1 can affect the stability of the major subgroup rhinoviruses.

Poliovirus and Rhinovirus Entry: Why the Differences?

The receptor binding sites of both poliovirus and the major subgroup human rhinoviruses are located in the canyon. Yet the interaction of CD155 with poliovirus induces conformational transitions in the virus that suffice for release of the RNA at the plasma membrane, whereas the interaction of ICAM-1 with many rhinoviruses only primes the viruses to release their genomes, with release occurring later in response to the low-pH trigger. The underlying basis for the difference between the entry modes of these viruses is likely due to differences in the molecular details of the interactions between these viruses and their respective receptors. As noted above, CD155 has a larger footprint on poliovirus particles than ICAM-1 has on rhinovirus particles. Indeed, CD155 binds to the walls of the surface protrusions that surround the poliovirus canyon, whereas ICAM-1 binds to deeper sites in the rhinovirus canyon. This difference may enable CD155 to sufficiently destabilize poliovirus particles so as to trigger poliovirus RNA entry at the cell surface.

What selective pressures might have been acting on these viruses to account for the differences in their interactions with their respective receptors? The major subgroup rhinoviruses use the most antibody-inaccessible receptor-binding sites yet known in the picornavirus family, presumably in response to selective pressure to escape

neutralization by antibodies. Poliovirus, as an acid-stable enteric virus, cannot utilize low pH to trigger uncoating. Consequently, it had to evolve a low-pH-independent uncoating mechanism. Doing so, it may have had to pay the price of using receptor-binding sites that are relatively (compared to the major subgroup rhinoviruses) unprotected from immune surveillance. This may explain why there are only three poliovirus serotypes, in contrast to the more than 100 major subgroup rhinovirus serotypes.

TRANSLATION

Translation: Part I

Since we are accustomed to thinking of transcription as a step that precedes translation, it might appear that we should take up picornavirus transcription before considering its translation. However, bear in mind that translation is a crucial first step in the replication of plus-strand RNA viruses because any virus-directed RNA synthesis (either to replicate the virus genome or to generate viral mRNA) requires production of the virus-specific, RNA-dependent RNA synthesis machinery. The parental plus-strand virus genome serves as the messenger RNA (mRNA) that encodes that machinery. Indeed, translation of the parental viral genome is generally the first biosynthetic event following uncoating of plus-strand RNA viruses (Box 6.7).

The earliest studies of poliovirus translation, carried out in the late 1960s, demonstrated that something most unusual was occurring. First, the virus-encoded proteins made in infected cells spanned a wide size range, the largest protein that was readily observed being 100 kilodaltons (kDa) in size. Second, the total molecular weights of those different proteins exceeded the coding capacity of

BOX 6.7

In Chapter 2 we discussed the rationale behind the Baltimore classification scheme, in which plus-strand RNA viruses are designated class IV viruses. An important distinction between the plus-strand RNA genomes of class IV viruses and the minus-strand RNA genomes of class V viruses is that the former have mRNA activity whereas the latter do not. Following uncoating, a first step in the replication cycle of class IV viruses is the translation of the parental genome, so as to generate the virus-encoded RNA polymerase/replicase activity required to replicate and transcribe the viral genome. Since the minus-sense genomes of class V viruses are not mRNAs, these viruses must bring the polymerase along with them, in order to begin their replication cycle.

the virus. Happily, pulse-chase experiments implied that some of the larger proteins actually might be precursors for several of the smaller ones. An early test of that premise was to compare **tryptic digests** generated from these proteins (Box 6.8). The experimental results demonstrated that one of the larger proteins, P1, indeed is cleaved to form the capsid proteins of the virus. Proteins such as P1, which are cleaved to yield two or more proteins, are referred to as **polyproteins**. (As discussed below, P1 initially is cleaved to generate VP1, VP3, and VP0. The last protein is later cleaved to yield VP2 and VP4 during the final maturation of progeny virus particles.)

The finding that the larger protein, P1, is a precursor that is processed to generate VP1, VP3, and VP0 implies that one or more proteases recognize specific sites on the P1 protein at which specific cleavages might occur. To test this hypothesis, infection was carried out in the presence of amino acid analogs (e.g., *p*-fluorophenylalanine [FPA]). The rationale for using these amino acid analogs was that they would be incorporated into new precursor proteins, thereby possibly impairing cleavage by sequence-specific proteases. Thus, the polyproteins might accumulate and the cleavage products might not appear. The amino acid analogs indeed prevented the appearance of some of the smaller proteins, including VP1, VP3, and VP0. Moreover, they caused the appearance of proteins that were larger than any that were detected in the absence of the analogs. Importantly, one of these larger proteins could well account for the entire coding capacity of the virus! (Consider the following. If poliovirus RNA were to encode any other **primary gene product**, i.e., one that is not a cleavage product of a larger precursor, in addition to this very large protein, then that other primary gene product must be one of the polypeptides made in the presence of the amino acid analogs. The combined molecular weights of the smallest

of those proteins and of the very large protein clearly exceed the coding capacity of the virus.)

In the late 1960s, a straightforward approach to determine whether a particular protein might be the cleavage product of a larger precursor polyprotein was to compare tryptic digests of these proteins. However, if amino acid analogs were used to cause the accumulation of the large precursor proteins, then this experimental approach would not be feasible since the analogs would impair experimental protease digestion. To circumvent this conundrum, David Baltimore and his colleagues used protease inhibitors (e.g., diisopropylfluorophosphate, which inhibits trypsin, chymotrypsin, thrombin, etc.) to cause the accumulation of the larger polypeptides. Tryptic fingerprints could now be generated from these large polypeptides, and from these fingerprints precursor-product relationships indeed could be established. Experimental findings from these several studies, taken together, demonstrated that the entire poliovirus genome is translated into one large *polyprotein* that is cleaved to yield the individual virus proteins.

As noted above, polypeptide P1 was shown to be the precursor for VP0, VP1, and VP3. In 1971, Baltimore and colleagues were able to map P1 on the viral genome and to map the individual capsid proteins within P1. They accomplished this using the drug pactamycin, which specifically inhibits only the initiation step of protein synthesis. Pactamycin allows any ribosomes that are already translating an mRNA to continue translation of that mRNA unimpeded. However, the drug does not permit new translation initiation events. If pactamycin is added to infected cell cultures and radioactive amino acids are subsequently added, then after a period of translation the radioactively labeled proteins will vary in their specific radioactivity. The closer the coding sequence for a protein is to the 3′ end of the genome, the greater will be its specific radioactivity. This is because translation occurs in the 5′-to-3′ direction on the positive-strand poliovirus genome (as well as on all mRNAs). Therefore, the most 3′ sequences will be traversed by the most ribosomes after the addition of the pactamycin and the labeled amino acids. (If this is not apparent, try drawing a picture that would demonstrate this fact.) Doing this experiment in the presence of FPA (which permitted the syntheses of only larger precursor polypeptides) demonstrated that the most 5′ region of the genome encodes P1. Doing the experiment in the absence of FPA demonstrated that the order of the capsid proteins within P1 is VP4, VP2, VP3, and VP1.

In 1981, Eckard Wimmer and colleagues succeeded in sequencing the entire poliovirus genome. Happily, the sequence they obtained was completely consistent with the earlier experimental results discussed above. The poliovirus

BOX 6.8

The rationale behind comparing tryptic digests is as follows. The proteins to be compared are cleaved into peptides by a protease, and the individual peptides generated from each protein are then separated from each other by two-dimensional polyacrylamide gel electrophoresis. The pattern of the peptides on the two-dimensional gel might be referred to as a "fingerprint." If a larger protein were the precursor of a smaller one, then some of the tryptic peptides generated from the smaller protein would correspond to a subset of tryptic peptides generated from the larger precursor.

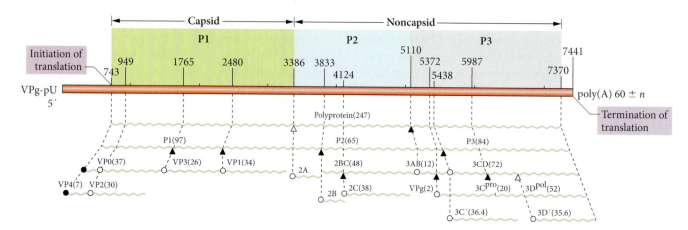

Figure 6.9 Gene organization and polypeptide processing of poliovirus. The viral genome (orange bar) is shown terminated at its 5′ end with the genome-linked protein VPg, and at the 3′ end with poly(A). Purple boxes indicate the sites of translation initiation (nucleotide 743) and translation termination (nucleotide 7370). The numbers above the viral RNA refer to the first nucleotide of the triplet encoding the N-terminal amino acid of each virus-specific protein. The coding region is divided into three regions, P1, P2, and P3, corresponding to the major cleavage products of the polyprotein. Individual polypeptides are shown as wavy lines. Numbers in parentheses are molecular masses of the polypeptides calculated from the amino acid sequences. Open circles indicate glycine in all cases, except for VP2, where the open circle is serine. The carboxy-terminal amino acid of 3CD is phenylalanine. Filled triangles, glutamine-glycine; open triangles, tyrosine-glycine. Polypeptides 3C′ and 3D′ are alternative cleavage products of 3CD. Filled circles indicate that the amino terminus is blocked with myristic acid. Modified from H. G. Krausslich and E. Wimmer, *Annu. Rev. Biochem.* 57:701–754, 1988, with permission.

genome is 7,433 nucleotides long, is polyadenylated at the 3′ terminus, and is covalently linked to a small protein (VPg) at its 5′ terminus. Importantly, the poliovirus genome contains only one open reading frame (ORF), which is 7,207 nucleotides long (89% of the genome) and codes only for the very large polyprotein noted above. (This protein was referred to as NCVP-00 in the early literature. NCVP stands for noncapsid viral protein. However, this terminology was abandoned 20 years ago.) This polyprotein is normally not seen in infected cells, since it is cleaved by proteases as specific cleavage sites are translated. That is, cleavage occurs while the polyprotein is still a ribosome-associated nascent polypeptide. (What experiment might you do to demonstrate that fact? Hint: It is possible to isolate polysomes, and from them the nascent polypeptides.)

The locations of the cleavage sites on the single poliovirus primary gene product and the N termini of the cleavage products were revealed by the sequence analysis (Fig. 6.9). With the single exception of VP2 (which is cleaved from VP0 as a final step in maturation), all viral polypeptides have glycine as their N-terminal amino acid. In 8 of 9 cases, the amino-terminal glycine of the cleavage product was preceded by a glutamine in the precursor polypeptide. In the ninth case, the preceding amino acid was tyrosine. These facts show that proteolytic processing occurs at Gln-Gly and Tyr-Gly pairs.

Processing of the precursor polypeptide is carried out by three virus-encoded proteases, 2A^{pro}, 3C^{pro}, and 3CD^{pro}.

The last protease is the immediate precursor for 3C^{pro} (Fig. 6.9). The proteases are active while still within the nascent polypeptide, from which they release themselves by autocatalytic cleavage. 2A^{pro} cleaves at a Tyr-Gly pair, whereas 3C^{pro} and 3CD^{pro} cleave at Gln-Gly pairs. (Other determinants of processing must exist in addition to these amino acid pairs, since cleavage does not occur at all of the Tyr-Gly and Gln-Gly sites in the precursor polyprotein.) After the proteases have been released by autoproteolytic cleavage, they are free to act in *trans*. 2A^{pro} carries out the cleavage that separates P1 from the remainder of the nascent polypeptide. 3CD^{pro} then generates VP0, VP1, and VP3. Interestingly, whereas 3CD^{pro} is required to process the Gln-Gly pairs within P1, 3C^{pro} and 3CD^{pro} are equally able to process each of the other cleavable Gln-Gly pairs. Importantly, the eventual cleavage of 3CD^{pro} activates the RNA polymerase activity of 3D^{pol}.

In the case of the nonstructural polyproteins, cleavage not only separates the individual end products of the precursor polyproteins but also activates their diverse functions in the process. Intriguingly, in some instances, the cleavage intermediates may themselves express functions that are distinct from their final cleavage products. One such example is that of 3CD and its cleavage products, 3C^{pro} and 3D^{pol}. The distinct functions of the precursor protein, 3CD, and the products, 3C^{pro} and 3D^{pol}, may serve to regulate the relative amounts of capsid proteins and progeny genomes that are made (see below). Such control would be a rather interesting solution in part to the impediment to gene regulation posed by the fact that the picornavirus mechanism of translation dictates that all portions of the genome must be translated with equal efficiencies. Consequently, regions encoding structural proteins (which are needed in large amounts), as well as regions encoding enzymes (which are needed in small amounts), are translated equally. Yet differences in the efficiencies of the various proteases and cleavage sites mean that not all proteins are produced in equal amounts.

The synthesis of multiple gene products via the proteolytic processing of a single precursor polyprotein would appear to be a most atypical and perhaps inefficient mode of translation. Why then might the picornaviruses have evolved this seemingly unusual mechanism? A reasonable guess is that it is the picornavirus solution to the problem of having to generate multiple gene products from a single-stranded RNA genome, in which all of the genes are colinear on a single plus-strand RNA molecule, a point discussed at length very shortly. However, translation of multiple proteins from a single polyprotein precursor is not the only atypical aspect of the picornavirus translation strategy.

I suspect that for at least some readers, the translation strategy of the picornaviruses, as considered thus far, might appear a bit bizarre. So, before continuing the story of picornavirus translation, it would be good for the sake of comparison to digress in order to consider the translation strategies of the positive-strand RNA phages, with the major purpose of appreciating why the translation strategies of these eukaryotic and prokaryotic viruses are so very different.

Translation: a Digression

The RNA Phages

The best known of the RNA phages are the several that infect *Escherichia coli*, in particular f2, MS2, R17, and Qβ.

Each of these phages contains a genome comprised of about 4,000 nucleotides that is packaged in an icosahedral capsid approximately 26 nm in diameter. Also, each phage contains a similar set of four genes, which have much the same arrangement on each of their plus-strand RNA genomes (Fig. 6.10).

Our major purpose in discussing the RNA phages at this time is to compare their mode of translation with that of the picornaviruses, with the aim of understanding why these prokaryotic and eukaryotic viruses differ in this regard. However, the replication cycles of the RNA phages are interesting in their own right. Moreover, during the 1960s these phages provided an important model system for asking fundamental questions regarding translation in general. Thus, we consider the RNA phages here in some detail before returning to the picornaviruses.

In principle, an RNA virus might need to encode only two proteins, an RNA-dependent RNA polymerase (replicase) and a coat protein. Each of these RNA phages also encodes a so-called "maturation" protein, also known as the "A protein." The phage capsid contains 180 copies of the coat protein. Sixty of these coat protein molecules constitute the pentamers at the vertices of the icosahedron. The other 120 constitute the hexamers, where the sides of the faces meet (for an example, see Fig. 2.8). The phage particle contains just one molecule of the A protein. The phage also encodes a fourth protein that plays a role in lysing the bacterial host cell. The arrangement of these genes on MS2 and Qβ is shown in Fig. 6.10. Note that the genes for the maturation, coat, and polymerase proteins are arranged similarly on these two viruses.

Despite the fact that the A protein is sometimes referred to as the "maturation" protein, its role, if any, in maturation is not clear. However, a notable feature of these RNA phages is that they infect only those bacteria that express male pili, and it is the A protein which facilitates

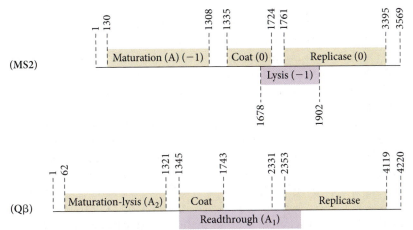

Figure 6.10 Genetic maps of bacteriophages MS2 and Qβ. The nucleotide numbers on the maps indicate the boundaries of the individual genes, as well as the RNA termini. Redrawn from J. A. Levy, H. Fraenkel-Conrat, and R. A. Owens, *Virology*, 3rd ed. (Prentice Hall, Englewood Cliffs, NJ, 1994).

binding of the phage to the sides of the male pilus of the *E. coli* host (Fig. 6.11). (Some bacteria express pili, which are tubules composed of a protein, pilin. Pili serve to move bacteria across a substrate and towards other bacteria. So-called conjugation or sex pili transfer DNA from a "male" cell to a "female" cell. Interestingly, male bacteria that are infected with the phage can transmit the infection to female bacteria. Thus, even bacteria have their sexually transmitted infections.)

After the phage attaches to the male pilus of its *E. coli* host, the phage RNA somehow enters the bacterium. The phage genome then must be translated to generate the phage polymerase, since the bacterial cell, like its eukaryotic counterpart, does not contain any enzyme activity that might use RNA as a template for RNA synthesis. Now events become interesting. In contrast to picornavirus genomes that contain one initiation site for translation, the phage RNA genomes contain three sites for the initiation of translation, one each for the A protein, the coat protein, and the polymerase. There were a variety of experimental

Figure 6.11 Micrograph of an F-pilus emerging from an *E. coli* cell that is covered with icosahedral MS2 phage particles. At the end of the pilus, a filamentous fd phage has attached itself. The thicker thread emerging at the right is a bacterial flagellum. Contributed by L. Caro; reprinted from J. A. Levy, H. Fraenkel-Conrat, and R. A. Owens, *Virology*, 3rd ed. (Prentice Hall, Englewood Cliffs, NJ, 1994), with permission.

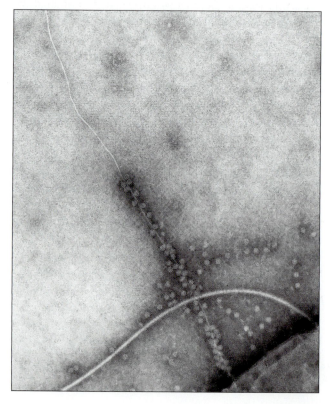

observations that led to this understanding. For example, translation of the A protein, coat protein, and polymerase genes each begins with an *N*-formyl methionine, indicating that each gene product is the result of a separate translation initiation event. In another experimental approach, the phage RNA was fragmented, and the fragments were allowed to bind to ribosomes in the presence of translation initiation factors. The ribosome-bound phage RNA fragments then were exposed to RNase. It was expected that the bound ribosomes would protect the translation initiation sites on the phage RNA segments from the RNase. Subsequent sequencing of the protected RNA sequences identified the three initiation sites. These sites then were located on the complete genome sequence.

Despite the fact that the phage's **polycistronic** RNA genome contains three translation initiation sites, only one of its three translation initiation sites is accessible to ribosomes at the start of translation. (The term "cistron" is rarely used now. Instead, the word "gene," or protein coding region, has taken on the same meaning as "cistron." If you have not already done so, see Box 2.5 for the origin of the cistron concept.) Contrary to what you may have expected, the translation initiation site for the A protein, at the 5′ end of the phage genome, is *not* the translation initiation site that is initially accessible. Instead, at the start of translation, ribosomes bind only to the internally located initiation site for the coat protein. Translation of the polymerase, which occurs next, depends on the prior initiation of the upstream coat protein coding region. The A protein, which is initiated from the most 5′ translation initiation site on the phage RNA, is actually the last protein to be made.

How might one explain these seeming anomalies? In particular, why are the initiation sites for the A protein and polymerase not available at the start of translation, and how might they subsequently become available? A possible explanation for this odd state of affairs was suggested by the experimental finding that translation of the polymerase depends on, and follows close on the heels of, initiation of the coat protein. Based on this finding, it was hypothesized that perhaps the phage RNA genome contains regions of secondary structure in which the translation initiation sites for the A protein and polymerase are embedded, in that way preventing those initiation sites from interacting with ribosomes. The hypothesis continues with the idea that the initially obstructed translation initiation site for the polymerase is exposed by translation of the upstream coat protein-encoding region. Translation of the polymerase then enables replication to begin, which, in turn, exposes the translation initiation site for the A protein.

Several lines of experimental evidence support these contentions. For example, as noted above, translation of the polymerase gene normally follows translation initiation of the coat protein gene. Thus, perhaps translation of the coat protein gene disrupts the secondary structure of the phage RNA to expose the translation initiation site for the polymerase. Also, partial denaturation of the phage RNA (e.g., by heating it or treating it with formaldehyde) allows the A protein and polymerase genes to be translated without prior translation of the coat protein gene.

Genetic evidence in support of the above hypotheses was provided by the experimental finding that an amber or chain-terminating mutation (Box 6.9) in the coat protein gene, at the position encoding the third amino acid (amber-3), not only prevented synthesis of coat protein but also impaired translation of the polymerase gene. In contrast, amber mutations at positions in the coat protein gene encoding amino acid 50 or greater (out of a total of 129 amino acids) did not block polymerase production. But how do these genetic studies support our hypothesis? (Congratulations if you already see the answer.) The amber-3 mutation causes translation of the coat gene to terminate before the ribosome has progressed far enough along the phage RNA to expose the initiation site for the polymerase (Fig. 6.12). This is the reason for the **polar effect** of the amber-3 mutation in the coat gene on translation of the polymerase gene (Box 6.9). In contrast, an amber mutation in the coat gene that encodes the 50th amino acid in the protein allows ribosomes to traverse far enough along the RNA to open up the secondary structure in which the translation initiation site for the polymerase is embedded. In support of this explanation, the amber-3 mutation in the coat gene did not affect translation of the polymerase if translation was carried out under conditions that denatured the RNA. Finally, the eventual sequencing of the phage genome confirmed that the initiation sites for the maturation and polymerase genes indeed are located within regions of secondary structure that might prevent ribosome binding.

The overall pattern of RNA phage translation is depicted in Fig. 6.12. Notice that the translation initiation sites for the A protein gene and the polymerase gene are embedded within "hairpins," making them inaccessible to ribosomes. Thus, the only open translation initiation site is that for the coat protein. Translation of the coat protein gene then disrupts the base pairing that obstructs the translation initiation site for the polymerase, thereby allowing translation of the polymerase to occur.

Next, we consider how the translation initiation site for the A protein is exposed. Translation of the polymerase enables the RNA to be replicated. The first step in that replication is the synthesis of a complementary minus-RNA strand that will serve as the template for transcribing progeny plus strands. Importantly, the translation initiation site for the A protein, at the 5′ end of the nascent plus strand, is briefly in its open conformation before the hairpin foldings can form. During that moment, translation initiation of the A protein can occur. (Bear in mind that bacterial cells do not have nuclei. Consequently, prokaryotic mRNAs can be transcribed and translated concurrently. This is referred to as coupled transcription/translation.) However, the hairpin structures quickly form so that only about one copy of the A protein is translated for every plus strand that is produced. Nevertheless, this stoichiometry is very efficient for the phage, since there is only one copy of the A protein, as well as one plus-sense strand, in the progeny phage particle. Note that this coordination of transcription of the plus-sense strand with translation of the A protein is possible because transcription occurs in the 5′-to-3′ direction and the translational machinery likewise moves on the nascent plus-strand RNA in the 5′-to-3′ direction. Thus, the two processes do not crash into each other.

Production of the polymerase, like that of the A protein, also is regulated, but by a completely different mechanism. Translation of the polymerase in fact is regulated by the coat protein, which serves as a translational repressor of the polymerase gene. As coat protein accumulates, it binds to the phage RNA near the translation initiation

BOX 6.9

A mutation that converts a codon for one amino acid into a codon for another amino acid is referred to as a **missense mutation**. Although the function of the protein might be altered by the missense mutation, a full-length protein is nevertheless generated. Amber (chain-terminating) mutations differ as follows. Of the 64 possible nucleotide triplets, three (UAG, UAA, and UGA) do not encode any amino acid. Conversion of any amino-acid-encoding triplet to one of these three triplets causes termination of translation at that point in the mRNA, followed by release of an incomplete polypeptide from the polysome. Such mutations are referred to as **nonsense, chain-terminating**, or **amber mutations**. Historically, the amber triplet was actually UAG. The UAA triplet was referred to as the ochre codon, and the UGA triplet was referred to as opal. However, with the recognition of the natural role of these triplets as termination signals in translation, the nonsense codons came to be called "Term" for "terminator" or stop codons. In prokaryotic systems, nonsense mutations may have a polar effect, that is, an effect on the expression of neighboring genes, which depends on their distance from the neighboring gene.

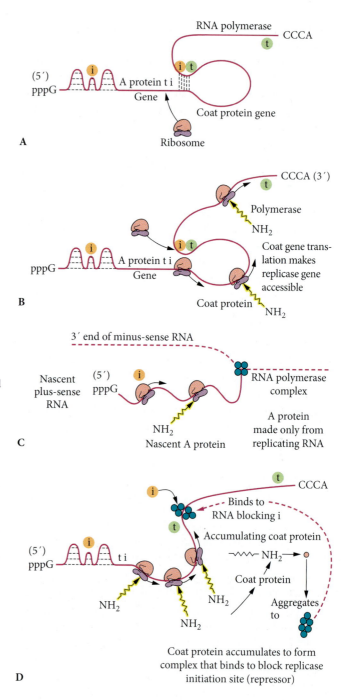

Figure 6.12 The pattern of RNA phage translation. The translation initiation sites of the A protein gene and the polymerase gene are embedded within regions of RNA secondary structure (shown schematically), making them inaccessible to ribosomes. (A) The only "open" translation initiation site is that for the coat protein gene. (B) Translation of the coat protein gene disrupts the base pairing that obscures the initiation site of the polymerase gene, thereby allowing translation of the polymerase gene to occur. (C) Translation of the polymerase enables the RNA to be replicated. The first step in that replication is the synthesis of a complementary minus-sense strand. Because the progeny plus-sense strand is synthesized in the 5′-to-3′ direction, the initiation site for the A protein is briefly in its open conformation, before the hairpin foldings can form. The drawing shows a ribosome bound to that open translation initiation site and a later position of that ribosome, with the associated nascent A protein. The hairpin structures quickly form so that only about one copy of the A protein is translated for every progeny plus-sense strand that is produced. It is possible for synthesis of the plus-sense strand to occur concomitantly with translation of the A protein, because transcription occurs in the 5′-to-3′ direction, and the translational machinery likewise moves on the nascent plus-strand RNA in the 5′-to-3′ direction. (D) As the coat protein accumulates, it binds to the RNA near the initiation site for the polymerase, thereby preventing its further translation. Modified from J. A. Levy, H. Fraenkel-Conrat, and R. A. Owens, *Virology*, 3rd ed. (Prentice-Hall, Englewood Cliffs, NJ, 1994), with permission.

site for the polymerase, thereby blocking it. Again, this provides for greater efficiency in the use of biosynthesis resources, since many fewer copies of the polymerase are needed than of the coat protein.

These RNA phages contain yet a fourth gene. MS2 contains the so-called lysis gene, and Qβ contains its A₁ gene. Interestingly, before the nucleotide sequences of these phage genomes were known, they did not seem to have

sufficient room to accommodate the additional genes. However, determination of the MS2 nucleotide sequence demonstrated that the lysis gene overlaps both the coat and polymerase genes and that it is read in the −1 reading frame, relative to the coat gene. The lysis gene is expressed when a ribosome that has been translating the coat gene undergoes a −1 **frameshift** upstream of the lysis gene. The ribosome then encounters a termination triplet in its new reading frame, followed by the initiation codon for the lysis protein.

As in the case of the MS2 lysis protein, the Qβ A_1 protein also is translated from an overlapping gene sequence, but by means of a different mechanism. In this instance, the translation initiation site for the Qβ A_1 gene is the same as that of its coat protein gene, and these two genes are translated in the same reading frame. The A_1 protein is produced when a ribosome **reads through** the weak UGA termination codon of the coat gene, thus producing A_1 instead of coat protein. (UGA is the weakest of the chain-terminating triplets [Box 6.9].)

Importantly, since the frameshift in the case of MS2 and the read-through in the case of Qβ are rare events, much less of the lysis and A_1 proteins is made than of coat protein. These unusual forms of regulation provide for a more efficient use of resources, since much less of the lysis and A_1 proteins is needed than of coat protein. Moreover, in the case of the MS2 lysis protein, it ensures that sufficient levels of that protein, as needed for lysis, will not accumulate until late in infection. Also, the use of overlapping gene sequences in the cases of these RNA phages provides yet additional examples of the extent to which viruses may go to economize the coding capacity of their genomes. Thus, the use of overlapping gene sequences allows for the differential expression and timing of phage gene products, as well as "genetic economy." (Ribosomal frameshift and read-through are translation strategies that are also found among plus-strand eukaryotic viruses, in particular the retroviruses. However, these phenomena are not seen in the picornaviruses.)

In summary, the RNA phage's mode of translation includes the following key points. (i) The phage genome contains three or four distinct translation initiation sites, and each gene product results from a separate translation initiation event. (ii) The internally located translation initiation site for the coat protein is the only initiation site that is immediately available for ribosome binding. (iii) The secondary structure in the RNA that obstructs the polymerase initiation site is disrupted by translation of the coat gene, thereby allowing initiation of translation of the polymerase gene. (iv) The secondary structure that obstructs the initiation site for the maturation protein (the A protein in the case of MS2) is made available to ribosome binding during synthesis of nascent progeny genomes. In this instance, the inhibitory conformation of the A protein initiation site cannot be established before translation of the A protein begins. Importantly, the A protein initiation site is exposed only long enough to permit one ribosome to enter the site, thereby coordinating the synthesis of one copy of the A protein with the synthesis of each progeny phage genome. (v) The coat protein regulates translation of the polymerase by acting as a translational repressor of the polymerase gene. (vi) The lysis gene of M2 phage and the A_1 gene of Qβ phage each overlap the coat and polymerase genes on their respective plus-strand RNA genomes. The lysis gene is translated as a result of a rare ribosome frameshift, whereas the A_1 gene is translated as a result of similarly rare ribosome read-through. Each of these unusual modes of translation results in the production of appropriate amounts of each protein. Taken together, these relatively simple viruses have evolved rather remarkable molecular mechanisms to ensure that each of their four protein products is produced in appropriate amounts.

Picornaviruses versus RNA Phages: Why the Differences?

The most fundamental difference between picornavirus translation and translation by RNA phages is as follows. Picornaviruses synthesize all of their individual proteins by proteolytic processing of a single polyprotein precursor molecule, which is the product of a single translation initiation event, in which nearly the entire viral genome is continuously translated from one end to the other. In contrast, the RNA phage genome contains multiple translation initiation sites, and the individual RNA phage gene products are the result of separate translation initiation events.

The difference in translation strategies between the picornaviruses and the RNA phages reflects the different modes of translation initiation in eukaryotic cells versus prokaryotic cells. In the case of eukaryotic cells, the vast majority of mRNAs begin with the 5′ 7-methylguanosine cap structure. A translation initiation complex, comprised of the 40S ribosomal subunit and several initiation factors, binds to the 5′ cap structure and traverses or "scans" the mRNA until it recognizes the first translation initiation codon, usually an AUG. At that point, the initiation factors are released, the 60S ribosomal subunit and several elongation factors attach, and translation begins. Because eukaryotic translation initiation complexes attach to an mRNA only at its 5′ cap structure and then scan for the first available initiation codon, generally only the most 5′ translation initiation site on a eukaryotic mRNA can be utilized. Thus, one possible strategy for a eukaryotic plus-strand RNA virus to generate multiple protein products, from a genome that encodes multiple products on a single RNA molecule, is to synthesize a single polyprotein precursor that is proteolytically processed to generate the individual protein products.

Prokaryotic translation, like eukaryotic translation, also does not simply begin at the 5′ end of the mRNA. However, the means by which prokaryotic ribosomes get to their initiation sites is very different from that used by their eukaryotic counterparts. Prokaryotic 30S ribosomal subunits bind to so-called "Shine-Dalgarno" sequences (named for their discoverers) on prokaryotic mRNAs. The Shine-Dalgarno sequences are complementary to sequences on the 16S RNA of the 30S ribosomal subunit, thereby properly aligning the ribosome on the mRNA so that it might recognize the associated downstream initiation site. Since Shine-Dalgarno sequences, unlike eukaryotic 5′ cap structures, can be located at multiple internal sites on an mRNA, multiple internal translation initiation sites can exist on prokaryotic mRNAs. Thus, **polycistronic** mRNAs that encode multiple proteins are common in prokaryotic cells, whereas eukaryotic mRNAs generally are **monocistronic**.

Translation: Part II

From the above discussion, one might have expected that picornavirus mRNAs would have 5′ cap structures at their 5′ ends. Well, picornavirus mRNAs indeed do *not* contain 5′ cap structures. Something that might have alerted you to this fact is that picornavirus genomes, which have mRNA activity, are covalently linked to VPg at their 5′ ends (Fig. 6.9). Instead, picornaviral plus-strand RNAs contain a single **internal ribosome entry site (IRES)**, which is characterized by regions of extensive secondary structure and is located upstream of the single AUG initiation codon on each picornaviral plus-strand RNA (Fig. 6.13). The 40S eukaryotic small ribosome subunit binds to the IRES and scans the mRNA until it encounters the first AUG initiation codon. Notice that the distances between the IRES

and the initiation codon vary among different picornaviruses (Fig. 6.13). Also, different picornavirus IRESs require different sets of translation initiation factors. The hepatitis A virus IRES requires all the same translation initiation factors as are needed for 5′ cap-mediated initiation, including eIF4E (see below). Other picornavirus IRESs require different subsets of these initiation factors (Box 6.10). (Interestingly, several attenuating mutations in the Sabin poliovirus vaccine strains are located on the left stem of loop V of the poliovirus IRES [Fig. 6.13].)

The existence of IRESs throws a partial kink into our above discussion, since there is no apparent reason why a eukaryotic mRNA might not be able to make use of multiple internally located IRESs, each located upstream of an ORF. Multiple IRESs on picornavirus genomes indeed might be desirable since they would enable the virus to regulate the expression of each gene independently (Box 6.10). Instead, the actual state of affairs is that since all picornavirus genomes contain only one ORF and all proteins are generated from a single polyprotein precursor, all genomic coding sequences are translated equally. Thus, despite the facts that there are variations in cleavage site efficiencies and that some cleavage intermediates have distinct functions, the picornavirus pattern of translation appears to be rather inefficient, since vastly more of the capsid proteins are needed than of replicase molecules and proteases. (On the other hand, it can be argued that there is more than one way to achieve efficiency. For instance, it may be more efficient to overproduce some proteins than to build into the genome all the regulatory elements required to fine-tune individual gene expression.

Figure 6.13 Schematic depiction of the secondary structures of the 5′ nontranslated regions of poliovirus (left panel) and EMCV (right panel). The IRES (shaded areas) enables eukaryotic ribosomes to bind directly to the internal site, without first scanning from the 5′ terminus. The poliovirus IRES is a type I IRES, which is found in the genomes of enteroviruses and rhinoviruses. The EMCV IRES is a type II IRES, which is found in the genomes of cardioviruses and aphthoviruses. The asterisk, by the left stem of poliovirus loop V, marks the location of several attenuating mutations in Sabin poliovirus vaccine strains. Modified from E. Wimmer and C. U. T. Helen, *Annu. Rev. Genet.* 27:353–443, 1993, with permission.

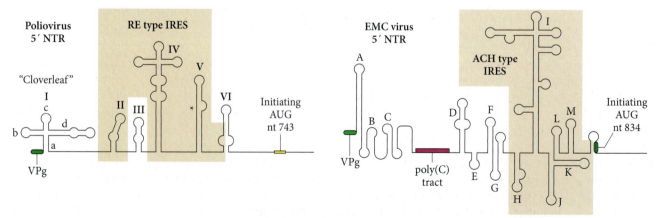

BOX 6.10

The insect picorna-like viruses share many features with mammalian picornaviruses, including the presence of a plus-strand RNA genome with a genome-linked protein at the 5′ end. However, sequencing of several insect picorna-like viruses shows that their genomes are organized differently from those of human picornaviruses. Specifically, their genomes contain two nonoverlapping ORFs, ORF1 and ORF2, which encode the nonstructural and structural proteins, respectively. Importantly, each ORF is preceded by it own IRES, and each ORF is translated into a single polyprotein, which is posttranslationally processed to give rise to the individual gene products. Furthermore, the intergenic IRES is actually more active than the IRES located at the 5′ end of the RNA, thereby explaining how these viruses produce vastly more of their structural than of their nonstructural proteins. At any rate, the genomes of these viruses provide examples of functionally dicistronic eukaryotic mRNAs, whose translation is controlled by two IRES elements.

The cricket paralysis virus is a member of the insect picorna-like viruses. Its internal IRES must promote translation by a novel mechanism, since it does not require any translation initiation proteins.

BOX 6.11

Has the idea of the following experiment occurred to you? What might be the properties of a picornavirus, say a poliovirus, which has been genetically engineered to contain a second IRES, for example, between the P1 and P2-P3 ORFs? Such experiments indeed have been done. In several, the second IRES was that from EMCV. These dicistronic poliovirus genomes produced infectious virus when transfected into cells. But for reasons that are not yet clear, they were genetically unstable, losing the second IRES after five serial passages. Might these dicistronic engineered polioviruses actually be less efficient than the wild-type virus? See the text. Note that the EMCV IRES is more efficient than the poliovirus IRES. Could the selection of IRESs in these experiments have been different so as to generate a more competitive virus?

Reference

Johansen, L. K., and C. D. Morrow. 2000. Inherent instability of poliovirus genomes containing two internal ribosome entry site (IRES) elements supports a role for the IRES in encapsidation. *J. Virol.* 74:8335–8342.

Alternatively, the IRES may have yet other functions [e.g., in RNA replication or packaging; see below], which might be compromised if there were a second IRES [Box 6.11].)

Despite the facts that several eukaryotic viruses are known to use IRESs and that even a few cellular mRNAs are known to contain an IRES, no mammalian mRNA is known to contain more than one IRES. An example of a cellular mRNA that contains an IRES is that which is translated into the endoplasmic reticulum chaperone protein BIP. A likely reason why BIP translation initiates using an IRES is that it enables this heat shock protein to be translated under conditions where 5′ cap-mediated initiation might be impaired by inactivation of cellular initiation factor eIF4F, as might happen during heat shock or growth arrest. The BIP IRES does not display any structural or sequence homology to any picornavirus IRES sequence, nor does any other cellular IRES. Also, the discovery of cellular IRES sequences raises the unanswered question of why there are so very few eukaryotic cellular mRNAs that contain multiple ORFs, as are seen in prokaryotes (Boxes 6.10 and 6.11).

Next, we ask why picornaviruses use IRESs, rather than 5′ cap structures, to initiate translation. One possibility is

that it provides for a mechanism by which picornaviruses can inhibit cellular protein synthesis, which indeed they do, presumably to completely subvert the host cell's translational machinery. Poliovirus achieves the shutdown of cellular translation by abolishing the ability of the cell to initiate translation of capped mRNAs (recall the reovirus example in this regard).

The mechanism by which poliovirus shuts down cellular translation is by using its 2A^{pro} to cleave the host translation initiation factor eIF4G, together with using its 3C^{pro} to cleave the poly(A)-binding protein (PABP). Initiation factor eIF4G is a component of the eIF4F translation initiation complex that recognizes the cap and recruits the 40S ribosomal subunit. Cleavage of eIF4G separates the cap-binding component of the eIF4F complex from the ribosome-recruiting component. The normal function of PABP is to bind to the poly(A) tail of eukaryotic mRNAs, in that way stimulating translation initiation by its interactions with several translation factors, including eIF4G. These interactions are thought to perhaps stabilize initiation complexes on capped mRNAs. Cleavage of PAPB removes its C-terminal domain, which interacts with the initiation factors.

Translation of uncapped poliovirus mRNA is unaffected by the cleavage of eIF4G, since that initiation factor is not needed for ribosomes to attach to the virus's IRES. Instead, the C-terminal fragment of eIF4G, which is generated from eIF4G by 2A^pro, facilitates ribosome attachment to the IRES. The 2A^pro of rhinoviruses and coxsackievirus likewise cleaves eIF4G during infection by those viruses, whereas EMCV affects yet another component of the cap recognition system.

Other experimental findings suggested that sequences or structures within the IRES are directly required for viral RNA replication, as well as for translation. For example, mutations in IRES, which do not affect its function in translation, can dramatically affect replication. Perhaps the IRES and cellular translation factors might facilitate interactions between the 5′ and 3′ ends of the RNA molecules that might be necessary to initiate RNA replication (see below).

TRANSCRIPTION AND GENOME REPLICATION

As stated above, and as discussed in Chapter 2, translation of the parental viral genome occurs early in the replication cycle of plus-strand RNA viruses, so as to generate the virus-encoded RNA polymerase/replicase activity that is required to replicate and transcribe the viral genome. Apropos that fact, picornavirus genomes contain a single ORF and, likewise, a single translation initiation site. Moreover, the picornavirus 3D^pol polymerase is encoded at the 3′ end of that ORF (Fig. 6.9). Consequently, the entire coding region of the picornavirus genome must be translated before transcription/replication can begin. This requires about 10 to 15 min. Yet, during the course of a single 8-h cycle of infection, a single input virus genome is replicated to yield about 50,000 progeny copies. (This number can be different for other picornaviruses.)

As has been implied in the preceding paragraph and elsewhere, picornavirus plus-strand RNA genomes are identical in base sequence to picornavirus mRNA molecules. Consequently, it is possible, if not likely, that the same mechanism that produces progeny viral genomes also produces viral mRNAs. (But if that is so, then do picornavirus mRNAs terminate in VPg? This issue is discussed later.)

Now, bear in mind that RNA replication is a two-step process, in which the first step is the transcription of a complementary template, from which the progeny genome might then be transcribed in the second step. Thus, in the case of plus-strand RNA viruses, can one assume that minus strands are synthesized by the same mechanism that synthesizes progeny plus-strand genomes? As a matter of principle, one should not make such an assumption. In the case of picornavirus replication, it is likely that plus-strand and minus-strand syntheses indeed are somewhat different processes. This is suggested by the as-yet-unexplained

fact that plus strands are synthesized in 100-fold-greater amounts than are minus strands (see below).

Picornavirus RNA synthesis is characterized by a number of distinctive features, which taken together set it apart from the replication of other plus-strand RNA viruses. These features are briefly enumerated here and are discussed in more detail below. All examples pertain to poliovirus, the prototype picornavirus.

First, poliovirus RNA synthesis depends on the activities of several viral and cellular proteins, in addition to the virus-encoded RNA-dependent RNA polymerase. These **accessory** proteins play multiple roles. Among these roles, (i) they target the replication complexes to appropriate sites in the cell; (ii) they recognize particular sequences on the RNA template so as to organize replication complexes; and (iii) one protein, VPg, functions as a primer (see below). Second, an important feature of poliovirus RNA synthesis is that viral precursor polyproteins, as well as their final cleavage products, may each express a distinct function in replication. This will be seen in the cases of the precursor polypeptides P3, 3AB, and 3CD and their respective cleavage products. Impressively, P3 itself is the precursor polyprotein for 3AB and 3CD, and 3AB and 3CD in turn are the precursors for VPg and 3C^pro plus 3D^pol, respectively, all of which play crucial roles in replication. Third, the 3CD intermediate polypeptide itself expresses multiple activities, functioning both as a protease and as a specific RNA-binding protein, with each of these activities playing a key role in replication. Importantly, the multiple activities of the intermediate precursor polypeptides expand the number of functional activities that can be expressed from the relatively small picornavirus genome. Fourth, bear in mind that replication factors must be able to recognize specific features on the RNA template. These features may include secondary structures, as well as particular nucleotide sequences. Fifth, poliovirus RNA synthesis is primed by a virus-encoded protein, VPg. Sixth, as described immediately below, viral RNA synthesis does not occur at random sites within the cytoplasm but instead occurs on specific membranous elements that are generated by virus-encoded **membrane-remodeling** activities.

As just noted, poliovirus RNA synthesis occurs on particular membranous elements, specifically, small cytoplasmic membranous vesicles. Poliovirus causes these cytoplasmic membranes to form by impairing the vesicle-mediated secretory pathway (i.e., that which goes from the endoplasmic reticulum to the Golgi apparatus, and then from the Golgi apparatus to the plasma membrane; this is the pathway that normally glycosylates intrinsic membrane proteins [see Chapter 8]). Poliovirus is not dependent on the activities of this cellular pathway, since there are no glycoproteins in nonenveloped poliovirus particles. (Some nonenveloped viruses do contain glycosylated capsid proteins and thus are still dependent on endoplasmic reticulum to Golgi body traffic.)

Viruses from a variety of families are known to affect the cell vesicle system and, consequently, glycoprotein trafficking. In the picornavirus instance, an assortment of virus-encoded proteins (e.g., 3AB, 2B, 2BC, 2C, and 3A) interact with, and localize at, intracellular membranes. Thus, the expression of some subset of these proteins may be responsible for the intracellular membrane remodeling that leads to disruption of the vesicle system and glycoprotein trafficking mediated by those vesicles. In one key experiment, the expression of recombinant 2BC or 2C caused the cytoplasm of transfected cells to develop vesicles of 50 to 350 nm in diameter, which resembled those found in poliovirus-infected cells. Moreover, both 2BC and 2C were associated with these vesicles.

What might be the advantage to the virus of membrane-associated transcription/replication? One advantage may be that it brings together the several factors that are necessary for virus RNA synthesis, concentrating them in two-dimensional arrays, which might then enhance the efficiency of their interactions. Another possible advantage of membrane-associated RNA synthesis to the virus is that it may concentrate progeny viral genomes and viral translation products at the same intracellular site, so as to enhance the efficiency of subsequent viral assembly. In these regards, most plus-strand RNA viruses carry out both genomic RNA synthesis and mRNA synthesis on cytoplasmic membranes, which in all instances originate from particular cellular membrane compartments.

The picornaviral plus-strand RNA molecule contains three key structural features that enable it to carry out its template function. These are (i) the **cloverleaf structure** at the 5′ end, (ii) an internal **stem-loop** referred to as the *cis-acting replication enhancer* (**cre**) **element**, and (iii) a **pseudoknot** at the 3′ end. (An RNA pseudoknot forms when a nucleotide sequence within the loop of a stem-loop is able to base pair with a nucleotide sequence outside the loop, as in Fig. 6.14.) The plus-strand template, with its 5′ cloverleaf structure, internal cre element, and 3′ pseudoknot, is depicted schematically in Fig. 6.15A. The cloverleaf is found in the first 108 nucleotides of the RNA molecule, upstream of the IRES (Fig. 6.13). The poliovirus cre element is located in a stem-loop structure in the 2C coding region. (The genomes of other picornaviruses also contain cre elements. That of rhinoviruses is located in the 2A coding region, those of Theiler's virus and mengovirus are found in the VP2 coding region, and that of FMDV is found in the 5′-noncoding region.) The roles of these features in picornaviral RNA synthesis are described below and depicted in Fig. 6.15.

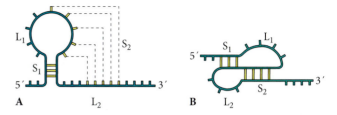

Figure 6.14 An RNA pseudoknot forms when a nucleotide sequence within the loop of a stem-loop is able to base pair with a nucleotide sequence outside the loop. Shown schematically are a stem-loop (A) and a pseudoknot (B).

A unique feature of poliovirus RNA synthesis is that it must be primed by a particular protein, the virus-encoded VPg protein. Regarding this fact, all DNA polymerases require a primer. In contrast, DNA-dependent RNA polymerases and RNA-dependent RNA polymerases generally do not require a primer. (The discovery of VPg covalently linked to the 5′ terminus of picornaviral genomic RNA was the first indication that picornaviral RNA synthesis might be primed by a protein. At present, there are two other known examples of RNA-dependent RNA polymerases that require a primer. These are the RNA polymerases of influenza virus and the bunyaviruses [Chapter 12]. In these two cases, the RNA primers are short-capped RNA segments, generated by cleavage of cellular mRNAs.)

Next, we consider the known and putative actions of the various viral and cellular proteins in poliovirus RNA synthesis. (After noting these activities, see Fig. 6.15 for a model of poliovirus RNA synthesis that takes them into account.) Early in the process, viral precursor polypeptide 3AB associates with cellular membrane elements via its hydrophobic domain. This is an important early step in viral RNA synthesis since 3AB secures both the primer, VPg, and the polymerase, 3D^{pol}, to the membranes. The 2C protein, which may also help secure the replication complex to the membrane, contains amino acid motifs with a high degree of homology with known nucleoside triphosphatase (NTPase)/helicase proteins. Moreover, 2C possesses adenosine triphosphatase (ATPase) and guanosine triphosphatase (GTPase) activities. Thus, it may serve as a helicase, perhaps separating the two strands of the replicative intermediate. 3CD is the precursor polyprotein that undergoes autocatalytic cleavage to generate 3C^{pro} and the RNA polymerase, 3D^{pol}. 3C^{pro} in turn cleaves 3AB to generate 3A and 3B (i.e., VPg). However, 3CD also binds to a stem-loop in the cloverleaf structure at the 5′ end of the viral plus strand. CD is then able to facilitate the attachment of the cellular poly(rC)-binding protein (PCbp) to a different stem-loop in the cloverleaf structure. (PCbp has several normal functions, including stabilizing mRNA and controlling both translational initiation and transcription.)

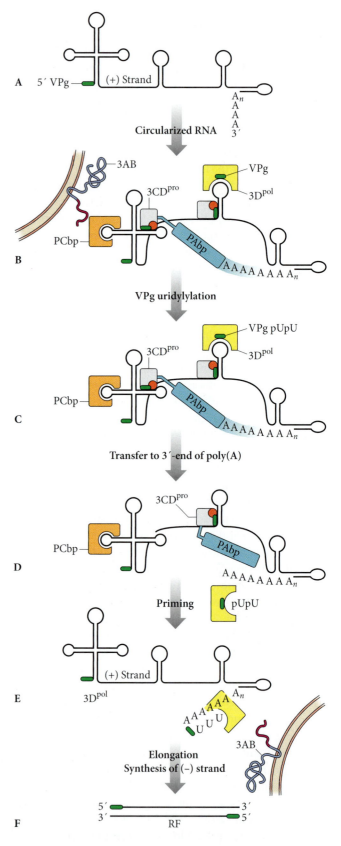

Figure 6.15 A model for poliovirus RNA synthesis. (A) Schematic of the plus-strand template, depicting the 5′ cloverleaf structure, the internal stem-loops known as cres, and the 3′ pseudoknot. (B) Since the input viral genome is a plus strand, the first product of viral RNA synthesis will be a minus strand. An important early step is for viral precursor polypeptide 3AB to associate with cellular membrane elements via its hydrophobic domain. Another early step is for virus protein 3CD to bind to a stem-loop in the cloverleaf structure at the 5′ end of viral plus strands. The cellular poly(rC)-binding protein (PCbp) binds to a different stem-loop in the cloverleaf structure, and the cellular poly(A)-binding protein (PAbp) binds to the poly(A) tail at the 3′ end of the viral RNA. Then, PAbp, which is bound to the 3′ poly(A) tail of the viral RNA, interacts with the complex of 3CD and PCbp at the 5′ end of the viral RNA, thereby bringing the two ends of the plus strand into apposition. The RNA-bound 3CD is recruited to the membrane vesicles by the membrane-bound 3AB. There, 3CD undergoes autocatalytic cleavage, to generate 3C^{pro} and 3D^{pol}, and 3AB is cleaved by 3C^{pro} to generate 3A and 3B (i.e., VPg). Next, the cre element comes into play by binding 3D^{pol}, 3CD, and VPg. (C) VPg now fulfills its role as the minus-strand primer as follows. First, VPg is uridylylated on a tyrosine residue by 3D^{pol}, using the sequence AAACA of cre as a primer. Cre-bound 3CD greatly stimulates this reaction. (D through F)

Figure continues

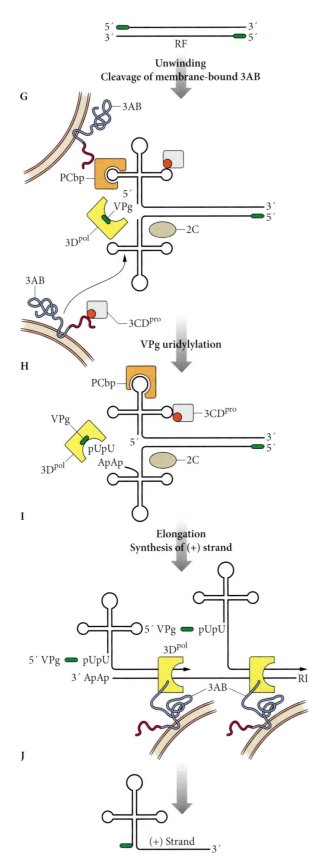

Figure 6.15 *(continued)* Next, the uridylylated VPg (i.e., VPg-pUpU), in association with the released 3D^pol^, is transferred to the poly(A) tail *at the 3′ end of the plus-strand RNA*, thereby being in position with 3D^pol^ to prime transcription of the complementary minus strand. Membrane-bound 3AB may play a role in stimulating the priming reaction. (G) Once minus-strand templates have been produced, synthesis of progeny plus strands can begin. As a first step, the viral 2C protein may separate the two strands of the replicative intermediate, itself binding to the cloverleaf structure that forms at the 3′ end of the *minus strand*. Also, PCbp and 3CD bind to the cloverleaf at the 5′ end of the *plus strand* and associate with membrane-bound 3AB, as they did earlier to initiate plus-strand synthesis above. (H and I) Also as above, membrane-bound 3AB is cleaved by 3CD to generate 3A and VPg, and 3CD autocatalytically generates 3C^pro^ and 3D^pol^. Moreover, VPg-pUpU is generated by 3D^pol^, but in this instance the poly(A) sequence at the 3′ end of the *minus strand* may be used as the template for the uridylylation of VPg, although this has not been resolved. (J) VPg-pUpU, in association with 3D^pol^, can then initiate plus-strand synthesis, perhaps stimulated once more by 3AB. Modified from S. J. Flint, L. W. Enquist, V. R. Racaniello, and A. M. Skalka, *Principles of Virology: Molecular Biology, Pathogenesis, and Control*, 2nd ed. (ASM Press, Washington, DC, 2004), with permission.

The complex of 3CD, PCbp, and the cloverleaf is necessary to initiate RNA synthesis. Note that each of 3CD, 3D^{pol}, 2C, and VPg binds to the cre element. cre-bound 3CD greatly stimulates the polyuridylylation of VPg, which is required for the initiation of both plus-strand and minus-strand RNA synthesis. VPg is polyuridylylated on a tyrosine residue by 3D^{pol}, using the cre AAACA sequence as the template for poly(U) synthesis (Fig. 6.15C).

Bearing in mind that synthesis of the minus strand must initiate from the 3′ end of the plus-strand template, the cellular PABP (see above) binds to the poly(A) tail at the 3′ end of the viral RNA. PABP might then interact with the complex of 3CD and PCBP2 at the 5′ end of the viral RNA, thereby bringing the two ends of the plus strand into apposition (Fig. 6.15B).

Fig. 6.15 depicts a model for poliovirus RNA synthesis that takes into account the activities of the viral and cellular proteins enumerated above. Models are conceptually useful but should not be taken as statements of fact. In this particular instance, despite the fact that the model takes into account experimental results from several research groups, these experiments have for the most part been carried out in vitro. Thus, the exact composition of the viral replication complexes as they occur in vivo is not yet entirely known, and the precise roles of individual viral and cellular proteins in viral RNA replication also are uncertain. (One exception is 3D^{Pol}, which clearly catalyzes the polymerization of viral RNA.) Moreover, some details in the model have never been proven and very well may be incorrect.

As noted above in a different context, viral plus-strand RNAs that might act as mRNA or progeny genomes are produced in 100-fold-greater amounts than minus strands during infection of cells. One possible explanation is that transcription of plus strands is more efficiently initiated than transcription of minus strands. Another possibility is that since plus strands serve as mRNA and also are packaged into progeny virus particles, fewer are available to serve as templates for minus-strand synthesis.

Since picornavirus genomes contain a VPg protein primer, and since viral mRNAs are likely generated by the same mechanism that generates viral genomes, we might ask whether functional picornaviral mRNA molecules likewise contain a covalently bound VPg. Interestingly, viral mRNA molecules isolated from polysomes mostly lack VPg. Instead, those mRNAs terminate with pUpUpA. A cellular enzyme of unknown function apparently removes the VPg from viral mRNAs. Note that there are no known cellular RNAs that terminate in VPg-like proteins. Yet removal of VPg is not necessary for plus strands to function as mRNA, since genomic RNA that contains VPg is an able mRNA, at least in vitro in cell-free translation systems. Moreover, some VPg-containing virus mRNA is found on polysomes late in infection. The last findings are perhaps not unexpected, considering that the IRES-mediated mode of translation initiation should not be impaired by the presence of VPg at the 5′ end of the mRNA.

Before moving on, consider the following fundamental feature of molecular biology, apropos the picornaviral replication cycle. RNA molecules are always read in the 3′-to-5′ direction when they are being transcribed, whereas they are translated reading in the 5′-to-3′ direction. Since (i) parental picornaviral RNA genomes must be translated before they can be replicated and (ii) parental and progeny plus-strand RNA molecules are templates for replication, as well as for translation, how do the replication complexes avoid colliding with ribosomes? This is another conundrum for which an answer is not yet in hand.

ASSEMBLY AND MATURATION

As described above, picornavirus capsids display icosahedral symmetry and are constructed from four nonidentical proteins, VP1, VP2, VP3, and VP4. These four proteins constitute a subunit structure called a protomer, which contains one copy of each protein. (Actually, at the time of capsid assembly, the protomer contains VP0, VP1, and VP3. VP0 subsequently undergoes cleavage to generate VP2 and VP4 during the final maturation of the virus particle [see below].)

The first step in the construction of picornavirus capsids is the assembly of pentamers, each of which is comprised of five protomers. (As noted just above, each of the protomers making up the pentamers initially consists of one copy each of VP0, VP1, and VP3.) The icosahedron is then built from 12 of those pentamers.

Whereas the capsid might appear to be a relatively simple assemblage of the 12 pentamers, its construction nevertheless involves specific interactions between the four distinct proteins that constitute each of the protomers. Moreover, these specific interactions occur between proteins within the same protomer, as well as between proteins in different protomers. Thus, we might wonder how those individual proteins might efficiently find each other and properly interact to form the protomers, especially in an intracellular environment in which the concentration of host proteins can be as high as that reached in protein crystals. As you might have guessed, the answer to this conundrum is in the strategy of picornavirus translation, in which VP1, VP2, VP3, and VP4 are generated by proteolytic cleavage from the P1 precursor protein (Fig. 6.9).

P1 is released from the nascent polyprotein precursor by 2A^{pro} (Fig. 6.9). Importantly, the amino acid sequences in P1 that correspond to VP2, VP4, VP1, and VP3 take on major features of the tertiary structures of the individual proteins, including formation of their β-barrel domains, while still covalently linked to each other on the nascent polyprotein (Fig. 6.16). Moreover, they also begin to make interprotein contacts before cleavage occurs. Thus, when P1 is cleaved twice by protease 3CD^{pro} to yield VP0, VP1, and VP3, these proteins have already begun to form protomers. Consequently, they do not then need to find each other in the protein-rich cytosol, where assembly might otherwise be hindered by nonspecific contacts with the great excess of cellular proteins. In addition, the cellular chaperone Hsp70 facilitates the conformational transitions leading to protomers. Once formed, the protomers assemble into pentamers, which are stabilized by extensive contacts between the individual proteins and by interactions involving the myristoyl residues at the five VP0 N termini within each pentamer. The great stability of the pentamers indeed causes their formation to be irreversible, thereby driving morphogenesis forward. Twelve pentamers then combine to form the protein shell.

It still is not certain whether the RNA genome enters into empty capsids or whether the pentamers form capsids around condensed RNA. The latter possibility is currently favored, since pentamers appear to interact with virus RNA to form virus particles, without empty capsids appearing first as precursors. In either case, there is experimental evidence suggesting that packaging is linked to ongoing viral RNA synthesis. For example, the drug L-guanidine, which inhibits initiation of synthesis of poliovirus plus and minus RNA strands, causes a simultaneous inhibition of RNA packaging. This experimental result is consistent with the finding that only newly synthesized RNA is encapsidated. Neither model explains how the viral RNA might be compacted to a concentration as high as 70%, as it exists in the capsid.

The cleavage of VP0 to yield VP2 and VP4 is the final step in the maturation of poliovirus particles. This step is believed to occur autocatalytically because the VP0 cleavage site is buried deep within the particle and thus is inaccessible to external proteases. Regardless, the cleavage of VP0 creates the C terminus of VP4 and the N terminus of VP3, which then are available to make the intra- and interpentameric contacts that further stabilize the extracellular virus particles. Moreover, this cleavage step is required for infectivity, since it primes the viruses to undergo the conformational transitions that enable release of the viral genomes from the particles into the cytosol, at the start of the next cycle of infection. Recall that these conformational transitions are triggered by particular interactions with the host cell at the start of the next cycle of infection.

It is rather remarkable that VP0 cleavage, the final step in the maturation of picornavirus particles, drives maturation by making it an irreversible process while stabilizing the extracellular virus particles and simultaneously priming them to undergo the triggered conformational transitions that lead to entry and uncoating at the start of the next replication cycle, a quite ingenious adaptation at the molecular level.

Figure 6.16 Assembly of the picornavirus structural unit or protomer from an intermediate polyprotein precursor. P1 is released from the nascent polyprotein by the action of 2A^{pro}. Importantly, the amino acid sequences in P1 that correspond to VP2, VP4, VP1, and VP3 assume major features of the tertiary structures of the individual proteins, including formation of their β-barrel domains, while still covalently linked to each other on the nascent polyprotein. Moreover, they also begin to make interprotein contacts before cleavage occurs. Thus, when P1 is cleaved twice by protease 3CD^{pro} to yield VP0, VP1, and VP3, these proteins have already begun to form protomers. Consequently, the individual proteins do not need to find each other in the protein-rich cytosol, where assembly might otherwise be hindered by nonspecific contacts with the great excess of cellular proteins. Modified from S. J. Flint, L. W. Enquist, V. R. Racaniello, and A. M. Skalka, *Principles of Virology: Molecular Biology, Pathogenesis, and Control*, 2nd ed. (ASM Press, Washington, DC, 2004), with permission.

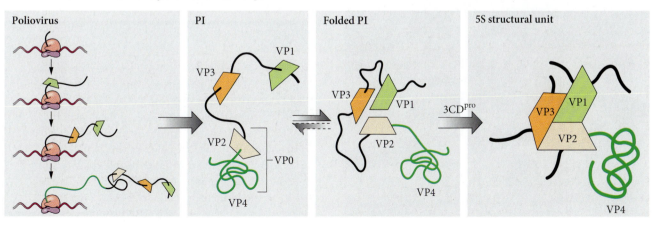

In the presence of L-guanidine, poliovirus RNA synthesis is blocked, empty shells accumulate, and, importantly, VP0 is not cleaved. The effect is reversible if the drug is removed. These experimental findings imply that viral RNA plays some role in the maturation cleavage. In addition, they are consistent with the less-favored assembly model, in which empty shells are precursors for virus particles. Nevertheless, empty capsids are not always seen in infected cells, and unlike pentamers, they show no tendency to associate with viral RNA in experimental systems.

MEDICAL ASPECTS

The *Picornaviridae* contain several medically important viruses. We begin with poliovirus.

Poliovirus

During the first half of the 20th century, paralytic poliomyelitis was one of the most feared diseases in the industrialized world. The author, who was a teenager when the polio vaccines first appeared in the mid-1950s, remembers well the panic that set in every summer of the prevaccine days, when the first neighbor or schoolmate was stricken. A visit to a hospital in those times was associated with shocking sights of victims in iron lungs. Indeed, before the polio vaccines, iron lungs filled the wards of children's hospitals (Fig. 6.17). (Even today, there are survivors of the polio epidemics of the early to mid-1950s who still remain confined to iron lungs. An impending crisis facing these individuals is that neither iron lungs nor spare parts for them have been made for years.)

Figure 6.17 Totally paralyzed poliomyelitis patients in mechanical respirators (iron lungs) during the epidemic in the United States before the advent of universal immunization (1955). From D. O. White and F. J. Fenner, *Medical Virology*, 4th ed. (Academic Press, San Diego, CA, 1994), with permission.

Even the emergence of acquired immunodeficiency syndrome (AIDS) in the early 1980s did not create the widespread fear generated by poliomyelitis. Yet despite the terror provoked by poliomyelitis, the average number of cases in the United States in the 1950s was "only" about 20,000 per year. (The peak year was 1952, when there were 57,879 cases.) For comparison, influenza, which we do not dread nearly as much, may kill 40,000 to 50,000 individuals yearly in the United States. A likely reason for the greater fear engendered by poliomyelitis was that the paralytic disease struck mainly adolescents and young adults, whereas influenza threatens mainly the elderly. It is not uncommon for individuals to be more emotionally invested in their children than in themselves or their parents (Box 6.12).

A key feature of poliovirus infections is that they tend to be mild and often clinically inapparent in infants and young children, while they are much more likely to lead to paralytic poliomyelitis in adolescents and young adults. A second important point is that poliovirus most often is contracted from contaminated food or water. Together, these facts help to explain the seeming paradox that paralytic poliomyelitis emerged as a major health problem with the development of modern sanitation practices. Prior to those developments, nearly everyone was infected as an infant or young child. Consequently, those infected individuals then had mild poliovirus infections, at least in part because of the passive immunity transferred to them by their previously infected mothers. Moreover, infection of those infants and children induced lifelong immunity. With the development of better sanitation, infection had a higher likelihood of not occurring until a later age in life, when passive immunity from maternal antibodies would have waned. When passive immunity was present, it would have delayed dissemination of the virus from the primary sites of infection in the oropharynx and gut (see below) to secondary sites in the central nervous system. This delay provided time for the development of active immunity that would have prevented dissemination to

BOX 6.12

The 1918 "Spanish Flu" epidemic was the most devastating epidemic in human history. During the single month of October 1918, that epidemic killed 196,000 people in the United States alone! Estimates of the total number of fatalities worldwide range between 20 million and 50 million. Yet unlike the polio epidemics, it did not cause any panic, and today most people either do not know what a huge impact the 1918 influenza outbreak had, or they need to be reminded of it. See Chapter 12.

the central nervous system, thereby preventing paralysis and death.

In the cases of many viruses, including poliovirus, the pattern of virus tropism and pathogenesis is largely determined by the sites in the host where the virus receptor is expressed. In the case of poliovirus, its receptor, CD155, is expressed mainly on anterior horn cells of the spinal cord, dorsal root ganglia, motor neurons, skeletal muscle cells, and lymphoid cells. The virus, which enters from the mouth, initially replicates in tonsils, lymph nodes of the neck, and the intestinal epithelium. The infection will be asymptomatic if it spreads no further (see above paragraph). This is the outcome more than 90% of the time. That is, the vast majority of poliovirus infections are clinically inconsequential. However, the virus may enter the bloodstream, from which it can invade the central nervous system. Muscles that are innervated by affected neurons may then undergo paralysis and subsequent atrophy. This outcome is generally rare, occurring in 0.1 to 2% of unprotected people infected with poliovirus. The extent of the paralysis and the muscles affected depend on the particular neurons that are destroyed by the virus.

A condition known as **postpolio syndrome** emerges in perhaps 20 to 40% of polio victims, years after the initial illness. It is characterized by further muscle atrophy, perhaps caused by the death of some of the fewer neurons that might have been available to innervate the affected muscle after the initial poliomyelitis episode (see Chapter 3).

In the 1950s, Jonas Salk developed the inactivated poliovirus vaccine, and Albert Sabin developed the live attenuated vaccine. The attenuated vaccine actually contains three different attenuated strains of poliovirus. Types 1 and 3 were attenuated by serially propagating them in cell culture until variants were obtained that displayed reduced neurovirulence in experimental animals. The attenuated type 2 strain was a naturally occurring variant. Each of these attenuated strains contains lesions that are located in the untranslated IRES.

The advantage of the *inactivated vaccine* is its greater safety. (While this is generally true, in the 1950s several vaccine stocks were incompletely inactivated, resulting in about 100 cases of vaccine-caused poliomyelitis. This episode is discussed further in Box 6.13.) A significant advantage of the *attenuated vaccine* is that it can be administered by the oral route, whereas the Salk inactivated vaccine needs to be injected. This advantage of the attenuated vaccine was particularly important in developing countries, where the infrastructure was often not in place to administer the inactivated vaccine. (In early trials, the Sabin attenuated vaccine was first administered either on a lump of sugar or in a sugary solution.) Also, oral delivery of a vaccine against an enteric virus has the advantage of mimicking the natural route of infection, thereby generating an immune response more like that evoked by a natural infection. (The attenuated vaccine works because the attenuated viruses replicate less well than the wild-type virus and very likely do not generate virus titers that might lead to infection of the central nervous system, while nevertheless raising up immunity.) Thus, while the Salk vaccine was effective at preventing most of the complications of polio, it was not as effective as the Sabin vaccine at impeding the initial, intestinal, infection by the wild-type virus. Consequently, an individual receiving the Salk vaccine might for a time become a carrier of the wild-type poliovirus. Moreover, the Sabin vaccine also had the advantage of producing lifelong immunity, with no need for subsequent booster shots. Finally, the live, attenuated virus could be transmitted naturally from its recipients to individuals who had not received the vaccine, thereby naturally transmitting the vaccine.

On the other hand, the last apparent advantage can also be an inherent problem. That is so because of the high rate of mutation of RNA viruses in general, which can result in reversion of an attenuated vaccine strain to virulence. Such revertant viruses indeed pose a threat to nonvaccinated individuals in the population. As recently as 2000-2001, the Dominican Republic and Haiti suffered an outbreak of poliomyelitis, encompassing 21 confirmed cases, which resulted from a single sample of the Sabin vaccine that was administered during the preceding year. In fact, the few cases of poliomyelitis that now occur in the West are vaccine related, resulting from rare reversions to virulence of the live vaccine virus. About 1 in 4 million live vaccine doses are estimated to lead to paralysis. In contrast, 1 in 100 infections of unprotected individuals with wild-type poliovirus lead to paralysis. Regardless of the overall risk-to-benefit ratio of the attenuated vaccine, these vaccine-related cases led to renewed interest in the inactivated vaccine, at least for the initial vaccinations, to be followed by the attenuated vaccine. Since 2000, the United States has been using the Salk vaccine exclusively.

The success of the poliovirus vaccines has led to the near disappearance of poliovirus from the Western world and indeed from much of the rest of the world as well. Wild-type poliovirus infections still occur in parts of Africa, the Middle East, and Asia. Yet such has been the success of the vaccines that poliovirus has realistically been targeted for elimination worldwide. Thus, the impact of Salk and Sabin indeed has been great (Box 6.13).

Ironically, the great success of the polio vaccines has generated some antivaccine sentiment. Since poliomyelitis essentially disappeared in countries where a strict vaccine regimen was followed, the risk of the vaccine is now perceived by many individuals in those countries to be

BOX 6.13

Jonas Salk and Albert Sabin

The fear evoked by poliovirus in the developed world ended abruptly when the Salk vaccine was announced in 1955. The response of the public was so great that Salk was hailed as a "miracle worker." Yet despite Salk's public acclaim, many of the great virologists of the time did not admire him for his accomplishment. Indeed, many resented his acclamation. One reason for resentment among virologists was that Salk's achievement was based on the earlier major breakthroughs of others, especially John Enders (recall that Enders and his colleagues developed methods for cultivating poliovirus in cell cultures [Chapter 1]). Some felt that Salk's work merely applied what already was known. Yet Enders did not express any resentment towards Salk and indeed acknowledged that scientists generally do build on the accomplishments of those who came before. Another cause for resentment was the publicity Salk had been getting on behalf of the private fund-raising that supported the vaccine work (in the pre-National Institutes of Health days of the 1950s). To boost donations, the March of Dimes promoted Salk as a celebrity, to a point that offended other researchers. Yet Salk actively tried to prevent getting all the credit for the vaccine, and he vigorously protested the Foundation's effort to have his name attached to it. Despite his efforts in that regard, people and the press referred to the vaccine as the "Salk vaccine." It just sounded and looked good. Yet another reason for resentment against Salk concerned the safety of the vaccine. Recall the early vaccine-related cases of poliomyelitis (see the text). These might have been caused by a mistaken understanding of the relationship between dosage of the inactivating agent (formaldehyde) and the surviving fraction of viable virus (see Box 9.5 for a discussion of this point). Nevertheless, the vaccine produced after 1955 was safe, and the development of that safe vaccine was realized because of Salk's conviction that it could be made. For reasons noted below, the Sabin live virus vaccine eventually won out over the Salk vaccine. Nevertheless, in countries where Salk's vaccine continued to be used, poliovirus has been virtually eradicated. And for reasons noted in the text, the United States has been using the Salk vaccine exclusively since 2000.

In 1963 Salk founded the prestigious Salk Institute for Biological Sciences in La Jolla, CA. At that time he was still somewhat alienated from other medical researchers. Thus, we can enjoy his remark "I couldn't possibly have become a member of this institute if I hadn't founded it myself." Jonas Salk died of congestive heart failure in 1995 at the age of 80. Today, he is one of the most venerated medical scientists ever.

And what of Albert Sabin, who developed the live polio vaccine? Sabin, born in 1906, immigrated to the United States in 1921, at least partly to escape the persecution of Jews in his native Poland. He obtained his medical education in the United States, and during World War II he served as a lieutenant colonel in the U.S. Army Medical Corps, helping to develop vaccines against dengue fever and Japanese encephalitis. Nevertheless, Sabin's major interest was polio, which he focused on after the war. But before developing the attenuated polio vaccine that bears his name, Sabin made other seminal contributions to poliovirus research. First, he provided evidence that the digestive tract is the poliovirus port of entry, rather than the respiratory tract as previously was thought, and that the virus later attacked nerve tissue. Second, Sabin established that poliomyelitis indeed was *rare* in those urban populations with the *lowest* standards of sanitation. Moreover, children in those populations had protective antibodies against poliovirus, despite never having shown signs of acute disease. These findings led to the proposition that poliovirus tends to be a mild infection in infants and young children, an age group more likely to be infected for the first time when living under conditions of low sanitation standards. In contrast, infection can be much more severe in teens and young adults, who are more likely to miss being infected earlier, as a result of living under high sanitation standards.

Although the Salk vaccine greatly reduced the number of poliomyelitis cases in the United States, Sabin believed that a "live" attenuated vaccine would be more effective than the killed vaccine developed by Salk. For that reason, Sabin persevered in his efforts to develop a live vaccine. The relative advantages of the inactivated and live poliovirus vaccines are enumerated in the text.

Interestingly, despite the fact that the World Health Organization (WHO) in 1957 invited Sabin to administer the attenuated vaccine to children in parts of Russia, Holland, Mexico, Chile, Sweden, and Japan, the U.S. Public Health Service did not approve testing of the vaccine in the United States until April 1960. By then, the Sabin vaccine had been administered to about 80 million people outside the United States. The delay in testing of the Sabin vaccine in the United States may have been due in part to the influence of advocates of the Salk vaccine. Disagreement between supporters of the rival Salk and Sabin vaccines indeed was intense. However, part of the reason for testing outside the

Box continues

United States was to ensure that the Sabin vaccine might not inadvertently be administered to people already protected by the Salk vaccine, thereby compromising the test results. Regardless, the U.S. Public Health Service did not endorse the Sabin vaccine for use in the United States until 1961. But it was the Sabin vaccine that actually eliminated polio from the United States. Yet the Salk vaccine continued to

have advocates, and a sharp rivalry between Salk and Sabin persisted thereafter.

Albert Sabin became president of the prestigious Weizmann Institute of Science in Israel but stepped down in November 1972 for health reasons. He passed away in 1993 at the age of 86.

---

greater than the risk of the virus itself. This is largely because the very rare vaccine-related cases, which occur in the absence of the natural disease, become major news events. Moreover, in the absence of natural infections, people become complacent and question the need for the vaccine.

The Nigerian states of Kano and Zamfara provide a recent example of the consequences of not maintaining vaccine compliance, and also of how politics and science can still be at cross-purposes in modern times. Rumors, spread in the region by politicians and clerics, claimed that the polio vaccine was tainted with human immunodeficiency virus and hormones, as part of a Western plot to cause infertility among Muslim women. These rumors resulted in the shutting down of the polio eradication program in those Nigerian states for 11 months. As a consequence of that noncompliance, there were 400 cases of paralytic poliomyelitis in Nigerian children in 2003, as well as 400 other cases in 10 neighboring countries.

Polio is still prevalent in four countries: Nigeria, India, Pakistan, and Afghanistan. Moreover, it is routinely found in a number of other countries, where it was reintroduced by travelers after many years of absence. In 2004, polio reemerged in Guinea and Mali, raising fears that a significant polio epidemic could yet hit west and central Africa.

Rhinoviruses: the Common Cold

Rhinoviruses are the most frequent cause of the common cold, and rhinovirus common colds indeed are common (i.e., widespread and frequent), with most of us acquiring at least one per year. This usually mild illness is transmitted by the respiratory route. Much research shows that a person is most likely to transmit rhinoviruses in the second to fourth day of infection, when the amount of virus in nasal secretions is highest, and there are studies which suggest that rhinovirus colds can be transmitted through the air. Nevertheless, hand washing is the simplest and most effective way to avoid getting rhinovirus colds. Not touching the nose or eyes is another. That is so because

the infection is most commonly transmitted as a result of touching respiratory secretions on an infected individual's skin or other environmental surface and then touching one's own eyes or nose. Nasal secretions can be transferred indirectly by shaking hands, or, for instance, by touching a doorknob or telephone that was touched by an infected person (Box 6.14).

Although ICAM-1, the receptor for the major rhinovirus subgroup, is present on most cells of the body, these viruses prefer the upper respiratory tract because they

BOX 6.14

Humans acquire more viral infections via the respiratory route than by any other means. Coronaviruses and influenza, measles, mumps, and chicken pox viruses are each examples of ubiquitous viruses that are transmitted by the respiratory route. Viruses evolved to exploit the respiratory route because it works so well for them. This is especially so in the modern world. We can practice good personal and public hygiene and exert some control over exposure to insect vectors. Yet we all need to breathe in the company of others, making these viruses very difficult to avoid (see Chapter 3). On the other hand (no pun intended), hand washing is the simplest and most effective way to avoid getting rhinovirus colds, since touching a contaminated object or surface and then going to one's eyes or nose is the major means of transmission of these diseases.

I thank Ellie Ehrenfeld for the following anecdote. An old Army study that one used to hear about found that when a curtain barricade was set up in an Army barracks, soldiers with flu on one side of the curtain readily spread the disease to those on the other side. However, the spread of rhinovirus infections was stopped. Rhinovirus infections then got the nickname "nose-picker's" disease.

grow best at 33°C. (The nasal mucosa is understandably cooler than more internal sites in the body.) Also, unlike many other picornaviruses, rhinoviruses are inactivated at low pH and thus cannot replicate in the gut. Consistent with the rhinovirus tropism for the upper respiratory tract, transmission generally occurs via aerosols produced by a sneeze or by touching contaminated objects or people and then rubbing one's face. A single droplet from an aerosol generated by a sneeze can contain as many as 100 million rhinovirus particles! And a single rhinovirus particle can initiate an infection. Clearly, the ability of the virus to induce sneezing serves it well by promoting its transmission.

Cold symptoms are usually no more serious than sneezing, followed by a runny nose and nasal obstruction, and perhaps a mild sore throat. The infection is self-limiting, and recovery is complete within 7 to 10 days. Interferon and other cytokines induced by the infection may be responsible for some of the symptoms (i.e., fluid leakage from capillaries to cause the runny nose), while also helping to control the infection. As noted above, the virus actually may subvert the action of the proinflammatory cytokines, since one of the cytokine effects is to upregulate expression of ICAM-1, which also is the receptor for the major subgroup rhinoviruses. Thus, activation of effectors of innate immunity (see below) may initially facilitate spread of the virus.

How the immune response to rhinoviruses mediates clearance of the infection is not well understood. Rhinovirus-specific IgA antibodies (Chapter 4) are produced at the mucosal surfaces, and there is evidence that rhinoviruses also elicit a cellular immune response in humans. In this regard, mucosa-associated lymphoid tissue is found throughout the mucosa of the respiratory tract, giving rise there to antibodies and cytotoxic T lymphocytes.

Despite the fact that rhinoviruses raise virus-specific antibodies, there is evidence suggesting that recovery from rhinovirus infections is not dependent on antibodies. First, rhinovirus infections are generally cleared before the appearance of virus-specific antibodies in serum and secretions. Second, individuals with an IgA deficiency and those with hypogammaglobulinemia appear to recover normally from rhinovirus infections. Thus, some sources state that the **innate immune system** may suffice to resolve rhinovirus infections. (For those who have not read Chapter 4, the innate immune system is the nonspecific branch of immunity, which is mediated by complement, professional phagocytes, cytokines, and natural killer cells. It acts soon after infection, before specific adaptive immune responses might develop.) However, in addition to **humoral immunity**, which is mediated by antibody-producing B lymphocytes, the **adaptive immune system** also includes **cell-mediated immunity**, which is mediated by cytotoxic T lymphocytes. Moreover, the cell-mediated immune response is generally the major means of recovery from virus infections, since cytotoxic T lymphocytes are the major effectors that destroy virus-producing cells (Chapter 4). Thus, many virologists and immunologists would not expect that a viral infection can be cleared solely by the innate immune system.

The IgA response at the mucosal surface dissipates quickly after resolution of the infection. For that reason, and because there are more than 100 rhinovirus serotypes, we have many episodes of rhinovirus colds in our lifetimes. These facts explain why rhinovirus common colds are not a good candidate for a vaccine. Moreover, it would probably not be worth exposing individuals to the potential risks of a vaccine when the illness itself poses little threat. As in the case of infection by many other viruses, rhinovirus infections often (50% of the time) produce no clinical symptoms. However, asymptomatic individuals produce virus and are able to transmit the infection.

Although rhinovirus-induced colds are generally mild and self-limiting, their overall impact, which includes missed work and school and the consequences of the all-too-common inappropriate use of antibiotics to treat this illness, can be quite significant. Moreover, rhinovirus colds may be predisposing factors to medical complications such as acute otitis media and sinusitis, and they also may exacerbate asthma in adults and children. Thus, there is a real need for means to prevent and treat the common cold. Yet there are several significant reasons why rhinovirus infections would not be a good candidate for a vaccine program, as enumerated above.

The general public frequently asks "Why can't 'they' cure the common cold?" It would be a worthwhile exercise to think through how you might answer that question as it pertains to virus infections in general, should you ever be asked. Also, it would be good to have an explanation ready for why antibiotics should not be used to treat a cold, except when there might be a rare bacterial complication, such as sinusitis or ear infections, or in the case of individuals who might be particularly vulnerable to secondary bacterial infections. The following points might be included in your response. First, antibiotics have no effect against viruses and they will not shorten a cold or lessen its symptoms. Second, the inappropriate use of antibiotics actually is counterproductive since it destroys the normal body flora, which plays several important natural roles, including preventing infections by bacterial pathogens and producing substances we need, such as vitamin K. Third, all antibiotics produce side effects, some of which are annoying to varying degrees (e.g., diarrhea and nausea), and some of which are potentially life threatening (e.g., anaphylactic shock). Fourth, and most important, misuse of antibiotics promotes the amplification of antibiotic-resistant bacteria, including those that cause devastating human disease. As a consequence, our limited armamentarium of antibiotics becomes progressively less effective against potentially life-threatening bacterial infections.

What then of anticold remedies? Although several agents that inhibit picornavirus replication in cell culture are presently undergoing clinical trials, to date no effective antiviral therapies have been approved for either the prevention or treatment of diseases caused by rhinovirus infection. Yet since the common cold remains the major reason patients visit their primary care physicians, the public expects that there should be agents available to lessen the effects of that illness.

Meanwhile, many over-the-counter options are available for relieving cold symptoms. Aspirin and acetaminophen (e.g., Tylenol) may relieve headache or fever, but the following important proviso should be borne in mind. Aspirin has been linked to the development of **Reye's syndrome** in children recovering from flu or chicken pox. Reye's syndrome is an acute encephalitis that may occur in children after a variety of acute febrile viral infections, including varicella (chicken pox, caused by varicella-zoster virus, a herpesvirus [Chapter 18]) and influenza. Although this complication is rare, its mortality rate may be as high as 40%. For this reason, the American Academy of Pediatrics recommends that children and teenagers should not be given aspirin or medicine containing aspirin when they have a viral illness such as a common cold. Also, nonprescription cold remedies do not shorten the length of a cold and may in fact be counterproductive. For example, nasal decongestants may initially provide relief, only to be followed by rebound congestion and worsening of symptoms. Other over-the-counter remedies may have other side effects, such as drowsiness, dizziness, insomnia, or upset stomach. Extensive research has failed to demonstrate any benefit of vitamin C therapy.

For the moment, hand washing remains the best means for preventing common colds. It is also helpful to keep your hands away from your eyes and nose. Avoiding close contact with people who have colds is also effective. But if you are really serious about not catching a cold, you might want to become a hermit.

Coxsackievirus and Echovirus

Most coxsackievirus and echovirus infections are mild. However, these viruses may cause more-serious illnesses, several of which are enumerated below. They are spread primarily by fecal contamination of food, water, and utensils. Most patients are children less than 10 years of age.

Aseptic (viral) meningitis is most often caused by echoviruses. But coxsackieviruses too may be responsible for some cases. Symptoms include fever, headache, stiffness of the neck, and muscular weakness. Nevertheless, viral meningitis is self-limiting and not nearly as dangerous as the bacterial meningitis caused, for example, by *Neisseria meningitidis* or *Haemophilus influenzae*.

Coxsackieviruses cause **sporadic** myocardial and pericardial infections that can be life threatening, particularly in newborns. (A sporadic disease is present only occasionally and is not epidemic.) Older children and adults often recover, but permanent heart damage may remain.

Herpangia (no, the name does *not* suggest herpesvirus involvement) is a throat infection involving coxsackievirus that occurs primarily in children and is self-limiting. Its characteristic symptoms are fever, sore throat, and difficulty in swallowing.

Other diseases caused by these viruses include pleurodynea (a muscular disease characterized by chest or abdominal pain that can be excruciating), and conjunctivitis.

Viral Hepatitis: Hepatitis A

Hepatitis generally refers to inflammatory disease involving the liver. Several different viruses, including cytomegalovirus (a herpesvirus) and yellow fever virus (a flavivirus), can cause *sporadic* disease of the liver. Four unrelated hepatitis viruses, hepatitis A, hepatitis B, hepatitis C, and hepatitis E viruses, cause **epidemic** viral hepatitis. These illnesses may occur in either acute or chronic forms. Together, the impact of the hepatitis viruses is enormous, as they are responsible for hundreds of thousands of fatal cases of acute hepatitis worldwide yearly. In addition, hepatitis B virus is responsible for over a million fatalities from liver cancer yearly. Hepatitis B, C, and E viruses are discussed in subsequent chapters. For the moment, we are concerned with hepatitis A virus, which is responsible for about 40% of acute cases of viral hepatitis.

As noted in the introduction, hepatitis A virus was originally included within the enterovirus genus. However, it is sufficiently different from the other picornaviruses that it is now classified as the only species within the genus *Hepatovirus*, the name reflecting its tropism for the liver. The distinguishing characteristics of hepatitis A virus include its unique stability at pH 1 and resistance to heat (56°C for 30 min). Also, in contrast to the other picornaviruses, hepatitis A virus is not cytolytic. In cell culture, it gives rise to stable persistent infections, with relatively small viral yields and little, if any, cell damage. Considering that hepatitis A virus is not cytolytic, one might ask how then does it cause disease? Importantly, liver pathology is not obvious until after virus replication has peaked and the immune response has been activated. Thus, hepatitis A virus-induced liver cell injury appears to be immunologically mediated, probably involving NK cells and cytotoxic T cells (Chapter 4).

Hepatitis A virus is spread by the fecal-oral route, entering the body by ingestion. It is believed to enter the blood circulation from either the oropharynx or the epithelial lining of the stomach, reaching the liver via the bloodstream. **Jaundice** appears at the time that antiviral antibody is detected. (Jaundice is a symptom of liver diseases in which the eyes, skin, and mucous membranes appear yellowish, due to the presence of bile pigments in the blood.) Earlier symptoms may include fever, malaise, nausea, abdominal pain, and lethargy. These symptoms and the liver pathology are similar to that caused by hepatitis B virus (Chapter 21). However, unlike hepatitis B virus, hepatitis A virus does not give rise to chronic liver disease, nor is it associated with liver cancer.

As stated above, hepatitis A virus is shed in the stool. Moreover, it is capable of surviving for many months in fresh or salt water. For these reasons, improperly treated sewerage can contaminate the water supply as well as marine life with hepatitis A virus. Filter-feeding shellfish, such as clams, oysters, and mussels, are readily contaminated with the virus, and thus, they are a source of infection if improperly cooked or eaten raw. Once the virus enters a community, it spreads rapidly. That is so because infected individuals shed virus well before symptoms appear, and 90% of children and 25 to 50% of infected adults have asymptomatic infections that nevertheless are productive.

As suggested by the percentages noted in the preceding paragraph, hepatitis A virus infections in children are generally much milder than infections in adults. Also, as might be expected from its mode of transmission, the incidence of hepatitis A virus infection is higher in the developing world than in the Western world. Nevertheless, 41 to 44% of the population of the United States has been infected with hepatitis A virus, as indicated by the presence of hepatitis A virus antibodies in their sera. In the United States, day care centers are a major source of contamination of children as well as their families. Recovery from hepatitis A virus infection is complete 99% of the time.

Considering the sources of hepatitis A virus, infection can be reduced by practices such as avoiding potentially contaminated water and food, especially uncooked shellfish. Proper hand washing, especially in day care centers, mental hospitals, and other care facilities, is likewise important.

A killed vaccine against hepatitis A virus, approved by the U.S. Food and Drug Administration, is available for use by individuals at high risk for infection, including those traveling to regions where infection is endemic. Since there is only one serotype of hepatitis A virus and the virus productively infects only humans, comprehensive vaccine programs against the virus would likely be successful.

The cost-effectiveness of routine childhood vaccination against hepatitis A virus in the United States was recently assessed by statistical modeling. This analysis concluded that

> "without childhood vaccination, the approximately 4 million children in the 2005 birth cohort would be expected over their lifetimes to have 199,000 hepatitis A virus infections, including 74,000 cases of acute hepatitis A and 82 deaths, resulting in 134 million dollars in hepatitis A-related medical costs and productivity losses... Compared with maintaining the levels of hepatitis A vaccination under the preexisting regional policy, routine vaccination at age 1 year would prevent an additional 112,000 infections..."

Consequently, in October 2005, the Advisory Committee on Immunization Practices recommended extending hepatitis A immunization to all U.S. children ages 12 to 23 months (see D. B. Rein et al., *Pediatrics* 119:e12–e21, 2007).

During the spring of 2007, there were several hepatitis A scares arising from reports that infected restaurant food service workers might be a source of transmission to patrons. This led to local proposals to institute mandatory vaccination for all food service workers, which, as might have been anticipated, led to some debate. On one hand, hepatitis A is not usually a serious infection. On the other hand, it can be dangerous for some individuals and it is highly infectious. In one episode in Los Angeles County, CA, an employee of the Wolfgang Puck Company was diagnosed with hepatitis A, after potentially infecting 3,000 people at 13 different events.

Preventive hepatitis A vaccines indeed are given to health workers and to school children. Yet the issue in the case of restaurant workers seems to be one of cost-to-benefit ratio. There is no question that the public would benefit, but the Hotel and Restaurant Workers Union argues that many workers might end up having to pay for the vaccinations themselves, which could be a hardship for some. But if the restaurant owners should have to pay, the cost may be significant for them as well. Thus, either way, compliance also might be an issue. At present, the debate goes on. (Bearing in mind the now huge federal expenditures for a variety of purposes, would you argue that the hepatitis A vaccine should be provided free, as a sound public health measure?)

Immune serum globulin is 80 to 90% effective at preventing hepatitis A virus illness, when given to nonvaccinated individuals within 2 weeks of exposure to the virus. Ig also may be used instead of, or in addition to, hepatitis A vaccine, to provide short-term protection for travelers going to regions where the virus is endemic.

Suggested Reading

Ren, R., and V. R. Racaniello. 1992. Poliovirus spreads from muscle to the central nervous system by neural pathways. *J. Infect. Dis.* **166:**747–752.

Flaviviruses

INTRODUCTION

The *Flaviviridae* are a family of plus-strand RNA viruses that includes several notable human pathogens. Among these are yellow fever virus, dengue hemorrhagic fever virus, Japanese encephalitis virus, St. Louis encephalitis virus, West Nile encephalitis virus, the several tick-borne encephalitis viruses (e.g., European and far eastern tick-borne encephalitis viruses), and hepatitis C virus. As their names imply, these viruses can cause serious, life-threatening diseases. Since flaviviruses are plus-strand RNA viruses, they are class IV viruses in the Baltimore classification scheme (Chapter 2). Interestingly, the name "flavivirus" is derived from the Latin word *flavus*, meaning yellow, signifying the jaundice caused by yellow fever virus.

Yellow fever virus is the prototype flavivirus and was the first virus shown to cause disease in humans. Moreover, it was the first virus that was found to be transmitted by an arthropod vector. As discussed below, the fact that yellow fever is not directly transmitted from one individual to another impeded the discovery of the virus. In any case, yellow fever virus, together with dengue and Japanese encephalitis viruses, is among the most dangerous viral pathogens in the developing world. West Nile virus, which causes fatal **encephalitis** (inflammation of the brain) in humans and domestic animals, has attracted much interest since 1999, when it suddenly emerged in the Western Hemisphere. The virus spread rapidly from its initial focus in New York City to essentially all 48 states of the continental United States, as well as to Canada and Mexico.

Before the molecular biology era, viruses were sometimes classified based upon their modes of transmission or patterns of pathogenesis. Thus, some viruses transmitted by the fecal/oral route were referred to as enteric viruses, while some transmitted by aerosols were referred to as respiratory viruses. Likewise, some viruses transmitted by an arthropod vector were referred to as arthropod-borne (arboviruses) (see Chapter 2).

Because medically important members of the *Flaviviridae* and the *Togaviridae* (Chapter 8) are transmitted by the bite of an arthropod (e.g., mosquitoes, ticks, and sand flies), these two virus families constituted the two original groups making up the arboviruses. But as we see here and in Chapter 8, the flaviviruses and togaviruses make up distinct virus families, based on more fundamental properties, which include their patterns of gene organization and expression. Nevertheless, even today, members of the *Flavivirus* genus are sometimes referred to as the Group B arboviruses to distinguish them from the Group A arboviruses, which are comprised of the *Alphavirus* genus of the *Togaviridae*.

Yet not all of the flaviviruses are arthropod-borne. One notable such exception is hepatitis C virus. Based on this fact, most of the known *Flaviviridae* are divided into the *Flavivirus* genus, comprised entirely of arboviruses, and *Hepatitis C virus*, which constitutes a separate genus.

All of the flaviviruses noted above are human pathogens. However, the family also includes the *Pestivirus* genus, which includes several important pathogens that infect both domestic and wild animals worldwide. The most important pestivirus is classical swine fever virus, which infects domestic pigs and wild boars. Other known pestiviruses include bovine viral diarrhea virus types 1 and 2 and border disease virus of sheep. Pestiviruses are transmitted for the most part by the large amount of virus that infected animals shed in the environment.

With the emergence of the molecular virology era, the flaviviruses were at first reclassified as a separate genus within the togavirus family. That was so because members of each of these two groups contain a plus-sense, single-stranded RNA genome and an enveloped icosahedral capsid. However, subsequent studies revealed that the molecular biology of the flaviviruses is more like that of the nonenveloped picornaviruses. Since the *Flaviviridae* differ from the togaviruses regarding their patterns of gene expression and from the picornaviruses regarding their envelope (and in other important ways as well; see below), they now constitute a separate family.

STRUCTURE AND ENTRY

The flavivirus envelope is a defining feature of the family and a factor that distinguishes these viruses from the "naked" picornaviruses (Chapter 6). Additionally, flaviviruses have capsids that are 37 to 50 nm in diameter, whereas picornavirus capsids are smaller, 27 nm in diameter. Although flaviviruses and picornaviruses each contain plus-strand RNA genomes, flavivirus genomes, at about 11,000 nucleotides in length, are larger than those of picornaviruses, which contain about 8,000 nucleotides. Also, members of the flavivirus genus of the *Flaviviridae* have genomic RNAs that contain a 5′ cap, whereas picornavirus

genomic RNAs have a VPg protein at their 5′ ends. (Recall that in the absence of a 5′ cap, picornavirus translation is driven by an **internal ribosomal entry site [IRES]**, resulting in cap-independent translation [Chapter 6].)

Recalling that *hepatitis C virus* constitutes a separate genus within the *Flaviviridae*, its genome, containing about 9,600 nucleotides, is smaller than the genomes of the other flaviviruses, which are typically about 11,000 nucleotides in length, as noted above. Moreover, hepatitis C virus plus-strand RNA genomes and the identical messenger RNAs (mRNAs) do not contain a 5′ cap. How then might hepatitis C virus genomes and mRNAs be translated? The answer is that hepatitis C virus translation, like that of the picornaviruses, is driven by an IRES. An IRES also has been identified in the RNA of classic swine fever virus, a member of the genus *Pestivirus*.

The flavivirus capsid is comprised of a single protein, the capsid or C polypeptide. Again, note the contrast to the picornaviruses, whose icosahedral capsids are comprised of four distinct capsid proteins (VP1, VP2, VP3, and VP4).

The flavivirus envelope contains two virus-encoded proteins. The major and larger one is the envelope (E) protein, and the smaller one is the membrane (M) protein. The function of the E protein is to bind to the receptor and to mediate envelope fusion to a cellular membrane. Since the E protein is the major envelope protein, it is also the major target on the virus for neutralizing antibodies. M protein is generated from a precursor polypeptide known as prM. There is experimental evidence suggesting that prM is a chaperone, which ensures that E folds and interacts properly during intracellular virion assembly and maturation. This function of prM is discussed further below.

The three-dimensional structure of dengue virus was solved using cryo-electron microscopy (cryo-EM) and three-dimensional image reconstruction. This approach involves taking hundreds of highly detailed, two-dimensional images, which are then processed by powerful computer programs to generate a detailed three-dimensional image (Box 7.1).

The computer-generated images show that 180 E proteins and 180 M proteins form a protective icosahedral shell around the particle. But what is interesting and almost unique in the case of these enveloped viruses is that this protein shell completely encloses the particle, such that the envelope is completely concealed from view. The togaviruses (Chapter 8) provide another example in which the envelope glycoproteins nearly completely conceal the virus envelope. In the case of most other enveloped viruses, at least some of the lipid bilayer is exposed at the virus surface.

It has been possible to determine the three-dimensional structures of a number of nonenveloped icosahedral viruses at near-atomic resolution because of their ability to form well-diffracting crystals for X-ray diffraction analysis. On the other hand, it generally has not been possible to generate crystals of enveloped viruses that might be useful for X-ray diffraction analysis. But, with advances in cryo-EM and powerful computer programs for three-dimensional image reconstruction, it is now possible to determine the structures of noncrystallizable enveloped viruses to 11-angstrom (Å) resolution or better. Even higher levels of resolution are possible when the X-ray structures of individual virus components can be solved and then fitted into the cryo-EM structure (see Chapter 18).

The core of the particle, which lies under the lipid bilayer, is comprised of the viral genomic RNA and multiple copies of the capsid (C) protein. However, the organization of the C protein within the nucleocapsid core is not apparent in the computer-generated images. In fact, the lack of a strong signal from the core suggests that it is poorly ordered or variable in structure. Moreover, neither the E nor the M proteins extend beyond the inner surface of the lipid envelope. This implies that neither of the envelope glycoproteins can influence the structure of the capsid, which is thought to be organized randomly relative to the outer glycoprotein shell.

Flaviviruses enter cells by receptor-mediated endocytosis and are transported via an endocytic vesicle to the endosomal compartment. Since flaviviruses are enveloped, we would expect membrane fusion to occur at some point in their entry pathway. Indeed, the mildly acidic pH of an endosomal compartment triggers the conformational transitions in the viral E proteins that promote membrane fusion and release of the nucleocapsid into the cytoplasm.

The flavivirus E protein, in its native state, is comprised primarily of β-sheets. This is noteworthy since it is in contrast to the well-studied and prototypical influenza virus fusion protein, which is comprised primarily of an α-helical stem domain (Chapter 12). Nevertheless, the structure of the flavivirus E protein is similar to that of togavirus fusion proteins (Chapter 8). Based on these structural distinctions, the fusion proteins of the flaviviruses and togaviruses are designated "class II fusion proteins," in contrast to the class I fusion proteins seen in influenza virus and in several other viruses as well (e.g., Ebola virus and human immunodeficiency virus [HIV]).

The E fusion protein in native flavivirus particles exists as an E/E homodimer. Low pH causes the homodimers to disassociate and re-form as fusion-active homotrimers, in which the hydrophobic fusion peptide is exposed in a position enabling it to insert into the endosomal membrane to promote fusion. The actual conformational rearrangements that virus fusion proteins undergo to promote membrane fusion are best understood in the case of influenza virus and are described in detail in Chapter 12.

REPLICATION

As in the example of the plus-strand picornaviruses (Chapter 6), upon entry into the cell, the flavivirus plus-strand RNA genome must be translated in order to generate the enzymatic machinery needed for viral transcription and replication. Also as in the picornavirus example, flavivirus plus-strand RNA genomes contain only one open reading frame and only one initiation site for translation. As then might be expected from the picornavirus precedent, each of the individual flavivirus proteins is generated by the proteolytic processing of a single precursor polyprotein. Moreover, a consequence of this mode of translation is that each of the flavivirus gene products is synthesized in equimolar amounts. This is a somewhat inefficient mode of gene expression since many more copies of the virus's structural proteins than of its replication proteins are needed. Yet this translation stratagem solves for the virus the problem of generating its individual gene products from a polycistronic, single-stranded, plus-sense RNA genome.

Another feature of flaviviruses reminiscent of the picornaviruses is that the genes encoding the structural or capsid proteins are clustered at the 5′ end of the genome. The nonstructural proteins, including a protease, a helicase, and a polymerase, are at the 3′ end of the plus-strand RNA (Fig. 7.1).

The flavivirus envelope contains the membrane-anchored M and E proteins, as noted above. The mechanism by which enveloped viruses acquire their envelope glycoproteins is discussed briefly in Chapter 2. General features of that process and aspects that are particular to the flaviviruses are discussed in detail below. For now, the process begins with the insertion of the flavivirus prM (the M protein precursor) and E proteins into the endoplasmic reticulum (ER) membrane, coincident with their translation.

ASSEMBLY AND RELEASE

Enveloped viruses acquire their envelopes as part and parcel of the budding of their nucleocapsids through a particular cellular membrane (see Fig. 2.9). In this regard,

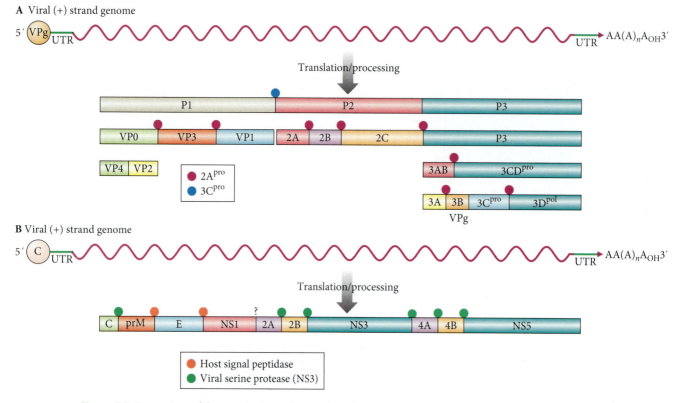

Figure 7.1 Comparison of the organization and expression of a picornavirus (poliovirus) genome (A) with a flavivirus genome (B). There is but one translation initiation site and one open reading frame on the plus-strand RNA genome in each instance. Processing of the flavivirus precursor polyprotein is carried out either by the host signal peptidase or by the viral protease, NS3. Adapted from S. J. Flint, L. W. Enquist, V. A. Racaniello, and A. M. Skalka, *Principles of Virology: Molecular Biology, Pathogenesis, and Control of Animal Viruses*, 2nd ed. (ASM Press, Washington, DC, 2004), with permission.

the cellular membranes from which viral envelopes are derived are generally modified by the presence of virus-encoded glycoproteins. In the cases of the flaviviruses, these proteins are the prM and E proteins, which become integral membrane proteins, as described below.

For any particular virus, the site of budding might be the plasma membrane, the nuclear membrane, the ER, or the Golgi apparatus. Since budding usually involves interactions between the proteins of the nucleocapsid and the cytoplasmic tails of the transmembrane envelope glycoproteins, the site of budding is generally determined by the targeting of the virus's envelope glycoproteins to a particular cellular membrane. In the case of the flaviviruses, since interactions between the envelope glycoproteins and the C protein may not be possible (see above), the site of budding may depend for the most part on the translation and processing of the flavivirus polyprotein, as described below. Regardless, immature flavivirus particles are believed to form when nucleocapsids bud into the lumen of the ER. These particles then traffic through the normal vesicle-mediated cellular secretory pathway: first

to the Golgi apparatus, and then to the plasma membrane, and then out of the cell.

The general process of targeting a protein to the ER is outlined in Fig. 7.2. Major features of this fundamental cellular pathway were discovered during the analysis of the translational strategy of the togaviruses (see Fig. 8.3). The sequence of events, as it occurs in the case of the flavivirus envelope glycoproteins, is as follows.

As the translation of the flavivirus nascent polyprotein ensues on a free ribosome in the cytoplasm, an internal **signal sequence**, located at the C terminus of the C protein, acts to target the still-ribosome-bound, nascent polyprotein to a **translocation complex** in the ER membrane. The ribosome binds to the translocation complex, and the signal sequence on the nascent polyprotein inserts into a translocation channel in the membrane. As the remainder of the nascent polyprotein, which contains the amino acid sequences of both prM and E, is translated, it crosses the membrane of the ER multiple times (Fig. 7.3). That is so because the prM protein has two transmembrane-spanning domains, which contain a hydrophobic **stop-transfer**

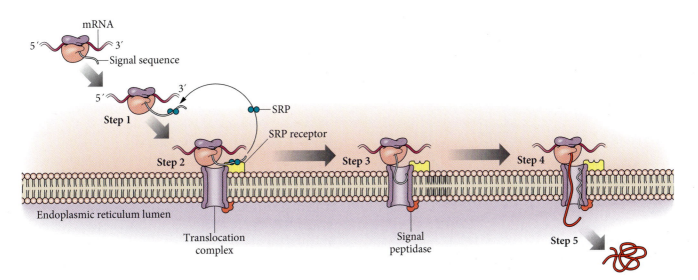

Figure 7.2 Targeting secretory proteins to the ER. Step 1: As the signal sequence emerges from the ribosome, it is recognized and bound by the signal recognition particle (SRP). Step 2: The SRP escorts the complex to the ER membrane, where it binds to the SRP receptor. Step 3: The SRP is released, the ribosome binds to a membrane translocation complex, and the signal sequence is inserted into a membrane channel. Step 4: Translation resumes, and the growing polypeptide chain is translocated across the membrane. Step 5: Cleavage of the signal sequence by signal peptidase releases the polypeptide into the lumen of the ER. If the protein were to contain a hydrophobic stop-transfer (transmembrane) domain (not shown), then it would become an integral membrane protein. If not, it would be released into the lumen of the ER. Adapted from G. M. Cooper, *The Cell* (ASM Press, Washington, DC, 1997), with permission.

sequence and a signal sequence. As a result, the E protein is also translocated into the lumen of the ER. After the indicated proteolytic cleavages (Fig. 7.3), the C protein is localized in the cytoplasm with the viral RNA, where it remains associated with the ER. On the luminal side of the ER, the prM and E proteins form a stable heterodimer. Figure 7.4 shows in more detail how a protein that spans a membrane multiple times might be generated.

A virally encoded protease cleaves the junction between C and prM, whereas the host signalase cleaves the junctions between prM and E and between E and NS1 (Fig. 7.3). Interestingly, signalase-mediated cleavage, even when it involves internal signal sequences, generally occurs cotranslationally. Thus, it is rather unique that a signalase-type cleavage of the flavivirus polyprotein is dependent on the catalytic activity of the subsequently translated, virally encoded cytosolic protease.

The interaction between prM and E in the ER lumen appears to promote the proper folding of E. Therefore, prM is said to be a chaperone for E. Moreover, prM continues to stabilize the conformation of E during the transport of immature virus particles through the secretory pathway. This is important because E must undergo pH-induced conformational transitions during the next cycle of infection that trigger membrane fusion and entry into the cytoplasm. Since immature virus particles pass through the acidic trans-Golgi network during their exit

from the cell, E must be prevented from undergoing the low-pH-induced transition during this passage. Protein prM ensures that E remains in its native conformation during transport through the secretory pathway. Shortly before virus release, the cellular protease furin converts the immature particles to mature viruses by cleaving prM to the shorter M protein. This cleavage will enable E to undergo the necessary low-pH-induced conformational transitions during the next cycle of infection.

Next, we consider the sorts of interactions that provide the driving force needed for budding. A reasonable hypothesis might contain the following steps. First, envelope glycoproteins acquire their proper membrane location, while nucleocapsids independently are assembling in the cytoplasm. Second, the ensuing specific interaction between the nucleocapsids and the cytosolic portions of the envelope proteins provides the driving force for enclosing the nucleocapsids within the envelopes and their budding through the membrane.

Although this is indeed the way in which a number of enveloped viruses acquire their envelopes and bud from the cell, it does not appear to be correct in the case of the flaviviruses. For example, recall that neither E nor M proteins extend beyond the inner surface of the lipid envelope. Thus, neither would appear to be able to directly interact with the capsid. Instead, it appears that lateral interactions between the viral transmembrane proteins

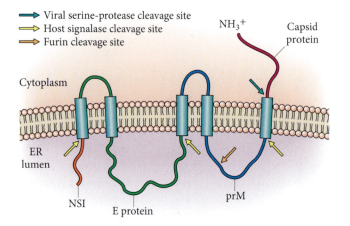

→ Viral serine-protease cleavage site
⇨ Host signalase cleavage site
⇨ Furin cleavage site

Cytoplasm

NH₃⁺   Capsid protein

ER lumen

NSI

E protein

prM

Figure 7.3 During translation of the flavivirus polyprotein, the envelope proteins are translocated and anchored in the ER membrane by various signal sequences and membrane anchor domains. The C protein contains a hydrophobic signal sequence at its C terminus, which translocates prM into the lumen of the ER from its site of synthesis at a translocation complex. The prM protein has two transmembrane-spanning domains, which contain a stop-transfer sequence and a signal sequence. As a result, the E protein is also translocated into the lumen of the ER. After the indicated proteolytic cleavages, the C protein and viral RNA are localized in the cytoplasm, where the capsid protein remains associated with the ER. On the luminal side of the ER, the prM and E proteins form a stable heterodimer within a few minutes of translation. The figure depicts the predicted orientation of the structural proteins across the ER membrane. Transmembrane helices are indicated by cylinders, arrows indicate the sites of posttranslational cleavage, and the cleavage sites of specific enzymes are indicated by different colors. NSI, nonstructural protein 1. Adapted from S. Mukhopadhyay, R. J. Kuhn, and M. G. Rossmann, *Nat. Rev. Microbiol.* 3:13–22, 2005, with permission.

provide that driving force for budding, without the involvement of the nucleocapsid core. That is, lateral interactions between the envelope proteins induce the local membrane curvature that leads to the budding and pinching off of the virus particle. Experimental evidence in support of that scheme includes the finding that expression of the cloned genes for proteins prM and E efficiently generates virus-like particles that are secreted from the cell. Importantly, these particles do not contain nucleocapsids. Moreover, similar empty particles also are a significant by-product of actual virus infection. Finally, unlike the cases of other enveloped viruses, nucleocapsids are rarely seen in flavivirus-infected cells, consistent with the premise that nucleocapsids assemble at the ER membrane, in a process that is somehow tightly coordinated with budding.

HISTORICAL INTERLUDE: IDENTIFICATION OF HEPATITIS C VIRUS

The identification of hepatitis A and B viruses and the development of sensitive assays for their detection led to optimism that the blood supply and blood transfusions might no longer be a source of viral hepatitis. However, it soon became clear that blood samples devoid of hepatitis A and B viruses could still transmit hepatitis. That is, non-A, non-B forms of infectious hepatitis must exist. Yet no infectious agent that might transmit hepatitis other than hepatitis A

Figure 7.4 Insertion of a protein that spans the membrane multiple times. In this example, an internal signal sequence results in insertion of the polypeptide chain with its amino terminus on the cytosolic side of the membrane. A stop-transfer sequence then causes the polypeptide chain to form a loop within the lumen of the ER, and translation continues in the cytosol. A second internal signal sequence triggers reinsertion of the polypeptide chain into the ER membrane, forming a loop in the cytosol. The process can be repeated many times, resulting in the insertion of proteins with multiple membrane-spanning regions. Adapted from G. M. Cooper, *The Cell* (ASM Press, Washington, DC, 1997), with permission.

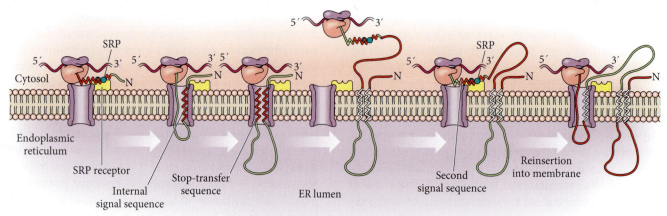

SRP

Cytosol

Endoplasmic reticulum

SRP receptor

Internal signal sequence

Stop-transfer sequence

ER lumen

Second signal sequence

Reinsertion into membrane

and B viruses was seen by EM, nor could such an agent be detected by growth in cell culture.

The mysterious non-A, non-B hepatitis agent, now known as hepatitis C virus, was finally identified in 1989 by a team of molecular biologists, using cutting-edge molecular biology techniques of the day (Choo et al., 1989). They began by obtaining a serum sample that was contaminated with an agent that caused non-A, non-B hepatitis when inoculated into chimpanzees. Based on the filtration characteristics of the agent and its sensitivity to chloroform, they believed that it might be an enveloped virus. Moreover, they reasoned that it might be possible to concentrate the agent from the plasma of chimpanzees that developed the disease, and they attempted to do so by ultracentrifugation. Total RNA and DNA then were extracted from the pellet, denatured, and reverse transcribed using random primers and the polymerase chain reaction (PCR) technique. The resulting complementary DNA (cDNA) was cloned into a bacteriophage λ (lambda) expression vector, and the resultant DNA library was infected into *Escherichia coli*. Individual *E. coli* colonies then were screened for production of a hepatitis-specific antigenic peptide by means of an immunoassay using serum from a human patient diagnosed with non-A, non-B hepatitis. About a million random clones from the library were screened before one was identified that encoded an antigen that was specifically reactive with sera from patients with non-A, non-B hepatitis virus infections. That DNA clone then was used as a hybridization probe to isolate a larger overlapping probe from the cDNA library, which then was used to identify a full-length (9.5-kilobase) RNA molecule that was plus-stranded with respect to the encoded hepatitis antigen. Importantly, there is no homology between that RNA molecule and host DNA. Instead, that RNA molecule was specifically present in individuals with non-A, non-B hepatitis infections. These experimental findings demonstrated that the DNA clone was derived from the genome of the non-A, non-B hepatitis agent. Subsequent sequencing of the full-length RNA molecule revealed that it had the characteristics of a flavivirus genome.

The events described above are notable since hepatitis C virus may have been the first virus identified by using molecular biology techniques rather than by using the standard virological procedures of the day. Such a radical approach was necessitated by the lack of a suitable cell culture system for hepatitis C virus.

By now, many new viruses have been identified in patient samples by using PCR-based procedures. Yet in many of these instances, while the viral genome has been sequenced and characterized, it has not been possible to cultivate and isolate an actual virus. In these cases, the new viruses may be referred to as "tentative" species, until an actual virus might be isolated.

WEST NILE VIRUS: AN EMERGING VIRUS

West Nile virus was isolated for the first time from a feverish adult female patient in the West Nile District of Uganda in 1937. However, it was not until a 1957 study of an outbreak of severe human **meningoencephalitis** (inflammation of the spinal cord and brain) in elderly patients in Israel that West Nile virus was recognized as a cause of inflammatory disease in the human central nervous system.

Interest in the West Nile virus suddenly increased in the United States when, in 1999, the virus made its first known appearance in the Western Hemisphere. The outbreak began abruptly in the late summer and early autumn of 1999 in New York City and surrounding areas with a spate of human encephalitis cases, whose occurrence and distribution were consistent with it being transmitted by an insect vector. That fact and analyses of patient blood samples led to the premise that a flavivirus might be the **etiologic** agent responsible for the disease in humans.

At the same time that humans in the New York City metropolitan area were experiencing this wave of encephalitis, unusual deaths were occurring among exotic captive birds at New York's Bronx Zoo, as well as among crows, blue jays, and other wild birds in the vicinity. An autopsy sample from one of the dead exotic birds (a flamingo) yielded flavivirus-like particles. Subsequent sequencing of the RNA extracted from those particles identified them as West Nile virus. Sequence analysis of virus isolates from several other infected species, including two fatal human cases, demonstrated that West Nile virus indeed was responsible for the human as well as the avian diseases. Isolation of the same virus from local mosquitoes supported the notion that the virus was transmitted by an arthropod vector.

Interestingly, the West Nile virus responsible for the New York City outbreak was most closely related to a West Nile virus isolated from a dead goose in Israel in 1998. Also, like the New York City isolates, the 1998 Israeli isolate was associated with increased pathogenicity for birds. This fact was notable because the ability to produce severe disease in birds is not a general characteristic of West Nile virus. Curiously, there were no reports of human cases during the Israeli **epizootic**. It was suggested that this might be due to widespread human immunity in Israel resulting from the longer presence of West Nile virus in the Middle East.

Before 1999, West Nile virus was commonly found in humans and birds and other vertebrates in Africa, West Asia, and the Middle East, but not in the Western Hemisphere. The occasional outbreaks that occurred in Europe are thought to have resulted from virus carried by migratory birds.

The geographic origin of the North American virus is not known, but genetically it most closely resembles strains found in the Middle East. (West Nile virus shows distinct genetic variation, providing evidence for at least two evolutionary lineages. Lineage 1 is seen in Africa, Asia, the Middle East, Europe, Oceania, and more recently in the Americas, whereas lineage 2 is restricted to Africa, where it causes the classical West Nile fever.) Also, it is not known for certain that the virus was not present in the United States before the summer of 1999. Regardless, since the initial outbreak of West Nile virus in New York City in 1999, the virus has rapidly spread to virtually all of the continental 48 states, as well as to Canada and Mexico. The apparently sudden emergence of West Nile virus in New York City and its subsequent rapid spread across all of North America emphasize the speed with which new, potentially dangerous pathogens can spread between population centers of the modern world.

The means by which West Nile virus was introduced into the Western Hemisphere is not known. Possibilities include travel by infected humans, unintentional introduction of virus-infected ticks or mosquitoes, or perhaps importation of infected birds or other domestic pets.

So what might be the future for West Nile virus in humans? Since epidemiological studies of West Nile virus have shown (i) that the virus is passed to humans only through the bite of an infected mosquito (see below) and (ii) that the virus is maintained in nature through a cycle involving avian hosts and mosquito vectors, the answer will likely be determined by the interrelationships between the virus, humans, mosquitoes, and birds. While West Nile virus emerged only recently in the Western Hemisphere, its record there nevertheless allows us to make plausible predictions for this corner of the world.

When West Nile virus first appeared in the New World, none of the indigenous birds had immunity, so the infection rapidly spread through the bird population. In time, the vast majority of birds became immune to the virus, leading to a decline in infected mosquitoes and human cases as well. However, as immune birds eventually die off and are replaced by new, susceptible birds, the epidemic cycle probably will begin again. In actuality, this is the track record that West Nile virus appears to be establishing in the Western Hemisphere.

Mosquitoes die with the first frost. How, then, does West Nile virus survive over the winter? The answer probably involves the role of birds in the cycle. The infection in most birds is not lethal. Thus, the majority of birds bitten by infected mosquitoes survive over the winter and pass the infection on to the new mosquitoes that hatch in the spring.

As discussed below, the most serious consequence of West Nile virus infection is fatal encephalitis in humans, as well as in horses and certain domestic and wild birds.

EPIDEMIOLOGY AND PATHOGENESIS

General Principles of Arthropod Transmission

For a virus to adopt the arbovirus lifestyle, it generally must be able to replicate both in an arthropod vector and in a vertebrate host. The arthropod vector is infected when it takes a blood meal from a **viremic** vertebrate host (i.e., one with virus in its bloodstream). The virus then replicates in the insect's gut and eventually spreads to the insect's salivary glands, from which it can be transmitted to a new vertebrate host during the arthropod's next blood meal. Transmission to the vertebrate host is efficient because the arthropod (usually a mosquito in these instances) actually regurgitates saliva into the victim's bloodstream.

The vertebrate host must develop sufficiently high levels of viremia to continue the cycle of infection. Humans are usually (but not always; see below) "dead-end" hosts for arboviruses because humans generally do not develop sufficient levels of viremia to enable efficient transmission to the arthropod vector. Instead, human infection usually occurs when the arthropod vector transmits the virus to humans from its normal animal or avian reservoir host. This leads to the following important general principle. An inherent consequence of the arbovirus lifestyle is that it provides a means for the virus to cross species barriers. Moreover, it is common for an arbovirus to be able to infect multiple species of mammals and birds that rarely, if ever, come together, since the insect vector may feed on each of them. However, only one or a very few of these vertebrate hosts act as a natural reservoir for the virus. Likewise, the virus also is usually specific for a particular arthropod host. The arthropod vector generally is not harmed by the virus, and it may continue to produce virus throughout its entire short lifetime. In contrast, infection of the usual vertebrate host is acute, generally followed by recovery and lifelong immunity.

What other factors are important for a virus to sustain the arbovirus lifestyle? We already mentioned the level of viremia in the vertebrate host. The duration of that viremia is important, too, as well as the population density and turnover rate of the vertebrate host. The last factor is important because infected members of the vertebrate population acquire immunity. Arthropod factors include their range of activity, because the arthropod must have a large enough range to be able to transmit the infection from one vertebrate host to another.

In tropical climates, where the arthropod is active year-round, if the above conditions are met, then the virus will flourish. In temperate climates, in addition to the factors noted above, the virus also must have a means for surviving over the winter. It may accomplish overwintering by **transovarial transmission** from one generation of arthropods to the next. Alternatively, the virus might migrate with its avian host and then return during the summer. Regardless, in temperate climates, arbovirus disease occurs during the summer, when the arthropod vector breeds and is able to mediate the transmission cycle.

Infection of the usual vertebrate host regularly results in recovery and immunity, as noted above. In contrast, serious disease becomes possible when a different, out-of-the-ordinary host is infected. As noted above, humans are generally not regular hosts for arboviruses. Consequently, either humans do not support sufficient levels of virus replication to efficiently transmit the virus to the vector or they may succumb before transmission might occur. Thus, while infection of humans is ordinarily a dead end for these viruses, it may yet result in very serious, life-threatening illness in humans.

Human exposure to arboviruses usually results from human activities that reflect recent changes either in lifestyle or in technology. For example, in our highly mobile society, humans may visit an area where a particular arbovirus is endemic. (In this regard, if there is an indigenous human population in the area, it may already express widespread immunity to resident arboviruses.) Other human activities that may result in exposure include wholesale movement into areas where arboviruses are prevalent (e.g., rain forests), deforestation, irrigation, and urbanization. The last two activities lead to an increase in arbovirus infections because of enhanced vector replication in accumulating water and sewerage. Indeed, the unplanned urbanization in developing countries that occurred during and after the Second World War, whereby millions of people congested in cities, resulted in major arbovirus epidemics. Finally, modern high-speed land and air travel can rapidly carry infected humans and other infected vertebrates and arthropods to geographic areas that were not previously exposed to the virus.

Infection, Dissemination, and Determinants of Pathogenesis

When the arthropod vector bites a vertebrate host, it regurgitates its virus-containing saliva into the vertebrate's bloodstream. The course of the subsequent initiation of the infection and its dissemination in the vertebrate host, particularly with regard to the viral tropism for particular cell types and tissues, influence the course of the resulting pathogenesis. The following examples of West Nile virus and dengue virus illustrate this point.

After West Nile virus enters into the vertebrate host, it is believed to initially replicate in Langerhans cells. These are cells of the monocyte/dendritic cell family that take up and process antigens in the epidermal layer of the skin (Chapter 4). Langerhans cells generally function as follows. After encountering an antigen, they migrate to those lymph nodes that drain their site of residence. Within those lymph nodes, the Langerhans cells differentiate into mature antigen-presenting dendritic cells. However, West Nile virus appears to undermine this trafficking pattern of Langerhans cells by using those cells as a means to disseminate to lymph nodes. The virus is then able to replicate in the lymphatic tissues, resulting in a primary viremia, which then leads to subsequent infection of peripheral organs such as the spleen and kidneys.

In animal model systems, West Nile virus is cleared from the blood and peripheral organs after about a week. But, the virus manages to infect the central nervous system in some of the animals. The neuropathology seen in these animals resembles that which occurs in West Nile virus-infected human patients. Injury occurs in the brain stem, hippocampus, and spinal cord.

Cells of the monocyte/dendritic cell lineage may also play a key role in dengue virus infections. That premise is consistent with the finding that dendritic cells are the most permissive cells known for dengue virus in the vertebrate host. Perhaps, then, those vertebrate cells are the initial target cells for dengue virus, as well as for West Nile virus. Macrophages, too, are important during early stages of dengue virus infection.

The receptor for dengue virus on dendritic cells appears to be the dendritic-cell-specific molecule DC-SIGN (an acronym derived from dendritic-cell-specific ICAM-grabbing non-integrin), which normally binds to ICAM-3 on naïve or resting T cells. (This interaction strengthens the adhesion of the dendritic cell to the T cell, and it is essential for dendritic-cell-induced T-cell activation [Chapter 4].) Yet DC-SIGN does not internalize with dengue virus, suggesting that it might act only as an attachment factor that promotes virus binding to an as yet unidentified cellular receptor, which then internalizes the virus (Box 7.2).

Bearing in mind that dendritic cells are key effectors of host immune defenses (they are arguably the most important of the professional antigen-presenting cells [Chapter 4]), we note again that West Nile virus and dengue virus subvert these important effectors of host immunity for their own benefit. In both instances they use dendritic cells to disseminate to lymph nodes, which in turn

Dendritic cells and DC-SIGN are also thought to play a key early role in HIV infection (Chapter 21). In addition, cells of the monocyte/macrophage lineage are reported to be target cells for various other microorganisms, including the hepatitis C virus, Ebola virus, human cytomegalovirus (a herpesvirus), *Leishmania* (a protozoan parasite), and *Mycobacterium tuberculosis*.

It might seem rather odd that a microbe would make its way in the world by infecting dendritic cells, since a major function of these cells is to phagocytose and kill microbes. However, the above cellular pathogens have evolved a variety of well-understood stratagems for surviving within a phagocytic cell. It is a safe bet that viral pathogens that subvert phagocytic cells have evolved their own means for avoiding being destroyed by those cells.

facilitates viremia, since the blood continually circulates through these lymphoid organs. Clinically, this viremia may then lead to invasion of the central nervous system (see below).

Another feature of dendritic cells, and of macrophages as well, that might influence the course of infection is that they express Fc receptors (Chapter 4). This fact accounts for the interesting phenomenon that infection, at least in the case of dengue virus, may actually be enhanced as much as 1,000-fold by the presence of *nonneutralizing* antibodies, since those antibodies enable virus particles to bind to Fc receptors and enter cells expressing those receptors. This phenomenon is referred to as immune enhancement or **antibody-dependent enhancement (ADE)**. It plays out in the case of dengue virus as follows.

There are four distinct dengue virus serotypes in circulation in tropical and subtropical areas, all of which generally cause uncomplicated dengue fever. However, dengue virus can also cause the more severe, life-threatening **dengue hemorrhagic fever**, which generally is seen during epidemics in communities that had earlier been infected with a dengue virus serotype different from the one now causing the more severe disease. The reason why the more serious dengue hemorrhagic fever generally is seen only in populations that earlier had been infected with a different serotype is as follows. Some antibodies against the earlier serotype bind to the current one but do not neutralize it. However, those nonneutralizing antibodies do facilitate binding and entry of the current virus into Fc-receptor-expressing dendritic cells and macrophages, which also are the cells in which the virus replicates best, as noted above. Interestingly, the presence of transplacentally acquired maternal antibodies against dengue virus also is associated with dengue hemorrhagic fever in infants. However, while ADE provides a satisfying explanation for dengue hemorrhagic fever, it is not the entire story (see below). (What experiment might you do to test whether antibody-dependent enhancement might be a factor during infection by a particular virus [Box 7.3]?)

The primary infection of macrophages and dendritic cells by dengue virus produces flu-like symptoms, including fever, headache, aches, and chills. As noted previously, these symptoms are due at least in part to the production of interferon (IFN) by these infected cells. This initial illness is considered to be relatively mild, and most infections do not progress beyond this stage. However, as noted above, after the virus replicates in dendritic cells and macrophages, viremia may reach sufficiently high levels for other organs, including the liver, vasculature, and brain, to be infected. Also, as noted above, the pattern of flavivirus pathogenesis is primarily determined by the tropism of the virus for other cells and tissues, after the initial infection of cells of the monocyte/macrophage lineage. (Neurotropic flaviviruses access the brain by first infecting the capillary endothelial cells that line the capillaries of the brain or choroid plexus. The tight junctions between these endothelial cells constitute the **blood-brain barrier** [Chapter 3].)

As discussed in Chapter 3, host immunity can be a two-edged sword, serving to protect the host but also an inadvertent cause of pathogenesis. Two specific examples of this point, apropos the flaviviruses, were already noted. First, nonneutralizing antibodies against a flavivirus can enhance infection of macrophages and other cells that express Fc receptors. Second, IFN, which activates the immune response, also causes the flu-like symptoms characteristic of the initial mild **systemic** illness (i.e., illness relating to the entire host, rather than any particular organ system). Importantly, the immune response, once activated, can be difficult to control. For example, hypersensitivity reactions such as delayed-type hypersensitivity (DTH) may also contribute to disease (Box 7.4).

Other relevant immunopathologic mechanisms include the formation of antibody complexes with viruses or virus antigens, which leads to infiltration by polymorphonuclear leukocytes, and later by macrophages, thereby intensifying inflammation. Finally, activation of complement can occur. When these reactions occur in the vasculature, they are liable to cause sufficient damage to result in hemorrhaging, such as in dengue hemorrhagic fever (see below).

BOX 7.3

ADE is not a general phenomenon, since antibodies produced in response to infection by a particular pathogen may well confer protection against serologically related pathogens. However, under conditions of antibody specificity where neutralization does not occur, cells that express Fc receptors, such as dendritic cells and macrophages, take up virus/antibody complexes more readily than they take up viruses not coated by antibody. In the dengue virus example noted in the text, it was demonstrated that preexisting antibodies induced by an earlier infection with a different dengue virus serotype did indeed bind to the new serotype but did not neutralize it.

How might we explain antibodies that bind to a virus but do not neutralize it? It is not entirely clear how neutralizing antibodies actually neutralize a virus, but it is generally thought that one way is for the antibody to impair the receptor-binding sites on the virus. Thus, in the above dengue virus instance, some antibodies against the first virus may recognize particular epitopes on both the first and second virus while not impairing binding of either virus (or neutralizing either virus by some other means). Moreover, neutralizing epitope(s) on the second virus may not be recognized by neutralizing antibodies against the first virus.

Whereas dengue hemorrhagic fever provides the most striking example of ADE in humans, the ADE phenomenon also has been observed in vivo in the case of at least two coronaviruses (Chapter 9): feline infectious peritonitis virus and the severe acute respiratory syndrome (SARS) human coronavirus. Moreover, it is an in vitro property of enveloped viruses in general.

As you already might have reasoned, ADE is thought to be a potentially significant impediment to the development of effective vaccines against those viruses for which it has been observed, including dengue virus and the SARS human coronavirus. Yet despite the finding that the SARS coronavirus enters B cells in an Fc-dependent manner in the presence of virus-specific antibodies, vaccinated animals showed no signs of enhanced lung pathology. Moreover, viral titers were greatly reduced in vaccinated animals, in comparison to nonvaccinated control animals.

ADE also has been observed in vitro in the case of the retroviruses, and it was potentially important clinically since it suggests that ADE might be a possible hindrance to the development of a vaccine against HIV, a retrovirus. But the following experimental findings imply that ADE might not be a cause for concern in this instance. A rare subset of HIV-infected individuals are said to be **nonprogressors**, since their infection does not lead to full-blown acquired immunodeficiency syndrome (AIDS). Yet the levels of experimental ADE seen in vitro were higher when the sera were obtained from these nonprogressors than when they were obtained from a group of rapid progressors. Thus, progression of HIV disease does not appear to correlate with ADE. Regardless, there are other, more serious obstacles to the development of a vaccine against HIV (see Chapter 21). (When a pathogen is decorated by antibodies [neutralizing or otherwise], it is said to be opsonized [Chapter 4]. It is then more readily phagocytosed by macrophages and dendritic cells, thereby usually facilitating its destruction. Dengue and several other viruses [below] survive their encounter with these cells and may actually replicate in them.)

Hemorrhagic Fever Viruses: Yellow Fever and Dengue Viruses

Yellow fever and dengue viruses are the two flaviviruses known to cause hemorrhagic fever. Infection by either virus can lead to bleeding from any and all body orifices. However, the hemorrhagic fevers caused by these two viruses are distinctly different from each other clinically. We consider yellow fever first.

Yellow fever is somewhat unique among the hemorrhagic fevers in that the liver is the major target organ. Consequently, the severe form of yellow fever virus infection is characterized by liver hemorrhaging and severe jaundice. Massive gastrointestinal hemorrhages ("black vomit") also may occur. Nevertheless, most yellow fever cases are mild. Interestingly, the name "yellow fever" does not have its origin in the yellowing of the skin and eyes that is characteristic of severe disease. Instead, it has its origin in the term "yellow jacket," which refers to the yellow flag that was flown in port to warn approaching ships of the presence of infectious disease. (But recall that the name for the flavivirus family derives from the Latin word *flavus*, meaning yellow, for the jaundice caused by yellow fever virus [Box 7.5].)

Based on its prevalence and impact, dengue virus is arguably the most medically important arbovirus in the world, causing nearly 100 million cases of dengue fever and 250,000 cases of dengue hemorrhagic fever worldwide. Uncomplicated dengue fever, also known as break-bone fever, is characterized by high fever, headaches, rash, and back and bone pain that lasts for up to a week. It is caused by the first exposure to dengue virus. Subsequent infection by another of the four related dengue virus serotypes

BOX 7.4

Prior exposure of an individual to an antigen induces the generation of a pool of T and B memory cells that are primed to react quickly and powerfully, should they reencounter the antigen (Chapter 4). DTH is an inflammatory response initiated by the activation of antigen-specific CD4 T cells in a previously sensitized individual. These activated T cells traffic to the infected site, where they release proinflammatory cytokines that recruit and activate antigen-nonspecific inflammatory leukocytes. Since these events occur over a period of 24 to 72 h, the phenomenon is referred to as delayed-type hypersensitivity. The clinical effects of DTH can last for several weeks and may indeed become chronic in certain instances. Other important examples of diseases in which DTH is a contributing factor in viral pathogenesis include smallpox and measles, where it is responsible for the pustular lesions in the former and the rash in the latter.

DTH may also involve CD8 T cells, with poison ivy providing a well-known example. Bear in mind that despite the contribution of DTH to pathology, DTH is nevertheless an inadvertent consequence of the crucial role of cell-mediated immunity in protecting the host against intracellular microbial pathogens.

---

can lead to the more severe dengue hemorrhagic fever, a consequence of ADE, as described above. In this instance, the presence of nonneutralizing antibodies facilitates virus uptake into dendritic cells and macrophages. These professional antigen-presenting cells in turn help to activate memory T cells, which then release inflammatory cytokines that are thought to act on the capillary endothelium, leading to internal bleeding and shock. However, only a small percentage of those previously infected individuals develop dengue hemorrhagic fever when infected by the new serotype. And while unsatisfying to most of us who prefer straightforward paradigms, the severe disease also appears to occur in individuals who were not infected earlier with a different dengue virus serotype. So we have here another example in which the determinants that predispose individuals to a severe virus infection are not yet entirely understood.

Yellow fever can largely be controlled in urban areas by severely reducing the level of the specific vector, in this instance *Aedes aegypti* mosquitoes. The vector control strategy works in this case because these mosquitoes have a very short flight range. Thus, it is necessary to control the vector population only in the immediate vicinity of human habitation. This can be accomplished by draining potential breeding sites such as swamps and ditches and destroying water-collecting objects such as discarded tires.

Humans can be a reservoir host for both yellow fever and dengue viruses. Thus, dengue and yellow fever viruses provide exceptions to the generalization that humans are a dead-end host for arboviruses.

An effective live vaccine against yellow fever, referred to as YF 17D, induces lifelong protective immunity in nearly 100% of vaccinated individuals. Interestingly, this vaccine was developed 70 years ago (Box 7.5) and has had a virtually unblemished safety record since its inception. It is regarded as one of the safest and most effective live-attenuated viral vaccines ever developed. YF 17D was generated from a yellow fever virus strain isolated from a patient in 1927. It was attenuated empirically by many passages of the virus in monkeys, mosquitoes, embryonic cell cultures, and embryonated eggs.

Despite its near-perfect safety record, like all vaccines, YF 17D has been linked to adverse effects. Rare cases of postvaccinal encephalitis (now called yellow fever vaccine-associated neurotropic adverse events) have been reported in very young infants, leading to the view that use of the vaccine is inadvisable in infants less than 9 months of age. More recently, rare but serious adverse vaccine effects that closely resemble the disease caused by wild-type yellow fever virus have been seen in the elderly. The reason these adverse effects in the elderly were not noted earlier may be because older individuals (particularly those 70 years or older) were not adequately represented in earlier vaccine trials. Why the elderly are at increased risk of serious yellow fever vaccine-associated adverse effects is not clear. However, a reasonable possibility is that it is because of immune deficits associated with aging. If so, this vaccine is better able to overcome these deficits than other vaccines, since its protective effect is not diminished with advancing age. The vaccine is recommended for the prevention of yellow fever in travelers to, and residents of, areas of tropical South America and Africa where the disease is endemic.

No vaccine against dengue virus is presently in use because of the potential risk of antibody-dependent enhancement, which might lead to dengue hemorrhagic fever upon subsequent challenge.

Encephalitis Viruses: Japanese Encephalitis, St. Louis Encephalitis, and West Nile Viruses

Japanese Encephalitis and St. Louis Encephalitis Viruses

Japanese encephalitis virus, which is prevalent throughout much of Asia, the Indian subcontinent, and islands

BOX 7.5

Yellow fever is one of the great plagues of human history. It originated in Africa, where it was endemic among Old World monkeys, its natural animal reservoir. It is believed that the yellow fever virus was brought from Africa to the New World by slave ships in the year 1596. But bearing in mind that yellow fever transmission requires an insect vector, in this case *Aedes aegypti* mosquitoes, how might the disease actually have been transported by the sailing ships of the day? The answer is that they inadvertently carried mosquito larvae in their water casks.

Of the numerous examples of the impact of infectious diseases on human history, here are a couple involving yellow fever. First, yellow fever may have been a factor in the decision of Napoleon to sell the Louisiana Territory to the United States in 1802. That is so because that disease destroyed 90% of his expeditionary force in the region. Also, thousands died of yellow fever during construction of the Panama Canal. Indeed, yellow fever caused the French to abandon their attempts to build the canal, enabling the United States to eventually take on the task.

In May 1900, at a time when neither the cause nor mode of transmission of yellow fever was known, U.S. Army Surgeon Major Walter Reed was appointed president of a medical board assigned to study infectious diseases in Cuba, in particular, yellow fever. A reason for American interest was that Cuba, at the time, was thought to be a major source of yellow fever epidemics in the United States, a belief that was said to have been a factor in the American annexation of Cuba.

When Reed began his inquiry, the prevailing hypothesis was that yellow fever might be caused by the bacterium *Bacillus icteroides*. However, Reed's board was unable to find any evidence in support of that notion.

Another hypothesis, advanced by Cuban physician Dr. Carlos Juan Finlay, was that the mosquito now known as *Aedes aegypti* is a vector for yellow fever. Reed embraced this idea because he noticed that people who cared for the patients with yellow fever had no increased risk of contracting the disease. So he concluded that people did not catch yellow fever from one another.

Reed, as president of the Board, is generally given major credit for solving the mystery of yellow fever transmission. Yet there were other heroes in this story. One was Finlay, who was the object of much ridicule for championing the mosquito hypothesis, at a time when there was not yet any

evidence that might support it. Acting Assistant Surgeon Major James Carroll was another hero. As a member of the Board, Carroll volunteered to be bitten, and he promptly developed yellow fever. Major Jesse Lazear, also a Board member, asked Private William Dean if he might be willing to be bitten. Dean consented, and he too contracted yellow fever. Fortunately, Dean and Carroll recovered. Not so for Lazear. After allowing himself to be bitten, he died after several days of delirium.

Lazear's martyrdom was not his only contribution to unraveling the mystery of yellow fever transmission. When Reed examined Lazear's notebook after his death, he found that it contained many crucial observations. For example, Lazear's carefully recorded findings demonstrated that in order for the mosquito to become infected it had to bite a yellow fever patient within the first 3 days of illness. Moreover, 12 days were then required before the virus could reach sufficiently high levels in the insect's salivary glands to be transmitted to a new human victim.

The above observations, taken together, convinced Reed and the Board that the mosquito hypothesis indeed was correct. Yet Reed knew that more extensive controlled experiments would be needed to convince the medical community. He directly supervised those experiments, which altogether involved 24 human volunteers, all of whom may rightly be considered heroes in the story.

The vector control measures that were instituted as a result of the work of Reed's Yellow Fever Board put an end to the yellow fever scourge in Cuba. Those measures were subsequently applied in Panama, where they enabled the United States to complete the Canal. Vector control worked in this instance because the female mosquito does not stray far from the source of her blood meal before laying her eggs. Thus, it is only necessary to control the vector population in the immediate area of human habitation.

Because of the work of the Yellow Fever Board, very few people living today have experienced this once dread disease. Yet the story does not end with the work of the Board. Its efforts were concerned with the means of transmission, not with the infectious agent itself. It was Carroll who in 1901 demonstrated that yellow fever is caused by a filterable virus. Finally, Max Theiler, at the Rockefeller Foundation's Division of Biological and Medical Research (although not a part of the Rockefeller Institute, it shared its facilities), developed a highly effective vaccine against yellow fever in the 1940s, well before the era of modern virology (see the text).

of the Pacific, is regarded as the world's most medically important encephalitic arbovirus. Its primary vector is the mosquito *Culex tritaeniorhynchus*, which breeds in irrigated rice fields and feeds on birds, pigs, and people. This niche works very well for the virus and against people, since pigs not only are the major domestic animal in much of Asia but also breed rapidly, thereby providing a constant supply of new, susceptible vertebrate hosts. Yet the human infection can be controlled by diligent human efforts. For example, the Japanese do an outstanding job of controlling human infection by draining rice fields when the mosquitoes are breeding and by keeping pigs away from areas of human habitation.

St. Louis encephalitis virus is the most frequent cause of severe virus encephalitis in the United States. Between 1964 and 1999, there were 4,478 reported cases, representing an average of 128 cases annually. However, in 1975, St. Louis encephalitis virus caused at least 1,791 cases of encephalitis, making up 42% of the total encephalitis cases reported that year to the Center for Disease Control (CDC). More-recent outbreaks have been much more limited in size, usually involving fewer than 30 cases. St. Louis encephalitis is an urban disease in the eastern and southwestern United States, where the major vector is *Culex pipiens*. It is a rural disease in the western United States, where the major vector is *Culex tarsalis*. Vertebrate hosts include birds and mammals.

Despite the capacity of these encephalitis viruses to cause serious aseptic (i.e., viral) meningitis (inflammation of the membrane around the brain and the spinal cord), as well as encephalitis, many more individuals are infected with these viruses than develop these severe diseases. Symptoms are most often no worse than fever, headache, and muscle soreness (myalgia).

West Nile Virus

West Nile virus caused a stir when it emerged in the Western Hemisphere in 1999 (see above). During the next 2 years, the CDC reported 149 cases of West Nile virus infections in the United States, including 18 deaths. (The CDC became the Centers for Disease Control and Prevention in 1980.) The subsequent case trend was as follows. In 2003 there were 9,862 cases, with 264 fatalities. In 2004 there were 2,539 cases, with 100 fatalities, and in 2005 there were 3,000 cases, with 264 deaths.

The above data show that many more humans have been infected with West Nile virus than have succumbed to it. Indeed, as in the cases of the other encephalitic flaviviruses, West Nile virus generally causes a mild disease in people, characterized by flu-like symptoms. The fever usually lasts only a few days and does not appear to cause any long-term effects. However, infection can result in more serious encephalitis, meningitis, and meningoencephalitis (inflammation of the brain and the membrane surrounding it).

Hepatitis C Virus

Hepatitis C virus is the most prevalent cause of non-A, non-B hepatitis. There are more than 4 million persistently infected carriers of hepatitis C virus in the United States and more than 170 million carriers worldwide! Indeed, it is a hallmark of this virus to establish persistent infection, doing so in about 85% of infected individuals. Clinically, persistent hepatitis C virus infection may lead to chronic hepatitis and hepatocellular carcinoma. The means by which hepatitis C virus evades host immunity in order to maintain persistent infection is considered below (Box 7.6).

Hepatitis C virus is most often transmitted by exposure to contaminated blood, in particular, by the sharing of syringes between intravenous drug users. (Most individuals in this group already are infected!) Still, the means of transmission in about 40% of cases is not known. Sexual transmission appears to be possible, and nosocomial transmission (i.e., transmission that occurs in a hospital, by a medical procedure) has an increasingly important role. Most recently, contaminated hospital equipment (e.g., kidney dialysis and computed tomography scanning machines) has been implicated. To date, humans are the only known host for hepatitis C virus.

It is not surprising that virus levels can become high in the blood, since large volumes of blood continuously circulate through the liver, the major site of hepatitis C virus infection. Even now, recipients of blood transfusions still are at some risk of being infected by hepatitis C virus. That is so despite sensitive antibody-based procedures for screening blood donors. The reasons are as follows.

BOX 7.6

Other cases of viral hepatitis are caused by **hepatitis E virus,** a plus-strand RNA virus, which is responsible for most of the epidemics of hepatitis in the developing world, but which also has become a concern in developed nations as well. Hepatitis E virus causes particularly severe liver disease in pregnant women, and unlike hepatitis B and C virus infections, hepatitis E virus infections do not become chronic. Much of the molecular biology of hepatitis E virus remains to be determined. Also, see the discussion of hepatitis D virus (the delta agent) in Chapter 22.

The safety of the blood supply depends mainly on selecting healthy donors and then effectively screening the blood. Donor selection is problematical because of the prevalence of asymptomatic hepatitis C virus carriers. Importantly, screening too remains a problem because antibodies, although present in the donor blood, may be at levels too low to detect. Moreover, by use of very sensitive PCR-based procedures, infected individuals who do not seroconvert even after several years of being infected have been identified. In the future, the routine use of PCR-based screening procedures may eliminate the threat of transmission via blood transfusion (Box 7.7).

Hepatitis C virus infection can have several clinical outcomes. In one scenario, the disease follows an acute course, with viremia lasting 4 to 6 months, followed by complete recovery. This is the type of infection seen in about 15% of cases. In another scenario, infection is asymptomatic and persistent. In these instances, it is possible that cirrhosis might develop after a period of 10 to 15 years, followed eventually by liver failure. This chronic course occurs with a frequency of about 70%. Finally, the remaining 15% of patients quickly develop severe cirrhosis.

Alcohol abuse is a major predisposing factor in the eventual development of cirrhosis in chronically infected individuals. Moreover, about 5% of chronic cases develop hepatocellular carcinoma, but only after about 30 years.

The mechanism by which hepatitis C virus causes this fatal cancer is not known with any certainty, but a likely possibility is as follows. Over the long course of chronic infection there is much destruction of liver cells. The host responds to the continual destruction of liver cells by extensively generating new liver cells to replace those that are destroyed. This vast proliferation of liver cells then provides a greatly increased opportunity for cancer-promoting mutations to occur during replication of cellular DNA. Consistent with this premise, hepatitis B and C viruses each give rise to chronic infections, and each induces hepatocellular carcinoma. In contrast, hepatitis A virus, which gives rise only to acute infections, does not induce hepatocellular carcinoma (see Chapter 3).

Persistent hepatitis C virus infections are extremely important clinically since they can lead to severe cirrhosis and fatal hepatocellular carcinoma. Thus, it is important to understand how this virus evades elimination, or even effective control, over the long course of persistent infection (see Chapter 3). HIV, a retrovirus, establishes and maintains persistent infection in part by generating a reverse transcript (i.e., a DNA copy) of its RNA genome, which it then inserts into the host DNA, a state in which it might remain latent essentially indefinitely. In addition, HIV has multiple stratagems for evading host immunity (see the Retroviruses, Chapters 20 and 21). Since hepatitis C virus does not encode a reverse transcriptase activity, it cannot establish latency or evade immune recognition by means of generating a DNA "provirus." On the other hand, several researchers report that immune responses are notably compromised in patients chronically infected with hepatitis C virus, as demonstrated, for example, by the increased susceptibility of those individuals to other pathogens. What little is known in these regards is briefly as follows.

The antibody response of chronically infected hepatitis C virus patients may be compromised, as suggested by the finding that some individuals do not produce detectable levels of virus-specific antibodies until as late as 4 months after the initial infection. Moreover, even after the appearance of virus-specific antibodies, the virus persists and disease progresses. This finding demonstrates that antibodies, even when present, are not sufficient to resolve or even control the infection. This is not entirely surprising, since cell-mediated immune responses are generally crucial for the control and resolution of virus infections. Indeed, other experimental findings show that CD8 cytotoxic T cells play a key role in controlling hepatitis C virus infections.

Why then do CD8 T cells fail to clear the hepatitis C virus infection in chronically infected individuals? A possible reason is that the magnitude of T-cell responses is significantly diminished in chronically infected individuals, in comparison to those who are acutely infected. One research group reported that the viral core protein in fact suppresses T-cell function. It was found to do so through its interaction with a complement receptor that is present on several professional antigen-presenting cell types, including B cells and macrophages. However, other research groups did not observe this inhibition, so the mechanism employed by hepatitis C virus to modulate the CD8 T-cell response is presently not known with certainty.

BOX 7.7

The single most important factor in lowering the incidence of posttransfusion hepatitis in the United States has been the elimination of the use of paid professional blood donors. Indeed, the establishment of an all-volunteer blood donor system at the U.S. National Institutes of Health clinical blood bank, together with routine screening for the hepatitis B virus surface antigen (Chapter 22), reduced the incidence of posttransfusion hepatitis of all types by 75%.

Hepatitis C virus shows considerable genetic heterogeneity. (As noted in earlier chapters, RNA viruses in general show a high level of genetic diversity. That is so because their RNA-dependent RNA polymerases lack the proofreading function of DNA polymerases.) However, the lack of an in vitro system in which to study hepatitis C virus has made it difficult to determine whether different hepatitis C virus genotypes in fact might constitute groups of distinct serotypes. On the one hand, it might be expected that there are distinct hepatitis C virus serotypes, since other flaviviruses manifest their genetic diversity in terms of their serotypes. But, on the other hand, there is evidence that immune "escape variants" of hepatitis C virus are not a necessary feature of virus persistence. For example, a study in a chimpanzee model of hepatitis C virus infection demonstrated that persistent infection could be established in the absence of the emergence of immune escape variants. Thus, the mechanisms used by hepatitis C virus to escape immune surveillance and to establish persistent infection remain to be clarified.

Regardless of the degree to which the nucleotide sequence diversity of hepatitis C virus might be clinically significant, that diversity is found throughout its genome. High levels of sequence variation are seen in hepatitis C virus genes that encode the envelope glycoproteins, as well as in genes encoding the nonstructural proteins. There is a higher degree of conservation in the 5′ untranslated region and in the gene that encodes the core (capsid) protein.

Hepatitis C virus isolates that have >90% sequence homology are said to be in the same genotype, and those displaying >20% variation are in different genotypes. By these criteria, six major hepatitis C virus genotypes and several subtypes are distributed worldwide. This categorization raises the question of whether these different genotypes vary with respect to their disease severity and response to therapy. Also, the issue remains whether the distinct serotypes might be an obstacle to the development of a vaccine against the virus.

Evidence from studies carried out in the United States, Europe, and Asia suggests that infection with genotype 1b poses the greatest risk of progression to cirrhosis and hepatocellular carcinoma. However, it is difficult to know this with certainty since other factors, such as the ages of the patients and the durations of their infections, might be influencing the data. Thus, the extent to which some hepatitis C virus genotypes might lead to a more severe course of infection is not yet resolved.

In contrast to the uncertainty of whether hepatitis C virus genotypes differ from each other regarding virulence, there is general agreement that genotypes do vary in their responses to antiviral therapy. First note that IFN-α, initially alone and then in combination with ribavirin (a guanosine analogue with a broad spectrum of activity against RNA and DNA viruses, including flaviviruses), has been the preferred therapy against hepatitis C virus (see below). Interestingly, patients infected with hepatitis C virus genotypes 1b and 4 respond to therapy significantly less well than patients infected with genotypes 2 and 3. The basis for the relative nonresponsiveness of genotypes 1b and 4 to therapy is not yet clear.

As just noted, the preferred therapy against hepatitis C virus is IFN-α, in combination with ribavirin. Previously, IFN-α alone had been the therapy of choice. However, the viral RNA in patient sera was reduced to undetectable levels in *only* about one-half of those individuals treated with IFN-α alone. Moreover, the majority of those patients relapsed once treatment was stopped. In contrast, viral RNA became undetectable in 80% of patients treated with IFN-α when administered in combination with ribavirin, and the rate of relapse in that patient group was significantly reduced as well (Box 7.8). Regardless, current therapies are expensive and are associated with significant side effects.

Several approaches are now under way to develop more effective therapies against hepatitis C virus. These approaches include new IFN-based drugs, new derivatives of ribavirin that are intended to cause lesser adverse side effects, immune modulators, and specific hepatitis C virus inhibitors. Regarding the last approach, virtually all steps of the hepatitis C virus replicative cycle can be targeted by specific inhibitors (Fig. 7.5). Major targets include the IRES on viral plus-sense RNAs (recall that in contrast to other flaviviruses, hepatitis C virus translation, like that of the picornaviruses, is driven by an IRES), the viral NS3 serine proteinase, which serves to process the viral polyprotein, and the viral RNA-dependent RNA polymerase. But there is concern that the multiple virus genotypes and high viral mutability may yet compromise the effectiveness of the drugs presently under development.

BOX 7.8

It is noteworthy that a few hepatitis C virus-infected individuals might have been cured by drug therapy, since hepatitis C virus infections may be the only virus infections that currently are curable, if only in a minority of cases. For the sake of comparison, see the discussion of HIV/AIDS (Chapter 21).

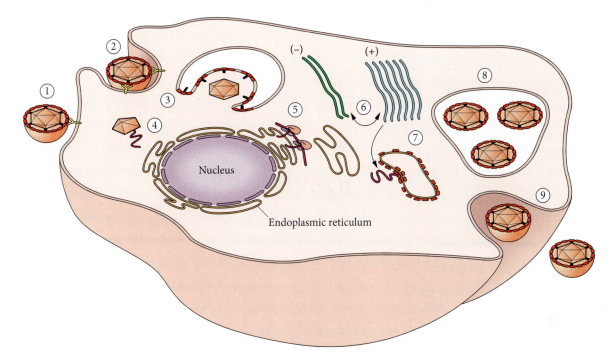

Figure 7.5 Virtually all steps of the hepatitis C virus life cycle can be targeted by specific inhibitors (when compounds already exist at the preclinical or clinical developmental stage, they are indicated in parentheses below): 1, virus binding to cellular receptor(s) (small-molecule inhibitors of cell attachment, monoclonal antibodies, and hyperimmune anti-hepatitis C virus immunoglobulins); 2, receptor-mediated endocytosis; 3, membrane fusion and nucleocapsid release; 4, nucleocapsid uncoating; 5, translation and polyprotein processing (IRES inhibitors, NS3 serine protease inhibitors, and NS2 zinc-dependent autoprotease inhibitors); 6, hepatitis C virus RNA replication (NS5B RNA-dependent RNA polymerase inhibitors, NS5A inhibitors, and inhibitors of replication complex formation); 7, virion formation and budding in intracellular vesicles; 8, virion transport and maturation; 9, virion release. (+), positive-strand RNA; (−), negative-strand RNA. Adapted from J. M. Pawlotsky, *Hepatology* **43:**S207–S220, 2006, with permission.

Vaccines against hepatitis C virus are being developed. However, bearing in mind that it is not yet clear whether or not genotypes might be recognized as serotypes, it remains possible that antigenic differences between genotypes and subtypes might result in a lack of cross-protection.

Suggested Reading

Choo, Q. L., G. Kuo, A. J. Weiner, L. R. Overby, D. W. Bradley, and M. Houghton. 1989. Isolation of a cDNA clone derived from a blood-borne non-A, non-B viral hepatitis genome. *Science* **244:**359–362.

Togaviruses

INTRODUCTION

The *Togaviridae* are a family of plus-strand RNA viruses. "*Toga*" is Latin for cloak, and here it refers to the fact that a lipid envelope surrounds all members of the *Togaviridae*. Yet this family name is not particularly informative, since other virus families are likewise enveloped. Since togaviruses are plus-strand RNA viruses, they are class IV viruses in the Baltimore classification scheme (Chapter 2).

The togavirus family contains three genera: the alphaviruses, rubivirus, and the arteriviruses. All of the alphaviruses are arthropod-borne or arboviruses (Box 8.1). The *Alphavirus* genus includes *Eastern equine encephalitis virus*, *Western equine encephalitis virus*, *Venezuelan equine encephalitis virus*, *Sindbis virus*, *Semliki Forest virus*, and *Chikungunya virus*.

The above three encephalitic alphaviruses are associated with mild systemic illness. Yet as in the case of the encephalitic flaviviruses (Chapter 7), infections with these alphaviruses can progress to severe encephalitis. Sindbis, Semliki Forest, and Chikungunya viruses are capable of causing arthralgia or joint disease, as well as fever. Chikungunya virus, which is endemic to Asia and Africa, is the only one of the alphaviruses for whom humans are known to be a natural host. All of these alphaviruses are transmitted to their vertebrate hosts by a mosquito vector.

Rubella virus, the cause of German measles, is the only member of the *Rubivirus* genus. Unlike the alphaviruses, rubella virus is transmitted from person to person by the respiratory route. Importantly, rubella virus is one of the very few viruses capable of crossing the human placenta, and infection of the developing fetus can result in serious developmental abnormalities.

The *Arterivirus* genus contains several animal pathogens, including *Equine arteritis virus* and *Lactase dehydrogenase-elevating virus* of mice. None of the arteriviruses are known to cause disease in people, and they are not discussed further.

Like the flaviviruses (Chapter 7), the togaviruses are enveloped, plus-strand RNA viruses. However, togaviruses are readily distinguished from flaviviruses by the organization of their genomes and by their pattern of gene expression (see below). Moreover, togavirus genomes, at 12,000 to 13,000 nucleotides in length, are somewhat larger than those of flaviviruses, at 10,000 to 11,000 nucleotides. Also, they are larger still than the genomes of the nonenveloped, plus-strand picornaviruses (7,200 to 8,400 nucleotides).

STRUCTURE AND ENTRY

Togavirus structure was determined by means of cryo-electron microscopy (cryo-EM) and three-dimensional image reconstruction. This approach involves taking hundreds of highly detailed, two-dimensional images, which then are processed by powerful computer programs to generate a three-dimensional image (Box 7.1). Additional detail, at atomic-level resolution, was attained by X-ray crystallographic examination of component structural proteins, which were then fitted into the cryo-EM structure.

Enveloped togavirus particles are 70 nm in diameter. The inner icosahedral capsid is comprised of 240 molecules of a single protein, the C protein. Molecules of the C protein are arranged as pentamers at the icosahedral vertices and as hexamers elsewhere in the capsid (see Fig. 2.8).

The outer lipid envelope closely adjoins the capsid. Embedded within the lipid bilayer, there are 240 copies of each of the two or three virally encoded glycoproteins, which are wound about each other to form the spikes at the viral surface (Fig. 8.1). These envelope glycoproteins are designated E1, E2, and E3. E1 and E2 are present on all togaviruses, and E3 is present on some but not all, depending on the particular togavirus. E1 and E2 each have one transmembrane helix that traverses the lipid bilayer. In contrast, E3, when present, lies completely outside the envelope. E1 is responsible for cell fusion, and E2 is primarily involved in receptor binding and cell entry. The function of E3 is unclear.

Reminiscent of the flaviviruses, and different from virtually all other enveloped viruses, the 240 copies of the togavirus E1 and E2 proteins nearly completely cover the virus surface, such that the lipid envelope is just barely exposed to the exterior. Moreover, like the flavivirus envelope proteins, the togavirus envelope glycoproteins too are arranged with icosahedral symmetry. But in contrast to the flavivirus envelope proteins, which do not traverse the lipid bilayer, the C termini of the togavirus E1 and E2 proteins indeed do traverse the lipid bilayer, and E2 interacts with the underlying capsid protein. Thus, unlike the flaviviruses, in which the capsid core is not in contact with the lipid bilayer or the glycoprotein shell, the togavirus E2 protein links the surface spikes to the internal capsid core. As a result, the togavirus icosahedral glycoprotein shell may very well influence the structure of the underlying capsid, or perhaps vice versa. (Recall that flavivirus capsids may lack a defined icosahedral structure [Chapter 7].)

The structures of the Sindbis virus E1 and C proteins have been solved at high resolution and have been fitted into the cryo-EM structure (Box 7.1). Thus, it is known that the transmembrane domains of E1 and E2 form a tight pair of α-helices within the lipid bilayer, and the cytoplasmic domain of E2 fits into a cleft in the C protein (Fig. 8.1). On the external side of the envelope, molecules of E1 lie at an angle of about 50° relative to the particle surface, and they interact to form a thin, icosahedral shell (the so-called "skirt") around the virus particle. The skirt virtually completely covers the envelope and provides support for the protruding spikes comprised of E2, and E3 as well in some instances.

Togaviruses acquire their spike-containing envelopes by budding from regions of the plasma membrane that contain the membrane-embedded envelope proteins. The interaction between the cytoplasmic tails of the E2 envelope glycoproteins and the C proteins of the capsids drives the budding process. Moreover, this interaction accounts for the close juxtaposition between the viral envelope and the capsid. (As described below, the transmembrane envelope glycoproteins are synthesized on the rough endoplasmic reticulum membranes and are transported to the plasma membrane by the vesicle-mediated secretory pathway.) At some point, the E2/E1 heterodimers trimerize to become the viral spikes. That is, the mature virus particle contains 240 E2/E1 heterodimers that are organized into the 80 trimeric spikes, each of which contains 3 E2/E1 heterodimers.

Togavirus entry begins with the binding of the E2 envelope glycoprotein to the receptor on the cell surface. E2 was identified as the viral attachment protein by genetic experiments involving Sindbis virus, in which mutations that affected virus binding were found to map in the gene

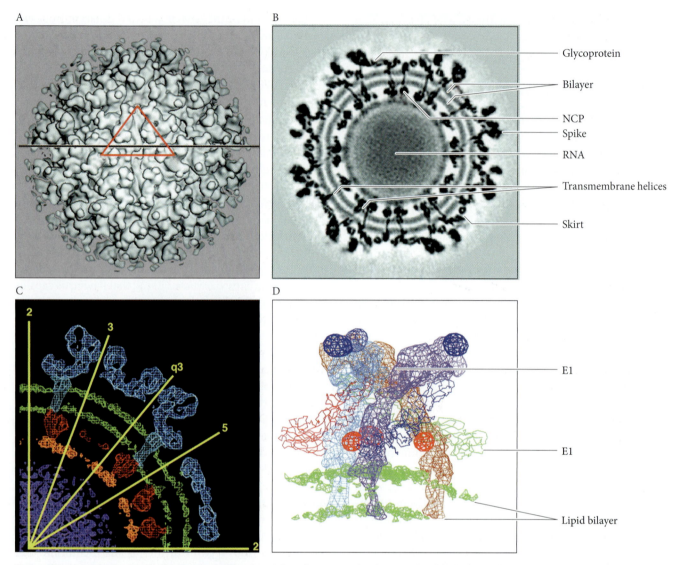

Figure 8.1 Cryo-EM density distribution for Sindbis virus. (A) Surface of the virus at 20-angstrom (Å) resolution. The outer surface is organized as an icosahedral shell, studded with 80 spikes, each built from three copies of each of the transmembrane glycoproteins, E1 and E2. The red triangle marks the boundary of an icosahedral asymmetric unit. (B) Cross section through the 11-Å map along the black line shown in panel A. Note the clearly defined lipid bilayer and the transmembrane domains of E1 and E2 crossing the membrane. Also notice that the spikes are connected by a thin external protein layer called the skirt. The nucleocapsid (NCP) is seen inside the membrane. (C) A central cross-section through the 11-Å resolution map, showing the glycoproteins (blue), the lipid bilayer (green), the nucleocapsid (red), the mixed RNA-protein region (orange), and the internal RNA (magenta). The orientation of the icosahedral (two-, three-, and fivefold) as well as quasithreefold (q3) axes is shown in yellow. (D) Structure of the Cα-backbones of three E1 molecules (colored magenta, dark blue, and green) and densities of three E2 molecules (colored light blue, purple, and brown) fitted into the cryo-EM map. The view is shown around a quasithreefold symmetry axis (q3 in panel C). The molecules are seen to form a complex trimeric spike. Adapted from W. Zhang et al., *J. Virol.* **76**:11645–11648, 2002, with permission.

for E2. (As might be expected, antibodies against the E2 protein have neutralizing activity against the virus.)

The laminin receptor serves as the togavirus receptor on some cell types. Laminin is a component of a type of extracellular matrix referred to as the basal lamina, which characteristically is found under epithelial cells. Other, less well characterized proteins serve as the receptor on other cell types.

After togaviruses bind, they are taken up into the cell by receptor-mediated endocytosis in clathrin-coated pits. They are then transported by endocytic vesicles to an endosomal compartment, in which the mild acidity triggers the viral spikes to undergo conformational transitions that promote membrane fusion and release of the virus core into the cytoplasm (Box 8.2). As seen in studies of Semliki Forest virus, these spike transitions include the disassociation of the E3/E2/E1 heterotrimers and the concomitant formation of E1 homotrimers (Figure 8.1). Since the fusion peptide on E1 is buried in the E3/E2/E1 heterotrimeric complexes, the low-pH-induced dissociation of the heterotrimers exposes the hydrophobic fusion peptide on E1, while also pointing it towards the cellular membrane, in that way enabling it to promote fusion of the viral lipid envelope with the endosomal membrane. Recall that generating the fusion-active form of the flavivirus E protein also involved a low-pH-induced transition to a trimer (Chapter 7). Also, the native conformation of the togavirus E1 protein, like that of the flavivirus E protein, is comprised primarily of β-sheets.

At several junctures in this book we underscore the importance of viral fusion proteins being in a metastable conformation, so that when appropriately triggered they might undergo the conformational rearrangements that promote membrane fusion. Viral fusion proteins often acquire their metastable conformation by first folding into their most stable conformation and then becoming metastable following proteolytic cleavage. Yet the togavirus E1 protein does not itself undergo proteolytic cleavage to

become metastable. Instead, as we see in more detail below, during maturation E1 is associated with the precursor of E2, called p62. The p62/E1 heterodimer cannot be triggered by low pH to undergo the conformational rearrangements that free up the E1 fusion peptide. However, the low-pH-induced conformational rearrangements that enable E1 to promote membrane fusion become possible after p62 is proteolytically cleaved to generate E2. (Interestingly, in order for both E1 and p62 to properly fold, they must fold in association with each other, in the endoplasmic reticulum [see below].)

Although alphaviruses appear to infect most vertebrate cells by receptor-mediated endocytosis, followed by low-pH-triggered fusion in an endosomal compartment, these viruses appear to be able to infect arthropod cells, as well as some vertebrate cells, by a low-pH-independent mechanism. Apropos these observations, there is no reason in principle why the fusion step in the entry of enveloped viruses could not occur at the plasma membrane in some instances, as well as from within an endosomal compartment in others. If togavirus fusion were to occur at the plasma membrane of some cell types, we might expect that the interaction of the virus with its receptor on those cells could trigger the conformational rearrangements that promote fusion. Note that influenza virus (Chapter 12) and human immunodeficiency virus (Chapter 21) provide important examples of enveloped viruses that enter by receptor-triggered fusion at the plasma membrane.

After fusion, either in endosomes or at the plasma membrane, the next crucial step is for the genomic RNA to be uncoated and released from the capsid. Despite the importance of this uncoating step, its mechanism is not yet clear. There is evidence from experiments done both in vitro and in vivo to suggest that uncoating is triggered by the interaction of the capsid with cytoplasmic ribosomes. Consistent with that premise, the site on the C protein that interacts with genomic RNA during encapsidation also appears to be able to bind to ribosomes by recognizing ribosomal RNA (rRNA).

If the ribosome-mediated togavirus uncoating mechanism indeed is correct, then it still must be explained how conditions at uncoating are different from those at assembly, so that both processes can occur at their appropriate times in the same cell. Actually, all viruses face a similar dilemma: how to generate at the end of one replication cycle a stable particle that readily disassembles at the start of the next. The answer, and an important general principle, is that virus assembly and disassembly usually occur under different conditions, in different cellular compartments.

Apropos the togaviruses, one possibility, as suggested by studies involving Sindbis virus, is that the capsid undergoes a conformational transition during budding that

BOX 8.2

Several lines of experimental evidence support the belief that the togavirus entry pathway proceeds via receptor-mediated endocytosis, followed by low-pH-induced membrane fusion in an endosomal compartment. First, EM shows that togaviruses are contained within small, coated vesicles soon after uptake, and shortly afterward they are found in endosomes. A second experimental approach uses the fact that dominant-negative mutant forms of the protein dynamin block the budding of clathrin-coated pits. Thus, the experimental finding that infection of cells by Sindbis virus is inhibited by expression of dominant-negative forms of dynamin supports the premise that Sindbis virus entry occurs via clathrin-coated pits and vesicles. Third, infection is impaired by lysosomotropic amines, which raise the endosomal pH. In addition, cell-free exposure of Semliki Forest virus to low pH induces the viral E3/E2/E1 heterotrimers to dissociate and form the fusion-active E1 homotrimers. Finally, Semliki Forest virus fuses with liposomes in a mildly acidic environment, demonstrating that the sole trigger for fusion is low pH.

primes it to uncoat during entry into a new host cell. An alternative hypothesis is that the capsid undergoes the crucial destabilizing conformational transition in response to the mildly acidic pH it encounters in the endosomes during entry. In this scenario, the capsid, as well as the envelope proteins, is affected by the low pH of the endosome during entry. In support of this hypothesis, exposure of purified Semliki Forest virus capsids to acidic pH in vitro leads to partial autoproteolysis within the N-terminal domain of the C protein. In addition, cleavage of the Semliki Forest virus C protein also has been detected in vivo, as the virus passes through the endosomal compartment. In either proposition, the final uncoating is believed to be triggered by ribosomes.

TRANSCRIPTION, TRANSLATION, AND GENOME REPLICATION

In general, plus-strand RNA viral genomes are translated on host ribosomes soon after being released from the virus particles. In that way, they are able to generate the enzymes needed for viral transcription and replication. We noted this crucial initial translation step when discussing the plus-strand picornaviruses and the flaviviruses (Chapters 6 and 7). Indeed, retroviruses are the only known exception to this generality among plus-strand RNA viruses (Chapter 21).

Still, there is an important difference between the organization of togavirus genomes and those of picornaviruses and flaviviruses, which distinguishes their respective patterns of translation in a significant and defining way. In the picornavirus and flavivirus cases in point, the genomic RNAs (and the identical messenger RNAs [mRNAs]) contain only one translation initiation site and only one corresponding open reading frame. Each of the picornavirus and flavivirus gene products is generated by the proteolytic processing of the single species of primary precursor polyprotein translated from their single open reading frame. (A primary protein is one that is not itself the cleavage product of a larger protein.) In contrast, togavirus genomic RNAs contain two translation initiation sites and two open reading frames, one open reading frame corresponding to each translation initiation site (Fig. 8.2). One togavirus translation initiation site is close to the 5′ end of the genomic RNA, and the other is located internally. As a consequence of this genomic organization, the togaviruses generate two distinct species of primary precursor polyproteins, which together are the source of all the individual togavirus gene products.

Togavirus genes, which encode the nonstructural proteins that catalyze transcription and replication of the genome, are translated from the translation initiation site near the 5′ end of the genome. Togavirus genes that encode the capsid and envelope proteins are translated from the internal translation initiation site. Thus, togavirus genes that encode nonstructural proteins are at the 5′ end of the genome, and the genes for the structural proteins are at the 3′ end. Since the structural genes of picornaviruses and flaviviruses are at the 5′ end of each of their respective genomes, togavirus genomes also differ from picornavirus and flavivirus genomes with respect to their gene order, as well as with respect to the numbers of their translation initiation sites and open reading frames.

The presence of two translation initiation sites on togavirus genomic RNAs and the single open reading frame downstream of each raises a new concern. That is, how is the togavirus internal translation initiation site able to function? This question arises because of the fact that translation of eukaryotic mRNAs initiates via a **ribosome scanning** process, in which the 40S ribosomal subunit, together with translation initiation factors, binds to the cap at the 5′ end of the mRNA. This initiation complex then scans along the mRNA until it encounters the first AUG translation initiation codon. The initiation factors are then released, enabling the 60S ribosomal subunit to associate with the 40S ribosomal subunit, so that translation can then begin. Like eukaryotic mRNAs in general, togavirus genomic RNAs are capped and do not contain internal ribosomal entry sites. Moreover, the ribosome dissociates from the mRNA when it encounters the termination triplets at the end of an open reading frame. Thus, the open reading frame that is closest to the 5′ cap of the genomic RNA is the only one that might be translated from that RNA. How then are the togaviral structural genes, which are downstream of the internal translational initiation site, expressed?

The answer to this conundrum is that togaviruses transcribe *two* distinct classes of capped mRNA (Fig. 8.2). One of these mRNAs is identical to the full-length (12-kilobase [kb]) genomic RNA. The other is a shorter (4-kb) mRNA that overlaps the 3′ end of the full-length mRNA. Importantly, the translation initiation site at the 5′ end of the larger mRNA, as well as the remainder of its corresponding open reading frame, is missing from the smaller mRNA. Instead, the translation initiation site that is internal on the larger full-length mRNA is at the 5′ end of the smaller mRNA. Thus, the translation initiation site, which is internal on the full-length RNA, can be used on the smaller RNA to generate the structural proteins. The larger and smaller RNAs are referred to as 49S and 26S RNAs, respectively. (S is an abbreviation for Svedberg, a unit of mass. It is a carryover from the days when the masses of macromolecules were determined by their sedimentation

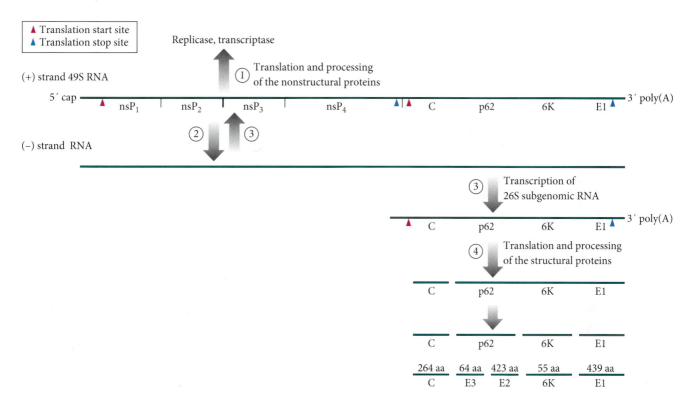

Figure 8.2 Pattern of Sindbis virus gene expression. The togavirus genomic RNA has two translation initiation sites and two translation termination sites. But initiation can begin only at the most 5′ translation initiation site. Thus, the first two-thirds of the togavirus genomic RNA is the only portion that can be translated (step 1). This portion of the genomic RNA encodes the components of the viral replicase/transcriptase. Once generated, these proteins transcribe the genomic RNA to produce a full-length minus-strand RNA copy (step 2). The minus strand then serves as the template to generate full-length plus strands and shorter subgenomic mRNAs (step 3). The internal translation initiation site on the genomic RNA is at the 5′ terminus of the subgenomic RNA, so that structural proteins can now be translated from the subgenomic RNA (step 4). The nonstructural proteins, as well as the structural ones, are each processed from polyprotein precursors.

velocities in sucrose gradients.) The 26S mRNA is also referred to as a **subgenomic** mRNA.

The actual steps in the transcription of the togavirus mRNAs are, in brief, as follows (Fig. 8.2). Once the genomic RNA has been translated to generate the viral replicase/polymerase, it is then transcribed into full-length negative-sense strands. These negative strands then serve as the templates for transcription of both the 49S and 26S mRNAs. Synthesis of the 49S mRNA initiates at the 3′ end of the negative strand. Synthesis of the 26S mRNA initiates at an internal *transcription* initiation site that lies at the junction between the nonstructural and structural gene clusters. The sequence of that initiation site is conserved among the togaviruses. The 26S mRNA does not serve as a template for minus-strand synthesis. Instead, all subgenomic mRNAs are transcribed from full-length minus strands.

The 49S and 26S mRNAs are each capped. Since cellular capping enzymes are localized in the nucleus, and since togavirus transcription occurs in the cytoplasm, the virus must encode its own capping activities. Indeed, the methyltransferase and guanylyltransferase capping activities are associated with the viral NSP1 protein (see below).

Notice that the viral replication enzymes are initially translated from the parental genomic RNA. In contrast, translation of the structural proteins requires at least partial translation of the parental genome, followed by the transcription of viral minus strands and then plus strands. Thus, togaviral gene expression has an "early" phase and a "late" phase, distinguishing it from the gene expression pattern of picornaviruses and flaviviruses. As we will see, *DNA* virus gene expression is characteristically divided into early and late phases. But the transition from the early phase to the late phase occurs much more quickly during a togavirus infection than during a DNA virus infection. (Is it clear why the viral replicase/polymerase genes must be in the reading frame at the 5′ end of the togavirus plus-strand RNA?)

Next, we consider in somewhat more detail the transcription and translation steps outlined above. The first of these steps is the translation of the nonstructural proteins from the genomic RNA. Translation of these proteins initiates from an AUG triplet close to the cap at the 5′ end of the genomic RNA. Translation then continues until the ribosome encounters three termination triplets that are located about two-thirds into the genome, just before the translation initiation codon for the structural genes (Fig. 8.2). The resulting translation product is a polyprotein that is co- and posttranslationally cleaved to generate four distinct proteins, NSP1, NSP2, NSP3, and NSP4. However, the initial translation products are the polyproteins P123 and P1234. There is a single weak UGA termination triplet between the coding sequences for NSP3 and NSP4, such that translation most often (90% of the time) terminates to yield P123 but sometimes continues to generate P1234.

At the time that P1234 is being proteolytically processed to generate NSP4, P123 and NSP4 rapidly form a complex that is believed to initiate the synthesis of 49S minus strands. This protein complex also can synthesize plus strands, using the 49S negative strands as templates. However, it does so inefficiently.

Efficient synthesis of plus strands occurs after the P123 precursor is cleaved to yield NSP1, NSP2, and NSP3. This set of cleavages also results in the shutoff of minus-strand synthesis. Importantly, the cleavages that generate NSP1, NSP2, and NSP3 are catalyzed in *trans* by the NSP2 segment of P123. That is, the NSP2 segment of one P123(4) polyprotein acts on another P123(4) polyprotein to generate the individual cleavage products. As a consequence of this cleavage in *trans*, the rate of cleavage accelerates as the concentration of P123 increases. Notice that a following consequence of this NSP2-mediated processing of P123(4) in *trans* is that synthesis of minus strands initially predominates, followed by the accumulation of both plus-strand and minus-strand RNA molecules. Synthesis of minus strands then diminishes, while plus-strand 49S and 26S RNAs accumulate. Thus, NSP2 plays a key role in regulating the switch from minus-strand to plus-strand synthesis.

Results from genetic experiments and sequencing studies imply that NSP4 contains the polymerase activity. The methyltransferase and guanylyltransferase capping activities are associated with NSP1, as noted above.

Synthesis of the 26S mRNAs enables the viral structural proteins to be translated. Translation of the 26S mRNA initiates at the single translation initiation site near the 5′ end of that mRNA molecule and proceeds uninterrupted to the end of the single 3,700-nucleotide open reading frame. However, there is a singularly important difference between the modes of translation of the nonstructural and the structural proteins. The precursor polypeptide of the nonstructural proteins is synthesized in its entirety on ribosomes that are free in the cytoplasm, and the individual nonstructural proteins generated from that precursor likewise are released into the cytoplasm. In contrast, the four structural proteins are not synthesized as one long precursor polypeptide, from which individual proteins are released into the cytoplasm. Instead, the pattern of translation of the structural proteins reflects the fact that two or more of these proteins are destined for inclusion in the viral envelope.

In Chapter 7, we detailed the pathways by which the flavivirus envelope glycoproteins first become integral membrane proteins and then are incorporated into virus particles. We now examine the particulars of how the togaviruses do the same.

Five structural proteins, C, E3, E2, 6K, and E1, are generated by the uninterrupted translation of the single open reading frame on the 26S mRNA (Fig. 8.2). The nucleotide sequence encoding the C protein is at the 5′ end of the 26S mRNA, and thus, it is the first sequence of the precursor polyprotein to be translated (Fig. 8.3). As in the case of the nonstructural proteins, translation of the C protein occurs on free cytoplasmic ribosomes. Then, almost immediately after the C protein is translated, it is autocatalytically cleaved from the nascent precursor polyprotein.

As translation of the nascent polyprotein continues, a signal sequence emerges, which is bound by a cellular signal recognition particle (SRP) (Fig. 7.2). The SRP temporarily blocks further translation of the 26S mRNA. Importantly, the SRP also targets the still-ribosome-bound nascent polyprotein to the rough endoplasmic reticulum, where the SRP binds to an SRP receptor on the endoplasmic reticulum membrane. The ribosome then binds to a protein **translocation complex** in the endoplasmic reticulum membrane, the SRP is released, and the signal sequence on the nascent polyprotein inserts into the translocation complex. Translation can now resume, and a polypeptide known as p62 (the precursor for proteins E2 and E3) is generated. Almost all of p62 passes through the translocation complex and into the lumen of the endoplasmic reticulum. However, a 22-amino-acid, hydrophobic **stop-transfer** sequence, which is close to the C terminus of the protein, anchors it in the endoplasmic reticulum membrane. When translation of P62 is complete, it is cleaved from the nascent polypeptide by a cellular **signal peptidase**. This cleavage leaves 31 amino acids of p62 extending into the cytoplasm, while also exposing a sequence on the ribosome-bound nascent polypeptide known as 6K, which is the signal sequence

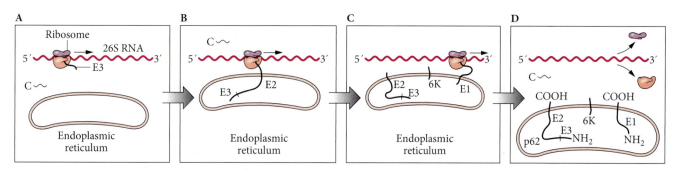

Figure 8.3 The 26S mRNA encodes a polyprotein that contains five structural proteins in the order C, E3, E2, 6K, and E1. (A) Translation of the C protein occurs on free cytoplasmic ribosomes. Then, almost immediately after the C protein is translated, it is autocatalytically cleaved from the nascent growing polyprotein. A signal sequence emerges on the nascent polyprotein, which targets the still-ribosome-bound nascent polyprotein to the rough endoplasmic reticulum (B). A polypeptide known as p62 (the precursor for proteins E2 and E3) is then generated. Almost all of p62 passes through the translocation complex in the ER membrane and into the lumen of the endoplasmic reticulum. However, a 22-amino-acid, hydrophobic stop-transfer sequence, which is close to the C terminus of the protein, anchors it in the endoplasmic reticulum membrane. (C) When translation of p62 is complete, it is cleaved from the nascent polypeptide by a cellular signal peptidase. This cleavage leaves 31 amino acids of p62 extending into the cytoplasm, while also exposing a sequence on the nascent polypeptide known as 6K, which is the signal sequence for protein E1. (D) Translation of protein E1 continues as it passes through a translocation complex, until its passage too is arrested by its hydrophobic stop-transfer sequence, leaving two amino acids extending into the cytoplasm. The E1 signal sequence, 6K, is cleaved from E1 by the signal peptidase.

for protein E1 (Fig. 8.3). Translation of protein E1 continues as it passes through a translocation complex, until its passage too is arrested by its hydrophobic stop-transfer sequence, leaving two amino acids extending into the cytoplasm. The E1 signal sequence, 6K, is cleaved from E1 by the signal peptidase.

The fates of p62 and E1 are considered below in the discussion of togavirus maturation. First, we consider what advantages the togaviruses might derive from making use of both a full-length and a subgenomic mRNA. After all, the picornaviruses and flaviviruses make their way in the world quite well, making use of but a single mRNA, a single translation initiation site, and a single open reading frame. We might begin our pursuit of an answer by bearing in mind the following facts. First, the full-length mRNAs are translated into the nonstructural enzymatic activities that the virus needs in least amounts. In contrast, the subgenomic mRNAs are translated into the structural proteins that the virus needs in greatest amounts. Second, the full-length mRNAs are identical to the viral genomic RNAs. Consequently, they are encapsidated as the pool of C proteins accumulates in the cell. In contrast, the subgenomic mRNAs are not encapsidated. Consequently, they continue to accumulate in the cytosol and be translated. Indeed, the ratio of 26S mRNAs to 49S mRNAs in the cytosol can become greater than 10 to 1. As a result, the togaviruses can make much more of their structural proteins than of their nonstructural proteins. This would appear to be a more efficient translation strategy than the ones of the picornaviruses and flaviviruses,

which necessarily must translate virtually all of their proteins in equimolar amounts.

ASSEMBLY AND MATURATION

Following the insertion of p62 and E1 into the endoplasmic reticulum membrane, these proteins then associate to form heterodimers that are transported by the normal vesicle-mediated secretory pathway to the Golgi body, where the portions of these proteins that eventually will face out from the virus envelope are glycosylated. From the Golgi body, the p62/E1 heterodimers are then transported to the plasma membrane. During this transport, p62 is cleaved by a cellular furin-like protease to yield E2 and E3. Note that E3 lies completely outside the membrane. E3 may or may not remain associated with the E2/E1 heterodimers, depending on the virus.

At the same time as the envelope proteins are being synthesized, processed, and transported to the plasma membrane, new 49S genomic RNAs are being enclosed within capsids comprised of C protein. Encapsidation is selective for full-length genomic RNAs, probably because those RNAs contain an encapsidation signal that is lacking on the 26S subgenomic RNAs. The encapsidation signal on the full-length genomic RNA is recognized by a sequence on the C protein that also recognizes rRNA, a fact that may be important during uncoating, as noted above.

The final stages of assembly begin with the interaction of the RNA-containing capsids with the envelope proteins at the plasma membrane. Recall that envelope glycoprotein

E2 has 33 amino acids on the cytoplasmic side of the bilayer. Experimental evidence from genetic and structural studies implies that the C-terminal cytoplasmic domains of E2 interact with the C protein during final assembly. Thus, the C termini of the E2 proteins might be thought of as comprising docking sites for the capsids at the plasma membrane. The interaction between the cytoplasmic tails of the E2 proteins and the capsids causes the plasma membrane to bend around the particle until the particle is completely enveloped within the membrane. At the completion of this process, the fully enveloped particles pinch off or bud from the cell surface.

Prior to, or perhaps at the time of, the envelopment process, three E3/E2/E1 heterotrimers trimerize to form each spike. Since there was initially a one-to-one correspondence between each of the 240 copies of the C protein and the E3/E2/E1 heterotrimers, the final virus contains 80 spikes, in which each spike consists of a trimer of E3/E2/E1 heterotrimers (Fig. 8.1). Recall that E1 and E2 are anchored to the lipid bilayer by hydrophobic stop-transfer sequences near their carboxy termini. In contrast, E3 lies completely outside the membrane. In some alphaviruses, including Sindbis virus (the prototype for the genus), protein E3 is released from the mature virus particles. Thus, the mature Sindbis virus particle is composed only of the E1, E2, and C proteins and the genomic RNA. In other viruses, such as Semliki Forest virus, E3 remains associated with the E2/E1 heterodimer.

EPIDEMIOLOGY AND PATHOGENESIS
Alphaviruses That Cause Encephalitis: Eastern, Western, and Venezuelan Equine Encephalitis Viruses

The alphaviruses, like the flaviviruses (Chapter 7), are prototypical arboviruses. This means that while these two virus groups have distinct patterns of genome organization and expression, they nevertheless display similar patterns of vector-mediated transmission. Thus, it would be good for the reader to review the general principles of arthropod-mediated transmission, seasonality, and dissemination discussed in Chapter 7.

Reminiscent of flavivirus infections, human exposure to alphaviruses generally results in asymptomatic infections. When illness does occur, it usually takes the form of a low-grade fever and flu-like symptoms, including chills, rash, and aches. However, three of these alphaviruses, eastern equine encephalitis virus, western equine encephalitis virus, and Venezuelan equine encephalitis virus, are known to cause potentially lethal encephalitis. Survivors may be left with permanent neurological damage, manifested by paralysis, mental retardation, blindness, and deafness. Yet these viruses are usually more of a problem to livestock than to humans. Nevertheless, urban outbreaks of arboviruses might occur if urban animals become reservoir hosts for these viruses.

Each of the encephalitic alphaviruses uses a different species of mosquito for its vector. Consequently, the range of each of these viruses is restricted to that of its specific vector. Western and eastern equine encephalitis viruses are found throughout the Americas, whereas Venezuelan equine encephalitis virus is found in tropical forests and marshes of Central and South America, and also in the Florida Everglades. Despite the designation of equine encephalitis viruses, wild birds are the usual vertebrate reservoir for eastern and western equine encephalitis viruses. Horses are an important vertebrate host for Venezuelan equine encephalitis virus during **epizootics** (outbreaks of disease involving multiple species of vertebrates), in which humans too are infected. In those instances, horses serve to amplify the virus. Little is known about the life cycle of this virus between these outbreaks.

Alphaviruses That Cause Arthritis: Chikungunya, Ross River, and Sindbis Viruses

Infections by Chikungunya, Ross River, and Sindbis viruses are noteworthy for the arthritis symptoms that accompany the usual flu-like fever and aches. Indeed, "*chikungunya*" is a colorful Swahili term meaning "that which bends up," referring to the arthritic symptoms caused by Chikungunya virus. The arthritis induced by these viruses affects mostly the small joints and usually resolves in a few weeks, but in the case of Chikungunya and Ross River viruses, it may persist for months or even years.

Chikungunya virus is the most prevalent of these viruses in humans. It is found throughout Africa, India, Southeast Asia, and the Philippines. Devastating chikungunya outbreaks have occurred in Africa and Asia, lasting for years at a time and affecting a majority of the population in some urban areas. There is concern that the virus could spread to the United States because of the appearance there of its vector, the *Aedes aegypti* mosquito. Ross River virus is found in Australia, and Sindbis virus is found in Africa, India, Malaysia, and northwestern Europe.

As in the cases of the encephalitic arboviruses and Chikungunya virus noted above, specific mosquitoes likewise serve as vectors for Ross River and Sindbis viruses. Vertebrate hosts for Chikungunya virus include humans and monkeys. For Ross River virus they include marsupials and rodents, and for Sindbis virus they include wild birds.

Rubella Virus

Rubella virus has the same structure and mode of replication as the alphaviruses. However, it is not an arbovirus.

Instead, rubella virus is spread by the respiratory route. Moreover, humans are its only known host. Rubella is one of the five classic childhood exanthems. The others are fifth disease (caused by a parvovirus [see Chapter 14]), measles, roseola, and chicken pox. German physicians were the first to distinguish rubella, which means "little red" in Latin, from measles and the other childhood exanthems. Hence its common name, **German measles**.

Rubella virus is highly transmissible by its respiratory route of infection, and it is endemic worldwide. The virus initially replicates in the respiratory tract and spreads to other parts of the body via the bloodstream. In prevaccine days, infection generally occurred in early childhood, usually producing a generally trivial illness, characterized by an exanthem or rash and swollen glands. Infection of adults may be more severe, and symptoms can include arthritis, decreased levels of blood platelets, and encephalopathy (degenerative disease of the brain). However, the major concern for rubella virus stems from the fact that a nonimmune woman who is infected during the first trimester of pregnancy runs the risk that the baby will suffer severe congenital abnormalities. (Recall that rubella virus is one of a small number of viruses that are known to cross the human placenta. Others include parvovirus B19, cytomegalovirus, and human immunodeficiency virus.) The rubella-associated congenital abnormalities include deafness, blindness, heart disease, mental retardation, and impaired growth (Box 8.3).

Prior to the development of rubella vaccines, rubella epidemics occurred at regular intervals of 6 to 9 years in temperate climates (Box 8.4). Consequently, only about 20% of women reached childbearing age without having been infected and acquiring immunity. The major concern involving rubella was that these nonimmune women might become infected during pregnancy. A particularly severe rubella epidemic struck the United States in 1964–1965, in which congenital rubella occurred in nearly 1% of all children born in some urban areas. Thus, it is understandable that the primary reason for rubella vaccination is to prevent infection of pregnant women. In actuality, this is accomplished by reducing the number of susceptible people in the population, since if infected,

BOX 8.3

In Chapter 1, we discussed how the development of Koch's postulates provided decisive criteria for identifying the causative agent of an infectious disease. We also noted that Koch put forth his postulates before viruses were identified and well before there was any understanding of their nature. Thus, one requirement of Koch's postulates, that the putative causative agent be grown in pure culture, meant that Koch's postulates could not strictly be satisfied for any virus. Nevertheless, a revised form of Koch's postulates can be applied to some, but not all, viruses.

Fortunately, several other empirical approaches have been devised to determine the cause of an infectious disease in which a virus might be involved. One approach is the so-called **case-control study**, where the goal is to identify trends that existed before the actual outbreak of the disease. Since case-control studies look back in time, they are also referred to as **retrospective studies**. Another approach is the so-called **cohort study**, in which a putative cause has been hypothesized. In this instance, one tracks a sample of the population that was exposed to the putative etiologic agent before the onset of disease. Since cohort studies look to the future, they are also referred to as **prospective studies**. In both retrospective and prospective studies, it is important to identify a suitable control group.

A perceptive Australian ophthalmologist, Norman McAlister Gregg, applied both retrospective and prospective approaches to identify rubella virus as the cause of the congenital defects (initially cataracts, but also deafness and heart disease) occurring in a large number of infant cases in Sydney, Australia, in 1940-1941. After observing the cases he encountered in his own practice, he wrote to other doctors in Australia to inquire into their experiences. They reported back a total of 78 other cases of congenital cataracts, 44 of which also had heart defects. Upon interviewing the mothers of these affected infants, Gregg deduced that the common factor might have been that the vast majority of them experienced rubella early in their pregnancies. Gregg then carried out a prospective study to test the hypothesis that there is a causal relationship between maternal rubella infection and congenital deformities. He identified a cohort or group of pregnant women who had experienced an exanthematous or rash-like illness during their current pregnancies and then monitored them to see if their babies displayed congenital defects. Comparing this group of babies with those born of mothers who had not experienced such infections confirmed the relationship between congenital defects and maternal exposure to rubella virus.

BOX 8.4

The following was edited from the Immunization Action Coalition website. More details on the rubella vaccine can be found there and elsewhere.

Three rubella vaccines were licensed in the United States in 1969. The currently used attenuated rubella vaccine was licensed in 1979, and the others were discontinued at that time. The vaccine is available singly, but it is recommended that it be given as part of the MMR vaccine, which also protects against measles and mumps, or as the MMRV vaccine, which additionally protects against varicella (chicken pox).

Vaccination is recommended by the Centers for Disease Control and Prevention (CDC), the American Academy of Pediatrics (AAP), and the American Academy of Family Physicians (AAFP), for all children and for adolescents and adults without documented evidence of immunity. Also, it is especially important to verify that all women of child-bearing age are immune to rubella before they get pregnant. The first dose of the MMR or MMRV vaccines should be given on or after the first birthday; the recommended range is from 12 to 15 months.

they might pass the infection on to a susceptible pregnant woman.

Children are a major source of rubella infection, since they are readily exposed to the virus in the crowded conditions existing in schools and day care centers. Consequently, children are the major targets for vaccination, usually receiving the live attenuated rubella virus vaccine, together with the measles and mumps vaccines (the **MMR** vaccine). Since rubella virus, like poliovirus, has no host other than humans, it is expected that it can be controlled by vaccination (Box 8.4).

Suggested Reading

Zhang, W., S. Mukhopadhyay, S. V. Pletnev, T. S. Baker, R. J. Kuhn, and M. G. Rossmann. 2002. Placement of the structural proteins in Sindbis virus. *J. Virol.* **76:**11645–11658.

Coronaviruses

INTRODUCTION

The *Coronaviridae* are a family of plus-strand RNA viruses that differ in several notable ways from other plus-strand RNA viruses. First, coronavirus RNA genomes, which contain 27,000 to 30,000 nucleotides, are considerably larger than the genomes of any other known plus-strand RNA viruses and, indeed, of other RNA viruses in general. For the sake of comparison, the genomes of the plus-strand RNA bacteriophages, picornaviruses, flaviviruses, and togaviruses contain 3,500 to 4,700, 7,200 to 8,400, 10,000 to 11,000, and 12,000 to 13,000 nucleotides, respectively. In fact, coronavirus RNA genomes are among the largest known RNA molecules of any kind. Second, coronaviruses contain helical nucleocapsids, a feature that also is unique among known plus-strand RNA viruses. Third, and most important perhaps, the coronaviruses express their plus-strand RNA genomes by a novel strategy whereby they generate a nested set of messenger RNAs (mRNAs) that have a common 3′ end. Each of the mRNAs of this nested set, except the smallest, contains multiple open reading frames (ORFs). However, only one ORF is translated from each of these mRNAs: the one at the 5′ end of the molecule.

Since coronaviruses are plus-strand RNA viruses, they are class IV viruses in the Baltimore classification scheme (Chapter 2). The mouse hepatitis virus (MHV) is the prototype coronavirus and molecular model for the family.

Coronavirus helical capsids are enclosed within lipid envelopes. Large, petal-like spikes radiate outwards from the envelopes, giving coronaviruses a corona-like appearance (i.e., the appearance of the sun, with its surrounding halo, seen only during a total eclipse) in electron micrographs (Fig. 9.1). Some sources also refer to the "crown-like" (corona) appearance of these viruses. Regardless of which meaning of corona might be implied, the appearance of these large spikes provides the basis for the family name.

Coronaviruses have been found in a number of vertebrates, including cats, dogs, horses, cattle, pigs, chickens, and turkeys, as well as humans.

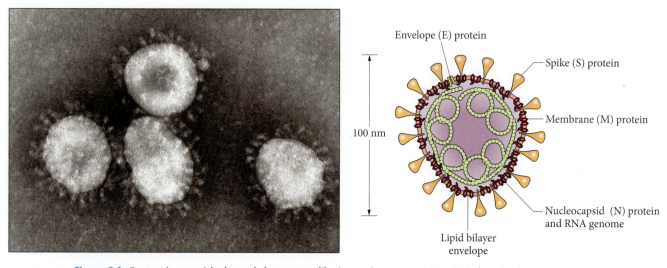

Figure 9.1 Coronavirus particles have a halo or crown-like (corona) appearance in negatively stained preparations. The schematic shows a coronavirus particle, with the minimal set of structural proteins. The viral nucleocapsid is a long, flexible helix comprised of the 30-kb-long genomic RNA and the nucleocapsid (N) protein. The large spikes are comprised of the S glycoprotein. The membrane (M) glycoprotein traverses the lipid envelope three times, and its cytoplasmic domain interacts with the nucleocapsid inside the particle. Some coronavirus envelopes also contain an HE glycoprotein (not shown). The micrograph is in the public domain. Schematic adapted from P. S. Masters, p. 193–292, *in* K. Maramorosch and A. J. Shatkin (ed.), *Advances in Virus Research*, vol. 66 (Academic Press, San Diego, CA, 2006), with permission.

Based on their nucleic acid sequences, coronaviruses have been sorted into three groups. Group 1 and group 2 coronaviruses infect mammals, including humans, whereas all of the group 3 coronaviruses infect only birds.

Coronavirus infections may lead to a variety of serious diseases that can affect the upper respiratory tract, gastrointestinal tract, liver, or central nervous system, depending on the particular coronavirus and host. However, coronaviruses are best known for the mild upper respiratory tract infections that they cause in humans. These infections are very similar to the common colds caused by the human rhinoviruses, and coronaviruses are acknowledged to be the second most frequent cause of common colds after the rhinoviruses.

Prior to 2003, only two coronavirus strains, HCoV-229E and HCoV-OC43, had been isolated from humans. For that reason, and because coronaviruses had been studied for decades, researchers in the field were taken by surprise in 2003, when the etiologic agent for the newly emerging severe acute respiratory syndrome (SARS) epidemic was found to be a "new" coronavirus. The discovery of the SARS coronavirus (SARS-CoV) prompted the search for other new coronaviruses in both humans and animals, leading to the discovery of two more new "human coronaviruses" (i.e., that infect humans): HCoV-NL63 in 2004 and HCoV-HKU1 in 2005. In contrast to the other human

coronaviruses, SARS-CoV causes a severe, life-threatening pneumonia in humans and is the most pathogenic human coronavirus yet known.

While SARS-CoV only recently emerged in the human population, HCoV-HKU1 and HCoV-NL63 have actually been circulating in humans for a long time. HCoV-HKU1 and HCoV-NL63 are each respiratory coronaviruses that are frequently associated with upper and lower respiratory tract infections. Infections by these two recently discovered human coronaviruses do not differ greatly from those caused by the "old" coronaviruses HCoV-229E and HCoV-OC43.

STRUCTURE

Coronavirus particles contain plus-strand RNA genomes that are 27,000 to 30,000 nucleotides in length. These genomes are both capped and polyadenylated, enabling them to be translated upon their release into the cytoplasm at the start of an infection cycle (see below). Within virus particles, coronavirus genomes are associated with the nucleocapsid (N) protein, in the form of helical nucleocapsids. Coronaviruses are unique in this regard, since they are the only known family of plus-strand RNA viruses that have helical nucleocapsids. The helical nucleocapsids are enclosed within a lipid envelope (Fig. 9.1), which contains the membrane (M), spike (S), and envelope (E) proteins. Some coronaviruses, including human coronavirus HCoV-OC43, contain a fourth envelope protein, the hemagglutinin esterase (HE). The functions of these proteins are discussed below.

Virus envelope glycoproteins generally contain a large domain, which faces outwards from the particle, and a short domain beneath the envelope. The coronavirus M protein differs from most other virus envelope glycoproteins in that only a short N-terminal domain is exposed on the exterior of the particle, while a large carboxy-terminal domain is exposed beneath the envelope, facing the interior. The reason for the atypical orientation of the coronavirus M protein is as follows. In contrast to coronaviruses, other enveloped viruses that contain helical nucleocapsids generally contain a major *internal* protein designated M for matrix. These matrix or M proteins mediate the association between the helical nucleocapsid and the viral envelope. Importantly, the M proteins of these other viruses usually associate with the lipid envelope entirely by virtue of their interactions with the cytoplasmic tails of the transmembrane envelope glycoproteins. That is, they generally have no hydrophobic transmembrane domain of their own that might anchor them to the envelope. In contrast to this more general situation, coronavirus nucleocapsids contact the coronavirus *transmembrane* M protein directly. Consequently, although the coronavirus M protein is a transmembrane protein, it nevertheless fulfills the function of a matrix protein for these viruses. This fact accounts for the large internal domain and small external domain of the coronavirus M protein. It is likely that the N protein of the nucleocapsid binds to the cytoplasmic domain of the M protein during coronavirus assembly and budding. The interaction between the N and M proteins was demonstrated experimentally using the mammalian (rather than the yeast) two-hybrid system, which has the advantage of detecting in vivo protein/protein interactions, which are more likely to be biologically relevant.

The S glycoprotein constitutes the large, prominent, 20-nm-long petal-shaped spikes that are seen in electron micrographs of coronavirus particles (Fig. 9.1). In contrast to the M protein, the S protein has a short carboxy-terminal cytoplasmic (i.e., internal) domain and two large external domains, consistent with its role in mediating virus binding and subsequent fusion of the virus envelope to a cellular membrane. Like the envelope glycoproteins that mediate entry of the flaviviruses and togaviruses (Chapters 7 and 8), the coronavirus S protein trimerizes within the endoplasmic reticulum membrane.

The E protein helps to facilitate virus assembly and morphogenesis, although it is not entirely clear how it does so. One possibility, as described in greater detail below, is that the E protein interacts with the M protein, thereby causing membrane curvature that promotes the budding process. In support of that notion, coexpression of the recombinant E and M proteins is sufficient to generate virus-like particles (VLPs). (Expression of the cloned genes for the flavivirus prM and E proteins similarly generates flavivirus VLPs [Chapter 7].) The E protein also forms cation-selective ion channels in lipid bilayer membranes. This ion channel-forming activity appears to plays a role in coronavirus replication, as shown by the ability of the Na$^+$ ion channel inhibitor hexamethylene amiloride to inhibit coronavirus replication in cultured cells.

ENTRY

As in other virus families, different coronaviruses use different cell surface molecules for their receptor. Also, different receptor specificities largely determine the various tissue tropisms and pathologies displayed by the different coronaviruses. For example, aminopeptidase N (APN), a ubiquitous enzyme present in a wide variety of human organs, tissues, and cell types (endothelial, epithelial, fibroblast, and leukocyte), is the primary receptor for several, but not all, of the group 1 coronaviruses. In contrast, the recently discovered HCoV-NL63 is a group 1 coronavirus that does not use APN for its receptor. Instead, HCoV-NL63, like SARS-CoV (a member of coronavirus group 2), uses angiotensin-converting enzyme 2 (ACE2) for its receptor. ACE2 is expressed in the human airway epithelia as well as the lung parenchyma (Box 9.1).

The envelopes of a minority of coronaviruses, including HCoV-OC43 and HCoV-HKU1 (see below), contain the HE-expressing glycoprotein (HE protein), which is thought to play an as yet uncertain role in entry. Since HE cleaves acetyl groups from the neuraminic acid-containing virus receptor, some coronavirus researchers suggest that HE may facilitate entry by releasing virus particles from the receptor. In support of that premise, the substrate preference of a particular coronavirus HE protein matches the sialic acid moieties of that virus's receptor. For example, the S and JHM strains of MHV bind to 5-*N*-acetyl-4-*O*-acetylneuraminic acid receptors, and these viruses encode HE proteins with sialate-4-*O*-acetylesterase activity. By similar reasoning, the HE activity might also promote release of the virus during its budding and egress from the cell.

The S protein mediates fusion for all coronaviruses, regardless of whether or not they express the HE protein. In order for the S protein to mediate fusion, in some cases it may need to be cleaved by cellular proteases to generate a heterodimer consisting of the receptor-binding S1 subunit, which lies completely outside the envelope, and the

BOX 9.1

Several interesting questions come to mind here. First, HCoV-NL63 is now known to be widespread in the human population, in which it is associated with moderate respiratory illness, primarily in children. Although HCoV-NL63 can give rise to lower respiratory tract infections, these infections are still generally much milder than those caused by SARS-CoV. But since HCoV-NL63 and SARS-CoV each use ACE2 for their receptor, how might we account for the much greater virulence of SARS-CoV? Second, both APN and ACE2 are zinc-binding, cell surface peptidases. Is there any significance to the fact that these two different zinc-binding, cell surface peptidases together serve as receptors for a sizeable number of coronaviruses, or is it merely coincidence?

Group 1 coronaviruses, which use APN for their receptor, include human coronavirus-229E (HCoV-229E), porcine respiratory coronavirus, feline infectious peritonitis virus, feline enteric coronavirus, TGEV, and canine coronavirus.

For the sake of completeness, carcinoembryonic antigen-cell adhesion molecule (CEACAM1) is the receptor for MHV, a group 2 coronavirus. However, other group 2 coronaviruses, including HCoV-OC43 and bovine coronavirus, use *N*-acetyl-9-*O*-acetylneuraminic acid for their receptor. But the cellular receptor for HCoV-HKU1, also a group 2 coronavirus, is not yet known.

time to determine the conformational rearrangements that the influenza virus hemagglutinin (HA) protein undergoes to promote fusion. Consequently, the influenza virus HA protein is the paradigm for membrane fusion proteins in general. Accordingly, the influenza virus fusion step is covered in detail in Chapter 12. Apropos the coronaviruses, the X-ray crystal structures of the native and rearranged MHV spike protein fusion core imply that the MHV fusion process is very similar to that of influenza virus and that of several other prominent enveloped viruses. These experimental findings are briefly reviewed in Box 9.2.

Does coronavirus membrane fusion occur at the plasma membrane or in an internal endocytic compartment? Taking SARS-CoV as an example, initial studies reported virus particles at various stages of fusion at the cell surface. These observations imply that the receptor triggers the conformational changes in the S protein that promote fusion. However, some later studies reported that SARS-CoV entry may be pH dependent, implying that the virus is taken up by endocytosis and that within a late endosomal compartment the low pH triggers the conformational changes in the S protein that lead to fusion. In these instances, the endosomal protease, cathepsin L, might also play a role in entry by cleaving the S protein, thereby priming it to mediate fusion. In this regard, recall that the S protein of some coronavirus strains is cleaved after the virus particle is taken up by endocytosis.

Since there is fairly strong experimental evidence that SARS-CoV can enter by direct membrane fusion, as well as evidence that the endocytic route too leads to infection, perhaps this virus is able to use multiple mechanisms to gain entry into host cells. The same might be true for other coronaviruses as well. This is the best one can say for now (Box 9.3).

GENOME ORGANIZATION AND EXPRESSION

Coronavirus mRNAs and Their Translation

Coronavirus plus-strand RNA genomes resemble eukaryotic mRNAs with respect to having both a 5′ cap and a 3′ poly(A) tail. However, they are vastly larger than most eukaryotic mRNAs, as they encode ≥20 individual proteins, while the generally monocistronic eukaryotic cellular mRNAs encode only 1 protein.

As in the case of all plus-strand RNA viruses (with the exception of the retroviruses), when coronavirus RNA genomes are released into the cytoplasm at the start of an infection cycle, they are translated on host ribosomes, so that they might produce the enzyme activities that are essential for transcription and replication of the viral RNA

membrane-anchored S2 subunit, which contains the fusion-active domain. In some instances, the S protein is cleaved by cellular proteases during transport of the maturing virus particle through the endoplasmic reticulum and the Golgi apparatus (see below). In other instances (e.g., the S protein of the MHV-2 strain and perhaps SARS-CoV), it is cleaved in an endosome after virus uptake by receptor-mediated endocytosis (see below). In either event, cleavage frees the hydrophobic fusion domain on the S2 subunit and causes the S protein to be in a metastable conformation, thereby priming it to undergo the conformational transitions that lead to membrane fusion and deposition of the viral nucleocapsid into the cytoplasm.

To understand how the S protein might promote fusion, it is necessary to know the conformational rearrangements that it undergoes to drive the fusion process. These conformational transitions were solved by X-ray diffraction analysis of crystals of the native and rearranged proteins. This experimental approach was employed for the first

BOX 9.2

The envelope glycoproteins of other enveloped viruses are generally posttranslationally cleaved into surface-bound and transmembrane subunits. The influenza virus HA glycoprotein is the best-studied such example (Chapter 12). Interestingly, the X-ray crystal structure of the fusion domain of the MHV S glycoprotein was recently found to resemble the low-pH-induced fusion-promoting conformation of the influenza virus HA glycoprotein (Xu et al., 2004). Moreover, it also resembles the structure of the fusion core of the HIV gp41 glycoprotein. This structure contains a centrally located "coiled coil," comprised of three α-helices. Each of the three monomers of the S protein, which make up each trimer, contributes one of the α-helices to the

coiled coil. The three α-helices are arranged about each other such that they splay apart at the base of the trimer. This structure and the membrane fusion process are discussed in detail in Chapter 12. For the moment, it is interesting that these experimental findings and others reported by Xu and coworkers (2004) suggest that coronavirus S proteins induce membrane fusion by a mechanism similar to those of influenza virus and HIV (Chapters 12 and 21).

Reference
Xu, Y., Y. Liu, Z. Lou, L. Qin, X. Li, Z. Bai, H. Pang, P. Tien, G. F. Gao, and Z. Rao. 2004. Structural basis for coronavirus-mediated membrane fusion. Crystal structure of mouse hepatitis virus spike protein fusion core. *J. Biol. Chem.* **279:**30514–30522.

Viral fusion proteins and models for membrane fusion. (**A**) Comparison of MHV fusion core with other viral fusion protein structures. The proteins under comparison include the simian virus 5 (SV5) F protein, the Ebola virus GP2 protein, the HIV gp41, the Maloney murine leukemia virus (MMLV) Env-TM protein, and low-pH-induced influenza virus HA protein. Top and side views are shown for the six fusion core structures. Notice the top view of the MHV fusion core structure, showing the threefold axis of the trimer. (**B**) Model for coronavirus-mediated membrane fusion. The first state is the native conformation of coronavirus spike protein on the surface of the viral membrane. It has been reported that the spike protein is trimeric in this conformation and about 200 Å (angstrom) in length, but the exact structure of the full-length protein remains unknown. The second state is the prehairpin state of the S2 subunit. After several conformational changes, the fusion peptide inserts into the cellular membrane with the aid of other regions of S protein, and possibly including the receptor. Although the internal fusion peptide is not exposed at the N terminus of S2, it could insert into part of the target membrane by means of some hydrophobic residues. This insertion would be stable enough to drive the membrane motion with the conformational changes of HR (heptad repeat) region 1, which is adjacent to the fusion peptide. The third state is conformational change and juxtaposition of the target and viral membranes. With the help of other regions of S protein, the HR1 and HR2 regions move together and facilitate juxtaposition of the cellular and viral membranes. The last state is the postfusion conformation. The coiled coil will reorient with its long axis parallel to the membrane surface. The fused cellular and viral membranes make it possible for subsequent viral infections. From Y. Xu et al., *J. Biol. Chem.* **279:**30514–30522, 2004, with permission.

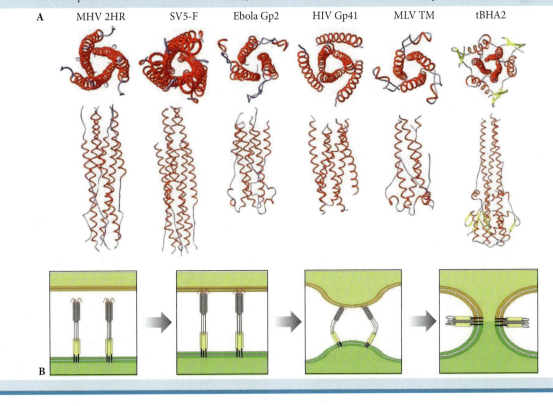

A MHV 2HR SV5-F Ebola Gp2 HIV Gp41 MLV TM tBHA2

B

BOX 9.3

Experiments to investigate the entry of a dangerous viral pathogen, such as SARS-CoV, might pose considerable risks to the experimenters, as well as to bystanders. It is possible to minimize experimental risks by using a **pseudotype virus**, instead of wild-type virus. Pseudotype virus particles contain the genome of one virus enclosed within the capsid or envelope of another. The capsid or envelope is from the virus of interest, since it dictates the entry pathway. However, entry is gauged by the expression of a reporter gene that is cloned into the genome of the other virus, in this instance, an avirulent one. Pseudotype virus particles can be generated by making use of the fact that viruses are assembled from pools of precursor components (recall the phenomenon of phenotypic mixing [Chapter 1]).

One widely used experimental approach is to generate pseudotype viruses that contain retroviral cores enclosed within the envelope of the virus of interest. In these instances, the retroviral genome is genetically engineered to express an easily detectable reporter gene, such as β-galactosidase, luciferase, or green fluorescent protein (GFP). This approach has been used successfully to study the entry of dangerous viruses from many families, including flaviviruses, filoviruses, rhabdoviruses, and bunyaviruses. In the example presented here, it was used by Simmons and coworkers to examine entry of SARS-CoV (Simmons et al., 2004).

Simmons and coworkers began by generating the pseudovirus. This was done by cotransfecting cells with two genetic constructs. One construct was a plasmid that expressed the SARS-CoV S glycoprotein. The other construct expressed the HIV polymerase and core proteins, and either a β-gal or luciferase reporter gene. The HIV cores generated by the HIV construct can express the reporter gene upon entry into a new host cell. However, the HIV construct does not express the HIV envelope glycoproteins, gp41 and gp120 (i.e., HIV too is an enveloped virus). Consequently, it cannot produce an infectious virus by itself. However, the SARS-CoV S glycoprotein, which is generated by the other

plasmid, can be incorporated into the pseudotype virus particles that contain the HIV cores. This occurs as follows. The SARS-CoV S glycoprotein is synthesized and transported to the plasma membrane by the usual mechanisms for synthesis and transport of transmembrane proteins. Consequently, when the HIV cores bud from the cell, they acquire an envelope that contains the SARS-CoV S glycoprotein, thereby generating the pseudovirus. Importantly, this pseudotype virus will bind to cells and enter them in a SARS-CoV-specific manner.

Using these pseudotype virus particles, the investigators observed that infection was inhibited by several lysosomotropic agents, suggesting that SARS-CoV envelope fusion is triggered by the acidity of an endosomal compartment. However, the S protein itself is not affected by low-pH treatment. A possible explanation for these seemingly contradictory experimental findings is that the interaction of the S protein with the receptor at the plasma membrane might prime the S protein to undergo a low-pH-induced transition. Priming of fusion by the receptor at the plasma membrane was seen earlier in the instance of some human rhinoviruses (Chapter 6).

Questions remain regarding coronavirus entry. Nevertheless, pseudotype viruses make it possible to study the entry pathways of dangerous viral pathogens and also to identify inhibitors of entry that might have therapeutic potential.

One last point: Given the makeup of the pseudotype virus in this instance, one might wonder whether it is any safer to work with than regular SARS-CoV. Well, it is. The major reason is that it is capable of only one replication cycle. Do you see why that is?

Reference
Simmons, G., J. D. Reeves, A. J. Rennekamp, S. M. Amberg, A. J. Piefer, and P. Bates. 2004. Characterization of severe acute respiratory syndrome-associated coronavirus (SARS-CoV) spike glycoprotein-mediated viral entry. *Proc. Natl. Acad. Sci. USA* **101**:4240–4245.

---

genome. This raises an issue that is central for all plus-strand RNA viruses but particularly crucial in the case of the coronaviruses, considering the exceptionally large size of their RNA genomes. Specifically, how might coronaviruses express each of their 20 or more individual gene products from their nonsegmented, plus-strand RNA genomes? We begin to address this issue by first considering the organization of coronavirus genomes.

The 14 to 16 nonstructural viral proteins involved in coronavirus genome replication, transcription, and proteolytic processing are encoded within the 5′-most proximal two-thirds of the viral genome, within ORF1 (Fig. 9.2 and 9.3). (As described below, ORF1 is actually comprised of two ORFs that partially overlap.) The structural proteins are encoded within the 3′-proximal one-third of the genome. Genes for the major structural proteins of all

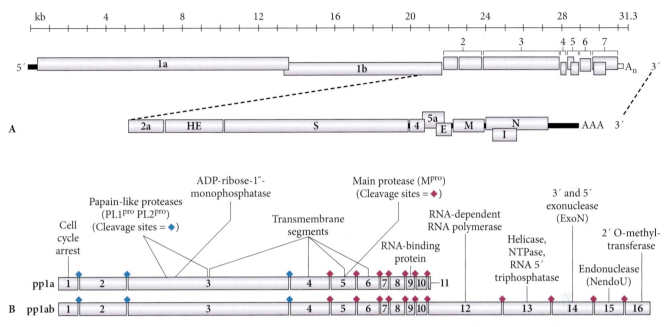

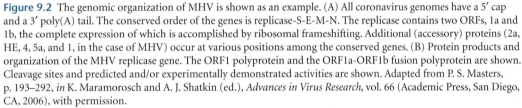

Figure 9.2 The genomic organization of MHV is shown as an example. (A) All coronavirus genomes have a 5′ cap and a 3′ poly(A) tail. The conserved order of the genes is replicase-S-E-M-N. The replicase contains two ORFs, 1a and 1b, the complete expression of which is accomplished by ribosomal frameshifting. Additional (accessory) proteins (2a, HE, 4, 5a, and 1, in the case of MHV) occur at various positions among the conserved genes. (B) Protein products and organization of the MHV replicase gene. The ORF1 polyprotein and the ORF1a-ORF1b fusion polyprotein are shown. Cleavage sites and predicted and/or experimentally demonstrated activities are shown. Adapted from P. S. Masters, p. 193–292, *in* K. Maramorosch and A. J. Shatkin (ed.), *Advances in Virus Research*, vol. 66 (Academic Press, San Diego, CA, 2006), with permission.

coronaviruses are arranged in the order S, E, M, and N. One might have postulated that this conserved gene order within the family provides some selective advantage to the coronaviruses. However, that does not appear to be the case. This was demonstrated by experiments in which viruses with engineered rearrangements of these genes were found to be as viable as standard virus.

In addition to the proteins involved in RNA synthesis and proteolytic processing and the four major structural proteins, coronaviruses also encode so-called accessory proteins. The coding regions for individual accessory proteins may lie anywhere in the genome, and the number of these may range from as few as one in the case of HCoV-NL63 to as many as eight in the case of SARS-CoV. The MHV accessory proteins are 2a, HE, 4, 5A, and I (Fig. 9.2). Some accessory proteins may be structural, as in the case of HE, which is the best characterized of the coronavirus accessory proteins. The functions of most of the others are not yet known. Thus far, all accessory proteins tested have been found to be nonessential for growth in cell culture, while some are necessary for growth and pathogenesis in animals.

Next, to better appreciate the uniqueness of the coronavirus pattern of gene expression, we briefly review the patterns of gene expression of the previously discussed

plus-strand RNA viruses: the picornaviruses, flaviviruses, and togaviruses.

The plus-strand RNA genomes of both the picornaviruses and flaviviruses contain only one translation initiation site (for details, see Chapter 6). Moreover, the single translation initiation sites on picornavirus and flavivirus genomes each correspond to the single ORF on those genomes. Also, the picornaviruses and flaviviruses each produce only one species of mRNA, which in each instance is identical to the genomic RNA.

In contrast to picornaviruses and flaviviruses, togavirus plus-strand RNA genomes contain two translation initiation sites, each of which corresponds to a single ORF. In order for togaviruses to express each of their two translation initiation sites, they generate two species of mRNA, one of which is identical to the genomic RNA, while the other is a subgenomic RNA. The translation initiation site that is internal on the larger genomic RNA is at the 5′ end of the subgenomic RNA, so that it is able to express its initiation function on that RNA.

Importantly, each of the above plus-strand RNA virus families translates each of its ORFs into a single large primary precursor polyprotein, and *all* of the individual proteins are generated by posttranslational cleavages of those

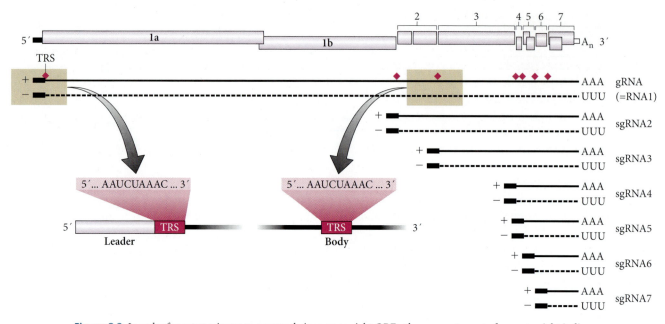

Figure 9.3 In order for coronaviruses to express their seven to eight ORFs, they generate a set of seven to eight individual mRNAs. The largest of those mRNA species is identical to the genomic RNA, and the six or seven others are subgenomic. The individual coronavirus mRNA species constitute a nested set, in which each mRNA has the same 3′ terminus. Shown is the nested set of both plus-sense and minus-sense strands produced during replication and transcription, using MHV as an example. (Inset) Details of the arrangement of leader and body copies of the TRS. Adapted from P. S. Masters, p. 193–292, *in* K. Maramorosch and A. J. Shatkin (ed.), *Advances in Virus Research*, vol. 66 (Academic Press, San Diego, CA, 2006), with permission.

precursor polyproteins. (Recall that a primary protein is one that is not itself the cleavage product of a larger protein.)

In contrast to the single translation initiation site on picornavirus and flavivirus genomes and the two translation initiation sites on togavirus genomes, coronavirus genomes contain *seven to eight* translation initiation sites, each of which corresponds to an ORF. Actually, the huge ORF1, which encodes the coronavirus replicase complex, contains two ORFs that overlap, and translation of the entire sequence depends on a programmed ribosomal frameshift.

In order for coronaviruses to express their seven to eight ORFs, they generate a set of seven to eight individual mRNAs. The largest of those mRNA species is identical to the genomic RNA, and the six or seven others are subgenomic. Importantly, the individual coronavirus mRNA species comprise a nested set, in which each mRNA has the same 3′ terminus (Fig. 9.3). That is, each mRNA species contains all of the gene sequences of the next smaller mRNA, plus one additional gene at its 5′ end. As might be expected, the only translation initiation site that is utilized on each mRNA is that which is closest to the 5′ terminus. (If it is not clear why that is so, recall how translation of eukaryotic mRNAs is initiated [see Chapter 6].) Thus, each of the coronavirus mRNA species, with the exception

of the smallest, is structurally polycistronic but functionally monocistronic.

Using this pattern of transcription, coronaviruses carry the togavirus strategy of drawing on a subgenomic mRNA to a new extreme.

With the exception of the largest coronavirus mRNA species, each of the coronavirus mRNAs is translated to generate a single primary protein product. (Regarding the huge ORF1, which encodes the coronavirus replicase complex, recall that it actually contains two ORFs that overlap and that translation of the entire sequence depends on a programmed ribosomal frameshift [see below].) The mechanism by which coronaviruses transcribe their multiple mRNAs is still poorly understood, but it is clearly unique, as will become clear. The overall scheme is as follows.

Coronavirus genomes, like the genomes of all plus-strand RNA viruses (excepting those of the retroviruses), are translated on cellular ribosomes upon being released into the cytoplasm. In that way, they can be translated into the RNA polymerase activities needed for transcription and replication. (Do you see why the coronavirus replication proteins *must* be translated from the most 5′ proximal ORF on the *genomic* RNA?)

Bearing in mind that the coronavirus genomic RNA, comprised of about 30,000 nucleotides, is among the largest known mature mRNAs, it is remarkable that the coronavirus polymerase/replicase is encoded by a nucleotide sequence that spans 70% of the genome. It is not known with certainty why the polymerase/replicase region should be comprised of such a long nucleotide sequence. However, it is probably, at least in part, because coronavirus replication and transcription involve a complex of virus-encoded enzymatic activities, which include an RNA-dependent RNA polymerase, a helicase, and protease activities, as well as a set of RNA-processing enzymes that are not needed by other RNA virus families. The viral replicase region encodes each of those activities (Box 9.4).

Another notable feature of the replicase region is that it actually is comprised of two overlapping ORFs. The 5′ end of the second ORF overlaps the 3′ end of the first ORF, and it is read in a −1 reading frame relative to the first ORF (Fig. 9.2). Consequently, the replicase region is translated into two large polyproteins from two overlapping open reading frames (ORF1a and ORF1b). ORF1 may be translated into a 495-kilodalton (kDa) polyprotein, or alternatively, ORF1a-ORF1b may be translated into an 803-kDa fusion polyprotein. The latter occurs as a result of a translational read-through by means of a −1 ribosomal frameshift. The resulting large polyproteins are cleaved co- and posttranslationally by three viral proteases (that are part of the replicase precursor polyprotein) to yield as many as 16 proteins.

Once the polymerase/replicase proteins have been produced, negative-sense copies of the genomic RNA can then be transcribed. These negative strands are used as templates for transcribing complete positive-sense copies of the genomic RNA. In addition, by mechanisms described below, the six to seven additional mRNAs are generated.

BOX 9.4

Coronavirus-encoded replicase activities include RNA-dependent RNA polymerase, RNA helicase, and protease activities, which are common to RNA viruses in general. However, the coronavirus replicase complex also includes a variety of RNA processing activities that generally are not used by other RNA viruses. These additional coronavirus activities may include sequence-specific endoribonuclease, 3′-to-5′ exoribonuclease, 2′-O-ribose methyltransferase, ADP ribose 1-phosphatase, and cyclic phosphodiesterase activities. Moreover, the replicase complex also contains an as-yet-undetermined number of cellular proteins. These proteins together express the set of enzymatic activities required for coronavirus transcription and replication.

Coronavirus Transcription

The seven to eight coronavirus mRNAs make up a nested set, in which each mRNA has the same 3′ terminus. Moreover, each of the coronavirus mRNAs has the same short 5′ leader sequence, of about 70 bases, which is identical to the 5′ end of the genomic RNA (Fig. 9.3). Any hypothesis to account for how the coronavirus mRNAs might be generated must account for these facts.

Perhaps the most plausible hypothesis to explain how coronaviruses generate their nested set of seven to eight mRNAs, which contain a common leader sequence and a common 3′ end, would be for them to make use of **alternative splicing**. In this regard, recall that the coding sequences of most eukaryotic mRNAs are interrupted by noncoding sequences called introns, which are spliced out of the pre-mRNA. Importantly, by means of alternative splicing, different mRNAs can be generated from the same pre-mRNA. Note that as many as 75% of eukaryotic pre-mRNAs may be alternatively spliced. Thus, alternative splicing is a well-established, straightforward means by which coronaviruses might generate their nested set of mRNAs.

Although alternative splicing might readily account for how the coronaviruses generate their set of nested mRNAs, the experimental evidence points against it. Instead, the evidence is consistent with either of two rather atypical mechanisms. But before we consider those mechanisms, note that if coronaviruses were to generate their nested set of mRNAs by means of alternative splicing, then they could not do so using cellular splicing enzymes since those enzymes are found exclusively in the nucleus. Thus, coronaviruses would have needed to encode their own splicing enzymes. At one time that prospect would have been attractive in helping to account for the large size of the coronavirus replicase region.

Key evidence *against* alternative splicing as the mechanism by which the nested set of coronavirus mRNAs is generated included the early experimental finding that the ultraviolet (UV) target size of each gene is proportional to its actual physical size. This meant that the production of each mRNA species did not require the transcription of upstream sequences, as would have been the case if these mRNAs were generated from a large pre-mRNA by alternative splicing. Target theory was a powerful experimental approach in the early days of molecular animal virology. Its theoretical basis and application to the issue of coronavirus transcription are explained in Box 9.5.

Since UV inactivation studies ruled out alternative splicing as the mechanism by which coronaviruses generate their nested set of mRNAs, other mechanisms needed to be considered. One of the leading mechanisms put forward is referred to as **leader-primed transcription during**

BOX 9.5

To understand the meaning of the key finding that the UV target size of each coronavirus gene is proportional to its physical size and how that finding eliminates the possibility that coronavirus mRNAs are generated by alternative splicing, one first needs to know how UV light affects nucleic acids. Additionally, it is helpful to have some insight into the mathematics of target theory. Fortunately, the latter is rather straightforward. First, note that UV target size determination is used only rarely today. Yet it is a powerful technique, and there are important references to it in the early virology literature, such as the case in point here. For those reasons, it is worthwhile to consider what the experimental approach entails and what it reveals.

In an actual experiment to determine the UV target size for the transcription of the individual coronavirus mRNA species, one can think of the polynucleotide template as a real physical target and of the UV light beam as being comprised of photon "bullets." Photon "hits" affect polynucleotides by causing adjacent pyrimidines, usually thymines in the case of DNA and uracils in the case of RNA, to covalently bond pairwise, giving rise to what are known as thymine or uracil dimers. The pyrimidine dimer is formed with a cis-syn cyclobutane linkage of the bases (see structures of thymine and thymine dimer, below).

The UV inactivation spectrum has a peak at 260 μm, showing that inactivating hits occur when photons of that energy directly strike the conjugated double bonds on pyrimidine rings. Thymine dimers can be extracted from UV-irradiated DNA, demonstrating that photon hits indeed induce adjacent pyrimidines to dimerize.

How might UV irradiation be used to determine the target size for the inactivation of expression of any particular viral gene? Importantly, transcription cannot proceed past a uracil dimer on the template RNA. Consequently, one means for obtaining experimental data is simply to infect cells with virus, irradiate with UV light for various periods of time, radiolabel with ^{32}P, incubate, and then extract and analyze the amount of each mRNA species by gel electrophoresis. The relative reduction in the amount of each mRNA species

as a function of the UV dosage is mathematically related to the size of the transcribed sequence. Specifically, at any UV dosage, the number of inactivating hits is directly proportional to the actual target size.

Presume for the moment that alternative splicing of a common pre-mRNA (i.e., one identical to the genomic RNA) generates the coronavirus subgenomic mRNA species. If that were the case, then the target size for inactivating transcription of each mRNA species would include the actual physical size of that mRNA, plus some or all of the sequence that is upstream from it on the template RNA. However, this was not the experimental result that was obtained. Instead, when the infected cells were irradiated late in infection, at 5 h, the target size for inactivating the transcription of each coronavirus poly(A)-containing mRNA species was directly proportional to the actual physical size of the particular mRNA (Yokomori et al., 1992). This is the expected experimental result if each mRNA species were transcribed independently of the gene sequences upstream of it. (If you are intrigued by this, try to account for the fact that early in infection, at 2.5 h, the majority of every mRNA species had a UV target size *equivalent* to that of genomic RNA! The likely explanation appears in the text.)

The mathematics of target theory is rather straightforward and requires no greater mathematical sophistication than college algebra and an understanding of limits. It is based on the Poisson distribution function, which is derived as a limiting condition of the familiar binomial probability distribution function, expressed in the following form:

$$P_n = N!p^n q^{N-n}/[(N-n)!n!]$$

In this expression, P_n is the probability of observing exactly n successes in N trials, when the probability of success in any one trial is p, and that of failure is $1 - p = q$. The Poisson distribution function is derived from the binomial probability distribution function by letting the number of trials N become very large and the probability of success in any one trial become very small. It has the form $P_n = x^n e^{-x}/n!$, where P_n is the probability of observing exactly n successes in N trials, when the average number of successes in N trials is x. It is applicable to the current analysis since the number of

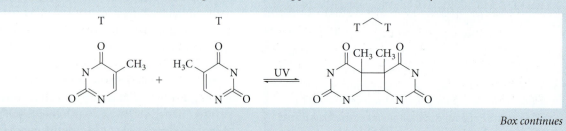

Box continues

BOX 9.5 (continued)

UV photons in the beam is very large and the probability of any one of them striking the RNA target to induce a uracil dimer is very small.

The P_0 term of the Poisson distribution function is especially useful since it represents the fraction of targets that have not been hit at any given dosage. Moreover, it is stated simply as $P_0 = e^{-x}$ (to obtain this expression, let $n = 0$ in the general equation). It also can be expressed in the form $lnP_0 = -x$. This equation says that the logarithm of the surviving fraction varies inversely and linearly with the average number of hits. Since the average number of hits x is directly proportional to the dosage, a plot of the logarithm of the surviving fraction (i.e., lnP_0) versus the dosage should yield a straight line. (In an actual experiment, the surviving fractions are represented by the actual amounts of each mRNA species produced at each UV dosage in the irradiated samples, relative to that in the nonirradiated samples.) Moreover, the slope of the line is directly proportional to the target size, since at any given UV dosage the number of hits on a target is directly proportional to its size. Thus, the fraction of survivors falls off more quickly for large targets than for small targets. Since the slopes of the inactivation curves directly reflect target sizes, relative target sizes can be derived from this approach. This fact provides the important advantage that the target size for inactivating transcription of each mRNA species can be related to the target size of the entire genome, where the latter is determined from the inactivation of infectivity. Since one can use other means to accurately determine the size of the genomic RNA in nucleotides, one can obtain very good estimates of the UV target size (in nucleotides) for the inactivation of transcription of each mRNA species. An elegant and instructive use of the target theory approach, as applied to the negative-strand vesicular stomatitis virus, can be found in an early paper written by Ball and White (1976).

The target theory approach also is valid when inactivating viruses by means other than UV irradiation, such as using formaldehyde. Importantly, from the expression $P_0 = e^{-x}$, one can see that the fraction of survivors (P_0) approaches but does not become zero as the dosage becomes very large. This is the point that Jonas Salk missed when preparing the early batches of his formaldehyde-inactivated poliovirus vaccine (Chapter 6). Based on the linear portion of his inactivation curves, he believed that he allowed for a sufficient margin of safety.

References

Ball, L. A., and C. N. White. 1976. Order of transcription of genes of vesicular stomatitis virus. *Proc. Natl. Acad. Sci. USA* 73:442–446.

Yokomori, K., L. R. Banner, and M. M. Lai. 1992. Coronavirus mRNA transcription: UV light transcriptional mapping studies suggest an early requirement for a genomic-length template. *J. Virol.* 66:4671–4678.

---

positive-strand synthesis (Fig. 9.4). This process begins with the transcription of full-length minus-strand intermediates. These minus-strand intermediates then are transcribed by a discontinuous process, in which the extreme 3′ end first is transcribed into the leader sequence. Then, the polymerase, with the bound leader sequence, "jumps" to one of the six or seven conserved transcription-regulating sequences (TRSs) that are complementary to the conserved intergenic sites present on the genomic RNA (Fig. 9.3) (see below). Transcription recommences at a different one of these sites to generate each subgenomic mRNA species. In addition to generating subgenomic mRNAs, the polymerase also copies the full-length negative-strand intermediates, which serve as templates for the transcription of full-length genomic RNAs.

Several important experimental findings support the leader-primed transcription model. First, free leader sequences were detected in MHV-infected cells, a finding now considered controversial by many in the field. Second,

an MHV temperature-sensitive mutant was isolated that transcribes leaders, but not mRNAs, at the restrictive temperature.

A second alternative mechanism is referred to as **discontinuous transcription during negative-strand synthesis** (Fig. 9.4). In this model, full-length plus-strand RNA molecules serve as templates for the discontinuous transcription of sub-genome-length minus strands. These sub-genome-length minus strands are produced when the polymerase stops at one of the conserved intergenic sequences on the genomic RNA. It then jumps with the bound nascent polynucleotide to the 5′ end of the positive-strand template to transcribe the leader (be aware that the template is always read in the 3′-to-5′ direction). The sub-genome-length minus strands then serve as the templates for transcription of subgenomic mRNAs.

Sadly for those who prefer definitive answers, several lines of experimental evidence also support discontinuous negative-strand synthesis. Importantly, infected cells

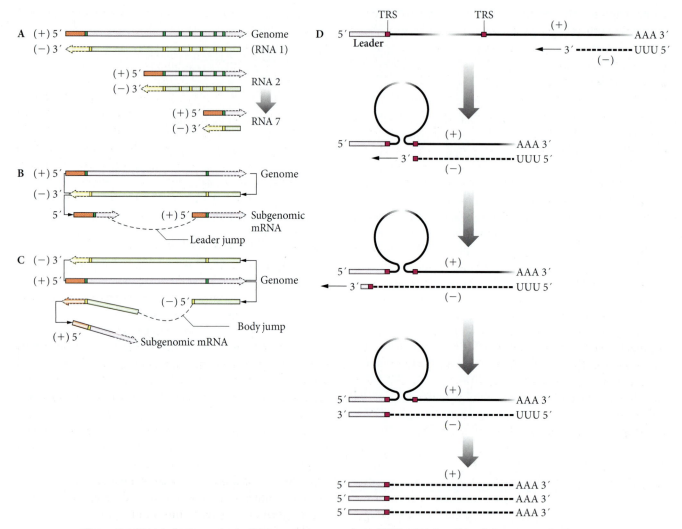

Figure 9.4 Models for the synthesis of the coronavirus nested set of RNAs. (A) Overview of plus-sense and minus-sense RNAs. (B) The process of leader-primed transcription during positive-strand synthesis begins with the transcription of full-length minus-strand intermediates. These minus strands then are copied by a discontinuous process, in which the extreme 3′ end first is transcribed into the leader sequence. Then, the polymerase, with the bound leader sequence, jumps to one of the six or seven conserved sites that are complementary to the conserved intergenic TRS sites present on the genomic RNA (Fig. 9.3). Transcription recommences at a different one of these sites to generate each subgenomic mRNA species. In addition to generating subgenomic mRNAs, the polymerase also copies the full-length negative-strand intermediates in order to produce full-length genomic RNAs. (C) The process of discontinuous transcription during negative-strand synthesis in which full-length plus-sense RNA molecules serve as templates for the discontinuous transcription of sub-genome-length minus strands. These sub-genome-length minus strands are produced when the polymerase stops at one of the conserved intergenic TRSs on the genomic RNA. It then jumps with the bound nascent polynucleotide to the end of the template to transcribe the leader. The sub-genome-length minus strands then serve as the templates for transcription of subgenomic mRNAs. (D) More detailed model for discontinuous transcription during negative-strand synthesis. Sub-genome-length minus strands are initiated at the 3′ end of the genomic RNA. Elongation continues up to a body copy of the TRS. A strand-switching event then occurs, pairing the newly transcribed negative-sense body TRS with the leader copy of the TRS, from which point transcription resumes. A complex of the plus-sense genomic RNA and the minus-sense sub-genome-length RNA then serves as the template for multiple plus-sense subgenomic mRNAs. Adapted from S. J. Flint, L. W. Enquist, V. A. Racaniello, and A. M. Skalka, *Principles of Virology: Molecular Biology, Pathogenesis, and Control of Animal Viruses,* 2nd ed. (ASM Press, Washington, DC, 2004), with permission, and P. S. Masters, p. 193–292, *in* K. Maramorosch and A. J. Shatkin (ed.), *Advances in Virus Research,* vol. 66 (Academic Press, San Diego, CA, 2006), with permission.

contain sub-genome-length minus strands that have poly(U) sequences at their 5′ ends and antileader sequences at their 3′ ends. Moreover, these sub-genome-length minus strands are transcriptionally active. That is, they serve as templates for the transcription of plus strands. In addition, infected cells contain equal amounts of these sub-genome-length minus strands and their corresponding subgenomic mRNAs. (Apropos these points, subgenomic mRNAs do not replicate when transfected into helper virus-infected cells. This experimental finding suggests that the sub-genome-length minus strands are not synthesized from subgenomic mRNA templates but instead are transcribed from the full-length viral genome, a conclusion consistent with the discontinuous negative-strand synthesis model.)

Actually, the leader-primed transcription model was put forward because of the initial failure of coronavirus researchers to detect sub-genome-length replication intermediates that contain plus strands and minus strands. These replication intermediates have now been detected in infected cells, and discontinuous negative-strand synthesis is currently considered the most likely transcription mechanism by most coronavirus researchers. However, since there is evidence in support of both the leader-primed model and the discontinuous negative-strand synthesis model, we should consider the possibility that *both* processes occur during coronavirus replication. In this regard, several experimental findings fit both models. For example, a highly conserved sequence, the TRS, which functions by some means in the transcription of each of the subgenomic mRNAs, precedes each ORF on the subgenomic mRNAs. This same conserved sequence also is found at the 3′ end of the genomic 5′ leader sequence. These facts are consistent with the premise that base pairing between the 3′ end of the leader sequence and one of the complementary regions on a full-length minus-sense strand plays an important role in coronavirus discontinuous transcription by the leader-primed mechanism (Fig. 9.4). They also are consistent with the premise that the intergenic sequence at the 3′ end of a nascent subgenomic minus strand may base pair with the complementary 3′ end of the leader sequence during discontinuous negative-strand synthesis (Fig. 9.4). In support of either possibility, in cells that are infected with two different strains of MHV, mRNAs that have the leader of one strain joined to the coding sequences of the other are seen. This experimental finding suggests that the polymerase, with its incomplete nascent RNA, may disassociate from its RNA template and reassociate with the complementary sequence on a different RNA template (see "Coronavirus recombination," next). The stability of the interaction between the leader and the complementary intergenic sequence was recently shown to be an important factor in the transcription of subgenomic mRNAs.

Now, we consider the pesky experimental finding noted above, that at early times in infection the majority of each mRNA species had a UV target size *equivalent* to that of the genomic RNA. A plausible explanation for this observation is that genome-length RNA molecules need to be made early so that they might serve as the template for the synthesis of subgenomic mRNAs by either (or both) of the two discontinuous transcription mechanisms described here. Recall that at late times, the UV target size for the inactivation of transcription of each mRNA species corresponded to its actual physical size.

Coronavirus Recombination

Coronavirus RNA genomes undergo a high frequency of recombination, as high as 25% for the entire coronavirus genome. This is noteworthy since the nonsegmented genomes of most other RNA viruses display levels of recombination ranging from low to undetectable.

The high frequency of coronavirus recombination may be a consequence of the unique means by which coronaviruses generate their RNAs, a process that involves discontinuous transcription and polymerase jumping. For instance, when the polymerase jumps, it might switch from its current template to a completely different template, from which it completes transcription, much as in copy choice models of DNA recombination. Coronavirus recombination is thought to occur primarily during minus-strand synthesis because plus-strand acceptor templates are in great excess over minus-strand templates.

The ability of coronaviruses to recombine at high frequency actually might play an important role in the survival of these viruses in nature. That is so because of the following two factors. First, coronaviruses have extraordinarily large RNA genomes. Second, RNA polymerases have a very high error frequency, due to their lack of a proofreading activity. Actually, the high error frequency of RNA polymerases places a theoretical upper limit on the size of RNA virus genomes at about 30 kilobases (kb), which, perhaps not coincidentally, is the size of a coronavirus genome. Larger genomes might not generate a sufficient number of viable progeny to ensure survival of the virus in nature. Coronaviruses might use their high frequency of genetic recombination to get around this threat to their viability.

The ability of coronaviruses to recombine at high frequency, together with their high mutation rate (which is a property of all RNA viruses), may also enable them to adapt to new hosts and ecological niches more readily than other RNA viruses. Recombination can also occur between different coronavirus strains, providing additional opportunities for these viruses to adapt to new niches (Box 9.6). For instance, consider an example involving the recently discovered human coronavirus HKU1, which shows that recombination indeed may give rise to new human coronavirus genotypes in nature. The story is as follows.

The discovery of SARS-CoV in 2003 prompted the search for new coronaviruses in both humans and animals, resulting in 2004 in the discovery of human coronavirus HCoV-NL63 and in 2005 of human coronavirus HCoV-HKU1 (Box 9.7). A 2006 study examined the complete genome sequences of 22 isolates of HCoV-HKU1, all of which were obtained from patients with respiratory tract infections. Analysis of these sequences showed that the 22 isolates could be sorted into three groups (genotype A, 13 isolates; genotype B, 3 isolates; and genotype C, 6 isolates). Interestingly, genotype C strains had *pol* gene sequences belonging to genotype A, whereas their S and N sequences belonged to genotype B. Multiple alignments of the nucleotide sequences of these strains confirmed that recombination between genotypes A and B gave rise to genotype C. (The site of recombination was localized to a sequence of 29 nucleotides in nonstructural protein [nsp] 16, just upstream of the stop codon of ORF1a.)

Coronavirus Reverse Genetics

Coronaviruses, like other RNA viruses, are not well suited for molecular genetic analysis, in part because it is more difficult to experimentally manipulate or alter RNA genomes than DNA genomes. With the advent of recombinant DNA technology in the 1970s, RNA virologists were able to reverse transcribe viral RNA genomes into complementary DNA copies (cDNAs), which then could be inserted into plasmids and propagated in bacterial hosts.

BOX 9.7

Prior to the SARS epidemic of 2003, a total of 19 coronaviruses were known. These included 2 human, 13 mammalian, and 4 avian coronaviruses. By 2006, 20 more coronaviruses had been discovered, including 3 human coronaviruses, 11 mammalian coronaviruses, and 6 avian coronaviruses. These new coronaviruses included SARS-like coronaviruses in Himalayan palm civets, in a raccoon dog from a live wild-animal market in mainland China, and in bats (see main text).

It is not unreasonable to suggest that the inherently high rate of coronavirus recombination among these diverse coronaviruses in nature might generate new coronavirus genotypes that can cross host species barriers, leading to **zoonotic outbreaks** (i.e., outbreaks transmitted from vertebrate animals to humans) with potentially devastating consequences. Influenza virus provides a precedent for such happenings, although the mechanism in the influenza virus instance is reassortment between the individual segments of its segmented genome (Chapter 12) rather than recombination per se, as in coronaviruses.

Apropos both influenza virus and coronaviruses, there is the intriguing speculation that human coronavirus-OC43 (HCoV-OC43) may be the result of a recombination between bovine coronavirus (BCV) and influenza C virus. In that way, HCoV-OC43 may have acquired the ability to bind to sialic acid, thereby enabling it to jump to human hosts (Zhang et al., 1992).

Reference
Zhang, X. M., K. G. Kousoulas, and J. Storz. 1992. The hemagglutinin/esterase gene of human coronavirus strain OC43: phylogenetic relationships to bovine and murine coronaviruses and influenza C virus. *Virology* **186**:318–323.

BOX 9.6

The terms "strain," "variant," and "mutant" are often used indiscriminately. All connote some dissimilarity from the canonical wild-type virus. Moreover, the term "wild type" is itself an arbitrary designation, which may refer to a laboratory-adapted form of the virus that does not accurately reflect the viruses actually circulating in nature. Bearing that proviso in mind, **strains** generally refer to different wild-type **isolates** of the same virus. Different isolates of the virus are categorized into strains based on shared nucleotide sequences. In contrast, **variants** display phenotypic differences from the canonical wild-type virus, but the genetic basis for the variation is not known.

In the bacterial host, the viral gene sequences could be transcribed into RNAs, which then could be transfected into eukaryotic host cells. Alternatively, the cDNAs might be transcribed in vitro and then transfected into the eukaryotic cells.

By generating the viral RNAs from cDNAs, it becomes possible to introduce any desired alteration into the cDNA, including deletions and point mutations, and then determine the effects of those mutations on function. This approach is referred to as **reverse genetics**, because the mutation is generated first and the effect on function is examined second. (In classical **forward genetics**, mutant organisms with a phenotype of interest are selected, and the mutated gene is then identified by standard molecular techniques. That is, forward genetics begins with a mutant phenotype, while reverse genetics begins with a gene sequence.) The most common method for introducing specific mutations is to first generate a synthetic oligonucleotide that bears the desired mutation and then use it as a primer for DNA synthesis.

Reverse genetics using cDNAs worked well for the 7.5-kb poliovirus genome but not for the approximately 30-kb coronavirus genomes. The large size of the coronavirus RNA genome was an obstacle, because (i) the low fidelity of the reverse transcription process precluded making an exact cDNA copy of so large an RNA molecule and (ii) the necessarily large bacterial plasmids, which contained a cDNA copy of the coronavirus genome, were unstable when propagated in *Escherichia coli*.

Before these difficulties could be overcome, a reverse genetics system for coronaviruses was developed, which made use of the high rate of coronavirus *RNA* recombination. In this experimental procedure, referred to as **targeted RNA recombination**, cells are infected with a coronavirus that has some characteristic that could be selected against, such as its tropism for cells of a particular host species. Then, a synthetic donor RNA, which bears the mutation of interest, as well as a spike gene that confers a different cell tropism, is introduced into those cells. One can then select for recombinants that might contain the marker of interest by selecting for viruses that contain the new cell tropism. As you might expect, this experimental approach worked best for mutations nearer to the 3′ end of the viral genome.

Although the targeted RNA recombination approach generated valuable findings, it was not as powerful an experimental approach as reverse genetics systems based on cDNA clones. But to use a cDNA-based approach, the two obstacles enumerated above first needed to be solved. The first obstacle, the low fidelity of reverse transcriptases, was surmounted by assembling full-length genomic cDNAs by ligating together smaller subcloned cDNAs. The second difficulty, the instability of bacterial plasmids containing a cDNA copy of the coronavirus genome, was overcome in a novel way by Almazán and colleagues in 2000. Their innovation was to transfect eukaryotic cells with a full-length cDNA copy of the coronavirus genome, which then was transcribed in situ by the eukaryotic RNA polymerase II. To accomplish this, the researchers placed the coronaviral cDNA under the control of the cytomegalovirus (a herpesvirus) immediate early promoter. The coronavirus used in this groundbreaking study was the transmissible gastroenteritis virus (TGEV), a major cause of enteric disease in pigs.

Almazán and colleagues also developed a novel two-step amplification system to generate the viral mRNA. In the first step, the transfected coronaviral cDNA genome was transcribed in the host cell nucleus (i.e., by RNA polymerase II, from the cytomegalovirus immediate early promoter). In the second step, the coronaviral RNA was amplified in the cytoplasm by the viral polymerase complex. In that way, the transfected, recombinant coronavirus cDNA generated a standard coronaviral mRNA pattern and indeed generated infectious coronavirus particles.

That nuclear transcription of the TGEV cDNAs actually gave rise to infectious virus was somewhat surprising since TGEV RNA contains multiple consensus splicing signals. These splicing signals are not recognized during natural infection because, under those circumstances, viral RNAs are exclusively in the cytoplasm while the cellular splicing enzymes are in the nucleus. Thus, it is not clear why the TGEV transcripts, generated in the nucleus from the cDNA, did not undergo significant splicing. (Bear in mind that splicing of primary RNA transcripts and export of the resulting mRNAs from the nucleus are generally coupled events.) Also, it is not clear why the viral genome contains these splicing signals at all. Splicing was not a concern when the cDNA transfection approach was used earlier in the analysis of influenza virus (Chapter 12) and hepatitis delta virus (Chapter 22), since these RNA viruses normally replicate in the nucleus.

What insights into coronavirus biology might have been learned from this unique reverse genetics approach? The coronaviral cDNA used in this initial study was derived from an attenuated TGEV isolate that replicated exclusively in the respiratory tract of pigs. However, the spike gene of this virus was experimentally replaced in the cDNA by the spike gene of an enteric isolate. The recombinant cDNA generated a fully virulent enteric virus. Thus, the pioneering initial experiments of Almazán and colleagues demonstrated that the spike protein alone is sufficient to determine the tropism and pathogenicity of TGEV, underscoring the power of the cDNA-based reverse genetics to coronavirus researchers.

Casais and coworkers (2001) reported a different clever cDNA-based reverse genetics strategy for coronaviruses. In this approach, a cDNA of the entire coronavirus genome (in this instance, avian infectious bronchitis virus) was generated by long-range reverse transcriptase PCR. This cDNA was then inserted into a unique restriction site in the vaccinia virus genome, in which it was much more stable than when in bacterial plasmids. Infectious RNA could be generated by in vitro transcription from the purified vaccinia virus DNA. Alternatively, it could be generated in vivo. To achieve transcription in vivo, the viral genomic cDNA was first ligated immediately downstream of a phage T7 RNA polymerase promoter, and then it was cloned directly into the vaccinia virus genome. Infectious coronaviral RNA was then generated after transfecting the recombinant vaccinia vector into primary chick kidney cells that were previously infected with a recombinant fowlpox virus that expressed the T7 RNA polymerase.

ASSEMBLY AND RELEASE

Coronaviruses, like most other plus-strand RNA viruses, carry out genomic RNA synthesis and mRNA synthesis on cytoplasmic membranes that originate from specific intracellular membrane compartments. By carrying out genome replication and transcription in association with membranous surfaces, viruses are potentially able to concentrate progeny viral genomes and viral translation products together, thereby enhancing the efficiency of subsequent viral assembly. Assembly and maturation at intracellular sites might provide yet another advantage for *enveloped* viruses (e.g., coronaviruses), since it may enable them to evade certain host immune defenses, which they otherwise might be vulnerable to if they were to undergo maturation at the plasma membrane.

Coronaviruses appear to induce the generation of the internal cellular membrane structures that support their replication complexes, since similar membranes are not seen in uninfected cells. The atypical coronavirus-induced membrane structures appear as double-membrane vesicles in thin-section electron micrographs (Fig. 9.5). In MHV infections, the viral replicase activities and the newly synthesized viral genomic and subgenomic RNAs are all detected in association with these internal double-membrane vesicles.

MHV likely generates these double-membrane vesicles from the endoplasmic reticulum and Golgi body, as suggested by immunofluorescence studies using organelle-specific markers. In the case of SARS-CoV, the double-membrane vesicles appear to originate from the endoplasmic reticulum.

The mechanism by which coronaviruses cause the double-membrane vesicles to form is not known with certainty, although there is evidence in the case of SARS-CoV that it might involve the effects of subunits of the viral replicase complex. These subunits are nonstructural proteins nsp3, nsp4, and nsp6. Each is encoded by ORF1a, and each contains a conspicuous hydrophobic domain. Nsp3 is known to express one of the viral protease activities. The activities of nsp4 and nsp6 are not known.

In the case of MHV, some experimental results showed that the virus-induced double-membrane vesicles contain the viral replicase subunits and also the viral integral membrane proteins, implying that these vesicles are where both genome replication and virus assembly take place.

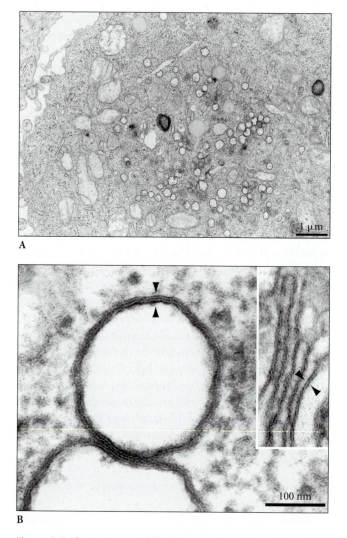

Figure 9.5 Electron micrographs showing membrane alterations in MHV-infected cells. (A) Multiple double-membrane vesicles in the cytoplasm of MHV-infected cells. The double-membrane vesicles were found either as separate entities or as small clusters of vesicles. (B) The vesicles range in size from 200 to 350 nm and consist of a double membrane. As seen here, the two bilayers often are fused into one tri-layer. Inset: For comparison, Golgi apparatus composed of a bilayer membrane. Arrowheads indicate the thickness of the membranes. The number of vesicles increases from 5 to 7 h postinfection, the time of peak viral RNA synthesis. Such vesicles were not found in mock-infected cells. From R. Gosert, A. Kanjanahaluethai, D. Egger, K. Bienz, and S. C. Baker, *J. Virol.* **76**:3697–3708, 2002, with permission.

However, other experimental results instead showed that while the MHV nonstructural replication proteins and replication complexes are localized at the double-membrane vesicles, structural proteins and virus assembly happen at the endoplasmic reticulum and Golgi body. In the case of SARS-CoV, too, replication complexes appeared to remain at the double-membrane vesicles, separate from the sites of virus assembly at the endoplasmic reticulum and Golgi body. So, for the moment at least, we presume that the three coronavirus envelope proteins, S, E, and M, and the nucleocapsid N protein, together with newly synthesized plus-strand virus genomes, congregate at endoplasmic reticulum and Golgi body membranes. The mechanism by which this sorting might be achieved is yet another part of the coronavirus replication story that remains to be clarified. Regardless, the broad outline of the coronaviral assembly pathway, as currently understood, is as follows.

The assembly process begins when multiple copies of the N protein bind to the newly synthesized viral genomes, thereby forming the helical nucleocapsids (Box 9.8). The C termini of the nucleocapsid-bound N proteins in turn bind to the carboxyl-terminal extensions of the M proteins that are exposed on the cytoplasmic surfaces of the endoplasmic reticulum and Golgi body membranes. In that way, the M protein mediates the condensation and packaging of the genome and budding. In addition, there is experimental evidence which suggests that the M protein might also specifically recognize the packaging signal on the genomic RNA (Box 9.8), thereby helping to selectively package genomic RNA into virus particles. Again, the means by which the S, E, and M proteins and the genomic RNA are targeted to the assembly sites is not yet known.

The E protein too is required for budding of the nucleocapsid into the membrane, and it is thought to perhaps alter membrane curvature as part of the budding process. In support of this notion, the E protein is required for the formation of coronavirus VLPs. Moreover, in promoting the formation of VLPs (and presumably virus particles as well), the E protein may interact directly with the M protein, as implied by the observation that coexpression of the E and M proteins from different coronavirus species does not support VLP formation. (Recall that the flavivirus E protein also was required to generate flavivirus VLPs. Moreover, lateral interactions between the flavivirus E and M proteins provide the driving force for the budding of those viruses, without the involvement of the nucleocapsid core [Chapter 7].) The S and HE proteins also interact in the plane of the membrane with the M protein, and they too associate with the budding virus particles.

The actual assembly and budding interactions occur at the cytoplasmic surfaces of the endoplasmic reticulum and the endoplasmic reticulum-Golgi intermediate compartment. The envelope proteins are glycosylated as the virus passes through the Golgi body at the same time as the virus matures morphologically. Viruses are then transported in membrane-bound vesicles to the plasma membrane, where fusion of the vesicles with the plasma membrane releases the viruses from the cell.

BOX 9.8

The molecular details of the assembly process for some helical viruses, such as the tobacco mosaic virus (TMV), are very well understood. In the TMV instance, copies of the single TMV coat protein repetitively interact with the RNA genome, while oligomerizing with one another. Doing so, they engage in identical interactions with the RNA genome (each coat protein molecule interacts with three nucleotides of the RNA genome) and with each other. The length of the helical capsid is determined by the length of the RNA genome.

In contrast to the well-understood assembly process of TMV, the molecular details of coronavirus helical packaging are not yet clear. To the point, neither the binding of the N protein to the RNA genome nor the oligomerization of N protein molecules is understood. But the following points are known. The N protein binds to the genomic RNA via both specific and nonspecific interactions. Little is known about the specific binding. In the case of MHV, the N protein binds specifically to a sequence, termed the packaging signal, located near the 3′ terminus of ORF1b in the viral genome, and this binding facilitates assembly of the genomic RNA into virus particles. Notice that the packaging signal is in a region of the genome that is not present in any of the subgenomic RNAs, perhaps accounting, at least in part, for why the genomic RNA is selectively encapsidated. The nonspecific binding is thought to involve the interaction of positively charged residues of N protein with the RNA. Structural studies of the N protein show that it contains an RNA binding domain near its N terminus and an oligomerization domain near its C terminus.

Note that the helical packaging of other important animal viruses, such as influenza virus, is also not well understood.

MEDICAL ASPECTS

Coronaviruses infect a variety of mammals and birds in nature. Yet most coronaviruses are species specific, their host range being limited by the availability of appropriate cellular receptors. Consequently, individual coronaviruses generally cause disease in only one particular host species (Box 9.9). The infection may affect various sites in the host, depending on the particular coronavirus-host interaction. Affected sites may include the upper respiratory tract, the gastrointestinal tract, and the liver and central nervous system, as in the examples of avian

The crossing of a virus from one host species to another does occur in nature, often with devastating consequences. The cross-species transmission of SARS-CoV from bats and civet cats to humans provides a prominent current example.

Interestingly, surveys of all known human pathogens (i.e., bacteria, fungi, and transmissible spongiform encephalopathies, as well as viruses) lead to the estimate that 58% are zoonotic in origin, with 73% of these pathogens considered to be emerging or reemerging. Furthermore, RNA viruses (e.g., Ebola virus, HIV, H5N1 influenza virus, H1N1 "swine flu" virus) are disproportionately represented among emerging pathogens, perhaps reflecting their much higher mutation rates. As discussed in the text of this chapter and in Chapters 12 and 21, the expansion of viruses into new hosts is not well understood. Although the ability of the jumping virus to use receptors in the new host is a necessary factor, other adaptations, so far poorly understood, are generally required as well.

Reference
Woolhouse, M. E., D. T. Haydon, and R. Antia. 2005. Emerging pathogens: the epidemiology and evolution of species jumps. *Trends Ecol. Evol.* **20**:238–244.

which is thought to have recently crossed from bats or civet cats to humans (see below), established human coronaviruses rarely cause lower respiratory tract disease in their human host.

Earlier (Chapter 6), we noted that there are more than 100 different serotypes of the human rhinoviruses circulating in the human population, accounting for the multiple rhinovirus common colds we contract during our lifetimes. In the case of the human coronaviruses, it was reported that there is no cross-immunity between the two old human coronavirus strains, HCoV-229E and HCoV-OC43. Moreover, new serologic variants of the human coronaviruses may continually arise by mutation. In addition, local immunoglobulin A immunity (Chapter 4) is relatively short-lived. Together, these facts may account for the multiple coronavirus colds we acquire throughout our lifetimes.

The recent SARS epidemic and the discovery of SARS-CoV in 2003 led to the search for other new coronaviruses in both humans and animals. As a result, two new human coronaviruses were discovered: HCoV-NL63 in 2004 and HCoV-HKU1 in 2005. As noted above, these so-called new human coronaviruses, unlike SARS-CoV, were not emerging viruses but instead had long been widespread in humans. Moreover, although infections with these viruses can be serious enough to require hospitalization, they generally do not differ greatly from those caused by the old coronaviruses, HCoV-229E and HCoV-OC43 (Box 9.10).

infectious bronchitis virus, TGEV of swine, and MHV, respectively.

Human coronaviruses are best known for the mild upper respiratory tract infections they cause, which are very similar to the common colds that are caused by the human rhinoviruses (Chapter 6). Data from serologic studies show that coronaviruses are responsible for 10 to 15% of all common colds, making them the second most frequent cause of the common cold after the rhinoviruses, which account for 30 to 40% of colds. (Other viruses associated with common colds include respiratory syncytial virus, parainfluenza virus, influenza virus, adenovirus, and enteroviruses.)

The minor differences between common colds caused by coronaviruses and those caused by rhinoviruses are as follows. Coronavirus colds are usually milder, with less-prominent sore throat and coughing. Also, coronavirus colds occur most often in winter and early spring, whereas rhinovirus infections peak in the early fall and late spring.

Although coronavirus common colds are generally mild, they can exacerbate preexisting chronic respiratory conditions such as asthma and bronchitis and on rare occasions may cause pneumonia. In contrast to SARS-CoV,

SARS

The recently identified SARS-CoV, which causes a severe life-threatening pneumonia in humans, is the most virulent "human" coronavirus yet known. The first signs of SARS are a high fever, sometimes associated with other symptoms such as chills, headache, and body aches. Between 10 and 20% of patients also experience diarrhea. Most patients go on to develop severe pneumonia, and as many as 40% require mechanical breathing assistance. The overall mortality rate for SARS is about 10%.

SARS affects relatively few children, but when it does, cases tend to be mild. In contrast, the mortality rate in the elderly can be as high as 50%. The SARS incubation period is typically 2 to 7 days, but incubation periods as long as 2 weeks have been reported. Patients appear to be contagious only when they have symptoms.

SARS-CoV may be transmitted by respiratory droplets from a cough or sneeze. However, SARS-CoV (possibly HCoV-HKU1 as well) is unique among known *human* coronaviruses in also being shed in feces, so that SARS may be spread by fecal contamination as well as by respiratory droplets. (In contrast to established human coronaviruses,

BOX 9.10

One research group isolated human coronavirus NL63 from an 8-month-old child suffering from pneumonia and subsequently detected the same virus in four additional children suffering from severe respiratory tract infections. A second research group isolated this virus from a 7-month-old child suffering from bronchiolitis and conjunctivitis.

Human coronavirus HKU1 was initially isolated from a pneumonia patient in Hong Kong. The research group that discovered HCoV-HKU1 subsequently detected it in 2.4% of patients (10 of 418) with community-acquired pneumonia. Nine of these patients were adults. For two of the patients the infection was fatal. In a subsequent study, 135 hospitalized patients were screened for HKU1, and 6 (5 children and 1 adult) were found to be positive. Three of these HKU1-positive individuals were admitted to the hospital for acute enteric disease. One was admitted for fever, otitis (inflammation of the ear, caused by infection), and febrile seizure. One was tested because of a failure to thrive, and one had a sample obtained as part of the exploration of X-linked agammaglobulinemia and hyperleukocytosis. These findings raise the possibility that human coronavirus HKU1 may be associated with both respiratory and enteric diseases, and possibly other conditions as well. Coronavirus particles have been seen in electron micrographs of stool samples from patients with diarrhea and gastroenteritis. Nevertheless, the connection between coronavirus infections and these symptoms in humans is not yet well-established.

many *animal* coronaviruses may be spread by fecal contamination as well as by the respiratory route.)

Some SARS-infected individuals appear to be SARS "superspreaders." One hypothesis to account for these superspreaders is that they are biologically contagious patients who become more efficient transmitters merely by going to places where conditions are particularly ripe for transmission, such as hospitals. Indeed, some health care professionals have advised against treating suspected SARS cases in the regular hospital setting, hoping to avoid the widespread transmission that occurs in hospitals.

Before the emergence of SARS, human coronaviruses were associated with relatively mild upper respiratory tract infections. Why then is SARS-CoV able to infect the lower respiratory tract in humans, with the potential of giving rise there to a life-threatening illness? Moreover, why is SARS unique among coronavirus infections in also being associated with gastrointestinal symptoms? The answers to these questions are not entirely known, but the following points do provide important insights.

The host range, tissue tropism, and virulence of animal coronaviruses are largely determined by the receptor-binding specificity of their S gene product. Thus, we might expect that the receptor-binding specificity of SARS-CoV largely accounts for the exceptional virulence of that virus in humans. Apropos that point, we noted above that SARS-CoV uses ACE2 for its receptor, a molecule that indeed is expressed in the human airway epithelia, the lung parenchyma, and on surface enterocytes of the small intestine. Moreover, ACE2 knockout mice are resistant to challenge with SARS-CoV, thereby providing experimental proof that ACE2 indeed plays a critical role in SARS-CoV infection in vivo.

Still, the use by SARS-CoV of ACE2 for its receptor cannot fully account for the extraordinary virulence of this virus, since HCoV-NL63, which likewise uses ACE2 for its receptor, gives rise to infections that are generally much milder than SARS-CoV infections. (NL63 infections appear to be common in childhood, as indicated by serology. However, in a small percentage of cases, especially involving young children, the elderly, and immunocompromised individuals, the virus can cause severe lower respiratory tract infections. HCoV-NL63 infection has also been associated with croup, an inflammatory condition of the larynx and trachea, especially in young children, that is manifest by a cough, hoarseness, and difficulty in breathing.)

Another important unanswered question is why some patients spontaneously recover from SARS while others develop severe life-threatening pneumonia. One possible explanation which might account in part for these individual differences is suggested by evidence that SARS-associated damage to the lungs is caused by immuno-pathologic mechanisms, rather than by virus cytopathic effects per se. The infiltration of macrophages and lymphocytes to the infected area and their release there of proinflammatory cytokines are regular features of the host response to infection (Chapter 4). Moreover, there are numerous precedents for such potentially destructive immune activity going awry, thereby leading to severe tissue damage. Apropos this premise, a recent study, carried out in a rat model, showed more severe acute lung injury in adult rats than in young rats, an age dependency of pathogenesis that likewise is seen in human infections. Moreover, higher levels of proinflammatory cytokines were seen after infection in adult rats than in young rats. Thus, it is plausible that host immune factors, which vary from one SARS patient to another, might account at least in part for differences in immunopathology.

Despite the gaps in our understanding of the SARS disease process, bear in mind that SARS is a very new disease and that SARS researchers have made enormous progress in a very short time. In particular, the etiologic agent was isolated and identified, diagnostic reagents were developed, and containment procedures were established. Achieving containment was a particularly impressive accomplishment, especially considering that SARS-CoV was initially able to spread internationally with striking ease and with no particular geographic predilection (see below). Yet containment was achieved, in no small measure due to the fact that the international response to the SARS outbreak was the most effective response to a major epidemic in history (see below).

The first known reference to SARS dates to February 2003, when the Chinese Ministry of Health announced the mysterious outbreak of an unexplained pneumonia in the Guangdong Province of southern China. (Indeed, the 2003 SARS epidemic is the first reported outbreak of a previously unknown emerging pathogen in the 21st century.) Chinese authorities reported a total of 305 cases of this atypical pneumonia, which included five deaths during the preceding 3 months. Yet the Chinese report of 305 cases was almost certainly an underestimate, as discussed below. The story of the 2003 SARS outbreak is as follows.

A Chinese doctor who had been treating SARS patients in Guangdong probably brought the disease to Hong Kong during that same February. He developed symptoms the first day of his stay in Hong Kong, was transferred to a hospital the following day, and succumbed the day after. However, during his brief stay at the Hotel Metropole in Hong Kong, he somehow infected at least 10 other guests. Eight of those infected guests were on the same floor as the doctor, and the two others were two and five flights up from the doctor. Those infected individuals subsequently boarded airplanes taking them to Singapore, Vietnam, Canada, and the United States. The epidemic that they spread lasted more than 100 days. The World Health Organization (WHO) estimated that there were 8,422 SARS cases during that period, resulting in 908 deaths. In all, 29 countries were affected. In the United States there were a total of eight confirmed SARS cases, none of which was fatal. Each of the infected individuals in the United States had shortly before traveled to an area where SARS transmission was occurring. Thus, the SARS outbreak of early 2003 was truly an epidemic of our modern global era, spread by air travel to at least three continents in a period of just a few weeks (Box 9.11).

The Vietnamese government responded to the SARS outbreak in their country by immediately quarantining the hospital in which the first infected individual was seen, while establishing infection control procedures at other hospitals (Box 9.11). Furthermore, it took the unexpected step on its part of issuing an international appeal for expert assistance, which it accepted from the U.S. Centers for Disease Control (CDC), the WHO, and Médecins sans Frontières (Doctors without Borders).

It is most interesting and important to compare the initial Chinese response to the 2003 SARS epidemic to that of the Vietnamese. The Chinese at first tried to cover up their SARS outbreak and then misrepresented the number of their cases. Indeed, there are reports that they had their SARS patients driven around in taxis to avoid being detected by WHO officials who came to visit their hospitals. News of the Chinese epidemic emerged in the outside world largely because scientists working in neighboring countries detected what the Chinese authorities knew about but tried to conceal.

By dealing with its outbreak openly and decisively, Vietnam was able to quickly contain its SARS outbreak. Had China acted more responsibly, many lives might have been saved worldwide. In addition, the quarantines that were instituted globally, the disruption of international travel, and the economic consequences that followed might have been much less extreme. International pressure finally caused China to take tough action against SARS in April 2003.

On 5 July 2003, the WHO announced that the cycle of human-to-human transmission of the initial SARS outbreak was broken and that the epidemic had come to an end. The fact that containment of the 2003 outbreak was achieved within 4 months of the first global alert is a tribute to effective public health policy and the united effort of the international community. Indeed, the global response to the SARS outbreak was the most effective response to a major epidemic in history. In further tribute to the power of good public health policy, note that containment was achieved without the benefit of a vaccine or critical diagnostic reagents. The importance of political commitment within a particular nation is underscored by comparing the responses of China and Vietnam to their respective SARS outbreaks. (Likewise, it is instructive to compare the responses of the governments of South Africa and Uganda to the acquired immunodeficiency syndrome [AIDS] epidemics in those nations [Chapter 21].)

Despite the eventual success of the world community in dealing with the 2003 SARS outbreak, the threat of new SARS outbreaks remained. In April 2004, the Chinese Ministry of Health reported several new SARS cases. And in contrast to its actions of the preceding year, this time China responded aggressively, quickly isolating patients who developed SARS, identifying their nearly 1,000 recent contacts, and sharing information with outside groups such as the WHO. (For an update on the current status of SARS, check the CDC SARS website.)

BOX 9.11

More than one-half of the first 60 SARS patients were health care workers. Indeed, with the exception of the patient who brought SARS to Vietnam, all of the patients who died in the subsequent Vietnamese outbreak were doctors and nurses. Of the many health care workers who went to heroic lengths in the fight against SARS, perhaps one, Dr. Carlo Urbani, can be singled out for special mention. His story is as follows.

 On 28 February 2003, the Vietnam French Hospital of Hanoi contacted the Hanoi office of the WHO concerning a patient, an American businessman, who seemed to be infected with what the Vietnamese doctors thought was an unusually bad case of the flu. Fearing the possibility of a potentially deadly avian influenza outbreak (see Chapter 12), hospital officials called on the WHO for help. Dr. Carlo Urbani, a specialist in infectious diseases, answered the call and quickly determined that the hospital was not facing influenza, but instead, a previously unknown disease. Thus, Urbani was the first to recognize SARS as a new, life-threatening, and possibly highly contagious disease entity. He immediately notified the WHO of his findings, setting in motion the most effective response to a major epidemic in history.

Without regard for his own safety, Urbani spent the next several days continually at the hospital, where he prepared patient specimens to be sent for analysis, and helped organize infection control procedures. One of the key procedures Urbani instituted was to establish an isolation ward that was kept under guard. As it became clear that the infection was highly contagious and deadly, Dr. Urbani worked closely with the hospital staff to maintain morale. Many on the hospital staff, like Urbani, decided not to leave the hospital, so as not to place their families or the community at risk. In doing so, they knowingly placed themselves at great personal risk.

Appreciating the danger posed by the new disease, Urbani undertook the difficult initiative of arranging a meeting between WHO officials and the Vietnamese Vice Minister of Health. Urbani was able to bring these parties together largely because of the strong trust he had built with Vietnamese authorities. This meeting resulted in the Vietnamese government taking extremely important measures to contain the outbreak in their country. First, they quarantined the Vietnam French Hospital of Hanoi, and second, they established infection control procedures at other hospitals. Third, the Vietnamese government took the extraordinary step on its part of issuing an international appeal for expert assistance. Specialists who answered the appeal came from the WHO, the CDC of the United States, and Médecins sans Frontières (Doctors without Borders). Urbani himself was president of the Italian chapter of Médecins Sans Frontières and was one of the individuals who accepted the 1999 Nobel Peace Prize on behalf of that courageous organization.

Sadly, Urbani began to develop symptoms of SARS during a March 11 flight to Bangkok to attend a conference there. He succumbed in a Bangkok hospital on 29 March 2003, not knowing that within weeks researchers working worldwide would isolate the SARS agent, sequence its genome, and identify it as a newly discovered coronavirus (see the text).

The SARS agent was isolated and identified within weeks of the initial 2003 outbreak. This achievement, in so short a time, was remarkable, especially considering the means by which it was accomplished. It happened as follows. The symptoms of SARS did not suggest any one particular cause. For that reason, and because of the severity of the disease, researchers went to extreme lengths, testing patient specimens for a broad range of viral, bacterial, chlamydial, and rickettsial agents that were known to target the lower respiratory tract. These agents included, among others, yersinia, mycoplasma, chlamydia, legionella, *Coxiella burnetii*, influenza viruses, paramyxoviruses, herpesviruses, and picornaviruses. None of these respiratory pathogens were consistently identified in the clinical specimens. However, a novel coronavirus was isolated from several SARS patients (Ksiazek et al., 2003). Isolation was made possible by the fortuitous fact that the virus replicated and caused cytopathic effects when inoculated into monkey kidney cell cultures. This was an unexpected stroke of good fortune, since the porcine epidemic diarrhea virus (isolated in China from pigs) was the only coronavirus (human or animal) that previously was shown to grow in such cells.

It is interesting that in the modern genomics era, SARS-CoV was identified using classic virology procedures. As noted above, a key step was to use standard cell culture procedures to isolate and amplify the pathogen from patient specimens. Then, using electron microscopy, a virus was detected in infected cell cultures that displayed

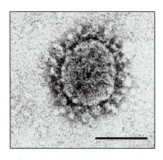

Figure 9.6 Electron micrograph of SARS-CoV grown in monkey kidney cell cultures, from the journal article reporting the identification of the SARS coronavirus. Note the internal helical nucleocapsid-like structure and club-shaped surface projections surrounding the periphery of the particle, typical of coronaviruses. Bar, 100 nm. From T. J. Ksiazek et al., *N. Engl. J. Med.* **348:**1953–1966, with permission.

characteristic coronavirus structural features (Fig. 9.6). Apropos the last step, it should be noted that for the purpose of pathogen identification, electron microscopy has the unique advantages of not requiring any specific reagents (e.g., specific antibodies or PCR primers) or prior knowledge of the pathogen. Subsequent molecular biological and immunologic studies confirmed that the isolated agent is indeed a novel coronavirus. Next, the SARS-CoV isolates that had been grown in cell cultures were found to cause lower respiratory tract disease when inoculated into monkeys. Together, these findings fulfilled Koch's postulates, as modified for highly virulent viral pathogens.

Although SARS-CoV was identified using classic virological procedures, the SARS epidemic was the first infectious disease outbreak in which virus researchers took full advantage of the powerful new techniques of the genomics era to analyze a new pathogen. Using these techniques, researchers determined the SARS-CoV genome sequence less than 1 month after the virus was first isolated and identified as the cause of SARS. Within the next 3 months, genome sequences of 20 independent clinical isolates of SARS-CoV were available in the GenBank database for comparison. Interestingly, the small differences between the genetic signatures of these viral isolates enabled them to be sorted into distinct groups that corresponded to contact source history and geography.

Several studies sought to trace the evolution of SARS-CoV during the 2003 epidemic. First, based on molecular epidemiological data, the epidemic was divided into three phases, designated early, middle, and late phases. The early phase refers to the first set of independent cases that were likely of zoonotic origin. The middle phase was characterized by extensive local transmission of the virus, while the late phase was the period of transmission of the infection to over 30 countries worldwide (Fig. 9.7). In one study, researchers found that only two major SARS genotypes

predominated during the early phase of the epidemic. Five of the early-phase isolates they examined contained a 29-nucleotide sequence that was missing from later, more virulent isolates. Importantly, those 29 nucleotides were present in coronavirus isolates from animals in a Shenzhen live-animal market. Four other early-phase isolates were characterized by a previously unreported 82-nucleotide deletion in the same region of the genome (*Orf8*). Moreover, an identical 82-nucleotide deletion was observed in coronaviruses isolated from farmed civets in Hubei Province, China. These findings provided support for the notion that early human infection by SARS-CoV had a zoonotic origin and that the virus evolved to become more virulent after its introduction into the human or intermediate host (see below). Since isolates that contained the 82-nucleotide deletion were viable, perhaps this region does not encode a protein but instead has a regulatory function.

As noted above, a second SARS outbreak, milder than the first, occurred in 2004 in the city of Guangzhou, China. Each of the four confirmed SARS patients of that episode had no obvious human-to-human contact history related to SARS, implying that their infections too most likely had a zoonotic origin. One study examined the nucleotide sequences of viruses obtained from these human patients, as well as from palm civets collected at the same period in the same region. Seventeen single-nucleotide variations were seen in the animals, 10 of which were located in the *S* gene. Of the single-nucleotide changes seen in the *S* gene of the animal isolates, seven also were seen in isolates from human patients of the 2004 outbreak. Moreover, one of these single nucleotide changes also was seen in a human isolate from the 2003 epidemic, and two of the single nucleotide changes were seen in isolates from both the 2003 and 2004 outbreaks.

Comparison of the SARS-CoV isolates from the 2003 and 2004 outbreaks reveals that the 2004 outbreak did not derive from the 2003 epidemic. Instead, each outbreak was an independent occurrence. Yet it is likely that in each instance the virus entered the human population from a common viral ancestor, which did not have the 29-nucleotide deletion characteristic of the most virulent forms of the 2003 isolates.

Were palm civets the source for the zoonotic transmission of SARS-CoV to humans? This was once considered a likely possibility. However, the answer is not known with certainty. On the one hand, phylogenetic analyses, together with some epidemiological data, support this prospect. On the other hand, recent studies show that SARS-CoV has only a limited distribution in palm civets. Moreover, SARS-CoV is capable of causing serious disease in these animals, also implying that they are an unlikely natural reservoir.

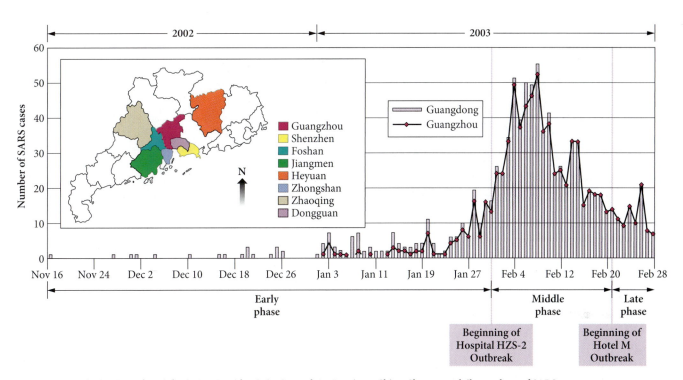

Figure 9.7 The triphasic SARS epidemic in Guangdong Province, China. Shown are daily numbers of SARS cases reported in Guangdong Province, in particular the city of Guangzhou. The early, middle, and late phases of the epidemic are defined in the main text. The map shows the geographical distribution of cases belonging to the early phase by administrative districts of Guangdong Province. Adapted from the Chinese SARS Molecular Epidemiology Consortium, *Science* **303**:1666–1669, 2004, with permission.

Many SARS researchers now believe that Chinese horseshoe bats may be the natural host for SARS-CoV. Bats are the known reservoir host for several other zoonotic viruses, including the recently emergent Hendra and Nipah viruses (Chapter 11), and bats and bat products are present in food and traditional medicine markets in southern China. Moreover, bats are a known natural host for coronaviruses that are closely related to those responsible for the SARS outbreak. Importantly, these bat SARS-like coronaviruses display greater genetic variability than the SARS-CoVs isolated from humans or from palm civets, in support of the premise that the virus was resident in bats before passing to humans and palm civets. Moreover, the genetic diversity of the bat SARS-like coronaviruses would be expected to increase the likelihood of variants that could cross the species barrier and cause outbreaks of disease in humans.

Still, the possibility remains that palm civets may have acted as intermediate hosts of the SARS-CoV in its transmission from bats to humans, perhaps after having been infected while in the live-animal markets. So, in summary, it is now believed that SARS-CoV succeeded in jumping from a zoonotic reservoir (probably bats) to the human population via one or more intermediate hosts (possibly including palm civets) and that rapid virus evolution played a key role in the adaptation of SARS-CoVs in the intermediate and human hosts.

Next, we ask which mutations are mainly responsible for enabling the cross-species transfer and adaptation of SARS-CoV to its human host. As noted above, the SARS-CoV S glycoprotein binds to the receptor ACE2. Sequencing analysis of hundreds of SARS-CoV genomes from humans and animals, during and after the 2003 epidemic, showed that the S protein underwent a high rate of evolution, which appeared to be critical for the transition from animal-to-human to human-to-human transmission. Then, in 2005, the structure of the receptor-binding domain of the SARS-CoV spike protein, in association with its ACE2 receptor, was solved (Li et al., 2005). The structure revealed that the spike protein is bound to the receptor via a loop that projects from a compact core within the receptor-binding domain. Fourteen residues on the loop contact 18 residues on human ACE2. Of these 14 residues on the loop, only 2 differ between the human and animal virus strains, but they cause a >1,000-fold difference in binding affinity to human ACE2 (Fig. 9.8). (Proviso: Adapting to

A possible attenuation mutation: L472P   From civet to human: K479N   From human to human: S487T

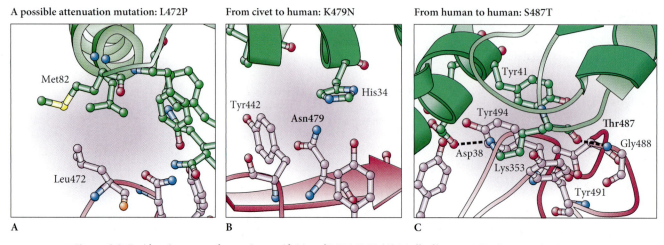

Figure 9.8 Residues important for species specificities of SARS-CoV. (A) Met[82] of human ACE2 is asparagine in rat ACE2, introducing a glycan that appears to interfere with infection of rat cells. (B) Asn[479] is present in most SARS-CoV sequences from human specimens. Lys[479], which is found in most sequences from palm civet specimens, would have steric and electrostatic interference from residues (e.g., His[34]) on the N-terminal helix of human ACE2. (C) Thr[487] appears to enhance human-to-human transmission of SARS-CoV. The methyl group of Thr[487] lies in a hydrophobic pocket at the ACE2/RBD interface. On rat and mouse ACE2, residue 353 is histidine, disfavoring viral binding. The dashed black lines indicate hydrogen bonds. Adapted from F. Li et al., *Science* **309**:1864–1868, 2005, with permission.

use the human ACE-2 protein for its receptor may not be the entire explanation for establishing the SARS-CoV host range, since there is evidence that ORF1 mutations also may have played a role in the expanding epidemic [see below].)

It is rather disturbing that a change in only two amino acids may be sufficient to enable an animal virus to cross a species barrier so that it can infect humans and do so with devastating consequences (Box 9.12). Yet this prospect applies to yet other potential emerging viruses. A pointed example was seen earlier in the case of the 1918 influenza pandemic, which may have resulted in as many as 50 million fatalities worldwide (Chapter 12). The 1918 influenza pandemic virus is believed to have crossed the species barrier from birds to humans as a consequence of a change of only one or two amino acids in the attachment protein (HA) of the virus. We might then ask why viruses do not constantly jump from one species to another. The answer is complex, in part because cross-species jumps require the opportunity as well as the capacity to infect the new host species. Human activities that may provide the opportunity for cross-species transfer include choosing to eat "exotic" animals and invasion of animal habitats. Still, at the molecular level, changes in receptor specificity alone may not be sufficient to enable a virus to thrive in its new host. Recall that whereas the 2003 SARS-CoV was particularly virulent, it was not efficiently

transmitted from one person to another, a factor that made early containment possible.

In view of the above, SARS-CoV, like West Nile virus (Chapter 7), should be considered to be an emerging virus only in the sense that it seems to have only recently found its way into the human population. In point of fact, it is highly unlikely that any emerging virus would have arisen as a "brand-new" virus. The notion that SARS-CoV only recently entered the human population is supported by retrospective serological surveys that did not detect any antibodies against SARS-CoV in the human population prior to the 2003 and 2004 outbreaks. Moreover, nucleotide sequence analyses show that SARS-CoV could not have arisen as a variant of any known human coronavirus.

What might we predict regarding future SARS outbreaks? The highly virulent SARS-CoV of the 2003 outbreak has not been seen either in humans or in animals since that outbreak was brought under control. However, milder forms of SARS-CoV have been detected in palm civets, and these may circulate in other animals as well, almost certainly including the natural reservoir host. Moreover, there have been sporadic cases of mild human infections. But bear in mind that despite extensive ongoing surveillance of SARS-CoV, one cannot predict when its more virulent form might reemerge to initiate another SARS epidemic in humans.

BOX 9.12

"Jumps" of an infectious viral agent from an animal host species to humans have been responsible for several of the most devastating epidemics in human history. HIV/AIDS is a relatively recent example, and the 1918 influenza pandemic, which killed as many as 50 million people worldwide, is another. Other major killer diseases, such as smallpox and measles, almost certainly emerged in humans as a result of the etiologic viral agent jumping from an animal host species to humans.

A critical determinant of whether an infectious agent will be able to spread through a new host population is its **reproduction number** (R_0), which is the average number of new individuals who are infected by each infected individual. An emerging pathogen can successfully establish itself in a new host only when its R_0 is greater than 1 in that host. If that should be the case, then there is a finite chance that a major epidemic will occur. If R_0 is less than 1, then outbreaks in the new host will be naturally self-limiting, even if the pathogen infects the new host on multiple independent occasions.

Ebola virus, which is exceptionally virulent in humans, is an example of an emerging virus that did not "take off" in the human population because its R_0 in humans is less than 1. Nevertheless, it is crucially important to take into account the possibility that the R_0 of an emerging pathogen can suddenly change. When the R_0 is close to 1 in the new host, even slight changes in its value can have potentially devastating consequences. Also note that changes in the host environment (e.g., population density) and behavior (e.g., sexual practices) can change R_0.

Of course, the ability of the pathogen to adapt to the new host is also an important determinant of whether it will be able to jump to, and establish itself in, the new host population. Apropos this point, although the few examples in this sidebar do not constitute an exhaustive list, nevertheless, all are RNA viruses, and it is true that RNA viruses are disproportionately represented among emerging viral pathogens. Also see Box 9.9.

Some public health experts have expressed optimism that a new outbreak of virulent SARS could quickly be brought under control by the same measures used to contain the first epidemic. Nevertheless, despite the dramatic success of the efforts to contain the first SARS outbreak, it would be extremely imprudent to take for granted that future outbreaks will be met with a similar level of success. One providential aspect of the 2003 SARS outbreak that helped make containment possible was that it occurred in places with well-developed health care systems. Moreover, virus transmission occurred along major routes of air travel, also enabling effective containment measures to be established. (Note that the WHO issued warnings against travelling to areas where risk of exposure to SARS was considered high [e.g., Toronto, Canada]. That in itself was an important incentive for governments to bring the infection in their nations under control.) The course of events would not have gone so well had the outbreak occurred in developing countries that do not have the capacity to deal with public health crises of that magnitude.

The relatively low transmissibility of SARS-CoV from human to human, in comparison to the transmissibility of, say, the 1918 pandemic influenza virus, was yet another factor that facilitated containment. But, it is possible that a future SARS outbreak virus might acquire the ability to efficiently spread through the human population,

dramatically changing the prospects for containing it (Box 9.12).

The fact that there is no way to know when, or if, SARS will reemerge in the human population is reason enough for us to learn more about the natural biology of SARS-CoV. Particularly important research goals include identifying with certainty its natural reservoir host and other possible intermediate host species. Since SARS is a respiratory infection, we might also ask whether SARS, like other virus-induced respiratory diseases, is seasonal. (If it were, and if its seasonality were to coincide with that of influenza, then how might this affect public health efforts?) Vaccines against SARS-CoV and new antiviral drugs are currently being developed. Still, as of this writing, there is no vaccine or antiviral drug effective against SARS-CoV. Thus, if SARS were to return, at present we still would need to rely on quarantine measures to contain the disease.

Here are a couple of additional thoughts to ponder regarding dangerous emerging viruses. First, in contrast to the approximately 900 individuals who were killed worldwide by the SARS outbreak of 2003, influenza virus regularly kills about 40,000 individuals yearly in the United States alone, even with vaccines at hand. Why then does the public generally become much more alarmed over emerging viruses such as the SARS coronavirus? Is such concern justified? Second, and in the same vein, how should our concern for major *established* human pathogens, such as influenza virus, hepatitis B virus, and now human immunodeficiency virus (HIV), measure up against our concern for *potential* bioterror agents? Bear in mind that one set of these threats is ongoing and real while the other is, for the present at least, speculative. Should that mean that the threat of bioterror agents is any less compelling?

Suggested Readings

Almazán, F., J. M. González, Z. Pénzes, A. Izeta, E. Calvo, J. Plana-Durán, and L. Enjuanes. 2000. Engineering the largest RNA virus genome as an infectious bacterial artificial chromosome. *Proc. Natl. Acad. Sci. USA* **97:**5516–5521.

Casais, R., V. Thiel, S. G. Siddell, D. Cavanagh, and P. Britton. 2001. Reverse genetics system for the avian coronavirus infectious bronchitis virus. *J. Virol.* **75:**12359–12369.

Ksiazek, T. G., D. Erdman, C. S. Goldsmith, S. R. Zaki, T. Peret, S. Emery, S. Tong, C. Urbani, J. A. Comer, W. Lim, P. E. Rollin, S. F. Dowell, A. E. Ling, C. D. Humphrey, W. J. Shieh, J. Guarner, C. D. Paddock, P. Rota, B. Fields, J. DeRisi, J. Y. Yang, N. Cox, J. M. Hughes, J. W. LeDuc, W. J. Bellini, L. J. Anderson, and the SARS Working Group. 2003. A novel coronavirus associated with Severe Acute Respiratory Syndrome. *N. Engl. J. Med.* **348:**1953–1966.

Li, F., W. Li, M. Farzan, and S. C. Harrison. 2005. Structure of SARS coronavirus spike receptor-binding domain complexed with receptor. *Science* **309:**1864–1868.

Rhabdoviruses

INTRODUCTION

The *Rhabdoviridae* contain minus-strand RNA genomes. That is, their single-stranded RNA genomes are complementary in base sequence to the messenger RNAs (mRNAs) that they generate. As such, rhabdoviruses are class V viruses in the Baltimore classification scheme (Chapter 2) and are the first minus-strand viruses that we consider in detail. The family derives its name from the Greek word *rhabdos*, meaning "rod," referring to the characteristic bullet shape of these viruses (Fig. 10.1).

Rhabdoviruses may be the most widely distributed virus family in nature. Indeed, there are more than 150 known rhabdoviruses, and these infect a range of vertebrate, invertebrate, and plant hosts. Two genera of the *Rhabdoviridae* infect vertebrates: the genus *Vesiculovirus* (named for vesicular stomatitis virus [VSV], the prototype rhabdovirus), and the genus *Lyssavirus*, which includes rabies virus and the rabies-like viruses.

Considering the number of rhabdoviruses in nature, it is perhaps remarkable that only one, rabies virus, is an important pathogen in humans. Moreover, in a sense, rabies virus is the deadliest virus known to man, since it kills essentially 100% of untreated, infected individuals. Rabies is also one of the oldest known diseases, having been recognized in Egypt before 2300 BCE, and in ancient Greece, where it was chronicled by Aristotle. Louis Pasteur developed a rabies virus vaccine in 1885, nearly two decades before viruses were recognized as unique microbial entities that are distinct from bacteria (Chapter 1). Before the rabies vaccine was developed, a bite from a rabid dog was invariably fatal.

VSV is the prototype and the most extensively studied rhabdovirus. It generally causes nonfatal diseases in domestic animals, including cattle, horses, and pigs. Although VSV infections are nonfatal, they nevertheless can have serious economic consequences for livestock owners, especially in the dairy industry. VSV rarely infects humans, but when it does it may cause influenza-like symptoms.

Rhabdovirus minus-strand RNA genomes are notably similar to the genomes of three other class V virus families that likewise package a

Figure 10.1 Preparation of rabies virus. The 11-kilobase genomic RNA is associated with 1,200 copies of the N protein, causing the ribonucleoprotein complex to take on its characteristic striations within the envelope of the virus particle. From D. E. White and F. J. Fenner, *Medical Virology*, 4th ed. (Academic Press, San Diego, CA, 1994), with permission.

monopartite (i.e., nonsegmented) minus-strand RNA. These other virus families are the paramyxoviruses, the filoviruses, and the bornaviruses. These distinct virus families do not share a common shape or genome size. Yet they appear to belong to a common "superfamily" or order based on other shared characteristics of their genomes and on their similar patterns of gene expression. Collectively, these four virus families are referred to as the *Mononegavirales*. (The order *Mononegavirales* is the first taxon above the family level to be recognized in virus taxonomy.)

Much of what is known about the patterns of gene expression that are common among the mononegaviruses was observed for the first time in studies of VSV. These shared features are enumerated below. But for the moment, note that not all minus-strand RNA viruses are mononegaviruses. For instance, the orthomyxoviruses (i.e., the influenza viruses), the arenaviruses, and the bunyaviruses each have **multipartite** (i.e., segmented) minus-strand genomes, comprised of two distinct gene segments in the arenaviruses, three in the bunyaviruses, and six to eight in the orthomyxoviruses.

Despite the diversity of minus-strand RNA viruses that infect vertebrates (including humans), plant rhabdoviruses are the only minus-strand RNA viruses known to infect plants, and no minus-strand RNA viruses of any sort are known to infect bacteria. What might we make of these facts? One possibility is that minus-strand viruses arose relatively recently in vertebrates. If that were the case, then we might expect that the minus-strand virus families might be related to each other. The existence of the order *Mononegavirales* is consistent with that possibility.

STRUCTURE

All minus-strand RNA viruses share the following characteristics. First, their minus-strand genomes are always associated with many copies of a nucleocapsid (N) protein and with fewer copies of the viral RNA polymerase molecules. Second, their nucleocapsids are always helically symmetric, reflecting the association of their minus-strand genomes with their N proteins. Third, the helical nucleocapsids of all minus-strand viruses are enclosed within lipid envelopes. (Viruses with plus-strand RNA genomes do not contain particle-associated polymerase activities, the retroviruses being the notable exception. This is a fundamental difference between class IV and class V viruses [Chapter 2]. Also, plus-strand RNA virus genomes are usually not coated with viral proteins in the virus particle, although the coronaviruses [Chapter 9] are an exception to that generalization.)

Rhabdovirus minus-strand RNA genomes contain about 11,000 nucleotides and are associated with about 1,200 copies of the N protein. Within the virus particles, these ribonucleoprotein complexes are visible as the characteristic striations seen in electron micrographs of the particles (Fig. 10.1).

As in the case of all other minus-strand RNA viruses, the rhabdovirus genomic RNA is also associated with the viral RNA polymerase, which in this instance is comprised of the large (L) protein and the P protein. The L protein is the catalytic subunit of the polymerase complex. Its large size reflects its several functions, which include ribonucleotide polymerization, capping functions, and polyadenylation of the viral mRNAs. The P protein derives its name from the fact that it is highly phosphorylated. It was formerly called NS, when it was believed to be nonstructural. It is an RNA-binding protein that functions to correctly position the polymerase complex on the template to initiate transcription. It also interacts with newly synthesized N proteins to promote the efficient encapsidation of progeny genomic RNA molecules. Later, during budding, the specific interactions of the L and P proteins with the lipid envelope are believed to facilitate the characteristic bullet-shaped appearance of the rhabdoviruses. The matrix or M protein is located between the helical nucleocapsids and the viral envelope. As in the case of other viral M proteins, it mediates the association between the nucleocapsid and the envelope.

Lastly, there is the transmembrane G protein (G for glycoprotein), which is located in the viral envelope in the form of homotrimeric complexes. These G protein trimers constitute the approximately 400 spikes that protrude from the surface of each virus particle. As you probably surmised, the G protein is the attachment protein for these viruses. In addition, it mediates membrane fusion after the virus has been taken up by receptor-mediated endocytosis. Not surprisingly, the major antigenic determinants of rhabdovirus particles are located on the G protein. Indeed, immunization with G protein alone can control VSV infections of agriculturally important livestock.

ENTRY

Rhabdovirus receptors are not yet known with certainty, although rabies virus appears to be able to bind the nicotinic acetylcholine neurotransmitter receptor, allowing it to access the central nervous system (see below). Regardless, rhabdoviruses enter cells by receptor-mediated endocytosis, using their G protein homotrimers as their attachment proteins and also to mediate low-pH-triggered fusion, which ultimately releases the nucleocapsid into the cytosol.

Interestingly, recent experimental results demonstrate that the virus envelope does not fuse with the limiting membranes of the endosome per se. Instead, the virus envelope fuses with membranes of the intraluminal vesicles that are a characteristic feature of organelles along the endocytic pathway (Fig. 10.2). Consequently, fusion releases the viral nucleocapsids into the lumens of these intraorganelle vesicles, rather than into the cytoplasm.

Apropos the above, the nomenclature describing the organelles of the endocytic pathway is somewhat confusing, in part because while all organelles of the endocytic pathway contain intraorganelle vesicles that arise by invagination of the organelle membrane, a particular organelle of the pathway, which traffics between early and late endosomes, is referred to specifically as a **multivesicular body** (**MVB**). MVBs bud from early endosomes and traffic to, and fuse with, late endosomes.

Fusion of VSV with the intraorganelle vesicles actually occurs within these MVBs, as triggered by their acidic pH. Then, in late endosomes, back fusion of the intraluminal vesicles with the endosomal membrane releases the viral nucleocapsids into the cytoplasm, so that infection might ensue. In the case of VSV, membrane fusion occurs at a very narrow pH range, between 6.2 and 5.8.

How might the virus benefit from this somewhat more complicated entry pathway, in which the nucleocapsid is first released into an intraluminal vesicle, which later fuses with a vesicle membrane to release the nucleocapsid into the cytosol? Perhaps it enables the virus to be safely transported across the cytoplasm, in particular through the cortical actin network that underlies the plasma membrane, while all the time in a milieu that is free of acid hydrolases. Interestingly, the anthrax toxin follows essentially the same pathway to reach the cytoplasm, perhaps for the same

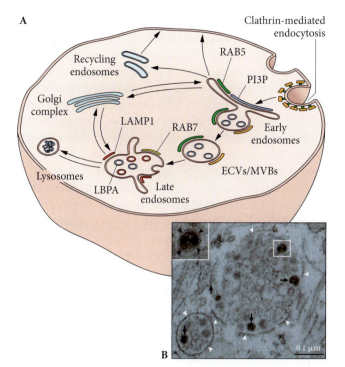

Figure 10.2 (A) Organization of the endocytic pathway and distribution of some important proteins and lipids in endosomal membranes. Note how the MVBs bud from the early endosomes and traffic to, and fuse with, late endosomes. (B) VSV enters cells by clathrin-mediated endocytosis and is subsequently routed down the endocytic pathway. Within MVBs, the low pH triggers the fusion of the viral envelope with the endosomal membranes, thereby releasing the nucleocapsid into the cytosol. However, this release may actually require two successive fusion events. First, the viral envelope fuses with the membranes of the intraluminal vesicles, which releases the nucleocapsid into the lumen of these vesicles, rather than into the cytoplasm. The nucleocapsid is then safely transported, in an environment free of hydrolases, through the cortical network to the perinuclear region of the cell, where the late endosomes mostly reside. Within this late-endosome compartment, back fusion of the luminal vesicle with the late endosomal membrane releases the nucleocapsid into the cytosol. The electron micrograph shows MVBs (the endosomal membrane is indicated by white arrowheads) that enclose intraluminal vesicles, which contain electron-dense structures that correspond to viral nucleocapsids (black arrows). The boxed area is shown at higher magnification in the inset and illustrates the clearly visible membrane (black arrowheads) of the endosomal vesicle that contains the nucleocapsid. Adapted from J. Gruenberg and F. G. van der Goot. *Nat. Rev. Mol. Cell Biol.* 7:495–504, 2006, with permission.

reason. It is possible (maybe even likely) that other enveloped viruses will be found to also use this entry pathway.

The molecular mechanism by which the rhabdovirus G proteins promote membrane fusion is not yet known. However, it almost certainly is different from the mechanism by which the fusion proteins of other enveloped viruses, including the coronaviruses, influenza virus, Ebola virus, and human immunodeficiency virus, promote fusion (Box 9.2). First, although rhabdovirus G proteins form homotrimers like the fusion proteins of other enveloped viruses, they do not form the trimeric coiled coils that characterize the fusion-competent conformations of the fusion proteins of these other enveloped viruses (see Chapter 12 in particular). Second, unlike the fusion proteins of these other viruses, rhabdovirus G proteins do not contain a nonpolar amino acid sequence at their termini, nor do they contain any other region that appears as a typical hydrophobic fusion peptide. Third, in contrast to the fusion proteins of most other viruses, rhabdovirus G proteins are not synthesized as fusion-incompetent precursors that are posttranslationally (and irreversibly) cleaved to generate the fusogenic protein. Instead, rhabdovirus G proteins display a pH-dependent equilibrium between the fusogenic and nonfusogenic conformational states. (A possible benefit to the rhabdoviruses of the reversibility of the fusogenic conformation of their G protein is that it enables those G proteins to be transported through the acidic compartments of the Golgi apparatus and then recover their native prefusion conformation at the cell surface, the site of virus budding.)

Fusion results in the release of the viral nucleocapsid into the host cell cytoplasm. That is, uncoating in the case of rhabdoviruses, and of minus-strand RNA viruses in general, consists of envelope removal. Intact helical ribonucleoprotein capsids, which contain the genomic RNA and the transcription/replication enzymes, are released into the cytoplasm, where replication then ensues. There is no further disassembly of the nucleocapsids. Instead, the intact nucleocapsids initiate replication.

GENOME ORGANIZATION, EXPRESSION, AND REPLICATION

The General Transcriptional Strategy of Viruses That Contain Negative-Sense RNA Genomes

We begin this section by restating the fundamental dilemma faced by minus-strand RNA viruses in general and how this dilemma affects their pattern of gene expression and replication.

There is no enzymatic machinery in an uninfected cell that can transcribe or replicate single-stranded RNA. Thus,

all RNA viruses, regardless of whether they have plus-strand or minus-strand genomes, must be able to provide their own transcriptase/replicase activities.

The genomes of plus-strand RNA viruses (class IV in the Baltimore scheme) are identical in nucleotide sequence to their mRNAs. Thus, the genomes of plus-strand RNA viruses can be translated directly upon their release into the cell and thereby generate the critical transcriptase/replicase activities as one of their translation products. However, the dilemma continues for the minus-strand RNA viruses because their minus-strand genomes do not serve as mRNAs. But if these viruses could somehow provide their transcriptase activity, then their predicament would be solved. That is so because (i) if the virus had its transcriptase at hand, it could then generate mRNAs that encode the transcriptase/replicase, as well as any other virus-encoded gene products; and (ii) these mRNAs can be translated by the cells' translational machinery (an important point developed at length in Chapter 2).

But how do minus-strand RNA viruses provide the needed transcriptase? The answer is that minus-strand RNA viruses bring the crucial transcriptase activity into the cell as an integral component of the virus particle. Those transcriptase molecules were produced during the preceding cycle of infection and were incorporated into the progeny virus particles as they matured. Note that the reoviruses (Chapter 5), which have double-stranded RNA genomes, face a similar dilemma and have devised a similar solution. Historically, VSV was the first RNA virus shown to contain a virus-encoded polymerase as an integral component of the virus particle. That pioneering study was carried out in 1970 by David Baltimore, Alice Huang (who is Baltimore's wife and herself a prominent scientist), and Martha Stampfer.

Gene Organization and Transcription

Rhabdovirus genomes, at about 11,000 nucleotides in length, encode five proteins, the N, P, M, G, and L proteins, the coding sequences of which reside in that order (in the 3′-to-5′ direction) on the minus-strand genome (Fig. 10.3). The rhabdovirus gene order was first established by ultraviolet (UV) inactivation studies (see below) and later confirmed by nucleotide sequencing. (The minus-strand RNA genomes of the four mononegavirus families

Figure 10.3 The genetic map of VSV. The 11-kilobase VSV genome encodes five proteins: the N, P, M, G, and L proteins. Since the genomic RNA serves as the template for mRNA synthesis, it is shown in the 3′-to-5′ orientation. The arrows indicate the intergenic sequences, where poly(A) sequences are added to the individual nascent mRNAs, by a stuttering process. l is a 47-nucleotide-long leader RNA (see the text).

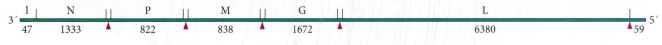

contain 5 to 10 genes. Yet it is interesting that those genes, which all the mononegaviruses contain in common, are similarly ordered on their genomes [for details, see Chapter 11]. That fact is consistent with the premise that these virus families evolved from a common ancestral virus.)

Rhabdovirus nucleocapsids do not fully disassemble upon their release into the cytoplasm. Instead, the intact ribonucleoprotein complex, with its associated N proteins and transcription/replication enzymes, generates all the viral transcription products. Indeed, mononegavirus genomes in general are never found as naked RNA molecules during the viral replication cycle. The rhabdovirus polymerase is initially located at the 3′ end of the minus-strand genome. During transcription, it must "worm" or "wiggle" under the N proteins that are spaced at regular intervals (15 to 20 nucleotides) throughout the length of the nucleocapsid (Box 10.1).

The RNA polymerase complex is comprised of the P and L proteins. The P protein serves as the RNA-binding subunit of the complex, while the L protein contains the catalytic activity. The P protein thus positions the L protein at appropriate sites on the RNA template to initiate RNA synthesis. Together, the L and P proteins generate

capped, polyadenylated mRNAs and also replicate the genomic RNA. (Rhabdoviruses must provide their own capping activities because the cellular capping enzymes are located in the nucleus.)

We might envision several possible stratagems by which a minus-strand RNA viral genome might be transcribed and then translated. For example, the entire minus-strand genome might be continuously transcribed to generate a polycistronic mRNA, which is then translated into a large precursor polyprotein that is proteolytically cleaved to generate the individual protein products. Alternatively, the large primary plus-strand RNA transcript might be alternatively spliced to generate a different mRNA species for translation into each protein. Yet another possibility is that transcription starts separately at each of the five genes, in that way generating a separate mRNA species for each protein. Each of these possibilities is eminently plausible. So which is correct?

Five distinct species of virus-specific mRNAs can be extracted from VSV-infected cells, and each of these mRNA species can be translated in vitro to generate one of the five individual virus proteins. These experimental findings seem to eliminate the possibility that the individual

BOX 10.1

We already accounted for why minus-strand RNA viruses, but not plus-strand RNA viruses, contain particle-associated polymerases. But why are minus-strand viral genomes always coated with proteins, such as the rhabdovirus N protein? Here is one possibility. The N protein, as a single-stranded RNA-binding protein, acts to keep the viral RNAs in a transcriptionally active conformation. This is important because minus-strand virus genomes must be ready to begin mRNA synthesis as a first step after they are released into the cell. In contrast, plus-strand RNA virus genomes are generally translated as the first step after they are released in the cell.

The ribonucleoprotein complexes of minus-strand RNA viruses serve as the templates for the synthesis of all plus-strand transcripts, including full-length plus strands that serve as templates for minus-strand synthesis and sub-genome-length mRNAs as well. Some nascent plus strands also associate with the N protein, which prevents them from being spliced and polyadenylated. This enables them to be transcribed into full-length minus-strand RNA genomes. Thus, the N protein regulates transcription versus replication, in addition to keeping the RNA molecules in a transcriptionally active conformation. Additionally, the N protein also promotes envelope acquisition and budding (see the text).

Since the N proteins of all minus-sense RNA viruses have most or all of these functions in common, a recent study considered the intriguing proposition that the N proteins of different minus-strand RNA virus families may each contain a similar conserved RNA-binding structural motif (Luo et al., 2007). Toward this end, the researchers compared the crystal structures of the N proteins of two rhabdoviruses, VSV and rabies virus, with those of influenza virus and Borna disease virus. (The bornaviruses are a fourth member of the *Mononegavirales* superfamily [Chapter 13].)

The topologies of the RNA-binding regions of the different representative N proteins were found to indeed be very similar. The RNA appears to fit into a cavity formed by distinct domains at opposite ends of the N protein. Structurally, these different N proteins contain at least five conserved α-helices in the N-terminal domain and three conserved α-helices in the C-terminal domain, which are connected by a central α-helix.

Reference
Luo, M., T. J. Green, X. Zhang, J. Tsao, and S. Qiu. 2007. Structural comparisons of the nucleoprotein from three negative strand RNA virus families. *Virol. J.* 4:72.

virus proteins are generated from a single precursor polyprotein. But how might the five distinct mRNA species be generated: by alternative splicing of a single primary RNA transcript or by separate transcription initiation of each mRNA species?

In principle, UV inactivation experiments should enable one to distinguish between these two plausible scenarios. The effect of UV irradiation on RNA is to induce the formation of uracil dimers, which block progression of the polymerase on the template (Box 9.5). If the VSV mRNA species were transcribed independently of each other, then the target size for inactivating transcription of each gene should reflect the actual physical size of the gene. In contrast, if the individual mRNA species were generated by alternative splicing of a large precursor RNA, then the target size for inactivating transcription of each gene would be the actual size of that gene, plus the actual size of the entire sequence that is upstream from it on the genome (see Box 9.5 for a discussion of target theory).

Early UV inactivation studies of Ball and White (see Box 9.5) demonstrated that the N protein gene, which is located at the 3' end of the VSV genome, is the only VSV gene that has a UV inactivation target size that corresponds to its actual physical size. The target sizes for inactivation of transcription of each of the other VSV genes correspond to the actual physical size of the particular gene, plus the physical size of the entire sequence that is 3' to that gene on the genomic RNA. For example, the UV target size for inactivating transcription of the P protein gene corresponds to the actual size of the P protein gene plus that of the N protein gene as well. The target size for inactivating transcription of the M protein gene corresponds to the actual size of the M protein gene plus the sum of the actual sizes of the genes encoding the N and P proteins. The target size of the G protein gene corresponds to the sum of the sizes of the genes for the N, P, M, and G proteins. (Do you see how these experimental findings also establish the order of the genes on the genomic RNA [Box 10.2]?)

The experimental findings from the UV inactivation study show that transcription of each gene is dependent on the transcription of the entire sequence that is upstream from it on the genomic RNA. While these findings are consistent with the premise that the individual mRNA species are generated by alternative splicing of a single large primary transcript, they leave open yet another possibility: that each gene is transcribed from a separate initiation event, with the proviso that transcription of the genes occurs in an obligatory sequence. Indeed, later molecular studies demonstrated that this somewhat contorted interpretation indeed is correct.

Interestingly, this mode of transcription is common among the mononegaviruses in general. The polymerase

BOX 10.2

Ball and White used a coupled in vitro transcription-translation system to determine UV target sizes. That is, transcription of gene sequences as a function of UV dosage was quantified by polyacrylamide gel analysis of the *protein* products generated by the coupled in vitro system. Sequences that are downstream of a hit cannot be transcribed and hence will not be translated into proteins. The researchers were careful to ascertain that their experimental conditions were such that the amount of protein syntheses was proportional to the amount of RNA synthesis.

always enters at a single 3'-proximal site and then separately transcribes each of the genes in an obligatory linear sequence. The polymerase cannot directly attach to an internal initiation site. Instead, it accesses each internal transcription initiation site by first transcribing each of the genes that are upstream from it.

Another notable feature of rhabdovirus transcription is that there is an apparent attenuation of approximately 30% at each of the gene junctions, resulting in a gradient of gene expression corresponding to the gene order (i.e., N > P > M > G > L). This experimental finding is entirely consistent with a transcription stratagem in which each gene is separately transcribed, but in an obligatory sequence along the minus-sense genomic template, with the polymerase initiating transcription at the 3' end of the template and reinitiating transcription only about 70% of the time at each successive intergenic boundary.

This gradient of gene expression, corresponding to the gene order, is also a feature of mononegavirus transcription in general. The gradient is physiologically significant because it enables these viruses to regulate the relative level of expression of each gene in accordance with its position on the genome. Apropos this point, recall that those genes that are common among all the mononegaviruses are similarly ordered on their genomes. In all instances, the gene for the N protein, which is needed in greatest amounts, is located at the extreme 3' end of the genome, a position from which it can be transcribed in greatest amounts. The P protein, a component of the polymerase complex, is next. Polymerases generally are needed in much lesser amounts than structural proteins. However, the P protein has other functions in addition to its major role in RNA synthesis. In particular, it interacts with newly synthesized N proteins to promote the efficient encapsidation of nascent progeny genomic RNAs. Indeed, the P protein is required in a precise molar ratio relative to the N protein to fulfill this function. The gene order and the related attenuation phenomenon ensure that these two proteins, and all the others as well, are produced in appropriate relative amounts. The M protein is next on the genome, followed by the genes for the glycoproteins, which vary in number among the *Mononegavirales*. The gene encoding the L protein, the actual polymerase, which is needed in least amounts, is always located at the 5′ end of the minus-strand genome. The mechanism by which attenuation is achieved at each intergenic boundary is not yet clear, but it is a most intriguing molecular adaptation. (See Wertz et al. [1998] for an analysis of the effects on VSV replication and pathogenesis of rearranging the position of the N gene.)

Next, we consider the highly conserved sequence at the junctions between the individual rhabdovirus genes, i.e., 3′...AUACUUUUUU G/CA U̲U̲G̲U̲C̲N̲N̲U̲A̲G̲...5′, in the instance of VSV. The first 11 nucleotides of the sequence correspond to the 3′ end of each gene. The poly(U) portion of that sequence is transcribed into the long poly(A) tail at the end of each mRNA by a special process that is described below. The underlined sequence is transcribed into the 5′ end of each mRNA. *G/CA* is not transcribed and thus is referred to as the intergenic sequence. Genetic engineering experiments demonstrated that this intergenic sequence is sufficient to direct the VSV to terminate synthesis of a polyadenylated upstream mRNA and initiate transcription of 5′-capped downstream mRNA. Not surprisingly, a remarkably similar gene junction sequence is present in the genomes of the paramyxovirus family of the mononegavirus superfamily (Chapter 11).

The following model for rhabdovirus transcription is consistent with all the experimental findings enumerated above. The gene that encodes the N protein, which is located at the 3′ end of the genomic RNA, is transcribed first. Transcription of the N gene begins with capping of the nascent mRNA. Transcription of the N gene continues until the polymerase complex reaches the conserved junction region that is present between the N and P genes. The poly(U) segment of the junction sequence causes the polymerase to "stutter," at which point it adds 100 to 300 adenine residues to the nascent mRNA, which then is completed and released. This process, by which the poly(A) tail is added, is referred to as **reiterative transcription**. About 30% of the time the polymerase too is released. If the polymerase should be released, then it is free to begin the process again at the 3′ end of the genomic RNA. About 70% of the time the polymerase reinitiates transcription at the start of the next gene on the RNA template (i.e., at the underlined portion of the junction sequence). The same sequence of events transpires at each of the intergenic

junctions. The gradient of transcription (N > P > M > G > L) is a consequence of the 30% probability of attenuation occurring at each intergenic junction sequence.

Replication

In order for rhabdoviruses to replicate their minus-strand RNA genomes, they need to transcribe full-length plus strands that can serve as templates for transcribing progeny minus-strand genomic RNAs. However, this is not a trivial matter, since, as you may have noted, the transcription stratagem described above generates only sub-genome-length mRNA species. How then are rhabdoviruses also able to generate full-length plus strands that will serve as intermediates in replication? The answer is as follows. As the sub-genome-length mRNAs are translated, the viral N protein accumulates and binds to the *nascent* mRNAs, doing so via its interaction with the P protein. When the N protein binds to the nascent plus strands, it prevents the bound polymerase from stuttering at the poly(U) segments of junction regions on the minus-strand template. In that way, the N protein prevents polyadenylation via reiterative transcription, as well as recognition of termination signals, thereby enabling full-length plus-strand transcripts to be generated, which then serve as templates for the synthesis of minus-strand genomic RNAs. The N protein also binds to the nascent minus-strand RNAs, thereby ensuring that full-length genomic RNAs too are generated. Moreover, the progeny genomic RNAs are produced as nucleocapsids, ready for packaging.

Notice that this mechanism, whereby newly translated N protein needs to bind to nascent plus-strand RNAs to prevent recognition of polyadenylation and termination signals, also ensures that *immediately* after entry (i.e., before new N protein is available) transcription would be favored over replication. N protein, when generated, then mediates the switchover from transcription to replication. Other mononegaviruses appear to use very similar mechanisms to mediate their switchover from synthesizing sub-genome-length mRNAs to synthesizing full-length plus strands. (The discovery that the N protein is required for the switchover from transcription to replication followed upon the early experimental observation that inhibiting protein synthesis with cycloheximide blocked replication, but not transcription.)

Here is one last detail concerning transcription and replication. The polymerase transcribes another subgenomic sequence in addition to the five capped and polyadenylated mRNA species. That sequence, referred to as the **leader**, is 47 nucleotides long, and it is neither capped nor polyadenylated (Fig. 10.3). It is transcribed from the extreme 3′ end of the genomic RNA. Despite its appellation, the leader is not contained in any of the mRNA species, and its role, if any, in transcription is not clear. Results from UV inactivation experiments show that the leader sequence does not contribute to the target size of immediately downstream genes. This implies that transcription of the mRNAs initiates directly at the first gene start sequence and does not require prior transcription of the leader sequence. Yet the leader is contained in the full-length positive-sense transcript, and therefore, it may play a role in replication.

ASSEMBLY AND RELEASE

The processes of transcription, translation, and genome replication continue, thereby generating sufficient levels of virus genomes and proteins for progeny viral nucleocapsids to assemble. Complete nucleocapsids are produced by the interactions of progeny genomic RNAs with the N, P, L, and M proteins. Molecules of the N protein polymerize on the nascent genomic RNA molecules during replication, such that neighboring N molecules form an extended network of interactions along the entire length of the RNA genome. The P protein too plays a role in generating the nucleocapsid by interacting with the N protein.

In the instances of at least some enveloped viruses, budding appears to depend on the interaction of the nucleocapsid with the envelope glycoproteins, as mediated by the matrix proteins. Despite that often-true scenario, a recombinant rabies virus mutant that lacked the gene encoding the G protein was nevertheless able to form enveloped, bullet-shaped, albeit spikeless virus-like particles (Box 10.3). However, while the G protein is not absolutely necessary for budding in this instance, it nevertheless enhances the efficiency of budding 10- to 30-fold when it is present.

On the other hand, the M protein is absolutely crucial for rhabdovirus budding, as shown by genetic experiments using temperature-sensitive mutants. Thus, the nucleocapsid

---

BOX 10.3

Structural proteins of several enveloped viruses, including the retroviral Gag and the Ebola virus VP40, as well as the rhabdovirus M proteins, are each by themselves able to bud from the cell surface in the form of lipid-enveloped virus-like particles. Moreover, each of these viral proteins contains a late-budding (L) domain, which is critical for efficient budding. In addition, each of these budding proteins engages in lateral interactions that promote the budding process.

Garoff et al. (1998) reviewed the data regarding budding of a number of enveloped viruses and then classified the budding strategies of these viruses into four distinct types,

as follows: type I, budding dependent on both capsid and spike proteins (alphavirus and hepadnavirus); type II, budding mediated by capsid or core protein only (retrovirus); type III, budding accomplished by membrane proteins only (coronavirus); and type IV, budding driven by matrix protein with the assistance of spikes and the ribonucleoprotein complex (rhabdovirus and possibly paramyxovirus and orthomyxovirus). These four categories are schematically depicted in the figure.

Reference
Garoff, H., R. Hewson, and D. J. E. Opstelten. 1998. Virus maturation by budding. *Microbiol. Mol. Biol. Rev.* **62:**1171–1190.

Viral proteins that drive budding. (I) Spike (red)- and NC (blue)-dependent budding of alphaviruses. (IIa) Gag protein-driven budding of a type C retrovirus. The membrane is shown in yellow, and the submembrane layer of Gag protein is depicted in blue. An RNA molecule (green) is also indicated, but it is unclear whether this is necessary for budding. (IIb) Budding of a type D retrovirus Gag particle. In this case, Gag molecules are assembled into a complete core in the cytoplasm prior to membrane attachment. (III) M (red, unfilled) and E (red, filled) membrane protein-driven budding of coronavirus. (IV) Rhabdovirus budding is depicted as an efficient (left) or inefficient (right) process depending on the presence of the spike proteins (red). The M protein layer below the membrane is shown in brown, and the RNP is depicted as a green helix with proteins (blue). Adapted from Garoff et al., 1998, with permission.

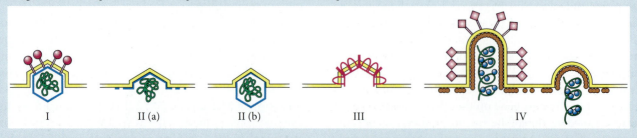

| I | II (a) | II (b) | III | IV |

and the M protein together can mediate budding in the absence of the G protein, although the process is considerably more efficient when the G protein is present.

Experiments in which baculovirus and vaccinia virus vectors were used to express the VSV M protein in insect and mammalian cells, respectively, demonstrated key ways by which the M protein might facilitate budding. In each instance, the M protein was found to bind to the plasma membrane and induce the budding of M-protein-containing vesicles from the cells. Thus, the M protein has an inherent budding activity, and it may be the major factor driving rhabdovirus budding. Interestingly, in the same studies, if the G protein was coexpressed with the M protein, then the G protein was *not* incorporated into the M-protein-containing vesicles. This experimental result might be explained if the nucleocapsid too needs to be present for the G protein to be incorporated into complete virus particles. In any case, together, these experimental findings support the notion that while the M protein plays a crucial role in budding, the nucleocapsid and G protein also collaborate in forming enveloped virus particles. Results of genetic experiments show that the cytoplasmic domain of the G protein is sufficient for it to support budding (Box 10.3), consistent with the premise that the G protein facilitates budding by interacting with a component of the nucleocapsid. Might it be the M protein?

Actually, the M protein has other roles in final virus assembly, in addition to its role in budding per se. First, it inhibits further viral transcription. Second, it interacts with the viral ribonucleoprotein core, thereby promoting its condensation into the nucleocapsid. Moreover, the M-protein-mediated condensation of the ribonucleoprotein at the bud site is thought to give rise to the bullet-shaped rhabdovirus morphology.

MEDICAL ASPECTS

Cytopathic Effects

Many viruses replicate in their host cells while causing little, if any, cell damage. Consequently, when infections with those viruses result in clinical symptoms (e.g., fever, nausea, aches, and tissue damage), those symptoms are most likely the result of host immune responses to the infection (Chapter 4).

On the other hand, some viruses directly damage and destroy the cells in which they replicate, and such virus-induced cytopathology might well contribute to the clinical symptoms caused by the infection. The effects of these cytopathic viruses on the cells they infect may also be seen in cell culture, where they are referred to as **cytopathic effects**. However, there is *no clear relationship* between cytopathic effects in cell culture and clinical symptoms of

infection. Moreover, so many changes occur in cells infected with cytopathic viruses that it is difficult to attribute cell death to any one particular effect of the virus. In at least some instances, cell death may result from the cumulative effects of a variety of insults inflicted by the virus (Box 10.4).

VSV is a classic cytopathic virus. Among its cytopathic effects, VSV inhibits cellular gene expression, impairs nucleocytoplasmic transport, and disrupts the cytoskeleton. The last effect causes cell rounding and detachment from the culture dishes. Multiple VSV factors lead to these changes. For example, the VSV leader RNA inhibits cellular transcription via its effect on RNA polymerase II and also inhibits cellular DNA synthesis. The N protein inhibits translation of cellular proteins and also impairs transport of proteins into and out of the nucleus. But it is the M protein which appears to cause the widest range of cytopathic effects. Expression of the M protein by itself inhibits all three host RNA polymerases, thereby effectively shutting down host gene expression. The inhibitory effect of the M protein on RNA polymerase II appears to be mediated via the inactivation of the cellular transcription factor TFIID. M protein also causes cytoskeletal (microtubule) disassembly. Recently, it was shown that a fraction of M protein binds to the nuclear membrane at nuclear pore complexes, thereby possibly contributing to the shutoff of host translation by inhibiting RNA export from the nucleus.

How might the above VSV-mediated cytopathic effects be of benefit to the virus? One possibility is that those cytopathic effects which affect host gene expression serve the virus by redirecting the biosynthetic resources of the cell for use by the virus. Other cytopathic effects might be of little or no benefit to the virus and might simply be epiphenomena associated with virus replication. Regardless, the relationship between these cytopathic effects in cell culture and clinical symptoms of infection is not clear.

BOX 10.4

How might you operationally define the moment of cell death? Those who study cytopathology caused by various toxic chemicals or lack of oxygen, for example, refer to cell death as the point where cell damage becomes irreversible. This serves as an operational definition in these instances, since up to a point, the toxic substance can be removed or the oxygen supply can be restored, followed by cell recovery. But what operational definition for cell death might be used in the case of a virus-infected cell?

VSV

As noted above, there are two genera of the *Rhabdoviridae* that infect vertebrates. These are the genus *Vesiculovirus* (named for vesicular stomatitis virus, the prototype rhabdovirus) and the genus *Lyssavirus*, which includes rabies virus and the rabies-like viruses. The genus *Vesiculovirus* includes some 35 serologically distinct viruses. Yet VSV is the only one of these that is known to infect humans, although it is primarily known as a pathogen of cattle, horses, and pigs. In these domestic species, VSV induces characteristic vesicular eruptions on the oral mucosa, for which the virus is named (i.e., vesicular stomatitis).

In humans, VSV induces influenza-like symptoms, including fever, chills, and muscle aches. Symptoms usually abate without complications within 7 to 10 days. VSV is transmitted to humans mainly during epizootic infections in cattle and horses. These human infections usually go unreported because of a lack of awareness of VSV. Usually, they are revealed by retrospective serological studies.

Rabies Virus

Rabies is responsible for 30,000 to 70,000 human deaths worldwide each year. Moreover, the mortality rate for clinical rabies is essentially 100%. This figure is in marked contrast to the mortality rates for virtually all other virus infections.

Rabies virus is generally transmitted to humans via the bite of a rabid animal. Indeed, rabies is the classic **zoonotic** infection. Dogs are the primary transmitters to humans, but many other species of wildlife can also transmit the infection. To the surprise of many, rabies is actually more common in cats than in dogs in the United States, since dogs are more likely to be vaccinated. Despite the large number of humans infected by rabies virus each year, the virus has never adapted to spread by direct human-to-human transmission. (Why might that be so? Is it perhaps because the virus is so well adapted for transmission via bites? See below.)

Rabies virus infection and clinical rabies (see below for the distinction) are more common in less developed areas of the world, particularly in areas with less developed public health systems. For example, rabies is estimated to kill about 25,000 humans annually in India, and dogs are believed to be the transmitter in 96% of those cases. In contrast, the frequency of rabies in the United States is currently about 1 case per year. The difference between India and the United States in these regards is largely due to extensive vaccination of dogs in the United States and their more limited contact with wildlife.

Rabies virus infects a variety of cell types in culture, as well as in vivo, reflecting the ability of the virus to bind nonspecifically to a variety of cell surface receptors (i.e.,

carbohydrates, phospholipids, and sialylated gangliosides). Rabies virus also binds specifically to the nicotinic acetylcholine receptors on neurons, thereby enabling it to invade the central nervous system (see below). The ability of rabies virus to infect a wide range of cell types is biologically significant since it enables the virus to disseminate in the host and to give rise to the unique pattern of rabies pathogenesis and transmission to new hosts, as described below.

When a rabid animal bites its victim, rabies virus-infected saliva is deposited deep into the victim's muscle tissue. The initial bite is followed by a so-called **incubation period** that lasts from several weeks to as long as several years. During most of this variably long incubation period, the virus is thought to remain close to the site of the bite, where it first entered the host. Eventually, virus produced in the initially infected muscle tissue invades motor or sensory neurons that innervate the muscle. Rabies virus can infect neural cells because it specifically binds to acetylcholine receptors that are present on those cells at neuromuscular synapses. Once in the neuron, rabies virus migrates along the axon (via the fast axonal transport system) towards the central nervous system, at a rate of about 50 to 100 mm per day. The virus somehow is able to cross the synaptic junctions between neurons, enabling it to eventually reach the brain.

Entry of the virus into the brain marks the end of the incubation period. Replication of the virus in the brain then gives rise to the **prodromal phase** of the disease, in which the first clinical signs of rabies emerge. Prodromal symptoms may include fever, malaise, headache, pain or itching at the site of the bite, gastrointestinal symptoms, and fatigue. The prodromal phase lasts from 2 to 10 days, after which the first neurological signs of rabies emerge. Hydrophobia (the fear of water), which occurs in one-half or fewer of all rabies victims, is one of the classic rabies symptoms. It is thought to be triggered by the pain associated with attempts to swallow water. Other neurological symptoms may include seizures, disorientation, and hallucination.

In the brain, rabies virus enters the **limbic system**, where it replicates extensively. The limbic system is comprised of a group of brain structures, including the hippocampus and parts of the cerebral cortex, and, significantly for the virus, the limbic system controls certain aspects of emotion and behavior. Consequently, rabies virus infection of the limbic system causes a loss of cortical control of behavior, resulting in an aggressive, vicious animal.

Importantly, the virus disseminates from the brain via afferent neurons to infect several highly innervated sites, including the adrenal cortex, the pancreas, the heart, and the retina and cornea. (Recently, a case of rabies transmission resulting from a corneal transplant was reported.) Most importantly, though, the virus also disseminates to the salivary glands, where it replicates furiously. These two factors, (i) rabies virus infection of the limbic system and (ii) its intense replication in the salivary glands, together result in a vicious animal, ideally suited to transmit its copiously infected saliva to a new host. (What better ways might there be to ensure transmission than a bite? These strike me as perhaps the most diabolical of all viral adaptations.)

The virus spreads further in the brain, causing more widespread neuronal dysfunction. The spread of the virus to the neocortex ultimately leads to paralysis, resulting in respiratory failure and death.

Despite the fact that rabies virus is highly immunogenic, neither humoral nor cell-mediated immune responses against rabies virus can be detected until after the virus reaches the central nervous system, after which symptoms inevitably develop. A possible explanation for the failure of the host to mount an immune response earlier is that too little virus is released from cells during the incubation period to trigger an immune response. Instead, most rabies viruses remain cell associated in the initially infected muscle tissue. This may be another diabolically effective adaptation of this virus, since during most of the incubation stage of rabies infection, the infection is sensitive to immune intervention (i.e., vaccine and hyperimmune immunoglobulin), which can prevent the otherwise fatal outcome.

Early immune intervention is thought to be effective because it neutralizes or inactivates the virus while it is still at the site of the bite, before it gains access to the nervous system, in that way accounting for the efficacy of postinfection vaccination (Box 10.5). Once the virus invades the brain, it is in an immunologically privileged site, so that immune intervention is no longer effective. Therefore, treatment after exposure to rabies virus is most urgent, even if the patient was bitten months earlier.

Despite the rabies vaccine, rabies remains a problem worldwide. More than 7 million patients each year in Asia alone receive postexposure treatment for rabies. The highest rate, 766 per 100,000 individuals per year, occurs in Vietnam. Yet, as in the case of many other infectious diseases, effective medical intervention is difficult or impossible in areas where medications are unaffordable and health facilities are remote. In South Africa in 2001, 26% of rabies vaccine treatment facilities had no vaccine in stock.

Important questions regarding rabies virus infections remain unanswered. To enumerate a few, first, it is not known how the virus crosses neuronal synapses to invade the brain. Second, it is not clear how the virus actually causes disease. Unlike VSV, rabies virus is not cytopathic in cultured cells. Moreover, rabies virus infection in humans

BOX 10.5

Pasteur developed his rabies vaccine in 1885, over a decade before viruses were identified as distinct microbial entities (Chapter 1). Yet Pasteur was able to empirically generate an attenuated rabies vaccine by applying the same principle that he previously used to prepare a vaccine against cholera. That is, he "aged" the microbe. First, he generated stocks of rabies virus by passing the virus in live rabbits. This was done by spinal cord inoculation, using a spinal cord extract of the previous passage as the inoculum. Spinal cord inoculation, rather than inoculation under the skin, was carried out because it was found to give rise to a shorter incubation period. Pasteur then incubated (i.e., aged) a rabies virus-infected spinal cord inside a glass flask. Pasteur measured the degree of attenuation on subsequent days of aging by preparing an extract from a sample of the spinal cord and observing its effect when injected into fresh rabbits. Within 2 weeks of the start of the aging process, the ability of the inocula to kill the rabbits was almost completely gone.

On 6 July 1885, a young French boy, Joseph Meister, who was bitten by a rabid dog, became the first recipient of the Pasteur rabies vaccine. Interestingly, Joseph, whose life was saved by Pasteur, later became gatekeeper of the famous Pasteur Institute in Paris.

Modern rabies vaccines are generally killed virus vaccines, prepared by chemically inactivating tissue culture lysates. These vaccines have fewer negative consequences than the earlier attenuated vaccines.

does not lead to the inflammatory reactions and cell degeneration that typically occur in the viral encephalopathies. Indeed, at the onset of illness, when evidence of neuronal dysfunction appears, there is little or no sign of histopathology. Thus, neuronal dysfunction, rather than neuronal death, may be responsible for the clinical features and fatal outcome in natural rabies. The underlying basis for that putative neuronal dysfunction is not known.

It also is not clear if rabies can be a chronic or latent infection in animals. This issue is fundamental to the biology of the virus, and it is clearly relevant to its epidemiology as well. On the one hand, rhabdoviruses in general (including rabies virus) have long been thought to give rise only to acute infections. On the other hand, antibody against rabies virus has been detected in apparently healthy wildlife species, including mongooses, skunks, raccoons, foxes, hyenas, jackals, fruit bats, vampire bats, insectivorous bats, and domestic dogs. (Recall that immune responses in humans are generally believed to occur only when symptoms appear.) These findings raise the intriguing question as to whether rabies virus might give rise to chronic or latent infections in some host species. Moreover, in India, a dog with no detectable antirabies antibodies was found to excrete rabies virus intermittently in its saliva over a period of 30 months. If this finding can be confirmed, it raises the question of whether rabies can be transmitted by asymptomatic animals.

Suggested Reading

Wertz, G. W., V. P. Perepelitsa, and L. A. Ball. 1998. Gene rearrangement attenuates expression and lethality of a nonsegmented negative strand RNA virus. *Proc. Natl. Acad. Sci. USA* 95:3501–3506.

11

Paramyxoviruses

INTRODUCTION

The *Paramyxoviridae*, like the rhabdoviruses of the previous chapter, contain nonsegmented, minus-strand RNA genomes. Moreover, paramyxoviruses and rhabdoviruses have similarly organized genomes and patterns of gene expression. For these reasons, the paramyxoviruses and rhabdoviruses belong to the "superfamily" or order *Mononegavirales* (Chapter 10). Yet there are notable differences between the paramyxoviruses and rhabdoviruses that justify their classification into distinct families, as detailed below (Box 11.1).

Several paramyxoviruses, including the well-known measles virus, mumps virus, parainfluenza virus, and respiratory syncytial virus, are important human pathogens. Each of these is introduced briefly here, as are a few others, and discussed in detail in the sections that follow.

Despite the availability of the measles vaccine, measles remains a potentially dangerous, even fatal disease, still causing 1 to 2 million deaths per year worldwide. Its introduction into unprotected populations can have devastating consequences.

Parainfluenza virus and respiratory syncytial virus each cause serious respiratory diseases, especially in children. Indeed, these two viruses alone are responsible for more than one-half of all cases of the croup, bronchiolitis, and pneumonia in infants.

Human metapneumovirus (HMPV), although discovered only in 2001, is now recognized as another important cause of serious upper and lower respiratory tract disease worldwide, especially in children. Despite its recent discovery, this virus is widespread in humans. Serological surveys showed that virtually all children in The Netherlands, Japan, and Israel are infected by 5 to 10 years of age.

Hendra virus is another newly discovered and highly pathogenic paramyxovirus that first was recognized in 1994. Four years later, in 1998, a closely related paramyxovirus, Nipah virus, emerged in Singapore and Malaysia, where it caused an outbreak of severe encephalitis. These viruses are transmitted to humans by infected animals (Box 11.2).

BOX 11.1

The paramyxoviruses also bear a superficial resemblance to the orthomyxoviruses (i.e., the influenza viruses [Chapter 12]). In fact, these two virus families once were grouped together in a family called the **myxoviruses**. That now-discarded designation derives from the Greek word "*myxa*," which means mucus, reflecting the fact that these viruses have an affinity for mucopolysaccharides and glycoproteins (in particular, sialic acid-containing receptors on cell surfaces), and consequently, they hemagglutinate red blood cells (Box 11.4).

The earlier coclassification of these families was also based on the facts that the envelope glycoproteins of each express neuraminidase activity and the genomes of each are single-stranded, minus-sense RNA molecules. Still, paramyxoviruses are distinguished from orthomyxoviruses by fundamental differences in their molecular biology. In particular, paramyxoviruses contain nonsegmented (monopartite) genomes (i.e., they are mononegaviruses), whereas orthomyxoviruses contain segmented (multipartite) genomes.

BOX 11.2

Discoveries of "new" medically important viruses, such as HMPV, underscore the need for medical and basic scientists to develop appropriate diagnostic tests and to characterize the natural history and epidemiology of each newly discovered virus. Likewise, medical providers need to remain current in their knowledge of pathogenic microorganisms, so that infections by these agents can be properly diagnosed and treated, and for effective public health measures to be implemented when necessary.

The natural history of HMPV is not yet known, but analysis of serological samples shows that it has been circulating in human populations since at least 1958. The recently discovered severe acute respiratory syndrome (SARS) coronavirus (Chapter 9), West Nile virus (Chapter 7), and HIV (Chapter 21) provide better-known examples of recently discovered, emerging viruses. In the cases of each of these more famous viruses, focused research efforts led to their identification and the characterization of their epidemiology.

Canine distemper virus (CDV) causes potentially fatal infections in a variety of domestic and wild animal species, even including marine mammals. It was once the leading cause of death in unvaccinated puppies, but widespread vaccination programs have dramatically reduced its incidence in dogs.

There is significant base sequence homology among all of the paramyxoviruses. Nevertheless, members of the family can be grouped into five genera that are distinguished by nucleotide sequence relatedness and by biological differences, such as differences in the activities of their envelope glycoproteins. The five genera of the *Paramyxoviridae* are as follows: *Morbillivirus* (e.g., measles virus, dolphin morbillivirus, CDV, and rinderpest virus), *Rubulavirus* (e.g., mumps virus, human parainfluenza virus types 2, 4A, and 4B, avian Newcastle disease virus, simian virus 5 [SV5], and Menangle virus), *Pneumovirus* (e.g., respiratory syncytial virus and HMPV), *Paramyxovirus* proper (e.g., human parainfluenza virus types 1 and 3 and Sendai virus of mice), and *Henipavirus* (Hendra and Nipah viruses).

STRUCTURE

Paramyxovirus genomes are approximately 15 to 18 kilobases (kb) in length (for comparison, rhabdovirus genomes are 11 kb in length). Like other minus-strand RNA viruses, paramyxoviruses contain helical nucleocapsids, comprised of the viral genome and associated proteins, which in this instance are the 2,600 molecules of the nucleocapsid protein (NP), 300 molecules of the polymerase phosphoprotein (P), and 50 molecules of the large (L) protein (Fig. 11.1). The turns of the coiled ribonucleoprotein complex are packed closely together. Yet all contacts between NP proteins are lateral. That is, there are no direct contacts between N subunits on adjacent turns of the helix. (If there were, it would be hard to envision how transcription and replication might transpire.)

The paramyxovirus NP protein serves several functions, including (i) controlling viral transcription versus replication, (ii) packaging the genomic RNA, and (iii) interacting with the M protein during virus assembly. In these respects, it is reminiscent of the rhabdovirus N protein. The P and L proteins constitute the viral RNA polymerase (Box 11.3). The L protein contains the actual polymerase activities (i.e., capping, polynucleotide polymerization, and polyadenylation), whereas the exact function of the P protein remains elusive. The L/P polymerase complexes initially associate with genomic RNA molecules at their 3′ end, but they eventually become distributed over the length of the genomic RNA, so that in association with the NP protein, they can promote its encapsidation. The

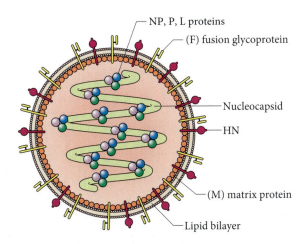

NP, P, L proteins

(F) fusion glycoprotein

Nucleocapsid

HN

(M) matrix protein

Lipid bilayer

Figure 11.1 Schematic drawing of a paramyxovirus. Note the lipid bilayer and the underlying matrix (M) protein. Inserted in the viral envelope are the HA/neuraminidase (HN) attachment glycoprotein and the fusion (F) glycoprotein. (Not all of the paramyxovirus attachment proteins contain HA and neuraminidase activities; see the text.) The helical nucleocapsid contains the 15-kb minus-strand genome, in association with the NP, P, and L proteins.

matrix (M) protein bridges the nucleocapsid to the lipid envelope.

Paramyxovirus envelopes contain two glycoproteins: a fusion (F) protein and an attachment protein. Notice that in this respect the paramyxoviruses differ from the rhabdoviruses, which use a single glycoprotein to mediate both attachment and fusion. (But as will be seen, attachment and entry of some paramyxoviruses [e.g., respiratory syncytial virus, a pneumovirus] may be mediated by the same protein.)

The activities of the paramyxovirus attachment proteins vary among the paramyxovirus genera. For example, the attachment proteins of the genera *Paramyxovirus* and *Rubulavirus* are designated HN for hemagglutinin/neuraminidase (Box 11.4). In contrast, the attachment proteins of the genus *Morbillivirus* are designated H, since they have hemagglutinin (HA) but not neuraminidase activity. The attachment proteins of the genus *Pneumovirus* are designated G (for glycoprotein), since they have neither HA nor neuraminidase activities. Like the fusion proteins of other enveloped virus families, paramyxovirus fusion proteins exist in the form of trimeric oligomers (see Chapter 12).

Viral neuraminidase activities, when present, cleave *N*-acetylneuraminic acid from glycoprotein receptors. The reasons why some viruses express this activity are as follows. Paramyxoviruses of the genera *Paramyxovirus* and *Rubulavirus*, which have neuraminidase-expressing HN proteins, bind to highly abundant, sialic acid-containing

molecules on the cell surface. Moreover, these highly abundant cell surface molecules are inadvertently incorporated into the viral envelopes at budding. Thus, paramyxoviruses and rubulaviruses express their neuraminidase (also called sialidase) to prevent virus self-aggregation during budding and also help to elute progeny virus particles from the cell surface. In contrast, measles virus, a member of the genus *Morbillivirus* that has HA but not neuraminidase activity, binds to CD46, a low-abundance cell surface protein, which is unlikely to be incorporated into viral envelopes in sufficient amounts to cause self-aggregation during budding. Thus, measles virus particles have little reason to express a neuraminidase, thereby accounting for their lacking that activity. (CD46 is also known as membrane cofactor protein. It is present on most cell types, protecting them against inappropriate complement activation [see Chapter 4].)

The attachment protein of the genus *Pneumovirus*, referred to as the G protein since it lacks both HA and neuraminidase activities, binds to as yet unknown receptors.

BOX 11.4

A variety of both enveloped and nonenveloped viruses have on their surfaces proteins that can bind to erythrocytes. Since multiple copies of these proteins are distributed over their surfaces, these viruses are able to bridge red blood cells together. If a sufficient number of such viruses are mixed with a suspension of red blood cells, then the cells and viruses can form a stable lattice, which remains in suspension. This phenomenon is referred to as **hemagglutination.**

One can measure the number of **hemagglutination units** (HAUs) in a virus sample as follows. A series of successive twofold serial dilutions of the virus are added to a fixed number of red blood cells in conical-bottom microtiter wells. In wells in which hemagglutination does not occur, the red blood cells settle out to form a readily visible red plug at the bottom of the well. The hemagglutination titer is given as the reciprocal of the highest dilution of the virus that still hemagglutinates the red blood cells.

The relationship between plaque-forming units (PFU) and HAUs for a particular virus can be determined empirically. Since hemagglutination assays can be carried out in several hours, one can use HA assays to get a quick estimate of PFU in samples of that virus. Note that HA assays do not yield usable data unless there is some minimal amount of virus

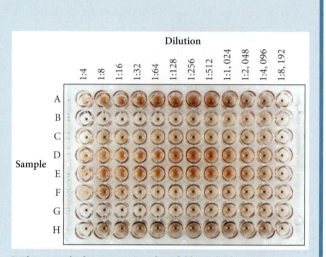

In the example shown, a series of twofold serial dilutions was prepared from different influenza virus samples, and an aliquot of each dilution was mixed with a suspension of chicken red blood cells and added to the wells. After 30 min at 4°C, the wells were photographed. Sample A causes hemagglutination through a dilution of 1:256 and has an HA titer of 256. From S. J. Flint, L. W. Enquist, R. M. Krug, V. R. Racaniello, and A. M. Skalka, *Principles of Virology: Molecular Biology, Pathogenesis, and Control,* 2nd ed. (ASM Press, Washington, DC), with permission.

present in the sample. For example, in the case of influenza virus, about 10^7 virus particles per ml are necessary for hemagglutination to occur.

ENTRY

When paramyxoviruses attach to their cell surface receptors, their fusion proteins undergo conformational rearrangements that promote envelope fusion with the plasma membrane. Thus, membrane fusion in these instances happens at neutral pH, rather than at the low pH found in endosomes. There are several plausible scenarios by which the conformational rearrangements of the paramyxovirus fusion proteins might be triggered at the cell surface. First, note that while different paramyxoviruses use different cell surface proteins for their receptors, in most instances the particular attachment protein of each virus is required to trigger membrane fusion mediated by the separate fusion protein of that same virus (i.e., the attachment and fusion proteins apparently must be from the same virus in order for them to act together to promote fusion). These experimental findings are consistent with the premise that fusion requires a **homotypic** interaction between the attachment and fusion proteins. Moreover, the HN or H

attachment proteins of at least some paramyxoviruses likely undergo their own conformational transitions, upon binding to their receptors. Thus, the changes in the attachment proteins might, in turn, elicit the conformational rearrangements of their homotypic F proteins. In other instances, the interaction of the F protein with a coreceptor might induce the transition to its fusion-active conformation.

The conformational transitions that viral fusion proteins go through to mediate membrane fusion are best understood in the case of the orthomyxoviruses, as discussed in detail in Chapter 12. For now, note that the fusion proteins of influenza virus, human immunodeficiency virus (HIV), and Ebola virus, as well as the fusion proteins of paramyxoviruses (e.g., measles virus, respiratory syncytial virus, Newcastle disease virus, and SV5), have all been found to form coiled-coil trimers. Thus, all of these viral fusion proteins might be expected to promote membrane fusion via the same sequence of conformational transitions.

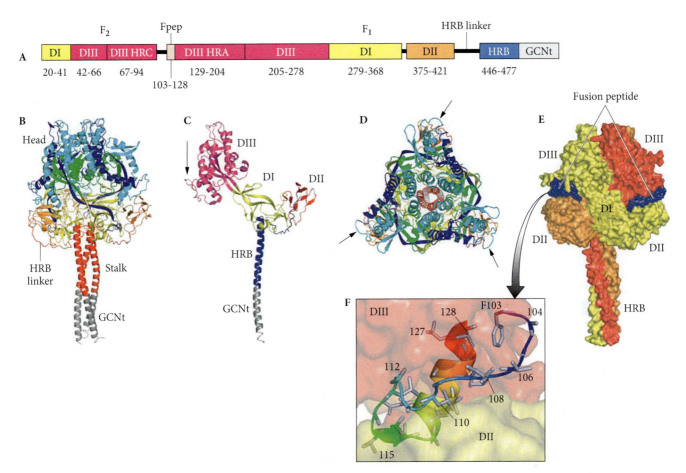

Figure 11.2 Structure of SV5 F-GCNt. (GCNt is a soluble, truncated SV5 F protein that does not contain a hydrophobic transmembrane domain. It was generated because the presence of the hydrophobic region would prevent crystallizing it, a step necessary to solving its structure. Since GCNt does not trimerize efficiently, the investigators appended an engineered, trimeric coiled-coil domain [GCNt] to HRB [heptad repeat B] to mimic the transmembrane domains.) (A) The F-GCNt domains. Important domains are highlighted in different colors, and their corresponding residue ranges are indicated. (B) Ribbon diagram of the F trimer, with each chain colored by residue number in a gradient from blue (N terminus) to red (C terminus). The head and stalk regions are indicated. (C) Ribbon diagram of one subunit of the F trimer, colored by domain. The domains are labeled, and the colors correspond to those in panel A. The arrow indicates the cleavage/activation site. (D) Top view of the trimer, colored as in panel B. Arrows indicate the cleavage/activation sites. (E) Surface representation of the F trimer, colored by subunit. The fusion peptide exposed surface is shown in blue. (F) Close-up view of the fusion peptide (residues 103 to 128). The peptide is folded back on itself with a small hydrophobic core and contains a mixture of an extended chain, a β-strand, and a C-terminal α-helix. The fusion peptide is sandwiched between two subunits of the trimer, between DII and DIII domains. Adapted from H.-S. Yin et al., *Nature* **439**:38–44, 2006, with permission.

The prefusion conformation of the SV5 fusion protein is as follows. Trimers of the protein form a globular head attached to a trimeric coiled-coil stalk, with the head oriented away from the viral envelope (Fig. 11.2). A large cavity is present at the base of the head, which contains the fusion peptides. The latter are sequestered between adjacent subunits, with cleavage sites exposed at the protein surface. (SV5 is a prototypic paramyxovirus not associated with any human disease.)

Viral F proteins in general are thought to facilitate membrane fusion by undergoing a sequence of conformational transitions that drive the fusion process. The first transition extends the hydrophobic fusion peptide from its sequestered location in the molecule upward and into the target membrane. Then, irreversible refolding of the fusion protein brings the target membrane into juxtaposition with the viral envelope. Class I fusion proteins generally are able to undergo these transitions because they initially assume a metastable conformation that subsequently undergoes discrete conformational rearrangements to a lower energy state. Acquisition of the metastable conformation generally requires an activation cleavage step (Box 11.5). The arrow in Fig. 11.2C indicates the cleavage/activation site in one subunit of the SV5 fusion protein. But see Box 11.5.

BOX 11.5

Paramyxovirus fusion proteins are categorized as class I fusion proteins because of the characteristic sequence of conformational transitions that they go through in mediating fusion. The best studied of the class I fusion proteins is the influenza virus HA protein, described in detail in Chapter 12. As noted there and elsewhere in the text, the prefusion conformations of class I fusion proteins are metastable, an important feature that enables them to undergo the triggered conformational rearrangements that promote fusion.

The metastable conformation of the fusion proteins is thought to be brought about by a posttranslational proteolytic cleavage event. However, recent experimental findings imply that protein cleavage may not be necessary for paramyxovirus F proteins to go through the sequence of conformational transitions that lead to membrane fusion. In particular, when the crystal structure of the *uncleaved* F protein of human parainfluenza virus 3 was solved, it was found to already be in the postfusion conformation, implying that proteolytic cleavage of that paramyxovirus fusion protein is not necessary to prime it to undergo the fusion-promoting conformational transitions. Perhaps cleavage is necessary in this instance only to enable the repositioning of the fusion peptide, so that it might insert into the target membrane. In support of this premise, the postfusion conformation of the cleaved protein, but not of the uncleaved protein, is able to interact with membranes.

Figure 11.3 depicts a model for SV5 F protein-mediated membrane fusion. Compare the prefusion and postfusion conformations of the viral F protein. Also, notice the open-stalk intermediate, which is probably followed by refolding of domain III (DIII), refolding of heptad repeat A into a coiled coil, and translocation of the fusion peptide towards the target cell membrane. (These domains are identified in Fig. 11.2.) Notice that the fusion peptides translocate through the globular head of the protein to its end. This step might well depend on activation cleavage, but it is not known with certainty whether cleavage is necessary to engender the metastable form of the protein (Box 11.5).

The attachment proteins of the genus *Pneumovirus* are designated G (for glycoprotein), since they have neither HA nor neuraminidase activities. Interestingly, the G protein of respiratory syncytial virus, a pneumovirus, is required for replication in the respiratory tract of mice, but not for replication in cell culture. Since the G protein of

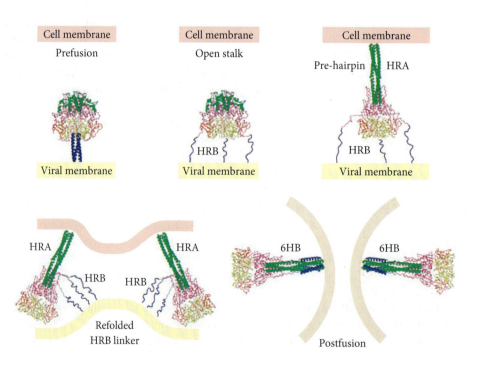

Figure 11.3 A model for paramyxovirus F-protein-mediated fusion. This particular example is based on the solution of the structure of the parainfluenza virus 5 F protein in its metastable, prefusion conformation. Notice the prefusion and postfusion conformations of the parainfluenza virus F protein. Also, notice the open-stalk intermediate, which is probably followed by refolding of domain III (DIII), refolding of heptad repeat A (HRA) into a coiled coil, and translocation of the fusion peptide towards the target cell membrane. Notice that the fusion peptides translocate through the globular head of the protein to its end. The formation of 6HB facilitates membrane fusion by stressing and juxtaposing the viral envelope and target membrane. Adapted from H.-S. Yin et al., *Nature* 439:38–44, 2006, with permission.

this virus is not required for replication in cell culture, it would appear that its F protein might be sufficient to mediate both attachment and fusion. That is, the respiratory syncytial virus G protein might not have a role in entry. The same might be the case for SH, a third respiratory syncytial virus envelope glycoprotein, which also is not required for replication in cell culture.

But why then might the respiratory syncytial virus G protein be required for viability in vivo? A possible explanation is as follows. The virus produces its G protein both as a complete transmembrane form and as an N-terminally truncated form that is secreted (sG). Recent experimental results suggest that sG downregulates the antiviral inflammatory response in infected lungs, in that way enhancing virus infection in vivo. The respiratory syncytial virus G protein helps the virus to subvert host immunity in other ways as well (see below and Box 11.16).

HMPV, like respiratory syncytial virus, is a pneumovirus, and its G protein too is not required for viability in cell culture but is required in vivo. These findings imply that the pneumovirus G protein serves purposes other than promoting F-protein-mediated membrane fusion. What that function might be in the case of HMPV is not known. (Might HMPV G protein deletion mutants have potential as an attenuated HMPV vaccine? See below and Box 11.16.)

In all instances, the intact paramyxovirus helical nucleocapsids, which contain the genomic RNA, the NP proteins, and the transcription/replication enzymes, are released into the cytoplasm. There is no further disassembly of the nucleocapsids during the ensuing replication cycle.

Syncytium Formation

Since paramyxovirus fusion proteins promote fusion at neutral pH, paramyxovirus-infected cells, which express viral glycoproteins at their surfaces, can fuse with adjacent, uninfected cells. Thus, paramyxovirus infection can give rise to **syncytia**. (A syncytium is a mass of cytoplasm within a single cell membrane that contains multiple nuclei, in this instance as the result of cellular fusion.) Indeed, formation of syncytia is a hallmark of paramyxovirus infection in cell culture. But do paramyxoviruses induce syncytium formation in vivo as well, and if so, how might syncytium formation benefit the virus? The answer to the first part of the question is that syncytia indeed can be found in tissues that have been infected by a variety of paramyxoviruses. (In the case of Nipah virus, an emergent paramyxovirus that causes fatal encephalitis in about 70% of infected individuals, the virus targets blood vessel endothelial cells, and syncytial

blood vessel endothelial cells are a characteristic feature of Nipah virus infection.) An answer to the second part of the question is that, in principle, being able to form syncytia could benefit the virus by enabling it to spread directly from cell to cell, thereby avoiding antiviral antibodies.

GENOME ORGANIZATION, EXPRESSION, AND REPLICATION
Genome Organization and Transcription

The key features of paramyxovirus gene expression are as follows. (i) All paramyxovirus transcription occurs on intact nucleocapsids (recall that paramyxovirus nucleocapsids contain the genomic RNA, the NP proteins, and the transcription/replication enzymes and that they remain intact throughout the entire viral replication cycle). (ii) Paramyxovirus polymerase complexes attach to the genomic RNA only at its 3′ end. (iii) The individual viral mRNA species are transcribed in an obligatory sequence, in the 3′-to-5′ direction on the genomic RNA template. (iv) Each of the mRNA species is capped, and a poly(A) sequence is added to each by **reiterative transcription** at the intergenic junction sequences. (v) Attenuation of transcription occurs at each intergenic junction sequence.

Note that the pattern of paramyxovirus gene expression is similar to that of the rhabdoviruses (Chapter 10) in each of the above respects. Apropos this point, the order of genes on paramyxovirus genomes is similar to the order of genes on rhabdovirus genomes. That is, genes having a similar function in these two virus families have similar positions on their respective genomes. This point is illustrated by comparing the schematic of the Sendai virus genome (Fig. 11.4) with that of a rhabdovirus genome (Fig. 10.3). The paramyxovirus NP protein and the rhabdovirus N protein have similar functions, and the gene encoding each is at the 3′ terminus of its respective genomic RNA. The next gene on both the paramyxovirus and rhabdovirus genomes is that which encodes the P protein. (The P genes of paramyxoviruses may encode yet other proteins, in addition to the P protein. This is the reason for the designation P/C/V in Fig. 11.4. See below.) The similarity in gene sequence between the two virus families continues with the M protein, followed by the envelope glycoproteins: two (e.g., F and HN) in the case of Sendai virus and one in the case of the rhabdoviruses. Finally, in each instance, the gene at the 3′ end of the genomic RNA is that encoding the L protein.

The conservation of gene order between the paramyxoviruses and rhabdoviruses, and indeed among the mononegaviruses in general, probably reflects the following.

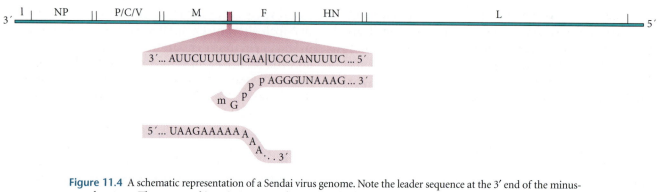

Figure 11.4 A schematic representation of a Sendai virus genome. Note the leader sequence at the 3′ end of the minus-strand genome. The conserved junction sequence between each of the individual genes is indicated. The first nine nucleotides of that sequence, beginning with the tetranucleotide AUUC, comprise the end (E) sequence of the upstream gene. That AUUC sequence signals the polyadenylation of the nascent mRNA, which occurs by reiterative transcription of the adjacent downstream poly(U) sequence, as shown. The start (S) sequence for the upstream gene begins with the tetranucleotide AGGG. Notice the cap on the mRNA initiated at that site. The intergenic (I) sequence, GAA, is not transcribed. The rhabdovirus consensus E sequence contains the tetranucleotide AUAC, rather than AUUC, and the rhabdovirus I sequence is GA, rather than GAA. There is no homology between the rhabdovirus and paramyxovirus S sequences, but in each family that sequence is 10 nucleotides long.

First, as noted above and in Chapter 10, transcription of mononegavirus genomes occurs in an obligatory sequence, in the 3′-to-5′ direction on the minus-strand template. Second, there is an attenuation of transcription at each of the gene junctions, resulting in the Sendai virus example of a gradient of gene expression corresponding to the gene order (i.e., NP > P > M > F > HN > L). Thus, a likely benefit to the virus of this transcription stratagem is that it results in closer-to-optimal relative levels of expression of each gene.

Next, we consider the nucleotide order of the conserved junction sequence between each of the paramyxovirus genes (Fig. 11.4). First, notice the tetranucleotide AUUC, which is present in the "end" portion of each junction sequence. (The end portion is so named because it specifies termination of transcription of the gene upstream of the junction.) These four nucleotides signal the polyadenylation of the nascent mRNA, which occurs by reiterative transcription of the adjacent downstream poly(U) sequence.

It is interesting to compare the junction sequence of the paramyxoviruses to that of the rhabdoviruses. While the end portion of the paramyxovirus junction sequence contains the tetranucleotide AUUC, the rhabdovirus end sequence instead contains the tetranucleotide AUAC (Fig. 11.4). That difference of just a single nucleotide is the only difference between the nine-nucleotide-long end sequences of the paramyxoviruses and rhabdoviruses. But in contrast to the similarity between the paramyxovirus and rhabdovirus end sequences, there is no homology between the "start" portions of their junction sequences. However, in both families, that sequence is 10 nucleotides long.

The conservation of the end sequence, rather than the initiation sequence, perhaps points to some special functional significance of the nucleotide order of the end sequence. Between the end sequence and the start sequence, there is the consensus "intergenic" sequence, which is GAA in the case of the paramyxoviruses (Fig. 11.4), whereas it is GA in the rhabdoviruses. These similarities between the paramyxovirus and rhabdovirus patterns of gene expression, gene arrangements, and gene junction sequences are all consistent with the possibility that these two families of the *Mononegavirales* derive from a common ancestor.

Before we consider some unique aspects of paramyxovirus gene expression, note that by convention, a paramyxovirus "gene" is defined as a sequence that encodes a single messenger RNA (mRNA), even if that mRNA contains more than one open reading frame and is translated into more than one protein. With that convention in mind, one difference between the patterns of gene expression of the rhabdoviruses versus the paramyxoviruses is that the paramyxoviruses have evolved several different mechanisms for expressing multiple proteins from the P gene. (This was the reason noted above for the designation P/C/V in Fig. 11.4.)

RNA editing is perhaps the most notable of the mechanisms used by paramyxoviruses to generate multiple proteins from their P genes. RNA editing was first described in the case of a trypanosomal mitochondrial mRNA in which U residues were added and deleted at multiple sites along the mRNA molecule. The enzymatic conversion of one base for another (e.g., a C to a U) is an additional form of RNA editing used by cells.

In the case of the paramyxoviruses, RNA editing involves the viral RNA polymerase occasionally inserting an extra G residue at a specific point in the open reading frame of the P mRNA. Because the reading frame is shifted downstream of the extra base, translation of the edited mRNA generates a protein that has the same N terminus as the P protein but a unique C terminus. This additional protein is the V protein, whose function is not yet known. Oddly, in the case of some paramyxoviruses, the V protein is translated from the unedited mRNA, while the P protein is translated from the edited mRNA. In some instances, a second G residue is inserted by RNA editing. This results in translation of the W protein. As in the instance of the V protein, the function of the W protein is not clear.

The mechanism of RNA editing used by the paramyxoviruses is somewhat reminiscent of the process of reiterative transcription, by which paramyxoviruses and rhabdoviruses add poly(A) tails to their nascent mRNAs. That is, the genomic RNA contains a poly(C) tract at the RNA editing site. The RNA polymerase is believed to "stutter" at that site, occasionally slipping backwards by one or two nucleotides on the template, thereby inserting one or two extra G residues.

Paramyxoviruses have yet other mechanisms to generate multiple proteins from their P gene. For example, they make use of multiple in-phase translation initiation sequences that are recognized with different levels of efficiency. Although this may seem to be at variance with the ribosomal scanning model for initiation of translation, there is no reason in principle why a scanning ribosome cannot occasionally bypass an initiation sequence that is in a less favored context. In the case of Sendai virus, this RNA editing stratagem results in the production of proteins Y1, Y2, and C′, in addition to P. (Recall a similar mechanism used by the reoviruses [Chapter 5], in which the S1 mRNA encodes two unrelated proteins, σ1 and σ1s. However, in that case the two initiation codons are not in the same reading frame. But the phage Qβ A$_1$ protein is generated by a leaky scanning mechanism, in which its coding sequence is in the same reading frame as the larger open reading frame in which it is embedded [Chapter 6].)

In no instance is the function of an additional P gene product known with certainty. There is some evidence suggesting that the V, W, and C proteins may serve to regulate transcription versus replication. However, none of these proteins appears to be crucial for replication, and none is universally present among all the paramyxoviruses (Box 11.6).

Replication

Replication requires that the minus-strand genomic RNA be transcribed into full-length plus strands, which then can serve as templates for the synthesis of progeny genomic RNAs. Since transcription involves the generation of sub-genome-length mRNAs, there must be some means

BOX 11.6

Here is another rather remarkable feature of paramyxovirus gene expression. Several of the paramyxoviruses, including Sendai virus, measles virus, and human parainfluenza virus type III, appear to replicate efficiently only when their genomes contain a precise multiple of six nucleotides, leading to the so-called **rule of six**. The explanation for the rule of six may be related to the facts that the intact nucleocapsid, rather than naked RNA, is the template for RNA synthesis and that each NP molecule is associated with a hexanucleotide in the nucleocapsid. In addition, the transcription start site for each gene tends to be conserved relative to the hexamer phasing. It is not clear why that should be so.

Interestingly, the remarkable rule of six is not applicable to all paramyxoviruses. For example, it does not apply to respiratory syncytial virus. This is all the more noteworthy since the respiratory syncytial virus genome contains 15,222 nucleotides, a number that *is* precisely divisible by 6. Yet when respiratory syncytial virus genomes that did not contain an exact multiple of six nucleotides were generated, those altered genomes were found to be transcribed as efficiently as standard-length genomes.

What might account for why some paramyxoviruses strictly adhere to the rule of six whereas others do not? Some researchers have suggested that the explanation is to be found in the NP and P proteins. Those viruses that adhere to the rule of six have larger NP and P proteins than respiratory syncytial virus. For example, measles virus and Sendai virus, which adhere to the rule of six, have NP and P proteins that contain 525 and 507 amino acids and 524 and 568 amino acids, respectively. In contrast, the respiratory syncytial virus NP and P proteins contain 391 and 241 amino acids, respectively.

Another intriguing fact is that those paramyxoviruses that obey the rule of six are the same paramyxoviruses that cotranscriptionally edit their mRNAs (see the text). The significance, if any, of that correlation is also not clear.

for switching over from transcribing mRNAs to generating full-length plus strands. Not surprisingly, paramyxoviruses accomplish this switchover in the same way as the rhabdoviruses. That is, as newly synthesized NP proteins accumulate in the cell, they bind to nascent plus strands, thereby impairing the ability of those nascent RNAs to undergo the "stuttering" necessary for them to respond to polyadenylation and termination signals on the genomic RNA. Note that the plus-sense, antigenomic RNA, like the genomic RNA, is present only in the form of ribonucleoprotein complexes. Importantly, there are no junctional editing signals on the plus-sense antigenome, so that it serves as the template for generating only full-length progeny genomes.

Paramyxovirus genomes, like rhabdovirus genomes, contain a sequence at their 3′ ends that is transcribed into a leader sequence, which appears at the 5′ end of full-length plus strands, but not on sub-genome-length mRNAs. The antileader sequence on the genomic RNAs contains the encapsidation signal for the packaging of progeny nucleocapsids into new virus particles. Its absence from the mRNAs explains why they, like plus strands, are not packaged into virus particles. Moreover, it explains why there is a termination and reinitiation sequence between the antileader sequence and the coding sequence for the first mRNA.

ASSEMBLY AND RELEASE

The assembly and release phases of the paramyxovirus replication cycle, like the earlier stages, resemble the corresponding phases of the rhabdovirus replication cycle. In brief, NP proteins associate with nascent genomic RNA molecules to initiate helical nucleocapsid assembly. P/L polymerase complexes also associate with each progeny helical nucleocapsid. The integral membrane glycoproteins, HN and F, are synthesized and transported to the plasma membrane by the cellular secretory pathway. The F protein, in at least some instances, is synthesized as a biologically inactive precursor that undergoes proteolytic cleavage in the *trans* Golgi apparatus. This step activates the fusion protein so that it can mediate the next entry stage.

Although the molecular mechanisms that mediate the paramyxovirus budding process per se are not well understood, the M protein is thought to play a major role in the process. The M protein of the rhabdoviruses also plays a pivotal role in the assembly and budding of members of that virus family (Chapter 10). The M protein of both the paramyxoviruses and the rhabdoviruses associates with the inner leaflet of the plasma membrane. Moreover, expression of the cloned M gene of at least some members of each of these virus families causes the membrane to curve and pinch off to generate virus-like particles that contain the respective M protein. And as expected, the M protein serves as the "bridge" between the nucleocapsids and the cytoplasmic tails of the envelope glycoproteins. The Sendai virus M protein appears to interact specifically with the cytoplasmic tails of both the HN and F glycoproteins during final virus assembly and budding.

MEDICAL ASPECTS

The paramyxovirus family contains a number of highly contagious and important human pathogens. Among these are measles virus, mumps virus, respiratory syncytial virus, the parainfluenza viruses, the recently discovered HMPV, and the newly emergent Nipah and Hendra viruses.

The paramyxoviruses in general are highly contagious, largely because they are transmitted primarily via the respiratory route. In that regard, the respiratory tract is the most common means by which viral infections are transmitted between humans. The reason the respiratory tract is a most attractive point of entry for a virus is that it has a large surface area comprised of a delicate mucosa that is in intimate contact with the vasculature. While these features enable the respiratory mucosa to carry out its principal role of oxygen and CO_2 exchange, they also make it an attractive site for viruses to initiate an infection (Chapter 3). (As you might have supposed, the respiratory tract is in turn notably well defended by immune effectors [Chapter 4].) Moreover, the respiratory route of transmission takes on an especially prominent role in the modern world, where good personal and public hygiene can protect us, at least to some extent, against infections that are transmitted by other routes (i.e., the fecal/oral route, intimate contact, and insect vectors). However, to avoid infections via the respiratory route, one would essentially need to be a hermit. Specific paramyxovirus infections and their associated disease syndromes are considered below.

Measles
Clinical Conditions
Before the introduction of measles vaccines and the global eradication program coordinated by the World Health Organization, death rates from measles may have been as high as 7 to 8 million children, worldwide, annually. Nevertheless, despite the development of an effective live-attenuated measles vaccine, there still are more than 30 million measles cases per year worldwide, of which more than 1 million are fatal. As you might expect, the vast majority of fatal measles cases occur in unvaccinated populations. In fact, in some unprotected groups, measles is the major cause of death in children less than 5 years of age.

Why might measles virus infections have the potential to be so virulent? The answer in part may be as follows. The rhinoviruses, coronaviruses, and influenza virus (Chapter 12), like measles virus, are transmitted by the respiratory route. However, unlike these other "respiratory" viruses, which remain in the respiratory tract, measles virus is able to disseminate widely throughout the body. One reason it can do so is that it can bind to several different receptors, including CD46, which is present on most cells of the body. Because measles virus can disseminate widely, serious complications from measles can occur in almost every organ system. Moreover, as described below, measles can lead to an immunosuppressed state that can last for as long as 6 months after the acute infection. During this time, individuals are susceptible to opportunistic infections, which in fact are the major cause of measles fatalities (Box 11.7).

Despite its virulence, measles is often regarded as no more than a common, relatively innocuous childhood disease. Indeed, prior to the availability of the vaccine, children in early school years regularly were deliberately exposed to the virus, either to get past this childhood "rite of passage" or because of awareness of the greater severity of measles in adults. Yet infection can be quite severe, even in infants and children, as described in more detail below.

Measles is one of the five classic childhood exanthems (i.e., eruptive or rash-like diseases). The others are rubella (caused by rubella virus, a togavirus), roseola (caused by human herpesvirus 6), fifth disease (caused by human parvovirus B19), and chicken pox (caused by varicella-zoster virus, a herpesvirus). The characteristic measles rash is caused by immune T cells that are targeted to measles virus-infected endothelial cells lining the small blood vessels under the skin. This same cell-mediated immune response likely contributes to other measles symptoms as well. Nevertheless, that immune response is crucial to controlling and eventually clearing the infection. (In view of the above, in individuals with severe deficiencies in cell-mediated immunity, measles is characterized by the absence of rash, severe complications, and a high fatality rate.)

The measles incubation period lasts 1 to 2 weeks, with the following key events taking place during this time. Measles virus initially infects the respiratory mucosa (Box 11.7), but only small amounts of transmissible virus are produced at this early stage. However, measles virus is able to infect dendritic cells associated with this mucosa. Infected dendritic cells traffic to lymph nodes, which then become secondary sites of infection, in that way giving rise to a massive secondary viremia. This viremia enables the virus to infect other tissues and to reinfect the respiratory

BOX 11.7

Measles virus pathogenesis is currently believed to result from a systemic infection, which follows a primary infection of the airway epithelium. Immune cells are thought to be a secondary target of the infection.

A brand new study implies a different scenario, in which lung lymphocytes are the first cells to be infected (Leonard et al., 2008). These primarily infected lymphocytes then transport the virus to the basolateral surface of the airway epithelium, which then sustains measles virus shedding into the airway lumen, a step that is crucial for transmission of the virus. Nevertheless, primary infection of the lymphocytes is sufficient to give rise to viremia and clinical measles.

This new view of measles virus pathogenesis is based on studies in which a mutant measles virus was generated, which has an attachment protein that no longer recognizes the measles virus receptor on airway linings but which still recognizes the measles virus receptor on lymphocytes. When rhesus monkeys were infected through the nose with this measles virus mutant, they showed clinical signs of measles infection, including rash and weight loss.

Importantly, they also developed viremia, but they did not shed virus into the airways. Thus, replication in the airway epithelium is necessary for virus shedding into the airway lumen, but it is not necessary for viremia and pathogenesis. Instead, virus replication in immune cells only can cause viremia and measles in monkeys.

Interestingly, in experiments in which wild-type measles virus was applied to polarized columnar epithelial cells that bore tight junctions, the virus was able to infect these cells only when applied basolaterally. As expected, the mutant virus could not infect these cells. Thus, the measles virus receptor on epithelial cells is probably a basolateral protein, which the virus accesses via infected lymphocytes. But, more importantly, infection of the airway epithelium may not be essential for systemic spread and virulence of measles virus, while it remains essential for transmission of the virus.

Reference
Leonard, V. H., P. L. Sinn, G. Hodge, T. Miest, P. Devaux, N. Oezguen, W. Braun, P. B. McCray, Jr., M. B. McChesney, and R. Cattaneo. 2008. Measles virus blind to its epithelial cell receptor remains virulent in rhesus monkeys but cannot cross the airway epithelium and is not shed. *J. Clin. Investig.* 118:2448–2458.

mucosa. The secondary attack on the respiratory mucosa is initiated by many more viruses than initiated the primary infection, resulting in the generation of many more infectious virus particles than were produced during the primary infection. See Chapter 3 for additional details.

The first measles symptoms to appear include high fever, cough, conjunctivitis, and coryza (cold-like symptoms). Importantly, this is the time when the patient is most infectious. Yet it is still not obvious that the patient has measles. Consequently, the patient may be freely interacting with others and spreading the virus.

Koplic's spots, which are lesions on the mucous membranes of the mouth, appear about 2 days later. Their characteristic appearance indicates, beyond a doubt, that measles is present. The Koplic's spots are soon followed by the distinctive measles rash. The patient usually is sickest on the day the rash appears.

The acute infection is also characterized by a general immunosuppression. This may be explained by several effects of measles virus. First, measles virus infects B cells and T cells, as well as macrophages, and in so doing arrests them in the G_1 phase of the cell cycle, thereby impairing their immunity functions. Second, measles virus attacks natural killer (NK) cells, leading to the downregulation of their lytic activity. Third, when measles virus binds to macrophages, its HA proteins cross-link CD46 molecules at the cell surface. This has the effect of suppressing production by the macrophage of interleukin-12 (IL-12), which, in turn, weakens the Th1 immune response (Chapter 4). Together, these actions of measles virus result in an immunosuppression that can last up to 6 months after an acute infection. This is of no mere academic interest, since the vast majority of the million or more children in Third World countries who die each year from measles infections actually succumb to opportunistic infections that occur during the measles-associated, transient immunosuppression (Box 11.8).

Some measles patients experience severe complications due to the direct effect of the virus per se. Moreover, these complications can occur in almost every organ system. The most common complications are pneumonia, croup, otitis media (an inflammation of the middle ear), and encephalitis. Measles is also a common cause of blindness in developing countries.

Pneumonia accounts for 60% of all measles deaths in developing countries (Box 11.9). In fact, measles-related pneumonias are more than twice as likely to be fatal than are severe pneumonias in which there is no measles involvement. Some of the measles pneumonia-related fatalities may be due to the direct effect of measles virus in the lung, but many, if not most, are related to a secondary bacterial or viral infection. Thus, the reasons for the high mortality rate of measles pneumonia include the immunosuppressive effects of measles infection (involvement of the lung in measles leads to severe disruption of lung immunity) and the likelihood of secondary bacterial or virus infection.

Otitis media is the most frequent complication of measles among children under 5 years of age in the United States, occurring in 14% of cases in that age group. It results from inflammation of the epithelial surface of the eustachian tube, which causes the eustachian tube to become obstructed, which in turn predisposes it to secondary bacterial infection. The fact that the diameter of the eustachian tube is small in young children likely accounts for why this complication occurs most often in that age group. Note that another paramyxovirus, respiratory syncytial virus (see below), is actually the most frequent cause of otitis media. Also, several bacterial pathogens (e.g., *Streptococcus pneumoniae*, *Haemophilus influenzae*, *Staphylococcus aureus*, and *Pseudomonas aeruginosa*) also are associated with otitis media.

The most relentless complications of measles virus infections occur within the central nervous system, in which

BOX 11.8

Interestingly, the first indications that a viral infection can impair host immunity were the incidents of measles patients who were not able to respond to a skin test for tuberculosis. Responsiveness to the test was partially recovered after the measles infection resolved. For history buffs, these observations were made by von Pirquet in 1908.

Reference
von Pirquet, C. 1908. Das Verhalten der kutanen Tuberkulin-Reaktion während der Masern. *Dtsch. Med. Wochenschr.* **34:** 1297–1300.

BOX 11.9

Malnutrition is an important contributing factor to severe measles pneumonia in developing countries. A vitamin A deficiency is thought to be the key deficit in this group of patients, since this deficiency causes the pulmonary epithelia to be more vulnerable to measles virus itself, as well as to secondary bacterial infection. For these reasons, vitamin A is prescribed to measles patients to reduce the likelihood of pneumonia and mortality. At present, an effort is under way in Africa to use vitamin A supplementation along with vaccination to reduce child mortality from measles.

three distinct forms of measles postinfection encephalitis are recognized. They are **acute disseminated encephalomyelitis** (ADEM), **measles inclusion body encephalitis** (MIBE), and **subacute sclerosing panencephalitis** (SSPE).

ADEM, which is the acute form of measles postinfection encephalitis, is considered to be the most dangerous complication of measles in the First World, and it is a principal reason for vaccination there against measles. ADEM occurs with an abrupt onset, about 5 to 6 days after the initial measles rash, in about 1 of 1,000 infected children, with a mortality rate of about 20%. Symptoms consist of a sudden recurrence of the fever, decreased consciousness, seizures, and other neurological signs. The neuropathology includes widespread perivascular demyelination (myelin sheaths around neurons are necessary for efficient nerve conduction in the central nervous system [Chapter 3]) and infiltration of mononuclear cells (cells that originate in the bone marrow, which differentiate into macrophages in the tissues). One-third of the patients who survive postinfection encephalitis have lifelong neurological **sequelae** (a sequela is a disorder caused by an earlier disease or injury in the same individual), including severe mental retardation, motor impairment, blindness, and sometimes hemiparesis (weakness on one side of the body).

How measles virus enters the central nervous system to cause acute encephalitis is not known. One possibility is that the virus invades the central nervous system by infecting T cells, which are able to cross the blood-brain barrier. Another possibility is that measles virus might infect brain endothelial cells. Regardless, sensitive reverse transcriptase polymerase chain reaction (PCR) in situ hybridization techniques show that the brain frequently becomes infected during acute fatal measles.

The mechanism by which measles virus induces demyelination in ADEM also is not known with certainty, although it is generally thought to be via immunopathological processes, rather than by direct viral cytopathic effects. The following scenario is often put forward to explain demyelination in ADEM. Infection causes brain myelin basic protein to appear in the patient's cerebrospinal fluid. This causes the patient's T cells to become reactive against myelin, resulting in autoimmune demyelination (Box 11.10).

MIBE, the second form of measles postinfection encephalitis noted above, is also referred to as subacute measles encephalitis, acute encephalitis of the delayed type, and immunosuppressive measles encephalitis. The last designation reflects the fact that MIBE occurs only in immunocompromised individuals. MIBE typically appears between 2 and 6 months after the acute measles infection and presents with seizures and an altered mental state. The mortality rate is greater than 75%, and as in the case

BOX 11.10

ADEM may follow vaccination against measles, mumps, and rubella (i.e., the combined MMR vaccine) in about 1 to 2 per million vaccine recipients, which is about 1,000-fold less than the risk of developing ADEM from an actual measles infection. A number of other viruses, as well as bacterial pathogens, are also thought to induce ADEM. Among these are influenza virus, mumps virus, rubella virus, varicella-zoster virus, Epstein-Barr virus, cytomegalovirus, herpes simplex virus, hepatitis A virus, coxsackievirus, and the bacteria *Mycoplasma pneumoniae*, *Borrelia burgdorferi*, and beta-hemolytic streptococci.

Several other antiviral vaccines have also been associated with ADEM. Consistent with the premise that viruses induce ADEM by causing myelin basic protein to become available for recognition by T cells, some vaccines, which later were shown to have been contaminated with host central nervous system tissue, gave rise to ADEM in as many as 1 in 600 vaccine recipients!

of ADEM, survivors have lifelong neurological sequelae. Interestingly, the measles viruses implicated in MIBE are usually replication defective. Replication-defective measles viruses are also implicated in the pathobiology of SSPE, the third measles virus postinfection encephalitis, which is discussed below under a separate subheading.

Notwithstanding the potential for measles virus infections to be highly virulent, most measles patients recover completely. They then enjoy lifelong immunity because of the following properties of measles virus. First, the major antigenic determinants on measles virus are on its envelope glycoproteins, HN and F. Second, although measles virus is an RNA virus and therefore generates mutations at a high rate, the parts of its envelope glycoproteins that are targets for neutralizing antibodies cannot mutate and still carry out their respective functions, namely, receptor binding and fusion. Consequently, only one serotype of measles virus circulates in the human population. Therefore, infection results in lifelong immunity, since there are no viable antigenic variants that might evade immunity.

SSPE

SSPE is one of only a few human demyelinating diseases known to be caused by a conventional virus (as opposed to prion diseases [Chapter 3]). Moreover, it is a paradigm of a kind of persistent infection that begins as an acute infection, which on rare occasions gives rise to late complications. Cases of SSPE have become rare indeed since the advent of the measles vaccine. Nevertheless, the disease is of great interest for what it might reveal about the pathogenesis of persistent infection in the brain and demyelinating disease in general (Box 11.11). For these reasons, SSPE is considered here under a separate subheading.

We begin by comparing SSPE with ADEM and MIBE. As noted above, ADEM appears 5 to 6 days after the initial measles rash and MIBE appears 2 to 6 months after the acute infection. In contrast to these forms of measles postinfection encephalitis, SSPE occurs between 5 and 10 years after the initial measles virus infection.

SSPE occurs in about 1 of every 1,000,000 measles cases, although there are regions where the incidence of SSPE is reported to be as high as 1 in every 25,000 measles cases. Apparent risk factors for SSPE include initial measles virus infection before the age of 2 years, large family, overcrowding in the home, and rural place of birth. The last three of these factors may increase risk because they contribute to a higher likelihood of measles infection before the age of 2.

Early in SSPE disease, children show a loss of attention span and neurological signs such as sudden muscle spasms. As the disease progresses, they slip into a vegetative state, which invariably ends in death.

Much remains to be explained about SSPE, but the following facts about the disease may be relevant to understanding its underlying basis. First, SSPE is characterized by exceptionally high titers of antimeasles neutralizing antibodies in both the cerebrospinal fluid and serum. In fact, antibodies against the N, F, and H proteins are present

BOX 11.11

Demyelinating diseases that are not associated with virus infections are known in both humans and animals. Examples in humans include adrenoleukodystrophy (a hereditary disorder of the nervous system in boys that affects the adrenal glands) and demyelination associated with vitamin deficiency and toxins. However, despite these examples, nearly all demyelinating diseases of known etiology, in both humans and animals, are viral in origin.

The two most prominent viral demyelinating diseases in humans are SSPE, caused by measles virus (discussed in this chapter), and progressive multifocal leukoencephalopathy, caused by JC virus (discussed in Chapter 15). Other important examples include HIV encephalitis and human T-cell lymphotropic virus type 1 (HTLV-1)-associated myelopathy (Chapter 21). With the exception of SSPE (see the text), in each of these instances, the etiologic agents clearly infect the central nervous system. In other demyelinating diseases, such as Guillain-Barré syndrome, which is associated with several different viruses (e.g., influenza virus, Epstein-Barr virus, and cytomegalovirus [Chapter 4]), and postinfection encephalitis, which also is associated with several viruses (e.g., measles, mumps, influenza, and vaccinia viruses), demyelination follows a systemic virus infection, but there is rarely any evidence of direct virus infection of the central nervous system. Demyelination in the latter set of instances is thought to be autoimmune mediated. On the other hand, many viruses, rabies virus for example, infect the central nervous system without causing demyelination.

Why do some viruses that invade the central nervous system cause demyelination while others do not? The tropism of the virus for a particular cell type is one factor. For example, JC virus targets and destroys oligodendrocytes, the progenitor cells for central nervous system myelin. In contrast, neurotropic viruses that do not cause demyelination rarely infect oligodendrocytes. For instance, rabies virus, which does not cause demyelination, primarily infects neurons. Nevertheless, in most instances, demyelination is thought to result primarily from immune responses, rather than the direct destruction of oligodendrocytes by the virus.

One last point about these known instances of virus-induced demyelination: in most instances, in both humans and animals, infections with demyelinating viruses only rarely lead to demyelinating disease. For example, SSPE occurs in only about 1 of every 1,000,000 measles cases. Also, most humans have lifelong JC virus persistent infections. Yet progressive multifocal leukoencephalopathy occurs only rarely, usually in association with immunosuppression (Chapters 15 and 21).

Multiple sclerosis is the most prominent demyelinating disease in humans, and although it has been studied since the late 19th century, its etiology is still not known. However, it is generally thought to be an autoimmune disorder, possibly triggered by a virus; this premise is consistent with the fact that most demyelinating diseases of known etiology, in both humans and animals, are viral in origin, as noted above. (For more on multiple sclerosis, see Chapter 4, Boxes 4.18 and 4.19.) Also note that our understanding of the pathogenesis of virus-induced demyelination derives for the most part from the study of animal models, such as that involving the picornavirus Theiler's murine encephalitis virus (Chapters 4 and 6).

in concentrations that are nearly 10-fold greater than seen during acute infection. On the other hand, there is actually very little, if any, virus in the brains of patients with SSPE, although nucleocapsids, their constituent proteins, and viral RNA can be detected. The levels of the M, F, and H proteins are much reduced, and the L protein is not detected at all (although some L protein presumably is made to sustain the low levels of transcription and genome replication).

Of the many issues concerning SSPE that remain unanswered, two major ones are as follows. First, the mechanisms that might turn an acute measles infection into a chronic one are not known. Second, the mechanisms underlying the demyelination seen in SSPE are not yet clear.

Several mechanisms have been proposed to account for the establishment and maintenance of the chronic measles virus infection that leads to SSPE. Some of these putative mechanisms postulate that defective viruses play a key role. For example, M protein mutants might have a limited ability to bud and spread, enabling them to persist in the host for long periods of time. Other mechanisms hypothesize that host immunity plays a key role in the establishment of the chronic infection, consistent with the exceptionally high antimeasles antibody titers seen in both the serum and the cerebrospinal fluid. The high antibody titers might be expected to limit virus spread. But in addition to their role in neutralizing extracellular virus, antimeasles antibodies also may be responsible in part for somehow downregulating measles transcription in neurons, as suggested by experimental results from studies of infected neurons in cell culture. Moreover, the neurons themselves have some property that inhibits measles transcription, even in the absence of antibodies. Finally, neurons do not generally express major histocompatibility complex class I molecules, which make neurons poor targets for cellular immune responses. These factors may conspire to ensure both the persistence of measles virus in the brain and the slow progression of SSPE.

Several mechanisms to account for the demyelination seen in SSPE likewise have been put forth. Most suggest that the demyelination results from immunopathological mechanisms, consistent with the unusually high antibody titers present in the cerebrospinal fluid and with the experimental finding that these antibodies and complement together are able to lyse brain cells from SSPE patients in vitro. (Antibody and complement together may lyse virus-infected cells when virus antigens are expressed at the cell surface [Chapter 4].) Moreover, CD4 and CD8 T cells are known to enter the infected brain, where they can release proinflammatory cytokines, such as tumor necrosis factor alpha and gamma interferon, which also can contribute to immunopathology. Additionally, if it is true that

the virus is slowly replicating in the brain, then direct cytopathic effects of the virus might contribute to the destruction of the myelin-generating oligodendroglial cells.

Interestingly, while it is clear that measles virus is the cause of SSPE, the widely believed assumption that measles virus is resident in the central nervous system between the acute infection and the onset of SSPE has never been verified. This is an important proviso, since most models to account for the SSPE disease process are based on the premise that the virus is present in the central nervous system. But in truth, the precise site of residence and replication of the virus during the "latency" period has never been demonstrated. Actually, there is precious little evidence that infectious measles virus is present in the central nervous system. Indeed, even after the onset of SSPE, the case for infectious virus in the brain is largely indirect, based on evidence such as the presence of measles **inclusion bodies** (Box 11.12) and the high antibody titers in the cerebrospinal fluid. Even at autopsy, infectious virus cannot be detected in brain samples. Thus, the possibility

---

BOX 11.12

Generally speaking, inclusion bodies are morphological features appearing in virus-infected cells that can be detected using light microscopy. Inclusion bodies seen in SSPE appear to contain accumulations of measles virus structural components. However, in other instances, inclusion bodies reflect degenerative cellular changes, such as the condensation and margination of chromatin seen in herpesvirus-infected cells.

Nerve cell from the substantia nigra (a region of the midbrain) of a child with MIBE, who was being treated for acute lymphoblastic leukemia. The cell has nuclear and cytoplasmic (arrows) inclusions. This is a light microscopy image, stained with hematoxylin and eosin. Magnification, ×1,760. From H. C. Drysdale et al., *J. Clin. Pathol.* **29**:865–872, 1976, with permission.)

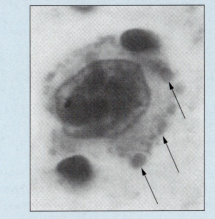

remains that measles virus might have sites of latency outside the central nervous system.

Where Did Measles Come from?

Measles virus is one of the most contagious pathogens known, with a reproduction number of 15 (see Box 9.12). One reason why measles virus is so contagious is that an infected individual can transmit the virus for a period of about 2 weeks, which is the time that elapses from just before the onset of symptoms to when the infection is cleared. At any rate, virtually every susceptible individual in an unprotected population is infected by early childhood. Moreover, each of those infected individuals either succumbs to measles or develops lifelong immunity. These facts would seem to create a dilemma for the virus. Specifically, how is measles virus able to keep finding new susceptible individuals to sustain itself in the population?

The answer to the above dilemma is that measles virus can survive only in populations that are large enough to continually generate a sufficient number of new susceptible individuals. Mathematical modeling by epidemiologists has led to the estimate that a community of at least 500,000 persons is necessary to sustain measles virus. This leads to the notable conclusion that measles virus became established in the human population relatively recently, since communities of adequate size to sustain measles virus did not exist before the last several thousand years, and the small communities that did exist had only occasional contact with other human groups. (Note that this line of argument leads to the more general conclusion that epidemic diseases in humans are a relatively recent development [Box 11.13].)

The above facts also explain why measles virus is not prevalent in other primates that express CD46. Their population densities are not nearly high enough to sustain this highly contagious virus. In addition, those facts explain why very young children are the main targets in populations in which the virus is established (i.e., they are the

BOX 11.13

In a marvelous book, *Guns, Germs, and Steel*, author Jared Diamond sets forth several key points relevant to the origin of "human" viruses. Historically, human groups that domesticated animals generally were the first to be exposed to zootic pathogens that would eventually adapt to humans (e.g., smallpox, influenza, and measles viruses). Domestication of large herds of animals likely began in the Middle East and spread to Europe and Asia. Consequently, Old World populations eventually developed widespread resistance to many of these "new" human pathogens. Thus, when Europeans first came into contact with groups in the Americas and Oceania, who had no previous exposure to Old World pathogens like measles virus, the consequences were devastating to the native populations. Indeed, the germs carried by these European newcomers played a far greater role in their conquests of native populations than any advantages in weapons or technology (see Chapter 19).

Domestication of large herd animals began in the Old World, and not elsewhere, simply because that is where these animals were available. Moreover, just as measles virus requires a large human population to sustain itself, its rinderpest progenitor (see the text) likewise requires a large herd to sustain itself.

It is estimated that more than one-half of all human infectious diseases, including influenza (from wild birds [Chapter 12]) and AIDS (from chimpanzees [Chapter 21]), originated in animals. From these examples, we see that infectious agents may be passed to us from wild animals as well as from domestic ones. And as we will see in the case of influenza, the pathogen might come to us by passing first from a wild animal to a domestic one, and then from the domestic one to us. Also, we should not go away with the idea that pathogens jump only from animals to humans, because they well may jump in the other direction as well. For example, humans have in fact transmitted measles virus to mountain gorillas and poliovirus to chimpanzees.

Finally, we consider the sequence of steps by which a virus which initially infects an animal exclusively becomes one that infects only humans. At first, the virus can be transmitted to humans only from its animal host. Then, the virus evolves so that it might be transmitted from human to human, but only inefficiently. At this stage, outbreaks in humans tend to be brief. With time, the virus evolves to sustain longer outbreaks in humans. It then may become so well adapted to the human host that it can no longer infect the original animal host. Such a sequence of events likely occurred in the case of measles virus, and in the case of HIV, which likely arose from the simian immunodeficiency virus, a virus of chimpanzees (Chapter 21).

Reference
Diamond, J. 1999. *Guns, Germs, and Steel.* W. W. Norton & Company, New York, NY.

only nonimmune individuals in the population). Finally, they account for the fact that epidemics recur in 1- to 3-year cycles in unvaccinated populations, since that is the time required for a sufficient number of new susceptible individuals to be born into the population (Box 11.14).

Since measles virus only recently became established within the human population, and since the virus has no known animal reservoir, where did it come from? Viruses do not emerge out of nowhere, nor do they leave fossils that might provide insights into their natural history. Nevertheless, the origin of measles virus can be determined with a fair amount of certainty. In general, animals are the source of new infectious agents that become established in humans (Box 11.13). The progenitor of measles virus is almost certainly "peste des petits ruminants" (the rinderpest virus), which infects cattle, sheep, and goats. That conclusion is based on the fact that rinderpest virus is the nearest known relative of measles virus, based on nucleotide sequence homology.

The transfer of rinderpest virus to humans likely occurred about 8,000 to 9,000 years ago, at the time humans were developing a nomadic existence that depended on the domestication of large herd animals. Moreover, the transfer probably occurred in the Fertile Crescent of the Middle East, one of the first areas where humans developed animal husbandry and, consequently, an area where the human population underwent a major expansion. Thus, measles virus subsequently spread from the Middle East throughout Europe and Asia, where it has been established for several thousand years. Moreover, as measles virus became established in humans, it diverged sufficiently from its rinderpest virus progenitor, such that it no longer can infect modern-day animals.

What then of rinderpest virus in its natural host, and in others as well? This close relative of measles virus gives rise to severe disease in domestic cattle, with death rates during outbreaks actually approaching 100%. Indeed, the term *Rinderpest* is adapted from German, meaning *cattle plague*. Such cattle plagues in fact were common in the past,

often precipitated by wars. Interestingly, the virus has also spread to wild ungulates in sub-Saharan Africa and is now considered a greater threat to the survival of these animals than human encroachment on their habitats.

Thanks to a rinderpest vaccine developed by Walter Plowright, there is now hope that rinderpest may join smallpox as an officially eradicated virus. The current Global Rinderpest Eradication Programme is administered by the U.N. Food and Agriculture Organization. If this program succeeds, rinderpest will become the first animal disease to be eliminated worldwide. Current hopes are to achieve this goal by the year 2010.

Measles Vaccine

A live attenuated measles vaccine has been in use since 1963, dramatically reducing the incidence of measles in those regions where it is rigorously used. For example, the number of measles cases in the United States declined from about 500,000 per year before 1963 to about 300 per year afterwards. John Enders developed the first attenuated live measles vaccine, which was further attenuated to produce the vaccine in use today.

The intent in the United States is to give the measles vaccine to all children at 2 years of age, in combination with mumps and rubella vaccines (MMR). The purpose of postponing vaccination until 2 years of age is to enable maternal antibody to completely disappear, as the presence of maternal antibody may interfere with the vaccine taking. (In some nations, where measles in infants younger than 9 months is a problem, a first dose of the vaccine is given at 6 to 7 months, with a second dose given in later infancy to be certain that seroconversion will have occurred.) Many states require revaccination for children entering grade school or junior high school. The booster is necessary because vaccine-induced immunity might wane, leaving the individual susceptible to infection at an age when the consequences might be considerably more serious. Also, a small minority of individuals do not develop immunity after the first dose of vaccine. The second dose provides another chance for immunity to develop in those individuals.

Unfortunately, the measles vaccine has not been available in much of the undeveloped world, which is the reason for most of the measles fatalities worldwide. Vaccination in the Third World is also hindered by the fact that unlike the oral poliovirus vaccine, the measles vaccine must be given by injection, creating a need for trained personnel to administer it.

Yet there are problems in the developed world as well, in particular, parental noncompliance with vaccine programs. For example, in the years between 1989 and 1991, there was a 15-fold increase in the number of measles cases

BOX 11.14

In general, infectious agents that give rise to epidemic diseases share three essential features. First, they are highly contagious, such that they spread quickly through the population. Second, they give rise to acute infections, such that those individuals who recover are then immune to reinfection. Third, they have no animal reservoir. Together, these three factors account for the phenomenon of periodic epidemics.

in the United States, largely because of poor compliance with vaccine programs. (Noncompliance was also a problem with polio vaccination programs [Chapter 6].)

Why is there a problem of noncompliance, when the benefits of vaccination appear to be so clear? When the likelihood of infection is not an immediate concern and there are concerns for vaccine safety (imagined as well as real), parents tend to become lax. But bearing in mind the far greater risk associated with primary measles virus infections, closing the immunization gap in young infants should be a high public health priority.

Most people regard the role of vaccination to be the protection of the individual vaccine recipient. Yet from a public health standpoint, **herd immunity** is more important than the immunity of individuals. Herd immunity refers to the immunity in the whole population, which comes about when a sufficient percentage of individuals have been vaccinated, so as to block the chain of transmission of the disease to unvaccinated individuals. Thus, we might ask what percentage of individuals in a population needs to be vaccinated in order to restrict or even eliminate the pathogen from the population. Well, for the MMR vaccine, the percentage of individuals that need to be vaccinated to establish and maintain herd immunity is 95%! At lesser levels, a reservoir of infected individuals exists in the population, maintaining the chance of future outbreaks or epidemics. For diseases such as measles, which have become very rare, the problem is one of maintaining enthusiasm for the vaccination program, especially when the risk of the vaccine is perceived to be greater than the risk of becoming infected. When compliance falls, herd immunity is diminished, and the health of the population as a whole is compromised (Box 11.15).

Mumps

Mumps virus causes an acute, non-life-threatening infection, characterized by a painful swelling of the salivary glands (*parotitis*) and fever. There is only one mumps virus serotype, and the virus has no known animal reservoir. These facts account for the effectiveness of the mumps vaccine, which is now administered as a component of the MMR vaccine. The introduction of the live attenuated mumps vaccine in 1967 led to a decline in the incidence of mumps in the United States from 76 cases per 100,000 individuals to 2 cases per 100,000 individuals. In prevaccine days, 90% of the population was infected by age 15.

Like other paramyxoviruses, mumps virus is highly contagious via the respiratory route, and an infected individual may be infectious for as long as 7 days before the onset of symptoms. Moreover, some infected individuals remain asymptomatic. Consequently, vaccination is the only way to prevent the spread of mumps.

Mumps virus initially infects the epithelial cells of the upper respiratory tract and then spreads to the parotid gland (a large salivary gland located over the jaw in front of the ear) and to other salivary glands as well. Then, reminiscent of measles virus, mumps virus too disseminates via the blood circulation to other organs, including the testes and ovaries, the pancreas, and the thyroid.

The central nervous system too may be infected, resulting in **meningitis** in as many as one-half of all symptomatic infections. (Meningitis is an inflammation of the meninges, the three protective membranes that surround the brain and spinal cord. These membranes are referred to as the dura mater, the arachnoid mater, and the pia mater.) Symptoms of meningitis may include severe headaches, vomiting, stiff neck, and fever. Indeed, mumps infection is the most frequent cause of meningitis. Nevertheless, mumps meningitis is far less life-threatening than bacterial meningitis.

Mumps is also associated with rarer, but much more serious, cases of **encephalitis**. (In contrast to meningitis, encephalitis refers to an inflammation of the brain per se, usually resulting from a virus infection.) Unilateral nerve deafness is another rare but important long-term consequence. Inflammation of the testes (*orchitis*) occurs in 25% of all mumps cases in postpubescent males and may lead to sterility. Inflammatory responses are responsible for most mumps syndromes.

Since the mumps vaccine is administered as a component of the MMR vaccine, a weakening of compliance with MMR vaccination programs has led to cases of mumps in unvaccinated adolescents and young adults, in whom the risk of complications is considerably higher than in children.

Parainfluenza Viruses

Parainfluenza viruses are ubiquitous human pathogens that most often cause relatively benign upper respiratory tract infections, from which recovery is generally complete within 24 h. However, these viruses are the major cause of **croup**, a serious condition in young children, characterized by acute obstruction of the larynx, resulting in coughing, hoarseness, and persistent stridor (harsh, high-pitched breathing sounds). Parainfluenza viruses also may cause pneumonia.

Since parainfluenza viruses are spread by the respiratory route, they too are highly contagious. However, unlike measles and mumps virus infections, parainfluenza virus infections rarely become viremic. Instead, parainfluenza viruses remain restricted to the upper respiratory tract. The antibody responses to localized mucosal infections are relatively short-lived, tending to wane in just a few months in humans. In comparison, systemic infections

BOX 11.15

Recent parental concerns over vaccination safety in Great Britain led to sufficiently low levels of vaccination compliance to actually result in measles virus outbreaks there. Parental concerns were initially based on anecdotal claims that measles itself is a cause of autism in children, a fear that was subsequently heightened by a 1998 report in the prestigious British journal *The Lancet* that the MMR vaccine might actually cause autism (Wakefield et al., 1998). The story takes on added interest because of a series of bizarre developments that led the authors of the *Lancet* paper to issue a retraction. The immediate cause of the retraction was a 2004 article in the *Sunday Times* of London, which reported that the newspaper discovered an undeclared conflict of interest relating to the 1998 *Lancet* paper. The newspaper disclosed that Wakefield, the lead author of the 1998 study, had accepted funding under Britain's legal-aid system to carry out an investigation into the possible link between the MMR vaccine and autism. Importantly, this effort was carried out on behalf of the Legal Aid Board to determine whether claims against the vaccine manufacturer by parents of autistic children might have any scientific merit.

The editors of *The Lancet* responded to the *Times* report with indignation directed against Wakefield. However, the sincerity of that indignation is somewhat suspect. Although Wakefield made no statement regarding a conflict of interest when submitting the paper to *The Lancet*, 2 months later he acknowledged in correspondence to the journal that he had evaluated a few children on behalf of the Legal Aid Board. There was no apparent reaction from the journal's editors at the time of that communication. However, note that Wakefield received £55,000 (US $103,000) for his effort on behalf of the Legal Aid Board, and that fact apparently was not disclosed to the journal at the time the paper was submitted. Regardless, Wakefield does not appear to have deliberately concealed a possible conflict of interest. Moreover, the authors of the paper clearly stated in their discussion, "We did not prove an association between measles, mumps and rubella vaccine and the syndrome described." Still, one might raise questions about the actions of both Wakefield and the journal editors. Regarding the latter, one can only wonder why so prestigious a journal as *The Lancet* would have published the 1998 paper in the first place, considering its dubious scientific merit and the effects it was bound to have on the public.

The 1998 study was weak on several counts. First, its conclusions were based on a very small sample size. More importantly, perhaps, the association between receipt of the vaccine and the onset of autism was made by the children's parents, who were not likely to be the most objective of observers, especially since at least some were looking for someone or something to blame. In actuality, the link between the MMR vaccine and autism was always highly speculative, and larger, better-controlled studies have not found any association between the two. Sadly, many worried parents are not convinced. Thus, many parents continue to keep their children from receiving the vaccine. (The MMR vaccine has also been blamed for autism in the United States. However, in the United States the guilty component is alleged to be a mercury-containing compound, thimerosal, used as a preservative.)

Here is another point that is apropos to the current discussion. For many professional scientists, maintaining their integrity and reputations in the face of conflicts of interest, whether potential or actual, remains a serious issue. For this reason, most are careful to disclose these conflicts, where they exist.

Despite the fact that careful scientific investigations have failed to establish links between vaccination and unintended consequences such as autism, many parents continue to consider vaccines to be unsafe. Some public health practitioners fault the mass media for this state of affairs. A case of autism, possibly resulting from a vaccine, makes for better press than the many thousands of serious illnesses that most certainly were prevented by the vaccine. What are your thoughts on these matters?

Reference
Wakefield, A. J., S. H. Murch, A. Anthony, J. Linnell, D. M. Casson, M. Malik, M. Berelowitz, A. P. Dhillon, M. A. Thomson, P. Harvey, A. Valentine, S. E. Davies, and J. A. Walker-Smith. 1998. Ileal-lymphoid-nodular hyperplasia, non-specific colitis, and pervasive developmental disorder in children. *Lancet* 351:637–641.

---

lead to high antibody levels that are sustained for decades. The consequence of this distinction for parainfluenza virus infections is that reinfection is common. However, reinfection disease is milder than primary disease, implying that some immunity may persist.

Respiratory Syncytial Virus

Respiratory syncytial virus infections are generally mild in healthy adults and older, healthy children, often resembling the common cold. However, in infants and young children, respiratory syncytial virus infections can be severe.

In fact, respiratory syncytial virus is the most important cause of fatal lower respiratory tract infections in infants and young children worldwide. In addition, this virus can cause potentially fatal respiratory tract infections in adult populations, in particular among the institutionalized elderly and immunocompromised individuals.

Like other paramyxoviruses discussed above, respiratory syncytial virus is highly contagious and has no known animal reservoir. Nearly 1% of all babies contract a respiratory syncytial virus infection that is severe enough to require hospitalization, and the infection is fatal in about 1% of those cases. Because of its virulence in newborns, great care must be taken to prevent the introduction of respiratory syncytial virus into nursery wards. This is a particularly important concern for intensive care nurseries, where the effects of the virus might be especially devastating (premature and low-birth-weight babies are very vulnerable). The virus spreads to virtually every infant in the affected ward. Special care should also be taken to prevent outbreaks among the elderly in nursing homes.

Since respiratory syncytial virus is so highly contagious, virtually everyone has been infected by 4 years of age. Indeed, most children are infected in the first 2 years of life. Thus, infections in older children and adults are usually reinfections, occurring in individuals who have some immunity. Indeed, natural immunity to respiratory syncytial virus does not prevent reinfection. Nevertheless, these reinfections tend to be milder, more like a cold, perhaps because of lingering immunity. However, reinfections in children often involve the middle ear, making respiratory syncytial virus the most frequent cause of otitis media (see measles virus, above). The great majority of infections occur in the winter.

Respiratory syncytial virus replicates in the mucous membranes of the nose and throat. In the very young and the very old, it also may invade the trachea, bronchia, and alveoli of the lower respiratory tract, thereby accounting for the more severe infections in those individuals. The development of disease results from local immunopathological mechanisms that damage the respiratory epithelium. (As its name implies, respiratory syncytial virus induces syncytia. Nevertheless, the pathologic effect of infection is due to immunologically mediated cell injury.) In those individuals in whom the infection has spread to the lower respiratory tract, the immunopathology leads to obstruction of the bronchia and bronchioles, a particularly dangerous situation in the very young, whose narrow airways are more likely to become restricted.

No vaccine is presently available against respiratory syncytial virus, although a variety of vaccines have been through clinical trials. Interestingly, trials of a formalin-inactivated vaccine in the 1960s showed that it actually resulted in more severe illness when vaccine recipients subsequently were challenged with wild-type respiratory syncytial virus. Animal studies suggested that the deleterious effect of the vaccine might have been the result of a heightened immunopathologic response to the challenge virus, possibly due to an imbalanced Th2 T-cell response to virus antigens (Box 11.16).

More recently, a **subunit vaccine** against respiratory syncytial virus, comprised of the viral F protein, was generated by genetic engineering techniques. Such subunit vaccines use only the parts of the infectious agent that yet may stimulate a strong immune response. The gene encoding the particular subunit of interest is isolated and transfected into bacteria or yeast cells, which then produce large quantities of the pathogen protein, which then is purified and used as the vaccine.

When evaluated for its effectiveness in humans, the F-protein-based respiratory syncytial virus subunit vaccine indeed was safe and immunogenic in adults. However, as in the case of the formalin-inactivated vaccine, the subunit vaccine evoked Th2-type responses in mice, suggesting that it might not be safe to administer to human infants. Indeed, protecting infants from respiratory syncytial virus is, at the time of this writing, still a major health issue that is yet to be solved. Very recent studies provide hope that the subunit vaccine can be made to evoke a more balanced Th1/Th2 response by modifying the formulation of the adjuvant, thereby making it safe for infants. (In immunology, an adjuvant is a substance used to enhance the adaptive immune response to an antigen.)

HMPV

Although HMPV was only recently discovered, in 2001, retrospective analyses of serological samples show that this virus has been circulating in human populations since at least 1958. (In general, **retrospective studies** look backward in time to examine exposure to suspected risk factors that might lead to a particular disease syndrome. They usually use medical records, saved specimens, and interviews with patients known to have had the disease.) Serological surveys also show that HMPV is widespread in humans worldwide, with virtually all individuals having been infected by 10 years of age. Perhaps even more remarkable, in view of its relatively recent discovery, HMPV is recognized as a serious human respiratory pathogen. In point of fact, the frequency of hospitalizations due to HMPV infections is comparable to that of some better-known respiratory virus infections, including those involving respiratory syncytial virus, parainfluenza virus, and influenza virus.

HMPV infections occur in adults as well as in children. Also, as we have noted in the case of other virus infections

BOX 11.16

Th1 helper T cells secrete cytokines (i.e., gamma interferon, IL-2, and tumor necrosis factor alpha) that favor a CTL response, whereas Th2 helper T cells secrete cytokines (i.e., IL-4, IL-5, and IL-10) that favor antibody production. Since viruses are intracellular pathogens, resolution of virus infections generally requires a predominantly Th1 response, whereas Th2 responses are typically needed to fight extracellular pathogens. However, both arms of adaptive immunity contribute to resolving virus infections, so that it is important that each be appropriately expressed. That is, virus infections are more effectively resolved when there is a balance of Th1 and Th2 responses that are appropriate to that particular infection (see Chapter 4). Actually, an excessive preponderance of either response can be damaging to the host. For example, a too-strong Th1 response can lead to dangerously high levels of Th1 proinflammatory cytokines, whereas a too-strong Th2 response can result in complement activation that results in lysis of bystander cells.

How might the inactivated respiratory syncytial virus vaccine have caused an immunopathologic immune response that is skewed towards Th2 upon challenge with wild-type virus? The answer is not entirely clear, but serological studies of respiratory syncytial virus-infected patients showed increased levels of immunoglobulin E (IgE) antibodies in their sera, as well as increased levels of Th2 cytokines. Importantly, Th2 cytokine IL-4 and IgE nonneutralizing antibodies are biomarkers of allergy to an environmental allergen, suggesting that an allergy-like condition can develop during respiratory syncytial virus infection.

A plausible premise, for which there is some experimental evidence, is that the soluble viral G protein (sG) is a superantigen. (Superantigens are proteins that can activate a very large fraction of helper T cells by nonspecifically binding to both the T-cell receptor and major histocompatibility complex class II molecules [see Chapter 4]. There are generally two sources of superantigens: virus proteins and toxins produced by bacteria. A well-known example of the latter is the superantigen generated by *Staphylococcus aureus*, which can cause toxic shock syndrome.) Since superantigens activate numerous different B-cell and T-cell clones, regardless of their specificities, this polyclonal activation can lead to autoimmunity. In addition, as a superantigen, sG also could bind to the Fc receptor on basophils, mast cells, and monocytes of the innate system, causing them to degranulate and release large amounts of the Th2 cytokines IL-4, IL-5, and IL-10 into the blood. The released IL-4 causes uncommitted Th0 cells to become Th2 cells and secrete the Th2 profile of cytokines. Moreover, there also is negative feedback at work, such that the IL-10 produced by the Th2 cells decreases the rate of proliferation of Th1 cells, thereby further skewing the immune response to an inappropriate Th2 one. Finally, if B cells undergo class switching in germinal centers containing Th2 cells that secrete IL-4 and IL-5, they tend to become producers of IgE antibodies, which mediate allergic reactions. That is, these B cells secrete allergy-promoting IgE antibodies, rather than antiviral IgG and IgA antibodies.

Based on the above, it was suggested that a respiratory syncytial virus strain engineered to contain a deletion of the superantigen domain of the *G* gene might be a candidate for vaccine development. However, does the experience with the subunit vaccine (see the text) diminish enthusiasm for this approach?

---

of the respiratory tract, the very young (particularly premature infants), the elderly, and patients with other conditions, such as asthma and cardiopulmonary disease, as well as those who are immunosuppressed, are predisposed to more severe HMPV infection. See Gray et al. (2006) for a recent large retrospective study, which showed that patients between 0.4 and 9 years of age and those receiving intensive care were more likely to show evidence of HMPV infection than their peers. These researchers also note that additional studies are needed to characterize the epidemiology of HMPV and the extent of HMPV disease in the general population.

The clinical syndromes seen in children hospitalized with HMPV infection included **pneumonitis** (inflammation of the air sacs in the lung) and **bronchiolitis** (inflammation of the bronchioles, i.e., the branches of a bronchial tree), whereas in elderly patients the most common syndromes were **bronchitis** (inflammation of one or more bronchi) and pneumonitis.

Several efforts are now under way to develop a vaccine against HMPV that might prevent severe disease in infants. One approach is to develop a live attenuated virus for intranasal immunization. That strategy is attractive because it has worked before to immunize against other respiratory viruses (Box 11.17). **Reverse genetics** provides an important means to developing such an attenuated vaccine. (Classical genetics is **forward genetics**. It begins with a phenotype and then seeks to find its genetic basis.

Since the mucosal surface of the respiratory tract is comprised of an extensive and delicate epithelial layer that is intimately contacted by the underlying vasculature, it has evolved to be extremely active immunologically. Thus, local exposure of the respiratory mucosa to an antigen can lead to very high levels of secretory IgA at the site of exposure. On the other hand, antibody responses to localized mucosal antigens tend to be relatively short-lived, as noted in the text.

In contrast, reverse genetics, which is made possible by modern DNA sequencing and site-directed mutagenesis, examines the phenotype that results from deliberately altering the known nucleotide sequence of a gene.) By using this approach, a candidate attenuated vaccine that contains a deletion of the HMPV gene encoding the G glycoprotein was engineered. The G protein was an attractive target, since altering it might attenuate the virus by impairing attachment, and perhaps by weakening its ability to modulate host immunity (see above). A potential advantage of this approach is that it may avoid the problem encountered with the candidate vaccine against respiratory syncytial virus, which actually caused a heightened immunopathologic response to the wild-type virus. That is so because there is no HMPV G glycoprotein in the vaccine that might predispose the vaccine to evoke that response (Box 11.16).

Another candidate vaccine was engineered to contain the avian metapneumovirus P protein gene, in place of the HMPV P protein gene. This may restrict the ability of the virus to replicate in human cells.

Hendra, Nipah, and Menangle Viruses

Hendra virus is named for the suburb of Brisbane, Australia, where that virus first emerged in 1994. The initial Hendra virus outbreak was associated with a respiratory tract disease that killed 14 racehorses and a horse trainer in a period of several weeks. The following year, the virus caused a second fatal human case, encephalitis in this instance, which may have been acquired while the patient was carrying out autopsies on the horses that had died the previous year.

Three years later, in 1997, another new paramyxovirus, the Menangle virus, emerged near Sydney, Australia. That virus caused a new syndrome in pigs, characterized by fetal death, stillbirths, and congenital malformations. Menangle virus also was implicated as the cause of an influenza-like illness in two humans who worked at Menangle virus-infected piggeries.

The following year, yet another new paramyxovirus, Nipah virus, emerged in Malaysia. It spread first in pigs, in which it caused few clinical signs, and then in humans, in which it caused encephalitis that killed over 100 individuals! Approximately 1 million pigs were slaughtered in an attempt to control the spread of the disease. That approach succeeded, largely because humans may be a "dead-end" host for this virus.

The ability of these three new paramyxoviruses to infect and cause potentially fatal diseases in a variety of host species, including humans, sets them apart from the other known paramyxoviruses. Moreover, Hendra and Nipah viruses appear to be more closely related to each other than they are to other paramyxoviruses. Both of these viruses have exceptionally large genomes, approximately 18 kb, compared to the more typical 15-kb genomes of other paramyxoviruses. In addition, they are more closely related based on nucleotide sequence homology, and antigenic cross-reactivity. Because of their unique biological and genetic features, Hendra and Nipah viruses have been classified into the new genus *Henipavirus*. In contrast, Menangle virus is a member of the genus *Rubulavirus*.

Serological evidence is consistent with the possibility that flying (pteropid) bats are the natural hosts of the Hendra, Nipah, and Menangle viruses, as is the fact that these bats do not become ill when infected by these viruses. Since these bats are distributed across an area encompassing northern, eastern, and southeastern regions of Australia, Indonesia, Malaysia, the Philippines, and some of the Pacific Islands, there is an ongoing risk of Hendra, Nipah, and Menangle viruses being reintroduced into the pig population.

In the case of Nipah virus, transmission to pigs may have been a consequence of the clearance of a tropical rainforest, carried out in order to build the new international airport for Kuala Lumpur. This habitat destruction is thought to have caused the indigenous bat population to move to trees around pig farms. (Ebola virus provides a better-known example of a virus that emerged as a consequence of human activity in rainforests [Chapter 13].)

Direct pig-to-pig transmission of Nipah virus apparently occurs, but there is no clear evidence for direct human-to-human transmission, meaning that humans may be a dead end for Nipah virus, as noted above. Regardless, human infections are severe; of the 269 known cases occurring in 1999, 108 were fatal.

Aside from pigs, it is not yet clear what other species might be sources of Nipah virus infection in nature. Also, it is not known how these viruses are transmitted from bats to animals or from animal to animal, but transmission appears to involve close contact with contaminated tissue or body fluids.

In February 2004, Nipah virus reemerged in two separate outbreaks in Bangladesh. Moreover, several features of these Bangladeshi outbreaks are particularly worrisome. First, unlike the virus that appeared in Malaysia in 1998, the virus in Bangladesh may be able to spread from person to person. The Malaysian victims were pig farmers, but there was no pig outbreak in Bangladesh, and there was no evidence to incriminate any other intermediate host. However, flying bats in Bangladesh tested positive for the virus, and it is conceivable that humans might have been directly infected from bat droppings. Second, the Bangladeshi virus seems to be significantly more lethal to humans than the Malaysian virus. The outbreak in Malaysia claimed the lives of 40% of those infected, whereas the mortality rate of the Bangladeshi outbreak was 74%! (What other factors may have contributed to the greater mortality rate of the Bangladeshi outbreak?)

CDV

CDV was for many years the most important virus affecting dogs. In particular, it once was the leading cause of death in unvaccinated puppies. However, widespread vaccination programs have dramatically reduced the incidence of CDV in domestic dogs. Still, CDV remains a deadly virus globally, killing numerous carnivorous species in the wild, including even African lions, spotted hyenas, and several species of seals. In the case of seals, a series of CDV epidemics over the last 10 years killed as many as 10,000 Caspian seals (the world's smallest seal), out of a total population of only 120,000.

As a paramyxovirus, CDV is efficiently transmitted via the respiratory route and was likely passed to the aforementioned carnivores by domestic and wild dogs. However, in order for the virus to successfully adapt to a new host, it likely would need to modify its attachment protein, so that it might efficiently recognize the version of its receptor in the new host. This premise was recently verified experimentally. The researchers found that the CDV gene encoding the HA underwent repeated changes at the sites involved in receptor recognition, while the rest of the protein showed very little variation among viruses isolated from different species.

CDV initially replicates in the lymphatic tissue and then spreads via the blood circulation to several organs, including the skin, the upper respiratory tract, and the central nervous system, enabling it to give rise to respiratory and neurological symptoms. In animals that survive the acute infection, the virus persists in the central nervous system, in which it causes demyelination due to degenerative changes in oligodendrocytes. The early stages of neurodegeneration may be due to the direct actions of the virus, while later stages may be due to immune-mediated mechanisms.

In humans, CDV infection is generally asymptomatic. Moreover, anyone who has been immunized against measles is usually protected against CDV as well.

Suggested Readings

Duke, T., and C. S. Mgone. 2003. Measles: not just another viral exanthema. *Lancet* **361**:763–773.

Gray, G. C., A. W. Capuano, S. F. Setterquist, et al. 2006. Multiyear study of human metapneumovirus infection at a large US Midwestern Medical Referral Center. *J. Clin. Virol.* **37**:269–276.

Yin, H. S., X. Wen, R. G. Paterson, R. A. Lamb, S. Theodore, and T. S. Jardetzky. 2006. Structure of the parainfluenza virus 5 F protein in its metastable, prefusion conformation. *Nature* **439**: 39–44.

Orthomyxoviruses

INTRODUCTION

The *Orthomyxoviridae* are a family of minus-strand RNA viruses that differ in the following major way from the minus-strand rhabdoviruses and paramyxoviruses considered in the two preceding chapters. Whereas the minus-strand genomes of the rhabdoviruses and paramyxoviruses are composed of a single, continuous RNA strand, orthomyxovirus genomes are divided into six to eight individual genomic RNA segments.

The orthomyxoviruses are similarly distinguished from the bornaviruses and filoviruses (Chapter 13), which, like the rhabdoviruses and paramyxoviruses, also contain nonsegmented, minus-strand genomes. Recall that the four families of nonsegmented, minus-strand viruses all display a common gene arrangement and strategy for gene expression. For that reason, they constitute a superfamily, referred to as the *Mononegavirales.*

But the orthomyxoviruses are not the only viruses that contain segmented, minus-strand RNA genomes, since the arenaviruses and bunyaviruses (Chapter 13) likewise have multipartite, minus-strand RNA genomes. But despite the fact that viruses of these three families have segmented, minus-strand genomes, they do not constitute a superfamily like the mononegaviruses. That is so because there are still other significant biological differences between them. For instance, arenavirus genomes and some bunyavirus genomes are actually **ambisense**. That is, their individual genomic segments are partially minus sense and partially plus sense. (Mother Nature indeed finds many ways to make a virus.) Also, in contrast to the six to eight RNA segments that make up orthomyxovirus genomes, arenavirus genomes are *bipartite*, and those of bunyaviruses are *tripartite*. This leads to the following key points. First, orthomyxoviruses contain the most highly segmented genomes of any known family of single-stranded RNA viruses. Second, the segmented nature of orthomyxovirus genomes has singularly important medical consequences, which are recounted below.

The orthomyxoviruses also differ from most other RNA virus families with respect to the fact that all of their RNA synthesis, both mRNA transcription and genomic replication, happens in the cell nucleus, whereas the replication cycle of most other RNA viruses occurs entirely in the cytoplasm. The reason why an RNA virus might be dependent on the cell nucleus for its RNA synthesis will be apparent shortly. For the moment, note that orthomyxoviruses are not unique among RNA viruses in their nuclear dependency, since the minus-strand bornaviruses (Chapter 13) and the plus-strand retroviruses (Chapter 20) also replicate in the nucleus.

The orthomyxovirus family name is derived by combining the Greek word "*orthos*," meaning "right," with the Greek word "*myxa*," which means "mucus," the latter in recognition of the fact that these viruses have an affinity for mucopolysaccharides, in particular, sialic acid-containing receptors on cell surfaces. The more familiar name "influenza" derives from early Italian astrologers who blamed the periodic flu epidemics on the *influenza*, i.e., influence, of the heavenly bodies (Box 12.1).

The family *Orthomyxoviridae* is subdivided into three genera: *Influenzavirus*, *Isavirus*, and *Thogotovirus*. Here we focus our attention on the influenza viruses because they constitute the largest genus of the orthomyxovirus family and, more importantly, because influenza viruses are among the most medically important pathogens that infect humans.

The most common, and also the most medically important, of the influenza viruses are those designated type A, which infect humans as well as many species of animals and birds. Actually, aquatic birds are the natural reservoir for influenza A viruses. Moreover, the avian reservoir plays a key role in the genesis of human epidemic and pandemic influenza strains, as described below.

The influenza B viruses can also be human pathogens. But an important distinction between the type A and type B viruses is that only the former may give rise to the occasional pandemics that can spread through the human population, with devastating consequences.

Little is known regarding the origin and possible reservoir of influenza B viruses. Indeed, until recently, influenza B virus was thought to be restricted to humans. However, in 2000, an influenza B virus was isolated from a naturally infected harbor seal. Interestingly, results of nucleotide sequence analyses and serology imply that the seal influenza B virus is closely related to type B strains that circulated 4 to 5 years earlier in humans. Moreover, retrospective serological analysis showed that this virus entered the seal population only sometime after 1995. Also, it was estimated that since 1995, between 0.5 and 2% of seals in the wild experienced influenza B virus infection.

Type C influenza viruses infect humans and, possibly, pigs. Influenza C virus infections differ clinically from those of types A and B, usually causing either a very mild respiratory illness or no symptoms at all. Moreover, type C influenza viruses do not cause epidemics and do not have the severe public health impact that influenza types A and B can have. Neither the natural reservoir of influenza C viruses nor their potential to infect other species is known.

The genus *Isavirus* presently consists of only one known member, the infectious salmon anemia virus, which contains eight RNA segments. This virus morphologically resembles influenza viruses but does not show significant sequence homology to them.

Until recently, the genus *Thogotovirus* was comprised of only two viruses: Thogoto virus, which contains six RNA segments, and Dhori virus, which contains seven RNA segments. Batken virus, which was isolated from mosquitoes and ticks and originally was thought to be a bunyavirus (Chapter 13), is now known to be a closely related variant of Dhori virus. Thus, it too is classified as a thogoto virus. Araguari virus, isolated from a marsupial, and which has six RNA segments, also is a candidate for inclusion in the genus *Thogotovirus*. Thogoto virus and Dhori virus, as well as Batken virus, are arboviruses, transmitted by ticks.

From this point on, we focus on the influenza viruses (i.e., viruses of the genus *Influenzavirus*). We begin by asking what biological features enable us to distinguish between influenza types A, B, and C. Although all influenza viruses contain helical nucleocapsids that are enclosed within lipid envelopes, types A, B, and C can be distinguished from each other serologically, by antigenic differences between their nucleocapsid (NP) and matrix (M) proteins. In addition, while influenza A and B viruses contain eight genomic RNA segments, influenza C viruses contain only seven. That difference between type C and

BOX 12.1

Recall that the orthomyxoviruses and the paramyxoviruses (Chapter 11) were once classified together in a family referred to as the *Myxoviridae*. The earlier coclassification was based on the fact that paramyxoviruses too have an affinity for mucopolysaccharides and, like orthomyxoviruses, they have envelope glycoproteins that express NA activity. Moreover, members of both families have minus-strand RNA genomes. But it now is recognized that the orthomyxoviruses and paramyxoviruses are not at all closely related. Most importantly, they differ in the organization and expression of their genomes, a key difference being that orthomyxovirus genomes are segmented, whereas paramyxovirus genomes are monopartite.

type A and B influenza viruses is explained by the fact that the cell attachment and neuraminidase (NA) activities of influenza A and B viruses are expressed by two distinct envelope glycoproteins, whereas type C viruses have a single multifunctional protein to carry out these activities. Also, since type C particles contain only a single glycoprotein species, their spikes are organized into orderly hexagonal arrays, a structural feature that distinguishes them from type A and B particles.

Type A influenza viruses are the most medically important members of the genus *Influenzavirus*, and for that reason, they also are the most extensively studied of the influenza viruses. Thus, from here on, we focus on influenza A viruses.

Influenza, together with smallpox, tuberculosis, malaria, plague, measles, and cholera, is historically one of the major infectious disease killers of mankind. (Currently, we might be inclined to add diseases caused by hepatitis B and C viruses and human immunodeficiency virus [HIV] to this inventory of major killers.) Even today, when effective flu vaccines are available, influenza viruses typically kill *nearly 40,000 individuals in the United States alone*, and over half a million worldwide, each year. The severity of influenza can vary from year to year. Indeed, the severest epidemic in human history was the 1918 influenza **pandemic**, which killed as many as *fifty million people* worldwide. In the United States there were an estimated 20 million cases (20% of the population at the time) and 850,000 dead. The reason for the severity of that epidemic is still not clear, nor is it clear when such an epidemic might recur. These issues are discussed in detail below.

STRUCTURE

Influenza virus particles contain a ribonucleoprotein core, enclosed within a lipid envelope. Most particles are spherical and about 80 to 100 nm in diameter, although filamentous forms also are seen (Fig. 12.1). These filamentous forms were once mistakenly believed to be noninfectious, empty particles. This misconception persisted largely because filamentous forms have been difficult to study, as they tend to be lost during passage of the virus in cell culture. Nevertheless, primary isolates of influenza virus actually are filamentous. Moreover, influenza viruses are thought to exist in the lungs of infected individuals mostly in the form of filamentous particles, which may be up to several micrometers in length. These particles indeed are infectious, although they internalize more slowly than spherical particles do. Like spherical particles, they are dependent on low pH for entry (see below). Still, because much more is known about the spherical forms, we discuss that form of the virus exclusively.

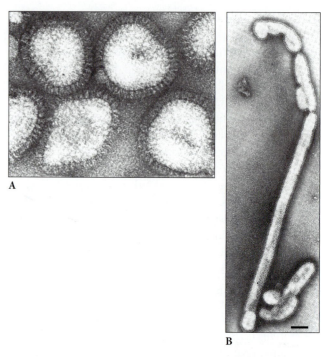

Figure 12.1 Electron micrographs of type A influenza virus particles. (A) Spherical. (B) Filamentous. From W. G. Laver, *Adv. Virus Res.* **18**:65, 1973, with permission.

A schematic diagram of a spherical influenza virus particle is shown in Fig. 12.2. The most striking morphological feature of these particles is the presence of approximately 600 spikes radiating out from the lipid envelope. These spikes are of two kinds. First, there are about 500 of the rod-shaped homotrimers of the hemagglutinin (HA) protein that are responsible for binding and fusion. Second, there are the approximately 100 of the mushroom-shaped tetramers of the NA protein that prevent virus clumping and promote release of progeny virus particles. As expected, these integral envelope proteins contain large external domains (the structure of HA is described in detail below).

A third integral envelope protein, M2, lacks a large external domain. This small (97-amino-acid) protein is a minor component of virus particles (20 to 60 molecules of M2 per virus particle) but plays a unique role in entry and is the target of several effective anti-influenza drugs (see below). M2 exists in the envelope in the form of tetramers.

As in the cases of other minus-strand viruses that have helical ribonucleoprotein cores, the influenza virus envelope is lined with a matrix (M1) protein, which connects the envelope to the ribonucleoprotein core. There are about 3,000 copies of M1 per virus particle. In addition, there is the nuclear export protein (NEP), so named for its role in facilitating the export of progeny virus genomic

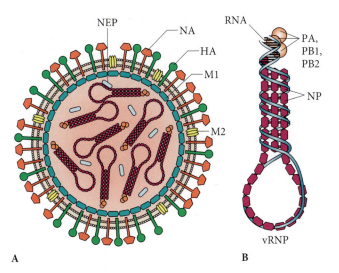

Figure 12.2 Schematic depiction of a spherical influenza virus particle (A) and a vRNP complex (B). The three envelope proteins of the virus are HA, NA, and the M2 protein. The matrix (M1) protein underlies the envelope, where it interacts with the cytoplasmic tails of the HA and NA proteins and with the vRNP complexes. The latter are comprised of the viral genomic RNA segments, complexed with the nucleoprotein (NP), the three polymerase proteins (PA, PB1, and PB2), and small amounts of the NEP, formerly called NS2. The 5′ and 3′ termini of the genomic RNA segments have complementary sequences that form a panhandle structure. The three polymerase proteins form a complex at the 5′ terminus of each RNA segment.

RNA segments from the nucleus (recall that all orthomyxovirus transcription and genomic replication occurs in the nucleus). NEP was previously known as NS2, based on the initial, erroneous belief that it is nonstructural.

The cores of A and B influenza viruses are comprised of eight distinct helical viral ribonucleoprotein (vRNP) segments. (Influenza C viruses have seven vRNP segments, since their genomes are comprised of seven distinct RNA segments, rather than the eight RNA segments that make up the influenza A and B virus genomes.) The explanation for this difference between these influenza types is as follows. The influenza virus spike glycoproteins (distinct from the short M2 envelope glycoproteins) are charged with three functions: binding to the receptor on the cell surface, mediating membrane fusion, and destroying the receptor. In type A and B influenza viruses, binding to the receptor and mediating membrane fusion are carried out by HA, while destroying the receptor is carried out by a second glycoprotein, NA. In influenza type C virus, a single glycoprotein, the HA-esterase fusion protein, carries out all three functions. Hence, type C influenza viruses do not have a separate genomic segment that encodes the NA activity.

Each vRNP segment contains a single molecule of genomic RNA and multiple copies of each of four distinct proteins. The NP protein is the most prevalent of the

proteins that associate with the vRNPs (about 1,000 copies per virus particle), serving to organize the genomic RNA segments into the helical vRNPs. In this respect, the influenza NP protein is analogous to the N and NP proteins of the rhabdoviruses and paramyxoviruses, respectively. The other three proteins associated with vRNPs are the three components of the viral polymerase, proteins PB1, PB2, and PA. There are approximately 30 to 60 copies of each of the polymerase component proteins per virus particle.

Each NP protein interacts with approximately 24 nucleotides of RNA. It is rather straightforward to envision the NP proteins ordered around the RNA. However, the RNA in the vRNPs is sensitive to ribonuclease (RNase), suggesting that the RNA segments may actually wrap around a core of NP proteins, instead of the NP proteins being ordered around the RNA. The crystal structure of the NP protein supports that premise (Fig. 12.3). First, the NP protein most commonly exists as a trimer, and each of the NP monomers has a deep RNA-binding pocket, which is located on the *outside* of the trimer. The surface of the pocket is made up of a large number of basic residues that likely function in RNA binding by interacting with the phosphodiester backbone of the RNA. Second, with the RNA bound to the pocket of each NP monomer at the outer periphery of the NP trimers, the distance between two neighboring RNA-binding sites in an NP trimer is approximately 70 angstrom (Å), which is consistent with the stoichiometry of one NP molecule per 24 nucleotides of RNA, as shown earlier. For the sake of comparison, the RNA-binding groove of the rhabdovirus N protein faces the interior of rhabdovirus N protein oligomeric complexes, consistent with the ability of the rhabdovirus N protein to protect the viral RNA from RNase digestion.

ENTRY

As a respiratory virus, influenza virus is spread by aerosolized droplets that are released by infected individuals. (Why might the impact of influenza be felt primarily during the winter months [Box 12.2]?) Influenza may also be spread through contact with fomites. (Fomites are otherwise harmless objects that may be contaminated with pathogenic microorganisms, hence the oft-repeated advice to wash your hands frequently. Is this effective?) The virus then initiates infection by binding to sialic acid residues on glycoproteins and glycolipids on cells of the respiratory epithelium. Binding is followed by receptor-mediated endocytosis, which delivers the virus to endosomes. The low pH within endosomes then triggers conformational rearrangements of the viral HA protein, which mediate the fusion of the viral envelope with the endosomal membrane, thereby releasing the vRNPs into the cytosol.

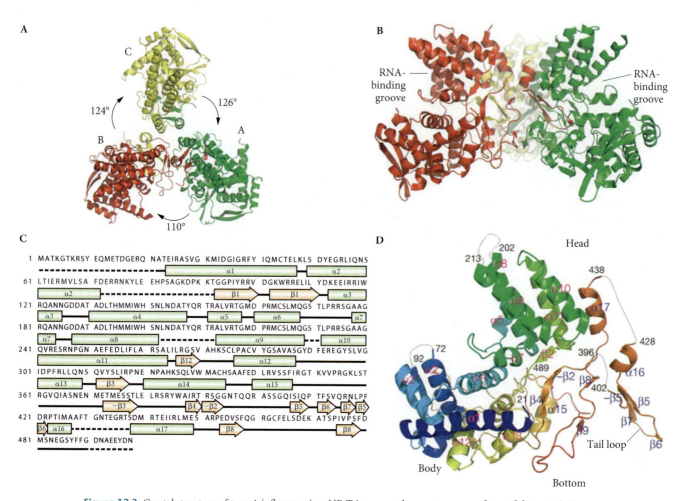

Figure 12.3 Crystal structure of type A influenza virus NP. Trimers are the most common form of the protein. In particular, note the RNA-binding grooves on the outside of the trimer. The surface of the grooves is made up of a large number of basic residues that are likely to function in RNA binding by interacting with the phosphodiester backbone. (A) NP trimer viewed along the noncrystallographic symmetry threefold axis, with three subunits shown in different colors. The rotation angles that relate the three subunits are marked. (B) Side view of an NP trimer, with the three subunits colored as in panel A. (C) NP secondary structure assignment (subunit B). α-Helices and β-strands are shown by rods and arrows, respectively. β-Strands with opposite signs form antiparallel β-strands. Residues preceding and succeeding disordered regions are marked and connected by dotted lines. (D) NP subunit B. The polypeptide is colored continuously from blue to red. Adapted from Q. Ye, R. M. Krug, and Y. J. Tao, *Nature* **444:**1078–1082, 2006, with permission.

The released vRNPs are next transported to the nucleus, which is the site for all influenza virus RNA synthesis, both transcription and replication. Why might influenza viruses carry out their transcription and replication in the nucleus, when all of the RNA viruses that we have discussed thus far (including plus-strand, minus-strand, and double-stranded RNA viruses) carry out their transcription and replication entirely in the cytoplasm? Why indeed should any RNA virus adapt to traffic to the nucleus, when there are no cellular enzymes there (or anywhere for that matter) that might transcribe or replicate their RNA genomes? Apropos the enzyme issue, influenza virus, like other minus-strand viruses, brings its own polymerase into the cell. The nuclear replication of influenza virus is explained below. For now, be aware that the monopartite, minus-strand bornaviruses (Chapter 13) and the plus-strand retroviruses (Chapter 20) also replicate in the nucleus. These are the only other RNA viruses that are known to do so.

Endocytosis and Intracellular Trafficking of Influenza Virus-Containing Endosomes

It recently became possible to track the entry of individual fluorescently labeled virus particles in real time by means

BOX 12.2

The impact of influenza is felt mainly during the winter months in temperate climates of both hemispheres, while no such seasonality is seen in tropical climates. Factors such as school attendance, crowding, and closed windows are generally thought to be responsible for the seasonal flu cycles so familiar to most of us. Also, seasonal fluctuations in immune responsiveness due to seasonal factors such as melatonin and vitamin D levels have also been suggested as factors. Nevertheless, little firm experimental evidence is available to show why influenza virus infections peak in the wintertime. Recent experiments using a guinea pig model have shown that both ambient air temperature and relative humidity affect the frequency of influenza virus transmission among guinea pigs. That is, both cold and dry conditions favor transmission.

Why might low relative humidity affect the frequency of transmission? A likely possibility is that breathing dry air could lead to damage to the epithelium or reduced mucociliary clearance. Yet the researchers do not believe that this was the reason for enhanced transmission, because they considered the time of exposure to the dry air (less than 72 h) to be too short to have that effect. Instead, they believe that the enhanced transmission is because aerosolized droplets of influenza virus are actually more stable in the air at low relative humidity (20 to 40%) than at intermediate relative humidity (50%). That is, droplets remain airborne for longer periods of time, thereby increasing the opportunity for transmission of the pathogens they carry.

Why might air temperature affect the frequency of transmission? Several possibilities might be envisioned, but a favored one is that in the cold, cilia in the respiratory tract beat more slowly, enabling the virus to persist longer in the respiratory tract, so that it might be more readily dispersed by a sneeze or a cough.

Be aware that other factors likely contribute to the periodicity of influenza epidemics. Also, if low temperature and low relative humidity are key determinants of influenza transmission in temperate climates, as seems to be the case, then it is not clear how influenza is maintained in tropical climates, where there is a high background of influenza activity throughout the year, on top of which epidemics occur at intermediate months between the influenza seasons in the temperate regions of the Northern and Southern hemispheres. However, estimating the influenza burden in tropical regions is considerably more difficult than in temperate regions (why might that be?), yet it is a matter of great importance for multiple reasons. Regardless, for most of us, it might be a good idea to turn up the thermostat in the winter and to get a humidifier.

P.S. Don't forget your flu shot!

References

Lowen, A. C., S. Mubareka, J. Steel, and P. Palese. 2007. Influenza virus transmission is dependent on relative humidity and temperature. *PLoS Pathog.* **3:**1470–1476.

Viboud, C., W. J. Alonso, and L. Simonsen. 2006. Influenza in tropical regions. *PLoS Med.* **3:**e89. [Epub ahead of print] doi:10.1371.

of fluorescence microscopy (Lakadamyali et al., 2003). This approach led to the observation that when influenza virus binds to the plasma membrane, it somehow induces formation of the very clathrin-coated pits that internalize it into the cell. These virus-induced coated pits form around the bound virus particles. This observation is intriguing since clathrin-coated pits are usually thought to form constitutively and continuously at the cell surface. Moreover, ligands usually associate with preexisting clathrin-coated pits at the cell surface, which then mature into clathrin-coated vesicles.

Observing virus entry in real time also revealed that intracellular transport of virus-containing endosomes occurs in three recognizably distinct stages. The first stage involves the slow movement of virus-containing endosomes at the cell periphery. This stage is mediated by actin-containing microfilaments, as demonstrated by the effects of drugs that specifically affect particular components of the cytoskeleton (e.g., the drugs cytochalasin D, which specifically disrupts actin filaments, and nocodozole, which specifically disrupts microtubules; these are standard reagents used in this experimental approach). In stage II, virus-containing endosomes move rapidly towards the cell nucleus. This stage is mediated by microtubules, again as shown by the differential effects of the inhibitors.

Transport of the virus-containing endosomes on microtubules is mediated by the **motor protein** dynein. This was shown by the blocking effect of microinjecting anti-dynein antibodies into the cells. (Motor proteins travel along microtubules. In so doing, they mediate the vectored transport of membrane-bounded structures within the cell. Their movement is driven by adenosine triphosphate [ATP] hydrolysis.)

During stage III the virus appears to move back and forth along the same path in the perinuclear area. Stage III movement also is mediated by microtubules. It was suggested that this back-and-forth motion results from the endocytic compartment coming under the influence of both minus-end- and plus-end-directed motor activities.

Bear in mind that during this entire time, the virus has been contained within an endosome. Thus, in the final stage of this process, the viral envelope must fuse with the endosomal membrane so that the vRNPs might be released into the cytosol. Interestingly, this fusion is seen to occur in the perinuclear region. This observation implies that the virus is released from late endosomes that are localized near the nucleus. Importantly, since envelope fusion is induced at low pH, this observation also implies that the acidification of endocytic cargo might usually occur in the perinuclear region, which challenges the general assumption that early endosomes at the cell periphery are the acidification sites for endocytic cargo. (These studies also underscore the fact that viruses have been valuable probes of cellular endocytic pathways.)

Note that viruses are known to use other endocytic pathways to enter cells, in addition to the one mediated by clathrin-coated pits and vesicles. For instance, the polyomavirus simian virus 40 (SV40) enters cells by a pathway that is mediated by caveolae (Chapter 15). Interestingly, influenza virus appears to enter cells via a second pathway, a pathway that is not well characterized, but which is independent of both clathrin and caveolae. Studies using fluorescence microscopy to follow the entry of individual influenza virus particles in real time showed that both of the influenza virus entry pathways result in fusion with intracellular membrane compartments and, presumably, release of the viral genome and productive infection.

Viral Membrane Fusion: the HA Protein

Membrane fusion, in which two separate lipid membranes fuse into a single continuous bilayer, is a fundamental cellular process. Cells use membrane fusion to transport cargo between specialized compartments or organelles and also for export of secreted molecules. Enveloped viruses use membrane fusion to release their genomes (Box 12.3).

Cells use assemblies of conserved protein families, such as SNAREs, to mediate membrane fusion. Binding of a particular v-SNARE on a vesicle membrane with a specific t-SNARE on a target membrane mediates vesicle-target recognition. Other fusion proteins, NSFs, and SNAPs, then bind to the SNARE complex to carry out fusion. In contrast, viruses use just a single envelope glycoprotein to mediate all the steps of viral membrane fusion. Yet despite this difference between cellular and viral membrane fusion, structural studies of the influenza virus HA protein and of the fusion proteins of some other enveloped viruses (e.g., SV5, HIV, and Ebola virus), as well as of the cellular protein complexes involved in the fusion of cellular vesicles with target membranes, reveal important

BOX 12.3

Vaccinia virus (Chapter 19) provides a single known exception to the generalization that *enveloped* viruses use membrane fusion to release their genomes into the cell. Vaccinia virus is an exception because one infectious form of that virus, the so-called extracellular virion, has two lipid envelopes surrounding the virus core. Consequently, a single fusion event will not deliver a naked viral core into the cell. Recently, a previously unknown vaccinia virus-induced nonfusogenic mechanism was shown to disrupt the outer viral membrane. This event is then followed by the fusion of the inner viral envelope with the plasma membrane and penetration of the virus core into the cytoplasm (Chapter 19).

similarities that are shared by all. (These similarities are illustrated in Box 12.4. But they will be appreciated better after you have read about the influenza HA. So read on, but do see Box 12.4 before leaving this subsection.) These shared structural features imply that these different viral and cellular fusion proteins all act via a similar mechanism. For that reason, and because the influenza HA protein is the best studied of the viral fusion proteins, HA-mediated membrane fusion is a paradigm for membrane fusion reactions in general. Consequently, influenza HA-mediated membrane fusion is discussed here in some detail.

The several examples of viral fusion proteins that we have discussed thus far were all seen to exist as trimers in the viral envelope, and in fact, this trimeric arrangement is typical. In addition, the membrane fusion activities of the viral fusion proteins generally needed to be primed by posttranslational cleavage. Apropos influenza, tryptase Clara, a cellular serine protease, is believed to carry out this priming of the HA protein of mammalian influenza virus strains. Importantly, expression of tryptase Clara is thought to be restricted to the nonciliated Clara cells of the bronchial and bronchiolar epithelia. Consequently, replication of mammalian influenza viruses may be restricted to epithelial cells of the respiratory tract that are in the immediate vicinity of the specialized Clara cells.

Tryptase Clara cleaves the HA precursor protein, referred to as HA_0, into two glycopeptides, HA1 (319 to 326 amino acids) and HA2 (221 to 222 amino acids). Thus, the HA homotrimer contains three identical subunits, each comprised of two glycopolypeptides, HA1 and HA2 (Fig. 12.4).

To understand how HA promotes fusion, it was necessary to determine the conformational transitions that HA undergoes, which drive the fusion process. Toward that end, X-ray crystal structures were determined for HA

BOX 12.4

The structures of the fusion proteins have been solved for several other viruses, including the retroviruses HIV, simian immunodeficiency virus, human T-cell leukemia virus, and Moloney murine leukemia virus; the paramyxoviruses SV5 and respiratory syncytial virus; and Ebola virus (a filovirus). Despite the fact that these viruses represent several distinct and quite unrelated families, their fusion proteins share key structural features. All contain a central core comprised of a triple-stranded coiled coil that is formed from N-terminal α-helices. This creates three grooves into which the three C-terminal α-helices pack antiparallel to the central core. In this way, some of these other virus fusion proteins generate the six-helix bundle seen after the final conformational transition of influenza HA2 (Fig. 12.5).

What then of cellular membrane fusion processes? These generally involve the formation of complexes between three groups of membrane receptor proteins called SNARES. One of these SNARE proteins, synaptobrevin, is on the vesicle membrane (i.e., synaptobrevin is a v-SNARE). The other two SNARE proteins, syntaxin and SNAP-25, are on the target membrane (i.e., they are t-SNARES). Importantly, these SNARES interact with each other to form a rod-shaped complex that is a coiled coil of four α-helices: one α-helix contributed by synaptobrevin, one by syntaxin, and two by SNAP-25. Do you see the similarities to the HA conformations? It continues.

Before the complex is assembled, the N terminus of syntaxin appears as a three-helix bundle. This helical bundle contains a groove that accommodates a fourth helix, provided by the syntaxin C terminus. SNAP-25 interacts with the C-terminal helix of syntaxin to form the t-SNARE complex. Note that SNAP-25 appears to be nonhelical before entering the complex. The resultant binary t-SNARE complex formed between syntaxin and SNAP-25 now contains a three-helical bundle into which the v-SNARE, synaptobrevin, appears to bind. Note that synaptobrevin too appears to be nonhelical before entering the complex. At the end of these steps, the SNARE complex has acquired the coiled-coil conformation that is comprised of four α-helices. Moreover, the membrane-anchoring regions of both the t- and v-SNARES, which are at their respective C termini, are at the same end of the complex. SNARE complexes are disassociated after fusion so that the individual SNARE proteins may be recycled.

Despite the fact that virus fusion proteins and cellular SNARE complexes differ in their subunit composition, they nevertheless share at least two key features. First, they each are α-helical rods. Second, in each instance the helices are oriented to bring the two membrane-associated domains

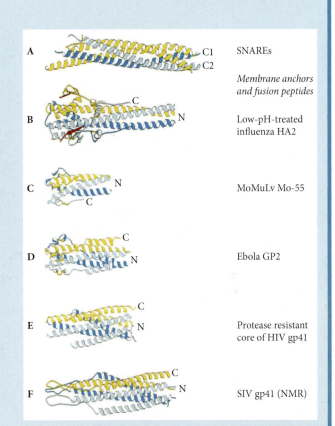

Rod-shaped α-helical bundles of the ectodomains of membrane fusion proteins, with the regions that insert in the participating membranes at one end (right side). (A) Recombinant synaptic fusion complex with the v-SNARE synaptobrevin ectodomain (light blue), t-SNARE syntaxin C-terminal ectodomain (dark blue), and the two helical regions of the t-SNARE SNAP-25B (yellow). C1 and C2 label the C termini that are inserted, prefusion, into the plasma membrane and vesicle membrane, respectively. (B) Low-pH-treated influenza virus HA after proteolytic cleavage at HA1 residue 27 and HA2 residue 38, which removes the bulk of HA1 (residues 28 to 328) and the fusion peptide region HA2 (residues 1 to 38). HA1 1 to 27 is in red. (C) Fifty-five-residue recombinant fragment of Moloney murine leukemia virus TM subunit. (D) Recombinant Ebola virus GP2. (E) Recombinant, proteolysis-resistant core of HIV type 1 gp41. (F) Recombinant SIV gp41. Adapted from J. J. Skehel and D. C. Wiley, *Cell* **95**:871–874, 1988, with permission.

together at the same end of the rod. These two points together strongly imply that viral fusion proteins and cellular SNARE complexes mediate membrane fusion by a similar mechanism: one in which the membranes that are to be fused are brought together by the juxtaposition of the membrane-anchoring domains that are at the ends of the fusion proteins, in the case of viruses, and at the ends of the SNARE complexes, in the case of the cell. In addition, these points highlight the functional significance of the low-pH-induced conformational transitions of HA.

Box continues

BOX 12.4 (continued)

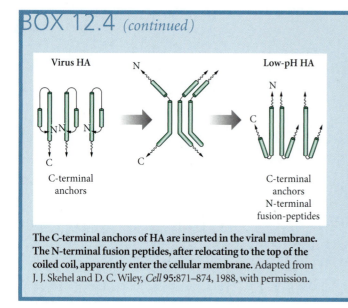

Virus HA   N   Low-pH HA

C-terminal
anchors

C-terminal
anchors
N-terminal
fusion-peptides

The C-terminal anchors of HA are inserted in the viral membrane. The N-terminal fusion peptides, after relocating to the top of the coiled coil, apparently enter the cellular membrane. Adapted from J. J. Skehel and D. C. Wiley, *Cell* 95:871–874, 1988, with permission.

In addition to the similarities in structure between the influenza HA complex and cellular SNARE complexes, there also may be similarities in the means by which their conformations are generated. For example, in each instance, initially nonhelical segments are recruited into the helical bundles. Moreover, the helix exchange step, which occurs when syntaxin and SNAP-25 form the t-SNARE complex, may be analogous to that which occurs in the pH-induced reorganization of the influenza virus HA. It was suggested that α-helical coiled coils may be particularly suited to carry out the transitions required for the juxtaposition of their molecular segments that lead to the fusion of the associated interacting membranes.

Reference
Skehel, J. J., and D. C. Wiley. 1988. Coiled coils in both intracellular vesicle and viral membrane fusion. *Cell* 95:871–874.

in its native conformation at neutral pH and also in its fusion-active conformation at low pH (Bullough et al., 1994) (Fig. 12.5). Before comparing these structures, first note that for both the native prefusion conformation and the fusion-active conformation, the structure had to be solved for an incomplete protein. The reasons were as follows. First, consider the native HA homotrimers. These can readily be isolated from purified virus particles by detergent extraction. Yet they are unsuitable for X-ray crystallographic structure analysis since their hydrophobic transmembrane domains cause them to aggregate in solution, thereby preventing crystallization. A soluble form of the native trimer, which can form crystals, was obtained by treatment with the protease bromelain, which cleaves HA close to its transmembrane domain (Fig. 12.4).

Second, consider the fusion-active conformation, which is induced by exposing the bromelain-generated native

trimers to low pH. As will be seen, when HA is in the prefusion, native conformation, the hydrophobic fusion peptides of the individual HA2 molecules are sequestered in the stalk of the trimer. In contrast, in the fusion-active conformation, the fusion peptides are thrust into an exposed position (see below). This exposure of the hydrophobic fusion peptides causes the trimers to aggregate in solution, again preventing crystal formation. Treatment with the protease thermolysin, which cleaves HA near the fusion peptide, solubilizes the fusion-activated HA trimers, thereby enabling them to be crystallized.

Despite the fact that the crystal structures obtained for both the native and fusion-active HA conformations are those of incomplete versions of the proteins, these incomplete forms nevertheless reveal the major conformational transitions that characterize the conversion of HA from

Figure 12.4 Schematic drawing of the primary structure of the 1968 Hong Kong influenza virus HA protein, showing the external domain (HA₁ plus HA₂ residues 185 to 211) and the cytoplasmic domain (HA₂ residues 212 to 221). The cleavage site between HA₁ and HA₂ is labeled "fusion activation." The signal sequence (removed by signalase), S-S bridges, carbohydrate attachment sites, and the fusion peptide (N terminus of HA₂) are shown. Adapted from B. N. Fields, D. M. Knipe, and P. M. Howley (ed.), *Fundamental Virology*, 3rd ed. (Lippincott-Raven, Philadelphia, PA, 1996), with permission.

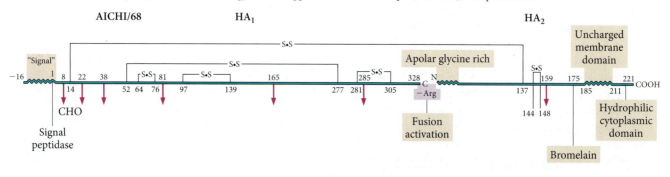

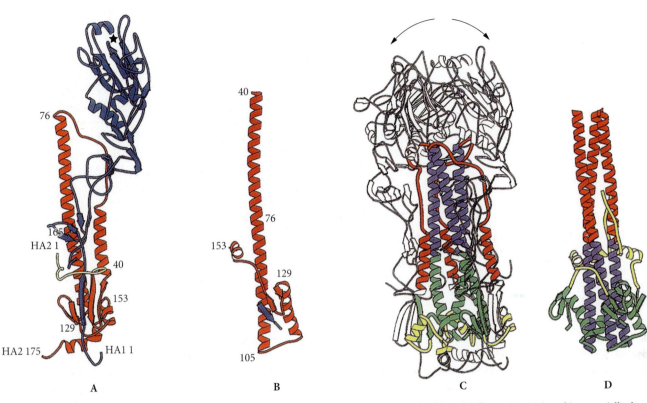

A   B   C   D

Figure 12.5 The folded structure of the influenza virus HA and its rearrangement when exposed to low pH. (A) The HA monomer. HA1 is in blue; HA2 is in red; the fusion peptide at the N terminus of HA2 is in yellow. Residue numbers in HA2 are shown at several key positions, in order to assist in visualizing the conformational change that occurs on fusion activation. The receptor-binding pocket in the β-barrel top domain of HA1 is indicated with a star. The viral membrane would be at the bottom of this figure. The transmembrane segment, which follows HA2 175 is not shown. (B) The HA2 monomer in the fusion-active form. The fragment shown is produced as the result of digesting the activated HA with thermolysin, removing most of HA1 and the fusion peptide of HA2. A short segment of HA1 (blue) remains bound to the rearranged HA2 as shown. Note the dramatic conformational change, in which residues 40 to 105 become a continuous α-helix. The introduction of a kink at residue 105 also causes the "bottom" part of the molecule to rotate upward. (C) The HA trimer.

The subunit to the right in this illustration is viewed in essentially the same orientation as the monomer in panel A. Parts of HA2 are colored: the short helix and loop (residues 40 to 75) are in red; the long helix from its N terminus to the point where the fusion peptide inserts into the long threefold axis (residues 76 to 105) is in violet; the remainder of the long helix (residues 106 to 128) and the β-sheet that follows (residues 129 to 140) are in green; and one of the C-terminal helices of HA2 (residues 141 to 162) is in yellow. The color scheme is designed to assist in visualizing the conformational change. Exposure to low pH initiates this change, causing the top domains of HA1 to move apart (arrow). (D) The fusion-active form of HA2. The tops of HA1 may be imagined to have spread to either side. The shaded segments correspond to the similarly colored segments in panel C. Modified from B. N. Fields, D. M. Knipe, and P. M. Howley (ed.), *Fundamental Virology*, 3rd ed. (Lippincott-Raven, Philadelphia, PA, 1996), with permission.

its native conformation to its fusion-active form. We begin by examining the structure of the native HA protein.

The HA1 subunits comprise the globular heads of the HA trimers, which cover the underlying HA2 subunits. This is most readily visualized by viewing the trimer structure, as depicted in Fig. 12.5C. Each HA1 subunit consists largely of an antiparallel β-sheet (Fig. 12.5A). A binding pocket for the sialic acid-containing receptor (indicated by a star in Fig. 12.5A) is located in the β-barrel "top" domain of each HA1 subunit, at the distal end of the trimer. Note that the receptor-binding pocket is approximately 135 Å from the viral envelope (see below). The HA1 and HA2 subunits of each HA monomer are linked together by a single disulfide bond between residues 14 and 137 on HA1 and HA2, respectively (Fig. 12.4). Each HA2 subunit is anchored at its carboxy terminus in the viral envelope.

The trimers of HA2 form a triple-stranded coiled coil of α-helices that extends 76 Å from the viral envelope. This is most readily visualized by examining both the HA monomer, as depicted in Fig. 12.5A, and the trimer in Fig. 12.5C. (The monomer in Fig. 12.5A is in the same orientation as that on the right in the trimer in Fig. 12.5C.) There is a loop region at the N terminus of the coiled-coil portion of each HA2 subunit (residues 55 to 75) that connects each α-helix of the coiled coil to a second α-helix, which runs antiparallel to the first α-helix (Fig. 12.5A).

The hydrophobic fusion peptide, comprised of about 20 nonpolar amino acids, is located at the N terminus of this second α-helix (yellow in Fig. 12.5A). In the trimers, the fusion peptide is buried in the interfaces of the coiled coils, about 35 Å from the viral envelope and 100 Å from the head of the molecule. This sequestering of the fusion peptide prevents the aggregation of virus particles, as well as premature interaction of HA with host cell membranes.

The structure of HA in its native conformation poses several enigmas. First, the receptor-binding sites in the heads of the trimer are 135 Å from the viral envelope. This distance between the viral envelope and the cellular target membrane is far too large to permit fusion between those membranes. Second, the fusion peptide at the N terminus of HA2 is still 100 Å from the receptor-binding pocket at the head of the trimer and at least that distance from the target membrane as well. How then might the fusion peptide interact with the target membrane and thereby promote fusion? These considerations lead to the conclusion we have been pointing towards: HA must undergo major conformational transitions to mediate fusion.

Before we consider the conformational transitions that HA undergoes to promote fusion, note that even in its native conformation, the HA molecule has already undergone a significant structural rearrangement. That is, following the proteolytic cleavage of HA_0 to generate HA1 and HA2, the N terminus of HA2, which contains the fusion peptide, moves 21 Å from the C terminus of HA1, with which it had been contiguous prior to cleavage. This proteolytic cleavage and rearrangement prime HA to undergo its transition to the fusion-active conformation, which is triggered when the endosomal pH drops to a value of 5 to 6.

The fusion-active HA conformation can be generated experimentally by incubating the isolated bromelain-treated HA trimers at pH 5. When the fusion-active form of the molecule was then crystallized and its X-ray structure was solved, it was clear that HA undergoes dramatic low-pH-triggered conformational transitions to assume the fusion-active form, with rearrangements occurring at each end of HA2. First, the loop that extended from residues 55 to 75 in the native HA and that had been folded back along the long α-helix is converted to an α-helix that is now part of a larger α-helical coiled coil that extends for at least 100 Å (Fig. 12.5; compare panels A and B). This first rearrangement has the effect of thrusting the fusion peptide upwards towards the target membrane, into which it now can insert. Second, residues 106 to 112, which were α-helical in the native structure, uncoil to form a loop. This enables residues 113 to 129, which remain α-helical, to pack up against the coiled coil, thereby permitting HA2 to fold back at the base.

Notice that the above two conformational rearrangements together result in an exchange of α-helices that pack up against the coiled coil. Importantly, the second transition results in an inversely bent structure in which the fusion peptide and transmembrane domain are now at the same end of the HA molecule, thereby bringing the viral envelope and the target membrane into close contact, while also imparting stress to those membranes (Box 12.4).

It would seem that the globular head domains of HA1 must rearrange in order to accommodate the conformational transitions of HA2. And so they do, as they separate to create space for the extended HA2 coiled coil. This was confirmed experimentally using a mutant in which the introduction of disulfide bonds between the HA1 domains blocked membrane fusion activity.

As for the membrane fusion process itself, bear in mind that membrane fusion is not a trivial process. First, the charge repulsion between the many phospholipid head groups and also between the sialic acid residues on cell surface glycolipids prevents membranes from spontaneously coming together. Second, membranes have no free edges that might initiate fusion. Thus, for fusion proteins to promote fusion, they must not only bring the interacting membranes into close apposition but also somehow destabilize the membranes to initiate the fusion reaction.

The process of membrane fusion, whether mediated by viral fusion proteins or cellular SNARE complexes, is believed to occur through an ordered sequence of steps. The first of these steps is the formation of a **hemifusion intermediate**, in which the outer (proximal) leaflets of the two interacting membranes fuse (Fig. 12.6). This step is followed by stalk formation (Fig. 12.6) and then by fusion of the inner (distal) leaflets, resulting in complete membrane fusion and pore formation. The hemifusion concept is based on theoretical considerations and is supported by experiments using model membrane systems, including experimental systems involving HA (Melikyan et al., 1995).

Figure 12.6 Transition states in membrane fusion. The monolayers are depicted as smooth and bendable sheets as described by the stalk hypothesis. Adapted from R. Jahn, T. Lang, and T. C. Südhof, *Cell* 112:519–533, 2003, with permission.

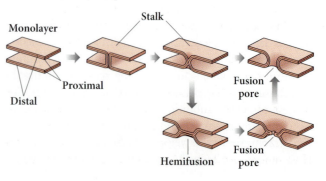

Despite all that is now known regarding the structure of fusion proteins and SNARE complexes and evidence in support of membrane fusion via hemifusion intermediates, there is no consensus regarding how fusion proteins actually go about mediating membrane fusion. In the most widely accepted model for HA-mediated fusion, the conformational transition, which generates the long trimeric coiled coil, thrusts the fusion peptide upwards towards the head of the molecule and into the target membrane. Importantly, as a result of that step, HA trimers are connected to the hydrophobic interiors of both the viral envelope and the target membrane. They are connected to the hydrophobic interior of the viral envelope via their transmembrane domains and to that of the target membrane via their fusion peptides. The second rearrangement of HA, in which the helix-to-loop transition allows the C-terminal end of the molecule to fold 180° up against the coiled coil, pulls the two membranes together. Moreover, it stresses and destabilizes the membranes, thereby perhaps initiating mixing of the membranes. In support of this model, an HA molecule was generated in which the extracellular domain is anchored to the viral envelope by a glycophosphatidylinositol linkage, rather than by the hydrophobic transmembrane domain. This altered form of HA was able to generate the hemifusion intermediate but could not generate complete membrane fusion. Presumably, the lipid linkage could not transmit sufficient stress to the membranes to complete the fusion process.

One objection to the above model is based on the following thermodynamic considerations. The conformational transition of HA that *releases* the most free energy is the formation of the extended coiled coil. Yet much of that free energy would be wasted if all it accomplished was the translocation of the fusion peptide to the target membrane. Indeed, most of the free energy *required* for fusion is needed to destabilize the interacting membranes. In the above model, this energy is provided by the second rearrangement, the helix-to-loop transition, followed by the helix turn. However, this second rearrangement releases less free energy than the first, since it is essentially a rearrangement of hydrophobic interactions.

An alternative model, which considers the above thermodynamic concerns, would have the fusion peptide insert into the *viral* envelope rather than into the target membrane. The subsequent transition of HA2 to the extended coiled coil would then exert a strong destabilizing stress on the viral envelope, while extending it in the direction of the target membrane. Note that during this process the target membrane would remain attached to HA via its receptor-binding globular head.

A third model proposes that fusion peptides insert simultaneously into both the viral envelope and the target

membrane, thereby stressing both of the interacting membranes. As in the second model, the subsequent formation of the extended coiled coil, rather than the helix-loop transition and the refolding of the adjacent helix, plays the major role in overcoming the energy barrier to fusion. Note that the major difference between these three models concerns whether the fusion peptides are inserted into the target membrane, the viral envelope, or both (Box 12.5).

Release of the Genome from the Envelope: the M2 Protein

In earlier chapters, we discussed the seeming paradox of how a virus is able to assemble its genome and other molecular subunits into an extremely stable structure and yet is also able to disassemble that structure and release its genome at the start of a new replication cycle in a new cell.

BOX 12.5

The mechanisms by which *nonenveloped* viruses breach membranes remain especially enigmatic, since there can be no membrane fusion event by which the capsid might pass into the cytoplasm. Recently, the conformational transitions of the major rotavirus (Chapter 5) entry protein, VP4, were determined using a combination of X-ray crystallography and electron cryomicroscopy (Dormitzer et al., 2004). As in the cases of the fusion proteins of enveloped viruses, the rotavirus VP4 membrane attack protein is primed for entry (i.e., membrane attack) by proteolytic cleavage, which in this instance occurs in the gut lumen of the host. Some as-yet-unknown entry-associated event is believed to then trigger the VP4 monomers to rearrange into the stable coiled-coil trimers seen in the crystal structure. This transition also unmasks the potential hydrophobic target membrane interaction peptide, which may insert into a host cell membrane. Moreover, VP4 then undergoes a third transition, in which each monomer of the trimer folds back on itself, bringing the target membrane-interaction peptide into apposition with the foot of VP4. The resulting tension on the cellular membrane may in turn create a breach that allows a subviral particle to enter the cytoplasm. Although this mechanism is somewhat speculative, the conformational transitions of the entry protein of these *nonenveloped* viruses bear remarkable similarities to those of the fusion proteins of many enveloped viruses (see the text).

Reference
Dormitzer, P. R., E. B. Nason, B. V. Prahad, and S. C. Harrison. 2004. Structural rearrangements in the membrane penetration protein of a non-enveloped virus. *Nature* **430**:1053–1058.

This paradox is well illustrated by influenza virus, as follows. During the final stages of an influenza virus replication cycle, the vRNPs associate with the M1 protein (see below). This association in turn bridges the vRNPs to regions of the plasma membrane that contain the HA and NA proteins. Mature virus particles then form by budding. Yet, as we have seen, during entry the influenza virus envelope fuses with an endosomal membrane, thereby releasing the vRNPs into the cytosol. But this release of the vRNPs requires that they dissociate from M1, the reverse of what occurs at maturation. Moreover, the disassociation of vRNPS from M1 also is necessary for the vRNPs to traffic to the nucleus, where all influenza virus transcription and genome replication take place. Release of the vRNPs from M1 is necessary for their targeting to the nucleus because the vRNPs contain nuclear localization signals (on the associated NP proteins) that are masked by M1 (see below). How then might the vRNPs be released at the appropriate time from their linkage to M1? (Another issue is how progeny vRNPs at later times might be exported from the nucleus so that they can be incorporated into progeny virus particles.)

The influenza virus M2 protein provides the solution to the dilemma spelled out above. The viral envelope contains relatively few (20 to 60) copies of the M2 protein. The individual M2 monomers are organized in the membrane as homotetramers, with an aqueous pore in the middle of the four subunits. In that configuration, the M2 homotetramers function as ion channels that are activated by low pH. Importantly, activation of the M2 ion channels occurs at a higher pH than envelope fusion and thus before envelope fusion takes place. When the M2 ion channel is activated, protons enter into the interiors of the virus particles, causing the vRNPs to disassociate from the M1 proteins. Thus, when the pH drops further and fusion occurs, the vRNPs can be released into the cytosol and traffic to the nucleus.

The first clue to the role of M2 during influenza virus infection actually came from investigations into the mode of action of the anti-influenza drug amantadine. Amantadine was already known to block an early step in the influenza virus replication cycle. Then, it was discovered that influenza virus mutations that conferred resistance to amantadine map in the M2 gene. Moreover, those amantadine-resistant mutations map in the nucleotide sequence encoding the M2 transmembrane domain. Since amantadine was already known to block an early step in the influenza virus replication cycle, these experimental findings pointed towards an early role for M2, in particular an activity of its transmembrane domain, perhaps in entry.

The M2 story takes an interesting and important twist with the later finding that when amantadine is added to cells in which the infection is already under way, then HA emerges at the cell surface in its low-pH, fusion-activated form. Since the fusion-active form of HA is usually induced only during entry, when the virus is exposed to acidic conditions in endosomes, the presence of amantadine during replication might have caused new HA proteins to be prematurely exposed to a low pH environment. But how might adding amantadine during viral replication cause HA proteins to be prematurely exposed to a low-pH environment? The answer is thought to be as follows. Since HA is a transmembrane protein, it is transported to the plasma membrane via the secretory pathway, which includes passage through the mildly acidic *trans* Golgi apparatus. Thus, unless the virus had some means for neutralizing the *trans* Golgi apparatus, HA would undergo premature conversion there to the fusion-active form, thereby preventing the formation of infectious virus. This line of argument led to the hypothesis that the M2 protein might function in the *trans* Golgi apparatus as a proton channel, in that way dissipating the proton gradient between the lumen of the *trans* Golgi apparatus and the cytosol. The result would be that new HA proteins could traffic via the secretory pathway to the plasma membrane in the native prefusion conformation (Box 12.6).

M2 ion channel activity was experimentally confirmed using the voltage-clamp procedure (used by neurophysiologists to measure ion currents across a neuronal membrane while holding the membrane voltage at a fixed level), as applied to *Xenopus laevis* oocytes and cultured mammalian cells that expressed recombinant M2 protein. Moreover, M2 ion channel activity was activated in these cells by low pH, and in addition, it was sensitive to amantadine. Finally, mutant M2 molecules that were resistant to amantadine were identified, and these amantadine-resistant mutations mapped in the sequence encoding the M2 transmembrane domains, completely consistent with the expectation that the ion channel pore is formed by the transmembrane domains of the tetramers (Fig. 12.7).

The above experimental findings show that the M2 ion channel plays a key role during influenza virus entry, and also later, during virus replication. Also, amantadine and its analog, **rimantadine**, are each used therapeutically to treat influenza. These two drugs and others that are active against influenza are discussed below.

Transport of vRNPs to the Cell Nucleus

As noted above, parental influenza vRNPs must traffic to the nucleus, since that is the site for all orthomyxovirus transcription and genomic replication. Why an RNA virus might need to carry out its transcription and replication in the nucleus is not at all obvious, and the reasons why influenza virus needs to do so will become clear shortly.

BOX 12.6

The influenza type A viruses that circulate in humans generally give rise to infections that are confined to the respiratory tract. These human viruses usually are less pathogenic in humans than avian strains belonging to subtypes H5 and H7, which can cause a more generalized infection in humans, as well as in avian hosts. As described in the text, differences in the dissemination and pathogenicity of these influenza virus types are determined at least in part by the variety of host proteases that might mediate cleavage activation of their HA proteins. As noted in the text, cleavage activation of the HA proteins of human type A influenza viruses is thought to be carried out by tryptase Clara, which is restricted to the Clara cells of the bronchial and bronchiolar epithelia, thereby

restricting replication of human influenza A viruses to the vicinity of these specialized cells in the respiratory tract. In contrast, cleavage of the HA proteins of the more highly pathogenic avian strains is carried out by more ubiquitous proteases, such as furin, enabling these strains to give rise to more-disseminated and hence more-severe infections.

Apropos the effect of adding amantadine at late times in infection, a premature conformational change in HA occurs only in the case of those influenza virus strains whose HA proteins are cleaved by furin (in the Golgi apparatus) and which have a pH-triggered conformational change at a pH close to that extant in the Golgi body.

For the moment, note that orthomyxoviruses are not unique among RNA viruses in their nuclear dependency, since bornavirus and retrovirus genomes also must enter the nucleus.

Because the nucleus is bounded by two lipid bilayers, macromolecules and macromolecular complexes cannot simply diffuse into, or out of, the nucleus. Instead, with few exceptions, the only means by which these entities

might cross the nuclear membranes is via **nuclear pore complexes**, and they can do so only if they contain short **nuclear localization signals** or **nuclear export signals**, which generally are fewer than 20 amino acids in length. Those signals are necessary and sufficient to target proteins and macromolecular complexes for transport into, or out of, the nucleus. As might be expected, influenza vRNPs contain nuclear localization signals, and they use

Figure 12.7 Interaction of amantadine with the transmembrane domain of the influenza A virus M2 ion channel. Amantadine is thought to exert its antiviral effect by blocking the virus-encoded M2 ion channel activity in infected cells. This inference is supported by genetic data and models of M2 protein, but the points in the virus life cycle critical for inhibition are still controversial. M2 protein is a tetramer with an aqueous pore in the middle of the four subunits. (A) Helical wheel representation of the transmembrane domain of M2 protein. The sequence Asp24-Arg45 of A/chicken/Germany/27 (H7N7) is given in single-letter code. Polar amino acids (Ser, Thr, and Asn) and charged amino acids (His and Glu) are found on the right side of the helical wheel, while nonpolar amino acids cluster on the left. The four locations of single amino acid changes in different amantadine-resistant mutants are boxed, and the substitutions most commonly observed are shown in parentheses. (B) Diagram of the putative interaction of amantadine with two diagonally located α-helices of the M2 tetramer in a lipid bilayer. The polar faces of the helices form the interior of the proposed ion channel. Adapted from A. J. Hay, *Semin. Virol.* **3:**21–30, 1992, with permission.

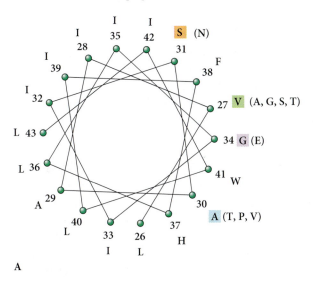

A

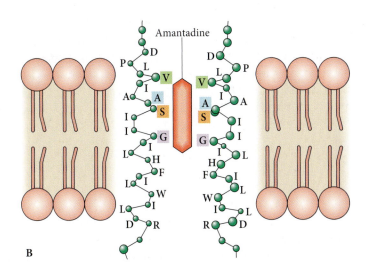

B

the same nuclear import machinery as that used by cellular proteins.

Macromolecules that contain nuclear localization signals bind to cytoplasmic nuclear localization signal receptors. The best-known of these receptors is importin-α, also known as karyopherin α. The complex of substrate and importin-α then binds to importin-β (karyopherin β), which facilitates docking of the substrate/importin complex at nuclear pore complexes. These nuclear pore complexes indeed are complex, being comprised of some 100 or more proteins, which are referred to as nucleoporins. Translocation into the nucleus also requires the activity of other proteins as well. The transport of cargo across the nuclear pore complex is illustrated schematically in Box 12.7.

BOX 12.7

The cytoplasmic and nuclear compartments are separated by the double nuclear membrane, which is studded with nuclear pore complexes. These are the only channels through the nuclear envelope. Thus, these nuclear pore complexes play a critical role, acting as the selective gateways that regulate macromolecular traffic into and out of the nucleus. Moreover, together with the nuclear membrane, they constitute the barrier that separates the contents of the nucleus from the cytoplasm.

In step 1 of nuclear entry, a nuclear localization signal (NLS) on a cytoplasmic cargo is bound by importin-α (karyopherin α). In step 2, importin-β binds to the cargo/importin-α complex and docks onto a nuclear pore complex. In step 3, additional cytoplasmic proteins, including the guanine nucleotide-binding protein Ran, in its guanosine diphosphate (GDP)-bound form (Ran-GDP), bind to the complex. These factors facilitate the transport of the cargo through the nuclear pore complex into the nucleus. The mechanism of this transport is not yet known, but it is thought that it may involve repeated cycles of docking and release of the cargo/importin complexes with specific components (importins) of the nuclear pore complex. Following import, the complexes are dissociated when Ran-GDP is converted to Ran-guanosine triphosphate (Ran-GTP) by Rcc1, a Ran-specific guanine nucleotide exchange protein. Export of cargo from the nucleus shares several mechanistic features with import. Like import, export is mediated by receptors (exportins) that, in this instance, recognize nuclear export signals and direct cargo to nuclear pore complexes. But Ran-GTP rather than Ran-GDP binds to the exportins to mediate export. In the cytoplasm, a Ran-GTPase-activating protein, RanGAP-1, converts Ran-GTP to Ran-GDP and RanBP1 and RanBP2 maintain Ran in the GDP-bound form. The combined actions of Rcc1 in the nucleus and RanGAP-1 in the cytoplasm maintain the Ran-GTP/Ran-GDP gradient that drives nuclear import and export. Modified from S. J. Flint, L. W. Enquist, R. M. Krug, V. R. Racaniello, and A. M. Skalka, *Principles of Virology: Molecular Biology, Pathogenesis, and Control,* 2nd ed. (ASM Press, Washington, DC), with permission.

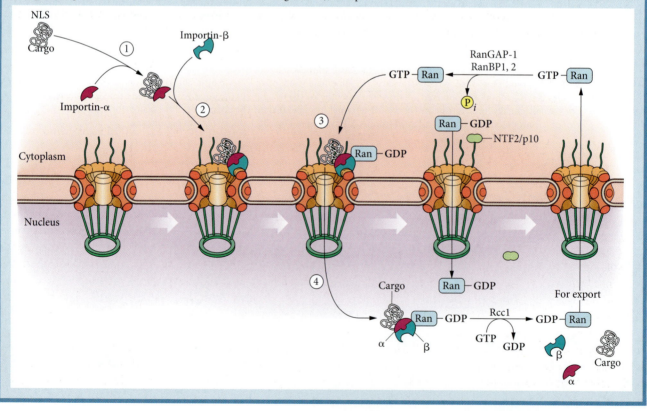

The nuclear localization signals on the vRNPs are contained on the NP proteins. "Naked" viral RNA, which has been stripped of NP, does not dock onto nuclear pore complexes and cannot enter the nucleus. Importin-α binds to the NP nuclear localization signals on NP, in that way initiating the sequence of steps leading to import of vRNPs into the nucleus, via nuclear pore complexes.

Interestingly, the diameters of the central channels of nuclear pore complexes are about 23 nm, while the diameters of the vRNPs are about 20 nm. Consequently, influenza vRNPs are among the largest entities that are known to enter the nucleus. An intriguing possibility is that influenza genomes are segmented mainly to enable the entire viral genome to pass through nuclear pores, since an unsegmented genome of the same total size might be too large to easily pass through nuclear pores.

Since all influenza virus RNA synthesis occurs in the nucleus, the RNA polymerase activities associated with the parental influenza virus particles also must be transported into the nucleus. This is a rather straightforward matter, since the polymerase proteins PB1, PB2, and PA traffic to the nucleus as components of the vRNPs. Later in infection, new polymerase proteins, which are synthesized on cytoplasmic ribosomes, must enter the nucleus to assemble with newly replicated RNA into new vRNPs. The same is true for the new NP, M1, and NS1 proteins. Each of these proteins contains a nuclear localization signal and enters the nucleus via the normal cellular nuclear protein import pathway. (Bear in mind that later in the replication cycle, there must be some means for reversing this transport so that new vRNPs might be transported from the nucleus to the sites of final virus assembly and release at the plasma membrane [another slant on our entry-exit paradox; see below].)

GENOME ORGANIZATION, TRANSCRIPTION, AND REPLICATION

Genome Organization and Transcription

The multipartite genomes of influenza A and B viruses each contain eight minus-strand RNA segments, whereas influenza C viruses contain seven genomic RNA segments, as explained above. The genomic organization of type A influenza viruses is shown schematically in Fig. 12.8. Each of genome segments 2, 3, 4, 5, and 6 encodes a single protein. In contrast, segments 1, 7, and 8 each encode two distinct proteins. As far as is known, no animal virus genomes other than those of the influenza viruses and the reoviruses are segmented to such an extent.

The first hint that influenza viruses might contain segmented genomes came from experiments in the early 1960s, in which extremely high rates of genetic recombination were unexpectedly observed when cells were coinfected with genetically distinct influenza virus strains. Such high rates of recombination are more consistent with genetic reassortment, rather than true breakage-and-joining recombination. By the late 1960s, the availability of polyacrylamide gel electrophoresis (PAGE) made it possible to confirm that influenza viruses in fact contain segmented genomes, since individual genome segments could readily be resolved on these gels (Box 12.8).

Several early experimental observations implied that influenza virus transcription would be different in important ways from that of other minus-strand and also plus-strand RNA viruses. First, influenza virus transcription in vivo requires the cell nucleus. (As noted above, influenza viruses are not unique among RNA viruses in this regard, since the minus-strand bornaviruses and the plus-strand retroviruses also require the cell nucleus for their replication. However, the nuclear dependency of an RNA virus was first seen in the case of influenza virus.) Second, influenza virus transcription in vivo is blocked by two drugs, actinomycin D and the fungal toxin α-amanitin, that would not be expected to affect transcription by an RNA virus. Actinomycin D inhibits transcription from a double-stranded DNA template, and α-amanitin specifically blocks transcription mediated by the host RNA polymerase II. Third, it was found that virus transcription in vitro is greatly stimulated by the dinucleotide AG, which is complementary to the two nucleotides, UC, at the 3′ end of each of the influenza virus genomic RNA segments. Moreover, the AG dinucleotide was found to be incorporated directly into the 5′ end of the new influenza virus transcripts. That is, the AG dinucleotide actually primed influenza virus transcription in vitro. (Before reading on, try to formulate a hypothesis that might incorporate all of these experimental findings into a model for influenza virus transcription.)

The first two observations are consistent with the premise that ongoing *host* transcription might be necessary for influenza transcription. The third finding suggests that influenza transcription in vivo might be primed by another RNA molecule. Taken together, they raise the possibility that cellular mRNAs somehow serve as primers for influenza virus transcription in vivo. This possibility was supported by the experimental finding that cellular mRNAs indeed are effective primers for influenza virus transcription in vitro, as catalyzed by the viral transcriptase.

The initial experiments that tested the priming hypothesis made use of a coupled transcription/translation system from rabbit reticulocyte lysates, in which globin mRNA was the predominant mRNA species present

(−) strand RNA
segments

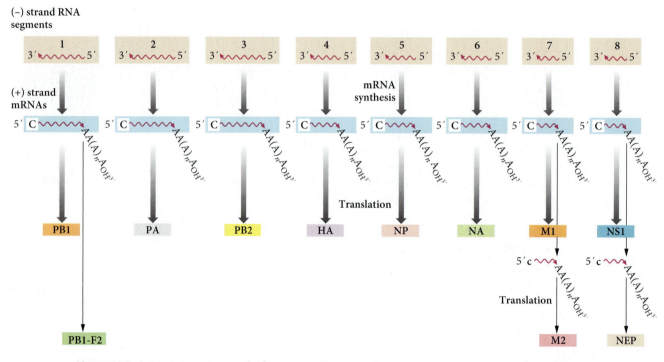

Figure 12.8 Genomic organization of influenza A virus, showing the protein-coding assignments of the individual genomic segments. Adapted from S. J. Flint, L. W. Enquist, R. M. Krug, V. R. Racaniello, and A. M. Skalka, *Principles of Virology: Molecular Biology, Pathogenesis, and Control*, 2nd ed. (ASM Press, Washington, DC, 2004), with permission.

(Plotch et al., 1979). As might be expected, other eukaryotic mRNAs, including growth hormone mRNA, and even retroviral and reoviral mRNAs, also are efficient primers of influenza virus transcription in the cell-free system.

Importantly, a 5′-methylated cap structure must be present on a cellular RNA for it to prime influenza virus transcription. Enzymatic or chemical removal of the caps from cellular mRNAs abolishes their priming activity, and recapping those RNAs restores their priming activity.

How might we account for such a bizarre process? First, note that influenza virus mRNAs must be capped in order to be efficiently translated. Second, since the influenza virus genome does not encode capping activities, influenza virus mRNAs acquire their caps by "stealing" them from cellular mRNAs. And the priming mechanism requires capped cellular mRNAs because this is the means by which influenza virus mRNAs acquire their caps. (In principle, how else might influenza virus mRNAs acquire their caps?) This was shown to indeed be the case in the ground-breaking and prescient study of Plotch and coworkers (1979). These investigators began by preparing cellular mRNAs that were labeled specifically in their caps. Then, when these cellular mRNAs were used in the in vitro priming reactions, the labeled caps were transferred to

each of the viral mRNA species. Also, analysis by PAGE showed that when influenza virus mRNAs were primed by cellular mRNAs, they were 10 to 13 nucleotides longer than when primed by dinucleotides. Thus, 10 to 13 nucleotides in addition to the cap also are transferred from cellular mRNAs to the influenza virus RNAs. This transfer of an additional 10 to 13 nucleotides to influenza virus RNAs also happens in vivo.

The process by which influenza virus mRNAs acquire their caps is referred to as **cap snatching**. In this process, capped cellular mRNAs are cleaved 10 to 13 nucleotides from their 5′ ends by a cap-dependent endonuclease activity associated with the virus polymerase complex. These capped RNA fragments then serve as primers for influenza virus transcription. These steps are described in more detail below. (Is cap snatching the reason why influenza virus transcription occurs in the nucleus? Couldn't cap snatching occur in the cytoplasm? See below.)

The sensitivity of influenza virus transcription to actinomycin D and α-amanitin can now be understood. As new cellular mRNAs mature, they are transported to the cytoplasm. Thus, the pool of capped cellular mRNAs in the nucleus continually turns over and must continually

BOX 12.8

Interestingly, the migration rates of the eight influenza virus RNA segments on PAGE gels varied from one influenza A virus strain to another, as seen in the figures in this sidebar. Likewise, the proteins of these different strains displayed different migration rates from one strain to another. Making use of these experimental findings, it was possible to correlate particular genes with particular RNA segments, that is, to assign particular genes to particular RNA segments. This was achieved by analyzing "reassortants" between two different influenza viruses. That is, the PAGE migration patterns

of the RNA and proteins of the reassortant strains were compared to that of each of the parental strains (Ritchey et al., 1976). A similar strategy was used to assign coding responsibilities to particular reovirus genomic RNA segments (Chapter 5).

Reference
Ritchey, M. B., P. Palese, and J. L. Schulman. 1976. Mapping of the influenza virus genome. III. Identification of genes coding for nucleoprotein, membrane protein, and nonstructural protein. *J. Virol.* 20:307–313.

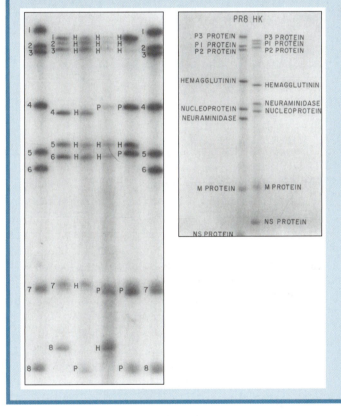

(Left) PAGE RNA analysis of influenza A/PR/8/34 and A/HK/8/68 viruses and three recombinant viruses derived from them. Lanes 1 and 6, PR8 virus; lane 2, HK virus; lane 3, recombinant virus number 1, identical to HK virus except for RNA 8; lane 4, recombinant virus number 2, identical to HK virus except for RNAs 4 and 7; lane 5, recombinant virus number 3, identical to HK virus except for RNAs 4, 6, 7, and 8. The letter P or H next to an RNA segment identifies it as a PR8 virus or HK virus gene, respectively. (Right) The RNAs of influenza A/PR/8/34 and A/HK/8/68 viruses were separated on a polyacrylamide gel as shown in the panel on the left. All eight RNA segments have been identified with respect to their gene products, as described above. Reproduced from P. Palese, *Cell* **10:**1–10, 1977, with permission.

be replenished to provide primers for influenza virus transcription. Recent experimental findings show that the cellular RNA polymerase II may play an additional role in the influenza virus replicative cycle, facilitating the export of most viral mRNAs from the nucleus (Box 12.9).

Next, we look at the specific steps of the influenza virus cap-snatching priming mechanism and transcription (Fig. 12.9). The influenza virus polymerase complex is comprised of the three P proteins, PB1, PB2, and PA. The PB1 subunit of the polymerase complex initiates transcription by binding specifically to a highly conserved

13-nucleotide-long sequence at the *5′ end* of each of the influenza virus genomic RNA segments.

But why would the polymerase initially bind to a site at the 5′ end of each genomic RNA segment? This question is raised because the RNA template is read in the 3′-to-5′ direction, and if the polymerase were to begin transcribing at the 5′ end of the genomic RNA, it would very soon run out of template. For this reason, it is believed that the polymerase remains fixed at the 5′ end of each genomic segment. Why then the polymerase binds to the 5′ end of each genomic RNA segment is explained shortly below.

BOX 12.9

While it is clear that blocking host RNA polymerase II at the start of infection leaves the virus without a source of cap donor mRNA molecules that might prime viral transcription, recent experiments of Paul Digard and coinvestigators show that this is not the only reason why the virus needs RNA polymerase II. These researchers found that the RNA polymerase II inhibitor DRB inhibits expression of viral mRNAs (with the exception of the NP and NS2 mRNAs) at the *posttranscriptional* level, specifically, at the level of nuclear export. That is, large quantities of mature viral mRNAs are generated in the presence of DRB, but they are not exported to the cytoplasm, where they can be translated. Thus, influenza A virus replication requires RNA polymerase II activity not just to provide capped mRNA fragments as primers of transcription but also to facilitate nuclear export of most viral mRNAs. Apropos these findings, other recent experimental results show that RNA polymerase II normally acts to link multiple transcriptional and posttranscriptional events that are involved in the synthesis, nuclear export, and translation of mature cellular mRNAs (see the text). Taken together, these studies underscore the fact that much remains to be learned about the interactions between influenza virus and cellular nuclear function.

Reference
Amorim, M. J., E. K. Read, R. M. Dalton, L. Medcalf, and P. Digard. 2007. Nuclear export of influenza A virus mRNAs requires ongoing RNA polymerase II activity. *Traffic* **8:**1–11.

The specific binding of PB1 to the conserved 5′ site on the genomic RNA template induces a conformational change in the PB2 subunit of the polymerase complex, which enables PB2 to bind to the cap of a cellular mRNA molecule. In the next step, the 12-nucleotide-long conserved nucleotide sequence at the *3′ end* of each of the genomic RNA segments binds to another site on PB1. This step induces an endonuclease activity of the polymerase complex, which cleaves the capped cellular mRNA 10 to 13 nucleotides from the cap structure. Earlier experimental results suggested that PB2 expresses this endonuclease activity. However, a more recent study found that the endonuclease activity actually is located in the PB1 subunit. Regardless, this endonucleolytic cleavage results in a mature transcription complex, in which the capped cellular RNA fragment is able to serve as the primer for influenza virus transcription.

Are you thinking that there is a requirement for homology between the cellular mRNA molecule which serves as the primer and the conserved sequence found at the 3′ ends of the influenza virus genomic RNA segments? That is a good thought, but that is not how it happens. Rather, the interaction between the primer RNA fragment and the viral genomic RNA segment is mediated entirely by the virus polymerase complex.

The first nucleotide added to the primer is a G residue, which is encoded by the penultimate C residue at the 3′ end of each genomic RNA segment. This step is followed by elongation, in which the genomic RNA segments pass through or over the polymerase complexes in the 3′-to-5′ direction, while the 5′ ends of the genomic RNA segments remain fixed on those complexes. The cap structure, which initially was bound to PB2, is released soon after the onset of elongation. The PB1 component of the polymerase complex catalyzes the addition of each nucleotide to the elongating viral mRNA chains. (Until recently, the PA protein had no known role in transcription, although it was known to have an active role during replication [see below]. However, results of a recent genetic analysis showed that PA is required for efficient cap-binding and for the cap-dependent endonuclease activity. Moreover, it is required for the polymerase complex to bind to the promoters on the viral minus-strand genomic RNAs.)

The polymerase complexes remain tightly bound to the 5′ end sequences of the minus-strand template RNA segments throughout the initiation and elongation steps of influenza virus transcription. This continual association between the polymerase complexes and the 5′ ends of the template RNAs raises the question of how the nucleotide sequences at the 3′ ends of the mRNA transcripts are generated. The answer is a bit odd. In point of fact, the final 15 to 22 nucleotides at the 5′ ends of the genomic RNA segments are *not* transcribed into the 3′ ends of the viral mRNAs. (What implications might this have apropos replication?) Another oddity is that each of the influenza virus mRNA species contains a poly(A) tail that is about 150 residues long. Yet there are no sequences on the minus-strand genomic RNA segments that might directly encode these poly(A) runs. These strange events are explained as follows.

The continual attachment of the polymerase complexes to the 5′ ends of the genomic RNA segments actually facilitates the polyadenylation of the viral transcripts in the

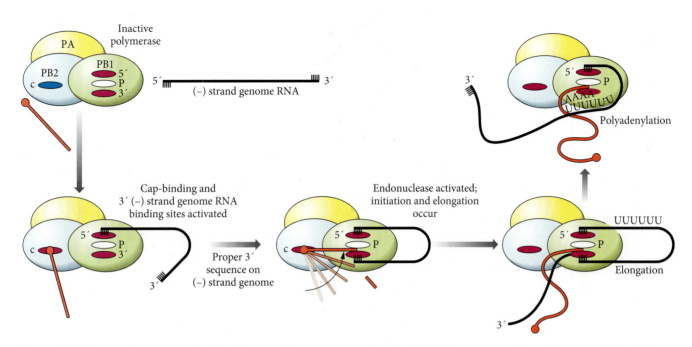

Figure 12.9 Influenza virus transcription. The influenza polymerase complex is comprised of the three P proteins, PB1, PB2, and PA. The PB1 subunit of the polymerase complex initiates transcription by binding specifically to a highly conserved 13-nucleotide-long sequence at the *5′ end* of each of the influenza virus genomic RNA segments. The specific binding of PB1 to the conserved 5′ site on the genomic RNA template induces a conformational change in the PB2 subunit of the polymerase complex, which enables PB2 to bind to the cap of a cellular mRNA molecule. In the next step, the 12-nucleotide-long conserved nucleotide sequence at the *3′ end* of each of the genomic RNA segments binds to another site on PB1. This step induces an endonuclease activity in the PB1 subunit, which cleaves the capped cellular mRNA 10 to 13 nucleotides from the cap structure. This endonucleolytic cleavage results in a mature transcription complex, in which the capped cellular RNA fragment is able to serve as the primer for influenza virus transcription. The first nucleotide added to the primer is a G residue, which is encoded by the penultimate C residue at the 3′ end of each genomic RNA segment. This step is followed by elongation, in which the genomic RNA segments pass through or over the polymerase complexes in the 3′ to 5′ direction, while the 5′ ends of the genomic RNA segments remain fixed on those complexes. The cap structure, which initially was bound to PB2, is released soon after the onset of elongation. The PB1 component of the polymerase complex catalyzes the addition of each nucleotide to the elongating viral mRNA chains. Results of a recent genetic analysis showed that PA is required for efficient cap-binding and for the cap-dependent endonuclease activity. The polymerase complexes remain tightly bound to the 5′ end sequences of the minus-strand template RNA segments throughout the initiation and elongation steps of influenza virus transcription. The continual attachment of the polymerase complexes to the 5′ ends of the genomic RNA segments facilitates the polyadenylation of the transcripts in the following way. As a genomic RNA segment "threads" past the polymerase complex in the 3′-to-5′ direction, the polymerase complex eventually encounters a sequence of 5 to 7 U residues that is adjacent to the terminal 15 to 22 nucleotides at the 5′ end of the genomic RNA segment. When that poly(U) sequence reaches the polymerase complex, the attachment of the polymerase complex to the 5′ end of the template RNA impairs further reading of the template in the 3′-to-5′ direction. This impairment favors the reiterative copying of the poly(U) sequence, which results in the addition of about 150 A residues to the 3′ ends of the nascent mRNAs. Adapted from D. M. Knipe et al. (ed.), *Fields Virology*, 4th ed. (Lippincott Williams & Wilkins, Philadelphia, PA, 2001); P. Rao et al., *EMBO J.* **22:**1188–1198, 2003; and S. R. Shih and R. M. Krug, *Virology* **226:**430–436, 1996, with permission.

following way. As a genomic RNA segment "threads" past the polymerase complex in the 3′-to-5′ direction, the polymerase complex eventually encounters a sequence of five to seven U residues that is adjacent to the terminal 15 to 22 nucleotides at the 5′ end of the genomic RNA segment. When that poly(U) sequence reaches the polymerase complex, the attachment of the polymerase complex to the 5′ end of the template RNA impairs further reading of the template in the 3′-to-5′ direction. This impairment favors the reiterative copying of the poly(U) sequence, which results in the addition of about 150 A residues to the 3′ ends of the nascent mRNAs. The rhabdoviruses (Chapter 10) and the paramyxoviruses (Chapter 11) also make use of the reiterative copying of a short poly(U) stretch to add poly(A) sequences to their mRNAs. However, in those instances the minus-strand template RNA is not bound to the polymerase complex at its 5′ end.

As noted above, five of the eight influenza A virus genomic RNA segments encode a single protein, while three genomic segments, 1, 7, and 8, each encode two proteins (Fig. 12.8). Genomic segment 7 encodes the matrix or

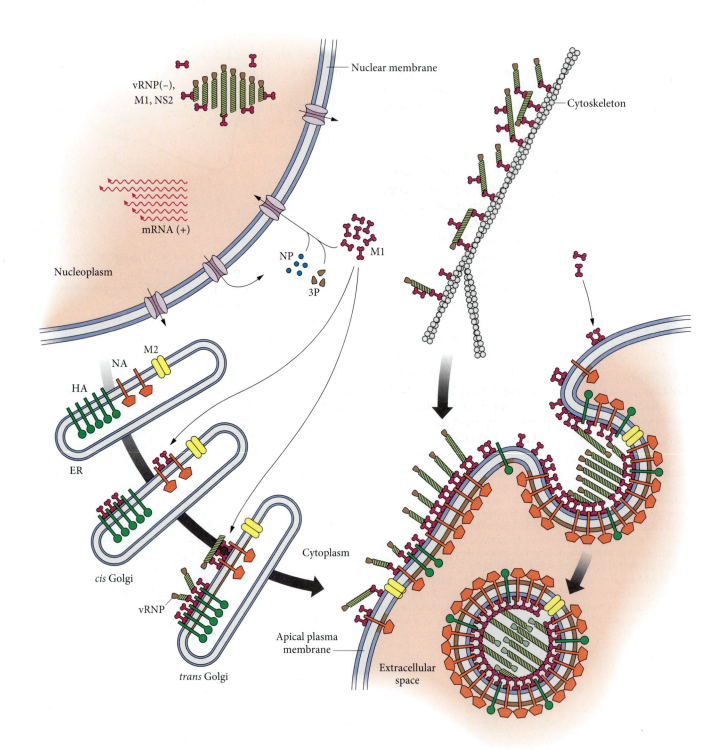

Figure 12.10 Schematic representation of influenza virus morphogenesis. For viral morphogenesis to occur, all the subviral components must be transported to the assembly site at the plasma membrane and interact with one another in an orderly manner. Both glycoproteins (HA and NA) use the exocytic pathway and are transported from the *trans*-Golgi network to the budding site, a specific region on the plasma membrane containing lipid rafts. Another glycoprotein (M2) is transported via the same route but does not require lipid rafts. The M1-vRNP complex, consisting of the viral genomic RNA, NP, NEP, 3P, and M1, is exported out of the nucleus and transported to the assembly site on the plasma membrane either via cytoskeleton elements or by piggybacking on the cytoplasmic tail of HA and NA. M1 binds to the cytoplasmic tail and transmembrane domain of HA and NA on the outer side, and the vRNP on the inner side. Finally, the plasma membrane bends at the assembly site containing glycoproteins and the M1-vRNP complexes, causing an outward membrane curvature. Eventually, fusion of the membranes leads to fission and pinching-off of the virus particle, releasing the enveloped progeny virus particle into the extracellular medium. Lipid raft microdomains in the membrane are shown in blue; nonraft regions are depicted in grey. Adapted from D. P. Nayak, E. K. Hui, and S. Barman, *Virus Res.* **106:**147–165, 2004, with permission.

M protein, as well as the M2 protein, while genomic segment 8 encodes NS1 and NEP. We noted above that NEP, formerly called NS2, mediates export of vRNPs from the nucleus. The actions of NS1 and NEP are described in detail below.

Genomic segments 7 and 8 employ similar mechanisms to generate multiple proteins. In each instance, the larger protein (M in the case of segment 7 and NS1 in the case of segment 8) is translated from an unspliced transcript, while the M2 and NEP proteins are each translated from spliced transcripts. Only about 10% of the M1 mRNA molecules are spliced to form M2 mRNA. This is to the advantage of the virus, since it requires many more copies of the M protein than of the M2 protein. Moreover, an excess of the M2 protein could impair the cellular secretory pathway, which would be disadvantageous to the virus.

The complete nucleotide sequence of influenza A virus was known by 1983, at which time it was believed that all of the viral gene products had been identified. Thus, it was unexpected when in 2001 a novel protein was identified, encoded by the +1 reading frame of the gene encoding the PB1 subunit of the viral polymerase. This protein, designated PB1-F2, is a proapoptotic polypeptide that specifically targets alveolar macrophages, in that way contributing to the ability of influenza virus to evade host immunity (see below).

Why Does Influenza Virus Transcription Really Occur in the Nucleus?

We recounted how influenza virus transcription is sensitive to the drugs actinomycin D and α-amanitin, the reason being that influenza virus transcription requires a continual supply of fresh capped cellular mRNA molecules, which are used by the virus to provide capped primers for its mRNAs. But is this the reason why influenza virus transcription occurs in the nucleus? The answer is, probably not. Here is why. There are lots of capped mRNA molecules in the cytoplasm, so that influenza virus should not have to carry out transcription in the nucleus for the sole purpose of cap snatching. This premise is supported by the example of the bunyaviruses. Bunyavirus transcription also is primed by capped cellular mRNAs (Chapter 13). Yet bunyaviruses replicate in the cytoplasm.

Now, recall that influenza viruses do not encode their own splicing activity. Thus, it is likely that influenza virus transcription occurs in the nucleus because the virus depends on the cellular splicing machinery, which is located exclusively in the nucleus. But then, what of the bunyaviruses? In contrast to the influenza viruses, bunyaviruses do not have spliced mRNAs and thus may replicate in the cytoplasm.

Now consider the following. Despite the fact that bunyaviruses, like influenza viruses, steal their mRNA caps from cellular mRNAs, bunyavirus transcription is not impaired by α-amanitin. How might that puzzle be explained? The answer is likely as follows. The sensitivity of influenza virus transcription to α-amanitin probably results from the fact that influenza virus transcription occurs in the nucleus, where capped cellular mRNA turns over rapidly (i.e., it is rapidly exported to the cytoplasm). In contrast, bunyavirus cap snatching occurs in the cytoplasm, where cellular mRNAs turn over much more slowly.

The bornaviruses (Chapter 13) provide yet another variation on this theme. Like influenza viruses, bornaviruses also carry out transcription in the nucleus. (In point of fact, the nonsegmented bornaviruses and the segmented influenza viruses are the only known nonretroid RNA viruses that replicate in the nucleus.) As in the case of influenza viruses, some bornavirus mRNA species likewise are spliced from precursor RNAs. However, rather than using a cap-stealing process such as used by influenza viruses, bornavirus mRNAs are capped by an activity associated with the virus-encoded polymerase. Why then do bornaviruses replicate in the nucleus? The answer is that they, like influenza virus, still depend on the nuclear splicing machinery to generate their spliced transcripts.

Replication

Influenza virus transcription generates mRNAs that are capped, polyadenylated, and prematurely terminated. However, in order to produce progeny viral genomic RNA segments, the viral polymerase complex must be able to first make noncapped, nonpolyadenylated, full-length plus strands, which can then serve as templates for producing minus-strand progeny genomic RNAs. Consequently, replication requires a means for switching from capped mRNA-primed initiation of plus strands to unprimed initiation of those strands. In addition, there must be a means to antiterminate at the short poly(U) sequences near the 5′ ends of the genomic template RNAs.

How might influenza virus accomplish the switchover from transcription to replication? In truth, the switchover from transcription to replication is not well understood, even after more than 2 decades of experimentally investigating this issue. Yet one reasonable hypothesis is that the transition is somehow initiated by the de novo synthesis of a particular viral protein. Consistent with that premise, in vivo transcription continues under conditions where protein synthesis is blocked, and impairing protein synthesis in fact does prevent the switchover to replication. What de novo factor might then be required to initiate the transition from transcription to replication? Biochemical as well as genetic experiments show that the influenza NP protein is required for the transition from transcription to replication. For example, mutants with temperature-sensitive lesions in the NP gene are able to synthesize mRNA, but not full-length, plus-strand, template RNA.

How then might production of NP protein promote the transition from transcription to replication? Recall that during transcription, the polymerase complex is bound to the 5′ end of the minus-strand template RNA and in that way causes the premature termination of transcription and the reiterative copying of the short poly(U) tract near the 5′ end of that RNA (see above). But when the concentration of newly synthesized NP protein becomes sufficiently high, this RNA-binding protein is believed to displace the polymerase complex from the 5′ end of the minus-strand template RNA. Consequently, the polymerase complex no longer impairs RNA synthesis beyond the poly(U) tract. Thus, there is no premature termination and no polyadenylation.

The action of the NP protein in freeing the 5′ end of the genomic RNA cannot be the sole basis for the transition of RNA synthesis from transcription to replication. That is so because there also must be some means for the polymerase complex to switch from its dependency on capped cellular mRNAs for priming RNA synthesis to being able to carry out unprimed initiation. It is likely that no additional newly synthesized viral protein is necessary for this aspect of the transition. This conjecture is based on the results of an in vivo experiment, in which the PB1, PB2, PA, and NP proteins were sufficient to replicate an artificial gene construct bracketed by the conserved sequences at the 3′ and 5′ ends of the influenza virus genomic RNA segments.

A recently proposed model to account for the switchover from transcription to replication hypothesizes that the polymerase complex synthesizes full-length plus strands, as well as mRNA molecules, throughout the replication cycle. However, the full-length plus strands initially are degraded until a sufficient amount of free NP protein and polymerase complexes have been produced to bind to them, thereby protecting them from being degraded. In support of this model, if viral polymerase complexes and NP protein are preexpressed in cells, then both full-length plus strands and viral mRNA molecules can be detected early in the viral infection cycle. Moreover, the researchers proposing this model found that the preexpressed polymerase complexes and NP proteins bind exclusively to nascent full-length plus strands. Still, this model does not account for how the polymerase complex might initiate the unprimed RNA synthesis that is required for replication.

PA and PB2 also have been implicated in the switchover of influenza virus RNA synthesis from transcription to replication, although there is no consensus among influenza researchers on this issue. Evidence for the involvement of PA and PB2 comes from studies of constructed mutations in the genomic RNA segments encoding PA and PB2. Some such mutations affect genomic replication but not transcription. The role of PA in replication is perhaps to facilitate the *unprimed* initiation of plus strands and minus strands by the polymerase complex. As for PB2, it is plausible that the mutations introduced into PB2 alter its recognition of a site on the RNA template required for replication but not transcription. Alternatively, since PB2 is known to interact in vitro with NP, perhaps these mutants fail to replicate because of an impaired ability to interact with NP. Then again, the mutations might affect the ability of PB2 to interact with a crucial cellular factor.

Indeed, it is likely that cellular factors do play a role in the switchover of influenza virus RNA synthesis from transcription to replication, and several potentially germane cellular proteins have been identified. A recent, comprehensive, proteomics-based screen identified 41 cellular proteins that copurified with influenza vRNPs, and four other cellular proteins that copurified with the viral polymerase complex were found. These cellular proteins included novel ones, such as importin-β3 and PARP-1, which were found in association with the polymerase complex, and nucleophosmin, which was associated with vRNPs.

PARP-1 normally acts as a DNA damage nick sensor that has been implicated in DNA repair. Nucleophosmin is an estrogen-regulated nucleolar protein that has been implicated in a number of cancers. It normally appears to affect centrosome duplication and genome stability but also appears to play a role in transcription mediated by RNA polymerase II. Interestingly, experimentally overexpressing nucleophosmin results in increased viral RNA polymerase activity. Other earlier experimental evidence suggested that a cellular splicing factor might facilitate formation of NP-RNA complexes. Still, it is not clear how any of these cellular factors might act in viral RNA synthesis or bring about the switchover to replication.

The next step in replication is the copying of the full-length, plus-sense template RNAs to make progeny genomic RNAs. This step also requires newly synthesized NP proteins, possibly because the single-stranded RNA-binding NP proteins prevent base pairing between template and the product, thereby maintaining the RNA in a single-stranded conformation.

The NS1 and PB1-F2 Proteins

Viral pathogens generally use a variety of strategies to evade host defenses (Chapter 4). A well-known and especially effective means by which influenza viruses evade the *acquired* immunity, which becomes widespread in the host population, is to undergo continuous antigenic change. This key feature of influenza virus is discussed in detail below. Influenza viruses are also able to evade *innate* antiviral defenses, in particular interferon (IFN)-mediated responses and apoptosis. In this regard, we consider two influenza virus gene products: first, the remarkable multifunctional NS1 protein that selectively inhibits intracellular signaling, IFN effectors, cellular mRNA processing, and nuclear export; and second, the recently discovered PB1-F2 protein that undermines host defenses by specifically targeting and destroying alveolar macrophages.

We begin by considering the mechanisms by which NS1 counters IFN. Recall that virus-infected cells are able to "sense" the presence of intracellular viral activity by means of specific receptors that recognize viral products, in particular, double-stranded RNA (Chapter 4). The Toll-like receptors TLR3, TLR7, and TLR8 have been thought to play a major role in that regard, doing so by transmitting a double-stranded RNA-triggered signal that activates transcription of IFN-α/β. Recent experiments that made use of mouse knockout cells and animals showed that the retinoic acid-induced gene I product (RIG-I) also is essential for activating transcription of IFN-α/β during infection by several RNA viruses, including paramyxoviruses, flaviviruses, and influenza viruses. Since the influenza virus NS1 protein was already known to inhibit transcriptional activation of IFN-β during infection, one study investigated whether NS1 might directly interact with RIG-I. The researchers found that NS1 indeed directly interacts with RIG-1, in that way inhibiting the RIG-I-mediated activation of IFN-β transcription.

NS1 can also impair IFN-mediated defenses by mechanisms that act downstream of IFN-β production. For instance, NS1 blocks the activation of RNase L, which is a key IFN effector (Box 12.10). This NS1 activity depends on a double-stranded RNA-binding domain at the N terminus of the NS1 protein. For that reason, and because activation of RNase L is absolutely dependent on double-stranded RNA-mediated activation of 2′-5′ oligo(A) synthetase (Box 12.10), it was suggested that when NS1 binds to double-stranded RNA, it prevents that double-stranded RNA from activating 2′-5′ oligo(A) synthetase. Regardless of whether that premise is correct, the importance of the double-stranded RNA-binding domain of NS1 was demonstrated by the experimental finding that mutating that site results in a highly attenuated virus, except in animals lacking innate antiviral defenses.

NS1 also impairs IFN-mediated antiviral defenses downstream of IFN production by blocking the activation of RNA-dependent protein kinase (PKR), another major effector of IFN-α/β (Box 12.10). Although PKR, like RNase L, is activated by double-stranded RNA, experimental findings suggest that NS1 blocks the activation of PKR by directly interacting with that protein. However, other experimental results imply that the NS1 double-stranded RNA binding domain is required for NS1 to block PKR activation, as in the case of the NS1-mediated impairment of RNase L activation.

BOX 12.10

The roles played by RNase L and PKR in the IFN response are described in Chapter 4. Beginning with RNase L, in brief, this nuclease degrades most viral and cellular RNAs. It is activated by a multistep process during a viral infection. In the first step, the presence of double-stranded RNA in a virus-infected cell induces the production and secretion of IFN-α/β by that cell. The IFN-α/β released by the infected cell then induces a 10- to 1,000-fold upregulation of RNase L in neighboring uninfected cells. However, the RNase L remains inactive in the neighboring bystander cells. But if a neighboring bystander cell should subsequently be infected, then double-stranded RNA that is produced within it will induce production of another cellular enzyme, 2′-5′-oligo(A) synthetase, which generates 2′-5′ oligomers of adenylic acid. This unusual polynucleotide then activates the previously upregulated RNase L, which then degrades most viral and cellular RNAs, thereby shutting down viral replication within the cell.

IFN-α/β released from virus-infected cells also upregulates PKR levels in neighboring bystander cells. But, as in the case of RNase L, PKR too remains inactive in uninfected cells. However, this serine/threonine protein kinase is activated if the cell should be infected. This occurs as follows. PKR has a double-stranded RNA-binding motif. When two molecules of PKR bind to the same molecule of double-stranded RNA, these two PKR molecules cross-phosphorylate each other, thereby activating each other's serine/threonine kinase activity. Thus, if a cell in the antiviral state is infected, the double-stranded RNA that is generated by the virus binds and activates PKR.

The activated PKR then acts by phosphorylating numerous target proteins, but the best-understood antivirus effect of PKR involves eIF2a, an initiation factor for protein synthesis. PKR phosphorylates eIF2a, thereby inactivating its ability to initiate translation, bringing about an abrupt shutdown of both viral and cellular protein synthesis in the infected cell.

What might we infer from the finding that influenza virus uses multiple mechanisms to undermine the host cell IFN-α/β-mediated antiviral defenses? For one, it demonstrates the importance to the host of this innate defense against virus infection; this notion is supported by the fact that mutants of influenza virus that are defective at blocking IFN responses are markedly attenuated, except in animals unable to express IFN-α/β. Equally important, IFN-α/β antiviral defenses are mediated by several independently acting effectors (e.g., RNase L and PKR), each of which needs to be blocked by the virus to completely subvert this innate host defense. The fact that the several influenza virus anti-IFN mechanisms are each mediated by NS1 is all the more remarkable. Still, the following proviso should be borne in mind. Notwithstanding the multiple means by which the influenza virus NS1 protein impairs the IFN-α/β response, hosts nevertheless remain able to mount an IFN-α/β response after influenza virus infection. However, they do so to a much lesser extent when infected by wild-type virus than when infected by an NS1 mutant, and as noted, replication of the mutant is therefore severely compromised. These findings also underscore the continuing coevolution of host defensive measures and pathogen countermeasures.

Other experimental findings show that the multifunctional NS1 protein also impedes the apoptosis or cell death program activated by virus infection (Chapter 4), doing so by binding to and activating phosphatidylinositol 3-kinase (PI3K) (Box 12.11). By that means, NS1 prevents premature induction of apoptosis, ensuring efficient virus replication. Thus, NS1 activates a cellular signaling pathway to block premature apoptosis while blocking other cellular

signaling pathways to undermine IFN-mediated defenses, doing each to ensure efficient virus replication.

Remarkably, NS1 has yet other roles to play. Before we recount these, first note that each of the IFN- and apoptosis-inhibiting activities of NS1 noted above happen in the cytoplasm. Thus, it is notable that NS1 also is active in the nucleus, where it inhibits host gene expression by blocking nuclear export of cellular mRNAs. Moreover, as you might now expect of this protein, it does so by multiple mechanisms, interacting separately with several different components of the cellular mRNA processing and export machinery.

So, how does NS1 block the nuclear export of cellular mRNA? As was noted above, it does so by several known mechanisms. In the first of these mechanisms, NS1 makes use of the fact that polyadenylation and splicing are prerequisites for the nuclear export of cellular mRNAs. Notably, NS1 blocks nuclear export of cellular mRNAs by blocking *both* the polyadenylation and splicing of mRNA. NS1 impairs polyadenylation of cellular mRNAs by interacting directly with the polyadenylation factors CPSF and PABII, and it blocks mRNA splicing by interacting with a putative cellular splicing factor, NS1-BP (acronym for NS1-binding protein; i.e., this cellular protein was identified and named by virtue of its binding to NS1 in a yeast interaction trap system). Second, the remarkable NS1 protein blocks nuclear export of mRNAs by interacting directly with components of the mRNA nuclear export machinery (see below and Box 12.12).

Since influenza virus transcription, as well as that of the host cell, occurs in the nucleus, you well may ask whether the NS1-mediated block of mRNA export, via its effect on polyadenylation factors, is specific for cellular mRNAs. The answer is yes. NS1 inhibits two of the proteins that are essential for the polyadenylation of cellular pre-mRNAs. Since influenza virus does not use the cellular polyadenylation machinery to generate the 3′ poly(A) tails of its mRNAs (see above), the nuclear export of cellular mRNAs, but not of viral mRNAs, is selectively inhibited by this mode of NS1 action.

Since influenza virus mRNAs, as well as cellular mRNAs, are exported by the nuclear export machinery, we also might ask whether the NS1-mediated block of mRNA export, via its inhibition of components of the mRNA nuclear export machinery, is selective against cellular mRNAs. And again, the answer is yes. NS1 blocks nuclear export of cellular mRNAs by interacting directly with components of the mRNA nuclear export machinery. Specifically, NS1 forms inhibitory complexes with NXF1/p15, E1B/AP5, and Rae1/mrnp41. The activities of these factors in nuclear export are normally as follows. Most fully processed mRNAs are exported to the cytoplasm by means of heterodimeric transport receptors, one of which

BOX 12.11

The cell has a variety of signaling pathways that promote cell survival by blocking apoptosis. Actually, most mammalian cells are programmed to undergo apoptosis unless they receive a survival signal from another cell. Such survival signals might be transmitted by polypeptide growth factors or cell contacts that signal via integrin receptors.

One of the major cell survival pathways is that mediated by the enzyme phosphatidylinositide (PI) 3-kinase, which phosphorylates the membrane-associated phospholipid phosphatidylinositol 4,5-bisphosphate (PIP_2) to generate PIP_3. A protein-serine/threonine kinase called Akt is recruited to the plasma membrane by binding to PIP_3. Akt is then phosphorylated and thereby activated by protein kinase PDK1, which also binds to PIP_3 at the plasma membrane. Akt then phosphorylates a number of proteins that regulate apoptosis, resulting in the inhibition of the apoptosis effectors caspase 9 and glycogen synthase-kinase 3β.

In the above Akt-mediated pathway, PI 3-kinase is normally activated by either protein-tyrosine kinases or G protein-coupled receptors. However, PI 3-kinase is directly activated during influenza virus infection by the direct binding of NS1.

BOX 12.12

The targeted export of RNA molecules across the nuclear membrane is a crucial cellular process, since cellular mRNA, ribosomal RNA (rRNA), and transfer RNA (tRNA) molecules are synthesized in the nucleus but function in the cytoplasm. Transport across the nuclear membrane of these classes of RNA molecules, like transport of ribosomal subunits and most proteins, is mediated by a large family of transport factors, known as **karyopherins**. Karyopherins that mediate nuclear import are called **importins**, whereas those that mediate nuclear export are referred to as **exportins**.

Exportins recognize, and bind to, cargos in the nucleus that contain a nuclear export signal. In the case of those cellular RNAs that are exported from the nucleus as nucleoprotein complexes, and in the case of influenza vRNPs as well, the nuclear export signals are on the protein components of the nucleoprotein complexes.

The small GTP-binding protein Ran, when in its GTP-bound form (i.e., Ran-GTP), facilitates the interactions between the cargo and the exportins (Box 12.7). Then, after the cargo/exportin/Ran-GTP complex is transported to the cytoplasm, a Ran-GTPase-activating protein, RanGAP-1, converts Ran-GTP to Ran-GDP, resulting in the release of the cargo from the export receptor. Importantly, the hydrolysis of Ran-GTP in the cytoplasm and the localization of the Ran GDP-GTP exchange factor in the nucleus maintain a Ran-GDP/Ran-GTP gradient across the nuclear

membrane, which is believed to provide the driving force for nuclear import and export.

Ribosomal subunits are exported by the karyopherin Crm1 through interactions with specific adaptor proteins that contain a leucine-rich nuclear export signal. In contrast, tRNAs are exported by the karyopherin exportin-t, which binds directly to an export signal that is part of the tRNA structure.

The export of mRNAs is unique. Unlike other export cargos, which have nuclear export signals that are recognized by members of the karyopherin family of export factors, mRNAs appear to be exported by members of the TAP/NXF (NXF is an acronym for nuclear export factor) family, which are not related to the karyopherins and are required specifically for mRNA export. (In the text, we describe how NS1 blocks nuclear export of particular mRNAs by forming inhibitory complexes with the heterodimeric transport receptor NXF1/p15.) Moreover, the newly synthesized mRNA primary transcripts must undergo capping, splicing, and polyadenylation before they may be exported. Also, some of the factors involved in these steps of mRNA processing also function in mRNA export, thereby ensuring that mRNA processing and export are tightly coupled. The specific mRNA-binding proteins and signals governing the interaction of the mRNA-protein complex with TAP/NFX are not yet entirely known, although some are noted in the text. See also Box 12.7.

---

is NXF1/p15 (Box 12.12). E1B/AP5 is believed to mediate the interaction between NXF1 and processed mRNAs. Rae1/mrnp41 forms a complex with NXF1, the mRNA, and the nucleoporin Nup98, which acts as a docking site for mRNA export factors. Moreover, Rae1/mrnp41 shuttles between the nucleus and the cytoplasm and, in so doing, is believed to mediate transport through the nuclear pore complex. However, viral RNAs are exported through the Crm1-mediated nuclear export pathway (see below and Box 12.12), which does not involve NXF1/p15 or Rae1.

The ability of NS1 to impair the nuclear export of cellular mRNAs may also contribute to the ability of influenza viruses to evade cellular immune responses. This conjecture is based on the following experimental observations. Cells that express mutant Nup98 or Rea1 preferentially retain in the nucleus mRNAs that encode immune-related functions (e.g., MHC-I and ICAM-1 [Chapter 4]). In contrast, mRNAs that encode "housekeeping" functions are not affected by Nup98 or Rea1 mutations. Since NS1 acts by forming inhibitory

complexes with Nup98 and Rae1, it might selectively impair the nuclear export of mRNAs that encode immune-related functions. (If that is so, it bears further testimony to the ability of viruses to "devise" rather amazing immune evasion stratagems, while also underscoring the remarkable range of activities of the influenza NS1 protein.) Note that the differential requirement for Nup98 and Rae1 for the export of particular mRNA species may reflect the observation that different classes of mRNAs preferentially bind specific subsets of RNA-binding proteins (Box 12.12).

The ability of NS1 to impair nuclear export of cellular mRNAs may also contribute to its ability to impair IFN-mediated defenses, as follows. Rae1, as well as Nup98, is induced in cells by IFN, supporting the contention that the nuclear export activities of Rae1 and Nup98 likely contribute to cellular antiviral defenses. Thus, since NS1 impairs IFN-mediated responses, it likewise moderates expression of both Nup98 and Rae1, in addition to forming inhibitory complexes with these proteins. Together, these effects of NS1 diminish both innate and adaptive immune responses, resulting in increased viral replication and virulence. For more on NS1, see Satterly et al. (2007).

Next, we consider the recently discovered influenza virus PB1-F2 protein, which is encoded by an alternate reading frame in the gene segment that encodes PB1. Like NS1, PB1-F2 also enhances influenza virus virulence in a mouse model. In particular, PB1-F2 undermines host defenses by specifically targeting and destroying alveolar macrophages, doing so by inducing those cells to undergo apoptosis. However, this effect of PB1-F2 does not merely impair host immune clearance of the virus. Importantly, as a consequence of targeting these alveolar macrophages, which are professional antigen-presenting cells (Chapter 4), PB1-F2 may also increase the likelihood of opportunistic secondary bacterial infection of the bronchia. This possibility is a matter of concern since secondary bacterial infections are, in fact, the major cause of mortality associated with type A influenza virus infections.

ASSEMBLY AND RELEASE
Assembly of vRNPs

The assembly of new vRNPs in the nuclei of infected cells and their subsequent nuclear export depend on a multifaceted set of specific protein-RNA and protein-protein interactions. First, since assembly of vRNPs and the transcription and replication of viral RNA take place in the nucleus, the viral RNA polymerase has to be imported into the nucleus. Not unexpectedly, then, nuclear localization signals are present in PB1, PB2, and PA, and all three polymerase subunits can independently enter the nucleus. However, PB1 and PA apparently must first form a complex in the cytoplasm to be efficiently transported into the nucleus. It is not yet clear whether PB2 enters the nucleus independently or in complex with PB1 and PA. Second, in the nucleus the polymerase complex associates with progeny genomic RNAs. The PB2 protein is bound to the PB1 protein, which is bound to the viral RNAs. Third, the PB2 protein interacts with the NP protein to initiate its binding to the viral RNAs. Fourth, the NP proteins cooperatively bind to the long viral genomic RNA molecules to generate the large vRNP structures. Finally, the M1 and NEP proteins bind, thereby enabling the vRNPs to interact with the nuclear export machinery (see Box 12.12 and below).

Two Issues Regarding Intracellular Targeting of Viral Components

Influenza virus assembly and release raise several interesting issues regarding the intracellular targeting of viral components. For example, progeny vRNPs must traffic *from* the nucleus to the plasma membrane, where they are incorporated into progeny virus particles. Yet, at an earlier stage of infection, parental vRNPs are transported *into* the nucleus, where they are transcribed and replicated. Moreover, the early transport of parental vRNPs to the nucleus is mediated by nuclear localization signals present on the NP proteins of the vRNPs. Yet the progeny vRNPs, which are assembled in the nucleus, likewise contain NP proteins which display nuclear localization signals. Indeed, those NP nuclear localization signals target newly made NP proteins from their site of synthesis in the cytoplasm to the nucleus, where they can assemble into progeny vRNPs. How then are progeny vRNPs exported from the nucleus to make possible virus assembly and release at the plasma membrane?

Next, consider that since influenza virus is a respiratory virus that infects the respiratory epithelium, it would be to the advantage of the virus if at the end of each replicative cycle it were able to target the release of progeny virus particles from the apical surface of the respiratory epithelium, since apical release would favor the spread of the infection in the epithelium, as well as the spread to new individuals. Indeed, influenza virus is released almost exclusively from the apical surface. How then might all the components of influenza virus particles be targeted to the apical surface of the respiratory epithelium for assembly and release? These two issues are addressed below.

Transport of vRNPs from Nucleus to Cytoplasm

The mechanism of vRNP export from the nucleus is only partly understood, yet some of its key features have been identified, in particular, the key roles played by the viral M1 and NEP proteins. (Recall that NEP is the acronym for "nuclear export protein.") Like all proteins, M1 and NEP are synthesized in the cytoplasm. Then, like all other proteins that are targeted to the nucleus, M1 and NEP contain nuclear localization signals. In the nucleus, new M1 proteins specifically bind to NP proteins that are associated with progeny vRNPs.

Importantly, the binding of M1 to the vRNPs blocks the nuclear localization signals on the associated NP proteins. In that way, M1 counteracts one impediment to the transport of progeny vRNPs from the nucleus. Apropos this fact, recall that during entry the vRNPs needed to be separated from the M1 proteins, so that they could be released into the cytosol and transported to the nucleus. (This separation was made possible by the action of the M2 ion channel.) The binding of M1 to the nascent vRNPs in the nucleus thus facilitates the opposite process of exporting the vRNPs from the nucleus.

Next, NEP proteins bind to the M1 proteins that now are attached to the vRNPs. NEP proteins contain a nuclear export signal that is believed to promote the export of the NEP-containing vRNPs from the nucleus. This export occurs through nuclear pore complexes and is mediated by specialized nuclear export receptor proteins. (See Box 12.12 for particulars regarding the molecular mechanisms that underlie the export of nuclear cargo in general and RNA species in particular.)

The nuclear export of influenza vRNPs, like that of ribosomal subunits, is mediated by the karyopherin Crm1/exportin1 (Box 12.12). Crm1 generally forms a ternary complex that includes the protein or cargo to be exported and Ran-GTP. Importantly, NEP is the adaptor protein that mediates the association of Crm1 with the vRNPs. This fact was demonstrated by the following experimental findings. First, NEP contains a leucine-rich nuclear export signal, which is characteristic of proteins exported by Crm1 (Box 12.12). Second, NEP interacts with Crm1, as demonstrated in a mammalian two-hybrid system. Third, by genetic analysis, the NEP leucine-rich nuclear export signal was shown to be essential for NEP to mediate the nuclear export of vRNPs. Fourth, using genetically engineered recombinant proteins, the NEP nuclear export signal was shown to mediate the nuclear export of heterologous proteins (Box 12.13).

Here are some final points. The nuclear export of influenza vRNPs depends on enough M1 protein accumulating in the nucleus to mask the nuclear localization signals on the NP proteins associated with the vRNPs. Since this does not occur until later stages of infection, progeny vRNPs will not be prematurely released from the nucleus. Also, M1 prevents the progeny vRNPs in cytoplasm from reentering the nucleus. Instead, once the progeny vRNPs are in the cytoplasm, the associated M1 proteins can guide them to the sites of virus assembly at the plasma membrane.

Apical Targeting of Virus Components and Final Virus Assembly

All influenza virus structural components are transported to the apical plasma membranes of the respiratory epithelium, where final virus assembly and release take place. The mechanisms underlying apical transport of influenza virus components are only partly understood, but special microdomains in the plasma membrane, known as **lipid rafts** (Box 12.14), appear to play a key role. These lipid rafts, which are highly enriched for glycosphingolipids and cholesterol, are preferentially localized in the apical membranes of polarized epithelial cells. They facilitate influenza virus apical targeting as follows. First, lipid raft domains form in the Golgi apparatus, and the influenza virus envelope glycoproteins HA and NA each preferentially partition into lipid raft domains during their transit through the *trans* Golgi body (Fig. 12.10). As in the case of other raft-associated proteins, HA and NA partition into lipid rafts because they contain particular sequences in their transmembrane and cytoplasmic domains that have a high affinity for rafts. Second, these Golgi body-derived lipid rafts are then targeted to the apical plasma membrane of polarized epithelial cells (Box 12.15).

M1 is not synthesized on membrane-bound polyribosomes and thus is not transported by the secretory pathway to the plasma membrane. Moreover, M1 does not preferentially partition into lipid rafts. How then does M1 localize at virus assembly sites? The answer is not yet known. However, M1 proteins contain specific sequences that cause them to bind to the cytoplasmic face of the plasma membrane, and the N terminus of M1 penetrates and interacts with the inner leaflet of that membrane. Consequently, M1 proteins, which bind to vRNPs in the nucleus, can anchor those vRNPs to the plasma membrane. Note that M1 proteins can associate with model membranes, even when HA and NA are not present in the membrane. However, M1 does not associate with lipid rafts or become incorporated into progeny virus particles, unless HA or NA is present in the membrane. This shows that the sorting of M1 proteins into progeny virus particles depends on their direct interaction with HA and NA, an interaction that occurs in lipid rafts. Thus, the sorting of M1 into lipid rafts and then into progeny virus particles may depend on sequences within the transmembrane and cytoplasmic domains of raft-associated HA and NA molecules. The interaction of M1 with HA and NA may involve the N terminus of M1, which penetrates the inner leaflet of the plasma membrane, as noted above.

BOX 12.13

Export of cellular mRNAs via the TAP/NXF family of export factors is generally coupled to the capping, splicing, and polyadenylation of those mRNAs. For that reason, unspliced HIV RNAs (Chapter 21), like influenza vRNPs, are exported by the Crm export factor. Crm mediates the export of HIV RNAs via its interaction with the HIV Rev protein, which binds to a *cis*-acting RNA sequence, the so-called Rev-responsive element that is present in unspliced and partially spliced HIV mRNAs. This is an important means by which HIV regulates its program of gene expression. The influenza virus NEP protein discussed in this chapter is reminiscent of the HIV Rev protein. In fact, the NEP nuclear export signal can functionally replace that of the HIV Rev protein.

BOX 12.14

Within the past several years, the view of the plasma membrane has changed from that of a homogeneous "sea" of lipids with embedded proteins to that of a variety of heterogeneous patches or domains comprised of more-defined lipids and proteins. One such membrane domain, referred to as a lipid raft, is created by lateral interactions in the membrane between glycosphingolipids and cholesterol. Lipid rafts are sometimes said to "float" in the plane of the membrane. They appear to be rather small (about 70 nm in diameter on the average) and very dynamic. Experimental results support the belief that their purposes are to support cell functions involving membrane traffic and signal transduction. The ability of lipid rafts to preferentially incorporate some proteins while excluding others is the basis for putative lipid raft functions (e.g., signal transduction). Particular proteins associate with lipid rafts because they contain sequences in their transmembrane and cytoplasmic domains that cause them to differentially partition into the raft.

Lipid rafts originate in the Golgi apparatus. Also, proteins that are transported through the secretory pathway may associate with lipid rafts during their transport through the Golgi body. The Golgi body-derived lipid rafts are then incorporated into transport vesicles that are specifically targeted to the apical membranes of epithelial and endothelial cells. Special raft-associated proteins, such as the annexins, have a role in that apical targeting.

Many of the proteins that specifically associate with lipid rafts are involved in intracellular signaling. Included among those signaling proteins are several protein tyrosine kinases and src family kinases. Finally, rafts play a significant role in the assembly and release of several other viruses in addition to influenza virus, in particular HIV (Chapter 21). A specialized kind of raft, called a caveola, mediates the entry of SV40 (Chapter 15).

In addition to those M1 proteins that bind directly to the plasma membrane, some M1 proteins also associate with HA in the Golgi body. Like those M1 molecules that interact with the envelope glycoproteins at the plasma membrane, those which do so in the Golgi apparatus may also be incorporated into progeny virus particles, as discussed below.

The M2 protein of type A influenza virus, like the M1 protein, also does not associate with lipid rafts, but it nevertheless is targeted to the apical plasma membrane, also by a mechanism that is not yet known. On the one hand, the apical targeting of M2 suggests that the protein may contain an apical targeting signal in its primary sequence.

On the other hand, the fact that M2 does not spontaneously associate with lipid rafts might account for why very low amounts of M2 (relative to HA) are present in virus particles, despite the fact that M2 is abundantly expressed at the apical membrane. Regardless, influenza virus envelopes are enriched in glycosphingolipids and cholesterol, a finding in agreement with their budding from lipid rafts of the apical plasma membrane.

The following model for influenza virus assembly incorporates the experimental findings enumerated above. Those M1 molecules that bind to lipid raft-associated HA in the Golgi apparatus traffic with those lipid rafts to the plasma membrane via the secretory pathway. Other M1 proteins, which may be components of M1/vRNP complexes and which bind directly to the inner leaflet of the plasma membrane, partition into the HA- and NA-containing lipid rafts at the plasma membrane via their interactions with the HA and NA proteins in the rafts. M1 interactions with HA and NA and M1-to-M1 interactions together are responsible for bringing vRNPs to the assembly sites. In this regard, recall that M1 is the most abundant protein in the virus particle. M2 is targeted to the assembly sites by an unknown mechanism, which might involve an apical targeting mechanism in its primary sequence (Fig. 12.10).

The cytoskeleton may well play a role in the targeting of M1/vRNP complexes to the virus assembly sites, consistent with the observation that the NP protein and cytoplasmic vRNPs are bound to the actin cytoskeleton. Moreover, lipid raft microdomains interact with, and are stabilized

BOX 12.15

Results of genetic studies showed that while the association of HA and NA with lipid rafts facilitated their apical sorting, the ability of HA and NA to associate with lipid rafts does not entirely account for their apical transport. For example, some HA and NA mutants, which sorted randomly, nevertheless showed a high degree of raft association. On the other hand, an HA mutant that was transported mainly to the apical plasma membrane did not show significant raft association. Thus, the determinants of HA and NA apical sorting are not entirely understood.

by, the underlying actin cytoskeleton. On the other hand, and inexplicably, drugs that disrupt the actin cytoskeleton do not affect the budding of spherical influenza virus particles, yet they do affect the budding of filamentous influenza virus particles.

Before moving on, it would be good to recap the several crucial actions of the influenza virus M1 protein during the viral replicative cycle. First, input M1 proteins must disassociate from parental vRNPs prior to the fusion step during entry. Second, new M1 proteins must enter the nucleus to associate with progeny vRNPs (an interaction that, by the way, blocks further transcription from those vRNPs). Third, M1 bound to vRNPs mediates the binding of NEP to the vRNPs, which facilitates their export to the cytoplasm. Fourth, M1 prevents the cytoplasmic vRNPs from reentering the nucleus. Fifth, M1 is a crucial factor in bringing together and assembling viral components (i.e., vRNPs, HA, and NA), and sixth, M1 provides the major driving force for virus budding at the plasma membrane. Apropos the last point, genetic analysis shows that neither HA nor NA is absolutely required for virus budding. On the other hand, there is no virus budding in the absence of M1. Moreover, recombinant M1 expressed alone can form virus-like particles in transfected cells. Thus, M1 is thought to provide the major driving force for influenza virus budding via its interaction with the inner leaflet of the lipid bilayer and via M1-M1 interactions, which cause the membrane bilayer to bend outward to initiate bud formation. (Recall that the M proteins of the rhabdoviruses [Chapter 10] and the paramyxoviruses [Chapter 11] play similar pivotal roles in the budding of their respective viruses.) Furthermore, as noted, M1 plays a key role in bringing together the vRNPs and envelope proteins at the assembly sites.

Encapsidating Eight Distinct Genomic Segments

An influenza virus particle must package at least one copy of each of its eight distinct vRNPs to be infectious. Two models have been proposed to explain how this might be accomplished. The first model postulates that each of the eight distinct vRNPs is specifically packaged into progeny virus particles. Phage Φ6, which contains a genome comprised of three different double-stranded RNA segments, provides a precedent for a specific packaging mechanism (Chapter 5). The second model proposes a nonspecific, random packaging of vRNPs, in which the assembly process does not distinguish between individual vRNPs. In this model, the assembly of virus particles with a complete set of genomic segments is determined entirely by chance.

It is somewhat difficult to envision how the "specific packaging" model might work in the case of influenza virus, since it would necessitate specific packaging signals in each vRNA segment. Moreover, even if those signals were identified, how might the budding process recognize that all eight specific vRNP segments are present prior to bud closure and virus release?

But is it in fact plausible that influenza virus might efficiently package complete virus genomes via the random packaging of individual genomic segments? The answer is yes, given the following proviso. If influenza viruses were able to randomly package *more* than eight genomic segments in each virus particle, then the fraction of particles that would be infectious could fall well within the ratios of infectious particles to total particles observed in virus preparations. For example, if influenza virus particles were to randomly package 12 vRNPs, then 10% of those particles would have a complete set of genomic segments and be infectious. Can you do the calculations to confirm that? As we have seen before in other instances, experimental results provide inconclusive evidence in support of each of the models described above.

Release: the Influenza Virus NA Protein

The final steps of the influenza virus replicative cycle include the budding of virus particles and their release from the host cell plasma membrane. The final step, release, requires the viral NA protein, which removes the terminal sialic acid residues from oligosaccharide chains on membrane glycoproteins. In the absence of NA activity, envelope HA proteins of the progeny virus particles remain bound to sialic acid residues on the host cell surface. Moreover, the HA and NA proteins in the envelopes of the budding virus particles themselves contain sialic acid-containing oligosaccharides, which were added during their transport through the secretory pathway. Thus, newly synthesized influenza virus particles would aggregate with each other, were it not for the activity of the NA protein. Aggregation of virus particles is in fact observed when the activity of the NA protein is experimentally impaired, for example by a temperature-sensitive mutation.

The anti-influenza drugs zanamivir and oseltamivir (Tamiflu) each act by inhibiting the NA protein. Note that amantadine and rimantadine, which inhibit the M2 ion channel activity (see above), are effective against influenza A virus but not against influenza B and C viruses. In contrast, zanamivir and oseltamivir are effective against influenza A and B viruses.

MEDICAL ASPECTS
Pathology and Clinical Syndromes

Human influenza viruses are transmitted primarily by the inhalation of aerosols. Consequently, they infect the epithelium of the respiratory tract. They rarely involve tissues other than in the lung.

The tropism of human influenza viruses for the respiratory epithelium has several medically significant consequences. Most importantly, perhaps, mucus-secreting, ciliated, and other epithelial cells are killed, thereby compromising the integrity of the respiratory epithelium, the physical barrier that is the first line of defense against respiratory pathogens (Chapters 3 and 4). Additionally, activity of the influenza virus NA protein cleaves sialic acid residues from the mucus, which has the effect of liquefying the mucus, which further enables opportunistic pathogens to gain access to the epithelium. Also, the influenza virus stratagem of targeting its release from the apical membranes of the epithelium facilitates cell-to-cell spread of the virus in the epithelium, which in turn results in the spreading of the pathology in the epithelium. One possible outcome is that influenza virus infection can give rise to primary influenza virus pneumonia (i.e., a pneumonia resulting from infection by that virus). However, primary influenza virus pneumonia is rare. More commonly, viral damage to the mucosa facilitates a secondary bacterial pneumonia, most often involving *Staphylococcus aureus*, *Haemophilus influenzae*, and *Streptococcus pneumoniae*. Secondary bacterial pneumonia is an important cause of fatalities in influenza virus infections, particularly in the elderly, the very young, the chronically ill, and the immunocompromised. Apical release of the virus also facilitates its transmission to new hosts.

The average incubation period for influenza is about 48 h, which is followed by the typical fever, cough, headache, muscular aches, sore throat, and conjunctivitis. The local symptoms result from damage to the epithelium, due in part to the inflammatory response at the mucosal membrane. The systemic symptoms are due to the action of proinflammatory cytokines, such as IFN-γ and tumor necrosis factor alpha (Chapter 4). Full recovery may take as long as a month.

The severity of an influenza virus infection generally depends on the virulence of the particular virus strain, the degree of immunity to that virus strain, and the age and health of the individual. Also, smokers are more likely than nonsmokers to have a more severe infection.

Reye's syndrome is an acute encephalitis that comes about as a complication of influenza A and B virus infections, as well as of varicella (chicken pox [Chapter 18]) and other virus infections. It occurs mainly, but not exclusively, in children between the ages of 6 and 11 years who are recovering from these infections, and it can be fatal if untreated. Its exact cause is not known, and there is no cure. Treatment is concerned primarily with protecting the brain from irreversible damage that might result from

its swelling. Importantly, giving children *aspirin* during a viral infection *increases their risk* of developing Reye's syndrome. (The U.S. Surgeon General, the U.S. Food and Drug Administration, the Centers for Disease Control and Prevention [CDC], and the American Academy of Pediatrics all recommend that aspirin [or any other salicylate compound] *not* be given to children under *19* years of age during episodes of fever-causing illnesses.)

Influenza is a serious infection. In the United States, influenza typically results in nearly 40,000 fatal cases per year and as many as 200,000 more cases that require hospitalization. Worldwide, influenza is responsible for an estimated 500,000 deaths per year. Erroneously, many people refer to any respiratory infection as the "flu." In truth, influenza is a serious illness. It is not the common cold!

The Flu Pandemic of 1918

A crucially important aspect of the influenza virus ecosystem is that antigenically novel influenza virus strains, against which there is little immunity in the human population, can emerge without warning. Moreover, these strains have the potential to cause devastating worldwide outbreaks that are referred to as **pandemics**. One such pandemic, the so-called **Spanish flu pandemic of 1918** (Box 12.16), may well have been the most devastating epidemic ever.

During the single month of October 1918, the 1918 pandemic killed 196,000 people in the United States alone. By the end of the winter of 1918–1919, 2 billion people around the world contracted influenza, and estimates of the total number of fatalities range between 20 million and 50 million. For the sake of perspective, that number of fatalities is about twice as many as would die of acquired immunodeficiency syndrome (AIDS) worldwide during the entire first 20 years of the AIDS epidemic. Moreover, the 1918 influenza epidemic killed more people in a single year than the 4-year bubonic plague that ravaged Europe

---

BOX 12.16

Despite the fact that the initial outbreak of the 1918 influenza pandemic is believed to have occurred in or around an Army camp in the United States in March 1918, by the fall it was being referred to as the Spanish influenza, probably because Spain, as a noncombatant in World War I, saw no need to impose wartime censorship on news of the epidemic. On the other hand, the United States and the major European combatants sought to suppress news of the epidemic to prevent public panic in time of war.

from 1347 to 1351. Note that influenza virus pandemics also occurred in 1957 (the Asian flu) and 1968 (the Hong Kong flu), but these were much less devastating than that of 1918. The numbers of deaths in the United States from those two more recent flu epidemics are estimated to be 70,000 and 50,000, respectively (see below) (Box 12.17).

Interestingly, the pandemic of 1918 began relatively benignly, with the first cases reported in early March. Within 4 months, the virus spread worldwide, and although millions of individuals became ill, few died. No one knew at the time how genetically changeable influenza virus can be. Indeed, influenza virus had not yet even been discovered, much less known to be the cause of the epidemic. Thus, the world was taken by surprise in August 1918, when the mild infection suddenly changed into something extraordinarily lethal. Outbreaks of the new variant occurred almost simultaneously in three locations, France, Sierra Leone, and Boston, and then spread worldwide. Indeed, this virus struck with a ferocity that stunned the medical professionals of the day.

Although many people who died during the pandemic succumbed to complications from pneumonia caused by opportunistic bacterial pathogens, the virus itself was quickly lethal in many individuals. Some patients had massively hemorrhaged lungs and were dying awash in their own blood. Indeed, their condition was more reminiscent of the pathology of the more recent Ebola virus outbreaks than of the fevers and aches typically associated with influenza virus.

The 1918 pandemic influenza virus strain was in fact unique in how quickly it could kill: literally overnight. There are anecdotes of people leaving for work in the

---

BOX 12.17

In 2003, Dr. Jeffery Taubenberger, of the Armed Forces Institute of Pathology in Rockville, MD, noted the following. "It is curious that the (1918) pandemic doesn't seem to be part of the cultural memory, at least in the United States, although it was a huge event with a huge impact. Everyone hears about the Black Death in the 1300s, yet here was an infectious disease only 85 years ago that killed 40 million people and for some reason we don't know about it."

It also is curious that whereas the 1918 influenza pandemic killed an extraordinarily large number of people, it did not cause any panic, mob rule, or disruption of law and order. With the possible exception of the recent winter of 2004-2005, when there were severe shortages of the flu vaccine in the United States (see the text and also below), and the spring of 2009, during an outbreak of swine influenza (Box 12.24), influenza has not appeared to cause alarm, even though it typically kills 35,000 to 40,000 individuals in the United States, and more than 500,000 individuals world-wide, each year. And, as dramatically demonstrated by the 1918 pandemic, many more individuals may be killed during more severe pandemic outbreaks.

We generally take for granted that influenza will appear every year, and afterwards we forget about it. On the other hand, recent history shows that we tend to fixate on new and emerging infectious agents, such as anthrax, Ebola virus, West Nile virus, and SARS, despite the fact that these more "exotic" infections currently pose much less of a threat to public health than influenza. Somehow, we feel more vulnerable to these new agents, even though we are not.

Perhaps this is caused in part by the fact that we tend to take influenza for granted, even confusing it with the much less severe common cold. Another factor may be the attention that the latest pathogen or outbreak receives from the media, which regard them as much more newsworthy than influenza. While there was greater public awareness of influenza during the winter of 2004-2005, this likely reflected the newsworthiness of the vaccine shortage at the time, which also was happening during a presidential election year. Perhaps we simply have become accustomed to the presence of influenza. Yet we did not become accustomed to poliovirus (Chapter 6). Perhaps that is because poliovirus struck the young, in whom we are most heavily invested emotionally. On the other hand, the public just may not appreciate the relative danger posed by influenza. Is it true that we spend a lot of time worrying about the wrong potential disasters? What are your thoughts on this issue?

Unfortunately, funding decisions of politicians are influenced by the fears of the public, which too commonly are focused on the new, regardless of their relative danger. This leads to the shifting of resources from the fight against infectious agents that cause the greatest mortality to ones that pose a perceived or remote threat. On the other hand, the emerging diseases have resulted in a greater public awareness of infectious diseases in general. The difficult goal is to order our priorities, and those of our policy makers, such that we are fighting the major killers of children and adults around the world, while maintaining our vigilance against new potential threats.

morning feeling fine, and then succumbing on the way. One story tells of four women in a bridge group playing together until 11:00 in the evening. By morning, three of them had died. The 1918 influenza pandemic highlights how dangerous influenza virus can be. In its most sinister form it is as deadly as almost any other virus known to man.

A particularly puzzling feature of the 1918 pandemic was that it tended to kill the hale and hearty, those individuals between the ages of 25 and 34, in the primes of their lives. In contrast, typical influenza epidemics cause the highest fatality rates in the elderly, the very young, the chronically ill, and people with weakened immunity.

The relatively lower mortality rates of the elderly during the 1918 pandemic might be explained by the chance that they acquired protective immunity from a previous infection by an influenza virus serologically related to the 1918 virus. An explanation that is often put forward to account for the higher mortality rate among individuals between the ages of 25 and 34 is that the pandemic occurred during the last year of World War I. Consequently, many of the individuals in the most susceptible group were being exposed to the crowded living conditions in army camps, which predisposed them to the secondary bacterial infections that led to many of the influenza fatalities of that preantibiotic era. On the other hand, the "crowded army camp theory" cannot explain why the same pattern of disease was seen in the populations of countries that did not participate in the war.

Another strange and probably significant (see below) aspect of the 1918 human pandemic was that a global influenza-like pandemic began almost simultaneously in pigs. Moreover, like the human pandemic, that flu-like illness in pigs was more intense than had ever been noted before in that species.

The 1918 influenza strain persisted into the 1920s. Then, it suddenly disappeared or perhaps lost its virulence. This is not clear. There are still other mysteries that remain. For example, retrospective serological studies showed that nearly 100% of the population was infected during the 1918 pandemic (a rather astonishing figure in its own right). Yet "only" 28% of the population actually became ill, and many fewer yet succumbed. (Bear in mind that the fatality rate of the 1918 influenza pandemic was still as much as 50-fold greater than the fatality rates that were seen during other influenza outbreaks.) It is not clear why a virus that was so infectious and so lethal caused illness in less than one-third of all infected individuals. Importantly, it is not known for certain why the 1918 pandemic occurred at all or why it was so virulent. (See below for current understanding of these crucial questions.) Nor is it known when a similar pandemic might occur again. Yet we are left with a vivid warning of what influenza virus

may hold in store. Given the current global population density and the speed with which a virus like the 1918 pandemic strain might spread in the modern world, as well as the increased numbers of elderly and immuno-compromised individuals, the emergence of such a virus today might be catastrophic (Box 12.18).

It was not possible to study the 1918 influenza virus at the time of the pandemic, since influenza virus was not yet known to exist, nor was there even a technology for cultivating a virus outside its host. Richard E. Shope, of the Rockefeller Institute (now the Rockefeller University), accomplished the first isolation of an influenza virus in 1930. Shope (a physician by training, who became an animal pathologist) isolated the virus from pigs when investigating the porcine infection that had been ongoing since 1918. (Shope was also the first researcher to isolate a papillomavirus, the cottontail rabbit papillomavirus [Chapter 16]. He went on to show that this virus is the cause of skin papillomas in its rabbit host, which was the first demonstration that a DNA virus might be tumorigenic.)

The pig pandemic seemed to parallel the human pandemic during the first fall and winter of 1918. Then, whereas the virulent human virus "disappeared," the pig flu continued to return each winter, sustained by new susceptible hogs. Interestingly, it would reemerge simultaneously everywhere throughout the hog population, rather than spread from place to place. Yet hogs pretty much stay put, except when on the way to the slaughterhouse. How then was it that the virus reemerged at the same time in virtually every pig pen in the country? The mysteries go on.

A human influenza virus was isolated by Smith, Andrewes, and Laidlaw 3 years after Shope isolated the pig virus. Interestingly, serological analyses of human survivors of the 1918 pandemic, which were done in the 1930s, demonstrated that antibodies in the blood of those individuals reacted better against the influenza virus circulating in pigs in the 1930s than against the influenza virus circulating in humans in the 1930s. (What possibilities do these findings suggest to you?) These findings led to the 1918 human virus being referred to as "swine flu." Results of more-recent molecular analyses (see below and Box 12.19) are consistent with the earlier serological analysis.

Origin of Epidemic and Pandemic Strains: Antigenic Drift and Antigenic Shift

We know from personal experience that aside from the irregularly occurring influenza pandemics, every winter is "flu season." Why then do we *not* have immunity against influenza virus from one year to the next, and why must a new vaccine be formulated each year? In addition, why do the irregular and unpredictable pandemics occur? The answers to these questions are related to the fact that

BOX 12.18

Influenza outbreaks occur yearly, usually during the winter, on all continents. Does the virus persist in some individuals in each region between outbreaks, and then emerge from some of those individuals to reseed the outbreak in that region the following winter? Alternatively, does the virus continue to circulate in each region between outbreaks, at a level too low to detect? Or, does the virus disappear from each region, only to reappear from elsewhere the following winter? Epidemiologists have been asking these questions for decades, and a definitive answer may finally be forthcoming.

Derek Smith and colleagues (Russell et al., 2008) recently provided evidence that from 2002 to 2007, yearly influenza H3N2 epidemics worldwide were seeded by viruses that originated in East and Southeast Asia. Note that H3N2 is the most prominent form of influenza virus currently circulating (see the text). The rationale for this study was as follows. If epidemic strains persist locally in a region and give rise to the next local epidemic, then local strains from any year should be more closely related to strains isolated from that region in other years than to nonlocal strains isolated the same year. Conversely, if epidemics were seeded by strains from outside the region, then local strains from any year should be more closely related to nonlocal strains of that year than to local strains from other years. With that in mind, Russell et al. generated an antigenic and genetic "map" of the HA genes of some 13,000 samples of H3N2 influenza virus samples isolated from six continents. Their analyses showed that there is a continuous region-wide overlap of temporary epidemics in East and Southeast Asia, which seed the yearly outbreaks elsewhere. Thus, the newly emerged H3N2 strains that first appear in East and Southeast Asia appear 6 to 9 months later in North America and Europe and, after another 6 to 9 months, they appear in South America. The authors note that this knowledge might lead to improvements in vaccine strain selection.

Reference
Russell, C. A., et al. 2008. The global circulation of seasonal influenza A (H3N2) viruses. *Science* **320**:340–345.

antigenic instability is a hallmark of influenza virus. Indeed, influenza virus seems to have an endless capacity to outpace the immunity in the population from one year to the next. (Antigenic instability is also a crucial factor in the course of an HIV infection [Chapter 21]. However, in the case of HIV, it plays out over the course of the chronic HIV infection in each infected individual rather than in the population as a whole.)

Influenza virus undergoes two types of genetic variation that affect its antigenicity. The first of these, called **antigenic drift**, is the reason why immunity does not persist from one flu season to the next and the influenza vaccine needs to be reformulated each year. The second type of genetic variation, called **antigenic shift**, accounts in large part for pandemics. Apropos both of these phenomena, antigenic determinants on influenza virus particles are located exclusively on the HA and NA envelope glycoproteins, and HA contains the major neutralizing antibody-binding sites.

We begin with antigenic drift, which is a continuous process. It is explained by the fact that the influenza virus RNA polymerase, like all RNA polymerases, lacks a proofreading function. Consequently, point mutations occur at a rate of about 10^{-4} per base per genome replication. And since replication requires synthesis of plus-strand templates as well as minus-strand progeny genomes, progeny genomes vary from the parental genome by more than one nucleotide, on the average. Consequently, the virus evolves so fast that it is able to generate antigenic variants that evade the antibodies that exist in the human population from the previous flu season. (Proviso: The measles virus polymerase, like the influenza virus polymerase, lacks a proofreading function. Yet for unknown reasons, measles virus is antigenically stable [Chapter 11]. Perhaps measles virus is under less selective pressure to change antigenically, even though both measles and influenza viruses are acute respiratory viruses.)

In contrast to antigenic drift, which is an ongoing process, antigenic shift is a sudden, major change in viral antigenicity. Consequently, few, if any, individuals in the host population will have immunity to the "shift" virus. Antigenic shifts arise because of two key features of influenza virus. The first of these is the segmented nature of the influenza virus genome. The second is the fact that influenza viruses infect several other host species in addition to humans. Birds and pigs are thought to be the most important of these other host species with respect to generating human pandemic shift strains. Actually, all influenza viruses that circulate in humans may have initially been avian viruses (see below).

The facts that (i) the influenza virus genome is segmented and (ii) the virus circulates in other mammalian as well as avian hosts might converge to generate pandemic shift strains as follows. A human is concurrently infected with both a human and a zoonotic influenza

BOX 12.19

In March 1997, Dr. Jeffery Taubenberger and his colleagues at the Armed Forces Institute of Pathology (AFIP) in Washington, DC, startled virologists when they reported the sequence of the HA gene of a 1918 pandemic virus (Taubenberger et al., 1997). Taubenberger was hired by the Institute to create a state-of-the-art molecular pathology laboratory. Towards that end, his unit, which included molecular biologist Ann Reid, developed new procedures to recover nucleic acids from tissue samples that were fixed in formaldehyde and paraffin. Pathologists routinely examine such fixed tissues, but molecular analysis of those fixed specimens had not been accomplished previously because the fixation can destroy nucleic acids.

Interestingly, Taubenberger's initial foray into influenza was not inspired by an interest in influenza per se. Instead, he wanted a means to showcase his Institute's new procedures, and also its vast collection of specimens that had been gathered over the past century. With those purposes in mind, Taubenberger and colleague Ann Reid put in a request for fixed tissue samples from soldiers who had succumbed during the 1918 flu pandemic. Taubenberger and Reid, expecting a long wait, were surprised when within a few seconds of receiving their request the Institute's automated recovery system successfully retrieved their samples. Those samples contained flecks of tissue from soldiers killed by the flu pandemic 80 years earlier. (The samples were taken by doctors who, at the time, had no knowledge of what might be causing the soldiers' illness.)

Taubenberger and Reid unsuccessfully screened several different patient specimens for influenza virus sequences, until they came to that of Private Roscoe Vaughn, who died in September 1918, during the peak of the pandemic. In Private Vaughn's cells they found small segments of influenza virus-like RNA. Their subsequent analysis generated the sequence of an H1 HA gene that was unlike any other that had been isolated after 1918. The closest known HA gene was that of the influenza virus that Shope isolated from pigs in 1930! The HA genes of at least five 1918 isolates have been sequenced to date.

Based on their initial success at generating the sequence of the 1918 HA gene, Taubenberger and his associates clearly wished to be able to reconstruct the complete genome sequence of the 1918 virus. However, the tiny quantities of tissue in the Institute's formaldehyde-fixed samples led them to doubt that they might ever be able to do so. Enter Dr. Johan Hultin, a retired pathologist, who unexpectedly provided a solution to their dilemma. Years earlier, in 1951, Hultin had unsuccessfully attempted to grow live influenza virus from Inuit victims of the 1918 pandemic, whose bodies were preserved in the Alaskan permafrost. Upon learning of the work of Taubenberger's group, Hultin was hopeful that the molecular biology procedures of that group might be able to reconstitute the genome of the 1918 virus from tissue samples of the frozen Alaskan victims. Thus, Hultin wrote to Taubenberger and succeeded in exciting his interest. Then, Hultin eagerly returned to Alaska in 1997 to obtain fresh specimens. He deliberately took tissue samples from a particularly obese woman, expecting that the combination of her fat and the permafrost might have preserved the viral genomes. Hultin's reasoning might indeed have saved the day, since his efforts enabled Taubenberger's group to generate the entire genome of the 1918 virus from the tissue samples he provided.

At the time of this writing (spring 2008), the AFIP was slated for closure by the U.S. Department of Defense as a cost-saving effort. The AFIP's host campus, the Walter Reed Army Medical Center, also was targeted for closure. The AFIP was founded in 1862 as a museum for specimens taken from American Civil War casualties. Over the years, the Institute's specimen collection became legendary, and it became known for its role in diagnosing difficult civilian as well as military cases. Moreover, its staff has included some of America's greatest pathologists. While its specimen collection may well remain intact, the value of that collection will be seriously compromised by the dispersal of the distinguished group of pathologists that oversees it.

Reference

Taubenberger, J. K., A. H. Reid, A. E. Krafft, K. R. Bijwaard, and T. G. Fanning. 1997. Initial genetic characterization of the 1918 "Spanish" influenza virus. *Science* 275:1793–1796.

---

virus. Next, if a cell in the human host were to be simultaneously infected with the human and the zoonotic influenza viruses, then reassortant influenza virus particles will be generated that contain genomic segments derived from each of the two strains (Fig. 12.11). A pandemic influenza strain could emerge if the human influenza virus suddenly were to acquire an HA genomic segment from the zoonotic strain, since few, if any, people will have immunity against the resulting reassortant virus. However, note the following two stipulations. First, the lack of widespread immunity to the shift virus does not necessarily account for its extreme virulence. Second, the shift virus will likely have to

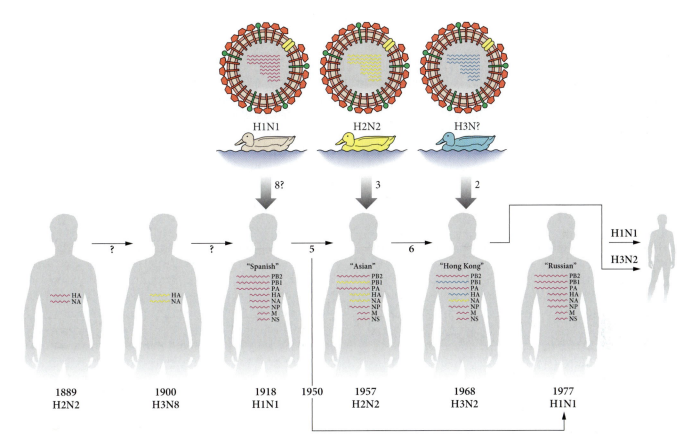

Figure 12.11 Postulated evolution of human influenza A viruses from 1889 to 1977. The appearance and transmission of distinct serotypes of influenza A virus in humans are shown. The upper part shows the nature of the avian influenza viruses that reassort with human viruses. The color of the genome segments represents a particular viral genotype. Segments of the predominant influenza virus genome and its gene products are indicated in each human silhouette for each year. The number next to each arrow indicates how many segments of the viral genome are known to have been transmitted. The earliest serology data we have are from 1889 and suggest that H2N2 was the predominant class in humans. In 1900 the predominant serotype was H3N8. No data exist for the other influenza virus genes present at these times, and these segments are not illustrated. Phylogenetic evidence is consistent with the appearance by 1918 of an influenza virus with eight distinct segments. This virus is thought to be of avian origin and is characterized by the H1N1 serotype (red). It was also found in pigs and was carried from North America to Europe by U.S. soldiers.

It gained notoriety as the cause of the catastrophic Spanish influenza pandemic of 1918. In 1957 the Asian pandemic was caused by a virus that acquired three genes (PB1, H2, and N2) from avian viruses infecting wild ducks (yellow) and retained five other genes from the circulating human strain (red). As the Asian strains appeared, the H1N1 strains disappeared from the human population. In 1968, the Hong Kong pandemic virus acquired two new genes (PB1 and H3) from the wild duck reservoir (blue) and kept six genes that were circulating in human viruses (red and yellow). The pandemic virus had a characteristic H3N2 serotype. After the appearance of this virus, the Asian H2N2 strain could no longer be detected in humans. In 1977 the Russian H1N1 strain that had circulated in humans in the 1950s reappeared and infected young adults and children. One theory was that this virus escaped from a laboratory. It has continued to cocirculate with H3N2 influenza viruses in the human population. Adapted from R. G. Webster and Y. Kawaoka, *Semin. Virol.* **5:**103–111, 1994, with permission.

undergo further changes before it might become highly infectious in humans. These points are discussed in greater detail below.

The source of the new HA of the pandemic shift virus is believed to be from the extensive pool of influenza viruses that infect wild birds (Fig. 12.11). (As discussed below, there are 16 known serologically distinct types of the influenza HA protein in nature, only three of which [HA1, HA2, and HA3] are found in human strains. There are nine known serologically distinct types of the NA protein.) This notion was initially based on the retrospective serological analysis of serum samples that were taken years earlier and that had been stored for other purposes. Importantly, its soundness is supported by more recent molecular analyses (Box 12.19; see also below).

Yet the process for generating pandemic shift viruses cannot be quite this simple. One reason is that influenza viruses tend to be host species specific. Indeed, avian influenza virus strains tend not to infect avian species other than their particular host, much less humans. Host species specificity is due in part to the fact that the HA proteins of a particular influenza virus tend to bind efficiently only to cells of the preferred host of that virus. Thus, the HA proteins of avian strains bind poorly to the sialic acid receptors in the human respiratory tract. However, it is now clear that a change of only a single amino acid in an avian influenza virus HA protein can enable a reassortant virus to bind to human cells, which is a prerequisite for the virus to replicate and disseminate efficiently in the human host and to spread between individuals (see below). Thus, a shift, followed by a short period of drift, might generate a new pandemic virus.

Until recently, there was no evidence that wholly avian influenza virus strains (i.e., strains in which all the genomic RNA segments are avian in origin) could replicate in humans. However, it recently became alarmingly clear that some avian influenza viruses indeed might replicate in humans (see below). Regardless, the early lack of evidence that avian strains might "leap" directly to humans led to the belief that the reassortment between avian and human strains *cannot* take place in humans but instead requires some other intermediary host species.

For several reasons, pigs were thought to be the intermediate host in which avian and human influenza virus strains might undergo nonselective reassortment to generate potential human pandemic strains. First, the epithelial lining of the respiratory tract of pigs contains receptors for known avian and human influenza virus strains, and human and avian influenza virus strains indeed can replicate in the porcine respiratory epithelium. Second, when pigs were experimentally coinfected with avian and human influenza virus strains, they in fact gave rise to a variety of reassortant viruses.

The route from birds and humans to pigs and then back to humans might seem a bit bizarre, but conditions actually exist to enable this to happen rather easily. The overall scenario is as follows. Pandemic influenza virus strains all originated in China, including even the misnamed Spanish flu of 1918. An important factor behind this Chinese origin is that wild ducks are the predominant reservoir for influenza virus, and China indeed has more ducks than people. These ducks are welcome on Chinese farms since they prey on many of the pests that attack the rice crops. Consequently, ducks are allowed to live in close proximity to people, as well as to farm animals, including pigs and chickens. The crowding together of humans, ducks, pigs, and chickens creates an ideal environment in which cross-infection and reassortment might occur. In this regard, note that influenza virus in ducks is spread by the fecal-oral route. Thus, ducks contaminate the water supply and, if that were not sufficient, pigs eat bird droppings. Moreover, influenza virus generally causes an asymptomatic or very mild gastrointestinal infection in ducks (but not in chickens, where it can be extremely lethal) that might go unnoticed, resulting in the acceptance

of virus-shedding ducks waddling through barnyards and kitchens, spreading the virus as they waddle along.

Our knowledge of antigenic shifts is based largely on retrospective serological analyses extending back to the human pandemic of 1889, which was followed by the pandemics that occurred in 1918, 1957, and 1968 (Fig. 12.11). But why were there "only" three influenza pandemics during the entire 20th century? Given our current understanding of how pandemic strains arise, we cannot explain why that was the case. We might be inclined to interpret the fact that there were only three influenza pandemics in the entire 20th century to mean that the generation of viable pandemic strains is a relatively rare event. If that is so, then we might consider ourselves fortunate.

Next, we consider the actual series of antigenic shifts that occurred during the 20th century, first noting that there are 16 serologically distinct subtypes of the HA protein that are presently known, all of which appear among the avian influenza virus strains. These HA serotypes are designated H1 through H16. In addition, there are nine known serologically distinct subtypes of NA, which are referred to as N1 through N9. Using antisera that are specific for these serologically distinct HA and NA subtypes, it was possible to carry out retrospective studies that reconstructed the evolution of human influenza A viruses throughout the 20th century (Fig. 12.11). These studies revealed the following.

Of the 16 serologically distinct subtypes of HA that are known in avian influenza viruses, only three, H1, H2, and H3, have circulated in humans during the last century. Specifically, only those avian viruses carrying H1, H2, or H3 have been able to transmit their HA type across the species barrier to establish themselves in humans. Moreover, and importantly, a pandemic resulted every time there was an antigenic shift (H3 to H1 in 1918, H1 to H2 in 1957, and H2 to H3 in 1968). Note that when a new shift virus appears, it immediately begins to undergo mutational adaptation (drift) that is almost certainly necessary for the virus to be able to cause a human pandemic. Antigenic drift continues, maintained by constant selective pressure to evade host immunity.

Recently, there were relatively small outbreaks in humans involving avian subtypes H5, H7, and H9. However, the inability of these viruses to easily spread from one human to another precluded major epidemics (see below).

Influenza B viruses undergo antigenic drift, but in contrast to influenza A viruses, they do not undergo antigenic shift. That difference is explained by the fact that influenza B viruses do not have any animal reservoir. Perhaps that is why influenza B viruses never cause pandemics. Also, influenza A and B viruses never reassort with each other.

In 1977, when an H3N2 subtype was the predominant influenza virus strain circulating in humans, an H1N1 strain suddenly showed up and appeared to be identical to a virus that was circulating in humans in the 1950s (Fig. 12.11). Because of the very great similarity between the 1977 H1N1 virus and the 1950s strain, the emergence of the 1977 strain appeared to be a reintroduction of an earlier strain rather than the result of a new shift. Also, this virus almost certainly had not been in continuous circulation, since it had not undergone sufficient antigenic drift to distinguish it from the 1950s strain. One speculation is that the 1977 virus escaped from a laboratory in which it had been stored. Regardless, the origin of the 1977 H1N1 strain is not known with any certainty. Also, in this instance, the emergence or reemergence of the H1N1 strain did not lead to the replacement of the H3N2 strain, and both continue to circulate in humans.

Interestingly, and in contrast to human influenza viruses, avian influenza viruses have not changed much in 80 years. A likely explanation is that influenza viruses exist in a state of "stasis" in their avian hosts, since birds indeed are the natural hosts for these viruses. In contrast, influenza virus might be considered an emerging virus in humans.

Each of the three influenza pandemics of the 20th century, which in each instance resulted from an antigenic shift, was more devastating than the regularly recurring winter epidemics. This was, of course, particularly so in the case of the 1918 pandemic. Nevertheless, the cumulative toll of the regularly occurring epidemics may have been greater than that of the sporadic pandemics. It is estimated that in the 20th century, drift viruses were responsible for between 40 million and 100 million fatalities worldwide. The upper end of these estimates far exceeds the number of influenza fatalities occurring during the shift virus pandemics. Moreover, in 1947, drift in the then-circulating H1N1 virus enabled it to evade the existing immunity in the population and generate an outbreak on a pandemic scale.

Jeffery Taubenberger and associates used polymerase chain reaction (PCR) technology to reconstruct the entire 1918 pandemic virus from RNA segments present in tissue samples of victims of the pandemic (Box 12.19). The sequence of that reconstructed virus is entirely consistent with the prospect that the 1918 pandemic virus came to humans directly from birds, a possibility consistent with the experimental results of others, both contemporary and earlier. However, these findings also leave open the chance that the virus went through some mammalian intermediary, where it underwent reassortment, followed by a period of drift. Pigs are considered a likely candidate for that mammalian intermediary, based on the following evidence. First, the HA protein of the 1918 human pandemic strain is most similar to the HA of the porcine strain isolated by Shope in 1930 (see above). Second, nearly simultaneous outbreaks occurred in humans and pigs in 1918. Third, both human and avian strains are able to infect pigs. And fourth, it was demonstrated experimentally that pigs are able to generate reassortants between human and avian strains. Note that the classic H1N1 swine flu, which so closely resembles the 1918 human pandemic strain, became widespread during the winter of 1918 not only in the United States but also in Europe and China. Thus, regardless of its origin, one and the same virus may have given rise to a pandemic in pigs as well as in humans.

Despite the evidence cited above, the role of pigs in the generation of the 1918 pandemic strain is by no means known with certainty. The reassortment leading to the antigenic shift might have occurred in some other mammalian species (e.g., horses). Humans and pigs might then have been infected independently, or humans might have passed the infection to pigs, rather than vice versa. Taubenberger favors the latter possibility because there are numerous instances in which human influenza viruses are known to have infected pigs, whereas (notwithstanding the 2009 swine influenza outbreak) swine influenza virus strains rarely have been detected in humans. Clearly, despite much recent progress, the true dynamics of the influenza virus ecosystem is only barely known.

Virulence of the 1918 Pandemic Influenza Virus: Opening Pandora's Box

After Taubenberger and coworkers successfully reconstructed the 1918 pandemic virus, they demonstrated that it is exceptionally virulent in mice. Indeed, no other influenza virus isolated from humans had ever shown such an astonishing level of virulence in mice. However, although those researchers were able to determine the sequences of all eight genomic segments of the 1918 pandemic virus, the sequences per se did not reveal the basis for the extraordinary virulence of the virus. But once the technology was developed to recover genomic sequences from samples of the 1918 virus, it became technologically feasible to identify the particular genes of the 1918 virus that might have accounted for its extreme virulence. That goal was realized by several different research groups, each working independently. Their experimental strategy was to first reverse transcribe the individual genomic segments of the 1918 pandemic virus, to likewise transcribe the individual genomic segments of a relatively benign human strain, and then to insert each of the resulting complementary DNAs (cDNAs) into individual plasmids. Reassortant viruses were then generated by microinjecting different combinations of these plasmids into cells in culture. These viruses could then be screened for their virulence in the mouse model (Box 12.20).

BOX 12.20

Textbooks, especially long, single-authored ones, have rather lengthy periods of gestation. When writing the first draft of this chapter in the spring of 2004 (it is currently the spring of 2009), the *potential* for reconstructing the entire 1918 virus or for generating reassortants between that influenza virus and relatively benign ones raised important concerns, which I first laid out 5 years ago. These experiments, which were conceivable in 2004, have since been successfully carried out. Nevertheless, the issues raised in 2004 remain relevant today.

On the one hand, such experiments have the potential to reveal the basis for the severe virulence of the 1918 virus. More importantly, perhaps, they have the potential to generate results that would enable virologists to recognize in a timely way the emergence of a new potential pandemic virus, thereby providing the knowledge and window of opportunity to develop strategies to cope with that threat.

On the other hand, one could raise important reasons for not doing these experiments. Most importantly, such experiments might give rise to a very lethal virus, one possibly as life threatening as the 1918 pandemic virus. In 2004, I noted that these viruses certainly would have to be very well secured. Yet the possibility that they might be accidentally released, even from the most secure facility, is not at all far-fetched. In this regard, in 2003 and 2004 the SARS virus "escaped" from three different Asian laboratories. Also, one frequently hears of concern over the possible release of smallpox virus from one of the few laboratories, particularly in Russia, that store samples of that virus. Moreover, while these experiments *might* be done safely in a very few laboratories in the United States and Europe, there is no global mechanism to ensure that they would be done safely elsewhere. So, in retrospect, what position might you have taken regarding these experiments?

With the technology at hand to identify the specific gene or genes of the 1918 virus that might have been the cause of its extreme virulence, what particular gene might have been the leading suspect? As you very well might have expected, for most researchers it was the gene encoding the HA envelope glycoprotein. One reason was as follows. As recounted above, the influenza virus HA protein needs to be activated by proteolytic cleavage into HA1 and HA2. Cleavage of the HA protein of contemporary human influenza viruses is thought to be mediated in vivo by tryptase Clara, which was thought to be expressed exclusively by the nonciliated Clara cells of the bronchial and bronchiolar epithelia. For that reason, replication of human influenza viruses was believed to be restricted to the immediate vicinity of those specialized cells of the respiratory tract epithelium (see below). In contrast, the HA cleavage sites of highly pathogenic avian influenza viruses are recognized by ubiquitous proteases, thereby enabling those viruses to disseminate throughout the entire lung and beyond the respiratory tract (Box 12.21).

In agreement with expectations, a reassortant virus that contained only the HA genomic segment of the 1918 pandemic virus was able to infect the entire mouse lung, presumably because its HA protein could be activated independently of tryptase Clara. Also as expected, this reassortant virus was much more virulent in the mouse model than any contemporary human influenza virus.

Another long-favored explanation to account for how the HA protein of the 1918 pandemic virus could have enhanced its virulence is that there would have been little, if any, immunity in the human population against the 1918 HA protein. However, recent research findings are not entirely consistent with that plausible premise. Actually, the 1918 HA protein caused enhanced immunopathology in the mouse model, particularly in combination with the 1918 NA protein. Taubenberger's group found that recombinant viruses that contained both the HA and NA proteins of the 1918 virus triggered higher than normal inflammatory and adaptive immune responses in the mouse model. Kawaoka and coworkers (Kobasa et al., 2004) reported a severe inflammatory response against an engineered virus that contained only the HA gene of the 1918 virus. Furthermore, they reported that this intense inflammatory response in mice, which was mediated by enhanced levels of alveolar macrophages, cytokines, and neutrophils, resulted in extreme hemorrhaging, reminiscent of the illness during the 1918 human pandemic.

Apropos the above findings in the mouse model, tissue samples of individuals who succumbed to the 1918 virus showed evidence of extreme lung pathology, which might well have resulted from severe inflammation and high levels of cytokines. On the other hand, experiments in the mouse model also showed that depleting neutrophils and alveolar macrophages results in increased death rates, demonstrating that these cells also play an important protective role. So, what might be the bottom line? At the moment, we might safely say that more research needs to be done to better understand the balance between protective immunity and immunopathology during infection with these viruses. Moreover, that understanding might lead to rational therapies based on modulating host inflammatory responses. The goal would be to maintain an immune

BOX 12.21

The protease sensitivity of an influenza HA_0 protein is determined by the amino acid sequence at its cleavage site, which is located in the uncleaved HA_0 molecule in the external loop that links HA1 to HA2. All influenza A viruses are cleaved at an arginine residue adjacent to a glycine residue, which becomes the N terminus of HA2. Two classes of proteases that can cleave HA_0 are known. The first is a lysine- or arginine-favoring trypsin-like protease (e.g., tryptase Clara) that is found in a limited number of cell types. The other is a ubiquitous subtilisin-like protease that recognizes a sequence of several arginine and lysine residues. With these facts in mind, contemporary *human* influenza H1N1 viruses need tryptase Clara to replicate in vivo. Moreover, these viruses require the exogenous addition of trypsin in order to carry out multiple cycles of replication in cell culture. In contrast, highly pathogenic avian influenza viruses can replicate in cell culture in the absence of added trypsin. Furthermore, the ability of an influenza virus to replicate in cell culture without added trypsin is a generally reliable predictor of its pathogenicity in mammals. Apropos these facts, the reconstructed 1918 pandemic virus is able to replicate in cell culture without added trypsin, consistent with the premise that its virulence might be due, at least in part, to its HA protein.

Highly pathogenic avian influenza A virus subtypes H5 and H7 have the so-called "multibasic" cleavage sites, which contain several arginine and lysine residues. All other type A influenza viruses have a "monobasic" cleavage site, which contains a single arginine or lysine residue. So, just when everything appeared so tidy, the HA protein of the 1918 H1N1 pandemic virus manifests the trypsin independence displayed by the multibasic cleavage sites of the highly pathogenic H5 and H7 avian influenza viruses, despite being monobasic.

HA subtypes H5 and H7 are cleaved by *endoproteases* during their transit through the secretory pathway. These endoproteases are present in many cell types, making possible the systemic infections caused by highly virulent avian strains. In contrast, tryptase Clara, which is believed to activate contemporary human influenza viruses, has been thought to be a secreted protease. Consequently, tryptase Clara-mediated activation cleavage was believed to occur extracellularly, either at the plasma membrane or after virus release from the cell. However, recent experimental results show that activation cleavage of H1N1 and H3N2 human viruses can be a cell-associated process, perhaps occurring soon after entry. These experiments were carried out by using primary human adenoid epithelial cell cultures, which contain a mixture of ciliated and Clara cell-like nonciliated secretory cells. In this respect, these cell cultures mimic the epithelium of the human respiratory tract. Bearing in mind that influenza virus infection occurs primarily in the ciliated cells of the epithelium, might the protease actually be produced in these cells, or, alternatively, is it acquired by those cells in a paracrine-like manner from the specialized Clara cells in their immediate vicinity? The answers are not yet known. But if the ciliated cells produce tryptase Clara, what then restricts the replication of contemporary human influenza viruses to the immediate vicinity of the Clara cells?

---

response that preserves its protective role while diminishing its immunopathological effects.

So, does the HA gene of the 1918 virus indeed account for its exceptional virulence? Yes, but only in part, as shown by several additional experimental findings. For example, a reassortant virus containing both the HA and NA genomic segments of the 1918 virus, with the remaining genes from a milder human H1N1 virus, was even more pathogenic than the reassortant virus containing only the HA genomic segment of the 1918 virus.

How might the NA gene of the 1918 virus have contributed to its virulence? One possibility is that the receptor-binding activity of HA and the NA activity work cooperatively at entry. If so, then each of these proteins might function more effectively when derived from the same virus, since the two proteins coevolved in that virus. This is a good thought, and perhaps true. However, a reassortant virus containing only the NA genomic segment of the 1918 virus also replicated in cell culture in the absence of exogenous trypsin (Box 12.21) and was more pathogenic in the mouse model than a milder contemporary human H1N1 virus. These findings imply that the NA protein itself contributed to the virulence of the 1918 virus. In addition, they raise the possibility that the capability of the 1918 HA protein to be activated in vitro in a trypsin-independent manner may be in part because of its NA protein, rather than exclusively because of the cleavage site of its HA protein.

Analysis of reassortant viruses in the mouse model also showed that the polymerase genes (i.e., those encoding PB1, PB2, and PA) of the 1918 virus likewise contributed significantly to its enhanced virulence (Tumpey et al., 2005). A straightforward explanation of these findings is that the polymerase of the 1918 virus may have enabled it to replicate faster than contemporary human H1N1 viruses, both in the lungs of the mouse model and, presumably, in the human respiratory tract. (Note: Searching on the PubMed website [www.pubmed.gov] for any of the articles cited here regarding the virulence of the 1918 pandemic virus will lead you to a plethora of current journal articles on this and related issues.)

The 1918 influenza virus PB1 protein may be particularly relevant, since the PB1 gene was transferred from an avian virus, together with the HA gene, to generate both the 1957 and 1968 pandemic viruses (Fig. 12.11). Taubenberger and coworkers recently showed that the PB1 protein of the 1918 pandemic virus and the avian influenza virus-derived PB1 proteins of the 1957 and 1968 pandemic viruses differ by only one specific amino acid from the consensus sequence of avian influenza virus PB1 proteins, suggesting that this change may be required for the avian PB1 to support replication in humans. Still, it is not yet understood how the avian PB1 per se, or the complete avian polymerase complex, might have contributed to the virulence of the pandemic viruses.

What of the 1918 influenza virus NS1 protein? Since the NS1 protein enables type A influenza viruses to evade IFN-mediated responses and apoptosis, we might suggest that the NS1 protein of the 1918 virus may have been unusually effective in those regards. However, experiments with reassortant viruses in the mouse model showed that the NS1 gene of the 1918 virus actually decreases influenza virus virulence, at least in mice. Still, the effect of NS1 might be host species specific. If so, then the NS1 protein of the 1918 virus might yet have contributed to the virulence of that virus in humans. Such speculations aside, experimental evidence to implicate a role in the extraordinary virulence of the 1918 pandemic virus exists only in the cases of the genes encoding the HA, NA, PB1, PB2, and PA proteins.

Before moving on, there is a most significant aspect of the 1918 HA protein to consider, an aspect that is relevant to the 1918 pandemic, as well as to the possible emergence of a future pandemic virus. Since the HA protein of the 1918 pandemic virus was almost certainly derived from an avian virus, some change in the binding specificity of that protein was necessary for the 1918 virus to efficiently infect humans. Thus, a key question is how many amino acids need to be changed in an avian influenza virus HA protein to enable it to bind efficiently to human receptors. In an experimental tour de force, Gamblin and coworkers (2004) demonstrated that a *single amino acid change* in an avian H1 HA protein might allow it to bind to a human receptor (Box 12.22).

The ability of the 1918 virus to acquire the capacity to efficiently infect human cells was a precondition for the 1918 pandemic. The fact that only a single amino acid

BOX 12.22

First, Gamblin and coworkers reconstructed the HA gene of the 1918 pandemic virus from 84 40-mer oligonucleotides (i.e., oligonucleotides that are 40 nucleotides long). Next, they cloned the reconstructed gene into a vaccinia virus vector, which produced the recombinant HA protein in CV-1 monkey kidney cells. The protein then was isolated after being cleaved by bromelain (see the text), and its crystal structure was determined. In addition, these researchers solved the crystal structures of the H1 HA proteins of the porcine strain isolated in 1930 and of a human strain isolated in 1934, each in association with its respective receptor.

Considering the fact that two of five isolates of the 1918 human pandemic strain have HA binding sites and, presumably, binding specificities that are identical to those of the 1930 swine strain, comparison of the latter with a representative avian H1 HA enabled these researchers to trace the binding specificity of the 1918 virus for human cells to a single amino acid change! The other three 1918 HA proteins differ at only one additional amino acid.

change in an avian influenza virus H1 HA can result in that capability is rather disturbing. Note that for H2 and H3 HA proteins, a minimum of two amino acid changes in receptor binding sites is necessary to shift from avian to human receptor-binding specificity. Importantly, there is a precedent for this change in the binding specificity of avian H2 and H3 HA proteins, as it occurred in 1957 and 1968, respectively (Fig. 12.11).

We end this subsection by underscoring the importance of understanding the basis for the extraordinary virulence of the 1918 pandemic virus, since that understanding will be crucially important in enabling virologists to recognize in a timely way the possible emergence of a new highly virulent pandemic virus and hopefully provide a basis and window of opportunity to develop strategies to cope with that threat.

Influenza Drugs and Vaccines

Effective anti-influenza drugs are presently available. For example, amantadine and rimantadine are effective against some type A influenza virus strains (but not against influenza virus types B and C), and zanamivir and oseltamivir (Tamiflu) are effective against influenza virus types A and B. Nevertheless, these drugs cannot be depended upon to protect the population as a whole against an influenza epidemic. The reasons are as follows. First, they are effective only if taken during the first 24 to 48 h *after* the onset of illness, a time when influenza cannot yet be distinguished from other respiratory infections. Second, these drugs cannot prevent the later host-induced immunopathogenic stages of the disease. Third, they cannot prevent the airborne spread of the virus. Thus, the best way to control the spread of influenza virus is by immunization.

Effective influenza vaccines are usually available every year. But because of antigenic drift, the antigenicity of influenza virus changes sufficiently from one year to the next to require a reformulation of the influenza vaccine each year (Box 12.23). To be able to correctly predict what influenza virus serotypes might prevail during the next flu season, the World Health Organization (WHO) coordinates a surveillance network of over 100 influenza centers, which routinely analyze the antigenic properties of the influenza virus strains that are circulating throughout the world. Techniques employed by these centers include hemagglutination inhibition, in which antibody samples raised against earlier strains are tested for their abilities to prevent hemagglutination by currently circulating strains (see Box 11.4). Additional information is obtained by sequencing portions of the HA genes of some of these strains. These sequences can be extremely valuable because of earlier singularly important studies that used X-ray crystallography to locate the limited subset of antigenic sites (epitopes) on the HA protein that bind *neutralizing* antibody. There are only five such sites altogether (Fig. 12.12). Antibodies that bind to other sites on HA do not neutralize the virus. Thus, attention can be focused on changes in the particular epitopes that bind neutralizing antibodies.

The WHO supervises surveillance continuously, worldwide, throughout the year. Information begins to be collected during the winter in the Southern Hemisphere (June and July), leaving vaccine manufacturers barely 6 months to get ready for the flu season in the Northern Hemisphere. Combined data from antigenic, molecular, and epidemiological approaches are used to select the strains from which each year's vaccine is formulated. Bear in mind that strain selection is done twice a year, once for each hemisphere.

Obviously, selecting the correct strains for the vaccine is crucially important. In that regard, the recent 2007-2008 vaccine may be the least effective vaccine since the 1997-1998 flu season, perhaps due to a poor match or to antigenic drift. A study carried out by the CDC showed that people vaccinated with the 2007-2008 vaccine were 44% less likely to get influenza than unvaccinated people. Typically, vaccine effectiveness is around 70%.

Since the influenza virus HA and NA proteins contain the major targets recognized by neutralizing antibodies against the virus, the inactivated human influenza vaccines in use today are generated from reassortant viruses that contain the HA and NA genomic segments of the selected target strains. The six remaining genomic segments are from strain PR8, a laboratory-adapted avirulent H1N1 strain (Box 12.8). (The M2, NP, M1, and polymerase proteins likely play a role in cellular immunity, but they are not part of the yearly evaluation process for seasonal influenza vaccines.) The vaccine is prepared by growing the vaccine viruses in embryonated chicken eggs. Viral HA and NA proteins are released from purified virus by detergent treatment and separated from other virus components that may be pyrogenic (fever inducing). The vaccine is said to be **trivalent**, since it is comprised of the selected circulating H1N1 and H3N2 strains (Fig. 12.11) and the currently circulating influenza B virus.

Clearly, the generation of influenza vaccines is unique in a number of ways. First, the vaccine is trivalent, and second, it needs to be reformulated twice every year, once in the Northern Hemisphere and once in the Southern Hemisphere. Reformulation is a time-consuming process, since the appropriate reassortant viruses must be generated each time the vaccine is reformulated. Consequently, the manufacturer has a very short window of opportunity to respond to epidemiological changes during the course of the flu season. Third, generating the vaccine requires a global enterprise, involving the WHO, numerous surveillance laboratories, regulatory and licensing agencies, and a network of vaccine manufacturers worldwide. Moreover, since the influenza vaccines are currently generated by the egg-based method (see below), a huge supply of embryonated eggs need to be produced and transported to vaccine manufacturers and then maintained, at each step under very carefully supervised conditions.

The unprecedented emergence in 1997 of a virulent *avian* H5N1 influenza virus in humans (see below) and the lag of several years before an effective vaccine against it could be developed struck home the fact that the then-current procedures for vaccine development would not have sufficed if that 1997 outbreak had led to a pandemic. Consequently, the 1997 experience led to new approaches to influenza vaccine development. These events are chronicled in the following two subsections.

Avian Influenza and Humans

Ever since 1889, the earliest year for which there is serological evidence, all of the influenza viruses that circulated in humans were of HA subtype H1, H2, or H3 (Fig. 12.11). Even in the three pandemic years in which antigenic shifts occurred, the only HA subtypes in humans were H1, H2, and H3.

BOX 12.23

Special circumstances led to flu vaccine shortages in the United States during the winter of 2004–2005, resulting in barely enough vaccine to protect even the highest-risk groups, such as the elderly and the immunologically compromised. These circumstances came about as follows. In October 2004, Chiron Corporation of Emoryville, CA, notified the CDC that none of its influenza vaccine would be available for distribution in the United States for the 2004–2005 influenza season. The reason was that the Medicines and Healthcare Products Regulatory Agency (MHRA) in the United Kingdom, where Chiron's influenza vaccine is produced, suspended the company's license to manufacture vaccine in its Liverpool facility for 3 months, because some of its 50,000,000 doses were contaminated with *Serratia marcescens*, a bacterium that can cause mildly pathogenic opportunistic infections. Since the United States obtains about one-half of its 100,000,000 doses of influenza vaccine from Chiron, the expected supply of vaccine in the United States would be reduced by about half. The remaining American supply of influenza vaccine is manufactured by Aventis Pasteur.

Although the above situation was unusual, in the opinion of many public health experts flawed public policy in the United States impedes having adequate vaccine supplies. For example, as exemplified by the flu vaccine shortage of 2004–2005, the potential pitfall of "putting all your eggs in *two* baskets" is that one or both of those baskets might break. (In this instance, the metaphorical basket might well have been filled with eggs, since, as noted in the text, influenza vaccines are currently prepared by growing the vaccine viruses in embryonated chicken eggs.)

In addition, some contend that government policies actually discourage vaccine development and production. It has been argued that under current conditions, the low returns on the investments by vaccine producers and their exposure to legal liabilities have resulted in less than a handful of potential producers remaining active in vaccine manufacture. Twenty-five different companies were producing vaccines in the 1970s, but by 2004 only five remained. Moreover, to maintain a profit margin, U.S. vaccine manufacturers continue to move their operations to lower-cost locations overseas. This compounds the safety issue by making it considerably more difficult for the FDA to carry out rigorous inspections of facilities. In point of fact, before the winter of 2004–2005, the FDA relied on assurances from Chiron that it was adequately addressing safety issues at its Liverpool plant. In contrast, British regulators relied on their first-hand inspections.

To meet the challenge posed by influenza, some public health experts have called for the government to play a significantly greater role, rather than leaving the challenge largely to a blind reliance on the free market. Suggestions include having the government subsidize vaccine manufacturers (note that the U.S. Government subsidizes the farm industry to maintain agricultural capacity) or guarantee to purchase a fixed number of vaccine doses, regardless of whether they are used (see below). Production of the flu vaccine poses particular risks on the part of the manufacturer, since the flu vaccine needs to be reformulated every year (see the text).

Government policies that are said to exacerbate the vaccine problem in general include those of the CDC, which as the largest purchaser of vaccines is able to use its buying clout to compel large discounts from the vaccine manufacturers. Also, in situations where no U.S.-made vaccine against a particular infectious disease is available (e.g., bacterial meningitis C), the FDA has refused to license well-tested vaccines that are routinely used in Canada and Europe. In addition, the FDA has removed vaccines from the market based on anecdotal but unverified *perceptions* of side effects. Finally, there is no policy in place that might protect vaccine producers from legal liability (e.g., legal compensation by the government) after they have fulfilled all of the accepted standards for vaccine testing and production. Indeed, liability concerns may have been the major reason why at least some vaccine manufacturers left the field.

Importantly, the above arguments represent one point of view. One might reasonably argue that these government practices serve to protect the public. What do you think?

When news of the flu vaccine shortage first broke in October 2004, alarm spread across the United States. To deal with a demand that far exceeded the supply, a policy was

From the *Omaha World Herald*, with permission.

Box continues

established to provide the vaccine only to those at the very highest risk. Then, remarkably, as a relatively few extra units of vaccine began to trickle in from other sources, demand for the vaccine dramatically diminished such that there actually was a surplus. Thus, even though restrictions were removed, in the end it appeared that many doses would go to waste. Among the reasons given to account for this strange sequence of events is that many had simply given up trying to obtain the vaccine when they could not get it during peak shortages. Yet I wonder if the public's behavior might be explained as follows. When people thought they could not obtain the vaccine, their desire for it increased dramatically. Then, when they thought that they could have it if they so wanted, they were once again willing to take their chances with the flu.

Bearing in mind that the only influenza viruses to have circulated in humans, at least since 1889, have had an H1, H2, or H3 HA subtype and that shifts of HA subtype have been associated with pandemics, it is notable that within the past decade or so, there have been several episodes of direct cross-species transmission of avian H5N1 influenza A viruses to humans. Moreover, avian influenza A viruses of other HA subtypes have also recently appeared in humans. However, the most alarming of these avian-to-human transmissions have involved subtype H5N1. The clinical manifestations of avian H5N1 influenza virus infections in humans have ranged from mild conjunctivitis to severe pneumonia, with multiorgan system failure. Importantly, these avian viruses have not yet adapted to be efficiently transmitted from one human to another.

The first documented cross-species transmission of avian H5N1 influenza to a human occurred in May of 1997, when a 3-year-old boy was admitted to the Queen Victoria Hospital in Kowloon, Hong Kong, with what turned out to be a fatal respiratory illness. Dr. Wilina Lim, Chief Virologist at the Hong Kong Department of Health, carried out the initial analysis of the young patient's blood sample. Using the reagents at hand from the WHO, Lim determined that the young boy had been infected with some form of influenza A virus. Yet using those reagents, Lim could not identify the particular subtype. Thus, Lim forwarded samples to the CDC in the United States and to England's Mill Hill, which are leading centers in the global flu surveillance network. In addition, she sent a sample to Jan De Jong, a virologist at the Dutch National Institutes of Health. De Jong was the first to determine that the virus specimen from the 3-year-old boy contained an H5 HA.

De Jong was astonished by his finding since an avian H5 HA subtype had never before been seen on an influenza virus in humans. De Jong at first thought that the virus was a contaminant. This seemed plausible, since poultry stalls with live chickens are ubiquitous in Hong Kong. However, based upon his subsequent inspection of Lim's Hong Kong laboratory and confirmatory analyses coming back from laboratories in Rotterdam, Atlanta, and Mill Hill, De Jong and others soon became convinced that Lim's finding was real.

To those who knew influenza, word of an H5 virus in a human instantly raised the specter of 1918. Moreover, the situation in 1997 was potentially even worse than that in 1918, since the new virus appeared to be a purely avian virus that was capable of infecting humans, and there was no immunity in the human population against such a virus.

Although there was some consolation in the fact that the new virus did not appear to be readily transmissible from one person to another, flu workers were well aware of how unpredictable influenza can be. Could this avian H5 virus eventually reassort with a human influenza virus or find some other means to adapt to humans? There indeed were precedents to indicate that it very well might.

From May until November there were no new cases, and the sense of impending crisis had largely abated. But on 8 November, Lim's laboratory, using new reagents supplied by the CDC, got a positive reading for an H5 virus in a sample from a 2-year-old child. The child was only mildly ill and recovered. The specimen was examined only because the child had a debilitating heart condition. Then, at the start of December, Lim identified an H5 virus as the cause of a fatal case of influenza in a 54-year-old man. Several days later Lim identified an H5 virus in two other severely ill patients. In short order, more cases of severe H5-related illnesses began turning up in Hong Kong.

It was reasonable to hypothesize that the ubiquitous Hong Kong live chicken stalls might have been the source of these human H5 cases. And, indeed, intensive research that December showed that 10% of the chickens in the Hong Kong markets carried the H5 virus. Ducks and geese carried the virus as well. As a result, on 28 December, Hong Kong officials initiated a territory-wide slaughter, in which all the birds in those markets were killed. On that same day, the last case in the 1997 outbreak of avian influenza

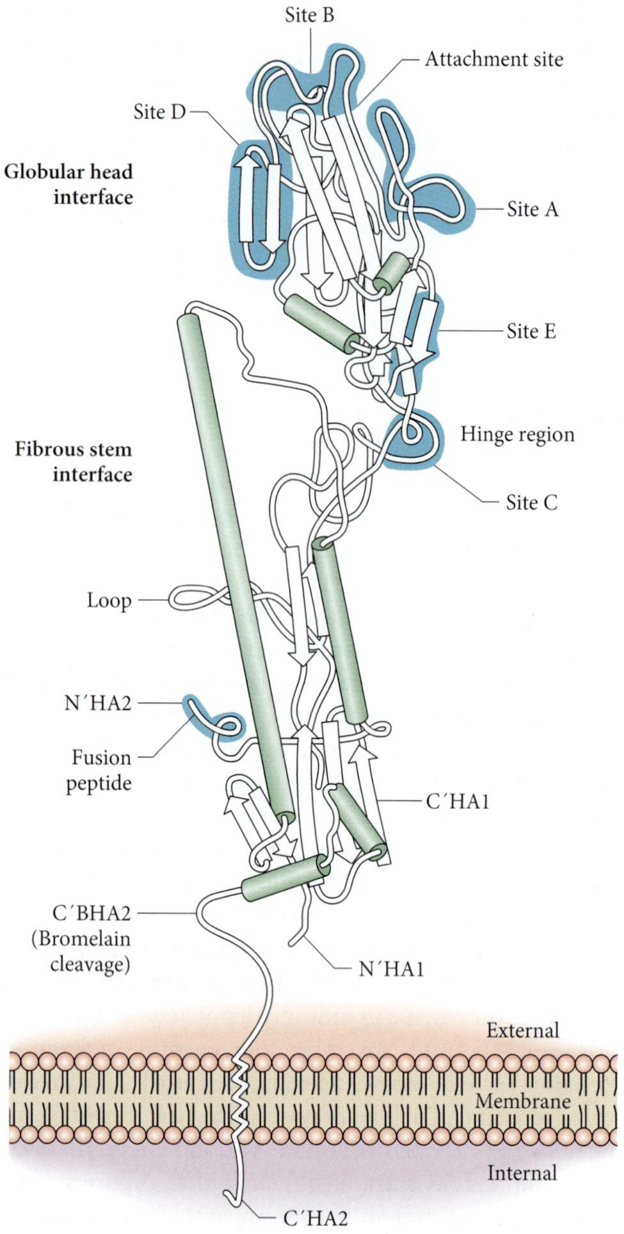

Figure 12.12 Monomer of the influenza virus HA protein, showing all of the antibody neutralization sites (shaded in blue), all of which are located in the globular head of HA1. Adapted from N. J. Dimock and A. J. Easton, *Introduction to Modern Virology*, 5th ed. (Blackwell Science, Oxford, United Kingdom, 2001), with permission.

spread the virus in their droppings. In fact, in February 2003, similar H5N1 viruses again struck in Hong Kong.

Meanwhile, thousands of miles away on poultry farms in The Netherlands, there was a sudden outbreak of an avian H7N7 virus, prompting the slaughter of more than 30 million birds and a ban on poultry exports. Importantly, this H7N7 outbreak in chickens coincided with 80 cases of confirmed H7N7 infections in humans (one of which was fatal). Infected individuals were mainly poultry workers who handled infected birds. In addition, an H9N2 subtype, which causes mild infections in domestic poultry, infected several humans that year in Hong Kong. These H7N7 and H9N2 viruses, like the H5N1 virus of 1997, were pure avian viruses, containing no human influenza virus sequences.

An outbreak of an H5N1 bird flu in humans occurred again in Asia in 2003-2004. This outbreak was particularly worrisome because, for the first time, there was evidence for direct human-to-human transmission. Yet, fortunately, there was no evidence that human-to-human transmission might be sustained. Instead, most human infections, like in the previous outbreak, resulted from direct contact with birds. The 2003-2004 outbreak led to 34 confirmed bird flu cases in humans, resulting in 23 deaths, a fatality rate of about 70%! About 100 million fowl were culled in response to that outbreak.

The H5N1 virus is now widespread in waterfowl in Southern China, Vietnam, Thailand, Indonesia, and several other Asian nations. Moreover, it appears to have evolved over the last several years to spread more easily among birds and to be more pathogenic in mice.

What might account for the elevated virulence of H5N1 avian influenza virus in mice? As expected, the avian H5 HA (as well as the avian H7 HA) can be processed by proteases that are more widespread than tryptase Clara in the mammalian host. (Recall that the tryptase Clara is believed to be restricted to the nonciliated Clara cells of the bronchial and bronchiolar epithelia in mice and humans.) The result is that viruses containing these avian HA proteins can replicate in the entire lung, and in other organs as well, thereby enabling them to cause more disseminated, potentially fatal infections.

As was the case with the 1918 H1N1 pandemic virus, the virulence of current H5N1 avian viruses also appears to be associated with their NA and polymerase genes, as well as with their HA gene. Moreover, several amino acid changes identified in the 1918 H1N1 pandemic virus, which may have contributed to its adaptation to growth in humans, have already been seen in the H5N1 virus now circulating in birds. Together, these findings heighten concern that the H5N1 virus may be close to becoming transmissible between humans.

in humans was confirmed. In all, there were 18 human cases, 6 of which were fatal.

At the end of the 1997 outbreak, several flu experts predicted that the avian H5N1 virus would continue to circulate in the environment, despite the slaughter of chickens in the markets to eliminate the reservoir. Their reasoning was based on the likelihood that the actual sources of the virus were wild ducks and geese, which would continue to

As of March 2006, the number of confirmed H5N1 infections in humans was greater than 200. Alarmingly, more than one-half of those infections were fatal. Yet it is possible, indeed likely, that the number of H5N1-infected humans may have been much greater still. That is, some percentage of the human infections may have been mild or asymptomatic and, consequently, may never have been recognized or reported. If that were so, then the fatality rate may have been considerably lower than the rate of more than 50% based on the number of confirmed H5N1 cases. But if that were so, then it raises the possibility that the H5N1 virus may be on the way to becoming capable of efficient direct human-to-human transmission. Although there have not yet been any confirmed instances of human-to-human transmission of the H5N1 virus, there have been a few instances of several infected individuals in the same family, suggesting that human-to-human transmission may be possible (Box 12.24).

Although there is concern over the likelihood that new bird flu outbreaks will occur in the future, the major worry among virologists is that the H5N1 virus might evolve or undergo reassortment to generate a virulent virus that is highly transmissible between humans, thereby creating conditions for a pandemic. Is there any means to estimate the likelihood that an avian influenza strain indeed might give rise to a human pandemic virus? The answer is yes. One way would be to experimentally generate reassortant viruses that contain various combinations of genomic segments from the emerging avian strain and from the currently circulating human strains. The reassortants could then be tested for their virulence in an animal model.

Experiments of this sort were carried out in the summer of 2006 by scientists at the CDC. A plasmid-based approach (see above) was used to construct hybrid viruses between the 1997 H5N1 strain and a seasonal human H3N2 strain. (Note that arguments for and against doing such experiments are the same as those for and against similar experiments to identify the basis for the virulence of the 1918 pandemic strain, as discussed in Box 12.20.) The most dangerous hybrids were expected to be those containing the H5N1 HA and NA proteins on the outside and H3N2 proteins on the inside. Surprisingly, when these reassortant viruses were tested in ferrets, which have respiratory tracts that are susceptible to human influenza viruses, they grew less well than the unmodified H5N1 virus and did not spread at all between the animals. So, these experimental results *suggest* that reassortants between avian H5N1 strains and prevailing human influenza strains might not generate pandemic viruses. But read on.

If an especially virulent reassortant virus had been generated by the above experimental approach, then it would

have served as a rather imposing warning that a pandemic might be imminent. But should we take comfort from the fact that in this instance the reassortant viruses were not especially virulent? Note that only one H5N1 strain was examined in this study. Moreover, the possibility remains that the avirulent hybrid viruses generated from this particular H5N1 strain (or from any other, for that matter) might yet evolve to become more transmissible and virulent in humans. So, can experiments of this sort be relied upon to predict when or where a pandemic outbreak might occur?

Preparedness for an Outbreak of Avian Influenza

Pandemics are hard to predict and difficult to prepare for because of (i) the complexity of factors responsible for influenza virus virulence and (ii) the genetic instability of the virus. Yet these aspects of the virus per se are only one part of the problem. Another relates to public health policies that cause vaccine manufacturers to have little incentive to develop and test vaccines against putative pandemic strains, since the pandemic, and consequently the market, may never materialize (Box 12.23). In addition, there have been technological hurdles to be overcome, as underscored by the experience of the 1997 H5N1 outbreak. Although that outbreak was successfully controlled by culling millions of birds in the Hong Kong markets, vaccinologists could not produce an effective vaccine against that H5N1 virus, even after several years of effort. Thus, it was distressingly clear that the vaccine technology of the day was not up to the task of rapidly responding to a newly emergent, potentially pandemic influenza virus. Since then, using new technologies, both Chiron and Aventis Pasteur have developed vaccines against H5N1 influenza. Moreover, the Aventis Pasteur vaccine was licensed by the U.S. Food and Drug Administration (FDA) in 2007 (see below). However, if an H5N1 virus were to undergo mutation, which may well be necessary to generate a pandemic virus, then currently stockpiled vaccines might no longer be effective against the altered virus. The bottom line is that even with new technologies, effective vaccines against the looming pandemic might be nonexistent initially (Box 12.25).

Another problem that hampered the response to the 1997 H5N1 outbreak was that there were no reagents generally available in hospitals and public health laboratories to identify influenza viruses of all subtypes. Thus, several months elapsed before the 1997 virus was even identified as being H5N1. In contrast, the causative agent of the 2003-2004 H5N1 outbreak was identified within hours of the admission of the first patients to the hospital. Not only were diagnostic reagents then in place, but there also was increased awareness on the part of health workers to

BOX 12.24

In the middle of last month (April, 2009), just before the copyedited Chapter 12 arrived from the publisher for me to proofread, a novel H1N1 influenza virus that originated in swine was identified in patients in the United States, Mexico, Canada, and elsewhere. (With every iteration of this chapter, there has been some new, potentially devastating outbreak in humans of avian, and now swine, influenza virus to comment on.) As of late May 2009, there were 642 confirmed cases in the United States of this H1N1 influenza virus, designated S-OIV. In eight of these cases the infection was fatal. However, since it is likely that the number of confirmed cases, as substantiated by reverse transcriptase PCR testing for S-OIV, is an underestimate of the actual number of cases, the fatality rate of 8 per 642 cases is very likely inflated. (See the report of the Novel Swine-Origin Influenza A [H1N1] Virus Investigation Team, 2009.)

Here are some relevant facts regarding the 2009 S-OIV outbreak, as of late May 2009. (i) S-OIV is a so-called triple-reassortant virus, since it contains genes from human, avian, and swine influenza viruses. Although triple-reassortant viruses have been known to circulate in pigs since 1998, S-OIV has a unique combination of gene segments that had not been seen earlier. (ii) Interestingly, S-OIV is not epidemic in pigs. Yet it manages to be transmitted from human to human, and indeed, the virus has been able to spread to 41 countries, as reported by the WHO. Apropos these points, culling of pigs has been carried out in one country as a control measure. However, this may have been ill-advised since S-OIV is not epidemic in pigs. Instead, people are spreading the virus. (iii) Bearing in mind that H1N1 influenza viruses have been circulating in humans from 1918 through 1957, and again from 1977 to the present (Fig. 12.11), it is not yet clear whether there is immunity in the human population against S-OIV. However, results of the 2009 study cited above are consistent with the possibility that individuals born before 1957, and thus exposed to the earlier H1N1 virus, may have some immunity to S-OIV. (iv) Clinical manifestations of the S-OIV infection have thus far largely resembled that of seasonal influenza. However, about 38% of the cases have involved vomiting or diarrhea, neither of which is typical of seasonal influenza. Also, as implied in point (iii) above, it is possible that older individuals (i.e., those born before 1957) tend to be spared or have milder infections. (v) S-OIV appears to be sensitive to the neuraminidase inhibitors oseltamivir and zanamivir (at least in the laboratory; the clinical effectiveness of antiviral treatment of S-OIV is not yet known). In contrast, during the 2008-2009 influenza season, almost all circulating human influenza H1N1 viruses were resistant to oseltamivir. The virus is not sensitive to amantadine or rimantadine.

Other key unanswered questions regarding S-OIV are as follows. (i) Will S-OIV become established in the human population? (ii) If it should, will it then generate seasonal antigenic variants? (iii) And will it adapt further to humans by mutation or reassortment, thereby becoming more transmissible or virulent, with more case fatalities? (Apropos these questions, bear in mind that the transmissibility, virulence, and drug sensitivity of influenza viruses in general may rapidly evolve (recall the precedent of the 1918 H1N1 pandemic strain). (iv) When might a vaccine be available? For updates on S-OIV, check the CDC website for the status of this swine H1N1 virus and for recommendations.

As of late May 2009, the S-OIV outbreak had not been notably more severe than seasonal influenza, although the general public would hardly have known that, considering the nonstop updates of new cases by the TV news media in the United States over the course of several weeks in April and May 2009, which created the impression that a killer pandemic was sweeping through the nation. Hardly ever was the S-OIV outbreak placed in the context of the fact that seasonal influenza typically kills nearly 40,000 individuals yearly in the United States. (See Box 12.17 for my remarks regarding public indifference to seasonal influenza.) And at a time when only about 100 Americans had been infected with the new virus (meaning that over 300,000,000 other Americans were not infected), Vice President Joe Biden told a national TV audience that he would tell members of his own family not to go anywhere where they might be in a confined place, such as on an airplane, in a subway, or a classroom (not bad advice if a 1918-type outbreak had been under way, but not good advice if the purpose was to reassure the public). At the same time, Biden's boss, President Barack Obama, in deference to the U.S. pork industry, began referring to this swine flu as "the H1N1 virus," a designation which accurately describes the 1918 pandemic virus, as well as a current seasonal influenza virus.

In fairness to Vice President Biden and others, the current swine flu outbreak, like other flu outbreaks before it, does have the *potential* to become a deadly pandemic. Indeed, as the virus spreads in the human population, its opportunities to generate mutants that might affect its transmissibility, immunity, and virulence will increase. It is to be hoped that this outbreak will be contained.

Reference
Novel Swine-Origin Influenza A (H1N1) Virus Investigation Team. 2009. Emergence of a novel swine-origin influenza (H1N1) virus in humans. *N. Engl. J. Med.* **360**:2605–2615.

the possibility that an avian influenza virus might infect humans.

One of the promising new approaches for developing influenza vaccines is via **plasmid-based reverse genetics**. In brief, the procedure is as follows. Certain so-called "qualified cells," which have been approved for the growth of human vaccines in vitro, are transfected with a set of eight plasmids, two of which express the HA and NA genes of the avian virus and six of which express the remaining genes (i.e., attenuating genes) from a cold-adapted attenuated virus. **Reverse genetics** is used to modify known virulence motifs in the HA or NA gene (e.g., the multibasic amino acid cleavage site motif in the HA gene; see above and Box 12.21). (The distinction between reverse genetics and forward or standard genetics is as follows. Reverse genetics seeks to identify possible phenotypes that may be derived from a specific DNA sequence, whereas forward genetics seeks to find the genetic basis of a phenotype.)

The plasmid-based reverse genetics approach was used to develop the H5N1 vaccine that was licensed in 2007 by the U.S. FDA, as mentioned above (see also below). In addition, a live, attenuated influenza A candidate vaccine was also developed using this approach. The engineered strains in this vaccine contain the two genes that encode the HA and NA proteins from human H3 and H1 viruses. The six genes that encode the internal proteins are derived from the cold-adapted attenuated virus. These vaccines have proven to be safe and effective in clinical trials.

The plasmid-based reverse genetics approach offers several significant advantages over the conventional egg-based method of vaccine production, especially with respect to the shorter time line needed to develop the vaccine.

In the conventional egg-based method, two influenza strains, which display the characteristics to be combined in the new vaccine strain, are injected into an embryonated egg, where their genome segments reassort "naturally." Each of the 256 possible reassortant viruses then have to be screened to find the one that displays the desired HA and NA antigenicity and that additionally is best able to replicate in the eggs. If it were desirable to modify the HA and NA virulence motifs of the vaccine strain, then it still would be necessary to reverse transcribe the corresponding influenza virus genome segments and insert them into plasmids to carry out the mutagenesis. In contrast, to generate a vaccine via the plasmid-based approach, only the HA and NA genes need to be cloned for each vaccine, since the plasmids encoding the genes from the attenuated donor strain are already available.

Growth of the vaccine in cell culture, rather than in embryonated eggs, offers other advantages, in addition to speed. For example, it avoids the possibility of the vaccine, even an attenuated one, killing the eggs. Destruction of the eggs may occur, particularly if the virulence motifs of the HA and NA proteins have not been modified by reverse genetics. (The embryonated eggs themselves are potentially susceptible to circulating avian influenza virus. Consequently, flocks dedicated for the purpose of vaccine production are raised under biosecurity regulations and are monitored by veterinarians.) Growth in cell culture also eliminates uncertainties of egg supply and is less costly than the standard egg-based procedure. However, large-scale production of influenza vaccines still needs to be carried out in eggs while cell culture-based technology continues to be developed.

Here are some additional, perhaps paradoxical concerns that were not enumerated above. None of the previous outbreaks of H5N1 virus has yet led to a pandemic. Thus, while new H5N1 outbreaks might well occur over the next several years, it may be that none of these future outbreaks will lead to a pandemic. Ironically, this is a cause for concern in its own right. First, developing vaccines is expensive. So, what might be the consequences with respect to investing in vaccine development, when the threats do not materialize? Most likely, in the absence of government cost guarantees, vaccine manufacturers might not be willing to respond to the next potential crisis, which could be the one that indeed results in a pandemic. Second, politicians and the public might eventually become complacent about the potential of future outbreaks. A third concern is relevant in the case of a live attenuated vaccine, which bears genes derived from an avian influenza virus. In that instance, there is the possibility that the vaccine virus itself might reassort with a prevailing human influenza virus, inadvertently producing a highly virulent or pandemic virus. Although we might take our chances regarding that prospect in the face of a potentially devastating pandemic, how might we feel, having been inoculated with the live vaccine, if the threatened pandemic did not materialize? For this last reason, other approaches to influenza vaccines, which do not leave the possibility of reassortment to generate a virulent virus, have been considered.

In one alternative approach, a vaccine against the 2003-2004 H5N1 virus was generated by expressing different portions of the H5 HA gene from a replication-incompetent adenoviral vector. Impressively, vaccine production was under way within 36 days of acquiring the virus H5 HA sequence. Moreover, the vaccine proved effective against a lethal dose of the H5N1 virus in a mouse model, as well as in chickens.

Yet another approach was to develop a **DNA-based vaccine** against the H5N1 virus. DNA-based vaccines contain only parts of the cloned reverse-transcribed viral genetic material. Once inside the body's cells, the DNA is transcribed and then translated to make proteins that act as a vaccine against the virus. Human trials of a DNA-based vaccine against H5N1 avian influenza were started at the U.S. National Institutes of Health (NIH) in December 2006. In this instance, the candidate vaccine went from the research bench into clinical trials in less than 6 months. The researchers at the NIH also showed that a DNA-based vaccine against the reconstructed 1918 pandemic virus protects against that virus in animal models. (DNA-based vaccines are thought to hold promise against other outbreak viruses, including HIV, Ebola virus, the severe acute respiratory syndrome [SARS] coronavirus, and West Nile virus.)

In April 2007 the FDA licensed an H5N1 vaccine developed by Aventis Pasteur, the first vaccine against an avian influenza to be licensed for use in humans in the United States. The vaccine was developed using the plasmid-based reverse genetics approach. The gene segments encoding the HA and NA proteins were from an H5N1 virus isolated from a Vietnamese patient in February 2004. The other six genomic segments were from the avirulent PR8 H1N1 strain. Also, the HA gene was modified to replace the stretch of six basic amino acids at the activation-cleavage site between HA1 and HA2, which is associated with high pathogenicity (Box 12.21). The vaccine virus was grown using the standard egg-based procedure. It then was purified by centrifugation, inactivated using formalin, and disrupted with detergent. A clinical trial conducted by the U.S. National Institute of Allergy and Infectious Diseases showed that the vaccine elicits an immune response against the H5N1 virus and is associated with mild side effects only, thus leading to its licensure. See Treanor et al. (2006) for details of the clinical trial.

Despite the availability of the Aventis Pasteur H5N1 vaccine and other promising efforts under way to develop vaccines against avian influenza viruses, the following proviso should be borne in mind. An emergent pandemic strain might have sufficiently different antigenic properties from that of any already stockpiled H5 vaccine such that the stockpiled vaccines might not be adequately protective. Thus, in the event of an H5N1 outbreak or pandemic, effective vaccines may be nonexistent (at least initially) or in short supply.

Might antivirals against avian influenza viruses possibly fill the protective gap until effective vaccines become available? Rimantadine and amantadine, which block the M2 protein ion channel, and oseltamivir (Tamiflu) and zanamivir, which block activity of the NA protein, are the antivirals used for treating human influenza A virus infections. Controlled clinical trials of the NA inhibitors, oseltamivir and zanamivir, against H5N1 avian viruses have not yet been carried out, but their efficacy in animal models is well documented. In some instances, treatment with oseltamivir appeared to be beneficial, but there also was at least one instance in which the H5N1 virus appeared to be resistant to the drug. Based on experience using NA inhibitors to treat human influenza virus infections, the timing of treatment is crucially important, with protection dropping rapidly if begun later than 60 h after infection. However, the window of opportunity is not known in the case of H5N1 viruses, and it may be longer than in the case of human influenza viruses.

Rimantadine and amantadine are not likely to be of use against avian influenza viruses because of widespread resistance of the avian influenza virus M2 protein to these drugs. Indeed, over 90% of isolates of H1 and H3 human subtypes are resistant to these drugs. Consequently, rimantadine and amantadine are now used only for prophylaxis, and only in those instances in which the circulating influenza strain is known to be susceptible to the drugs.

Suggested Readings

Bullough, P. A., F. M. Hughson, J. J. Skehel, and D. C. Wiley. 1994. The structure of influenza hemagglutinin at the pH of membrane fusion. *Nature* 371:37–43.

Gamblin, S. J., L. F. Haire, R. J. Russell, D. J. Stevens, B. Xiao, B. Y. Ha, N. Vasisht, D. A. Steinhauer, R. S. Daniels, A. Elliot, D. C. Wiley, and J. J. Skehel. 2004. The structure and receptor binding properties of the 1918 influenza hemagglutinin. *Science* 303:1838–1842.

Kobasa, D., A. Takada, K. Shinya, M. Hatta, P. Halfmann, S. Theriault, H. Suzuki, H. Nishimura, K. Mitamura, N. Sugaya, T. Usui, T. Murata, Y. Maeda, S. Watanabe, M. Suresh, T. Suzuki, Y. Suzuki, H. Feldmann, and Y. Kawaoka. 2004. Enhanced virulence of influenza A viruses with the haemagglutinin of the 1918 pandemic virus. *Nature* 431:703–707.

Lakadamyali, M., M. J. Rust, H. P. Babcock, and X. Zhuang. 2003. Visualizing infection of individual influenza viruses. *Proc. Natl. Acad. Sci. USA* **100:**9280–9285.

Melikyan, G. B., J. M. White, and F. S. Cohen. 1995. GPI-anchored influenza hemagglutinin induces hemifusion to both red blood cell and planar bilayer membranes. *J. Cell Biol.* **131:**679–691.

Plotch, S. J., M. Bouloy, and R. M. Krug. 1979. Transfer of 5′-terminal cap of globin mRNA to influenza viral complementary RNA during transcription in vitro. *Proc. Natl. Acad. Sci. USA* **76:**1618–1622.

Satterly, N., P.-L. Tsai, J. V. Jan van Deursen, D. R. Nussenzveig, Y. Wang, A. Paula, P. A. Faria, A. Levay, D. E. Levy, and B. M. A. Fontoura. 2007. Influenza virus targets the mRNA export machinery and the nuclear pore complex. *Proc. Natl. Acad. Sci. USA* **104:**1853–1858.

Treanor, J. J., J. D. Campbell, K. M. Zangwill, T. Rowe, and M. Wolff. 2006. Safety and immunogenicity of an inactivated subvirion influenza A (H5N1) vaccine. *N. Engl. J. Med.* **354:**1343–1351.

Tumpey, T. M., C. F. Basler, P. V. Aguilar, H. Zeng, A. Solórzano, D. E. Swayne, N. J. Cox, J. M. Katz, J. K. Taubenberger, P. Palese, and A. García Sastre. 2005. Characterization of the reconstructed 1918 Spanish influenza pandemic virus. *Science* **310:**77–80.

Miscellaneous RNA Viruses

INTRODUCTION

Seven distinct, single-stranded RNA virus families are discussed in this chapter. Although each of these seven virus families is described much more briefly than those families receiving full-chapter treatments, each of the families covered here includes clinically important members, some of which are attracting considerable attention as emerging viruses. Also, the molecular biology of each of these families is unique in at least one respect.

The caliciviruses, astroviruses, and hepatitis E-like viruses are plus-strand RNA viruses and thus are class IV viruses in the Baltimore classification scheme. In contrast, the filoviruses and bornaviruses are minus-strand RNA viruses and thus recapitulate the replication strategy of class V viruses in the Baltimore scheme. Moreover, the filovirus and bornavirus genomes are nonsegmented and are organized like the previously considered rhabdovirus and paramyxovirus genomes (Chapters 10 and 11). Thus, these four virus families together constitute the superfamily or order *Mononegavirales*. Nevertheless, the bornaviruses differ from other members of the mononegavirus superfamily by replicating in the nucleus rather than the cytoplasm. Ebola virus, a filovirus, is perhaps the most infamous of the emerging viruses.

The arenaviruses and bunyaviruses are nominally minus-strand viruses but differ from the mononegaviruses in several defining respects. First, they contain segmented genomes, which are bipartite in the case of the arenaviruses and tripartite in the case of the bunyaviruses. Second, the bipartite arenavirus genomes differ in a major way from the genomes of all the RNA virus families discussed in earlier chapters. In point of fact, arenaviruses do not have conventional single-stranded RNA genomes of either the plus sense or minus sense. In actuality, one end of each of the individual arenavirus genomic RNA segments is of the plus sense, while the other end is of the minus sense. Consequently, these viruses are said to be **ambisense** and cannot be perfectly classified as either class IV or class V viruses in the Baltimore classification scheme. Interestingly, some, but not all, of the bunyaviruses also contain ambisense genomic segments.

(See Box 14.1 for a brief description of the circoviruses, which are very small viruses that contain ambisense, circular, single-stranded DNA genomes.)

We begin our consideration of these seven virus families with the arenaviruses and bunyaviruses, describing their curious genomic organizations and replication strategies. We then move on to the bornaviruses, which have recently been attracting attention as a possible cause of neuropsychiatric disorders in humans.

Next, we come to the plus-strand caliciviruses, astroviruses, and hepatitis E-like viruses. Each of these plus-strand virus families somewhat resembles the plus-strand togaviruses (Chapter 8), since like the togaviruses, they translate their nonstructural proteins from a genomic-length messenger RNA (mRNA) and their structural proteins from a subgenomic mRNA. One of the caliciviruses, the Norwalk virus, is responsible for about 40% of all outbreaks of gastroenteritis among adults in the United States. This virus has recently gained notoriety as the cause of several highly publicized outbreaks of gastroenteritis on cruise ships. The human astroviruses are a major cause of acute viral gastroenteritis in young children worldwide. Hepatitis E-like viruses cause an acute form of hepatitis that resembles the disease caused by hepatitis A virus (Chapter 6).

We conclude this chapter with the filoviruses, a family that includes the Ebola and Marburg viruses, two emerging **hemorrhagic fever viruses** that have generated much interest since their discoveries. But before moving on, note that the majority of known hemorrhagic fever viruses are members of the arenavirus, bunyavirus, and filovirus families discussed here, as well as of the flaviviruses discussed in Chapter 7. Infection by any one of the hemorrhagic fever viruses can be life threatening. Moreover, several of these viruses, including Ebola and Marburg viruses, have been discussed for their potential use as deadly biological weapons.

ARENAVIRUSES

The arenaviruses are enveloped single-stranded RNA viruses. They derive their name from their sandy appearance in electron micrographs (*arena* is the Latin word for sand), which is due to the presence of cellular ribosomes in the virus particles, which resemble grains of sand in electron micrographs (Fig. 13.1). Interestingly, these ribosomes retain their functionality but appear to play no role in the life cycle of the virus.

Lymphocytic choriomeningitis virus (LCMV) is the prototype arenavirus. LCMV is not a hemorrhagic fever virus, but the family includes several members which are, including Lassa fever, Junin, and Machupo viruses.

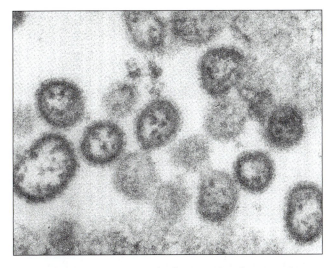

Figure 13.1 Electron micrograph of an arenavirus (courtesy of R. C. Williams). From J. A. Levy, H. Fraenkel-Conrat, and R. A. Owens, *Virology*, 3rd ed. (Prentice-Hall, Upper Saddle River, NJ, 1994), with permission.

The more famous Ebola virus, which is a filovirus, is the most virulent of the known hemorrhagic fever viruses. However, the arenavirus Lassa fever virus is much more prevalent in the world, making Lassa fever the best known of the hemorrhagic fevers caused by an arenavirus.

Viral hemorrhagic fevers in general are characterized by widespread hemorrhaging from the body's epithelial surfaces, including the internal mucosa, as well as the skin. Taking Lassa fever as an example, clinical illness begins with flu-like and gastrointestinal symptoms. Bleeding, organ failure, and shock may occur in the late stages of the disease, with death occurring after an average period of about 2 weeks following the onset of symptoms.

Most of the hemorrhagic fever viruses are found in either tropical Africa (e.g., Lassa fever virus) or South America (e.g., Junin and Machupo viruses). Rodents appear to be the natural hosts for these viruses, and as is usually the case in the natural host, infections of rodents by these viruses tend to be benign. Additionally, transmission in nature is neonatal, thereby resulting in immune tolerance. Moreover, as one might expect from the mode of transmission, infections in rodents are also persistent. Consistent with the above, infections in cell culture are noncytocidal.

The above facts raise the possibility that pathogenesis of arenavirus hemorrhagic fevers in humans might be immunomediated. Still, little is known about the mechanisms underlying the pathogenesis of these infections in humans (Box 13.1). Regarding the possibility that pathogenesis might be mediated by immunopathological mechanisms, neutralizing antibodies typically appear only late in infection, after recovery in survivors, and not at all in most fatal

BOX 13.1

Lassa fever virus is the most common of the hemorrhagic arenaviruses. Nevertheless, the pathogenesis of Lassa fever is not well understood. This is in part because Lassa fever occurs predominantly in the "Mano River Union (MRU) countries" of Sierra Leone, Liberia, and Guinea, which have experienced extensive civil unrest during the past 2 decades, thereby seriously impeding research on human infections. Moreover, in some regions, it is dangerous to conduct postmortem examination of patients who died of the disease, because of certain African cultural taboos against the manipulation of corpses. Thus, much of what is known of Lassa fever pathogenesis is based on limited data from humans (Walker et al., 1982) and extensive experimental observations from primate models of Lassa fever.

Reference
Walker, D. H., J. B. McCormick, K. M. Johnson, P. A. Webb, G. Komba-Kono, L. H. Elliott, and J. J. Gardner. 1982. Pathologic and virologic study of fatal Lassa fever in man. *Am. J. Pathol.* 107:349–356.

cases, thus implying that the pathogenesis of arenavirus hemorrhagic fevers is not caused by the humoral immune response. On the other hand, severe disease appears to correlate with the activity of the effectors of innate immunity, specifically, macrophages and dendritic cells and the proinflammatory cytokines they release. In animal models, following inoculation with the virus, dendritic cells and macrophages appear to be the initial sites of Lassa fever virus replication. The virus might then replicate locally in the tissues and in regional lymph nodes and then disseminate through the lymph and blood to a wide variety of organs and tissues, including the liver, spleen, endothelium, lymph nodes, kidney, adrenal gland, pancreas, placenta, uterus, breast, and gonads. This widely disseminated infection might then give rise to a **septic shock**-like process, in which proinflammatory cytokines released at these many infected sites cause blood vessels to become leaky, so that blood fluid escapes into the surrounding tissues, resulting in a system-wide drop in blood pressure. (Septic shock occurs during widely disseminated infections in which the inflammatory response can become excessive.) However, actual hemorrhaging in the instance of Lassa fever appears to result from Lassa fever virus-induced release of a soluble mediator that impairs platelet aggregation.

Chronically infected rodents shed the virus in urine, feces, and saliva. Humans can be infected by inhaling virus-containing aerosols, by contact with contaminated objects (fomites), and by consuming contaminated food. Lassa fever virus can also be transmitted from human to human, especially in hospital settings, with fatality rates there as high as 20%.

Lassa Fever Virus: an Early Emerging Virus

Lassa fever virus first emerged in 1969, causing quite a stir at the time. Indeed, its appearance marked one of the first recognized outbreaks of a so-called emerging virus. The episode began with the death of a nurse at a mission in Lassa, Nigeria. Then, a nurse attending the first victim also died. A third nurse, who assisted at the autopsy of the first nurse, became seriously ill but recovered after being evacuated to Yale University Medical School in New Haven, CT, where she received intensive care. Virologists at Yale succeeded in isolating the virus from their African patient. One of the Yale virologists then became infected but recovered following transfusion with immune plasma from the African patient. However, a Yale laboratory technician also was infected and died.

The natural reservoir for Lassa fever virus is the West African mouse, *Mastomys natalensis*, which lives in or near human habitats, contaminating human dwellings with its urine. The virus gives rise to an asymptomatic, chronic infection in its rodent host, as noted above.

Lassa fever virus is thus far unique among arenaviruses in its capacity to be transmitted from one human to another. Fortunately, however, human-to-human transmission is inefficient, and the virus remains confined to the West African habitat of its natural *Mastomys* host. Yet serological surveys show that the virus is endemic there, with millions of humans having been infected. There are an estimated 300,000 new human infections yearly. Approximately 20% of these infections result in clinical illness, with a case-fatality rate that may be greater than 15% under some circumstances. Lassa fever virus is thought to be responsible for about 5,000 fatalities yearly in West Africa. Urbanization, which brings together large numbers of people and also facilitates propagation of *Mastomys*, is likely a factor underlying the spread of Lassa fever virus in the West African human population.

LCMV and Immunopathology

LCMV, the prototype arenavirus, is widely distributed worldwide. In the United States, its reservoirs are the house mouse *Mus musculus* and hamsters. Note that the popularity of hamsters as pets has led to many instances of LCMV disease in humans. Notwithstanding the name of this virus, the typical clinical syndrome in humans is a flu-like fever, not meningeal illness. But if meningitis should occur, it is usually subacute. LCMV is not a hemorrhagic fever virus.

Until the early 1980s, LCMV infection of mice was the most extensively studied of all persistent virus infections. Most importantly, some of these early studies of the LCMV mouse model were the first to demonstrate the contribution of host immunity to the pathology of virus diseases. An early key observation, which in fact led to the immunopathology paradigm, was that the virulence of an LCMV infection is highly dependent on the age of the mouse at the time of infection. Mice infected in utero or as newborns demonstrated no signs of illness while continuing to produce large amounts of the virus throughout their lives. In contrast, LCMV-infected adult mice developed encephalitis in about 10 days to 2 weeks. Mice infected in utero or neonatally were found to be immunologically tolerant to the virus (Chapter 4), thereby implying a connection between an immune response to the virus and pathogenesis.

The link between LCMV pathogenesis and the host immune response was confirmed by the experimental finding that LCMV disease could be prevented in LCMV-infected adult mice by immunosuppressive treatments such as X-irradiation or neonatal thymectomy prior to the infection. Moreover, transfer of immune lymphocytes to immunosuppressed LCMV-infected mice led to lesions and death from classic LCMV disease. Recipients of these lymphocytes did not produce antiviral antibodies, implying that cytotoxic T cells might be responsible for the pathology. Subsequent experiments involving adoptive transfer of specific T-cell subtypes into gene knockout mice demonstrated that CD8 cytotoxic T cells indeed are responsible for the tissue damage in infected mice. These experimental findings are further supported by the observation that T-cell-mediated cytotoxicity is greatly impaired in perforin-deficient infected mice (perforin is a key effector of cytotoxic T lymphocyte-mediated cytotoxicity [Chapter 4]). (Yet bear in mind that these cytotoxic T cells also are crucial mediators of virus clearance in vivo. Immunopathology is the price we sometimes pay for the benefits of our prodigious immune defenses.)

The Ambisense Arenavirus Genome

Next, we consider the organization of arenavirus genomes and their patterns of transcription and replication. Arenavirus genomes are comprised of two single-stranded RNA molecules, designated L and S, which are approximately 7,200 and 3,400 bases long, respectively. Sequence analysis of these genomic RNA segments demonstrated that they are neither conventional plus-strand nor minus-strand RNA molecules. Instead, one region of each genomic segment is of the plus sense, and the other region, which does not overlap the first, is of the minus sense. Because each of the genome segments contains a region of plus sense and

a region of minus sense, these genomic segments are said to be ambisense. The plus-sense and minus-sense regions of each genomic RNA segment are separated from each other by an intergenic noncoding stem-loop structure (Fig. 13.2).

The S segment encodes the major structural proteins of the virus, which include the NP internal nucleoprotein, and the components of the spikes, GP1 and GP2. The L segment encodes the viral RNA-dependent RNA polymerase and the Z protein, a zinc-binding protein that acts as a matrix protein. There are two open reading frames (ORFs) on each genomic segment, for a total of four

Figure 13.2 Schematic representation of the coding arrangements for the genomic L and S RNA segments. The stem-loop structures depict the intergenic noncoding regions. The lower part of the figure summarizes the transcription and replication events for the S genomic RNA, and this general scheme also applies to the genomic L RNA segment. As a consequence of the ambisense coding arrangement, GP mRNA (and Z mRNA) can be transcribed only after the initiation of genomic RNA replication. The S-derived subgenomic mRNAs contain a short stretch of nontemplated nucleotides and a cap (designated "x" and "c," respectively, in the diagram) at the 5′ ends and terminate at the 3′ ends at multiple sites within the intergenic region. Adapted from B. N. Fields, D. M. Knipe, and P. M. Howley (ed.), *Fundamental Virology*, 3rd ed. (Lippincott-Raven, Philadelphia, Pennsylvania, 1996), with permission.

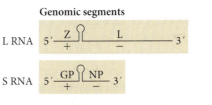

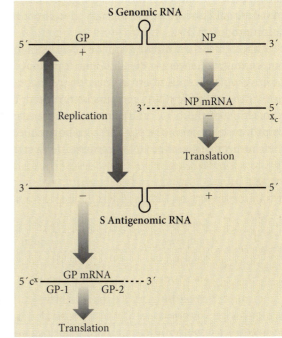

ORFs. The S segment is able to encode NP, GP1, and GP2, since the last two proteins are derived by posttranslational cleavage of the precursor polyprotein GP-C.

How does the ambisense nature of arenavirus genome segments play out during transcription and replication? Consider first the S genomic RNA segment. After entry and uncoating, a virus particle-associated RNA polymerase transcribes the negative-sense 3′ end of the S genomic RNA segment into a subgenomic complementary mRNA that is translated into the NP protein (Fig. 13.2). Likewise, the viral polymerase transcribes the negative-sense 3′ end of the L genomic RNA segment into a subgenomic complementary mRNA that is translated into the L protein. These transcripts terminate within a short stem-loop configuration in the intergenic region of the genomic RNA segment, suggesting that this element plays a role in transcription termination. The subgenomic mRNAs, which are translated into the NP and L proteins, are the only viral RNAs produced during the first few hours of infection.

Translation and accumulation of the NP and L proteins are necessary and sufficient to enable the genomic RNA segments to be copied into full-length antigenomic RNA templates. The role of the NP protein in the process is to overcome the block to copying imposed by the stem-loop structure in the genomic RNA segments, thereby permitting synthesis of full-length antigenomic segments. Note that the 3′ ends of each of these antigenomic templates are of the minus sense. Thus, the 3′ ends of these antigenomic S and L segments can be transcribed into subgenomic complementary mRNA molecules that are translated into the GP and Z proteins, respectively. As noted above, the GP1 and GP2 proteins are generated by posttranslational cleavage of the primary glycoprotein translation product, GP-C.

Although the ambisense genomic RNA segments contain coding sequences for the GP-C and Z proteins at their positive-sense 5′ ends, there is no evidence that any portion of the genomic RNA segments is translated. Instead, translation of the GP-C and Z proteins depends on the prior transcription of the NP and L coding sequences on their respective genomic RNA segments, followed by the translation of those proteins, the generation of full-length antigenomic RNA segments, and the transcription of mRNAs from the 3′ ends of those full-length antigenomic RNAs. This strategy may impart a method of temporal regulation for the translation of the envelope glycoproteins, which are produced as late proteins, only after genome replication.

The synthesis of the full-length complementary copies of both of the ambisense genomic RNA segments requires read-through of the transcription termination signals. It is likely that the single-stranded RNA-binding NP protein plays a role in regulating relative levels of transcription versus replication, by blocking termination at the intergenic sequences.

BUNYAVIRUSES

The bunyaviruses are one of the largest animal virus families, containing more than 300 known members. These are grouped into five genera (Table 13.1), four of which contain viruses that are pathogenic to humans. Human pathogens include Rift Valley fever virus, Crimean-Congo virus, and Hantaan virus, each of which is a hemorrhagic fever virus. Most human infections with these hemorrhagic fever viruses tend to be mild. However, in those instances in which patients develop hemorrhagic symptoms, the case-fatality rate is about 50%.

California encephalitis virus, La Crosse virus, and Sin Nombre virus ("no-name virus") are pathogenic human bunyaviruses that do not give rise to hemorrhagic fevers. Instead, the first two of these viruses give rise to febrile illness and encephalitis. The third, Sin Nombre virus, causes the so-called hantavirus pulmonary syndrome, a severe condition that ends in pulmonary edema, respiratory failure, and death within days.

Most of the bunyaviruses are spread by insect vectors, such as mosquitoes, ticks, fleas, thrips, and flies. Infected females of many of these insects can transmit the virus transovarially to their eggs, thereby enabling the virus to persist in its arthropod hosts from one generation to the next. Humans are infected when bitten by an infected arthropod.

Table 13.1 The bunyaviruses[a]

Genus	Vector	Examples
Bunyavirus	Mosquito	La Crosse encephalitis virus, Bunyamwera virus
Nirovirus	Tick	Dugbe virus, Nairobi sheep disease virus
Phlebovirus	Sandfly	Rift Valley fever virus, Uukuniemi virus
Hantavirus	Rodent	Hantaan virus, Sin Nombre virus
Tospovirus	Thrip	Tomato spotted wilt virus

[a]From E. K. Wagner and M. J. Hewlett, *Basic Virology*, 2nd ed. (Blackwell Publishing, Malden, MA, 2004).

Sin Nombre and Hantaan Viruses

While most bunyaviruses are arthropod-borne, members of the genus *Hantavirus* (Table 13.1), which includes Hantaan virus and Sin Nombre virus, are spread by rodents, their natural hosts. They are transmitted from rodent to rodent by urine and saliva. Humans are infected by inhalation of aerosolized rodent urine or by direct contact with their excreta.

Sin Nombre virus first emerged in May 1993, giving rise at the time to an outbreak of hantavirus pulmonary syndrome on the Navajo Indian Reservation in the Four Corners area of the American Southwest (i.e., where Utah, Colorado, Arizona, and New Mexico converge). To determine the cause of the outbreak, investigators from the Centers for Disease Control and Prevention (CDC) screened patients for antibodies to a wide variety of infectious agents. This effort led to the identification of a previously unknown hantavirus, the Sin Nombre virus, which also was found in many deer mice trapped in the area. The outbreak in humans was attributed to a rise in the population of the deer mouse vector, caused by an unusually abundant rainfall in that area during the previous winter, which led to plentiful amounts of a favored mouse food, pinion nuts. Consequently, there were more frequent incursions of mice into human habitations. Such episodes may have occurred earlier, as suggested by Navajo folklore that warns against approaching rodents after periods of unusual weather.

Another notable bunyavirus outbreak occurred earlier, during the Korean War of 1950-1952, when thousands of United Nations soldiers serving in Korea developed a feverish disease that was also characterized by hemorrhaging, kidney failure, and shock, with a case-fatality rate of 5 to 10%. Interestingly, the virus responsible for this outbreak was not identified until 1978, when it was isolated from a Korean rodent, *Apodemus agrarius*, and was found to be a previously unrecognized bunyavirus that was named Hantaan virus (after the Hantaan River in South Korea). It is the prototype of the *Hantavirus* genus, which also includes Sin Nombre and Muerto Canyon viruses, which are endemic to the American southwest.

The Tripartite Bunyavirus Genome

All bunyaviruses contain segmented, tripartite, negative-strand RNA genomes. Like the genomes of other minus-strand RNA viruses, each of the bunyaviral genomic RNA segments exists as a helically symmetric viral ribonucleic protein (vRNP), in association with molecules of a nucleocapsid (N) protein and a polymerase (L) protein. Also like other minus-strand RNA viruses, bunyavirus particles are enveloped. Their envelopes contain two glycoproteins, G1 and G2.

The bunyavirus replication stratagem is by and large the same as that of other minus-strand RNA viruses. When bunyavirus vRNPs are released into the cytoplasm, they are transcribed by the viral, vRNP-associated RNA polymerase complexes. Viral mRNAs are transcribed from each of the three genomic RNA segments (Fig. 13.3). The largest bunyavirus genomic RNA segment encodes the transcriptase (L protein). The middle-sized genomic RNA segment encodes the envelope glycoproteins G1 and G2. In some genera it also encodes a nonstructural protein, NS_M, which not all bunyaviruses express. The smallest genomic RNA segment encodes the N protein, and in some genera it also encodes another nonstructural protein, NS_S, which may be involved in transcription and replication.

As in the case of influenza viruses (Chapter 12), bunyaviruses acquire caps for their mRNAs by "snatching" them from cellular mRNAs. In this regard, the bunyavirus polymerase, like the influenza virus polymerase, has an endonuclease activity that mediates cap snatching by cleaving

Figure 13.3 When bunyavirus vRNPs are released into the cytoplasm, they are transcribed by the viral, vRNP-associated, RNA polymerase complexes. Viral mRNAs are transcribed from each of the three genomic RNA segments. The example depicted is of La Crosse virus. The largest (L) bunyavirus genomic RNA segment encodes the transcriptase/polymerase (L protein). The middle-sized (M) genomic RNA segment encodes the envelope glycoproteins G1 and G2. In some genera it also encodes a nonstructural protein, NS_M, which not all bunyaviruses express. These individual proteins are generated by cleavage of a precursor polyprotein. The smallest genomic RNA segment encodes the N protein, and in some genera it also encodes another nonstructural protein, NS_S, which may be involved in transcription and replication. This is possible because the plus-sense transcript of the S genomic RNA segment contains two partially overlapping, out-of-phase reading frames. Bunyaviruses, like influenza viruses (Chapter 12), acquire caps for their mRNAs by a cap-snatching process. Adapted from E. K. Wagner, M. J. Hewlett, D. C. Bloom, and D. Camerini, *Basic Virology*, 3rd ed. (Blackwell Publishing, Malden, MA), 2008, with permission.

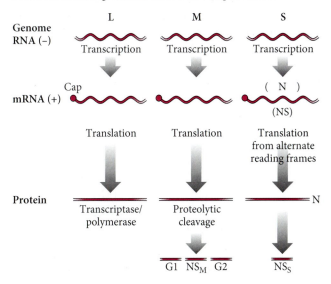

12 to 15 nucleotides from the ends of host cell mRNAs. However, while influenza virus mRNA synthesis occurs in the nucleus, bunyaviral mRNA synthesis occurs in the cytoplasm. The explanation for this distinction is as follows. Influenza virus RNA synthesis is believed to occur in the nucleus so that the virus might use the splicing machinery of the host cell to generate spliced mRNAs (Chapter 12). In contrast, bunyaviral mRNAs are not spliced, and bunyaviruses have no other need for nuclear activities.

In the *Phlebovirus* and *Tospovirus* genera of the *Bunyaviridae* (Table 13.1), the smallest genomic RNA segment is actually ambisense, and its expression follows the same strategy as in the transcription of the arenavirus genomic RNAs (Fig. 13.4). The N protein is translated from an mRNA that is transcribed from the negative-sense 3′ end of the genomic S segment. The NS_S protein is translated from an mRNA that is transcribed from the negative-sense 3′ end of a full-length complementary copy of the genomic S segment. The *Bunyavirus* genus of the *Bunyaviridae* also expresses an NS_S protein. In that instance, NS_S is translated from the 3′ end of the N gene, but in a different reading frame (Fig. 13.3).

The GP1 and GP2 envelope glycoproteins are generated by proteolytic cleavage of a precursor polyprotein. This cleavage usually occurs in the endoplasmic reticulum, where these transmembrane glycoproteins are translated.

Figure 13.4 In the *Phlebovirus* and *Tospovirus* genera, the smallest genomic RNA segment, which encodes the N and NS_S proteins, actually is ambisense, and its expression follows the same strategy as in the transcription of the arenavirus genomic RNAs (Fig. 13.2). (1 and 2) The N protein is translated from an mRNA that is transcribed from the negative-sense 3′ end of the genomic S segment. (3 through 5) The NS_S protein is translated from an mRNA that is transcribed from the negative-sense 3′ end of a full-length complementary copy (vcRNA) of the genomic S segment. Notice that the 3′ minus-sense sequence of the vRNA is separated from the 5′ plus-sense sequence by an intergenic sequence (orange). The mRNAs contain a cap, plus 12 to 15 additional nucleotides (blue), acquired from cellular mRNAs by a cap-snatching mechanism. Adapted from D. O. White and F. J. Fenner, *Medical Virology*, 4th ed. (Academic Press, San Diego, CA), 1994.

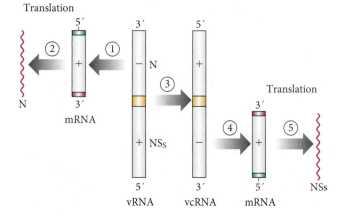

They then accumulate in the Golgi membranes. Progeny nucleocapsids bud into these Golgi membranes to generate progeny virus particles, which are transported in vesicles from the Golgi to the plasma membrane, where vesicle fusion with the plasma membrane releases the virus from the cell.

BORNAVIRUSES

Borna disease virus (BDV) is, thus far, the only member of the bornavirus family. So, then, what sets it apart from the already established virus families? Based on the facts that (i) it is a nonsegmented, minus-strand RNA virus and (ii) its genomic organization is similar to that of the nonsegmented, minus-strand paramyxoviruses, rhabdoviruses, and filoviruses, it is classified as a member of the order *Mononegavirales*. Nevertheless, the BDV manner of replication differs from that of the other members of the mononegavirus superfamily in the following way. Each of the other mononegaviruses replicates in the cytoplasm. In contrast, transcription and replication of bornavirus genomes occur entirely in the nucleus. After bornavirus particles enter into the cell, vRNPs are released and traffic to the nucleus, where transcription and replication of the genomic RNA take place. Bornaviruses are the only known animal viruses with nonsegmented, minus-strand, RNA genomes that replicate in the nucleus.

BDV and Human Neuropsychiatric Disease

BDV was first identified as a pathogen that causes fatal central nervous system disease (T-cell-mediated meningoencephalitis) in horses and sheep. This virus was then found to be highly neurotropic and potentially fatal in a wide variety of experimentally infected other animals, including chickens, rabbits, cattle, dogs, and monkeys. The virus disseminates to the brain by axonal transport and transneuronal transmission (Chapter 3). Recently, the white-toothed shrew (*Crocidura leucodon*) was identified as a BDV reservoir species in an area in Switzerland where Borna disease is endemic. As might be expected, BDV persists in these shrews without giving rise to any obvious disease.

BDV has been attracting attention because of evidence that it causes emotional disturbances in animals, which some researchers believe bear a resemblance to some neuropsychiatric conditions seen in human patients. Moreover, results of serological analysis suggest that BDV might infect humans. For these reasons, there is the intriguing possibility that BDV might be a cause of neuropsychiatric disorders in humans.

Yet, whether BDV actually contributes to human neuropsychiatric disorders remains controversial. Moreover, because of inconsistent findings from serological and polymerase chain reaction (PCR)-based analyses, it is not at all clear how prevalent BDV might be in the human population (Box 13.2). However, since 1996, several different seroepidemiological studies have reported an increased BDV seroprevalence in human psychiatric patients. In addition, BDV antigens and RNA have been reported in human brain samples collected at autopsy from individuals with a history of mental disorders. On the other hand, similar findings also have been made in clinical samples from patients with brain tumors and from brain tissue of some apparently healthy control subjects.

BOX 13.2

While reverse transcriptase PCR is exceptionally sensitive, its use is fraught with the danger of false-positive results, due in fact to the very high sensitivity of PCR-based procedures. For instance, false positives can result from even the slightest inadvertent laboratory contamination. Regarding how this proviso might apply to the issue of BDV in humans, the putative human-derived BDV sequences initially reported were virtually identical to the animal BDV strains being handled in the same laboratory. This was so despite the fact that the geographic location and time of isolation of the human-derived and laboratory strains varied considerably. These facts cast doubt on these PCR-based findings and highlight the importance of maintaining scrupulous standards when carrying out PCR analyses.

Several more-recent and more carefully designed PCR-based studies failed to detect BDV RNA in human tissues. Thus, the case for BDV in humans is now based mainly on serological evidence. Unfortunately, the diagnostic value of anti-BDV antibodies was initially compromised by the fact that these antibodies recognize BDV antigen with only low avidity, and such low-avidity antibodies commonly are cross-reactive to a similar extent with unrelated antigens. However, the recent development of a peptide array-based screening test allowed the further characterization of these anti-BDV antibodies. It revealed the presence of small amounts of BDV-reactive antibodies in crude human sera that specifically recognized various epitopes of three major BDV proteins. Importantly, upon purification, these epitope-specific antibodies bound to BDV antigen with high avidity, in support of the premise that their presence in human sera reflects an actual human BDV infection. At any rate, most BDV researchers consider the jury to be out regarding whether or not BDV infects humans.

In 2000, BDV was isolated for the first time from a human brain, in this instance, from a schizophrenic patient at autopsy. A pathological examination of the brain revealed mild inflammatory changes in the hippocampus. Regardless, whether BDV is a cause of human mental disorders is still an open question.

BDV-infected Lewis rats are the preferred animal model of Borna disease. Experimental infection of normal, immunocompetent, adult rats with BDV leads to extensive inflammation in the brain and significant neuronal damage, which is thought to be mediated by cytotoxic T cells. Evidence in that regard is as follows. First, the appearance of T cells in the brain correlates with onset of disease symptoms. Second, experimentally depleting cytotoxic T lymphocytes prevents neuronal damage. Interestingly, these T cells do not eliminate the virus from the brain. The inflammation eventually does subside, but the cortex remains in a severely damaged state in surviving animals, which display a condition known as hydrocephalus ex vacuo, in which cerebrospinal fluid replaces lost brain tissue.

In the mouse model, susceptibility to BDV-induced neurological disease depends strongly on the genetic background of the animals, despite the facts that BDV replicates to similar levels in the brains of all mouse strains and, as in the rat model, an inflammatory reaction is seen in the brains of all mouse strains. How then might we explain the difference between the susceptibilities of the different mouse strains to BDV-induced neurological disease? Both susceptible and resistant mouse strains produce antibodies to BDV antigens, but only in susceptible animals does the cellular arm of the immune system recognize viral antigens in the brain. This appears to be because BDV replication is restricted to immunoprivileged sites in the brains of resistant mice, while in susceptible animals replication takes place in peripheral sites as well, where the cellular arm of the adaptive immune response might then be activated. This conjecture was supported by the experimental finding that peripheral immunization of persistently infected, healthy, resistant mice with a recombinant vaccinia virus expressing the BDV nucleoprotein results in neuronal degeneration.

The experimental findings in vivo described above are consistent with the fact that BDV is not cytopathic to cells in culture, in which it establishes persistent infections. In point of fact, there are no readily apparent differences between the viabilities of infected and uninfected cells in culture. Thus, as you might have predicted, laboratory infections of immunologically impaired rats do not show signs of encephalitis or other pathology. However, infection of immunotolerant newborn rats impairs development of their central nervous systems, implying that BDV can affect cellular function in the absence of immunopathology.

Also, severe neurobehavioral disorders have been observed in naturally and experimentally infected animals that did not show any signs of encephalitis or other immunopathology.

Bornavirus Genomes and Their Nuclear Transcription and Replication

The BDV genome is about 8.9 kb in length, and it is organized like the genomes of the other mononegaviruses (Fig. 13.5). It contains at least six ORFs. The first ORF at the 3′ end of the negative-sense genomic RNA encodes the nucleoprotein (N). Next, there is the phosphoprotein (P), then the matrix protein (M), followed by the envelope glycoprotein (G), and finally the RNA-dependent RNA polymerase (compare to the rhabdoviruses [Fig. 10.3] and the paramyxoviruses [Fig. 11.4]). The sixth ORF overlaps the P ORF and encodes a protein called p10, which also is known as X.

Bornavirus gene expression is largely similar to that of the other mononegaviruses. Yet all bornavirus RNA synthesis, including both mRNA transcription and genomic replication, occurs in the nucleus. In that regard, bornaviruses are the only family of the mononegavirus superfamily that replicates in the nucleus, and together with the orthomyxoviruses, they are the only nonretroid RNA viruses that are known to do so. The other mononegaviruses and all plus-strand RNA viruses other than the retroviruses replicate in the cytoplasm.

The reason why bornaviruses replicate in the nucleus is similar to the reason that orthomyxoviruses do so. As in the case of the orthomyxoviruses, some bornavirus mRNA species are spliced from precursor RNAs. Also like the orthomyxoviruses, bornaviruses depend on the cellular splicing machinery, which is located only in the nucleus, to mediate this splicing. Note that bornaviruses do not use a cap-snatching process, such as used by orthomyxoviruses, to cap their mRNAs. Instead, bornavirus mRNAs are capped by an activity associated with its virus-encoded polymerase.

The bornavirus transcripts contain two splice donor sites (SD1 and SD2) and three splice acceptor sites (SA1, SA2, and SA3) (Fig. 13.5). The splice donor and splice acceptor sites act together to generate mRNAs that may lack one or two of the three different introns. Transcripts that

Figure 13.5 Genomic organization and transcription map of BDV. N, nucleoprotein p38/p40N; X, protein X p10; P, phosphoprotein p24/p16; M, matrix protein gp18; G, envelope glycoprotein gp94; L, polymerase protein p190; S1 through S3, transcription initiation sites; T1 through T4 and t6, polyadenylation/termination sites; SA1 through SA3, splice acceptor sites; SD1 and SD2, splice donor sites; ESS, exon splicing suppressor. The positions of the sites are given for the antigenome in parentheses. Positions of the introns (I, II, and III) are also indicated. Adapted from K. Tomonaga, T. Kobayashi, and K. Ikuta, *Microbes Infect.* 4:491–500, 2002, with permission.

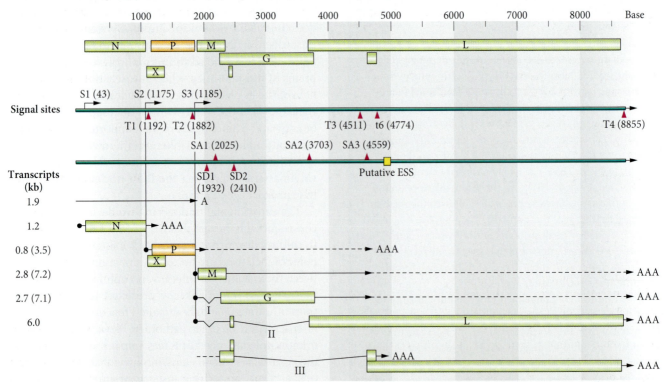

retain intron I are translated into the M protein. Those that retain intron II are translated into the G protein. Those that lack introns I and II are translated into the L protein. Transcripts that lack intron III are believed to be translated into novel, as-yet-unidentified proteins.

Since bornaviruses replicate in the nucleus, parental bornaviral vRNPs must be able to traffic to the nucleus after viral entry and uncoating. Moreover, progeny vRNPs must be able to traffic from the nucleus, where they are generated, to sites of virus assembly at later times in the replication cycle. As you might have expected, this nuclear transport activity is mediated by the nuclear localization and nuclear export signals present on viral proteins such as the N and P proteins (see the example of influenza virus [Chapter 12]). Indeed, the N and P proteins each contain nuclear localization signals, and the N protein also contains a nuclear export signal. (The X [p10] protein, which is the most recently identified BDV gene product, also localizes in the nuclei of infected cells and also contains a nuclear localization signal, albeit a nonconventional one. The role of p10 in the BDV replicative cycle is not yet known, but its nuclear localization would be consistent with a possible role in viral RNA synthesis, splicing, or the intracellular trafficking of vRNPs.)

What underlies the switch from nuclear import of vRNPs to their nuclear export? One possibility is that the P protein plays a role in controlling nuclear import versus export, since the P protein binds to a site on the N protein that overlaps its nuclear export signal. Thus, relative levels of P versus N may control nuclear retention versus export. In addition, note that there are two isoforms of the N protein. They differ in the location of their translation start sites, which are separated by a stretch of 13 amino acids. The shorter version lacks the nuclear localization signal present at the N terminus of the larger version. The shorter version might enter the nucleus in a complex with other viral proteins, incorporate into progeny vRNPs, and change the ratio of nuclear export signals to nuclear localization signals on those vRNPs. But note that these mechanisms remain speculative.

CALICIVIRUSES

The caliciviruses are the first of the three families of plus-strand RNA viruses discussed in this chapter. Their plus-strand RNA genomes are comprised of approximately 7,500 nucleotides and are contained within 27-nm, naked (nonenveloped), icosahedral capsids, which consist of 60 trimers of a single capsid protein. The family name is derived from the cup-shaped depressions (*calyx*, cup) on the surfaces of virus particles, as seen in electron micrographs (Fig. 13.6A). As expected, the plus-sense, single-stranded,

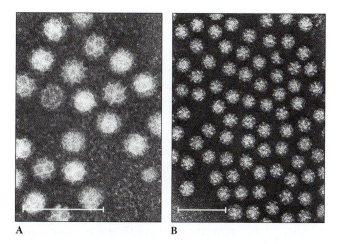

Figure 13.6 Negatively stained electron micrographs of virus particles of *Caliciviridae* (A) and *Astroviridae* (B). Bars, 100 nm. (Panel A micrograph is courtesy of Drs. M. Szymanski and P. J. Middleton; panel B micrograph is courtesy of Dr. D. Snodgrass.) From D. O. White and F. J. Fenner, *Medical Virology*, 4th ed. (Academic Press, San Diego, CA, 1994), with permission.

calicivirus genomic RNA is infectious. It is polyadenylated at its 3′ end, but it is not capped at its 5′ end. Instead, like the picornaviruses (Chapter 6), the 5′ terminus is bound to a VPg-like protein.

Norwalk Virus (Norovirus)

Norwalk virus, recently renamed norovirus, is the best-known calicivirus. Its name reflects its discovery in stool samples taken from adults during an outbreak of acute gastroenteritis in Norwalk, OH. Notably, about 40% of all outbreaks of gastroenteritis among adults in the United States are caused by Norwalk virus. The virus is cytocidal in cell culture, and it replicates extensively in intestinal epithelial cells. As a result, infection in vivo compromises the function of the intestinal brush border, thereby impairing the proper adsorption of water and nutrients across the epithelium. Consequently, after an incubation period of 24 to 48 h, infection leads to diarrhea, abdominal cramps, vomiting, nausea, myalgia, and mild fever, which usually resolve within a day or so.

Norwalk virus is transmitted primarily by the fecal-oral route. Most outbreaks in the United States have been traced to an infected food handler. Caliciviruses are inactivated by heating. Consequently, raw foods (e.g., salads) pose the greatest risk. Shellfish concentrate these viruses from contaminated water and thus pose a serious risk when inadequately cooked or served raw.

Outbreaks occur year-round in developed countries, often originating in schools, day care centers, resorts, hospitals, nursing homes, and restaurants. Over the past several years, there have been several highly publicized episodes in which Norwalk virus outbreaks affected hundreds of individuals during outbreaks on cruise ships, including the Queen Elizabeth II, and on the maiden voyage of Cunard's Queen Victoria. In November 2006, Norwalk virus struck 700 victims on Carnival's Liberty, one of the largest outbreaks ever, according to the CDC.

Calicivirus Replication

Plus-strand calicivirus genomes contain three major ORFs: ORF1, ORF2, and ORF3 (Fig. 13.7A). ORF1 encodes a polyprotein, which is proteolytically cleaved to produce all of the nonstructural proteins, including VPg (see below), the 3C-protease, and the 3D polymerase/replicase. ORF2 encodes the major capsid structural protein, and ORF3 encodes a minor structural protein. Importantly,

Figure 13.7 (A) Genomic organization and translation strategy of the feline calicivirus. The caliciviruses use genome-length mRNA molecules to translate the nonstructural proteins encoded by ORF1, while the structural proteins expressed by ORF2 and ORF3 are translated from a single abundant species of subgenomic RNA. Translation of ORF3 requires a termination/reinitiation event. Both RNAs are covalently linked to protein VPg at their 5′ termini. (B) The astrovirus coding strategy. ORFs 1a and 1b encode the nonstructural proteins. ORF2 encodes the capsid protein. Nucleotide positions indicate the overlap of ORF1b with ORF1a and ORF2. ORF1b is translated as a result of a −1 ribosomal frame shift, following translation of ORF1a, to yield an ORF1a/ORF1b fusion protein. Like the caliciviruses, astroviruses use a full-length mRNA to translate the nonstructural proteins and a subgenomic mRNA to encode the capsid protein. (C) Genomic organization of hepatitis E virus. The genome contains three overlapping reading frames: ORF1, ORF2, and ORF3. ORF1 encodes a polyprotein that is proteolytically processed to yield nonstructural replication proteins, including the RNA-dependent RNA polymerase and a helicase, as well as a protease for processing the precursor polyprotein. ORF2, which overlaps ORF1, encodes the capsid protein. This protein contains a signal sequence and three probable glycosylation sites. ORF3, which overlaps ORF1 and ORF2, encodes a small protein that associates with the liver cell cytoskeleton, where it is thought to facilitate virus assembly. Like the caliciviruses and astroviruses, the hepatitis E-like viruses use a full-length mRNA to translate the nonstructural proteins and a subgenomic mRNA to encode the capsid protein.

the RNA replicase not only synthesizes more full-length RNA molecules but also generates a single species of bicistronic, subgenomic mRNA (Fig. 13.7A).

The caliciviruses use genome-length mRNA molecules to encode their nonstructural proteins, while the structural proteins expressed by ORF2 and ORF3 are translated from the single abundant species of subgenomic RNA. In these respects, the caliciviruses somewhat resemble the plus-strand togaviruses (Chapter 8), which likewise use genome-length mRNA molecules to translate a polyprotein that is cleaved to yield the nonstructural proteins while using an abundant subgenomic mRNA to express the capsid protein from a second ORF. However, note that the calicivirus pattern of gene expression yet differs from that of the togaviruses, since the calicivirus subgenomic mRNA contains two ORFs while that of the togaviruses contains one ORF. Interestingly, the calicivirus ORF3, which is located at the 3′ end of the subgenomic RNA, overlaps ORF2 by four nucleotides in the case of the feline calicivirus and by eight nucleotides in the case of the rabbit calicivirus. Translation of ORF3 requires a termination/reinitiation process that is not yet well understood.

The caliciviruses also resemble the plus-strand picornaviruses (Chapter 6) in that the plus-strand RNA genomes of both families are not capped but instead are covalently linked at their 5′ ends to a viral protein called VPg. Recall that the picornavirus VPg protein plays a key role in genome replication, while translation on non-capped picornavirus mRNAs is initiated at internal ribosomal entry sites. In contrast to the picornaviruses, calicivirus genomes do not have an internal ribosomal entry site at which ribosomes might bind. Instead, the calicivirus VPg protein appears to function in translation initiation on calicivirus mRNAs through unique protein-protein interactions with the cellular translation machinery. As might then be expected, the calicivirus VPg protein is present on the subgenomic mRNA, as well as on the genomic RNA. Also, genomic RNA molecules are infectious if the VPg is replaced by a conventional cap, implying that VPg functions essentially as a cap.

At the structural level, caliciviruses, like picornaviruses, are naked, whereas the togaviruses are enveloped. Thus, caliciviruses have properties of both picornaviruses and togaviruses and might be considered an intermediate between those two families. Yet they have at least one defining feature, thus far unique among animal viruses: a VPg protein that may function in translation initiation.

ASTROVIRUSES

Astroviruses were once considered a genus within the calicivirus family. Like the caliciviruses, they contain plus-strand

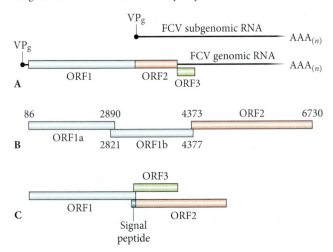

RNA genomes. But they feature a genome organization and pattern of gene expression that differ sufficiently from those of other nonenveloped, plus-strand RNA viruses to justify assignment to a separate virus family, the *Astroviridae*. Moreover, they present a unique star-shaped appearance in electron micrographs, thereby accounting for the family name (Fig. 13.6B).

Astroviruses were first identified by electron microscopy in stool samples of children with diarrhea. They are now well established as a cause of viral gastroenteritis. In fact, they are associated with 2 to 10% of all pediatric cases of diarrhea. However, they typically cause a less severe gastroenteritis than that caused by other enteric viruses, rotaviruses in particular.

Astrovirus genomes are about 6,800 nucleotides in length and are organized into three ORFs, referred to as ORF1a, ORF1b, and ORF2 (Fig. 13.7B). ORF1a and ORF1b are located at the 5′ end of the genome, where they encode nonstructural proteins. ORF2, located at the 3′ end of the genome, encodes the structural proteins. Whereas ORF1a encodes a variety of nonstructural proteins, including a protease, ORF1b encodes the viral RNA-dependent RNA polymerase. Actually, little is known about the role of most of the ORF1a products.

As in the case of other plus-strand viruses, upon infection, the nonstructural astroviral proteins are translated from the 5′ end of the genomic RNA. But how then are ORF1a and ORF1b translated from the same RNA molecule? Actually, ORF1b is translated as a result of a −1 ribosomal frameshift, which occurs near the end of ORF1a, in that way yielding an ORF1a/ORF1b fusion protein. The products of both ORF1a and ORF1b then mature by proteolysis.

In a stratagem reminiscent of the togaviruses (Chapter 8), the newly translated astroviral replication proteins then transcribe full-length minus-strand RNAs, which serve as the templates for the transcription of new genomic and subgenomic RNAs. The astrovirus subgenomic mRNAs, which contain the complete ORF2, generate large amounts of the structural proteins.

Interestingly, the N-terminal portion of the ORF1a polyprotein contains a functional nuclear localization signal that causes proteins expressed from ORF1a to accumulate in the nucleus. The reasons for this nuclear localization are unclear. In contrast, the products of ORF1b remain in the cytoplasm. Since nuclear involvement is rare among nonenveloped RNA viruses, it might be considered to be another distinguishing feature of astroviruses.

HEPATITIS E-LIKE VIRUSES

Hepatitis E-like viruses originally were classified as caliciviruses but were later found to be more similar to togaviruses than to caliciviruses. They are now known to be different enough from togaviruses to merit their own family.

Hepatitis E virus is the family prototype. It causes a disease similar to that caused by hepatitis A virus (Chapter 6). In that regard, hepatitis E virus infection gives rise to acute disease only. It does not lead to chronic hepatitis, cirrhosis, cancer, or a carrier state. Moreover, like hepatitis A virus, it is spread by the fecal-oral route. Hepatitis E virus is found worldwide, but it is most troublesome in developing countries. The mortality rate for hepatitis E is about 1 to 2%. However, infection is especially severe in pregnant women, in which the fatality rate is about 20% (Box 13.3).

Hepatitis E virus particles are small, nonenveloped icosahedrons that contain plus-strand RNA genomes 7.5 kb in length. These genomes contain three overlapping ORFs: ORF1, ORF2, and ORF3 (Fig. 13.7C). ORF1 encodes a polyprotein that is proteolytically processed to yield nonstructural replication proteins, including the RNA-dependent RNA polymerase and a helicase, as well as a protease for processing the precursor polyprotein. ORF2, which overlaps ORF1, encodes the capsid protein,

---

BOX 13.3

The events surrounding the original discovery of hepatitis E virus are bizarre enough to be worthy of mention. Mikhail S. Balayan, of the Russian Academy of Medical Sciences, Moscow, while investigating an outbreak of hepatitis in Tashkent, *volunteered himself to drink the pooled filtrate of stool samples from the patients*. His efforts were not for naught, since he indeed contracted hepatitis as he presumably desired. He then isolated a 32-nm virus particle from his own feces, which he inoculated into monkeys, in which it produced a hepatitis-like illness. By means of immunoelectron microscopy, using immune sera from the original patients, Balayan identified in the stools of these monkeys a virus that appeared to be identical to the virus in the original patient samples.

Balayan states in the original report, "Hepatitis E virus (HEV) was first identified in the excreta of an experimentally infected *human volunteer* and further confirmed by similar findings in clinical specimens from patients with acute jaundice disease different from hepatitis A and B." Were all of these steps really necessary to satisfy Koch's postulates?

Reference
Balayan, M. S. 1993. Hepatitis E virus infection in Europe: regional situation regarding laboratory diagnosis and epidemiology. *Clin. Diagn. Virol.* 1:1–9.

pORF2. This protein contains a signal sequence and three probable glycosylation sites. Thus, it is likely synthesized in the endoplasmic reticulum and may be transported to the plasma membrane via the secretory pathway. ORF3, which overlaps ORF1 and ORF2, encodes a small 123-amino-acid protein that associates with the liver cell cytoskeleton and forms complexes with the viral RNA and pORF2. It is thought to perhaps anchor the assembly of pORF2 and progeny viral genomes into virus particles at cytoskeletal sites, in that way facilitating virus assembly.

Since transfected full-length hepatitis E virus genomes are infectious, it is thought that ORF1 of the genomic RNA is translated immediately upon entry into cells to produce the enzymes responsible for viral RNA synthesis. Since the viral polymerase is required for the expression of ORF2 and ORF3, it is thought that a bicistronic subgenomic mRNA that encodes both the ORF2 and ORF3 proteins needs to be transcribed. Notice that the arrangement of hepatitis E virus genomes is distinct from that of the caliciviruses, which likewise use full-length mRNA molecules to encode the nonstructural proteins and a separate subgenomic mRNA to encode the capsid protein.

FILOVIRUSES

The filovirus family thus far contains only the Marburg and Ebola viruses. Like the rhabdoviruses, paramyxoviruses, and bornaviruses, filoviruses are members of the mononegavirus superfamily. That is, their minus-strand genomes are nonsegmented and organized like the genomes of other members of the superfamily.

Structure and Replication

Filovirus particles are filamentous and **pleomorphic**. That is, the filaments take on several shapes, including long filamentous rods, sometimes with branches. Other particles are U shaped, or S shaped, or circular (Fig. 13.8). The virus

Figure 13.8 *Filoviridae.* Negatively stained preparation of Ebola virus particles. Bar, 100 nm. (Courtesy of Dr. F. A. Murphy.) From D. O. White and F. J. Fenner, *Medical Virology*, 4th ed. (Academic Press, San Diego, CA, 1994), with permission.

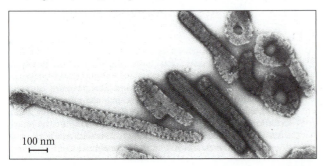

100 nm

filaments vary in length from 800 nm to as long as 1,400 nm. But they have a uniform diameter of 80 nm, and they are enveloped.

The filovirus genome encodes seven proteins. They are GP (the envelope glycoprotein), NP (the nucleoprotein), M (the matrix protein), L (the RNA-dependent RNA polymerase), P (probably a component of the polymerase), a second minor nucleoprotein, and a second matrix or membrane-associated protein. Each of these proteins is contained in virus particles. The filovirus genome is about 19 kb in length.

The filovirus replication strategy is not completely understood. However, the gene order and intergenic sequences are reminiscent of those seen in the other mononegaviruses. Thus, filovirus replication is thought to be similar to that of rhabdoviruses and paramyxoviruses.

Marburg and Ebola Viruses

Marburg and Ebola viruses each cause severe hemorrhagic fever in humans as well as in nonhuman primates. Onset of disease is sudden in each instance, with fatality rates ranging from 25% to as high as 90%, depending on the virus strain.

Marburg virus was first discovered in 1967, when laboratory workers at the Behring Works factory in Marburg, Germany, developed hemorrhagic fever after handling tissues from infected African green monkeys. The workers were engaged at the time in producing vaccines that were grown in kidney cells from the monkeys. For this purpose, the factory regularly imported monkeys from Uganda. In 1967, one shipment of several hundred monkeys contained two or three animals that were harboring the virus. However, the virus spread from those few infected monkeys to others, and then, it suddenly spread to humans. That initial outbreak included a total of 31 human cases, of which seven were fatal. Since then, Marburg virus has caused sporadic outbreaks of fatal hemorrhagic fevers in sub-Saharan Africa, most notably in Zimbabwe, Kenya, and South Africa.

Ebola virus, which was named after a river in Zaire (now the Democratic Republic of the Congo), was first recognized in 1976 after outbreaks in Zaire and Sudan. Altogether, those outbreaks resulted in about 550 cases, leading to 430 fatalities, a case-fatality rate of almost 80%! Afterwards, there were small to midsize sporadic outbreaks of Ebola virus in Africa (Table 13.2). However, the virus did not garner widespread attention until after the outbreaks in Zaire and Gabon in 1995 and 1996, respectively, and the release of the books and movie noted below.

Marburg virus and Ebola virus indeed cause the most severe viral hemorrhagic fevers yet known. Each is disseminated through the blood, and each replicates in many

Table 13.2 Ebola hemorrhagic fever (from the CDC web site); known cases and outbreaks are shown in chronological order

Year	Ebola subtype	Country	No. of human cases	% of deaths among cases	Situation
1976	Ebola-Zaire	Zaire (Democratic Republic of the Congo [DRC])	318	88	Occurred in Yambuku and surrounding area. Disease was spread by close personal contact and by use of contaminated needles and syringes in hospitals/clinics. This outbreak was the first recognition of the disease.
1976	Ebola-Sudan	Sudan	284	53	Occurred in Nzara, Maridi, and the surrounding area. Disease was spread mainly through close personal contact within hospitals. Many medical care personnel were infected.
1976	Ebola-Sudan	England	1	0	Laboratory infection by accidental stick with contaminated needle.
1977	Ebola-Zaire	Zaire	1	100	Noted retrospectively in the village of Tandala.
1979	Ebola-Sudan	Sudan	34	65	Occurred in Nzara. Recurrent outbreak at the same site as the 1976 Sudan epidemic.
1989	Ebola-Reston	USA	0	0	Ebola-Reston virus was introduced into quarantine facilities in Virginia, Texas, and Pennsylvania by monkeys imported from the Philippines. Four humans developed antibodies to Ebola-Reston virus but did not become ill.
1990	Ebola-Reston	USA	0	0	Ebola-Reston virus was introduced once again into quarantine facilities in Virginia and Texas by monkeys imported from the Philippines. Four humans developed antibodies but did not get sick.
1992	Ebola-Reston	Italy	0	0	Ebola-Reston virus was introduced into quarantine facilities in Sienna by monkeys imported from the same export facility in the Philippines that was involved in the episodes in the United States. No humans were infected.
1994	Ebola-Zaire	Gabon	49	59	Occurred in Mékouka and other gold-mining camps deep in the rain forest. Initially thought to be yellow fever; identified as Ebola hemorrhagic fever in 1995.
1994	Ebola-Ivory Coast	Ivory Coast	1	0	Scientist became ill after conducting an autopsy on a wild chimpanzee in the Tai Forest. The patient was treated in Switzerland.
1995	Ebola-Zaire	Democratic Republic of the Congo (formerly Zaire)	315	81	Occurred in Kikwit and surrounding area. Traced to index case-patient who worked in forest adjoining the city. Epidemic spread through families and hospitals.
1996	Ebola-Zaire	Gabon	31	68	Occurred in Mayibout area. A chimpanzee found dead in the forest was eaten by people hunting for food. Nineteen people who were involved in the butchery of the animal became ill; other cases occurred in family members.
1996	Ebola-Zaire	Gabon	60	75	Occurred in Booué area with transport of patients to Libreville. Index case-patient was a hunter who lived in a forest camp. Disease was spread by close contact with infected persons. A dead chimpanzee found in the forest at the time was determined to be infected.
1996	Ebola-Zaire	South Africa	2	50	A medical professional traveled from Gabon to Johannesburg, South Africa, after having treated Ebola virus-infected patients and thus having been exposed to the virus. He was hospitalized, and a nurse who took care of him became infected and died.
1996	Ebola-Reston	USA	0	0	Ebola-Reston virus was introduced into a quarantine facility in Texas by monkeys imported from the Philippines. No human infections were identified.
1996	Ebola-Reston	Philippines	0	0	Ebola-Reston virus was identified in a monkey export facility in the Philippines. No human infections were identified.
2000-2001	Ebola-Sudan	Uganda	425	53	Occurred in Gulu, Masindi, and Mbarara districts of Uganda. The three most important risks associated with Ebola virus infection were attending funerals of Ebola hemorrhagic fever case-patients, having contact with case-patients in one's family, and providing medical care to Ebola case-patients without using adequate personal protective measures.
2001-2002	Ebola-Zaire	Gabon and The Republic of the Congo	122	79	Outbreak occurred over the border of Gabon and the Republic of the Congo. Additional information is currently available on the WHO web site.

organs, including the liver, spleen, kidney, lungs, lymph nodes, and adrenal glands. Viral replication leads to the breakdown of endothelial cells, resulting in vascular injury. The ensuing hemorrhaging, which is widespread (especially in the gastrointestinal tract), leads to edema and shock.

Ebola virus was, and may still be, the most prominent emerging virus in the minds of many individuals. This is largely because it was featured in the popular press and the media, in particular in the movie *Outbreak*, based on the book of the same name by Robin Cook, and in the book *The Hot Zone* by Richard Preston. Is Ebola virus the most likely candidate for a real-life "Andromeda strain" (Box 13.4)? We will consider that issue shortly. But first, Richard Preston, in his book *The Hot Zone*, describes the progression of a lethal Ebola virus infection as follows.

> "Ebola . . . triggers a creeping, spotty necrosis that spreads through all the internal organs. The liver bulges up and turns yellow, begins to liquefy and then it cracks apart . . . The kidneys become jammed with blood clots and dead cells, and cease functioning. As the kidneys fail, the blood becomes toxic with urine. The spleen turns into a single large hard blood clot the size of a baseball. The intestines may fill completely with blood. The lining of the gut dies and sloughs off into the bowels and is defecated along with large amounts of blood."

As noted above, fatality rates for Ebola virus infections range from 25% to as high as 90%, depending on the particular virus strain. Death occurs within 1 to 3 weeks after the onset of infection. For comparison, note that the fatality rate for human immunodeficiency virus (HIV) is higher still. That virus kills well over 90% of untreated infected individuals, but it does so only after a period of 10 years on the average. While neither of these infamous viruses is easily transmitted from one human to another, Ebola is the more readily transmissible of the two. In contrast to Ebola virus, HIV is not spread by aerosols, and one does not need to wear a biological "space suit" while working with HIV-infected blood.

Four subtypes of Ebola virus have been identified, three of which are known to cause severe hemorrhagic disease in humans. These are Ebola-Zaire, Ebola-Sudan, and Ebola-Ivory Coast. Ninety percent of the Ebola-Zaire cases and 50% of the Ebola-Sudan cases resulted in death.

The fourth Ebola virus subtype, Ebola-Reston, caused an episode that initially was quite frightening. It happened as follows. Several shipments of cynomolgus monkeys, imported from the Philippines to the United States in 1989 and 1990, were alarmingly found to be infected with a filovirus that was serologically indistinguishable from the highly virulent Ebola-Zaire virus. The outbreak in the monkeys occurred while they were being held at the Reston Quarantine Facility in Reston, VA, just a short distance from Washington, DC. The monkeys were being held at that facility because imported monkeys must be held in quarantine for 1 month before they can be shipped anywhere else in the United States. When some of these monkeys became ill and died, there were fears that lethal infections might spread to humans. Fear was heightened when serological analysis could not distinguish between the Reston virus and the Ebola-Zaire virus. Yet no humans became ill, despite the fact that 14% of humans who had close contact with the monkeys later tested positive for filovirus antibodies. Moreover, subsequent screening of thousands of cynomolgus monkeys imported into the United States in 1989 from the Philippines and Indonesia demonstrated that about 12% had antibodies to the Ebola-Reston virus. Inexplicably, and contrary to initial expectations, Ebola-Reston virus has not caused disease in humans. In fact, no case of Ebola hemorrhagic fever in humans has ever been reported in the United States. Another enigmatic fact is that the Ebola-Reston virus, which was isolated from Asiatic monkeys, is virtually indistinguishable from the African Ebola-Zaire virus. Perhaps the Ebola-Zaire virus somehow spread to Asia at some recent earlier time. If so, then African Ebola virus may be spreading in the world.

The outcome of the Reston Ebola virus outbreak in monkeys must be considered hugely providential, since the potential existed for a crisis of horrific proportions. The Ebola-Reston virus appeared to spread among monkeys by the respiratory route. If that virus had been pathogenic to humans, consider the likely consequences of an outbreak

BOX 13.4

During the Spring 2008 semester, while discussing Ebola virus in the context of emerging viruses, I asked my Virology class whether Ebola virus might be the real-life "Andromeda strain." To my surprise, none of my students knew what I was referring to, an indication of how fleeting popular culture can be. So, *The Andromeda Strain* was a best-selling novel by Michael Crichton that was faithfully converted into a 1971 cinematic box office hit. The story tells of a military satellite that returns to Earth with a deadly alien virus that threatens to destroy all life on Earth. A *small* group of scientists is assembled at a secret (of course) underground laboratory in Nevada (where else?) to figure out how to prevent worldwide disaster. The movie was already technologically dated when it ran in 1971, and it hardly reflected how virologists actually work. Yet it was a hit, and the public later worried that Ebola virus might be the real Andromeda strain.

in the Capital region of the United States, which then would probably have spread throughout the densely populated East Coast, and likely further still.

The natural reservoirs of Ebola virus are not yet known. One thought is that bats that are native to the areas where Ebola virus is found may be carriers of the virus. This belief is also consistent with the fact that the virus can replicate in some bat species. How then might Ebola virus be transmitted from its natural zoonotic host to humans? The answer is not known. But once a human is infected, human-to-human transmission then occurs by close personal contact or by contact with an infected individual's body fluids.

Ebola virus does have a limited ability to spread via small-particle aerosols in the laboratory. Nevertheless, airborne spread among humans is not known to occur. These facts lead to the important conclusion that when close contact between infected persons and uninfected individuals is controlled, the spread of Ebola virus can be contained. Indeed, isolation of infected individuals and the wearing of protective clothing by health care workers are the best known means for controlling an outbreak. Unfortunately, in poor areas of the developing world, such procedures cannot always be practiced. For example, in financially impoverished African health care facilities, health care providers often attend to patients without the protection of a mask, gown, or gloves. Consequently, health care providers have been especially at risk of becoming infected themselves. The reuse of unsterilized syringes and other medical equipment in these facilities also is an important factor in the spread of disease.

The incubation period for Ebola hemorrhagic fever ranges from 2 to 21 days. The first signs of illness occur without warning. They include fever, headache, joint and muscle aches, sore throat, and weakness. These symptoms are followed by diarrhea, vomiting, stomach pain, and severe hemorrhaging, which leads to prostration and in many instances death.

It is not known why some people recover from Ebola hemorrhagic fever while others do not. It is possible that survivors mount a more effective immune response to the virus, since patients who die usually do not show evidence of an effective immune response. However, even survivors have no detectable neutralizing antibodies.

Is Ebola Virus the "Andromeda Strain?"

The popular press and media may have led one to conclude that Ebola virus is the real-life "Andromeda strain" (Box 13.4). However, contrary to what might be presumed, and perhaps paradoxically, the very hypervirulence of Ebola virus argues against its emergence as the Andromeda strain. First, since Ebola virus is so lethal, infected humans generally die before they can pass on the infection. Second, and related to the first argument, is the fact that Ebola virus is not easily transmitted from human to human. Third, diseased individuals are usually bleeding from the nose and gums, giving ample warning of their condition and probably not inspiring the intimate contact generally needed for passing on the infection. Fourth, Ebola virus gives rise only to acute infections in humans. There are no persistently infected human carriers. Together, these features account for the fact that quarantine measures are an effective strategy for containing Ebola virus outbreaks. (Would quarantine be an effective strategy against HIV and acquired immunodeficiency syndrome [AIDS]? Might HIV be the Andromeda strain? See Chapter 21.)

DNA VIRUSES: SINGLE STRANDED

Parvoviruses

INTRODUCTION

The *Parvoviridae* are the only viruses that are known to have *linear*, single-stranded, DNA genomes. In contrast, other single-stranded DNA viruses, such as the *Escherichia coli* phage φX174, have *circular*, single-stranded, DNA genomes. In any case, all single-stranded DNA viruses are class II viruses in the Baltimore classification scheme (Chapter 2).

Parvovirus genomes are quite small, containing only about 5,000 nucleotides, and their icosahedral capsids are small as well, only 18 to 26 nm in diameter. For comparison, polyomaviruses, which are the smallest double-stranded DNA viruses, have icosahedral capsids that are 45 nm in diameter. Yet, polyomavirus genomes are no longer than those of parvoviruses. Polyomavirus particles are larger because they package double-stranded genomes, which also are associated with cellular histones (Chapter 15). Regardless, parvoviruses are among the smallest known viruses. Reflecting that fact, the family derives its name from the Latin word *parvo*, meaning "small" (Box 14.1).

The parvovirus family is comprised of two subfamilies: the subfamily *Parvovirinae*, which infects vertebrates, and the subfamily *Densovirinae*, which infects insects. The subfamily *Parvovirinae* is further subdivided into five genera: the parvoviruses, dependoviruses, erythroviruses, amdoviruses, and betaparvoviruses.

Parvovirus B19, which is the only parvovirus known to be a pathogen in humans, is an erythrovirus. B19 is associated with a common, but usually mild, rash-like (the jargon terminology is *exanthematous* or *eruptive*) disease of children, referred to as fifth disease (see below). In adults, B19 is associated with a similar disease and with a more serious, potentially life-threatening disease called **aplastic crisis**, which is characterized by the loss of erythroblasts (red blood cell precursors) from the bone marrow. However, aplastic crisis is generally seen only in individuals manifesting any of several types of chronic hemolytic anemia (Box 14.2).

The life cycle of the dependoviruses is quite remarkable in that their replication is dependent on the presence, in the same cell, of a totally

BOX 14.1

Members of the family *Circoviridae* are smaller than even the parvoviruses. Circoviruses, which like parvoviruses also infect vertebrates, contain *circular*, single-stranded DNA genomes that comprise only about 2,000 nucleotides, and their nonenveloped, icosahedral capsids are only 17 nm in diameter. Given their small size, circoviruses are able to encode only two proteins: Rep and Capsid. Theoretically, this is the minimal number of proteins that might be required to replicate a virus.

Since circoviruses express only a replicase and a capsid protein, might circoviruses then be largely reliant on host cell proteins? Indeed, they are, but so are viruses that contain much larger genomes.

Despite the small size of circovirus genomes, they nevertheless display a noteworthy amount of complexity. For example, the coding regions of *rep* and *capsid* are opposite in polarity. That is, the circovirus single-stranded, circular genomes are **ambisense** (Chapter 13), and consequently, circovirus proteins are encoded both by the genomic viral DNA and by the complementary antigenomic strand of the replication intermediate that is synthesized in the host. Also, *rep* and *capsid* are separated by an intergenic region of approximately 80 nucleotides, which contains the *ori* and a pair of inverted repeat sequences.

unrelated **helper virus**, usually an adenovirus or a herpesvirus, but perhaps a papillomavirus as well. Since dependoviruses were first discovered in throat and fecal samples from patients who were infected with an adenovirus, members of the genus are sometimes referred to as **adeno-associated viruses** (AAVs). A distinctive feature, which is characteristic of the entire parvovirus family, concerns the mechanism by which their linear single-stranded DNA

BOX 14.2

What fundamental biological characteristic of B19 might its name reflect? Here is the answer. In 1975, while carrying out routine screening of blood donors in Australia, Yvonne Cossart found a parvovirus-like antigen and corresponding parvovirus-like particles in the sera of nine healthy blood donors and two patients. Because the original blood sample in which the parvovirus-like particles were detected was coded "B19," this became the name of the first parvovirus detected in humans.

genomes are replicated. More details about these points are given shortly.

A few words about terminology: from here on, the term "parvoviruses" refers to the entire family. Members of the subfamily *Parvovirinae* that replicate independently are referred to collectively as **autonomous parvoviruses**. The densoviruses are autonomous and are not considered further.

PARVOVIRUS STRUCTURE, BINDING, AND ENTRY

Parvovirus capsids contain 60 protein subunits. This detail is noteworthy because 60 is the least number of subunits required to construct an icosahedral capsid (in which there are 3 subunits for each of the 20 facets of the icosahedron [Chapter 2]). Consequently, parvoviruses are among the structurally simplest viruses that display icosahedral symmetry. On the other hand, rather than being composed of a single species of protein subunit, parvovirus capsids are composed of two to four different capsid proteins, referred to as VP1 to VP4, depending on the particular parvovirus species.

One of the parvovirus capsid proteins is usually predominant. For example, in the case of the AAVs, the capsid consists of three viral proteins, VP1 (90 kilodaltons [kDa]), VP2 (72 kDa), and VP3 (60 kDa), at a ratio of about 1:1:10, respectively, such that VP3 constitutes about 90% of the capsid. These proteins are related since they are encoded by in-frame, overlapping gene sequences (see Fig. 14.6). That is, the entire coding sequence for VP3 is contained in the coding sequence for VP2, whose coding sequence is, in turn, entirely contained within VP1. Since all three capsid proteins are translated in the same reading frame, they contain a common C-terminal domain of about 530 amino acids, which appears to be essential for cell binding. (For the sake of comparison, recall that picornavirus capsids are composed of equal numbers of four proteins, VP1, VP2, VP3, and VP4, which are generated by proteolytic cleavage of a precursor polyprotein [Chapter 6].)

The unique N terminus of the VP1 minor coat protein contains a lipolytic enzyme, phospholipase A$_2$ (PLA$_2$), which is sequestered within the capsid structure. The significance of this enzymatic activity of VP1 and its sequestering within the capsid structure will be apparent shortly.

The core of the C-terminal domain, which is common to each of the parvovirus capsid proteins, is composed of an eight-stranded, antiparallel β-barrel motif (Chapter 6). Much of the surface structure of parvovirus particles consists of large loops inserted between the β strands of the β-barrel. These loops take the form of (i) projections at, or surrounding, the icosahedral threefold axes and

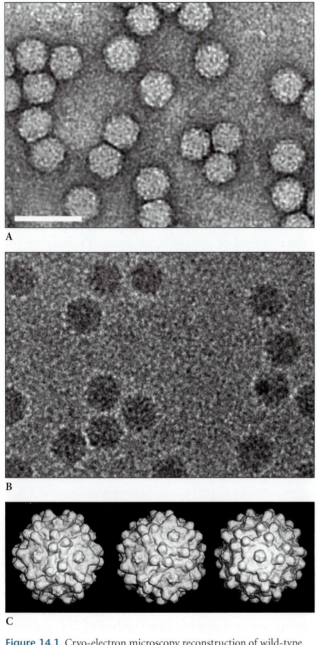

Figure 14.1 Cryo-electron microscopy reconstruction of wild-type AAV4. Negatively stained (A) and frozen (B) micrographs of wild-type AAV4 particles are shown. Bar, 50 nm. (C) Surface-shaded cryo-electron microscopy reconstructed AAV4 images viewed down the icosahedral twofold (left), threefold (center), and fivefold (right) axes. Notice the projections at, or surrounding, the icosahedral threefold axes and the depressions at the twofold and around the fivefold axes. From E. Padron et al., *J. Virol.* **79**:5047–5058, 2005, with permission.

(ii) depressions at the twofold and around the fivefold axes (Fig. 14.1).

Several of the AAVs bind to cell surface heparin sulfate **proteoglycans** (proteoglycans are proteins that are linked to a type of polysaccharide called a glycosaminoglycan), using those molecules as a low-affinity **primary receptor**.

Some parvoviruses then use integrins as their **secondary receptor** (explained immediately below), while others use the fibroblast growth factor receptor for that purpose.

So, what is meant by primary and secondary receptors? In brief, many viruses enter cells after binding to a single receptor. In the case of other viruses, including parvoviruses, the situation is more complex. In the case of these other viruses, the particle attaches to its primary receptor, but entry is triggered by subsequent interaction of the virus with a secondary or **coreceptor**. This type of virus-receptor interaction is considered in detail in the chapters on the adenoviruses and human immunodeficiency virus (HIV). In those two instances, the interaction of the virus with the secondary receptor is known to trigger conformational rearrangements in the virus particle that facilitate entry. It is not known how the interaction of parvoviruses with their secondary receptors promotes parvovirus entry.

After parvoviruses bind, they are taken into the cell by receptor-mediated endocytosis. Next, since parvoviruses replicate in the nucleus, they somehow must escape from the endosomal compartment in order to traffic to the nucleus. In that regard, all viruses must cross at least one cellular membrane to complete their entry process. However, parvoviruses, like all viruses that replicate in the nucleus, need to cross two cellular membranes. In the case of the parvoviruses, the first membrane they need to cross is that of the endosomal/lysosomal compartment, and the second is the nuclear membrane.

As seen in earlier chapters, enveloped viruses may enter the cytoplasm via fusion of their envelopes either with the plasma membrane or with an intercellular endosomal or lysosomal membrane. In contrast, nonenveloped viruses must use some other strategy to enter the cytosol, and in general, these alternative strategies are poorly understood. So, by what means might parvoviruses carry out membrane penetration?

Recent experimental findings imply that parvoviruses may use a previously unrecognized enzymatic mechanism to penetrate and cross endosomal membranes. Specifically, parvoviruses appear to escape from the endosome by means of a lipolytic enzyme, PLA_2, which is expressed at the N terminus of the parvovirus VP1 minor coat protein. As noted above, this enzymatic region of VP1 is normally sequestered within the capsid structure. However, it is exposed in the endosomal/lysosomal compartment.

Exposure of PLA_2 might be triggered by the low pH of the late endosomal/lysosomal compartment, but that is not known with certainty. Regardless, the importance of the VP1 PLA_2 activity is demonstrated by the experimental finding that a single amino acid substitution in the active site of the VP1 PLA_2, which inactivates its enzymatic activity, prevents infectivity. (For the sake of comparison, membrane penetration of the nonenveloped reoviruses [Chapter 5] is mediated by the reovirus μ1 protein, which has an endosome-disrupting activity that depends on its hydrophobic N-terminal myristoyl group.)

Incoming parvovirus particles undergo several other structural rearrangements of note, which also are presumably triggered by the low pH of the endosomal/lysosomal compartment. For instance, the N terminus of VP1 contains three nuclear localization signals, which are exposed concomitant with exposure of the PLA2 catalytic domain. Moreover, two of these nuclear localization signals also are present at the N terminus of VP2, and they too are exposed in the endosomal compartment. In addition, the viral genome may be at least partially exposed.

The molecular organization of the parvovirus particles that escape from the endosome is not yet clear. In any case, the parvovirus genome must be imported into the nucleus. In this regard, bear in mind that (i) nuclear import requires that the entering cargo expresses a nuclear localization signal and (ii) nuclear localization signals are expressed on proteins. (The nuclear import pathway is illustrated schematically in Box 12.7.) Thus, viral genomes are imported into the nucleus through nuclear pore complexes in association with viral proteins that express the requisite nuclear localization signals. As then might be predicted, results from genetic experiments show that the parvovirus VP1 and VP2 nuclear localization signals are required for the initiation of infection. (During replication per se, these nuclear localization signals serve to promote the import of the newly synthesized VP1 and VP2 molecules into the nucleus, where virus assembly takes place.)

Viral nuclear entry is discussed in greater detail in other chapters, in particular, Chapters 12, 15, and 17. In some instances (e.g., the orthomyxoviruses [Chapter 12]), the virus must undergo extensive disassembly before interacting with the nuclear pore complex, whereas in other instances most disassembly may occur at the nuclear pore complex (e.g., the adenoviruses [Chapter 17]).

Apropos the above, to the best of my knowledge, there is not yet any definitive evidence that intact capsids of any virus can enter the nucleus to initiate infection. However, there is experimental evidence in support of the possibility that intact parvovirus particles indeed might do so. First, note that the largest macromolecular complexes known to be imported into the nucleus via nuclear pore complexes are 39 nm in diameter. Since parvovirus icosahedral capsids are 18 to 26 nm in diameter, based on size constraints alone, they might well be able to cross the nuclear pore while still more or less intact. Experimental evidence in support of that possibility includes the finding that microinjecting several capsid-specific antibodies into the cell nucleus blocks AAV2 infection completely (Box 14.3).

BOX 14.3

Regarding the possibility that intact parvovirus particles might pass through nuclear pore complexes to initiate infection, consider the following. We previously underscored the seeming paradox that while viruses assemble stable particles at the end of one infectious cycle, these stable particles nevertheless must disassemble when triggered to do so, at the start of the next infectious cycle. The explanation offered for this incongruity is that disassembly during early stages of infection generally occurs in cellular compartments with conditions that are different from those in the compartments in which assembly and maturation occur at later times. Therefore, if parvovirus disassembly were to occur in the nucleus, it would raise the intriguing question of how both disassembly and assembly of parvovirus capsids could each occur in the nucleus. Perhaps the explanation involves the triggered partial disassembly or rearrangement of parvovirus particles, as evidenced by the surfacing of the VP1 and VP2 nuclear localization signals, which occurs before the particles enter the nucleus. These structural transitions might leave the virus particles primed to undergo further disassembly and genome release in the nucleus. Alternatively (or additionally), parvovirus disassembly and assembly might occur within different subcompartments of the nucleus. Still, at present, it is not clear exactly when and how the DNA is released from parvovirus particles.

PARVOVIRUS GENOMES

All parvoviruses have linear, single-stranded DNA genomes that contain about 5,000 nucleotides. Autonomous parvoviruses package genomes that are of the minus sense only. In contrast, dependoviruses package equal numbers of plus-strand and minus-strand genomes. The reason for this difference between autonomous parvoviruses and dependoviruses is explained later. At any rate, dependoviruses are infectious regardless of whether they contain plus-strand or minus-strand genomes. (In either case, the proviso remains that productive dependovirus infection requires coinfection with a helper virus.)

Importantly, all parvovirus genomes have **palindromic** sequences at both their 5′ and 3′ ends (Fig. 14.2). (Palindromes are sequences that in the double-stranded form appear unchanged when rotated 180°. For example, GATGCATC is a palindrome.) Note that a palindromic, single-stranded sequence can fold back on itself to form a hydrogen-bonded duplex. Early studies (reviewed below) demonstrated that parvovirus genomes indeed terminate

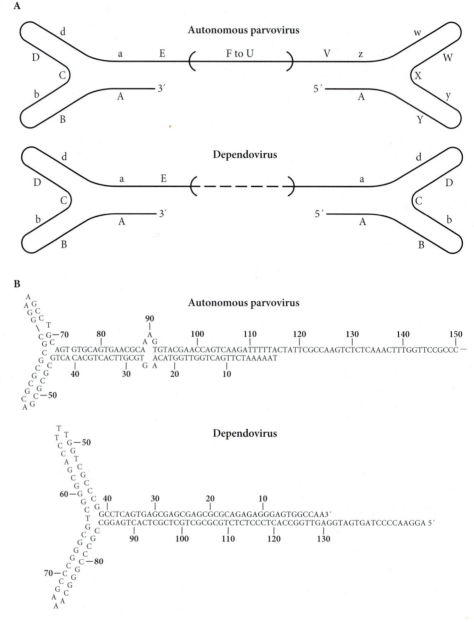

Figure 14.2 (A) Schematic depiction of the genomic structures of autonomous parvoviruses and dependoviruses. Complementary sequences are indicated by capital and lowercase letters. Notice that while the end terminal sequences of the autonomous parvoviruses are unrelated to each other, those of the dependoviruses are present as inverted complementary repeats. Also, the palindromic sequences at each end of dependovirus genomes appear in each orientation in equal frequency. In the case of the autonomous parvoviruses, while the 5′ palindromic sequence occurs in either orientation, the orientation of the 3′ palindromic sequence is fixed. (B) The actual nucleotide sequence at the 3′ end of the Kilham rat virus, an autonomous parvovirus, and the AAV, a dependovirus. Notice the bilobed structure in each instance. Adapted from J. A. Levy, H. Fraenkel-Conrat, and R. A. Owens, *Virology*, 3rd ed. (Prentice Hall, Upper Saddle River, NJ), 1994.

in hydrogen-bonded duplexes that were initially interpreted as "hairpins" (Box 14.4).

The genomes of a number of autonomous parvoviruses and dependoviruses have by now been completely sequenced. This sequence analysis shows that the hydrogen-bonded structures at the ends of some parvovirus genomes are more complex than simple hairpins, instead being comprised of two internal palindromes that are flanked by sequences that together constitute a palindrome (Fig. 14.2). This more elaborate combination of palindromic sequences forms bilobed structures rather than simple hairpins. As is discussed below, these hydrogen-bonded

sequences play a crucial role in the replication of parvovirus genomes.

Although the genomic ends of both autonomous parvoviruses and dependoviruses end in palindromic sequences, the genomic ends of these parvovirus groups differ as follows. First, there is a difference in the relationship of the two ends to each other. The palindromic sequences at each end of autonomous parvovirus genomes are completely unrelated to each other (Fig. 14.2). In contrast, the palindromic sequences at the ends of dependovirus genomes are present as inverted, complementary sequences. Second, the ends of the dependovirus and autonomous

BOX 14.4

It is an important general truth that the genomes of animal DNA viruses are not simple linear DNA molecules. In the current instance, single-stranded parvovirus genomes terminate in palindromic sequences that fold back to form hydrogen-bonded duplexes. Circoviruses (Box 14.1) contain covalently closed, circular, single-stranded DNA genomes. All the other known animal DNA viruses contain double-stranded genomes. The genomes of the polyomaviruses (Chapter 15) and papillomaviruses (Chapter 16) are covalently closed circles. Adenovirus genomes (Chapter 17) are linear, but end in covalently bound terminal proteins. Poxvirus genomes (Chapter 19) also are linear, but at each end, the two strands are covalently connected, such that these genomes are continuous polynucleotide chains. Herpesvirus genomes (Chapter 18) contain extensive terminal and internal duplications, and although they are linear in virus particles, they circularize upon infection. Hepadnavirus genomes (Chapter 22) are circular, although not covalently closed in the virus particle.

Note that cellular chromosomal DNA ends in telomeres, which enable the cell to regenerate the ends of its DNA, which would otherwise be foreshortened at each replication by the requirement for priming. All known viral DNA genomes, in contrast to genomic cellular DNA, do not contain telomeres. As will be seen, the various genomic arrangements of the DNA viruses likely exist to solve a common problem, i.e., generating the 5′ ends of progeny DNA molecules.

parvovirus genomes differ concerning their orientation. The palindromic sequences at each end of dependovirus genomes appear in each orientation, in equal frequency. In the case of the autonomous parvoviruses, whereas the 5′ palindromic sequence occurs in either orientation, the orientation of the 3′ palindromic sequence is fixed. The significance of these facts is discussed shortly.

The organization and expression of parvovirus genomes are discussed in detail below. For now, note the following. Approximately one-half of each parvovirus genome encodes replication proteins. These are designated NS1 and NS2 (NS for nonstructural) in the case of the autonomous parvovirus MVM (minute virus of mice) and Rep in the case of AAV2 (see Fig. 14.6). In AAV2, transcriptional promoters P5 and P19 initiate transcription at different points of the *rep* open reading frame, to generate messenger RNAs (mRNAs) encoding Rep proteins Rep78 and Rep68.

Moreover, an intron at the 3′ end of this open reading frame allows expression of short versions of the two full-length Rep proteins. These shorter Rep proteins are known as Rep52 and Rep40. Rep78/68 acts primarily to initiate replication, whereas Rep52/40 serves primarily to encapsidate progeny viral genomes. NS1 serves as the sole initiator of MVM replication and is functionally equivalent to Rep78/68. (In some examples that follow we refer to Rep78/68, and in others to NS1. Regardless, these proteins are functionally comparable.) NS2 proteins actually are a series of alternatively spliced, multifunctional molecules. Each of the parvoviral initiator proteins expresses a site-specific, single-stranded, nuclease activity and a **helicase** activity. (DNA helicases are enzymes that catalyze the unwinding of DNA by disrupting the hydrogen bonding between paired bases as they progress along the DNA strand in a polar fashion.) The remaining one-half of the parvovirus genome encodes the capsid proteins.

REPLICATION OF PARVOVIRUS SINGLE-STRANDED DNA

In most chapters, the pattern of viral genome organization and transcription is discussed in detail before considering the mechanism by which the viral genome is replicated. The reasons for doing so are as follows. First, the strategy of viral gene expression is the key feature that underlies the classification of a virus family in the Baltimore classification scheme (Chapter 2), which is the principal basis for the organization of this book. Second, most viruses need to express at least a portion of their genomes before they can replicate their genomes. Regarding this second point, parvoviruses manifest the following distinctive feature. Replication of parvovirus genomes depends on the generation of double-stranded DNA replicative intermediates, which have the dual function of also serving as templates for transcription.

Considering that parvovirus genomes are comprised of single-stranded DNA molecules, do these viruses need to encode their own DNA polymerase? Well, it may be somewhat of a surprise to some readers that cellular enzymes (aided by one or two virally encoded proteins) are responsible for replicating parvovirus single-stranded DNA genomes. On the other hand, bear in mind that DNA replication is a complex process, involving a number of different activities. Since parvovirus genomes are far too small to encode all of the activities needed to carry out the replication of their genomes, we can appreciate why parvoviruses depend upon cellular enzymes to mediate their replication. Moreover, knowing that parvovirus replication is mediated by the cellular DNA replication machinery enables us to appreciate why parvovirus replication occurs in the nucleus. Another reason parvoviruses replicate in the nucleus is that after the parvovirus genome is converted to a double-stranded replicative intermediate, its transcription is carried out by cellular RNA polymerase II (pol II). All pol II transcription occurs in the nucleus.

Like the single-stranded DNA-containing parvoviruses, double-stranded DNA viruses of the polyomavirus, papillomavirus, and adenovirus families also use cellular DNA replication enzymes to mediate their replication. Consequently, all of these viruses have to deal with the fact that the cellular enzymes for DNA metabolism are efficiently expressed only in cells that are in the S phase of the cell cycle. The polyomaviruses, papillomaviruses, and adenoviruses deal with this impediment to their replication by expressing functions that induce "resting" cells to enter into an S phase. In contrast, the parvoviruses do not encode any such mitogenic function. Consequently, parvovirus replication is "on hold" until the host cell enters S phase of its own accord. (Notice that it might not be useful for parvoviruses to encode a mitogenic function, since parvovirus genomes cannot be expressed before being converted to double-stranded replicative intermediates, a process that itself depends on the cell being in S phase.) The dependoviruses are even more fastidious, since they also depend on the presence of a helper adenovirus or herpesvirus to create an intracellular milieu that is conducive to their replication.

The next problem that parvoviruses face when replicating their genomes is one that all linear DNA molecules must solve. That is, how to generate their 5' ends. This problem arises because of the following facts. First, DNA polymerases can add a deoxynucleoside triphosphate to the 3' end of a DNA strand only if (i) that 3' deoxynucleoside triphosphate is hydrogen-bonded to its complementary strand and (ii) there is an exposed, unpaired template. For these reasons, DNA synthesis must be primed, usually by means of short RNA primers. (The reason that RNA priming can fill this role is that RNA synthesis does not need to be primed.) Second, DNA polymerases add new bases only in the 5'-to-3' direction. Because of these facts, both viral and cellular genomes must come up with mechanisms for generating the extreme 5' end of each DNA strand (Box 14.4).

Why might this complication to DNA replication exist? A good guess is that it exists to provide a proofreading step, which ensures that correct nucleotides are incorporated with very high fidelity. If incorporation of correct bases were dependent only on base pairing, then incorrect bases would be incorporated with a frequency of about 1 in 10^4. (Notice that this is about the frequency at which RNA polymerases incorporate an incorrect base.) Actually, the fidelity of DNA replication is much greater, with

an error frequency of 1 incorrect base per 10^9 to 10^{10} bases! This is partly because the DNA polymerase complex contains an exonuclease activity that recognizes mismatches in the helix and excises mismatched bases. To accomplish this proofreading function, DNA polymerases must add bases to a growing 3' end only when it is hydrogen-bonded in a helix. The bottom line is that DNA polymerases require both a 3' OH terminus and an unpaired template in order to add a nucleotide.

The cell solves the problem of generating the 5' ends of its DNA by means of special short repeated sequences at the ends of chromosomes called telomeres, which are replicated by means of a special enzyme called telomerase. This enzyme is able to synthesize the DNA ends by using an RNA template that is associated with the enzyme. The parvoviruses solve the end generation problem by means of the unique configurations at the ends of their single-stranded DNA genomes, which actually enables them to self-prime their replication. (No virus is known to use telomerase to solve its "ends" problem. Why do you suppose that no virus exploits this cellular machinery?)

A General Model: the "Rolling Hairpin"

Peter Tattersall and David Ward (1976) discovered the unique nature of the ends of parvovirus genomes and provided the following model for parvovirus genomic replication, which is based on their experimental observations. Most importantly, Tattersall and Ward reported that the 5' and 3' ends of the MVM genome are hydrogen-bonded, hairpin structures, represented schematically by the sequences ABa and EFe, respectively (Fig. 14.3). Tattersall and Ward deduced that the ends of the MVM genome are hairpins, based on the experimental finding that those ends could be isolated as hydrogen-bonded fragments after the genomes were treated with the single-strand-specific S1 nuclease. As noted above, later sequencing of parvoviral genomes revealed that the ends of some parvovirus genomes assume a more complicated bilobed configuration (Fig. 14.2). In contrast to the ends of the genome, the remaining 95% of the genome is S1 sensitive and therefore single stranded (represented by D in Fig. 14.3).

Tattersall and Ward proposed that the free 3' end of a hairpin could, in principle, serve as a primer for parvoviral DNA synthesis. Indeed, the first step in the replication model is for the 3' end of the molecule to prime the synthesis of polynucleotide sequence d, which is complementary to single-stranded sequence D (Fig. 14.3, step i). This step is referred to as "gap filling" because it fills in the single-stranded gap. It is analogous to the synthesis of the "leading strand" during the replication of double-stranded DNA.

Replication continues with the synthesis of the sequence complementary to aBA (Fig. 14.3, steps ii and iii).

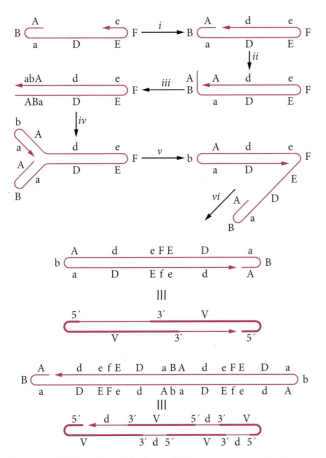

Figure 14.3 The rolling-hairpin model for parvovirus replicative DNA synthesis. Step *i*, gap-fill synthesis; step *ii*, displacement; step *iii*, gap-fill synthesis to complete genomic hairpin; step *iv*, enzymatic rearrangement of terminal palindrome to give rabbit-eared structure; step *v*, displacement synthesis; and step *vi*, gap-fill synthesis. The molecule is now essentially the same structure as the first intermediate between steps *i* and *ii*, and repetition of steps *ii* and *iii* generates the dimer genomic hairpin. Repetition of steps *iv*, *v*, and *vi* generates the tetramer, and so on for larger multimers. The progeny DNA strands, V, are indicated by heavy lines. Modified from P. Tattersall and D. Ward, *Nature* **263**:106–109, 1976, with permission.

This step is referred to as "displacement synthesis" because it displaces sequence aBA from the 5′ terminal hairpin.

A central feature of the model is that the stretched-out palindrome sequences (i.e., abA and ABa) re-form the hairpins to generate the "rabbit-eared" structure depicted in Fig. 14.3 (step *iv*). This rearrangement is referred to as "hairpin transfer." Importantly, it creates a new base-paired 3′ end, which can serve as the primer for continued synthesis of additional linear sequences. Also, notice that hairpin transfer repositions the base-paired 3′ end in what is effectively a strand-switching maneuver. Also, in the case of MVM, the helicase activity of NS1 is needed to "melt" and reconfigure the duplexes into the rabbit-eared structures (more on this point below). At any rate, a new round of displacement synthesis can follow, to yield a duplex

that contains the parental genome plus a complete progeny genome (Fig. 14.3, steps *v* and *vi*). The duplex structure at the end of step *vi* is capable of further replication via displacement synthesis, hairpin transfer, and gap filling, to generate the tetramer shown at the bottom of Fig. 14.3.

This replication model is supported by several early experimental observations. First, the gap-filling step (Fig. 14.3, step *i*) can be catalyzed in vitro by several DNA polymerases. Second, there is no evidence for RNA priming when the virus is replicating in vivo, nor are there Okazaki fragments or any other evidence for discontinuous lagging-strand synthesis. Third, some of the duplex parvoviral genomes formed in infected cells are **concatemers**, containing as many as 10 genome equivalents. Fourth, after being denatured (i.e., heated to melting temperature) and then annealed (i.e., slowly cooled), many of these genomes generate hydrogen-bonded duplexes with one-hit kinetics. That is, they anneal monomolecularly, implying that the complementary strands are covalently linked (Box 14.4).

Before moving on, we need to elaborate on a detail of the rolling-hairpin model depicted in Fig. 14.3. Following the first gap-filling step, in which cellular enzymes convert the input parvoviral DNA into a duplex intermediate (Fig. 14.3, step *i*), the growing DNA strand is not initially able to displace the hairpin against which it abuts (ABa in the figure), so that displacement synthesis cannot yet occur. That is so because, first, the gap-filling step necessarily precedes expression of NS1, which is encoded within d/D, and second, displacement requires the NS1 helicase activity. Consequently, the 3′ end of the new DNA sequence is instead ligated to the 5′ end of the hairpin, creating a covalently closed duplex molecule. That duplex structure then serves as a template for transcription of NS1 mRNA. When NS1 then becomes available, its nicking activity introduces a nick into the replication origin, which is comprised of sequences in the hairpin (ABa) and the new DNA segment that is ligated to it. This nicking reaction creates a new base-paired 3′ end that can prime DNA synthesis, and since the NS1 helicase activity is now in place, the hairpin can be displaced and copied. Notice that the NS1-mediated nicking step is analogous to the nicking step in rolling-circle replication, which creates a 3′ end that functions as the primer for continuous unidirectional synthesis about the circular template (Box 14.5).

The rolling-hairpin model accounts for the replicative intermediates that are found in infected cells. Yet we have not yet accounted for the structures of the genomes found in virus particles. Importantly, we have not explained why dependoviruses package equal numbers of plus-strand and minus-strand genomes whereas some autonomous parvoviruses package only minus-strand genomes. This issue is explained in the next section. But first, we must

BOX 14.5

The model for parvovirus DNA replication described in the text has been dubbed "rolling-hairpin" synthesis, by analogy with the **rolling-circle** mechanism that replicates circular genomes. *E. coli* phage φX174 provides an example of rolling-circle replication of a single-stranded circular genome.

In both the rolling-hairpin and rolling-circle processes, replication proceeds via duplex replicative intermediates. (A) In the case of φX174, generation of the replicative intermediate requires RNA-primed DNA synthesis. After the circular φX174 replicative intermediates have been formed, a 3′ end is created by a break in one of the strands. That new 3′ end can then function as the primer for continuous, unidirectional synthesis about the circular template. As the 3′ end is extended, the strand containing the 5′ end is displaced, resulting in the generation of single-stranded, linear DNA. Genome-length pieces then are cut off from the linear DNA product and circularized. In contrast to the rolling circle, in the rolling hairpin, a unidirectional replication fork shuttles back and forth along the linear template, changing direction every time a new terminal palindrome is synthesized and rearranged.

(B) The rolling-circle process is also able to generate double-stranded genomes. Bacteriophage λ (lambda) and herpes simplex viruses are examples of prokaryotic and eukaryotic viruses, respectively, that replicate in this way. In each of these instances, the first step is for their linear double-stranded DNA genomes to circularize. Then, in each instance there is an initial phase of theta (θ) replication (described in detail in Chapter 15), which is followed by a rolling-circle mode of replication. Double-stranded progeny genomes are generated by the rolling-circle mechanism as follows.

As in the case of φX174, one of the daughter strands is synthesized continuously around the circle. Synthesis on this strand is analogous to leading-strand synthesis. However, in instances where a double-stranded genome is being replicated, the displaced strand also serves as a template. Bearing in mind that the polymerase complex can incorporate nucleotides only in the 5′-to-3′ direction and requires a free 3′ OH end, synthesis on the displaced strand is discontinuous and is analogous to lagging-strand synthesis. The process generates a concatemeric double-stranded tail, from which unit-length genomes are excised. Modified from J. A. Levy, H. Fraenkel-Conrat, and R. A. Owens, *Virology*, 3rd ed. (Prentice Hall, Upper Saddle River, NJ), 1994, with permission.

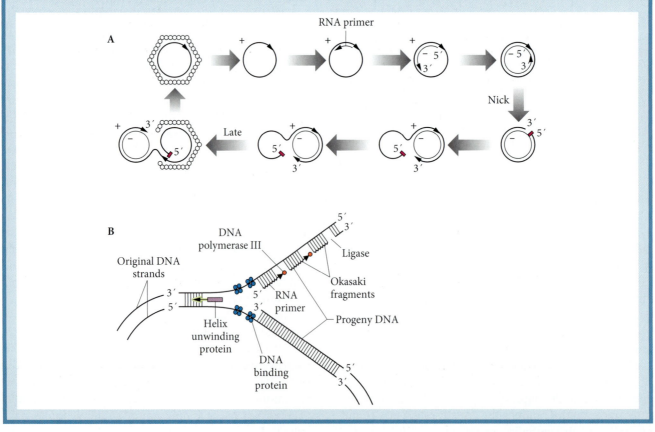

account for the following two features of the parvovirus genomes.

Progeny dependovirus genomes contain terminal palindromes in which the original sequences and their inverted complements are present in equal numbers at each end. A process called "terminal resolution" is believed to be responsible for generating equal numbers of termini in both orientations in vivo. In this process, the early dependovirus proteins Rep68 and Rep78 bind specifically to the (GAGC)$_3$ repeated sequence in the stem region of the terminal palindrome and introduce a nick 20 bases away at a site called the terminal resolution site or *trs* (Fig. 14.4). This creates a base-paired 3' hydroxy end that can prime a displacement synthesis that generates a new terminal palindrome in the original orientation, while simultaneously inverting the original palindrome. Since this process inverts the original terminal sequence every time a new one is generated, progeny genomes contain equal numbers of termini in each orientation. This process has been demonstrated to occur in an in vitro model system.

The terminal palindrome at one end of a dependovirus genome is the inverted complement of the sequence at the other end. In contrast, the palindromes at the ends of the autonomous parvovirus genomes are distinct, unrelated sequences. In the case of the dependoviruses, inversion of the terminal sequences occurs at each end of the genome by terminal resolution as described in the preceding paragraph. In contrast, inversion occurs only at the 5' ends of autonomous parvovirus genomes, by a process that may be similar to terminal resolution. A somewhat more complicated process termed "junction resolution" is believed to generate 3' termini that remain in a single orientation. This process is described in detail by Cotmore and Tattersall (2006).

ENCAPSIDATION OF PROGENY DNA

We continue deviating from our customary order of describing the events of a viral replication cycle by describing parvovirus encapsidation before considering parvovirus gene expression. We do so because in the case of the parvoviruses, encapsidation is closely tied to the process of DNA replication just described. We begin by noting that assembly of AAV particles occurs in two separate phases.

Figure 14.4 Terminal resolution reaction for AAV. *D* and *d* denote sequences that are contained within the terminal repeat, but not within the hairpin. The image of the Rep protein indicates that the active form is believed to be a dimer. In the expanded box, the sequence protected from DNase I digestion by Rep in an isolated linear origin is indicated by underlining. The terminal resolution site is denoted *trs*, and the vertical arrow indicates the position of the Rep-induced nicking and 5'-attachment site. Compare to the steps depicted in Fig. 14.3. Adapted from S. F. Cotmore and P. Tattersall, Parvoviruses, chapter 29, p. 593–608, *in* M. L. DePamphilis (ed.), *DNA Replication and Human Disease* (Cold Spring Harbor Press, Cold Spring Harbor, NY), 2006, with permission.

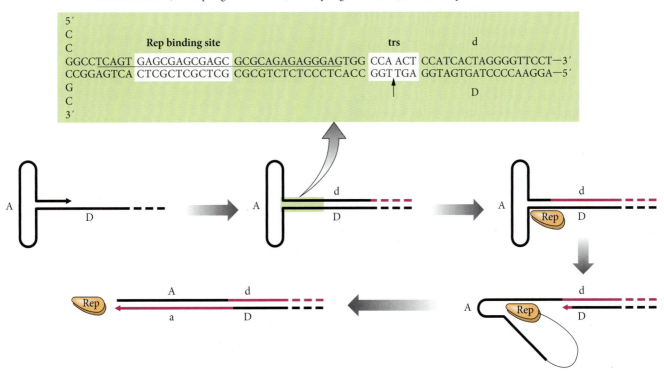

In the first phase, the three capsid proteins, VP1, VP2, and VP3, assemble into empty capsids in the nucleus. In the second phase, single-stranded genomic DNA is inserted into the preformed empty capsids. The main issues are (i) how individual parvovirus genomes might be generated from the duplex concatemers that act as replicative intermediates and (ii) how they might be enclosed within the capsids.

It once was believed that empty parvovirus capsids are filled by a process of strand displacement synthesis from the double-stranded concatemers. This was thought to occur as follows. First, single-strand nicks are introduced into the double-strand concatemers, at the 5′ ends of the individual genome sequences (Fig. 14.5A). This step would create free 3′ ends in the duplexes, which can then serve as primers for the displacement synthesis of progeny DNA molecules. Encapsidation of the displaced strands would then occur, concurrent with their displacement. Note that this model leads to the prediction that the viral DNA enters the capsid in the 5′-to-3′ direction.

More recently, in 2001, Kleinschmidt and colleagues showed that AAV2 single-stranded genomes actually are packaged in the 3′-to-5′ direction, contrary to the expectations of the earlier model. Other recent experimental observations now suggest a plausible new model. The relevant observations are noted first, followed by the model.

Here are the relevant findings. First, as noted above, AAV2 encodes two small polypeptides, Rep40 and Rep52, which express a helicase activity that is required to translocate single-stranded DNA genomes into preformed empty capsids. Second, Rep52/40 helicase complexes processively scan in a 3′-to-5′ direction, along a single DNA strand. Thus, the direction of the processivity of the Rep52/40 helicase is the same as the direction in which the DNA is encapsidated. Third, Rep52/40 complexes also express an adenosine triphosphatase (ATPase) activity, and the energy derived from ATP hydrolysis drives the helicase activity of those complexes. Fourth, Rep52/40 complexes are able to bind to empty capsids as well as to DNA replicative intermediates.

Here then is the new model, which resembles the earlier model, to the extent that progeny single strands are generated from the high-molecular-weight replicative forms and are then encapsidated. First, the viral initiator nuclease Rep78/68 (in the case of the dependoviruses) or NS1 (in the case of the autonomous parvoviruses) introduces site-specific, single-strand nicks into the duplex, concatemeric, replicative forms (Fig. 14.5B). Then, in the case of AAV2, Rep52/40 complexes concurrently bind to replicative intermediates and to empty capsids at a single capsid **portal**. (A portal is a unique site on a preassembled empty capsid, through which a viral genome can enter.) The

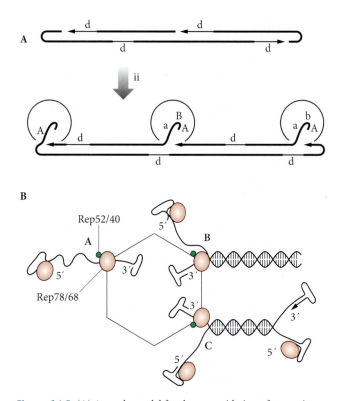

Figure 14.5 (A) An early model for the encapsidation of parvovirus genomes involving strand displacement synthesis from the double-stranded concatemers. First, single-strand nicks are introduced into the double-strand concatemers, at the 5′ ends of the individual genome sequences. This step creates free 3′ ends in the duplexes, which can then serve as primers for the displacement synthesis of progeny DNA molecules (boldface). Encapsidation of the displaced strands would then occur, concurrent with their displacement. (B) Model for the involvement of Rep helicases as genome packaging motors. Most helicases bind to single-stranded DNA adjacent to duplex regions and then proceed to unwind the two strands, moving along the bound single strand, under the consumption of ATP. The AAV packaging complex represents an immobilized helicase complex, composed of large and small Rep proteins, on the capsid surface. Both single-stranded and double-stranded genomes are able to bind to the packaging complex via interaction with the large Rep proteins, which bind sequence specifically to both ends and covalently to the 5′ end. The genome is then translocated through the packaging complex and into the capsid either as a single-stranded molecule using the initial "scanning" function before the first duplexed base pairs are encountered (labeled A in the figure) or by unwinding a double-stranded dimer or multimer genome on the capsid surface at the same time that it inserts the 3′ end of one of the single strands into the capsid (labeled B). The hypothesis that dimer and multimer genomes can act as substrates for packaging would also require that the packaging reaction is terminated, most probably by the nicking activity of the large Rep proteins. A situation in which the palindrome-containing termini of a double-stranded monomer not engaged in the packaging reaction open up to allow a further round of strand displacement replication to occur during encapsidation is shown (labeled as C). This situation would result in "collision" of the packaging and replication complexes and premature strand displacement, which could either promote or inhibit encapsidation, depending on whether single-stranded or double-stranded genomes are the preferred substrates for the packaging complex. Modified from P. Tattersall and D. Ward, *Nature* **263**:106–109, 1976, and J. A. King et al., *EMBO J.* **20**:3282–3291, 2001, with permission.

Rep52/40 helicase then acts as a molecular motor to "pump" single-stranded DNA into the capsid through the portal. The energy is provided by ATP hydrolases. (Note that a specialized portal through which DNA might enter the capsid has not yet been detected for AAV. Also, recall that the entire Rep52 and Rep40 coding regions are contained entirely within those of Rep78 and Rep68, respectively [Fig. 14.6]. Yet the larger proteins catalyze the packaging reaction up to 1,000-fold less efficiently than the smaller ones. Why might that be so? Also, whereas NS1, like Rep78/68, has ATPase and helicase activities, NS2, unlike Rep52/40, does not have these activities. Actually, much less is known about the encapsidation mechanism of the autonomous parvoviruses [Box 14.6].)

It is not yet known whether newly displaced single strands are released before being encapsidated or if packaging occurs concurrently with their displacement (Fig. 14.5B). Regardless, viral replication occurs in two distinct

Figure 14.6 Genome of AAV, which is the best-characterized parvovirus. Note the inverted terminal repeats (ITRs) at the ends of the genome, the three promoters for the initiation of transcription (P5, P19, and P40), and the single polyadenylation site. Each of the three viral promoters generates at least two mRNAs that are related by differential splicing. Parvovirus genomes consist of two functional regions: a region encoding the Rep proteins and a region encoding the capsid proteins. Transcriptional promoters P5 and P19 initiate transcription at different points of the *rep* open reading frame, to generate mRNAs encoding proteins Rep78 and Rep68. Moreover, an intron at the 3′ end of this open reading frame allows expression of short versions of the two full-length Rep proteins, known as Rep52 and Rep40. A common in-phase DNA sequence encodes the three AAV capsid proteins. VP2 and VP3 are translated from the same mRNA species, with the initiation of VP3 occurring at an in-phase, internal AUG codon in the VP2 open reading frame. Adapted from R. M. Linden and K. Berns, p. 68–84, *in* S. Faisst and J. Rommelaere (ed.), *Parvoviruses: from Molecular Biology to Pathology and Therapeutic Uses*, vol. 4, *Contributions to Microbiology* (S. Karger, Basel, Switzerland, 2000), with permission.

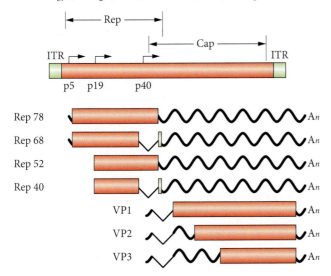

phases. In the first phase, the input viral genome is amplified by the rolling-hairpin mechanism, thereby generating high-molecular-weight, duplex replicative forms. In the second phase, individual viral genomes are excised, displaced, and encapsidated.

We return now to an issue discussed earlier: dependoviruses package equal numbers of plus-strand and minus-strand genomes, whereas some autonomous parvoviruses package only minus-strand genomes. This difference is thought to be explained as follows. Recall that the terminal hairpins serve as replication origins, where, during "terminal resolution," a nick is produced in one strand of the DNA in a sequence-specific manner to create a base-paired 3′ nucleotide that can prime unidirectional strand displacement DNA synthesis (Fig. 14.4). Importantly, the ability to package positive strands maps to the viral hairpins, mainly to the nick site of OriR (the right-end replication origin). (This was demonstrated by experiments in which genomic chimeras were created using viruses that encapsidate strands of both polarities and viruses that package predominantly minus strands and also by experiments using viral mutants.) In the case of the dependoviruses, the palindromic sequence at the other end of the genome is the inverted complementary sequence and therefore does not contain the high-efficiency nick site that can prime displacement synthesis. However, since each terminal palindrome undergoes a "flip-flop" during each round of dependovirus replication, the high-efficiency nick site is always present at one end on both plus-strand and minus-strand dependovirus genomes. Consequently, both plus-strand and minus-strand dependovirus genomes are packaged. In contrast, a high-efficiency nick sequence on some autonomous parvovirus genomes is located only on the 3′ terminal palindrome, which does not undergo a flip-flop and is present only on negative-sense genomes. Consequently, only the negative-sense genomes are packaged by these autonomous parvoviruses. Note that there is no integral "packaging signal" per se, and only those parvoviruses that have different termini at their left and right ends preferentially encapsidate negative-strand genomes.

GENOME ORGANIZATION AND EXPRESSION

The genome of AAV, which is the best-characterized parvovirus, is depicted schematically in Fig. 14.6. First, note the inverted terminal repeats at the ends of the genome, which are characteristic of dependoviruses. Next, note that the genome contains three promoters for the initiation of transcription, P5, P19, and P40, which are located at map units 5, 19, and 40, respectively. (There are 100 map units in all.) Also, there is only a single polyadenylation site.

All parvoviral mRNAs are transcribed in the nucleus from the double-stranded replicative intermediates, using the cellular RNA pol II enzyme, which initiates transcription at each of the three viral promoters. In addition, each of the three viral promoters generates at least two mRNAs that are related by differential splicing. An example of this fact was noted above in the case of the AAV2 Rep proteins.

Parvovirus genomes may be thought of as consisting of two functional regions: a region encoding the Rep proteins and a region encoding the capsid proteins. The AAV Rep proteins, which are the translation products of mRNAs transcribed from promoters P5 and P19, are then regarded as early proteins. That is so because their expression activates promoter P40, the promoter for the genes encoding the capsid proteins, VP1, VP2, and VP3. The particular functions of the four AAV Rep proteins in replication (Rep78 and Rep68) and encapsidation (Rep52 and Rep40) were noted above. (As will be seen in subsequent chapters, the genomes of DNA viruses are typically organized into early and late regions and the expression of these distinct regions is temporally regulated.)

A common in-phase DNA sequence encodes the three AAV capsid proteins (Fig. 14.6). This gives rise to two interesting aspects of AAV capsid protein expression. First, VP2 and VP3 are translated from the same mRNA species, with the initiation of VP3 translation occurring at an in-phase, internal AUG codon in the VP2 open reading frame. This is so despite the fact that on eukaryotic mRNAs the most 5′ terminal translation initiation site is generally the only one that is functional. Second, despite being translated from an internal AUG codon in the VP2 open reading

frame, VP3 is the major AAV capsid protein, and more VP3 than VP2 is translated from their common mRNA. The explanation for these anomalies is that translation of VP2 initiates from an unconventional ACG codon, which ribosomes read through more often than not.

The early and late genes of autonomous parvoviruses are organized like those of AAV. Nevertheless, the autonomous parvoviruses and the dependoviruses have differing patterns of transcription. For example, the autonomous parvovirus MVM uses only two promoters: one for the early region and one for the late region. But, like AAV, it uses only one polyadenylation site at the end of the genome. B19, another autonomous parvovirus, uses only a single promoter. However, B19 uses two polyadenylation sites, one of which is in the middle of the genome.

THE DEPENDOVIRUS LIFE CYCLE

The dependovirus life cycle exemplifies one of the most unique of all virus-host interactions. Specifically, dependovirus replication requires that the host cell be simultaneously infected with any of several unrelated helper viruses. Adenoviruses and several of the herpesviruses (e.g., herpes simplex viruses I and II, cytomegalovirus, and pseudorabies virus) all provide complete helper activity, and papillomaviruses perhaps do so as well.

You may very well be wondering how likely it is that a dependovirus might be present in a cell that is concurrently infected with a helper virus, and as you might expect, the answer is, not very likely at all. How then do dependoviruses survive in nature if they must depend on helpers, which only rarely are present simultaneously? Congratulations if you have already surmised that dependoviruses are able to establish latent infections during which they await rescue by a helper virus. Indeed, the ability to establish latent infection is a key feature of the dependovirus life cycle.

The AAVs (for adeno-associated viruses) are so named because they were originally identified in throat and fecal samples from patients who were infected with an adenovirus. The occurrence of AAV latency and the activation of latent AAVs by adenoviruses were discovered only later. The discovery happened when, for the purpose of vaccine production, primary cultures of African green monkey kidney cells were infected with adenoviruses. AAV then appeared in 20% of the lots of African green monkey kidney cells and also in 1 to 2% of human embryonic kidney cell lots. These findings implied that in the absence of a helper virus, dependovirus genomes can remain latent until they are rescued by infection with a helper virus.

Other experiments showed that when tissue culture cells are infected with AAV, in the absence of helper virus, the virus assumes its latent state, in which it is able to persist for over 150 passages. ("Passage" refers to the splitting of a cell culture to permit its continued propagation.) Importantly, the latent AAV could be rescued at any time by superinfection of the cell cultures with an adenovirus.

Taken together, the above experimental findings demonstrate that (i) latent infection with AAV is common in nature, (ii) this virus readily persists for extended periods in its latent state, and (iii) its helper-dependent replication is an effective strategy for it to sustain itself in vivo. That is, when rescue occurs, the dependovirus replicates and can be transmitted to new individuals, thereby enabling the cycle to repeat. It also is important that while AAV is in its latent state in vivo, it avoids immunological detection.

What might be the mechanism of action by which several unrelated helper viruses rescue the latent AAV genome? The answer is not entirely clear. However, the following facts suggest a plausible explanation. Several of the adenovirus early gene products are able by themselves to provide helper activity (e.g., E1A, E1B, E2A, and E4). Moreover, expression of each of these adenovirus gene products results in transactivation of AAV gene expression. Yet none of these adenovirus products appear to interact directly with the AAV genome. Additionally, unrelated herpesviruses and perhaps papillomaviruses also provide helper activity. Also, several genotoxic substances or treatments (e.g., several chemical carcinogens and ultraviolet [UV] irradiation) can make cells permissive for AAV, in the complete absence of helper virus. Since these different viral helper activities and genotoxic treatments enhance the expression of certain cellular genes, it is reasonable to suggest that they induce a poorly defined intracellular milieu that is conducive to AAV rescue and replication. Regardless, all in all, dependoviruses display a rather bizarre form of parasitism, in which viruses exploit totally unrelated viruses to ensure their reproductive success.

Integration of Dependovirus Genomes

For more than 15 years, researchers believed that when dependovirus infection occurs in the absence of helper virus, then dependovirus genomes integrate into the cellular genome, where they were thought to remain latent until being "rescued" by infection with a helper virus. This belief was based on an early finding that the AAV genome integrates in a site-specific manner into human chromosome 19 *of cells growing in culture*. However, when human tissue samples from tonsils/adenoids, spleens, and lungs were directly examined, the majority of wild-type AAV DNA was found to exist as circular, double-stranded, extrachromosomal plasmids. Nevertheless, the researchers who carried out this study suggested that more human

samples need to be studied before concluding that site-specific integration of AAV genomes is restricted to cells growing in culture.

Even if site-specific integration of AAV genomes were only an in vitro phenomenon, it still would be noteworthy for several reasons. First, the ability of AAV genomes to undergo site-specific integration, even if only in vitro, is thus far unique among eukaryotic DNA virus genomes. Indeed, the AAV genome may be the only known example of an exogenous DNA of any sort that integrates at a specific site in the human genome. (Note that any exogenous DNA introduced into a mammalian cell can integrate nonspecifically by nonhomologous recombination.) Second, because of the unique site specificity of AAV integration into the human genome and because of the nonpathogenic nature of AAV, this virus has been attracting attention for use as a vector for human gene therapy.

What might be the mechanism that mediates the site-specific integration of the AAV genome into human chromosome 19? The answer is not entirely clear, but the following experimental findings bear on this issue. First, the single-stranded AAV genome is converted to a double-stranded form soon after infection. Then, the viral P5 promoter is activated and Rep78/68 proteins are generated. However, only low levels of these proteins are produced in the absence of helper virus, and the expression of other AAV genes is repressed. Under these conditions, the Rep proteins mediate integration of the AAV genome, rather than its replication. Apropos these facts, the target sequence for integration on human chromosome 19 was identified as a 16-nucleotide-long sequence that closely resembles the Rep78/68 binding sequence on the viral genome (Fig. 14.4). Moreover, data from a cell-free system show that Rep78/68 actually can form a complex both with that cellular sequence and with the viral Rep78/68 binding site, introducing nicks at those sites, as a prerequisite to integration.

Two models have been proposed to explain how Rep78/68 might facilitate the joining of the viral and cellular DNA sequences. In the first model, Rep78/68 is thought to mediate strand switching during DNA replication, thereby creating links between the viral and cellular DNAs. In fact, data from a cell-free system show that Rep78/68 actually initiates DNA synthesis at the target sequence on chromosome 19, presumably after introducing a nick there. Thus, this model proposes that a multimeric form of Rep78/68 binds to both the viral Rep binding site and the cellular target sequence, in that way positioning the viral genome for integration. Then, Rep78/68 nicks the target DNA and initiates displacement DNA synthesis on it. As DNA synthesis at the target site ensues, a series of DNA template switches (copy choice) links the viral DNA to the chromosomal DNA. Downstream template switches complete the integration process (see Linden et al., 1996). In the second model, Rep78/68 is thought to be involved in directly ligating the nicked viral and cellular sequences.

Interestingly, in most latently infected cells, the integrated viral DNA is present as a tandem array of several genome equivalents. This means that some genomic replication must have occurred at some point, presumably prior to or during integration. It is possible that integration of tandem repeated AAV genomes serves the virus by increasing the likelihood that intact complete genomes might be released upon rescue.

As noted above, it is not clear how infection by the helper virus leads to activation of the latent virus. Nevertheless, infection by the helper virus does result in activation of the AAV Rep genes, which in turn activates transcription from all three AAV promoters, resulting in rescue of the integrated AAV genome and replication. The AAV terminal repeats are required for rescue.

MEDICAL ASPECTS

Parvovirus B19 is the only member of the parvovirus family known to cause disease in humans. It is usually transmitted via respiratory secretions and close personal contact, and it is highly contagious. Consequently, most adults have been infected.

B19 is an autonomous parvovirus. Yet, like other autonomous parvoviruses, it still requires cellular functions that are efficiently expressed only when cells are in the S phase of the cell cycle. Moreover, since parvoviruses do not express any mitogenic activity, B19 has a preference for rapidly dividing tissue, in particular, hematopoietic cells. This preference of B19 for hematopoietic cells is a key determinant of its virulence, as will soon be seen.

Studies of human volunteers who were experimentally infected with B19 have been carried out. Interestingly, these studies were deemed feasible because B19 was once thought to cause only relatively mild illness in people. Importantly, these studies showed that B19 selectively infects erythroblasts (red blood cell progenitors) and that B19 disease results from the direct killing of these cells and also from the immune response to the infection.

In children, B19 causes a well-known disease with the medical name *erythema infectiosum*. This condition is typified by mild fever and rash, which actually result from host immune responses. *Erythema infectiosum* is more commonly known as "fifth disease," because it was the fifth of the childhood **exanthems** (i.e., rash diseases) for which the causative agent was identified (in this instance, in the 1980s). (The four previously recognized childhood exanthems are varicella [chicken pox], rubella [German measles], roseola [caused by human herpesvirus 6], and measles.)

B19 causes a similar disease in adults, but in adults, it is often accompanied by **arthralgia** (i.e., joint pain). In one study, 40% of B19-infected adults were affected by this complication. Arthralgia is much less common in children, being seen as a complication in less than 10% of childhood cases.

Rare, but more serious, B19 illnesses occur in people who are immunocompromised (e.g., acquired immunodeficiency syndrome [AIDS] patients), anemic, or pregnant. For example, **aplastic crisis**, which occurs in people with chronic hemolytic anemia (i.e., sickle-cell anemia), is the most life-threatening complication of B19 infection. This disease is characterized by the sudden disappearance of erythroblasts from the bone marrow, which, together with the underlying anemia, can lead to a rapid drop in **hematocrit** (the percentage of a blood sample that consists of red blood cells) and potentially life-threatening anemia.

Interestingly, B19 is one of a very small group of viruses that can be transmitted from the mother to the fetus across the placenta. Others are rubella virus, HIV, and the human cytomegalovirus (one of the herpesviruses). About 30% of women who are infected with B19 during pregnancy pass the virus on to the fetus. Usually, these fetal infections cause no permanent harm. But in less than 10% of intrauterine B19 infections, the outcome is death of the fetus and spontaneous abortion, resulting from severe anemia in the fetus that can lead to congestive heart failure. In contrast to rubella virus, B19 is not known to cause fetal deformities. Fetal loss, rather than birth defects, is the major concern of infection during pregnancy.

Many adults have antibodies against one or more AAV serotypes, implying that those individuals have repeatedly been exposed to these viruses. Nevertheless, AAVs are not associated with any illness in people.

Unlike the small double-stranded DNA viruses of the polyomavirus and papillomavirus families, there is no evidence that AAVs might act as tumor viruses. On the one hand, this might be expected, since AAVs do not encode mitogenic activities, such as those encoded by the polyomaviruses and papillomaviruses, which inactivate cellular tumor-suppressor proteins p53 and Rb (Chapters 15 and 16). On the other hand, like the slowly transforming retroviruses, which also do not express transforming activities but instead transform by means of a *cis* effect of their integrated genomes (Chapter 20), AAVs too integrate their DNA, at least in vitro.

The human bocavirus (HBoV) is a recently discovered human parvovirus that is closely related to B19. It was initially detected in 2005 by random polymerase chain reaction (PCR) amplification of pooled respiratory samples from hospitalized children in Sweden, followed by the sequencing of the PCR products, which revealed the presence of the new virus. Specific primer sequences were then developed for the selective detection of the virus. By use of these primers, HBoV has since been identified in Europe, the United Kingdom, the United States, Canada, Africa, Asia, and Australia. Although HBoV appears to be relatively common (about as common as influenza virus and more common than coronaviruses), the clinical consequences of HBoV infection are not yet clear. Results from a U.S. study suggest that infection might be associated with lower respiratory tract disease. However, in many cases there was coinfection with another respiratory virus, so the exact role of HBoV in human disease is not known.

Suggested Readings

Cotmore, S. F., and P. Tattersall. 2006. Parvoviruses, p. 593–608. *In* M. L. DePamphilis (ed.), *DNA Replication and Human Disease.* Cold Spring Harbor Press, Cold Spring Harbor, NY.

King, J. A., R. Dubielzig, D. Grimm, and A. J. Kleinschmidt. 2001. DNA helicase-mediated packaging of adeno-associated virus type 2 genomes into preformed capsids. *EMBO J.* **20:** 3282–3291.

Linden, R. M., P. Ward, C. Giraud, E. Winocour, and K. I. Berns. 1996. Site-specific integration by adeno-associated virus. *Proc. Natl. Acad. Sci. USA* **93:**11288–11294.

Tattersall, P., and D. C. Ward. 1976. Rolling hairpin model for replication of parvovirus and linear chromosomal DNA. *Nature* **263:**106–109.

Polyomaviruses

INTRODUCTION

The *Polyomaviridae* are a family of small, nonenveloped, icosahedral, double-stranded DNA viruses. As such, they are class I viruses in the Baltimore classification scheme and the first family of class I viruses discussed in this text. They and the *Papillomaviridae* (Chapter 16) formerly were classified as distinct genera within a family referred to as the *Papovaviridae*. However, with the determination of the nucleotide sequences of these viruses, it was apparent that the *Polyomaviridae* and *Papillomaviridae* are not at all related and instead constitute distinct virus families. Thus, the term "papovaviruses" is rarely used today. Nevertheless, for reasons that soon will be apparent, the polyomaviruses, the papillomaviruses, and the larger adenoviruses are collectively referred to as the "small DNA tumor viruses."

Polyomaviruses that infect monkeys, humans, mice, hamsters, rats, and parakeets are known. Each of the known polyomaviruses is highly host species specific and somewhat tissue specific as well within its preferred host. The best known and most intensively studied of the polyomaviruses is **simian virus 40 (SV40)**, which naturally infects Asian rhesus macaques. **Polyomavirus of mice** also has been studied extensively. These two viruses have generated widespread interest because of their abilities to induce tumors in laboratory animals and **neoplastic transformation** of cells in culture (Box 15.1). Moreover, whether SV40 might be circulating in humans and causing cancer and other illnesses in the human population is a matter of current interest and controversy.

Humans are the natural host for **JC virus** (JCV) and **BK virus** (BKV). Moreover, each of these polyomaviruses is widespread in the human population. JCV is attracting increasing attention as the sole cause of the human neurodegenerative disease **progressive multifocal leukoencephalopathy** (PML [Chapter 3]). PML is a disease of immunocompromised individuals, and as such, it is the cause of death in 5% of all acquired immunodeficiency syndrome (AIDS) patients. Also, there is mounting evidence that both JCV and BKV may be associated with cancer in humans and that BKV causes kidney nephropathy. In that regard, BKV is now believed to be

Neoplastic transformation, or simply cell transformation, refers to the conversion of a "normal" cell in culture into one displaying the characteristics of a tumor cell growing in culture. (Is any animal cell growing in culture truly normal?) Major differences between normal and transformed cells include the following. First, normal epithelial cells and fibroblasts generally stop dividing in cell culture when they have formed a confluent monolayer; this phenomenon is referred to as "contact inhibition." In contrast, transformed cells pile up and grow in multiple layers. Second, normal fibroblasts and epithelial cells must adhere to a solid surface in order to replicate. In contrast, their transformed counterparts are able to replicate when suspended in an agar-containing medium. This characteristic of transformed cells is the basis for a useful quantitative assay for cell transformation, which also is useful for obtaining clonal isolates of transformed cells. Importantly, studies of transformed cells in culture have produced insights into the molecular events underlying cancer that hardly could have been obtained by studies in animals alone.

a major cause of graft loss following kidney transplantation. JCV and BKV each acquired their names from the initials of the human patients from which they first were isolated.

Remarkably, some 35 years after the discoveries of JCV and BKV in 1971, possible third and fourth human polyomaviruses were independently discovered in 2007. At present, these new viruses are known as the KI and WU polyomaviruses, reflecting their discoveries at the Karolinska Institute and Washington University, respectively. Nothing is yet known regarding the prevalence of these viruses in humans or whether they might be associated with any disease. Then, less than a year later, yet another new polyomavirus was discovered in humans, named the Merkel cell virus because of its association with a rare human skin cancer, Merkel cell carcinoma.

The mouse polyomavirus (referred to earlier simply as polyomavirus) was the first member of the family to be discovered. Its discovery was fortuitous, occurring in 1953 while Ludwig Gross was carrying out experiments in which he was injecting mice with filterable extracts from other mice that had "spontaneously" developed leukemia. The injected mice likewise developed leukemia, demonstrating that leukemia could be induced by a filterable agent. Surprisingly, however, some injected mice developed parotid tumors as well as leukemias. These experimental results suggested that the extracts contained a second tumorigenic agent, in addition to a leukemia virus

(which years later would be shown to be a retrovirus [Chapter 20]). Gross indeed was able to separate the parotid tumor agent from the leukemia virus by ultracentrifugation. Stewart and Eddy in 1957 demonstrated that the parotid tumor agent is able to cause tumors at multiple sites and proposed the name "polyomavirus" to reflect that fact.

SV40 was discovered by Sweet and Hilleman in 1960 as a frequent contaminant of rhesus macaque kidney cell cultures that were being used at the time to grow poliovirus and adenovirus vaccines. Then, in 1962, it was reported that SV40 induces sarcomas when inoculated into newborn hamsters. In addition, SV40 was found to induce neoplastic transformation of cells in culture. However, by the time of those discoveries, SV40 had *inadvertently been administered to hundreds of millions of humans* via both the live (Sabin) and inactivated (Salk) poliovirus vaccines (Chapter 5), fulfilling one of the most dread nightmares of vaccinologists! (Note that about one in 10,000 SV40 particles survived the formaldehyde treatment used to inactivate poliovirus in the inactivated vaccine.)

The unintended widespread exposure of humans to SV40 was a *potential* public health threat of major proportions. The consequences of that exposure are discussed below.

Because SV40 and polyomavirus can induce neoplastic transformation of cells in culture and tumors in laboratory animals but yet are relatively simple viruses that could easily be studied in cell culture, they have been the most intensively studied and, consequently, the best understood of the *Polyomaviridae*. As you may now suppose, the initial purpose of studying these viruses was to identify the mechanisms by which they induce neoplasia (Box 15.2). However, early studies with SV40 in particular, while driven by the desire to elucidate the molecular biology of cancer, also resulted in major and sometimes unplanned advances in fundamental eukaryotic cell and molecular biology. These advances included the identification of the role of enhancers in eukaryotic transcription, the phenomenon of alternative splicing of messenger RNA (mRNA) precursors, identification of steps in the initiation of eukaryotic DNA replication, and the identification of protein nuclear localization signals.

In addition, and as hoped, studies of SV40 indeed resulted in important insights into the molecular basis of cancer. In particular, they led to the identification of the key cellular tumor suppressor protein p53. Moreover, studies of the effects of the SV40 LT protein (see below) on the activity of p53 and on the activity of the pRb tumor suppressor protein led to singularly important insights into the roles of these proteins in ensuring that cellular replication is carefully regulated to meet the needs of the organism as a whole. Moreover, the SV40 genome

James Watson, in his now classic *Molecular Biology of the Gene* (W. A. Benjamin, Inc., New York, NY, 1965), writes the following:

> Until recent work with polyoma and RSV (Rous sarcoma virus), the search for chemical differences between normal and cancer cells resembled the search for a needle in a haystack. There may be more than a million different genes in a mammalian cell, of which we now can assign functions to perhaps 1,000. Thus, the odds are that a biochemist studying cancer is looking for changes in molecules that have not been discovered.

It now is known that mammalian cells contain many fewer genes than the million suggested by Watson in 1965. Regardless, his statement reflects the mind-set of the time. From it, we can appreciate the importance to the virologists and cancer researchers of the 1960s of the discovery of oncogenic viruses containing only five or so genes. These viruses raised the possibility of studying the molecular biology of cancer in "a straightforward, rational manner."

was the first eukaryotic viral genome to be sequenced. That effort was prompted by the desire to better understand the workings of the viral origin of DNA replication, its enhancer and promoter elements, and the organization and expression of the viral genome in general. More recently, SV40 has been serving as a marker for a novel endocytic pathway that it uses to mediate its entry into the cell. That pathway uses caveolae rather than clathrin-coated pits. Recently, SV40 also was found to encode a micro-RNA (miRNA), thereby providing the first known example of a functional virus-encoded miRNA. The exploitation by SV40 of the host RNA interference (RNAi) machinery presents a prototype for this brand new virus-host interaction.

Considering that SV40 is the most intensively studied of the polyomaviruses and that it has led to key insights into eukaryotic cellular and molecular biology, we will use SV40 as a paradigm for the polyomaviruses in general. The closest known relative of SV40, as shown by DNA homology, is BKV of humans.

STRUCTURE

Polyomavirus capsids are icosahedrally symmetric structures that are about 500 angstroms (Å) in diameter. They contain 360 copies of the major capsid protein VP1, all of which are arranged as pentamers. Each VP1 pentamer also

contains one copy of either VP2 or VP3, which are the minor capsid proteins. The capsids enclose the viral genome, which is a closed, supercoiled, circular, double-stranded DNA molecule of about 5,200 base pairs (bp). The genome is complexed with cellular histones that are organized into about 25 nucleosomes. (Nucleosomes are the basic structural units of chromatin, consisting of DNA wrapped around a histone core.) Thus, the viral genome is referred to as a minichromosome. Papillomavirus genomes likewise are arranged as minichromosomes (Chapter 16).

Polyomaviral icosahedral capsids are organized and built differently from the capsids of most other icosahedral viruses. To appreciate these differences, we briefly review some features of the icosahedral building plan (Chapter 2). Recall that an icosahedron is a regular 20-sided polygon, in which each face is an equilateral triangle. Moreover, icosahedra have 12 vertices, which are the points where five of the triangular faces come together. The simplest icosahedral viruses contain 3 protein subunits per face and thus 60 subunits in the entire capsid (see Fig. 2.8).

An icosahedron that contained only 60 subunits would be too small to contain the SV40 minichromosome with its 25 nucleosomes. To solve this space problem, polyomaviruses do the same as other icosahedral viruses: they add additional protein subunits to the capsid. In almost all instances, this is done according to strict rules discovered by Casper and Klug. Essentially, each of the 20 triangular faces of the icosahedron is divided into smaller triangles, each of which contains three subunits. Thus, the basic icosahedral building plan is maintained, and icosahedral capsids always contain a number of subunits that is a multiple of 60. The structure may no longer look like an icosahedron per se. However, it always retains the fivefold, threefold, and twofold axes of symmetry that are characteristic of icosahedra (Fig. 2.8).

The larger icosahedrally symmetric capsid shown in Fig. 2.8C and D represents the structure of a togavirus (e.g., Sindbis virus and Semliki Forest virus). Looking carefully at the arrangements of subunits in that structure, notice that the subunits are arranged as pentamers, and also as hexamers. Sixty subunits make up the 12 pentamers that comprise each of the 12 vertices of the icosahedrally symmetric capsid. All of the remaining subunits (180 in this instance) are arranged as hexamers. In cases where pentamers and hexamers are comprised of the same protein subunit, the interactions between subunits in pentamers cannot be exactly identical to their interactions in the hexamers. Yet they can be very similar. Moreover, the interactions between subunits of adjacent pentamers and hexamers cannot be identical to the interactions between subunits of adjacent hexamers, although they too can be very similar. For these reasons, Casper and Klug introduced the concept of **quasiequivalent bonding**, in which each of the subunits can be bound to its neighbors in similar but not exactly identical ways (e.g., head-to-head and tail-to-tail interactions [Fig. 2.8E]). Importantly, all subunits interact via noncovalent bonds and, therefore, are arranged to make maximal contact.

Still looking at the conventional icosahedrally symmetric capsid in Fig. 2.8, notice that there is fivefold rotational symmetry around the pentamers and sixfold (actually it is threefold) rotational symmetry around the hexamers. Now, considering that the SV40 capsid is constructed entirely of pentamers (Fig. 15.1) and that it must likewise display the characteristic rotational axis of symmetry of an icosahedron, how can it establish sixfold (actually threefold) symmetry around a pentamer? Also, bear in mind that it is impossible to fit a pentamer into a hexavalent surrounding without violating the principle of quasiequivalence.

The explanation for these conundrums can be found in the actual structures of the polyomaviruses. Whereas the pentamers and hexamers of other icosahedral viruses interact across complementary surfaces, the pentamers of the polyomaviruses are joined together by the extended arms of the C termini of their VP1 subunits (Fig. 15.1). Five arms extend from each pentamer and insert into the neighboring pentamers in three distinct kinds of interactions. Also, VP1 is the only protein of SV40 that has cysteines. One of these cysteines, C104, forms interchain disulfide bonds, which link VP1 subunits in adjacent pentamers. This means of interaction between pentamers affords great stability as well as flexibility in how a given subunit of one pentamer might interact with a subunit of a neighboring pentamer. That is, alternative arrangements of the long C-terminal arm of VP1 allow different contacts between pentamers, all of which are strong.

Interestingly, at the time the X-ray structure of SV40 was solved, it was suggested that a chaperone might be needed to correctly establish these three distinct sorts of interpentameric VP1 interactions. In that regard, recall that chaperones bind to other proteins that are in prenative folding states and assist them in reaching a correctly folded

Figure 15.1 (A) In contrast to typical icosahedral capsids, which are constructed from subunits organized as pentamers and hexamers (Fig. 2.8), polyomavirus capsids are assembled entirely from pentamers. Nevertheless, like other complex icosahedral capsids, polyomavirus capsids display sixfold as well as fivefold rotational symmetry. The 12 five-coordinated pentamers, which are located at the vertices of the icosahedral structure, are shown in white, whereas the 60 six-coordinated pentamers are in color. (B) In order to establish sixfold as well as fivefold rotational symmetry about a pentamer, the pentamers of the polyomaviruses are joined together by the extended arms of the C termini of their VP1 subunits. Pentamers and their subunits are color coded as in panel A. Five arms extend from each pentamer and insert into the neighboring pentamers in three distinct kinds of interactions. This is in contrast to most other icosahedral viruses, in which the pentamers and hexamers interact across complementary surfaces. From R. C. Liddington et al., *Nature* **354**:278–284, 1991, with permission.

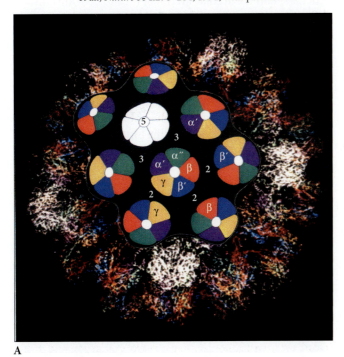

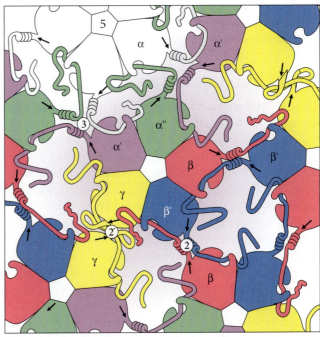

A

B

and functional conformation, both as individual proteins and as components of multimolecular complexes. Indeed, it now appears that SV40 itself may provide a chaperone activity to facilitate virus assembly (and for other purposes as well; see below).

Each VP1 pentamer contains a cavity at its center, which contains a molecule of either VP2 or VP3, inserted in a hairpin-like manner. These VP2 and VP3 molecules are anchored in the cavity by hydrophobic interactions involving their C termini. In that regard, VP2 and VP3 share 234 identical amino acids at their C termini. That is so because translation of VP3 initiates from an internal AUG codon within the VP2 coding sequence and uses the same translational reading frame (see below).

VP2 contains about 100 additional residues at its N terminus, which is myristylated. Recall that the poliovirus VP4 protein likewise is myristylated and that the poliovirus VP4 myristyl group is believed to play a role in enabling the poliovirus genome to traverse the cell membrane (Chapter 6). Indeed, myristylated residues are found in the capsid proteins of a variety of other nonenveloped virus families and are believed to play a role in virus entry in each instance. Similarly, the myristyl group on polyomaviral VP2 molecules is believed to play a role at some point in polyomavirus entry. But for the polyomavirus VP2 myristyl group to play a role in entry, it must be exposed or released from the virus particle (see below). The function of VP3 is not clear. Perhaps it is there because of insufficient room in the capsid to accommodate a larger VP2 at each pentamer, or perhaps it too plays a role in entry, or perhaps in release, as suggested by recent experimental findings (see below).

The structural features of the SV40 genome, referred to as a "minichromosome," are discussed below.

ENTRY AND UNCOATING

The SV40 infectious cycle begins when the virus binds to major histocompatibility complex class I (MHC-I) molecules (Chapter 4), which the virus uses for its receptor. Although SV40 specifically binds to MHC-I molecules, these molecules do not internalize with the virus (Box 15.3). This implies that SV40 uses a secondary receptor to mediate its uptake into the cell. In that regard, SV40 recently was found to inefficiently infect a cell line deficient in ganglioside biosynthesis. When cells of this line were preincubated with ganglioside GM1, infection became more efficient. Perhaps SV40 uses GM1 as a secondary receptor or coreceptor to deliver the virus into the cell (see below).

Endocytosis, as mediated by clathrin-coated pits, is the major means by which animal cells take up ligands

from the extracellular spaces. For this reason, perhaps, viruses commonly exploit the clathrin-mediated endocytic pathway to gain entry into cells. However, it is now clear that animal cells have other means for internalizing membrane-bound ligands. SV40 enters the host cell by one of these alternative endocytic pathways. Specifically, SV40 uses **caveolae**, rather than clathrin-coated pits, to mediate entry. Caveolae are small (70- to 100-nm) invaginations of the plasma membrane that are distinguished from clathrin-coated pits by their size and distinctive flask-like shape (Fig. 15.2). Also, in thin-section electron micrographs, caveolae lack the visible coats that are characteristic of clathrin-coated pits. Figure 15.2 is an early electron micrograph showing SV40 entering in tight-fitting noncoated invaginations of the plasma membrane that later were shown to have the biochemical characteristics of caveolae. Compare Fig. 15.2 to Fig. 2.10.

Figure 15.2 Particles of SV40 are seen entering cells in tight-fitting, noncoated plasma membrane invaginations that were later shown to have the biochemical characteristics of caveolae. SV40 was the first infectious agent shown to enter cells by a caveola-mediated pathway. From G. G. Maul et al., *J. Virol.* **28:**936–944, 1978, with permission.

Caveolae form in special microdomains of the plasma membrane referred to as **lipid rafts**. Importantly, the lipid raft concept is based on the notion that the plasma membrane is not homogeneous in its composition. Instead, the plasma membrane is comprised of distinct domains, which have different compositions and functions. Lipid rafts are one such domain. A principal property of lipid rafts is that they are enriched in sphingolipids and cholesterol. Caveola formation is triggered by expression of the host protein caveolin-1, which is an integral membrane protein that localizes in lipid rafts. Thus, caveolae may be thought of as a special type of lipid raft that contains caveolin-1 and that is invaginated as a consequence of the presence of caveolin-1.

The cellular functions of caveolae are not yet entirely clear, but they have been implicated in the uptake of cholesterol and raft-associated receptors. Also, they are thought to play a role in organizing signal transduction pathways. SV40 was the first microorganism shown to invade cells via caveolae (Anderson et al., 1996). Moreover, SV40 is now one of the most extensively used ligands for tracking caveola-mediated endocytosis. For those reasons, we look at the caveola-mediated SV40 entry pathway in some detail.

After SV40 binds to the plasma membrane, it diffuses in the plane of the membrane until it encounters a caveola. The virus then transmits a signal that causes it to be enclosed within preexisting caveolae. This was demonstrated by the finding that pharmacologically blocking the SV40-induced transmembrane signal causes virus particles to accumulate at the annuli or "mouths" of the caveolae. The signal appears to be transmitted from the lipid raft membrane regions that surround the annuli of the caveolae, as suggested by the finding that SV40 signal transmission is prevented by pharmacologic agents that specifically disrupt lipid rafts. (Recalling that SV40 is not initially associated with lipid rafts or caveolae, note that GM1 is localized in caveolae and lipid rafts, consistent with the likelihood that this ganglioside is not a primary receptor for SV40.)

RNA interference was used to identify the protein kinases that might be upregulated by the SV40-induced signal (Pelkmans et al., 2005). That is, genes that are predicted to encode kinases were systematically silenced to evaluate their roles in SV40 entry. Experiments were carried out in parallel with vesicular stomatitis virus, a rhabdovirus (Chapter 10), which enters by clathrin-coated pit-mediated endocytosis. In all, 590 putative kinase genes were screened. The silencing of 80 of these kinase genes specifically affected SV40 infections! Vesicular stomatitis virus entry was specifically affected by the silencing of 92 additional kinase genes. Thirty-six other kinases were involved in the entry of both viruses (Box 15.4).

BOX 15.4

The numbers of kinases involved in virus entry are far larger than what most experts initially would have predicted. Why might so many kinases be necessary? Bear in mind that endocytic pathways in general involve multiple steps, including enclosing the ligand within the caveolae or clathrin-coated pits, the pinching-off process, reorganization of the cytoskeleton (see below), and the sorting, tethering, and transport of the ligand (the virus in this instance) along specific points to its final destination. Thus, the answer to our question may come from a more complete understanding of each of the multiple steps of the entry pathway.

Earlier, the SV40 signal was found to upregulate the **primary response** genes c-*myc*, c-*jun*, and c-*cis*. These genes are referred to as "primary response" genes because the inducing signals directly upregulate their expression, without any intervening protein synthesis. Their expression plays a role in the progression of cells into the S phase of the cell cycle. For that reason, upregulation of primary response genes by the viral signal might prime cells to support virus replication (see below). However, that possibility has not been verified experimentally.

Apropos the activities of caveola during SV40 entry, note that caveolae generally are static at the cell surface and that their internalization occurs only when it is induced. Moreover, induction of caveola internalization appears to be local. That is, when cells are exposed to SV40, those caveolae that are devoid of virus particles do not internalize. Still other studies suggest that SV40 particles can induce the de novo formation of caveolae around themselves and that these caveolae then internalize. These features of caveola dynamics are in contrast to those of clathrin-coated pits, which appear to form and internalize constitutively.

Other recent studies suggest that the generally static caveolae are held in place by the cortical actin skeleton beneath the plasma membrane. Thus, it is interesting that one consequence of SV40 signal transmission may be the local depolymerization of the actin cytoskeleton to enable the pinched-off caveolar vesicles to pass deeper into the cytoplasm. At the same time that the actin cytoskeleton is locally breaking up, actin is recruited to the virus-containing caveolae, forming so-called actin "tails." It was noted that these actin tails resemble similar entities that play a role in the intracellular mobility of bacterial pathogens of the genera *Listeria*, *Shigella*, and *Rickettsia*.

SV40 signaling also recruits dynamin to sites of virus internalization. Dynamin is a guanosine triphosphatase (GTPase) best known for its role in clathrin-mediated endocytosis, in which it assembles into rings around the necks of invaginated clathrin-coated pits. It leads to the constriction of those necks and the pinching off of clathrin-coated vesicles. Studies using pharmacologic inhibitors and dominant-negative cell mutants demonstrated that dynamin recruitment, tyrosine phosphorylation, and actin reorganization are all necessary for caveola-mediated SV40 infection.

The possibility that the ganglioside GM1 might serve as a coreceptor for SV40 was suggested above. GM1 preferentially associates with lipid raft domains of the plasma membrane. Thus, SV40 may become associated with GM1 when the virus diffuses to a caveola/raft. (Other experimental results suggest that SV40 may induce tight-fitting membrane invaginations to form via the multivalent binding of its VP1 pentamers to cell surface GM1.)

After the virus-loaded caveolae pinch off from the plasma membrane, they become the primary endocytic vesicles, which deliver the virus to a new class of endosomes that have been named "caveosomes." Caveosomes contain caveolin-1 but thus far are not known to contain recognized endosomal markers, nor are they acidic like late endosomes and lysosomes. Furthermore, caveosomes do not contain markers for either the Golgi complex or the endoplasmic reticulum (ER). Importantly, caveosomes do not accumulate ligands taken up by clathrin-coated pits. It is noteworthy that studies of SV40 entry led to the identification of this potentially novel and important organelle (Pelkmans et al., 2001).

SV40 is sorted from caveosomes into caveolin-1-free vesicles that deliver it along microtubule tracks to the ER, which is a very unusual destination for a ligand that is taken up by endocytosis. Indeed, the cholera and Shiga toxins were previously the best-known examples of ligands that traffic from the cell surface to the ER.

The next step in SV40 entry is equally unusual. SV40 undergoes at least partial disassembly in the ER, in which the internal capsid protein VP2 becomes susceptible to immunofluorescent staining using antibodies specific for VP2. Recall how the disassembly and uncoating of other nonenveloped viruses is triggered by the virus-receptor interaction at the plasma membrane or by the acidic conditions within an endosome or lysosome. What then might trigger SV40 disassembly in the nonacidic environment of the ER? The answer may be known by the time this book is published. For now, it is reasonable to suggest that since a chaperone activity is necessary to assemble SV40, a chaperone might likewise be required for the reverse process of disassembly. As you are likely aware, the ER contains various chaperones (e.g., protein disulfide isomerase, BiP, and Hsc70), one or more of which might do the job. (Hsc70 is the constitutively expressed form of the 70-kilodalton [kDa] heat shock proteins [HSPs],

whereas Hsp70 is the inducible form. Hsp70 is induced in response to heat shock, ischemia, and other stresses.)

The premise that ER chaperones might facilitate the disassembly of SV40 is supported by recent studies of the related mouse polyomavirus, which, like SV40, also traffics to the ER. First, ERp29, an ER chaperone that is structurally related to protein disulfide isomerase, was shown to act in the ER to expose the carboxy-terminal arm of mouse polyomavirus VP1. Then, a second research group reported that Hsc70 might continue the mouse polyomavirus uncoating process, as implied by the finding that this chaperone coimmunoprecipitated with the viral VP1 protein 3 h after infection of mouse cells. Moreover, Hsc70 chaperones were able to disassemble mouse polyomavirus particles in vitro.

More recently, the ER chaperones, ERp57 and protein disulfide isomerase, were shown to cause the release of VP1 pentamers from the fivefold vertices of SV40 capsids, at least in vitro.

The SV40 minichromosome must somehow pass from the ER to the nucleus, which is the site of all polyomavirus transcription, minichromosome replication, and assembly of progeny virus particles. Little is known about how SV40 minichromosomes might make their way from the ER to the nucleus. One working hypothesis is as follows. After virus particles partially disassemble in the ER, VP2 or perhaps VP3 that is exposed or released from particles creates pores in the ER membrane through which the minichromosome might pass into the cytoplasm. Then, transport of the minichromosome to the nucleus might be mediated by molecules of VP2 or VP3 that remain associated with the minichromosome. The following experimental findings are consistent with this model. First, VP2 and VP3 each efficiently integrates posttranslationally into the ER membrane. Indeed, at the time of the final editing of this chapter (May 2009), SV40 VP2 and VP3 were the only known proteins (either viral or cellular) that posttranslationally insert into the ER membrane. When VP3 does so, it appears to acquire a multimembrane spanning topology. Moreover, expression of VP2 and VP3 in bacteria compromises the ability of the bacteria to exclude membrane-impermeative agents (Box 15.5). Second, VP2 and VP3 each possess a DNA-binding domain to provide a link to the minichromosome, and each possesses a nuclear localization signal that might target the VP2/VP3/minichromosome complex to the nucleus. Third, when antibodies against VP2 and VP3 are microinjected into the cytosol, they block infection by SV40. Fourth, microinjected minichromosomes do not enter the nucleus unless VP2 and VP3 are present as well. Still, the exact nature of the disassembly intermediate that crosses the ER membrane is not known with certainty. (Current experimental results from the author's laboratory imply that the SV40

BOX 15.5

Here is a puzzler. The **viroporin** abilities of SV40 VP2 and VP3 might facilitate viral entry, and perhaps exit as well (see the text). However, if these proteins were to disrupt the plasma membrane during viral replication, the result would likely be premature host cell death. How then might the membrane-disrupting activities of VP2 and VP3 be regulated during infection? One possibility is that the major capsid protein, VP1, regulates the process, as follows. The onset of VP1 synthesis precedes that of VP2 and VP3 by about 12 h. This gives VP1 ample time to generate an excess of pentamers that would then be available to bind the newly synthesized VP2 and VP3, thereby preventing their insertion into membranes. As these VP2- and VP3-associated VP1 pentamers then assemble into particles (see the text), the remaining VP2 and VP3 could then insert into the plasma membrane, perhaps facilitating virus release. Expression of VP2 and VP3 indeed is necessary for virus release.

More recently, SV40 was found to express a new "later" protein termed VP4 that may facilitate cell lysis and the release of progeny virus. This is suggested by the experimental finding that coexpression of recombinant VP4 and VP3 in bacteria, but not their individual expression, results in bacterial cell lysis. During SV40 infection of monkey kidney cells, VP4 is expressed coincident with cell lysis, about 24 h after VP1 synthesis commences (see the text and Daniels et al., 2007, and references therein).

minichromosome is exposed either concurrent with, or subsequent to, the translocation of the disassembly intermediates from the ER into the cytosol. Moreover, SV40 genomes released into the cytosol remain associated with VP2 and/or VP3. Exposed VP2 and VP3 sequences on the disassembly intermediates might facilitate their translocation across the ER membrane and their subsequent trafficking to the nucleus.)

You might now be wondering whether other polyomaviruses, like SV40, also enter via a caveola-dependent pathway. The answer for BKV is yes, for mouse polyomavirus the answer is sometimes, and for JCV, which enters via a clathrin-dependent pathway, the answer is no. Mouse polyomavirus, which can enter by means of caveolae, also appears to enter by non-clathrin-, non-caveola-derived, uncoated vesicles, the exact nature of which is still not clear. In this regard, recent experiments demonstrated that SV40 also can infect caveolin-1-deficient cells. Under those circumstances, the virus associates with lipid rafts and then enters via indentations of the plasma membrane

that progressively enclose the virus. Reminiscent of caveola-mediated SV40 entry, the generation of these endocytic vesicles is triggered by virus-induced signals that involve activation of tyrosine kinases. The resulting endocytic vesicles deliver SV40 to organelles that resemble caveosomes, except that they lack caveolin-1 in these caveolin-1-deficient cells. The virus then is delivered to the ER by microtubule-dependent vesicular transport, just as in caveolin-1-expressing cells.

Each of the polyomaviruses uses a different specific receptor. As noted above, SV40 binds to MHC-I molecules. Mouse polyomavirus binds to oligosaccharides with terminal α-2,3-linked sialic acid residues. The JCV receptor recently was found to be the 5-HT2A serotonin receptor, which is present on glia, among other cells, accounting for the neurotropism of JCV. Incidentally, 5-HT2A also is the receptor for the psychoactive drug LSD. Mouse polyomavirus and JCV, like SV40, each induce a signal that promotes virus entry. BKV likely does the same.

GENOME ORGANIZATION AND EXPRESSION

Genome Organization

The SV40 genome contains about 5,240 bp in a covalently closed superhelical DNA molecule. It is referred to as a **minichromosome** because, like cellular chromatin, it is associated with cellular histones H2A, H2B, H3, and H4, which are assembled into 24 to 26 **nucleosomes** along the DNA molecule. In that regard, eukaryotic chromosomes typically contain about twice as much protein as DNA. The major proteins of eukaryotic chromatin are the histones H1, H2A, H2B, H3, and H4, which are organized with the DNA into nucleosomes. Each nucleosome contains a sequence of 146 bp of DNA, which enfolds the nucleosome core, which is comprised of two molecules each of H2A, H2B, H3, and H4. A molecule of H1 is associated with the nucleosome, where the DNA first enwraps and then leaves the nucleosome (Fig. 15.3). Adjacent nucleosomes are separated by about 200 bp. Minichromosomes isolated from SV40 particles do not contain histone H1. However, newly synthesized minichromosomes from infected cells do contain H1. It is not known how or why packaged minichromosomes do not contain H1. (In Box 14.1 we noted that none of the animal DNA viruses contains a simple linear DNA genome. The varieties of forms that those DNA genomes might take were briefly described in that sidebar. The polyomaviruses and papillomaviruses are the two virus families that contain covalently closed, circular, double-stranded DNA genomes. Nevertheless, polyomavirus genomes are thus far unique among animal DNA virus genomes in being associated with cellular histones.)

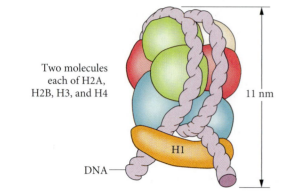

Two molecules
each of H2A,
H2B, H3, and H4

11 nm

H1

DNA

Figure 15.3 The SV40 genome is referred to as a "minichromosome" because, like cellular chromatin, it is associated with cellular histones H2A, H2B, H3, and H4, which are the major proteins of eukaryotic chromatin. These histones are assembled into 24 to 26 nucleosomes along the SV40 DNA molecule. Each nucleosome contains a sequence of 146 bp of DNA, which enfolds the nucleosome core, comprised of two molecules each of H2A, H2B, H3, and H4. A molecule of H1 is associated with the nucleosome, whereby the DNA first enwraps and then leaves the nucleosome. Adjacent nucleosomes are separated by about 200 bp. Note that minichromosomes isolated from SV40 particles do not contain histone H1. However, newly synthesized minichromosomes from infected cells do contain H1. It is not known how or why packaged minichromosomes do not contain H1. Adapted from G. M. Cooper, *The Cell: a Molecular Approach*, Sinauer Associates, Inc., Sunderland, MA, 1997.

The SV40 genome is divided into an **early region** that is expressed before viral DNA replication and afterwards and a **late region** that is expressed efficiently only after the onset of viral DNA replication (Fig. 15.4). The early region encodes the large T (LT) and small T (ST) antigens. In addition, the early region encodes a 17kT antigen, about which little is known. (LT and ST are referred to as tumor or T antigens, reflecting the fact that they can be detected using antiserum from animals bearing tumors induced by these viruses or by antiserum from animals inoculated with cells transformed by them.)

As will be seen, LT is a particularly fascinating protein. It contains only 708 amino acids. Yet it affects viral replication in numerous ways via its rather notable variety of functional domains (Fig. 15.5). These domains enable SV40 to form specific multimeric complexes with both itself and several different cellular proteins. In addition, LT contains a DNA-binding domain, which enables the protein to interact with specific sequences in the unique viral regulatory region (see below). As a result of these specific interactions, LT orchestrates both transcription and replication of the viral genome. Moreover, the ability of SV40 LT to specifically interact with particular host proteins (i.e., p53 and pRb) underlies the ability of the virus to induce neoplastic transformation. In addition, LT contains a nuclear localization signal, the discovery of which

led to the identification of cellular nuclear membrane receptors that recognize proteins bound for the nucleus.

The late region encodes the capsid proteins VP1, VP2, and VP3, as well as a small, 7.9-kDa protein referred to as the **agnoprotein**. Much more is said below about the capsid proteins than about the agnoprotein. However, the circumstances leading to the discovery of the latter are interesting and worth noting. When the SV40 genome was sequenced, it was revealed that there is an open reading frame in the section that encodes the late mRNA leader, at the 5′ end of the late region. The finding of this small open reading frame then led to a search for a protein that it might encode, which indeed led to the identification of a protein with a molecular mass of 7.9 kDa that is translated at late times in infection. It was named agnoprotein because its function was unknown at the time of its discovery.

Recently, SV40 was found to express a fourth late protein, termed VP4, which like VP3 also initiates synthesis from a downstream AUG start codon within the VP2 transcript. VP4 may interact with VP3 to facilitate cell lysis and release of progeny virus (Box 15.5).

The early and late regions of all polyomavirus genomes are separated by a **regulatory region** of about 500 bp in length, which is recognized by LT and by cellular proteins that regulate transcription and replication. The SV40 and mouse polyomavirus regulatory regions are depicted in Fig. 15.6. The regulatory region also includes the **origin (ori) of DNA replication**. The sequences regulating transcription are located close to *ori* and include the promoter/enhancer for the early region and the promoter for the late region. The components of the regulatory region and their functions are described in greater detail below.

Notice that transcription is bidirectional from initiation sites near *ori* and extends about halfway around the genome in each direction (Fig. 15.4). Consequently, opposite strands of the DNA genome serve as templates for transcription of early and late mRNAs. (What benefit to the virus might accrue from this atypical pattern of genomic organization? A possible explanation is suggested by the fact that polyomaviruses are small viruses, containing small genomes with limited coding capacity. Their bidirectional mode of transcription, from a common regulatory region, minimizes genomic space that needs to serve as regulatory and punctuation sequences.)

The Regulatory Region

Key features of the SV40 regulatory region are shown in Fig. 15.6. We begin with *ori*, which is comprised of the *ori* **core**, and the two flanking auxiliary sequences, *aux-1* and *aux-2*. The *ori* core is the minimal *cis*-acting sequence required for SV40 DNA replication, whereas *aux-1* and *aux-2* augment *ori* core activity. The SV40 *ori* core is 64 bp

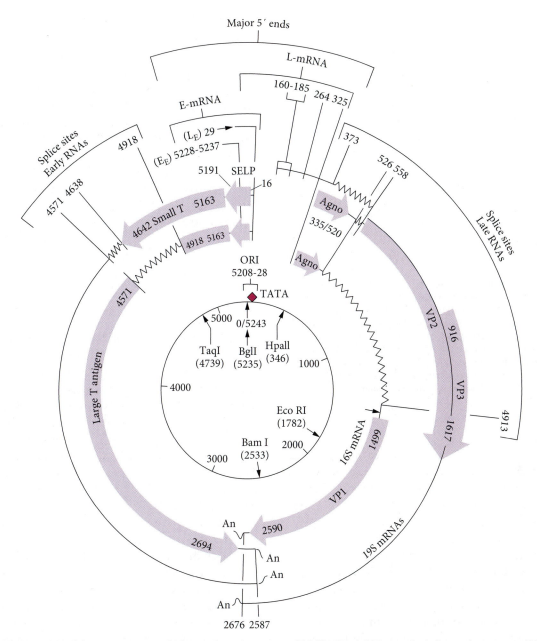

Figure 15.4 Organization of the SV40 genome, which contains a single origin of DNA replication *(ori)* located in a regulatory region between the early and late transcription start sites. E_E and L_E indicate early-early and late-early transcription start sites, respectively. The early region extends counterclockwise, and the late region extends clockwise from the regulatory region. By means of alternative splicing, the early region encodes two proteins, LT and ST. (The regions encoding all viral proteins are shaded. Introns are indicated by zigzag lines.) By means of alternative splicing, the late region likewise generates multiple mRNAs. The larger of these mRNAs encodes the minor capsid proteins VP2 and VP3, in which VP3 initiates synthesis from a downstream AUG start codon within the VP2 transcript. The smaller of the late transcripts encodes the major capsid protein, VP1. Notice that the region encoding the C termini of VP2 and VP3 overlaps the region encoding the N terminus of VP1, which is read in a different reading frame. Recently, a fourth SV40 late protein, VP4, was identified, which like VP3 initiates synthesis from a downstream AUG start codon within the VP2 transcript (see the text). The section that encodes the late mRNA leader contains a small open reading frame, which encodes the agnoprotein. The early regions of all polyomaviruses encode a minimum of two proteins, and the late regions encode at least three structural proteins. The BglI restriction endonuclease makes a single cut on the SV40 genome, within *ori*. By convention, SV40 nucleotides are numbered in a clockwise direction from the unique BglI restriction site. Other restriction endonuclease cleavage sites also are indicated. Restriction endonuclease digestion analysis was used to generate transcription maps of the SV40 genome before the era of DNA sequencing. Adapted from B. E. Griffen and Y. Ito, DNA viruses: polyomavirus, p. 1.105–1.113. *In* S. J. O'Brien (ed.), *Genetic Maps: Locus Maps of Complex Genomes* (Cold Spring Harbor Laboratory Press, Cold Spring Harbor, NY, 1993), with permission.

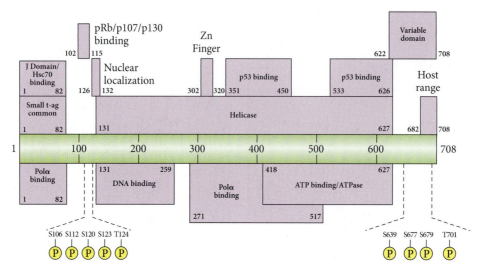

Figure 15.5 The SV40 large T (LT) antigen is a multidomain, multifunctional protein that can induce cellular proliferation and neoplastic transformation and also act as an efficient molecular machine to unwind duplex SV40 DNA for DNA replication. The numbers (1 through 708) correspond to the amino acid residues of the LT protein. Amino acids 1 through 82 are encoded by the first exon of the *LT* gene and are common to LT and ST (see Fig. 15.4). This region of LT contains a sequence necessary for binding to the DNA polymerase α-primase complex (Pol α) and to the cellular chaperone, Hsc70. It is referred to as the J domain because it shares sequences and functional properties with the DnaJ protein of *E. coli* (see the text). The DnaJ domain interacts with Hsc70 to remodel the pRB-E2F complex, thereby promoting cell growth. The pRb/p107/p130 binding region, which is required for binding the Rb tumor suppressor protein and the Rb-related proteins p107 and p130, and the nuclear localization signal are indicated. Expression of the full LT helicase activity requires the span of amino acids 131 through 627. Actually, the functional helicase domain contains three separable structural domains, which include the "Zn finger" (the region that binds zinc ions), the "ATP binding/ATPase" domain that participates in ATP binding and hydrolysis, and a domain containing residues 346 to 414 and 549 to 627 (not indicated). The "DNA binding" region is also contained within the span of amino acids encompassing the helicase domain. It is the minimal region required for binding to the SV40 ori (Fig. 15.4). The "p53 binding" regions and a segment required for binding to Pol α are also located within the long helicase domain. The functions of the C-terminal segment (residues 628 to 708) are not well defined, but residues 682 to 708 play a role in host range determination. Mutations in this portion of LT affect the ability of SV40 to replicate in some African green monkey kidney lines. Moreover, this region, sometimes referred to as the "variable domain," contains amino acid differences between SV40 strains. The circles indicate sites on LT that undergo phosphorylation; S indicates a serine residue, and T indicates a threonine residue. From A. R. Stewart et al., *Virology* **221**:355–361, 1996, with permission.

in length, as defined by mutations on either side of it that still allow DNA replication. At its center it contains a palindrome that is 27 bp in length (Fig. 15.6 and 15.7). Note that the SV40 *ori* is typical of eukaryotic origins of DNA replication in general, which usually contain a core component that specifies where replication begins and one or more auxiliary components that enhance replication under particular conditions (Box 15.6).

Both *aux-1* and *ori* core contain strong binding sites for LT. These are LT binding sites I and II, respectively, the strongest of which is site I. Aux-2 also contains an LT binding site, site III, which is much weaker than sites I and II. Next, notice the 21-bp GC-rich repeat sequences that partially overlap *aux-2*. These GC-rich repeats contain binding sites for transcription factors, and in fact, these sites are an important component of the SV40 early promoter. Thus, *aux-2* is required for early gene promoter activity, as well as for maximum *ori* activity.

As mentioned above, LT binding to specific sites in the regulatory region orchestrates viral transcription as well as replication. This important aspect of the SV40 life cycle is considered in detail below. For now, note that binding of LT to sites I and II negatively regulates expression of the SV40 early region, while LT binding to site II also is a key step in the initiation of viral DNA replication. The role of LT binding site III is not known, but as noted, the affinity of LT for site III is much less than its affinities for sites I and II.

At very early times after infection, early transcription initiates from multiple start sites clustered at EE (acronym for early-early). At somewhat later times, early transcription initiates from multiple start sites clustered at LE (for late-early).

Both the **TATA** box and the 21-bp repeats are components of the SV40 early promoter (Fig. 15.6). The **TATA box-like sequence** (i.e., TATTTAT), also referred to as a Hogness-Goldberg box, is located about 30 bp upstream from the EE start sites. In fact, most eukaryotic genes transcribed by RNA polymerase II (pol II) contain a TATA box that is positioned 25 to 30 bp upstream of the transcription start site. Interestingly, analysis of point and

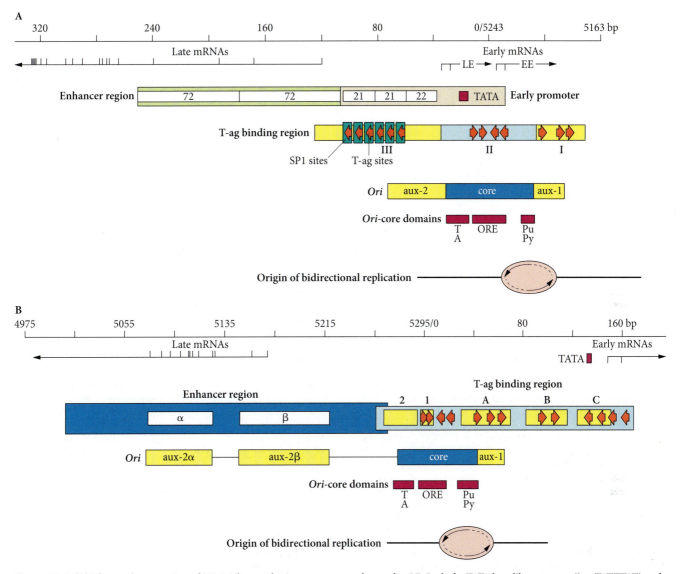

Figure 15.6 (A) The regulatory region of SV40. The numbering scheme is the same as in Fig. 15.4 (although the orientation is reversed). The 64-bp-long origin core (*Ori*-core) is the minimal *cis*-acting sequence required for SV40 DNA replication. The auxiliary sequences, aux-1 and aux-2, which flank the *ori* core, augment *ori* core activity. Within the *ori* core there is a TA tract with a strand bias for Ts and As, a palindromic region that serves as an origin recognition element, and a region with a purine (Pu)/pyrimidine(Py) strand bias, which is the site at which bidirectional DNA replication is initiated, as indicated by the bubble of replicating DNA, in which the two strands of the template are unwound (also see Fig. 15.8). The LT protein binding sites I, II, and III are shown. The arrowheads indicate the location and direction of the pentanucleotides 5′GAGGC3′, which are involved directly in LT binding. At very early times after infection, early transcription initiates from multiple start sites clustered at EE. At somewhat later times, early transcription initiates from multiple start sites clustered at LE. Both the TATA box-like sequence (i.e., TATTTAT) and the 21-bp repeats, which overlap aux-2, are components of the SV40 early promoter. The 21-bp repeat region contains six repeats of the sequence CCGCCC, which binds the transcription factor SP1. Further upstream of the TATA box, there are two copies of a 72-bp-long element, which function as enhancers of early transcription and are necessary for maximal expression of both the EE and LE start sites. Note that strains of polyomaviruses that are isolated from nature tend to contain only one copy of the enhancer element. Duplication of this region appears to occur during cultivation in vitro. (B) The mouse polyomavirus regulatory region is shown for comparison. Notice that it contains functional elements that are similar to those comprising the SV40 regulatory region but somewhat differently arranged. Figure courtesy of Dr. Mel DePamphilis, National Institute of Child Health and Human Development, National Institutes of Health, Bethesda, MD.

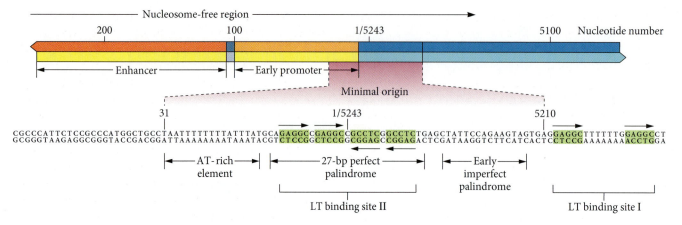

Figure 15.7 The SV40 origin of DNA replication. The 64-bp-long *ori* core is the minimal origin necessary for SV40 DNA replication. At its center there is a perfect palindrome that is 27 bp in length. Notice that this perfect palindrome overlaps LT binding site II (the major LT binding sites), which is comprised of 5′G(A/C)GGC3′ sequences (also see Fig. 15.6 and 15.9). The early promoter, which contains SP1 binding sites, and the enhancer augment SV40 DNA replication. They are thought to do so by helping to recruit essential replication factors to the origin, or perhaps by remodeling the replication origin (e.g., by creating a nucleosome-free region). Adapted from S. J. Flint, L. W. Enquist, R. M. Krug, V. R. Racaniello, and A. M. Skalka, *Principles of Virology: Molecular Biology, Pathogenesis, and Control*, 2nd ed. (ASM Press, Washington, DC, 2004), with permission.

deletion mutants mapping in this region of the SV40 genome demonstrated for the first time that TATA box elements act to position transcription initiation to sites that are about 30 nucleotides downstream of the TATA box. However, as described below, SV40 early transcription initiates at different sites at different times, implying that factors in addition to the TATA box must act to determine these start sites.

The **21-bp repeats** are located upstream of the SV40 TATA box-like elements, overlapping *aux-2*. (One of these 21-bp repeats is actually 22 bp long, but the overall sequence is still referred to as the 21-bp repeats.) This region contains six repeats of the sequence CCGCCC, which binds the transcription factor SP1.

Further upstream of the TATA box, there are two copies of a 72-bp-long element, which function as **enhancers** of early transcription and are necessary for maximal expression of both the EE and LE start sites. In point of fact, analyses of these SV40 sequences led to the discovery of enhancer elements and were the model for their subsequent analysis. Enhancer sequences can have a powerful effect on the levels of transcription from promoters that are located at sites distant from them. Moreover, the enhancer can act on a promoter that is either upstream or downstream from it. Furthermore, the effect of the enhancer is not dependent on its orientation with respect to the promoter. As might be expected, enhancers contain binding sites for multiple transcription factors. Note that strains of polyomaviruses that are isolated from their natural hosts tend to contain only one copy of the enhancer element. Duplication of this region appears to occur during cultivation in vitro.

Comparison of the SV40 and mouse polyomavirus regulatory regions shows that they contain similar functional elements that are somewhat differently arranged in the two instances (Fig. 15.6).

T Antigen in Temporal Regulation of Transcription

All SV40 templates are transcribed by the cellular RNA pol II system. That is one reason why SV40 has served as an important model for the study of mammalian gene

BOX 15.6

By analogy with transcription, one might think of the *ori* core as the promoter and of the auxiliary sequences as the enhancer. However, this analogy is not perfect, in part because SV40 *aux-1* and *aux-2* must lie proximal to opposite ends of *ori* core and in their proper orientations, whereas transcriptional enhancers can lie distal to either end of a promoter and in either orientation (see below). Still, it is interesting apropos this imperfect analogy that auxiliary sequences, including *aux-2* of SV40 and the mouse polyomavirus *aux-2α* and *aux-2β*, may consist of transcription factor binding sites (see below). Also, the mouse polyomavirus *aux-2α* and *aux-2β* indeed are components of that virus's transcriptional enhancer. However, the SV40 *aux-1* and *aux-2* are not components of the SV40 enhancer. Regardless, the SV40 and mouse polyomavirus examples, in which there is partial overlap of viral replication origins and transcription factor binding sites, show that transcription factors may participate in replication.

expression. In addition, the SV40 genome is small, and it is organized into nucleosomes, as noted above.

The following is a very brief overview of the pattern and regulation of SV40 gene expression, which is followed below by a more detailed account of possible underlying mechanisms and supporting experimental findings. As noted above, at early times after infection, early region transcription initiates from multiple start sites clustered at EE. At these times, transcription of the late region is inefficient at best. Then, coincident with the onset of viral DNA replication, there are two changes in this pattern of early transcription: (i) initiation of early region transcription shifts to multiple start sites clustered at LE, and (ii) efficient expression of the late region commences. Importantly, LT plays a crucial role in orchestrating this pattern of viral gene expression and replication, since LT is necessary for both the onset of viral DNA replication and the concurrent shift from EE to LE start sites, as well as for the beginning of efficient expression of the late region.

Bearing in mind that early region transcription and late region transcription are controlled from the same bidirectional promoter-regulatory region, what might be the mechanisms that underlie the LT-mediated temporally regulated changes in the pattern of SV40 gene expression? As we look into the relevant experimental findings, we will see that this issue remains to be better understood. However, it is clear that these switches are regulated by several cooperating mechanisms. For example, whereas LT affects the expression of all the SV40 promoters, it can affect transcription negatively (e.g., in the case of the EE promoter), as well as positively (e.g., in the case of the late promoter). (The remainder of this section reviews in some detail current understanding of how the SV40 LT protein orchestrates viral gene expression. For many readers this attention to detail may be excessive. However, note that in large part our incomplete understanding of the temporal regulation of SV40 gene expression reflects the fact that living systems [even relatively simple virus ones] may be far more complex than our working hypotheses might suppose. In spite of that fact, or perhaps because of it, I hope that most readers will persevere through the following elements of the story. Moreover, doing so will also provide examples of several reasonable hypotheses, some of which were ultimately at odds with experimental findings, a not infrequent outcome of the scientific endeavor.)

The regulatory activity of LT was first shown with respect to the SV40 early promoter. Initial studies made use of SV40 mutants that encode a temperature-sensitive LT protein. At their restrictive temperature, these mutants overproduce LT mRNA and, consequently, LT protein as well, thereby showing that LT autoregulates its own expression. Other experimental results showed that LT-mediated regulation of early region transcription requires the direct interaction of LT with the viral DNA. For example, transcription from an SV40 genome, which contains an intact early promoter but lacks LT binding site I or II, or both sites, was poorly inhibited by LT. In fact, maximum autoregulation of LT expression requires that LT binding sites I and II each be intact. Thus, both of these LT binding sites play a role in autoregulation. How might LT binding lead to autoregulation? As you may have supposed, when LT is bound to its binding sites, then RNA polymerase is blocked from binding to the early promoter.

The LT-mediated down-regulation of early region expression occurs at the same time as the LT-mediated activation of late region expression. That is so because the onset of each of these respective LT-mediated processes is dependent on the accumulation of LT in the cell. Remarkably, earlier in infection, when LT levels are low and LT binding sites are not yet occupied, LT stimulates rather than represses expression of its own promoter. Moreover, this LT-mediated early upregulation of the early promoter occurs by a mechanism that does not involve direct binding of LT to the viral genome. The nature of that mechanism is not entirely clear, but there is evidence that it involves an induction of, or effect on, cellular regulatory proteins, which then mediate the process. A somewhat similar process, involving an LT effect on cellular regulatory proteins, is thought to mediate the activation of the late promoter, as described below. First, we consider the mechanisms by which initiation of early transcription switches from EE to LE start sites.

We begin our discussion of the switch from EE to LE start sites by asking why the virus might use two differently regulated promoters for the expression of the same transcripts. Generally speaking, such a strategy might enable an organism to fine-tune gene expression to particular physiological or developmental conditions. Here, it enables the virus to fine-tune the amount of early region expression to its changing needs during the viral replication cycle. In this regard, it is important that the EE promoter is stronger than the LE promoter. Thus, the stronger EE promoter enables the virus to efficiently express the early region before the onset of viral DNA replication, when the template copy number is low. However, once viral DNA replication has begun, which gives rise to many more templates, the continued transcription of the early region from the highly efficient EE start sites might be wasteful, or perhaps even deleterious. Hence, a shift to the less efficient LE start sites could ensure the continued production of LT protein at levels appropriate to support the late stages of the replication cycle. This conjecture is consistent with the fact that the shift from EE to LE start sites requires both LT and the onset of viral DNA replication, as noted above. Now, we move on to the actual mechanism of the shift.

An intact LT binding site II, which incidentally also forms the *ori* core, is a crucial component of the switch from EE to LE start sites. This was indicated by genetic analysis in which mutations affecting LT binding site II prevented the switch. In contrast, mutations that affect LT binding sites I and III have no effect on this switch.

Now, consider the following. Since LT binding site II is at the *ori* core, site II mutants also are defective for DNA replication. Thus, does LT binding per se cause the switch from EE to LE start sites, or is DNA replication the crucial factor that triggers the switch? This question was answered by a clever experiment in which replication of the site II mutants was restored by inserting a functional origin at a site away from the early promoter region. When this was done and replication was allowed to proceed, levels of transcription from LE start sites were comparable to wild-type levels. Thus, replication per se, rather than LT binding, induces the switch from EE to LE start sites.

The following model was put forward to account for the above experimental findings. Early in infection, when LT levels are low, early transcription initiates from strong EE start sites. As LT levels increase, LT binds to site II, thereby depressing transcription from the strong EE start sites, while also initiating replication of the viral DNA (see below for the role of LT in initiating DNA replication). A key element of the model is that while LT binding to sites I and II impairs transcription from EE start sites, replication increases the copy number of transcription templates, thereby enabling increased transcription from inefficient LE start sites. (Can you envision other means by which DNA replication might activate transcription from LE start sites? Several might come to mind.) This model is consistent with the experimental finding that low levels of transcription occur from the inefficient LE start sites, even in the absence of DNA replication.

The 72-bp repeat elements, which constitute the enhancer regions, are also necessary for maximal expression of both the EE and LE start sites. However, that is so only for templates that are *not* replicating. In the case of replicating templates, the enhancer region is needed only for synthesis from EE start sites. Thus, replication appears to make the LE start sites independent of the viral enhancer. Perhaps the enhancer and replication each act to maximally open the chromatin structure to enable more efficient transcription from LE start sites. (The following experimental findings are not crucial to our main story, but they may provide food for thought for some readers. The six GC-rich repeats in the 21-bp repeat region contain binding sites for transcription factors and are promoter elements for transcription from both the EE and LE start sites. Nevertheless, these GC repeats play distinct roles in the expression of the EE versus the LE start sites. For example, mutations that affect the GC-rich repeat closest to the TATA box significantly decrease the activity of the EE start sites, whereas they increase that of the LE start sites. In contrast, mutations affecting the four GC repeats nearest to the 72-bp repeats decrease the activities of both the EE and LE start sites.)

Next, we consider activation of the late promoter, which, like the EE-to-LE switch, also occurs concurrently with the onset of viral DNA replication. Interestingly, LT can affect late promoter activity by multiple mechanisms, one of which does not involve direct binding of LT to the viral DNA. The latter mechanism is thought to involve effects of LT on cellular transcription factors. (Recall that very early in infection, when LT levels are low and LT binding sites are not yet occupied, LT can stimulate expression of the early promoter by a mechanism that also does not involve direct binding of LT to the viral genome. The nature of that mechanism too is thought to involve an LT-mediated induction of, or effect on, cellular regulatory proteins.)

Since LT binding to the viral DNA is necessary to initiate viral DNA replication, and since LT-mediated upregulation of the late promoter can occur in the absence of LT binding to the viral DNA, we would expect that the upregulation of the late promoter can occur, at least in part, in the absence of viral DNA replication. This indeed is so, as shown by the following experimental findings. First, LT can activate late genes in the presence of inhibitors of viral DNA replication. Second, LT can activate late genes even when *ori* is deleted from the DNA. However, the induction of late gene expression is most efficient when viral DNA replication is able to proceed. Clearly, replication amplifies the template copy number, accounting at least in part for the positive effect of replication on expression of the late region. In addition, some researchers have proposed that replication might titrate or sequester cellular factors that act as sequence-specific repressors of the late promoter. This possibility is discussed further below. First, to account for the replication-independent component of the LT-mediated upregulation of the late promoter, we consider the possibility that LT directly affects cellular factors that might facilitate late region transcription.

What cellular factors might LT interact with in order to activate the late promoter? The answer is not known with certainty, but several plausible candidates have been identified. One of these, the transcription-enhancing factor 1 (TEF-1), is a promising contender for the following reasons. First, SV40 LT interacts directly with TEF-1, and second, the late promoter region indeed contains TEF-1 binding sites. Third, mutations at all three late promoter TEF-1 binding sites prevent LT-mediated activation of the promoter, while not significantly affecting basal levels of late promoter activity. Fourth, the interaction of LT with TEF-1 might account for the fact that LT is a promiscuous activator of cellular promoters that do not contain LT binding sites. This is so because the presence of TEF-1 sites is sufficient for these heterologous promoters to be inducible by LT. (Have you been presuming that the interaction of LT with TEF-1 activates a transcription-enhancing function of this protein? See below for a completely different take on the role of TEF-1 in the regulation of the SV40 late promoter.)

Although the preceding experimental findings strongly imply that TEF-1 has a role in the LT-mediated upregulation of the late promoter, the interaction of LT with TEF-1 is probably not sufficient for LT to activate late gene expression. This deduction is implied by the discovery of deletion mutants of LT that are impaired in their ability to activate late transcription but still retain their ability to bind to TEF-1. (Incidentally, these LT mutants, which were unable to activate the late promoter, nevertheless retained their ability to down-regulate the early promoter. What conclusions might you draw from that finding?)

The above experimental results raise the possibility that LT may need to interact with one or more cellular factors, in addition to TEF-1, in order to activate late transcription. What then might these additional factors be? Reasonable candidates include activator protein Sp1 (which binds to the SV40 21-bp repeat region, as noted above), the TATA-binding protein (TBP), and RNA pol II itself. Indeed, LT is able to bind each of these factors. Moreover, the interaction of LT with each of these factors is specific, as shown by the fact that LT does not bind at all to numerous other transcription factors. However, the interaction of LT with TBP appears to be the only one of these interactions that has been well investigated.

But here is an apparent paradox. Although LT appears to bind specifically to TBP, the TATA box is a component of the early promoter, not of the late promoter. Despite this fact (or perhaps even because of it), the interaction of LT with TBP was put forward as an element of late promoter activation. The reasons are as follows. Transcription initiation in mammalian cells involves the ordered assembly of a functional preinitiation complex, which is usually comprised of RNA pol II and a number of general initiation factors. The latter include TFIID, which itself is comprised of several proteins, one of which is TBP. TFIID starts off the assembly of a functional preinitiation complex by interacting specifically with the TATA sequence. But what then of the SV40 late promoter, which does not contain a TATA sequence? One hypothesis for how the interaction of LT with TBP might activate the "TATA-less"

SV40 late promoter is that LT might bind concurrently to the late promoter and to the preinitiation complex, in that way bringing the preinitiation complex to the TATA-less late promoter. In this model, LT would interact with the preinitiation complex via its interaction with TBP. This model is attractive because it might explain the promiscuous activation by LT of many promoters, since it now is known that TBP is required for transcription of TATA-less as well as TATA-containing promoters. Another possibility is that LT might stabilize the preinitiation complex by binding concurrently to TBP and other transcription factors. This notion is attractive because it accounts for the fact that LT does not need to bind directly to the DNA in order to activate late transcription. However, there are other experimental findings that conflict with each of these models. For example, LT can transactivate late transcription even when it is restricted to the cytoplasm (i.e., by a mutation in its nuclear localization signal). Also, LT does not appear to be able to bind concurrently to more than one of the transcription factors enumerated above. These provisos led to the hypothesis that transactivation by LT mainly results from its putative ability to remove, or better yet, to prevent the binding of factors that inhibit the formation of preinitiation complexes. An attraction of the latter variant of this hypothesis is that it explains the ability of LT to transactivate late transcription even when restricted to the cytoplasm, from where it might sequester limiting amounts of the putative inhibitory factors.

So, we now ask whether there is any experimental evidence in support of the premise that LT upregulates the late promoter by sequestering factors that might inhibit late promoter activity. In fact, the answer is yes. Several research groups provided evidence that the transcription factor TEF-1, discussed above, actually might be an inhibitor rather than an activator of the late promoter and that LT interacts with TEF-1 to relieve that repression. First, LT binding to TEF-1 prevents TEF-1 from binding to DNA, and in so doing, LT activates transcription from the late promoter, at least in vitro. Second, late start sites also may be activated by mutations in the TEF-1 binding sites which are near the late promoter. Finally, LT mutants that could not bind to DNA but retained the ability to bind to TEF-1 were found. These mutant LT molecules indeed were able to activate the late promoter. Together, these experimental findings imply that LT can activate the late promoter by sequestering the putative inhibitor protein, TEF-1, a process that does not require LT binding to the late promoter.

TEF-1 is not the only cellular protein that might negatively regulate the SV40 late promoter, and the sequestering of these factors by LT may not be the only means by which LT relieves their repressive effect. Three proteins isolated from HeLa cell extracts, referred to collectively as

IBP-s (for initiator-binding proteins of SV40), each associate with high affinity to the SV40 late promoter, and in so doing, these IBP-s in fact repress transcription from the promoter (at least in a cell-free system). SV40 mutant genomes that lack the IBP-s binding sites are not repressed by the IBP-s in vitro. Moreover, these mutants overproduce late mRNA at early times after transfection into monkey cells, while also underproducing early transcripts.

The IBP-s were subsequently found to be comprised of members of the steroid/thyroid hormone receptor superfamily. Importantly, binding of IBP-s to the SV40 late promoter prevented preinitiation complexes from assembling at the promoter. Also, the thyroid hormone L-triiodothyronine (T3) can reverse this repression of the late promoter in transfected monkey cells. These novel experimental results imply that the SV40 late promoter could be regulated by steroid hormones. That is, the SV40 late promoter may contain a thyroid hormone response element. (Thyroid hormones are involved in regulating differentiation, growth, and metabolism in a variety of tissues of all vertebrates. Their effects are mediated by specific thyroid hormone receptors that are located intracellularly, rather than at the cell surface. Hydrophobic thyroid hormones diffuse across the plasma membrane and attach to their specific intracellular receptors, which may be bound to DNA on specific sequences referred to as thyroid hormone response elements or thyroid response elements, which are located in promoters of target genes. Ligand binding to a thyroid hormone receptor may either activate or repress expression of responsive genes, depending on the particular ligand/receptor complex.)

Why might SV40 mutants lacking IBP-s binding sites underproduce early mRNA? One possibility is that since such mutant late promoters are no longer impeded by IBP-s, they might compete with the early promoter for limiting amounts of *trans*-activating factors necessary for early and for late mRNA synthesis. But more to the point, what might be the mechanism by which IBP-s-mediated repression of the late promoter is relieved during the replication cycle of the virus? The answer is not yet known, but one suggestion is that the LT-mediated replication of viral DNA to high copy number might titrate away the IBP-s proteins. This possibility might account for that fraction of late promoter upregulation that is dependent on replication. However, the general significance of the late promoter IBP-s binding sites to the viral life cycle is not clear, since they seem to matter in some cells but not at all in others.

Clearly, the means by which the single SV40 LT protein orchestrates the pattern of viral gene expression during the course of infection may be quite complex and remains to be better understood. Nevertheless, the multiple activities that LT employs to coordinate these events are remarkable. Bear this in mind as we consider additional roles of the very same LT protein in the replication of the viral genome, in the maturation of the virus particle, and in its effects on the growth state of the host cell.

Before we leave the issue of the temporal regulation of SV40 gene expression, note that a recent study demonstrated that an SV40 sequence, located in the untranslated region that is 3′ to the polyadenylation cleavage site of the late pre-mRNA, encodes a miRNA that is complementary to early mRNAs. This miRNA marks early mRNAs for degradation. This study and miRNAs in general are considered in more detail later (see "SV40 in Humans," below).

Splicing Pattern of Viral mRNA

Studies involving SV40 transcription were among the very first to demonstrate the phenomenon of **splicing of mRNA precursor RNAs**. (The discovery of mRNA splicing actually came about during analysis of adenovirus mRNA structure, as described in Chapter 17. Philip Sharp and Richard Roberts directed separate research groups that independently reported their findings in 1977. The next year, Sharp and coworker Arnold Berk reported their findings regarding SV40, as described here. Sharp and Roberts were awarded the Nobel Prize in 1993 for their independent discovery of mRNA splicing.) The discovery that SV40 precursor mRNAs are spliced came about because of several paradoxical observations concerning the organization of SV40 DNA sequences that encode LT and ST. First, about one-third of the way into the SV40 early region, there are two UAA chain termination triplets in each of the three reading frames. Second, neither of the sets of open reading frames on either side of these chain termination triplets is large enough to encode LT. Third, deletion mutations encompassing the chain termination triplets do not prevent production of wild-type LT but do prevent expression of functional ST. Fourth, LT and ST contain a common N terminus but distinct C termini.

Arnold Berk and Phil Sharp resolved these paradoxes by demonstrating that LT is translated from two open reading frames, one on each side of the site containing the chain termination triplets in each of the three reading frames (Berk et al., 1978). That is, the site containing these chain termination triplets is spliced out of a precursor RNA molecule to generate the LT mRNA. Stated somewhat differently, LT is encoded in two separate regions of the viral genome, which are brought together by the phenomenon of mRNA splicing (Fig. 15.4). ST and 17kT are translated from mRNAs generated by alternative splicing of the early precursor RNA. The late primary transcript likewise is differentially spliced for translation into the different capsid proteins (see below).

Why did SV40 evolve this complex pattern of transcription, in which its early and late primary transcripts are spliced in different ways to yield different proteins? One thought is that it enables the virus to maximize the efficiency of its relatively small genome. The phenomenon of RNA splicing actually was first discovered as an event in adenovirus transcription, as noted above. Shortly afterwards it was seen for SV40, as described here, and is now known to occur throughout the eukaryotic kingdom.

Looking more carefully at splicing events for SV40 early pre-mRNA, splicing occurs at either of two 5′ splice sites to form mRNAs encoding either LT or ST (Fig. 15.4). As noted above, LT and ST have a common N terminus but differ at their C termini. The mRNA encoding ST actually is longer than the mRNA encoding LT. The reason is that the ST mRNA contains a termination triplet that is spliced out of the LT mRNA. The still-mysterious 17kT (not shown in the figure) likewise has the same N-terminal sequences as ST and LT but yet a third C terminus, as a result of a different 3′ splice site in its mRNA.

As in the case of SV40, mouse polyomavirus encodes an ST molecule and an LT molecule. However, both the mouse and hamster polyomaviruses encode yet another T antigen, the middle T (MT) antigen (see below). Each of these polyomaviral T antigens is generated by the translation of a differentially spliced mRNA.

Next, we consider expression of the SV40 late region. In contrast to the means by which the overlapping early-region-encoded T antigens are translated, late proteins VP2 and VP3 are translated from the same mRNA molecules (Fig. 15.4). In this instance, VP3 is translated by means of a leaky scanning mechanism, in which the VP2 translation initiation site is occasionally bypassed by the ribosome. This process results in the two distinct proteins that have a common C terminus.

Interestingly, the sequence encoding the N terminus of VP1 overlaps the sequence encoding the C termini of VP2 and VP3. Moreover, the sequence encoding VP1 is translated in a different reading frame from the sequence encoding VP2 and VP3, and from a differently spliced mRNA (Fig. 15.4). (Consider for a moment that this particular region of the SV40 genome encodes three different proteins in two different reading frames.) As noted above, the SV40 agnoprotein is encoded by the leader region of some species of late mRNA.

DNA REPLICATION

Our consideration of polyomavirus DNA replication begins with the following important points, which also are relevant to DNA replication in the cases of the papillomaviruses (Chapter 16) and adenoviruses (Chapter 17). Because

of the complex nature of DNA replication, as imposed by the requirements for priming, bidirectional replication at the replication fork, the need to unwind the helix, and so forth, DNA replication requires multiple enzyme activities. Since DNA replication requires multiple enzyme activities, and because viruses have limited coding capacity, all DNA viruses except the poxviruses make extensive use of cellular DNA replication activities. Yet, invariably, the replication of DNA viruses requires the expression of at least one viral protein. Why might that be so?

Now, bearing in mind that the polyomaviruses, papillomaviruses, and adenoviruses are highly dependent on cellular DNA replication activities, note that differentiated cells in vivo rarely divide. Instead, these cells are usually in a nondividing state termed G_0. Importantly, such cells contain very low levels of the enzymes and substrates needed for DNA replication. Because polyomaviruses (and papillomaviruses and adenoviruses as well) are dependent on those enzymes and substrates for their replication, they would not be able to replicate efficiently unless they were able to induce resting cells to enter into the cell cycle. In the case of SV40, the LT protein induces cells to bypass the complex circuits that regulate exit from G_0. The means by which LT does so is described in detail below (see "T Antigens and Neoplasia"). For the moment, note that the ability of the polyomaviruses to induce cell cycling underlies their ability to induce neoplastic transformation of cells in culture and to cause tumors in experimental animals.

As noted above, SV40 has provided an important model for studying fundamental features of eukaryotic DNA replication. The usefulness of SV40 in this regard is due to several facts. First, as just noted, SV40 makes extensive use of the host cell DNA replication machinery. Second, a single SV40 protein, LT, orchestrates the entire SV40 DNA replication process in monkey cell-free extracts. As described below, LT directs the initiation of viral DNA replication by specifically binding to the SV40 *ori* at GAGGC sequences, where it assembles into a double hexamer, which then recruits several cellular DNA replication proteins, including cellular replication protein A, topoisomerase I, and polymerase-α/primase, to form a replication initiation complex.

The circular double-stranded SV40 genome is replicated bidirectionally. That is, replication proceeds from *ori* in each direction around the circle (Fig. 15.8). This was demonstrated rather dramatically by making use of the fact that the EcoRI restriction endonuclease makes a single cut on the SV40 genome. Each end of the expanding replication "bubble" can be seen to approach its nearest end of the EcoRI-cleaved viral genome (Fig. 15.8). Note that by having a circular DNA genome, SV40 avoids the

A

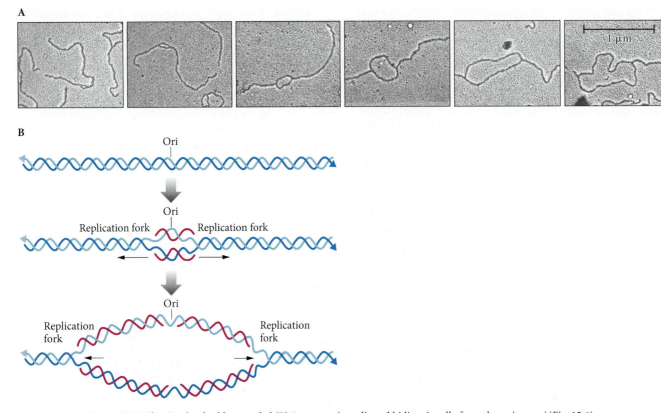

B

1 μm

Figure 15.8 The circular double-stranded SV40 genome is replicated bidirectionally from the unique *ori* (Fig. 15.6). This was demonstrated by making use of the fact that the EcoRI restriction endonuclease makes a single cut on the SV40 genome. (A) The electron micrographs are arranged to show how each end of the expanding replication bubble progressively approaches each corresponding end of the EcoRI-cleaved viral genome. From G. C. Fareed et al., *J. Virol.* **10**:484–491, 1972, with permission. (B) A graphic illustration of the events shown in panel A. Adapted from S. J. Flint, L. W. Enquist, R. M. Krug, V. R. Racaniello, and A. M. Skalka, *Principles of Virology: Molecular Biology, Pathogenesis, and Control*, 2nd ed. (ASM Press, Washington, DC, 2004), with permission.

"end replication problem" of linear DNA genomes (see Chapter 14 for a discussion of this issue).

The functions of LT and of several of the cellular proteins that participate in replicating the SV40 genome are as follows. The LT double hexamer acts as a helicase to unwind each of the two replication forks. Topoisomerase I is a critical enzyme that acts to release the topological stress created by DNA unwinding. It does so by nicking and religating the DNA ahead of the replication fork. Replication protein A is a single-stranded DNA-binding protein. As such, it binds to the free single-stranded DNA generated by the LT helicase activity, thereby stabilizing the unwound DNA and enabling the DNA polymerase-α/primase to lay down the RNA primer and then extend it with a short stretch of DNA. LT remains associated with the leading strands of the DNA during elongation, continuing to unwind the template at the replication forks. (Readers interested in a detailed review of these events might see Simmons et al., 2004).

The cellular replication machinery carries out all subsequent steps on both the leading and lagging DNA strands. Additional cellular proteins include polymerase-δ, PCNA, and RF-C. In all, about 11 cellular proteins are used to fully replicate the viral DNA.

Surprisingly little is known of the actual way in which these proteins interact in DNA replication, although recent experimental results demonstrated a direct interaction between LT and topoisomerase I. More precisely, two regions of LT bind to two regions on topoisomerase I. Also, topoisomerase I must be present at the time the initiation complex is forming in order for LT to promote initiation. Importantly, the interaction between topoisomerase I and LT causes the unwinding and nicking to be specific to *ori*. Bear these new points in mind when considering all the activities and interactions of the 708-amino-acid LT protein.

The ability of SV40 LT to initiate viral DNA replication is regulated by specific phosphorylations of serine and threonine residues located in clusters at each end of the LT molecule (Fig. 15.5). Phosphorylation of one residue, Thr124, is particularly important because it facilitates the binding of LT to its major *ori* binding site, site II (Fig. 15.7).

Now, here is an interesting point. The phosphorylation of LT and hence its ability to initiate DNA replication are dependent on the activity of a **cyclin-dependent kinase**, cdc2, which is activated at the time the cell enters S phase. (The cyclin-dependent kinases [Cdks] are activated by the **cyclins**, a family of proteins that control progression through the distinct phases of the eukaryotic cell cycle [see below].) This cell cycle dependence of cdc2 kinase activation and the phosphorylation of LT explain in part why viral DNA replication must wait until the start of the first S phase following infection. Regulation of the replication

activity of LT in this way is beneficial to the virus because it prevents abortive attempts at replication before the cell has entered S phase and thus before necessary replication factors that might support replication are in place. This is rather clever on the part of the virus.

Electron microscopy of LT hexamers shows that they are planar rings containing a central hole through which the DNA presumably passes (Fig. 15.9). Assembly of hexamers begins when a single LT monomer binds to a specific pentanucleotide sequence (GAGGC) at the origin (Fig. 15.7). The bound LT monomer then associates with other monomers

Figure 15.9 LT and the unwinding of the SV40 origin. (A) In vitro, in the presence of ATP but in the absence of DNA, LT assembles into hexamer complexes. These are active as helicases when a partially single-stranded (3′) entry site exists on the DNA substrate. In electron micrographs, these hexamers appear as propeller-shaped structures with a large central hole. From M. C. San Martin, C. Gruss, and J. M. Carazo, *J. Mol. Biol.* **268:**15–20, 1997, with permission. (B) Assembly of hexamers begins in vivo when a single LT monomer binds to a specific pentanucleotide sequence (GAGGC) at the origin (also see Fig. 15.7). The bound LT monomer then associates with other monomers to form a single hexamer. The second hexamer then forms from individual monomers over a second GAGGC pentanucleotide sequence. When the double hexamer forms over the origin, 8 bp of DNA at a flanking early palindromic (EP) region is melted, thereby initiating DNA synthesis. LT also recruits several cellular DNA replication proteins, including topoisomerase I, polymerase-α/primase, and cellular replication protein A (RpA), a single-stranded-DNA-binding protein that stabilizes the unwound DNA. The initiation of DNA replication is then mediated by the complex of RpA, DNA polymerase α/primase, and LT. The drawing depicts the two hexamers moving apart, but in actuality, they are thought to remain together. Adapted from S. J. Flint, L. W. Enquist, R. M. Krug, V. R. Racaniello, and A. M. Skalka, *Principles of Virology: Molecular Biology, Pathogenesis, and Control*, 2nd ed. (ASM Press, Washington, DC, 2004), with permission.

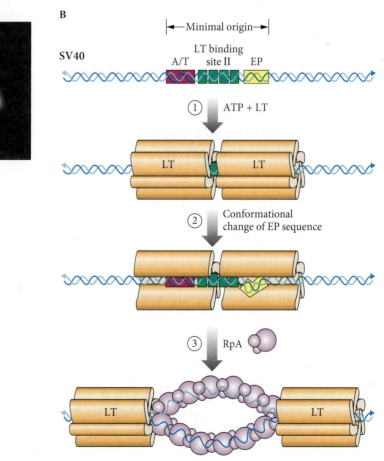

to form a single hexamer. The second hexamer then forms from individual monomers over a second GAGGC penta-nucleotide sequence. When the double hexamer forms over the origin, 8 bp of DNA at a flanking early palindromic region (EP in Fig. 15.9) are melted, thereby initiating DNA synthesis.

The helicase activity of the double LT hexamers is powered by the binding and hydrolysis of ATP. As a consequence of the cycle of ATP binding and hydrolysis, the LT hexamers exist in three nucleotide-binding states: ATP bound, ADP bound, and nucleotide free. Recently, high-resolution X-ray diffraction structures of LT hexamers were determined for each of these three distinct nucleotide binding states (Gai et al., 2004).

As hoped, these X-ray structures provided insights into how conformational changes in the LT hexamers, which result from ATP binding and hydrolysis, might initiate DNA "melting" and replication. It is interesting that these conformational changes affect the angles and orientations between the individual LT monomers rather than the conformations of the individual monomers per se. That is, LT binding and hydrolysis lead to changes in the molecular arrangement between the individual LT monomers within the hexamer, not to changes within the individual monomers. These conformational changes within the hexamer are most apparent in the size and shape of the central channel and in the contacts between the monomers (Fig. 15.10).

The sequence of molecular rearrangements of the hexamers caused by ATP binding and hydrolysis appears to be as follows. When ATP binds, residues on the neighboring monomer B move to the ATP site on monomer A to interact with that ATP. This reorientation results in a closure

Figure 15.10 The helicase activity of the double LT hexamers is powered by the binding of ATP and its hydrolysis. Consequently, the LT hexamers exist in three nucleotide-binding states. These are the ATP-bound state (A), the ADP-bound state (B), and the nucleotide-free state (C). Each of the LT monomers is shown in a different color. The six ATPs in panel A and the six ADPs in panel B are shown in pink. To provide a clearer view of the ATP binding cleft at the interfaces between the LT monomers, the N-terminal D1 domains of the LT monomers are not shown. (D) A close-up view of the cleft between two neighboring LT monomers (in green and cyan) of the ATP-free structure. (E) The same view of the cleft between two LT neighbors in two different ATP-bound states: the ATP-free structure (in green and cyan) and the ATP-bound structure (in gold). The bound ATP and Mg^{2+} are shown in purple. Note the narrowing of the cleft when ATP is bound. This reorientation results in a closure and twisting of the hexamer, which create an "iris"-like motion in it, which is thought to promote the translocation of the DNA (also see Fig. 15.11). From Gai et al., *Cell* **119**:47–60, 2004, with permission.

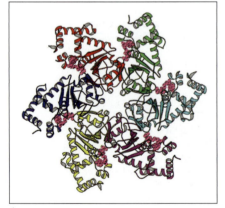

A

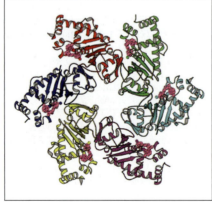

B

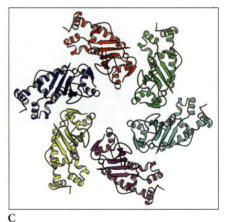

C

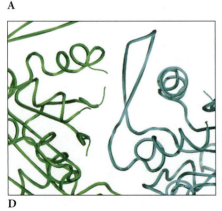

D

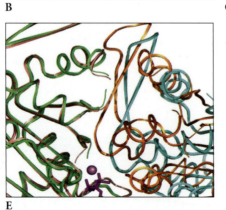

E

and twisting of the hexamer, which create an "iris"-like motion in it. Additionally, each of the LT monomers has a unique β-hairpin structure on the central channel surface. The conformational changes induced in the hexamer by ATP binding cause these β hairpins to move longitudinally along the central channel, perhaps serving as a motor to draw the DNA into the central channel of the double hexamer (Fig. 15.11). When ATP binds, the β hairpins move about 17 Å inward toward the center of the double hexamer, drawing about 5 or 6 bp of the double-stranded DNA into the central channel and thereby promoting strand separation. Single-stranded DNA is thought to exit from side channels in the hexamers. ATP hydrolysis and ADP release cause the hexamers to reassume their original conformation, so that the cycle might repeat.

The N terminus of LT (and of ST as well) contains a so-called **DnaJ domain** (Fig. 15.5), which is necessary for SV40 DNA replication. The LT DnaJ domain resembles the J domain of the DnaJ family of host chaperones, which specifically interact with, and modulate, other chaperones. Actually, the DnaJ class of chaperones might be thought of as cochaperones, since they promote the interaction of heat shock 70 (Hsc70) family chaperones with their substrates and stimulate their ATPase activity. DnaJ chaperones, including LT, interact with Hsc70 chaperones via their J domain.

The need for the LT DnaJ domain in DNA replication was shown by the experimental finding that mutations mapping in the DnaJ domain are defective for DNA replication. Moreover, a recombinant chimeric T antigen, in which the LT DnaJ domain is replaced with a human Hsj1 J domain, is able to support SV40 DNA replication.

The role of the LT DnaJ domain in SV40 DNA replication is not yet clear. One thought is that the LT DnaJ domain recruits Hsc70, which is needed to properly arrange components of the replication complex to carry out replication. Indeed, a precedent for that model is provided by λ (lambda) phage, which uses the *Escherichia coli* DnaK (the prokaryotic analog of Hsc70) and DnaJ proteins to free up the DnaB helicase activity, which is otherwise inhibited in the preinitiation complex. This results in unwinding of the λ phage DNA template and initiation of its replication. (Note that the SV40 LT DnaJ domain also acts in virus assembly and tumor genesis [see below].)

LATE PROTEINS AND ASSEMBLY

VP1 is the major structural protein of the icosahedral polyomavirus capsids. Monomers of VP1 assemble into pentamers in the cytoplasm, and each pentamer also binds

Figure 15.11 A model depicting the coupling of the conformational rearrangements of LT hexamers to the translocation and unwinding of double-stranded DNA. (A) An LT double hexamer, with two single-stranded DNA loops coming out from side channels. The interaction of ATP with LT hexamers causes a closure and twisting of the hexamers, resulting in an "iris"-like motion in the hexamers (Fig. 15.10). Additionally, each of the LT monomers has a unique β-hairpin structure on the central channel surface, as represented by the red bars within the helicase domain. The conformational changes induced in the hexamers by ATP binding also cause these β hairpins to move longitudinally along the central channel, perhaps serving as a motor to draw the DNA into the central channel of the double hexamer. When ATP binds, the β hairpins move about 17 Å inward toward the center of the double hexamer, drawing about 5 or 6 bp of the double-stranded DNA into the central channel and in that way promoting strand separation. Single-stranded DNA is thought to exit from side channels in the hexamers. ATP hydrolysis and ADP release cause the hexamers to reassume their original conformation, so that the cycle might repeat. (B) An LT hexamer corresponding to the left half of the LT double hexamer in panel A in the ATP-free state. (C) The β hairpins move when ATP binds, thereby drawing the DNA into the helicase for unwinding. (D) The β hairpins move about halfway back to the ATP-free position after ATP hydrolysis. (E) The β hairpins move back to the ATP-free position upon ADP release, ready to repeat the cycle. The colored dots on the DNA are position markers for translocation. Adapted from D. Gai et al., *Cell* **119**:47–60, 2004, with permission.

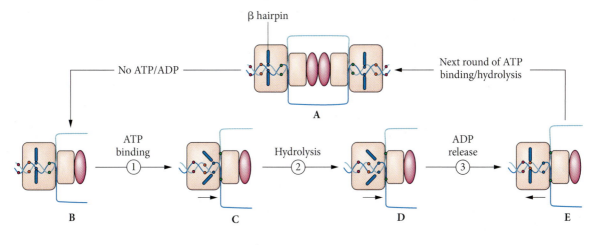

a monomer of either VP2 or VP3 in its central fivefold cavity. These VP1 pentamers, with their associated monomer of either VP2 or VP3, are transported into the nucleus, where they assemble into virus particles. Each capsid is comprised of 72 pentamers. One pentamer is located at each of the 12 vertices of the icosahedron. Each of these 12 vertex pentamers is surrounded by five other pentamers, whereas each of the remaining 60 pentamers is surrounded by six other pentamers (Fig. 2.8).

The presence of VP1 pentamers in both pentameric and hexameric arrays in polyomavirus capsids is a distinguishing structural characteristic of these viruses. It requires that there be multiple kinds of nonidentical contacts between VP1 molecules that reside in separate pentamers, thereby violating the principle of quasiequivalence. In actuality, polyomavirus VP1 molecules display three distinct modes of interpentameric interactions. These different interactions are possible because they involve the flexible N-terminal and C-terminal arms of the VP1 molecule, rather than the face-to-face contacts that characterize the interactions between pentamers and *hexamers* of most other icosahedral viruses. The C termini of VP1 molecules interdigitate into the VP1 molecules of neighboring pentamers, and as noted above, they do so in three different ways (Fig. 15.1).

The fact that polyomaviral interpentameric bonding involves three distinct and specific complex molecular interactions between identical VP1 subunits implies that these interactions need to be regulated during assembly. This leads to the conjecture that chaperones might be needed to regulate polyomavirus assembly. Consistent with that premise, mutations in the LT DnaJ domain impair capsid assembly. In addition, Hsc70 binds to the C termini of newly synthesized VP1 molecules in the cytosol and then cotranslocates with them to the nucleus, where virus assembly then ensues.

The need for chaperone activity during SV40 assembly is underscored by the fact that VP1 has the intrinsic tendency to spontaneously assemble in vitro into polymorphous capsid-like structures, a process that is triggered by the addition of calcium. In contrast, assembly in vivo generates virus particles of uniform structure. (Apropos the above, chaperones generally promote the proper folding of proteins. In addition, they also prevent the aberrant aggregation of proteins that might occur while they are still in an incompletely folded configuration.)

All polyomaviruses undergo assembly in the nucleus. LT (which is synthesized early), like VP1 (which is synthesized late), has a nuclear localization signal that targets it to the nucleus. Thus, LT already is in place to trigger virus assembly upon the arrival of VP1 pentamers in the nucleus. (The first nuclear localization to be identified was in fact that of SV40 LT. Proof that the LT nuclear localization

signal indeed acts as an autonomous signal capable of specifying nuclear location was provided by transferring it to two normally cytoplasmic proteins, beta-galactosidase and pyruvate kinase, causing them to be transported to the nucleus.)

Actually, the benefits of the convergence of VP1 pentamers with LT in the nucleus may be more impressive yet, since there also is a need for factors that might couple capsid assembly to encapsidation of the viral minichromosome. Since LT binds to the minichromosome, while also functioning as a cochaperone during capsid assembly, LT also might be a factor that couples encapsidation to final virus assembly. However, note that neither LT nor ST is found in virus particles.

SV40 DNA itself contains a packaging signal, which is located in a segment of the regulatory region that also contains binding sites for several transcription factors. VP2 and VP3 are thought to recognize the packaging signal indirectly by binding to the particular transcription factors SP1 and AP2.

The enigmatic SV40 agnoprotein, which like LT is not found in virus particles, also might play a role in virus assembly. This premise is based on several experimental observations. First, mutation of the agnoprotein results in the slow accumulation of progeny virus particles. Second, a missense mutation in the SV40 VP1 protein resulted in small plaques and reduced virus yields. Apropos the agnoprotein, this VP1 mutant gave rise to six independently isolated **pseudorevertants**, all of which mapped to the same serine residue in the agnoprotein. (A pseudorevertant is a revertant in which a mutation that affects a second gene compensates for the defect of the original mutation.) Moreover, in each of these second-site revertants, that particular serine was replaced with a hydrophobic amino acid. The simplest interpretation of these experimental findings is as follows. First, VP1 and the agnoprotein interact in a specific way during virus assembly. Second, the VP1 mutation impaired that interaction, and third, the compensating mutation in the agnoprotein somehow restored it.

Other experimental results, from studies also involving agnoprotein mutants, imply that the agnoprotein is involved in localizing VP1 to the nucleus. A recent study suggests that the agnoprotein may disturb the nuclear envelope to promote the egress of progeny virus particles from the nucleus.

The mechanism by which progeny virus particles are released at the end of the replication cycle is not known with certainty. As just noted, there is experimental evidence that the agnoprotein may promote nuclear release by destabilizing the nuclear envelope, perhaps doing so via its interactions with the **nuclear lamina** (a meshwork of intermediate membrane filaments that lies under the inner nuclear membrane, providing structural support to the nucleus). Results of other studies suggest a role for the nuclear enzyme **poly(ADP-ribose) polymerase (PARP)**, which normally functions in DNA damage surveillance and repair and in the decision between apoptosis and necrosis. SV40 VP1 and VP3 each bind to PARP. In addition, VP3, but not VP1, stimulates PARP activity. Moreover, stimulation of PARP activity by VP3 leads the infected cell to a necrotic pathway, characterized by the loss of membrane integrity. In that way, stimulation of PARP might facilitate release of mature SV40 particles from the cells.

It also is possible that virus release might depend on the regulated viroporin activities of VP2 and VP3 (Box 15.5). In addition, SV40 recently was found to encode a previously unknown 15-kDa protein, VP4, which like VP3 initiates synthesis from a downstream AUG start codon within the VP2 transcript. VP4 is expressed about 24 h after the onset of VP1 expression, immediately before cell lysis. In the absence of VP4 expression, cell lysis is delayed. While VP2 and VP3 are able to permeabilize host cell membranes, they may do so only in a nonlethal way in the absence of the new 15-kDa protein. The experimental data also suggest that VP4 may interact with VP3 to change the physical properties of the latter, resulting in its enhanced lytic activity (Daniels et al., 2007).

T ANTIGENS AND NEOPLASIA

The first role of SV40 LT (and ST as well) is to prepare cells to support viral replication. In this regard, differentiated cells in vivo rarely divide and usually are in a nondividing "resting" state termed G_0. Such cells contain very low levels of the enzymes and substrates required for DNA replication, transcription, and translation. Because polyomaviruses are dependent on those enzymes and substrates for their replication, they would not be able to replicate efficiently unless they were able to induce resting cells to leave G_0 and enter the cell cycle. The ability of polyomavirus T antigens to induce cell cycling underlies their ability to induce neoplastic transformation of cells in culture and tumors in experimental animals. The possibility that these viruses might be etiologic agents of human cancer is discussed later in the chapter.

Before getting on with our discussion of the T antigens in neoplastic transformation by the polyomaviruses, has the following occurred to you? On the one hand, polyomaviruses can induce neoplastic transformation because of their ability to induce cell cycle progression, which is necessary for efficient virus replication. On the other hand, polyomavirus replication is cytolytic, invariably resulting in cell death. So here then is the conundrum. Since

a transformed cell is one that proliferates under conditions where its normal counterpart does not, and since dead cells do not proliferate, how might the cytolytic polyomaviruses transform cells? The answer is that cell transformation generally occurs only in **susceptible** cells that support transcription of the viral early region T-antigen genes but do not express one or more factors that are necessary for the virus to complete its replication cycle and kill the cell (Fig. 15.12). (Some definitions might be in order here. Virologists sometimes make a distinction between cells that are **permissive** for a particular virus and

Figure 15.12 SV40 replication versus transformation. On the one hand, SV40 and other polyomaviruses can induce neoplastic transformation because of their ability to induce cell cycle progression, which is necessary for efficient virus replication. On the other hand, the replication of all polyomaviruses is cytolytic, invariably resulting in cell death. Since a transformed cell is one that proliferates under conditions where its normal counterpart does not, and since dead cells do not proliferate, polyomaviruses can only transform cells in which lytic replication is not possible. For a cell to be transformed by a polyomavirus, it must support transcription of the viral early region T-antigen genes but must not express one or more factors that are necessary for the virus to complete its replication cycle and kill the cell. Adapted from G. M. Cooper, *The Cell: a Molecular Approach*, Sinauer Associates, Inc., Sunderland, MA, 1997.

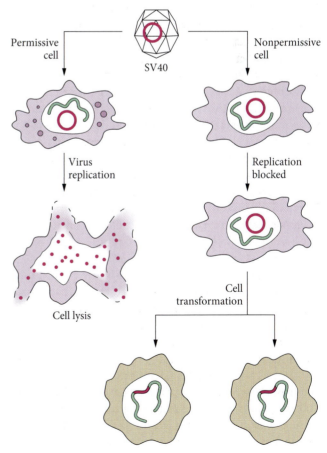

those that are merely *susceptible* to that virus. A permissive cell expresses all the factors that are necessary for the virus to complete its replication cycle. In contrast, a susceptible cell is one that expresses a receptor that the virus can engage to facilitate its entry. What cellular factor then determines permissiveness for the polyomaviruses? In the case of SV40, permissiveness largely depends on the species of origin of the cellular DNA polymerase α-primase. This enzyme efficiently forms complexes with SV40 LT [an interaction that is required for SV40 DNA replication] only when it is produced in certain cells of monkey or human origin.) Consequently, cell transformation is not an event associated with polyomavirus replication. Instead, polyomaviruses transform nonpermissive susceptible cells as an inadvertent consequence of their replication strategy in permissive cells.

Since polyomavirus-transformed cells do not produce progeny polyomaviruses, cell transformation is of little use to these viruses. Yet regardless of that stipulation, neoplastic transformation induced by polyomaviruses provided an important model for tumorigenesis, revealing crucial facts that are recounted below.

It is now about 50 years since the discovery that polyomaviruses can induce neoplastic transformation. Very soon after those initial discoveries, two alternative plausible general schemes were put forward to account for viral transforming activity. In the first, viral transforming activities were hypothesized to induce a permanent change in the physiological state or phenotype of the cell. These changes were imagined to be like those leading to irreversible differentiation of normal cells, but in these instances they result in neoplastic transformation. An implicit feature of this model is that after the initial induction of the transformed cell phenotype, viral gene expression would not be necessary to maintain that phenotype. This scheme is referred to as a **hit-and-run** mechanism, reflecting the key premise that the continued presence of the viral genome would not be required to maintain the transformed cell phenotype. In the alternative scheme, maintenance of the transformed cell phenotype requires the continued expression of viral transforming genes.

It was crucially important to distinguish between the above alternative possibilities as a first step to understanding transformation by these viruses. Before reading on, can you think of an experiment that might distinguish between these different pathways to transformation? The actual experimental approach (ca. 1965) was as follows. First, it was presumed that since the polyomaviral genome is small, any viral function that might be required for cell transformation would necessarily be required by the virus for its replication as well (i.e., viruses usually do not carry excess "genetic baggage"). This presumption implies that

viral mutants that are conditionally defective for cell transformation should be present among viral mutants that are conditionally defective for replication. Thus, mutants of mouse polyomavirus that were temperature sensitive for replication in permissive cells were generated and isolated. Then, these mutants were individually screened for their ability to transform nonpermissive cells at the respective temperatures that were permissive and restrictive for replication in permissive cells. Indeed, a mouse polyomavirus mutant was identified that was temperature sensitive for transformation of nonpermissive cells (the nonselected characteristic), as well as temperature sensitive for replication in permissive cells (the selected characteristic). Finding this mutant confirmed the premise that viral cell-transforming functions are *necessary* for replication. Importantly, this temperature-sensitive mutant could be used to transform nonpermissive cells at the mutant's permissive temperature. These transformed cells continued to express the transformed cell phenotype when they were incubated at the permissive temperature. However, the cells reacquired the normal cell phenotype when they were incubated at the mutant's restrictive temperature. Thus, maintenance of the transformed cell phenotype requires the *continuous activity* of at least one polyomaviral gene.

Here is another conundrum. Since one or more polyomaviral genes need to be expressed continually to maintain the transformed cell phenotype, and since only nonpermissive cells, which do not replicate polyomaviral genomes, are transformed by these viruses, how do viral genes persist in these nonpermissive transformed cells? The answer is that viral gene persistence in these cells results from the **integration** of the viral genome into the cellular chromosomal DNA. However, this integration is a chance rare event that is not a part of the viral life cycle in permissive cells. Consequently, the efficiency of cell transformation by polyomaviruses is very low.

Now, we are ready to consider the molecular mechanisms that underlie neoplastic transformation as induced by the polyomavirus T antigens. To set the stage, we briefly note that SV40 LT induces cells to bypass the complex circuits that regulate exit from G_0 by abolishing the normal functions of two particular types of **tumor suppressor proteins**. These are (i) **p53** and (ii) members of the **pRb** family of proteins. The latter includes the retinoblastoma tumor suppressor protein pRb and the related proteins p130 and p107. These facts are discussed in detail below. But for now we can assert that the study of the transforming activities of the polyomaviral T antigens, in particular SV40 LT, resulted in singularly important advances towards the understanding of cellular growth control and the origins of human cancer. Of particular significance, those studies resulted in the discovery of the crucially important p53 protein. In point of fact, p53 was first discovered in association with LT in SV40-transformed cells (Fig. 15.5). Moreover, the roles of p53 in regulating cell proliferation were first understood in the context of how LT expression brings about neoplastic transformation. The critical role of p53 in maintaining normal cellular replication, as described below, is underscored by the fact that lesions at the p53 locus are among the most frequently occurring mutations in human cancer, being present in about two-thirds of all human cancers! For reasons that soon will be apparent, the loss or inactivation of p53 function strongly predisposes a cell to accumulate additional mutations that may further its progression to the neoplastic state.

As noted above, SV40 LT also interacts specifically with the tumor suppressor protein pRb, thereby abolishing its normal role in regulating cellular proliferation (Fig. 15.5). (Considering the specific interactions of LT with p53 and pRb, once more we are reminded of the numerous activities of this remarkable 708-amino-acid protein.) Consequently, studies of the LT-pRb interaction likewise contributed significantly to our understanding of cellular growth control and neoplasia. The cellular gene encoding pRb also is mutated in many human cancers.

Next, we consider how pRb and p53 regulate normal cell proliferation. Beginning with pRb, note the following key points. First, cell cycle progression into S phase depends on the expression of genes that respond to transcription factor E2F. Second, pRb is located in the nucleus, where it may bind to transcription factor E2F, which in turn is bound to specific E2F-responsive promoters (Fig. 15.13). Third, when pRb is bound to E2F, the pRb-E2F complexes act as inhibitors rather than as activators of E2F-responsive genes. Fourth, pRb can be bound to E2F only when pRb is in a hypophosphorylated state. Fifth, pRb is hypophosphorylated only in nondividing cells; and sixth, E2F-responsive genes are activated by the phosphorylation of pRb.

Bearing the above points in mind, pRb-regulated, E2F-responsive genes normally are activated by the action of Cdks (Fig. 15.13). As noted above, the cyclins are a family of proteins that control progression through the distinct phases of the eukaryotic cell cycle. They act by the effect of each on a specific Cdk.

Cyclin D, which is synthesized in response to stimulation by growth factors, interacts with Cdk4 to phosphorylate pRb. When pRb is thus phosphorylated, it is released from E2F, which then is able to promote expression of the E2F-responsive growth-stimulatory genes. In this way, the regulation of genes under the control of pRb is managed by the availability of growth factors. Cyclin D itself rapidly turns over, so that its continued effect normally requires the continued presence of growth factors. (As you may

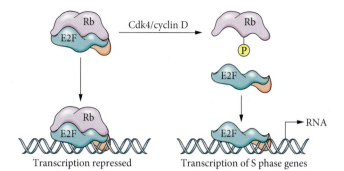

Figure 15.13 The Rb protein in the regulation of cell proliferation. Cell cycle progression into S phase depends on the expression of genes that respond to transcription factor E2F. However, when pRb is bound to E2F on E2F-responsive promoters, the pRb/E2F complexes act as inhibitors rather than as activators of E2F-responsive genes. Note that E2F binds to its target sequences in either the presence or the absence of pRb. However, pRb can be bound to E2F only when pRb is hypophosphorylated. The phosphorylation of pRb and hence its E2F-binding activity are regulated as cells move through the cell cycle. Specifically, as cells pass through the restriction point in G_1, cyclin D interacts with the cyclin-dependent kinase Cdk4 to phosphorylate pRb. When pRb is so phosphorylated, it is released from E2F, which then is able to promote expression of the E2F-responsive growth-stimulatory genes. Adapted from G. M. Cooper, *The Cell: a Molecular Approach*, Sinauer Associates, Inc., Sunderland, MA, 1997.

have supposed, mutations that lead to unregulated expression of cyclin D can promote the development of human cancers, including lymphomas and breast cancer.)

Next, we consider p53, which is a crucial component of the surveillance mechanism that prevents cells from undergoing unscheduled or potentially disastrous cell divisions. The latter might occur if a cell were to sustain DNA damage and then were to attempt to replicate before repairing that damage. However, such potentially disastrous circumstances cause the cell to accumulate p53, which results either in cell cycle arrest to provide an opportunity to repair the damage or in **apoptosis** if the damage is not repaired or if the cell enters an inappropriate or unscheduled S phase. (Apoptosis is a cell death program that is regulated by a variety of signals from within and from outside the cell. It may be triggered by inappropriate progression of the cell cycle, DNA damage, normal development, and also antiviral immune defenses [Chapter 4].) From the point of view of the host, cell suicide by apoptosis is preferable to the generation of daughter cells that might contain genetic lesions capable of leading to cancer. Cell suicide also is preferable to aberrant cell proliferation that likewise might lead to cancer. (Moreover, immune-mediated apoptosis is a most effective means for bringing virus replication to a halt within an infected cell [see Chapter 4].)

Now, we are ready to consider how SV40 LT undermines the normal regulatory functions of pRb and p53.

We begin with pRb. When SV40 LT binds to *hypo*phosphorylated pRb in resting cells, it disrupts pRb-E2F complexes. As a result, E2F is then free to activate E2F-responsive promoters in these cells. Consequently, these cells can then reenter into the cell cycle.

Since the interaction of LT with pRb activates cells to reenter the cell cycle, why must LT also inactivate p53? The answer is as follows. The interaction of LT with pRb, which induces cells to enter S phase, is an event outside the normal cell cycle. Therefore, it would trigger apoptosis were it not for the fact that LT also binds and inactivates p53. In addition, SV40 LT contains a Bcl-2-like domain in its C-terminal region. This is important because host Bcl-2 proteins negatively regulate apoptosis in mammalian cells.

Before the discovery of the LT DnaJ activity, LT was thought to disrupt the function of pRB and p53 simply by "soaking up" the available pools of these molecules. (Recall that LT contains specific binding sites for pRb and p53 [Fig. 15.5].) However, it now is known that the LT-mediated interference with the activities of pRB and p53 is more complicated, requiring the DnaJ domain of LT and its association with Hsc70. How then might the complex of the LT DnaJ domain with Hsc70 interfere with the activity of pRb? Actually, Hsc70, in association with the LT DnaJ domain, appears to disrupt the pRB-E2F complexes. Moreover, the energy for these disruptions is likely supplied by Hsc70-mediated ATP hydrolysis, an activity of Hsc70 that is triggered by the LT DnaJ domain.

This model is consistent with the fact that disrupting multimolecular complexes is a normal Hsc70 function. Moreover, it is supported by the experimental finding that the LT DnaJ domain is required in *cis* with the LT pRb-binding domain to upregulate expression of E2F-dependent genes. That is, the LT DnaJ domain must act on pRb-E2F complexes that are bound by that same molecule of LT. Note on Fig. 15.5 the separate locations on LT for the DnaJ domain (amino acids 1 to 82) and for the domain that binds to pRb (amino acids 102 to 115). A distinct third region of LT contains the p53 binding sites (amino acids 351 to 450 and 533 to 626).

DnaJ domains thus far have been found in the LT and ST proteins of all members of the *Polyomaviridae*. Recall that the chaperone activity of the SV40 LT DnaJ domain also is required for DNA replication and to organize the SV40 capsid during virus assembly.

It is noteworthy, indeed remarkable, that neoplastic transformation induced by two other unrelated DNA virus families, the *Papillomaviridae* (Chapter 16) and the *Adenoviridae* (Chapter 17), also results from inactivation of the tumor suppressor proteins pRb and p53. The E7 protein of oncogenic papillomaviruses and the E1A protein of adenoviruses inactivate pRb. The papillomavirus

E6 protein and the adenovirus E1B protein inactivate p53. Moreover, in each instance the ability of these viral proteins to bind to these host tumor suppressor proteins correlated with the ability of the respective viruses to induce neoplastic transformation.

Considering the above, we now can state that neoplastic transformation by the small DNA tumor viruses (i.e., the polyomaviruses, papillomaviruses, and adenoviruses) generally results from the inactivation of tumor suppressor gene products, which negatively regulate cell growth. Moreover, as in the precedents of the small DNA tumor viruses, human herpesvirus 8 (HHV-8), which is associated with Kaposi's sarcoma in AIDS patients (Chapters 18 and 21), likewise expresses viral genes that inactivate p53 and pRb. Moreover, expression of these HHV-8 genes has been implicated in HHV-8 oncogenesis.

Importantly, the mechanism of cell transformation by the DNA tumor viruses (i.e., the inactivation of tumor suppressor genes) is in marked contrast to that of the oncogenic retroviruses, which transform cells by activating mitogenic signal transduction cascades (Chapter 20). This difference between the DNA tumor viruses and the oncogenic retroviruses underscores the importance of the discovery that the DNA tumor viruses inactivate the functions of cellular growth-suppressing proteins, which indeed led to an understanding of the roles of these proteins in normal cellular growth control. (But see the discussion of the mouse polyomavirus middle T [MT] antigen below.)

Although transformation by the polyomaviruses and transformation by the papillomaviruses have the same underlying molecular basis, it is important to note the following key distinction regarding transformation by these viruses. Neoplastic transformation by the polyomaviruses is for the most part a laboratory artifact, having little, if any, relevance to cancer in nature, including cancer in humans. (That said, the issue of whether SV40 might cause cancer in humans is controversial and is explored in detail below.) In contrast, papillomaviruses indeed are the etiologic agents of cervical carcinoma, the second leading cause of deaths due to cancer in women (Box 15.7).

Bearing in mind the multiple replication- and transformation-related activities of LT enumerated above, what then might ST do? Apropos that question, the SV40 ST protein contains only 174 amino acids. Moreover, 82 of those amino acids are identical to those at the N terminus of LT (Fig. 15.4). The remaining C-terminal 92 amino acids of ST are unique to that protein. Note that ST contains the DnaJ domain but lacks the pRb and p53 binding regions (Fig. 15.5). Considering the duplication of ST sequences in LT, it may not be surprising that ST is not absolutely required for SV40 replication in cultured cells,

BOX 15.7

The finding that the polyomaviruses, papillomaviruses, and adenoviruses all induce neoplastic transformation by means of their abilities to inactivate the tumor suppressor proteins pRb and p53 has given rise to a compelling paradigm regarding neoplastic transformation in vitro and the regulation of normal cell proliferation in vivo. Moreover, lesions in the cellular genes encoding these proteins, particularly p53, are associated with naturally occurring neoplasms. What then are we to make of the fact that these "DNA tumor viruses," with the exception of a few high-risk serotypes of the human papillomavirus (Chapter 16), are *not* generally associated with neoplasia in vivo?

There is no straightforward answer to this question, but the following points should be borne in mind. The progression of a normal cell to a tumor cell and then to a fully malignant one involves a progression of mutagenic events, rather than any single event. These events include a breakdown of the molecular mechanisms that control cell proliferation. They also include a breakdown of the safeguard systems that act to repair damaged DNA and that initiate apoptosis when the DNA damage is extensive or leads to unrestricted cell proliferation. In addition, continued enlargement of a tumor requires the growth of new blood vessels within the tumor, which in turn requires the inappropriate expression of angiogenic factors by the tumor. Moreover, for the tumor to metastasize, it must inappropriately express enzymes that break down surrounding tissue. Finally, in the case of virus-associated cancers (and also lymphomas and leukemias), the emergence of an immunodeficiency is often a precondition to the emergence of cancer.

Considering the above, viruses that interfere with cellular growth-regulating and safeguard systems may well decrease the number of cellular genes that must be mutated for a cancer to emerge. Yet other cellular mutations are still necessary to generate a cancer. Thus, even in the case of the high-risk papillomaviruses (Chapter 16), the frequency of infected females far exceeds the frequency of cervical carcinoma cases. As you might have expected, the high-risk strains are the ones most efficient at inactivating p53 and pRb. Note that HIV facilitates carcinogenesis by immunosuppression (Chapter 21). For yet a different take on this intriguing issue, see Box 17.8.

---

as indicated by the fact that virus mutants that contain deletions of nucleotide sequences unique to ST are viable. Nevertheless, the rate of replication, virus yield, and plaque size of SV40 generally are reduced in infections with ST mutants. Thus, ST expression is necessary for efficient virus replication.

SV40 ST also is required to induce cell transformation. But that is so only under certain circumstances and in certain cells. For example, ST may be required to transform quiescent cells, and ST is almost always required to transform human cells. Yet SV40 deletion mutants that contain lesions only in sequences unique to ST generated tumors when injected into hamsters, although it took them longer to do so.

Since ST is required to induce transformation under certain circumstances, we conclude that LT is not sufficient to induce transformation under all circumstances. While this may appear somewhat surprising, it perhaps could have been expected. LT may be sufficient to affect transformation in cells in which growth is regulated primarily by pRb and p53. However, the fact that the viral functions required for transformation depend on the particular host cell and its growth conditions likely reflects the more general truth that there is no single common straightforward pathway to neoplasia for all cells under all conditions. Regarding that, transformation of a normal cell into a neoplastic one requires a progression of changes, and consequently, different cells and cell lines may be at different stages of that progression when infected (Box 15.7). The experimental findings with SV40 specifically underscore the fact that the impairment of the tumor suppressor proteins p53 and pRb is not sufficient to bring about cell transformation under all circumstances. We also need to remind ourselves that our tidy models often do not adequately reflect Mother Nature's intricacies.

The cellular targets and actions of LT are well characterized, as described above. What then are the specific cellular targets of ST? The primary activity of ST is thought to be impairing the down-regulation of signal transduction pathways that stimulate cell proliferation. These signal pathways generally involve a cascade of protein kinase activities, in which each kinase in the pathway phosphorylates and thereby activates the kinase downstream of it in the pathway. In normal cells, these mitogenic signals are expressed transiently, being regulated in part by the activity of **protein phosphatase 2A** (pp2A), one of the major serine/threonine phosphatases in the cell. ST binds to pp2A, thereby inhibiting its activity. This prevents pp2A from mediating the down-regulation of mitogen-activated protein (MAP) kinases, which normally are inactivated by dephosphorylation of serine or threonine residues. Amino acids 97 through 103 of ST, which are not present in LT, mediate its interaction with pp2A.

Recent studies show that ST may ultimately exert its oncogenic potential by means of its effect on the cellular Myc protein. Myc is a nuclear transcription factor that is regulated by signal pathways that are triggered by mitogenic stimuli. Inappropriate expression of even small amounts of Myc can lead to cellular transformation. Mitogenic signals affect Myc by enhancing its stability, doing so by phosphorylating its Ser 62 residue. Myc activity is negatively regulated by pp2A, which dephosphorylates the Myc Ser 62. Consequently, ST, which inactivates pp2A, exerts its oncogenic potential, at least in part, by preventing dephosphorylation of Myc, thereby resulting in its stabilization. (One might marvel at the number of biological activities packed into the SV40 early region. In this and the following instances, the unique functions of ST are encoded by an intron of the transcript for the multifunctional LT protein.)

Other recent studies showed that ST also activates the phosphatidylinositol 3-kinase (PI3K) signaling pathway, which plays a key role in promoting cell growth and survival. Consistent with the premise that this activity of ST is relevant to transformation, constitutive PI3K signaling functionally mimics and replaces ST in the transformation of human cells. The primary consequence of PI3K activation is the conversion of phosphatidylinositol-4,5-diphosphate (PIP2) to phosphatidylinositol-3,4,5-triphosphate (PIP3) in the plasma membrane, which then functions as a second messenger to activate downstream pathways that involve a serine/threonine kinase called Akt, and other proteins as well. PI3K and Akt are known to be aberrantly activated in breast and ovarian cancers, underscoring the importance of inappropriate PI3K signaling in human malignancies. Although ST clearly perturbs PI3K signaling, ST has not yet been shown to directly interact with PI3K. Thus, ST may target the PI3K pathway at other levels, possibly at Akt. Interestingly, regarding the study noted in the preceding paragraph, which implicated Myc as a target of ST, PI3K and Myc are known to cooperate in the transformation of human cells. Specifically, Myc-induced apoptosis can be suppressed by activation of PI3K/Akt.

Current studies also show that both ST and LT are required for SV40 to induce human mesothelial cells to overexpress **Notch-1**, a key cellular regulatory gene that promotes cell cycle progression and is required for the continued proliferation of SV40-transformed human mesothelial cells. Consequently, these cells provide an important example in which cell transformation cannot be accounted for solely by the LT-mediated inhibition of pRb and p53. It will be important to understand how ST contributes to the upregulation of Notch-1 expression in human mesothelial cells, since these cells are at least 1,000-fold more susceptible to transformation by SV40 than are human fibroblasts. Moreover, how SV40 might transform human mesothelial cells is potentially important clinically because SV40 has been implicated as an etiologic agent of human mesotheliomas in vivo (see below).

Earlier experimental results demonstrated that ST decreases the level of the Cdk inhibitor p27^{KIP1} whereas LT decreases the level of the Cdk inhibitor p21^{WAF1}. The differential effects of ST and LT on these separate cellular targets may explain, at least in part, why ST and LT are each required for the transformation of certain cell types. LT is believed to cause a decline in p21 levels via its inactivation of p53. Indeed, p21 was first identified as a p53-responsive gene product. In contrast, the mechanism by which ST affects the level of p27 is not yet known. However, since the stability of p27 is affected by its phosphorylation, perhaps ST affects p27 via its effect on pp2A.

Next, we consider the third SV40 early gene product, the 17kT antigen. The SV40 17kT antigen consists of the first 131 amino acids of LT and four unique carboxy-terminal amino acids that are encoded by a different reading frame. This protein has not been well studied. However, rat fibroblast cells can be transformed by transfection with an expression vector containing the cloned 17kT antigen gene, raising interesting questions about how 17kT might do so. These questions are answered in part as follows.

Both 17kT and ST contain the same amino-terminal sequence as LT. Thus, each of these T antigens contains the DnaJ domain (amino acids 1 to 82 [Fig. 15.5]) in its amino terminus and, like LT, each can stimulate the ATPase activity of Hsc70. Moreover, 17kT contains the Rb-binding domain in addition to the DnaJ domain. Like LT, 17kT reduces levels of the pRb-related protein p130 and stimulates cell cycle progression of quiescent fibroblasts, two LT activities that are disrupted by DnaJ mutations. As then might be expected, LT DnaJ mutants can be rescued by plasmids that express 17kT.

Since the functions of 17kT appear to be redundant to some of the functions of LT and ST as well, we might ask why the virus encodes the smaller 17kT protein. One possibility is that when functional domains of LT such as the p53- and DNA-binding domains are engaged, the functions of the LT N-terminal region are not available. Thus, 17kT can provide N-terminal functions such as inactivating pRb, thereby freeing up those LT functions that interact with the cell's DNA replication machinery and p53. In that regard, note that since 17kT contains the first 131 amino acids of LT, it contains, unlike ST, both the DnaJ and pRb-binding domains, which are required in *cis* to up-regulate expression of E2F-dependent genes (Fig. 15.5). In contrast, ST contains its unique pp2A-inactivating activity.

Interestingly, human fibroblast cells can be induced to progress through S phase by transfection with a complementary DNA (cDNA) construct that expresses only ST. However, that is the case only if both the ST DnaJ and pp2A-binding domains are intact. In addition, despite the fact that constructs of ST, which contain mutations in either the DnaJ domain or the pp2A-binding domain, are each totally inactive in the above growth induction assay, these mutant constructs complement each other in *trans* to induce cell cycle progression. These experimental results imply that the DnaJ and pp2A-binding domains play distinct roles in inducing cell cycle progression. Yet the roles of the DnaJ domains of ST and 17kT in the SV40 replication cycle and in transformation have not been well enough studied, and a more thorough accounting of their effects may be revealed by future analysis. These roles may well be considerably more extensive and complex than might be suggested by the preceding discussion.

Now we come to the **MT antigen** of the rodent polyomaviruses. Like their ST and LT antigens, the MT protein of the rodent polyomaviruses is translated from an alternatively spliced early mRNA molecule (Fig. 15.14).

As we noted above, the DNA tumor viruses, which include members of the *Polyomaviridae*, *Papillomaviridae*, and *Adenoviridae*, generally encode proteins that inactivate both p53 and pRb. However, the mouse and hamster polyomaviruses are clear exceptions to that generalization. Although their LT proteins bind and inactivate pRb, they do not encode any protein that directly affects p53. Perhaps reflecting that fact, the MT protein of the rodent polyomaviruses is their primary transforming protein. Note that there is no counterpart of the rodent polyomavirus *mt* gene in the genomes of SV40, JCV, and BKV. Also note that MT expresses no enzymatic activity of its own and exerts its effects via its interaction with certain key cellular proteins.

MT is a membrane-associated protein that acts by binding to, and activating, Src family tyrosine kinases.

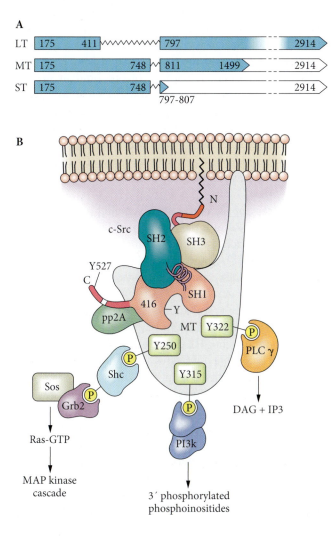

A

B

Figure 15.14 The MT protein of rodent polyomaviruses. (A) Like the LT and ST proteins, the MT protein of the mouse and hamster polyomaviruses is translated from an alternatively spliced early mRNA molecule. The shaded regions in the drawing correspond to the coding regions of the mouse polyomavirus LT, MT, and ST mRNAs, and the wavy lines correspond to the introns. The nucleotides are numbered from the center of the unique origin of DNA replication. (B) Signal mediators with which the Src/MT complex interacts to activate cellular mitogenic signaling pathways. But note that MT must first bind to pp2A before it may bind to Src. Then, the MT/pp2A complexes are thought to bind to the small percentage of Src molecules that already are in the activated conformation, in that way trapping those Src molecules in the activated state. Src-bound MT is itself a substrate for the activated Src kinase. As a result, MT becomes phosphorylated on specific tyrosine residues. This enables certain signaling molecules, which contain so-called SH2 domains, to now bind to phosphotyrosine residues on the Src-associated MT. That is so because SH2 domains bind to specific short amino acid sequences that contain phosphotyrosine residues. In this way, Src-bound MT is able to bring these signaling molecules into apposition with the activated Src family kinase. As a result, these signaling molecules too are phosphorylated, in that way activated by the Src kinase, so that they too might then transmit signals into the cell. PI3K and phospholipase C-γ1 (PLC-γ1) are two signal mediators that contain SH2 phosphotyrosine-binding motifs and that consequently associate with Src-phosphorylated MT. Phosphorylated MT also binds the adaptor protein Shc, which likewise contains an SH2 domain. This results in the phosphorylation of Shc and activation of the mitogenic Ras/MAP kinase signaling pathway. Adapted from S. J. Flint, L. W. Enquist, R. M. Krug, V. R. Racaniello, and A. M. Skalka, *Principles of Virology: Molecular Biology, Pathogenesis, and Control*, 2nd ed. (ASM Press, Washington, DC, 2004), with permission.

The latter proteins are found at the plasma membrane, in association with signaling receptors that possess no intrinsic tyrosine kinase activity of their own. When these receptors are activated by their respective ligands, they transmit signals into the cell via their associated Src family kinases. Examples of such receptors include cytokine receptors, and also the B- and T-cell receptors (Chapter 4). Mouse polyomavirus MT interacts with three Src family kinases, src, yes, and fyn, and activates the first two. Hamster polyomavirus MT associates with, and activates, only fyn. Interestingly, members of the herpesvirus family (Chapter 18) and the rodent polyomaviruses encode proteins that directly interact with Src family kinases. But the mouse polyomavirus MT protein was the first such viral gene product to be discovered (Box 15.8).

How might MT activate Src family kinases? The answer is not known with certainty. However, it appears that MT must bind to pp2A in order for MT to interact with Src. Now, note that the enzymatic SH1 (acronym for Src homology 1) domain of Src is normally maintained in an inactive conformation by the interaction of the Src SH2 domain with a phosphorylated tyrosine residue (Y527) located near the C terminus of the protein. Consequently,

the dephosphorylation of Y527 is one means of activating Src (Fig. 15.15). Together, the above points led to the suggestion that the pp2A moiety of the MT/pp2A complex might express a tyrosine phosphatase activity, in addition to its serine/threonine phosphatase activity, which might dephosphorylate Y527 on the Src protein, thereby enabling the SH1 kinase domain of Src to assume its active conformation. However, contrary to that attractive hypothesis, the phosphatase domain of pp2A does not need to be intact for the MT/pp2A complex to activate Src. Moreover, it is possible that the MT/pp2A complex binds only to already dephosphorylated or otherwise activated Src molecules. In support of that premise, only about 10% of Src molecules are bound to MT molecules in polyomavirus-infected or transformed cells, despite the fact that many more MT molecules are available for binding. A model consistent with these experimental findings is that the MT/pp2A complex binds only to the small percentage of Src molecules that might already be in the activated conformation, in that way trapping them in the activated state.

Importantly, Src-bound MT is itself a substrate for the activated Src kinase. As a result, MT becomes phosphorylated on specific tyrosine residues. This enables certain signaling molecules, which contain the so-called SH2 domain, to now bind to the Src-associated MT. That is so because SH2 domains, like that of Src, bind to specific short amino acid sequences that contain phosphotyrosine residues. In this way, Src-bound MT is able to bring these signaling molecules into apposition with the activated Src family kinase. As a result, these signaling molecules too are phosphorylated and thereby activated by the Src kinase, so that they might then transmit a mitogenic signal into the cell.

PI3K (see above) is one such signaling molecule that associates with mouse polyomavirus MT, after the latter has been phosphorylated by the Src family kinase. Phospholipase C-γ (gamma) 1 (PLC-γ1) is another SH2-containing signaling molecule activated by the Src/MT complex. Upon activation, PLC-γ1 generates the signal mediators diacylglycerol and inositol triphosphate. Phosphorylated MT also binds the adaptor protein Shc, which likewise contains an SH2 domain. The binding of Shc to the Src/MT complex activates the mitogenic Ras/MAP kinase signaling pathway (Fig. 15.14).

Why might the MT/Src complex activate so many cellular signaling pathways? One possibility is that activating these multiple signaling pathways might be needed to ensure both cell proliferation and cell survival. Consider the following experimental results. Expressing a constitutively active version of Src in the mammary glands of transgenic mice only rarely leads to tumor genesis. In contrast, coexpression of Src and MT leads to full malignancy.

BOX 15.8

It is highly noteworthy that the prototype of the Src family kinases, Src, was discovered as a gene product of a retrovirus, the Rous sarcoma virus (see Chapters 1 and 20). Expression of the Rous sarcoma virus *src* gene is sufficient for that virus to induce tumors. In addition, Src was the first protein ever found to possess tyrosine kinase activity. Until then, only serine/threonine kinases were known. Indeed, the discovery of Src led to an appreciation of the key role of tyrosine kinases in intracellular signaling pathways.

In a study of singular importance, Harold Varmus, Michael Bishop, and their coworkers demonstrated that the retroviral *src* gene actually originated from a gene present in normal chicken cells. This breakthrough finding led to the discovery that numerous other retroviral **oncogenes** have their counterparts in the genomes of normal cells. The aberrant expression of these genes, such as when under the control of a viral promoter, can lead to cancer. These issues are considered in detail in Chapter 20.

Reference
Stehelin, D., H. E. Varmus, M. Bishop, and P. K. Vogt. 1976. DNA related to the transforming gene(s) of avian sarcoma viruses is present in normal avian DNA. *Nature* **260**:170–173.

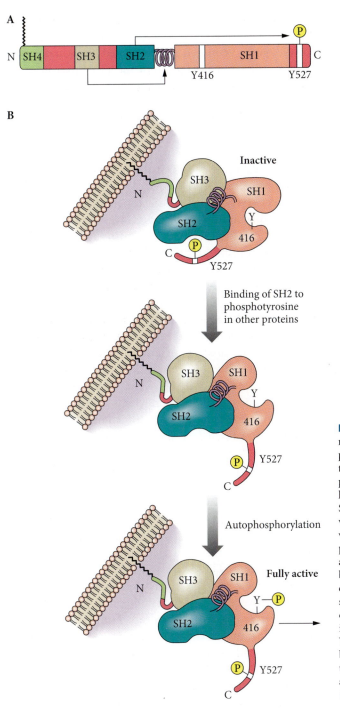

Figure 15.15 Regulation of the cellular Src tyrosine kinase. (A) The arrows denote the intramolecular interactions of the Src protein in its repressed state. The interaction between phosphorylated Y527, located near the C terminus of the protein, and the SH2 domain near the center of the protein facilitates an interaction between the SH3 domain and the helix located between SH2 and SH1. The latter interaction forces the enzymatic SH1 kinase domain into an inactive conformation. (B) Src can be activated by several means. For example, as depicted here, Src can be activated by the binding of its SH2 domain to a phosphotyrosine on another protein. This would displace the phosphorylated Src Y527, thereby enabling the SH1 domain to assume its active conformation. Src also may be activated by the binding of its SH3 domain to a proline-rich sequence on another protein, which likewise would enable the SH1 domain to assume its active conformation. In addition, Src might be activated by the dephosphorylation of Y527. Once Src is released from its inhibited state, its active conformation might be stabilized by the phosphorylation of Y416, as shown. Similarly, when the mouse polyomavirus MT protein binds to activated Src, it might stabilize Src in its active state (see the text). Adapted from S. J. Flint, L. W. Enquist, R. M. Krug, V. R. Racaniello, and A. M. Skalka, *Principles of Virology: Molecular Biology, Pathogenesis, and Control*, 2nd ed. (ASM Press, Washington, DC, 2004), with permission.

Furthermore, mutations affecting MT tyrosine phosphorylation sites that impair the binding of either Shc or PI3K result in dramatic reductions of the transforming activity of the MT protein, showing that each of these MT interactions is important for cell transformation (Fig. 15.14). Moreover, mammary epithelial cells expressing MT mutants defective in recruiting PI3K are also highly apoptotic, implying that recruitment of PI3K by MT affects cell survival.

Unlike the hamster and mouse polyomaviruses, neither SV40 and the human polyomaviruses nor the papillomaviruses and adenoviruses encode proteins that associate with, or activate, Src family tyrosine kinases. Instead, these viruses all encode proteins that transform cells by inactivating p53 and pRb. The facts that (i) the LT proteins of the rodent polyomaviruses do not inactivate p53 and (ii) the rodent polyomavirus MT protein, which operates through an interaction with Src, is the principal rodent polyomavirus transforming protein underscore the point that the viral inactivation of p53 and pRb, although a powerful paradigm, is not the only means by which DNA viruses might induce cell proliferation and neoplastic transformation. In this regard, it is noteworthy that two members of the herpesvirus family, Epstein-Barr virus and herpesvirus saimiri, are also known to encode proteins that, like MT, interact with Src family kinases (Chapter 18). Furthermore, it is interesting that the bovine papillomavirus E5 protein transforms cells by binding and activating the platelet-derived growth factor receptor (which expresses an intrinsic tyrosine kinase activity when activated), thereby causing continual receptor signaling. (Nevertheless, full transformation in this instance also requires expression of the bovine papillomavirus E6 and E7 proteins, which inactivate p53 and pRb, respectively [Chapter 16].)

Still, given the commonality of strategies used by SV40, the human polyomaviruses, the papillomaviruses, and the adenoviruses for inducing cell proliferation while ensuring cell survival, it is somewhat surprising that the rodent polyomaviruses have developed an alternative way to achieve the same result, which does not involve inactivating p53. However, a mechanism recently was proposed by which an MT-induced, Src-mediated signal indeed might induce the expression of a natural regulator of p53. If rodent polyomaviruses were to inactivate p53 via this MT-mediated indirect route, then each of the known polyomaviruses, papillomaviruses, and adenoviruses would induce neoplasia by inactivating the same key cellular targets: the tumor suppressor proteins p53 and pRb.

For the most part, the above discussion of the polyomaviral T antigens in transformation reflects well-established phenomena that are compatible with our current understanding of cellular growth control and neoplasia. Yet active investigation in this area is ongoing, and new effects of these proteins are continually being uncovered. The meaning of some of these effects is not entirely clear, in part because of the intricacies of the pathways that are influenced by the T antigens and their overlapping and interactive nature. Thus, despite the major paradigms that emerged from the analyses of the T antigens, it is almost certain that future work will lead to deeper levels of understanding. The same might of course be said of many other issues considered in this text. Yet in the case of cancer, it may be worth reiterating that a few simple paradigms are not likely to explain all the molecular and cellular changes that underlie this disease.

Before moving on, we note that in addition to driving resting cells into S phase to support replication of rodent polyomavirus genomes, MT-activated Src-kinases also promote the phosphorylation of the VP1 capsid protein, which is necessary for assembly of rodent polyomavirus particles.

SV40 IN ITS NATURAL HOST

The natural host for SV40 is the Asian rhesus macaque (*Macaca mulatta*). And, under captive conditions, SV40 may infect several related species, including the cynomolgus macaque and the African green monkey. SV40 infections of the natural host are similar to those of the other polyomaviruses in their natural hosts. That is, infections of the natural hosts are persistent and usually asymptomatic. SV40 is present in the kidney in vivo and shed in the urine. In addition, SV40-infected hosts may be viremic. Susceptible animals can be infected by the oral, respiratory, and subcutaneous routes, but the urine is thought to be the primary means of transmission among these hosts.

Consistent with the fact that SV40 infections of rhesus macaques generally are persistent and asymptomatic, SV40 infections of rhesus kidney cells in vitro give rise to long-term stable persistent infections, which display few obvious cytopathic effects from the infection. The lack of readily evident SV40 cytopathology in these cell cultures accounts for the fact that SV40 was not immediately noticed in the rhesus kidney cell cultures used to propagate the early poliovirus vaccine stocks. Yet how can these cultures produce a lytic virus while not displaying obvious cytopathology? The answer is that whereas all cells in the culture harbor the virus, only a small percentage produce virus at any moment. These virus-producing cells indeed are killed, and new virus-producing cells emerge to maintain the productive infection. Such an in vitro persistent infection is referred to as a **carrier culture**. The factors regulating virus production in a carrier culture are not entirely clear.

Monkeys infected with the simian immunodeficiency virus (SIV), a retrovirus like human immunodeficiency virus (HIV), develop simian AIDS. When then infected with SV40, these SIV-afflicted monkeys provide a useful model of human polyomavirus infections in human AIDS patients. That said, it is noteworthy that SV40 can cause widely disseminated infections in monkeys suffering from simian AIDS. In these animals, SV40 is detected in brain, lung, kidney, lymph nodes, and spleen. Viral DNA also has been detected in circulating peripheral blood mononuclear cells. In addition, SV40 has been found to be associated with PML in these animals. PML is better known as the demyelinating disease that is induced by JCV in immunocompromised humans, and which is seen as a late complication in human AIDS patients (Chapters 3 and 20; see also below). Together, these different findings show that (i) SV40 may disseminate in the host by the hematogenous route, (ii) it is neurotropic in addition to being kidney-tropic, and (iii) SV40 can become an opportunistic pathogen in immunocompromised hosts.

SV40 IN HUMANS

The prospect that SV40 might be a pathogen in humans is given more attention here than many virologists would likely deem appropriate. However, this is an item that particularly intrigues me because the human population was inadvertently and widely exposed to SV40 via vaccines that the public commonly regarded as godsends (Chapter 6). The subsequent discovery of SV40 as a contaminant in these vaccines and, in particular, the finding that it is an oncogenic virus were shocking moments in the history of vaccinology, fulfilling the most dreaded of fears, the presence of a cryptic tumor virus in a vaccine already administered to hundreds of millions of humans worldwide. Here are the background and current state of affairs regarding this issue.

Unintended exposure of millions of individuals to SV40 via the contaminated polio vaccines posed a *potential* public health crisis of immense proportions. Yet even with the intense concern stemming from those episodes, and after 4 decades of research, it still is not clear whether SV40 is an infectious agent in humans or an agent of human disease. Epidemiological analysis in the 1960s did not indicate any increased incidence of cancer in recipients of the contaminated poliovirus vaccines, thus somewhat quieting early concerns. However, bear in mind that cancer is a disease that may take decades to emerge. (In this regard, see the discussions of papillomavirus-induced cervical cancer and hepatitis B virus-induced hepatocellular carcinoma in Chapters 16 and 18, respectively. Also, see Box 15.7.) In addition, these studies generally had important limitations. For example, sample sizes were often too small to generate statistically significant results, and it was sometimes difficult to define suitable control populations. Small sample sizes were a particular concern in the case of rare childhood tumors (see below).

Interest in the possible relationship between SV40 and human cancer was rekindled by experimental results obtained using modern, extremely sensitive polymerase chain reaction (PCR)-based procedures. These studies, carried out in several independent laboratories, detected SV40 DNA sequences in human mesotheliomas, osteosarcomas, non-Hodgkin's lymphomas, and childhood brain tumors (choroid plexus and ependymomas). These findings were particularly alarming to some researchers

because the four tumor types (i.e., pleural mesothelioma, bone, lymphoma, and brain) in which SV40 was found in humans are the same tumor types that are induced experimentally by SV40 in hamsters. Also, many of the SV40-positive human subjects in these studies were never exposed to SV40-contaminated vaccines. Provided that the experimental findings from these PCR-based studies are correct, they imply that SV40 indeed may be causing cancer in humans. Moreover, they imply that SV40 is presently circulating in the human population, spreading by horizontal transmission from one individual to another. (For a detailed review, see Vilchez et al., 2004.)

Despite the research findings enumerated above, the issue of SV40 in humans remains controversial because other studies, from other research groups using similar PCR procedures, could not detect SV40 DNA in human tissues. In addition, newer, more sensitive serologic procedures could not provide evidence for SV40 infection in humans.

Given the possibly immense significance of a virus with proven neoplastic capability that might be circulating in the human population, we well might ask why the issues of whether SV40 is now circulating in the human population and whether it is a pathogen in humans *remain unresolved* to this day. Several answers to this question were put forward by Robert Garcea and Michael Imperiale, two researchers who have been active in this area (Garcea and Imperiale, 2003). One set of reasons is technical, concerning the possible generation of erroneous results, particularly stemming from the use of PCR-based procedures. First, PCR detection of SV40 sequences is difficult, requiring many cycles (40 to 60) of amplification. One upshot is that DNA extraction procedures are critical for success, since they determine whether there will be adequate amounts of suitable DNA or RNA for PCR analysis. Second, because of the great sensitivity of PCR, great care must be taken to avoid false positives due to inadvertent contamination of samples. To prevent contamination, an ever-present possibility, some researchers strongly recommend that tissue processing and subsequent PCR analysis be carried out in separate facilities. Also, it would be good if positive controls could be processed in yet another facility. Regarding the issue of controls, it may in fact be difficult to obtain proper control tissues. This issue is noted as a separate point below. Here we note that in one study, when a mesothelioma was microdissected, SV40 sequences were detected in the tumor but not in adjacent normal tissue, thereby providing an important internal control and enhancing confidence in the result. Third, the widespread prevalence of the related polyomaviruses JCV and BKV in the human population leaves the possibility that these or other polyomaviruses, rather than SV40, are being detected by PCR-based procedures in some instances. With that proviso in mind, several researchers took the extra step of confirming the presence of SV40 sequences by direct sequencing of the PCR-amplified products. That practice and the development of more-specific PCR primers should increase the reliability of these procedures. Fourth, serological studies have not provided critical results because of immune cross-reactivity between SV40 and the ubiquitous human polyomaviruses JCV and BKV. That is, antibodies against JCV and BKV in sample sera might be cross-reactive with SV40, thereby giving rise to false-positive results. On the other hand, even in cases where infection with SV40 may have occurred, cross-reactive responses to JCV or BKV may have developed (due to SV40-induced activation of BKV- or JCV-specific memory cells), rather than specific responses against SV40, thereby possibly giving rise to false-negative results.

A fifth reason for controversy is somewhat theoretical. It is raised by experimental results which imply that the SV40 genome is not present in all cells of tumors that show evidence of SV40 involvement. The presence of tumor cells not containing SV40 was implied indirectly by quantitative PCR procedures that did not find enough SV40 DNA in the tumors for every cell in the tumors to have contained an SV40 genome. Also, immunohistochemical staining of tumor biopsy samples for LT often did not detect LT in all cells of the tumors. These experimental findings lead to controversy because they are at odds with the widely held paradigm that the continued expression of polyomaviral genes is necessary for maintenance of the transformed cell phenotype, as related above. Yet there are several plausible explanations for these discrepancies between experimental observations and theory. The simplest ones involve technical issues affecting the sensitivity of the procedures. Then, the paradigm itself might not be universally true. Cancers result from a complex multistage process (see Boxes 15.7 and 17.8), and in some instances, the virus may play a necessary role only at particular points in the oncogenic process. For example, the progressive acquisition of tumorigenic mutations in individual cells of the tumor, together with the possible death of tumor cells expressing virus genes (possibly by immunoselection acting against those cells), might lead to a preponderance of cells in the tumor in which the virus genome is no longer expressed, or perhaps lost. (Note that chromosome instability is a characteristic feature of SV40-transformed cells.) Alternatively, tumor cells might be recruited from surrounding normal tissue by a paracrine mechanism. For example, LT-expressing tumor cells might secrete growth factors that cause neighboring cells to express neoplastic characteristics. Studies involving murine and canine cells show that SV40 LT expression indeed promotes the secretion of the insulin-like growth factor, the hepatocyte growth factor, and the vascular endothelial growth factor.

Three other reasons for controversy may be noted. Hence, sixth, despite the fact that SV40 DNA sequences have been detected in a variety of human cancers, the small sample sizes and lack of control groups have made it difficult to establish statistically significant relationships between SV40 in the human population and human cancer. Seventh, the prevalence of SV40 in the human population very well may vary from one geographical location to another. Moreover, it may be important to examine additional potential target tissues. Finally, eighth, there is the possibility that SV40 is *not* a causative agent of human tumors but instead is present in tumors only because tumor cells are more permissive than normal cells for SV40. That is, SV40 might be a "passenger" virus in the tumors.

Although there is no consensus as to whether SV40 is a tumor virus in humans, several key facts cannot easily be denied. First, SV40 can transform a variety of human cells in culture and induce tumors in laboratory animals. Second, the types of tumors that SV40 induces in laboratory animals are the same as those that are most often found to contain SV40 DNA in human cancers (at least as reported by those researchers who found SV40 DNA in human cancers). Third, infectious SV40 indeed was isolated from a primary brain cancer of a 4-year-old child. Fourth, LT/pRb and LT/p53 complexes were identified in human brain tumors, consistent with current understanding of how SV40 induces neoplasia. Fifth, despite the fact that JCV and BKV are ubiquitous in humans and are known to be oncogenic in rodents, SV40 has been detected much more frequently than either JCV or BKV in human cancers. (Recent findings have implicated both JCV and BKV as etiologic agents of cancer in humans [see below]. Yet it makes sense that SV40 *might* be a more common cause of human cancer than either JCV or BKV since humans are the natural host for JCV and BKV, whereas if SV40 indeed is present in humans, it is likely a much more recent arrival there.) Notwithstanding the force of these arguments, bear in mind that impressive evidence against a role for SV40 in human cancer also has been presented. If these issues are to be resolved to everyone's satisfaction, it will be necessary to carry out carefully controlled studies, involving statistically significant sample sizes from a variety of geographic locations, in which the most rigorous experimental procedures are scrupulously adhered to.

SV40 also has been incriminated as a cause of human kidney disease. One study, using PCR-based procedures, found SV40 DNA sequences in kidney biopsy samples from 56% of patients with focal segmental glomerulosclerosis, a kidney syndrome that has been recognized since the 1980s. Also, SV40 DNA sequences were detected in the transplanted kidneys of immunosuppressed pediatric renal transplant recipients. What is more, SV40 DNA was detected in an indigenous kidney of a young adult lung transplant recipient, who also displayed polyomavirus nephropathy (nephropathy refers to disease of the kidney; see below with respect to BKV). These findings and others appear to demonstrate that the kidney is a potential site for SV40 in humans, as well as in monkeys. Moreover, they imply that immunocompromised patients are at risk for SV40-induced disease. Regarding the latter point, note that in its natural host (the rhesus macaque), SV40 is known to induce disease almost exclusively when the animal is in an immunocompromised state, such as that induced by SIV (see above). Yet be aware that the same provisos apply to the issue of whether SV40 is an etiologic agent of human kidney disease as apply to the issue of SV40 in human cancer.

Since there is no consensus as to whether SV40 actually is circulating in humans, any discussion of how SV40 might be transmitted between humans is speculative. Yet JCV and BKV provide precedents for how SV40 transmission might occur in the putative human host. Like SV40, both JCV and BKV target the urinary tract. Yet JCV and BKV each have been found in tonsillar tissue, suggesting that they may spread by the respiratory route, or by hand-to-mouth contact. They then might disseminate to the kidneys via the blood circulation or the lymphatic system. Also, inoculation of SV40 into pregnant hamsters can lead to tumors in the offspring, demonstrating that maternal-infant transmission is a possible route for SV40 to spread. This mode of transmission may account for the presence of SV40 in rare childhood ependymomas and choroid plexus tumors. How might SV40 cross from the mother into the fetal circulation? One possibility is suggested by the SV40 caveola-mediated entry process. Caveolae have been implicated in the transcytosis of ligands across vascular endothelial cells. Perhaps SV40 crosses placental trophoblasts (Chapter 3) by transcytosis (Fig. 15.16).

The systemic spread of SV40 in the host, its persistence in the urogenital system, and its possible expression in certain tumors all imply that SV40 should elicit a detectable immune response in humans, provided of course that it is circulating in the human population. Yet it has been difficult to obtain reliable serological data in the absence of a dependable serological assay for SV40, one that might clearly distinguish it from JCV and BKV. Thus, very little is known about the human immune response to SV40 infection. However, humans recently were shown to mount a cytotoxic T-cell response against cells expressing LT. Thus, it is not clear how either the virus or LT-expressing tumors might persist in the human host.

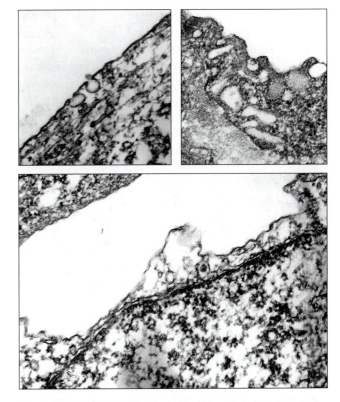

Figure 15.16 SV40 was introduced into a human umbilical cord vein in vitro. Shown are three images of SV40 entering into the endothelium of the vein. (Top left) The virus is seen at the annulus of a caveola. (Top right) The virus is seen internalizing into a caveola. (Bottom) The virus is enclosed within a caveola vesicle that appears to be in contact with both the plasma membrane and the nuclear membrane. From L. C. Norkin, unpublished.

A recent report (Sullivan et al., 2005) provides at least a partial answer as to how SV40 might avoid immune clearance in its natural host, as well as in putative infections of humans. This study demonstrated that an SV40 DNA sequence, in the untranslated region 3′ to the polyadenylation cleavage site of the late pre-mRNA, encodes an miRNA. (Noncoding miRNAs, which are about 22 nucleotides in length, are found in all metazoa. At least some of these miRNAs regulate gene expression. In plants they are known to do so by base-pairing with their target mRNAs, thereby marking them for degradation by the RNAi machinery. Other mechanisms also operate in animals.) The SV40 miRNA accumulates at late times in infection, consistent with it being processed from a late transcript. It is perfectly complementary to early mRNAs, in agreement with the premise that it marks early mRNAs for degradation, in that way resulting in a down-regulation of T-antigen expression at late times in infection.

To determine whether expression of the SV40 miRNA indeed might be functionally relevant to the SV40 replicative cycle, an SV40 mutant was generated in which the miRNA coding sequence was altered by site-specific mutagenesis. As predicted, the miRNA mutant produced excess levels of LT and ST. Nevertheless, viral replication in cell culture was not affected by the mutation. These results show that excess T antigens produced late in infection do not impair SV40 replication, at least in vitro. But what might be the effect of the mutation in vivo, where the T antigens are well known as targets for cytotoxic T lymphocytes (CTLs [Chapter 4])? With that question in mind, the researchers asked whether cells infected with the miRNA mutants might be more susceptible to CTL-mediated attack than cells infected with wild-type virus. Indeed, this was the case, demonstrating that the down-regulation of early region expression caused by the SV40 miRNA does reduce the susceptibility of SV40-infected cells to attack by CTLs. (Note that the SV40 miRNA was the first known instance of a functional virus-encoded miRNA. Moreover, it is the first example of the exploitation by a virus of the host RNAi machinery.)

JCV AND BKV IN HUMANS

BKV and JCV are widespread in the human population, with most individuals having been infected by the age of 15 years. As noted above, transmission may occur via the respiratory route, an efficient means of transmission that is consistent with the widespread prevalence of BKV and JCV in humans. BKV and JCV infections of humans generally are benign. Moreover, these viruses persist indefinitely in the kidneys of infected individuals, usually without causing any serious disease.

A likely reason why BKV and JCV infections are generally benign is that virus replication is usually restricted by host immune responses. This premise is supported by the following observations. First, the usually "latent" BKV and JCV infections may become active in immunocompromised individuals (e.g., AIDS patients, transplant recipients on immunosuppressive therapy, and patients with certain malignancies). Virus particles appear in the blood and urine of these individuals. Indeed, BKV and JCV are found in urine samples from as many as 60% of immunosuppressed patients. Second, reactivation of BKV in immunocompromised individuals may lead to potentially severe urinary tract infections, while reactivation of JCV may lead to infection of the central nervous system, with fatal consequences, as described below. Yet reactivation of these viruses also may occur naturally during pregnancy, with no apparent consequences to the mother or fetus. Some specific outcomes resulting from reactivation of these viruses are as follows.

Reactivation of JCV in immunocompromised patients may lead to progressive multifocal leukoencephalopathy (PML), a deadly subacute or slow demyelinating disease of the central nervous system for which there is no treatment. The term "leukoencephalopathy" refers to the fact that lesions are seen only in the white matter. In actuality, PML was the first demyelinating disease in humans that was found to be caused by a conventional virus (i.e., in contrast to the transmissible spongiform encephalopathies, which are caused by prions [Chapter 3]).

One research group proposes that JCV enters the central nervous system via infected B cells. (JCV-infected B cells were detected in the systemic circulation of 30 to 50% of individuals with a variety of immunosuppressive conditions, whereas JCV-infected B cells are detected in less than 5% of healthy individuals.) Once in the central nervous system, the virus has a tropism for oligodendrocytes (the myelin-generating cells in the brain) and astrocytes (star-shaped glial cells that support and nourish nerve cells in the brain) (Fig. 3.10). Lytic JC virus infection of oligodendrocytes causes the demyelination characteristic of PML lesions.

PML develops only in adults, usually in the fifth to seventh decades of life. Patients display multiple symptoms, including weakness, loss of coordination, cognitive impairment, speech abnormalities, and sensory loss, including visual impairment. Once under way, PML is relentlessly progressive, with paralysis and death generally occurring 3 to 6 months after the onset of symptoms.

Prior to the emergence of HIV and the AIDS pandemic in the early 1980s, PML was considered to be a very rare disease, with a total of "only" several hundred cases reported from its discovery in 1958 as a new disease entity, up to the early 1980s. (PML was recognized as a new disease by Astrom and colleagues, based on the nuclear inclusions that they observed in enlarged oligodendrocytes at the margins of the lesions. These neuropathological observations foreshadowed the subsequent visualization of a viral agent by electron microscopy and its eventual isolation.) However, the AIDS pandemic vastly increased the number of patients developing PML. One reason, as you might have expected, is that PML is associated with an immunosuppressed state, and AIDS patients display a most severe form of immunosuppression. Moreover, there are a large number of AIDS patients. And it is estimated that PML develops in as many as 5% of all AIDS patients. Indeed, 1% of HIV-infected patients present with PML as their first clinical manifestation of AIDS. Considering that in 2002 there were 330,000 individuals in the United States alone who were living with AIDS, the impact of PML is now considerable. It is estimated that there are 7,000 to 8,000 new PML cases per year in the AIDS population in the United States.

Most AIDS/PML patients do not survive more than 6 months after the onset of PML symptoms. Nevertheless, highly active antiretroviral therapy (HAART [Chapter 21]) can prolong the life of these AIDS/PML patients, presumably because it allows partial recovery of immune function that might help to control the JCV infection. Indeed, with the availability of HAART, AIDS/PML patients now have a better prognosis than PML patients with other predisposing conditions. (For the same reason that the PML course of AIDS/PML patients is abated by HAART, a renal transplant recipient who developed PML while on immunosuppressive therapy underwent remission of his PML when immunosuppression was discontinued. Unfortunately, terminating this patient's immunosuppression increased the risk of transplant failure, and the patient had to return to hemodialysis.)

Note that the frequency of PML is much greater in AIDS patients than in individuals with other predisposing immunosuppressive conditions. Why then is AIDS a condition that so uniquely inclines patients to developing PML? The answer is not known with certainty. One possibility is that HIV infection causes an immunosuppression that qualitatively or quantitatively is more severe in ways that predispose patients to reactivation of JCV and, consequently, PML. Another is that HIV infection may compromise the blood-brain barrier to facilitate the entry of JCV into the brain. A further suggestion is that an HIV gene product (e.g., Tat [Chapter 21]) may transactivate the JCV genome. Inflammatory cytokines and chemokines, generated locally in response to HIV infection, also might transactivate JCV (Box 15.9).

Can SV40 in primates provide a useful animal model of PML? Perhaps it can. First, SV40 has 69% homology with JCV. Second, SV40, like JCV, usually gives rise to persistent yet harmless infections in its natural host, in this instance the rhesus macaque. Third, SIV, a relative of HIV, gives rise to an AIDS-like illness in rhesus macaques (Chapter 21). In addition, SIV-induced immunosuppression leads to reactivation of latent SV40 in SV40-infected rhesus monkeys, and about 3% of these monkeys develop a PML-like illness. No doubt there are differences between this "simian PML" and PML in humans. Nevertheless, there is hope that the simian PML model may provide valuable insights into the incurable human disease.

BKV has emerged as the principal etiologic agent of **polyomavirus nephropathy** (PVN), a major cause of kidney transplant failure. PVN occurs with a frequency of 1 to 18% in renal transplant recipients, usually within the first year. Like other polyomavirus syndromes, PVN is associated with immunosuppression, in this instance, rigorous immunosuppressive therapy to prevent rejection of the graft. Thus, reducing the immunosuppressive regimen in PVN patients may restore control of the BKV infection. However, that intervention needs to be balanced against the risk of immune

rejection of the transplant. Thus, it may be wise to routinely monitor renal transplant recipients for the presence of BKV in their urine (an early risk factor for PVN), since altering the immunosuppressive regimen is more effective against PVN the earlier it is begun. In these regards, recall the examples in which natalizumab therapy was withdrawn from multiple sclerosis patients who developed PML and the effect of HAART in AIDS/PML patients.

BKV also has been implicated in hemorrhagic cystitis (a hemorrhagic inflammation of the bladder), a well-known complication following bone marrow transplants.

Recently, BKV and JCV were implicated in the development of a variety of human tumors. For example, BKV and JCV DNA sequences were detected in human mesotheliomas and colorectal carcinomas that arose in immunocompromised patients. One study, involving pediatric patients, detected JCV DNA sequences in 5 of 18 ependymomas and in 1 of 5 choroid plexus papillomas. (Recall that SV40 DNA sequences likewise were detected in childhood choroid plexus tumors and ependymomas.) However, just as there is controversy regarding whether SV40 causes cancer in humans, there is no consensus yet regarding whether JCV and BKV are human tumor viruses. One obstacle to resolving this uncertainty is that BKV and JCV are ubiquitous in the human population. This fact makes it difficult to rule out the possibility that they might be passenger viruses in human tumors. Moreover, since PCR-based techniques were used to detect JCV and BKV DNA sequences in human tumors, the same reservations regarding PCR-based detection noted above in the case of SV40 and human cancer apply to JCV and BKV as well. Yet bear in mind that JCV and BKV have proven oncogenic potential in the laboratory. So, at the moment, the full clinical consequences of polyomavirus infections of humans remain unknown.

KI, WU, AND MERKEL CELL HUMAN POLYOMAVIRUSES

BKV and JCV were each discovered in 1971. Subsequently, 35 years elapsed before the independent and virtually simultaneous discoveries of the next known human polyomaviruses, provisionally named the KI and WU polyomaviruses. Why might so much time have passed between the discoveries of JCV and BKV and the concurrent discoveries of these two new human polyomaviruses? The answer is that the new discoveries were driven by new technologies, specifically, the generation of random libraries from patient samples that could be analyzed by high-throughput DNA sequencing.

The KI and WU polyomaviruses are about 67% homologous to each other and appear to be each other's

BOX 15.9

In a recent PML-related incident, the drug natalizumab (Tysabri), which is used to treat multiple sclerosis, appeared to cause PML in 2 of about 3,000 natalizumab-treated multiple sclerosis patients. In a third case, the patient was being treated with natalizumab for Crohn's disease, in which the drug decreases the rate of remission.

The implication that natalizumab might cause PML in multiple sclerosis patients was extremely disquieting because this drug had produced the most dramatic reduction of multiple sclerosis symptoms yet seen. Moreover, it was shown to reduce the frequency of multiple sclerosis relapses by 67%. Regardless, because of these cases of natalizumab-associated PML, the drug's manufacturer, Biogen Idec, voluntarily removed it from the market. This episode is worth considering in some detail because it provides an example of how hundreds of millions of dollars of drug research and development and a promising approach to treating a serious human illness may virtually disappear down the drain. Moreover, it raises important social issues.

Why do a "mere" three cases of PML in 3,000 treated individuals point to natalizumab as the cause? Consider that the incidence of PML is generally only about 3 cases per million individuals per year. Moreover, these cases almost always are seen in individuals showing clear generalized immunosuppression. Since PML is otherwise not associated with multiple sclerosis, these facts strongly imply that the natalizumab-associated PML cases likely were caused by exposure to the drug.

How does natalizumab work and why might it have caused PML? First, as we noted in Chapter 4, multiple sclerosis results from an autoimmune attack on the patient's myelin basic protein. Second, and consistent with the first point, despite the fact that the central nervous system is often thought of as an immunologically privileged site, leukocytes do cross the blood-brain barrier during various pathological conditions in the central nervous system. Third, natalizumab is a humanized monoclonal antibody against α4β1 integrin, which is a protein on the surfaces of lymphocytes that interacts with its ligand, VCAM-1, on vascular endothelial cells. Since VCAM-1 is up-regulated locally on vascular endothelial cells by proinflammatory cytokines coming from an inflamed site, this interaction enables lymphocytes to home in on inflamed sites, where they cross the blood vessel endothelium to get to the infected tissue. Together, these facts imply that natalizumab acts against multiple sclerosis by binding to α4β1 integrin on the surfaces of lymphocytes, thereby preventing the trafficking of lymphocytes into the brain. This possibility is consistent with the earlier finding in animal models that antibodies against α4β1 integrin inhibit or even reverse experimental autoimmune encephalitis (Chapter 4).

One might be tempted to conclude from the above findings that natalizumab causes PML because it impairs immune surveillance at sites of latent JCV infection, and possibly in the brain as well. However, that conclusion might be premature since natalizumab-treated patients did not show any tendency toward acquiring other opportunistic infections or cancer. Thus, the drug's inadvertent effect appears to be specific to PML. One hypothesis offered to explain this puzzle is that natalizumab induces the release of B cells from the bone marrow, some of which may harbor latent JC virus. Regardless, given the efficacy of natalizumab in the treatment of multiple sclerosis, it is singularly important to understand the means by which it might inadvertently cause PML.

Incidentally, one patient recovered from PML after natalizumab treatment was discontinued. On the one hand, that outcome provides hope that natalizumab might yet be used to safely treat multiple sclerosis patients, if those patients could be monitored for conditions that might predispose them to PML, such as measuring blood levels of JCV. On the other hand, two patients did not recover after treatment was stopped. Also, it is not entirely clear how predictive JCV blood levels are of the risk of developing PML. Furthermore, the drug remains active in patients for as long as 3 months after administration is stopped. Consequently, stopping treatment after JCV viremia is detected may be too late to prevent PML from developing. While awaiting clarification of these issues, if you were a physician, would you use a drug that is effective against a nonfatal but potentially crippling illness (multiple sclerosis), if there were a 1 in 3,000 chance that it might induce a fatal disease (PML)? If you were a multiple sclerosis patient, would you want the drug?

PML has also been seen in patients treated with monoclonal antibodies against two other lymphocyte surface antigens. One of these drugs, rituximab (Rituxan), which is comarketed by Biogen Idec and Genentech, is a humanized monoclonal antibody against CD20 (a protein of unknown function expressed on B cells), used in the treatment of B-cell non-Hodgkin's lymphoma, B-cell leukemia, and some autoimmune diseases (e.g., rheumatoid arthritis and systemic lupus erythematosus). Two patients receiving rituximab died of PML. Both were being treated for lupus. Although rituximab has been under investigation for use in

Box continues

multiple sclerosis, at this time the drug is approved for use only in non-Hodgkin's lymphoma and rheumatoid arthritis. Understanding how treatment with these agents might predispose patients to PML may provide new insights into the biology of this disease.

Postscript: On 5 June 2006, after reevaluating the effectiveness and safety of natalizumab, the U.S. Food and Drug Administration (FDA) reapproved the drug for use in patients with relapsing forms of multiple sclerosis, but only under certain conditions. These included a restricted distribution program known as the TOUCH Prescribing Program ("TOUCH" stands for "Tysabri Outreach: Unified Commitment to Health"), in which patients must be periodically evaluated for PML while being treated with natalizumab. Note that the treated patients who earlier acquired PML had also been exposed to immunosuppressants, which may have been a contributing factor to their developing PML. Still, there were not enough cases to rule out the possibility that natalizumab treatment alone might lead to PML. Regardless, based on the known association of PML with immunosuppression, TOUCH protocols recommend against using natalizumab to treat immunosuppressed patients. Likewise, patients treated with natalizumab should not concurrently use immunosuppressants or immunomodulators. Incidentally, the cost for natalizumab treatment is approximately $30,000 per patient per annum.

---

closest known relative. Each virus varies significantly from JCV and BKV, especially in their late regions, where they display only 15 to 27% homology to the previously discovered human polyomaviruses.

The pathogenicity, prevalence, and modes of transmission of the KI and WU polyomaviruses are not yet known. However, the presence of KI and WU viruses in respiratory secretions and their absence from urine samples suggest that their mode of transmission may be different from that of JCV and BKV, a possibility consistent with the sequence divergence of their capsid proteins. (Proviso: The very existence of the KI and WU viruses is implied by PCR-based procedures. To date, virus particles have not been isolated, nor is it certain that humans are the primary host for the KI and WU viruses.)

Since the LT proteins of the KI and WU polyomaviruses contain binding sites for pRb and p53, they have the transforming *potential* of other polyomaviruses. Moreover, like other polyomaviruses, it is possible that they may cause severe disease in immunocompromised individuals. Thus, these viruses well may be human pathogens. However, as thoughtfully noted by the discoverers of the KI polyomavirus, it is difficult to establish associations of persisting viruses with particular diseases since these viruses are often discovered outside their symptomatic context.

Within a year of the discovery of the KI and WU viruses, in January 2008, a third new human polyomavirus was discovered, named the Merkel cell polyomavirus (MCV) because of its possible association with Merkel cell carcinoma, a rare aggressive human skin cancer. The incidence of this uncommon cancer tripled between 1986 and 2001 (corresponding to the emergence of AIDS), such that there are now about 1,500 cases of Merkel cell carcinoma diagnosed in the United States each year. In this regard, it is interesting to compare Merkel cell carcinoma with Kaposi's sarcoma, a cancer associated with HHV-8 and AIDS (Chapters 18 and 21). Before the arrival of transplant surgery and the emergence of AIDS, both Kaposi's sarcoma and Merkel cell carcinoma generally affected people 65 years of age and older. Now both cancers occur much more commonly among transplant recipients on immunosuppressive therapies and AIDS patients (Chapter 21); these facts are consistent with the premise that Merkel cell carcinoma, like Kaposi's sarcoma, has an infectious origin. Interestingly, MCV was discovered by husband and wife researchers Patrick Moore and Yuan Chang, who also discovered HHV-8 in 1994. The technique used to identify MCV, digital transcriptome subtraction, eliminates known human molecular sequences from a sample, leaving unknown or nonhuman sequences that can then be used to detect possible infectious agents.

MCV has a genome of 5,387 bp. Notably, its genomic sequence is most homologous to that of the mouse polyomavirus, rather than with SV40 or the other human polyomaviruses, which are more similar to SV40.

What is the evidence that MCV is an etiologic agent of Merkel cell carcinoma? MCV genomic sequences were detected by PCR analysis in 8 of 10 Merkel cell tumors but not in any of 59 control tissues, including nine skin samples. Similarly, only 4 of 25 additional skin and non-Merkel cell carcinoma skin tumor samples, from both immunocompetent and immunosuppressed patients, tested positive for MCV sequences. Interestingly, in six of the eight MCV-positive tumors, the viral DNA was integrated into the tumor cell genomes, in a pattern implying that integration of the viral DNA preceded clonal expansion of the tumor cells.

Together, these findings are consistent with the premise that MCV indeed plays a role in the genesis of Merkel cell carcinomas. Still, the association of the virus with the tumor is not yet proof that the virus is the etiologic agent. Nevertheless, these findings are intriguing to researchers in the field because although polyomaviruses have long been suspected of causing human cancers, no polyomavirus has yet unequivocally been proven to do so, and MCV may be the best candidate yet for a human polyomavirus tumor agent. Still, much work remains to be done, including determining the distribution of the virus in the human population and ascertaining its role in Merkel cell carcinoma and in other human cancers (e.g., Hodgkin and non-Hodgkin lymphomas) and in human disease in general.

SV40-BASED GENE DELIVERY VECTORS

A goal of gene therapy is to transmit a gene to a patient who either is lacking that gene or expresses a defective version of it. Since viruses by their very nature are gene delivery vehicles, recombinant viruses engineered to carry particular host genes have offered a promising approach to gene therapy. Adenoviruses (Chapter 17), herpesviruses (Chapter 18), and retroviruses (Chapters 20 and 21) currently are showing potential for use as vectors for gene therapy.

Since studies of the polyomaviruses in general, and of SV40 in particular, have led to key advances in eukaryotic cell and molecular biology, as well as to crucial insights into neoplasia, it is fitting that progress also has been made over the past several years towards developing SV40-based gene delivery vectors that might have therapeutic potential in the treatment of human diseases. These SV40 vectors generally are assembled in vitro from purified DNA and cell-free extracts that contain recombinant SV40 capsid proteins. Packaged plasmids may be as large as 17.6 kilobases (kb), since they do not contain nucleosomes. These in vitro-assembled **virus-like particles** resemble SV40 structurally and are capable of delivering functional genes into a variety of tissues, including bone marrow and liver, each of which is a potentially important target for therapy.

T-antigen genes are deliberately not present on the engineered plasmid that is packaged within the virus-like particles, thereby ensuring that they are not tumorigenic. In general, special care needs to be taken when engineering a viral vector to ensure that it does not have the potential to do harm (Chapter 17). Thus, the vector needs to be disabled in some way so as not to express a potentially pathogenic gene product (a neoplastic one in the case of SV40) and to prevent replication that might lead to cell destruction.

Recently, to demonstrate the *potential* of SV40-based vectors for liver gene therapy, a vector that expressed a luciferase-tagged LT was constructed. To target the liver, mice were injected in the tail vein with stocks of the vector, and in vivo expression of luciferase activity was monitored over time. Injection of the vector led to efficient gene delivery to the liver, in which transgene expression occurred in hepatocytes for at least 107 days. There was a transient inflammatory response that resolved completely within several days. Low levels of anti-SV40 antibodies were detected, but there were no indications of a cellular immune response against either the transgene or the viral proteins (Arad et al., 2005).

Suggested Readings

Anderson, H. A., Y. Chen, and L. C. Norkin. 1996. Bound simian virus 40 translocates to caveolin-enriched membrane domains, and its entry is inhibited by drugs that selectively disrupt caveolae. *Mol. Biol. Cell* 7:1825–1834.

Arad, U., E. Zeira, M. A. El-Latif, S. Mukherjee, L. Mitchell, O. Pappo, E. Galun, and A. Oppenheim. 2005. Liver-targeted gene therapy by SV40-based vectors using the hydrodynamic injection method. *Hum. Gene Ther.* 16:361–371.

Berk, A. J., and P. A. Sharp. 1978. Spliced early mRNAs of simian virus 40. *Proc. Natl. Acad. Sci. USA* 75:1274–1278.

Daniels, R., D. Sadowicz, and D. N. Hebert. 2007. A very late viral protein triggers the lytic release of SV40. *PLoS Pathog.* 3(7):e98.

Gai, D., R. Zhao, D. Li, C. V. Finkielstein, and X. S. Chen. 2004. Mechanisms of conformational change for a replicative hexameric helicase of SV40 large tumor antigen. *Cell* 119:47–60.

Garcea, R. L., and M. J. Imperiale. 2003. Simian virus 40 infection of humans. *J. Virol.* 77:5039–5045.

Pelkmans, L., E. Fava, H. Grabner, M. Hannus, B. Habermann, E. Krausz, and M. Zerial. 2005. Genome-wide analysis of human kinases in clathrin- and caveolae/raft-mediated endocytosis. *Nature* 436:78–86.

Pelkmans, L., J. Kartenbeck, and A. Helenius. 2001. Caveolar endocytosis of simian virus 40 reveals a new two-step vesicular-transport pathway to the ER. *Nat. Cell. Biol.* 3:473–483.

Simmons, D. T., D. Gai, R. Parsons, A. Debes, and R. Roy. 2004. Assembly of the replication initiation complex on SV40 origin DNA. *Nucleic Acids Res.* 32:1103–1112.

Sullivan, C. S., A. T. Grundhoff, S. Tevethia, J. M. Pipas, and D. Ganem. 2005. SV40-encoded microRNAs regulate viral gene expression and reduce susceptibility to cytotoxic T cells. *Nature* 435:682–686.

Vilchez, R. A., and J. S. Butel. 2004. Emergent human pathogen simian virus 40 and its role in cancer. *Clin. Microbiol. Rev.* 17:495–508.

Papillomaviruses

INTRODUCTION

The *Papillomaviridae* are a family of small, double-stranded DNA viruses that superficially resemble the *Polyomaviridae* (Chapter 15). Members of each of these virus families contain small, circular, double-stranded DNA genomes that are arranged in association with cellular histones as "minichromosomes." Moreover, the genomes of both polyomaviruses and papillomaviruses are enclosed within small, nonenveloped, icosahedral capsids. Because of these and other superficial similarities between the *Papillomaviridae* and *Polyomaviridae*, they formerly were classified as distinct genera within a family referred to as the *Papovaviridae*. Yet the papillomaviruses are slightly larger than the polyomaviruses. Their genomes are 8,000 base pairs (bp) long, versus 5,000 bp for the polyomaviruses. Likewise, their capsids are 55 nm in diameter, versus 45 nm for the polyomaviruses. More importantly, comparing the genome organization of the papillomaviruses with that of the polyomaviruses shows clearly that these two virus groups are not related. Still, one may refer to the papillomaviruses and the polyomaviruses, as well as the larger adenoviruses (Chapter 17), collectively as the "small DNA tumor viruses."

Papillomaviruses are widespread in nature, infecting a variety of mammalian and bird species. Yet each papillomavirus type is usually specific for only one host species. Our major concern in this chapter is with the **human papillomaviruses (HPVs)**. The bovine papillomavirus 1 (BPV-1) also comes up, since this virus too has been widely studied.

Different HPV types are classified based on the nucleotide sequences of specific regions of their genomes, and thus they are referred to as **genotypes**. This practice has been followed because serologic reagents that might distinguish between HPV **serotypes** are not generally available. Currently, *more than 100 different HPV genotypes* that specifically infect humans are known.

The HPVs are particularly intriguing for several reasons. Most importantly, some HPV genotypes that infect the anogenital region, the so-called "high-risk" types, are the etiologic agents of cervical carcinoma, a cancer second in frequency only to breast cancer in women. As such, cervical carcinoma is the *second leading cause of cancer death in women* (approximately 500,000 new cases per year, and about half that many deaths).

Yet most HPV types that infect the anogenital region (the so-called "low-risk" types) cause "only" benign genital warts, which are referred to as condylomas. (A benign growth, such as a wart [moles and polyps are other examples], replicates the same cells that it started with but produces too many of them. Since the daughter cells are like the initial cell, the wart is not malignant. That is, it does not metastasize to form new foci. Yet even though a wart is benign, it still is referred to as a neoplasm.) Although these genital warts or condylomas are nonmalignant, they are persistent and may be irritating and occasionally painful. Note that benign condylomas are not precursors of malignant carcinomas. Instead, the genesis of malignant carcinomas is a unique process (see below). Importantly, regardless of whether particular papillomavirus-induced lesions are malignant or benign, HPVs are a *frequent* cause of sexually transmitted disease in humans.

It is noteworthy, and somewhat puzzling, that at least 80% of vaginal HPV infections appear to be transient and do not result in any cervical lesions. Even infections with high-risk HPV genotypes only rarely give rise to malignant disease (see Box 15.7). Thus, the relatively high incidence of cervical carcinoma (relative to most other cancers in women) reflects the even greater prevalence of HPV infections in the human population. Indeed, more than 10% of asymptomatic young women, with cytologically normal cervical epithelia, already have an ongoing infection with a high-risk HPV genotype. Moreover, it is estimated that the risk of contracting a genital HPV infection of either the high-risk or low-risk type during one's lifetime is 80%. Indeed, HPVs are the most common sexually transmitted infection in humans.

Those HPVs that infect the mucosal epithelia, including that of the female genital tract, are referred to as **mucosal HPVs**. Those HPVs that infect the skin are referred to as **cutaneous HPVs**. Although mucosal HPVs that infect the anogenital mucosa are of particular concern because some are associated with cervical carcinoma, the cutaneous HPVs too are clinically important since they cause common warts or **papillomas**. These are benign, self-limited proliferations of skin cells that typically resolve spontaneously, possibly because of immune responses (see below). Interestingly, each of the cutaneous HPVs has a specific tropism for a particular site on the body. For example, HPV-1 causes plantar warts (on the soles of the feet), whereas HPV genotypes 2, 4, and 7 cause warts on the hands (Boxes 16.1 and 16.2).

Papillomaviruses are also distinguished by their rather unusual lifestyle, which differs from that of many other viruses. Specifically, they can only initiate infection in vivo in the undifferentiated basal cells of an epithelium. But then, they can complete their replication only in those cells that subsequently differentiate and move up towards the surface of the epithelium (Fig. 16.1). This papillomavirus lifestyle facilitates both virus persistence and transmission, as described below. Moreover, the viral adaptations that enable this lifestyle underlie the virus's neoplastic potential. (The human cytomegalovirus, a herpesvirus [Chapter 18], provides another example of a virus that uses cell differentiation as a means to trigger its productive phase.)

STRUCTURE

Papillomaviruses have nonenveloped icosahedral capsids that are about 55 nm in diameter and that are comprised of two structural proteins: the major capsid protein, L1, and the minor capsid protein, L2. The capsid architecture is comprised entirely of L1 molecules that are assembled into pentamers, a total of 72 pentamers per capsid.

Polyomaviruses too are assembled entirely from pentamers that are comprised of their major capsid protein, VP1 (Chapter 15). Nevertheless, icosahedral capsids are more commonly assembled from subunits that are arranged as hexamers, as well as pentamers (Fig. 2.8). In these more typical icosahedral capsids, each of the 12 vertices of the capsid is comprised of a pentamer, and all of the remaining subunits are typically arranged as hexamers. Consequently, in those capsids in which pentamers and hexamers are comprised of the same protein subunit, the interactions between subunits in the pentamers cannot be exactly identical to their interactions in the hexamers. Yet they can be very similar. Moreover, the interactions between subunits of adjacent pentamers and hexamers cannot be identical to the interactions between subunits of adjacent hexamers, although they too can be very similar. For these reasons, Casper and Klug introduced the concept of **quasiequivalent bonding**, in which each of the subunits can be bound to its neighbors in similar but not exactly identical ways (e.g., head-to-head and tail-to-tail interactions [Fig. 2.8E]). Importantly, the subunits in these capsids interact via noncovalent bonds, and therefore, they are arranged to make maximum contact.

In the cases of the papillomaviruses and the polyomaviruses, the situation is more complicated because some pentamers need to be surrounded by six other pentamers to create the characteristic icosahedral axes of symmetry. Each of the 12 pentamers that are located at the vertices of the icosahedron makes contact with 5 other pentamers, whereas each of the 60 other pentamers makes contact with 6 other pentamers (Fig. 15.1 and 16.2). But how might a pentamer be surrounded by six other pentamers? The X-ray structure provided the answer in the case of the

BOX 16.1

The first clue that cervical cancer might be caused by a transmissible agent was the 1842 observation by a doctor in Florence, Italy, that uterine (cervical?) cancer developed in the prostitutes and married women of Florence but was rare in the presumably celibate nuns who lived outside the city. The good doctor might himself have hypothesized that cervical cancer is a sexually transmitted disease. However, he was thrown off by the fact that the nuns had a higher than typical rate of breast cancer, which he attributed to their corsets being too tight.

In 1907, G. Ciuffo, an Italian researcher, became the first to demonstrate that human warts indeed have a viral etiology when he showed that warts could be transmitted by cell-free filtrates. Towards this end, Ciuffo inoculated himself, personally!

In 1933, Richard E. Shope, of the Rockefeller Institute, became the first researcher to isolate a papillomavirus, the cottontail rabbit papillomavirus. He then established that this virus is the cause of skin papillomas in its rabbit host. In fact, this was the first demonstration that a DNA virus might be oncogenic. Recall that Shope also accomplished the first isolation of an influenza virus in 1930 (Chapter 12).

Yet it was not until a 1954 study, in which the wives of American servicemen returning from Korea were found to be the hapless recipients of genital warts (transmitted by their philandering husbands), that the medical community was convinced that genital warts can be sexually transmitted.

Harald zur Hausen and coworkers were responsible for the next major breakthrough, demonstrating in 1983 and 1984 that HPV-16 and HPV-18 (which now are recognized as high-risk HPVs) are found in cervical cancers at high frequency. These observations provided the first suggestion that HPVs might be the causative agents of these malignancies. Subsequent epidemiological studies confirmed this association. Importantly, the correlation of virtually all cases of cervical carcinoma (>95%) with only a small subset of HPV genotypes provides compelling evidence for the viral etiology of these malignancies. It is noteworthy that prior to these findings, it was widely thought that herpesvirus infections might be the cause of cervical carcinoma, based on the observation that many women afflicted with cervical carcinoma had a history of genital herpes.

Historically, it is interesting that virus involvement in animal cancer, as demonstrated by Shope in 1933, and by others even earlier still, was not seen as being relevant to human cancer. Indeed, the experience of Peyton Rous, whose 1911 seminal discovery of virus-induced sarcomas in chickens was not recognized by the Nobel Committee for 66 years, provides a dramatic example of this disconnect (see below and Chapters 1 and 20). In actuality, it took a major paradigm shift in the late 1970s for some viruses to be recognized as tumor viruses in humans (Box 16.2). In the case of the papillomaviruses, the World Health Organization did not officially recognize that HPV-16 and HPV-18 are carcinogenic in humans until 1995, more than 60 years after Rous first demonstrated the viral etiology of cancer in chickens. Interestingly, Peyton Rous also played a significant role in characterizing the nature of the skin lesions induced by the rabbit papillomavirus, demonstrating that they are benign tumors, which nevertheless might progress to skin cancers in rabbits (see the text).

polyomaviruses. In order to establish sixfold as well as fivefold rotational axes of symmetry about a pentamer, the pentamers of the polyomaviruses are joined together by the extended arms of the C termini of their VP1 subunits (Fig. 15.1). Five arms extend from each pentamer and insert into the neighboring pentamers in three distinct kinds of interactions. Moreover, interchain disulfide bonds link VP1 molecules in adjacent pentamers. An "invading arm" model likewise has been proposed to account for the papillomavirus capsid architecture. This is in contrast to most other icosahedral viruses, in which the pentamers and hexamers interact noncovalently across complementary surfaces (Fig. 2.8E).

The ratio of L1 to L2 molecules in the BPV-1 capsid is about 30:1. This ratio, together with three-dimensional image reconstructions, suggests that the L2 minor capsid protein is located in the centers of the 12 pentamers that constitute the vertices of the icosahedral capsids. Sequences close to the N terminus of the L2 protein are displayed on the surfaces of BPV-1 capsids.

As in the case of polyomavirus genomes, the papillomavirus genomes are circular and are associated with cellular histones, thereby forming chromatin-like complexes.

ENTRY AND UNCOATING

Since the outer layer of the skin is comprised of dead cells, cutaneous HPV genotypes generally require a break or puncture of the skin in order to access living cells of the underlying germinal stratum of the epidermis (Fig. 3.1

BOX 16.2

The family *Papillomaviridae* is the first virus family in this book that includes members that are undeniable agents of human cancer. With that in mind, we well might ask what percentage of human cancers have a viral etiology. In Chapter 15, we noted that it is not yet clear whether the polyomavirus SV40 causes cancer in humans. However, hepatitis C virus, a flavivirus, is clearly an etiologic agent for hepatocellular carcinoma (Chapter 7). Other viruses that are indisputable agents of human cancer include the herpesvirus Epstein-Barr virus, which causes Burkitt's lymphoma and nasopharyngeal carcinoma (Chapter 18); human herpesvirus 8, which is a cause of Kaposi's sarcoma, the most frequent cancer seen in acquired immunodeficiency syndrome (AIDS) patients (Chapters 18 and 21); the retrovirus human T-lymphotropic virus 1, which induces adult T-cell leukemia (Chapter 21); and hepatitis B virus, a hepadnavirus that also causes hepatocellular carcinoma (Chapter 22). Together, viruses are thought to account for

about 20% of all human cancers and a similar percentage of all deaths due to cancer. Importantly, the realization that a particular cancer has a viral etiology opens up the possibility of rational strategies for prevention (i.e., vaccines) and therapies against those cancers. As will be seen, these possibilities have been and are being realized in the case of the human papillomaviruses.

Note that the frequency of individuals infected with these oncogenic viruses is vastly greater than the frequency of individuals in which infection results in cancer. This point is discussed in Box 15.7.

Remarkably, bacterial and parasitic infections also may lead to cancer. For example, *Helicobacter pylori* infections may lead to stomach cancer, and *Schistosoma*, *Opisthorchis*, and *Clonorchis* have been linked to rectum and bladder cancers in areas of Northern Africa and Southeast Asia, where they are prevalent.

and 16.1). Papillomavirus replication then tends to remain localized, since there are no blood or lymphatic vessels at that level of the skin that might spread the infection. These localized foci of infection may lead to the generation of warts.

Mucosal HPV genotypes, which infect the mucosal epithelia of the female genital tract, are usually transmitted sexually. Infections in these instances also remain localized, restricted to the urogenital epithelia, sometimes giving rise to genital warts or, in the case of infections with a high-risk HPV genotype, to cervical carcinoma. (Since mucosal HPV genotypes are transmitted primarily by the sexual route, men as well as women are infected with mucosal HPVs. As in the case of women, infection in men rarely leads to any clinical consequences, but infection in men can lead to genital warts and to penile cancer or anal cancer. See below.)

It has been difficult to investigate papillomavirus entry into cells, as well as papillomavirus biology in general, since these viruses cannot be propagated in standard cell cultures. That is so because the papillomavirus replication cycle depends on the biology of the intact stratified squamous epithelium. One consequence of that dependency is that it has been difficult to grow sufficient amounts of virus to investigate entry. That being the case, so-called pseudovirions or virus-like particles (VLPs), which are assembled in vitro from papillomavirus capsid proteins, have been used to study papillomavirus entry.

Some researchers point out that when VLPs are used as a model to study papillomavirus entry, there is the underlying *assumption* that the entry process for VLPs reflects the entry pathway of the actual virus. Regardless of this proviso, experiments using VLPs have led to progress in the characterization of papillomavirus entry. First, they led to the identification of heparin sulfate proteoglycans as key molecules on the cell surface that facilitate papillomavirus binding. Second, VLPs, and probably virus particles as well, enter by a clathrin-dependent endocytic pathway that requires actin filaments. Moreover, acidified endosomes and microtubules likewise play a role in the entry process. Third, the papillomavirus L1 protein mediates binding to cells. Fourth, the L2 protein is not necessary for initial binding to cells. The last conclusion is implied by the finding that DNA-free VLPs consisting of L1 only, or consisting of L1 and L2, are equally effective at competing with virus particles for binding and infection. Fifth, L2 indeed plays a role in the papillomavirus entry pathway, although a postbinding one. This conjecture is implied by the finding that antibody to L2 can neutralize papillomaviruses while not blocking binding. Since there is evidence that L2 also may bind to cells, it was suggested that L2 may interact with a secondary receptor.

Other recent experimental findings suggest that the L2 protein also may play a role in releasing the viral genome into the cytoplasm. Regarding that possibility, all viruses must transport their genomes across at least one cellular

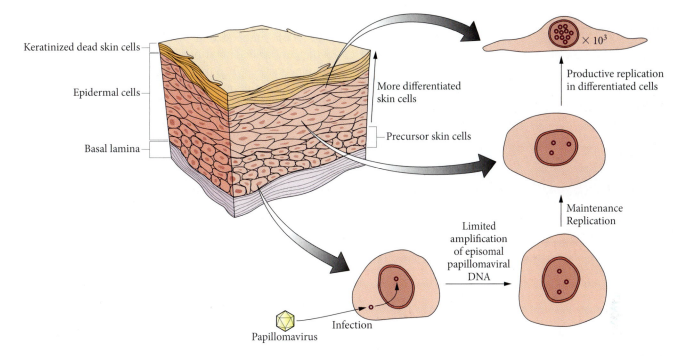

Keratinized dead skin cells

Epidermal cells

Basal lamina

More differentiated skin cells

Precursor skin cells

Infection

Papillomavirus

Limited amplification of episomal papillomaviral DNA

Maintenance Replication

Productive replication in differentiated cells

$\times 10^3$

Figure 16.1 Papillomaviruses depend on wounds (e.g., cuts and abrasions), which enable them to infect undifferentiated basal cells of the stratified epithelium, the only cells of the normal stratified epithelium that are actively dividing. In these cells, viral genomes replicate more frequently than the cellular genome, in that way amplifying the viral genome copy number. This initial phase of viral replication is referred to as the establishment phase. Following that, there is a transition to a maintenance phase, in which the viral genomes replicate only once per cell cycle, on average, to keep the viral copy number constant as the basal cells divide. Finally, as infected cells go through their final stages of differentiation in the outer layers of the epithelium, the virus life cycle switches to its productive phase. Viral DNA amplification is triggered, along with late transcription, and thousands of progeny virus particles are generated from each of a small number of terminally differentiated cells. Thus, the papillomavirus replicative cycle is directly regulated by the differentiation state of the host cell within the stratified squamous epithelium. Adapted from S. J. Flint, L. W. Enquist, R. M. Krug, V. R. Racaniello, and A. M. Skalka, *Principles of Virology: Molecular Biology, Pathogenesis, and Control*, 2nd ed. (ASM Press, Washington, DC, 2004), with permission.

membrane, so that their genomes might then traffic to the sites of their expression and replication. In the case of the papillomaviruses that site is the nucleus.

Apropos the above, enveloped viruses may transport their capsid or genome across a cellular membrane by means of fusion between their envelope and the target cellular membrane. However, envelope fusion is not possible in the case of nonenveloped viruses and, consequently, the mechanisms by which nonenveloped viruses transport their genomes across cellular membranes are generally not as well understood as the mechanisms used by enveloped viruses.

One of the better-understood nonenveloped virus entry mechanisms is that of poliovirus, in which receptor engagement triggers the release of the poliovirus VP4 protein, which contains a hydrophobic domain that is thought to form pores in the membrane through which the genome might pass (Chapter 6). In contrast, nonenveloped reoviruses use the low pH of an acidic endosomal compartment to trigger a conformational rearrangement of

the reovirus capsid that exposes a hydrophobic domain on the reovirus μ1 protein that enables viral cores to cross the endosomal membrane into the cytosol (Chapter 5).

Returning to the papillomavirus L2 protein, the following recent experimental evidence suggests that L2 facilitates the release of the viral genome from endosomes into the cytosol. First, HPV VLPs, which are comprised only of L1 and DNA, uncoat their DNA in endosomes. This demonstrates that L2 is not necessary for uptake of the virus into the cell, or its transport to endosomes, or even the release of the viral DNA into the endosomal compartment. But in the absence of L2, the viral DNA remains entrapped within endosomes, and infection is aborted.

How might L2 mediate the translocation of the viral genome into the cytosol? L2 appears to do so by virtue of a hydrophobic domain that is believed to generate channels through the vesicular membrane, through which the DNA then might pass. This premise is supported by the experimental finding that a 23-amino-acid hydrophobic sequence at the C terminus of L2 is necessary for the release

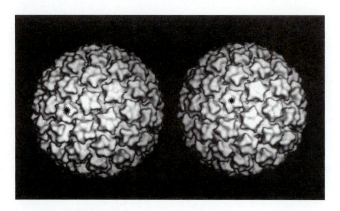

Figure 16.2 Papillomavirus icosahedral capsids are constructed from the major capsid protein L1, which is organized entirely as pentamers. Individual pentamers are surrounded by either five other pentamers (starred, left) or six other pentamers (starred, right). The image is a reconstruction generated by cryo-electron microscopy analysis of particle structures. With only a little effort it can be seen in three dimensions. Place the edge of your hand between the images, and your nose up against your hand. From T. S. Baker et al., *Biophys. J.* **60**:1445–1456, 1991, with permission.

of viral genomes from endosomal compartments. Moreover, a 23-amino-acid peptide containing this sequence itself displays membrane-disrupting activity in a pH-dependent manner. Also, the 23-amino-acid peptide and the wild-type L2 protein, but not C-terminal mutants of L2, integrated into cellular membranes. Finally, L2 also accompanies the viral genome to the nucleus. Regarding that, L2 has two nuclear localization signals and a DNA-binding domain (Kamper et al., 2006).

Experimental results from other very recent studies, which used BPV VLPs, imply that the entry pathways of some papillomaviruses, like the entry pathways of simian virus 40 (SV40) and mouse polyoma virus (Chapter 15), involve trafficking to the endoplasmic reticulum (ER). Interestingly, trafficking of BPV to the ER also is dependent on the virus's L2 protein and its interaction with the ER snare protein, syntaxin 18.

GENOME ORGANIZATION AND EXPRESSION

Papillomavirus gene expression, like papillomavirus entry, has been challenging to study because of the inability of these viruses to replicate in standard cell cultures. (Recall that the papillomavirus life cycle depends on the differentiation of infected cells within a stratified squamous epithelium.) To surmount this obstacle, an alternative cell culture system that largely mimics the histological and physiological characteristics of a normal stratified squamous epithelium was developed. It is referred to as an **organotypic raft culture** or, more commonly, as a raft culture.

Raft cultures contain a dermal equivalent comprised of fibroblasts embedded in a collagen matrix, on top of which are seeded keratinocytes from the foreskins of human neonates. When this assemblage of cells is cultured at an air-medium interface, the keratinocytes proliferate, differentiate, and stratify into an "epithelium" that resembles a normal stratified squamous epithelium. By using HPV-infected raft cultures and experimental approaches of modern molecular biology (e.g., complementary DNA [cDNA] cloning and sequencing, RNA protection/S1 nuclease analysis, and primer extension), it has been possible to analyze the pattern of gene expression of a number of HPV genotypes.

From such studies, it now is clear that the genome organization of the papillomaviruses and the pattern of papillomavirus gene expression are highly conserved among the human and animal papillomavirus genotypes and quite distinct from those of the polyomaviruses. Nevertheless, there are differences in the details between different papillomaviruses, so that what follows is not necessarily entirely the case for all HPV genotypes. Nevertheless, it is intended to reflect the overall fundamental nature of the HPV pattern of gene expression.

Noting that BPV has for many years served as a model to study the biology of HPV, a schematic of the circular BPV genome is shown in Fig. 16.3. It contains eight open reading frames, encoding the early proteins E1, E2, E5, E6, and E7, which are expressed in the undifferentiated and partially differentiated cells of the squamous epithelium, and late proteins L1 and L2, which are expressed in keratinocytes undergoing terminal differentiation (Fig. 16.4). E4, which overlaps E2 in a different translational reading frame, now is known to be expressed primarily as a late gene (see below). Thus, there is overlap on papillomavirus genomes between early and late genes. This is in contrast to polyomavirus genomes, which contain distinct early and late regions. Also in contrast to polyomavirus genomes, all papillomavirus open reading frames are organized in the same orientation (clockwise in Fig. 16.3). That is, only one strand of the papillomavirus genome is transcribed, whereas polyomavirus early and late genes are transcribed from separate strands and in opposite orientations (Chapter 15).

Papillomavirus gene maps from some earlier sources depict an E3 gene. The reason is that the sequence analysis revealed an out-of-phase reading frame within the E2 coding sequence. That out-of-phase reading frame within E2 was designated E3. However, no protein has ever been found that might be encoded by the E3 open reading frame, and the reading frame was not present in most papillomavirus genomes, suggesting that E3 is not a bona fide viral gene.

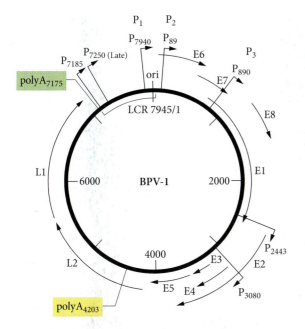

Figure 16.3 A schematic of the circular genome of BPV-1. It contains eight open reading frames, which encode early proteins E1, E2, E5, E6, and E7 and late proteins L1 and L2. E4, which overlaps E2 in a different translational reading frame, actually is expressed primarily as a late gene. The E3 open reading frame is not known to encode any protein, and it is absent from most papillomavirus genomes. The E4 open reading frame is embedded out of phase within the longer E2 open reading frame. Likewise, the E8 open reading frame is embedded out of phase within the E1 open reading frame. The RNA polymerase II promoters are designated P_{89}, P_{890}, etc. (P1, P2, and P3 are the major promoters of the low-risk human papillomaviruses. P1 is located just upstream of the E6 open reading frame, P2 is located within the E6 open reading frame, and P3 is within the E7 open reading frame. In addition, several minor promoters have been identified as well.) The sites of mRNA polyadenylation and the origin of DNA replication are indicated. Compare this to the genome of the polyomavirus SV40, in which early and late genes are expressed in different orientations and in which there is no overlap between early and late genes (Fig. 15.4). Adapted from N. J. Dimmock, A. J. Easton, and K.N. Leppard, *Introduction to Modern Virology*, 5th ed. (Blackwell Science Ltd., Oxford, 2001), with permission.

Other general features of the papillomavirus genome organization and expression include the following. First, as noted above, the E4 open reading frame is embedded within the longer E2 open reading frame and is read in a different reading frame. Likewise, the E8 open reading frame is embedded out-of-phase within the E1 open reading frame.

The ends of HPV transcripts were identified by primer extension and by RNA protection/S1 nuclease analysis. The results of those experiments implied the existence of three distinct major promoters. These are referred to as P1, P2, and P3 in the case of the low-risk HPVs. They have other designations in the cases of other papillomaviruses. P1 is located just upstream of the E6 open reading frame. The second promoter, P2, was found within the E6 open

reading frame. The third promoter, P3, is within the E7 open reading frame. In addition, several minor promoters have been identified as well. Thus, papillomavirus promoters are found in several regions of their genomes. This is another feature of the papillomaviruses that differs from the polyomaviruses, in which all promoters are found in the single regulatory region.

Sequence analysis of papillomavirus genomes and of their transcripts shows that both early and late primary transcripts are extensively spliced. Indeed, primary transcripts undergo alternative splicing between several combinations of distinct donor and acceptor sites (Fig. 16.5). Splicing is closely associated with polyadenylation. Thus, note that the early and late regions are separated by the early polyadenylation signal (sometimes designated A_E) and all transcripts are polyadenylated either at A_E at the end of the early region or at the polyadenylation signal (A_L) at the end of the late region.

In addition to being spliced, most papillomavirus messenger RNAs (mRNAs) contain multiple open reading frames (Fig. 16.5). Moreover, multiple open reading frames on the same mRNA may be available for translation. That is, these mRNAs may be polycistronic. For instance, the mRNAs that encode E1 or E4, and also E5, may have the potential to initiate translation at the downstream E5 translation initiation site, as well as at the upstream E1 translation initiation site. However, it is not yet clear exactly which proteins are translated from some of the polycistronic transcripts.

There are several species of mRNA that encode a protein designated E1^E4, which is comprised of the first 5 amino acids of E1 and the last 85 amino acids of E4 (Fig. 16.5). One such mRNA that also contains a downstream sequence encoding L1 is particularly interesting. That is so because this is the only L1-expressing transcript generated by at least some HPV genotypes. Consequently, in the case of those HPVs, all L1 molecules need to be translated from an L1 translation initiation site that is downstream from the E1^E4 translation initiation site on this bicistronic mRNA. This mRNA produces 17-fold more E1^E4 protein than L1 protein. This is consistent with the ribosomal scanning model of translation, in which ribosomes usually initiate translation at the first available AUG, but in which they are able to initiate or reinitiate translation at a downstream start site, but only at low efficiency. (The role of E1^E4 is discussed in detail below.)

One of the intriguing aspects of the papillomavirus life cycle is that the activities of particular viral promoters, and of polyadenylation sites and splice sites as well, are dependent on the differentiation state of the host cell. These facts are illustrated by the expression of the mRNAs

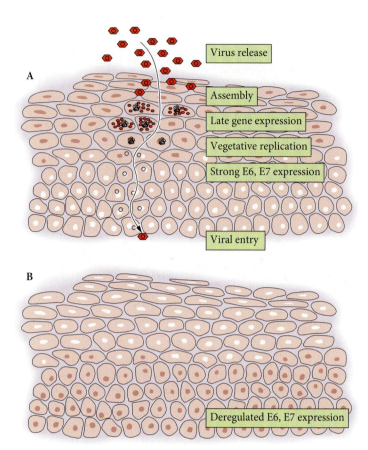

Figure 16.4 (A) Papillomavirus production is strongly coupled to the differentiation program of the infected epithelium. The figure shows strong expression of E6/E7 and of late viral proteins and virus particles only in differentiated epithelial cells. Nuclei expressing E6/E7 are indicated in dark pink. Viral genomes are indicated as black circles, and viral capsid proteins are indicated in red. Virus particles are depicted schematically as black circles in red, icosahedral structures. (B) Schematic representation of deregulated E6/E7 expression in (para)basal cells associated with cell transformation and failure of the viral life cycle. Adapted from P. J. F. Snijders et al., *J. Pathol.* **208**:152–164, 2006, with permission.

that encode E1^E4. First, note that there are several such mRNAs, each of which is under the control of the P3 promoter (P_{742} in Fig. 16.5). One of these mRNAs is the bicistronic E1^E4/L1 mRNA noted above. Another is also a bicistronic mRNA, which encodes E1^E4 and E5. Yet another encodes E1^E4 and L2. Each of these mRNAs generates E1^E4, both in cell culture and in a cell-free system. Importantly, the activity of the P3 promoter depends on the differentiation state of the cell, such that P3 becomes fully active only in terminally differentiated keratinocytes in the upper levels of the squamous epithelium. For that reason, and also because the P3 promoter generates mRNAs that encode L1 and L2, it is sometimes referred to as a late promoter. However, the P3 promoter also is active, although at much lower efficiency, at the earliest times in the viral life cycle in the basal and parabasal cells of the epithelium. Nevertheless, the P3 promoter and the late open reading frames under its control are expressed much more efficiently in the middle and upper thirds of the epithelium, in which the E1^E4/E5 transcript constitutes the bulk of the viral mRNA that is present.

The fact that the papillomavirus P3 promoter is regulated by cell differentiation is particularly noteworthy, since this feature distinguishes it from the late promoters of the polyomaviruses and of other DNA viruses as well. In the cases of those other DNA viruses, their late promoters are activated by viral DNA replication.

Now, bear in mind the following two points. First, the P3 promoter is active at early times in the viral life cycle, although inefficiently, as just noted. Second, *pre*-mRNAs, which are transcribed under the control of P3 and which contain either the L1 or L2 open reading frame, indeed are present in the nuclei of less differentiated cells. Why then are L1 and L2 produced only in differentiated keratinocytes in the upper layers of the epithelium? The answer is that the polyadenylated spliced bicistronic mRNAs that might be translated into L1 (or L2) are generated only in the differentiated upper strata of the squamous epithelium. Thus, the viral polyadenylation and splice sites and viral promoter activity are each regulated by the differentiation state of the cell, leading to the following important conclusion.

The transition from the early stage of the papillomavirus life cycle to the late stage is not achieved by the turning on of a distinct late promoter dedicated to late functions. Instead, it is accomplished by the upregulation of the late P3 promoter and by increased read-through beyond the early polyadenylation site, thereby generating transcripts

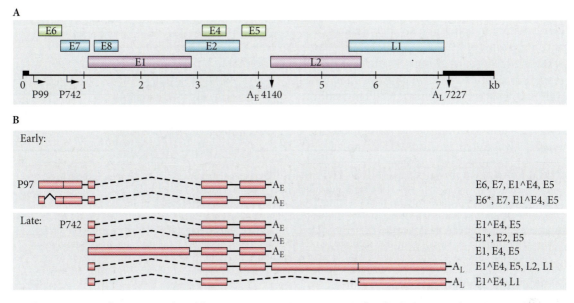

Figure 16.5 Papillomavirus early and late primary transcripts are extensively spliced. Shown are the genome and pattern of transcription of the high-risk human papillomavirus HPV31. Primary transcripts undergo alternative splicing between several combinations of distinct donor and acceptor sites, and splicing is closely associated with polyadenylation. Notice that the early and late regions are separated by the early polyadenylation signal (pAE) and all transcripts are polyadenylated either at pAE at the end of the early region or at the polyadenylation signal (pAL) at the end of the late region. Also notice that most papillomavirus mRNAs contain multiple open reading frames, and multiple open reading frames on the same mRNA may be available for translation. Adapted from K. M. Spink and L. A. Laimins, *J. Virol.* **79:**4918–4926, 2005, with permission.

that contain the L1 and L2 open reading frames. Moreover, these end results are brought about by cellular differentiation. In addition, multiple splicing and polyadenylation options lead to distinct late transcripts that might be translated into different combinations of viral proteins (Fig. 16.5). Some late transcripts extend continuously to the late region polyadenylation site, and these transcripts might generate some combination of E1^E4, E5, L2, and L1, depending on where they were spliced. Other transcripts might terminate at the early polyadenylation site and generate some combination of E1, E2, E1^E4, and E5, again depending on where they were spliced. Recall that translation can initiate at downstream start sites, but only at low efficiency, and by means that are not yet known.

Recapping the above, late viral gene expression is dependent on (i) a late-stage viral promoter, (ii) late-stage-specific polyadenylation sites, and (iii) late-stage-specific splicing sites. Moreover, the activity of each of these viral factors is dependent on the differentiation state of the host cell. How then might the differentiation state of the cell affect the activities of these disparate viral elements? Emerging results tell a complex and still poorly understood story of how transcription, polyadenylation, and splicing of papillomavirus transcripts might be regulated by the host cell.

One possibility that might account for the linkage between the activity of the P3 promoter and cellular differentiation is that P3 promoter activity is dependent on cellular transcription factors that are produced efficiently only when epidermal cells differentiate into keratinocytes. This premise has not yet been verified experimentally. However, it is plausible and consistent with experimental findings in the case of the HPV early promoter, discussed next.

The HPV early promoter (P1 in Fig. 16.3), which is responsible for the expression of E6, E7, and other viral genes, likewise is regulated by keratinocyte differentiation, and the mechanism underlying the linkage between the activity of that promoter and keratinocyte differentiation has been characterized. Expression of the E6/E7 promoter is regulated by a *cis*-acting enhancer element, which is located upstream of the E6/E7 promoter in the **upstream regulatory region** (**URR**) of the viral genome (Fig. 16.6). The URR also is known as the **long control region** or **LCR** (Fig. 16.3). Notice that the URR is situated between the end of the late gene *L1* and the start of the early gene *E6*, and also encompasses the *ori*. Importantly, a domain of the URR enhancer, which regulates expression of E6/E7, is affected by keratinocyte differentiation. Thus, it is referred to as the keratinocyte-specific enhancer (KE) domain. The KE domain is affected by keratinocyte differentiation

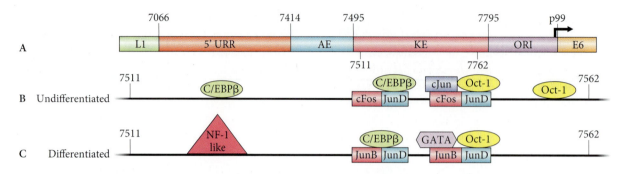

Figure 16.6 The E6/E7 early promoter (known as P99 for HPV31) is regulated by *cis*-acting elements located in the URR of the viral genome. (A) URR is located between the end of *L1* and the start of *E6* and can be subdivided into several functional domains. These include a 5′ URR domain, an auxiliary enhancer (AE) domain, an epithelial cell-specific KE domain, *ori*, and the P99 promoter. The region of the KE domain, between nucleotides 7511 and 7762, is expanded to underscore the differential occupancy of the same *cis* regulatory elements by alternative transcription factors under undifferentiated (B) and differentiated (C) cellular conditions. The differential recruitment of transcription factors to the same *cis* elements under different states of cellular differentiation may be a means by which expression of E6/E7 is linked to keratinocyte differentiation. Modified from E. Sen et al., *J. Virol.* **78**:612–629, 2004, with permission.

because its activity depends on the binding of cellular transcription factors that are produced upon keratinocyte differentiation.

Although KE is the main transcriptional regulator of E6/E7 expression, the activity of the E6/E7 promoter also is affected by the auxiliary enhancer (AE) domain within the URR (Fig. 16.6). As discussed below, the URR also interacts with viral transcription factors, in particular the viral E2 protein, which negatively regulates the E6/E7 promoter. (Notice that the URR contains no open reading frames but instead contains numerous regulatory sequence elements, which bind various transcription factors, and the *ori*. In these ways the papillomavirus URR functionally resembles the promoter/origin region of the polyomaviruses [Chapter 15]. Nevertheless, recall that in contrast to the papillomavirus URR, the polyomavirus regulatory region controls early and late genes that are expressed in different orientations.)

Next, we consider the linkage between the selection of splice sites and cellular differentiation. The selection of splice sites generally depends on *cis*-acting exon recognition elements, now known as exonic splicing enhancers (ESEs) and exonic splicing suppressors or silencers (ESSs). Alternative splicing of the papillomavirus transcripts is regulated by at least five viral *cis* elements: three ESEs and two ESSs. These five *cis* elements act coordinately to determine which 3′ splice site might be used. Importantly, multiple *cellular* splicing factors interact with the viral ESEs and ESSs to select the alternative 3′ splice sites. Some of these cellular splicing factors are differentially expressed in accordance with the stage of cell differentiation.

Now, we consider how selection of the papillomavirus polyadenylation sites might be linked to cellular differentiation. First, note that nearly all viral early pre-mRNAs are polyadenylated at the early polyadenylation site. Thus, the early-to-late switch might require (i) that the late polyadenylation site be released from whatever block it may be under and (ii) the suppression of the early polyadenylation site. Also, notice that the early polyadenylation site lies within the coding region for the L2 protein (Fig. 16.3). Consequently, the early polyadenylation site is spliced out of the L1 pre-mRNA, showing that splicing and polyadenylation can be competing and mutually exclusive events. (How might the early polyadenylation site be avoided during generation of L2 mRNA? Sorry, but the answer is not yet known.)

Returning to the principal question, how might selection of the papillomavirus polyadenylation sites be linked to cellular differentiation? Recent experimental results show that polyadenylation of HPV-16 early transcripts is positively regulated by two separate *cis*-acting elements that flank the HPV-16 early polyadenylation site. One of these elements contains a motif that binds hnRNP H, a protein family involved in the regulation of alternative splicing and polyadenylation. Interestingly, hnRNP H levels are highest in the lower levels of the cervical epithelium, where early viral transcription predominates. Likewise, studies on the regulation of late gene expression revealed the presence of a *cis*-acting negative regulatory element in the 3′ untranslated region of the viral genome. Later studies demonstrated that this element negatively regulates the late polyadenylation site through its interaction with cellular factors involved in polyadenylation.

Earlier studies suggested that *cis*-acting elements on the viral RNAs also might affect RNA stability and RNA export from the nucleus. However, more-recent experimental results suggest that these viral elements instead affect polyadenylation or splicing, as discussed above. Regardless, the regulation of papillomavirus gene expression is complex, reflecting its dependency on a variety of *cis*-acting viral RNA elements as well as on a variety of host cell factors that are expressed in conjunction with the cell differentiation program. (For a recent review of developments in the area of papillomavirus gene expression and posttranscriptional regulation, see Zheng and Baker, 2006.)

Before moving on, we well might ask why the papillomavirus pattern of gene expression is so complex, using multiple promoters, multiple splice donor and acceptor sites, and a variety of polycistronic mRNAs. Plausibly, the use of multiple promoters might enable papillomaviruses to fine-tune the levels of the different proteins they express during the different stages of their complex replicative cycle in the different strata of the epithelium. In addition, alternative splice sites make it possible for these viruses to coordinately express different proteins from the same promoter. But then, we still need to account for why the papillomaviruses evolved to make their pattern of gene expression dependent on host cell differentiation. Can you propose any advantages the virus might derive from this unique lifestyle? See below.

REPLICATION

Replication of circular papillomavirus genomes is reminiscent of that of the polyomaviruses (Chapter 15). In each instance, a circular viral genome is replicated bidirectionally from a single origin of DNA replication (see Fig. 15.8). However, there is experimental evidence that papillomavirus replication may switch to a rolling-circle mechanism during the late productive phase of the viral replicative cycle (see below). Regardless, by having circular DNA genomes, these viruses avoid the "end replication" problem of linear DNA genomes (see Chapter 14, and particularly Box 14.3).

Papillomavirus DNA replication depends upon the viral E1 and E2 proteins, the only viral proteins required for viral DNA replication. The E1 protein serves as an initiator that binds specifically to the viral *ori*, which is located in the LCR of the viral genome, as noted above. The E2 protein serves as an accessory replication factor by forming a complex with E1, in that way enhancing DNA binding by E1. E1 expresses a DNA helicase activity that is thought to unwind the advancing replication forks. In addition, E1 interacts with several cellular proteins required for DNA replication. Indeed, the rest of the replication machinery is provided by the host cell. Together, the roles of the papillomavirus E1 and E2 proteins in genomic replication functionally resemble that of the polyomaviral LT proteins. (Interestingly, and somewhat reminiscent of the

SV40 LT, E1 and E2 initially bind to *ori* cooperatively as homodimers. As ATP is hydrolyzed [presumably by the E1 ATPase activity], E2 dimers are replaced by additional molecules of E1, until a larger E1 structure, thought to be a ring-like hexamer, is assembled on single-stranded DNA [see Fig. 15.9].)

The papillomavirus E5, E6, and E7 proteins facilitate viral DNA replication indirectly by creating an intracellular milieu conducive to DNA replication. In this way, the actions of these three other papillomavirus early proteins also are reminiscent of the polyomaviral T proteins. These similarities will become more evident when we consider the molecular mechanisms by which E6 and E7 exert their effects. For the moment, note that expression of the viral E6 and E7 proteins (and E5 in some instances) induce the proliferation of epithelial cells. The development of warts and, in the case of high-risk HPV strains, of cancer is a consequence of E6 and E7 expression.

In addition to its function in viral DNA replication, the papillomavirus E2 protein appears to have further but incompletely understood roles in the viral life cycle. First, E2 may serve as a DNA-binding transcription factor, consistent with the experimental observation that low doses of E2 induce a weak activation of transcription of the *E6* and *E7* genes. However, the biological significance of this observation is called into question by the finding that an E2 mutant, which is inactive in the transcription regulatory function but has normal DNA replication activity, displays normal growth properties.

On the other hand, higher doses of E2 act to repress transcription of *E6* and *E7*. E2 exerts this repressive effect by binding to the same sites on the DNA that initiate DNA replication. When E2 binds to these sites, which are proximal to the E6/E7 promoter, it sterically hinders the binding of cellular transcription factors such as SP1 and TFIID, which are essential for the transcription of *E6* and *E7*. Thus, the E2 protein serves to activate DNA replication, while simultaneously repressing transcription of the E6 and E7 genes. (As discussed below, *E6* and *E7* are oncogenes, and their continued expression is necessary to maintain the neoplastic cell phenotype. Therefore, E2 must be inactivated in cervical carcinoma cells. Experimentally expressing E2 in cervical carcinoma cells results in cellular senescence and apoptosis [see below].)

E2 plays yet another role in papillomavirus replication, serving to facilitate viral genome segregation during cell division. (The significance of this fact will be apparent shortly.) It does so by tethering viral genomes to mitotic chromosomes.

As noted several times above, the HPV life cycle is directly regulated by the differentiation state of the host cell within the stratified squamous epithelium. The means by which cellular differentiation regulates the pattern of viral gene expression was also discussed above. Now, we note that papillomaviruses also switch between different modes of DNA replication, in response to the differentiation state of the cell. These different modes are as follows. After initial viral entry into the undifferentiated basal cells of the epithelium, the viral genomes replicate more frequently than the cellular genome, in that way amplifying the viral genome copy number. This initial phase of viral replication is referred to as the **establishment phase**. Following that, there is a transition to a **maintenance phase**, in which the viral genomes replicate only once per cell cycle, on average, to keep the viral copy number constant. Finally, during the **productive phase**, high levels of viral DNA replication occur in the more terminally differentiated cells of the epithelium, which leads to thousands of viral genome copies per cell. How cellular differentiation regulates these switches in the viral replication mode is not fully understood. Individual stages in the viral replicative cycle are considered in somewhat more detail below.

First, papillomaviruses depend on wounds (e.g., cuts and abrasions) to the stratified epithelium in order to infect cells in the basal layer (Fig. 16.1 and 16.4). Importantly, these basal cells are the only cells of the normal stratified epithelium that are actively dividing.

After the viral genome is uncoated in a basal cell, it translocates to the nucleus, where it is established as an extrachromosomal plasmid. In these infected basal cells, viral genomes are replicated until there are about 100 or so genomes per cell. This is the establishment phase, as noted above. Since the circular viral genomes are present as extrachromosomal plasmids, they replicate independently of the cellular chromosomal DNA. They do not hop in and out of cellular chromosomes.

Following the initial establishment phase, the viral genomes replicate only in conjunction with cellular DNA replication. That is, the viral DNA replicates only when the basal cell genome replicates. As will be seen, this scheme enables the virus to maintain its genomes in the basal layer of the epithelium and thereby sustain a persistent infection. Consequently, this period is sometimes referred to as the maintenance phase. Importantly, the virus does not produce its capsid proteins in these basal cells, and no progeny virus particles are yet generated.

When basal cells replicate, one daughter cell remains behind as a basal cell, whereas the other daughter cell migrates away from the basal layer and begins to differentiate. Although the virus initially infects a basal cell, in which it replicates its DNA as described above, virus production occurs in cells only as they become highly differentiated in the outermost layers of the epithelium. In this way, a reservoir of viral genomes is stably maintained in the basal layer, during which time the basal layer continually generates daughter cells that differentiate and produce progeny viruses.

The time required for a daughter basal cell to transit through the layers of the epithelium and undergo differentiation is generally one to several weeks, depending on the particular epithelium. After this time, the cells desquamate (i.e., flake off) from the epithelium. These sloughed-off cells carry with them high concentrations of virus particles that are able to survive in the environment for relatively long periods, until they might infect a new host. This complex pattern of replication provides the papillomaviruses with several important advantages (see below).

The mechanism responsible for the switchover from the initial establishment phase of viral DNA replication to the maintenance phase is not yet known. One thought is that during the maintenance phase, the viral *ori* acts like a cellular origin of DNA replication. In that way viral genomes replicate concurrently with the cellular genome and the viral DNA is sustained at an approximately constant copy number per cell. At these times, early viral proteins are synthesized at low levels, and late genes are not expressed at all.

Recall that E1 and E2 are the only viral proteins that are needed to replicate the viral genome. This is so during the maintenance phase, as well as during other phases. The P1 promoter, which is immediately upstream of the E6 open reading frame, may regulate expression of each of the early genes. Importantly, P1 is regulated by keratinocyte differentiation, such that during differentiation of papillomavirus-infected keratinocytes, E1 and E2 protein concentrations appear to increase. These results are consistent with the notion that E1 and E2 levels coordinate the transition to the productive phase of the replication cycle (Box 16.3).

It is not yet known with certainty just when and where in the epithelium the E6 and E7 oncoproteins are first expressed. Some researchers believe that they are expressed either at low levels or not at all in the basal cells, since if they were, then those cells might be expected to be in a continued state of proliferation, and the likelihood of infection leading to cancer would be increased considerably (see below). Still, most researchers in the field believe that E6 and E7 are expressed in all layers of the epithelium, though perhaps at a slightly reduced level in the basal cells.

As infected cells go through their final stages of differentiation in the outer layers of the epithelium, the virus life cycle switches from its nonproductive early phase to its productive late phase. Viral DNA amplification is triggered along with late transcription, and thousands of progeny virus particles are generated from each of a small number of terminally differentiated cells (Fig. 16.4).

BOX 16.3

As is often the case, living systems are considerably more complex than our rather straightforward models might lead us to believe. A case in point is that by means of alternative splicing, papillomaviruses express a protein, E8^E2C, that fuses the peptide encoded by the small E8 open reading frame to the carboxy terminus of E2. Moreover, E8^E2C transcripts are present throughout the complete replication cycle. Importantly, mutation of E8^E2C results in runaway viral DNA replication in basal cells. Contrary to what might have been expected, this leads to the inability to stably maintain viral genomes in the basal cells and, consequently, over the long term in keratinocytes as well. Thus, E8^E2C appears to play a key role in regulating viral DNA replication, affecting both the establishment and maintenance of the persistent infection.

But how might E8^E2C have a positive effect on replication during the establishment phase and a negative effect during the maintenance phase? Not surprisingly, the answer is not clear. E8^E2C lacks the amino-terminal domain of E2

responsible for activation of transcription and DNA replication but retains the carboxy-terminal E2 domain that mediates specific DNA recognition and dimerization among E2 proteins. Thus, E8^E2C might act by competing with E2 at E2 DNA-binding sites or possibly by forming heterodimers with E2 that are nonfunctional for replication. Yet there are several remaining enigmas. One is that E8^E2C is needed for the long-term maintenance of papillomavirus extrachromosomal genomes, despite the fact that the protein is a negative regulator of their replication. A second enigma is that while E8^E2C is thought to act as an antagonist of E2, either by displacement of E2 molecules from their binding sites or through formation of inactive heterodimers, recent experimental findings suggest that neither of these mechanisms can be the entire story. For example, genetic studies showed that the E8 domain also contributed significantly to the inhibition of replication. Thus, even with the introduction of E8^E2C, the papillomavirus copy number control mechanism is hardly understood.

During the early establishment and maintenance phases, and perhaps during the entire replicative cycle, papillomavirus genomes are replicated much like polyomaviral genomes, that is, by bidirectional replication of **theta structures**. However, there is experimental evidence of a switch to a **rolling-circle** mode of replication in differentiated epithelial cells. (The rolling-circle mode of replication is described in detail in Chapter 14; see Box 14.5.) For example, during the late phase, replicating viral DNA forms characteristic of the theta mode of replication were not detected. Instead, replicating forms had only a single replication fork, characteristic of rolling-circle replication (see Box 14.5). Moreover, the sizes of the replicating DNA molecules, as indicated by two-dimensional gel analysis, were consistent with a rolling-circle model of DNA replication.

Why might there be a switch to the rolling-circle mode of replication? In the theta mode, an initiation event is required for each round of replication. In contrast, in the rolling-circle mode, a single initiation event can generate multiple progeny genomes. Thus, the rolling-circle mode may be inherently more efficient than the theta mode. This might facilitate productive replication in the growth-arrested, fully differentiated keratinocytes, in which conditions for initiating replication would usually be less than favorable. However, evidence that there is a fundamental switch in the mode of HPV-16 DNA replication in differentiated

cervical epithelial cells is not widely accepted among papillomavirus researchers. Moreover, as will be seen, the E6 and E7 proteins make for a more replication-supportive milieu in these differentiated keratinocytes.

Some of the papillomavirus factors that are known to facilitate productive replication in differentiating epithelial cells were discussed above and are briefly summarized here. In addition, several factors, not previously noted, are enumerated as well. However, bear in mind that our list of factors is incomplete, since not all are known. So, our partial list includes the following. (i) The P1 promoter is upregulated in differentiated cells. Consequently, the E1 and E2 proteins, which sustain papillomavirus DNA replication, likewise are expressed at high levels in differentiated cells. The resulting amplification of the viral genome copy number in turn significantly enhances the levels of late gene expression. (ii) The P3 promoter likewise is upregulated in differentiating cells. (iii) There is increased readthrough of the early polyadenylation site of P3-regulated transcripts, resulting in mRNAs that encode L1 and L2. (iv) Production of the BPV L1 protein requires recognition of a 3′ donor splice site that is not recognized in incompletely differentiated cells. (v) Polyadenylation of BPV late mRNAs only occurs in fully differentiated keratinocytes. (vi) The viral regulatory region is hypermethylated in undifferentiated epithelial cells and hypomethylated in differentiated cells, suggesting that viral gene expression might be regulated by methylation of the LCR sequence. (However, the premise that hypo- and hypermethylated genomes play a role in differentiation-dependent replication is still controversial.) (vii) Differentiating cells may contain cellular factors that stabilize late viral mRNAs.

Before moving on, we may once again ask why the papillomaviruses might have evolved such a complex life cycle, one so intimately coupled to the differentiation program of epithelial cells. We cannot know the answer with certainty, but the papillomavirus strategy provides these viruses with several conceivable advantages. For example, production of virus in the differentiated cells that have migrated to the surface of the epithelium facilitates transmission of the infection to new individuals. At the same time, restriction of late protein synthesis and virus production to the terminally differentiated keratinocytes at the outermost layers of the epithelium places these virus-producing cells beyond the reach of immune attack and prevents viruses from being exposed to neutralizing antibodies. Indeed, the papillomaviruses are quite "ingenious" at avoiding host immunity, as further indicated by their limited and noncytolytic replication in the basal layer of the epithelium, which precludes a robust immune response (Chapter 4). Meanwhile, since the basal cells are self-renewing, by preventing virus production in those cells while stably maintaining a reservoir of viral genomes in them, papillomaviruses are able to establish a long-term source of virus for transmission to new individuals. HPV infections that might initially target the differentiated, more superficial cells of the epithelium would likely be transient, since the viral genomes would be lost as the infected cells are shed during the terminal differentiation process.

RELEASE AND TRANSMISSION

Release presents a somewhat unique challenge for the papillomaviruses. That is so because during late stages of differentiation, keratinocytes develop a cornified cell envelope comprised of a dense matrix of proteins and lipids, which are covalently cross-linked by transglutaminases expressed in those cells. These cornified cells provide the host with a tough barrier against water loss, mechanical damage, and virus infection. Yet papillomaviruses must somehow be released across the tough cornified envelopes of differentiated keratinocytes, in order for these viruses to be transmitted to new individuals. Bearing in mind that papillomaviruses do not induce cell lysis, how then might this release occur? The answer is not known with certainty, but HPVs do appear to induce changes in the thickness, morphology, composition, and strength of the cornified cell envelopes of infected keratinocytes, by mechanisms that are not known with certainty, and it is reasonable to propose that these alterations might facilitate virus release.

There is experimental evidence which suggests that the E1^E4 protein may cause the changes in the cornified keratinocyte cell envelopes noted above. First, E1^E4 is localized at the cornified cell envelopes and copurifies with them. Second, E1^E4 begins to be profusely expressed during the late stages of the virus life cycle in the differentiated keratinocytes (see above). Third, expressing E1^E4 from a retrovirus vector in differentiated keratinocytes results in morphologically altered cornified cell envelopes that are similar to those seen in HPV-infected genital epithelium. Together, these experimental findings support the premise that the papillomavirus E1^E4 protein is responsible for the morphological changes in the cornified cell envelopes of infected differentiated keratinocytes and are consistent with the possibility that E1^E4 might facilitate the release of HPV particles from these cells. (E1^E4 has several other effects on host cells, including inducing cell cycle arrest in G_2 and causing the reorganization of the keratin intermediate-filament network [keratins act to structure the cytoplasm and to resist external stresses]. The significance of these other E1^E4 activities in the virus life cycle is presently unknown.)

As noted above, several spliced mRNAs are known to encode E1^E4, including a monocistronic one that encodes E1^E4 only and a bicistronic one that encodes E1^E4 and L1 as well. Also, whereas E1 and E4 are encoded in the early region, the mRNAs encoding the hybrid E1^E4 protein are generated under the control of the P3 promoter, which is greatly upregulated in differentiated cells. Consequently, E1^E4 is detected primarily late in infection, in differentiated cells that also contain the L1 major capsid protein. For these reasons, it may be appropriate to think of E1^E4 as a late gene product. Also, nearly all cells in natural lesions, such as condylomas and premalignant tumors of the cervix, which contain detectable L1, also contain detectable E1^E4 (see below).

The restriction of the papillomavirus productive phase to the differentiated keratinocytes of the outermost layers of the epithelium would appear to facilitate transmission from one individual to another as noted above. However, it is not that straightforward, since extracellular virus is seen only rarely in HPV lesions; this is an important fact, since large numbers of virus particles likely are required to initiate an infection. So, how then is the infection transmitted? First, the infected keratinocytes desquamate (i.e., flake off), becoming concentrated packages of virus particles, which might efficiently initiate infection if there were a means to release the virus from them. This might happen, for example, if they were broken open by the mechanical stress of sexual activity. Release of virus from these cells also might be facilitated by the E1^E4-induced changes in the thickness and durability of the corneal envelopes. It is believed that HPVs might then gain access to the basal layer of the recipient's stratified squamous epithelia through wounds or abrasions.

The E6 and E7 proteins also may indirectly facilitate transmission of the virus. The mitogenic activities of these oncoproteins greatly amplify the number of virus-producing cells in the epithelium. That is, whereas normal human epithelium usually contains only 8 to 10 layers of cells, condylomas may contain as many as 100 layers.

Regarding the actual spread of papillomaviruses, in particular the genital HPV genotypes, a variety of studies demonstrate that these are transmitted primarily by the sexual route. Thus, while we discuss genital HPV types mainly with respect to infections in women (i.e., because of the link to cervical carcinoma), these viruses also infect the genital areas of men as well. As in the case of women, men infected with HPV of any type usually do not develop any symptoms or health problems. But men infected with low-risk types may develop genital warts, and infection with high-risk types can lead to penile cancer or anal cancer. Yet the incidence of HPV-associated penile cancer is relatively rare compared to the incidence of cervical carcinoma (see below).

Likelihood of infection in both women and men correlates with age at first intercourse, the number of sexual partners, and the likelihood that one or more of these partners was an HPV carrier as estimated by their sexual behaviors. Early studies of the male role in transmission were based on questionnaires that addressed their sexual behavior, while later studies assessed the presence of HPV DNA in various penile tissues.

It is possible that women might tend to be long-term carriers of genital HPV types, whereas men might tend to be only transiently infected. If that is so, then men who might have passed on their HPV infections might not have been recorded as positive when subsequently screened in epidemiologic studies. If that is so, then the data correlating transmission to the female and the likelihood that a male partner was infected may underestimate the actual correlation. Interestingly, uncircumcised men were about 3 times more likely to harbor HPV DNA in their penile tissue than circumcised men and were correspondingly more likely to transmit HPV to their female partners. Also, there is evidence that HPV might be present in the semen of some infected males.

Nonsexual transmission of genital HPV strains may also be possible. For instance, infected mothers might transmit their infections to their children via normal handling.

CELL TRANSFORMATION AND ONCOGENESIS: CERVICAL CARCINOMA

As noted above, *more than 99% of cervical cancer biopsy samples* show evidence of infection with a high-risk HPV, in particular HPV-16. This correlation between cervical carcinoma and the presence of one of a small subset of genital HPV types provides compelling evidence that these high-risk HPVs are the etiologic agents of cervical carcinoma in humans. Low-risk mucosal genotypes, such as HPV-6 and HPV-11, rarely cause invasive cancers. Instead, these HPVs induce benign neoplastic genital lesions called condylomas. Some cutaneous HPVs, particularly HPV-5 and HPV-8, like the high-risk mucosal HPVs that are associated with cervical carcinoma, also may be classified as high-risk genotypes, since they induce *epidermodysplasia verruciformis*, a skin lesion that has a tendency to become malignant. Since papillomaviruses are significant agents of neoplastic disease in humans, this topic is explored at some length in this chapter (Box 16.4).

Papillomavirus-induced *warts* occur as an inadvertent and unnecessary feature of the replicative cycles of some papillomaviruses. (Warts are believed to result from the replication of usually nondividing cells, which maintain extrachromosomal viral genomes. This process is distinct from the genesis of carcinomas. See below.) This linkage between papillomavirus replication and neoplasia is in marked contrast to neoplasias induced by the polyomaviruses, which are mainly laboratory artifacts and a dead end for the polyomaviruses with respect to their replication and survival in nature (Chapter 15). However, while papillomavirus-induced warts are coincident with the replication cycle of some papillomaviruses, the progression of high-risk HPV-induced genital lesions to malignant carcinomas is antithetical to the life cycle of those viruses. That is so because HPV infection is abortive in high-grade (i.e., advanced) premalignant lesions (see below) and in carcinomas. That is the case mainly because HPV replication is dependent on the differentiated state of the host cell, and these advanced premalignant and malignant lesions are characterized by the loss of epithelial differentiation. In addition, the viral genome generally integrates into the cellular DNA in these high-grade lesions, a state from which it cannot replicate. Thus, it may seem ironic that high-risk HPVs actually are generated in clinically unapparent infections and in low-grade squamous intraepithelial lesions (SILs; see below), while they are not generated in cervical carcinomas, which are the major clinical consequence of infection with these types.

Molecular Mechanisms

Next, we examine the molecular mechanisms by which papillomaviruses induce neoplasia. A particular concern

BOX 16.4

Another historic interlude

In Chapter 1, we recounted how Peyton Rous, in 1911, demonstrated that a chicken tumor could be transmitted by a cell-free filtrate of the tumor. In that way, Rous demonstrated that some chicken tumors, which previously were thought to be "spontaneous," actually are induced by a virus. The virus in that instance came to be known as the Rous sarcoma virus. Its name reflects the fact that it not only induces the tumor but also determines its form. The Rous sarcoma virus is a retrovirus. Its biology, which is very different from that of the DNA tumor viruses, is described in detail in Chapter 20.

Rous spent the next several years unsuccessfully trying to demonstrate that mouse cancers too might be caused by transmissible agents, finally leaving off work with tumors in

1915. However, Rous returned to the field in 1934, encouraged to do so by Richard Shope, who was Rous's friend and colleague at the Rockefeller Institute. Shope had just recently discovered that a virus is responsible for the giant warts that often are seen on the skin of wild rabbits in the southwestern United States. What was the nature of those warts? Were they indeed real tumors? Rous was intrigued by these issues and studied those warts, finding them to indeed be tumors, although benign ones, which nevertheless might progress to skin cancers in rabbits. (Note that genital warts in humans are distinct from cervical carcinomas and generally do not progress to malignant lesions. The genesis of cervical carcinomas in humans is described in the text.) This explains how Rous, together with Shope, was among the first researchers to demonstrate that some papillomavirus infections have a natural tendency to progress to malignancy.

is to account for the differences between the low-risk and high-risk HPV genotypes which enable the latter to induce malignant disease. We begin by noting that the greater ability of high-risk HPVs (compared to that of low-risk HPVs) to induce malignancies in vivo correlates with their greater ability to induce neoplastic transformation of cells in culture. This is important, since it provides the rationale for comparative studies of cell transformation by high-risk and low-risk HPV types, as an experimental approach to understand the molecular basis underlying the capacity of the high-risk genotypes to induce cervical carcinoma.

We also note that papillomaviruses and polyomaviruses alike have a need to create an intracellular milieu that might support their replication. In the case of papillomaviruses, this need is closely associated with the fact that they undergo productive replication only in differentiated keratinocytes. (Recall that the establishment and maintenance phases of the infection occur in the basal layers of the epithelium, where they are characterized by only low levels of viral activity. Viral genomes eventually are replicated [as extrachromosomal plasmids] in conjunction within the division of the basal cell, in that way maintaining the viral copy number at a constant level in the basal layer of the epithelium. Vegetative amplification of the viral DNA and the assembly of progeny virus particles occur later, in concert with the terminal differentiation of the epithelial cells into keratinocytes.) However, differentiated keratinocytes do not divide and thus are in a state ill suited to support virus replication. Expression of

the papillomavirus oncogenes creates a replication-supportive milieu within cells, supporting the productive phase of the infection. (In contrast to the above, recall that warts are thought to result from the viral stimulation of cell growth within the basal layers of the epithelium. Warts are not thought to be precursors of malignant carcinomas. The genesis of the latter is described below.)

With the above points in mind, recall that SV40 LT induces cells to bypass the complex circuits that regulate exit from G_0 by abolishing the normal functions of tumor suppressor proteins p53 and pRb and other members of the pRb family. Remarkably, the papillomaviruses likewise inactivate p53 and pRb to cause resting cells to exit from G_0. However, whereas SV40 uses separate domains on the single LT protein to inactivate both p53 and pRb, papillomaviruses use separate proteins to inactive p53 and pRb: E6 for p53 and E7 for pRb.

It is all the more remarkable that adenoviruses too induce cell cycle progression and neoplastic transformation by inactivating p53 and pRb (Chapter 17). Thus, members of three unrelated DNA virus families have evolved to target the same cellular proteins in order to facilitate virus replication. Yet bear in mind that cell transformation by the polyomaviruses, high-risk HPVs, and adenoviruses is not required for their replication. Instead, it occurs as a rare inadvertent consequence of their need to induce resting cells to enter the cell cycle. For the sake of comparison, see the oncogenic retroviruses (Chapter 20), which transform cells via completely different mechanisms. Moreover, although transformation by the polyomaviruses, papillomaviruses, and adenoviruses occurs as a rare accident that is not crucial to their replication, it nevertheless reflects viral activities that promote viral replication. In contrast, transformation by the retroviruses bears no relevance whatsoever to their replication.

Experiments in vitro demonstrated that the *continuous* expression of the HPV E6 and E7 proteins is *required* to maintain the continued proliferation of cervical carcinoma cell lines (see below). These experimental findings are reminiscent of those which demonstrated that the continued expression of SV40 LT likewise is required to maintain the transformed phenotype of SV40-transformed cells (Chapter 15).

Since all HPVs need to induce cell cycling, how might we account for the greater ability of high-risk HPVs to induce cell transformation and neoplasia? One reason is that E7 proteins of high-risk genotypes are more efficient at inactivating pRb than are E7 proteins from low-risk genotypes. That is so because E7 of high-risk genotypes is much better able to bind to pRb than E7 of low-risk genotypes. Moreover, E7 from high-risk genotypes inactivates pRb not only by sequestering it but also by marking it for degradation via the ubiquitin/proteosome pathway, doing so by binding both pRb and a subunit of the proteosomal complex. Likewise, E6 from high-risk genotypes targets p53 for ubiquitination and rapid proteosomal degradation by recognizing both p53 and a component of the ubiquitin/ proteosomal pathway. In contrast, the E6 proteins of low-risk strains bind to p53 only weakly, if at all, and these interactions do not lead to p53 degradation.

So, recapping the above, E6 proteins of low-risk mucosal strains, such as HPV-6 and HPV-11, bind to p53 only weakly, if at all. Moreover, these interactions do not lead to p53 degradation. Likewise, E6 proteins from cutaneous HPVs also bind p53 only weakly, and the E6 proteins of those HPV genotypes likewise are incapable of inducing p53 degradation. Still, even the cutaneous HPVs and low-risk mucosal HPVs have a need to prevent apoptosis, which otherwise would result from the aberrant cell cycling induced by their E7 proteins. How then might cutaneous HPVs and low-risk mucosal HPVs deal with p53 so as to prevent apoptosis? Also, bear in mind that high-risk HPV infections are rather common, yet they only rarely lead to malignancies. What might these facts suggest? Clearly, while expression of E6 and E7 by high-risk HPVs is necessary to cause malignant disease, expression of those proteins is not sufficient to do so. But do these facts also imply that whereas differences between E6 and E7 of high-risk and low-risk HPV types, with respect to their interactions with p53 and pRb, are important in the establishment and progression of cervical cancer, they cannot alone account for why high-risk HPV types cause malignancies, while other HPV

genotypes do not? If so, what might these other factors be? The answers to these questions are not known with certainty, but bear these questions in mind as you read on, probing for possible answers. But first, a cautionary note is in order.

In case we take comfort in believing that the inactivation of p53 and pRb provides a universal paradigm to explain neoplastic transformation by the DNA tumor viruses, be aware that Mother Nature seems to take pleasure in jarring us from our self-satisfaction. A case in point is that as in the example of the SV40 T proteins, the papillomavirus E6 and E7 proteins have several other known targets in the cell. Although the biological significance of some of these other interactions is not yet known, there are two other effects of E6 that likely are important to the neoplastic process. These are (i) the induction of telomerase activity and (ii) the interaction of high-risk E6 proteins with a few proteins that contain a PDZ domain. (PDZ domains are protein-binding motifs, and their most general function may be to localize their ligands to appropriate membrane domains. In polarized epithelial cells, particular PDZ proteins clearly localize at distinct apical, basal, and junctional membrane domains.) The E6-mediated induction of telomerase prevents the erosion of the chromosomal termini that otherwise would occur as a result of sustained cell proliferation. The interaction of high-risk E6 proteins with proteins containing PDZ domains may lead to the inactivation of additional tumor suppressor proteins.

The effects of E6 on telomerase activity and on PDZ proteins are discussed in greater length below. But first, is it possible that yet other papillomavirus gene products may promote neoplasia? The answer indeed is yes. For example, the transforming activity of BPV is based not on its E6 and E7 proteins but instead on its E5 protein, which has a mechanism of action distinctly different from the mechanisms of E6 and E7. The 44-amino-acid BPV E5 protein binds to the *transmembrane* domain of the platelet-derived growth factor receptor. Doing so, it transforms cells by activating that receptor, which then expresses its tyrosine kinase activity. Moreover, the BPV E5 protein may be able to similarly activate the epidermal growth factor (EGF) receptor. Activation of these receptors stimulates downstream mitogenic signaling pathways, such as the mitogen-activated protein kinase cascade (Box 16.5).

E5 may act in another way as well. When growth factors bind to the extracellular portions of their cell surface receptors, they induce receptor dimerization and signal transduction. In addition, they also cause the activated receptors to be internalized, a step leading to their degradation in acidified endosomes. This degradative process serves to ensure that signals are propagated only in response to

Two points are notable in this sidebar. First, retroviruses commonly induce cell proliferation by activating membrane receptors (Chapter 20). Yet in most such instances, the means by which retroviruses do so is quite different from that of the BPV E5 protein. However, a retrovirus, the Friend erythroleukemia virus, encodes a dimeric transmembrane glycoprotein, gp55, which although larger than the BPV E5 protein, transforms cells via a similar mechanism, inducing cell proliferation by constitutively activating the erythropoietin receptor.

Second, the difference in transforming mechanisms between the human and bovine papillomaviruses is somewhat reminiscent of the difference in transforming mechanisms between SV40 and the rodent polyomaviruses (Chapter 15). Recall that the primary transforming protein of the rodent polyomaviruses is their MT protein rather than LT. Also, whereas SV40 LT acts by inactivating p53 and pRb, the rodent polyomavirus MT acts by its effect on Src family kinases, which leads to the activation of mitogenic signaling cascades.

current growth factor conditions. The E5 protein binds to a component of the cellular vacuolar adenosine triphosphatase (ATPase), the enzyme responsible for acidifying endosomes. As a consequence, endosomal acidification is impaired and a significant fraction of internalized receptors avoid degradation and, instead, are recycled to the cell surface to be further stimulated by E5 or by their natural ligands.

The above describes the effects of the BPV E5 protein. What then of the E5 proteins of the high-risk HPVs? Might they play a role in human malignancies? Recent experimental results support this possibility. For example, expression of the cloned HPV-16 E5 gene is sufficient to transform cells in culture (as indicated by cell growth in soft agar and accelerated growth in low-serum media). Moreover, transgenic mice that express the HPV-16 E5 gene developed epithelial tumors, showing that expression of the E5 protein of a high-risk HPV is tumorigenic in vivo. Importantly, the ability of E5 to induce epithelial tumors in these mice was impaired when they were crossed to mice carrying an allele of the EGF receptor that produces a dominant-negative receptor. These experimental results strongly support the premise that the HPV-16 E5 protein influences tumor genesis in vivo, doing so at least in part via its interaction with the EGF receptor, presumably by activating a mitogenic signaling pathway.

Another recent study assessed the role of the HPV-16 E5 protein in the life cycle of that virus. The E5 protein was not needed during the establishment and maintenance phases of infection in the basal epithelial cells, during which the viral genome is replicated and then maintained as a low-copy-number extrachromosomal plasmid. In contrast, E5 expression did enhance the productive phase of the viral life cycle in differentiated keratinocytes. Apparently, E5 somehow acts to more effectively reprogram the differentiated cells to support viral DNA replication. In that regard, E5 has been shown to cooperate with E7 to stimulate the proliferation of human and mouse cells in culture. E5 also enhances the cell immortalization activities of E6 and E7 (see below). Interestingly, other studies showed that, unlike E6 and E7, E5 often is not expressed in cervical cancers. Thus, unlike E6 and E7, E5 is not necessary for the maintenance of the malignant phenotype.

Evidence that the high-risk HPV E5 protein affects tumor genesis is consistent with the notion that expression of high-risk HPV E6 and E7 is insufficient to cause malignancy, despite the ability of these proteins to impair cell cycle control. The truth of that premise should have already been evident from the fact that only a small percentage of women infected with a high-risk HPV ever develop cervical carcinoma. In most instances of infection with a high-risk HPV, cervical lesions appear only as inconspicuous, so-called low-grade **cervical intraepithelial neoplasias** (CINs), also referred to as low-grade **squamous intraepithelial lesions** (SILs, see below). Furthermore, most of these low-grade lesions are cleared within 6 to 12 months after their appearance, probably as a result of host immune responses to the lesion.

A small percentage of low-grade CINs persist and may eventually progress to higher-grade CINs, in time becoming malignant carcinomas. However, they generally do so only after 12 to 15 years of persistent infection. Thus, progression to the malignant cell phenotype must involve a succession of other cellular changes, which happen over a period measured in years. We begin our consideration of these changes by focusing on one that involves the virus genome. Specifically, the virus genome ceases to be maintained as an extrachromosomal plasmid and, instead, integrates into the cellular genome.

Why might the transition of the viral genome from its extrachromosomal state to an integrated one be relevant to neoplasia? Experiments in cell culture demonstrate that cells with an integrated HPV genome have a growth advantage over cells in which the viral genome is maintained as an extrachromosomal plasmid. Interestingly, this is so despite the fact that cells with extrachromosomal viral genomes may contain many more copies of the viral genome than cells with integrated genomes. For example, in one study, cell clones containing *only* between 3 and 60 integrated viral genomes per cell outgrew cell clones containing as many as 1,000 stably maintained, extrachromosomal viral genomes per cell.

What might account for the growth advantage of cells that contain integrated viral genomes only? The answer may be that these cells actually express higher levels of the viral E6 and E7 proteins, since these are the only viral **oncoproteins** (proteins that cause transformation or tumor genesis) that continue to be expressed in these carcinoma cells. Why then do cells with a few integrated genomes produce more of the viral E6 and E7 proteins than cells with many more genomes, all of which are extrachromosomal? First, recall that the E2 protein can act as a transcriptional repressor of the E6/E7 promoter. Second, the E2 gene generally is disrupted by integration in HPV cervical carcinoma cells. Thus, a likely overall explanation is that integration of viral genomes disrupts the E2 gene, leading to a loss of the E2 transcriptional repressor activity and, consequently, deregulated expression of E6 and E7. This premise is supported by the experimental observation that reexpressing E2 in cervical cancer cell lines by introducing a plasmid that constitutively expresses E2 leads to growth suppression. Since the continued expression of E6 and E7 is necessary to maintain the continued proliferation of cervical carcinoma cells, and since expression of E2 in those cells would repress expression of E6 and E7, we can account for at least one consequence of integration: the disruption of the E2 gene and the resulting enhanced expression of E6 and E7.

The cellular genome too must change before an HPV-induced cervical lesion can develop into an invasive carcinoma. This is evident from the fact that progression of high-risk HPV-positive cervical lesions to malignant carcinomas generally is a slow process, which in fact occurs only rarely. That is, the lesions are *not* initially malignant and only rarely become so, despite the fact that the viral oncogenes are expressed within them. Thus, the most likely reason for the slow progression to malignancy is that malignancy in general requires the accumulation of cellular mutations, while at the same time the rate of spontaneous mutagenesis in human cells is relatively low. However, and importantly, expression of the high-risk HPV E6 and E7 oncoproteins dramatically enhances the occurrence of cellular mutations.

Expression of the viral E6 and E7 proteins increases the frequency of cellular mutagenic events at least in part because they inactivate the p53 and pRb tumor suppressor pathways. When cell proliferation is stimulated by E7, under conditions in which E6 prevents p53 from arresting the replication of damaged DNA, the setting is then ripe for the accumulation of mutations. Moreover, the normal p53-mediated mitotic checkpoints are not operating to prevent tetraploid cells from inappropriately reentering the cell cycle. Thus, additional oncogenic events may eventually occur, such as the emergence of aneuploid cells and cells with other chromosomal abnormalities.

As just noted, the progression of low-grade cervical lesions to malignant carcinomas requires the accumulation of cellular genetic changes. Nevertheless, and interestingly, these cellular mutations are not sufficient to maintain the malignant cell phenotype, since malignancy may also require the continued expression of the viral E6 and E7 proteins. This was implied by experiments in which the viral oncogenes were silenced by introducing a vector that constitutively expresses the BPV E2 protein. Thus, the acquisition of secondary cellular genetic changes and the continued expression of the viral oncogenes are necessary to establish and maintain the malignant state.

E6 expresses yet another effect relevant to neoplasia, which was noted only briefly above: the activation of **telomerase**. (We discussed telomerase in Chapter 14. In short, since DNA synthesis must be primed, there is a need for a special mechanism to generate the 5′ ends of daughter DNA strands; this need is fulfilled by telomerase.) Here is how E6 activation of telomerase plays out in the development of carcinomas. First, primary cultures of mammalian cells generally are capable of only a limited number of cell divisions before entering a state known as **senescence**, a nonproliferative state that is nonresponsive to growth factors. Senescent cells generally do not express telomerase and thus contain shorter telomeres than the original primary cells. Cells may avoid becoming senescent if telomerase remains active and if pRb is inactivated. Cells in which senescence never occurs are said to have become **immortalized**. Since E6 induces telomerase and E7 inactivates pRb, expression of the papillomavirus oncogenes can induce the immortalization of keratinocytes in vivo, a necessary step in the progression of those cells to a malignant state. Indeed, results of genetic studies demonstrate that the activation of telomerase by E6 is essential for keratinocyte immortalization. Moreover, telomerase activity is detected in exfoliated cervical cells from high-risk HPV infections, but *not* from low-risk HPV infections. Not surprisingly then, experimentally induced E2 expression in cervical carcinoma cells, which shuts down E6 and E7, induces cellular senescence.

Since expression of E2 in cervical carcinoma cells impairs expression of both E6 and E7, thereby leading to cellular senescence, it would be interesting to ask what might be the consequences of shutting down expression of either E6 or E7, but not both, in cervical carcinoma cells. This issue is addressed in some detail in Box 16.6. The bottom line is that the *continuous inactivation* of p53 by E6 and that of pRb by E7 are each necessary to maintain the proliferative phenotype of cervical carcinoma cells. In

BOX 16.6

As noted in the text, expression of E2 in cervical carcinoma cells results in a shutdown of E6 and E7 and the occurrence of cellular senescence. What then might be the consequences of shutting down either E6 or E7, but not both, in these cells? This experiment is technically feasible (do you see how?) and indeed has been done in Dan DiMaio's laboratory. If you follow this story you will almost certainly better appreciate the intricacies of the effects of pRB and p53. Here then are the results. Suppressing only E6 (which blocks p53) results in a complex outcome in which about two-thirds of the cells become senescent, while the remaining cells continue to proliferate, with many of these cells then undergoing apoptosis after about a week. The apoptosis-inducing effect of repressing E6 occurs only in cells that continue to express E7. A likely explanation of this experimental finding is that the continuing inactivation of pRb by E7 causes inappropriate cell divisions (i.e., cell divisions in which normal checkpoints in the cell cycle are not passed). Inappropriate cell divisions would normally activate p53-regulated apoptosis, which can occur in this instance in the absence of the p53-repressing effect of E6. Still, it was not clear from these studies why suppressing only E6 induces senescence in most cells. (As noted in the text, E6 induces telomerase, and senescent cells do not express telomerase. But continue reading.)

Suppressing only E7, which then enables pRb to block cell proliferation, leads to cellular senescence rather than to apoptosis. This finding is consistent with earlier reports that a growth-suppressing effect of pRb is sufficient to induce senescence in normal fibroblasts. Moreover, introduction of an exogenous pRb gene can induce senescence in human cancer cells. (These cancer cells are derived from "spontaneous" human cancers, which presumably contain a mutant *pRb* gene.)

Together, these experiments show that apoptosis is induced only when E6 alone is suppressed. They also show that repressing either E6 or E7 can lead to cell senescence. Moreover, it is likely that E6 and E7 act in separate ways to block cell senescence. It is true that inactivating telomerase can induce senescence. However, blocking E6-mediated activation of telomerase may not account for the effect of repressing E6 in the above experiments, as indicated by the fact that similar experimental results were obtained in cervical carcinoma cells engineered to express greatly elevated telomerase activity.

A subsequent study from the DiMaio laboratory demonstrated that p53 activity is necessary for both senescence and apoptosis when E6 alone is suppressed. However, p53 is not necessary for senescence to occur when both E6 and E7 are suppressed. Two conclusions may be drawn from these findings. First, there is a p53-mediated pathway to senescence and apoptosis that is activated by suppressing E6. Second, inactivating E7 can induce senescence via a p53-independent pathway, consistent with the earlier conclusion that E6 and E7 act in separate ways to block cell senescence.

In yet another report from this research group, cell senescence induced by inactivating E7 was shown to require the resultant activation of pRb, which is consistent with other research findings noted above. Thus, there is a pRb-mediated pathway to senescence that is activated when E7 is suppressed. Still, it was noted that E7, like the SV40 LT, is a modular protein that may have other cellular targets in addition to Rb family members. Thus, further studies may reveal other E7 targets, in addition to pRb, that may be involved in these phenomena. Regardless, the continuous inactivation of p53 by E6 and of pRb by E7 is necessary to maintain the neoplastic phenotype of cervical carcinoma cells.

Reference

DeFilippis, R. A., E. C. Goodwin, L. Wu, and D. DiMaio. 2003. Endogenous human papillomavirus E6 and E7 proteins differentially regulate proliferation, senescence, and apoptosis in HeLa cervical carcinoma cells. *J. Virol.* 77:1551–1563.

---

addition, the inactivation of E2 is a crucial step towards the genesis of cervical carcinomas.

Repressing expression of either E6 or E7 can lead to cellular senescence, and repressing only E6 can lead to apoptosis (see Box 16.6 for current understanding of these effects). These experimental findings further underscore the significance of the inactivation of *E2* in the development of cervical carcinomas, since it results in *derepressing both* the *E6* and *E7* oncogenes. Note that E2 expression can lead to apoptosis in cells devoid of any other papillomavirus genes, perhaps even further underscoring the inactivation of E2 as a crucial step in HPV-mediated carcinogenesis as well as the complexity of HPV malignancy, discussed further below.

Experimental evidence noted above implied that E6 and E7 may promote carcinogenesis in ways other than merely inactivating the cellular pRb and p53 tumor-suppressing proteins. This notion is supported further by the fact that there are *E6* mutants unable to inactivate p53 but which nevertheless retain the ability to transform cells. Moreover, there are *E6* mutants that are able to target p53 for degradation but are yet unable to induce transformation. Thus, E6 apparently must interact with other cellular targets, in addition to p53, to affect neoplasia. Moreover, E6 proteins that are not able to inactivate p53 may yet be able to affect cell transformation via their interaction with other cellular targets. (Might these experimental findings be relevant to the question of how E6 from low-risk HPV types induces release of cells from G_0 growth arrest?)

Earlier, we briefly noted that a group of cellular proteins that contain a PDZ domain may be important targets for the E6 proteins of high-risk HPVs. Evidence supporting that premise includes the following. First, the carboxy termini of E6 proteins of high-risk HPV strains, but not of low-risk strains, contain a binding motif for the PDZ domain of several PDZ domain-containing cellular proteins. Second, the PDZ-binding motif of E6 proteins of high-risk HPV genotypes indeed is essential for transformation of rodent cells in vitro and for inducing tumors in the skin of transgenic mice. (Interestingly, a mutant E6 protein that does not contain the PDZ-binding motif but retains the ability to inactivate p53 is not able to induce tumors in these transgenic mice. This mutant continues to synthesize DNA after exposure to DNA-damaging ionizing radiation. Thus, the E6 PDZ-binding motif is not required to block the p53-mediated arrest of cellular DNA synthesis induced in response to DNA damage, although it is required to effect neoplasia.) Third, Dlg and Scribble, which have been identified as tumor suppressor proteins in *Drosophila*, are two of the PDZ domain-containing proteins that are bound by E6 of high-risk HPVs.

More-recent experiments that also made use of the transgenic mouse model demonstrated that the interaction of E6 with PDZ domain proteins correlates with a step involved in the formation of benign tumors, but not with the progression of those tumors to malignant cancers. Thus, some property of E6, other than its interaction with PDZ domain proteins, accounts for its role in the final progression to malignancy. This other interaction well may be the E6-mediated inactivation of p53, since mice deficient for p53 show enhanced rates and frequency of progression to malignancy, but no increase in the initiation of tumor genesis (Simonson et al., 2005).

Stages of Cancer Development

Having considered the molecular events underlying cervical carcinoma, we next look into the cytopathological stages in the development of these cancers. First, note that most cervical carcinomas develop from precursor non-cancerous (i.e., noninvasive) lesions that are referred to as CINs (also known as SILs, as noted above). Pathologists grade or classify these lesions based on how atypical their constituent cells are and also on how far they extend from the parabasal layer of the epithelium up through the entire thickness of the epithelium (Fig. 16.7). Many of the CIN 1 and CIN 2 lesions, which are only mildly **dysplastic** (i.e., unusual in their growth or development), are *not* thought to be precursors of cervical cancers. Rather, the viral expression patterns within those lesions suggest instead that they are sites of active viral replication. That being the case, the cells within those lesions already have differentiated and have lost the ability to divide, and moreover, they are destined to be shed. Consequently, those low-grade lesions are not likely to progress further towards becoming carcinomas. Consistent with that premise, similar low-grade lesions appear during infection with low-risk HPV genotypes, which pose a negligible risk of causing carcinoma. Thus, the histology of these low-grade lesions is believed to reflect the cytopathological effect of productive viral infection, rather than a premalignant stage that usually progresses to cancer.

What then is the source of the carcinomas? The apparent answer is that within the proliferating basal and parabasal cells of only some CIN 2 and CIN 3 lesions, there is a dramatic increase in the expression of the papillomavirus E6 and E7 genes, which results in the proliferation of dysplastic cells (Fig. 16.7). Unfortunately, the mechanism underlying this upregulation of E6 and E7 expression in the basal and parabasal cells is not yet known. Perhaps the rare integration of the viral genome into that of the host cell explains this upregulation of E6 and E7, at least in part (see above). The disruption of E2, concomitant with integration, also would prevent productive infection in those cells (Box 16.7).

Host immune responses against the low-grade CIN lesions appear to be one reason why those lesions only rarely persist and progress to malignant carcinomas. This is implied by the greater incidence of lesion persistence and progression to malignancy in immunocompromised women. In addition, both humoral and cellular immune responses against HPV antigens have been detected in individuals undergoing lesion regression. Individual differences in human leukocyte antigen (HLA) haplotype (Chapter 4) may be a factor that determines which low-grade lesions escape from immune surveillance, since certain HLA alleles correlate with a decreased risk of cervical carcinoma.

Tissue Microenvironment

This section ends with a brief consideration of the influence of the **tissue microenvironment** on HPV infection and carcinogenesis in vivo, a topic that likely will receive

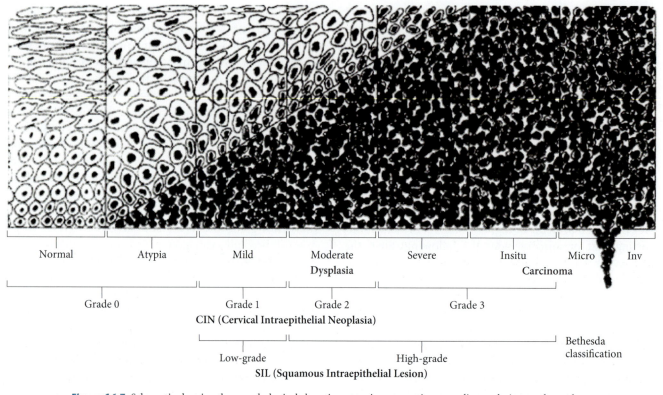

Figure 16.7 Schematic showing the morphological alterations seen in consecutive premalignant lesions and correlation of the CIN classification with the SIL classification. Many of the grade 1 and grade 2 lesions, which are only mildly dysplastic, are not thought to be precursors of cervical cancers. Rather, the viral expression patterns within those lesions suggest instead that they are sites of active viral replication. The apparent source of the carcinomas are some grade 2 and grade 3 lesions, in which there is a dramatic increase in the expression of the papillomavirus E6 and E7 genes within the proliferating basal and parabasal cells. From P. J. F. Snijders et al., *J. Pathol.* **208**:152–164, 2006, with permission.

increasing attention in the future. Most (87%) CINs and cervical carcinomas develop within a specific microenvironment of the cervix, the so-called **transformation zone**, where the single-layered epithelium of the glandular endocervix progressively changes into the mature multilayered squamous epithelium of the exocervix. Therefore, factors specific to the milieu of the transformation zone appear to favor the development of CINs and carcinomas. Those factors are not known with certainty, but several possibilities are noted in the following paragraphs (Box 16.8).

BOX 16.7

CIN 3 lesions progress to both **severe dysplasia** and so-called **carcinoma-in situ**, terms that reflect the increasing atypia and spread of the lesion into the upper layers of the epithelium. Importantly, cervical carcinoma-in situ is treatable. But if untreated, these lesions can progress to frank malignancy, at which point there is only limited treatment success. Pap smear screening programs play an important role in identifying women with early-stage precancerous lesions, which still are treatable.

BOX 16.8

Men too are commonly infected with high-risk HPVs. After all, that is how women get infected. But then, why is the incidence of HPV-associated penile cancer relatively rare compared to the incidence of cervical carcinoma? Perhaps it is because there is no penile equivalent of the cervical transformation zone. If that is the reason, it further underscores the importance of the tissue microenvironment in HPV-induced cervical carcinoma.

A usual explanation for the increased susceptibility of the transformation zone to HPV infection and associated neoplasia is that the virus can more easily access the basal epithelial cells of that thinner epithelium. Yet note that HPV infections are equally common in the cervix and the vagina, but while cervical cancer is the second most common cancer in women worldwide, vaginal cancer is remarkably rare. Thus, it is likely that other characteristics of the transformation zone also make it conducive to CIN development and cervical cancer. For example, some research groups report that the density of Langerhans cells (a specialized type of dendritic cell found in epithelia [Chapter 4]) may be reduced in the transformation zone, relative to the exocervix. Since Langerhans cells are important for immunosurveillance in the squamous epithelium, their reduced density in the transformation zone might result in reduced stimulation of T helper (Th) cells.

One recent study reported that the density of Langerhans cells actually does undergo an increase in squamous intraepithelial lesions. However, the cytokines that are then produced within the microenvironment of the transition zone are more typical of a Th2 response than a Th1 response, despite the fact that the latter is more appropriate for tumor immunity. (In Chapter 4 we noted that when Th cells are activated by dendritic cells, they orchestrate the adaptive immune response by secreting particular cytokines. Th1 cytokines, such as interleukin-2, favor a cytotoxic T-cell response, which is appropriate for tumor immunity. Th2 cytokines, such as interleukin-10, favor the activation of antibody-producing B cells, which is an inappropriate response against tumors.)

Another possible factor predisposing the transformation zone to neoplasia is that it is particularly sensitive to the carcinogenic effect of estrogen. Furthermore, the transformation zone also was reported to lack factors that downregulate expression of E6 and E7. Moreover, the neoplasia that occurs within the transformation zone often is associated with inflammation and production of proinflammatory cytokines like tumor necrosis factor alpha (TNF-α). Although this cytokine favors a cytotoxic T-cell response, it also can be a risk factor for the generation of cancer. For example, TNF-α (as well as other cytokines generated from the tumor) may induce tumor-associated macrophages to produce a metalloprotease that enhances invasiveness of malignant cells.

TREATMENT AND PREVENTION

This section begins with a consideration of the prevention and treatment of benign warts. It then continues with the recent development of a vaccine that indeed can prevent papovavirus-induced cervical carcinomas.

Since warts usually regress spontaneously, why might it be important to have effective therapies? The reason is that regression may occur only after an interval of months to years. Moreover, in the interim, these warts may be painful or cosmetically unsightly. It also is desirable to remove them to prevent their spread. Fortunately, removal may be accomplished by cryotherapy, by electrocautery, or chemically.

Currently, the best way to avoid transmission of HPVs is to avoid contact with infected tissue or surfaces. The use of condoms may help to prevent sexual transmission of HPV. However, protection by condoms *may be limited* because these viruses are highly prevalent in the sexually active population, and virus-infected cells are present at external genital sites, as well as on and in the genitalia.

HPV Vaccine To Prevent Cervical Cancer

Why might it have been thought that a vaccine would protect against cervical carcinoma? The intent of this question may not be immediately apparent, since cervical carcinoma is most commonly caused by a high-risk HPV. Yet vaccines generally neither prevent nor cure virus infections. Rather, they enable the host to bring the infection under control before symptoms might arise (see Chapter 4). But cervical carcinoma may develop after years of virus persistence, despite a continual host immune response against the infection during that entire period. Nevertheless, immune surveillance likely plays a key role in protecting the host against the malignant progression of HPV-induced lesions. This assertion is strongly supported by the fact that the human immune system naturally clears many HPV infections over time and also by the increased incidence of HPV-associated neoplasms in immunosuppressed patients (see above). For that reason, and also because cervical carcinoma is generally associated with HPV infection, work has been under way to develop a vaccine that might yet protect against infection, or, less ambitiously but still crucially important, protect against progression of infection to malignancy. This work led to a prophylactic vaccine, Gardasil (marketed by Merck & Co. Inc.), which was approved by the U.S. Food and Drug Administration in June 2006 for administration to girls and women 9 to 26 years of age.

In clinical studies, Gardasil was seen to effectively prevent persistent infection by the two HPV strains that cause 70% of cervical cancers (i.e., HPV-16 and HPV-18) and by two other strains that cause 90% of genital warts. Despite the fact that the vaccine provides substantial protection against genital warts, it is being promoted as a protection against cervical cancer, rather than as a vaccine against sexually transmitted HPV disease in general.

Gardasil consists of VLPs that are comprised of the L1 protein of the constituent HPV genotypes (see above).

Clinical trials indicate that 99% of vaccinated individuals mount an antibody response in which the anti-HPV antibody titer is >50-fold greater than that resulting from natural infection! However, these antibodies are specific for the genotypes in the vaccine, which currently contains HPV-16 and HPV-18 but no other known high-risk types.

Considering the prevalence of cervical carcinoma and the efficacy of the HPV vaccine, vaccine usage could in principle diminish the incidence of cancer in women worldwide by as much as 15%. Still, the vaccine will not provide any protection to those girls or young women who have been exposed to the vaccine strains before receiving the vaccine. Since more than one-half of sexually active adults will be infected by HPV at some point in their lives, and since the high-risk HPVs are fairly common, many health experts want girls to be vaccinated before becoming sexually active (Box 16.9). But since the vaccine cannot fight an existing HPV infection, some experts recommend cancer screening as a better preventive measure than vaccination for women who have begun sexual activity. Also, since there are many more HPV strains than the four in the vaccine, all women are advised to have regular Pap smears.

Since Gardasil will not protect women who already are infected, attempts are under way to develop a therapeutic vaccine. Unlike Gardasil, which raises a humoral immune response against major HPV capsid proteins, the prospective therapeutic vaccine targets the viral oncoproteins E6 and E7, which are expressed in cancer cells. It is hoped that the therapeutic vaccine will induce a cell-mediated immune response (Chapter 4) that might destroy cancer cells in cervical carcinoma patients. Such vaccines are effective at promoting the destruction of transplanted E7-expressing tumor cells in mice, but for reasons that are not yet clear they have not shown promise in human trials. Perhaps human patients are tolerant of the tumor antigens, or perhaps they do not present those antigens via the major histocompatibility complex proteins (Chapter 4).

OROPHARYNGEAL CANCER

Ever since Harald zur Hausen first reported isolating HPV DNA in cervical cancer lesions in the early 1980s, evidence has been mounting that HPV-16 and, to a lesser extent, HPV-18 are involved in oropharyngeal cancers, in addition to their involvement in cervical carcinomas. Results of more recent studies support the assertion that HPVs are an etiologic agent of oropharyngeal cancers. For example, a 2008, PCR-based epidemiological analysis found that up to 60% of oropharyngeal cancer cases in the United States from 1998 to 2003 were associated with HPV infection.

Note that at earlier times, oropharyngeal cancers were generally seen in individuals in their 50s or 60s, and in most cases these cancers were associated with smoking or heavy alcohol consumption. Consequently, the more recent correlation between oropharyngeal cancers and HPV infection may help to account for the facts that while the total number of oropharyngeal cancers in the United States is in decline, the rate among Americans in the 20-to-49 age group more than doubled between 1975 and 2005. Moreover, results of patient surveys show that while the risk factors for HPV-negative tumors included several decades of heavy smoking or drinking, the risk of developing HPV-positive oral cancer was instead significantly correlated with having multiple oral sex partners. There was no correlation between HPV-negative cancers and multiple oral sex partners.

BOX 16.9

Gardasil is being targeted at girls and young women because it is most effective when given to females before they engage in sexual activity. Since the median age at which girls first have sex is 15 years, and since the vaccine has no effect against vaccine-sensitive strains if administered after exposure, some public health experts have argued for making the vaccine mandatory for school attendance.

Not unexpectedly, socially conservative groups in the United States have rallied against adopting this policy, arguing that it would increase premarital sex and that a child's parents or guardian should make the decision regarding the vaccine. A spokesperson for Focus on the Family, a Christian advocacy organization based in Colorado Springs, CO, stated the following: "We can prevent it (papillomavirus infection) by the best public health method, and that's not having sex before marriage." Presenting a different view, a cervical cancer survivor and mother, whose therapy included surgical removal of her uterus, stated, "This vaccine should not lead to an argument about when girls have sex. It's about saving the lives of women in their child-bearing years, letting them have children or take care of the children they already have."

Others have objected to Gardasil because it protects against a sexually transmitted disease, whereas other vaccines mandated for school attendance usually are meant to prevent diseases easily spread through casual contact, such as measles and mumps. Is that a relevant basis for not mandating use of the vaccine? How do you feel about this issue? Also, see Chapter 21.

Indirect evidence that the HPV E6 and E7 proteins might be involved in the HPV-associated orophoryngeal cancers was provided by the experimental finding that the tumor suppressor proteins of HPV-negative oral cancers often show incapacitating mutations, perhaps caused by long-term exposure to carcinogens in tobacco and alcohol, while such mutations are rare in HPV-associated oral cancers. Moreover, other recent studies show that suppressing E6 and E7 expression in HPV-positive oral cancer cells reactivated the cells' tumor suppressor genes, thus providing more direct evidence for the involvement of E6 and E7.

As a consequence of the above findings, public health experts are considering whether to recommend HPV vaccines for use in men. A recently completed study, sponsored by Merck, the developer of Gardasil, found that their vaccine was able to prevent HPV infection of male genitals, and a National Institutes of Health clinical study is under way to determine whether the vaccine can prevent oral HPV infection. Meanwhile, Merck is seeking to get Gardasil approved for use in men.

Suggested Readings

Kamper, N., P. M. Day, T. Nowak, H. C. Selinka, J. Bolscher, L. Hilbig, J. T. Schiller, and M. Sapp. 2006. A membrane-destabilizing peptide in capsid protein L2 is required for egress of papillomavirus genomes from endosomes. *J. Virol.* **80:**759–768.

Simonson, S. J., M. J. Difilippantonio, and P. F. Lambert. 2005. Two distinct activities contribute to human papillomavirus 16 E6's oncogenic potential. *Cancer Res.* **65:**8266–8273.

Zheng, Z.-M., and C. C. Baker. 2006. Papillomavirus genome structure, expression, and post-transcriptional regulation. *Front. Biosci.* **11:**226–302.

Adenoviruses

INTRODUCTION

The *Adenoviridae* are a family of viruses containing double-stranded DNA genomes that range from 20×10^6 to 25×10^6 daltons (Da) in molecular mass, about four to six times larger than the double-stranded DNA genomes of the polyomaviruses and papillomaviruses. Moreover, adenovirus genomes are linear, in contrast to the circular genomes of the polyomaviruses and papillomaviruses. Furthermore, each 5′ end of linear adenovirus genomes is covalently bound to a so-called terminal protein. Adenovirus particles are nonenveloped icosahedrons that are between 70 and 100 nm in diameter.

Adenovirus serotypes are classified according to their resistance to neutralization by antisera against other known adenovirus serotypes. (The major targets for neutralizing antibodies on adenovirus particles are epitopes found on the viral hexon protein and on the terminal knob of the fiber protein [see below].) On this basis, about 100 adenovirus serotypes are known, at least 47 of which infect humans. The first human adenoviruses discovered (see below) were serotypes 1 through 7, which also are the most common.

The endemic (i.e., common) adenoviruses, which include serotypes Ad1, Ad2, Ad3, Ad5, Ad6, and several others, together infect more than 80% of the human population early in life. In contrast to these endemic serotypes, which infect mainly children, the remaining adenovirus serotypes seem to have an epidemic epidemiology and can infect anybody who has not been infected with them previously.

While most adenovirus infections are asymptomatic (a feature that actually enhances the spread of these viruses), they are associated with several distinct clinical syndromes, including feverish colds and also a pertussis-like illness (i.e., like whooping cough) in children and adults. Adenoviruses are also the most common cause of viral infections of the conjunctiva, giving rise to conjunctivitis (also known as "pinkeye"). In addition, adenoviruses cause pharyngitis (inflammation of the pharynx), which often is accompanied by conjunctivitis; this syndrome is known as pharyngoconjunctival fever. Adenoviruses also cause a set of symptoms known as **acute respiratory tract disease syndrome** (ARDS). This syndrome is seen primarily among military recruits and is characterized by fever, cough, pharyngitis, and cervical adenitis (enlarged, inflamed, and tender lymph nodes in the neck). Moreover, adenoviruses are the major cause of acute gastric enteritis; 15% of all such hospitalized enteritis cases have an adenovirus etiology. The serotypes responsible for this gastrointestinal condition rarely cause respiratory tract infections (Box 17.1).

It is commonly believed that adenoviruses, like polyomaviruses and papillomaviruses, persist in infected individuals and that reactivation of latent adenovirus infections in immunocompromised children and adults can lead to serious disease. The discovery of the adenoviruses did in fact result from their apparent tendency to persist in infected individuals. It happened as follows. In 1953, Robert J. Huebner, then chief of the Laboratory of Infectious Diseases at the National Institute of Allergy and Infectious Diseases, and coworker Wallace P. Rowe were exploring means for isolating viruses responsible for common colds. Since chick embryos, which had been useful for propagating influenza virus, did not appear to support replication of cold viruses, Huebner and Rowe wondered whether adenoids might provide a suitable culture medium for the isolation of cold viruses. This thought was prompted by the fact that all viruses which inhabit the breathing passages have access to these lymphoid tissues that reside in the back of the throat.

Huebner and Rowe then proceeded to collect adenoids that had been removed from children in nearby Washington hospitals and placed them in culture medium. For the first week or so the adenoids did well in culture, putting out new sheets of cells along the walls of the culture vessels. But in subsequent weeks the majority of cultures began to show signs of cytopathology, followed by their complete destruction.

A hypothesis put forward to explain these findings was that the adenoids, prior to their removal, had been naturally infected in vivo with a virus that was able to establish a latent infection. Then, upon removing the adenoids and culturing them in the absence of host immunity, reactivation of the *latent* infections occurred, resulting in unrestrained virus replication and concomitant cell death.

To test the premise that the cell degeneration seen in the adenoid cultures was due to a virus, other cell cultures were exposed to the media in which the adenoids had been cultured. In almost all instances, the recipient cultures underwent a similar destruction. Moreover, Huebner and Rowe were able to passage the cytopathic agent through 17 subcultures over a 69-day period. In addition, they were able to passage their **adenoid degeneration agent** through a wide variety of human and animal tissue cultures. Inoculating rabbits with the agent resulted in the generation of antiserum that could then protect adenoid cell cultures (Box 17.2).

Together, the above studies appeared to demonstrate that a potentially cytopathic infectious agent was latent in the healthy adenoids of 9 of 10 healthy children. But bear in mind the proviso in Box 17.2.

At this point in our story, Huebner and Rowe had a virus with no associated disease. This changed when they isolated the virus from the postnasal drippings of a child ill with a fever and runny nose. Subsequently, Huebner and a technician working with the cultures developed conjunctivitis. The same virus then was isolated from the infected eyes of Huebner and the technician. The virus then was isolated from hospital staff members and their patients, all of whom had a fever, inflamed pharynx, and an eye infection. Huebner and Rowe also found the virus in several groups of individuals suffering from a

---

BOX 17.1

For reasons that are not clear, military recruits are more susceptible to ARDS than the adult population in general. A frequently offered explanation suggests that it is because of the close living conditions in military barracks. However, crowding may not be the reason, because college students of the same age, living in similarly crowded dormitories, are not similarly susceptible to adenovirus infections. Virologists at the Centers for Disease Control and Prevention suggest that a reason might be that the virus lives for prolonged periods of time on dry surfaces and the military insists on reusing pillows in their basic training facilities.

BOX 17.2

As noted in the text, in early studies adenoviruses were isolated from the adenoids and tonsils of 90% of all children. More-recent studies found adenovirus DNA in 75 to 80% of tonsil samples. Importantly, very few of those samples contained replicating virus, and the others did not yield replicating virus even after extensive culture in vitro. These findings imply that it may be incorrect to refer to adenovirus infections as "latent." Indeed, no one has yet demonstrated reactivation of silent virus from human tissues. It was suggested that laboratory contamination may have been the reason that adenoviruses were found in so many samples in those early studies.

kind of feverish cold, in one instance traceable to a swimming pool.

One month after Huebner and coworkers reported their findings, Maurice Hilleman, who was then at the Walter Reed Army Institute of Research, reported the isolation of a new virus from Army recruits suffering from feverish colds at a military camp. (Recall that Hilleman isolated simian virus 40 [SV40] in 1960 [Chapter 16].) Hilleman grew his isolates in cultures established from human placentas. Hilleman then swapped his specimens for those isolated in Huebner's laboratory. Serological analyses then established that the two research groups had independently isolated the same virus. It took several years and the intervention of a committee chaired by John Enders before a name for the new virus was agreed upon. "Adenovirus" was chosen, since that was the tissue from which the first isolates were obtained.

Despite the early finding that adenoviruses could be isolated from the adenoids and tonsils of 90% of all children, it soon became clear that these viruses are not responsible for the common cold. Indeed, adenoviruses are responsible for only a small minority of all common respiratory illnesses. Yet adenoviruses are responsible for a much higher percentage of serious respiratory illness in children under 6 years of age and in military recruits.

Huebner, Rowe, their collaborators, and also Hilleman went on to develop successful adenovirus vaccines that were especially useful when administered to military recruits. Indeed, from the military's point of view, the results of vaccination were sensational. Only a decade earlier, during World War II, more than 4 million U.S. recruits were hospitalized during training with adenovirus respiratory illnesses. One-half of these individuals required hospitalization for an average of 10 days. The Hilleman vaccine resulted in a 98% reduction in incidence of respiratory infection (Box 17.3).

In 1962, J. J. Trentin and colleagues made the seminal observation that human adenoviruses *induce malignant cancers* when inoculated into newborn hamsters. Theirs was the first demonstration that a human virus might lead to neoplasia. In addition, human adenoviruses induce neoplastic transformation of hamster, rat, and rabbit cells in culture. Yet adenoviruses have *not* been associated with *human cancer*, despite extensive searches for such a linkage. Regardless, that initial finding led to adenoviruses serving as an important model for probing the underlying mechanisms of carcinogenesis. As in the case of the polyomaviruses (Chapter 15), initial interest in the adenoviruses as a model system for the study of oncogenesis led to major fundamental contributions in the areas of eukaryotic cellular and molecular biology, particularly regarding gene expression, DNA replication, cell cycle control, and

cell growth regulation. The splicing of eukaryotic premessenger RNAs (pre-mRNAs) was observed for the first time in an adenovirus system, which was a contribution of particular importance to eukaryotic molecular biology.

STRUCTURE

Adenovirus particles are nonenveloped icosahedrons that are between 70 and 100 nm in diameter, not including the fibers that vary in length from 12 to 32 nm, depending on the serotype. The outer shell is composed of 252 multiprotein subunits, of which 240 are arranged as hexons and 12 are arranged as pentons (Fig. 17.1). (Icosahedral adenovirus capsids are more conventional than polyomavirus

---

BOX 17.3

Wyeth Laboratories produced an oral adenovirus vaccine against Ad4 and Ad7 that was licensed for use only by the military. Production began in 1980 and ended in 1995, and the last of that vaccine was used in 1999. Wyeth stopped production because its facilities needed renovation, which the Department of Defense (DoD) declined to underwrite. In 2004, a spokesman for the DoD stated publicly that the vaccine had been dropped by the military due to a military study which erroneously found that the vaccine was no longer needed. Regardless, the problem of adenovirus infection in basic training resumed with the termination of the vaccine program.

In April 2000, an expert committee of the Institute of Medicine of the National Academies convened to advise the Army on the management of natural infectious disease threats to its personnel. Among the findings in the Committee's report, issued in January 2002, were the following: (i) "that the adenovirus vaccine is urgently needed to control the epidemic respiratory disease that has caused much morbidity among recruits in the past, and now once again threatens the health and even the lives of military trainees; since *acute pulmonary infection due to adenovirus is a nearly unique occupational risk of the military trainee*, it is imperative that DoD take rapid and effective action to once more eliminate this preventable disease"; and (ii) "that the DoD not only evaluate the cause(s) underlying this serious procurement system failure, but also make a clear commitment to the changes necessary to prevent similar breakdowns in the future."

The Pentagon now states that the development of a new vaccine is a top priority.

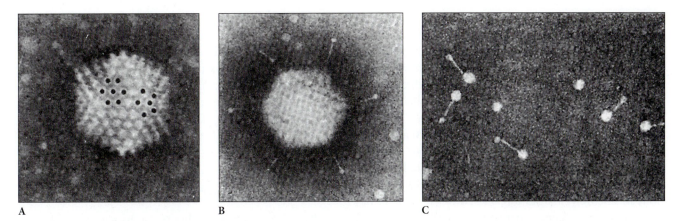

Figure 17.1 (A) Adenovirus particles are nonenveloped icosahedrons that are between 70 and 100 nm in diameter. The outer shell is composed of 252 subunits, of which 240 are arranged as hexons and 12 are arranged as pentons. Pentons are located at each of the 12 vertices of the icosahedron, and each is surrounded by five hexons. Each of the hexons is a center of threefold rotational symmetry. One of the 12 pentons that is surrounded by 5 hexons is marked by dots, as is one of the hexons that is surrounded by 6 hexons. (B) Six of the 12 fibers that are present on each virus particle are shown projecting from penton capsomeres containing penton base and fiber. (C) Free penton capsomers, containing penton base and fiber (×285,000). From R. G. Valentine and H. G. Pereira, *J. Mol. Biol.* **13:**13–20, 1965, with permission.

and papillomavirus capsids, which are built entirely from pentamers [Chapters 15 and 16]. Nevertheless, as will be seen shortly, adenovirus icosahedrons are not entirely conventional, since multiprotein subunits *arranged* as hexons are not actually hexons per se.) As you might expect, pentons are located at each of the 12 vertices of the icosahedron (that is, at the positions of fivefold rotational symmetry), and each penton is then surrounded by five hexons. Each of the hexons is a center of threefold rotational symmetry. Hexons that surround pentons necessarily occupy a bonding environment different from that of hexons that are surrounded entirely by other hexons. Hence, formation of adenovirus capsids depends on nonequivalent contacts between subunits.

The fibers that project from each of the 12 pentons are a striking feature of adenovirus particles. Each of the fibers terminates in a distal knob. As you might have supposed, the knobs bind to cell surface receptors. As noted above, neutralization by type-specific antibodies results from antibody binding to epitopes on the hexon protein and on the terminal knob of the fibers.

The three major adenovirus structural proteins are polypeptides II, III, and IV, which comprise the hexon, the penton, and the fiber, respectively. Looking more carefully at the arrangement of these three major structural proteins, each of the hexons is actually a *trimer* of viral polypeptide II, the most abundant protein in the outer shell. That is, each hexon is comprised of only three molecules of viral polypeptide II (see Fig. 17.5). How then does a hexon manifest its sixfold positions? The answer is as

follows. Each molecule of polypeptide II contains two β-barrel domains. (Recall that the capsid proteins of the picornaviruses and the VP1 protein of the polyomaviruses likewise are built on the β-barrel plan [Chapters 6 and 15].) The two β barrels, which are located at the base of polypeptide II, are similar in structure. Thus, the individual "hexons," each comprised of a trimer of polypeptide II, exhibit pseudohexagonal symmetry. Importantly, the shape of the β-barrel domain facilitates close packing within and between hexons. (See Chapter 6 for an explanation of how the β-barrel domain facilitates close packing between subunits. This advantage might account for the fact that the β-barrel motif is highly conserved among the capsid proteins of a variety of virus families, despite the fact that there is little, if any, amino acid sequence identity within this motif from one virus family to another. Do these facts imply that these various virus families that make use of the β-barrel motif are descended from a common ancestor, or might they suggest evolutionary convergence?)

The penton protein is polypeptide III. A pentamer comprised of five molecules of polypeptide III is referred to as a "penton base." Thus, a total of 780 proteins make up the framework of the icosahedral shell (12 × 5 in pentons plus 3 × 240 in hexons). The fiber protein is itself a trimer of polypeptide IV. The amino-terminal 40 residues of each polypeptide IV are rooted in the penton base, and each of the carboxy-terminal 180 residues contributes to the formation of the terminal bulb.

In addition to the three major capsid proteins (the hexon, penton base, and fiber proteins), the capsid also contains four additional minor proteins: proteins IIIa, VI, VIII, and IX. The structure and locations within the virus particle of the four minor capsid proteins, in particular proteins IIIa and VI, are less well characterized than those of the major capsid proteins. This is regrettable because proteins IIIa and VI play important roles in the adenovirus life cycle. Protein IIIa participates in virus assembly and maturation, while protein VI may participate in endosomal disruption after cell entry. Thus, not knowing their precise location in the particle has limited our understanding of how they carry out their respective functions in assembly, maturation, and entry. (Protein VI has additional important functions, including regulating hexon import into the nucleus during particle assembly; see below.)

Cryo-electron microscopy and single-particle image reconstruction were recently used to determine the structure of adenovirus particles to a resolution of 6 angstroms (Å) (Saban et al., 2006). Based on this analysis, minor capsid protein IIIa appears to be located below the penton base, on the inner capsid surface, where it may associate with both hexon and penton proteins in the vertex area, thereby perhaps enabling it to play roles in both assembly and disassembly of particles. Protein VI and protein VIII also appear to be located on the inner capsid surface. Protein IX, which is known to stabilize the hexon lattice, appears to be located on the surface, at the edges of the icosahedral facets.

The core of the virus particle contains five additional proteins (proteins V, VII, and μ; the so-called terminal protein; and about 10 copies of the adenovirus cysteine protease p23). The core also contains the linear double-stranded DNA viral genome, which is condensed by its noncovalent association with proteins V, VII, and μ. Protein VII is the major of the three DNA-binding proteins, with about 835 copies per virus particle. All are arginine rich and basic. Polypeptide V additionally may form a layer around the core, connecting it to the inner surface of the capsid via interactions with protein VI and either core protein VII or the DNA, or both.

One copy of the terminal protein is covalently linked to the 5' end of each DNA strand via a phosphodiester bond between a serine residue on the protein and the 5' hydroxyl of the terminal deoxycytosine on the DNA. As will be seen, a precursor of the terminal protein serves as a primer for DNA replication. The genome also associates with about 10 copies of the cysteine protease p23. (Notice that there is no polypeptide I. The "protein" originally designated polypeptide I turned out to be an aggregate of smaller molecules.)

The molecular masses of adenovirus genomes range from 20×10^6 to 25×10^6 Da. As noted above, the association of the genomes with the core proteins appears to facilitate a condensed chromatin-like structure that can be folded into the capsid interior. Ad2 was the first adenovirus serotype for which a complete genome sequence was determined (a total of 35,937 base pairs [bp]), confirming that adenovirus genomes contain inverted terminal repeats ranging in size from 100 to 140 bp, depending on the serotype. The repeats likely play a role in DNA replication (see below).

ENTRY AND UNCOATING

The adenoviruses, like other respiratory viruses, enter the respiratory tract in the form of aerosolized droplets generated by a cough or sneeze, or via saliva. Also, because the virus is stable in the environment, infection can be transmitted by touching a contaminated object (fomite) and then touching the face. The advantages provided by the respiratory tract as a site of viral entry were discussed in Chapter 3.

Many of the adenoviruses, as well as group B coxsackieviruses (Chapter 6), share a common cell surface receptor, termed "CAR," an acronym for "coxsackievirus and adenovirus receptor" (Fig. 6.7). CAR is a type-1 transmembrane protein which, in humans, contains 346 amino acids. Based on its amino acid sequence, CAR is classified as a member of the immunoglobulin superfamily of proteins. Adenovirus type 2 also can bind to major histocompatibility complex class I (MHC-I) molecules, and Ad5 can bind to the $\alpha_M\beta_2$ integrin.

Binding to CAR, or MHC-I, or $\alpha_M\beta_2$ is not sufficient to promote adenovirus entry. Instead, most adenoviruses also need to bind to a coreceptor, identified as the vitronectin-binding α_v integrins $\alpha_v\beta_3$ and $\alpha_v\beta_5$ (Fig. 6.7). Indeed, the interaction of the adenoviruses with their receptor and coreceptor provides a classic example of this variety of virus-cell surface interaction, in which a virus must use multiple cell surface components to facilitate its entry. It occurs as follows.

First, the knob domain on the protruding fiber binds to the membrane-distal, amino-terminal immunoglobulin domain on CAR (Fig. 6.7). Second, the penton base interacts with the coreceptor, α_v integrin. The interaction of adenovirus particles with the α_v integrins is mediated by a conserved arginine-glycine-aspartic acid (RGD) sequence on the penton base. RGD is a signature recognition sequence for the interaction of natural ligands with their integrin receptors. Thus, the molecular basis for the interaction of the adenovirus penton base with the α_v integrins mimics that of the natural ligand, vitronectin, which likewise binds via its RGD sequence. Earlier (Chapter 6), we noted that the foot-and-mouth disease virus (a picornavirus) likewise binds to an integrin via an RGD sequence on that virus's VP1 protein.

Bearing in mind that the interaction of the fiber protein with CAR (or MHC-I or $\alpha_M\beta_2$ integrin) is not sufficient to facilitate cellular entry, how then might the interaction of the penton base with the coreceptor, α_v integrin, promote entry? The answer is that when the CAR-docked particles bind to the coreceptor, a variety of cell responses, including endocytosis, are triggered. Thus, while CAR, MHC-I, or $\alpha_M\beta_2$ integrin serves to capture the virus, the coreceptor, α_v integrin, serves to trigger endocytosis, at least in part via the signals it transmits. In this regard, recall that integrins are cell surface α/β heterodimeric glycoproteins that have a variety of functions, including cell-to-cell and cell-to-matrix adhesion and the triggering of signal transduction pathways.

Now, consider that in polarized respiratory epithelial cells, which are the target cells for the respiratory adenoviruses, CAR is localized in tight junctions between the cells and along the basolateral membranes. Also, the integrins, which are involved in cell-to-cell adhesions and in cell-to-matrix adhesions, likewise are found between the cells and along the basolateral membrane. How then are adenoviruses (and coxsackieviruses as well) able to gain access to their receptors and coreceptors? One possibility is that incoming adenovirus particles might bind to specialized nonpolarized epithelial cells that express CAR on the apical membrane, and which then might transfer the virus to the basal surface, perhaps by transcytosis. (Please forgive a speculation on top of a speculation, but this possibility is consistent with the experimental finding that engineering CAR to be expressed on the apical surface of *polarized* human airway epithelial cells is sufficient for adenovirus infection in vitro [Box 17.4].) Alternatively, lesions in the epithelium may temporarily expose CAR to allow viral docking. The latter alternative might account for how the virus also accesses the basolateral integrins. Still, with no experimental data to provide a definite answer, the question remains an open one.

Following the interactions of the adenoviral fiber proteins and penton bases with CAR and the α_v integrins, respectively, the virus is internalized via receptor-mediated endocytosis (Fig. 17.2). This endocytosis is clathrin mediated and requires the dynamin guanosine triphosphatase (GTPase) (which is involved in membrane-severing events as well as in vesicular trafficking) and the clathrin adaptor protein AP2 complex (which binds to defined endocytic motifs in cargo proteins). Most likely, the adenovirus particles are endocytosed while in association with their integrin coreceptors.

As we have noted before, the entry pathway of nonenveloped viruses is particularly enigmatic due to the fact that in every instance the virus particle or its genome must cross at least one cellular membrane, and membrane fusion (as occurs in the case of enveloped viruses) is not an option. Thus, whereas virus escape across a cellular

---

BOX 17.4

Epithelial cells are organized into sheets, in which the cells are bound together by **tight junctions**, which seal the space between the cells and form a barrier between the exterior or apical surface of the epithelium and its basolateral surface, which faces the basal lamina and the underlying tissue. The tight junctions also prevent the diffusion of membrane proteins between the apical and basal domains, thereby enabling the two domains to differ in their protein content. Under appropriate cell culture conditions, epithelial cells form polarized sheets of cells in vitro.

Following their generation in polarized human airway epithelia, progeny adenovirus particles are released from the basolateral surface. How then might progeny virus particles escape to the apical surface of the epithelium? The fact that they do so is significant, since the emergence of the virus at the apical surface enables transmission of the infection. With apologies once again for indulging in speculation, the answer to our conundrum might be as follows. Molecules of CAR, located on adjacent cells, form homodimers, thereby serving as cell-cell adhesion molecules. Since the adenoviral fiber protein docks to the same site on CAR as the one that mediates CAR dimerization, the adenovirus fibers might disrupt the intercellular **adherens junctions** which constitute the epithelial barrier. There indeed is evidence that the fiber protein plays a role in facilitating virus escape to the apical surface. For instance, fiber protein, like progeny virus particles, is released from the basolateral membrane during viral replication and, importantly (perhaps), adherens junctions are disrupted by the protein and the virus is able to travel between the epithelial cells to escape at the apical surface. Thus, the fiber protein may act at two important steps in the adenovirus replicative cycle: initially during entry and later to promote escape across the epithelia.

Reference
Walters, R., P. Freimuth, T. O. Moninger, I. Ganske, J. Joseph Zabner, and M. J. Welsh. 2002. Adenovirus fiber disrupts CAR-mediated intercellular adhesion allowing virus escape. *Cell* **110:**789–799.

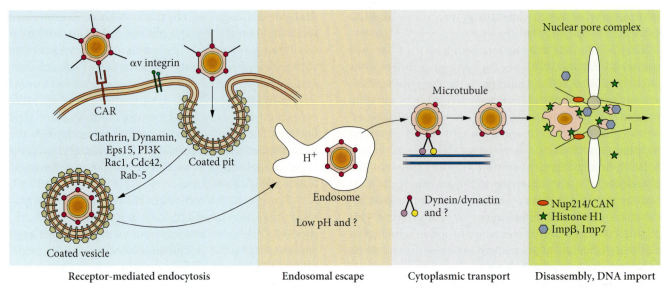

Figure 17.2 Infectious entry pathway of Ad2 and Ad5 into epithelial cells. Ad fibers bind to CAR and locally activate α_v integrins, which triggers clathrin-mediated viral endocytosis. As indicated in the figure, this stage requires dynamin and several other cellular factors. The virus is then delivered to a slightly acidic intracellular compartment, escapes to the cytosol, and is transported to the nucleus by a dynein/dynactin-dependent microtubule-mediated process. The virus then docks at a nuclear pore complex, and it disassembles by recruiting the nuclear histone H1 and the H1 import factors importin β and importin 7. Adapted from O. Meir and U. F. Greber, *J. Gene Med.* **6:**S152–S163, 2004, with permission.

membrane is a crucial stage of infection for these viruses, this stage is less well understood than other stages of their replicative cycles. Yet enough is known about this step and subsequent steps of adenovirus entry to make for a rather good story.

Adenovirus entry into the cytosol and then the nucleus is made possible by a series of regulated conformational changes that occur in the virus particles. These conformational changes are triggered at specific stages of the entry pathway and culminate in the entry of the uncoated viral genome into the nucleus.

The first conformational change that the infecting virus particle undergoes actually occurs at the cell surface, where loss of the fiber protein is triggered by the interaction of the virus particle with the receptor. Following clathrin-mediated uptake, adenovirus particles are then transferred to a slightly acidic intracellular endosomal compartment, in which the virus begins to undergo additional conformational changes that are triggered by the low pH (Fig. 17.2). The critical role played by the endosomal acidic milieu is demonstrated by the fact that transport of adenovirus particles into the cytosol is inhibited by **lysosomotropic** agents and the fact that adenovirus-mediated disruption of endosomal membranes in vitro requires an acidic milieu. (Lysosomotropic agents [e.g., chloroquine] are neutral weak bases that diffuse into all cellular compartments. However, they accumulate in acidic compartments since they are protonated there and thus become unable to diffuse back out. Consequently, they raise the pH of acidic compartments.)

The low-pH-induced conformational changes that the virus particle undergoes in the endosome include the release of protein IIIa, protein VIII, the penton base, the stabilizing protein IX, and some of the hexon proteins. Moreover, the reducing environment of the endosome activates the cysteine protease p23, which previously was inactive in the core of the virus particle. The activated protease p23 cleaves protein VI, thereby activating it. Activated protein VI plays an important role in entry, as is described below. Interestingly, the p23 cleavage site on protein VI may be exposed by the interaction of the penton base with the α_v integrin coreceptor. (Recall that the triggering of virus conformational changes during entry is possible because viruses commonly assemble particles that are metastable, doing so by the postassembly proteolytic processing of one or more capsid proteins. Thus, we might anticipate that such a step occurs during the adenovirus replicative cycle.)

At this point in entry, adenovirus particles are composed of the viral genome, the core proteins, and a small amount of the hexon protein. It appears that peripentonal hexons and their associated underlying proteins are preferentially dissociated from the capsids, along with the penton bases and fiber proteins. Thus, capsid disassembly thus far appears to be localized largely to the icosahedral vertices of the particles.

Studies of an adenovirus mutant called ts1 (for temperature sensitive 1) have shed light on subsequent steps in entry. ts1 is capable of binding to cells and being endocytosed with the same efficiency as wild-type virus. However, ts1 is unable to penetrate the endosomal membrane. The basis for this ts1 deficiency is as follows. ts1, when propagated at nonpermissive temperature, differs from wild-type virus in having only two or three copies of the p23 protease, rather than the approximately 10 copies present in wild-type virus particles. As a consequence of their p23 deficiency, ts1 mutants contain uncleaved preproteins, and thus, they are retained in an immature state. But how does this fact affect their entry?

Recent studies comparing entry of the ts1 mutant with that of wild-type viruses demonstrated that wild-type but not ts1 virus particles are able to permeabilize model membranes (Wiethoff et al., 2005). Moreover, low-pH-induced conformational rearrangements of the wild-type capsids, but not of ts1 capsids, lead to the release of a factor that permeabilizes membranes. Importantly, the membrane-permeabilizing factor was identified as protein VI.

The identification of protein VI as the permeabilizing factor was accomplished as follows. Dissociated capsid proteins were separated from partially uncoated virus particles by centrifugation through a step gradient. Immunodepletion of the free proteins with specific antisera then showed that the majority of the membrane lytic activity could be removed with antisera against protein VI.

Recalling that the p23 protease is needed to activate protein VI, p23 does so by proteolytically cleaving the precursor protein VI, such that an amphipathic α-helix becomes the N-terminal sequence on the cleaved protein. This enables protein VI to interact with the endosomal membrane to manifest its membrane-permeabilizing activity.

Protein VI also needs to be released from the capsid to efficiently promote membrane permeabilization. Disruption of membranes by protein VI is independent of pH, implying that low-pH conditions are required for activation of protein VI and virus disassembly but not for the disruption of endosomal membranes per se (Box 17.5).

Entering adenoviral particles, which by now are devoid of most of their components, except for their genomes, core proteins, and remaining hexon protein, now undergo microtubule-dependent and dynein/dynactin-dependent transport to the nucleus, where they dock to the nuclear pore complex receptor filament protein CAN/Nup214. These filaments extend out from the ring assembly on the cytoplasmic side of the nuclear pore complex (Fig. 12.10). (Dynein is a motor protein that is part of a larger protein

---

BOX 17.5

Apropos the means by which nonenveloped adenoviruses use low pH to trigger the sequence of conformational changes that promote permeabilization of endosomal membranes, it would be good to recall the general principle that the membrane-disrupting activity of viruses in general is shielded from the environment until needed, so as to avoid nonproductive interactions with cell membranes that might occur at the wrong time or place.

In this regard, recall the chain of events whereby the membrane-disrupting protein VI is activated and released from the parental adenovirus particle. It begins with the interaction of the penton base with an α_v integrin at the cell surface, followed by activation of protease p23 by the reducing environment of the endosome. The p23 cleavage site on protein VI may have been exposed by the interaction of the penton base with the α_v integrin coreceptor. Regardless, the proteolytic activation of protein VI by the activated p23 was dependent on the reducing environment of the endosome. In addition, the endosomal acidity also triggers the release of the penton base and some of the hexon proteins, thereby enabling the release of the activated protein VI from the particles, which then permeabilizes the endosomal membrane to enable the release of partially uncoated viral particles into the cytosol; all-in-all, a rather intriguing set of viral adaptations.

---

complex, dynactin, which mediates movement of cargo along microtubules. Microtubule-mediated adenovirus transport may be stimulated by virus-induced signals that upregulate the protein kinase A and mitogen-activated protein kinase cascades.)

The capsid undergoes its final disassembly at the nuclear pore complex, and the viral genome then enters the nucleus (Fig. 17.2). Final capsid disassembly is necessary prior to nuclear entry, since adenoviral capsids, at 70 to 100 nm in diameter, are considerably larger than the largest entities (39 nm in diameter) that are known to be imported via nuclear pore complexes.

Interestingly, the mechanisms by which the partially disassembled adenovirus capsids dock at nuclear pore complexes and then transport their genomes into the nucleus are distinctly different from the mechanisms that usually mediate the binding of cellular macromolecules and macromolecular complexes at nuclear pore complexes and their ensuing import into the nucleus. As illustrated in Box 12.7, proteins targeted to the nucleus typically are recognized by cytoplasmic receptors of the importin/karyopherin family, which deliver the cargo to the cytoplasmic face of the nuclear pore complex. Transport of the substrate into the nucleus then requires additional cytosolic factors, including the guanine nucleotide-binding protein Ran. In contrast, as shown by Urs Greber and co-workers, adenoviral particles bind to the nuclear pore complex filament protein CAN/Nup214, independently of any additional cytoplasmic factors, including Ran. Thus, adenoviruses may have evolved an intrinsic ability to dock at nuclear pore complexes. The docked adenoviral particles then trap the nuclear histone H1. This step is followed by the binding of the H1 import factors Impβ and Imp7, together with additional cytoplasmic factors, such as the chaperone Hsc70. These factors trigger final disassembly of the remaining capsid at the nuclear pore and promote import of the viral genome into the nucleus.

Some final points are notable. (i) The hexon coat protein is the major protein on the surface of the partially disassembled adenovirus particles at the time they dock at nuclear pore complexes. Thus, the hexon protein may be the principal nuclear targeting determinant on these particles. (ii) The nuclear pore complex filament protein, CAN/Nup214, appears to also function as a docking site for the import of the U1 small nuclear ribonucleoproteins, which help to mediate mRNA splicing. (iii) The interaction of histone H1 with the partially disassembled adenovirus particles is *insensitive* to nucleases, implying that H1 proteins actually interact with the hexon protein rather than with the viral DNA. (iv) Some hexon protein is imported into the nucleus at final disassembly, possibly guiding the viral DNA in. Yet the hexon proteins do not contain any apparent nuclear localization signal. Thus, they could rely on binding to a cellular protein that does possess a nuclear localization signal, presumably histone H1 in this instance. Moreover, binding of H1 to the capsids is necessary for final disassembly. This fact was demonstrated by the experimental finding that microinjecting anti-H1 antibodies into the nucleus was as effective at blocking final disassembly as antibodies that blocked capsid binding to nuclear pore complexes. (v) Immunofluorescence microscopy demonstrated that Hsc70 associates with capsids after their release from endosomes. This interaction appears to be necessary for the import of viral DNA into the nucleus, since nuclear entry is blocked by introducing anti-Hsc70 antibodies into cells. Hsc70 might help to disassemble the capsid at the nuclear pore complex.

The following scenario for final capsid disassembly and nuclear import of the viral genome is consistent with all the experimental findings enumerated above. Histone H1 binds to partially disassembled virus particles that are docked at nuclear pore complexes. This is followed by the binding of the H1 import factors Impβ and Imp7, which recognize H1 on the docked capsids. The bound H1 import factors then transport into the nucleus the H1-hexon complexes that are most proximal to the nuclear pore complex, thereby creating the opportunity for the DNA to follow. Hsc70 may facilitate these events, possibly by promoting capsid disassembly or by stabilizing disassembled capsids at nuclear pore complexes until genomes might enter the nucleus.

Summing up this intriguing entry pathway, recall that virus disassembly occurs in a regulated series of steps that are elicited at specific stages of entry. It begins with the release of the fiber proteins, which is triggered by the interaction of the capsid with the receptor at the plasma membrane. It is followed by conformational changes induced by the low pH of the endosomal compartment. These changes include activation of the p23 protease, which leads to further conformational changes, including activation and release of the membrane-permeabilizing protein VI. Finally, interactions with histone H1 and H1 import factors at nuclear pore complexes lead to final capsid disassembly and genome import into the nucleus.

GENOME ORGANIZATION AND EXPRESSION

All adenovirus genomes examined to date appear to be organized similarly (Fig. 17.3). They are linear and contain as many as 36,000 bp. Moreover, they contain a terminal protein (molecular mass, 55 kDa) covalently attached to each 5′ end. In addition, adenoviral genomes contain short terminal repeats, each of which contains an identical origin of DNA replication. Adenoviral genomes also contain a *cis*-acting packaging sequence that initiates the interaction of the DNA with encapsidating proteins. The packaging signal overlaps the early enhancers I and II, which are close to the origin of replication. Adenovirus enhancer I specifically stimulates expression of the E1A transcription unit, and enhancer II stimulates transcription of all transcription units (see below; see also Fig. 17.6).

Earlier, we noted how the polyomaviruses and papillomaviruses process their primary transcripts in multiple ways, to assemble several different messages from a limited amount of primary genetic information. In the case of the adenovirus, we see the potential advantages of differential mRNA splicing realized to a far greater extent than seen in the cases of either the polyomaviruses or papillomaviruses.

The pattern of transcription of the adenoviruses is depicted in Fig. 17.3. Arrowheads indicate the direction of transcription along the rightward (r-strand) strand and the leftward (l-strand) strand of the DNA. Solid vertical

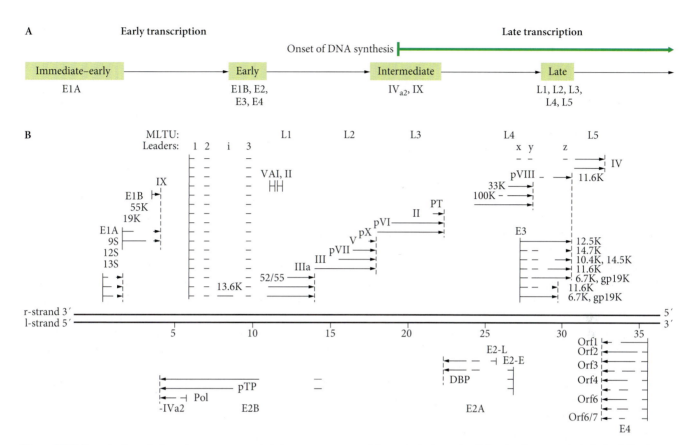

Figure 17.3 Organization of adenovirus genomes and their pattern of transcription. (A) Adenovirus gene expression is temporally regulated into immediate-early, early, intermediate, and late phases. The genome regions expressed during each of the temporal phases are indicated. (B) The linear genome is numbered in kilobase pairs. Both DNA strands contain ORFs, and both strands serve as templates for transcription. Individual mRNAs are indicated as solid lines, with introns indicated as gaps. Arrowheads indicate the direction of transcription along the rightward strand (r-strand) and the leftward strand (l-strand) of the DNA. Solid vertical lines indicate RNA polymerase II (Pol II) promoters. Broken vertical lines indicate polyadenylation sites. Notice that *rightward* adenoviral transcripts overlap *leftward* transcripts. Also, whereas adenoviruses encode more than 40 different proteins, the sequences encoding these proteins are organized into only eight transcription units. Other details to note include the following. Each of the five early region transcription units (E1A, E1B, E2, E3, and E4) has its own promoter, and the E2 and E3 transcription units each contain two alternative polyadenylation sites. The major late region produces five families of mRNAs, L1 through L5. Each of the late mRNA families is generated from the same promoter, and each is processed from the same pre-mRNA. The families are distinguished on the basis of which of the five polyadenylation sites in the pre-mRNA is used. That is, all members of a late mRNA family terminate at the same polyadenylation site. All late transcripts contain the same 5′ tripartite leader sequence. Major proteins are pII, hexon; pIII, penton; pIV, fiber; pVII, core; PT, cystein protease p23; DBP, DNA-binding protein; pTP, terminal protein precursor; Pol, DNA polymerase. Adapted from K. Leppard, *Seminars in Virology* **8**:301–307, 1998 (copyright Academic Press).

lines indicate RNA polymerase II (Pol II) promoters. Broken vertical lines indicate polyadenylation sites.

Major points to note are as follows.

1. Both strands of adenoviral genomes contain open reading frames (ORFs), and both strands serve as templates for transcription. In this respect, adenoviral transcription resembles that of the polyomaviruses, but not that of the papillomaviruses. However, in contrast to what occurs in the polyomaviruses, *rightward* adenoviral transcripts overlap *leftward* transcripts.

2. Adenoviruses encode more than 40 different proteins. However, the sequences encoding these proteins are organized into only eight RNA polymerase II (Pol II) transcription units, as explained below.

3. Expression of adenovirus genes, like those of polyomaviruses and papillomaviruses, is temporally regulated. That is, there is an early phase of transcription that precedes DNA replication, and a late phase that follows DNA replication. But in contrast to the polyomaviruses and papillomaviruses, the early phase of adenovirus transcription actually comprises two distinct transcriptional phases, which are expressed as follows. The E1A transcription unit is expressed first, before the other early genes (Fig. 17.3). For that reason, its expression is referred to as "immediate-early."

In addition, E1A expression differs from that of all other adenoviral transcription units in that it does not require any prior viral protein synthesis. That is, E1A genes, unlike other adenoviral genes, are transcribed in infected cells when pharmacological inhibitors are present to block protein synthesis. In contrast, expression of the remaining viral genes requires prior E1A expression. The initial expression of adenovirus genes, like those of polyomaviruses, is controlled by enhancers that interact efficiently with cellular transcription factors.

4. The switchover to the late transcriptional phase of the adenovirus replication cycle differs from that of both the polyomaviruses and papillomaviruses. In the case of the polyomaviruses, late gene expression appears to be *completely* dependent on early gene expression and subsequent viral DNA replication. In the case of the papillomaviruses, expression of the P3 late promoter is regulated by cell differentiation, such that while P3 is active at early times, it becomes fully active only in terminally differentiated keratinocytes. Moreover, the complete set of P3 transcripts is generated only in terminally differentiated keratinocytes. (See Chapter 16 for details.) As in the case of the papillomaviruses, but in contrast to the polyomaviruses, adenovirus late genes also are expressed at early times, but at very low levels only. But in contrast to the papillomaviruses, adenovirus DNA synthesis is necessary, and perhaps sufficient, for the switchover to the increased rate of initiation from the major late promoter (Fig. 17.3). (Adenovirus late genes are those genes that attain their greatest rate of expression following viral DNA replication.) And, as noted below, only replicated viral DNA molecules are able to generate all of the major late transcripts.

5. The small IVa2 and IX genes form an intermediate class, rather than being true late genes. At late times, the IVa2 protein binds to a sequence in the major late promoter and, in conjunction with a cellular protein, stimulates transcription from that promoter at least 20-fold. Notice that the protein IX gene is completely contained within the early E1B transcription unit (Fig. 17.3). Consequently, transcription from the pIX promoter is initially inhibited by transcription from the upstream E1B promoter. However, as viral DNA replication ensues, the concentration of E1B promoters exceeds that of the cellular factors that promote E1B transcription, thereby making the pIX promoter available to initiate transcription of the pIX gene.

6. Each of the five early region transcription units (E1A, E1B, E2, E3, and E4) has its own promoter. Within each transcription unit, alternative splice sites are used, which enables each of the early transcription units to encode multiple distinct proteins. Moreover, notice that two of the early region transcription units, E2 and E3, each contain two alternative polyadenylation sites. This enables each of these transcription units to transcribe two families of mRNAs, each of which is differentially spliced.

7. The major late region produces five families of mRNAs, L1 through L5. Each of the late mRNA families is generated from the same promoter, and each is processed from the same pre-mRNA. The families are distinguished on the basis of which of the five polyadenylation sites in the pre-mRNA is used. That is, all members of a late mRNA family terminate at the same polyadenylation site. The major late region is transcribed as follows. As just noted, the major late promoter produces only one primary transcript. That transcript encompasses the entire genomic sequence from the promoter to the right end of the genome. Again, note that the major late region contains five polyadenylation sites, and the members of each family of late mRNAs share the same polyadenylation site (Fig. 17.3). The first processing step is for the primary transcript to be cleaved at one of the five late polyadenylation sites, a step that actually occurs while the RNA is still a nascent transcript. Each polyadenylation site is selected with approximately equal frequency. Then, the transcript is polyadenylated, and a 7-methyl Gppp cap is added to its 5′ end. Importantly, the random selection of the polyadenylation site determines which of the five mRNA families the final mRNA will belong to.

8. A common 5′ tripartite leader sequence is generated on all late transcripts by the splicing together of the sequences labeled 1, 2, and 3 in Fig. 17.3. This leader is then spliced to one of the main body sequences contained in the family of mRNAs that terminate in the selected polyadenylation site. This differential splicing within the family determines which polypeptide will be translated, since only the 5′ proximal ORF in a eukaryotic "polycistronic" mRNA can usually be translated (Boxes 17.6 and 17.7).

9. Several instances of regulated adenoviral gene expression were noted above. Among these examples, gene expression is temporally regulated, giving rise to the immediate-early, early, intermediate, and late phases of transcription. Moreover, the means by which adenoviral genes are organized into transcription units gives rise to coordinate control of gene expression within transcription units. In addition, there is the intermediate IVa2 protein, which stimulates transcription from the late promoter, and the IX gene, whose

BOX 17.6

The discovery of mRNA splicing actually took place during the detailed analysis of adenovirus mRNA structure. Moreover, the discovery was made independently, and virtually simultaneously, by two different research groups (Berget et al., 1977; and Chow et al., 1977).

The first unexpected finding was that several different transcripts encoded by the major late region have the same capped 11-nucleotide sequence at their 5′ ends. Hybridization studies with restriction fragments, which previously had been mapped on the genome, demonstrated that these 5′ terminal sequences are not encoded adjacent to the main bodies of the mRNAs. However, the most dramatic finding came when the mRNA encoding the hexon protein was hybridized to the intact complementary strand of viral DNA. The resulting heteroduplex structure contained three short DNA loops, labeled A, B, and C in the figure, which is actually an accurate color rendering of a real electron micrograph of a heteroduplex between the hexon mRNA and the viral DNA (Berget et al., 1977). These loops correspond to the introns removed from the mRNA by splicing. Other major late mRNAs generated similar heteroduplexes when hybridized to the DNA.

Within the next several months it was shown that pre-mRNA splicing is not an oddity associated with viral

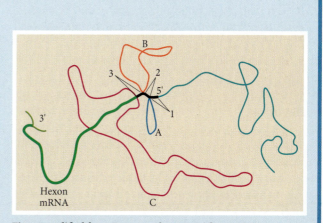

Figure modified from Berget et al., 1977, with permission.

molecular biology but instead is characteristic of eukaryotic transcription in general. Philip Sharp and Richard Roberts shared the 1993 Nobel Prize in physiology or medicine for their independent discovery of mRNA splicing.

References

Berget, S. M., C. Moore, and P. A. Sharp. 1977. Spliced segments at the 5′ terminus of adenovirus 2 late mRNA. *Proc. Natl. Acad. Sci. USA* 74:3172–3174.

Chow, L. T., R. E. Gelinas, T. R. Broker, and R. J. Roberts. 1977. An amazing sequence arrangement at the 5′ ends of adenovirus 2 messenger RNA. *Cell* 12:1–8.

---

expression is dependent on DNA replication. Additional examples of adenoviral regulation of gene expression are as follows.

Recall that the major late adenovirus promoter is in fact active prior to DNA synthesis, but at a relatively low level. During this early period, the L1 polyadenylation site is preferentially selected, and consequently, the production of L1 family proteins is favored. Only replicated viral DNA viral molecules are able to generate all of the major late transcripts. Accordingly, viral DNA synthesis is necessary, and perhaps sufficient, for the switchover at late times to random selection of polyadenylation sites. However, the mechanism responsible is not entirely understood.

In addition to making use of different polyadenylation sites to alter the levels of expression of different mRNA families, adenoviruses also are able to regulate pre-mRNA splicing to alter the proportion of mRNAs produced within an mRNA family. For example, the relative levels of different mRNAs produced from the L1 pre-mRNA change with the transition from

the early to the late phases of infection. The only L1 mRNA produced at early times is that encoding the 52- to 55-kDa protein. Interestingly, a viral E4 protein acts to relieve the inhibition of the other L1 splice site at later times. It does so by inducing the dephosphorylation of cellular SR proteins, thereby relieving the block to the splicing that would generate the mRNA for the 13.6-kDa protein. (Cellular SR proteins act as inhibitors or activators of splicing, depending on the pre-mRNA sites to which they bind. Dephosphorylated SR proteins are unable to repress recognition of the splice site that generates the mRNA for the 13.6-kDa protein.)

The E2 transcription unit encodes three proteins: the viral DNA polymerase, the DNA-binding protein (DBP), and the terminal protein (Fig. 17.3). The precursor of the terminal protein serves as a primer for viral DNA replication (see below). Of the three proteins encoded by the E2 region, DBP is needed in the greatest amounts. That is so because of its role in DNA replication, in which it coats the displaced strand of the

BOX 17.7

Why might the adenoviruses have evolved such a complex mechanism for late transcription, in which they use elaborate alternative splicing and polyadenylation at multiple sites in the pre-mRNA? We cannot know for certain. However, one apparent benefit is that this strategy of alternative polyadenylation sites and differential splicing enables the major late region to encode at least 15 different proteins from a common promoter. On the other hand, it appears somewhat wasteful since each of the late mRNAs, which are between 1 and 5 kbp long, is processed from a precursor of up to 28 kbp in length.

Yet it is typical for eukaryotic primary transcripts to extend beyond the polyadenylation site. The 3′ end of the mRNA is then generated by a specific cleavage of the pre-mRNA. Poly(A) is then added to the new 3′ terminus, and the cleaved downstream sequences are degraded. The signal for polyadenylation is the sequence 5′AAUAAA3′, which is located 10 to 30 nucleotides upstream from the cleavage site. Also, additional necessary polyadenylation sequences are present on the other side of the cleavage site. Splicing and capping of the pre-mRNA occurs at about the time it is polyadenylated, and both the capping and polyadenylation activities are associated with RNA Pol II.

Bearing in mind the above description of eukaryotic mRNA processing in general, it appears that the adenoviruses have elaborated upon a process already in place in the cell. That elaboration enables adenoviruses to express many proteins from a minimum of gene space, while also achieving coordinate control of gene expression.

The advantages of using multiple splice sites to generate many proteins from a limited gene space likewise apply to the organization and expression of the early genes. The five early genes generate more than 30 proteins from only five promoters. Moreover, the E1A, E1B, E2, and E3 transcription families each encode a set of polypeptides with related functions. Grouping related functions under the control of a single transcriptional control element should enable the virus to simultaneously express related functions as they are needed.

---

template DNA molecule (see below). Accordingly, the virus differentially expresses polypeptide DBP within the E2 transcription unit by preferentially using the particular polyadenylation site on the long E2 pre-mRNA, which results in the mRNA that encodes DBP.

Another example of regulation, in this instance involving expression of the intermediate phase IVa2 mRNA, depends on an indirect effect of DNA replication. Specifically, transcription of the IVa2 gene is apparently controlled by a repressor that is present in limited amounts, and which is titrated by DNA replication. The story of IVa2 is better still, since IVa2 itself is a sequence-specific activator of transcription. In cooperation with a cellular protein, IVa2 stimulates transcription from the major late promoter at least 20-fold, as noted in point 5 above. In addition, recall how transcription of the intermediate phase IX mRNA also is coupled to DNA replication.

10. Adenoviruses encode two short RNAs, VAI and VAII, which, unexpectedly, are transcribed by RNA Pol III. Since RNA Pol II is the RNA polymerase responsible for transcribing cellular mRNAs, it is not surprising that many DNA viruses that replicate in eukaryotic cells make use of cellular RNA Pol II to carry out their transcription. In contrast, since Pol III normally transcribes transfer RNA (tRNA) genes and 5S ribosomal

RNA (rRNA), the use of RNA Pol III by adenoviruses was hardly expected. Indeed, the adenovirus VA RNAs were the first viral RNAs found to be transcribed by Pol III. Yet other viruses (e.g., Epstein-Barr virus) have since been found to also encode RNA Pol III transcripts. The RNA Pol III transcripts VAI and VAII are shown by paired vertical lines in Fig. 17.3.

E1A AND E1B PROTEINS IN REPLICATION AND NEOPLASIA

Genetic studies showed that the E1A and E1B transcription units *each need to be expressed* for adenoviruses to induce permanent cell transformation. Expression of E1A alone leads to *abortively* transformed cells that cannot be stably maintained in cell culture. These experimental findings were among the first indications that multiple genetic changes are necessary to bring about neoplastic cell transformation in general.

Importantly, expression of the E1A and E1B transcription units likewise is *necessary* for virus replication. The fact that adenovirus transforming genes also are necessary for adenovirus replication might be expected since transformation is not a usual feature of the adenoviral life cycle and size-restricted virus genomes do not carry unneeded genetic baggage. The transforming genes of the polyomaviruses and the papillomaviruses likewise are required for the replication of those viruses (Chapters 15 and 16).

How then do the adenovirus transforming functions play out in the normal replicative cycle of these viruses? As in the case of the polyomaviruses and papillomaviruses, the crucial problem that the adenoviruses solve with their transforming functions is as follows. Differentiated cells in vivo rarely divide and usually are in the G_0, nondividing state. Such cells contain very low levels of the enzymes and substrates required for DNA replication. Because adenoviruses, like polyomaviruses and papillomaviruses, are dependent on those cellular enzymes and substrates for their replication, they would not be able to replicate efficiently unless they were able to induce resting cells to enter into the cell cycle.

By what means do the adenovirus E1A and E1B proteins induce resting cells to divide, and as an inadvertent consequence of that ability, bring about cell transformation? If you are thinking that their mechanism of inducing transformation has something to do with inactivating p53 and pRb (Chapters 15 and 16), then, to be sure, you are correct. Indeed, we commented earlier on the remarkable fact that whereas the polyomaviruses, the papillomaviruses, and the adenoviruses are three distinct unrelated virus families, they have nevertheless evolved similar means for inducing resting cells to enter the cell cycle, in each instance also enabling these viruses to induce neoplastic transformation of cells in culture, and tumors in experimental animals. The adenovirus E1A and E1B proteins inactivate pRb and p53, respectively.

The roles of the pRb family of proteins and of p53 in regulating cellular replication were discussed in Chapter 15 (see Fig. 15.13). Regarding pRb, in brief, tumor suppressor protein pRb binds to, and thereby inactivates, transcription factor E2F, which is necessary for expression of various genes whose products are necessary for resting cells to enter into S phase. The E2F-pRb complexes have the same DNA-binding specificity as E2F alone, but the presence of pRb in the complex represses transcription. When pRb is released from E2F, then cells that are in the G_0, or resting, state are able to progress into S phase. In the normal course of the cell cycle, pRb is inactivated by the action of a cyclin-dependent kinase. Specifically, the Cdk4/cyclin D complex phosphorylates pRb, resulting in its dissociation from E2F, thereby activating expression of E2F target genes.

Whereas pRb regulates entry of cells into S phase, p53 is a key component of the surveillance mechanism that comes into play if a cell should bypass normal check points in the cell cycle and then undergo an inappropriate unchecked cell division. If a cell were to embark on an unscheduled S phase, then cell cycle checkpoint proteins would activate p53. Also, p53 prevents a cell from undergoing potentially ruinous cell divisions, such as might occur if a cell were to sustain DNA damage and then were to attempt to replicate before repairing that damage. Such potentially disastrous situations cause the cell to accumulate p53, which results either in cell cycle arrest in order to repair the damage or in apoptosis if the damage is not repaired or if the cell enters an inappropriate or unscheduled S phase. From the point of view of the host, cell suicide is preferable to the generation of daughter cells that might contain genetic lesions capable of leading to cancer.

In looking at the interaction of E1A proteins with pRb in somewhat more detail, bear in mind that E2F-pRb complexes have the same DNA-binding specificity as E2F alone but that the presence of pRb in the complex represses transcription from E2F-dependent promoters. Adenovirus E1A proteins bind to the same site on pRb as E2F, thereby sequestering pRb and displacing it from E2F. Consequently, the now free E2F can promote cell cycle progression to S phase. Interestingly, although the polyomavirus LT proteins, the papillomavirus E7 protein, and the adenovirus E1A protein are unrelated, nevertheless, each interacts with pRb via a common Leu-X-Cys-X-Glu motif.

In addition to promoting cell cycle progression, so that cells might make the cellular enzymes and substrates needed for DNA synthesis, E2F also is a transcription factor for the viral E2 transcription unit. Free E2F binds to the viral E2 promoter, thus enabling expression of the E2 family of mRNAs. (Indeed, E2F was originally identified as a cellular factor that binds to the adenovirus E2 promoter, thereby accounting for it being named "E2F.") Recall that the adenovirus E2 transcription unit encodes three viral proteins: the viral DNA polymerase, DBP, and the terminal protein. The precursor of the last serves as the primer for viral DNA replication (see below).

Next, we consider the biology of the E1A and E1B proteins in somewhat more detail. Beginning with the E1A proteins, the E1A transcription unit makes use of differential splicing to produce two mRNAs that are translated into two proteins with sedimentation coefficients of 12S and 13S (Fig. 17.3). The difference between these two proteins is that the 12S protein lacks the 49 amino acids that constitute the internal segment, CR3 (acronym for "conserved region 3"), of the 13S protein. Despite this difference, the two E1A proteins interact in the same way with pRb, and consequently, they are functionally equivalent with respect to transformation. However, the 13S protein also stimulates adenovirus early gene transcription via its CR3 domain. This effect of E1A is expressed before DNA synthesis, which in turn fully activates expression of intermediate and late genes. Interestingly, the E1A 13S protein does not bind to DNA. Instead, it stimulates early transcription by an indirect mechanism, in which it interacts with several different cellular transcription factors.

The adenovirus E1B proteins are expressed to counter the apoptosis-inducing effect of E1A. (As noted above, since E1A expression induces cells to bypass normal checkpoints in the cell cycle, E1A expression alone would cause activation of p53, an effect that would lead to apoptosis.) By means of differential splicing, the E1B transcription unit encodes two antiapoptotic proteins: one with a molecular mass of 55 kDa and the other with a molecular mass of 19 kDa (Fig. 17.3).

The E1B 55-kDa protein blocks p53-mediated apoptosis by directly binding to p53. This plays out as follows. Apoptosis induced through the p53 pathway is mediated by the caspases, which are the actual effectors of apoptosis (see Chapter 4). Moreover, the ability of p53 to induce activation of the caspases appears to depend on its function as a transcriptional activator. Thus, it is interesting that the binding of the E1B 55-kDa protein to p53 does not disrupt the DNA-binding activity of p53. Rather, the binding tethers the transcription-repressing effect of the 55-kDa protein to p53-responsive genes.

The E1B 55-kDa protein has yet another means by which it undermines p53, but only in combination with the E4 ORF6 protein. These two proteins appear to form a complex with each other that binds to p53, thereby blocking its function as a transcriptional activator. But, in addition, the complex also causes p53 to be degraded at an accelerated rate. While the E1B 55-kDa protein and the E4 ORF6 protein could each bind independently to p53 and reduce its transcriptional activation activity, the accelerated degradation of p53 activity depends on the complex between the E1B 55-kDa and the E4 ORF6 proteins. This was demonstrated using an E4 *orf6* mutant, which encodes a protein that fails to associate with the E1B 55-kDa protein but still interacts with p53.

The E1B 19-kDa protein acts to block apoptosis differently from, and independently of, the E1B 55-kDa protein. First note that the cellular Bcl-2 and Bax proteins serve to regulate apoptosis; Bcl-2 is a negative regulator, and Bax is a positive regulator (Fig. 4.25). The life-or-death fate of the cell is determined by the balance of these proapoptotic and antiapoptotic factors. The remarkable adenovirus E1B 19-kDa protein not only upregulates transcription of host Bcl-2 to negatively regulate apoptosis but also binds to Bax. The latter binding activity prevents Bax from activating the caspase cascade, a rather diabolical dual activity.

Interestingly, adenovirus mutants defective at blocking apoptosis originally were discovered because the DNA of cells infected with E1B mutants was unstable. The cells died prematurely, and plaque sizes and viral yields were reduced.

We return briefly to E1A, whose mitogenic effect creates the need for E1B. Are you surprised to now be told that E1A also acts to block p53 function? That it does so is true, but its mode of action is very different from that of the E1B proteins. E1A blocks p53 as follows. In Chapter 15 we noted that the p53 protein also oversees cell cycle progression by stimulating transcription of genes that encode proteins which regulate cyclin-dependent kinases. p21 is one such p53-regulated protein. Induction of p21 transcription by p53 depends on the transcriptional coactivators p300 and Creb-binding protein. E1A binds to these coactivators, thereby blocking the p53-induced upregulation of p21. Note that E1A is acting here to impair the ability of p53 to block cell cycle progression per se rather than to block the p53-mediated induction of apoptosis. The E1B 55-kDa protein, which transcriptionally represses p53-responsive genes (as noted above), might have a similar effect.

Considering the array of measures the adenoviruses take to undermine p53 function, we might ask whether these various measures are redundant. It may seem so, as implied by the finding that expression of the E4 ORF6 protein, which binds to p53, is sufficient to transform rodent cells in cooperation with E1A proteins. However, when all the proteins that act to subvert p53 are present, they might act together to ensure that viral replication is not impeded by the onset of apoptosis or cell cycle arrest. Another explanation that is frequently offered is that this redundancy enables the virus to deal with differences among different host cells and conditions at the time of infection. (Note that while transformation is not relevant to the natural life cycle of the adenoviruses, the concurrent expression of the adenovirus transforming proteins ensures the survival of transformed cells [Box 17.8].)

Interestingly, it is now known that the adenovirus E4 gene encodes a PP2A-inactivating protein, reminiscent of the polyomavirus ST protein, once again underscoring the similar mechanisms that these unrelated viruses have independently evolved to affect cell cycle progression.

SHUTOFF OF HOST PROTEIN SYNTHESIS

Adenoviruses disrupt cellular gene expression at late times by means of several diverse mechanisms. First, the export of cellular mRNA to the cytoplasm is selectively impaired by a mechanism that is not yet clear but requires the cooperative action of the viral E1B 55-kDa and E4 ORF6 proteins, which form a complex with each other as noted above. (Recall that the E1B 55-kDa and E4 ORF6 proteins also act together to accelerate the breakdown of p53.)

Second, the action of the adenoviral L4 100-kDa protein causes translation to switch from a cap-dependent to a cap-independent mode. It does so as follows. The cellular eIF4F translation initiation complex plays several crucial roles during translation initiation, including cap

BOX 17.8

After the discovery about 50 years ago that the polyomaviruses can transform cells, two alternative plausible hypotheses were put forward to account for viral oncogenesis. In the first of these hypotheses, the initiation and maintenance of the transformed cell phenotype were said to require the continued expression of viral transforming genes. In the second, the so-called **"hit-and-run" hypothesis**, the virus was said to induce a permanent change in the physiological state or phenotype of the cell, such that the continued presence of the viral genome would not be required to maintain the transformed cell phenotype. Experiments in which cells in culture were transformed by polyomaviruses containing temperature-sensitive lesions in transforming genes demonstrated that the continued expression of polyomavirus transforming genes indeed is necessary to maintain cell transformation (Chapter 15). (Note that experimental results with these mutants also showed that the viral transforming genes are crucial for virus replication.)

As described in the text, expression of the adenovirus E1A and E1B gene products can lead to cell transformation, in which the transformed cells permanently express the viral transforming genes, consistent with the classical notion of viral oncogenesis. Other experimental results demonstrated that the adenovirus E4 region encodes two new oncogenic proteins, the products of E4 ORF3 and E4 ORF6, which can cooperate with the viral E1A proteins to mediate cell transformation (see the text). But surprisingly, cells transformed by E1A and either E4ORF3 or E4ORF6 did not express the viral E4 gene products, and only a few expressed E1A proteins. In actuality, the majority of these cells appeared to have lost the E1A and E4 DNA sequences. Indeed, the majority of transformed cell lines lacked any detectable viral DNA sequences whatsoever, consistent with a hit-and-run

mechanism of cell transformation. (Interestingly, the E1A and E4 genes are retained and expressed in transformed cells when E1B is expressed as well. In the absence of E1B, the E1A and E4 ORF3 or E4 ORF6 genes apparently cooperate to initiate transformation and then are lost from the cells, which continue to express the transformed phenotype. Also, the E4 ORF6 protein forms a complex with the E1B 55-kDa protein, as described in the text. That complex binds to p53, causing it to be degraded at an accelerated rate.)

How might the transient expression of the viral oncogenes have led to cell transformation? The answer is not yet known, but the process is thought to perhaps be due to the mutagenic effects of the viral transforming genes. In fact, the mutagenic effect of E1A is already apparent within 11 h after infection.

These experimental findings are extremely significant because they raise the possibility that naturally occurring tumors, which lack any detectable virus-specific molecules, may yet have a viral etiology. If that is the case, then viruses may have a greater impact on human cancer than presently realized. Yet there is also the following conundrum. Adenoviruses have never been convincingly associated with cancer in humans because none of the extensive screens found a correlation between human cancer and the presence of adenoviral DNA. So, if adenoviruses (or any other viruses) were responsible for cancer in humans, as mediated by a hit-and-run mechanism, how might this be demonstrated?

Reference
Nevels, M., B. Täuber, T. Spruss, H. Wolf, and T. Dobner. 2001. "Hit-and-run" transformation by adenovirus oncogenes. *J. Virol.* 75:3089–3094.

recognition and recruiting the 40S ribosomal subunit. The activity of the eIF4F complex can be regulated in the cell by phosphorylation of the eIF4E subunit, which is the actual cap-binding protein. Phosphorylation of eIF4E leads to its increased association with the other components of the eIF4F complex and higher levels of translation of capped mRNAs. The adenovirus L4 100-kDa protein appears to impair phosphorylation of eIF4E as follows. eIF4E is phosphorylated in vivo by protein kinase MnK1 when each is bound to eIF4G, which is also a component of the eIF4F initiation complex. The L4 100-kDa protein binds to eIF4G at or near the site normally occupied by MnK1, thus displacing MnK1 from eIF4F

complexes, in that way preventing phosphorylation of eIF4E late in infection.

Adenoviral mRNAs, although capped, continue to be translated, probably because their tripartite leaders (see above) can recruit ribosomes in a cap-independent manner, somewhat reminiscent of the picornavirus internal ribosomal entry site sequences (Chapter 6). (In Chapter 6 we noted that poliovirus likewise achieves the shutdown of cellular translation by abolishing the ability of cells to initiate translation of capped mRNAs. The underlying mechanism in that instance is for the poliovirus 2A^{pro} protein to mediate cleavage of eIF4G. Translation of uncapped poliovirus mRNA is unaffected by the cleavage of eIF4G, since that initiation factor is not needed for ribosomes to attach to the internal ribosomal entry site used by poliovirus.)

What purpose does it serve the virus to shut down host protein synthesis at late times? One possibility is that the host cell's limited biosynthetic capacity is directed entirely to serving the needs of the virus. More importantly, perhaps, it may facilitate final virus assembly. The argument is as follows. Viral structural components must find each other for final virus assembly in an intracellular environment in which the concentration of cellular proteins may be as high as that attained in protein crystals (i.e., 20 to 40 mg/ml). In that milieu, many viral structural subunits may inadvertently interact with irrelevant cellular proteins, thereby impairing assembly.

DNA REPLICATION

DNA replication always needs to be primed, thereby creating the problem of how to generate the 5′ ends of the daughter DNA strands (Chapter 14). Cells solve the "end replication" problem by means of special short repeated sequences at the ends of their chromosomes called telomeres, which are replicated by means of a special enzyme called telomerase (Chapter 14). In contrast, viruses have evolved a variety of solutions to the end replication problem. Bacteriophage λ (lambda) and herpes simplex viruses circularize their linear double-stranded genomes, which then replicate via a "rolling-circle" mechanism (Chapters 14 and 18). Parvoviruses, which have single-stranded DNA genomes, solve the end replication problem by means of a "rolling-hairpin" mode of replication, in which the free 3′ end of the hairpin structure serves as the primer (Chapter 14). In contrast, polyomaviruses (Chapter 15) and papillomaviruses (Chapter 16) solve the end replication problem by having circular genomes.

Adenoviruses solve the end replication problem by using an 80-kDa precursor of the 50-kDa terminal protein to prime DNA replication. The 80-kDa precursor of the terminal protein, designated pTP, is incorporated into the progeny DNA strands during priming, as described below. It is proteolytically cleaved to the smaller terminal protein during final virus assembly and maturation, thereby explaining how a copy of the 50-kDa terminal protein comes to be linked to the 5′ end of each viral DNA strand.

Bear in mind that adenovirus genomes contain an inverted terminal repetition, each copy of which contains the adenovirus *ori* sequence. Priming and ensuing replication thus occur as follows (Fig. 17.4). A complex, consisting of pTP and the virally encoded DNA polymerase (Pol), binds specifically to the core regions of the viral origins of replication. Both of these early viral proteins are encoded by the E2 transcription unit (Fig. 17.3). The binding of the pTP/Pol complexes to the *ori* sequences is facilitated by two cellular *transcription* factors, Nf-1 and Oct-1. These transcription factors promote *replication* by specifically binding to sequences adjacent to the *ori* core and then binding the pTP/Pol complexes. In this way, Nf-1 and Oct-1 bring the viral replication proteins to the *ori* core. The bound viral polymerase then initiates DNA synthesis by reading the guanine moiety at the 3′ end of the template strand. Doing so, it adds a cytosine residue to a serine side chain of the pTP priming protein. Synthesis then proceeds in the 5′-to-3′ direction along the DNA template. In this way, the terminal protein becomes linked to the DNA via a phosphodiester bond between a serine residue on the terminal protein and the 5′ hydroxyl of the terminal deoxycytosine on the DNA. As the template strand is read, the complementary strand is displaced from the parental duplex and coated by DBP, the single-stranded DNA-binding protein, which also is encoded by the E2 transcription unit. This protein is necessary for elongation since it acts to unwind the duplex structure. A cellular topoisomerase acts to relieve overwinding of the double-stranded DNA template.

Considering that each terminus of a linear adenovirus genome contains the same sequence, that each of these repeated sequences is inverted with respect to the other, and that each repeat contains the *ori*, we might correctly surmise that initiation may occur independently and concurrently at each end of the genome. Indeed, that is so. On those viral genomes in which replication does initiate at each end, the template duplexes fall apart when the replication complexes meet. Each replication complex then may continue along its template strand to completion.

When initiation occurs at only one end of the viral genome, the complementary parental strand is completely displaced by the replication process. What then happens to the displaced complementary strand? Since the viral genome has inverted terminal repeated sequences, the repeat sequences at each end of the displaced strand can base-pair to form a "panhandle" structure (Fig. 17.4). Since the short double-stranded DNA segment comprising the panhandle resembles the end of the genome and contains the *ori* sequence, it can serve as an *ori* to initiate a new cycle of protein priming and continuous DNA synthesis, as described above (Fig. 17.4, steps 3 and 4).

As a consequence of the unique adenovirus initiation process, adenovirus DNA replication is always continuous, analogous to leading-strand synthesis during replication of cellular double-stranded DNA; there are no short Okazaki fragments. Moreover, since the virus uses pTP to prime replication, there is no RNA primer complementary to viral sequences that would need to be excised and somehow replaced by a corresponding DNA sequence.

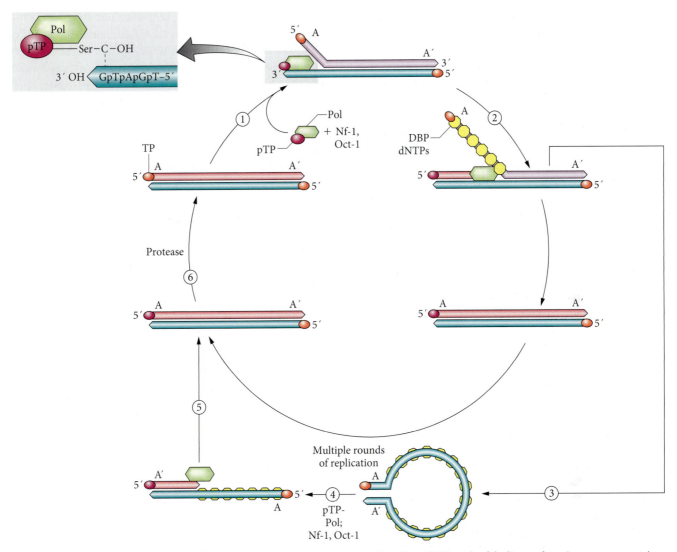

Figure 17.4 Replication of adenovirus DNA. Adenovirus genomes contain an inverted terminal repetition, each copy of which contains an *ori* sequence. (Step 1) Priming and ensuing replication begin with the binding of a complex, consisting of pTP and the virally encoded DNA polymerase (Pol), to the *ori* sequences, as facilitated by two cellular transcription factors, Nf-1 and Oct-1, which specifically bind to sequences adjacent to the *ori* core, as well as to the pTP/Pol complexes. The bound viral polymerase then initiates DNA synthesis by reading the guanine moiety at the 3′ end of the template strand. Doing so, it adds a cytosine residue to a serine side chain of the pTP priming protein. (Step 2) Synthesis then proceeds in the 5′-to-3′ direction along the DNA template. In this way, the terminal protein becomes linked to the DNA via a phosphodiester bond between a serine residue on the terminal protein and the 5′ hydroxyl of the terminal deoxycytosine on the DNA. As the template strand is read, the complementary strand is displaced from the parental duplex and coated by the single-stranded DBP, which helps to unwind the duplex structure. A cellular topoisomerase relieves the overwinding of the double-stranded DNA template. (Step 3) The ends of the linear adenovirus genome contain an inverted terminal repeated sequence, which contains the *ori*. Thus, the repeat sequences at each end of the displaced strand can base-pair to form a "panhandle" structure, in which the short double-stranded DNA segment comprising the panhandle contains the *ori* sequence and resembles the end of a linear double-stranded genome. (Steps 4 and 5) Consequently, it can serve as an *ori* to initiate a new cycle of protein priming and continuous DNA synthesis. If replication were to initiate independently and concurrently at each end of the genome in step 1, then the template duplexes would fall apart when the replication complexes met, and each replication complex might then continue along its template strand to completion. (Step 6) pTP is proteolytically cleaved to the smaller terminal protein during final virus assembly and maturation. Adapted from S. J. Flint, L. W. Enquist, R. M. Krug, V. R. Racaniello, and A. M. Skalka, *Principles of Virology: Molecular Biology, Pathogenesis, and Control*, 2nd ed. (ASM Press, Washington, DC, 2004), with permission.

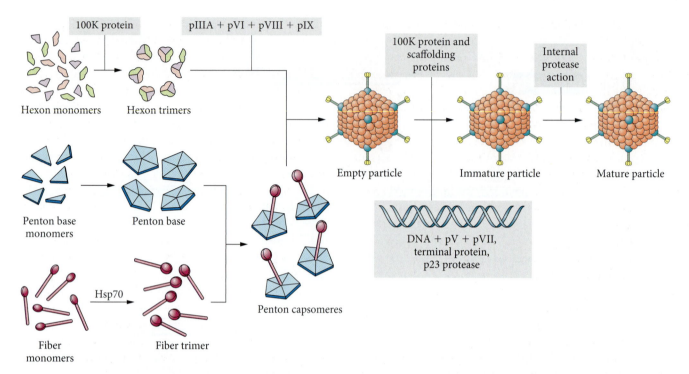

Figure 17.5 Assembly of adenoviruses. Hexons, pentons, and fibers embody the major adenovirus structural components. As in the example of the T-even bacteriophages (Chapter 2), each of these major adenoviral structural components is assembled via a separate pathway. The individual pathways converge only at final assembly. Assembly of hexons depends on the direct interactions of the hexon protein (polypeptide II) with the 100-kDa viral L4 chaperone or scaffolding protein, which participates in the construction of the capsid but is removed as the job is completed. The fibers, which are trimers comprised of three monomers of polypeptide IV, likewise assemble with the aid of a chaperone, in this instance the cellular Hsp70. Meanwhile, five molecules of polypeptide III assemble to form each of the penton bases. Then, the amino-terminal 40 residues of each the three monomers of polypeptide IV, which make up each fiber, are embedded into a penton base. Immature virus particles then form, which do not yet contain the viral genome or the core proteins but do contain the viral L1 52- and 55-kDa scaffolding proteins, which are thought to play a key role in facilitating the encapsidation of the viral DNA. These scaffolding proteins are eliminated as final assembly proceeds; this is likely facilitated by proteolytic degradation. Late in the assembly process, proteolytic cleavage also generates capsid proteins IIIa, Tp, VI, VII, VIII, and μ from their precursors, which are present in immature capsids. These cleavage reactions stabilize the particle and enable it to be infectious. They are carried out by the L3-encoded protease, which requires the genomic DNA as a cofactor. Modified from N. J. Dimmock, A. J. Easton, and K. N. Leppard, *Modern Virology*, 5th ed. (Blackwell Science Ltd., Oxford, United Kingdom, 2001), with permission.

Two final points regarding adenovirus DNA replication are to be considered. First, using a protein to prime replication seems to be a relatively simple means for solving the end replication problem. Indeed, as a consequence of the relatively simple adenovirus DNA-priming mechanism, adenovirus DNA replication requires many fewer proteins than are needed to replicate the DNA of polyomaviruses (Chapter 15). Yet using a protein to prime DNA replication is a solution to the end replication problem that is known to occur only in the cases of the adenoviruses and hepadnaviruses (Chapter 22). (Do you have any thoughts as to why cells and other viruses might not avail themselves of this means for initiating DNA replication? Before reading further, give this question a moment's consideration.)

Second, we noted in Chapter 14 that a likely reason that cellular DNA replication needs to be primed is that priming is intrinsic to the proofreading mechanism, which is part and parcel of cellular DNA replication. Since the polyomaviruses and papillomaviruses use cellular DNA polymerase to replicate their DNA, replication in those instances is likely to be highly accurate. What then of the fidelity of adenovirus DNA synthesis, since these larger viruses encode their own DNA polymerase and use a protein primer to initiate DNA replication? Although the answer is not known with certainty, the adenovirus DNA polymerase does possess a built-in 3′-to-5′ exonuclease activity that excises mismatched bases from duplex DNA molecules.

ASSEMBLY

Adenovirus assembly takes place in the nucleus. Thus, since viral structural proteins are synthesized in the cytosol, they must be transported to the nucleus, where they are incorporated into nascent progeny virus particles. It is likely that the viral structural proteins enter the nucleus via the usual nuclear pore-mediated nuclear import pathway (Box 12.7).

As described above, the outer shells of adenovirus particles are composed of 252 multiprotein subunits, of which 240 are in the form of hexons and 12 are arranged as pentons (Fig. 17.1). In addition, fiber structures project from each of the pentons. These hexons, pentons, and fibers constitute the major adenovirus structural components. As in the example of the T-even bacteriophages (Chapter 2), each of these major adenoviral structural components is assembled via a separate pathway. The individual pathways converge only at final assembly (Fig. 17.5).

As explained above, each of the hexon subunits actually is a trimer of polypeptide II, rather than a hexamer. The hexon subunits nevertheless manifest pseudohexagonal symmetry because each of its three constituent molecules of polypeptide II contains two similarly shaped β-barrel domains at the base of the hexon. In Chapter 6, we noted how the β-barrel shape of picornavirus capsid proteins facilitates their close packing in picornavirus capsids. Likewise, the shape of the β-barrel domains of polypeptide II facilitates close packing within and between adenovirus hexons.

Interestingly, the assembly of hexons is dependent on the direct interactions of polypeptide II with the L4 100-kDa protein. (Recall that the L4 100-kDa protein also mediates the inhibition of cellular protein synthesis.) In the cytosol, the L4 100-kDa protein acts as a chaperone or **scaffolding protein** to mediate the trimerization of polypeptide II to form hexons. Moreover, the L4 100-kDa protein facilitates hexon nuclear transport and acts again as a scaffolding protein in the nucleus, where it aids capsid assembly. "Scaffolding proteins" are so termed because they participate in the construction of the capsid but are removed as the job is completed.

The fibers, which are trimers comprised of three monomers of polypeptide IV, likewise assemble with the aid of a chaperone, in this instance the cellular Hsp70. Meanwhile, five molecules of polypeptide III assemble to form each of the penton bases. Then, the amino-terminal 40 residues of each of the three monomers of polypeptide IV, which make up each fiber, are embedded into a penton base.

Next, immature virus particles form. These particles do not yet contain the viral genome or the core proteins but do contain the viral L1 52- and 55-kDa scaffolding proteins. These two proteins are differentially phosphorylated forms of the same translation product. They are thought to play a primary role in facilitating the encapsidation of the viral DNA rather than in the assembly of the capsid per se. This belief is based on the experimental finding that L1 52- and 55-kDa mutants make empty capsids.

Elimination of the scaffolding proteins occurs as final assembly proceeds and likely is aided by proteolytic degradation. Moreover, late in the assembly process, proteolytic cleavage also generates capsid proteins IIIa, Tp, VI, VII, VIII, and μ from their precursors, which are present in immature capsids. These cleavage reactions stabilize the particle and enable it to be infectious. They are carried out by the L3-encoded protease, which requires the genomic DNA as a cofactor.

Packaging of the adenoviral DNA is mediated by a packaging signal located near the left end of the genome, just upstream of the E1A transcription unit (Fig. 17.3 and 17.6). Notice that the packaging signal is comprised of repeats of short sequences, which overlap the early enhancers I and II and are close to the origin of replication. (Adenovirus enhancer I specifically stimulates expression of the E1A transcription unit, and enhancer II stimulates transcription of all transcription units.) Importantly, this positioning of the packaging signal is essential for encapsidation. Also, it causes the end of the genome with the packaging signal to enter the capsid first. The covalently attached terminal protein has no role in encapsidation. (The SV40 packaging signal likewise is located in that virus's regulatory region, which contains its *ori*, enhancer, and early and late promoters. These common features of the packaging signals of the unrelated polyomaviruses

Figure 17.6 Packaging of the adenoviral DNA is mediated by a packaging signal located near the left end of the genome, between the left inverted terminal repeated sequence (ITR) and the E1A transcription unit (Fig. 17.3). The packaging signal is comprised of repeats of short sequences, which overlap the early enhancers I and II and are close to the origin of replication. Adapted from S. J. Flint, L. W. Enquist, R. M. Krug, V. R. Racaniello, and A. M. Skalka, *Principles of Virology: Molecular Biology, Pathogenesis, and Control*, 2nd ed. (ASM Press, Washington, DC, 2004), with permission.

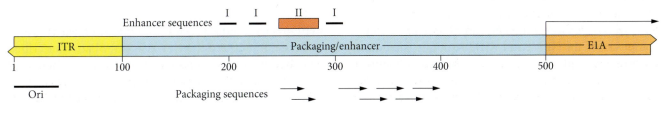

and adenoviruses suggest that they embody some functional significance. What might that be? Read further.)

Although assembly appears to begin with the generation of empty capsids, capsid formation does not occur if the viral genomes contain a defective packaging signal. This implies that capsid formation may be triggered by an interaction of the pentons and hexons with the viral DNA, in particular with the packaging signal. Since the end of the DNA with the packaging signal is the first to enter the capsid, one possibility is that proteins that interact with the packaging signal, and perhaps with the *ori*/regulatory region, might play a role in triggering assembly and facilitating packaging.

At the very least, we expect that there should be one or more proteins that recognize the packaging sequence. One such protein is the L1 52/55-kDa protein, which facilitates encapsidation, as noted above. The IVa2 protein also is needed for DNA packaging. IVa2 interacts with the L1 52/55-kDa protein and also binds to the packaging sequence.

The mechanism of adenovirus release from cells is not yet known.

EVASION OF HOST DEFENSES

Adenoviruses have multiple means by which they undermine host immune defenses. We consider adenoviral countermeasures against both innate and adaptive immune defenses, since the effectors of innate immunity provide the first tier of immune defenses against invading pathogens, whereas specific adaptive immune responses are usually needed to resolve the infection (Chapter 4).

When the innate immune system senses the presence of an invader, local macrophages and natural killer (NK) cells secrete chemical effectors called cytokines, which help the host to check the infection (Chapter 4). Multifunctional tumor necrosis factor (TNF) is one of the major cytokines produced by the innate response. One of its effects is to induce apoptosis when it binds to TNF receptors on virus-infected cells. Likewise, the Fas ligand on cytotoxic T cells (CTLs) (which are mediators of adaptive immunity) initiates apoptosis when it binds to Fas on virus-infected target cells. To counter these host defenses, the E3 region of many adenoviruses encodes small proteins that inhibit the ability of both the TNF receptor and Fas to transmit extrinsic apoptosis-inducing signals into the target cell. For example, the E3-encoded receptor internalization and degradation (RID) protein, which is composed of the RIDα and RIDβ subunits (previously referred to as the E3 10.4- and 14.5-kDa proteins), induces the endocytosis and subsequent lysosomal degradation of Fas, in that way preventing the apoptosis-activating

signals that Fas transmits (Fig. 4.25). An additional E3 protein, the 6.7-kDa protein, which is expressed on the cell surface, forms a complex with RID that down-modulates two cell surface receptors in the TNF receptor superfamily, namely, TRAIL receptor 1 and TRAIL receptor 2. (TRAIL is an acronym for "TNF-related apoptosis-inducing ligand.") Moreover, RID prevents the mobilization of phospholipase A2, which results in reducing the secretion of proinflammatory cytokines.

Here are a couple of points to note before moving on. First, there are two cellular signaling pathways to apoptosis. One is the extrinsic pathway that is triggered when a proapoptotic ligand (e.g., TNF-α or Fas ligand) binds to its receptor (e.g., TNF-α receptor or Fas) at the cell surface. The other is the intrinsic pathway that is triggered in response to internal cues, such as DNA damage or inappropriate cell divisions. With that distinction in mind, the means by which E3 proteins block the apoptosis that would result from *extrinsic* signals transmitted by host cytokines and T cells are in addition to the ways discussed above by which E1A, E1B, and E4 proteins block apoptosis that would result from the *intrinsic* mitogenic activity of E1A. Also, the E1B 19-kDa protein, as a Bcl-2 homologue (see above), blocks apoptosis induction via extrinsic pathways (e.g., TNF-activated cell death), as well as by the intrinsic p53-mediated pathway. Second, as will be seen below, E3 proteins act in yet additional ways to block cell-mediated adaptive immune responses. The impressive varieties of ways by which E3-encoded proteins act to prevent apoptosis are depicted in Fig. 4.25 and 4.26.

Adenoviruses also counter the antiviral effects of the alpha/beta interferons (IFN-α/β), which, like macrophages and NK cells, are effectors of innate immunity. Interestingly, in contrast to other adenovirus stratagems to undermine host defenses, all of which involve virus-encoded proteins, the anti-IFN stratagem involves a short RNA segment, VAI (VA stands for "virus-associated"). VAI acts by forming a double-stranded duplex that binds to the double-stranded RNA-binding site on PKr, a major enzyme effector of the IFN-induced antiviral state (Chapter 4). This binding is thought to block activation of PKr by authentic double-stranded RNA. The biological significance of VAI RNA is shown by adenovirus mutants which are defective for VAI RNA and consequently generate reduced viral yields. (Activated PKr impairs viral and cellular translation by phosphorylating translation initiation factor eIF2α, thereby diminishing viral yields [Chapter 4].) Interestingly, VAI and VAII are transcribed by RNA Pol III, as described above. (The VAI and VAII RNA Pol III transcripts are marked by paired vertical lines in Fig. 17.3.)

Next, we direct our attention to adenoviral stratagems for undermining T-cell-mediated defenses. Such defenses are especially important in regulating infections by pathogens that establish persistent or latent infections, such as adenoviruses and herpesviruses (Chapter 18). (See the proviso below and Box 17.2 concerning the issue of whether adenoviruses actually establish clinical latency.) Thus, it might be expected that adenoviruses succeed in nature at least in part because of their ability to undermine cell-mediated immune defenses. Indeed, that is the case, and one means by which they do so is via the E3-mediated down-regulation of Fas, as described above. In addition, they also impair the MHC-I antigen presentation pathway (Chapter 4). (CD8 killer T cells specifically attack target cells that display foreign peptides presented by MHC-I molecules.) Adenoviruses have at least two ways by which they impair antigen presentation via the MHC-I pathway (Fig. 4.24). First, E1A proteins block transcription from the MHC-I promoter. Second, the E3 19-kDa glycoprotein (gp19) contains at its C terminus a dilysine motif, which is a classic endoplasmic reticulum recycling/retention determinant. The gp19 protein binds to MHC-I molecules, thereby causing them to be recycled to, or retained in, the endoplasmic reticulum. In addition, this protein also binds to TAP (Chapter 4), preventing its association with MHC-I molecules and, consequently, MHC-I presentation of viral peptides.

CLINICAL SYNDROMES

The endemic adenovirus serotypes include Ad1, Ad2, AD3, Ad5, Ad6, and several others. These serotypes infect mainly children, in whom they are a significant cause of respiratory disease, as well as serious infections of the gastrointestinal tract and the eye. Infections with the endemic adenovirus serotypes indeed are common, as more than 80% of the human population is infected early in life, with the peak incidence of infection occurring between the ages of 6 months and 5 years.

In contrast to the endemic adenovirus serotypes, which infect mainly children, the remaining adenovirus serotypes seem to have an epidemic epidemiology and can infect anybody who has not been infected with them previously. Some of these adenoviruses are a particular problem for the military, as noted above.

The clinical syndromes resulting from adenovirus infections reflect the pattern of virus entry and dissemination in the host. Adenoviruses can enter the host in the form of aerosolized droplets generated by a cough or sneeze, or via saliva, or via the fecal-oral route. Also, since adenoviruses are resistant to drying, they can be spread by inanimate objects, such as towels. Adenoviruses also are resistant to chlorine, and thus, they may be spread in swimming pools. Consequently, adenoviruses can infect the respiratory and alimentary tracts and the conjunctiva. Most enteric adenovirus infections, and some respiratory infections, are subclinical. However, viremia may occur following local replication of the virus, resulting in spread to visceral organs, a relatively uncommon occurrence in immunocompetent individuals.

Adenoviruses once were mistakenly believed to cause the common cold, which is now known to be caused mainly by rhinoviruses and coronaviruses. Yet adenoviruses too may cause cold-like symptoms. In addition, they cause **pharyngitis**, particularly in young children. This illness is characterized by fever, cough, nasal congestion, tonsillitis, and inflamed throat. Adenoviruses also cause laryngitis, bronchiolitis, and **croup** (a condition characterized by acute obstruction of the larynx, resulting in loud coughing, hoarseness, and stridor [a harsh high-pitched wheezing sound caused by obstruction of the air passages]). Adenoviruses also may cause a disease in children that resembles the whooping cough caused by the bacterium *Bordetella pertussis* (Latin for "severe cough"). It is characterized by classic whooping cough paroxysms (i.e., a sequence of repetitive coughs followed by an aspiratory whoop). The adenovirus illness is prolonged and constitutes a true viral pneumonia.

Adenoviruses are one of several viruses (e.g., influenza virus, herpesviruses, and hantaviruses) that cause ARDS, which is a life-threatening condition characterized by fever, cough, pharyngitis, and inflamed lymph nodes of the neck (cervical adenitis). Strangely, adenoviruses give rise to ARDS mainly in military recruits; the condition is rare in nonmilitary settings (Boxes 17.1 and 17.3). Adenovirus ARDS is caused primarily by serotypes 4 and 7. The latter serotype is the most pathogenic human adenovirus.

Any of the common adenovirus serotypes may cause **pneumonia** in young children. Adenovirus pneumonia is often severe and may be fatal. ARDS in military recruits also occasionally progresses to pneumonia. As might be expected, adenovirus pneumonia may be life threatening in acquired immunodeficiency syndrome (AIDS) patients and organ transplant recipients.

Adenoviruses may cause gastrointestinal infections as well as respiratory infections, despite the fact that the alimentary tract presents a harsh environment to most would-be pathogens (Chapter 3). Adenoviruses can infect the alimentary tract because they are resistant to its low pH, gastrointestinal proteases, and the emulsifying action of bile detergents, which generally make the alimentary tract an uninviting environment for microbes. But even so, it is not clear how adenoviruses breach the immune system's defensive barriers that protect the gastrointestinal tract. At any rate, adenoviruses are a major cause of acute **viral**

gastroenteritis, accounting for 15% of all hospitalized gastroenteritis patients. Adenovirus serotypes 40, 41, and 42 may cause epidemic gastroenteritis in children. Interestingly, these adenovirus serotypes do not appear to cause fever or respiratory disease, as they appear to have a tropism for the alimentary tract.

Ocular diseases caused by adenoviruses include **pharyngoconjunctival fever** and **epidemic keratoconjunctivitis**. The former disease occurs either sporadically or in outbreaks that may be traced to a common source, such as a swimming pool. It involves a systemic infection associated with a mild to painful sore throat and high fever, as well as conjunctivitis. It is usually caused by Ad3 and occasionally by Ad4 or Ad7. A distinguishing feature of the latter condition is corneal involvement. Interestingly, epidemics often initiate in ophthalmology outpatient clinics by direct contact with contaminated instruments. This condition is most commonly caused by Ad8 and Ad19. Irritation of the eye, as might be caused by a foreign body (e.g., dust or debris), is an apparent risk factor. Indeed, the disease, once known as "shipyard eye," may be an occupational hazard for certain industrial workers. A striking outbreak of epidemic keratoconjunctivitis occurred among shipyard workers at Pearl Harbor, in which there were more than 10,000 cases during 1941 and 1942.

Adenoviruses also have been associated with urogenital infections, causing cervicitis, urethritis, and cystitis (inflammation of the bladder). Adenoviruses commonly establish persistent infections in the kidney, and virus may be shed in the urine for months to years.

Apropos the above, adenoviruses are often said to persist indefinitely as latent infections in lymphoid tissue, adenoids, and tonsils (but see the proviso at the end of this paragraph). Importantly, reactivated adenovirus infections might give rise to disease in immunocompromised adults and children. Such immunocompromised individuals include organ transplant recipients and AIDS patients. The mechanism of adenovirus latency is not known, but host immunity is believed to play a role in its maintenance. For example, adenovirus types 34 and 35 originally were isolated from renal transplant recipients on immunosuppressive regimens, and they often are detected in AIDS patients. Importantly, adenovirus infections in immunocompromised patients may lead to pneumonia, hepatitis, and other symptoms of more generalized infection and indeed can be life threatening. (Proviso: The issue of whether adenoviruses may establish latency following initial infection is not well resolved. Adenoviral DNA indeed has been detected in peripheral blood monocytes, lungs, tonsils, and the upper airways. However, these samples did not contain replicating virus, even after extensive cell culture in vitro. Thus, it is unclear what percentage of adenovirus infections are due to reactivation of latent virus as opposed to primary infection. Also, see Box 17.2.)

The discovery that adenoviruses induce neoplastic transformation of some cell types in culture and malignant tumors in newborn rodents raised concern that adenoviruses might play a role in human cancer. However, it soon appeared that this is not the case since adenovirus DNA sequences have not been detected in human cancers. But see Box 17.8.

RECOMBINANT ADENOVIRUSES

Gene Therapy

The goal of **gene therapy** is to transmit a gene to a patient who either is lacking that gene or expresses a defective version of it. Viruses have been evaluated extensively for their potential as therapeutic gene delivery vectors, since they are gene delivery systems by their very nature.

When engineering a viral vector, the gene of interest is inserted into the viral chromosome, in association with a suitable promoter that might mediate its expression. In some instances, a viral sequence must be deleted to make room on the viral chromosome for the transduced gene and its promoter, as well as to impair the potential virulence of the vector. As medium-sized viruses, adenoviruses are able to accommodate large segments of foreign DNA, without the need for deleting a portion of their genomes. But if need be, **adenovectors** can be made replication defective, usually by deleting an E1 gene (see below).

Adenovirus gene delivery vectors were initially developed to treat **cystic fibrosis**, a lethal genetic disease affecting an epithelial cell ion channel protein, the cystic fibrosis transmembrane conductance regulator. Indeed, gene therapy for cystic fibrosis has been a major goal of gene therapy research for more than a decade, and adenovirus vectors have been the delivery systems of choice, since they naturally target the airway epithelium.

Unfortunately, although adenovirus vectors have been highly effective in transferring the cystic fibrosis transmembrane conductance regulator gene in cell culture, the efficiency of gene transfer in animals and humans has been low, possibly because the airway epithelium has evolved to avoid the uptake of potential pathogens, including adenoviruses. Yet experimental studies currently (2008) are in progress to develop optimal protocols for using adenovirus gene delivery vectors to treat cystic fibrosis. This is so notwithstanding the disastrous outcomes of some clinical trials involving viral gene vectors (Box 17.9). Indeed, 240 clinical trials involving recombinant adenoviruses were initiated between 1989 and 2004. Thus, the examples noted above and those that follow are but a small fraction of the total that involved adenovectors. These trials

BOX 17.9

Despite the promise of using viral vectors for gene therapy, several high-profile incidents, beginning in 1999, nearly brought an end to these efforts. One such case, involving an adenovirus vector, occurred during a 1999 clinical trial at the University of Pennsylvania. This tragic incident concerned teenager Jesse Gelsinger, who suffered from ornithine transcarbamylase deficiency, a rare metabolic disorder that is generally fatal. However, since Gelsinger's cells were a mosaic for the condition, his illness could be controlled by a low-protein diet and a drug regimen. Thus, the medical investigators involved in this incident knew that Gelsinger did not actually need the gene therapy. However, Gelsinger answered the call for volunteers in the hope that success might free him from his dependence on his demanding drug regimen. Ironically, the purpose of this clinical trial was to determine the vector dosage that might be high enough to get the ornithine transcarbamylase gene to work, but low enough so as not to cause serious side effects.

To the great dismay of all involved, following the initiation of gene therapy, Gelsinger quickly developed jaundice, kidney failure, lung failure, and brain death. Gelsinger's tragic death was the first in humans that was related to gene therapy. Its cause was not entirely clear, but it probably was not due to the gene therapy per se. Instead, the vector dosage may have been high enough to trigger an overwhelming inflammatory reaction (Chapter 4). Prior to this incident, adenovirus vectors had been used in one-fourth of all gene therapy trials. Moreover, in this particular study, the vector previously had been tested in mice, monkeys, baboons, and one human, all at the same dosage as the one Gelsinger received. The major side effect seen earlier at that dosage was a mild liver inflammation that disappeared on its own. However, a 20-fold-higher dosage caused severe, fatal liver inflammation in three monkeys, which perhaps should have been viewed as a warning.

This episode led to extensive discussion of the issues that ought to be considered in the scientific and ethical review of clinical research proposals, which must occur before clinical studies involving human subjects can be carried out. One obvious issue is to weigh the risks faced by the test subjects against the potential benefits to society. Also, since potential financial rewards of successful clinical outcomes can be enormous, care must be taken that they not cloud the judgment of the researchers. Note that there were no indications whatsoever that the judgments of the researchers in the Gelsinger incident were affected by self-interested considerations.

---

included cancer therapies and prophylactic human immunodeficiency virus (HIV) vaccines (as described below), in addition to therapies for genetic deficiencies.

Despite the scope of the efforts noted above, no gene therapies of any sort have yet been approved for use in the United States or Europe. Moreover, it has become extremely difficult to undertake pilot gene therapy trials in the United States and Europe, since these studies require biotechnology and pharmaceutical companies that are able and willing to invest the enormous sums of capital needed to bring a gene therapy from conception to fruition. To do so requires assembling large teams of basic scientists and physician scientists, as well as lawyers who know how to navigate the complex regulatory constraints that have been imposed. And, bear in mind the great odds against a successful outcome. (See Chapter 20 for an account of an initially promising gene therapy trial using a retrovirus-based vector that ultimately had an unfortunate outcome.)

Cancer Therapies

A promising strategy for treating cancer, which is based on gene therapy, aims to correct specific genetic lesions that lead to carcinogenesis. For example, more than 50% of all human cancers contain mutations in the p53 tumor suppressor gene. For this reason, most gene therapies for treating cancer have been aimed at restoring p53 function. This anticancer strategy indeed is supported by studies in vitro, which show that introducing wild-type p53 into human cancer cells which lack functional p53 induces the cells to undergo apoptosis. Importantly, expression of the recombinant p53 gene need only be transient, since apoptotic cell death occurs within 24 h of wild-type p53 expression both in vitro and in vivo. A noteworthy advantage of this gene therapy approach is that it avoids the toxic side effects of conventional chemotherapy.

Apropos the subject of this chapter, the most extensively studied p53-based anticancer therapeutic agents are adenovectors that express the wild-type p53 tumor suppressor gene. Phase I and phase II clinical trials of such adenovirus vectors for cancer therapy indeed have been conducted in the United States. One trial involved a replication-defective adenovector, in which the viral E1 gene was replaced with a wild-type p53 gene driven by a cytomegalovirus promoter. The vector was injected directly into the tumors. The phase I study included 25 patients with advanced nonsquamous cell lung carcinomas that previously were resistant to conventional treatment. In these individuals, the adenovector was safe and stabilized the disease for as long as 24 months. Moreover, the tumors regressed in two of the patients. In the phase II study, 12 of 17 patients showed no evidence of tumors 3 months after treatment ended. This negative rate of 70% compares very favorably indeed with that of 17% reported for patients receiving conventional chemotherapy combined with radiation therapy. The success rate for the adenovector therapy might have been greater still if it too had been combined with radiation therapy. Promising results also were seen in phase I and phase II trials in which a p53 adenovector was administered to patients with bronchoalveolar cell lung cancer. Generally, these adenovector therapies have shown little toxicity. Serious adverse effects occurred in fewer than 5% of treated patients.

It hardly seems likely that all the cells in the tumor can be transduced by the procedure described above. Thus, considering the high success rates achieved with this stratagem, perhaps it is not necessary to transduce all the cells in the tumor to effect a cure. If that is so, how then might the cures be explained? One possibility is that a "bystander effect" might be at work to destroy nontransduced cells in the tumor. Such a bystander effect might be explained by several plausible mechanisms. For example, local cell death might cause the innate immune response to release antitumor cytokines such as TNF-α, or it might initiate long-lived T-cell-mediated immunity to tumor cells, or it might activate NK cells that express antitumor activity, as described below.

Another adenovector-based gene transfer approach to treating cancer uses vectors that express the tumoricidal cytokine TNF-α or TRAIL (see above). Yet another gene therapy approach is based on the finding that pancreatic cancer cells are particularly sensitive to IFN-α gene transfer. IFN-α expression has a direct cytotoxic effect on pancreatic cancer cells and also elicits an indirect antitumor effect. In animal experiments, the indirect effect led to regression of tumors at sites distant from the injected tumor. How might that indirect effect be explained? Well, it was due at least in part to the ability of IFN-α to stimulate NK cells (Chapter 4). NK cells are now being found to play a primary role in destroying cancer cells in the body, which they recognize by the presence of certain molecules that identify them as tumor cells. NK cells are readily activated by IFN.

Note that retroviral vectors have been developed for similar purposes (see Chapter 20). However, adenoviral vectors have several advantages over retrovirus vectors. They transduce nondividing as well as dividing cells, whereas retroviruses can only transduce dividing cells. Moreover, adenovectors achieve higher expression levels and can be produced to higher titers and on a larger scale.

Another adenovirus-based approach to antitumor therapy is not founded on gene therapy in the strictest sense. Instead, it is a **targeted therapy approach**, based on the premise that cytolytic adenoviruses which have been engineered to contain a deletion in the E1B region should be able to replicate and thereby cause cellular lysis, but only in cells lacking functional p53. (Why are E1B mutants able to replicate only in cells lacking p53 function? Recall that E1A expression induces inappropriate cell cycling, which would trigger p53-mediated apoptosis in the absence of E1B.) Thus, these E1B mutants should selectively replicate in the tumor cells, selectively releasing large amounts of virus that can infect and cause lysis of other cells in the tumor. E1B-defective adenoviruses indeed were antitumorigenic when administered to nude mice that carried transplanted human tumors. Moreover, an E1B-defective adenovirus has been developed for clinical application by Onyx Pharmaceuticals. The Onyx virus indeed preferentially destroyed a broad spectrum of p53-deficient human tumor cells that were transplanted into nude mice. Moreover, the tumor cytotoxicity was found to result from lytic virus replication, as expected. These promising experimental results provide hope that this targeted therapy approach may be clinically effective.

Considering the promise of the targeted therapy approach, other targeting strategies have been developed to target adenovectors to other particular target cells. One approach to directing the viral vector to particular cells or tissues is to specifically alter the cell-surface-binding specificity of the vector. This can be done by generating a hybrid protein that links an antibody Fab fragment specific for the adenovirus fiber protein to a ligand that specifically binds to a receptor on the target cell. For example, one such hybrid protein contains an adenovirus-specific Fab fragment linked to the epidermal growth factor (EGF). This protein causes the adenovector to preferentially attach to certain cancer cells that overexpress the EGF receptor. For example, the A431 epidermoid carcinoma cell line greatly overexpresses the EGF receptor. Adding the hybrid proteins to an adenovector enhanced the transduction efficiency of A431 cells 16-fold, compared to infection with the untreated adenovector.

Currently, there is great interest in **antiangiogenic** approaches to cancer therapy. Antiangiogenesis stratagems are based on the fact that when cancerous tumors become larger than 1 to 2 mm, they must promote the generation of new blood vessels (angiogenesis) to sustain tumor viability and further tumor growth. In point of fact, one of the most notable characteristics of tumor growth is the absolute requirement for the tumor to increase its blood supply through angiogenesis.

Tumor cells promote angiogenesis by secreting angiogenic factors such as the vascular endothelial growth factor (VEGF). VEGF in turn binds to receptors on endothelial cells of the local vasculature, causing it to generate new vessels which grow in the direction of the VEGF source. As you might then expect, there is a correlation between VEGF expression by a cancer and its malignancy potential.

Normally, angiogenesis is regulated by the balance between angiogenesis activators and angiogenesis inhibitors. Endostatin is the most effective of the known angiogenesis inhibitors at blocking tumor-induced angiogenesis. For this reason, endostatin has been widely used in antiangiogenesis-based cancer therapies. However, cancer therapies based on soluble angiogenesis inhibitors are hampered by the difficulty of sustaining high enough levels of the agent locally to control new vessel formation and thus tumor growth. Nonviral DNA vectors and replication-deficient viral vectors are similarly hindered, due mainly to low transduction efficiency and poor distribution throughout the solid tumor mass.

In the hope of improving on these endostatin-based therapies, replication-competent recombinant adenovectors, which express the gene for endostatin, were constructed and then evaluated in nude mice for their potential efficacy against malignant tumors. In one study, the recombinant endostatin-adenovector was injected into established human lung carcinoma xenografts (i.e., grafts between species), and a low dose of the chemotherapeutic agent gemcitabine was administered intraperitoneally. This combination treatment produced no ill side effects and resulted in a marked decrease in microvessel density around the tumors and in their diminished growth. It also inhibited the generation of new tumors.

As noted above, tumors promote angiogenesis by secreting the proangiogenic factor VEGF. Thus, another antiangiogenesis-based approach to cancer therapy aims to nullify VEGF. One such stratagem involved a soluble form of the VEGF receptor which directly sequesters VEGF, and indeed it inhibited the mitogenic response to VEGF in cell culture. However, the soluble protein is not an effective cancer therapy, since, as in the endostatin instance, it has not been possible to sustain levels in vivo that are high enough to effectively inhibit the generation of new tumor vasculature.

Still, in recognition of the potential of an anti-VEGF-based cancer therapy, an attempt was made to surmount the obstacle associated with using the soluble VEGF receptor. Specifically, an adenovector that carries an expression cassette for the soluble VEGF receptor was developed. Additionally, this vector contained an E1B deletion that restricted its replication to p53-deficient tumor cells (see above). Thus, the advantage of the adenovector over the soluble form of the VEGF receptor is that the vector might replicate specifically within the tumor and thus locally amplify the initial gene dose, so that clinically effective levels of the soluble VEGF receptor might be generated in the vicinity.

This vector indeed caused a distinct reduction of tumor growth in nude mice that carried a transplanted subcutaneous human tumor, and consequently, it increased long-term survival rates. Moreover, in support of the rationale on which it was based, it was a more effective antitumor agent both in vitro and in vivo than the original adenovector from which it was derived and which contained only the E1B lesion. In addition, the vector was even more effective when administered together with the chemotherapeutic agent 5-fluorouracil. (For the sake of completeness, note that other successful antiangiogenesis-based strategies to cancer therapy have been developed. For example, Genentech developed a humanized monoclonal antibody against VEGF that has proven to be effective against a variety of cancers in clinical trials.)

Despite the promise of some of the virus-based therapeutic stratagems enumerated above, no gene therapies of any sort, including those involving recombinant viruses, have yet been approved for use in the United States or Europe. Yet the Chinese company Shenzhen SiBono GenTech has developed and is now marketing an adenoviral vector that expresses wild-type p53 to treat squamous cell cancers of the head and neck. The Chinese approach is based upon much earlier work carried out in the United States and elsewhere.

Vaccines: HIV and Influenza Virus

During the past few decades, adenovirus-based vector systems have been developed as potential vaccines against many infectious diseases. In particular, replication-defective recombinant adenoviruses have been used successfully to vaccinate against pseudorabies virus in mice and pigs, foot-and-mouth disease virus in pigs, and bovine herpesvirus in rats.

Recent attempts at developing an **HIV vaccine** used adenovectors in the hope of eliciting an effective CTL response against HIV. The rationale for specifically activating a CTL response is that CTLs play a primary role in controlling most virus infections. Moreover, HIV characteristically establishes a persistent infection, and CTLs are particularly important in controlling persistent virus infections. The rationale behind using a viral vector to induce the CTL response is that CTL responses are induced against *intrinsic* antigenic peptides that are presented via the MHC-I antigen presentation pathway (Chapter 4).

A CTL-mediated immune response against HIV indeed was elicited in nonhuman primates immunized with

Has the following occurred to you? Considering that adenoviruses are prevalent in the human population, how might the efficacy of an adenovector be affected if the recipient already has immunity to the vector virus? To answer this question, the adenovector-based anti-HIV vaccine was administered to monkeys that either expressed or did not express adenovirus antibodies. In those monkeys that previously were exposed to adenovirus, the adenovector induced an anti-Gag response that was diminished but not entirely abolished. These findings are consistent with the notion that immunity does not block infection per se, a factor that works against the prospect of an effective HIV vaccine (Chapter 21).

replication-incompetent adenovectors that expressed the HIV Gag or Env gene (Chapter 21). Importantly, this CTL response controlled viremia and prevented destruction of CD4 T cells (the primary target cells of HIV) following challenge with related simian AIDS viruses. However, for reasons discussed in Chapter 21, it may be impossible to develop an AIDS vaccine that might prevent AIDS in humans. (Vaccinia virus vectors that expressed the HIV Gag were developed for the same purpose as these adenovectors. Although the vaccinia virus vectors were effective in animal trials, they were less so than the adenovectors [Box 17.10].)

As discussed in Chapter 12, recent outbreaks of avian H5N1 influenza viruses have raised concern over a possible H5N1 pandemic in humans. Since there might be little warning of such a pandemic, it is important that procedures be in place to develop and produce an effective H5N1 vaccine more quickly than is possible using current egg-based procedures (Chapter 12 and below). Moreover, it is important that such a vaccine have broad specificity against a variety of H5N1 strains, since the vaccine might not precisely match the pandemic strain. Apropos this point, current influenza vaccines, based on inactivated

virus, are strain specific. At this time (spring 2009), there also is concern over an H1N1 swine influenza pandemic, and the above points are also applicable to H1N1 swine influenza virus strains. With these concerns in mind, consider the following.

A recombinant adenovector that encoded the hemagglutinin (HA) gene from a swine H3N2 influenza virus was generated. The vector then was evaluated to determine whether it might induce protective immunity against a lethal challenge with a homologous H3N2 influenza virus strain and a heterologous H1N1 influenza virus strain. As expected, mice inoculated with the recombinant adenovector generated *antibodies* that neutralized the homologous virus but not the heterologous virus. Interestingly, however, immunized mice were yet protected against a lethal challenge with the heterologous virus. Since the vaccine did not induce neutralizing antibodies against the heterologous virus, nor did it prevent initial infection per se, the experimental results suggest that cross-protective cell-mediated immunity may have protected the vaccinated mice from the lethal challenge with the heterologous virus. In this regard, it is a well-known and well-understood fact that CTLs are less specific than antibodies (Chapter 4).

All influenza vaccines thus far approved for use in humans are prepared by growing the influenza viruses in embryonated chicken eggs. Virus then must be purified and treated with detergent to release the HA and neuraminidase proteins from the viral envelope. The HA and neuraminidase proteins then must be separated from the nucleoproteins, since these are pyrogenic (fever-inducing) molecules. Overall, the process is expensive and slow. Adenovectors have the potential to improve upon the currently inefficient means of producing influenza vaccines.

Suggested Readings

Saban, S. D., M. Silvestry, G. R. Nemerow, and P. L. Stewart. 2006. Visualization of α-helices in a 6-angstrom resolution cryoelectron microscopy structure of adenovirus allows refinement of capsid protein assignments. *J. Virol.* **80:**12049–12059.

Wiethoff, C. M., H. Wodrich, L. Gerace, and G. R. Nemerow. 2005. Adenovirus protein VI mediates membrane disruption following capsid disassembly. *J. Virol.* **79:**1992–2000.

18

Herpesviruses

INTRODUCTION

The *Herpesviridae* are a family of large, structurally complex viruses that contain double-stranded DNA genomes. Of note, the family includes several medically important viruses that are responsible for a number of common human ailments of varying levels of severity. In addition, the herpesviruses have a particularly interesting lifestyle, in which they establish latent infections that may "spontaneously" reactivate over the lifetime of an infected individual.

The ability of herpesviruses to establish lifelong latent infections has several important consequences. First, it enables infected individuals to be lifelong reservoirs of the virus. Second, it figures heavily in the nature of herpesvirus disease syndromes.

The eight known human herpesviruses are herpes simplex virus types 1 and 2 (HSV-1 and HSV-2), varicella-zoster virus (VZV), Epstein-Barr virus (EBV), human cytomegalovirus (HCMV), human herpesvirus 6 (HHV-6), human herpesvirus 7 (HHV-7), and human herpesvirus 8 (HHV-8). Together, these viruses cause cold sores and fever blisters, chicken pox and shingles, venereal disease, birth defects, brain and eye infections, mononucleosis, and several different cancers in humans. Indeed, herpesviruses were the first viruses to be associated with human malignancies. Importantly, herpesvirus infections are ubiquitous in the human population. They also are responsible for various animal diseases, including cancers. Yet in humans and animals, most infections are mild or show no signs of clinical disease.

The name "herpes" has its origin in a Greek word meaning "to creep," in reference to the spreading skin lesions associated with HSV (see below). Consistent with the fact that herpesviruses have circulated in humans since antiquity (see below), cold sores, as caused by HSV-1, have been well recognized for more than two millennia. Indeed, the Greek physician Hippocrates (460 BCE to 370 BCE) described them, and another Greek, Herodotus (484 BCE to 425 BCE), known as the "Father of History" in Western culture, noted the association between these lesions and fever.

Shakespeare too was no doubt aware of recurrent herpes lesions. If you can make sense of Shakespearean English, note that in *Romeo and Juliet* the Bard informs us that Queen Mab uttered "O'er ladies lips, who straight on kisses dream, which oft the angry Mab with blisters plagues, because their breaths with sweetmeats tainted are." Yet, while these early observers were unquestionably brilliant, Lowenstein was the first to recognize the viral etiology of herpes lesions in 1919, when he demonstrated that an inoculum from a human cold sore could produce lesions on a rabbit cornea that were similar to ones seen in human eye infections.

The ability of all known herpesviruses to establish latent infections in their natural hosts is a defining feature of this virus family (see Chapter 3). The initial herpesvirus infection, which is referred to as the **primary infection**, is acute and generally occurs in a cell type or tissue in which virus replication is efficient. The host counters with an immune response that may *appear* to have resolved the infection. Indeed, the host immune response does provide protection against reinfection. However, the immune response is not able to actually clear the herpesvirus from the host. Instead, the herpesvirus establishes a latent infection in a cell type or tissue, which is usually different from that of the primary infection, and in which the virus cannot replicate efficiently. The virus can then persist indefinitely in that tissue. Importantly, the productive infection might periodically reactivate to enable transmission to new susceptible individuals.

As an example of the above, HSV-1 establishes a primary infection in mucosal epithelia and a latent infection in the ganglia of the sensory nerve which innervates that particular epithelium (e.g., the ganglia of the trigeminal cranial nerve in the case of oral HSV-1 infections). When the latent virus reactivates, it can descend down the nerve to productively reinfect the epithelium, resulting in a recurrence of lesions at the site of the primary infection. Recurrent infections are usually less severe than the primary infection and may even be unapparent (because of immune memory responses). In any case, an infected individual can be a reservoir of the virus for life, as noted above. Historically, it was interesting, and initially paradoxical, that recurrent HSV infections occurred only in individuals who expressed anti-HSV antibodies. The paradox was resolved with the understanding of herpesvirus latency and reactivation.

Herpesviruses are large viruses, approximately 200 nm in diameter. Their double-stranded DNA genomes are linear and range in size from about 90×10^6 daltons (Da) to about 150×10^6 Da. Compare that to polyomavirus genomes (Chapter 15), which are about 3.2×10^6 Da, or to the larger adenovirus genomes that range from 20×10^6 to 25×10^6 Da. In terms of base pairs, the HSV-1 genome contains about 152,000 bp, compared to 5,200 bp for the polyomaviruses. Herpesvirus genomes encode about 70 to 165 different proteins, in comparison to the polyomavirus genomes, which encode 5 major proteins.

The herpesvirus DNA core is contained within an icosahedral capsid, which in turn is enclosed within a glycoprotein-containing lipid envelope. Notice that the herpesviruses are the first *enveloped* DNA animal viruses thus far mentioned in this book. However, they are not alone in that regard. The poxviruses (Chapter 19) and the hepatitis B viruses (Chapter 22) are enveloped as well. In addition, there is the African swine fever virus, the sole member of the family called *Asfarviridae* (Box 18.1). Between the capsid and the envelope there is a layer called the **tegument**, which contains virally encoded enzymes and transcription factors (see below).

The eight presently recognized human herpesviruses fall into three subfamilies: alpha, beta, and gamma. The alphaherpesviruses, which include HSV-1, HSV-2, and VZV, are characterized by their rapid reproductive cycle and, importantly, their ability to invade sensory neurons and establish latent infections in sensory ganglia. HSV-1 and HSV-2 are responsible for recurrent facial and genital **herpetic lesions**, whereas VZV causes chicken pox (varicella) and shingles (herpes zoster; a painful recurrence of a latent VZV infection that was manifest as chicken pox earlier in the patient's life). Whereas HSV-1 and HSV-2 infections may be painful, they generally are not life threatening. However, an infection that disseminates to the brain is potentially lethal, as is infection of an immunocompromised person or newborn. Moreover, HSV-1 infection of the eye can lead to blindness if the virus should spread to a deeper layer of the cornea.

BOX 18.1

African swine fever virus (ASFV) causes the most severe viral disease known in pigs and for that reason poses a serious threat to swine industries worldwide. Of note, it is the only DNA virus that can be classified as an arbovirus. ASFV is spread by ticks, and because of the long-distance transport of livestock and of their resident infected ticks, this virus is now established globally. ASFV is a very large virus, with a linear, double-stranded DNA genome comprised of 170,000 to 190,000 bp, depending on the isolate. Interestingly, there is no protective immunity in formerly infected pigs, which presumably explains why efforts to produce a vaccine against this important animal pathogen have so far not succeeded.

The beta- and gammaherpesviruses are lymphotropic. The two groups are distinguished from each other and from the alphaherpesviruses on the basis of their genome organization and pattern of replication. While HSV-1 is often taken to be the prototype of the family in general, HCMV is the prototype of the betaherpesviruses. Congenital HCMV disease is the most common congenital infection in humans, occurring in an average of 1% of all live births in the United States. This translates to 30,000 to 40,000 congenitally infected infants born each year in the United States. The risk for serious birth defects is especially high in infants born of mothers who were infected during pregnancy. This virus also is responsible for a form of mononucleosis in young adults. HCMV infection can be life threatening in individuals with impaired immune function (e.g., acquired immunodeficiency syndrome [AIDS] patients and immunologically suppressed organ transplant recipients). The other two betaherpesviruses, HHV-6 and HHV-7, generally cause mild early-childhood illnesses.

The two gammaherpesviruses, EBV and HHV-8, are notable (i) for their ability to establish latent infections of lymphoid cells and (ii) as the etiologic agents, or at least cofactors, of cancers in humans. Thus, while EBV is best known as the agent responsible for infectious mononucleosis, it also is associated with two malignant proliferative diseases, Burkitt's lymphoma and nasopharyngeal carcinoma. In addition, EBV has been associated with other human malignancies, including some T-cell lymphomas, Hodgkin's disease, posttransplant lymphoproliferative disease (PTLD) (Box 18.2), and some gastric carcinomas. HHV-8, also known as Kaposi's sarcoma-associated herpesvirus, is now known to be the cause of Kaposi's sarcoma, one of the characteristic opportunistic diseases associated with AIDS. The AIDS-related malignancies are distinctly different from their counterparts in non-AIDS patients in that nearly one-half of all cases of AIDS-related cancers are associated with either EBV or HHV-8. See Box 16.2 for more on the viral etiology of human cancers.

Importantly, immunologic dysfunction, as occurs in AIDS patients and in those undergoing organ transplants (Box 18.2), is a predisposing factor for herpesvirus-induced cancers, which can be aggressive and often fatal. Clearly, immunity acts to regulate herpesvirus latency, yet not every immunosuppressed patient develops tumors, perhaps in part because the viruses have evolved patterns of gene expression that prevent infected cells from proliferating out of control. (See Box 15.7 for more on why cancer does not develop in every individual infected with a tumor virus.)

STRUCTURE

Herpesvirus particles are comprised of a rigid icosahedral capsid, which is approximately 125 nm in diameter and 15 nm thick and is surrounded by a lipid membrane envelope (Fig. 18.1). The material between the envelope and the surface of the capsid is referred to as the tegument. Within the capsid, the virus DNA is folded into a highly condensed form. Yet, in contrast to the DNA genomes of the polyomaviruses, papillomaviruses, and adenoviruses, herpesvirus genomes are not associated with any DNA-binding proteins. The DNA-binding proteins of the polyomaviruses, papillomaviruses, and adenoviruses neutralize the sugar-phosphate backbones of their respective DNA genomes, thereby facilitating the tight packing of those genomes within their respective capsids. It is not clear how herpesvirus genomes might be neutralized, so as to facilitate their encapsidation. One possibility is that the DNA is packaged in association with small, positively charged molecules such as spermine.

Since HSV-1 is the prototypical herpesvirus, this virus and its proteins are used to exemplify herpesvirus structure.

BOX 18.2

PTLDs are among the most serious complications occurring after organ transplantation. The term PTLD actually refers to a wide range of abnormal lymphoproliferative conditions, ranging from a benign self-limited form to an aggressive, widely disseminated disease. Approximately 90% of these growths are of B-cell origin (see the text), and 90 to 95% of them contain EBV. Patients with PTLD appear to have a more aggressive clinical course, are less responsive to conventional treatments for lymphoma, and have poorer outcomes than immunocompetent individuals who develop malignant lymphomas. Importantly, EBV infection (primary or reactivation) is the main etiologic factor of PDV. The intensity of immunosuppression is another key underlying factor.

Figure 18.1 Electron micrograph of a frozen, hydrated HSV-1 particle showing the surface glycoproteins, lipid envelope, tegument, and icosahedral nucleocapsid. From F. J. Rixon, *Semin. Virol.* 4:135–144, 1993, with permission.

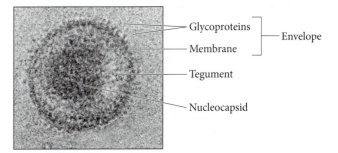

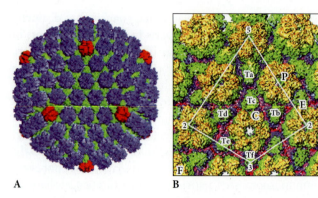

Figure 18.2 Three-dimensional reconstruction of HSV-1. (A) Shaded surface view of the HSV-1 capsid with 12 pentons (red), 150 hexons (blue), and 320 triplexes (green). (B) A radially color-coded surface representation of the outer capsid surface of the 8.5-Å (angstrom) reconstruction. An asymmetric unit, defined by the icosahedral fivefold, threefold, and twofold axes (indicated by the labels 5, 3, and 2), is enclosed by the white line. A penton (labeled 5), the three types of hexon (designated P, E, and C), and the six types of triplex (labeled T_a, T_b, T_c, T_d, T_e, and T_f) are shown. Modified from Z. H. Zhou et al., *Science* 288:877–880, 2000, with permission.

The HSV-1 icosahedral shell is composed of 12 pentons and 150 hexons (Fig. 18.2). (With respect to being assembled from pentons and hexons, herpesviruses capsids are like those of adenoviruses [Chapter 17] but differ from those of polyomaviruses [Chapter 15] and papillomaviruses [Chapter 16], which are assembled entirely from pentons.) As you might have expected, based on the precedent of other icosahedral viruses that are assembled from pentons and hexons, each of the 12 pentons in the herpesvirus capsid is located at a vertex of the icosahedron. The 150 hexons are located on the capsid edges and faces. Notice that there are three classes of hexons. Those designated type C are located within a face of the icosahedron. Those designated types P and E have differing locations along an edge. (These distinctions can be readily discerned by examining Fig. 18.2B, bearing in mind that the pentamers [in red] are at the vertices of the icosahedron.)

Each of the 150 hexons is comprised of six molecules of VP5, the major capsid protein (encoded by UL19), and six molecules of VP26 (encoded by UL35). (The designation "UL" refers to the position of the gene on the map; see below.) The latter proteins form a ring on top of the VP5 subunits. Eleven of the 12 pentons also are assembled from monomers of VP5, whereas the 12th "penton" is a cylindrical structure, called the **portal**, which actually is assembled from 12 monomers of the UL6 gene product. The portal has an axial channel through which DNA enters the capsid during assembly. It presumably also provides the channel through which the DNA leaves the capsid during the early stages of infection.

The capsids of adenoviruses and herpesviruses might seem to be similar, insofar as they each are icosahedrons assembled from hexons and pentons. Still, herpesvirus particles are considerably more complex than adenovirus particles, with their additional complexity manifested by their additional layers of proteins, the tegument, and the lipid envelope. Indeed, over one-half of the 80 or more different proteins encoded by herpesviruses are components of virus particles.

The following description of HSV-1 further indicates the structural complexity of the herpesviruses as a group. Six molecules of VP26, as well as of VP5, are associated with each of the hexons as was just noted. Also, so-called triplexes, which are heterotrivalent structures composed of proteins encoded by UL18 and UL38, lie above the VP5 icosahedral shell, where they connect the pentons and hexons in groups of three (Fig. 18.2B). Notice that there are six types of triplex (T_a through T_f in Fig. 18.2), differing with respect to their location in the capsid. Moreover, the tegument contains at least 15 different proteins, including enzymes (e.g., a functional ubiquitin-specific protease embedded in the large tegument protein) and transcription factors (Box 18.3). The envelope contains a dozen or more viral proteins and glycoproteins that promote binding and entry. Yet despite this complexity, and with the exception of the portal protein (the UL6 gene product), the subunits of the icosahedral capsid per se are comprised entirely of VP5, with the connections between VP5 monomers making up the floor of the capsid shell.

BOX 18.3

While the tegument transcription factor helps to initiate the infection (see the text), the function of the HSV-1 ubiquitin-specific protease is not yet clear. However, this activity is conserved across all members of the *Herpesviridae*, implying that it has an essential function, at least in vivo. In Marek's disease virus, a tumorigenic alphaherpesvirus that infects chickens, the ubiquitin-specific protease appears to be required to establish persistence in an immunocompetent host and also to contribute to malignant outgrowths, while Marek's disease virus mutants defective in this activity show only moderate impairment of replication in vitro. Apropos these findings, deubiquitinating enzymes have been associated with the control of antigen presentation via MHC-I molecules and with oncogenesis.

ENTRY AND UNCOATING

Much more is now known of how enveloped RNA viruses enter cells than of how enveloped DNA viruses (i.e., herpesviruses, poxviruses, hepadnaviruses, and African swine fever virus) accomplish the same. Nevertheless, clear distinctions between the entry mechanisms of the enveloped RNA viruses and those of the herpesviruses are recognized.

For the sake of comparison, we briefly review the entry mechanisms of the enveloped RNA viruses, before considering that of the herpesviruses. Enveloped RNA viruses generally have a single transmembrane glycoprotein, which undergoes a conformational transition that is triggered either by receptor recognition at the plasma membrane (e.g., human immunodeficiency virus [HIV] [Chapter 21]) or by low pH in an endosomal compartment (e.g., influenza virus [Chapter 12]). These conformational transitions of the envelope glycoprotein lead to the insertion of the viral fusion peptide into the plasma membrane or into the membrane of an endocytic vesicle. In either case, insertion of the fusion peptide and a subsequent conformational rearrangement of the fusion protein result in membrane fusion and the release of the viral core into the cytoplasm. Some enveloped RNA viruses, including members of the paramyxovirus family (Chapter 11), use a variation on this theme, in which separate proteins mediate attachment and fusion.

The herpesvirus fusion machinery is somewhat more complex, making use of as many as five envelope glycoproteins to mediate binding and entry. Three of these herpesvirus glycoproteins, gB, gH, and gL, which are believed to constitute the core fusion complex, are conserved among all human herpesviruses, suggesting that all herpesviruses share a common mechanism by which they undergo binding and fusion.

The preceding details about herpesvirus entry may suffice for many readers. Still, I urge all to at least scan the following.

Most herpesviruses make their first contact with host cells by interacting with **glycosaminoglycans** on the cell surface (Fig. 18.3). Glycosaminoglycans are polysaccharides that consist of repeating units of particular disaccharides. For most herpesviruses, the glycosaminoglycan of choice is heparan sulfate, which is mainly comprised of a disaccharide of N-acetylglucosamine linked to glucuronic acid. (The related glycosaminoglycan heparin is mainly comprised of the disaccharide of 2-deoxy-2-sulfamido-α-D-glucopyranosyl-6-O-sulfate linked to iduronic acid. Problems arise in classifying glycosaminoglycans that contain both "heparan sulfate-like" and "heparin-like" sequences.) The sugars usually are modified by the addition of sulfate groups, and the glycosaminoglycans are commonly linked to proteins to form proteoglycans. Some proteoglycans are components of the extracellular matrix, while others are cell surface proteins that generally function in cell adhesion.

Viral glycoproteins gB and gC can mediate the binding of herpesviruses to cell surface glycosaminoglycans. gC is not essential for entry per se, nor does it trigger the irreversible fusion-inducing conformational changes that lead to entry. Instead, the initial binding event promoted by gC can enable the virus to seek out receptors that might trigger entry. Therefore, heparan sulfate is thought of as a *binding receptor* rather than an *entry receptor*.

After herpesvirus particles attach to the binding receptor, entry is thought to be triggered by the interaction of other virus glycoproteins, which recognize a specific entry receptor for that virus. For example, in the case of HSV, the interaction of gD with one of several different entry receptors can trigger fusion, which is mediated by the core fusion complex, comprised of gB, gH, and gL (Fig. 18.3; Box 18.4). Protein gB of some herpesviruses forms trimers that appear as prominent spikes on virus particles, while gH and gL form heterodimers. Other envelope proteins on the viral surface may enable different herpesviruses to target their favored cells and tissues.

gB is the most highly conserved component of the herpesvirus entry machinery, and the recent determination of its crystal structure (Fig. 18.4) revealed that its structure is strikingly similar to that of the G protein of vesicular stomatitis virus (VSV). This experimental finding is particularly interesting since VSV is a minus-strand RNA virus (a rhabdovirus [Chapter 10]) and its G protein is its sole envelope glycoprotein, serving as both an attachment protein and a fusogen. But despite the similarity of the herpesvirus gB protein to the VSV G protein, gB does not have an obvious fusion peptide, nor does gHgL. Nevertheless, the unexpected similarity of gB to the VSV G protein suggests that gB too is a fusion effector and that it accomplishes fusion through an extensive conformational rearrangement.

The conformational rearrangement of gB is triggered, at least in part, by the rearrangement of gD, as induced by its binding to the receptor (Box 18.4). In contrast, the VSV G protein takes on its fusion-active conformation in response to low pH. The rearrangement of gD is not believed to participate in fusion per se. Also, the precise functions of gB in entry and of gH/gL too are not known with certainty. But all are required for entry. One premise is that gH/gL hold gB in its prefusion conformation and that the triggered conformational rearrangement of gD enables it to interrupt this inhibitory contact.

The HSV entry receptors include HVEM (acronym for herpesvirus entry mediator), which is a member of the

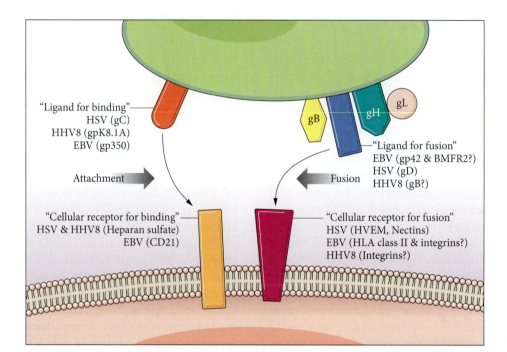

Figure 18.3 Participants in herpesvirus entry and virus-induced cell fusion. For both alphaherpesviruses and gammaherpesviruses, binding to cells can be mediated by a virus particle glycoprotein that is not essential for entry. The binding receptors are heparan sulfate for HSV gC and HHV-8 K8.1A and CD21 (a complement component, also known as 3d [Chapter 4]) for EBV gp350 (in the case of B cells). Entry requires interaction of a viral ligand with another cell surface receptor. For HSV, the viral gD protein is the ligand for several cell surface receptors, any one of which can mediate entry. They include herpesvirus entry mediator (HVEM), a member of the TNF receptor family; nectin-1 and nectin-2, two members of the immunoglobulin superfamily; and specific sites in heparan sulfate generated by certain isoforms of 3-*O*-sulfotransferases. For EBV entry into B cells, gp42 binds to human leukocyte antigen (HLA) class II molecules. Note that gp42 is not required for EBV infection of epithelial cells. The EBV entry receptors in epithelial cells have not yet been identified but could include integrins. The viral ligands could be gH and/or BMRF2. For HHV-8 entry, gB can bind to one of the integrins. Any one of these interactions of a viral ligand with an entry receptor is thought to activate the fusion activity of gB and gH-gL. Adapted from P. G. Spear and R. Longnecker, *J. Virol.* 77:10179–10185, 2003, with permission.

tumor necrosis factor (TNF) receptor family (Chapter 4); nectin-1 and nectin-2, which are members of the immunoglobulin family of proteins; and a specifically modified heparan sulfate.

HVEM is expressed on a variety of cell types, including epithelial cells, which are the initial cells targeted by HSV in humans. The natural ligands for HVEM are LIGHT and lymphotoxin-α, both of which are members of the TNF family of cytokines. LIGHT regulates T-cell immune responses, while lymphotoxin-α mediates a large variety of inflammatory, immunostimulatory, and antiviral responses.

The nectins too are expressed on a variety of cell types, including epithelial cells and neurons. These immunoglobulin family proteins appear to be involved in cell-to-cell adhesions and are structurally similar to the poliovirus receptor (Chapter 6). The fact that nectins are expressed on neural cells is relevant to the HSV lifestyle,

as may be apparent from what is discussed above and noted again in the next paragraph.

HSV characteristically infects two different cell types during the course of an infection. The virus initially infects epithelial cells, in which the infection is acute. However, during this acute phase, HSV invades neurons that innervate the epithelium. Importantly, regarding the HSV lifestyle, the virus establishes true latency in the neurons. As discussed in more detail below, latent virus in neurons can spontaneously reactivate and reinfect the epithelium. Thus, an infected individual can transmit the virus over a lifetime.

Many integral membrane proteins of epithelial cells are largely restricted to either the apical or basal surface of a polarized epithelium. For that reason, the role of a particular HSV entry receptor during infection in vivo is determined by whether it is located on the apical or basal surface of the epithelium. For example, HVEM is available

BOX 18.4

Krummenacher and coworkers (2005) determined the structure of the HSV gD protein, and based on that structure and the facts that are indicated below, they proposed a mechanism by which gD might be activated by receptor-binding and the means by which it might then promote entry.

First, here are the facts. HSV-1 attaches to a cell through an interaction of gC with a binding receptor (a heparan sulfate proteoglycan), followed by an interaction of gD with one of three entry receptors: a nectin, HVEM, or a specifically modified heparan sulfate (see the text). Additionally, the recent solutions of the crystal structures of the ectodomain of gD, both unbound and in complex with the ectodomain of HVEM, reveal that binding of gD to the entry receptor causes gD to undergo a conformational change in which a C-terminal segment of the ectodomain polypeptide chain is released from a strong intramolecular contact. In this regard, gD is dimeric in the virus envelope, and it was

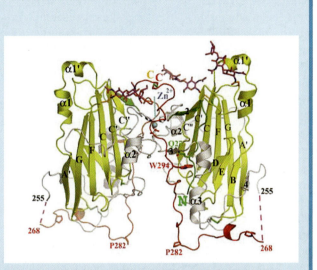

Disulfide-linked dimer of gD (23–306)$_{307C}$. A Zn^{2+} ion, trapped at the dimer interface, and the disulfide bonds are shown. The red dotted lines represent a disordered part of the C-terminal region. N and C termini and important amino acids are indicated. Figure from C. Krummenacher et al., 2005, with permission.

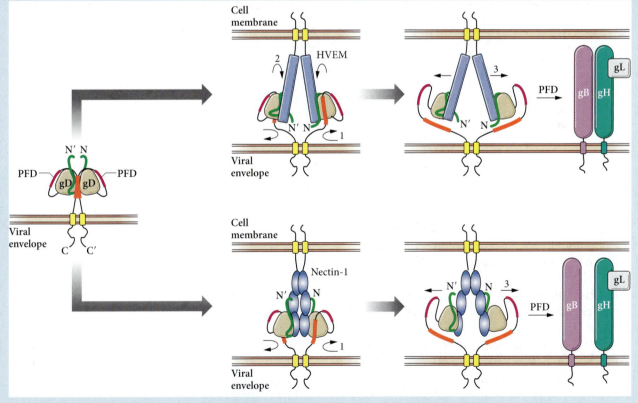

Proposed mechanism for receptor-mediated activation of HSV gD. Envelope gD is shown, as a dimer, in its unbound state as well as during interaction with HVEM (top) or nectin-1 (bottom). Conformational changes are chronologically indicated by numbered arrows: 1, displacement of the C terminus; 2, folding of the gD N terminus in the case of HVEM binding; and 3, exposure of the profusion domain (PFD), a region shown to be important in a post-receptor-binding step of the entry process. The N terminus of gD is shown in green, and the PFD (residues 260 to 285) is shown in red. How exactly gD might trigger fusion after being activated by receptor binding is not known, but a likely possibility is that the "activated" gD core interacts with gB or gHgL to activate the fusion complex. Figure from C. Krummenacher et al., 2005, with permission.

Box continues

suggested that dimerization could stabilize the inactive conformation of the C terminus. The liberated C-terminal segment of gD may then interact with gB or the gH/gL complex to trigger molecular rearrangements in these fusion components.

Reference
C. Krummenacher, V. M. Supekar, J. C. Whitbeck, E. Lazear, S. A. Connolly, R. J. Eisenberg, G. H. Cohen, D. C. Wiley, and A. Carfi. 2005. Structure of unliganded HSV gD reveals a mechanism for receptor-mediated activation of virus entry. *EMBO J.* **24:** 4144–4153.

to gD on the exterior surface of the intact epithelium and thus may facilitate the initial infection. In contrast, since the nectins are present in **adherens junctions** (i.e., the protein complexes that occur at cell-cell junctions in epithelial tissues), they generally are not initially accessible to the HSV gD protein. However, the nectins may well play a role in the cell-to-cell spread of HSV within the epithelium. Moreover, they may play a role in neuroinvasion, since they are present on neuronal cells as well as the epithelium. Thus, HSV might use one set of receptors for

Figure 18.4 (A) Domain architecture of gB. Domains observed in the crystal structure are highlighted in different colors, and their corresponding first residue positions are shown. (B) Ribbon diagram of a single gB protomer. The domains are rendered in colors corresponding to those shown in panel A. Labeled residues indicate the limits of individual domains and the disordered loop in domain I. Residues Arg[661] to Thr[669] of the shown protomer are in gray because they belong to domain III of a neighboring protomer. Residues Arg[661] to Thr[669] of the other neighboring protomer are included here and shown in yellow, because they contribute to a sheet in domain III of the shown protomer. Disordered segments are shown as dots of appropriate color. Disulfides are shown in ball-and-stick representation. Cysteines are numbered according to the scheme shown in panel A. (C) gB trimer. Protomer A is the same as in panel B. Protomer B is shown in white, and protomer C is shown in gray. (D) Accessible surface area representation of gB trimer. Adapted from E. E. Heldwein, H. Lou, F. C. Bender, G. H. Cohen, R. J. Eisenberg, and S. C. Harrison, *Science* **313:**217–220, 2006, with permission.

A

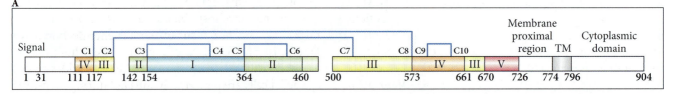

B C D

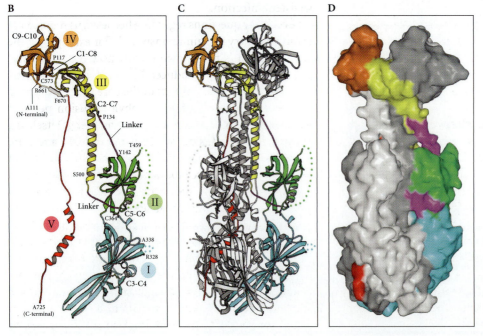

initial entry into the epithelium and others for cell-to-cell spread within the epithelium and entry into neuronal cells.

EBV, like HSV, is an important human pathogen, and like HSV, the EBV life cycle also involves two different cell types: in this case, epithelial cells and B lymphocytes. The site of primary EBV infection is thought to be the mucosal epithelium of the nasopharynx. From there, the virus may then establish a latent infection in B lymphocytes. Actually, EBV has been referred to as the "definitive lymphocyte pathogen," in recognition of the clinical consequences of EBV infection of B cells (see below).

The core fusion machinery of EBV, like that of all herpesviruses, consists of the three conserved glycoproteins gB, gH, and gL. However, the EBV attachment proteins are uniquely its own. EBV binds to B cells via its major envelope glycoproteins, gp350 and gp220 (Fig. 18.3). These proteins bind to the CR2 complement receptor (also known as CD21), a protein that is plentiful on B cells. This interaction between gp350/220 and CD21 is analogous to the interaction between the HSV gC protein and heparan sulfate. That is, it promotes initial binding but not membrane fusion. EBV fusion with B cells is then promoted by EBV glycoprotein gp42, which interacts with major histocompatibility complex class II (MHC-II) proteins, a group of molecules that also are plentiful on B cells (Chapter 4). (The other human gammaherpesvirus, HHV-8, also expresses gp42.) The gp42 interaction with MHC-II is believed to trigger a conformational transition in the gB, gH, gL core fusion complex, which then mediates membrane fusion. Notice that the role of the EBV gp42 protein is analogous to that of the HSV gD protein. Yet there is no sequence homology between these two fusion-triggering proteins.

The fact that EBV makes use of MHC-II molecules for its entry receptor on B cells and the fact that humans are highly polymorphic for MHC loci (Chapter 4) may help to account for why this virus preferentially causes disease (e.g., Hodgkin's disease; see below) in humans with particular MHC-II haplotypes.

EBV entry into epithelial cells must be different from its entry into B cells, since epithelial cells generally do not express either complement receptors or MHC-II molecules. Consequently, EBV does not use gp350/220 and gp42 to initiate entry into epithelial cells. Instead, the EBV gH protein, as part of a gHgL complex, interacts directly with an as-yet-unknown receptor in the epithelial cell membrane. In addition, the EBV BMRF2 protein contains an RGD (Arg-Gly-Asp) motif, which is the common motif by which integrin-binding ligands interact with integrins. Thus, BMRF2 is thought to facilitate EBV binding to the basal surface of the epithelium, where extracellular matrix-binding integrins are located.

Here is a most interesting twist, and one with a rather novel resolution. As just noted, the interaction of the EBV gp42 protein with the entry receptor activates the core fusion complex when the virus is infecting B cells. The twist is that gp42 actually impairs the fusion complex when it is present on virus particles trying to infect epithelial cells. The resolution is that EBV particles that infect B cells have gp42 on their surface, while those that infect epithelial cells are deficient in the protein. So then, how does EBV generate particles that either contain or are deficient for gp42 in vivo, and how might this state of affairs benefit the virus?

The answer to the first part of the question is that EBV particles generated in B cells have a shortage of gp42 because this protein interacts *intracellularly* with MHC-II molecules expressed in B cells, and this interaction targets the gp42/MHC-II complexes to an intracellular degradative pathway. In contrast, EBV particles made in epithelial cells, which do not usually express MHC-II molecules, lose none of their gp42. Thus, EBV particles made in B cells have an enhanced tropism for epithelial cells, whereas particles made in epithelial cells have an enhanced tropism for B cells. As for the second part of the question, as a consequence of this stratagem, EBV is able to shuttle between epithelial cells and B cells. This shuttling may very well play a role in efficient initiation of infection of new individuals, since B cells are the primary EBV reservoir in the old host, while epithelial cells are the initially infected cells in the new host. Moreover, this shifting tropism might facilitate the establishment of latent infection in the B cells. In addition, it might aid in replenishing the infected B-cell pool, since there is increasing evidence that epithelial cells are periodically infected during long-term persistent infection.

Several key questions regarding herpesvirus membrane fusion and entry remain unanswered. For example, in the case of HSV, it is not known exactly how the conformational change in gD, as induced by receptor binding, activates conformational changes in the core fusion complex. Moreover, gB does not have an obvious fusion peptide, nor does gHgL. Also, it is not known whether gD itself is a component of the fusion mechanism per se. In addition, might there be yet other receptors on the cell surface that interact directly with components of the HSV core fusion complex, or is the gD-receptor interaction entirely responsible for activating the fusion complex? (In this regard, the binding/fusion events in the case of HHV-8, which like EBV is a gammaherpesvirus, require only gB, gH, and gL, implying that HHV-8 fusion is activated by the direct interaction of one or more proteins of the core fusion complex with the cell surface. Indeed, the HHV-8 gB protein can bind to heparan sulfate, a reaction not reported for the HSV gB protein. Also, EBV appears to bind to epithelial cells but not B cells via its gH protein.) Furthermore, it is puzzling that under certain conditions, the HSV envelope fuses with the plasma membrane at physiological pH. Yet under other conditions, the virus is taken up by endocytosis, followed by viral envelope fusion with an endosomal membrane. The biological relevance of entry via endocytosis is not clear, since most evidence indicates that productive infection results from fusion at the plasma membrane. On the other hand, the lifestyle of the herpesviruses generally involves infection of multiple cell types. Thus, do different entry pathways mediate herpesvirus entry into different host cell types? Some differences in this regard were noted above, in the cases of HSV-1 and EBV. But might there be others?

Here is yet another basic question: Why do herpesviruses require three or more envelope glycoproteins to mediate binding/fusion, while influenza virus binding/fusion (Chapter 12) is mediated by the influenza virus hemagglutinin protein alone? Perhaps it is because influenza virus fusion occurs in endosomes, as triggered by the low pH of the endosomal compartment. In contrast, other viruses, such as the paramyxoviruses (Chapter 11) and HIV (Chapter 21), which undergo fusion at the plasma membrane, may require multiple viral factors to mediate binding/fusion. That may be so, since there is no low-pH trigger at the cell surface to activate the process. Regardless, the known fusion-active conformations of viral fusion proteins, of enveloped RNA viruses for the most part, but now including the HSV-1 gB protein, show key similarities (e.g., the coiled-coil structure), implying that these distinct viral fusion proteins mediate fusion via similar mechanisms. On the other hand, unlike the fusion proteins of the enveloped RNA viruses, gB does not have an obvious fusion peptide.

Following fusion of the viral envelope with the plasma membrane, the capsid and some of the surrounding tegument are transported via microtubules towards the nucleus. As noted above, the tegument contains at least 15 different proteins, including enzymes and transcription factors that promote the infection. The virus-associated host shutoff protein (VHS), which is the product of the UL41 gene, is one of the most prominent of the tegument proteins. As the capsid traffics to the nucleus, VHS is released into the cytosol, where it causes disruption of polysomes and degrades host mRNAs. In that way, VHS helps to redirect the cell from supporting host protein synthesis to supporting viral protein synthesis. In addition, it regulates the levels of different viral mRNAs, so as to facilitate the sequential expression of different viral genes.

The incoming herpesvirus capsids next dock at nuclear pore complexes. Although herpesvirus capsids are now naked (i.e., stripped of their envelopes), they are nevertheless larger than adenovirus particles, which likewise dock at nuclear pore complexes (Chapter 17). Since adenovirus particles are too large to pass through nuclear pore complexes, herpesvirus capsids must be too large as well. Indeed, parental herpesvirus capsids are not seen in the nucleus, but neither are they seen to disassemble in the cytosol. How then is the herpesvirus genome delivered into the nucleus? Not surprisingly, and reminiscent of the adenoviruses, the now-naked but largely intact herpesvirus particles "inject" their genomes across the nuclear pore into the nucleus (Fig. 18.5).

Figure 18.5 Electron micrographs of pseudorabies virus (porcine herpesvirus) capsids juxtaposed to nuclear pores, in different stages of DNA release. Invariably, nucleocapsids at nuclear pores are oriented with one vertex pointing to the central granulum of the pore, at a distance of 40 nm. Importantly, viral DNA appears to be released from morphologically intact nucleocapsids via the juxtaposed vertex region through the nuclear pore into the nucleoplasm. During this stage of infection, neither disassembly of nucleocapsids nor localization of nucleocapsids in the nucleus could be detected. Bars are 200 nm. From H. Granzow, F. Weiland, A. Jöns, B. G. Klupp, A. Karger, and T. C. Mettenleiter, *J. Virol.* **71:**2072–2082, 1997, with permission.

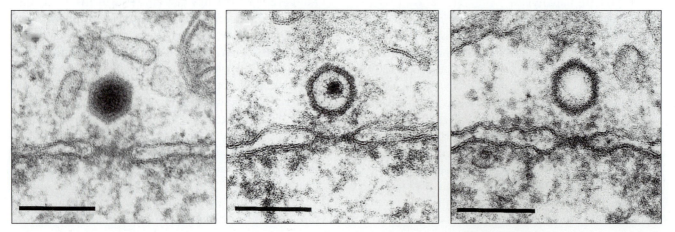

Experiments in vitro, using isolated capsids and purified nuclei, show that docking of herpesvirus capsids at nuclear pore complexes and genome entry can be blocked by antibodies against the nuclear pore complex and importin-β (see Box 12.7). These experimental results show that transport of herpesvirus genomes into the nucleus indeed depends on the specific docking of capsids to nuclear pore complexes. Moreover, they imply that herpesvirus capsids somehow interact with nuclear pore complexes to affect genome release and transport into the nucleus.

The HSV-1 α-TIF (alpha-trans-inducing factor) protein, a structural component of the tegument, accompanies the virus genome into the nucleus, where it enhances transcription of viral immediate-early genes (see below). Following transport of the viral genome and the α-TIF protein into the nucleus, the now-empty capsids are released from the nuclear pore complexes and back into the cytosol (Box 18.5).

GENOME ORGANIZATION AND EXPRESSION

Herpesvirus genomes vary considerably in size, ranging from 70 to 165 genes. Thus, there are considerable differences in the particular genes that different herpesviruses might express during the normal course of events in vivo. Nevertheless, all herpesvirus genomes share common organizational features and patterns of expression. Once again we use HSV-1, the prototype herpesvirus, as our example. Like all herpesviruses, the HSV-1 genome contains a terminally redundant, direct repeated sequence, depicted as XY (Fig. 18.6). Importantly, this terminal repeated sequence, together with adjacent sequences, often is repeated in an inverted orientation within the genome. This can be demonstrated by denaturing the DNA and allowing isolated single strands to self-anneal. In the case of HSV-1, several different forms of self-annealed strands are generated, which together show the following. The terminal sequence at the left end of the genome (depicted as XYABC/X'Y'A'B'C' in Fig. 18.6) is separated from its internal inverted repeat by a relatively long unique sequence, designated U_L (U for "unique" and L for "long"). The terminal sequence at the right end of the genome (depicted as DEFXY/D'E'F'X'Y') is separated from its internal inverted repeat by a relatively short unique sequence, designated U_S. All herpesvirus genomes contain the direct repeats. They then have the same or other repeated sequences at internal sites. The repetitive sequences contain several interesting elements, including immediate-early genes (see below) that mobilize the subsequent stages in the herpesvirus cycle of gene expression, the promoter for the LAT RNAs that act to establish latent infection, and DNA packaging signals.

Another somewhat unique feature of herpesvirus genomes is that while they are linear in the virus particle, they circularize upon release into the cell nucleus, a step which is necessary for genomic replication to occur (see below). This is the reason why the HSV-1 genetic and transcription map is depicted as a circle (Fig. 18.7; Box 18.6).

Also, some herpesviruses contain multiple origins of DNA replication. It is thought that the different origins have different functions in the herpesvirus life cycle. For example, one or more origins might support vegetative replication, whereas another might support the maintenance of episomal viral genomes in latently infected cells, while a third may support reactivation from the latent state. In the case of EBV, replication of the latent viral genome is mediated by *oriP*, whereas lytic genomic replication is mediated by *oriLyt*. (During latency, the EBV genome is replicated once per cell cycle, when the infected cell replicates its genome.) HSV-1 contains three origins of DNA replication: one copy of *oriL*, located in the center of the U_L region, and two copies of *oriS*, each of which is located in a repeat flanking the U_S region of the genome (Fig. 18.6 and 18.7). The particular functions in vivo of the individual HSV-1 *ori* sequences are not known, but they can substitute functionally for one another in vitro.

BOX 18.5

The mechanism by which herpesvirus genomes (and adenovirus genomes as well [Chapter 17]) are transported into the nucleus (i.e., from largely intact particles docked at nuclear pore complexes) is different from the way that polyomavirus genomes (Chapter 15) enter the nucleus. In the latter case, the virus particles undergo partial disassembly in the ER, prior to transport of their genomes into the nucleus. And, as noted in Chapter 15, current work in the author's laboratory shows that the simian virus 40 genome is at least partially exposed concomitant with, or subsequent to, its release from the ER into the cytosol, but before it arrives at the nucleus. Moreover, entry of polyomavirus genomes into the nucleus is facilitated by virus capsid proteins that contain nuclear localization signals. Also, in contrast to the genomes of polyomaviruses, papillomaviruses, and adenoviruses, herpesvirus genomes are not associated with histones, or histone-like proteins, that might enter the nucleus with the genome. Note that the smaller DNA viruses do not contain portal complexes.

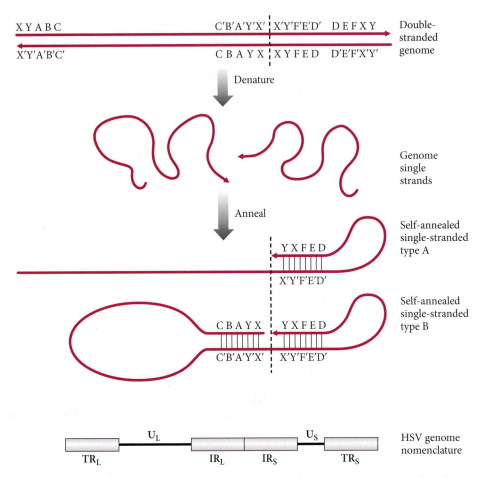

Figure 18.6 Some features of herpesvirus genomes, as illustrated by HSV-1. Note the terminally redundant, direct repeated sequence, depicted as XY. This terminal repeated sequence, together with adjacent sequences, often is repeated in an inverted orientation within the genome. This can be demonstrated by denaturing the DNA and allowing isolated single strands to self-anneal, as depicted in the figure. Again, taking HSV-1 as an example, the several different forms of self-annealed strands that are generated show the following. The terminal sequence at the left end of the genome (depicted as XYABC/X′Y′A′B′C′) is separated from its internal inverted repeat by a relatively long unique sequence, designated U_L. The terminal sequence at the right end of the genome (depicted as DEFXY/D′E′F′X′Y′) is separated from its internal inverted repeat by a relatively short unique sequence, designated U_S. All herpesvirus genomes contain the direct repeats. They then have the same or other repeated sequences at internal sites. The repetitive sequences contain several interesting elements, including immediate-early genes (see the text) that mobilize the cycle of gene expression during productive infection, the promoter for the LAT RNAs that act to establish latent infection, and DNA packaging signals. Herpesvirus genomes also contain multiple origins of DNA replication. HSV-1 contains three origins of DNA replication: one copy of *oriL*, located in the center of the U_L region, and two copies of *oriS*, each of which is located in a repeat flanking the U_S region of the genome. Adapted from N. J. Dimmock, A. K. Easton, and K. N. Leppard, *Introduction to Modern Virology*, 5th ed. (Blackwell Science Ltd., Oxford, United Kingdom, 2001), with permission.

As described below, herpesviruses replicate their genomes via a rolling-circle mechanism (Box 14.5) that produces "head-to-tail" concatemers. These concatemers are then cleaved to produce unit-length viral genomes that are packaged in progeny virus particles. The cleavage/packaging signals on the genome are the "a" sequences found in R_S and R_L (Fig. 18.7).

The pattern of HSV-1 gene expression has been studied more extensively than that of any other herpesvirus.

Importantly, the organization of HSV-1 genes and their pattern of expression reflect the unique HSV-1 life cycle. (The same might be said for any of the human herpesviruses.) HSV-1 infection begins with lytic replication in the epithelium. Before this acute phase of the infection is resolved by host immune effectors, the virus invades the endings of the sensory neurons innervating the affected area and then makes its way up to the ganglia of the nerve,

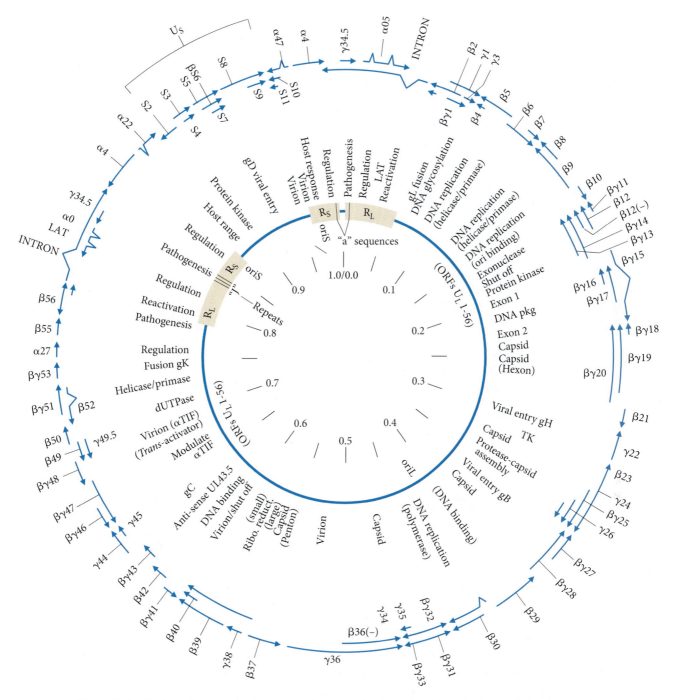

Figure 18.7 HSV-1 genetic and transcription map. Details are given in the text. Most often, transcripts are controlled by their own specific promoters and splicing is uncommon. The time of expression of the various transcripts is roughly divided into immediate early (α), early (β), late ($\beta\gamma$), and strict late (γ). This in turn is based on whether the transcripts are expressed in the absence of protein synthesis (α), before viral DNA replication and shutoff following this (β), before viral DNA replication but reaching maximal levels following this ($\beta\gamma$), or only following viral DNA replication (γ). Adapted from E. K. Wagner, M. J. Hewlett, D. C. Bloom, and D. Camerini, *Basic Virology*, 3rd ed. (Blackwell Publishing, Malden, MA, 2008), with permission.

Bacteriophage λ likewise has a genome that is linear in the virus particle but circularizes upon release into the bacterial cell (Chapter 2). But here is an odd variation. The T-even phages have circular genetic maps, but their genomes are nevertheless quite linear at all times. The explanation of this conundrum is that the ends of T-even phage genomes are circularly permuted. That is, if their genetic information is represented by ABCDEFGH, then a **circular permutation** would generate molecules ABCDEFGH, BCDEFGHA, CDEFGHAB, DEFGHABC, EFGHABCD, and so on.

But how does the population of circularly permuted phage genomes arise in the first place? The answer is that these phage genomes also are terminally redundant (e.g., ABCDEFGHA, BCDEFGHAB, and so on). That being the case, during replication, recombination can occur between the terminal redundant regions of two daughter genomes, giving rise to molecular concatemer. Genome-sized molecules are then randomly excised from these concatemers, thereby generating a population of terminally redundant, circularly permuted genomes.

where it establishes latent infection. The latent infection may reactivate several times per year in some individuals, resulting in lytic reinfection of the area of the epithelium innervated by the latently infected nerve.

Interestingly, only about one-half of the 70 or so HSV-1 genes are essential for virus replication in cell culture. The remaining HSV-1 genes are sometimes referred to as "nonessential." But, bearing in mind that the necessarily limited size of virus genomes dictates that viruses tend not to accumulate unnecessary genetic baggage, what might be the functions of the nonessential HSV-1 genes? The answer is that the needs of HSV-1 are very much different in vivo from what they are in vitro (as you well may have expected). The genes that are essential for HSV-1 replication in vitro are the same genes that are needed for the lytic phase of infection in vivo. Since the productive replication phase is similar among the different herpesviruses, these essential genes are largely conserved among all herpesviruses. That being said, herpesvirus genes that are dispensable in vitro likely enable the different herpesviruses to evade host immunity and to establish latent infections in specific target cells and tissues in the immunocompetent host.

Since the different herpesviruses use different cell types and tissues in which to establish latency, and since these different cell types and tissues present different conditions

for establishing latency and evading host immunity, the "dispensable" genes are more divergent among the different herpesviruses than are the essential genes. The lifelong persistence of the herpesviruses and the ability of the latent viruses to reactivate and reestablish the productive infection also imply that a significant fraction of herpesvirus genes serve regulatory functions, as well as functions that might undermine host immune responses.

The expression of herpesvirus genes is temporally regulated. That is, early genes are expressed before DNA replication, and late genes, for the most part, are expressed afterwards. The expression of polyomavirus, papillomavirus, and adenovirus genomes likewise is temporally regulated. Yet the dependence of the switchover from early to late gene expression and its dependence on DNA replication are not as clearly marked in the case of the herpesviruses as in the cases of the smaller DNA viruses. Actually, HSV-1 late-gene expression accelerates upon the onset of DNA replication but occurs even if DNA synthesis is blocked. Moreover, even some early genes continue to be transcribed at late times, although at much lower rates.

Most herpesvirus genes have their own promoter and termination site, although adjacent, similarly oriented genes can share the same poly(A) site. Also, herpesviruses make little use of splicing primary transcripts. Regarding the latter point, only 4 of the 70 or so HSV-1 genes are known to contain introns. However, spliced messenger RNAs (mRNAs) play key roles in the HSV-1 life cycle. For instance, three of the spliced HSV-1 transcripts are expressed at immediate-early times after infection (see below). One of the products of a spliced mRNA, α27 (ICP27 in ICP [infected-cell polypeptide] nomenclature), impairs the spliceosome, thereby causing a drop in the levels of spliced mRNAs in the infected cell. This impairment of the spliceosome favors expression of viral genes, since viral transcripts are unspliced for the most part.

A family of spliced transcripts, and a rather unique one at that, appears to play a key role in both the establishment and the reactivation of HSV-1 latency. These are HSV-1 **latency-associated transcripts (LATs)**, which are present in the nuclei of latently infected sensory neurons. In point of fact, the LATs are comprised of spliced introns from an RNA precursor molecule that actually is transcribed from a site on the genome, on a strand that is complementary to that encoding the α0 mRNA, an immediate-early gene transcript.

A unique feature of the major 2-kilobase (kb) LAT is that it is a stable intron, spliced from a much less stable primary transcript. (Generally, it is the intron that is unstable.) As introns, the LATs do not encode any protein. Nevertheless, they appear to play several roles in latency.

The LATs and the α27-mediated impairment of the splice-osome noted above are a couple of rather fascinating viral adaptations that we have not noted before. These are each discussed further below.

The EBV latent cycle EBNA (NA is an acronym for nuclear antigen) genes provide a spectacular exception to the single promoter/no splicing paradigm. During latent EBV infection of human B lymphocytes, six EBNAs (EBNA1, 2, 3A, 3B, 3C, and LP) are expressed from a single complex transcriptional unit that spans nearly 100 kb of the viral genome. The six EBNAs are expressed from their single transcriptional unit by means of alternative splicing, alternative polyadenylation sites, and distinct, sequential usage of three promoters (Cp, Wp, and Qp).

Regardless of the above special exceptions, the pattern of herpesvirus gene expression mostly differs from the transcription patterns seen in the cases of the smaller polyomaviruses, papillomaviruses, and adenoviruses, which all make extensive use of shared transcription initiation sites and splicing of primary transcripts. What purpose might there be to the alternative pattern of herpesvirus gene expression, which involves 70 or more separate transcription units? Perhaps it provides the potential for more finely tuned regulation of gene expression, in keeping with the more complex herpesvirus life cycle.

Herpesviruses, like adenoviruses, express immediate-early genes. The defining feature of herpesvirus and adenovirus immediate-early genes, which distinguishes them from "standard" early genes of those viruses, is that immediate-early genes are expressed independently of any prior virus protein synthesis, whereas expression of the standard early genes depends on the prior expression of immediate-early genes. The immediate-early, early, and late herpesvirus genes are also referred to as α, β, and γ genes, respectively. Herpesvirus genes are not localized together along the viral genome by these categories, but instead members of the different classes of genes are interspersed, as in the adenovirus genome.

The herpesvirus immediate-early or α-gene products have regulatory functions, which are reminiscent of the adenovirus immediate-early E1A-gene products. They have other functions as well, including ones that are dispensable in vitro but important in vivo. For example, α-gene products regulate virus latency and subvert host immunity. As in the cases of the early and late genes of the smaller DNA viruses, the herpesvirus early β genes encode proteins required for DNA replication, and late γ genes produce the structural proteins needed for generating progeny virus particles.

HSV-1 gene expression during lytic infection is triggered by the UL48 gene product. This raises the interesting question of how a viral gene product might trigger viral gene expression, if viral gene expression itself needs to be triggered by a viral gene product. The answer is as follows. The UL48 gene product is a virus particle-associated protein called α-TIF (noted above), also known as **VP16**. Note that VP16 is actually a γ-gene product that is produced during the late phase of infection and packaged into progeny virus particles as a component of the tegument. Importantly, VP16 has a very effective transcription activation domain that functions soon after entry during the next round of infection. It can do so since VP16 accompanies the HSV-1 genome into the nucleus, where it amplifies immediate-early transcription. Notice how this stratagem completes the gene expression cycle, from immediate-early gene expression to early gene expression, to late gene expression, and then back to immediate-early gene expression. All in all, this is an effective means by which the virus ensures the timely activation of a new transcription cycle upon infecting a new host cell.

VP16 does not bind directly to a viral promoter. Instead, it interacts with viral promoters via its association with a ubiquitous cellular DNA-binding transcription factor, octamer-binding protein 1 (Oct1), so named because it recognizes a consensus DNA sequence termed the octamer motif. This motif is present on particular cellular enhancers, as well as on HSV-1 enhancers that are associated with α-gene promoters. Although the octamer motif is present on cellular as well as viral enhancers, the virus uses a rather remarkable means to ensure that the transcription activation activity of the few input VP16 molecules is restricted only to Oct1 molecules that are bound to viral enhancers. The virus does so by means of special cis-acting sequences which flank the octamer sites of the viral enhancers. These sequences, 5′TAATGARAT3′, in which R is a purine, are referred to as GARAT elements. GARAT elements do not flank cellular enhancers, and they are not necessary for Oct1 binding per se. But when Oct1 binds to a viral octamer sequence that is flanked by GARAT, it undergoes a conformational alteration that is a prerequisite for VP16 to then bind to it. In that way, the viral GARAT sequences dictate selectivity in response to VP16.

Interestingly, VP16 itself must undergo a conformational alteration before it might stably bind to Oct1. That conformational change in VP16 is induced when it binds to another cellular protein, Hcf (for host cell factor). Thus, VP16, in conjunction with Oct1 and Hcf, activates herpesvirus immediate-early promoters (Fig. 18.8; Box 18.7).

Why might HSV-1 have evolved this indirect means for activating transcription? The answer can only be surmised. However, since Hcf is a protein that plays a role in cell cycle progression in uninfected cells, a reasonable possibility is that it enables the virus to initiate replication only in cells that express a milieu that might optimally

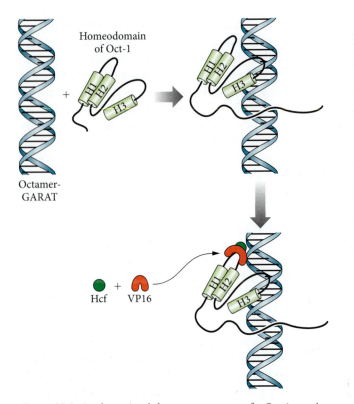

Homeodomain
of Oct-1

Octamer-
GARAT

Hcf + VP16

Figure 18.8 Conformational changes are necessary for Oct-1 protein-dependent recruitment of HSV-1 late protein, VP16, to HSV-1 promoters. VP16 is a transcription factor that binds to viral promoters via its interaction with a ubiquitous cellular DNA-binding transcription factor, Oct1, which recognizes a consensus DNA sequence termed the octamer motif. This motif is present on particular cellular enhancers, as well as on HSV-1 enhancers that are associated with α-gene promoters. Although the octamer motif is present on cellular as well as viral enhancers, VP16 binds only to Oct1 that is bound to viral enhancers. VP16 can discriminate between viral and cellular promoters by means of a *cis*-acting sequence that flanks the octamer sites of the viral enhancers. These sequences, 5′TAATGARAT3′, in which R is a purine, are referred to as GARAT elements. GARAT elements do not flank cellular enhancers, and they are not necessary for Oct1 binding per se. But when Oct1 binds to an octamer sequence that is flanked by GARAT, it undergoes a conformational alteration that is a prerequisite for VP16 to then bind to it. In that way, the viral GARAT sequence dictates selectivity in response to VP16. VP16 itself must undergo a conformational alteration before it might stably bind to Oct1. That conformational change in VP16 is induced when it binds to another cellular protein, Hcf (for host cell factor). Thus, VP16, in conjunction with Oct1 and Hcf, activates herpesvirus immediate-early promoters. Adapted from S. J. Flint, L. W. Enquist, R. M. Krug, V. R. Racaniello, and A. M. Skalka, *Principles of Virology: Molecular Biology, Pathogenesis, and Control*, 2nd ed. (ASM Press, Washington, DC, 2004), with permission.

support virus replication. Moreover, since the lytic pattern of HSV-1 transcription is dependent on the presence of Oct1 and Hcf, then under conditions where these proteins are absent or not available (see below), there can be no transcription of viral α genes and no cascade of lytic transcription (see the following paragraph). In addition,

VP16 is the prototype of a family of transcriptional activators that have an activation domain referred to as an **acid blob**. The origin of this term is as follows. While DNA-binding domains of transcriptional activator proteins have well-defined domains, the distinct activator domains appear to be comprised of conformationally ill-defined polypeptides that function irrespective of sequence. That is, these activating domains clearly do not form compact globular forms, with distinct tertiary structures. However, the activity of these domains is critically dependent on the presence of a sufficient excess of acidic residues (aspartate and glutamate) clustered within or about them. Thus, these nearly shapeless acidic activation domains are referred to as acid blobs or "negative noodles." VP16 has an intensely acidic region within its 78-amino-acid-long carboxy terminus, which is essential for its activation activity. As also shown by VP16, the DNA-binding domain of the activator does not need to be covalently linked to the activating module (i.e., VP16 has no DNA binding domain of its own, but instead binds to the DNA only in association with Oct-1; see the text).

there is no activation of Oct1-dependent cellular genes. Such a state of affairs might promote viral latency. (As will be seen, there is considerably more to the establishment of latency than the absence of Oct1 and Hcf, and as might be expected, external factors can trigger expression of Hcf and of α genes, leading to reactivation of latent virus [see below].)

The potent transcription activation activity of VP16 sets in motion a cascade of viral gene expression, beginning with the activation of α genes, which in turn trigger expression of β genes. One or more β genes then inhibit expression of α genes and, moreover, upregulate expression of γ genes. One or more of the γ genes then inhibit expression of β genes. Thus, genes that are expressed in one phase of the infection activate genes that are expressed in the next phase and downregulate genes expressed in the previous phase. The overall result is an economy of viral gene activity during the productive cycle. Again, the cycle comes full circle when VP16, a γ gene, activates expression of α genes at the start of the next round of infection.

The HSV-1 α genes include α4, α27, α47, α0, and α22 (Box 18.8). The HSV-1 α4 and α27 gene products activate transcription of HSV-1 β genes, thereby leading to viral DNA replication. The effect of these HSV-1 α gene products is reminiscent of the adenovirus E1A gene products,

A word about nomenclature. HSV genes or transcriptional units generally have names such as UL54. Such names have the value of indicating the region of the genome where the particular gene might be found. UL54 encodes the α27 protein, also known as ICP27. Some herpesvirus experts suggested that I use the ICP nomenclature throughout, but I prefer not to since the designation α27 tells us that we are dealing with an immediate-early protein. In some instances, particularly where the ICP nomenclature appears widely in the journal literature, I have tried to include both designations. In cases where I use the ICP nomenclature only, the immediate-early, early, or late classification is given as well.

which likewise activate expression of early genes. Interestingly, α4 is located in the R_S regions of the viral genome, and α27 is located in the U_L region (Fig. 18.5 and 18.6). Thus, the HSV-1 genome contains two copies of α4 and one copy of α27.

The α0 gene (located in R_L) encodes the α0 protein, which, like the α4 and α27 gene products, is a transcriptional regulator. Yet each of these α-gene transcriptional regulatory proteins appears to act via a different mechanism. The α4 protein seems to upregulate transcription by interacting with basal transcription factor complexes located at viral promoters. In contrast, and as noted above, α27 (ICP27) inhibits precursor mRNA splicing by interfering with the spliceosome, thereby favoring translation of viral proteins, which are unspliced for the most part. The α0 protein may cause a change in nuclear organization, leading to more efficient transcription.

The α22 gene product is nonessential in vitro. It appears to play a role in the posttranscriptional processing of some transcripts and might enhance the range of cells that the virus may infect in vivo. Other nonessential HSV-1 gene products include the α47 (ICP47) protein, which inhibits antigen presentation by the MHC-I antigen presentation pathway (see below). Another nonessential α-gene product, ICP34.5, may play a role in reactivating latent virus in neuronal cells.

The pattern of HSV-1 gene expression in latently infected neurons is distinctly different from the pattern described above for productive infection of epithelial cells. In brief, most of the viral genome is shut down during latency. However, the so-called "latency-associated" transcripts are expressed from a single latent-phase promoter that is located in both copies of R_L. The LATs actually are comprised of spliced introns, transcribed from an RNA precursor molecule encoded by a site on the genome, on a strand that is complementary to that encoding the α0 mRNA (an immediate-early gene transcript; see below). As then expected, LATs and the precursor RNA molecules from which they are derived do not encode any proteins. The pattern of gene expression during establishment and during maintenance of latency and the presumed mechanism of action of the LATs are described in greater detail below.

DNA REPLICATION

Herpesvirus β genes are activated by α-gene products, as described above. As the β-gene products accumulate, they support high levels of herpesvirus DNA replication. The expression of seven HSV-1 β genes, each of which encodes a protein that has a function in DNA synthesis, is necessary and sufficient to replicate HSV-1 DNA. These HSV-1 β genes include UL30, which encodes the viral DNA polymerase; UL42, which encodes a processivity factor; UL29, which encodes a single-stranded DNA-binding protein; UL9, which encodes the *ori*-binding protein; and UL5, UL8, and UL52, which encode the helicase-primase complex. (In contrast to the HSV-1 example, recall that the small DNA viruses [i.e., the polyomaviruses and papillomaviruses] use the host cell DNA polymerase and associated cellular activities to replicate their genomes. Although the polyomavirus LT protein and the papillomavirus E1 and E2 proteins do participate in DNA replication, they do so in accessory roles only. The larger adenoviruses encode their own DNA polymerase but are still dependent on cellular factors to replicate their genomes.)

Other HSV-1 early gene products play an indirect role in viral DNA replication, for example, by increasing the pool of deoxyribonucleotide substrates. These other viral genes generally are nonessential for HSV-1 replication in vitro and in vivo. However, their expression is necessary for HSV-1 to manifest its full pathogenesis in vivo, perhaps because pathogenesis depends on the ability of the virus to replicate efficiently in some particular cell types (Box 18.9).

The linear herpesvirus genomes must circularize in order to replicate. Bacteriophage λ (lambda) also has a double-stranded DNA genome, which circularizes in order to replicate (Chapter 14). Herpesvirus genomes (and those of bacteriophage λ as well) circularize very early in infection. Circularization is mediated through complementary single-nucleotide 3′ overhangs at the DNA ends and probably requires the action of a protein to bring the two ends together.

Once the linear herpesvirus genome has circularized, DNA replication appears to proceed in two phases. First,

The herpesviruses and the poxviruses (Chapter 19), which constitute the two families of large animal DNA viruses, each encode their own DNA replication machinery, thereby making their DNA replication independent of the cellular DNA replication machinery. This independence also allows these viruses to shut down cellular DNA synthesis, which further serves the viruses by making available to them the entire cellular pool of nucleotides. It also may be vital to the ability of at least some herpesviruses to establish reactivating latent infections. For example, the neurotropic herpesviruses, HSV and VZV, establish latency in "resting" neurons that do not enter into S phase.

there is an initial phase of **theta (θ) replication** (as typified by the polyomaviruses [Chapter 15]), initiated at one or more of the replication origins, which is then followed by a **rolling-circle** mode of replication. (The rolling-circle phase of bacteriophage λ DNA replication likewise is preceded by a phase in which λ genomes replicate via theta intermediates. Theta replication, which generates only circular progeny genomes, is the sole means by which polyomavirus genomes are replicated [Chapter 15].)

The rolling-circle process begins with the nicking of one of the uninterrupted circular strands. The resulting 3′ end then functions as the primer for continuous unidirectional synthesis about the circular template. Synthesis on this strand is analogous to leading-strand synthesis during conventional DNA replication. Synthesis on the displaced strand is discontinuous and is analogous to lagging strand synthesis (see Box 14.5). It generates a concatemeric double-stranded tail from which unit-length genomes can be excised.

Recall that the HSV-1 genome contains three *ori* sequences: one in each of the R_S repeats and one in U_L. Interestingly, virus-encoded *ori*-binding proteins can bind to, and initiate replication at, any one of the *ori* sequences.

EBV too has a clearly defined phase of replication via theta structures. During that phase, EBV uses a single viral protein to organize replication complexes, the rest of which are comprised of cellular proteins. The T antigen of the polyomavirus simian virus 40 (Chapter 15) has a similar role in the replication of that smaller DNA virus. During the subsequent rolling-circle phase of EBV replication, the virus uses additional viral proteins, as enumerated above for HSV-1.

Herpesvirus DNA polymerases, like the cellular enzyme, must prime discontinuous replication by using short RNA primers. RNA priming exists as a feature of the proofreading function of DNA polymerase complexes (Chapter 14). To accomplish this proofreading function, DNA polymerases, including that of the herpesviruses, express a 3′-to-5′ exonuclease activity, which excises erroneously incorporated nucleotides. Bearing the above facts in mind, we can appreciate why the herpesviruses might replicate their genomes via theta intermediates and rolling circles. If herpesviruses were to replicate their genomes as linear DNA molecules, they would then have to solve the problem of generating the 5′ ends of their daughter DNA strands; this problem exists because linear DNA synthesis must be primed with RNA primers. Cells solve this problem by means of telomerase, but no virus is known to generate its genomic ends by that means. By replicating their linear, double-stranded DNA genomes via first circularizing, and then proceeding by theta intermediates and rolling circles, herpesviruses solve the problem of how to generate the 5′ ends of their daughter DNA strands (Box 18.10; see also Box 14.4).

ASSEMBLY, MATURATION, AND RELEASE
Assembly of Procapsids

As in the cases of the smaller DNA viruses, herpesvirus capsid assembly occurs entirely in the nucleus. Thus, herpesvirus

Since the herpesvirus-encoded DNA polymerase expresses an exonuclease activity, does it then display the same level of fidelity as the cellular polymerase? Apparently it does not. Nucleotide incorporation by the HSV-1 DNA polymerase is less faithful than it is in the case of the cellular DNA polymerases, despite the exonuclease activity of the viral enzyme. Nevertheless, the exonuclease activities of herpesvirus DNA polymerases are biologically relevant. Moreover, they are clinically relevant as well, as illustrated by the HCMV polymerase. The HCMV polymerase is the target of the currently licensed systemic anti-HCMV drugs ganciclovir (GCV), foscarnet (FOS), and cidofovir (CDV). As might be expected, the emergence of drug-resistant mutants has impeded therapy in some instances. For example, mutations in the exonuclease domain of the HCMV polymerase are found relatively frequently in association with GCV and CDV resistance. One such mutation, found in a clinical specimen, was inserted into a laboratory strain of HCMV. This resulted in a mildly growth-impaired recombinant virus that displayed resistance to GCV and CDV, as well as an increased spontaneous mutation rate during propagation in vitro.

proteins that participate in capsid assembly must be transported to the nucleus after having been synthesized in the cytoplasm. Transport of these virus proteins into the nucleus presumably occurs via the normal cellular pathway of nuclear protein import.

Empty herpesvirus capsids, termed procapsids, are the precursors of mature virus particles. These herpesvirus procapsids are assembled with the aid of an internal protein scaffold (reminiscent of the assembly of adenovirus capsid precursors [Chapter 17]). The HSV-1 scaffolding proteins are encoded by a pair of overlapping genes, UL26 and UL26.5 (Fig. 18.9). The open reading frame of UL26 extends in frame from the N terminus of the UL26.5 open reading frame. UL26.5 has its own promoter, which is embedded in the coding sequences of the UL26 open reading frame. The two proteins thus have the same C-terminal end, which contains the oligomerization and capsid protein-binding domains. The unique N-terminal sequence of the UL26 gene product, which is the larger of these two proteins, encodes a serine protease domain, the role of which is described below.

Figure 18.9 The HSV-1 UL26 and UL26.5 gene products. The UL26 open reading frame is indicated below its location in the prototype orientation of the HSV-1 genome (shown with a thin line representing unique long [U$_L$] and unique short [U$_S$] regions bounded by terminal repeated sequences, represented by thick bars). UL26 and UL26.5 mRNA transcripts and protein products are also depicted. Primary translation products are represented by pink boxes. Light orange boxes represent proteolysis products. Vertical lines indicate the sites of proteolytic processing of the scaffold proteins. Numbers indicate amino acid residues at the N and C termini of the polypeptides or the sites of proteolytic processing. Adapted from A. K. Sheaffer, W. W. Newcomb, J. C. Brown, M. Gao, S. K. Weller, and D. J. Tenney, *J. Virol.* **74:** 6838–6848, 2000, with permission.

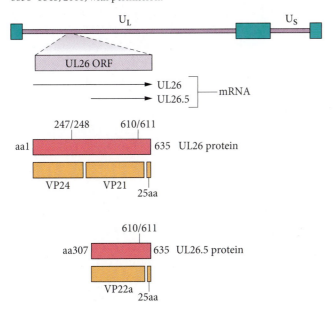

The UL26 and UL26.5 gene products are first incorporated into the developing procapsids (see below) in their native unprocessed forms. As the capsid matures, the UL26 protein autocatalytically cleaves itself to release the minor scaffold proteins VP24 and VP21, the more abundant UL26.5 protein, and the major scaffold protein VP22a (Fig. 18.9). Moreover, the protease activity removes 25 amino acids from the common C termini of VP21 and VP22a. This 25-amino-acid sequence contains the binding region for VP5, the major capsid protein. The latter cleavages are essential for the maturation of the capsid. If they are prevented, then the scaffolding proteins are retained within the capsid, and encapsidation of the genome cannot occur. The scaffolding proteins are thought to be released from the procapsid as the DNA enters.

Before considering the actual assembly of the procapsid, we first briefly review some features of herpesvirus capsid architecture. All of the 150 hexons, which are found on the edges and faces of the capsid, are comprised of the major capsid protein, VP5. Each of these hexons is also associated with six molecules of VP26. Eleven of the 12 pentons, which are located at the vertices of the icosahedral shell, are like the hexons comprised of VP5. The 12th vertex is the unique portal structure, comprised of 12 molecules of the UL6 gene product. The portal is a cylindrical structure, with an axial channel through which the DNA enters into the procapsid. The UL18 and UL38 gene products also play a major role in capsid assembly. They make up the triplexes, which are heterotrivalent structures that lie above the VP5 icosahedral shell, where they connect the pentons and hexons in groups of three.

Bearing the above facts in mind, capsid assembly begins with the assembly of small oligomers that contain one or a few molecules each of VP5 and scaffolding proteins, such that the scaffolding proteins form a core that is internal to the shell proteins. Assembly is thought to proceed with the small assemblages of VP5 and the scaffolding proteins adding on to the edges of the growing capsid. The triplexes then act to secure those assemblages in place.

The structural transformation of the fragile HSV-1 procapsid into a stable mature capsid was analyzed by cryo-electron microscopy (Heymann et al., 2003). Key features are as follows. In the fragile procapsid, interactions between pentamers and hexons are first mediated by the surrounding triplexes (Fig. 18.10). Then, as the capsid matures, gaps between the capsomers (i.e., pentamers and hexons) disappear because the floor domains of the VP5 subunits of each capsomer rotate to make contact with neighboring capsomers. A continuous network of interactions between VP5 molecules is established, thereby forming the floor of the capsid. In the process, the axial channel through each hexon constricts, in so doing sealing off the

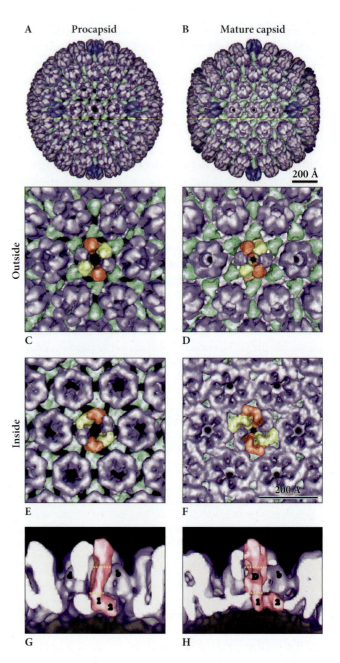

Figure 18.10 Molecular anatomy of the HSV-1 capsid end states. (A and B) The spherical procapsid (A) and the polyhedral mature capsid (B) are shown. The particles are color coded to distinguish the various components. VP5 hexons are light blue, pentons are dark blue, and triplexes are green. The particles are viewed along a twofold axis of symmetry (Fig. 18.2). (C through F) An area centered on an E hexon at a twofold axis of symmetry for the procapsid (C and E) and the mature capsid (D and F) and for both the outer surfaces (C and D) and the inner surfaces (E and F). The three quasiequivalent pairs of VP5 subunits in the E hexon are yellow, red, and blue, respectively. They surround the axial channel through this hexon. Other VP5 subunits are blue, as in panels A and B. Note the regularization and compaction of the protruding domains. Compare panels C and D (these domains protrude from the outer surface of the procapsid; these portions of the VP5 subunits are colored red, yellow, and blue in the E hexon in panel C). Note the constriction of the transhexon channel (compare panels C and D with panels E and F) and the rotation of the innermost ("floor") domains of VP5 to form the continuous floor (compare panels E and F; six floor domains form the red, yellow, and blue ring at the center of panel E). (G and H) Cross-sectional views through a hexon in the procapsid (G) and mature state (H), showing the three domains of the VP5 subunit separated by dotted lines: the outer (top), the intermediate, and the floor (bottom). In each case, a single VP5 subunit is delineated in pink. The drawbridge domain (marked D in panel H) is a subdomain of the intermediate domain and is retracted against the rest of this domain in the procapsid state. Two subdomains of the floor domain are denoted 1 and 2. Their changes in position between panels G and H illustrate the rotation undergone by the floor domain in maturation. From J. B. Heymann, N. Cheng, W. W. Newcomb, B. L. Trus, J. C. Brown, and A. C. Steven, *Nat. Struct. Biol.* **10:**334–341, 2003, with permission.

capsid. Moreover, the orientation of the triplexes changes, so that they transform from being mere structural elements of the procapsid to becoming molecular clamps that overlie the capsomers, binding them together in the stable mature capsid. Furthermore, the scaffolding proteins are cleaved and lost.

It is not yet entirely clear how the portal is incorporated into the growing structure. Part of the mystery concerns the fact that one, and only one, portal complex is present in the midst of the 12 otherwise geometrically identical vertices. Regardless, eventually, the closed spherical procapsid is complete (Box 18.11).

Encapsidation of DNA

Progeny viral genomes are generated by rolling-circle replication, as described above. As such, progeny genomes are initially contained within concatemers that comprise the linear tails of the actively replicating rolling-circle replication complexes. Consequently, progeny genomes must be cleaved from the rolling-circle replication complexes before they can be enclosed within preassembled procapsid structures. For this purpose, herpesvirus genomes contain so-called "*a*" sequences, which are located in R$_L$ and R$_S$ on the viral genomes (Fig. 18.7). Each *a* sequence contains within it two short segments, *pac1* and *pac2*, which are necessary

BOX 18.11

The analyses described in this sidebar strike me as especially intriguing. One has to read further into it to appreciate why that is so and why it is located here in the text.

Based on an extensive amount of sequence data and detailed analyses of capsid structures, it is now clear that mammalian and avian herpesviruses are descended from a common ancestral virus. A fascinating corollary to emerge from these analyses is that the individual herpesviruses have been coevolving with their hosts. For example, considering the mammalian herpesviruses, it appears that there existed, before the mammalian radiation of 80 to 60 million years ago, an ancestral herpesvirus which gave rise to the variety of mammalian herpesviruses present today and that speciations within sublineages (*Alpha-*, *Beta-* and *Gammaherpesvirinae*) took place in the last 80 million years. Importantly, the branching pattern of the viral evolutionary tree is congruent with that of the corresponding host lineages, implying coevolutionary development of virus and host lineages.

Interesting, no doubt, but why introduce this story here? At a certain point, DNA and protein sequences become too distant to imply phylogenetic relationships. However, such relationships can still be inferred from detailed structural considerations. The bottom line is that aspects of capsid structure now suggest an evolutionary link between the herpesviruses and the tailed, or T4-like, bacteriophages!

Such a possible link between the eukaryote-infecting herpesviruses and the prokaryote-infecting T4-like bacteriophages

was suggested earlier, based on parallels in their capsid assembly pathways. This possibility was reinforced by detailed analyses of their respective portal complexes, through which DNA enters the capsid, and by a low level of similarity in amino acid sequences of capsid scaffold-processing proteases from both these eukaryotic and prokaryotic viruses. More recently, improvements in computer-generated capsid reconstructions now enable comparisons at the level of protein secondary structure, which reveal previously unsuspected key similarities between the major capsid protein, VP5, of HSV-1 and analogous bacteriophage proteins.

Taken together, these experimental findings and others reported in the original journal articles imply that the entire capsid-packaging machinery of modern-day herpesviruses and tailed bacteriophages is of ancient origin, having been passed down to these contemporary viruses from a common progenitor, from a time close to, or predating, the emergence of vertebrates in the Cambrian Period, 570 to 505 million years ago! Thus, the common elements of capsid architecture seen in the herpesviruses and T4-like viruses may well have arisen at a very early stage in the evolution of life.

References

McGeoch, D. J., S. Cook, A. Dolan, F. E. Jamieson, and E. A. R. Telford. 1995. Molecular phylogeny and evolutionary timescale for the family of mammalian herpesviruses. *J. Mol. Biol.* **247**:443–458.

McGeoch, D. J., F. J. Rixon, and A. J. Davison. 2000. Topics in herpesvirus genomics and evolution. *Virus Res.* **117**:90–104.

---

for cleavage and packaging. Viral maturation/encapsidation proteins in the procapsid recognize the *pac* sequences and cleave the DNA concatemers precisely at the genome termini, while simultaneously inserting the free genome end into the procapsid. Importantly, this process ensures that complete genomes, which contain all repeated sequences, are packaged. At least seven herpesvirus gene products are necessary for the specific cleavage and packaging of herpesvirus genomes. None of these gene products is necessary for DNA replication per se.

The precise mechanism by which the HSV-1 portal promotes DNA encapsidation is not yet known. However, the herpesvirus portal is now known to resemble architecturally the bacteriophage portals, which are used by many large double-stranded DNA bacteriophages as the docking site for DNA packaging and as the channel for DNA entry and exit. (Bacteriophage portals also are

referred to as "connectors," since they also serve as the attachment site for the phage tail [Box 18.11].) Thus, the herpesvirus DNA encapsidation mechanism may well be similar to that of the tailed bacteriophages.

Unfortunately, encapsidation in the case of the tailed bacteriophages is also not yet understood. However, one very plausible mechanism involves a ratcheting process involving the portal proteins, which is driven by adenosine triphosphate (ATP) hydrolysis (Fig. 18.11).

Final Assembly and Release

Before herpesvirus capsids exit from the nucleus, they appear to become surrounded by the primary tegument protein, encoded by UL31 in the case of HSV-1. The pathway by which the mature herpesvirus capsids then exit from the nucleus almost certainly begins with their budding through the inner nuclear membrane. Interestingly,

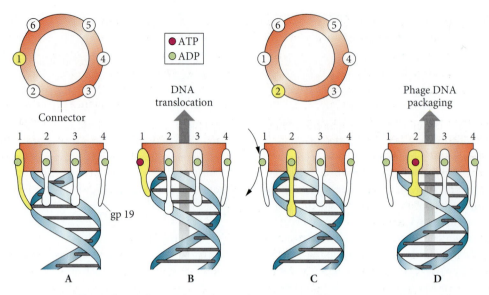

Figure 18.11 The ratchet mechanism for packaging phage DNA is as follows. (A) Six molecules of the packaging protein are bound to the phage portal (connector). When one of the packaging proteins is bound to the sugar-phosphate backbone of the DNA duplex, the DNA translocation adenosine triphosphatase (ATPase) is activated. (B) When ATP is hydrolyzed, a conformational change is induced in the packaging protein, forcing the DNA into the head. (C) When a new ATP binds to this packaging protein, the packaging protein disassociates from the DNA and returns to the original conformation. (D) At this moment, the next packaging protein binds to the neighboring sugar-phosphate backbone. When this cycle proceeds six times, six molecules of ATP are hydrolyzed, and one turn of the DNA is translocated into the head. Modified from H. Fujisawa and M. Morita, *Genes Cells* 2:537–545, 1997, with permission.

herpesviruses are the only enveloped viruses that are known to bud from the nucleus.

Bearing in mind that the tegument is topologically analogous to the matrix of an enveloped RNA virus, the UL31 protein facilitates the budding of the capsid through the inner nuclear membrane. The HSV-1 particle also acquires (but only temporarily) the UL34-encoded phosphorylated membrane protein, which has been inserted into the inner nuclear membrane.

What happens next is not yet clear. One possibility is that the particle enters the cytoplasm by fusing with the outer nuclear membrane, losing its initial (primary) envelope in the process. The UL31 and UL34 proteins, which are absent from mature virus particles, may be lost in this process.

If the HSV-1 release pathway were to involve transport of virus particles into the cytoplasm as suggested above, then that is where the maturing particles likely would acquire additional tegument proteins, including α-TIF (i.e., VP16) and VHS. The capsids then could be transported to the plasma membrane, where they would be reenveloped by the modified plasma membrane, which by now would contain the viral envelope glycoproteins. (In this regard, and with respect to the possibilities enumerated below, the HSV-1 glycoproteins, like host glycoproteins, are synthesized at the endoplasmic reticulum [ER] and are transported by the normal cellular secretory pathway through the Golgi body and trans-Golgi network [TGN] before reaching their final destination.) Reenvelopment at the plasma membrane would then constitute the final maturation step. It may be mediated in part by α-TIF and VHS. If reenvelopment indeed were to occur at the plasma membrane, the process would concurrently release the mature virus particle from the cell.

Another possible final envelopment and release pathway might be as follows. After the enveloped capsids fuse with the outer nuclear membrane (thereby losing the initial envelope), they might then undergo reenvelopment by budding into an intracellular compartment of the secretory pathway, such as the ER, the Golgi body, or the TGN. From those locations, secretory vesicles might transport the reenveloped capsids to the plasma membrane, where they would be released from the cell by fusion of the secretory vesicle with the plasma membrane. Budding of HSV-1 capsids into the Golgi body and into the TGN indeed has been observed in electron micrographs. Note that a reenvelopment step is a feature of both this model and the one above. In the current model, reenvelopment occurs at an internal membrane, whereas in the above model it occurs at the plasma membrane. Reenvelopment at an intracellular membrane is actually favored by some herpesvirus researchers, in part because of an experimental finding that the lipid composition of extracellular mature virus particles resembles that of the TGN/Golgi body. In this model, the virus would need to target its envelope glycoproteins to the appropriate internal membrane.

Since the space between the two nuclear membranes is continuous with the lumen of the ER, it also is possible that the capsid might migrate to the ER lumen after crossing the inner nuclear membrane. The still-enveloped capsid then could be released from the ER and from the cell via the secretory pathway, without a reenvelopment step. In that event, the virus would have to acquire the envelope glycoproteins before leaving the ER, perhaps acquiring them at the inner nuclear membrane. Alternatively, the capsid might be released into the cytosol by fusing with the ER membrane, losing its primary membrane in the process. The capsid then might undergo reenvelopment by budding into the downstream Golgi or TGN compartments, with release then occurring via the secretory pathway. Otherwise, the capsid might bud through the plasma membrane, as in the first scenario hypothesized above.

Which of the above plausible scenarios might actually be correct? The answer would largely depend on the experimental verification of the specific intracellular targeting of the envelope glycoproteins and of the proteins comprising the tegument. Regrettably, relatively few studies have attempted to establish the precise cellular locations of those viral proteins during acute infection. Yet even if those locations were known, they might well include all the compartments of the secretory pathway and thus not point to any clear answer. Consequently, much of what actually is known comes from analysis by electron microscopy. Unfortunately, the results from electron microscopy analysis support specific steps in each of the above final maturation/exit models. That is, each pathway would have the virus interacting with several different cellular membranes, making it difficult to make a critical distinction between models. Moreover, analysis by electron microscopy provides only a static view of events, with no indication of their direction or sequence. At any rate, the currently favored model would have the capsid acquiring a primary envelope by budding through the inner nuclear membrane. The enveloped capsids then would fuse with the outer nuclear membrane, thereby generating naked cytosolic capsids. These capsids then would undergo reenvelopment by budding into an intracellular compartment of the secretory pathway, from which they would be transported by secretory vesicles to the plasma membrane, where they would be released from the cell by fusion of the vesicle with the plasma membrane.

Despite the general preference for the model in which final budding occurs at an internal cellular membrane rather than at the plasma membrane, there still is no direct proof to substantiate it, as noted above. Regardless, we still might ask what advantages the virus might gain by adopting such a convoluted exit pathway. A likely advantage is that by specifically targeting its envelope glycoproteins to an internal membrane rather than the plasma membrane, the virus can be somewhat more cryptic to the host's immune system, which is a most important consideration in the case of the herpesviruses.

LATENT INFECTION AND IMMUNE EVASION

The capacity of the human herpesviruses to establish latent infections that might "spontaneously" reactivate is a hallmark of the group. Moreover, because human herpesviruses can establish lifelong latency, with the potential for recurrent productive infection, the herpesvirus-infected individual can be a lifelong source of infection to individuals in the population who are not yet infected. (This point has already been noted, but repetition may be justified by its importance, one reason for which is as follows.) This attribute of the human herpesviruses enabled them to be sustained in small human groups such as existed in preagricultural, preindustrial times. For comparison, recall that measles virus, which gives rise only to acute infections, can survive only in populations that contain 500,000 or more individuals (Chapter 11). That is so because smaller populations do not generate a sufficient number of new susceptible individuals to sustain that virus. This fact led to the conclusion that measles virus became established in the human population relatively recently, since communities large enough to sustain measles virus did not exist before the last several thousand years. By a similar line of argument, and supported by the fact that herpesvirus infections generally are not life threatening in humans, we conclude that herpesviruses have long been present in humans. What is more, host-specific herpesviruses have been found in virtually every vertebrate species in which they have been sought, suggesting that herpesviruses have been thriving in their respective hosts for millennia. (If you have not already done so, now would be a good time to see Box 18.11.)

A second basic truth about the herpesviruses, which might be inferred from fundamental principles, is as follows. Pathogens that persist in infected hosts *must* evolve mechanisms by which they evade their hosts' antiviral immune mechanisms. Since latency is a characteristic feature of the herpesvirus life cycle, we are not surprised that these viruses indeed have developed numerous stratagems to undermine host immunity. Indeed, herpesviruses are able to impair both innate and adaptive immune responses at almost every level, including recognition of viral epitopes, presentation of viral peptides by MHC-I and MHC-II molecules, the recruitment of immune effector cells, complement activation, and apoptosis. In fact, some researchers estimate that as many as one-half of the 80 or more herpesvirus genes may function to subvert host immune defenses. The cases in point that follow are only representative examples of a much greater number expressed by each of the human herpesviruses. But before moving on, considering that herpesviruses have coexisted with us for millennia, might we actually derive any benefit from this association? The answer may well be yes, as discussed in Box 18.12.

HSV

The initial (or primary) lytic HSV-1 infection occurs in mucosal epithelial cells. During this stage, the virus invades the termini of sensory neurons that innervate the infected epithelial site. The virus is then transported to neuronal nuclei in the sensory ganglia, where it establishes latency. Latent HSV genomes persist in the neural cells as nonintegrated, circular DNA molecules. Since neural cells do not divide, the latent HSV genome can persist for the life of the neuron, without itself dividing. (To the best of my knowledge, HSV-1, HSV-2, and VZV are the only viruses known to establish latent infection in terminally differentiated, nondividing neuronal cells.)

BOX 18.12

Latent herpesviruses are generally thought of as having a parasitic relationship with the host, since they leave the infected individual at risk for subsequent viral reactivation and disease. But Barton et al. (2007) now show that mice that are latently infected with either murine gammaherpesvirus 68 or murine cytomegalovirus (which genetically resemble EBV and HCMV, respectively) may indeed derive a benefit from the latent infection, as they are resistant to infection with the bacterial pathogens *Listeria monocytogenes* and *Yersinia pestis*.

An initial key experimental finding was that macrophages from latently infected mice were uniformly activated, displaying upregulation of surface MHC-II molecules, as well as other indicators of macrophage activation (Chapter 4). This led the researchers to ask whether herpesvirus latency might be associated with host resistance to infection by other microbial pathogens. Thus, they found that macrophages from the latently infected mice indeed are particularly bactericidal, killing *L. monocytogenes* (which generally can survive in macrophages) rapidly after uptake, whereas macrophages from mock-infected mice showed no such protective activity.

To test whether this protection might be more general, latently infected and mock-infected mice were challenged with *Y. pestis*, an extracellular bacterium and the causative agent of bubonic plague. Again, the latently infected but not the mock-infected control animals were protected. Moreover, acute herpesvirus infection did not affect susceptibility to these pathogens. Additionally, a mutant herpesvirus that is unable to establish latency was not able to induce protection against *L. monocytogenes* in mice, nor did it stimulate macrophage activation, thereby confirming the importance

of latency in this protection. However, this protection was not universal, since latently infected and control mice were equally susceptible to West Nile virus.

Latency-induced protection is not antigen specific but instead involves elevated levels of the cytokines IFN-γ and TNF-α, which are soluble mediators of macrophage activation (Chapter 4). Thus, latency results in an upregulation of the basal activation state of innate immune defenses against subsequent infections. This activation of mediators of innate immunity was sufficient to mediate protection, as shown by the fact that in vivo depletion of CD4 and CD8 T cells (key mediators of adaptive immune responses [Chapter 4]) prior to challenge with *L. monocytogenes* had no effect. Plausibly, the chronic, low-level presentation of viral antigens, as a result of viral reactivation, results in the prolonged secretion of IFN-γ and TNF-α.

There are not yet any studies to show whether herpesvirus latency in humans affects susceptibility to secondary bacterial infection. However, the researchers note that the observation of latency-induced cross protection, as induced by two subfamilies of the *Herpesviridae*, suggests that this is a general aspect of the herpesvirus-host relationship, as they then state, "Thus, whereas the immune evasion capabilities and lifelong persistence of herpesviruses are commonly viewed as solely pathogenic, our data suggest that latency is a symbiotic relationship with immune benefits for the host."

Apropos the above, see Box 4.16.

Reference
Barton, E. S., D. W. White, J. S. Cathelyn, K. A. Brett-McClellan, M. Engle, M. S. Diamond, V. L. Miller, and H. W. Virgin IV. 2007. Herpesvirus latency confers symbiotic protection from bacterial infection. *Nature* **447**:326–329.

Recently, by use of sensitive polymerase chain reaction (PCR)-based procedures, low levels of transcripts of several HSV-1 lytic genes were detected in latently infected ganglia in a mouse model. Prior to those findings, it was thought that only latency-associated functions were expressed in latently infected cells. Nevertheless, the LATs, referred to above, are the only abundant HSV-1 RNAs expressed in these ganglia. Moreover, mutational analyses and transfection assays show that the HSV LATs play a key role in regulating latency. The LATs comprise two RNAs, of 1.5 and 2 kb in size, which actually are generated as spliced introns from an RNA precursor molecule transcribed from the genome strand complementary to that encoding the α0 mRNA. The unique latent-phase-active promoter for the LATs is present in each copy of R_L (Fig. 18.7). See Preston (2000) for a detailed review of HSV-1 latency.

The LATs and the RNAs from which they are derived do not encode any proteins, and they are retained within the nuclei of latently infected neurons. Their mode of action is not known with certainty, but one plausible notion is that they may act to interfere with production of the α0 protein by an antisense mechanism (see the preceding paragraph for the relationship between the LATs and the α0 coding sequence). As a consequence of impairing α0 expression, the transcriptional cascade that underlies productive infection would be restricted. Also worthy of note is the fact that whereas introns generally are unstable, LATs are stable and accumulate to high concentrations in the nuclei of latently infected neural cells, in which they are retained, as noted.

There also is experimental evidence suggesting that the LATs play a role in viral reactivation while also expressing antiapoptotic, anti-interferon (anti-IFN), and proneuronal survival activities. Regarding inhibition of apoptosis and neuronal survival, about 25% of the neurons of a trigeminal ganglion are killed when infected by wild-type HSV-1, while more than one-half are killed when infected by HSV-1 lacking *LAT*. The ability of LATs to promote neuronal survival results at least in part from their ability to block apoptosis, which cells commonly call into play to prevent production of infectious virus. This was shown by introducing LATs into cultured cells and then exposing them to apoptosis-inducing agents.

How might HSV-1 LATs impair apoptosis? Nigel Fraser and coworkers (Gupta et al., 2006) reported that the *LAT* gene contains an inverted repeat sequence that might be seen by the host cell as a micro-RNA (miRNA) precursor. (miRNAs regulate gene expression either by degrading specific mRNAs or by inhibiting their translation into protein.) Fraser and coworkers found that the *LAT*-encoded miRNA, miR-LAT, promotes the degradation of mRNAs encoding two host proteins. One of the affected

host mRNAs is that encoding the transforming growth factor β, a protein that can induce cell death. The other affected mRNA is that encoding SMAD3, which is a mediator of the signaling induced by transforming growth factor β.

Next, we ask by what mechanism efficient HSV-1 transcription is restricted to the LATs during latent infection. The answer appears to be as follows. During lytic infection, HSV-1 DNA is in the form of transcriptionally active euchromatin. (Euchromatin is the noncondensed, transcriptionally active form of chromatin, in contrast to heterochromatin, which is the transcriptionally inactive, condensed form of chromatin.) But during the establishment of HSV-1 latency in trigeminal ganglion neurons, lytic-gene promoters happen to be complexed with histones, as they become embedded in transcriptionally inactive heterochromatin. However, the LAT promoter and 5′ exon remain in transcriptionally active, histone-free euchromatin (Wang et al., 2005). Further analysis shows that the *LAT* locus is bounded by short DNA sequences that serve to insulate it from this chromosome remodeling.

Experiments that compare LAT⁺ and LAT⁻ viruses show that LAT expression itself increases the extent to which viral lytic-gene promoters are impaired by chromatin remodeling. In this regard, histone modifications, particularly methylation and acetylation, play an important role in determining whether DNA is in the euchromatin or heterochromatin state. Thus, LAT expression also causes an increase in the amount of the dimethyl lysine 9 form of histone H3 (a component of heterochromatin), while reducing the amount of the dimethyl lysine 4 form of histone H3 (a component of euchromatin) on viral lytic-gene promoters (Wang et al., 2005). Apparently, the HSV LATs manipulate the cellular histone modification machinery to repress viral lytic-gene expression, in addition to their other effects enumerated above. Interestingly, emotional stress (a known activator of latent HSV-1; see below) can lead to a reversal of this chromosome remodeling, thereby possibly opening up the rest of the genome to transcription. Also, LAT expression itself is turned off following stress, consistent with a role for LATs in the maintenance of viral latency.

If expression of HSV-1 LATs were a sufficient condition for the establishment of HSV-1 latency, then we might expect to find latent infection in the epithelium as well as in neuronal cells. Since latency is not observed in the epithelium, what might be special about neurons to predispose them to latent HSV-1 infection? The answer in part may be as follows. Productive infection depends on the action of the HSV-1 VP16 protein (α-TIF), a component of the tegument that enters the nucleus at entry (see above). In the nucleus, VP16 is a powerful transcriptional activator of α-gene expression. However, it can serve this purpose only in association with two cellular proteins, Oct1 and Hcf (Fig. 18.8; see above also). In neuronal cells, Hcf is found only in the cytoplasm, thereby precluding the ability of VP16 to trigger the transcriptional cascade in neuronal cells that underlies lytic infection in epithelial cells.

If latency is to play a significant role in the life cycle of the herpesviruses, then there must be some natural means to reactivate the latent infection, thereby giving the virus an opportunity to infect new, immunologically naïve individuals in the population. HSV-1 is known to be reactivated by a variety of factors. These include stress, trauma, nerve damage, the menstrual cycle, steroids, and ultraviolet (UV) radiation (e.g., exposure to intense sunlight). Stimuli that cause reactivation of HSV-1 from latency also cause the relocation of Hcf to the nuclei of neuronal cells. This experimental finding provides support for the above assertion that the cytoplasmic localization of Hcf in neuronal cells is a factor predisposing those cells to latent rather than lytic HSV-1 infection. However, much remains to be learned about the mechanisms that trigger reactivation.

HSV-1 latency in neuronal cells is a significant adaptation of the virus, because neurons provide the latent virus with a site in which to reside that is somewhat immunologically privileged. The immunologically privileged status of neurons is due in part to the fact that they express only low levels of MHC-I molecules. But the peripheral nervous system is still accessed by lymphocytes, as well as by antibodies and complement. Still, the virus manages to remain "stealthy" in neuronal cells by generating negligible amounts of proteins that might be recognized by immune effectors.

While neuronal cells, by their nature, express only low levels of MHC-I molecules, that is not so in the case of the mucosal epithelial cells that support the primary HSV-1 infection. However, HSV-1 is able to suppress the MHC-I antigen presentation pathway in epithelial cells, so that the virus might evade cell-mediated immune responses while undergoing vegetative replication.

Importantly, the ability of HSV-1 to impair the MHC-I-mediated antigen presentation pathway (as well as other immune effectors) in epithelial cells facilitates the establishment of latency in neuronal cells. That is so because it enables the virus to sustain the primary infection of the mucosal membranes for a longer time, thereby providing a greater opportunity for the virus to invade the local innervating neuronal cells. Thus, the ability of HSV-1 to evade host immune defenses during the acute phase of the infection and its ability to do the same during latency as well are each key to the HSV-1 lifestyle, and we continue our discussion of HSV-1 immune evasion stratagems.

Bearing in mind that cell-mediated immune defenses are principal means by which the mammalian host resolves a primary virus infection (and controls and resolves reactivating infections), HSV-1 has evolved two distinct mechanisms by which it suppresses the MHC-I antigen presentation pathway. In the first, the HSV-1 VHS protein, which is contained in the tegument, impairs host protein synthesis, in that way nonspecifically shutting down antigen presentation at several steps. (Recall that as the capsid traffics to the nucleus, VHS is released into the cytosol, where it causes disruption of polysomes and degrades host mRNAs.) It is thought that VHS might degrade the mRNAs of IFN-stimulated genes, including those encoding MHC-I components. However, *vhs* mutants are yet able to block the MHC-I antigen presentation pathway, implying that there is a second mechanism by which HSV-1 blocks antigen presentation. In that second mechanism, the α47 gene product, ICP47, binds to subunits of the TAP transporter on its cytoplasmic side, thereby blocking import of peptides into the ER, where they otherwise would associate with MHC-I molecules (see Fig. 4.17 and 4.24). Interestingly, the HCMV US6 protein likewise blocks peptide transport by jamming TAP, but it does so from the ER lumen. It is noteworthy that the evasion of cell-mediated immune defenses is sufficiently important to the establishment and maintenance of herpesvirus latency that all of the herpesviruses studied to date have one or more means for impairing the MHC-I antigen presentation pathway.

Memory CD8 T cells, which are in place after the acute infection has been resolved (Chapter 4), constitute a ready line of defense against future HSV-1 challenge. Thus, when the latent HSV-1 infection reactivates and the virus reenters the epithelium, it then must contend with the host's powerful immunological memory. Under these circumstances, as well as during the primary infection, the virus likely benefits from expression of its ICP47 protein, since the ICP47-mediated inhibition of the MHC-I antigen presentation pathway may buy the reactivated virus additional time to transmit the infection to new susceptible individuals. In the epithelium, the effects of ICP47 eventually are offset by the IFN-γ secreted by the natural killer (NK) cells and the CD4 helper T cells that enter the lesions. Also, cells in which MHC-I expression is down-regulated at the cell surface are susceptible to attack by NK cells (Box 18.13). NK cells are discussed further below and in Chapter 4.

Under the conditions of a reactivated infection, where virus replication is subject to the immunological memory component of the adaptive immune response (Chapter 4), virus replication would be particularly sensitive to the induction of IFN, specifically IFN-α and IFN-β that are produced by virus-infected cells. However, HSV-1 is able

BOX 18.13

It would be most useful to have an in vivo mouse model in which to evaluate the role of ICP47 in HSV-1 latency. Unfortunately, HSV-1 ICP47 binds very poorly to the murine TAP and thus only weakly inhibits antigen presentation by MHC-I molecules in mice. To address this limitation, recombinant HSV-1 was generated to express either the murine cytomegalovirus protein m152 or HCMV US11 (see the text), each of which effectively inhibits antigen presentation by murine MHC-I molecules. The recombinant viruses indeed prevented CD8 T cells from killing infected cells in vitro. Importantly, the recombinant viruses replicated to higher titers in the central nervous systems of experimentally infected mice and induced paralysis more frequently than a mock HSV-1 recombinant. The differences between the recombinant viruses and the control HSV-1 were not observed in mice that were deficient for MHC-I expression or in mice in which CD8 T cells were depleted. These results confirm that the increased virulence of the recombinant viruses was due to inhibition of antigen presentation to CD8 T cells, which likely resulted in an increased virus presence in the central nervous system.

Reference

Orr, M. T., K. H. Edelmann, J. Vieira, L. Corey, D. H. Raulet, and C. B. Wilson. 2005. Inhibition of MHC-I is a virulence factor in herpes simplex virus infection of mice. *PLoS Pathog.* 1:e7.

to counter the IFN-α/β response by producing two γ-gene products, ICP34.5 and US11.

Before we consider how ICP34.5 and US11 hamper the IFN-α/β response, first recall that IFN-α/β do not necessarily protect the virus-infected cells in which they are first induced. Instead, the IFN molecules secreted by virus-infected cells have a more important effect on as-yet-uninfected neighboring cells. It is in these cells that IFN-α/β actually induce the antiviral state that helps to contain the infection, doing so by inducing the upregulation of expression of more than 100 distinct IFN-stimulated genes. The double-stranded RNA-dependent protein kinase, PKR, is one of the key products of these IFN-stimulated genes. However, the PKR that is upregulated in response to IFN-α/β remains inactive unless and until the cell might be infected. The primary activator for PKR in virus-infected cells is the viral double-stranded RNA that many viruses inadvertently produce during infection. More recently, a human protein called PACT was found to also activate PKR in virus-infected cells. When PKR is activated, it blocks virus infection by phosphorylating

eukaryotic initiation factor-2α (eIF2-α), thereby shutting down protein synthesis in the infected cell. (See Chapter 4 for a discussion of IFNs.)

ICP34.5 acts by directly binding to, and activating, the host protein phosphatase 1α, which then dephosphorylates eIF2-α, thereby preventing the IFN-α/β-induced shutdown of translation and the resulting apoptosis. See Box 4.6 for an experiment in a mouse model confirming the role of ICP34.5 in vivo.

US11 is produced late in the HSV-1 lytic cycle and is packaged into the viral tegument. How this late HSV-1 protein blocks the IFN-α/β-induced antiviral state is not entirely clear. But before considering how it might do so, we first need to explain how double-stranded RNA and PACT each activate PKR. Double-stranded RNA activates PKR as follows. PKR contains two double-stranded RNA-binding motifs (dsRBMs). One of these motifs, dsRBM2, binds to the kinase domain on the nonactivated PKR protein, via an intramolecular interaction that keeps PKR in a closed, inactive conformation. When double-stranded RNA binds to the dsRBMs, it thus induces a conformational change of PKR, leading to its activation. Considering PACT next, it activates PKR, by an RNA-independent mechanism, by directly binding to the same motif in the PKR kinase domain that specifically interacts with dsRBM2. In that way, PACT releases PKR from its inactive conformation.

Since US11 can bind to PKR in an RNA-independent manner, by doing so it may well block the activation of PKR by double-stranded RNA. But US11 can also block the activation of PKR by PACT. In this regard, US11 binds to PACT, as well as to PKR. However, it needs to bind only to PKR to inhibit PKR activation by PACT. But contrary to expectations, the binding of US11 to PKR does not affect the interaction of PKR with PACT. However, by binding to PKR, US11 may yet block the conformational change of PKR induced by PACT binding.

Remarkably, US11 has yet another means by which it impairs the IFN-α/β-induced antiviral state; in this instance it involves ribonuclease L (RNase L), another key effector of the IFN-α/β-induced antiviral state (Chapter 4). RNase L is a nuclease that degrades most viral and cellular RNAs, and, like PKR, it too is activated by a multistep process. First, IFN-α/β induces the upregulation of RNase L, as well as the upregulation of another cellular enzyme, 2′-5′-oligo(A) synthetase. But, as in the case of PKR, these cellular enzymes remain inactive until the IFN-stimulated cell might be infected. Should the cell be infected, then the double-stranded RNA that is generated by the virus activates the 2′-5′-oligo(A) synthetase, which then produces 2′,5′ oligomers of adenylic acid, which is an unusual polynucleotide that binds to, and activates,

RNase L. Getting back to US11, this HSV-1 protein blocks the activation of RNase L by blocking the activation of 2′-5′-oligo(A) synthetase. It is not clear how US11 blocks the activation of 2′-5′-oligo(A) synthetase. However, US11 has a dsRBM domain, which is necessary for it to block the activation of 2′-5′-oligo(A) synthetase, suggesting a mechanism that might, in part, result from the sequestering of double-stranded RNA produced during infection. At any rate, the HSV-1 US11 gene product is able to counteract RNase L, in addition to PKR and PACT.

HSV-1 also encodes two envelope glycoproteins, gC and gE, that together counteract complement pathways and antibodies. Glycoprotein gC inhibits complement activation by binding to C3b, an interaction that in turn prevents the assembly of membrane attack complexes (Chapter 4). Glycoprotein gE blocks antibody-mediated host defenses by acting as a receptor for the Fc portion of antibodies, thereby blocking Fc-mediated functions (Chapter 4). The binding of the Fc portions (rather than the Fab portions) of HSV-1-specific antibodies to the plasma membranes of infected cells and to the envelopes of free viruses could protect the former against antibody-dependent cell cytotoxicity (Chapter 4) and the latter against neutralization.

Although HSV-1 is able to undermine IFN-, complement-, and antibody-mediated immune defenses, keep in mind that cell-mediated immune defenses are the arm of the immune system that is primarily responsible for resolving primary virus infections and for the control and resolution of reactivating infections. This is demonstrated in animal model systems and in immunosuppressed human patients as well. For example, HSV-1 reactivates more frequently in AIDS patients and in organ transplant recipients undergoing immunosuppressive treatment (in whom cell-mediated immune defenses are severely compromised) than in immunocompetent individuals. Indeed, the eight human herpesviruses all pose an increased threat to AIDS patients (Box 18.14). Severe Kaposi's sarcoma in AIDS patients provides a dramatic example of this point, as described below. Cell-mediated defenses are crucial to resolving primary infections and recurrences, since they are required to kill productively infected cells. Antibodies can help to prevent the spread of the virus, but they cannot resolve the infection because of the ability of the virus to spread directly from cell to cell. Thus, we might better appreciate the importance to the herpesviruses of their ability to impair the MHC-I antigen presentation pathway.

The belief that latent HSV-1 infection is "silent" led to the premise that latently infected neurons are largely invisible to the immune system. Recently, with the emerging evidence that some lytic genes indeed are expressed (although at very low levels) in latently infected neurons, evidence of a persisting immune response within latently

BOX 18.14

HHV-6 and HHV-7 are thought of as benign betaherpesviruses that generally cause mild early-childhood illnesses. However, high concentrations of the HHV-6 genome have been detected in association with both progressive multifocal leukoencephalopathy (Chapter 15) and multiple sclerosis lesions in immunologically impaired individuals (see the text), raising the possibility that HHV-6 activation might play a role in the pathogenesis of these demyelinative diseases. HHV-7 has been associated with cytomegalovirus "CMV disease" in renal transplant recipients, suggesting that it could play a part in that syndrome.

infected ganglia also emerged. For instance, in mouse and rabbit models, as well as in humans, the persistence of inflammatory cells and cytokines has been observed in the absence of infectious virus. Moreover, low levels of both specific and nonspecific T cells infiltrate the ganglia during routine immunosurveillance.

Some research reports state that the inflammatory response eventually subsides within latently infected ganglia. Yet it is possible, if not likely, that activated CD8 T cells remain in the ganglia, where they might play a role in maintaining HSV-1 latency. This notion is consistent with results from a mouse model that showed that CD8 T cells act to diminish the virus presence in the central nervous system. See Box 18.13 for details.

Importantly, there is experimental evidence that CD8 T cells may block viral reactivation in neuronal cells by a means that is *nonlethal* to the neuronal cells, an effect that might be mediated by the release of IFN-γ by the CD8 T cells. This cytokine is known to be a powerful inducer of an antiviral state that is different from that induced by IFN-α/β (Chapter 4). But the mechanism by which IFN-γ inhibits replication of HSV-1 is not yet clear and may vary in different cell types. (IFN-γ is secreted mainly by T lymphocytes and NK cells and perhaps by antigen-presenting cells as well, while IFN-α/β are secreted by virus-infected cells [Chapter 4]. Genetic studies involving mice that have targeted mutations of either IFN-γ or IFN-α/β receptors show that signaling pathways induced by both types of IFNs play an important role in inhibiting viral replication.)

The pool of HSV-1-specific CD8 T cells in the latently infected trigeminal ganglia is likely maintained by the intermittent exposure of these T cells to virus antigens expressed there. This premise is consistent with an experimental observation in a latently infected mouse model, which showed that cytokine production in the trigeminal ganglia is inhibited by treatment with the antiherpes drug acyclovir (see below). Yet much remains to be learned about the maintenance of an HSV-1-specific T-cell response and its role in maintaining latent infection. Interestingly, in a mouse model, virtually all of the HSV-1-specific CD8 T cells in the latently infected trigeminal ganglia were specific for a single epitope on viral glycoprotein B.

EBV

EBV, like HSV-1, is both ubiquitous in the human population and extremely efficient at establishing latent infection. Indeed, more than 90% of the human population worldwide carries EBV as a latent virus.

Primary EBV infection occurs in the mucosal epithelium of the nasopharynx. The virus then infects B cells in the underlying lymphoid tissue, in which it establishes latency. Because of the omnipresence of EBV in the human population, it has been referred to as "the ultimate B lymphocyte parasite."

As currently understood, the pathway to EBV latency begins when the virus infects *naïve*, resting B cells. (B and T cells that have never been activated are referred to as being naïve or virgin [see below and Chapter 4].) Actually, EBV can indiscriminately infect any resting B cell, but for reasons that will become clear (hopefully), latency is established in those B cells that are naïve. In B cells, the virus expresses only a subset of its genes (referred to as the **growth-promoting program**). This restricted expression of EBV genes induces infected, resting B cells to proliferate and become activated large lymphocytes or **lymphoblasts**, which exhibit an increased rate of RNA and protein synthesis.

In the normal course of events, resting, small B lymphocytes are activated and proliferate when their B-cell receptors are stimulated by their cognate antigen (Chapter 4). Upon receiving that signal, the resting B cell then becomes a lymphoblast. Antigen-activated lymphoblasts migrate to the follicles of secondary lymphoid organs, where they form **germinal centers**. (Germinal centers are sites in lymphoid organs where there is intense B-cell proliferation, selection, and maturation. It is where B cells become short-lived antibody-producing plasma cells and long-lived memory B cells.)

As in the case of normally activated B cells, EBV-activated B cells likewise migrate to lymph nodes, where they too form germinal centers and where some differentiate into long-lived memory B cells. Thus, the EBV growth-promoting program may have an effect on the B cell that is equivalent to normal B-cell stimulation by antigen, at least with respect to the needs of the virus. Importantly, the end result is that EBV can persist in the peripheral blood in long-lived memory B cells. These cells proliferate slowly and are not a pathogenic threat to immunocompetent carriers. (Normal memory B cells undergo a slow rate of homeostatic proliferation.)

Why does EBV go through this circuitous pathway of infecting naïve B cells to get to the memory B-cell compartment, when it might have infected memory B cells directly? Perhaps it is because infection and activation of memory cells would cause them to then become short-lived plasma cells (Chapter 4). In contrast, activation of naïve B cells leads to long-lived memory cells. Thus, EBV subverts the normal biology of B-cell differentiation to establish latency in the long-lived memory B-cell subset. This premise is consistent with the experimental observation that just as antigen-driven B-cell activation/differentiation occurs in the secondary lymphoid organs rather than in the peripheral circulation, so do the key steps in establishing EBV latency.

A minor proviso to the above argument is that whereas there is experimental evidence that memory B cells are long-lived, other evidence suggests that they are short-lived unless restimulated by their cognate antigen. But leave it to the virus to find a way out of this potential conundrum. As described below, EBV continuously expresses its so-called "LMP2" protein within latently infected B cells. LMP2 constitutively transmits a signal that mimics the signal transmitted by the B-cell receptor (Chapter 4). Thus, LMP2 expression may act to continuously maintain the latently infected memory B cell population, independent of antigenic stimulation.

Next, we consider the pattern of EBV gene expression during the establishment and maintenance of latency, noting that it becomes more restricted in the latently infected memory B cells than it was at the time of initial B cell infection. The highly restricted pattern of EBV gene expression in the memory B cells enables the virus to be considerably less visible to the host's immune system. However, this is not the sole means by which EBV evades host immunity (see below).

In the first stage of B-cell infection, sometimes referred to as **stage 3 latency**, the virus expresses as many as 11 different genes. This pattern of transcription is referred to as the **growth program**. The 11 genes expressed during this stage comprise all of the known latency-associated EBV genes. They include (i) the six nuclear proteins, referred to as the EBV nuclear antigens or EBNAs, i.e., EBNA-1, EBNA-2, EBNA-3A, EBNA-3B, EBNA-3C, and EBNA-LP (LP for latent protein); (ii) three integral latent membrane proteins or LMPs, i.e., LMP1, LMP2A, and LMP2B; and (iii) two small nonpolyadenylated EBV-encoded RNAs or EBERs, i.e., EBER1 and EBER2 (Fig. 18.12). Six of these viral latency proteins, EBNA-1, EBNA-2, EBNA-3A, EBNA-3C, EBNA-LP, and LMP1, are absolutely required for activating B cells. The specific functions of some of these latency factors are described below. Interestingly, the genes expressed during lytic infection by EBV (a gammaherpesvirus) are similar to those expressed during lytic infection by HSV-1

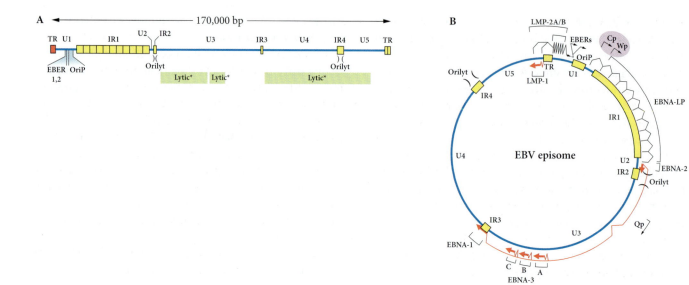

Figure 18.12 The EBV genome and latency transcripts. (A) The 170,000-bp EBV genome contains five regions of unique sequence, separated by regions of DNA containing numerous repeated sequence elements. The IR1 region contains a number of copies of a basic repeat element. The virus has two origins of replication, the *oriP*, which mediates latent genome replication and segregation in concert with cellular chromosome replication, and two copies of the *oriLyt*, which functions during productive infection for vegetative DNA replication. Three regions of the genome contain genes homologous to those expressed during the productive replication of the alphaherpesviruses. The structures of these genes are similar in the two classes of herpesviruses, with each gene being expressed under the control of its own cognate promoter. (B) During the latent phase of infection, the genome circularizes and becomes a histone-associated minichromosome. A number of latent transcripts are expressed; one family, the EBNAs, are derived by alternative splicing of a very large precursor transcript which extends from one of two promoters (Cp and Wp) rightward to the end of the IR3 region. The LMP transcripts are expressed from their own cognate promoters, and LMP-2A/B are also derived by alternate splicing of a small precursor RNA. The EBERs are expressed from RNA polymerase III promoters. Adapted from E. K. Wagner, M. J. Hewlett, D. C. Bloom, and D. Camerini, *Basic Virology*, 3rd ed. (Blackwell Publishing, Malden, MA, 2008), with permission.

(an alphaherpesvirus). In contrast, the EBV genes expressed in latently infected B cells have no homology to the genes of other herpesviruses, including HSV-1. How might you explain this? See above.

During stage 3 latency, EBV genomes circularize and then undergo limited replication, as mediated by OriP (origin of plasmid replication; so named because it maintains the viral genome as an *episome* in latently infected B cells). (Strictly speaking, a plasmid is an extrachromosomal DNA molecule that is capable of replicating independently of the chromosomal DNA, whereas an episome can exist either autonomously or as part of a chromosome. Despite the fact that the EBV genome is not known to persist as a part of a cellular chromosome, the journal literature often refers to the extrachromosomal EBV genome as an episome.)

Following the limited amplification of the viral genome (see below) and activation of the B cell, the activated B cell migrates to a lymphoid follicle, and the pattern of viral gene expression is then downregulated to the so-called **default program**, in which only EBNA-1, LMP1, and LMP2 are expressed. Expression of LMP2 is sufficient to cause the B cell to form a germinal center (see above) in the follicle.

Together, LMP1 and LMP2 can drive the B cell to undergo antibody isotype switching and affinity maturation (see Chapter 4), which are defining germinal center events. LMP1 also triggers resulting memory cells to leave the germinal center and enter the circulation.

The pattern of EBV gene expression may change again, such that latently infected memory B cells express only EBNA-1. This can be important to long-term EBV persistence, since EBNA-1 has properties that enable it to avoid being presented by the MHC-I antigen presentation pathway (see below). The EBNA-1-only stage is referred to as **type I latency**.

During type I latency, EBV genomes are stably maintained by being replicated just once per cell cycle, concurrent with cellular DNA replication in S phase. Note that an important distinction between EBV latency in B cells and HSV latency in neuronal cells is that viral genomes replicate in the former instance, but not in the latter. That is so because B cells may divide whereas neuronal cells do not. (Proviso: Cells displaying the type I latency transcription pattern are seen in Burkitt's lymphoma cells [see below] but have been difficult to detect in other EBV-infected individuals.)

Whereas EBV DNA synthesis during lytic infection of nasopharyngeal mucosal cells is mediated by the viral DNA polymerase, the limited EBV DNA replication at all stages of latency in B cells is instead mediated by the cellular DNA polymerase. In fact, EBV DNA replication in a latently infected B cell requires only one viral protein, EBNA-1.

EBNA-1 is a DNA-binding protein that binds to *oriP*, the latency-specific origin of replication. Moreover, EBNA-1 that is bound to *oriP* enables the viral genome to replicate in synchrony with cellular chromosomes. EBNA-1 also binds to other sites on the viral genome and to host chromosomes as well. In so doing, EBNA-1 tethers viral genomes to host chromosomes, so that the replicated viral genomes can be segregated to each of the daughter cells upon each cell division. The end result is that 10 to 50 copies of the viral genome are maintained per cell, on average.

As we have seen, a key feature of the EBV latency stratagem is to promote the maturation of naïve B cells into memory B cells, presumably because memory B cells are long-lived, perhaps even immortalized. Thus, EBV creates a pool of possibly immortalized memory B cells that can proliferate indefinitely and in which the virus can maintain the latent infection. For the sake of comparison, recall that the smaller DNA viruses (i.e., the polyomaviruses, papillomaviruses, and adenoviruses) transform normal cells into immortalized neoplastic cells by inactivating pRb and p53 (Chapters 15, 16, and 17). Another important distinction between B-cell immortalization induced by EBV and immortalization induced by the smaller DNA viruses is that while the immortalization of B cells is a crucial feature of the EBV lifestyle, immortalization by the smaller DNA viruses is usually a rare, unintended event that is not known to further the aims of the virus.

EBV induces the immortalization of B cells as follows. In the germinal centers, where B cells normally mature into antibody-secreting plasma cells and memory cells, the EBV latency-associated membrane antigens, LMP1 and LMP2, are thought to mimic signals usually provided via CD40 and the B-cell receptor, respectively. This is important because the normal role for CD40 and the B-cell receptor is to signal B-cell activation and maturation (Chapter 4). Consequently, LMP1 and LMP2 provide the necessary signals, in the absence of the actual cognate antigen and T-cell help, to stimulate B cells to proliferate and differentiate into memory B cells. (This particular adaptation of EBV strikes me as one of the more amazing features of its rather extraordinary latency stratagem.) EBNA-2 plays a key role in the process, since it transcriptionally upregulates LMP1 and LMP2. The mechanisms of action of LMP1 and LMP2 are as follows.

The EBV LMP1 protein, which contains six hydrophobic transmembrane domains, becomes an integral protein of the host cell plasma membrane. At the cell surface, LMP1 functions as a constitutive signal transducer, because its six transmembrane domains cause it to form clusters in the plane of the plasma membrane, in the absence of any ligand. The cytoplasmic carboxy termini of the clustered LMP1 molecules then interact with, and activate, cellular proteins that are mediators of intracellular signaling pathways. Accordingly, LMP1 is able to constitutively transmit signals which mimic cellular signals that normally induce cell proliferation. One of the cellular proteins through which LMP1 acts is involved in signaling pathways induced by TNF (see Chapter 4), and some of the effects of LMP1 are thus due to the activation of NF-κB (Fig. 18.13; also see Chapter 4).

LMP1 has also been shown to protect B cells from apoptosis; this effect is mediated in part through LMP1 signals that upregulate transcription of Bcl-2, a cellular regulatory protein that can block apoptosis (Chapter 4). LMP1 also may prevent apoptosis by interfering with the transmission of apoptotic signals from the cell surface.

LMP2, like LMP1, also contains multiple transmembrane domains that likely enable it to form clusters in the plasma membrane and consequently transmit signals in the absence of any ligand. One such LMP2-mediated signal mimics B-cell-receptor-mediated signaling that promotes the maturation of the naïve B cell into an immortalized memory B cell, as described above. (Recall that LMP1 and LMP2 together provide all the cues necessary to cause an infected B cell to become a memory B cell.) However, the effects of LMP2 may be more extensive yet, acting also to regulate the latent infection. This occurs as follows. LMP2 cytoplasmic tails bind to, and thereby sequester, the Src family tyrosine kinases Fyn and Lyn. Consequently, a significant fraction of these cytoplasmic signal mediators are not available to other signaling receptors. This affects the maintenance of EBV latency, since it impairs B-cell-receptor-mediated signaling that can reactivate the latent EBV infection in memory B cells (see below).

LMP2 has yet another way to impede signaling mediated via Fyn and Lyn. It results from the fact that the clustering of LMP2 causes the bound Fyn and Lyn kinases to be cross-phosphorylated at negative regulatory sites. Src kinases that are phosphorylated at these sites cannot transmit signals, since their kinase domains are then maintained in an inactive conformation (Fig. 15.15 and Fig. 20.15).

Once EBV latency has been established in memory B cells, EBNA-1 and LMP2 are the only EBV proteins that can be detected. Importantly, this highly restricted pattern of EBV gene expression minimizes the likelihood that the latently infected cells will be recognized by host immune effectors. This is significant because people carrying latent EBV in their B cells indeed maintain CD8 cytotoxic

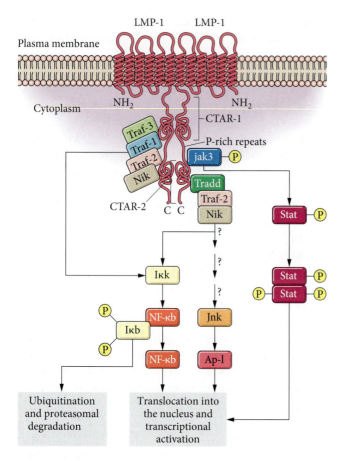

Figure 18.13 The EBV LMP1 protein contains six hydrophobic transmembrane domains, which cause LMP1 to become an integral plasma membrane protein. It becomes a constitutive signal transducer because its six transmembrane domains cause it to form clusters in the plane of the plasma membrane, in the absence of any ligand. This clustering is depicted in the figure for two LMP1 molecules. The cytoplasmic carboxy termini of the clustered LMP1 molecules then are able to interact with, and activate, cellular proteins that are mediators of cellular cytoplasmic signaling. Accordingly, LMP1 is able to constitutively transmit signals that mimic cellular signals that normally induce cell proliferation. As shown, members of the TNF receptor-associated protein family (Trafs) associate with the LMP1 C-terminal activation regions, and some of the effects of LMP1 are thus due to its signal-mediated activation of NF-κB. Also note that LMP1 can activate the JAK/STAT signaling pathway (Chapter 4). Adapted from S. J. Flint, L. W. Enquist, R. M. Krug, V. R. Racaniello, and A. M. Skalka, *Principles of Virology: Molecular Biology, Pathogenesis, and Control,* 2nd ed. (ASM Press, Washington, DC, 2004), with permission.

T lymphocytes (CTLs) specific for EBV-infected cells. Thus, downregulation of latent EBV gene expression to a pattern in which only EBNA-1 and LMP2 are expressed or to an EBNA-1-only pattern of expression, as occurs in type I latency, may be an important mechanism that EBV uses to evade an EBV-specific CTL response.

The stealth of the latent EBV infection is enhanced further by the notable property of EBNA-1 of being intrinsically resistant to proteasomal degradation. Consequently,

EBNA-1 is resistant to being presented by the MHC-I antigen presentation pathway. The upshot is that the endogenous expression of EBNA-1 in latently infected B cells will not cause those cells to be recognized and killed by EBNA-1-specific CD8 T cells. The mechanism by which EBNA-1 resists proteasomal degradation is the same as that used by long-lived cellular proteins that likewise avoid proteasomal degradation. These proteins contain a glycine-alanine repeat domain that confers this resistance. This inhibitory sequence, when transferred to any other protein, will confer resistance to proteasomal degradation.

Despite the immune-evasion tactics manifested by EBV, the cell-mediated immune response nevertheless controls the latent EBV infection in vivo. In support of that premise, EBV-specific CTLs can prevent the outgrowth of EBV-infected B cells in tissue culture. More to the point, immunosuppression in vivo associated with organ transplantation or AIDS can result in uncontrolled proliferation of EBV-infected cells. (In light of these facts, adoptive immunotherapy with EBV-stimulated T cells has been suggested as an effective treatment for EBV-associated posttransplant lymphoproliferative syndrome [a life-threatening B-cell proliferative disease; see below].) The antigen specificity of the CTLs that control EBV infection in vivo is not yet known.

In order for latency to play its key role in the EBV life cycle, the virus must be able to at times reactivate from its latent state. The switch from latency to a productive phase is controlled by two EBV immediate-early proteins, Zta (*BZLF1*, Zta, ZEBRA) and Rta (*BRLF1*), which are coordinately expressed from a common bicistronic RNA transcript.

Zta is a DNA-binding protein that recruits cellular transcription factors to multiple sites on the viral genome, thereby functioning as an essential activator of EBV early gene transcription. In addition, Zta binds directly to the EBV origin for *lytic* replication (*ori*Lyt) and recruits the virus-encoded DNA primase and polymerase processivity factors that are essential for productive DNA replication. In fact, Zta is the only EBV-encoded protein known to specifically recognize *ori*Lyt. In so doing, it ensures a switch from *ori*P-dependent genomic replication to *ori*Lyt-dependent genomic replication.

Rta is known to promote the translation of Zta from the common bicistronic mRNA that encodes both proteins and also to autostimulate its own synthesis. Moreover, Rta activates transcription of early and late lytic-cycle genes. Thus, Rta, in conjunction with Zta, induces the transcriptional cascade that culminates in productive infection. Rta is also necessary for viral DNA replication, but unlike Zta, Rta is not known to bind directly to *ori*Lyt.

Here is yet another means by which Zta promotes reactivation of the latent infection. During latency, the EBV genome is largely silenced by cell-driven methylation of

CpG motifs. In the switch to the lytic cycle, this epigenetic silencing is overturned, in part by the activation of the viral *BRLF1* gene by Zta.

Rta and Zta are each necessary for lytic EBV infection of epithelial cells, as well as for reactivation of the latent infection in B lymphocytes. As then might be expected, Zta and Rta are not generated in latently infected cells, as long as the infection remains latent in the cell, since their expression there would be sufficient to activate productive infection.

What then triggers expression of Zta and Rta in latently infected B cells, which in turn reactivates the latent virus? First, expression of Zta and Rta is controlled by several cellular DNA-binding proteins that stimulate or repress transcription of the Zta/Rta bicistronic mRNA. The action of many of the cellular proteins that control Zta and Rta expression are regulated by extracellular signals. Consequently, EBV can be reactivated from its latent state by any extracellular signal that leads to expression of Zta and Rta, such as cross-linking of the B-cell receptor. That is, extracellular signals that cause latently infected B cells to differentiate into antibody-secreting plasma cells will cause the latent virus to reactivate, a fact that is clearly relevant to the life history of EBV. A second relevant phenomenon is that once memory B cells are induced to become antibody-secreting plasma cells, they migrate to the mucosal epithelia, including the salivary gland mucosal epithelium. Together, these facts make it possible for EBV to spread via the saliva. In addition, EBV that is shed from the B cells is likely amplified by reinfecting mucosal epithelial cells of the nasopharynx, thereby increasing the viral titer in the saliva.

Upon being reactivated, EBV (like HSV) would be sensitive to IFN-α/β. As then might be expected, EBV has evolved a means to counter the host IFN response. EBV does so by expressing a set of small virus-encoded RNAs, the EBERs. Like the adenovirus VA RNAs (Chapter 17), which also block the IFN-α/β response, the EBERs are highly structured, noncoding RNAs that are transcribed by the cellular RNA polymerase III. However, in contrast to the adenovirus VA RNAs, which act by binding to PKR, in that way blocking its activation by double-stranded RNA, the EBERs do not appear to directly bind to PKR. Nor are they able to block the phosphorylation of PKR. Thus, the EBERs appear to inhibit the effects of IFN-α/β downstream of PKR, by an as-yet-unknown mechanism.

The EBERs help EBV to evade host defenses by yet other mechanisms, which involve (i) enhancing cellular expression of interleukin-10 (IL-10), by an as-yet-unknown means, and (ii) augmenting the elevated levels of cellular IL-10 by expressing a virus-encoded IL-10 homologue. How do these EBV activities subvert antiviral host defenses? The answer is as follows. When helper T cells reach an infected site, the costimulatory signals they receive at the site determine whether they will secrete the T helper 1 (Th1) or Th2 subset of cytokines (see Chapter 4). For example, activated macrophages responding to a viral (or bacterial) infection secrete IL-12, which causes helper T cells arriving at the site to secrete the Th1 subset of cytokines. In contrast, helper T cells at a site of parasitic infection will find themselves in a cytokine environment dominated by IL-4, which will cause them to secrete a Th2 cytokine profile. The Th1 cytokine profile, which is associated with virus infections, includes IFN-γ, IL-2, and TNF. These cytokines are needed to defend against the virus and likely promote the expansion of the pool of virus-specific, CD8 cytotoxic T cells. The Th2 cytokine profile includes IL-4, IL-5, and IL-10, which are appropriate for defending against invading parasites. Importantly, IFN-γ made by Th1 cells decreases the rate of proliferation of Th2 cells, whereas IL-10 produced by Th2 cells decreases the proliferation of Th1 cells. Thus, via the EBER-induced upregulation of IL-10 and expression of its IL-10 homologue, EBV is able to bias the host immune response towards a Th2 direction, which is inappropriate for defending against a viral infection. Thus, EBV manages to manipulate the Th1/Th2 differentiation pathway to its advantage. (Provisos: First, although the Th1/Th2 paradigm is valuable for educational purposes, it is likely an oversimplification of the actual state of affairs in vivo, since the diversity of T-cell responses is more varied than merely the strict Th1 versus Th2 distinction. Second, while it is clear that EBV encodes an IL-10 homologue, it is not yet entirely clear that EBV actually upregulates cellular IL-10 expression.)

EBV displays still another immune evasion stratagem. Recall that the EBV lytic-phase protein, gp42, serves as the viral attachment protein that binds to MHC-II molecules on B cells. Recent experimental results demonstrate that lytically infected cells also generate a soluble form of gp42, in addition to the membrane-bound form. The soluble form of gp42 is generated in the ER by proteolytic cleavage of the membrane-bound form. Importantly, soluble gp42 is secreted, raising the possibility that it might bind to surface MHC-II molecules, thereby impeding MHC-II-restricted antigen presentation to T cells.

HCMV

HCMV appears to establish latency in monocyte-derived dendritic cell precursors that circulate in the blood. It is not clear why HCMV latency is restricted to these particular blood cells, and it is possible that there may be as-yet-undiscovered other sites of HCMV latency in vivo. Additionally, it is not known whether the virus expresses specific RNAs associated with, or necessary for, latent infection. The low frequency of latently infected cells in vivo has been one impediment to resolving this important issue.

The mechanism of HCMV reactivation within dendritic cell progenitors is not entirely understood. However, it is known that the differentiation of dendritic cell progenitors into mature dendritic cells results in induction of HCMV immediate-early genes and, consequently, in reactivation of the latent virus. Moreover, as in the case of HSV-1 (see above), it is now known that chromosome remodeling is a key factor underlying the reactivation of a latent HCMV. The details are as follows. The major immediate-early enhancer/promoter of latent HCMV genomes in dendritic cell precursors, in vivo, is in the form of condensed, transcriptionally inactive heterochromatin. But when these dendritic cell progenitors differentiate into mature dendritic cells, specific chromatin remodeling occurs, whereby the HCMV major immediate-early enhancer/promoter goes to a form of open, transcriptionally active euchromatin. In this course of events, the association of the HCMV immediate-early promoter with heterochromatin protein 1 (indicative of closed, transcriptionally inactive chromatin) is replaced by its association

with acetylated histone H4 (indicative of open, transcriptionally active euchromatin), and viral lytic genes are then expressed. Thus, the periodic differentiation of dendritic cell precursors into mature dendritic cells could lead to sporadic episodes of reactivation of the latent HCMV infection, thereby providing a means by which HCMV might be maintained and spread within a human population. (Interestingly, free HCMV cannot be detected in the blood of latently infected healthy carriers. Thus, transmission of HCMV via blood transfusion can be prevented by leukocyte depletion of the blood.)

HCMV is a champion, even among herpesviruses, at evading host immunity. Some of its means for doing so are enumerated in Table 18.1, and these selected examples are described in the remainder of this subsection. In particular, note the multiple ways by which HCMV blocks the MHC-I and MHC-II antigen presentation pathways. HCMV encodes at least four gene products that inhibit those pathways: US2, US3, US6, and US11. The HCMV US3 protein causes peptide-loaded MHC-I molecules to be retained in the ER and thus prevents their transport through the Golgi apparatus and then to the cell surface (Fig. 4.24). The HCMV US2 and US11 proteins appear to

Table 18.1 A sampling of human cytomegalovirus immune evasion strategies

Strategy	Virus product	Mechanism
MHC-I antigen presentation pathway	US6	Inhibits TAP
	US3	Retains peptide-loaded MHC-I molecules in ER
	US2	Causes proteolysis of MHC-I heavy chain
	US11	Causes proteolysis of MHC-I heavy chain
	UL111a	IL-10 homologue; impairs Th1 cytokine profile
MHC-II antigen presentation pathway	US2	Causes proteolysis of class II DR-α chains
	US3	May target Ii
NK cells	UL18	Decoy for NK cell inhibitory receptor
	UL42	Decoy for NK cell inhibitory receptor
Apoptosis	IE1	Inhibits TNF-α-induced apoptosis
	IE2	Blocks caspase 8 of the FasL/Fas pathway
	vMIA	Encoded by the UL37 gene; inhibits mitochondrial megapore activation like Bcl
	vICA	Encoded by UL36; inhibits extrinsic apoptosis induced via TNFR1, Fas, or Trail; inhibits caspase 8
	UL144	TNF-α receptor homolog
Complement	CD55, CD59	Intrinsic membrane proteins, accelerate breakdown of C3bBb
IFN	TRS1	Evasion of PKR-mediated protein shutoff
	IRS1	Evasion of PKR-mediated protein shutoff
Chemokines	US28	Chemokine receptor homologue
	UL146	Chemokine receptor agonist
IgG-mediated immunity	gp34	Fc receptor
	gp68	Fc receptor

bind directly to MHC-I molecules in the ER, causing them to be shed from the ER into the cytoplasm, where they are then degraded. The HCMV US6 protein blocks the TAP transporter from the luminal side of the ER. (This is reminiscent of the HSV ICP47 protein, which binds to the cytoplasmic side of the TAP transporter [see above].) The HCMV US2 protein also targets MHC-II α chains for degradation, while the US3 protein also destabilizes MHC-II molecules, perhaps via an effect on the invariant chain (Chapter 4). HCMV, like EBV, also encodes an IL-10 homologue, which inhibits the Th1 cytokine response that would favor cell-mediated immunity (see above).

NK cells, of the innate immune system, provide a host response against the ability of some viruses to downregulate MHC-I expression and do so by preferentially killing cells that express low levels of MHC-I molecules on their surface (Chapter 4). As a countermeasure to this host defense, HCMV produces *two* MHC-I molecule homologues, UL18 and UL42, which are believed to engage the NK cell inhibitory receptor, thereby blocking the NK cell attack on HCMV-infected cells. In support of that premise, small-interfering-RNA-mediated knockdown of UL42 mRNA expression in HCMV-infected cells resulted in increased sensitivity to lysis by NK cells. The significance of NK cell-mediated defenses in vivo is underscored by the fact that humans who lack NK cells experience frequent and severe HCMV infections.

Why might HCMV have evolved such a redundancy of functions that impair the MHC-I antigen presentation pathway? One possibility is that different viral gene products might work better against particular MHC-I isoforms or in different cell types. Another plausible premise is that the independent effects of these multiple HCMV gene products might act additively to more thoroughly impair the MHC-I antigen presentation pathway.

Why might HCMV impair the MHC-II as well as the MHC-I antigen presentation pathway? Here is one possibility. HCMV establishes latency in dendritic cell precursors, which are progenitors of professional antigen-presenting cells. Next, reactivation of the latent HCMV infection occurs when those progenitor cells differentiate into professional antigen-presenting cells. In professional antigen-presenting cells, endogenous antigens can make their way into MHC-II loading compartments for presentation to CD4 helper T cells. Thus, by preventing MHC-II-mediated antigen presentation, HCMV escapes detection by CD4 T cells that produce antiviral cytokines. Moreover, dendritic cells normally activate CD8 killer T cells via the MHC-I-mediated cross-presentation pathway (Chapter 4). Thus, in theory at least, HCMV-infected dendritic cells also might activate CD8 killer T cells via direct MHC-I-mediated priming. Regardless, the evolution of the multiple HCMV countermeasures to cell-mediated immune responses underscores the importance to persistent viruses of evading these host responses.

HCMV, like the other herpesviruses, encodes several proteins that interfere with apoptosis. One of the HCMV antiapoptotic gene products, UL44, is a TNF-α receptor homologue that acts by intercepting TNF-α. Its effect is based in part on the fact that the cellular TNF-α receptor, like Fas, is a cell death-inducing cell surface receptor. Moreover, there is evidence that HCMV encodes another protein that reduces the levels of the cellular TNF-α receptor in the plasma membrane. In addition, HCMV immediate-early protein IE86 also inhibits TNF-α-induced apoptosis, although by an as-yet-unknown mechanism. Furthermore, IE86 also modulates the function of p53, also by an unknown mechanism.

vMIA is another HCMV-encoded antiapoptosis protein. It functions in a way analogous to cellular Bcl-2 proteins, which normally regulate apoptosis by sequestering proapoptotic proteins, such as Bax. The latter protein acts by causing the release of cytochrome *c* from mitochondria, which in turn triggers activation of a caspase cascade. (Caspases are a family of proteases, all of which cleave after aspartate residues. They are sequentially activated and are effectors of all apoptosis pathways [see below].) Interestingly, the virus-encoded vMIA protein has no homology to the cellular Bcl-2 family proteins. Nevertheless, like Bcl-2, vMIA sequesters Bax at mitochondria.

VICA is yet another HCMV-encoded cell death suppressor. It associates with the caspase 8 precursor, pro-caspase-8, thereby blocking its recruitment to the death-inducing signaling complex; this step normally precedes the autocatalytic activation of caspase 8 and is part of the sequence of events triggered when a death receptor is activated at the plasma membrane.

Once more, note the impressive redundancy of stratagems by which HCMV targets particular host defenses, in this instance, apoptosis. A possible explanation for these multiple viral countermeasures against apoptosis might be as follows. There are two cellular pathways to apoptosis. First, there is the *extrinsic pathway* that is triggered when a proapoptotic ligand (e.g., TNF-α and FasL) binds to its receptor (e.g., TNF-α receptor and Fas) at the cell surface. Second, there is the *intrinsic pathway* that is triggered in response to internal cues, such as DNA damage. Each pathway is mediated by a convergent caspase cascade. In the intrinsic pathway, p53 is activated and goes to the nucleus while Bax goes to the mitochondria to trigger activation of the caspase cascade. Consequently, the multiple HCMV antiapoptotic countermeasures cover all the bases by impeding steps of both the extrinsic and intrinsic apoptotic pathways.

HCMV acquires resistance to complement by incorporating into its envelope the cellular complement control proteins CD55 and CD59. These intrinsic membrane proteins normally prevent the assembly of complement components into membrane attack complexes on host cells, doing so by accelerating the breakdown of C3bBb (the C3 convertase of the alternative complement pathway; see Chapter 4). (Compare the HCMV means for counteracting complement to that of HSV, described above. That comparison provides another interesting example of how even related viruses "devise" distinctly different solutions to a common problem.)

Since HCMV, like all herpesviruses, encodes gene products on both of its DNA strands, it produces complementary transcripts that can inadvertently form double-stranded RNA, making it susceptible to inhibition by IFN-α/β. Not surprisingly then, HCMV encodes two proteins, TRS1 and IRS1, which are involved in the viral evasion of PKR-mediated protein shutoff.

Remarkably, HCMV contains genes that encode 14 putative G-protein-coupled receptors, which together take up about 7% of its genome. G-protein-coupled receptors make up the largest family of receptor proteins in mammals, in which they play important roles in many physiological and pathological processes. They are activated by a variety of ligands, including hormones, neurotransmitters, and, importantly in the current context, chemokines, which induce migration and activation of macrophages and lymphocytes (Chapter 4). Considering the extent to which the HCMV genome is dedicated to encoding these proteins, it is safe to assume that they serve an important, if not yet understood, role in HCMV biology. One of the HCMV-encoded, G-protein-coupled receptor homologues, US28, actually functions as a signal-transmitting chemokine receptor. It is thought to perhaps induce chemotaxis of cells that might promote viral dissemination within an infected human host. An alternative means for achieving a similar result is illustrated by the HCMV UL146 protein, which is a secreted chemokine mimic that binds to, and activates, the IL-8 receptor. In these ways, HCMV can subvert chemokines and chemokine receptors to serve its own purposes.

We conclude this survey of HCMV immune-evasion stratagems by noting that the virus encodes two transmembrane receptors, gp34 and gp68, each of which binds the Fc portion of immunoglobulins. Thus, expression of these HCMV proteins at the cell surface could protect infected cells against antibody-dependent cell cytotoxicity (Chapter 4).

CLINICAL SYNDROMES

Despite differences between the human herpesviruses in their cell and tissue tropism and molecular mechanisms displayed during infection, these viruses present several common features that help us to understand how they are sustained, spread, and cause disease in human populations. We already considered their capacity to establish lifelong latent infections that may sporadically reactivate. In addition, unlike those viruses that are spread by the respiratory or fecal-oral route, the spread of most herpesviruses depends on close, even intimate, human contact. Since we crave and even need these contacts, they provide an effective means for viral transmission. This means of transmission and the capacity of latent infections to periodically reactivate combine to form a most successful stratagem for these viruses to persist and spread in the population (see Chapter 3).

For the sake of comparison, we take a moment to compare human infection by herpesviruses HSV-1, HSV-2, and EBV with human infection by HIV (Chapter 21). This comparison is informative because these herpesviruses and HIV each establish lifelong persistent infections in humans, and each of these viruses is spread by intimate personal contact. Yet there are important differences between how herpesviruses and HIV interact with humans. (HIV infection is described in detail in Chapter 21. Only a couple of general points are noted here.) First, the human herpesviruses are vastly more efficient than HIV at infecting humans. Second, and more important in the current context, HIV mounts a nonstop, aggressive attack on the human immune system that eventually incapacitates it. In contrast, the herpesviruses usually spend their time attempting to remain unnoticed by the immune system, rather than destroying it. As a consequence of the unremitting HIV attack on the immune system, untreated HIV infections are almost always fatal. Death in these instances results from the opportunistic infections that the immunocompromised AIDS patient is susceptible to, rather than from HIV per se. In contrast to HIV infections that are eventually fatal in most untreated patients, herpesvirus infections generally are mild. Moreover, since host immunity plays an important role in regulating primary as well as latent herpesvirus infections, these infections pose a serious threat mainly to individuals whose immunity has been compromised (e.g., AIDS patients).

Each of the human herpesviruses occupies a specific niche within the host. Yet in the case of some herpesviruses, such as HSV-1 and HSV-2, the heparan sulfate-containing glycosaminoglycans to which they each bind are present on almost all cells in the body. What then might prevent the spread of these viruses from their particular niches to become a systemic infection? The answer may be related in part to the fact that glycosaminoglycans are covalently linked to proteins to form proteoglycans, some of which are components of the extracellular matrix, while others are cell surface proteins. Consequently, the spread of free herpesviruses (at least in the case of HSV-1 and HSV-2) may be impaired by the virus binding to the extensive extracellular matrix surrounding the cells. IFN, NK cells, and adaptive immune responses also may act to prevent dissemination of the infection. (However, bear in mind that T-cell responses are absolutely necessary to resolve a disease episode resulting either from a primary or from a reactivated infection.)

HSV

HSV-1 and HSV-2 each initially infect and replicate in mucosal epithelial cells. These viruses then establish latency in neuronal cells that innervate the initially infected mucosa. HSV-1 is known to primarily infect the oral mucosa. Less commonly, it can give rise to clinically important infections of the eye. Oral infections are usually transmitted by saliva, typically during kissing, or by the sharing of saliva-contaminated items. HSV-1 infection of the eye likely results from rubbing the eye with saliva-contaminated fingers and most often occurs on the cornea (the normally clear dome that covers the front part of the eye).

HSV-2 is generally associated with genital or anal infections, and in either instance transmission is sexual.

However, the distinctions between the tropisms of HSV-1 and HSV-2 are increasingly blurred by the practice of oral sex. Indeed, HSV-1 is now thought to be responsible for about 10% of genital infections, and HSV-2 may be responsible for as many as 40% of oral infections (Box 18.15). HSV-2 may also be transmitted from infected mothers to newborns during birth, in those instances giving rise to severe, often fatal neurological disease. Infections of newborns are especially severe because of the immaturity of their immune systems.

HSV-1 and HSV-2 are each ubiquitous in humans. Recent epidemiologic data state that 5 of every 12 adults in the United States are infected with HSV-2, with as many as 1 million newly infected individuals each year. The incidence of HSV-1 is greater still. Not surprisingly, infection with HSV-2 usually occurs later in life than infection with HSV-1, since HSV-2 infections correlate more with becoming sexually active.

The primary HSV-1 infection can be mild or unapparent, and when disease does occur, it is generally benign. The classic HSV-1 lesions begin as clear vesicles that rupture to form ulcers and then crusted lesions (Fig. 18.14). The lesions may be distributed throughout the mouth, including on the lips, tongue, palate, and gums.

Recurrences of HSV infections are almost invariably reactivations of an ongoing latent infection. That is so because immunity induced by the primary infection is long-lasting. For the same reason, the cold sores of recurrent

BOX 18.15

Despite the increasing frequency of oral HSV-2 infections, the vast majority of HSV-2 infections are genital, at least according to one study. A reasonable explanation is that most individuals first acquire an oral HSV-1 infection, which may then provide some immunity against oral infection with HSV-2. Yet the prevalence of oral HSV-2 infections may actually be higher than current estimates, since an oral HSV-2 infection (unlike an oral HSV-1 infection or a genital HSV-2 infection) rarely reactivates and thus goes undetected.

Now, suppose an individual acquires a genital HSV-1 infection through oral sex. How likely is it that the HSV-1-infected individual might then transmit the genital HSV-1 infection via genital sex? The answer is that the likelihood that the genital HSV-1 infection might be spread by genital sex is less than the likelihood of its transmission via oral sex. That is so because HSV-1, like HSV-2, reactivates and sheds less often outside its usual preferred site. Regardless, it is clear that HSV-1 can be spread by both oral and genital sex.

Reference
Cherpes, T. L., L. A. Meyn, and S. L. Hillier. 2005. Cunnilingus and vaginal intercourse are risk factors for herpes simplex virus type 1 acquisition in women. *Sex. Transm. Dis.* **32:**84–89.

Figure 18.14 (A) Primary herpes gingivostomatitis. (B) HSV establishes latent infection and can recur from the trigeminal ganglia. From P. R. Murray, K. S. Rosenthal, G. S. Kobayashi, and M. A. Pfaller, *Medical Microbiology*, 4th ed. (Mosby, Inc., St. Louis, MO, 2002), with permission.

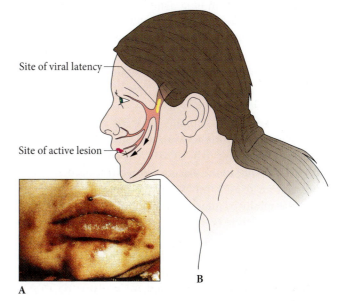

Site of viral latency

Site of active lesion

A

B

herpes labialis are usually even less severe than the generally benign lesions of the primary infection. On the other hand, individuals may experience recurrent cold sores and fever blisters, even when the primary infection was clinically unapparent.

HSV-1 infections of the eye are referred to as **herpetic keratitis**. These ocular infections can be superficial, involving only the top layer of the corneal epithelium. In that case, there is usually no permanent damage. But infection of the deeper layers can lead to corneal scarring and blindness.

Herpes gladiatorum is an infection of any region of the skin that is transmitted through cuts or abrasions. It is found among wrestlers (hence its name) and rugby players. **Herpetic whitlow** is an infection of the finger that is transmitted through a break in the skin or, more commonly, through a torn cuticle. It is an occupational risk among health care workers exposed to infected oropharyngeal secretions and can easily be prevented by use of gloves. Interestingly, children with oral HSV-1 lesions may autoinoculate themselves for herpetic whitlow by finger or thumb sucking. In adults, herpetic whitlow most often involves HSV-2, due to autoinoculation from genital herpes. HSV-1 may also infect the upper respiratory tract, giving rise to **herpes pharyngitis**, or sore throats. **Herpes encephalitis**, usually associated with HSV-1, is the most common sporadic (i.e., nonepidemic) viral encephalitis. Symptoms may include seizures and focal neurological defects. The condition may be fatal even with treatment, and most survivors suffer permanent neurological **sequelae** (a disorder caused by a preceding disease or injury in the same individual).

Primary genital HSV-2 infections are often asymptomatic, but they can give rise to painful lesions, as well as to fever, malaise, and muscle aches, in both males and females. **HSV meningitis** is a neurological complication of a genital HSV-2 infection, which usually resolves on its own. It occurs in about 10% of all cases of genital infection, and its typical features include headache, fever, stiff neck, and an abnormal level of lymphocytes in the cerebrospinal fluid.

HSV infection of neonates happens most commonly during passage through an infected birth canal. However, some babies are indeed born with herpes lesions. That fact and the presence in the placenta of necrotic herpes lesions and viral inclusions imply that HSV may be transmitted transplacentally in utero. Neonatal infection is associated more often with HSV-2 than with HSV-1, as might be expected. Importantly, since cell-mediated immunity is not yet developed in newborns, the infection is able to disseminate to the liver, lungs, and central nervous system, resulting in neurological damage and, perhaps, death.

The risk of a fetus being infected in utero with HSV-1 strongly depends on whether the infection in the mother is a recurrent one or a primary one. If virus present in the birth canal is from a recurrent infection, the risk of neonatal herpes is 3 to 4%. It rises to 30 to 40% during a primary HSV infection. (How might you account for those facts?)

Recurrences of genital HSV-2 infections are more frequent than recurrences of oropharyngeal HSV-1 infections. Importantly, some individuals may experience primary or recurrent HSV infections that are asymptomatic. Thus, these individuals may transmit the infection, despite showing no signs that a productive infection is under way. Such individuals are important lifelong vectors for transmission of HSV.

EBV

The primary lytic EBV infection takes place in the mucosal epithelium of the nasopharynx and salivary glands. While the lytic infection of the mucosal epithelium is in progress, the virus establishes latent infection in B cells that infiltrate the local lymphatic tissue (Chapter 4). The presence of virus in the saliva of latently infected but asymptomatic individuals (see below) may lead to transmission during kissing, hence the colloquial term "kissing disease" for **infectious mononucleosis.**

In point of fact, kissing is the major means for EBV transmission that leads to "mono." Consequently, most individuals in the United States contract the disease as teenagers or young adults. But consider that more than 70% of the population in the United States is infected with EBV by 30 years of age. Moreover, while the highest rates of infectious mononucleosis are seen in persons between 10 and 19 years of age, there are but a "mere" six to eight cases per 1,000 persons in that age group per year. Clearly, then, EBV infection in the great majority of instances is unapparent or very mild. But why then is infectious mononucleosis most often seen in adolescents and young adults? The likely answer is that most individuals are infected with the ubiquitous EBV as young children, and childhood infection with EBV is generally asymptomatic. In contrast, the same virus gives rise to mononucleosis in those individuals who are first infected as adolescents and adults. This is not the first instance we encountered in which age is a factor contributing to the severity of a viral infection. (Poliovirus is another example that readily comes to mind [Chapter 6].) Possible reasons for the age dependence of the clinical course of EBV infections are discussed below. Meanwhile, in developing countries, where hygiene is poorer, almost all individuals are infected in the first year or two of life. Consequently, the glandular fever of infectious mononucleosis is virtually unknown in the Third World.

Interestingly, EBV was actually seen for the first time in biopsy specimens from **Burkitt's lymphoma** patients. This B-cell neoplasm, which affects children in certain parts of East Africa and New Guinea, was originally described in detail by Dennis Burkitt in 1958. As a result of Burkitt's efforts, the cancer bearing his name came to be recognized as a distinct clinical entity.

The discovery of EBV happened as follows. In 1961, Tony Epstein met Burkitt in London, at the occasion of a talk given by Burkitt on children's cancers in tropical Africa. During his talk, Burkitt proposed that the lymphoma he characterized might have a viral etiology. His premise was based on the geographic distribution of cases, which was similar to that of yellow fever, which is caused by a flavivirus (Chapter 7). When the two researchers met, Burkitt agreed to send Epstein frozen specimens for him to analyze. Then, in 1964, Tony Epstein and Yvonne Barr, working together with Bert Achong, discovered EBV by electron microscopy of cells that they cultured from Burkitt's lymphoma tissue. According to its discoverers, the virus was named Epstein-Barr-(Achong) virus (EBV) by Werner and Gertrude Henle, whose own research demonstrated that EBV is a distinctly different virus from the related HSV. EBV was linked to infectious mononucleosis when serum from a recovering mononucleosis patient was accidentally found to react with Burkitt's lymphoma cells. This discovery was followed by the finding that EBV is associated with other disorders as well, including **nasopharyngeal carcinoma** and **Hodgkin's disease**.

The EBV-associated diseases enumerated above, i.e., Burkitt's lymphoma, nasopharyngeal carcinoma, and Hodgkin's disease, underscore the fact that EBV may give rise to lymphomas and carcinomas, as well as to infectious mononucleosis. Yet, bear in mind that the vast majority of EBV infections are clinically unapparent or very mild. These facts lead us to ask (i) what might determine whether an EBV infection may lead to one of these more severe clinical outcomes and (ii) what might determine whether the more severe outcome might be infectious mononucleosis (usually a nonfatal, self-limiting lymphoproliferative disease) versus the usually fatal Burkitt's lymphoma or nasopharyngeal carcinoma? (Thorley-Lawson [2005] makes this point somewhat differently, noting that "… EBV has the potential to be highly pathogenic yet rarely manifests this potential.")

The strength of the immune response is one determinant of the clinical outcome. An overactive immune response can give rise to infectious mononucleosis. Alternatively, a weak immune response may lead to lymphoma. To see how the strength of the immune response might affect the clinical outcome, we begin by considering infectious mononucleosis.

In instances of infectious mononucleosis, EBV stimulates B-cell proliferation. Yet the actual **lymphocytosis** (i.e., increase in mononuclear cells) associated with this disease actually results from the activation and proliferation of EBV-specific T cells and also of NK cells. The activated lymphocytes are rather atypical in appearance and are referred to as **Downey cells**. By the second week of infection they may constitute up to 80% of the total white blood cell count. This extremely active T-cell response results in the swollen lymph nodes, liver, and spleen that are characteristic of later stages of the disease. The major complaint of "mono" patients is their constant fatigue. Infectious mononucleosis is rarely fatal, but it may lead to serious complications, including neurological disorders and rupture of the spleen.

Even after recovery from infectious mononucleosis, EBV persists in about one memory B cell per 10^5 to 10^6 B lymphocytes, for the lifetime of the individual. Moreover, the virus may be reactivated in a latently infected B cell whenever the B cell might be activated (see above and Chapter 4). Reactivation leads to shedding of virus into the saliva, so that an asymptomatic recovered infectious mononucleosis patient remains a lifelong reservoir and vector for the virus (as does one who experienced an asymptomatic EBV infection earlier in life).

Burkitt's lymphoma, nasopharyngeal carcinoma, and Hodgkin's disease are considered below. But first, with respect to the fact that the vast majority of EBV infections are clinically unapparent or very mild, recall that most individuals are infected as young children. As such, they may experience a much milder course of EBV infection since their immune responses are weaker than those of adolescents and adults. In addition, adolescents are likely to transmit a larger inoculum during heavy kissing than might be transmitted by adults to infants. (What of maternal antibodies? Might these be a factor?)

Burkitt's lymphoma is a malignant B-cell lymphoma of the jaw and face that is most prevalent in children living in tropical Africa and New Guinea (Fig. 18.15). Actually, Burkitt's lymphoma is the most common childhood cancer in equatorial Africa. Yet bearing in mind once more that EBV is ubiquitous in the human population worldwide, the geographic restriction of Burkitt's lymphoma suggests that the disease may be associated with a similarly geographically restricted cofactor (a somewhat different notion from that originally proposed by Burkitt). Moreover, the age distribution of the disease (i.e., the younger the child, the more likely it is that the lymphoma will develop) is consistent with the notion that the immune status of the infected individual is important. Together, these details are consistent with the premise that other infections may affect B-cell development so as to

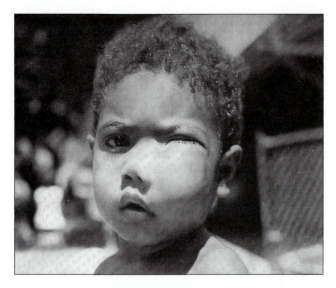

Figure 18.15 Characteristic facial tumor in a child from Papua New Guinea with Burkitt's lymphoma. (Courtesy of Dr. J. Biddulph.) From D. O. White and F. J. Fenner, *Medical Virology*, 4th ed. (Academic Press, San Diego, CA, 1994), with permission.

facilitate the carcinogenic outcome of the EBV infection. The malarial parasite, *Plasmodium falciparum*, which has a fitting geographical distribution and also is chronically immunosuppressive, is often discussed as a possible cofactor in Burkitt's lymphoma, but that relationship is still being debated (Box 18.16). *P. falciparum* is not required for Burkitt's lymphoma, since sporadic cases not associated with malaria occur outside the regions of endemicity. Regardless, notice that diseases caused by EBV may occur either from an overactive immune response, as in the case of infectious mononucleosis, or from compromised immune system control of the EBV infection, as seen in Burkitt's lymphoma, and perhaps Hodgkin's disease as well (see below).

Most Burkitt's lymphoma cells contain multiple free copies of the EBV genome. Importantly, these genomes display a latency I pattern of transcription, in which they express only EBERs (see above). This highly restricted pattern of EBV gene expression in Burkitt's lymphoma cells

BOX 18.16

Patients with Burkitt's lymphoma are less likely to carry the sickle cell trait than the population as a whole. Recall that the sickle cell trait provides protection against *P. falciparum* malaria. This might account for why individuals carrying the sickle cell trait might be less likely to develop Burkitt's lymphoma. Also, experiments in animal models show that there is a synergistic action between other potentially oncogenic viruses and a plasmodium.

raises the issue of how the virus might cause and maintain neoplasia. Apropos this issue, note that Burkitt's lymphoma cells also contain a characteristic chromosomal translocation, in which the *c-myc* **proto-oncogene**, located on chromosome 8, is translocated to chromosome 14 (or more rarely to chromosome 2 or 22), where it comes under the transcriptional control of immunoglobulin gene enhancers. (Proto-oncogenes are normal cell genes that may be oncogenic when inappropriately expressed. See Chapter 20.) Expression of *c-myc* is normally tightly regulated, such that Myc is produced only during a short period in the G_1 phase of the cell cycle. However, the translocated *c-myc* gene is expressed throughout the cell cycle, resulting in unregulated proliferation of B cells. Since *c-myc* encodes a transcription factor that drives cell proliferation in several other human cancers, this translocation likely is a critical step leading to Burkitt's lymphoma. Moreover, it appears to be sufficient to cause Burkitt's lymphoma, as demonstrated by Burkitt's lymphoma patients who are EBV-negative, but who carry the *c-myc* translocation. Moreover, experiments involving mice show that a similar translocation causes B-cell lymphomas in the mouse model. Thus, the principal carcinogenic role of EBV in Burkitt's lymphoma may be to facilitate the translocation of the *c-myc* proto-oncogene to an immunoglobulin gene locus.

George Klein in 1981 proposed the following overall sequence of events to account for how EBV infection leads to Burkitt's lymphoma. First, during the first stage of B-cell infection, EBV stimulates B-cell proliferation, thereby enhancing the probability of chromosomal damage. Second, an environmental factor, presumably infection by *P. falciparum*, impairs the ability of host T-cell responses to control proliferation of the immortalized B cells. Finally, translocation of the *c-myc* oncogene leads to its constitutive expression, resulting in neoplasia.

More-recent experimental results suggest that LMP1 may promote lesions in the cellular genome by repressing DNA repair. In addition, it now appears that the EBNA-1 protein and the noncoding EBER RNAs might have a role by mediating partial protection from the apoptotic response associated with unrestricted *c-myc*-driven cell growth, as well as the apoptotic response triggered by nonproductive rearrangements of the B-cell receptor (Chapter 4). Note that the above is likely an oversimplified description of a very complex process, some steps of which are still somewhat controversial.

EBV infection is also associated with two other human malignancies, nasopharyngeal carcinoma and Hodgkin's disease. The former is a generally rare cancer in the Western world (1 per 100,000 individuals) but one that is endemic in the densely populated regions of Southern China, Taiwan, and other regions, where it accounts for

20% of all cancers. It also is endemic in individuals from those populations who migrated to other parts of the world. The conspicuous restriction of nasopharyngeal carcinoma to people of Cantonese origin suggests a predisposing genetic or cultural (e.g., dietary) factor. Because of its high incidence and frequently fatal outcome, nasopharyngeal carcinoma is considered a significant health problem in regions where it is endemic. As in Burkitt's lymphoma, nasopharyngeal carcinoma cells contain multiple copies of episomal EBV genomes. But in contrast to Burkitt's lymphoma, in which tumor cells are derived from B lymphocytes, tumor cells in nasopharyngeal carcinoma have an epithelial origin.

Hodgkin's disease is a complex lymphoid tumor that is associated with severe primary EBV infection. Approximately 40% of Hodgkin's tumors in the United States contain EBV genomes, and the frequency approaches 80% in developing countries; EBV genomes are present in the Hodgkin's tumors of 100% of AIDS patients who present with this malignancy. In addition, LMP1 and LMP2 have also been detected in Hodgkin's tumors. Together, these facts would seem to provide virtually unequivocal evidence linking EBV to Hodgkin's disease.

How might severe EBV infection act as a risk factor for Hodgkin's disease? The answer is not known. But taking the liberty to speculate more than a tad, consider that the infectious mononucleosis seen in teenagers and young adults, as opposed to the milder infection seen in children, is characterized by a profound disruption of immune function (e.g., substantial numbers of virus-infected memory B cells [≥50%], a striking proliferation of T cells, and the production of a broad range of nonspecific antibodies due to the ability of the virus to express genes that activate naïve B cells as though they were responding to antigen). Thus, infectious mononucleosis underscores the fact that EBV infection may severely compromise immune function. This point is consistent with the possibility that Hodgkin's disease might be caused in part by an EBV infection leading to the disruption of immune function. In fact, both Burkitt's lymphoma and Hodgkin's disease (each of which is a B cell lymphoma) are associated with immune system dysfunction. On the other hand, failure of the immune system to control the EBV infection is not likely to be the whole story, since not every immunosuppressed individual has these cancers, despite the fact that these individuals may harbor more than 10^7 EBV-infected cells. So, knowing that humans are the natural host for EBV and that viruses tend to give rise to mild infections in their natural hosts, what goes wrong in an EBV infection that leads to fatal neoplasia? The answer to this question is not yet clear. However, it is worthwhile to approach it from a different point of view, noted below, first noting that EBV is an efficient inducer of

transformation in vitro and then asking why the incidence of EBV-associated cancers is as low as it is.

Apropos the above examples of human cancers that are linked to EBV, note that EBV was indeed the first virus to be implicated in human cancer. Moreover, this virus is an efficient inducer of neoplastic transformation in vitro. On the other hand, since EBV may give rise to lifelong persistent infections in >90% of the human population, why is the incidence of EBV-associated cancer as low as it is? A possible explanation might involve the fact that the virus persists in "resting" memory B cells rather than in proliferating cells (see above). In the memory cells, there is no expression of those EBV latency genes that drive cell proliferation. Instead, EBV uses its cell growth-promoting functions only to gain access to the memory compartment. Having accomplished that aim, the virus then blocks expression of its potentially oncogenic functions. Thus, something must upset the regulated EBV latency transcription pattern in order for neoplasia to emerge. In addition, if B cells in which the growth program is reactivated are to proliferate, then the cytotoxic T-cell response against those cells also must be impaired. The latter point accounts for the increased risk of B-cell lymphoproliferative diseases in AIDS patients and in others with impaired cell-mediated immunity. Relevant to this discussion, the polyomaviruses, papillomaviruses, and adenoviruses all induce neoplastic transformation in cell culture, yet with the exception of a few high-risk serotypes of the human papillomavirus (Chapter 16), these viruses are *not* associated with neoplasia in vivo. This seeming incongruity is discussed in Boxes 15.7 and 17.8, and some of the points raised there are relevant to EBV as well. Before moving on, we should appreciate that much remains to be learned about the molecular mechanisms that regulate the pattern of EBV transcription during the various stages of the establishment and maintenance of latency. That knowledge might well provide important insights into how EBV causes cancers and possibly other diseases in humans. For more on these issues, see Thorley-Lawson (2005).

EBV-associated PTLD is an often-fatal complication that occurs in transplant recipients on immunosuppressive regimens. It is a condition characterized by uncontrolled lymphoid growth, usually involving B cells, and it may involve other organs as well. PTLD may arise at any time after the transplant has been performed. However, most cases occur within the first 2 years after the transplant. The standard treatment for PTLD has been to reduce immunosuppression. However, newer therapies are being evaluated, in particular the use of rituximab, a humanized chimeric anti-CD20 monoclonal antibody. The rationale for this approach is based on the facts that most

PTLD cases involve B cells in origin and that cell surface CD20 is a fairly specific B-cell marker. Another promising therapy has been to use EBV-specific CTLs derived from EBV-seropositive blood donors.

Oral hairy leukoplakia (OHL) is an EBV-associated disease resulting from a productive EBV infection of the oral mucosa. Its name refers to the characteristic white grooved lesions that usually appear along the sides of the tongue. OHL is known as one of the first opportunistic infections that occurs in AIDS patients, and indeed it may be diagnostic of AIDS. Importantly, however, OHL can affect patients who are HIV negative. The condition itself is not life threatening, although it is indicative of immune deterioration in AIDS patients. OHL was first described in 1984, and there has been a marked decrease in its frequency in the era of potent antiretroviral therapy (Chapter 21). Accordingly, when AIDS patients present with it while on antiretroviral therapy, it is an indication that their drug regimen is failing.

EBV has also been implicated in multiple sclerosis, an important autoimmune disease of the central nervous system that results in severe neurological disability. The cause of multiple sclerosis is still unknown, and while several infectious agents have been proposed as the potential etiologic agent, no virus has as yet been shown unequivocally to be its cause (see Boxes 4.18 and 4.19). Yet at present, there is more compelling evidence linking EBV to multiple sclerosis than there is for any other infectious agent. For example, seroepidemiological studies consistently show that multiple sclerosis patients are more likely to be seropositive for EBV than are control subjects, and, importantly, these high seroprevalence rates for EBV in multiple sclerosis patients are not seen for other viruses.

A number of studies have been carried out to identify the mechanism by which EBV *might* trigger multiple sclerosis. In one such study, the EBV DNA polymerase was detected in the cerebrospinal fluid from a multiple sclerosis patient. Moreover, CD4 T cells specific for the EBV DNA polymerase were present in the cerebrospinal fluid from the same patient. Bearing in mind that multiple sclerosis is an autoimmune disease characterized by a T-cell-mediated attack against myelin sheaths, it is noteworthy that the EBV DNA polymerase-specific CD4 T cells were able to cross-recognize a myelin basic protein peptide. These cross-reactive EBV-specific CD4 T cells could be involved in the pathogenesis of multiple sclerosis, since activated T cells can gain entry into the brain compartment and therefore target myelin basic protein in the central nervous system (see HHV-6, below).

HCMV

Approximately 50% of the human population in developed countries is infected with HCMV. Moreover, as many as 10% of all persons may be shedding HCMV at any one time, consistent with the notion that asymptomatic recurrences happen throughout life. HCMV is known to be transmitted by the oral and sexual routes, as well as by blood transfusions and tissue transplantation. Importantly, HCMV may also be transmitted in utero, across the placenta. In fact, congenital HCMV is *the* most prevalent viral infection known to be transmitted in utero. (Rubella virus, HIV, and parvovirus B19 are the other viruses most often associated with transplacental infections in humans.)

Regarding sexual transmission of HCMV, the virus is present from time to time in cervical secretions and semen of infected individuals. Moreover, transmission by kissing and sexual activity likely accounts for the fact that the frequency of HCMV-infected individuals rises from 10 to 15% in early adolescence and to 50% by 35 years of age. Some of the individuals infected as young adults may develop a disease reminiscent of EBV infectious mononucleosis, although with less-severe pharyngitis and lymphadenopathy.

Apropos the fact that congenital HCMV is the most common viral infection known to be transmitted in utero, between 0.5 and 2.5% of all live newborns are infected with HCMV. Importantly, about 10% of those infected newborns show clinical signs of disease, with hearing loss and mental retardation being the most common clinical consequences of congenital HCMV infection. Furthermore, some infected babies with abnormalities are stillborn or die shortly afterwards. In others, progressive damage may occur after birth. Indeed, as many as 90% of infants who show clinical signs of infection at birth will have neurologic abnormalities later in life.

Notwithstanding the significant clinical impact of congenital HCMV infection, note that the number of newborns with HCMV-associated birth defects is much smaller than the number actually infected in utero. The explanation may be related to the fact that most congenital HCMV infections result from reactivation of latent maternal infections rather than from primary infections. (Reactivation of a pregnant female's latent HCMV infection may occur at any time during pregnancy, with the likelihood rising as term approaches. Reactivation in these instances may be triggered by hormonal changes and altered immune status.) Reactivation of an ongoing persistent infection generally is uneventful to both mother and fetus. (Why might that be? See the following paragraph.) In contrast, 10 to 15% of those fetuses infected as a result of a primary maternal infection are born with birth defects. (Another question to ponder before reading on is why HCMV-associated birth defects are fivefold more common in developed countries than in Third World countries.)

Actually, more newborns are infected with HCMV during passage through the birth canal or via the mother's milk than are infected transplacentally. The relatively high frequency of postnatal infections of newborns is explained by the facts that (i) about 10% of women shed HCMV from the cervix at delivery and (ii) 10 to 20% of nursing mothers shed HCMV in their milk. As in the instances of transplacental infection, most postnatal HCMV infections of the newborn result from reactivations in the mother rather than from primary infections. Reactivating infections are relatively benign in the mother and in the baby as well, regardless of whether infection of the latter occurred in utero or postnatally. That is so because preexisting maternal antibodies in the latently infected mother protect both mother and fetus or baby.

HCMV has been implicated as a risk factor in the development of **atherosclerosis**, a widespread and potentially fatal condition that is more commonly linked to genetic and other risk factors (e.g., cholesterol, diabetes, and obesity). Evidence for a link between HCMV and atherosclerosis includes epidemiological data and the presence of virus, viral antigens, or viral nucleic acid in atherosclerotic lesions. How HCMV *might* contribute to atherosclerosis is not entirely clear. Results of one study suggest that the artery itself may be a site of HCMV latency. Other experimental data show that the risk for progressive atherosclerosis is specifically increased in patients with an inflammatory response to HCMV. Still other data suggest that atherosclerosis might result from an HCMV-induced autoimmune response. Apropos that notion, two HCMV proteins, UL122 and US28, show sequence similarity with molecules normally expressed on the surfaces of endothelial cells, and antibodies against US28 and UL122 are able to induce endothelial cell damage through cross-reaction with normally expressed cell surface molecules.

In addition to the above clinical conditions, there is now substantial evidence for HCMV involvement in **transplant arterial disease**, in which HCMV-induced vasculopathy can lead to graft rejection. The risks are greatest when the transplant donor is HCMV positive and the recipient is HCMV negative. Interestingly, HCMV prophylaxis (e.g., with ganciclovir; see below) can reduce the risk of acute graft rejection in these cases.

HCMV, like the other human herpesviruses, is an important pathogen in immunocompromised individuals. In point of fact, HCMV is the most lethal of the human herpesviruses in AIDS patients, as well as in other immunocompromised people. HCMV-associated diseases seen in immunocompromised individuals include **pneumonitis** (inflammation of the air sacs in the lungs, usually caused by a virus), which, if not treated, can be fatal. In addition, HCMV causes **retinitis** in as many as 15% of AIDS patients and **colitis** in as many as 10% of these patients.

Despite the above, HCMV is generally relatively benign in immunocompetent adults.

VZV

VZV is responsible for two well-known human diseases. One of these, **varicella** or **chicken pox**, is a virtually universal disease of childhood. The other, **herpes zoster** or **shingles**, is a late complication of an earlier episode of chicken pox. As expected, herpes zoster results from reactivation of the latent VZV infection. It is a painful, disabling disease that is seen primarily in the elderly and in patients who are immunocompromised.

Whereas HSV and EBV are spread by intimate contact, VZV is spread mainly by the respiratory route. Accordingly, primary VZV infection occurs in the mucosal epithelia of the oropharynx and the respiratory tract. From there, the virus enters the bloodstream and the lymphatic system. Viral replication then occurs in monocytes and capillary endothelial cells, giving rise to a secondary viremia, which leads to virus replication in epithelial cells of the skin and the resulting characteristic chicken pox rash. Then, reminiscent of HSV, VZV ascends the axons of the sensory nerves that innervate the infected skin areas, becoming latent in their ganglia (usually dorsal root and cranial ganglia).

Chicken pox is one of the five classic childhood exanthemas (i.e., eruptive diseases). The others are measles, rubella, roseola (caused by HHV-6; see below), and fifth disease (caused by human parvovirus B19). Although chicken pox is generally symptomatic, it is not life threatening, at least not in children. However, a primary infection can be much more severe in adults. Chicken pox is characterized by fever and, of course, the distinctive rash, which appears about 14 days after the initial infection. Successive crops of lesions emerge over the next 3 to 5 days.

Herpes zoster (shingles) occurs when VZV, which is latent in a ganglion, reactivates and descends the axon to reinfect the area of the skin innervated by that axon. The patient experiences severe pain in that area of the skin, which is followed by the local appearance of chicken pox-like lesions. Herpes zoster can lead to a chronic painful condition known as **postherpetic neuralgia**, which occurs in as many as 30% of individuals over 65 years of age who experience herpes zoster.

Herpes zoster affects mainly the elderly because of the declining efficiency of their immune responses. About 1% of individuals between 50 and 60 years in age are affected by herpes zoster per annum, with the frequency of herpes zoster increasing rapidly with age, such that most individuals will have experienced at least one episode by age 80, a fact likely reflecting the importance of cell-mediated immunity in controlling VZV latency. Herpes zoster also can be triggered by injury to the spinal cord. But the facts that (i) herpes zoster affects mainly the elderly and (ii) even in that population, the frequency of herpes zoster recurrences are relatively rare in comparison to the frequency of HSV recurrences imply that VZV latency is more tightly controlled than HSV latency.

As in the case of other herpesvirus infections, either primary or reactivated VZV infection can be very severe in immunocompromised individuals, such as in AIDS patients and transplant recipients. Defects in cell-mediated immunity in those individuals might allow the virus to disseminate to the brain, liver, or lungs, with potentially fatal consequences.

HHV-6

HHV-6 was originally isolated in 1986 from immunocompromised patients, some of whom were also infected with HIV. As its name implies, HHV-6 was the sixth human herpesvirus to be discovered, and it was the first new human herpesvirus to be discovered after the discovery of EBV about 20 years earlier. The discovery of HHV-6 was soon followed by the discovery of HHV-7 and HHV-8 (see below).

Two variants of HHV-6 are known: HHV-6A and HHV-6B. The latter is a known cause of roseola in children (see below), whereas the former has been isolated mainly from adults, in particular, from patients with AIDS. The two HHV-6 variants are closely related, with most of their proteins showing >90% amino acid sequence identity. The differences between their biological properties, clinical aspects, and distribution remain largely to be determined.

HHV-6 is believed to be spread by saliva, since it is present there in infected individuals. But whatever the mechanism, HHV-6 transmission is efficient, since most infants are infected by 2 to 3 years of age. Infection at this early age suggests transmission within the family, most likely from the mother, at a time when the protection afforded the child by maternal antibodies is waning. Serologic surveys show that more than 95% of the world's population is seropositive for HHV-6, consistent with the premise that primary infection is almost certainly followed by lifelong persistence of the virus.

It is not entirely clear which cells in the host are most important for supporting the productive and latent phases of an HHV-6 infection. The acute phase is thought to involve the salivary glands, as suggested by the presence of HHV-6 DNA in that tissue and by the presence of infectious virus in the saliva. Lymphocytes, macrophages, endothelial cells, and epithelial cells are among the cell types that support productive HHV-6 infection in vitro.

But CD4 T cells are the peripheral blood cells that are most readily infected with HHV-6 in cell culture, suggesting that these cells may be the primary producers of HHV-6 in the blood. The identity of the receptor for HHV-6 on CD4 T cells is not yet known, but it is not CD4 (which HIV subverts for use as its receptor). HHV-6 can also replicate in glial cells in vitro. This experimental finding is consistent with indications that HHV-6 is neurotropic in vivo (see below). The virus may establish latency in T lymphocytes and monocytes.

HHV-6 is linked with **roseola** (also known as **exanthema subitum** and **sixth disease**; the latter name is because the disease is caused by HHV-6). Roseola is one of the five classic childhood exanthemas (see the discussion of chicken pox above). It is characterized by the rapid onset of fever that lasts for only a day or two, followed by a generalized rash in some, but not all, patients that also lasts a day or two.

Acute HHV-6 infections in children are usually self-limiting. However, they can lead to serious complications. For example, HHV-6 is associated with neurologic disease, causing as many as one-third of all febrile seizures in children up to 2 years in age. More-serious incidences of HHV-6-associated encephalitis have been reported in both immunocompetent and immunosuppressed adults, as well as in children. Importantly, these findings demonstrate that HHV-6 is neuroinvasive. In this regard, HHV-6 DNA has been detected in brain tissue from apparently healthy adults, suggesting that the virus also may persist in the central nervous system without clinical consequences.

Numerous studies have linked HHV-6 (as well as EBV [see above]) to multiple sclerosis. One such study reported evidence of HHV-6 infections in brain tissue (taken at autopsy) in over 70% of multiple sclerosis patients. Moreover, these sites of HHV-6 infection corresponded to sites of myelin destruction, in agreement with the known ability of HHV-6 to infect and destroy myelin-producing oligodendroglial cells. In contrast, there was no evidence of HHV-6 in control samples taken from normal brains and from those of patients who succumbed to other neurological diseases. Analysis of blood samples revealed HHV-6 viremia in 56% of multiple sclerosis patients and in none of 61 healthy control subjects. Yet, as noted above, when considering a possible role for EBV in multiple sclerosis, no virus has yet been shown unequivocally to cause this disease. On the other hand, considering the number of different viruses that have been implicated as the etiologic agent of multiple sclerosis, is it possible that this disease can be triggered by any of several distinct viruses? This issue is discussed at some length in Boxes 4.18 and 4.19.

Cell-mediated immunity plays a major role in resolving lytic HHV-6 infections and in controlling latent infections.

Consequently, latent HHV-6 is more likely to reactivate in AIDS patients and in other immunosuppressed individuals. When HHV-6 does reactivate, it may cause severe opportunistic diseases, including encephalitis, pneumonitis, hepatitis, and retinitis, and also bone marrow graft failure.

Evidence from clinical case studies and from experimental model systems suggests that HHV-6 itself may act as an immunosuppressive agent. For example, HHV-6 was able to cause thymocyte depletion in a mouse model, and there is a report of a fatal immunodeficiency in an HIV-negative child that was thought to result from a disseminated HHV-6 infection.

Interestingly, it was suggested that HHV-6 infection may promote the progression of AIDS; this inference was based on the experimental finding that HHV-6 can superinfect cells that are already latently infected with HIV and transactivate the HIV long terminal repeat (Chapter 21). In doing so, HHV-6 can upregulate HIV gene expression, leading to productive HIV infection. Moreover, HHV-6 upregulates expression of CD4, the HIV receptor, in CD8 T cells, making these cells susceptible to HIV in coinfected patients.

HHV-6 DNA sequences have been detected in a variety of human cancers, including Hodgkin's disease. Considering that HHV-6 might be immunosuppressive, HHV-6-induced immunologic disruption alone might be a risk factor for this tumor. Regardless, as in the case of EBV, the means by which HHV-6 might promote human cancer is not yet known with certainty.

HHV-6 also has been considered as a possible cause of **chronic fatigue syndrome**, although conflicting data leave this issue unresolved. One study reported increased levels of anti-HHV-6 antibodies in people with chronic fatigue syndrome, but these findings were not confirmed in several other studies.

Figure 18.16 Kaposi's sarcoma in a young man infected with HIV-1. Note the distribution of the lesions, suggesting lymphatic involvement. Photo courtesy of P. Volberding. From J. A. Levy, *HIV and the Pathogenesis of AIDS*, 2nd ed. (ASM Press, Washington, DC, 1998), with permission.

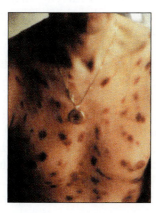

HHV-7

HHV-7 was initially isolated from an AIDS patient, who incidentally was also infected with HHV-6. (AIDS patients were among those from whom HHV-6 originally was isolated, as noted above.) HHV-7 is not yet associated with any disease and is thus referred to as an "orphan virus." Yet there is recent evidence that HHV-7, like HHV-6, may be a cause of roseola.

HHV-8

HHV-8 was discovered by Yuan Chang and Patrick Moore in 1994 in biopsy samples from **Kaposi's sarcoma** tumors. This malignant connective tissue tumor is one of the characteristic opportunistic diseases associated with AIDS and, indeed, is diagnostic of AIDS in HIV-positive patients (Fig. 18.16). In fact, Kaposi's sarcoma is the most common AIDS-related cancer, and HHV-8 is present in all Kaposi's sarcoma lesions. (Be sure to see Box 18.17.)

BOX 18.17

In 1981, 34 cases of an unusually aggressive form of Kaposi's sarcoma suddenly were seen in Haitian immigrants to the United States. At this time, there were coincident reports of a new disease causing opportunistic infections (in particular, *Pneumocystis* pneumonia and toxoplasmosis) in homosexual men. Shortly afterwards, aggressive Kaposi's sarcoma was also seen in 26 homosexual men who lived in New York City and California. Michael Gottlieb, who originally identified the new disease causing opportunistic infections in homosexual men, found that individuals with that syndrome, as well as those with aggressive Kaposi's sarcoma, all suffered from an affliction (that eventually would be called AIDS) that singled out their CD4 T lymphocytes for destruction.

A Hungarian physician, Moritz Kaposi, initially described Kaposi's sarcoma in 1872. It had been very rare (less than 1 case per 100,000 people per year in the United States prior to the AIDS epidemic) and had not been seen previously in Haiti. In fact, before the AIDS epidemic, Kaposi's sarcoma was seen mainly in elderly Italian and Jewish men and very rarely in elderly women. Moreover, in these individuals, the lesions developed slowly, in contrast to the much more aggressive form of the disease initially seen in the Haitian men.

HIV was discovered in 1983 as the cause of AIDS. Yet the discovery of HHV-8 did not happen until more than 1 decade later. HHV-8 is the most recent human herpesvirus to be discovered and among the most recent human cancer-associated viruses to be discovered.

HHV-8 is also associated with two B-cell disorders: **primary effusion lymphoma** and **multicentric Castleman's disease**. The latter is a rare disease, characterized by hyperproliferation of B cells, leading to noncancerous tumors that may develop in lymph node tissue.

The oncogenic potential of HHV-8 likely depends on its genes encoding homologues of cellular proteins that regulate cell growth (e.g., v-cyclins) and apoptosis (e.g., v-Bcl-2). And like most of the beta- and gammaherpesviruses, HHV-8 encodes a homologue of a cellular chemokine receptor, in this instance a G-protein-coupled receptor (GPCR) that can bind a number of chemokines. However, the HHV-8-encoded GPCR is constitutively activated without chemokine binding. Thus, it constitutively activates multiple signaling pathways, resulting in cell proliferation in the absence of added ligands. In addition, the HHV-8-encoded GPCR causes cells to secrete vascular endothelial cell growth factor, thereby promoting angiogenesis. This is important because the proliferation of new blood vessels is essential for tumor progression. Indeed, the malignant potential of a cancer correlates with its level of vascular endothelial cell growth factor expression (Chapter 17).

While HHV-8 is necessary for the development of AIDS-related Kaposi's sarcoma, bear in mind that immunosuppression too is an important risk factor for this cancer. Yet even in people with AIDS, HHV-8 infection is not sufficient to cause Kaposi's sarcoma, since many more AIDS patients are infected with this virus than develop the disease. (Note how we keep encountering the apparent paradox that whereas different herpesviruses have the ability to cause severe life-threatening diseases, it is only rarely that they actually do so.) In that regard, also consider the following enigma. Kaposi's sarcoma affects about 20% of HIV-infected homosexual men, but *only* about 2% of HIV-infected heterosexual men and HIV-infected women. No other HIV/AIDS-associated opportunistic infection so selectively targets a single segment of the population as does Kaposi's sarcoma in homosexual men, suggesting that other epidemiological factors might be determinants for the development of this tumor. Do you have any thoughts on what these might be?

Kaposi's sarcoma in AIDS patients is generally a much more aggressive disease than the "classic" or sporadic Kaposi's sarcoma, which occurs mainly in elderly men living in Mediterranean countries such as Italy. In AIDS patients, Kaposi's sarcoma can involve the gut, lungs, liver, spleen, lymph nodes, the hard and soft palates, and other internal organs, as well as the skin. Moreover, it is difficult to manage, it can be painful and have an enormous emotional impact, and it can be fatal. In contrast, the "classic" or sporadic Kaposi's sarcoma generally affects only the lower extremities.

The introduction of highly active antiretroviral therapy (Chapter 21), which involves a combination of multiple antiretroviral drugs, has led to a dramatic decline in the incidence of Kaposi's sarcoma in those HIV-infected individuals receiving the therapy. In areas where highly active antiretroviral therapy has not been available, such as sub-Saharan Africa, Kaposi's sarcoma remains a major problem for HIV/AIDS patients.

In contrast to the other human herpesviruses, HHV-8 does not appear to be especially ubiquitous in the human population. Its prevalence ranges from less than 5% of the population in the United States and Northern Europe to 50% in some regions of Africa. The geographic incidence of classic or sporadic cases of Kaposi's sarcoma coincides with the geographic distribution of HHV-8. HHV-8 infection is generally benign in otherwise healthy individuals.

TREATMENT AND PREVENTION
Antiviral Drugs

Although there are only a few effective antiviral drugs in general and no treatments are available for most viruses, several drugs are useful for treating herpesvirus infections. Thus, we might ask what unique feature of the herpesviruses made it possible to develop effective drug therapies against them. Mainly, it is that herpesviruses encode several of their own enzymes to support their DNA synthesis rather than using the corresponding cellular enzymes. Accordingly, it has been possible to find nucleoside analogues that preferentially affect the virally encoded enzymes (Box 18.10).

Acyclovir [9-(2-hydroxyethoxy) methyl guanine], a guanosine analogue, is the prototype and most-prescribed anti-HSV drug. Its development was in fact a major step in the development of rational antiviral therapies. Acyclovir is an effective inhibitor of herpesvirus DNA replication, while having low toxicity to uninfected host cells. The drug's specificity for herpesvirus DNA synthesis is explained as follows. The virus-encoded thymidine kinase phosphorylates acyclovir, thereby enabling the now-activated drug to be a substrate for the viral DNA polymerase (Fig. 18.17). When the drug is then incorporated into the viral DNA, it blocks further elongation, since it lacks the 3'-OH group of the sugar ring. Since there is no enzyme in uninfected cells that might activate acyclovir, the drug has no effect on uninfected cells.

As might be expected, herpesviruses generate acyclovir-resistant mutants. And consistent with the mode of action of acyclovir, these mutants map in the viral thymidine kinase and DNA polymerase genes. In the first case, the mutants are unable to phosphorylate acyclovir, and in the second case, they are unable to incorporate the

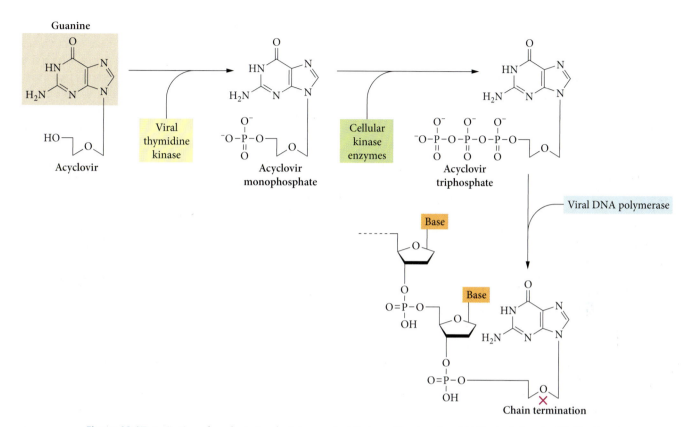

Figure 18.17 Activation of acyclovir. Acyclovir is a product that must be phosphorylated in the infected cell before it can inhibit viral DNA replication. The viral thymidine kinase, but not the cellular kinase, adds one phosphate to the 5′-hydroxyl group of acyclovir. The monophosphate compound is a substrate for cellular kinases that convert the monophosphate compound to acyclovir triphosphate. The triphosphate compound is recognized by the viral DNA polymerase and incorporated into viral DNA. As acyclovir has no 3′-hydroxyl group, the growing DNA chain is terminated and viral replication ceases. Adapted from D. R. Harper, *Molecular Virology* (Bios Scientific Publishers Ltd., Oxford, United Kingdom, 1994), with permission.

phosphorylated drug into DNA. The emergence of drug-resistant mutants may be particularly troublesome in cases where long-term therapy might be required to control a latent infection. (AIDS therapy presents the same problem, for the same reason [Chapter 21].) Fortunately, acyclovir-resistant strains are generally less virulent, although they still may be devastating to AIDS patients. Acyclovir can be administered by injection, as a topical cream, or as an ingestible capsule.

Chemical modifications to acyclovir have led to **valacyclovir**, **penciclovir**, and **famciclovir**, which act in a manner similar to that of acyclovir (Fig. 18.18). The advantage of the newer drugs is that they are taken up more efficiently than acyclovir after oral ingestion. Each is approved by the U.S. Food and Drug Administration (FDA) for treating HSV.

Ganciclovir [9-(1,3-dihydroxy-2-propoxy)methyl guanine] is another guanosine analogue. It resembles acyclovir and is active against the DNA polymerases of HSV,

HCMV, and HHV-6. Resistance mutations map in the viral DNA polymerase gene.

Foscarnet (phosphonoformate) is a pyrophosphate analogue that acts against herpesvirus DNA polymerases by mimicking the pyrophosphate portion of nucleoside triphosphates. Once again, resistance mutations map in the viral DNA polymerase gene. This simple molecule is the only nonnucleoside inhibitor of herpesviruses and has been used to treat infections in which acyclovir- and ganciclovir-resistant mutants have emerged. Unfortunately, some acyclovir-resistant mutants are resistant to foscarnet also. (Note that foscarnet also is effective against hepatitis B virus [Chapter 22] and HIV [Chapter 21].)

The drugs described above and others have been approved by the FDA for treating specific herpesvirus infections as follows. For treating HSV-1 and HSV-2, approved drugs include acyclovir, penciclovir, valacyclovir, and famciclovir. Adenosine arabinoside, iododeoxyuridine, and trifluridine also have been approved, but they

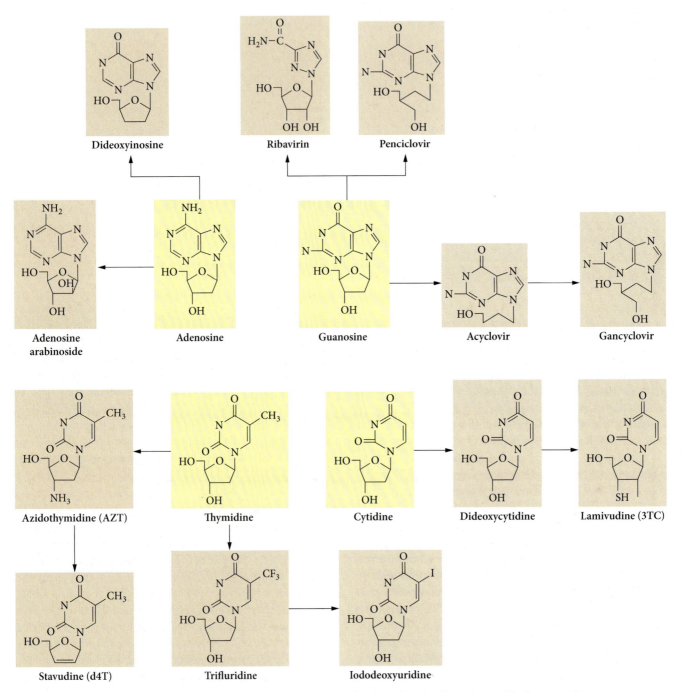

Figure 18.18 Many well-known antiviral compounds are nucleoside and nucleotide analogues. The four natural deoxynucleosides are highlighted in yellow. Adapted from E. De Clercq, *Nat. Rev. Drug Discov.* 1:13–25, 2002, with permission.

are less effective. But trifluridine, like acyclovir and penciclovir, is useful as topical treatment for HSV lesions on the eyelids and cornea (keratitis). Approved drugs for treating HCMV include ganciclovir and foscarnet. Approved drugs for treating VZV include acyclovir, famciclovir, and valacyclovir. (The VZV DNA polymerase is much less sensitive to acyclovir than is the HSV enzyme.) VZV immunoglobulin and zoster immune plasma also may be used to treat VZV. HHV-6 can be treated with cidofovir (a nucleoside phosphonate analogue), ganciclovir, and foscarnet. Note that new developments in herpesvirus therapy are occurring continuously. But, also note the following provisos.

The above drugs can *shorten the duration* of primary and recurrent episodes of some herpesvirus infections. Yet *no drug treatment can cure* a latent infection. This truth holds for HIV as well as for the herpesviruses. On the other hand, drugs have been used with some success to prevent herpesvirus as well as HIV infections. While on the topic of prevention, condoms are *not* fully protective against genital HSV infection but certainly are better than nothing (e.g., see Wald et al., 2005). Since HCMV is spread largely by sexual activity, with semen being a major vector for transmission, condom usage would limit the spread of that virus as well.

It seems appropriate to at least mention some of the over-the-counter treatments for cold sores and genital ulcers (e.g., Abreva and lysine) that are widely advertised as disease course-shortening compounds. Much is available on the World Wide Web from the manufacturers and dispensers of these products. However, for a reliable and thorough account of herpes treatment options, read the *Sexually Transmitted Diseases Treatment Guidelines* on The Centers for Disease Control and Prevention (CDC) website. The following is a sampling from that source:

> Antiviral chemotherapy offers clinical benefits to the majority of symptomatic patients and is the mainstay of management. Counseling regarding the natural history of genital herpes, sexual and perinatal transmission, and methods to reduce transmission is integral to clinical management.
>
> Systemic antiviral drugs can partially control the signs and symptoms of herpes episodes when used to treat first clinical and recurrent episodes, or when used as daily suppressive therapy. However, these drugs neither eradicate latent virus nor affect the risk, frequency, or severity of recurrences after the drug is discontinued. Randomized trials have indicated that three antiviral medications provide clinical benefit for genital herpes: acyclovir, valacyclovir, and famciclovir. Valacyclovir is the valine ester of acyclovir and has enhanced absorption after oral administration. Famciclovir also has high oral bioavailability. Topical therapy with antiviral drugs offers minimal clinical benefit, and its use is discouraged.

Vaccines

VZV is the first human herpesvirus for which a live attenuated vaccine was developed. In one study, this vaccine was found to reduce the number of reported verified varicella cases by 85%. Among older adults, the VZV vaccine reduced the incidence of herpes zoster (shingles) by slightly more than 50% percent (Box 18.18).

BOX 18.18

On 25 May 2006, the FDA approved the first vaccine against adult shingles, and the following October, the vaccine advisory panel of the CDC voted to make shingles vaccination routine for all Americans 60 years and older.

The "shingles vaccine," known as Zostavax, is actually a boosted dose of the chicken pox vaccine currently given to children. In adults age 60 and older, it can prevent shingles in slightly more than one-half of the vaccine recipients and dramatically reduce the severity of shingles in many others. Without vaccination, about 20% of people who have had chicken pox eventually will get shingles. A person who lives to be 85 has a 50% chance of getting shingles.

Unfortunately, as many as 10% of older patients will not be able to avail themselves of this vaccine because of weakened immune systems due to cancer therapy, organ transplants, AIDS, or other causes. The vaccine contains a live attenuated VZV that could overwhelm the immune systems of those patients.

Despite the apparent efficacy of the VZV vaccine, optimism that other effective herpesvirus vaccines might be developed is tempered by the following considerations. Taking HSV as an example, live, attenuated HSV vaccines are under development. However, since HSV disseminates rapidly to infect neurons, there is concern that the immune response elicited by the vaccine may not be strong enough to completely protect neurons from infection. Optimism for the potential effectiveness of HSV vaccines is also diminished by the fact that natural repeated exposure to HSV and clinical recurrences do not elicit immune responses capable of completely preventing reactivations of the endogenous virus. Thus, the effectiveness of HSV vaccines (and those for EBV as well; see below) may depend on their abilities to block the initial infection; this is a tall order. (A similar argument can be made regarding the prospects for an effective AIDS vaccine [Chapter 21].) Note that while some candidate HSV vaccines are highly effective in mouse or guinea pig models, they are not nearly as effective in humans. This may well be because the human virus is better adapted to evade human immune responses than those of mice and guinea pigs as well as to initiate infection in humans.

Next, considering EBV, this virus was found to establish latency in B cells of a mouse model, almost immediately after the initial infection. Moreover, even inhibiting lytic EBV replication at the site of infection (using a replication-deficient mutant of EBV) did not affect establishment of the latent infection. These experimental findings underscore the involvement of B cells at the earliest times of EBV infection and suggest that extensive lytic replication of EBV is not crucial for the establishment of latent infection.

The observations described above regarding EBV call attention to the fact that prophylactic vaccination strategies that might be developed against EBV, and perhaps other herpesviruses as well, must target the virus at the site of infection. That is so because it will be necessary to prevent early infection of cells that harbor latent virus. It remains to be seen whether vaccination, which mimics the immune response to natural EBV or HSV infection, will be able to reduce infection or disease from an exogenous viral challenge. Nonetheless, the success of the VZV vaccine provides some reason for optimism.

At present, there is no effective treatment or vaccine available for EBV, nor is there a vaccine for HCMV.

Suggested Readings

Gupta, A., J. J. Gartner, P. Sethupathy, A. G. Hatzigeorgiou, and N. W. Fraser. 2006. Anti-apoptotic function of a microRNA encoded by the HSV-1 latency-associated transcript. *Nature* 442:82–85.

Heymann, J. B., N. Cheng, W. W. Newcomb, B. L. Trus, J. C. Brown, and A. C. Steven. 2003. Dynamics of herpes simplex virus capsid maturation visualized by time-lapse cryo-electron microscopy. *Nat. Struct. Biol.* 10:334–341.

Preston, C. M. 2000. Repression of viral transcription during herpes simplex virus latency. *J. Gen. Virol.* 81:1–19.

Thorley-Lawson, D. A. 2005. EBV the prototypical human tumor virus—just how bad is it? *J. Allergy Clin. Immunol.* 116:251–261.

Wald, A., A. G. Langenberg, E. Krantz, J. M. Douglas, Jr., H. H. Handsfield, R. P. DiCarlo, A. A. Adimora, A. E. Izu, R. A. Morrow, and L. Corey. 2005. The relationship between condom use and herpes simplex virus acquisition. *Ann. Intern. Med.* 143:707–713.

Wang, Q. Y., C. Zhou, K. E. Johnson, R. C. Colgrove, D. M. Coen, and D. M. Knipe. 2005. Herpesviral latency-associated transcript gene promotes assembly of heterochromatin on viral lytic-gene promoters in latent infection. *Proc. Natl. Acad. Sci. USA* 102:16055–16059.

Poxviruses

INTRODUCTION

The *Poxviridae* are the largest and most complex of the animal viruses. Their double-stranded DNA genomes range from 130 to 360 kilobases (kb) in size. Thus, the largest poxvirus genomes are more than twice the size of the herpes simplex virus (HSV) genome and about 70-fold larger than a polyomavirus genome. Moreover, the morphologically complex poxvirus particles, which display neither icosahedral nor helical symmetry, can be as large as 330 by 280 nm, making them just barely discernible by light microscopy. For comparison, the icosahedral polyomaviruses are about 50 nm in diameter. Poxviruses are found in animals throughout the vertebrate kingdom and in insects as well.

Variola virus was the most medically significant of the poxviruses, since it gave rise to smallpox, one of the great scourges of humankind. In fact, smallpox was once reputed to have killed, crippled, or maimed nearly 1/10 of all individuals who ever have lived, making it by that reckoning the most devastating of all human diseases. Smallpox is believed to have emerged in human populations in about 10,000 BCE, and outbreaks of this now-eradicated disease are known to have occurred periodically for thousands of years. It is estimated that smallpox killed 400,000 Europeans each year during the 18th century. In late 18th century England, just before the advent of smallpox vaccination (see Chapter 1), smallpox was responsible for 7 to 12% of all human deaths and was the cause of death in one-third of all children. During the 20th century, prior to its eradication, it is estimated that smallpox killed between 300 million and 500 million people. As recently as 1967, the World Health Organization (WHO) estimated that there were 15 million smallpox cases worldwide, 2 million of which were fatal.

Famous individuals who survived smallpox but were permanently scarred by it included Elizabeth I of England in 1562 (her reason for applying heavy makeup may have been to hide the pockmarks), Abraham Lincoln in 1863 (Box 19.1), and Joseph Stalin, who had his photographs retouched to make his pockmarks less apparent. George Washington, too, contracted and recovered from the disease.

521

BOX 19.1

A little-known fact is that Abraham Lincoln was actually suffering from smallpox prodrome while delivering the Gettysburg Address. During the train trip from Washington, DC, to Gettysburg, PA, on 18 November 1863, the day before the speech, Lincoln told John Hay, his private secretary and assistant, that he felt weak. Shortly after delivering the Gettysburg Address on November 19, Lincoln developed a high fever and severe headache, and within a week, his skin erupted with scarlet blisters. Although Lincoln's doctors diagnosed the president with a mild form of the disease, contemporary researchers suggest that Lincoln's physicians knew otherwise but wanted to reassure the public that their president was not gravely ill. The subsequent history of the United States would certainly have been very much different had Lincoln succumbed to smallpox at that time.

Lincoln's famed sense of humor remained in evidence during his potentially fatal brush with smallpox. His physician informed Lincoln of his diagnosis while the President was interviewing an office seeker. After the office seeker heard the physician advise Lincoln that the disease is highly contagious, he made excuses and left immediately, causing Lincoln to remark, "There is one good thing about this. Now I have something I can give everybody."

BOX 19.2

A singularly important factor in the success of the smallpox vaccination program was that variola virus does not have any animal virus reservoir in which it might have persisted in nature. Nevertheless, it was not possible to vaccinate a sufficient percentage of individuals in some densely populated tropical regions, where the disease was endemic, to achieve **herd immunity** (refers to the immunity in the whole population, which comes about when a sufficient percentage of individuals have been vaccinated [Chapter 11].) Success in those regions was achieved instead by combining surveillance with containment. That is, infected individuals were actively sought out and isolated, and their contacts were then isolated and vaccinated.

Note that the surveillance and containment strategy worked in the case of smallpox, because infected individuals were not infectious during the prodromal stage. This is not so in the case of measles virus (Chapter 11), which is nevertheless considered as a target for eradication. Poliovirus (Chapter 6), which also may eventually be eradicated, is also somewhat problematical because it gives rise to a preponderance of subclinical infections.

Smallpox was exceptionally devastating to indigenous native peoples, who were first exposed to the virus by European explorers and conquerors. In point of fact, smallpox played a far greater role than guns in the European conquests of the Americas and of other native peoples as well. These points are discussed further below and in Box 19.20.

Following a successful worldwide vaccination program, variola virus was officially declared to be eradicated in 1977, with the last documented smallpox case occurring that year in Somalia. The last known case of smallpox in the United States was reported in 1949. Smallpox was, in fact, the first infectious disease to be controlled by vaccination, and its eradication is regarded as one of medicine's greatest triumphs (Box 19.2). After the eradication of variola virus, routine smallpox vaccination was stopped, since it was no longer necessary for prevention and the risk associated with vaccination now clearly exceeded the risk of acquiring a natural infection. (As recounted in Chapter 1, the smallpox vaccine was indeed the first vaccine ever developed. Remarkably, it was developed in 1796, 70 years

before the acceptance of the germ theory of disease and a century before the recognition of viruses as distinct microbial entities.)

Because of the success of the smallpox eradication program, reference stocks of variola virus, which were maintained in WHO laboratories, were destroyed. Today, stocks of the virus are known to exist only at the Centers for Disease Control and Prevention (CDC) in the United States and at the Institute of Virus Preparations in Siberia, Russia.

Current concern regarding smallpox is that variola virus stocks might fall into the hands of groups that would use it as a bioterror agent. For that reason, many concerned individuals have argued that the last remaining variola virus stocks should be destroyed. This issue is discussed later in the chapter.

Here is a conundrum of some historical interest. The original smallpox vaccine contained cowpox virus, as recounted in Chapter 1. Yet, somehow, vaccinia virus came to replace cowpox virus in the vaccine (see Chapter 1). Vaccinia virus is closely related to variola virus, as they share the same overall genome organization and many of their genes show >90% homology. Regardless, it is not clear how vaccinia virus came to replace cowpox virus in the vaccine, and vaccinia virus is now considered to be the prototype poxvirus. Yet neither the origin nor the natural host of vaccinia virus is presently known (Box 19.3).

Molluscum contagiosum virus is the only poxvirus other than variola virus that is known to be specific for humans. Like its more famous relative, it too is pathogenic in humans, in which it induces a chronic rash that is characterized by wart-like skin lesions. Actually, all poxvirus diseases in humans involve lesions of the skin.

Yaba virus, a primate poxvirus, causes benign tumors in macaques, patas, and baboons. These Yaba virus-induced tumors spontaneously regress within several weeks to months. Interestingly, Yaba virus induces similar tumors in humans, as seen in the case of a laboratory attendant who was accidentally infected while handling affected monkeys and also in experimentally infected human volunteers.

The poxviruses have evolved a mode of replication that is unique among DNA viruses. To the point, their entire replication cycle occurs exclusively within the cytoplasm, largely independent of host nuclear function. This mode of virus replication requires that the infecting poxvirus particles contain the enzymes needed to initiate transcription of their DNA genomes. In this respect, the poxviruses are reminiscent of the minus-sense, single-stranded RNA viruses and of the double-stranded RNA viruses (i.e., the reoviruses), which also must contain an RNA polymerase (albeit an RNA-dependent one) within the virus particle. Interestingly, the first virus-encoded RNA polymerase to be discovered was that of vaccinia virus, not of an RNA virus, as one might have otherwise thought.

The fact that poxvirus replication occurs entirely outside of the nucleus may have several important consequences for these DNA viruses. Clearly, it necessitates the large size of poxvirus genomes, which is required so that they might encode their own DNA replication machinery as well as their own RNA polymerase; enzymes for capping, methylating, and polyadenylating their messenger RNAs (mRNAs); and initiation proteins that recognize specific viral promoters. Also, the cytoplasmic replication site of the poxviruses might restrict their lifestyle possibilities. For example, poxvirus infections are generally acute, while other double-stranded DNA viruses, all of which can establish chronic, persistent, or latent infections, replicate in the nucleus.

Despite the possible drawbacks of poxvirus replication in the cytoplasm, it may yet benefit these viruses by ensuring that they will be able to replicate, regardless of the growth state of the host cell. Moreover, it enables the poxviruses to shut down host macromolecular synthesis, thereby diverting the cell's biosynthetic capacity to support the needs of the virus (Box 19.4).

STRUCTURE

The morphology of poxvirus particles is rather unique, since it does not display any structural features that might correspond to either icosahedral or helical symmetry. Instead, when observed by electron microscopy, poxvirus cores may appear as smooth, homogenous rectangles of about 250 to 300 nm by 200 to 250 nm, enclosed within a 30-nm membrane (Fig. 19.1). Poxvirus cores also may appear to be dumbbell shaped, with so-called lateral bodies filling the space between the concavities of the core and the external membrane. The concavities and lateral bodies can be seen best in Fig. 19.1A, B, and K. In panel B, the inner and external membrane layers are marked by large arrows, and a lateral body is labeled by a star. Spikes (also referred to as the palisade layer) are marked by an arrowhead in the figure. The concavities and lateral bodies were once thought to be artifacts of the fixation procedure but now are believed to reflect actual structural features. The external membranes of vaccine virus particles are discussed below.

Different sections through the virus particles provide different impressions of the relationship between the core and the external membrane (compare Fig. 19.1, panels C through E, with panels F through H and I through K). Also, notice the tube-like structure within the core. Rupturing vaccinia virus particles can release tubules that are 30 to 40 nm in diameter and that may have a helical substructure. These tubules may be the tube-like structures seen in virus particles, but that is not yet known with certainty. There is evidence that tubules are comprised of proteins, in association with supercoiled DNA, which leads to the suggestion that the vaccinia virus genome exists within the cores as an organized DNA/protein complex. However, this contention is still controversial.

Figure 19.2 depicts a schematic three-dimensional model of a vaccinia virus particle, in which the sections

BOX 19.3

During the smallpox eradication campaign, it was believed that vaccine strains could not survive in nature and that wild-type vaccinia virus had become extinct. However, over the past several decades, several genetically distinct isolates of vaccinia virus were isolated during outbreaks in humans and cattle in Brazil and elsewhere. Based on the diversity of the Brazilian isolates, it was suggested that an ancestral vaccinia virus existed in Brazil prior to the beginning of the WHO smallpox eradication vaccination campaigns and that these viruses continue to circulate.

BOX 19.4

The poxviruses, with their large genomes and near independence of the host transcriptional and replication machinery, challenge our general conception of a virus. Yet the recently discovered mimivirus, *Acanthamoeba polyphaga*, which is larger than 400 nm and contains a genome that is 1,200 kbp in length and a correspondingly immense genetic repertoire (911 putative protein-coding genes, in comparison to the 200 of vaccinia virus), further defies our concept of viruses as relatively uncomplicated entities that extensively exploit the biosynthetic capacities of the host cell.

Mimivirus was first identified growing in amoebas, and because it stained gram positive, it was originally mistaken for a bacterium. A remarkable part of the story is that the mimivirus was first isolated from a cooling tower in Bradford, United Kingdom, during the investigation of a pneumonia outbreak. In fact, a high level of seroconversion to the mimivirus was seen at the time in patients with pneumonia. Actual mimivirus infection of humans was confirmed when a technician from the laboratory that isolated the virus contracted acute pneumonia and seroconverted to the mimivirus! The technician regularly performed Western blot assays, including ones that documented infection with the mimivirus.

But the story becomes still more interesting. Recently (August 2008), Didier Raoult and colleagues found a new strain of mimivirus that was even larger than the first mimivirus, in this instance, from a water cooling tower in Paris (La Scola et al., 2008). Because of its size, the researchers dubbed the new virus "mamavirus." More remarkably still, mamavirus had a small virus attached to it, which contained only 20 genes. The researchers appropriately named this **satellite virus** Sputnik. More amazingly still, Sputnik "infects" both mamavirus and mimivirus. To be more precise, Sputnik cannot multiply in uninfected amoebae, but it grows rapidly, after an eclipse phase, in the giant virus factories found in amoebae coinfected with either mimivirus or mamavirus. Sputnik is deleterious to the "host viruses," causing them to produce abortive forms and defective capsids. The investigators coined a new term for a virus like Sputnik: **virophage**.

References

La Scola, B., C. Desnues, I. Pagnier, C. Robert, L. Barrassi, G. Fournous, M. Merchat, M. Suzan-Monti, P. Forterre, E. Koonin, and D. Raoult. 2008. The virophage as a unique parasite of the giant mimivirus. *Nature* **455**:100–104.

Raoult, D., P. Renesto, and P. Brouqui. 2006. Laboratory infection of a technician by mimivirus. *Ann. Intern. Med.* **144**:702–703.

---

through the model in three different orientations correspond to the sections through the actual virus particles shown in Fig. 19.1. Notice how the tube, which has a total length of about 250 nm, is folded upon itself into three continuous segments. In viewing Fig. 19.2, bear in mind that it is a *model* that is meant to clarify *possible* interrelationships between virus components.

The issue of poxvirus structure is more complicated still, since there are multiple forms of infectious vaccinia virus particles, each of which plays a distinct role in the viral life cycle. These different morphological forms are generated at different stages of the viral maturation process, as described below.

The most abundant of the poxvirus structural forms is called an **intracellular mature virion** (IMV), also referred to as a mature virion or MV, which is the type depicted in Fig. 19.1E, H, and K and in the model shown in Fig. 19.2. (It is my particular fetish to avoid the term "virion," which refers to an infectious virus particle. Throughout the text I have made an effort to use the term "virus particle" rather than virion. But in the case of the poxviruses, convention cannot be denied.) IMVs consist of the nucleoprotein core, surrounded by a lipid envelope that contains more

than a dozen viral proteins. IMVs are released from cells when they lyse and are thought to mediate viral spread from host to host. (IMVs are also the infectious units of the smallpox vaccine. The reasons are noted below [see "Assembly, Maturation, and Release"].)

A second major form of infectious vaccinia virus particle is a type called an **intracellular enveloped virion** (IEV), also referred to as a wrapped virion or WV. IEVs differ from IMVs by containing two additional membranes, which are derived from the *trans*-Golgi network or from an endosomal membrane by an ill-defined wrapping process during maturation (see below). Prior to wrapping, these additional membranes are modified to contain at least seven other viral proteins.

IEVs exit the cell to become **extracellular virions** (EVs). The exit process is believed to involve fusion of the outermost membrane of the IEV with the plasma membrane (Fig. 19.3). Thus, upon exiting the cell, IEVs lose their outermost membrane but retain an additional membrane that is not present on IMVs. Other electron microscopic images suggest that some IMVs exit the cell by budding directly through the plasma membrane, thereby obtaining an IEV-like extra membrane.

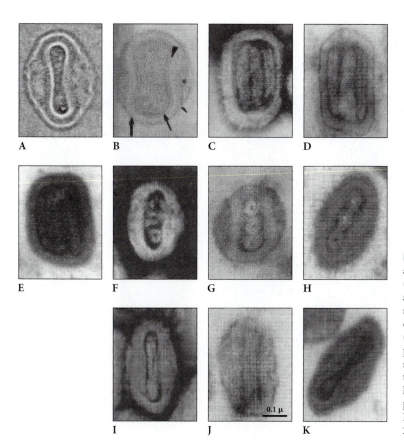

A   B   C   D

E   F   G   H

I   J   K

0.1 μ

Figure 19.1 Vaccinia virus. (A) Cryosection; (B) a uranyl acetate-stained whole-mount cryo-electron micrograph; (C, F, and I) three different projections of a phosphotungstic acid-stained whole-mount preparation; (D, G, and J) sections of uranyl acid-stained purified virus in three different perpendicular planes (scale bar in panel J, 0.1 μm); (E, H, and K) sections of IMVs (see below) in three different perpendicular planes. The concavities and lateral bodies are seen best in panels A, B, and K. In panel B, the inner and external membrane layers are marked by large arrows, and a lateral body is labeled by a star. Spikes (also referred to as the palisade layer) are marked by an arrowhead. From R. Condit, N. Moussatche, and P. Traktman, *Adv. Virus Res.* **66:**36–124, 2006, with permission.

Figure 19.4 provides a summary of vaccinia virus morphogenesis, in which the relationships between IMVs (or MVs), IEVs (or WVs), and EVs are depicted. IVs refer to immature virions, and IVNs are IVs that contain nucleoids. The steps leading to the formation of IVs and IVNs are described below (see "Assembly, Maturation, and Release").

EVs may remain cell associated as depicted in Fig. 19.3, or they may become unattached. (In the journal literature, attached EVs are referred to as **cell-associated enveloped virions** [CEVs], and those that dissociate from the cell are called **extracellular enveloped virions** [EEVs].) In the case of several strains of vaccinia virus, the vast majority of EVs are the cell-associated form, even at late times in infection. In the example of the cell-associated form depicted in Fig. 19.3, notice how the outer wrapping membrane of the IEV has fused to the plasma membrane. Both the cell-associated EVs and extracellular EVs contain the same outer membrane. They are thought to mediate cell-to-cell spread between adjacent cells and long-range spread, respectively.

As might be expected of viruses that encode 200 or so proteins, poxvirus particles themselves contain a large number of different virus-encoded polypeptides. Indeed, vaccinia virus particles contain about 80 or so different virus-encoded proteins. Of these, about 40 are nonenzymatic virus components and about 20 are envelope proteins.

Others are viral DNA-binding proteins, which are thought to play a role in condensing and packaging the viral genome within cores. Four of the latter proteins, encoded by vaccinia virus genes F17, L4, A3, and A10, comprise about 70% of the viral core by weight. Vaccinia virus particles also contain 24 different host proteins, the most abundant of which are cytoskeletal proteins, heat shock proteins, and proteins involved in translation.

Consistent with their cytoplasmic replication cycle, poxvirus particles also contain all of the enzymes and factors necessary for early-gene transcription. In all, poxvirus particles contain about 25 virus-encoded enzymes and factors, including DNA-dependent RNA polymerases and RNA processing enzymes, all of which are located in the viral cores.

ENTRY AND UNCOATING

IMVs and EVs are each infectious. Bearing in mind that EVs are enwrapped within an additional outer membrane, which contains proteins different from those in the IMV envelope, EVs and IMVs cannot enter cells by exactly the same mechanism. Although the existence of two forms of infectious vaccinia virus particles has complicated the analysis of poxvirus entry into host cells, recent progress has made it possible to separately describe the entry of IMVs and EVs.

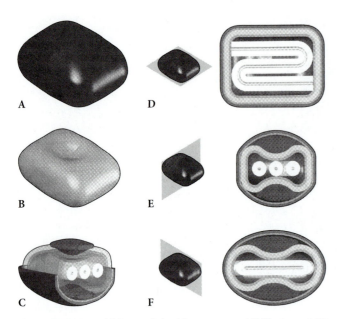

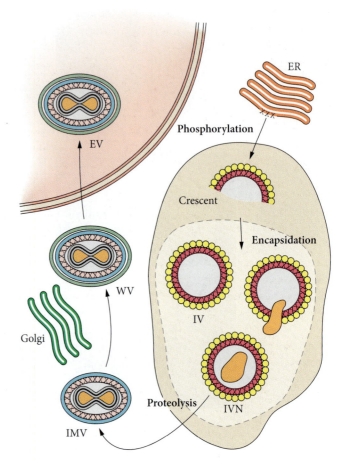

Figure 19.2 A model for vaccinia virion structure. (A) The intact MV. No attempt has been made to represent surface tubule elements. (B) The virion core. (C) Cutaway view. The membrane has been removed from the upper half of the virion, the near end has been removed, and the core wall has been rendered transparent, thus revealing the multiple layers, the concavities in the core, the lateral bodies, and the tubular internal structure. (D, E, and F) Sections through the virion in three different perpendicular planes. The sections correspond to sections shown in Fig. 19.1. See the text for details. (Model courtesy of Michel Moussatche. A dynamic three-dimensional model is available online at http://www.vacciniamodel.com/.) From R. Condit, N. Moussatche, and P. Traktman, *Adv. Virus Res.* **66**:36–124, 2006, with permission.

IMVs

Binding of vaccinia IMVs is thought to be mediated by the viral A27, D8, and H3 envelope proteins. These proteins bind to cell surface **glycosaminoglycans** (i.e., polysaccharides

Figure 19.3 Transmission electron micrographs of the IMV (left panel), IEV (middle panel), and EV (right panel) forms of vaccinia virus. (Left panel) The arrow points to the single membrane of IMV. (Middle panel) One arrow points to the IMV membrane, and two point to the outer wrapping membranes. (Right panel) One arrow points to the outer wrapping membrane that has fused with the plasma membrane, and the others point to the IMV membrane and the remaining EV wrapper. Electron microscopic images were kindly provided by Andrea S. Weisberg. From B. Moss, *Virology* **344**:48–54, 2006, with permission.

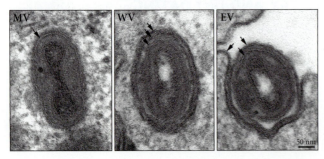

Figure 19.4 Vaccinia virus morphogenesis summary. Following the arrows in sequence from the upper right, several integral viral membrane proteins (black Xs) are made in the endoplasmic reticulum (ER, red lines) and transported to viral factories (sand-colored area) along with endoplasmic reticulum-derived lipid to be assembled into crescents which contain a lipid bilayer (red line) and the membrane proteins, scaffolded on a honeycomb structure composed of the D13 protein (yellow circles). Crescent formation is controlled by phosphorylation. The crescents mature to IVs and IVNs, accompanied by encapsidation of the genome. The dotted line surrounding IV, IVN, and encapsidation signifies that the order of these events is uncertain. Metamorphosis to IMVs is accompanied by loss of the D13 scaffold, proteolysis, further addition of membrane proteins (blue line), and movement of particles outside factories. IMVs acquire Golgi apparatus-derived membranes (Golgi, green lines) to become IEVs (WVs) and are exocytosed through the plasma membrane (EV) to become EVs. Adapted from R. Condit, N. Moussatche, and P. Traktman, *Adv. Virus Res.* **66**:36–124, 2006, with permission.

that consist of repeating units of particular disaccharides; see Chapter 18). For example, the D8 protein was found to bind to the glycosaminoglycan chondroitin sulfate, and the A27 and H3 proteins were found to bind to the glycosaminoglycan heparan sulfate. However, more recent studies found that soluble glycosaminoglycans affected IMV infectivity only partially in some cells, and not at all in others, suggesting that glycosaminoglycans may not be the only poxvirus receptors. Thus, the specific cellular receptors for poxviruses are not yet known with certainty.

Enveloped viruses generally penetrate cells by either of two mechanisms. In each of these mechanisms, virus attachment to cells is followed by the activation of a viral fusion protein, which then facilitates fusion of the viral envelope with a cellular membrane. The key difference between the two mechanisms concerns whether the viral envelope fuses with the plasma membrane or with an internal cellular membrane. So, in the first mechanism, the viral fusion protein is activated by the binding of the virus to its receptor at the cell surface, and the viral envelope then fuses with the plasma membrane at neutral pH. In the second mechanism, the virus binds to its receptor on the plasma membrane, and it is then taken up into the cell by receptor-mediated endocytosis. Next, the viral fusion protein is activated by the low pH of the endosomal compartment, triggering fusion of the viral envelope with an internal endosomal membrane. In either event, the viral core is released into the cytoplasm.

So, which of the above mechanisms mediates the entry of vaccinia virus IMVs? Earlier experiments found that vaccinia IMVs could enter cells by fusion with the plasma membrane at neutral pH. Moreover, these early studies did not find the entry of vaccinia IMVs to be inhibited by drugs that raise the pH of endosomal compartments. Also, low-pH conditions did not enhance IMV entry at the plasma membrane. Together, these early experimental findings argued against entry by endocytosis. But, lest matters be too straightforward for us, more recent experiments found that entry of IMVs indeed is accelerated by brief low-pH treatment and, additionally, is significantly reduced by inhibitors of endosomal acidification (Townsley et al., 2006). In these more recent experiments, some virus particles were seen to fuse with the plasma membrane at neutral pH. However, fusion efficiency at the plasma membrane increased 30-fold after low-pH treatment.

Although the more recent study reports that vaccinia virus IMVs enter cells more efficiently via the low-pH-dependent endosomal pathway, both the earlier and more recent studies found that these particles could enter cells by fusion with the plasma membrane at neutral pH. Thus, it is possible that the virus makes use of fusion at the plasma membrane, as well as fusion from within an endosomal compartment, to enter cells. But why would the virus have adapted to be able to enter cells by both pathways? Possibly, being able to use multiple pathways to enter cells enables the virus to enter more cell types (Box 19.5). In addition, IMVs and EVs may have different requirements for a low-pH environment during entry (see below).

The vaccinia virus A27 protein was once thought to mediate fusion. This notion was implied by the experimental finding that a monoclonal antibody against the A27 protein

> ## BOX 19.5
>
> Vaccinia virus can induce **syncytium formation** (i.e., cell fusion). This phenomenon occurs when cells with still-bound progeny EVs, or adsorbed IMVs, are briefly acidified and then returned to neutral pH. The mechanism of syncytium formation and its possible role, if any, in the vaccinia virus life cycle are not known. But based on the precedents of other viruses, one might expect that vaccinia virus-induced syncytium formation very likely reflects the fusion reaction that occurs during entry through acidified endosomes. That being so, if vaccinia virus were to enter cells exclusively by direct fusion with the plasma membrane at neutral pH, then its ability to induce syncytia after low-pH treatment would have no obvious explanation.

blocks uncoating and infectivity, while not impairing virus attachment per se. Moreover, anti-A27 monoclonal antibodies also block vaccinia virus-induced syncytium formation (Box 19.5). However, more-recent studies found that A27 deletion mutants are able to enter cells and also induce syncytia. Such experiments, using null mutants, provide a critical "gold standard" test, a test that A27 did not pass. (What might account for the earlier experimental findings, which suggested that A27 mediates fusion?)

The gold standard described above recently was met by several other vaccinia virus proteins. These include the A21, A28, H2, and L5 proteins, each of which is located in the IMV envelope. Each is a nonglycosylated late protein that contains a single transmembrane domain, a hydrophobic sequence at or near its N terminus, and cystine residues that form two to four intramolecular disulfide bonds. Null mutants that affect each of these proteins produce noninfectious IMVs that appear morphologically normal and attach to cells but cannot penetrate into the cytoplasm. These null mutants were also unable to induce syncytium formation, showing that A21, A28, H2, and L5 indeed are part of the viral fusion machinery.

Results of other recent experiments show that A21, A28, H2, and L5, as well as four additional putative envelope proteins (A16, G3, G9, and J5), form a stable fusion/entry complex, and *each* of these proteins is essential for uncoating and infectivity but not for initial attachment. Moreover, studies using individual null mutants show that *each* protein is necessary to assemble the stable complexes, suggesting that these proteins engage each other within a specific interaction network. Together, these findings imply that each of the eight proteins of the vaccinia virus fusion/entry complex has a nonredundant function.

Thus far, no cellular receptor proteins that might be necessary to trigger vaccinia virus fusion have been identified.

Recall that the herpesviruses (Chapter 18) likewise assemble a fusion complex from multiple membrane proteins (designated gB, gH, and gL in the case of HSV). In contrast, enveloped RNA viruses generally have but a single transmembrane glycoprotein to mediate membrane fusion, which occurs either at the plasma membrane (e.g., human immunodeficiency virus [HIV] [Chapter 21]) or in an endosomal compartment (e.g., influenza virus [Chapter 12]). At any rate, the large number of virus proteins involved in vaccinia virus entry suggests that the poxvirus fusion mechanism may be markedly different from that of other, less complex viruses.

While specific receptors for poxviruses have not yet been identified, there is evidence that entry of IMVs involves plasma membrane **lipid raft** domains. (Membrane microdomains called lipid rafts form because of the tendency of sphingolipids and cholesterol to segregate and pack tightly together in membranes; see Chapter 15.) For example, the cholesterol-depleting, raft-disrupting drug methyl-β-cyclodextrin inhibited IMV entry and uncoating, while not affecting IMV attachment. In addition, several IMV envelope proteins, including A27 and D8, copurified with cell membrane lipid raft fractions that were isolated from cells immediately after virus infection. Other experimental findings suggest that IMV binding and insertion into the lipid rafts might be initial steps in the fusion process. Lipid rafts might play yet another role in IMV entry, and postentry as well, as follows.

The IMV entry process, but not that of EV entry, activates mitogenic cell signaling pathways that involve the signaling molecules Rac, mitogen-activated protein kinase/extracellular signal-regulated kinase, protein kinase A, and protein kinase C. The triggering of cellular signaling pathways by IMVs may be related to their entry via lipid rafts, since lipid rafts are thought to function as plasma membrane domains that organize cell signaling pathways that emanate from the cell surface. These signaling pathways are normally activated by ligand-receptor interactions or, in the case of vaccinia virus, when triggered by pathogen binding.

It is not yet known whether the IMV-triggered signals might play a role in IMV entry. However, the entry of several other viruses, including vesicular stomatitis virus and SV40, is known to be signal-dependent (Box 15.4). Interestingly, the mitogenic signal transmitted by the binding of vaccinia IMVs appears to enhance virus multiplication postentry, perhaps by inducing an intracellular milieu that is more conducive to viral replication.

EVs

Recall that IEVs are derived from IMVs via their acquisition of two additional membranes (which occurs by means of the enwrapping process, described below). One of those additional membranes is subsequently lost when the IEV exits the cell by membrane fusion, becoming an EV in the process. But one additional membrane remains on the EV. Because of the additional EV membrane, entry of EVs has been more difficult to understand than entry of IMVs.

Infectious entry of EVs cannot simply happen by the fusion of their outermost membrane with the plasma membrane, since that would result in internalized IMVs that are still enclosed within the IMV membrane. That is, EVs would still have to shed an additional membrane in order for their cores to enter the cytoplasm. To further complicate matters, there is no evidence that even the outermost membrane of EVs fuses with the plasma membrane. In addition, despite the fact that the EV outer membrane contains a different set of proteins from those of the IMV envelope, the same proteins required for IMV entry are also required for the cell-to-cell spread of EVs. Consequently, it is likely that vaccinia EVs use the same fusion mechanism as IMVs, one involving only the IMV membrane. But how can that be the case if the fusion complexes present in the IMV membrane lie under, and are obscured by, the outer EV membrane? Apparently, the IMV membrane must somehow be exposed on EVs prior to EV entry. But it has generally been thought that enveloped viruses shed their lipid envelopes exclusively by membrane fusion. Thus, to resolve the enigma of the entry of poxvirus EVs, we consider the possibility that this generalization is not universally true, with poxvirus EVs providing an exception. With that possibility in mind, one current model has EVs taken up by endocytosis, followed by the disruption of the outer membrane within an endosomal compartment, followed in turn by fusion of the IMV envelope with the endosomal membrane, thereby releasing the core into the cytoplasm. In support of this model, experimental results have indeed been reported which show that EV entry requires one or more low-pH-dependent steps. In the first step, a low-pH environment induces disruption of the EV outer membrane, and in the second step, the IMVs released from the EVs display a low-pH-dependent fusogenic activity. The latter finding is consistent with the recent experimental results noted above, which show that IMVs may enter primarily by a low-pH-triggered fusion step. At any rate, nearly all reports are in agreement with the premise that only the IMV membrane is fusogenic.

Interestingly, other recent experimental results (obtained by immunoelectron microscopy) imply another totally new and unique mechanism by which vaccinia virus EVs might enter cells. In this mechanism, the outer membranes of vaccinia virus EVs are disrupted at neutral pH, at the point where they bind to the cell. This enables the IMV within the EV membrane to enter the cell by fusion with the plasma membrane (Law et al., 2006). This ligand-induced, nonfusogenic rupturing of the EV membrane, which exposes the IMV membrane, requires the EEV A34 and B5 proteins and cellular glycosaminoglycans. The molecular mechanism by which the EV outer membrane is disrupted remains to be determined. In view of these experimental findings, we might suppose that those EV proteins that are exclusive to the outer EV membrane play a unique role in the infectivity of these particles.

Since the outer membrane proteins of EVs play a role in binding to cells, and since these proteins are present on EVs but not on IMVs, then EVs and IMVs very likely use different cell surface receptors. Three independent lines of experimentation demonstrated that this premise is in fact correct. First, IMVs and EVs show no correlation in their efficiency of binding to different cell lines. Second, treating the cell surface with different proteases differentially affects the binding and infectivity of IMVs and EVs. Third, the binding of a monoclonal antibody (against an unknown cell surface epitope) to cells prevents IMV binding but not EV binding.

Irrespective of the mechanism by which poxvirus cores gain access to the cytoplasm, once there, they are transported on microtubules from the cell periphery to more central regions of the cell, where the replication cycle might then continue.

GENOME ORGANIZATION AND EXPRESSION

Vaccinia virus, the prototypical poxvirus, has served as the model for poxvirus gene organization and expression. Thus, the following features of poxvirus genome organization and expression are known primarily from studies of vaccinia virus.

Poxvirus genomes are linear, double-stranded DNA molecules that range between 130 and 360 kb in size. The vaccinia virus genome, which is about 200 kb in length, encodes about 200 proteins. Coding sequences are closely packed on the genome and are present on each of its DNA strands. Genes encoding structural proteins and essential enzymes are clustered within the central 120 kb or so. These replication genes are highly conserved within the poxvirus family. Genes that affect immune evasion and virulence (but which are largely nonessential for virus replication in cell culture) are found at the ends of the genome. In contrast to the highly conserved, centrally located replication genes, these flanking genes are divergent, being specific to the particular poxvirus type. (As might be expected, these large DNA viruses have evolved numerous mechanisms to evade the host immune response to infection [see below]. The conservation of poxvirus replication

genes, but not of genes that might be involved in immune evasion, is reminiscent of the herpesviruses, in which so-called "essential genes," which encode replication functions, are likewise conserved, but in which so-called nonessential genes, which largely encode immune evasion functions, are not conserved [Chapter 18].)

Poxvirus genomes are characterized by the following singularly unique feature. The two DNA strands of poxvirus genomes are covalently connected at each end, such that a poxvirus double-stranded DNA genome constitutes a continuous polynucleotide chain. In addition to this distinctive attribute, there are inverted terminal sequences at each end of the genome. Figure 19.5 shows a portion of the terminal repetition on the genome of a vaccinia virus strain. The lengths of the poxvirus terminal repetitions vary by poxvirus type and strain. They are 10 kb or more in the case of vaccinia virus DNA and as short as 725 bp in the case of variola virus DNA. Notice that the terminal regions are AT rich and do not perfectly base pair. The terminal repetition exists in two complementary forms. Why do you think that should be? More importantly, what purpose might be served by the unique poxvirus genomic configuration? See Box 14.4 and below.

The poxviruses are the *only* known animal DNA viruses that replicate *entirely* in the cytoplasm. Their genomes never enter the nucleus, where all cellular nucleic acid synthesis (excepting mitochondrial) otherwise occurs. As a result, poxviruses have no access to the cellular transcription and DNA replication machinery, and thus, they must be independent of that machinery. Consistent with their nuclear independence, poxviruses can replicate in experimentally enucleated cells.

In view of the above, not only does a poxvirus particle need to encode the enzymes needed for transcribing its genome, but also a poxvirus *particle* itself must contain those very enzymes. (This point was noted above. Do you see why it is so?) These poxvirus-encoded enzymes include an RNA polymerase; enzymes for capping, methylating, and polyadenylating its mRNAs; and initiation proteins that recognize specific viral promoters. On the other hand, poxviruses do not encode splicing activities, which is fine by the viruses, since poxvirus mRNAs are not spliced; a good thing too, since cellular splicing enzymes are in the nucleus.

The poxviruses also encode their own DNA replication enzymes, including a DNA polymerase, helicase, and ligase. However, these replication activities are not found in the particle but instead are the products of viral early genes.

The poxvirus RNA polymerase is responsible for transcribing all poxvirus gene classes during the phased cycle of poxvirus gene expression (see below). It is a multicomponent complex, containing as many as nine different

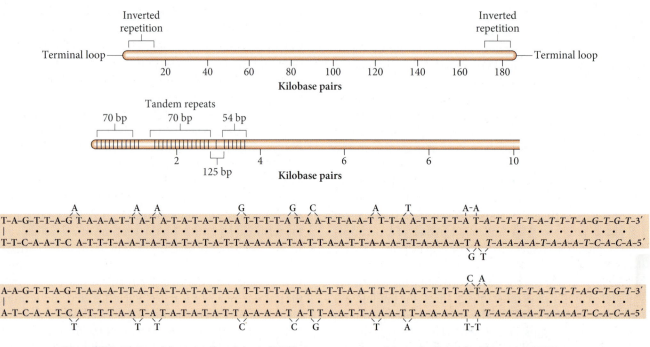

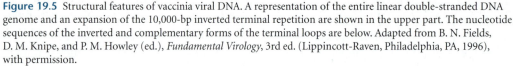

Figure 19.5 Structural features of vaccinia viral DNA. A representation of the entire linear double-stranded DNA genome and an expansion of the 10,000-bp inverted terminal repetition are shown in the upper part. The nucleotide sequences of the inverted and complementary forms of the terminal loops are below. Adapted from B. N. Fields, D. M. Knipe, and P. M. Howley (ed.), *Fundamental Virology*, 3rd ed. (Lippincott-Raven, Philadelphia, PA, 1996), with permission.

subunits. The basic "core" enzyme is comprised of eight subunits, which is the form of the enzyme used for intermediate and late gene transcription (see below). In contrast, the virus particle-associated form of the enzyme, which is used for early transcription, contains the core enzyme in association with Rap94, the viral transcription specificity factor, which is needed for both the initiation and termination of early transcripts. Interestingly, the poxvirus RNA polymerase complex is remarkably similar to the cellular polymerase II enzyme, with several of its subunits showing notable homology to subunits of the cellular enzyme.

Reminiscent of the adenoviruses (Chapter 17) and herpesviruses (Chapter 18), the poxvirus pattern of gene expression involves three distinct temporal phases, referred to as early, intermediate, and late. Early genes encode proteins involved in viral DNA replication, evasion of host defenses, and triggering the expression of intermediate-phase genes. Postreplicative intermediate- and late-phase genes encode products that participate in virus morphogenesis and assembly. In general, all large, double-stranded DNA viruses, including bacteriophages, exhibit a similar cascade of transcription stages.

Transcription of the early genes, which comprise about one-half of all the poxvirus genes, initiates upon release of the viral cores into the cytoplasm. Interestingly, the early genes are transcribed from within the confines of the viral

cores. What is more, early mRNAs are capped and polyadenylated within cores and then extruded through pores in the core surface. Early mRNAs can be detected in the cytoplasm within 20 min of synchronized infection. (Recall the manner of transcription of reovirus double-stranded RNA genomes, which also happens within viral cores [Chapter 5].)

As noted above, in order for transcription to occur within the poxvirus cores, it is necessary that the viral RNA polymerase be contained within the cores. In addition, the capping activities and two necessary virus-encoded, early-gene-specific transcription factors, VETF and RAP94, are also contained within the virus core (Box 19.6). (Do you see that the cytoplasmic poxviruses would need to bring their RNA polymerase and transcription factors into the cell, even if transcription did not occur within the parental virus cores?) VETF binds specifically to early promoters. RAP94, which is bound to the core RNA polymerase, binds to VETF. The net effect is to recruit the RAP94/RNA polymerase complexes to the early viral promoters, so that initiation complexes can then form. VETF also expresses a DNA-dependent ATPase activity that might enable VETF to disassociate from the promoter, so that the polymerase might move processively down the template to carry out elongation.

Despite the fact that VETF and RAP94 are early-gene-specific transcription factors, these proteins are

BOX 19.6

An active poxvirus transcription/mRNA-processing complex, which is able to initiate and terminate transcription in vitro, can be isolated from vaccinia virus particles. This complex, which likely carries out early transcription and processing in vivo, is comprised of the nine-subunit polymerase complex (i.e., the core enzyme in association with Rap94), VETF (viral early transcription factor), and both subunits of the capping enzyme. VETF recruits the polymerase to the template. The viral capping enzyme is multifunctional, as it also provides the viral termination factor, which is required for the termination of vaccinia virus early transcripts. The viral capping reactions, which are catalyzed entirely by the viral capping enzyme, occur in the same three steps as occur during the capping of cellular mRNAs (Fig. 5.10). First, the triphosphate at the end of the primary transcript is converted to a diphosphate by the RNA triphosphatase activity of the viral enzyme. Second, a guanosine monophosphate (GMP) cap is added by the RNA guanyltransferase activity. Third, the guanyl residue is methylated by the RNA (guanine-7-) methyltransferase.

The poxvirus RNA polymerase is reminiscent of the host enzyme in its use of additional proteins to recognize specific promoters. Still, it may be more appropriate to think of RAP94 as a subunit of the "early" form of the RNA polymerase, rather than as a transcription factor per se. Apropos that notion, RAP94 remains tightly associated with the polymerase throughout transcription. This is important because RAP94 is required to terminate early transcripts, as well as to initiate them (see the text).

independent particle-targeting signal. Thus, VETF and RAP94 may have roles in particle assembly, as well as in transcription. Moreover, these experimental findings are consistent with the thought that it is more appropriate to think of RAP94 as a subunit of the polymerase complex than as a transcription factor (Box 19.6).

Despite certain similarities between the poxviral and cellular RNA polymerases (Box 19.6), the viral RNA polymerase displays at least one key feature that distinguishes it from the cellular RNA polymerase II enzyme. The cellular RNA polymerase generates primary transcripts that extend far beyond the ends of the mature mRNAs. These primary transcripts, which do not terminate at discrete sites, are then processed to produce mature mRNAs with their distinct polyadenylated termini. In contrast, the poxvirus RNA polymerase terminates transcription of early mRNAs at distinct sites that are 20 to 50 bp downstream of the sequence TTTTTNT on the template strand of the DNA. (Actually, the UUUUUNU sequence in the nascent mRNA is the termination signal that is sensed by the polymerase.) Moreover, the multifunctional viral mRNA capping enzyme is also the viral transcription termination factor, which is carried along by the RNA polymerase as it traverses the template. When the termination signal appears on the nascent mRNA, the capping enzyme somehow causes the RNA polymerase to cease transcription and release the template. It is not clear whether incorporating two dissimilar functions into the capping protein provides some particular advantage other than economy. Note that this site-specific termination of poxvirus mRNAs appears to apply only to early transcripts, as the intermediate and late transcripts have heterogeneous ends. Thus, poxviruses use at least two different mechanisms of mRNA 3'-end formation.

Interestingly, RAP94, which is required to initiate transcription of early genes, is (like the capping enzyme/viral termination factor) also required for terminating early-gene transcripts. Apropos that fact, the late polymerase does not respond to the UUUUUNU termination signal on the nascent mRNA. Thus, the early enzyme complex, which contains RAP94, but not the late polymerase, which lacks RAP94, retains a "memory" of the promoter at which it initiated. That is possible because RAP94 remains associated with the core enzyme of the early polymerase throughout transcription. In contrast, the late enzyme, which lacks RAP94, ignores the early termination signal. Termination of late transcription occurs by an as-yet-unknown mechanism.

Nascent early viral mRNAs are modified by the viral capping activity when they are about 30 nucleotides long. Once the viral mRNAs are mature, they are translated by the cellular translation machinery.

synthesized only during the late phase of the replicative cycle. Importantly, these late proteins are incorporated into progeny virus particles, so that they might be available to function at the start of the next cycle of infection. (Note the similarity of this poxvirus stratagem to that of HSV-1, which brings its late α-TIF [VP16] protein into the cell as a component of the virus particle [Chapter 18].)

When RAP94 expression is repressed, defective virus particles that appear normal but are lacking RNA polymerase are generated. These defective particles do contain VETF, which is bound to the DNA, presumably at an early promoter. Together, these results are consistent with the interesting likelihood that the transcription proteins, including RAP94, are packaged as a multiprotein complex, recruited to the nascent particle by VETF, thereby obviating the need for each protein of the complex to have an

The poxvirus transcriptional cascade begins upon release of the viral cores into the cytoplasm, as noted above. What is more, early-gene products, which are translated from mRNAs generated from within the viral cores, actually accomplish the complete uncoating of the cores. In so doing, these early-gene products release the viral genomes into the cytoplasm. An early experimental finding that complete uncoating is prevented by inhibitors of protein synthesis first suggested that viral gene expression might be required to facilitate uncoating. Protein synthesis inhibitors also greatly increase the accumulation of early transcripts, consistent with the possibility that genome release correlates with the end of the early phase of gene expression.

Other poxvirus early-gene products help the infection to evade host defenses (see below). Yet other viral early genes encode proteins that resemble host growth factors. These viral proteins are secreted from infected cells and act to stimulate the growth of neighboring cells. Interestingly, these viral genes may be derived from host genes. Moreover, their products play a role in the formation of smallpox pustules. Importantly, some poxvirus early proteins activate the *intermediate phase* of gene expression. They do so by acting as transcriptional regulatory proteins, interacting with the viral RNA polymerase, so as to redirect it to intermediate-gene promoters.

Still other poxvirus early proteins replicate the viral genome (see below). Importantly, poxvirus DNA replication *is also necessary* to trigger *intermediate-gene* expression. Thus, the phased sequence of poxvirus gene expression is punctuated by viral DNA synthesis. Note that poxviruses are like other DNA viruses in this regard. Yet bear in mind that unlike other DNA viruses, the transcriptional and replication machineries used by poxviruses are viral in origin.

Why might poxvirus DNA replication be necessary to trigger intermediate gene expression? One possibility is that particular virus proteins are bound to, and obstruct, intermediate-gene promoters on parental viral genomes. These obstructing proteins might then be stripped from the intermediate gene promoters during viral DNA replication. That possibility is consistent with the experimental finding that deproteinizing the DNA from purified virus particles enables the transcription of intermediate genes in vitro. In Chapter 2, under the subheading "Regulated Gene Expression," we discussed how a virus might benefit from a phased sequence of gene expression.

Transcription of poxvirus intermediate genes appears to also require the synthesis of new viral RNA polymerase complexes. This fact correctly implies that all components of the viral RNA polymerase are transcribed from early promoters. Nonetheless, the genes encoding the RNA polymerase core subunits continue to be expressed throughout the replication cycle. But why might new RNA polymerase

molecules be necessary at later stages of the replication cycle? One reasonable possibility is that the core-associated polymerase molecules are inactivated by the final uncoating process. Another reasonable possibility concerns the fact that the core-associated enzyme is complexed with RAP94, an early promoter specificity factor or early polymerase subunit (see above and Box 19.6). Since intermediate and late genes cannot be transcribed efficiently by RNA polymerase complexes associated with RAP94, it may be advantageous to the virus to generate new RNA polymerase molecules that are not complexed with RAP94. This is readily doable by the virus, since RAP94 mRNA is actually transcribed from a late promoter, which is not accessible at early times. Transcription of RAP94 from a late promoter is also important with respect to the fact that it enables RAP94-bound polymerase to be made at late times and to be incorporated into progeny virus particles at final assembly, so as to be available at the early stage of the next cycle of infection, for the transcription of early genes (see above).

Despite the fact that poxvirus transcription occurs entirely in the cytoplasm, it is not entirely independent of the host cell. A poxvirus early-gene product somehow causes a cellular nuclear protein, Vitf2, to appear in the cytoplasm, where it too has a role in redirecting virus transcription from early to intermediate genes. The possible dependency of poxvirus intermediate-gene transcription on cellular Vitf2 was unexpected because poxvirus transcription was thought to be self-contained. That belief was driven in part by experimental results obtained using enucleated cells. One possibility that is consistent with all previous experimental observations is that infection activates a cryptic *cytoplasmic* Vitf2, rather than causing the relocation of an active nuclear one. (Vitf is an acronym for vaccinia virus intermediate transcription factor. Vitf1 and another protein, Vitf2, were initially isolated from the cytoplasm of vaccinia virus-infected cells. It came as a surprise when Vitf1 and Vitf2 could also be isolated from the nuclei of uninfected cells.)

The precise role of Vitf2 in poxvirus transcription is yet another unknown. Moreover, the normal role of Vitf2 in the cell is also not known. This protein is comprised of two subunits, one of which has RNA-binding and putative helicase activities, which might play a role in RNA elongation. There is little information regarding the other subunit. A more interesting question for a virologist is why a virus whose transcription is otherwise autonomous might depend on a single cellular protein. A possible answer relates to the fact that Vitf2 is most highly expressed in replicating cells. Thus, activating or mobilizing Vitf2, which is present only at low levels in resting cells, might be part and parcel to enabling the virus to create an intracellular milieu that is more favorable to virus replication (Box 19.7).

BOX 19.7

Although much of the evidence is compelling, it has not yet been proven that vaccinia virus replication is indeed dependent on Vitf2. But we still might ask the following: aside from the possible dependency of vaccinia virus replication on Vitf2, is that replication truly independent of nuclear factors? Other experimental results imply that poxviruses can recruit cellular nuclear proteins to viral replication complexes in the cytoplasm, by a process that can be blocked by inhibitors of transport through nuclear pores. Consequently, some researchers have suggested that poxvirus genomes, despite their very large size, may not encode all the proteins required for their replication and that nuclear proteins that might function in poxvirus nucleic acid metabolism may yet be revealed. See the discussion of VLTF-X and associated proteins in the text.

Just as early-stage gene products activate expression of intermediate-stage genes, intermediate-gene products include transcriptional regulatory proteins that activate expression of late genes. Thus, whereas transcription of early-stage genes is mediated entirely by proteins carried into the cell in virus cores, transcription of intermediate-stage genes and late-stage genes requires the products of genes expressed in each of the respective preceding stages.

Recall that transcription of poxvirus late genes requires new RNA polymerase molecules that are encoded by early genes, as well as transcription factors encoded by intermediate-stage genes. An additional late-gene transcription factor, VLTF-X, initially thought to be a virus gene product, was later identified in uninfected cells. VLTF-X appears to specifically bind to late viral promoters, perhaps forming a nucleation site for virus-encoded factors, which then recruit the viral RNA polymerase to late transcription initiation sites. At any rate, the discovery that VLTF-X is a cellular protein, present in both cytoplasmic and nuclear extracts of uninfected mammalian cells, implies that poxvirus transcription is less autonomous than previously thought. Moreover, in a recent study, two proteins, hnRNP A2 and RBM3, were each found to copurify with VLTF-X in uninfected HeLa cell extracts. Additionally, each of these proteins independently stimulated viral late-gene transcription in vitro. The hnRNP A2 protein is known to be an RNA-binding and single-stranded DNA-binding protein.

Late-gene products include viral structural proteins, as well as nonstructural proteins that play a role in morphogenesis. In addition, they include the RNA polymerase and other transcription factors that need to be in place in progeny virus particles to initiate the next cycle

of infection. (Actually, the genes encoding the polymerase core subunits are expressed throughout the infection, as noted above.)

DNA REPLICATION

Vaccinia virus DNA replication begins between 1 and 2 h postinfection in synchronously infected cells and eventually generates about 10,000 progeny viral genomes per infected cell, about one-half of which are incorporated into virus particles.

It is notable that poxvirus DNA replication in general is distinguished from that of all other known DNA viruses, as follows. First, it happens entirely in the cytoplasm, and second, poxviruses encode all the factors necessary to mediate replication of their double-stranded DNA genomes. Poxvirus-encoded DNA replication factors include a DNA polymerase, a thymidine kinase, a thymidylate kinase (which phosphorylates ribosylthymine monophosphate [TMP]), a ribonucleotide reductase (which reduces ribonucleotide diphosphates to deoxyribonucleotide diphosphates), an adenosine triphosphate (ATP)-dependent DNA ligase, a deoxyribonuclease (DNase) (which has nicking activity), and a topoisomerase (which catalyzes the reversible breakage and rejoining of DNA strands). Other factors are required as well, although their roles or mechanisms of action are not as well understood. These other players include a DNA processivity factor, a protein kinase, a DNA-dependent ATPase, and a uridyl DNA glycosylase. The uridyl DNA glycosylase forms a complex with the processivity factor. The ATPase is thought to be a helicase, although no one has yet demonstrated that.

Perhaps you have already given some thought as to why the two DNA strands of poxvirus genomes are covalently connected at each end, such that these double-stranded DNA genomes actually comprise continuous polynucleotide chains. If so, then you very likely came to the following plausible reason. As was noted in several earlier chapters, DNA replication, unlike RNA synthesis, cannot simply initiate de novo. Instead, nucleotides, which are added complementary to the template strand, may be added only to the free 3′ end of a preexisting polynucleotide chain, a requirement necessitated by the proofreading function intrinsic to all DNA polymerases. As a consequence of this fact, there are only two known mechanisms for initiating and completing DNA synthesis. The first mechanism involves RNA primers and a replication fork. The RNA primers are used both in the initiation and elongation of the lagging strand. This is the mechanism used by the cell and also by the polyomaviruses, papillomaviruses, and herpesviruses. The second mechanism involves strand displacement. This is the mechanism used by the adenoviruses (Chapter 17), in which the primer (a protein in this instance) is present on the viral genome. It also is the mechanism used by the parvoviruses (Chapter 14) to replicate their single-stranded DNA genomes. However, in the parvovirus instance, a free 3′ end, which is present on the DNA in a hairpin configuration, serves as the primer. The poxviruses likewise make use of strand displacement, and they too use a free 3′ end for their primer, which they are thought to generate by nicking their continuous, double-stranded DNA genome (see below). Significantly, since DNA synthesis must be primed (usually by the free 3′ end of a short RNA primer), there is the problem of how progeny genomes might acquire the 5′ ends of their DNA strands. While cells make use of telomerase for this purpose, no virus is known to use telomerase. Thus, the fact that poxvirus genomes are actually continuous polynucleotide chains enables these viruses to adapt a DNA replication mechanism by which they can generate the 5′ ends of their progeny genomes, while also satisfying the requirement for a free hydrogen-bonded 3′ nucleotide to serve as the primer for nucleotide polymerization (see below).

With the above points in mind, a model for poxvirus DNA replication was put forward to account for the following: (i) the linear poxvirus genome is comprised of a continuous polynucleotide strand; (ii) concatemers appear to be generated during the replication process; and (iii) there is no evidence for an *ori* sequence on the viral DNA. The first step in the model is the introduction of a nick, which generates a free hydrogen-bonded 3′ nucleotide to serve as the primer for nucleotide polymerization. The nick is thought to occur within one of the inverted repeats, near one or both ends of the genome (Fig. 19.6).

The left branch in Fig. 19.6 depicts the sequence of events that might transpire when a single nick is introduced into a parental DNA molecule. Polymerization is shown to proceed until the end of the template is reached, while at the same time the template strand is displaced. (Changes in the sedimentation of the parental DNA provided early experimental support for the nicking, self-priming, and strand displacement steps. In addition, results of labeling studies suggested that replication begins near the ends of the genome.)

Next, notice how the terminal repeats then enable the daughter sequence (and the displaced template strand as well) to fold back, thereby creating a base-paired 3′ end on the daughter sequence that can prime the copying of the rest of the genome. This copying would result in the generation of a concatemer (in this instance, two unit-length genomes that are covalently joined), which can be resolved to generate two genomes. However, polynucleotide synthesis might continue all the way to the end of the template. New "rabbit ears" then might form, and the entire process might repeat, generating larger concatemers.

Concatemeric forms of poxvirus DNA are found to accumulate when late-gene expression is prevented. If late-gene expression is allowed to ensue, then the concatemers are seen to be resolved. These experimental findings show that concatemers are intermediates in poxvirus DNA replication and, moreover, that late-gene products are needed to resolve the concatemers. However, note that vaccinia virus concatemeric intermediates are resolved independently of (and prior to) DNA encapsidation (see below).

Recalling that no DNA virus is known to make use of telomerase to regenerate the 5′ ends of its DNA genome, the model outlined above provides a means by which poxviruses can generate the 5′ ends of their progeny genomes. Also, notice how similar this model is to the rolling-hairpin model depicted for parvovirus DNA replication (Fig. 14.3).

The right branch in Fig. 19.6 depicts what might transpire when a nick is introduced at each end of the poxvirus genome. Initiation might then occur at each end of the molecule. When the two replication complexes run into each other, the two halves fall apart, and the polymerase complex on each half then might complete the replication by running up against the base-paired 5′ end. In the example shown, two genomes are generated in the absence of concatemer formation, but the process can continue with concatemer formation. Do you see how? Regardless, genome replication can happen, irrespective of whether one or two nicks are introduced into the DNA.

Interestingly, cells infected with several different poxviruses can support the origin-independent, cytoplasmic replication of any circular DNA molecule, via the generation of concatemeric intermediates. Moreover, experiments using temperature-sensitive vaccinia virus mutants demonstrated that each of the five proteins known to be required for viral genome replication also is required for plasmid replication in these experimental systems. Since these results could be obtained with several different plasmids, it was suggested that poxvirus-mediated plasmid replication might initiate nonspecifically in these instances, perhaps at random nicks in the DNA. What is more, replication of these circular plasmids lacking specific poxvirus DNA sequences occurred in cytoplasmic **viral factories** (see below).

ASSEMBLY, MATURATION, AND RELEASE

The putative steps in the assembly and maturation of vaccinia virus particles are surmised mainly from electron microscopy analysis, in which the viral structural intermediates that are seen in infected cells are organized into a plausible sequence (Fig. 19.7). In some experiments,

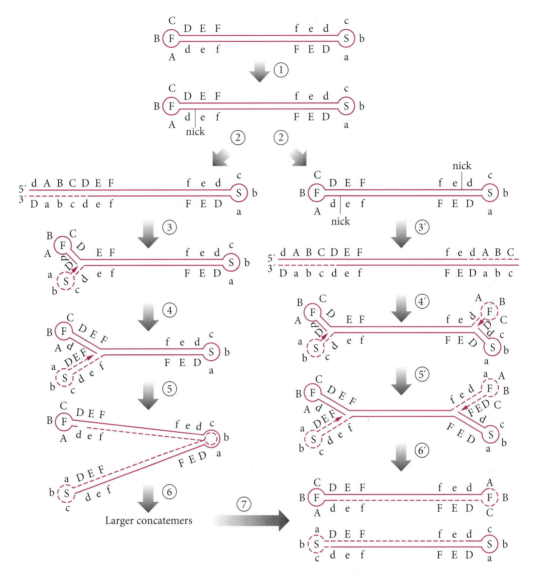

Figure 19.6 Self-priming model for vaccinia virus DNA replication. *F* and *S* refer to the difference in electrophoretic mobilities of the two alternative inverted and complementary hairpin sequences present at the ends of the vaccinia virus genome. The scheme on the left assumes that the replication of a single DNA molecule is initiated at one end and continues to the other end to form a concatemer, whereas on the right initiation occurs at both ends without concatemer formation. Compare to Fig. 14.3. Adapted from B. N. Fields, D. M. Knipe, and P. M. Howley (ed.), *Fundamental Virology*, 3rd ed. (Lippincott-Raven, Philadelphia, PA, 1996), with permission.

cells are infected with a collection of viral mutants as well as with wild-type virus. The final intermediate structures generated by the mutants help to identify the roles of individual viral gene products in the assembly and maturation process. (R. S. Edgar and W. B. Wood pioneered this experimental approach in their 1965 analysis of the assembly pathway of bacteriophage T4 [Fig. 2.5].)

The first recognizable virus structures become visible in the cytoplasm in so-called viral factories, which are largely devoid of cellular organelles. These early viral structures materialize as crescent-shaped membrane segments that are covered with spikes on their convex side, and they appear to engulf electron-dense "viroplasm." In this regard, the first and all other viral assembly steps appear to take place in association with membranes. The crescent-shaped membrane segments are comprised of virus-encoded proteins and host-derived lipids. IV particles result from the completion and closure of the aforementioned membrane about the viroplasm (Fig. 19.4). The vaccinia virus D13 protein is thought to form a scaffold, which confers the curvature on the crescent membrane and on the IV membrane as well.

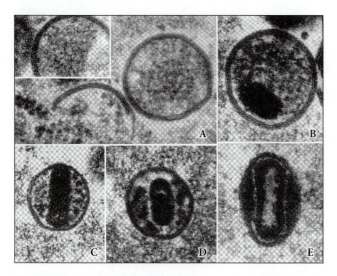

Figure 19.7 Vaccinia virus morphogenesis intermediates. (A) Crescents; (B) IV; (C) IVN; (D) structures intermediate between IV and IMV; (E) IMV. From R. Condit, N. Moussatche, and P. Traktman, *Adv. Virus Res.* **66:**36–124, 2006, with permission.

Viral DNA replication occurs at the same cellular sites (i.e., viral factories) where the above assembly reactions are taking place, and nascent DNA molecules are inserted into the IVs before they are completed. The mechanism by which the viral DNA is enclosed within the maturing particles is not yet understood. The vaccinia A32 and I6 proteins are the only virus proteins that have been implicated directly in the DNA packaging process. A32 possesses an ATPase activity that might provide the energy needed for DNA translocation into the maturing virus particle. However, the precise roles of A32 and I6 are not known. Unlike the case of encapsidation of HSV-1 DNA, where resolution of the concatemeric DNA intermediates is coupled to encapsidation (Chapter 18), vaccinia virus concatemeric intermediates are resolved independently from DNA encapsidation, in an earlier step. Following DNA encapsidation, IVs then undergo several additional maturation steps, including the proteolytic processing of precursor proteins, to give rise to IMVs (see above) (Fig. 19.7).

As noted above, IMVs comprise the majority of infectious progeny virus particles that are associated with infected cells. Most of the IMVs are released when the cell lyses. However, before cell lysis occurs, some IMVs are transported from the viral factories, via microtubules, to sites near the microtubule organizing center (MTOC) of the cell. There, they are enclosed by two additional membranes, which are derived from the *trans*-Golgi or endosomal cisternae (Fig. 19.4). In that way, these IMVs become IEVs (see above). Contrary to what you may have expected, IMVs do not bud into the Golgi cisternae or the endosomal compartment in this membrane-acquiring

process. Instead, IMVs induce engulfment by these membranes in a process referred to as "wrapping."

What determines that the IEVs acquire their two additional membranes from the *trans*-Golgi or endosomal cisternae, rather than from the plasma membrane? More generally, what determines which particular cellular membrane will be the site for a specific step in virus assembly or maturation? As you likely expected, the particular membrane is usually determined by specific targeting sequences on viral envelope proteins. (As noted in earlier chapters, viruses use the same protein-targeting sequences as used by the cell, and they subvert the cellular systems that normally sort proteins to each of the intracellular sites addressed by those targeting sequences.) In the case of the maturation of an IMV into an IEV, Golgi or endosomal membranes are modified to contain at least seven viral proteins, each of which has an endosomal targeting sequence. Of these viral proteins, only B5 and F13 are required for wrapping of IMVs to form IEVs, as demonstrated by experiments with mutant viruses. It is now known that F13 associates posttranslationally with the Golgi and endosomal membranes and that this association is peripheral rather than transmembrane and is mediated by palmitoylation of the protein.

Once IEVs are formed, they are transported on microtubules away from the MTOC towards the cell periphery. Then, they are thought to exit the cell by fusion of their outermost membrane with the plasma membrane (Fig. 19.3 and 19.4; see above). Some released EVs remain at the cell surface as CEVs, while others are released as EEVs.

Regardless of whether IEVs become CEVs or EEVs, they lose their outer membrane upon exiting the cell but retain an additional membrane that is not present on IMVs. Why then do IMVs not acquire the additional membrane by simply budding directly through the plasma membrane? A reasonable answer is that budding through the plasma membrane would require the virus to target envelope proteins to the plasma membrane, rather than to internal membranes. Doing so might cause the infected cell to be more readily visible to host immune defenses, before the maximal number of progeny virus particles might be generated and released from the infected cell. But this seemingly credible scenario is made somewhat less plausible by the fact that some electron microscope images suggest that some IMVs indeed might exit the cell by budding directly through the plasma membrane.

As noted above, IMVs and EVs are each infectious, but each plays a different role in dissemination and spread of the virus. IMVs are thought to mediate virus spread from host to host; this role takes advantage of their ability to survive outside the host for prolonged periods. (Because of their robustness, IMVs are able to resist freeze-thawing and desiccation, which is a reason why IMVs are used as the infectious units of the smallpox vaccine.) In contrast, the fragile CEVs and EEVs are thought to mediate dissemination within the host: cell-to-cell spread in the case of CEVs and long-range spread in the case of EEVs. CEVs and EEVs are suited to these roles because their outer membrane makes them more resistant to neutralization by antibodies.

The final steps in the mechanism adapted by CEVs for carrying out cell-to-cell spread are truly remarkable. The overall process occurs as follows. First, IEVs move from the MTOC-containing perinuclear area to the cell surface via microtubules and a kinesin-associated motor. Two vaccinia virus proteins, F12 and A36, which are present on IEVs but not on CEVs and EEVs, also are involved in this movement. (F12 and A36 deletion mutants form IEVs, but their movement to the cell surface is severely restricted, and they remain intracellular.)

Next, upon reaching the inner surface of the plasma membrane, IEVs then exit the cell by fusion of their outermost membrane with the plasma membrane (Fig. 19.3). The next steps are the remarkable ones. The EVs that remain attached at the cell surface (i.e., the CEVs) do so because they induce the reorganization of the actin cytoskeleton just beneath the site of fusion. The numbers of stress fibers under the fusion sites are seen to decrease, and in their place the CEVs induce the formation of "actin tails," which remain in contact with the CEVs at the surface (Fig. 19.8).

The viral A36 protein, which is required for transport of IEVs to the cell periphery, also is needed for inducing the formation of the actin tails. The A36 protein acts at this stage by recruiting cellular **Src family protein kinases**, which are activated locally by the newly emergent CEVs. (Src family protein kinases are associated at the plasma membrane with signaling receptors that possess no intrinsic tyrosine kinase activity of their own. When these receptors are activated by their respective ligands, they transmit signals into the cell via their associated Src family kinases. The Src family kinases are discussed briefly in Chapter 15 and in detail in Chapter 20.) This results in Src-mediated phosphorylation of tyrosine residues 112 and 132 of A36 and activation of a signaling cascade that acts to stimulate the actin polymerization beneath the virus. The growing actin tails then propel the CEVs on actin-filled membranous protrusions towards neighboring uninfected cells, thereby facilitating direct cell-to-cell spread of the virus (Fig. 19.8; Box 19.8).

Why do some EVs generate actin tails to be retained at the cell surface as CEVs, while others are released as EEVs? The answer is not yet known. However, the relative amount of CEVs versus that of EEVs is known to be influenced by the cell type and virus strain. Several strains of vaccinia

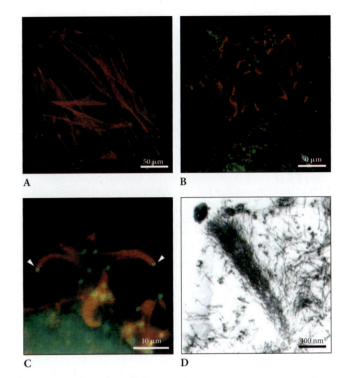

Figure 19.8 Actin tails in HeLa cells infected with vaccinia virus. (A through C) Immunofluorescence images. (A) Uninfected control HeLa cell showing a typical actin stress fiber staining. Scale bar, 50 μm. (B) Infected cells showing intracellular actin tails (red) and viral particles (green). Scale bar, 50 μm. (C) Viral particles projecting from the cell surface on actin tails show a clear overlap between actin (white arrowheads). Scale bar, 10 μm. (D) Electron micrograph of an actin tail decorated with S1 myosin showing a direct contact between actin filaments and the virus particle. Scale bar, 400 nm. From S. Cudmore et al., *Nature* 378:636–638, 1995, with permission.

virus generate many more CEVs than EEVs, even at late times in infection.

This generation of actin tails by vaccinia virus CEVs and their dissemination to neighboring cells are in marked contrast to budding viruses, such as influenza virus and the paramyxoviruses, which encode an envelope-associated neuraminidase activity to ensure their efficient release. Thus, vaccinia virus does not encode a neuraminidase activity that might promote its release. Instead, vaccinia EVs tend to be retained at the cell surface by their association with the virus-induced actin tail. Perhaps those EVs that are released fail to induce actin tails in time to ensure their retention. Or perhaps EEVs are generated by the rare direct budding of IMVs rather than via the formation of IEVs.

Recalling that vaccinia virus induces infected cells to form syncytia, what role, if any, might syncytia play in spread of the virus? An examination of vaccinia virus plaques reveals that syncytia are formed mainly at the centers of the plaques and late in the formation of the plaque. The most straightforward interpretation of this observation

The movement of vaccinia virus on actin tails is reminiscent of a process used by a small group of bacteria that includes *Listeria*, *Shigella*, and *Rickettsia*, which, like viruses, have adapted an intracellular lifestyle. Concomitant with that intracellular lifestyle, these bacteria have evolved a mechanism for spreading from cell to cell by inducing the formation of actin tails at one of their ends. Polymerization of the actin filaments at one end of the bacterium drives it to the periphery of the host cell. There, the forming actin tails cause plasma membrane protrusions to form, which extend into neighboring cells. The neighboring cells then are thought to phagocytose the invading membrane protrusions. The bacteria escape from the resulting phagocytic vacuoles by secreting a pore-forming toxin. This process of bacterial dissemination, which is sometimes referred to as "rocket motility," enables these crafty microbes to spread without being exposed to antibodies.

is that syncytia form between cells that are each already infected, rather than between infected and uninfected cells. If that is the case, then syncytia are not likely to contribute significantly to viral spread. For a thorough review of much of the above, see Condit et al., 2006.

IMMUNE EVASION

Considering the toll exacted by the smallpox scourge, we might well expect that variola virus brought multiple strategies into play to evade host immune defenses. Despite the medical importance of those putative variola virus immune evasion strategies, we have little direct knowledge of them. Our ignorance in this area is explained in part by the fact that humans are the only known natural host for variola virus, leaving us without animal models of variola virus infection. It also is explained in no small part by the fact that variola virus cannot be routinely studied because of the inherent danger to the experimenter, and also because of the threat of inadvertent virus escape from the laboratory. Thus, much of what is known concerning variola virus immune evasion strategies is inferred from studies of the less virulent vaccinia virus, the prototypical poxvirus. Moreover, the veracity of vaccinia virus models is compromised by the fact that the natural host of vaccinia virus is not known. Consequently, we cannot know how representative vaccinia virus animal models are even of vaccinia virus in its natural host, much less of variola virus in humans.

Despite the above provisos, the identification and analysis of vaccinia virus genes that play a role in immune

evasion (particularly those genes that are homologous to variola virus genes) have provided important insights. From such studies, we now know that poxviruses display a broad array of immune evasion strategies that target key players of innate and adaptive immune responses (see Fig. 4.7; see also Jha and Kotwal, 2003). As we consider the examples that follow, make particular note of the multiple ways in which poxviruses employ homologs of host immune effector and regulatory proteins as a means to evade host defenses.

Countermeasures against Innate Defenses

Vaccinia virus employs two known mechanisms for evading **complement**. First, vaccinia virus, and variola virus as well, expresses **complement control proteins** (CCPs), which were discovered as homologs of known human and mouse CCPs (Chapter 4). The vaccinia CCP (known as VCP) is a *soluble* 27-kilodalton (kDa) protein that is homologous to human-encoded complement regulators. VCP inhibits both the classical and the alternative complement pathways. It does so by binding to the cellular C3b and C4b complement proteins, which are the C3 convertases of the alternative and classical complement pathways, respectively (Chapter 4; Fig. 4.6 and 4.7). By binding to these complement proteins, the viral CCPs accelerate their decay. Second, like human cytomegalovirus (Chapter 18), vaccinia virus incorporates *cellular* CCPs CD46, CD55, and CD59 into its envelope, thereby enabling vaccinia virus particles to resist complement attack (Boxes 19.9 and 19.10).

The poxviruses have evolved a more extensive array of countermeasures against the **interferons (IFNs)** than have any other family of viruses. Among these poxvirus-encoded IFN countermeasures are the so-called **viroceptors**, which are soluble homologs of cellular proinflammatory cytokine receptors. These viral molecules generally contain only the extracellular ligand-binding domain of the corresponding cellular protein. Thus, they act as decoy receptors that sequester the ligand, while being unable to transmit the signals that these ligands trigger when they bind to their normal receptors on the plasma membrane. Vaccinia viroceptors bind all three IFNs (i.e., alpha, beta, and gamma interferons [IFN-α, -β, and -γ]) in the extracellular environment, effectively quashing their activities at the infected site (Box 19.11).

Another means by which poxviruses undermine the host IFN response is illustrated by the vaccinia virus E3 protein. First, by way of review, RNA-activated protein kinase (**PKR**) and 2′-5′-oligoadenylate synthetase (2′-5′OAS) are two cellular enzymes that are induced by IFN-α/β and then activated in response to double-stranded RNA (Chapter 4). Activated 2′-5′OAS in turn activates the

BOX 19.9

The following experimental results demonstrate the clinical significance of poxvirus complement evasion strategies. If mice are infected either with cowpox virus mutants that lack the cowpox CCP gene or with wild-type virus, the findings are as follows. In mice infected with the mutant viruses, in comparison to infections with wild-type virus there is an enhanced infiltration of mononuclear cells to infected sites. This finding is explained by the fact that the wild-type viral CCP, but not the mutant CCP, down-modulates the generation of complement-derived chemotactic cytokines. These cytokines cause immune effector cells to hone in on the infected site (see Chapter 4). Thus, the net effect of the wild-type viral CCP is to weaken the inflammatory response. This not only diminishes immune recognition of the infection but also contributes to diminished destruction of uninfected tissue (immunopathology) at the infected sites. The latter effect may be advantageous to the virus because it preserves the viral habitat. As seen from examples in the text, modulating inflammation per se is another stratagem employed by poxviruses to evade host immunity (Box 19.10).

BOX 19.10

Interestingly, the ability of the vaccinia virus CCP to temper inflammation suggests that the CCP protein may have potential as a therapeutic agent for the treatment of inflammatory diseases. Indeed, the vaccinia virus CCP recently was shown to have promise as a therapeutic agent in a rat mild-injury model, in which it prevented cell death and inflammation. Moreover, the vaccinia virus CCP protein reduced joint destruction in a mouse model of arthritis.

The myxoma virus MT-7 protein (Box 19.11) and the IL-18BP homolog encoded by vaccinia virus and several other poxviruses (Box 19.12) likewise have potential as therapeutic agents for the treatment of inflammatory diseases.

BOX 19.11

The myxoma virus M-T7 protein is a viroceptor for IFN-γ. When European rabbits were infected with a myxoma virus mutant lacking the M-T7 gene, there was a marked increase in the infiltration of inflammatory cells into infected lesions, in comparison to infections with wild-type virus. (Myxoma virus is responsible for myxomatosis in European rabbits.) These experimental findings demonstrate that poxviruses can modulate the infiltration of inflammatory cells by reducing the effects of the IFNs, as well as by blocking the generation of complement-derived chemotactic cytokines (Box 19.9). Each of these inflammatory response-obstructing stratagems benefits the virus by diminishing immune recognition of the infection and by preserving cells that might support virus replication.

cellular endoribonuclease, **RNase L**. Activated PKR and RNase L are two key effectors of the IFN-mediated antiviral state. Activated PKR phosphorylates the α subunit of the eukaryotic translation initiation factor eIF-2 (eIF-2α), leading to a nonspecific inhibition of protein synthesis

and, consequently, of virus replication as well. Activated RNase L cleaves cellular and viral RNAs, thereby also blocking virus replication. Bearing this in mind, the vaccinia virus E3 protein undermines the IFN response as follows. The C-terminal domain of E3 binds to double-stranded RNA, which it thereby sequesters, in that way inhibiting the activation of PKR and 2′-5′OAS and thus of RNase L as well. Other experimental results suggest that the N-terminal half of E3 interacts directly with the protein kinase domain of PKR.

For good measure, the vaccinia virus K3 protein and the myxoma virus M156 protein provide examples of yet another poxvirus countermeasure against IFN, one that like the E3 protein also acts at the level of PKR. K3 and M156 each have sequence similarity to the cellular translation initiation factor eIF-2α. Thus, each binds to PKR, and in so doing each inhibits PKR-mediated phosphorylation of the cellular eIF-2α. Notice again how the virus makes use of virus-encoded homologs of cellular proteins.

Recalling that the IFNs are **cytokines** (i.e., soluble substances secreted by cells that affect other cells), poxviruses display strategies that affect yet other cytokines of innate immunity. One remarkable example is provided by several poxviruses, including vaccinia virus. It involves a virus-encoded protein that is highly homologous to the cellular interleukin-18-binding protein (IL-18BP), whose natural role is as follows. IL-18, to which IL-18BP binds, is a proinflammatory cytokine that is an inducer of IFN-γ. Thus, IL-18 is an important factor in host defense against infection. But, as a proinflammatory cytokine, IL-18 is also thought to contribute to the pathogenesis of chronic inflammatory diseases. IL-18BP comes into play to modulate the proinflammatory effects of IL-18. Indeed, an imbalance between IL-18 and extracellular IL-18BP is thought to contribute to a variety of disease conditions, including inflammatory bowel disease, Crohn's disease, and ulcerative colitis. The poxvirus-encoded IL-18BP homolog likewise impedes the IL-18-mediated induction of IFN-γ and inflammation. Thus, the host IL-18BP normally serves to modulate inflammation to prevent immunopathology, while the viral homolog of the protein serves as an immune evasion factor. The poxvirus IL-18BP homolog also impedes the activation of NF-κB, an important common mediator of signaling pathways that activate innate immunity responses, including inflammation and apoptosis (Chapter 4). See Box 19.12 for points that are relevant to the preceding facts.

Other poxvirus viroceptors bind **tumor necrosis factor** (TNF), which, like IFN-γ, is a major component of the Th1 cytokine profile; this is the ideal cytokine response for coordinating an immune defense against a virus infection (Box 19.12; Chapter 4). One role of TNF is to activate **natural killer (NK) cells**. Consequently, poxvirus-encoded viroceptors that bind TNF impair NK-cell-mediated responses to poxvirus infections. The interactions of poxviruses with NK cells are discussed further in the subsection that follows. Historically, the poxvirus-encoded TNF receptors were the first viroceptors to be discovered.

Two vaccinia virus-encoded proteins, A46 and A52, inhibit signaling by **Toll-like receptors** (Chapter 4). In brief, Toll-like receptors recognize common molecular patterns that are found on a variety of invading pathogens or their products but that are not generally seen in an uninfected host. Each individual Toll-like receptor recognizes a particular pathogen-associated molecular pattern. When activated, Toll-like receptors usually use an NF-κB-mediated signaling pathway to activate expression of genes that encode proinflammatory cytokines and T-cell costimulatory molecules. (NF-kB is a major transcription factor that activates genes involved in both **innate** and **adaptive immune responses**.) Importantly, Toll-like receptors can recognize viral pathogens early in an infection and thus play a key role in initiating early innate immune responses, including inflammation.

The respective mechanisms by which A46 and A52 inhibit signaling by Toll-like receptors are interesting, especially since they are so different from each other. Beginning with A46, note that signaling by Toll-like receptors is mediated by the interaction of their common **TIR domain** with particular **adaptor** molecules (Box 19.13). The vaccinia virus A46 protein contains a region that is homologous to the TIR domain. Consequently, A46 acts by sequestering a TIR adaptor protein, in this instance, myeloid differentiation factor 88 (MyD88), which is the prototypical TIR adaptor protein. MyD88 mediates the activation of NF-κB by every known Toll-like receptor, except Toll-like receptor 3.

In contrast to the vaccinia virus A46 protein, its A52 protein has no obvious homology to any cellular protein. It inhibits signaling by Toll-like receptors as follows. Recruitment of MyD88 to Toll-like receptors activates so-called IL-1 receptor-associated kinases (IRAKs), which in turn bind TNF-associated factor 6 (TRAF6). The recruitment and activation of these signal mediators lead to the downstream activation of NF-κB. The A52 protein binds to both IRAK2 and TRAF6, thereby disrupting signaling complexes containing those proteins (Boxes 19.13, 19.14, and 19.15).

Poxviruses, like several other viruses, have evolved means to impair **apoptosis**. One of the best-studied of these poxvirus antiapoptosis strategies involves the so-called cytokine response modifying, or Crm, proteins.

BOX 19.12

IFN-γ indirectly affects T cells by inducing macrophages to secrete IL-12. This cytokine in turn induces T cells to secrete the so-called Th1 cytokine profile, which includes TNF and IL-2, as well as IFN-γ, and which comprises the ideal set of cytokines for coordinating defenses against a viral attack (Chapters 4 and 18). Thus, by blocking the induction of IFN-γ, IL-18BP effectively suppresses expression of the Th1 cytokine response.

Also, as in the case of the poxvirus-encoded CCPs, the poxvirus-encoded IL-18BPs and the myxomavirus M-T7 protein are thought to have excellent potential as anti-inflammatory therapeutic agents (Box 19.10). In this regard, the cellular IL-18BP is known to block the development of atherosclerotic lesions in mouse models.

BOX 19.13

First, we explain the origin of the acronym TIR: Toll-like receptors and IL-1 receptors each play key roles in activating host immune responses. The Toll/IL-1 receptor (TIR) domain is the common intracellular adaptor-protein-binding domain of the two families of receptors. Second, the association of a particular adaptor protein with an activated receptor is the first step in the transmission of the intracellular signal. Because Toll/IL-1 receptors share the conserved TIR domain, most of them can activate NF-κB through a common signaling pathway.

BOX 19.14

When expressed in mammalian cells, the vaccinia A46 protein partially inhibited IL-1-mediated activation of NF-κB, while A52 strongly inhibited it. Importantly, vaccinia virus mutants that lack either the A46 or A52 gene are attenuated in a mouse model, demonstrating the importance of these genes as vaccinia virus virulence factors.

Historically, it is interesting that Toll-like receptors were initially thought to be involved only in the response to bacterial infections. However, the discovery of vaccinia virus countermeasures against Toll-like receptors provided early evidence for the involvement of Toll-like receptors in the response against virus infections as well.

The CrmA protein encoded by vaccinia, cowpox, and myxoma viruses acts as follows. Recall that apoptosis can be triggered by either Fas or TNF, and in either event, apoptosis is then mediated by a family of cysteine proteases termed **caspases** (Chapter 4). All caspases are synthesized as inactive precursors that are activated by proteolytic cleavage, usually by another caspase. That is,

BOX 19.15

Many viruses are known to affect NF-κB-mediated signaling pathways. However, while poxviruses tend to inhibit these pathways, other viruses actually activate them. The latter viruses include Epstein-Barr virus, influenza virus, HIV, and hepatitis B virus, among others. Poxviruses almost certainly inactivate NF-κB pathways to evade host defenses. But why then do other viruses activate these pathways? Apparently, they do so to create an intracellular milieu that is more conducive to supporting viral biosynthetic events and, in some instances, to activate particular viral or cellular genes. In an important example, activation of NF-κB by HIV leads to the transcriptional upregulation of critical HIV early genes (see Chapter 21), while in other instances NF-κB activation leads to the activation of cellular genes not involved in host defense but which instead augment virus replication. An advantage to a virus of achieving these ends via the activation of NF-κB is that NF-κB activation occurs very quickly after exposure to an inducer, since it does not require protein synthesis. Importantly, activation of NF-κB in these instances is not mediated by Toll-like receptors, since that would result in consequences that likely would not serve viral purposes.

apoptosis inducers cause apoptosis by setting in motion a cleavage cascade that involves members of the caspase family, and other cellular proteins as well. CrmA inhibits apoptosis by interacting directly with caspase 1 (ICE) and caspase 8.

Other poxvirus-encoded Crm proteins inhibit apoptosis by yet other mechanisms. For example, the cowpox virus CrmB and CrmD proteins and the homologous myxoma virus T2 protein are soluble homologs of the cellular TNF receptor. These viral homologs are secreted from infected cells. Then, in the extracellular space, they act as decoy receptors that sequester TNF, one of the key inducers of apoptosis, as noted above. Importantly, myxoma virus mutants lacking the T2 gene are less pathogenic in rabbits, demonstrating that the ability of poxviruses to impair apoptosis is due to a poxvirus virulence factor in vivo. Recently, polymerase chain reaction (PCR) technology was used to identify sequences on the variola virus and monkeypox virus genomes that encode CrmB homologs. Thus, this antiapoptotic stratagem may be widespread among the poxviruses.

Importantly, there are two functionally distinct cellular pathways to apoptosis. First, as just noted, there is the **extrinsic pathway**, which is triggered when one of the proapoptotic ligands, TNF-α, FasL, or TRAIL, binds to its respective receptor at the cell surface. Activation of these receptors in turn activates membrane-proximal activator caspases (caspase 8 and caspase 10), which then set in motion the effector caspase cascade. Second, there is the **intrinsic pathway**, which is triggered in response to internal cues, such as DNA damage or inappropriate cell divisions. The intrinsic pathway, which is independent of caspase 8, requires the opening of the mitochondrial permeability transition pore complex, which enables the release of cytochrome *c*, a key signal protein of the mitochondrial apoptosis pathway. Note that the extrinsic and intrinsic pathways converge to activate the same effector caspase cascade. By now you likely correctly surmised that poxviruses have also evolved countermeasures against the proapoptotic stimuli that induce apoptosis through a mitochondrion-dependent but caspase-8-independent pathway (Box 19.16).

Countermeasures against Adaptive Defenses

The recognition and destruction of virus-infected cells by **cytotoxic T lymphocytes** (CTLs) are generally essential for the mammalian host to resolve a virus infection. Yet poxviruses as a family provide only a few examples of viral countermeasures against CTL-mediated antiviral defenses. Why might that be? One suggested possibility is that countermeasures against cell-mediated immunity are most important in the case of viruses, such as the herpesviruses

BOX 19.16

Poxviruses directly affect the mitochondrion-dependent apoptosis pathway by blocking the "opening" of the mitochondrial permeability transition pore. Normally, the mitochondrial permeability transition pore is opened when intrinsic proapoptotic signals activate the cellular Bak and Bax proteins, which then form oligomers in the outer mitochondrial membrane, leading to the release of cytochrome c and activation of the caspase cascade.

Bak and Bax are members of the so-called Bcl-2 family of proteins. Some family members, like Bcl-2 itself, are antiapoptotic, whereas others, like Bak and Bax, are proapoptotic. Bcl-2 family members act at mitochondria. Whether or not apoptosis occurs is determined by the balance of activities of the pro- and antiapoptotic Bcl-2 family members.

The myxoma virus antiapoptotic M11L protein binds to Bax and prevents the conformational change to Bax necessary for its activation at the mitochondrial membrane. Since Bak and Bax are normally regulated by Bcl-2, one might have expected that M11L is homologous to Bcl-2. But contrary to that expectation, M11L has no obvious sequence homology with Bcl-2, and its mode of binding to Bax is not yet known.

For the sake of comparison, note that the adenovirus E1B 19K protein is a Bcl-2 homolog that interacts with Bak and Bax to prevent their oligomerization. Likewise, the BHRF1 protein of Epstein-Barr virus, a gammaherpesvirus, is also a homolog of Bcl-2.

---

we begin this subsection by considering the countermeasures employed by molluscum contagiosum virus against CTLs. Then, we consider the possibility that other poxviruses too express countermeasures against CTL-mediated immune defenses.

CTLs recognize virus-infected cells by means of their T-cell receptors, which interact specifically with antigenic peptides that are presented at the surface of the infected cell by major histocompatibility complex class I (MHC-I) molecules. The T-cell receptor recognizes the complex comprised of both the antigenic peptide and the MHC-I molecule (Chapter 4). Molluscum contagiosum virus evades CTL-mediated immunity by impairing the MHC-I antigen presentation pathway. Not surprisingly, perhaps, it does so by exploiting a recurrent poxvirus stratagem, i.e., by using viral homologs of cellular proteins, in this instance, an MHC-I homolog (Box 19.17).

Molluscum contagiosum virus is thus far the only poxvirus that is known to express homologs of MHC-I molecules. And perhaps the resultant ability of this virus to impair the MHC-I antigen presentation pathway is relevant to its aforementioned ability to persist in its human host for several months or longer. However, other poxviruses, which give rise only to acute infections, are also able to impair MHC-I-mediated immune defenses, by yet other mechanisms. For instance, a myxoma virus-encoded protein, referred to as MV-LAP, causes the ubiquitination of MHC-I molecules and their consequent degradation by proteosomes. (The ubiquitin/proteosome pathway is the major pathway for the *selective* degradation of proteins in eukaryotic cells. The other major pathway for protein degradation in eukaryotic cells involves proteolysis within lysosomes. Recall that the high-risk papillomaviruses use the ubiquitin/proteosome pathway to degrade cellular tumor suppressor proteins [Chapter 16].) MV-LAP is expressed early in the myxoma virus replicative cycle. Why might that be expected? See Chapter 4 if the answer is not apparent. Also see Box 19.18.

MV-LAP is a significant virulence factor for myxoma virus in vivo, as shown by experiments in which rabbits were infected either with an MV-ΔLAP myxoma virus mutant or with wild-type virus. Rabbits infected with the mutant virus had fewer secondary cutaneous **myxomas** and milder respiratory infections than those infected with the wild-type virus. (Myxomas are benign tumors, usually embedded in connective tissue.) MV-ΔLAP-infected rabbits also had a high rate of recovery (over 70%), compared to a generally fatal outcome after infection with wild-type myxoma virus. Curiously, the ΔLAP deletion did not affect the ability of the mutant virus to replicate in vivo.

Vaccinia virus too may express CTL evasion strategies, since there are now several reports in the journal literature

---

and HIV, which establish lifelong persistent infections as an important aspect of their lifestyle. In contrast, poxviruses, which generally give rise to acute infections only, may not have as critical a need to impair CTL-mediated immune defenses. (The thinking here is that while the innate immune system can effectively act against an infection within the first several hours, the adaptive immune system is not effective until the fourth day after infection or later, time enough for an acute infection to be transmitted to new susceptible individuals. Yet variola virus-infected individuals were not contagious until the onset of fever, which occurred between days 12 and 16 of the infection [see below].)

But what then of molluscum contagiosum virus, a poxvirus that has the capacity to persist in humans for several months or longer? The answer is that molluscum contagiosum virus indeed impairs CTL-mediated immune defenses, consistent with the above generalization. Thus,

BOX 19.17

How might the molluscum contagiosum virus-encoded MHC-I homolog impair CTL-mediated immune defenses? Several possibilities may come to mind. (Stop for a moment to consider what these might be.) However, three key experimental findings lead towards a particular mechanism. First, rather than being transported to the plasma membrane like cellular MHC-I molecules, the poxvirus-encoded MHC-I homolog is retained as a transmembrane protein in the membranes of the endoplasmic reticulum and the Golgi apparatus. Second, molluscum contagiosum virus-infected keratinocytes do not display β2-microglobulin at the cell surface. Apropos this point, recall that β2-microglobulin is the invariable light chain that is common to all MHC-I molecules (Chapter 4). Importantly, MHC-I heavy chains must associate with β2-microglobulin in the endoplasmic reticulum in order to present peptides at the cell surface. Third, the molluscum contagiosum virus-encoded MHC-I homolog forms complexes with β2-microglobulin in the endoplasmic reticulum. Together, these experimental findings

point towards a mechanism whereby the molluscum contagiosum virus-encoded MHC-I homolog acts by sequestering β2-microglobulin in the endoplasmic reticulum and the Golgi apparatus.

For the sake of comparison, several herpesviruses, in particular human cytomegalovirus, likewise encode MHC-I homologs (Chapter 18). However, in contrast to the molluscum contagiosum virus-encoded MHC-I homolog, the herpesvirus proteins appear to function primarily as decoys that activate the NK-cell inhibitory receptors, thereby protecting herpesvirus-infected cells from being lysed by NK cells. Apropos this point, recall that NK cells preferentially kill target cells that express low levels of MHC-I molecules at the cell surface, as a defensive countermeasure against those viruses that attempt to evade CTL-mediated attack by downregulating MHC-I expression. The molluscum contagiosum virus MHC-I homolog does not appear to act in this way since it is not present at the cell surface.

that this virus downregulates MHC-I expression. But other than the fact that this effect appears to depend on a vaccinia virus early-gene product, there is not yet any information regarding possible underlying mechanisms.

Actually, studies bearing on the issue of vaccinia virus evasion of CTL-mediated defenses have been concerned mainly with the observed increased sensitivity of vaccinia virus-infected cells to lysis by NK cells. Apropos this point, recall that NK cells, which are a component of innate immunity, preferentially kill target cells that express low levels of MHC-I molecules at the cell surface. As noted above, this is an NK-cell-mediated countermeasure against viruses that attempt to evade CTL-mediated attack by downregulating MHC-I expression. Yet ever-resourceful

vaccinia virus may in turn counter this activity of NK cells by actually infecting those cells. (Proviso: Recent experimental results showed that the impaired expression of MHC-I molecules by vaccinia virus-infected cells may not be essential to altering the susceptibility of those cells to NK-cell-mediated lysis. These findings underscore the fact that the mechanisms used by NK cells to distinguish between healthy and virus-infected cells remain to be better understood.)

Results of another recent study demonstrated that vaccinia virus can also impair the MHC-II-mediated antigen presentation pathway. (Recall that "professional" antigen-presenting cells present antigen to CD4 helper T cells via the MHC-II-mediated antigen-presentation pathway, in that way activating the CD4 T cells, which then can participate in activating CTLs [Chapter 4].) The effect of the virus on the MHC-II-mediated antigen presentation pathway was noted within 1 h of infection. The total levels of MHC-II proteins in the cell and at the cell surface were not affected. Instead, vaccinia virus infection appeared to directly interfere with the binding of antigenic peptides to MHC-II molecules.

BOX 19.18

Cellular ubiquitin ligases transfer ubiquitin groups to their target proteins via a so-called C_4HC_3 domain that is critical for their ubiquitin ligase activity. Likewise, MV-LAP contains a C_4HC_3 domain, by which it transfers ubiquitin groups to the cytoplasmic tails of its targets, i.e., MHC-I molecules. Interestingly, that same C_4HC_3 domain is present in herpesvirus proteins that serve herpesviruses in the same way that MV-LAP serves molluscum contagiosum virus.

CLINICAL SYNDROMES

Smallpox

Smallpox was declared to be eradicated worldwide in 1977. Before then, mortality rates associated with smallpox were as high as 40%. But why then did more than 60% of infected individuals survive their highly virulent smallpox infections, while others succumbed? One possibility is as follows. A pathogen as widespread and virulent as variola virus must have strongly selected for host alleles that confer less virulent disease, or even resistance. Experimental results from a mouse model involving ectromelia virus, which is closely related to variola virus and causes a disease in mice similar to smallpox in humans, are consistent with this premise. Resistant strains of mice indeed arise, and four genetic loci associated with resistance have been identified (Box 19.19).

Before its eradication, smallpox was a highly contagious disease (although less so than measles and influenza), most often transmitted by inhalation of droplets of saliva. Since this route of transmission required prolonged face-to-face contact with an infected individual, the greatest risk of infection occurred between members of the same household and between others having close contacts with infected persons. Smallpox was also spread via direct contact with scabs or infected bodily fluids or by contact with dried virus on clothes, blankets, and other materials (Box 19.20).

Variola virus initially infected the upper respiratory tract, consistent with its transmission by the respiratory route. During this early stage of infection of the upper respiratory tract, patients were still symptom free and not yet infectious. Virus that was generated in the upper respiratory tract would then disseminate via the lymphatic system to regional lymph nodes, from which it passed into the blood circulation, enabling the infection to spread to the spleen, additional lymph nodes, liver, and other organs, and finally appearing in the skin.

The first symptoms would generally appear between days 12 and 16. They included fever as high as 104°F, malaise, head and body aches, and, in some cases, vomiting. This phase, called the **prodrome**, lasted for 2 to 4 days. It was followed by the appearance of the characteristic smallpox rash and "pox."

The rash would appear first on the tongue and in the mouth. There, it would develop into pocks that eventually broke open to release large amounts of virus into the mouth and throat. Consequently, this was the time when an infected person was most contagious (see above regarding transmission by inhalation of droplets of saliva). Generally, infected individuals were not contagious until the onset of fever, which occurred between days 12 and 16, as noted above.

As the pocks in the mouth were breaking down, the rash would begin to spread over the entire body, appearing first on the face and then spreading to the arms and legs and, finally, to the hands and feet. The process of rash spreading was generally complete within 24 h of its onset. If the patient were to recover, the fever would usually subside at that time, and the patient would begin to feel somewhat better. The infection is thought to have always been acute. That is, either the patient would both recover and be immune to subsequent reinfection, or the patient would succumb. Resolution of the infection required cell-mediated immune responses.

During development of the rash and "pox," the latter took on the appearance of raised bumps that eventually came to have a depression in the center that was said to resemble a bellybutton. This was the major distinguishing characteristic of the smallpox lesion (Fig. 19.9). Note that these lesions were due to hemorrhaging of the small vessels of the dermis.

By the end of the second week after the rash appeared, most of the lesions would have formed a scab. Then, the scabs would begin to fall off. The shedding of scabs was usually complete by 3 weeks after the first appearance of the rash. An infected person remained contagious until the last smallpox scab fell off.

Fortunately for the community as a whole, high fever and fatigue often limited the movement and contacts of contagious individuals. Moreover, the highly visible pustules provided a means to identify and quarantine infectious individuals.

Molluscum Contagiosum

Molluscum contagiosum virus is the only poxvirus other than variola virus that is known to be specific for humans. It typically proliferates within the follicular epithelium

---

BOX 19.19

About 10% of Europeans express an allele that confers resistance to HIV (see Chapter 21). Specifically, those HIV-resistant individuals produce a defective form of the chemokine receptor CCR5. The wild-type CCR5 protein is used by HIV as a coreceptor. Some researchers have suggested that the defective *ccr5* allele exists at its high level in the human population because it conferred resistance to bubonic plague, caused by the bacterium *Yersinia pestis*. More recently, it has been argued that the defective *ccr5* allele was selected in response to variola virus. Whereas *Y. pestis* is a bacterium that replicates outside cells and does not appear to use CCR5 in infection, there is some evidence that variola virus may use chemokine receptors like CCR5 to enter cells. Also, variola virus may well have placed greater selective stress on the human population than did *Y. pestis*.

BOX 19.20

Lord Jeffrey Amherst (for whom my hometown in Massachusetts is named) was the commanding general of British forces in North America during the final battles of the so-called French and Indian war (1754–1763). Amherst's significant military victories over the French forces played a key role in securing British control over all of North America. Nevertheless, Amherst's current reputation is tarnished by the belief that he deliberately gave smallpox-infected blankets to Native American Indians, thereby causing a deadly epidemic among them. This episode actually occurred after the war, when hostility between the British and the Native Americans led to the 1763 Indian uprising known as Pontiac's Rebellion. To end the rebellion, Amherst *suggested* giving smallpox-infected blankets as "gifts" to the attacking Indians, as was clear from Amherst's own letters to a subordinate. Although Amherst is usually regarded as the villain of this episode, there is historical evidence that, unbeknownst to Amherst, the British commander at Fort Pitt had already independently attempted this very tactic, sending smallpox-infected blankets and handkerchiefs to the Delaware Indians surrounding the fort. Regardless of who actually perpetrated this episode, it is the first known example of *deliberate* germ warfare in North America.

In his marvelous book, *Guns, Germs, and Steel*, Jared Diamond describes how smallpox gave the Spaniards an inadvertent but decisive advantage in their conquest of the Aztecs, beginning in 1519. Cortez came to Mexico with only 600 individuals, yet he defeated an Aztec empire of several millions. One infected slave, arriving in Mexico from Spanish Cuba, is thought to have transmitted smallpox to the Aztecs, who had no prior exposure to the infection. As a result, most of the Indians and their leaders were killed by the Old World germ. That is, the vast majority of Aztecs were killed by smallpox rather than by Spanish guns and swords. Indeed, 3.5 million Aztecs succumbed to smallpox in a mere 2 years! The Spaniards, in contrast, were largely immune or had genetic resistance and were spared. (The issue of genetic resistance to smallpox is considered in the text and in Box 19.19.)

Pizarro, in 1532, with a mere 168 men (and only 12 guns, which were the slow-to-load, inaccurate harquebuses of the day) conquered the Inca Empire of millions. However, Pizarro, like Cortez, "benefited" from smallpox, which arrived in Peru by chance at around 1526, killing much of the Inca population and leaving it dispirited and fragmented.

Smallpox was the major culprit among the infectious diseases that were endemic in Europe and that devastated the indigenous peoples of the New World. The other most important of these infectious diseases included measles, influenza, typhus, and bubonic plague. Estimates of how many Native Americans actually were living in the Americas when Columbus arrived and of how many actually succumbed to smallpox and other Old World diseases vary considerably. That is so because of the use of arbitrary formulas, unreliable historical sources, and premises that in some instances are driven by "politically correct" ideologies. Nevertheless, the harm racked up by Europeans transmitting these diseases to Native American populations was unquestionably enormous. Moreover, this devastation continued into the 20th century, particularly affecting the Alaskan Inuit peoples, as well as the native populations of Australia, New Guinea, and Africa.

Reference
Diamond, J. 1999. *Guns, Germs, and Steel: the Fates of Human Societies.* W. W. Norton & Company, New York, NY.

(Fig. 3.1). Consequently, like other poxviruses, molluscum contagiosum virus causes skin lesions. But in marked contrast to smallpox lesions, which were pustular (i.e., derived from vesicles subsequently infiltrated by large numbers of lymphocytes, as may occur when viruses spread from capillaries to superficial layers of the epidermis), molluscum contagiosum lesions are wart-like and proliferative. These lesions project up above the adjacent skin, appearing as visible tumors, and each contains huge amounts of poxvirus particles (Box 19.21).

Most molluscum contagiosum lesions are seen on the trunk, genitals, and upper extremities, where they are generally clustered in groups of 5 to 20. The time that elapses between initial infection and the appearance of the lesions ranges between 14 and 50 days. The lesions usually resolve within between 2 and 12 months, presumably because of host immune responses. But some lesions may persist for as long as 2 years. For this reason, molluscum contagiosum disease is considered to be chronic.

Individuals most frequently infected with molluscum contagiosum virus include preadolescent children, sexually promiscuous adults, participants in sports that involve skin-to-skin contact (e.g., wrestling), and individuals with impaired cellular immunity. Molluscum contagiosum disease occurs worldwide, but it is particularly prevalent in certain regions. For example, 4.5% of the individuals of a

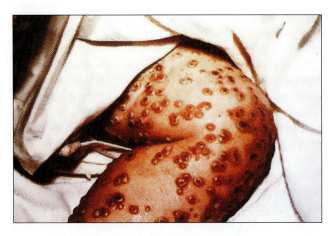

Figure 19.9 Characteristic smallpox lesions in a young smallpox victim. From S. J. Flint, L. W. Enquist, R. M. Krug, V. R. Racaniello, and A. M. Skalka, *Principles of Virology: Molecular Biology, Pathogenesis, and Control*, 2nd ed. (ASM Press, Washington, DC), with permission.

village in Fiji bore molluscum contagiosum lesions. High levels of molluscum contagiosum also have been seen in Papua New Guinea and Zaire (now the Democratic Republic of the Congo).

Vaccinia Virus and the Smallpox Vaccine

The smallpox vaccination protocol consisted of scratching live vaccinia virus into the patients' skin. The vaccinated individual then would be observed for the development of pustules, which generally appeared by 6 to 10 days after vaccination. Their appearance would confirm a "take."

BOX 19.21

At present, molluscum contagiosum virus is being associated with an increasing number of incidents of skin tumors in immunocompromised patients. In this regard, molluscum contagiosum virus is one of three poxviruses that are known to be tumorigenic. The others are Shope fibroma virus of rabbits and Yaba monkey tumor virus. Historically, the discovery of the Shope fibroma virus was first reported by Richard E. Shope in 1932. Recall that Shope, in 1930, was the first researcher to isolate an influenza virus (Chapter 12) and in 1933 was the first to isolate a papillomavirus: the cottontail rabbit papillomavirus (Chapter 16).

It is not known how poxviruses induce tumors. They do not encode known oncogenes. Many poxviruses do encode homologs of cellular growth factors (e.g., epidermal growth factor), but expression of those viral proteins does not necessarily lead to tumorigenesis.

No vaccine is free of risk. At times before smallpox was eradicated, when vaccination against the disease was routine, there were about 7 to 9 deaths per year in the United States that were attributed to vaccination per se. In addition to those fatal outcomes, there also were nonfatal adverse responses to the vaccine, which ranged from mild and self-limited to severe and life threatening. Serious vaccine-related complications included encephalitis, general disseminated disease, and progressive vaccinia. The latter condition occurred principally in individuals with defective cell-mediated immunity. Indeed, the live smallpox vaccine poses an unacceptable risk to any immunocompromised individual. Children with eczema also were at high risk (Box 19.22).

As smallpox eradication was in its final stages, the general risk of vaccination began to exceed the risk of smallpox infection. Then, when eradication was finally realized, routine vaccination of the general public could no longer be justified, and its practice came to an end in most countries. Importantly, recall that the absence of a nonhuman reservoir for variola virus was a key facilitating factor in the eradication of smallpox. The possibility that vaccination may have conferred lifelong immunity may have been another important factor.

The smallpox vaccine that currently is stockpiled in the United States is live vaccinia virus that was grown in the skin of calves. Although it is now more than 30 years old, it is still believed to be effective.

Antipoxvirus drug therapies are currently limited to cidofovir. This drug inhibits variola virus and other orthopoxviruses in vitro and is effective in mice that have been infected with lethal doses of cowpox virus. Nevertheless, while cidofovir is licensed to treat some herpesvirus infections, including human herpesvirus 6 and retinitis caused by cytomegalovirus in HIV-infected individuals, it is not licensed to treat complications associated with the smallpox vaccine. Consequently, it is used "off-label" for this purpose. (By off-label, one means as prescribed by a physician to treat a condition for which it has not been specifically approved.) Alternatively, it may be used via a special protocol called an Investigational New Drug (IND) protocol. (Check the Centers for Disease Control and Prevention website for the current status of research in these areas. The following is from a current [2009] posting. "However, its [cidofovir] use for the treatment of vaccinia adverse reactions is restricted under an Investigational New Drug (IND) protocol. According to the IND protocol, cidofovir would only be used when VIG [vaccinia immune globulin] was not efficacious. Healthcare providers considering use of cidofovir for the treatment of vaccinia adverse reactions should consult with HHS/CDC.")

BOX 19.22

Live vaccine agents have the potential to spread from vaccinated individuals to nonvaccinated individuals in a community. In fact, in the case of the smallpox vaccine, each vaccinated individual develops a pox at the site of inoculation that can release the vaccine virus for as long as 21 days.

In the past, the long-drawn-out ability of a vaccinated individual to transmit the live smallpox vaccine virus was an important positive attribute of the vaccine, since it helped to spread immunity in the population. However, in the contemporary world, in which the acquired immunodeficiency syndrome (AIDS) pandemic has given rise to large numbers of severely immunocompromised individuals and in which organ transplant recipients on immunosuppressive regimens are increasingly common, reintroducing routine smallpox vaccination as a preparedness measure against a potential bioterrorist attack (see the text) is not an option to be taken lightly, since immunocompromised individuals may be put at severe risk from those receiving the current live smallpox vaccine. So, then, should public health policy prevent the vaccine from being available to immunocompetent individuals who are willing to accept the risk to themselves?

What then of developing a new, safer vaccine? The main problem in this regard is that before any crisis against which the new vaccine might be called upon (such as a bioterrorist attack), the effectiveness of the new vaccine would not be known, since it cannot be tested on humans, nor can it be tested in animals, since none is known to contract smallpox.

On 24 January 2003, the U.S. Department of Health and Human Services put into operation a preparedness program in which smallpox vaccine was administered to federal, state, and local individuals who might be first responders in the event of a bioterror attack. As part of this preparedness program, the CDC established a surveillance protocol to monitor for adverse reactions resulting from smallpox vaccination. The complete text of that protocol is available online (Casey et al., 2006).

Reference
Casey, C., C. Vellozzi, G. T. Mootrey, L. E. Chapman, M. McCauley, M. H. Roper, I. Damon, and D. L. Swerdlow. 2006. Surveillance guidelines for smallpox vaccine (vaccinia) adverse reactions. *MMWR Recomm. Rep.* 2006 55(RR-1):1–16. (Placing the title in your Web browser will bring up the manuscript.)

SMALLPOX AND BIOTERRORISM

Biological warfare is hardly a new concept. In fact, there is evidence of its having been practiced as early as the 6th century BCE, when the Assyrians were reputed to have poisoned enemy wells with a fungus that they believed would make the enemy delusional. The Greek historian Thucydides reported between 431 and 404 BCE that during the siege of Athens by the Spartans in the Peloponnesian War, the Spartans deliberately "poisoned" Athenian wells, causing a devastating epidemic that killed thousands of Athenians. The Mongols and Turks are known to have used infected animal carcasses to poison wells in medieval Europe and to have catapulted diseased corpses into besieged cities. During this era, Europeans too catapulted bubonic plague victims, and their excrement, over castle walls. The last known instance of plague corpses being used as a bioweapon occurred in 1710, when Russian forces flung plague-infected corpses over the city walls of Revel (now Tallinn), the capital city and main seaport of Estonia. See Box 19.20 for an account of a 1763 episode in which British forces in North America used smallpox-infected blankets against Native American Indians. During the Second World War, nations either experimented with, or actually used, biological warfare agents, and during the Cold War, both the United States and the Soviet Union developed and stockpiled biological warfare agents. (For a fascinating review, see T. J. Johnson, *A History of Biological Warfare from 300 B.C.E. to the Present*, available at www.aarc.org/resources/biological/history.asp.)

Despite the use of bioweapons over the past 3 millennia, many contemporary military experts believe that biological weapons would be of little use on a modern battlefield. That is so because unlike nuclear, chemical, and even conventional weapons, bioweapons would not immediately stop an advancing army. Moreover, once released, the spread of a bioagent would be virtually impossible to control. In addition, its use would invite retaliation in kind. Thus, in the contemporary world, the major concern over bioweapons is their potential use by terrorists and rogue states.

Of the many pathogenic microorganisms that might be used as bioterror agents, including those causing plague, tularemia, botulism, and tuberculosis, most biological warfare experts believe that smallpox and *Bacillus anthracis* pose the greatest threats. That is so because they are the only ones of these potential biowarfare agents that are easy to produce and dispense as well as being deadly. (The manufacture of a lethal anthrax aerosol [i.e., weaponized anthrax] would be beyond the ability of groups not having access to advanced biotechnology. However, with sufficient funding, individuals or groups may be able to acquire weaponized anthrax. The Japanese terrorist group Aum Shinrikyo was able to acquire weaponized anthrax, which they successfully dispersed in Tokyo. Fortunately, the strain turned out to be nonvirulent.)

When dispersed as an aerosol, anthrax has a case fatality rate of more than 80%. In contrast, smallpox has a case fatality rate of "only" about 30%. Nevertheless, smallpox may be the more effective bioterror agent for the following reasons. Anthrax-infected individuals simply die, whereas smallpox-infected individuals can infect many others, potentially killing many more victims than anthrax. Additionally, smallpox is difficult to diagnose until the variola infection is well under way in an individual. That is so because its incubation period is long and it initially presents with flu-like symptoms (see above). Moreover, antibiotics are not effective against viruses, so that effective treatment against smallpox is virtually impossible. Finally,

in 1980, 3 years after smallpox was considered to have been eradicated, the World Health Assembly recommended an end to routine vaccination, and most countries complied. Thus, most individuals alive today have never been vaccinated against smallpox, and it is not known whether those who received vaccinations 25 or more years ago are still protected. As a consequence of all of these factors, a smallpox outbreak would almost certainly severely overstress any public health system.

Not surprisingly, an infectious bioterror agent does not have to cause an epidemic to cause widespread panic and disruption. A 2001 episode in the United States, in which *B. anthracis* was disseminated to particular targets via the postal system, provides an example of this fact. This incident occurred over a period of several weeks, beginning in September 2001. Letters that contained anthrax spores were mailed to several news media offices and to two U.S. Senators. Five individuals were killed and 17 others were infected. The perpetrators were never identified, although

BOX 19.23

Numerous procedures and standards have been compiled by the NIH to ensure against accidents within, and release of a pathogen from, a Biosafety Level 4 facility. The following account highlights only a small sample of these procedures and standards. It is taken verbatim from the complete set of NIH guidelines.

(1) Access to the laboratory is strictly controlled by the laboratory director. (2) The facility is either in a separate building or in a controlled area within a building, which is completely isolated from all other areas of the building. (3) There are two models for Biosafety Level 4 laboratories: (A) the Cabinet Laboratory where all handling of the agent is performed in a Class III Biological Safety Cabinet, and (B) the Suit Laboratory where personnel wear a protective suit. Biosafety Level-4 laboratories may be based on either model or a combination of both models in the same facility. If a combination is used, each type must meet all the requirements identified for that type. (4) Within work areas of the facility, all activities are confined to Class III biological safety cabinets, or Class II biological safety cabinets used with one-piece positive pressure personnel suits ventilated by a life support system. (5) The Biosafety Level 4 laboratory has special engineering and design features to prevent microorganisms from being disseminated into the environment. (6) Personnel enter and leave the laboratory only through the clothing change and shower rooms. They take a decontaminating shower each time they leave the laboratory. Personnel use the airlocks to enter or leave the laboratory only in an emergency. (7) Personal clothing is removed in the outer clothing change room and kept there. Complete laboratory clothing, including undergarments, pants and shirts or jumpsuits, shoes, and gloves, is provided and used by all personnel entering the laboratory.

When leaving the laboratory and before proceeding into the shower area, personnel remove their laboratory clothing in the inner change room. Soiled clothing is autoclaved before laundering. (8) Supplies and materials needed in the facility are brought in by way of the double-doored autoclave, fumigation chamber, or airlock, which is appropriately decontaminated between each use. After securing the outer doors, personnel within the facility retrieve the materials by opening the interior doors of the autoclave, fumigation chamber, or airlock. These doors are secured after materials are brought into the facility. (9) Practical and effective protocols for emergency situations are established.

Once again, the preceding are just a few of the extensive precautions in place in a Biosafety Level 4 facility.

(A) **A Biosafety Level 4 containment facility at the Canadian Science Center for Human and Animal Health in Winnipeg, Canada.** (B) Ebola infections being carried out in a biosafety cabinet under Biosafety Level 4 conditions.

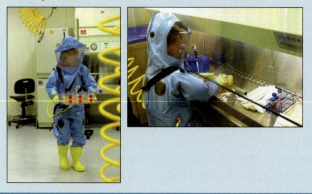

at the time this was being written (August 2008) a key suspect, an anthrax researcher at a U.S. Army laboratory, committed suicide before being charged.

At any rate, the potential use of a bioterror agent is a very real concern, with no apparent solution. Current best options for dealing with the threat of a bioterror attack include vigilance and preparedness efforts, such as vaccinating first responders. Also, it would be prudent to carry on research and development efforts towards the development of new, more effective drugs and diagnostics, etc. Is reintroducing the *routine* use of the current vaccine, or perhaps a new vaccine, the answer to the bioterror threat of smallpox? This question and other related matters are discussed in Box 19.22 and below.

Reintroducing routine smallpox vaccination as part of a preparedness effort may not be advisable because the potential benefits of doing so might not outweigh the risks of adverse events (Box 19.22). But what about using the smallpox vaccine in response to an actual smallpox bioterror attack? The CDC Advisory Committee on Immunization Practices (ACIP) in fact advised against vaccinating the general public in the event of such an attack. Instead, the ACIP recommended the so-called **ring vaccination strategy**, in which direct contacts (as well as household contacts of contacts) of diagnosed cases are identified and vaccinated. This strategy is based in part on empirical findings that vaccination within 3 to 4 days of contact with an infectious individual can prevent infection of the contactor or lessen the severity of the disease. In any event, the alertness of health workers and their ability to diagnose quickly would be crucial to the success of the ring vaccination strategy. The ACIP also recommended that those who would be caring for patients at smallpox isolation and care centers should be vaccinated.

Here is yet another issue related to the threat of bioterrorism: what is to be done with existing stocks of variola virus? As a result of the success of the smallpox eradication program, reference stocks of variola virus that were maintained in WHO laboratories were destroyed. Currently, stocks of the virus are known to exist in only two public health laboratories: one in the United States at the CDC and one at Russia's State Research Center of Virology and Biotechnology (VECTOR).

Because of concern that this remaining virus might somehow fall into the hands of terrorists or inadvertently escape from these laboratories, many have argued that these last remaining variola virus stocks should be destroyed. The case for destroying these stocks is clear. Why then might one argue for maintaining them? One reason given is that someday they might be needed for the development of vaccines and antiviral drugs that might be more effective than current options against variola virus. Because of the possible use of variola virus by terrorists, the WHO in fact does oversee live variola virus research at the above-mentioned laboratories, with the purpose of developing accurate diagnostic tests and therapeutics and the means for assessing their effectiveness. These efforts are carried out under the most rigorous conditions for biocontainment (i.e., Biosafety Level 4 [Box 19.23]). What other reasons for maintaining variola virus stocks come to mind?

Suggested Readings

Condit, R. C., N. Moussatche, and P. Traktman. 2006. In a nutshell: structure and assembly of the vaccinia virion. *Adv. Virus Res.* **66**:31–124.

Jha, P., and G. J. Kotwal. 2003. Vaccinia complement control protein: multi-functional protein and a potential wonder drug. *J. Biosci.* **28**:265–271.

Law, M., G. C. Carter, K. L. Roberts, M. Hollinshead, and G. L. Smith. 2006. Ligand-induced and nonfusogenic dissolution of a viral membrane. *Proc. Natl. Acad. Sci. USA* **103**:5989–5994.

Townsley, A. C., A. S. Weisberg, T. R. Wagenaar, and B. Moss. 2006. Vaccinia virus entry into cells via a low-pH-dependent endosomal pathway. *J. Virol.* **80**:8899–8908.

III

Retroviruses: Oncogenes and Cancer, HIV and AIDS

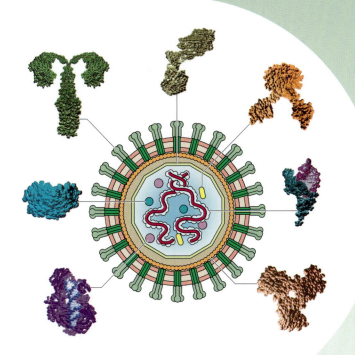

The Retroviruses: the RNA Tumor Viruses

INTRODUCTION

The *Retroviridae* have a rather extraordinary replication strategy that differs in major ways from that of all other known plus-strand RNA viruses. The genomes of other plus-strand RNA viruses typically function as messenger RNAs (mRNAs) upon their release from the virus particle, being directly translated on cellular ribosomes and in that way generating the virus-encoded, RNA-dependent transcriptase/replicase activities needed to initiate the virus replication cycle.

Parental retroviral plus-strand RNA genomes do not serve as mRNAs. Instead, parental retroviral RNA genomes are *reverse transcribed* by a virus-encoded, particle-associated reverse transcriptase enzyme, a process that actually replaces the retroviral RNA genome with a double-stranded DNA copy of it. The DNA genome is then integrated into the host genome, from which it serves as the template for the transcription of all viral mRNA molecules, as well as all progeny plus-strand RNA genomes. Moreover, transcription, which generates retroviral mRNAs and progeny viral genomes, is carried out by the cellular polymerase II (Pol II) RNA polymerase.

The prototype retrovirus is the Rous sarcoma virus (RSV) of chickens, discovered in 1911 by Peyton Rous. The subsequent discoveries in the 1950s of mouse mammary tumor virus and the Gross murine leukemia virus, the first known oncogenic mammalian retroviruses, demonstrated that Rous's original observations of a transmissible malignancy are not restricted to a particular cancer in chickens.

Since retroviruses were initially discovered as filterable agents, able to cause malignant solid tumors and leukemias, they were originally known as **RNA tumor viruses**, then as **oncornaviruses**, and in some instances as **leukoviruses**. They acquired the more general **retrovirus** designation after the discovery of the remarkable reverse transcription step in their replication cycle. While all other known plus-strand RNA viruses are class IV viruses in the Baltimore classification scheme, retroviruses occupy a separate niche in that scheme, as the class VI viruses.

In 1980, Robert Gallo identified the first known human retrovirus, the human T-lymphotropic virus (also known as the human T-cell leukemia virus). In 1984, research groups headed by Gallo and Luc Montagnier discovered the human immunodeficiency virus (HIV), a retrovirus identified as the causal agent of acquired immunodeficiency syndrome (AIDS) (Chapter 21). Retroviruses are now known to be present in all eukaryotes. Moreover, there is evidence for retroviruses, or retrovirus-like genetic elements, even in bacteria.

Retroviruses are now grouped into seven subclasses, based on their morphologies and genetic relatedness. Five of these subclasses are comprised of the so-called RNA tumor viruses or oncornaviruses, regardless of whether or not they are actually oncogenic. In fact, most are not. (As explained below, the ability of retroviruses to cause tumors in vivo or to transform cells in culture is not relevant to their replication or lifestyle.) Well-known examples of oncornaviruses include Rous sarcoma virus, murine leukemia virus (MLV), mouse mammary tumor virus, and avian leukosis virus (ALV).

The two remaining retrovirus subclasses are the **lentiviruses** and the **spumaviruses**. The lentiviruses (*lenti* is Latin for slow) are so named for the slow courses of the diseases they cause. HIV is the most famous of the lentiviruses. The spumaviruses (*spuma* is Greek for foamy) are so named for the vacuolar, soapsuds-like or foamy appearance of cells infected with these viruses. For that reason, these viruses also are referred to as **foamy viruses**.

It is not necessary for us to recount the morphological and genomic characteristics of each of the five RNA tumor virus subclasses, since their common features are much more significant than their differences. Instead, it is more important to note the differences between the RNA tumor viruses as a group and the lentiviruses and spumaviruses. All of the RNA tumor viruses are **simple retroviruses**, whereas the lentiviruses and spumaviruses are **complex retroviruses**.

The genomes of the simple retroviruses most often consist of only the *gag*, *pol*, and *env* genes, each of which is crucial for replication. In some instances, a simple retrovirus genome may also contain a tumor-inducing **oncogene**, which is irrelevant for replication, as mentioned above (Fig. 20.1). The genomes of the complex retroviruses likewise contain *gag*, *pol*, and *env* genes, but those genomes also contain nonstructural regulatory genes (e.g., *tat*, *rev*, and *nef* in the case of HIV) that are not present in simple retrovirus genomes (Fig. 20.1). Also, and importantly, the simple retroviruses encode only unspliced and singly spliced mRNAs, whereas the complex retroviruses also encode multiply spliced mRNAs. The latter are translated into regulatory proteins that may play key roles in replication and pathogenesis (see Chapter 21).

The remainder of this chapter is mainly concerned with the simple retroviruses (i.e., the RNA tumor viruses). It also considers the spumaviruses, but only briefly, because these viruses are not as well characterized as either the RNA tumor viruses or the lentiviruses. Still, the spumaviruses are noteworthy because they employ a replication strategy that differs in a fundamental way from that of all other retroviruses. Chapter 21 is mainly devoted to HIV/AIDS and discusses selected other lentivirus infections as well.

Before we consider the simple retroviruses in detail, note that *all* members of the retrovirus family employ a replication strategy that includes the following essential and distinctive features. First, the reverse transcription step, in which the viral plus-strand RNA genome is replaced by a double-stranded DNA copy, is the key distinguishing feature of the retrovirus replication strategy. Second, reverse transcription of the viral plus-strand RNA genome is catalyzed exclusively by the virus-encoded reverse transcriptase enzyme. Third, the retroviral reverse transcriptase is incorporated into the virus particle at the time of particle assembly and thus is available to reverse transcribe the parental virus genome at the start of the next cycle of infection. In marked contrast to the retroviruses, the Baltimore scheme class IV plus-strand RNA viruses do not contain any enzyme in the virus particle. Instead, when the genomes of the class IV viruses are released into the newly infected cell, they serve as mRNAs that are translated into the viral transcriptase/replicase activities, which then mediate virus transcription and replication. Consequently, fourth, the retroviruses are the only plus-strand RNA viruses in which the parental genomic RNA does not serve as an mRNA. Fifth, the DNA copy of the retrovirus genome is integrated into the cellular genome. This integration event is a crucial step in the retroviral replication cycle, since, as a sixth feature, all retroviral mRNAs and all retrovirus progeny genomes are generated by transcription of the integrated retroviral genomes. (Note that this integration event enables HIV genomes to be stably maintained in latently infected CD4 T cells, which is a key factor in the clinical course of AIDS [Chapter 21].) Seventh, all retrovirus mRNAs and progeny RNA genomes are transcribed from the integrated retrovirus genomes by the cellular Pol II RNA polymerase. In point of fact, the retroviruses are the only RNA viruses that are transcribed by the cellular transcription machinery. Eighth, as in the case of all DNA polymerases, the retroviral RNA-dependent DNA polymerase (reverse transcriptase) requires a primer. Yet the retrovirus reverse transcription process is unique with respect to how it meets its requirement for a primer. Specifically, retroviruses use cellular transfer RNAs (tRNAs) to prime reverse transcription.

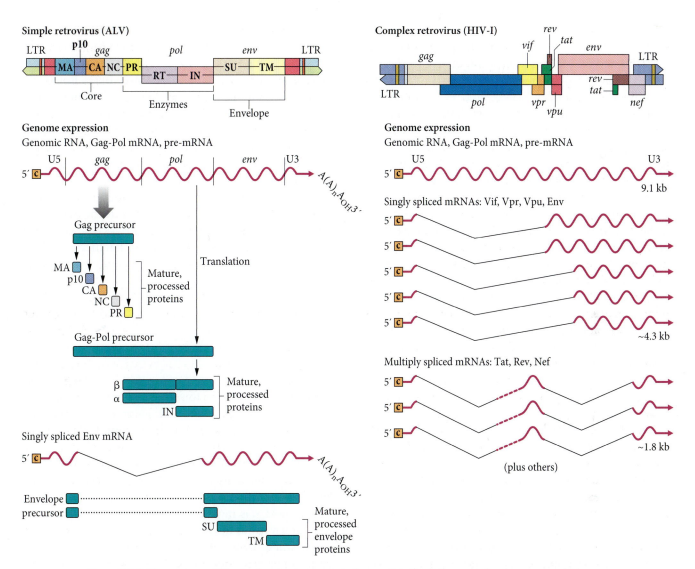

Figure 20.1 The genomes of the simple retroviruses most often consist of only the *gag*, *pol*, and *env* genes, which are crucial for replication. In some instances, a simple retrovirus genome may also contain an oncogene, which is irrelevant for replication. The simple retrovirus genome depicted is the **provirus** of ALV. (The provirus is the DNA copy of the RNA genome, which contains the LTR sequences that are generated during reverse transcription [see the text].) The genomes of the complex retroviruses, such as that of the HIV type 1 provirus depicted, likewise contain *gag*, *pol*, and *env* genes but, in addition, contain nonstructural regulatory genes (e.g., *tat*, *rev*, and *nef*, in the case of HIV) not present in the simple retroviruses. Also, and importantly, notice that the simple retroviruses encode only unspliced and singly spliced mRNAs, whereas the complex retroviruses also encode multiply spliced mRNAs. The latter are translated into the regulatory proteins that may play key roles in replication and pathogenesis. From S. J. Flint, L. W. Enquist, R. M. Krug, V. R. Racaniello, and A. M. Skalka, *Principles of Virology: Molecular Biology, Pathogenesis, and Control*, 2nd ed. (ASM Press, Washington, DC, 2004), with permission.

A cellular tRNA is base paired near the 5′ end of the genomic RNA molecules. As in the case of the retroviral reverse transcriptase, the tRNA primer is incorporated into progeny virus particles at the time of assembly. Ninth, the retroviruses are the only known viruses that can productively infect a cell while simultaneously transforming it. (Recall that the DNA tumor viruses invariably kill cells in which they replicate. A cell may be transformed by a DNA tumor virus only if the productive infection should abort.) Tenth, the retroviruses differ from all other known viruses in having a diploid, single-stranded RNA genome.

The earlier discovery that retroviruses can cause a variety of cancers in mammals as well as in birds led to intense interest in these viruses. The more recent discovery that

HIV is the etiologic agent of AIDS further heightened interest in the retroviruses. Consequently, the retrovirus family has received far more attention than any other virus family, and retrovirus researchers were awarded Nobel Prizes on four separate occasions. First, Peyton Rous was awarded a Nobel Prize in 1966 for having demonstrated in 1911 that a malignant sarcoma of chickens could be transmitted by a filterable extract. (The preceding sentence does not contain a misprint. See below and Box 1.5.) Next, Howard Temin and David Baltimore were awarded Nobel Prizes in 1975 for having independently discovered in 1970 the retrovirus reverse transcriptase. This discovery largely confirmed Temin's iconoclastic hypothesis that the retrovirus RNA genome might be "reverse transcribed" to generate a DNA copy. Next, Michael Bishop and Harold Varmus were awarded Nobel Prizes in 1989 for demonstrating that retrovirus oncogenes have their counterparts (proto-oncogenes) in normal cells. Then, in 2008, the Nobel Committee awarded the prize in Physiology or Medicine to Luc Montagnier and Françoise Barré-Sinoussi for the discovery of HIV (Chapter 21).

The experimental findings of Bishop, Varmus, and others led to major breakthroughs in the understanding of cancer. In particular, these researchers showed that cancer can result from the inappropriate expression or mutation of specific cellular genes. In addition, the discovery of the retrovirus reverse transcriptase made possible much of modern recombinant DNA technology and genetic engineering. That is so because RNA cannot directly be cloned. However, it can be extracted from cells, copied into DNA by the retroviral reverse transcriptase, and then cloned; this process is referred to as complementary DNA (cDNA) or expression cloning. And the discovery that AIDS is caused by HIV led to the development of a blood test that made the world's blood supply safe from that virus and of a diagnostic to identify HIV-infected individuals, while also bringing about the use of antiretroviral drugs for treating HIV-infected individuals, dramatically reducing morbidity and death.

A HISTORICAL INTERLUDE

Peyton Rous's 1911 report of the virus bearing his name was the first to document a filterable, infectious agent that causes solid tumors. In this account, Rous stated,

> A transmissible sarcoma of the chicken has been under observation in this laboratory for the past fourteen months, and it has assumed of late a special interest because of its extreme malignancy and a tendency to widespread metastasis. In a careful study of the growth, tests have been made to determine whether it can be transmitted by a filtrate free of the tumor cells .... Small quantities of a

cell-free filtrate have sufficed to transmit the growth to susceptible fowl.

Present-day students may be amazed that Rous's extremely novel findings generated little interest at the time they were reported. But medical researchers of the day did not see the relevance of a transmissible cancer in chickens to human malignancies, and they regarded Rous's findings as merely a curiosity. Rous's unique observations were not appreciated until the 1950s, when it was shown that related viruses could likewise cause tumors in mammals.

In Howard Temin's 1975 Nobel lecture, he had this to say regarding the lack of impact of Rous's findings when they were first reported, and indeed for the next 40 or more years afterwards as well:

> Although Rous and his associates carried out many experiments with Rous sarcoma virus, as the virus is now called, and had many prophetic insights into its behavior, they and other biologists of that time did not have the scientific concepts or the technical tools to exploit his discovery.

Howard Temin was a towering and heroic figure in the history of retrovirology and of modern molecular biology in general. Thus, we continue with this historical interlude to recount some notable aspects of his research career.

Howard Temin began working on Rous sarcoma virus in the 1950s, as a graduate student in Renato Dulbecco's laboratory at Caltech (see Box 1.11). At that time, the retrovirus field was in its infancy, and nothing whatsoever was known about the molecular events leading to cancer. Temin's first contribution to retrovirology was achieved under the supervision of Harry Rubin, who was then a postdoctoral fellow in Dulbecco's laboratory. It happened somewhat fortuitously, as follows. Temin had actually been doing his graduate research in another laboratory, looking into the embryology of the innkeeper worm, *Urechis caupo*. But he was also a laboratory assistant in the General Biology course, and as such, he was dispatched to Dulbecco's laboratory to obtain some fertilized chicken eggs. Harry Rubin supplied the chicken eggs. But he also told Temin about the chicken sarcoma viruses that were being studied in the Dulbecco laboratory. Rubin had earlier transformed a normal chicken cell with a single Rous sarcoma virus particle and then showed that the initially transformed cell produced hundreds more transformed daughter cells in a week's time. During their discussion, Rubin suggested to Temin that he (Temin) might try to develop a quantitative tissue culture assay for Rous sarcoma virus based on that observation. So Temin soon switched from embryology to virology and indeed developed an **in vitro focus-forming assay** for Rous sarcoma virus.

The assay developed by Temin is analogous in principle to a plaque assay. Importantly, it made it possible to quantify the titer of **transforming units** in a virus sample. The assay is truly quantitative, since the number of foci of morphologically transformed cells that appear in an infected cell culture is proportional to the concentration of the virus in the inoculum. In fact, Temin's assay was the first quantitative assay for viral transformation in general and the first quantitative assay for a retrovirus. Its development was an impressive achievement at the time because cell culture was still in its infancy. But more importantly, it opened up the study of retroviruses in cell culture. This, in turn, made it possible to study the molecular biology of the retroviruses, which led others to make the eventual connection between viral carcinogenesis and the genetic basis of cancer. But Temin had only just begun. His more important contributions were still to come. They happened as follows.

Rous sarcoma virus is not cytocidal. Consequently, it could stably transform chicken cells while at the same time replicating in them. From Temin's point of view in the late 1950s, these facts were consistent with the attractive but still untested premise that virus-mediated transformation results from the continuous presence and expression of virus genes in the transformed cells. But how then to account for the fact that Rous sarcoma virus could likewise stably transform rat cells, which do not support replication of the avian Rous sarcoma virus? To do so, either one must abandon the paradigm that continued viral gene expression is necessary to maintain the transformed cell phenotype or one must somehow explain how a viral RNA genome might persist in a line of cells that do not support replication of the virus.

At this early time in the history of retrovirology, it was discovered that bacteriophage λ (lambda) could stably integrate its DNA genome into the chromosome of the host bacterium, a state in which the temperate bacteriophage genome might stably persist for many cell generations (Chapter 2). Did the example of the temperate bacteriophages influence the radical hypothesis that Temin was about to put forward? That issue is discussed below. In any case, Temin proposed that the Rous sarcoma virus genome likewise persists in eukaryotic host cells in a similar integrated state. Temin realized, of course, that such an option would not be possible unless the Rous sarcoma virus RNA genome might somehow be copied into a DNA form. Accordingly, that is precisely what Temin hypothesized. In 1964, Temin's premise became known as the **provirus hypothesis**. It explicitly proposed that Rous sarcoma virus generates a DNA copy of the viral RNA genome, which is then integrated into the cellular DNA and stably maintained in that state in clones of transformed cells.

Temin's hypothesis was iconoclastic in the extreme. As a matter of fact, it was put forward at precisely the time that the "central dogma of molecular biology" was taking hold. The central dogma, which was expounded by Francis Crick in 1958, maintained that information in biological systems always "flows" from DNA to RNA and then to protein. The dogma, which became the foundation of molecular biology, had taken such a strong hold on thinking at the time that Temin's proposal was regarded as the scientific equivalent of heresy. Consequently, Temin had to fight a long, lonely battle against the criticism and ridicule that his hypothesis generated.

Yet it now seems odd that the dogma should have exercised such a strong influence, since RNA viruses already were known and clearly were an exception. Apropos that point, Temin noted the following in his 1975 Nobel lecture. "Studies with the newly discovered RNA bacteriophage and with animal RNA viruses, especially using the antibiotic actinomycin D, indicated that RNA viruses transferred their information from RNA to RNA and from RNA to protein and that DNA was not directly involved in the replication of these RNA viruses." (But note that when non-retroid RNA viruses replicate, information does not flow in a direction opposite to that of the central dogma.) Regardless, for almost a decade, Temin's provirus hypothesis was either ignored or else scoffed at.

Temin was a serious researcher, and he indeed carried out experiments to test his radical hypothesis. Moreover, his experiments generated results that were consistent with his proposal. However, his experimental results, which were limited by the technology of the day, could not meet the standard of compelling proof that was demanded to verify a hypothesis that challenged such strongly entrenched beliefs as enunciated by the central dogma.

Temin's early experimental findings included the following. First, he reported that actinomycin D, which inhibits transcription from a DNA template, impaired replication of Rous sarcoma virus in cultured chicken cells. Yet while this experimental finding is in agreement with the hypothesis that there is a DNA stage in the life cycle of the virus, the result also might be explained if viral replication were dependent on the expression of particular cellular genes. Second, Temin showed that inhibitors of DNA synthesis blocked Rous sarcoma virus infection. However, one could argue that cellular DNA synthesis rather than viral DNA synthesis is necessary for infection, perhaps as part and parcel of establishing an intracellular milieu conducive to viral replication. Third, Temin indeed carried out nucleic acid hybridization experiments to demonstrate the presence of viral DNA in Rous sarcoma virus-transformed rat cells. Once more, Temin's experimental results supported his proposal. But while critical polymerase chain reaction (PCR)-based experiments might readily be carried out today, the cutting-edge technology of the 1960s could not generate data compelling enough to have convinced Temin's critics.

In Temin's Nobel lecture, he had the following to say about how his experimental findings were received by his colleagues:

> Based on the results of these experiments, I proposed the DNA provirus hypothesis at a meeting in the Spring of 1964—the RNA of infecting Rous sarcoma virus acts as a template for the synthesis of viral DNA, the provirus, which acts as a template for the synthesis of progeny Rous sarcoma virus RNA.... At this meeting and for the next 6 years this hypothesis was essentially ignored.

The major breakthrough, which led to widespread acceptance of the provirus hypothesis, came in 1970, when Temin and David Baltimore independently, and at the same time, showed that the crucial enzyme reverse transcriptase is present in RNA tumor virus particles. In recognition of their independent discoveries, Temin and Baltimore were awarded Nobel Prizes in Physiology or Medicine in 1975. Renato Dulbecco shared in that award for his work on the DNA tumor viruses. (Dulbecco was a pioneer in the development of animal virology [Chapter 1; Box 1.11]. One of Dulbecco's findings, which is relevant in the current context, is that transformation by the polyomaviruses is associated with the stable integration of their DNA genomes into the host cell genome [Chapter 15].)

Temin and Baltimore had somewhat different reasons for carrying out their breakthrough experiments. Temin, whose entire research career was dedicated to studying RNA tumor viruses, primarily yearned to validate his provirus hypothesis. In 1969, in a study that never was reported in its entirety, Temin and coworker Satoshi Mizutani did experiments demonstrating that the alleged Rous sarcoma virus DNA synthesis could occur in the absence of any de novo protein synthesis. This experimental finding implied (to believers at least) that the DNA polymerase activity that catalyzed the synthesis of the viral DNA was present before infection. Thus, Temin and Mizutani sought the putative polymerase in the virus particle.

Baltimore had earlier been studying other virus-specific enzymes, ones that transcribe/copy RNA from RNA. With his wife, Alice Huang, and other colleagues, he wanted to explain why the single-stranded RNA genome of vesicular stomatitis virus is not infectious while the single-stranded RNA genome of poliovirus is infectious. The lack of infectivity of the vesicular stomatitis virus genomic RNA led Baltimore, Huang, and coworkers to hypothesize that this virus had an obligate requirement for a particle-associated polymerase. Several of their earlier findings also pointed towards this hypothesis. First, the poliovirus RNA genome

functioned as an mRNA after it was released from the virus particle into the cytoplasm. In contrast, they found that vesicular stomatitis virus genomic RNA is complementary to the mRNAs that encode that virus's proteins. Translation of the poliovirus parental genomic RNA generates the viral transcriptase necessary to initiate viral transcription/replication. Thus, to understand how the initial stages of a vesicular stomatitis virus infection might transpire, they looked for, and found, an RNA-dependent transcriptase activity within vesicular stomatitis virus particles. (Note that vaccinia virus [a DNA virus] had already provided a precedent for a virus particle-associated RNA polymerase [Chapter 19].) From this sort of perspective, Baltimore sought after a reverse transcriptase activity in retrovirus particles.

During his Nobel lecture, Baltimore reflected on how his earlier experiences might have led him to his key retrovirus discovery:

> If we look back to virology books of 15 years ago, we find no appreciation yet for the variety of viral genetic systems used by RNA viruses. Since then, the various systems have come into focus, the last to be recognized being that of the retroviruses ("RNA tumor viruses"). As each new genetic system was discovered, it was often the identification of an RNA or a DNA polymerase that could be responsible for the synthesis of virus-specific nucleic acids that gave the most convincing evidence for the existence of the new system....

Baltimore spoke after Temin at the Nobel ceremony. The following item from Baltimore's lecture is cited here for his comments about his co-award recipient:

> In his Nobel lecture, Howard Temin has outlined how he was led to postulate a DNA intermediate in the growth of RNA tumor viruses. Although his logic was persuasive and seems in retrospect to have been flawless, in 1970 there were few advocates and many skeptics. Luckily, I had no experience in the field and so no axe to grind—I also had enormous respect for Howard dating back to my high school days when he had been the guru of the Summer School I attended at the Jackson Laboratory in Maine. So I decided to hedge my bets—I would look for either an RNA or a DNA polymerase in virions of RNA tumor viruses....

Temin died at the age of 59 from lung cancer. He was never a smoker, and his lung cancer was an adenocarcinoma, a type not linked to smoking. In fact, Temin was a zealous crusader against smoking. Even at the Nobel Prize ceremony in his honor, he reprimanded smokers in the audience.

After Temin's death, Baltimore wrote the following:

> Ten years in the scientific wilderness is a long time; few have had to bear the silence of their colleagues for so long. I can remember meetings in the 1960s when Howard would present his latest data supporting the provirus

notion only to be greeted by either skeptical questions or quiet, polite disbelief. Howard's conviction that there had to be a provirus never seemed to waver over the whole decade. He knew he was right—and he was—but what fortitude it took to keep looking for the experiment that would show it! My first reaction when I realized that I had seen the reverse transcriptase was to call Howard, because I so much wanted him to know that he was vindicated in his commitment to the idea of a provirus. But he had already found out for himself.

Howard Temin and David Baltimore published their independently discovered, singularly important finding in back-to-back papers in the British journal *Nature* (Baltimore, 1970; Temin and Mizutani, 1970).

In 1972, Hill and Hillova provided further proof that RNA tumor virus genomes persist in transformed cells in the form of a DNA copy. They did so by demonstrating that purified DNA from a Rous sarcoma virus-transformed cell could produce infectious virus when transfected into a normal cell.

Indisputable evidence of reverse transcription led to the RNA tumor viruses becoming known as the retroviruses. More importantly, it led to the realization that the central dogma need not always be valid. Other viruses, such as the hepadnaviruses (Chapter 22), were later found to also make use of reverse transcription. Indeed, the phenomenon of reverse transcription is more general still, applying also to cellular genetic constituents known as **retroelements** (see below), which may comprise 10% or more of the mammalian genome. Some of these retroelements may be replication-defective retroviruses. Alternatively, some may be intermediate stages in the evolutionary process that gave rise to the retroviruses, as proposed by Temin in his **protovirus theory**. (The protovirus theory hypothesizes that a primordial, cellular, reverse transcriptase-like enzyme copied cellular RNA molecules into DNA molecules that inserted into the cellular DNA. These integrated protovirus sequences were theorized to eventually acquire other features that enabled them to become retroelements and then viruses.) Reverse transcription is also a key feature of the cellular process whereby the host enzyme **telomerase** maintains the ends of chromosomes (Chapter 14). Additionally, the reverse transcriptase enzyme is now a critical tool of modern molecular biology, enabling researchers to make a DNA copy of any mRNA molecule, which then may be cloned and studied under ideal experimental conditions. Finally, knowledge of the reverse transcription step in the retrovirus life cycle led to a more complete understanding of retroviral persistent infections, particularly as regards HIV infection and the pathogenesis of AIDS (Chapter 21).

Before moving on, we return to an issue raised above: did the discovery of bacteriophage lysogeny prompt Temin to put forward the provirus hypothesis? Here is Baltimore's take on this question:

> Although the pregnant analogy to known lysogenic bacteriophage might have guided Howard, people who were at Caltech at that time assure me that Howard was unlikely to have arrived at the notion of a DNA intermediate through this route. Apparently, the influence of Max Delbrück—who was totally committed to the study of lytic phages and did not really believe in the importance of phage lysogeny—was so great that there was little discussion of lysogeny at Caltech then. Furthermore, Howard has minimized the importance of lysogeny as a precursor to his concepts. Therefore, he must have arrived at the concept of a DNA intermediate simply from the persuasive power of such a concept to explain the properties of the transformed state. He was particularly influenced by the morphological difference between cells transformed by particular Rous sarcoma virus variants, which he felt had to mean that the viral genome continued forever to affect the transformed cell.

Temin, in his Nobel lecture, indeed cites the mutant studies to which Baltimore refers as a key factor in the genesis of the provirus hypothesis.

STRUCTURE

Rous sarcoma virus, the prototypical simple retrovirus, is enveloped and spherical (Fig. 20.2). Individual Rous sarcoma particles can vary somewhat in size, but they have a mean diameter of 127 nm. The virus envelope contains two proteins: the transmembrane (TM) and surface (SU) glycoproteins, which extend out from the envelope. TM and SU are generated by proteolytic cleavage of a precursor protein encoded by the *env* gene (Fig. 20.1). After the precursor protein is cleaved, SU yet remains covalently linked to TM in the particle via disulfide bonds.

The viral core is comprised of the matrix (MA) protein, the capsid (CA) protein, and the nucleoprotein (NC). As in the case of the TM and SU proteins, which are cleavage products of the polyprotein encoded by the *env* gene, the MA, CA, and NC proteins are cleavage products of the polyprotein encoded by the *gag* gene. (Interestingly, retrovirus capsids are actually assembled from molecules of the Gag precursor polyprotein. The Gag polyproteins do not undergo cleavage until the time of final virus maturation. This is an important feature of retrovirus assembly that is discussed in greater detail below. Also, there are two NC proteins among the Rous sarcoma virus cleavage products: one of 19 kilodaltons [kDa] and the other of 12 kDa. See below.) The "Gag" designation stands for group-specific antigen. This name derives from the early finding that retrovirus capsid proteins induce antibodies that are cross-reactive between members of a retrovirus

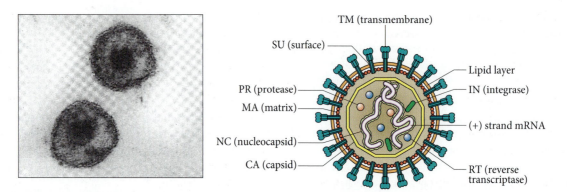

Figure 20.2 (Left) Electron micrograph of negatively stained Rous sarcoma virus particles. The method of preparation for electron microscopy (staining of thin sections) did not preserve the envelope glycoproteins for visualization. (Right) Diagram of an ALV particle, depicting the locations of the component proteins, the diploid RNA genome, and the envelope. From S. J. Flint, L. W. Enquist, R. M. Krug, V. R. Racaniello, and A. M. Skalka, *Principles of Virology: Molecular Biology, Pathogenesis, and Control*, 2nd ed. (ASM Press, Washington, DC, 2004), with permission.

group (e.g., the avian retroviruses and the murine retroviruses). The *gag* gene of other retroviruses may encode yet additional internal proteins, of unknown function in some instances.

Retroviral cores also contain 50 to 100 molecules of the virus-encoded protease (Pro) and reverse transcriptase (Pol). Pol is comprised of the polymerase itself, plus ribonuclease H (**RNase H**), and the **integrase** (IN). RNase H is a special RNase that degrades the RNA component of RNA/DNA heteroduplexes. It plays a crucial role in converting (an accurate word choice; see below) the RNA genome into a DNA copy. The IN is the viral enzyme that inserts the reverse-transcribed DNA copy of the retroviral RNA genome into the cellular chromosomal DNA. Three forms of the reverse transcriptase are found in Rous sarcoma virus-infected cells: the 63-kDa α protein, which contains the polymerase and RNase H domains; the 95-kDa β protein, which contains the polymerase, RNase, and IN domains; and αβ heterodimers, which are the most abundant form of the reverse transcriptase.

The MA protein is located between the envelope and the capsid. The latter is assembled from the CA protein. The remaining internal proteins are found within the capsid. The NC proteins are bound to the genomic RNA; there is about one copy of the NC protein for every 10 nucleotides. In this regard, the retroviral plus-strand RNA genomes resemble the nucleoprotein complexes characteristic of the minus-strand RNA viruses, rather than the "naked" RNA genomes of more conventional plus-strand RNA viruses. (Single-stranded DNA-binding proteins typically play a role in DNA replication. The single-stranded RNA-binding NC protein is thought to participate in

reverse transcription.) Pol likewise is associated with the RNA genome. The retroviral core can appear to be isometric with distinct facets, while some cores appear to have a smooth, continuous surface. Cores also vary in size. Larger enveloped particles generally have larger cores.

Retrovirus particles contain two identical copies of their positive-strand RNA genome within each core. That is, they are *biochemically* diploid. Each of the identical strands of the Rous sarcoma virus genome contains about 7.4 kilobases (kb). The individual RNA strands are hydrogen bonded to each other at multiple points along their lengths, while the most stable interactions occur between short sequences near their 5′ end. The reason for the diploid genome is not clear, since one copy is sufficient for the entire process of reverse transcription.

In addition to containing the diploid genome and enzymes that catalyze reverse transcription, retrovirus cores also enclose an assortment of about 100 tRNA molecules. Out of this random mixture of tRNA molecules, one particular one (tryptophan tRNA in the case of Rous sarcoma virus) is base paired to each copy of the genome, in position to serve as a primer for reverse transcription. Other small cellular RNA molecules are also present within cores, including 5S ribosomal RNA (rRNA), 7S RNA, and other RNAs that may be breakdown products of cellular mRNA molecules. Interestingly, the mechanism by which these cellular RNAs are incorporated into retroviral particles is not known. Moreover, the exact amount of cellular RNA in retrovirus particles is not known, but it is thought to not exceed 40% of the total packaged RNA. Taking that upper estimate for the amount of cellular RNA in the cores and taking into account the volume of the proteins in the core, it appears that the remaining space still far exceeds that needed to enclose the diploid viral genome.

ENTRY

The simple retroviruses were once thought to use a single cell surface receptor to mediate both binding and entry. However, newer experimental findings show that feline leukemia virus and several other simple retroviruses may require two distinct cell surface proteins to initiate infection: one to mediate binding and another to trigger fusion leading to entry. (HIV and related complex retroviruses also use two host proteins for these separate purposes [Chapter 21].)

Those retrovirus receptors which mediate both binding and entry are referred to as "classical" receptors. For many of the retroviruses, these classical receptors are so-called multipass transmembrane proteins (i.e., proteins whose polypeptide chains traverse the lipid bilayer multiple times) that are involved in transport of small molecules (e.g., amino acids or ions).

All retroviral Env proteins contain an N-terminal SU subunit, which is involved in receptor binding, and a C-terminal TM subunit, which anchors the protein to the viral envelope. The TM subunit also carries out membrane fusion, by means of its N-terminal hydrophobic fusion peptide. The overall structure of a retrovirus Env protein is reminiscent of the influenza virus hemagglutinin protein (see Chapter 12; Box 12.4).

Like the attachment and fusion proteins of other enveloped viruses, the retroviral SU and TM subunits are in a metastable conformation before binding to their receptor. This metastable state is important because it enables these proteins to go through the conformational transitions that promote fusion and entry. The metastable condition of the retroviral SU and TM (like that of the analogous proteins of most other enveloped viruses) results from their being generated by the cleavage of a precursor polyprotein. Whereas the precursor polyprotein may very well fold to its most stable conformation, the individual cleavage products will be in a metastable conformation following cleavage of the precursor.

The conformational changes that the retroviral SU and TM subunits undergo to mediate fusion are, in some instances, triggered by a unique sequence of steps. Recall that viral fusion proteins generally undergo entry-promoting conformational rearrangements that are triggered either by pH-independent interactions with the receptor at the cell surface (e.g., as in the case of the paramyxoviruses) or by the low-pH milieu of an intracellular endosomal compartment (e.g., as in the case of influenza virus). In contrast, the mechanism that activates Rous sarcoma virus

fusion proteins and those of its close relatives in the avian sarcoma-leukosis virus (ASLV) group occurs via a novel multistep process that incorporates features of both the pH-independent and pH-dependent virus entry pathways. This happens as follows.

The pH-independent steps occur when Env binds to the receptor. This interaction triggers a conformational change in the SU subunit, in which its N and C termini interact, in that way causing SU to dissociate from TM, thereby freeing the latter subunit to rearrange into an extended conformation (Fig. 20.3). As a consequence of these conformational transitions, the TM fusion peptide is exposed in the vicinity of the cell membrane, into which it then inserts. Notice that the TM protein has heptad repeat sequences near its N-terminal end and also at the envelope-proximal region and that these repeat sequences are exposed in the receptor-primed, extended conformation (Fig. 20.3).

Rous sarcoma virus and retroviruses of the ASLV group are next taken up by endocytosis, a process which transports them to an endosome. The low pH of the endosome then causes TM to undergo a second conformational transition, in which the heptad repeats "zipper back," folding into six-helical bundles. As in the case of other viral fusion proteins (for example, the influenza virus hemagglutinin [Chapter 12]), this conformational transition brings the fusion peptide, which now is inserted into the cellular target membrane, into close apposition with the transmembrane domain that traverses the viral envelope. Thus, the viral envelope is brought near the cellular target membrane, while both membranes are stressed, thereby facilitating fusion (which is thought to occur via a **hemifusion intermediate** step [Fig. 20.3; Chapter 12]). (The entry process of the avian retroviruses described above is not shared by all of the simple retroviruses. Some, such as the murine retroviruses, fuse with the plasma membrane at neutral pH.)

The intermediate, extended conformation of TM is a key feature of the above retroviral entry pathway, since it prepares TM to undergo the subsequent low-pH-dependent conformational transition, in which the exposed heptad repeats interact to promote fusion. But what sort of experimental evidence might support this sequence of events? The following is one experimental approach that has been applied to other viruses, in addition to the retroviruses. Soluble peptides that contain the heptad repeat were found to bind to TM when they were added while the virus was still at the cell surface. Moreover, when the heptad repeat peptides were added before low-pH treatment, they prevented low-pH-induced fusion and infection. The most likely explanation for the effect of the heptad repeat peptide is that it sterically hinders the low-pH-induced

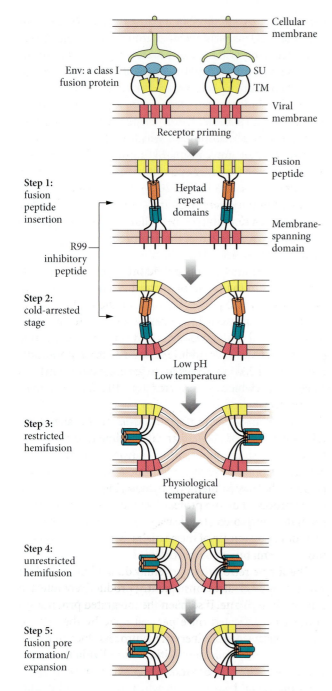

Env: a class I fusion protein

SU

TM

Cellular membrane

Viral membrane

Receptor priming

Step 1:
fusion peptide insertion

Fusion peptide

Heptad repeat domains

Membrane-spanning domain

R99 inhibitory peptide

Step 2:
cold-arrested stage

Low pH
Low temperature

Step 3:
restricted hemifusion

Physiological temperature

Step 4:
unrestricted hemifusion

Step 5:
fusion pore formation/ expansion

Figure 20.3 ASLV Env-membrane fusion driven by receptor and low pH. Following receptor interaction at the cell surface, the fusion peptide of ASLV Env is exposed and inserted into the target membrane (Step 1). The receptor-primed form of ASLV Env then forms a stable pre-six-helical bundle intermediate (Steps 1 and 2). In this state, ASLV Env can be triggered by low pH at nonphysiological temperatures (Step 2). Upon addition of low pH, ASLV Env folds into its tight six-helical bundle, promoting first a restricted hemifusion intermediate (Step 3), an unrestricted hemifusion intermediate (Step 4), and finally formation and expansion of the fusion pore (Step 5) that allow delivery of the viral core into the cytoplasm. (TM, in the exposed, extended conformation, can bind the heptad repeat peptides that inhibit fusion and entry; see the text.) From R. J. O. Barnard, D. Elleder, and J. A. T. Young, *Virology* **344**:25–29, 2006, with permission.

refolding of TM and the interactions between the heptad repeats that generate the six-helix bundle, thereby blocking fusion and infection. (Similar experimental evidence shows that the fusion proteins of other viruses [e.g., influenza virus and Ebola virus] likewise transiently exist in a comparable intermediate, extended conformation prior to fusion. Thus, peptides analogous to the retroviral heptad repeat peptide may have potential as antiviral therapeutic agents.)

THE REPLICATION CYCLE: A BRIEF OVERVIEW

In preceding chapters, we generally considered the organization and expression of viral genomes before considering their replication. However, we break that mold here because in the case of the retroviruses, gene expression and genomic replication each depend on the prior process of reverse transcription. Yet this deviation from our usual format is really more apparent than real because reverse transcription per se does not actually replicate the viral genome. Instead, the viral RNA genome is degraded during reverse transcription, so that the net result is the replacement of the RNA genome with a DNA copy of it. But while *reverse transcription is not replicative*, it is crucial to replication since the DNA copy of the RNA genome serves as the template for productive genome replication. In fact, genome replication and expression each occur by virtually the same process, i.e., transcription of the integrated provirus by the cellular Pol II RNA polymerase. Because of the uniqueness of this aspect of the retroviral replication cycle, a brief overview of that cycle is offered here, in order to provide a perspective for the more detailed description of the organization, expression, and replication of retroviral genomes in the sections that follow.

The entry process described in the preceding section results in the release of the retroviral core into the cytoplasm, where the viral RNA genome is then reverse transcribed by the core-associated reverse transcriptase. That is, reverse transcription occurs within the viral cores. The product of reverse transcription is a double-stranded DNA copy of the viral genome, which is then transported to the nucleus and integrated into the cellular chromosomal DNA. In at least some instances, migration of the viral DNA into the nucleus requires that the cell be dividing, so that the nuclear membrane might break down, providing access to the nuclear interior.

What happens next is no less remarkable than the preceding reverse transcription and integration steps. The integrated viral DNA is transcribed by the cellular Pol II RNA polymerase to generate *all* viral mRNAs and *all* progeny viral genomes. Note that the full-length viral RNA molecules generated in this way are identical to progeny genomic RNAs. But unlike genomic RNA molecules, they are active as mRNA molecules unless incorporated into particles as progeny genomes (Box 20.1). A *singularly distinctive* upshot of the unique means by which retroviruses generate progeny viral genomes (i.e., via transcription of a single integrated provirus) is that retroviral replication *does not have the characteristic exponential growth phase* that is typical of almost all other viruses.

The fact that the cellular Pol II RNA polymerase catalyzes all retrovirus transcription and replication, using the integrated provirus as the template, creates a special dilemma for the virus, which follows from the following two facts. First, the Pol II RNA polymerase initiates transcription at particular promoters. Second, the Pol II promoters themselves are not transcribed by the Pol II RNA polymerase. Thus, if the promoter used by the Pol II RNA polymerase to transcribe the proviral DNA was present in the viral RNA genome, then that promoter could not be replicated and perpetuated as a part of the genome. Thus, the promoter used to initiate transcription of progeny RNA genomes must not lie within the viral genome. But what then serves as the promoter for the transcription and replication of the retroviral provirus DNA? The answer is that the promoter is generated during the remarkable reverse transcription process (see below), such that it is present on the proviral DNA but outside the coding sequence for the viral genome. Remarkably, however, the promoter is generated from information contained entirely within the genomic RNA. (Can you envision other ways by which retroviruses might have solved the dilemma posed by the need to have a promoter outside the genomic RNA? See below.)

Transcription of the integrated provirus generates full-length viral RNAs, some of which become encapsidated progeny genomes. However, some of these plus-strand, full-length RNA molecules serve as mRNAs. In that role, they are translated into the Gag and Gag-Pol polypeptide precursor proteins. In addition, some of these RNA molecules are spliced in the nucleus to generate shorter mRNA molecules, which are translated into the Env precursor polyprotein.

Env is incorporated into the plasma membrane of the cell, while Gag and Gag-Pol are assembling into capsids in the cytoplasm. Immature virus particles then are assembled at the plasma membrane and are released by budding through the plasma membrane, acquiring their envelopes in the process. Pol, the protease, and the Gag proteins are generated by proteolytic cleavage of the precursor polypeptides upon release of the particles from the cell, thereby producing mature virus particles.

The simple retroviruses generally do not kill the cells in which they replicate. Consequently, productively infected cells might replicate. If so, then the integrated proviral genomes are replicated once per cell cycle by the cellular DNA replication machinery. This process does not contribute to virus replication within an individual infected cell per se. Rather, it expands the population of productively infected cells.

BOX 20.1

The purified RNA genomes of plus-strand RNA viruses are generally infectious when transfected into permissive cells. That is so because those plus-strand genomes function as mRNAs, which are translated to generate the crucial RNA transcriptase/replicase proteins needed to transcribe and replicate the viral genomes.

Like the RNA genomes of other plus-strand viruses, retrovirus plus-strand genomes similarly have the typical features of a eukaryotic mRNA molecule, including a methylated 5′-terminal cap and a 3′-terminal poly(A) tail. Then, why do retroviral plus-sense RNA genomes *not function as mRNAs*, and why are they *not infectious*? The following two points are pertinent to the first part of this question. First, the original RNA genome is degraded during reverse transcription. Second, the seclusion of the retroviral genome within the core, prior to the reverse transcription event that destroys it, precludes its availability to ribosomes.

You would be correct if you argued that the preceding two points do not necessarily rule out the purified retroviral genome being infectious. But even if the retrovirus RNA genome were not secluded prior to being degraded, it still would not be infectious because it does not encode an RNA polymerase that might replicate and transcribe it. Reverse transcription, which is a necessary antecedent of transcription, requires the particle-associated reverse transcriptase, as well as other core proteins.

GENOME ORGANIZATION

The retrovirus plus-strand RNA genome has the typical features of a eukaryotic mRNA molecule, including a methylated 5′-terminal cap and a 3′-terminal poly(A) tail. The sequence order on the genomic RNA is 5′-R, U5, *gag*, *pol*, *env*, U3, R-3′. R represents a short (16- to 65-nucleotide) sequence present at each end of the genome, whereas U5 and U3 represent unique sequences at the 5′ and 3′ ends, respectively (Fig. 20.4).

Importantly, during the process of reverse transcription (see below), the **long terminal repeat** (LTR) sequence, comprised of U3, R, and U5, is generated at each end of the proviral DNA (Fig. 20.4). This creates the viral promoter outside the actual sequence that encodes the viral genome. Thus, when the DNA provirus is integrated into the cellular DNA and transcribed by the cellular RNA Pol II enzyme, full-length copies of the viral genome can be transcribed (i.e., the promoter itself is not transcribed; see above). The LTRs also contain the viral enhancers, which ensure high levels of transcription from the integrated provirus promoter. In addition, they contain transcription termination/polyadenylation signals. (In answer

to a question posed above, the virus might have solved its promoter needs by means of a site-specific integration mechanism that positioned the provirus next to a suitable cellular promoter. However, generating the LTRs allows transcription of the provirus at all integration sites. If the virus did not generate the LTRs but instead integrated at a cellular promoter, what other issues might it have to deal with?)

The genomes of all replication-competent retroviruses contain the *gag*, *pol*, and *env* genes. Oncogenic retroviruses may contain an additional gene, the oncogene (*onc*), which is responsible for transformation. Note that the oncogene is not required for replication. Moreover, in most instances, acquisition of an oncogene results in partial deletion of a structural gene. In those instances, the oncogenic retroviruses are defective and require a helper virus in order to replicate. In some well-known examples, portions of the envelope gene are missing from the replication-defective, oncogenic retrovirus, which can yet be propagated in the presence of a helper virus that provides the envelope protein.

REVERSE TRANSCRIPTION

Reverse transcription of retroviral genomic RNA occurs upon release of the virus core into the cytoplasm. The *entire* process of reverse transcription takes place *within the viral core*, where it is catalyzed by the particle-associated reverse transcriptase.

The sequence of steps constituting the current model for the reverse transcription of the retroviral RNA genome was pieced together by characterizing the intermediates generated in infected cells and also from the intermediates generated by isolated virus particles in cell-free systems. Reverse transcription in cell-free systems, as well as in vivo, occurs within viral cores. Moreover, identical reverse transcription intermediates are generated under both conditions. In either case, the process requires only (i) the integrity of the core structure; (ii) the viral genome; (iii) the reverse transcriptase, which actually expresses several activities (see below); and (iv) precursor deoxyribonucleoside triphosphate molecules. In cell-free systems, the process begins virtually immediately upon the addition of the precursor deoxyribonucleoside triphosphates to the viral cores, implying that retrovirus cores in vivo are poised to begin reverse transcription upon being released into the cytosol.

The reverse transcription process requires four distinct enzymatic activities: an RNA-dependent DNA polymerase, a DNA-dependent DNA polymerase, a DNA helicase, and an RNase H. Remarkably, each of these four activities is expressed by the *pol* gene product. Moreover, the first

Figure 20.4 A comparison of the simple retrovirus plus-strand RNA genome with the DNA provirus. The retrovirus RNA genome has the typical features of a eukaryotic mRNA molecule, including a methylated 5′-terminal cap and a 3′-terminal poly(A) tail. The sequence order on the genomic RNA is 5′-R, U5, *gag*, *pol*, *env*, U3, R-3′. The location of the oncogene (*onc*), when present, is indicated. R represents a short (16- to 65-nucleotide) sequence present at each end of the genome, whereas U5 and U3 represent unique sequences at the 5′ and 3′ ends, respectively. During the process of reverse transcription, the LTR sequence, comprised of U3, R, and U5, is generated at each end of the provirus, thereby creating the viral promoter outside the sequence that encodes the viral genome. The LTRs also contain the viral enhancers, which ensure high levels of transcription from the integrated provirus promoter. In addition, they contain transcription termination/poly-adenylation signals. The linear provirus is converted to a noncovalently closed circular intermediate before being integrated into the cellular genome. Adapted by J. Leong from R. Weiss, N. Teich, H. Varmus, and J. Coffin, *RNA Tumor Viruses*. Cold Spring Harbor Press, Cold Spring Harbor, NY, 1982, with permission.

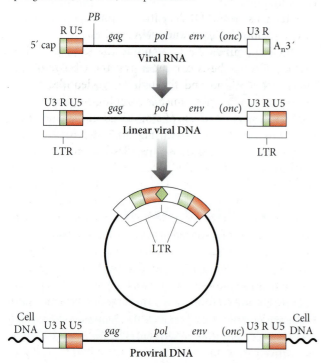

three of these activities are expressed by a single domain of the reverse transcriptase protein, with RNase H occupying a separate domain. The reverse transcriptase molecule contains sequence motifs that are present in the catalytic domains of other RNA and DNA polymerases, implying that it catalyzes nucleic acid polymerization reactions by molecular mechanisms similar to those used by other nucleic acid polymerases.

DNA synthesis in general needs to be primed, and that which occurs during reverse transcription is not an exception. But the priming that occurs during retroviral reverse transcription is still unique, since retroviral reverse transcriptases use particular tRNA molecules for their primers. For example, the Rous sarcoma virus reverse transcriptase uses tryptophan tRNA for its primer. The tRNA is partially unwound and hydrogen bonded to the RNA template via 18 or more of its 3′-terminal nucleotides, at a site on the viral genome referred to as the **primer-binding site** (pbs), which is located near the 5′ end of the genome (Fig. 20.5, step 1.) (What do you make of the location of the pbs? How far can transcription proceed before reaching the end of the template? Before reading on, what do you suppose might happen next?)

Bear in mind that the requirement for a tRNA primer and the location of the primer-binding site dictate that the genome cannot simply be reverse transcribed from one end to the other. Thus, as you follow the reverse transcription process, note the means by which the ends of the provirus DNA are generated. Moreover, watch closely how the LTRs are generated. This feature of the reverse transcription process is important, since it creates a promoter that is outside the coding sequence for the genome itself, so that complete progeny genomes and full-length viral mRNAs might be transcribed. Also, observe how the LTRs are created from sequences encoded entirely by the viral genome; all in all, this represents a rather remarkable adaptation at the molecular level.

Priming near the 5′ end of the plus-strand RNA genome leads to the synthesis of a short (approximately 100-nucleotide-long) segment of minus-strand DNA, before the reverse transcriptase reaches the end of the template (Fig. 20.5, steps 1 and 2). When the reverse transcriptase reaches the end of the template, it faces a bottleneck, so to speak, and the resulting intermediate thus tends to accumulate during reverse transcription (in cell-free systems, that is; see below). For this reason, this intermediate is referred to as **strong stop DNA**. Now, recall the RNase H activity noted above. As reverse transcription proceeds along the genomic RNA template, the RNase H activity hydrolyzes the template about 18 nucleotides behind the DNA synthesis.

The next step is for the reverse transcriptase to engage the 3′ end of the same template strand (Box 20.2). The reverse transcriptase is often said to "jump" from one template sequence to another. This terminology is somewhat misleading, since enzymes (including reverse transcriptase) do not have such Olympian abilities. Instead, it is likely that the reverse transcriptase acts to bring both ends of the genome into proximity to ease this template exchange. The RNase H also eases the so-called jump by digesting the copied end of the template, thereby exposing the R sequence on the strong stop DNA, so that it can base pair with the complementary R sequence at the 3′ end of the genome. Interestingly, the strong stop DNA does not accumulate in infected cells, implying that the jump, or template exchange, is an efficient process in vivo.

Once this first jump has taken place, reverse transcription resumes and the rest of the RNA genome is reverse transcribed. Importantly, notice how this first jump and the ensuing reverse transcription have generated an LTR.

As the reverse transcriptase generates the minus-strand DNA in the preceding step, the associated RNase H activity hydrolyzes the remainder of the genomic RNA template. However, a short (11-nucleotide-long) RNase H-resistant polypurine tract (ppt) near the 3′ end of the genome is spared (Fig. 20.5, step 4). This sparing is perhaps due to incomplete hydrogen bonding within the sequence. Regardless, sparing of ppt is important, since this short RNA sequence serves to prime the synthesis of the plus-strand DNA, as catalyzed by the DNA-dependent DNA polymerase activity of the reverse transcriptase protein (Fig. 20.5, step 5).

After plus-strand DNA synthesis has been primed, as described above, plus-strand DNA synthesis proceeds, using the new minus DNA strand as the template. Plus-strand DNA synthesis continues past the U5 site on the minus DNA strand and through 18 nucleotides of the tRNA primer sequence, which is complementary to pbs. It finally terminates because of a modified nucleotide in the tRNA that cannot be copied (Fig. 20.5, step 6). The pbs-complementary sequence on the tRNA primer is hydrolyzed by RNase H, thereby releasing the remainder of the tRNA. In this way, a double-stranded LTR sequence is generated at one end of the proviral DNA, flanked by a sequence on the positive strand that is complementary to pbs. (In the case of some retroviruses [e.g., HIV and ASLVs], but not others [e.g., MLV], additional internal RNA sequences, which are partially spared from RNase H digestion, also may prime positive-strand DNA synthesis [step 7].)

At this point in the reverse transcription process, there are complementary single-stranded pbs sequences at each end of the replicative intermediate (Fig. 20.5, step 8). These sequences may hydrogen bond to each other, resulting in

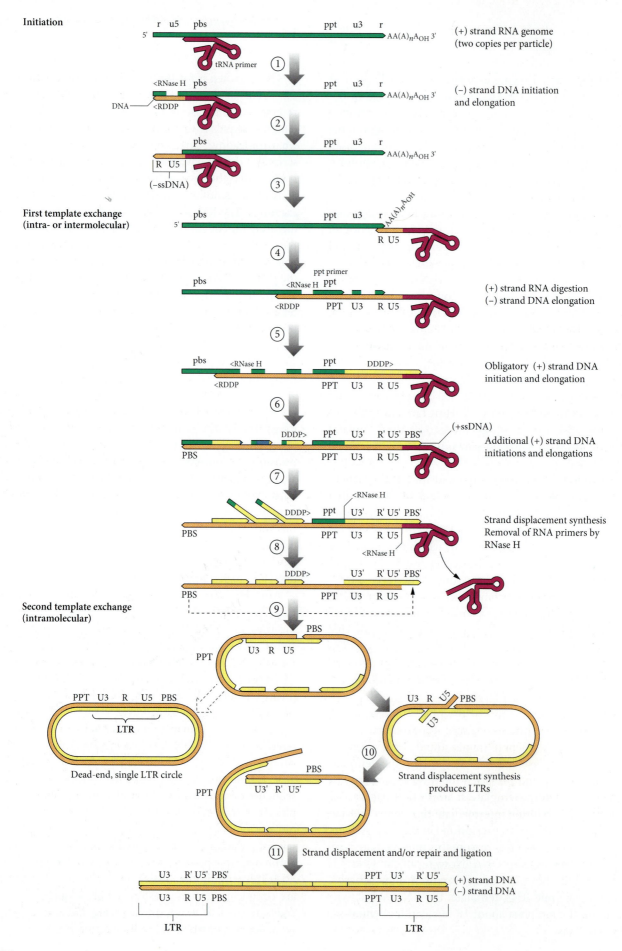

Initiation

r u5 pbs   ppt u3 r
5'   AA(A)$_n$A$_{OH}$ 3'   (+) strand RNA genome
tRNA primer   (two copies per particle)

①

<RNase H pbs   ppt u3 r
                                          AA(A)$_n$A$_{OH}$ 3'   (−) strand DNA initiation
DNA — <RDDP   and elongation

②

pbs   ppt u3 r
                                          AA(A)$_n$A$_{OH}$ 3'
R U5
(−ssDNA)

③

First template exchange
(intra- or intermolecular)

pbs   ppt u3 r   AA(A)$_n$A$_{OH}$
5'
                                          R U5

④

ppt primer
pbs   <RNase H ppt   (+) strand RNA digestion
                                          (−) strand DNA elongation
                 <RDDP   PPT U3 R U5

⑤

pbs   <RNase H   ppt   DDDP> Obligatory (+) strand DNA
                                          initiation and elongation
         <RDDP   PPT U3 R U5

⑥

                 DDDP> PPT U3' R' U5' PBS' (+ssDNA)
PBS   PPT U3 R U5   Additional (+) strand DNA
                                            initiations and elongations

⑦

                           <RNase H
                 DDDP> PPT U3' R' U5' PBS'   Strand displacement synthesis
PBS   PPT U3 R U5   Removal of RNA primers by
                           <RNase H   RNase H

⑧

                 DDDP>   U3' R' U5' PBS'
PBS   PPT U3 R U5

Second template exchange
(intramolecular)

⑨

                    PBS
              U3 R U5
PPT

PPT U3 R U5 PBS   U3 R U5 PBS
                                        U3
         LTR   Strand displacement synthesis
                              ⑩   produces LTRs
Dead-end, single LTR circle
                         PBS
              PPT   U3' R' U5'

⑪ Strand displacement and/or repair and ligation

U3 R' U5' PBS'   PPT U3' R' U5'   (+) strand DNA
U3 R U5 PBS   PPT U3 R U5   (−) strand DNA

LTR   LTR

BOX 20.2

The polymerase might also engage the 3′ end of the second copy of the RNA genome present in the core. The necessary transfer of the reverse transcription process to the second copy of the RNA genome would provide a rationale for retroviruses having diploid genomes. However, Temin showed that the whole reverse transcription process can be completed on a single template strand. Indeed, the function of the second genome remains unclear.

a second template exchange event and a circular intermediate structure (step 9). Cellular enzymes may repair this intermediate structure to generate a closed circle, which contains only one LTR; this represents a dead end for the virus (step 9, left branch).

But an alternative path leads to a biologically active provirus, which contains the LTR sequence at each end. This outcome, which is mediated by the helicase and DNA-dependent DNA polymerase activities of the reverse transcriptase, occurs as follows. When the complementary single-stranded pbs sequences at each end of the replicative intermediate hydrogen bond to each other (Fig. 20.5, step 9), displacement synthesis ensues at the 3′ end of each DNA strand of the noncovalently closed circle; this

outcome generates a linear, double-stranded proviral DNA molecule, with an LTR at each end (Fig. 20.5, step 9, right branch; step 10).

Once again, note that the reverse transcription process outlined above does not replicate the viral genome. Instead, it replaces the parental RNA genome with a single linear, double-stranded DNA copy, which contains the LTR sequence at each end. This molecule enters the nucleus, as part of a preintegration complex that also contains the reverse transcriptase and the IN, NC, and CA proteins. The IN protein then catalyzes the insertion event that establishes the LTR-containing DNA genome as the integrated provirus.

Before moving on to the integration process per se, two additional issues need to be addressed. The first is the means by which retroviral preintegration complexes enter the nucleus. Nuclear entry of these complexes is not well understood, since they are too large to pass through nuclear pores. In some instances, such as the MLV, nuclear entry seems to depend on the nuclear membrane breaking down at mitosis. Consistent with that notion, MLV is able to productively infect only replicating cells. In addition, pharmacologic inhibitors of mitosis delay MLV nuclear entry and integration. Also, the MLV provirus segregates into only one daughter cell at the first cell division after infection, suggesting that integration occurs only after replication of the cellular DNA. Some retrovirus researchers believe that simple retroviral preintegration complexes in general cannot enter an intact nucleus. Yet preintegration

Figure 20.5 Current model of the reverse transcription process. Retroviral reverse transcription is primed by particular tRNA molecules, such as tryptophan tRNA in the case of Rous sarcoma virus. The tRNA is partially unwound and hydrogen bonded to the RNA template at a site on the viral genome referred to as the pbs, located near the 5′ end of the genome (step 1). Priming near the 5′ end of the genome leads to the synthesis of a short (approximately 100-nucleotide-long) segment of minus-strand DNA, before the reverse transcriptase reaches the end of the template (steps 1 and 2). As reverse transcription proceeds down the genomic RNA template, the RNase H activity hydrolyzes the template about 18 nucleotides behind polynucleotide synthesis. The reverse transcriptase then engages the 3′ end of the same template strand (step 3). (It also might engage the 3′ end of the second copy of the RNA genome, although the whole reverse transcription process can be completed on a single template strand.) The RNase H eases the jump of the reverse transcriptase from one end of the template strand to the other by digesting the copied end of the template, thereby exposing the R sequence there, so that it can base pair with the complementary R sequence at the 3′ end of the genome. Reverse transcription resumes, and the rest of the genome is copied (step 4). Moreover, an LTR has been generated. As the reverse transcriptase generates the minus-strand DNA, the associated RNase H activity hydrolyzes the remainder of the genomic template, while sparing a short (11-nucleotide-long) RNase H-resistant polypurine tract (ppt) near the 3′ end of the genome. The ppt segment then primes the synthesis of the plus-strand DNA, as catalyzed by the DNA-dependent DNA polymerase activity of the reverse transcriptase protein (step 5). Plus-strand DNA synthesis progresses through the U5 site on the minus DNA strand and through 18 nucleotides of the tRNA primer sequence, which is complementary to pbs. It finally terminates because of a modified nucleotide in the tRNA that cannot be copied (step 6). The pbs-complementary sequence on the tRNA primer is hydrolyzed by RNase H, thereby releasing the remainder of the tRNA. In this way, a double-stranded LTR sequence is generated at one end of the proviral DNA, flanked by a sequence on the positive strand that is complementary to pbs. In the case of some retroviruses (e.g., HIV and ASLV) but not others (e.g., MLV), additional internal RNA sequences, which are partially spared from RNase H digestion, also may prime positive-strand DNA synthesis (steps 7 and 8). At this point in the process, there are complementary single-stranded pbs sequences at each end of the replicative intermediate. These sequences may hydrogen bond to each other, resulting in a second template exchange event and a circular intermediate structure (step 9). Cellular enzymes may repair this intermediate structure to generate a closed circle with only one LTR, which consequently is biologically inactive (step 9, left branch). But an alternative outcome, which may be catalyzed by the helicase and DNA-dependent DNA polymerase activities of the reverse transcriptase, is for displacement synthesis to ensue at the 3′ end of each DNA strand of the noncovalently closed circle, an outcome that generates a linear, double-stranded proviral DNA molecule, with an LTR at each end (step 9, right branch; step 10). Modified from R. A. Katz and A. M. Skalka, *Annu. Rev. Biochem.* **63**:133–173, 1994, with permission.

complexes of HIV, a complex retrovirus, do enter the nuclei of nondividing cells, by mechanisms not yet known.

Second, the fidelity of retroviral reverse transcription is quite low. As you may have correctly surmised, the reason for the low fidelity of reverse transcription is that the reverse transcriptase lacks the editing activity of other DNA polymerases (i.e., the activity that removes mispaired bases from newly replicated DNA). Misincorporation rates of retroviral reverse transcriptases may actually be as high as 1 per 10^4 nucleotides, in comparison to 1 per 10^7 to 1 per 10^{11} nucleotides for the cellular DNA polymerase. Since retroviral genomes are about 10^4 nucleotides in length, each individual retroviral provirus will contain, on the average, approximately one new mutation per replication cycle. Moreover, the Pol II RNA polymerase, which generates progeny viral genomes by transcribing the integrated provirus, likewise does not have a proofreading activity.

The facts enumerated in the preceding paragraph take on great significance in the case of HIV, since they explain why HIV generates mutants at a fierce rate, thereby vastly complicating efforts by the host immune system to contain the infection, as well as efforts to develop a vaccine and long-term anti-HIV therapies. Moreover, the ability of HIV to generate mutants at a fast clip is an important aspect of its life cycle, since it facilitates dissemination of that virus in its human host and its spread from one individual to another. (How mutagenesis affects HIV dissemination and spread is not obvious and is explained in Chapter 21.)

INTEGRATION

Integration is a crucial step in the retroviral life cycle, since the integrated proviral DNA is the template for the transcription of all viral mRNAs and all progeny viral genomes. The integration process is catalyzed by the particle-associated viral IN protein, which is encoded by the 3′ end of the *pol* gene (Fig. 20.1). IN, the viral protease, and the reverse transcriptase are incorporated into progeny virus particles during virus assembly, as segments of the Gag-Pol precursor polyprotein. The individual proteins are released from Gag-Pol by the action of the viral protease at the time of virus maturation (see below).

We begin our consideration of the mechanism by which IN promotes integration of the provirus by examining the nucleotide sequences making up the joints between the integrated proviral DNA and the cellular genome, paying particular attention to the changes that occur in the proviral and cellular sequences as a result of the integration process (Fig. 20.6). First, notice that the

unintegrated viral DNA contains a short (4 to 6 base pairs [bp]), inverted, terminal repeat sequence at each of its blunt ends and that the terminal two nucleotides present in those inverted terminal repeats are deleted by the integration process. Second, notice that the target site on the cellular DNA is duplicated, such that it is present on each side of the integrated proviral DNA.

All integrated retroviral proviruses examined to date terminate with a 5′-TG and a 3′-CA. In contrast to the uniformity in the termini of integrated proviruses, irrespective of the particular retrovirus, the length of the flanking host cell duplication does vary with each particular retrovirus. The variation in the length of the flanking cellular sequence as a function of the particular retrovirus led to the realization that integration is catalyzed by a virus-encoded activity.

Figure 20.7 depicts a straightforward model for retroviral integration, which accounts for the above experimental observations. In the first step, IN removes two nucleotides from the 3′ terminus of each retroviral DNA strand. These processed 3′ termini then carry out an IN-catalyzed attack on the cellular DNA, such that each of the retroviral 3′ termini is joined to a strand of the cellular DNA. This attack is nonspecific with respect to the host nucleotide sequence, but the distance between the resulting joints is dependent on the IN protein of the particular virus. The intermediate structure is then resolved and repaired by cellular enzymes, thereby generating the integrated provirus, which lacks the terminal two nucleotides of the unintegrated DNA and is flanked by the duplicated cellular sequence. Cellular proteins that are known to be involved in the repair of DNA damage have been implicated in the final stages of retrovirus integration.

Features of the nucleotide sequences across the joints between the retroviral provirus and the cellular DNA bear striking similarities to features of the joint sequences seen between a variety of prokaryotic transposable genetic elements and the genomes of their bacterial host cells. For example, inverted terminal repeat sequences are found at the ends of prokaryotic transposable genetic elements, as well as at the ends of integrated retroviral proviruses. Moreover, a duplication of a cellular sequence is found on each side of prokaryotic transposable genetic elements, as observed in the case of integrated retroviral proviruses. These similarities lead to the provocative notion that prokaryotic insertion sequences and retroviruses may have evolved from a common molecular ancestor that existed before the divergence of eukaryotes and prokaryotes (see below). A corollary premise is that retroviral integration and prokaryotic DNA transposition occur via similar mechanisms.

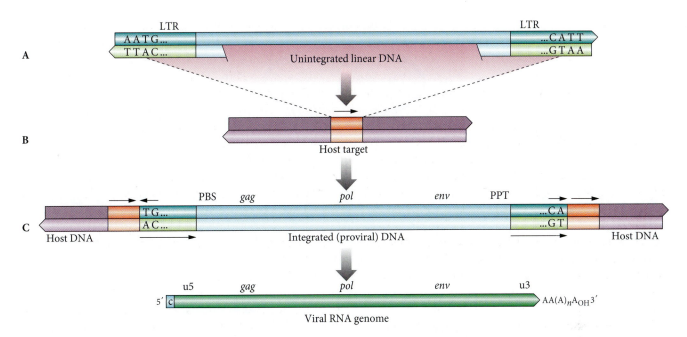

Figure 20.6 Nucleotide sequences across the joints between the integrated proviral DNA and the cellular genome. (A) The unintegrated viral DNA contains a short (4- to 6-bp) inverted terminal repeat sequence at each of its blunt ends. The example shown is ALV. (B) The nonspecific target site on the host genome. (C) The integrated provirus. The terminal two nucleotides, present in the inverted terminal repeats prior to integration, are deleted upon integration. Also, the target site on the cellular DNA is duplicated, such that it is present on each side of the integrated proviral DNA. Adapted from S. J. Flint, L. W. Enquist, R. M. Krug, V. R. Racaniello, and A. M. Skalka, *Principles of Virology: Molecular Biology, Pathogenesis, and Control*, 2nd ed. (ASM Press, Washington, DC, 2004), with permission.

GENE EXPRESSION AND REPLICATION

Retroviral LTRs, which are generated during reverse transcription, contain transcriptional promoters and also enhancers, which vastly stimulate transcription from the retroviral promoter. The enhancer and promoter regions in the U3 region of an avian retrovirus LTR are drawn to scale in Fig. 20.8. The figure depicts the individual elements of the promoter and enhancer and the soluble transcription factors that bind to each.

Simple retroviruses do not encode soluble transcription factors of their own. (HIV and other complex retroviruses do encode transcription factors. The HIV Tat protein is one such transcriptional regulator [Fig. 20.1] [see Chapter 21].) Instead, transcription of simple retroviruses is controlled by cellular transcription factors, as depicted in Fig. 20.8. The LTRs also contain transcription termination sequences, polyadenylation signals, and sequences that are transcribed into ribosomal binding sites.

The transcription pattern of simple retroviruses is exemplified here by Rous sarcoma virus. The cellular Pol II RNA polymerase transcribes the integrated proviral DNA to generate an unspliced 38S transcript and two spliced transcripts: one of about 28S and the other of about 21S.

Some of the full-length, unspliced 38S transcripts are incorporated into nascent virus particles as progeny viral genomes, while others serve as mRNAs that are translated into Gag and Gag-Pol (Fig. 20.1; see also below). The larger of the Rous sarcoma virus spliced transcripts is translated into the *env* gene product. The smaller spliced transcript is translated into the *onc* gene product, which is *not* required for virus replication. (The smaller, spliced mRNA is generated only by those rare, simple retroviruses that like Rous sarcoma virus contain an oncogene, and each of these retroviruses is a special case. The nature of the retroviral *onc* gene products and their means of action are discussed below.)

Each of the viral RNAs is capped by cellular capping enzymes. Also, each of the viral RNAs contains the same 5′ leader sequence. That is, the common 5′ leader sequence is present on the unspliced 38S transcript, and it is joined to the coding sequences of the 28S and 21S transcripts by the splicing process. (The polyomaviruses, papillomaviruses, and adenoviruses provided similar examples of alternative splicing.) Polyadenylation occurs posttranscriptionally, after the pre-mRNA has been spliced by cellular enzymes.

The efficiency of retrovirus replication actually depends rather strictly on the relative amounts of spliced

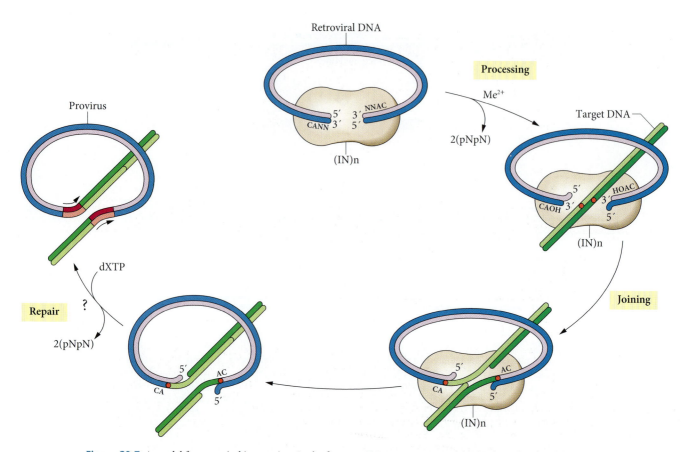

Figure 20.7 A model for retroviral integration. In the first step, IN removes two nucleotides from the 3′ terminus of each retroviral DNA strand. These processed 3′ termini then carry out an IN-catalyzed attack on the cellular DNA, such that each of the retroviral 3′ termini is joined to a strand of the cellular DNA. This attack is nonspecific with respect to the host nucleotide sequence, but the distance between the resulting joints is dependent on the particular virus IN protein. The intermediate structure then is resolved and repaired by cellular enzymes, thereby generating the integrated provirus DNA, which lacks the original terminal two nucleotides and is flanked by the duplicated cellular sequence. The small red circles represent the phosphodiester bonds cleaved in the joining reaction. Modified from N. D. Grindley and A. E. Leschziner, *Cell* **83**:1063–1066, 1995, with permission.

and unspliced mRNAs that are transcribed from the integrated provirus. Consider the case of ALV, which transcribes only one spliced mRNA species (i.e., that which is translated into the *env* gene product). The splice site on the ALV primary transcript is normally recognized about one-third of the time. But a mutation that affects the splicing signal causes the ratio of unspliced mRNAs to spliced mRNAs to go from 2:1 to 1:2 and also results in virtually undetectable viral growth.

In addition to having to properly balance the ratio of unspliced to spliced transcripts, retroviruses also must deal with the issue of exporting their *unspliced* RNAs from the nucleus. This is an important concern because eukaryotic primary transcripts, or pre-mRNAs, generally contain introns, and eukaryotic cells have a mechanism in place to ensure that pre-mRNAs do not exit the nucleus prematurely, before splicing is completed.

The nuclear export mechanism for cellular mRNAs and the process that ensures against the premature export of unspliced mRNAs are not entirely clear. But it is known that pre-mRNAs must undergo several processing steps, including 5′ capping, splicing, 3′ end cleavage, and polyadenylation, before they might be exported. Moreover, there is evidence that particular components of the splicing machinery may recruit mRNA export factors to the mRNA/spliceosome complex, in that way coupling splicing to export.

Thus, bearing in mind that (i) mRNAs normally must undergo splicing before they may be exported from the nucleus and (ii) export of mRNAs may normally be coupled to the splicing machinery, how might retroviruses ensure the export of their unspliced RNAs? Their solution is as follows. Unspliced retroviral RNAs themselves contain specific nucleotide sequences that promote their export.

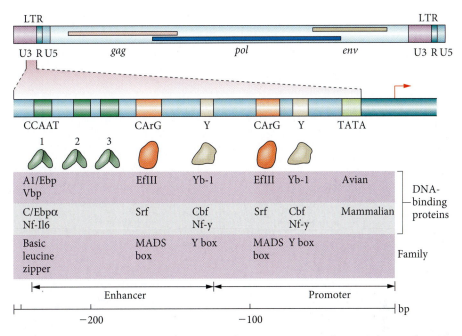

Figure 20.8 The transcriptional control region of an avian retrovirus. The proviral DNA of an ALV is shown at the top. The enhancer and promoter present in the U3 regions of the LTRs are drawn to scale below. Each of the multiple CCAAT, CArG, and Y-box sequences present in this transcriptional control region, which are required for maximally efficient transcription, is recognized by both avian and mammalian proteins. All those recognizing the CCAAT enhancer sequences are members of the basic "leucine zipper" family of proteins. (Such proteins contain a leucine zipper dimerization motif next to their DNA-binding domains. Dimerization is necessary for DNA binding.) The chick protein EfIII that binds to the CArG sequences appears to be the avian homolog of mammalian serum response factor (Srf). This protein plays an important role in the activation of transcription in response to serum growth factors, hence its name. These proteins are members of a large widespread family defined by a conserved sequence motif within the DNA-binding domain (the MADS box) and named for four of the originally identified members. From S. J. Flint, L. W. Enquist, R. M. Krug, V. R. Racaniello, and A. M. Skalka, *Principles of Virology: Molecular Biology, Pathogenesis, and Control,* 2nd ed. (ASM Press, Washington, DC, 2004), with permission.

Since these sequences function independently of any viral protein, they are referred to as **constitutive transport elements** (CTEs). The retroviral CTEs interact with regular cellular RNA export factors, as shown by the finding that the export of unspliced retroviral transcripts competes with the export of mature, spliced cellular mRNAs (Box 20.3).

Next, we consider the translation of the Rous sarcoma virus mRNAs. The unspliced 38S retroviral mRNA is translated into two primary translation products (Fig. 20.1). The major one of these is the 76-kDa Gag protein, i.e., the translation product of the *gag* gene, which is the polyprotein precursor of the NC, MA, and CA proteins. The minor translation product of the 38S mRNA is the 180-kDa Gag-Pol protein, i.e., the fusion product of the *gag* and *pol* genes.

Earlier, it was thought that the 180-kDa protein might result from an occasional translational read-through from the *gag* gene into the *pol* gene. This was an attractive hypothesis because it explained why many fewer copies of the reverse transcriptase than structural proteins are generated. However, sequence analysis of the Rous sarcoma virus genome revealed that *pol* actually overlaps a short sequence at the 3′ end of *gag*. Moreover, *pol* is in a different (−1) reading frame relative to *gag*. These experimental findings favor a mechanism in which ribosomes occasionally (i.e., about 5% of the time) undergo a −1 frameshift before reaching the *gag* stop codon, thereby enabling the 180-kDa Gag-Pol fusion protein to be generated. Historically, this instance of ribosomal frameshifting by Rous sarcoma virus is the first known example of the phenomenon.

Two features of the Rous sarcoma virus 38S mRNA are known to augment ribosomal frameshifting. The first is a so-called "slippery" homology sequence, which is comprised of the heptanucleotide AAAUUUA. The second is a

Despite the finding that retroviral CTEs interact with regular cellular RNA export factors, the CTE-mediated retroviral RNA export pathway is not entirely identical to the pathway that exports mature cellular mRNAs. In particular, the CTE-mediated pathway does not involve GTP-Ran, which normally mediates nuclear export (Box 12.7).

While the CTE-mediated nuclear export pathway is not completely understood, it is known to involve a cellular protein, Tap, which binds specifically to the retroviral CTE. (Tap refers here to the cellular mRNA export factor, Tip-associated protein, rather than to the transporter associated with antigen processing [Chapter 4].) Tap also facilitates the nuclear export of cellular mRNAs, but rather than binding directly to cellular mRNAs, Tap instead binds to components of the splicing complex that is associated with cellular mRNAs. Thus, the simple retroviruses appear to be able to export their unspliced mRNAs via a novel Tap-mediated RNA export pathway that does not couple nuclear export to the splicing machinery and splicing and does not involve GTP-Ran.

so-called "pseudoknot," a tertiary RNA structure (Fig. 6.14) located several nucleotides downstream of the slippery homology sequence. The pseudoknot is thought to cause ribosomes to pause at the slippery homology sequence, thereby increasing the likelihood that the −1 ribosomal frameshift will occur there.

In contrast to the ribosomal frameshift strategy used by Rous sarcoma virus to generate its Gag-Pol precursor polyprotein, MLV adapted a strategy in which its *gag* and *pol* genes are in the same reading frame but separated by an amber (UAG) chain termination codon. Approximately 4 to 10% of ribosomes bypass the chain termination triplet on the 38S mRNA to generate the MLV Gag-Pol polyprotein. Translational suppression of the MLV UAG codon occurs when that triplet is misread by a Gln tRNA.

Just as the Rous sarcoma virus ribosomal frameshifting strategy is enhanced by particular sequences on the 38S mRNA, the MLV translational read-through strategy is likewise augmented by particular nucleotide sequences in its corresponding unspliced transcript. In the MLV instance, read-through is augmented by a purine-rich sequence downstream of the UAG codon and by a pseudoknot further downstream in the mRNA. As in the case of ribosomal frameshifting described above, the pseudoknot is thought to cause the ribosome to pause; in this instance,

this step favors the misincorporation of a Gln tRNA at the chain termination site.

On the one hand, the distinct stratagems used by Rous sarcoma virus and MLV to fine-tune the relative levels of replication enzymes to structural proteins provide yet another illustration of how even closely related viruses may evolve different solutions to a common problem. On the other hand, the use of common determinants for each of these mechanisms (i.e., a polypurine sequence and a downstream pseudoknot) suggests that one of these mechanisms might have evolved from the other, in at least one branch of the retrovirus family tree.

ASSEMBLY, MATURATION, AND RELEASE
Gag and Gag-Pol

Nascent retrovirus particles are actually assembled from intact Gag and Gag-Pol precursor polyproteins (Fig. 20.9). In fact, Gag and Gag-Pol are not cleaved to generate the MA, CA, NC, reverse transcriptase, and IN proteins until after particle budding and release. Why might retroviruses postpone maturation cleavage until so late in the replication cycle? This strategy of postponing proteolysis was likely adapted because it makes for the efficient incorporation of each of the individual Gag and Gag-Pol cleavage products into progeny virus particles. That is, these proteins do not have to first come together in the cytoplasm by random association. Moreover, assembly is not impaired by the likelihood that the individual Gag cleavage proteins might make nonspecific contacts with the great excess of cellular proteins in the cytosol. Also, in conjunction with the mechanisms that determine the ideal relative amounts of Gag and Gag-Pol (see above), this process ensures that the individual proteins are combined in the nascent particle in their proper relative amounts.

Earlier, we saw that the plus-strand picornaviruses also exploit a strategy of assembling virus particles from precursor polyproteins (Chapter 6). The individual picornavirus capsid proteins VP1, VP2, VP3, and VP4 are incorporated into assembling virus particles as components of the picornavirus P1 precursor polyprotein. Importantly, amino acid sequences in P1 that correspond to VP1, VP2, VP3, and VP4 take on major features of the tertiary structures of the individual proteins, while still covalently linked to each other in the polyprotein. Moreover, they also begin to make interprotein contacts before assembling nascent virus particles. Thus, at the time of cleavage, these polyproteins have already given shape to the protomers that form the capsid.

The fact that picornavirus capsids are assembled from precursor polyproteins is recalled here to underscore the

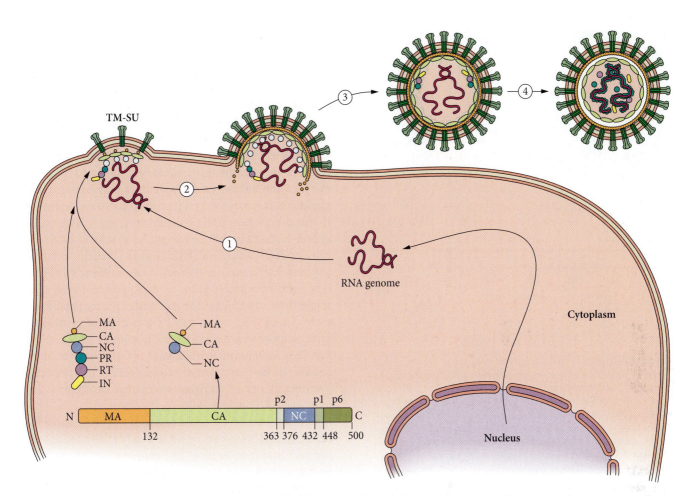

Figure 20.9 Nascent retrovirus particles are actually assembled from intact Gag and Gag-Pol precursor polyproteins, which are not cleaved to generate the MA, CA, NC, reverse transcriptase, and IN proteins until after particle budding and release. The map of the Gag polyprotein depicted is that of HIV, and the folded Gag and Gag-Pol polyproteins are shown as well. Note that the individual Gag proteins are positioned in the same order within the precursor polyprotein as they are in when they are within the layers of the virus particle. The association of Gag molecules with the plasma membrane, with one another, and with the RNA genome (the latter via interactions with the NC segment of Gag) initiates assembly at the plasma membrane (step 1). In some instances, the MA segment also interacts specifically with the cytoplasmic tail of the TM segment of the envelope glycoprotein. Assembly continues with the incorporation of additional molecules of Gag and Gag-Pol (step 2). Eventually, the membrane around the nascent particle fuses, releasing a still-noninfectious virus particle (step 3). Cleavage of Gag and Gag-Pol by the viral protease produces mature, infectious virus particles (step 4). Modified from S. J. Flint, L. W. Enquist, R. M. Krug, V. R. Racaniello, and A. M. Skalka, *Principles of Virology: Molecular Biology, Pathogenesis, and Control*, 2nd ed. (ASM Press, Washington, DC, 2004), with permission.

point that this stratagem may be used even more elegantly by the retroviruses than by the picornaviruses. This conjecture is based on the following. In contrast to the non-enveloped picornavirus particles, with their "naked" plus-strand genomes enclosed within a one-protein-layer-thick capsid, retrovirus particles contain three protein layers beneath an envelope. First there is the MA protein layer, then the CA protein layer, and finally the NC protein-coated genome (Fig. 20.10). Importantly, the individual Gag proteins are positioned in the same order within the polyprotein as they are found within the layers of the virus particle. Consequently, despite the fact that the maturation

cleavage of Gag results in a reorganization of the internal structure of the virus particle required for its infectivity, the individual Gag proteins are arranged within the precursor polyprotein in such a way that they are, by design, in their correct layer within the virus particle at final virus maturation. Moreover, as we will see, the Gag polyprotein has plasma membrane localization and RNA-binding signals that likewise are positioned to correctly orient MA with respect to the lipid envelope and to properly position NC with respect to the genome. These signals also facilitate delivery of the Gag-bound RNA genome to the virus assembly sites.

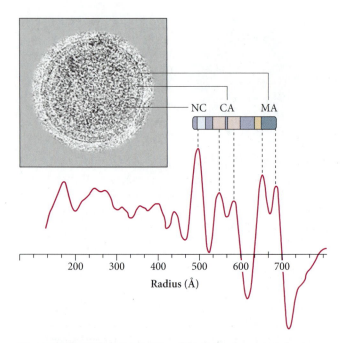

Figure 20.10 Radial organization of the Gag polyprotein in immature HIV particles. The inset is a cryo-electron micrograph of a viruslike particle assembled from the HIV Gag protein. It is actually reconstructed from radial density measurements of digitalized images of the particles. The plot (red line) shows density as a function of distance from the particle center, in angstroms. The organization of the Gag polyprotein is depicted alongside the micrograph. From S. J. Flint, L. W. Enquist, R. M. Krug, V. R. Racaniello, and A. M. Skalka, *Principles of Virology: Molecular Biology, Pathogenesis, and Control*, 2nd ed. (ASM Press, Washington, DC, 2004), with permission.

The above principles apply to the Gag-Pol polyprotein as well as to the Gag polyprotein. Indeed, the arrangement of individual protein sequences within Gag-Pol ensures the proper positioning of the reverse transcriptase and IN within the interior of the virus core. However, virus particles cannot be constructed from Gag-Pol alone, probably because there would not be enough room within the core to accommodate the C-terminal Pol sequence present on each Gag-Pol polyprotein. Thus, we can begin to appreciate the importance of the mechanisms that ensure that the proper relative amounts of Gag and Gag-Pol are generated (see above).

Nascent virus particles also must incorporate the tRNA molecules that serve as primers of reverse transcription. These tRNAs are brought into the assembling virus particles by specifically binding to the reverse transcriptase component of the Gag-Pol polyprotein and by base pairing with the pbs sequence on the viral genome.

Here is yet another remarkable adaptation of the retroviruses. Bearing in mind the advantages these viruses gain by assembling progeny virus particles from intact Gag and Gag-Pol polyproteins and the fact that the protease is

itself a component of the Gag and Gag-Pol precursors (Fig. 20.1), any premature cleavage of these polyproteins, before their incorporation into nascent virus particles, would be wasteful for the virus. How then is premature cleavage of Gag and Gag-Pol prevented? The answer is essentially the fact that the active protease is a dimer. Consequently, a concentration of protease monomers sufficient to lead to efficient dimerization is attained only within the confines of assembling particles. But since the protease exists as a monomer on the precursor polyproteins, how might it become free so that it might dimerize? The answer to this conundrum is not entirely clear, but a likely explanation is that a small percentage of precursor polypeptides undergo autoproteolytic cleavage, thereby freeing a few protease monomers within the assembling virus particle. Efficient dimerization of these protease molecules might then occur within the confines of the particle. These early dimers might then initiate a cascade of proteolytic processing within the particle. (In support of the above premises, mutations that increase the efficiency of autoproteolysis lead to impaired production of virus particles, presumably due to increased premature cleavage of Gag and Gag-Pol.)

Although proteolytic processing of Gag to generate the MA, CA, and NC proteins occurs only after final assembly and budding, each of the components of the still-uncleaved Gag polyprotein nonetheless has a particular role to play during assembly. Some of these roles are as follows. (i) As in the assembly of many other enveloped viruses, retrovirus assembly takes place concomitantly with the binding of structural proteins to a cellular membrane. The MA segment of Gag, which is located at the N terminus of the polyprotein, contains signals that target Gag proteins to plasma membrane assembly sites. Moreover, the association of Gag with the cytoplasmic face of the plasma membrane is facilitated by a myristate (a hydrophobic, 14-carbon, saturated fatty acid) covalently linked to the N-terminal glycine residue of the MA domain of most known Gag proteins. The sequences on the MA component of Gag specifying myristoylation and targeting to the plasma membrane are crucial for viral assembly. Moreover, in some cases, the MA segment also interacts specifically with the cytoplasmic tail of the TM segment of the envelope glycoprotein. (ii) The Gag proteins of several retroviruses are known to bind to positive-end-directed microtubule motors of the kinesin family, suggesting that Gag and Gag-RNA complexes specifically harness the microtubule transport system to traffic to the plasma membrane. (iii) Sequences on CA, as well as on NC and MA, are required for the intermolecular interactions between Gag polyprotein molecules during virus assembly. Interestingly, these association sequences are also necessary to transport Gag to the plasma membrane, since alterations of those sequences cause the polyprotein to accumulate in the cytoplasm. (iv) The NC segment of Gag contains an RNA-binding domain that specifically recognizes unspliced retroviral RNA molecules. This sequence facilitates the delivery of progeny viral genomes to virus assembly sites at the plasma membrane. The genomic RNA in turn contains a *cis*-acting packaging signal termed **Ψ (psi)** in the 5′ leader region of the RNA. The RNA binding sites on the NC component of Gag ensure that genome packaging is coordinated with assembly of the protein shell. This mode of assembly, in which the protein subunits of the virus particle assemble in association with the genome, is said to be "concerted." It is in contrast to "sequential" assembly, in which the viral genome is inserted into a preformed shell.

When considering the size of Gag and Gag-Pol and the number of specific intramolecular and intermolecular interactions that these polyproteins go through during folding, assembly, and final maturation, one might expect that chaperones act to ensure the productive folding and interactions of Gag and Gag-Pol. Indeed, that appears to be the case, as implied, for example, by studies of the Mason-Pfizer monkey virus (M-PMV). These M-PMV studies were prompted by the finding that the carboxy terminus of the M-PMV Gag protein yields a small (36-amino-acid-long) protein called p4. Genetic studies seeking to identify the role of the M-PMV p4 protein in virus replication showed that the p4 domain both stabilizes Gag and facilitates capsid assembly. Returning to the notion that a chaperone facilitates the proper folding and interactions of Gag and Gag-Pol, screening by yeast two-hybrid analysis showed that the p4 domain interacts with TRiC, a cellular chaperone that is involved in the folding of numerous cellular proteins and protein complexes.

The retroviral genomic RNA itself appears to play a crucial role in virus assembly, apparently by acting as scaffolding within the particle. Moreover, the RNA also seems to help maintain the structural integrity of the mature virus particle. In fact, retroviral cores can be disrupted by treatment with RNase (Box 20.4).

In the cases of MLV and of HIV (a complex retrovirus), their ψ packaging sequences overlap intronic sequences on their full-length transcripts. Thus, in those viruses, only the unspliced RNAs contain the complete ψ sequence. This would appear to be a rather clever adaptation of the virus to account for the selective encapsidation of full-length, unspliced RNA genomes. However, in the cases of some other simple retroviruses, such as ALV, the ψ packaging sequence is located in the 5′ leader sequence, upstream of any introns. Thus, ψ is present on both unspliced and spliced RNAs. Yet in the case of these viruses,

BOX 20.4

Despite the fact that the genomic RNA contains a *cis*-acting packaging signal, any RNA appears to be able to be packaged and fulfill the scaffolding and stabilizing roles. Several experimental findings together illustrate these points. For example, in a cell-free system, while the assembly of recombinant Gag proteins into capsid-like structures is dependent on the addition of RNA, the RNA requirement can be satisfied by the addition of cellular RNA. In another example, when cells were transfected with MLV genomes that contained a deletion in the viral RNA packaging signal, progeny particles that contained cellular mRNA in place of viral RNA were generated.

A corollary conclusion from the above experimental findings is that since nascent retrovirus particles appear to readily package nonretroviral RNA, the ψ packaging signal on retroviral genomic RNAs must play an important role in enhancing the specificity of RNA packaging. Apropos that point, retroviral genomic RNA comprises much less than 1% of the total cytoplasmic RNA within an infected cell. Yet due to the selective encapsidation of RNA molecules containing the ψ packaging signal, the vast majority of progeny virus particles do contain authentic genomic RNA. Interestingly, the majority of "cellular" RNA that is incorporated by mistake into retrovirus particles is virus-like RNA transcribed from *endogenous retroviruses* (see the text). The endogenous retrovirus-encoded RNA carries viral packaging signals and can compete with viral genomic RNA for packaging into virus particles. Some "genuine" cellular RNAs can be incorporated into virus particles, but only rarely.

unspliced RNAs are selectively encapsidated by some as-yet-unknown mechanism.

Here is another unexplained aspect of retrovirus assembly. While the NC protein binds specifically to the ψ sequence, and although ψ itself is >300 nucleotides in length, there are >2,000 individual genome-bound copies of NC within each retroviral core. Thus, all but a small percentage of these NC molecules are apparently bound nonspecifically to other sequences in the genomic RNA.

Env and Budding

Viral envelope glycoproteins are generally synthesized in association with the rough endoplasmic reticulum and are transported by normal cellular pathways to their final destinations, which are dictated by their particular targeting signals (Box 2.5; Fig. 8.3). In the case of the retroviruses, the *env* primary gene products contain a targeting signal that directs them to the plasma membrane. Like other polypeptides destined to become transmembrane proteins, retroviral Env proteins pass through the Golgi apparatus, en route to their final destination at the plasma membrane. During their transport through the Golgi apparatus, retroviral Env polyproteins are cleaved by furin family proteases to generate the TM and SU subunits, which in some instances are joined via disulfide bonds and in other instances remain noncovalently associated.

Two distinct mechanisms by which enveloped viruses acquire their envelopes and bud are known. The first of these is referred to as a **concerted process**, in which capsid assembly, envelope acquisition, and budding all take place concurrently (Fig. 20.9). The second mechanism might be referred to as a **sequential process**, in which the capsid or core assembles in the cell interior, independent of any cellular membrane and envelope glycoproteins. Then, the assembled core is transported to a particular cellular membrane, which has been modified to contain the viral envelope proteins. The interaction of the assembled capsid or core with the modified membrane then leads the particle to acquire its envelope, bud, and be released. (The terms "concerted" and "sequential" were used differently above, to distinguish between DNA packaging coincident with, versus subsequent to, capsid assembly.)

Some retroviruses acquire their envelopes and bud via a concerted process, while others do so via a sequential process. For example, HIV uses a concerted process. Its Gag polyproteins concurrently associate with each other and with the Env-modified cytoplasmic face of the plasma membrane, so that both assembly and budding occur in tandem (Fig. 20.9). In contrast, the mouse mammary tumor virus uses a sequential process, in which complete cores are assembled in the cytoplasm before they associate with the modified plasma membrane and bud. In each instance, the MA, CA, and NC components of the Gag precursor polyprotein carry out their respective roles as detailed above.

Interestingly, in the case of those retroviruses that assemble via a concerted process, the MA segment of Gag must associate with Env-modified membrane in order for Gag polyproteins to interact with one another. That is, assembly and envelope acquisition *must* be concerted in these instances. Apparently, association of Gag with the plasma membrane causes it to adopt an assembly-competent conformation. This feature of Gag may be an adaptation that ensures against premature, nonproductive interactions between Gag polyproteins in the cell interior.

Above, we noted several advantages that the retroviruses gain by postponing the cleavages of Gag and Gag-Pol until after assembly and budding. We add here that postassembly cleavage ensures an irreversible, forward direction to the assembly process. Importantly, it also enables these viruses to solve a conundrum discussed in several earlier chapters: how to assemble at the end of one replication cycle a virus particle that is stable but is yet able to undergo triggered disassembly at the start of the next cycle. From thermodynamic considerations, we would expect that the capsid should assemble into its most stable conformation, and indeed it does. However, the structure becomes metastable upon the specific proteolytic cleavages of Gag and Gag-Pol at the time of budding and release. At the start of the next cycle of infection, the release of the capsid structure from its metastable conformation provides the driving force for its partial disassembly. Likewise, the proteolytic cleavage of Env enables SU and TM to undergo the conformational transitions that facilitate fusion.

Because virus maturation via budding at the plasma membrane is not deleterious to the host cell, oncogenic simple retroviruses are able to establish persistent, productive infections within a cell, while also transforming it. Thus, the noncytocidal nature of these viruses is one factor that enables them to cause tumors in their natural hosts. Retrovirus-induced oncogenesis is the topic of the next section.

ONCOGENESIS

Before beginning our discussion of retroviral oncogenesis, we briefly reiterate the insights into neoplasia that were gained from studies of the DNA tumor viruses. Most importantly, studies involving the DNA tumor viruses led to a clearer view of the cellular mechanisms that *negatively* regulate cellular replication, particularly with respect to processes affected by the key cellular **tumor suppressor proteins**, pRb and p53 (Chapters 15, 16, and 17). In brief, pRb family proteins negatively regulate entry of cells into the cell cycle, while p53 activates apoptosis in cells that attempt to divide without having appropriately passed earlier checkpoints in the cell cycle. And, the DNA tumor viruses (in particular, the polyomaviruses, papillomaviruses, and adenoviruses) affect transformation by inactivating these tumor suppressor proteins.

While studies of the DNA tumor viruses led to fundamental insights into processes that negatively regulate cell proliferation, studies involving retroviruses led to breakthroughs of exceptional and far-reaching importance into the opposite face of cellular replication control: the signaling pathways that *positively* govern cellular replication. Moreover, those studies led to an understanding of how mutation or inappropriate expression of genes encoding components of those pathways may result in cancer. (Note that mutations in genes encoding tumor suppressor proteins may also lead to cancer. See Chapter 15.)

Also note the following distinction between the DNA tumor viruses and the RNA tumor viruses. The DNA tumor viruses include members of the polyomavirus, papillomavirus, adenovirus, herpesvirus, poxvirus, and hepadnavirus families. In contrast, all RNA tumor viruses are members of the retrovirus family. However, and importantly, the retroviruses fall into two distinct groups regarding their capacity to induce tumors in vivo and neoplastic transformation of cells in culture. Members of the first of these retrovirus groups, typified by Rous sarcoma virus, cause malignancies in nearly 100% of infected animals within days of infection, and they efficiently transform cells in culture. These rapidly transforming retroviruses express an oncogene, and they are the most efficient inducers of neoplasia known. For reasons that soon will become clear, they are referred to as **oncogene-transducing oncogenic retroviruses.**

Members of the second of these retrovirus groups, typified by the avian, murine, and feline leukemia viruses, rarely induce tumors in vivo, and then only after an incubation period of weeks to months. Moreover, they virtually never induce transformation of cells in cell culture. These viruses do not carry an oncogene and are referred to as **oncogene-deficient oncogenic retroviruses.** As you might surmise, the oncogene-transducing and oncogene-deficient oncogenic retroviruses induce tumors by quite distinct mechanisms.

Next, consider that "only" about 20% of human cancers are caused by viruses and that the great majority of those cancers are caused by DNA viruses (i.e., papillomaviruses, herpesviruses, and hepadnaviruses [Box 16.2]). Why then have studies of retroviral tumor genesis and transformation so profoundly enhanced our understanding of cancer? The answer is that the study of retroviruses led to the singularly important discovery that retroviral oncogenes in fact are related to, and almost certainly derived from, cellular genes. Moreover, these findings led to the equally important discovery that mutation or inappropriate expression of one or more of the cellular analogs of retroviral oncogenes is a factor in the development of virtually all human cancers. These developments involved studies of the oncogene-transducing oncogenic retroviruses, and they happened as follows.

Oncogene-Transducing Oncogenic Retroviruses

In the late 1960s, it was found that after exposing Rous sarcoma virus to γ (gamma) or ultraviolet (UV) irradiation, there were surviving viruses that could no longer transform chicken fibroblasts but could still replicate. These experimental findings demonstrated that the ability of Rous sarcoma virus to transform cells is not required for the viability of the virus. This is a key part of our story,

which we will return to momentarily. Also note that the isolation of nontransforming mutants of Rous sarcoma virus defined for the first time a retroviral oncogene, which was dubbed *src* for its ability to induce sarcomas (i.e., cancers of connective tissue).

The isolation of temperature-sensitive *src* mutants next enabled researchers to answer a fundamental question regarding retrovirus-induced neoplastic transformation: is the continued expression of a retrovirus oncogene necessary for retrovirus-transformed cells to continue displaying the transformed cell phenotype (as predicted by Howard Temin; see above), or does the oncogene need to be expressed only transiently to lock in transformation? The experiment that resolved this issue was carried out as follows. Cells were transformed by the mutant virus at the transformation-permissive temperature of its *src* mutation. When these cells were then maintained at that temperature, they continued to express the transformed cell phenotype. But when these cells were shifted to the nonpermissive temperature of the *src* mutation, then the cells regained the normal cell phenotype. That is, transformation was reversible when *src* expression was impaired. In control experiments, cells transformed by wild-type Rous sarcoma virus expressed the transformed cell phenotype when propagated at either temperature. Thus, the continued expression of the *src* gene function is necessary to maintain the transformed cell phenotype.

The finding that *src* expression is not necessary for Rous sarcoma virus replication may not be particularly surprising to many contemporary readers who already know how oncogene-transducing oncogenic retroviruses acquire their oncogenes. But it was exciting to researchers of the day, since the *src* gene is present in standard strains of Rous sarcoma virus and viruses generally do not carry excess "genetic baggage." So researchers of the day asked the following key questions: why is *src* superfluous for Rous sarcoma virus replication, and, more intriguingly apropos our story, why is *src* present in standard strains of Rous sarcoma virus if it is not necessary for replication? Several different experimental observations led to a resolution of these issues, as follows.

One key experimental finding was that transforming retroviruses (that now would be referred to as oncogene-transducing oncogenic retroviruses) could be isolated after inoculating animals with nontransforming retroviruses (that now would be referred to as oncogene-deficient retroviruses). In one such experiment, 150 mice were inoculated with the oncogene-deficient Abelson murine leukemia virus. One of the inoculated mice subsequently developed a lymphoma (a solid cancer of lymphoid tissue). Next, a highly oncogenic variant of the Abelson virus was isolated from the tumor itself. Importantly, this variant contained a brand new oncogene termed *abl*. This experimental observation and other, similar ones raised the possibility that retroviral oncogenes might be acquired from the host cell. But bear in mind that other possible explanations still remained. For example, the virus with the new oncogene might have been an endogenous virus that was reactivated by infection with the Abelson virus.

The next major breakthrough was the demonstration that the Rous sarcoma virus *src* gene, in point of fact, has its counterpart in the genomes of normal chicken cells. This crucial discovery came to pass as follows. Harold Varmus came to Michael Bishop's laboratory (at the University of California School of Medicine, San Francisco) as a postdoctoral fellow in 1970, at a time when it already was believed that cancer likely has an underlying genetic basis; this premise is strongly implied by the fact that when a cancer cell divides, each of its daughter cells is likewise a cancer cell. Varmus and Bishop had the insight to see the *possibility* that cancer-causing genes (oncogenes?) might in fact be necessary parts of the cell's normal genetic makeup and that cancer might occur when these genes are mutated or otherwise improperly expressed. With those thoughts in mind, Varmus and Bishop set out to test the hypothesis that *src* might actually be present in the normal chicken genome.

To test their hypothesis, Varmus and Bishop needed to generate a probe that might detect the putative cellular *src* gene. They could not use the complete Rous sarcoma virus genome for this purpose, because the complete virus genome might possibly have detected endogenous retrovirus sequences within the cellular genome, rather than a cellular *src* gene per se. So, they needed to generate a more specific probe.

In the days before recombinant DNA procedures, Varmus and Bishop cleverly generated their *src* probe by making use of a transformation-defective mutant of Rous sarcoma virus, isolated earlier by Peter Vogt. The important feature of this mutant was that its *src* gene was deleted. (Before reading on, can you see how Varmus and Bishop might have used Vogt's *src* deletion mutant to isolate their *src* probe?) Varmus and Bishop generated their probe by first using reverse transcriptase to make a radioactively labeled, single-stranded DNA copy of the entire standard Rous sarcoma virus genome, which contains the *src* gene. This cDNA was then fragmented and annealed to an excess of genomes of the *src* deletion mutant. The only DNA fragments that did not anneal were those containing only *src* sequences. These single-stranded DNA fragments could be separated from the annealed product and used as the *src* nucleic acid hybridization probe.

Using their cDNA probe, Varmus and Bishop were able to demonstrate the presence of *src* not only in the genomes

of normal chicken cells but also in the genomes of many other vertebrates as well, including humans (Stehelin et al., 1976). These experimental findings led to the remarkable conclusions that the cellular *src* gene was present early in vertebrate evolution and that it has remained conserved to this day.

More experiments of this kind demonstrated that other highly oncogenic retroviruses contain oncogenes of their own, which are distinct from *src* and which have their own counterparts in normal cell genomes. Indeed, each of the known retroviral oncogenes corresponds to a gene present in a normal cellular genome, and each of these retroviral oncogenes appears to be derived from a cellular genome (Box 20.5).

Direct evidence for the involvement of *cellular* oncogenes in human cancers was provided shortly afterwards by the experiments of Robert Weinberg, Geoffrey Cooper, and others, in which oncogenes were cloned directly from "spontaneous" human cancers and then were shown to be present in normal cells as well. To appreciate just how impressive these achievements were at the time, bear in mind that this pioneering work was done at the very start of the recombinant DNA era, before the development of the more powerful experimental procedures available to contemporary researchers.

Perhaps the most straightforward of the experimental stratagems devised at the time was one in which "normal" mouse cells in culture were transfected with fragmented DNA from a human tumor: a bladder carcinoma in this instance. The few mouse cells in the culture that received the transforming gene from the human tumor gave rise to foci of neoplastically transformed cells that were readily visible in the cell culture. DNA was extracted from clones of those transformed mouse cells. That DNA was then fragmented and used for a second cycle of transfection into fresh mouse cells. The second transfection was carried out because transformed mouse cells arising from it would have incorporated much less human DNA than the mouse cells transformed by the first transfection. Indeed, cells transformed by the second transfection would likely have incorporated only the human oncogene and short associated sequences. Thus, the second transfection cycle was carried out to further isolate the human oncogene from other human DNA sequences. A λ phage genomic library was then made from the DNA extracted from a mouse cell clone carrying the human oncogene. Next, this λ library was expressed in *Escherichia coli*, and the resulting plaques were transferred to a filter paper replica, which then was probed with human *alu* sequences. (*alu* sequences are short, approximately 300-bp, repetitive sequences that are dispersed throughout the human genome at about a million different sites. They derive their name from the

BOX 20.5

JSRV, a simple retrovirus, does not express a host-derived oncogene, but it nevertheless can induce lung cancer in immunodeficient mice, in as little as 10 days. Note that this is a much shorter time to oncogenesis than is typically found in the case of oncogene-deficient oncogenic retroviruses (see the text).

So, how might JSRV induce neoplasia? In contrast to known retroviral oncogenes, which have their counterparts in cellular genes that positively regulate cell growth, the oncogenic activity of JSRV is *due to the viral Env glycoprotein*. When the JSRV Env is expressed from a recombinant vector, it transforms cells in culture and induces tumors in vivo, doing so by activating known mitogenic signaling pathways, via as-yet-unknown triggering mechanisms (Wootton et al., 2005). These experimental findings provide a rare example of an oncogenic viral *structural* protein. Although their processes are less dramatic, the Env proteins of the murine mammary tumor virus and the avian hemangioma virus are also oncogenic (Alian et al., 2000; Ross et al., 2006).

Since the activities expressed by a virus gene generally provide the virus with some advantage, what benefit, if any, might JSRV derive from the transforming activity of its Env protein? Miller and coworkers suggest that since the primary effect of the JSRV Env protein in JSRV-infected sheep is to cause proliferation of infected lung epithelial cells, which in turn causes increased secretion of lung fluids, the benefit to the virus may be to facilitate its transmission via enhanced fluid secretions and aerosol production. These secretions are indeed full of virus. Moreover, since all progeny viral genomes are generated by RNA Pol II-mediated transcription of the integrated proviral DNA, the JSRV Env protein gives virus production a boost by causing infected cells to proliferate, thereby further enhancing transmission.

The example provided by JSRV and the other examples noted here provide a new paradigm in retrovirus biology. Still, the vast majority of known retroviral oncogenes are derived from the host genome.

References
Alian, A., D. Sela-Donenfeld, A. Panet, and A. Eldor. 2000. Avian hemangioma retrovirus induces cell proliferation via the envelope (*env*) gene. *Virology* **276**:161–168.

Ross, S. R., J. W. Schmidt, E. Katz, L. Cappelli, S. Hultine, P. Gimotty, and J. G. Monroe. 2006. An immunoreceptor tyrosine activation motif in the mouse mammary tumor virus envelope protein plays a role in virus-induced mammary tumors. *J. Virol.* **80**:9000–9008.

Wootton, S. K., C. L. Halbert, and A. D. Miller. 2005. Sheep retrovirus structural protein induces lung tumours. *Nature* **434**:904–907.

---

fact that they are recognized at a single site by the AluI restriction endonuclease. Most importantly, *alu* is a uniquely human sequence.) By using the human *alu* sequence as a probe for human DNA sequences, a single phage in the library, which carried the human oncogene, was isolated. The cloned oncogene was then used as a probe to demonstrate that the oncogene indeed is present in normal human cells (Shih and Weinberg, 1982).

The oncogene isolated from the bladder carcinoma, as described in the preceding paragraph, was also isolated from several different human bladder carcinomas. That oncogene is now known as *ras*, reflecting the fact that it was initially discovered in the genomes of several rapidly transforming *rat sarcoma* viruses. But importantly, *ras* is different from the oncogenes isolated from several other human tumor types. Indeed, beginning with these initial discoveries, *more than 100 or so different oncogenes* have been isolated directly from bladder, breast, colon, lung, and other cancers.

What then became of isolating oncogenes directly from retroviruses? For a while, whenever a new retrovirus was isolated, there was the exciting possibility that it might lead to the discovery of a new cellular oncogene. However, that approach eventually became exhausted, and more oncogenes have since been found by directly isolating them from normal cells and tumors.

Further work showed that the normal cellular counterparts of viral oncogenes are commonly mutated in "spontaneous" human cancers (i.e., the 80% or so of human cancers that are not induced by viruses). That is, cancer may arise as a consequence of mutations in cellular genes that are related or identical to retroviral oncogenes. Bishop coined the term **proto-oncogene** for these cellular genes in their normal state. Regarding the **Ras** protein, 15% or more of all human tumors have a *ras* mutation that leads to unregulated Ras activity (see below). In yet other tumors, the *ras* gene has been amplified (Box 20.6).

Why would cells harbor such potentially dangerous genes? Bishop and Varmus correctly postulated that these genes normally regulate cell growth. They are proto-oncogenes in their normal state, but they may become oncogenes when mutated or inappropriately expressed, as in the *ras* example. What then are the normal activities of the various proto-oncogenes, and what sorts of lesions cause them to become oncogenes?

We begin to answer these questions by examining the cellular signaling pathways that normally respond to growth-stimulatory signals (Fig. 20.11). Our reason for doing so is that the various retroviral oncogenes are analogs of cellular proto-oncogenes, which encode proteins involved at many different steps of these pathways. Indeed, there are examples of proto-oncogene products that act as signal mediators at the plasma membrane, in the cytoplasm, and in the nucleus (Fig. 20.12). Some cases in point are discussed below, and others are enumerated in Table 20.1. Note that in many instances, the gene encoding a particular intracellular signaling molecule was initially discovered as a retroviral oncogene, and only afterwards was it shown to be a cellular gene, that is, a proto-oncogene. So, what are the components of these mitogenic signaling pathways?

Cellular growth and differentiation are positively regulated by extracellular peptide growth factors, such as the epidermal growth factor (EGF). These growth factors trigger mitogenic signals when they bind to the extracellular domains of their specific transmembrane receptors. This happens as follows. Most growth factor receptors consist of a single polypeptide that contains an intrinsic **protein-tyrosine kinase** activity located on its cytoplasmic tail. When growth factors bind, they cause their receptor molecules to dimerize, in that way enabling the individual receptor monomers to cross-phosphorylate each other (Fig. 20.13). This cross-phosphorylation event creates phosphotyrosine-binding sites on the cytoplasmic tail of the receptor. Downstream signal mediators, which have so-called SH2 domains (see below), bind to those phosphotyrosine-binding sites, in that way setting off the cascade of molecular interactions that constitute the

signal. Subsequent mediators of these signaling pathways are enumerated below. But first, note the following.

Some signaling receptors, such as cytokine receptors and integrins, do not have an intrinsic protein-tyrosine kinase activity. In order for these receptors to transmit signals into the cell, they associate with an intracellular **nonreceptor protein-tyrosine kinase**. Like growth factor receptors, cytokine receptors also dimerize when they bind their ligands. The associated nonreceptor protein-tyrosine kinases are then able to cross-phosphorylate each other, resulting in their activation (as described below). The activated nonreceptor protein-tyrosine kinases then phosphorylate the receptor polypeptide per se, thereby creating phosphotyrosine-binding sites for downstream signaling molecules, as in the case of growth factor receptors (Fig. 20.14).

Cellular Src and Abl are examples of nonreceptor protein-tyrosine kinases. Src, in point of fact, was the first known nonreceptor protein-tyrosine kinase. Moreover, the discoveries of the biological role of Src and of its part in revealing receptor tyrosine phosphorylation were major steps forward in the understanding of cell signaling and the regulation of cell growth. How these breakthroughs happened and why their impact was so significant are recounted at length in Box 20.7.

Next, we consider subsequent steps of mitogenic signal transmission. So-called **adaptor proteins** bind via their SH2 domains (SH is an acronym for Src homology; see below) to specific phosphotyrosine-containing sequences on activated receptors. These adaptor proteins are so named because they link the activated receptor to downstream mediators of the signal pathway.

Grb2 is a particularly important example of an adaptor protein. A complex of Grb2 and the guanine nucleotide exchange factor SOS binds to a phosphotyrosine-containing sequence in the activated receptor, doing so via the SH2 domain in Grb2 (Fig. 20.13). This interaction brings SOS to the plasma membrane, where it can stimulate GDP$\rightarrow$GTP exchange on Ras (a small guanosine triphosphate [GTP]-binding protein), thereby activating Ras. (Recall that the *ras* oncogene was initially discovered in the genomes of several rapidly transforming rat sarcoma viruses and was later isolated directly from several human bladder carcinomas.) The activated Ras/GTP complex then binds to, and activates, Raf, a serine/threonine protein kinase. Activated Raf then triggers the MAP (mitogen-activated protein) kinase cascade by activating MEK (MAP kinase kinase), also a serine/threonine kinase (Fig. 20.11). Activated MEK in turn activates the serine/threonine kinase MAPK (for mitogen-activated protein kinase; also known as ERK, for extracellular signal-regulated kinase). The signal is then translocated to the nucleus, where it activates

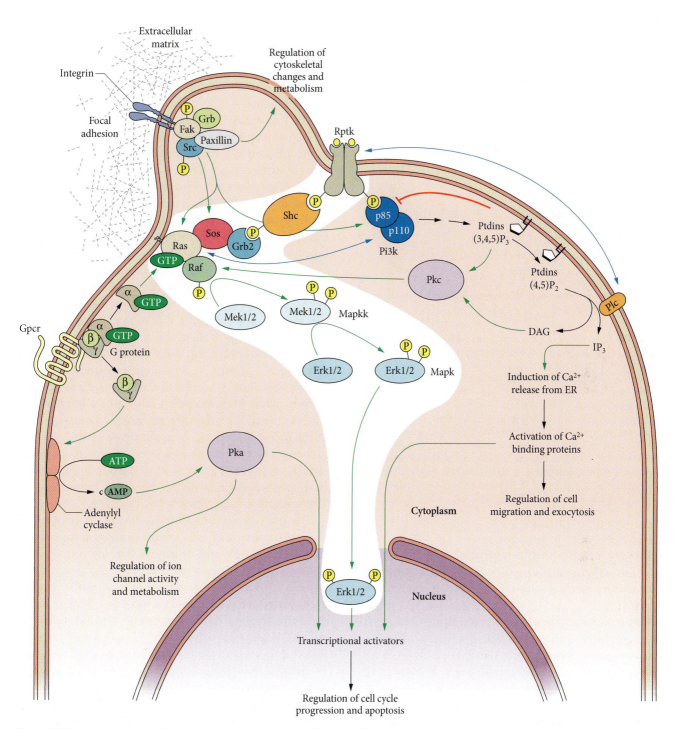

Figure 20.11 Some of the intracellular signaling pathways that normally respond to growth-stimulatory signals and their interconnections. The MAP kinase pathway, which extends from growth factor receptors at the cell surface to transcription factors in the nucleus, is a key mitogenic pathway. Ligand binding to the receptor induces receptor dimerization and autophosphorylation of tyrosine residues on the receptor, as mediated by an intrinsic receptor protein-tyrosine kinase activity. This creates phosphotyrosine-containing sites on the receptor, which serve as binding sites for downstream signaling molecules, such as Shc and Grb2. The binding of these adaptor proteins enables them to bring other signaling mediators to the activated receptor, beginning with the GTP-binding protein, Ras. Inactive GDP-Ras is activated by Sos, doing so by inducing the exchange of the Ras-bound GDP with GTP. Ras, in its now-active GTP-bound state, interacts with, and activates, the serine/threonine kinase activity of Raf, which in turn triggers the MAP kinase cascade by activating MAP kinase kinase (MEK). MEK then activates the serine/threonine kinase MAPK (for mitogen-activated protein kinase; also known as ERK, for extracellular signal-regulated kinase). The signal is then translocated to the nucleus, where it activates transcription factors that turn on expression of genes involved in cell growth and proliferation. Other mitogenic intracellular signaling pathways, which intersect the MAP kinase pathway, are also diagrammed. Note the signaling pathway emanating from focal adhesions and the association of the nonreceptor protein-tyrosine kinases Src and Fak with the focal adhesion. When integrins bind to the extracellular matrix, a tyrosine residue on Fak is autophosphorylated. The Src SH2 domain then binds to the phosphorylated tyrosine residue on Fak, thereby activating Src. The activated Src then phosphorylates additional sites on Fak, in that way creating binding sites for the Grb2-Sos complex and consequently triggering the MAP kinase pathway. Modified from S. J. Flint, L. W. Enquist, R. M. Krug, V. R. Racaniello, and A. M. Skalka, *Principles of Virology: Molecular Biology, Pathogenesis, and Control*, 2nd ed. (ASM Press, Washington, DC, 2004), with permission.

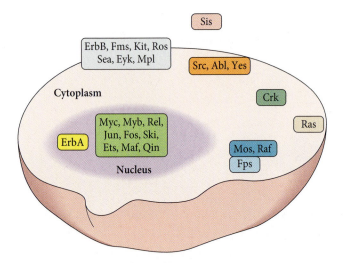

Figure 20.12 Schematic of sites of action of oncoproteins. Oncoproteins function as components of mitogenic signaling chains. The products of important retroviral oncogenes are shown grouped into seven categories: growth factors (Sis); integral plasma membrane proteins with the structure of growth factor receptors that are also tyrosine-specific protein kinases (ErbB, Fms, Kit, Ros, Sea, Eyk, and Mpl); membrane-bound nonreceptor tyrosine kinases (Src, Abl, and Yes); linker-adaptor proteins (Crk); GTPases (Ras); cytosolic serine/threonine (Mos and Raf) and tyrosine-protein kinases (Fps); transcription factors (Jun, Fos, Rel, Myb, Myc, Ets, Ski, Qin, and Maf); and hormone receptors (ErbA). Adapted from J. R. Nevins and P. K. Vogt, p. 276, *in* B. N. Fields, D. M. Knipe, and P. M. Howley (ed.), *Fundamental Virology*, 3rd ed. (Lippincott-Raven, Philadelphia, PA, 1996), with permission.

transcription factors that turn on expression of genes involved in cell growth and proliferation. Other mitogenic intracellular signaling pathways also are diagrammed in Fig. 20.11.

Retroviral oncogenes encode proteins involved in virtually every step of the mitogenic signal transduction cascade outlined above and other signaling pathways as well (Table 20.1). But if retroviral oncogenes are merely captured cellular proto-oncogenes, why does their expression from the viral genome cause neoplastic transformation in cell culture and cancer in vivo? Several possibilities may come to mind. Those which are recounted here are known to actually facilitate neoplasia in vivo. They are as follows. First, viral oncogenes are transcribed under the control of the viral promoter and enhancer elements. Under these conditions, expression of these genes is not regulated by extracellular signals and checkpoints of the cell cycle that normally would regulate their expression. Second, the highly efficient viral enhancer and promoter elements may lead to extremely elevated levels of the viral oncogene products in the cell. Third, the recombination process that incorporates the cellular proto-oncogene into the retroviral genome might lead to alterations in a regulatory domain of the proto-oncogene product, resulting in its

constitutive activity. Specific real examples of these possibilities follow.

The simian sarcoma virus (SSV) *sis* oncogene encodes the actual platelet-derived growth factor (Table 20.1). *sis* is oncogenic in SSV-transformed cells because it is constitutively expressed under control of the viral promoter/enhancer elements, resulting in the continual "autostimulation" of the MAP kinase signaling cascade. That is, the virally generated *sis* gene product is providing the transformed cell with its own intrinsic growth factor, a type of cell stimulation referred to as **autocrine**.

In contrast to the SSV *sis* oncogene, which encodes a soluble growth factor, the avian erythroblastosis virus (AEV) *erbB* oncogene encodes a truncated form of an actual growth factor receptor, that of the EGF. Also in contrast to the SSV Sis protein, which is oncogenic because it is continuously produced, the AEV-encoded ErbB protein is oncogenic because it is missing regulatory sequences and consequently is thought to transform cells as a constitutively active EGF receptor (Table 20.1). Interestingly, whereas the AEV *erbB* oncogene encodes a mutated form of the EGF receptor that is constitutively active, *naturally* occurring human breast and ovarian cancers express a normal *erbB* gene but do so at elevated levels. (You may have been wondering whether there is an *erbA* oncogene. There is, and it is described below.)

The nonreceptor protein-tyrosine kinases encoded by the retroviral *src* and *abl* oncogenes are tumorigenic because, like the AEV *erbB* oncogene, they too contain deletions that affect a regulatory domain in the proteins they encode. The functional domains of the *cellular* Src protein are depicted in Fig. 20.15. SH1 is the kinase domain of Src. Note the tyrosine residues at amino acid positions 416 and 527. Phosphorylation of Tyr527 negatively regulates Src. It does so because phosphorylated Tyr527 binds to the Src **SH2 domain**, thereby bringing the Src **SH3 domain** into contact with the polyproline helix located between SH1 and SH2. This interaction between SH3 and the polyproline helix distorts SH1, and in so doing, it blocks its kinase activity. (Recall that adaptor proteins such as Grb2 bind via their SH2 domains to specific phosphotyrosine-containing sequences on activated receptors.)

In contrast to the normal cellular Src protein, the retroviral Src protein is not subject to autoinhibition via the phosphorylated Tyr527 but instead is constitutively active. That is so because the region of the viral oncogene encoding the C-terminal regulatory sequence of Src was deleted, presumably at the time the cellular *src* was incorporated into the viral genome.

Cellular Src can normally be activated by signals that trigger tyrosine phosphorylation of an unrelated target protein. That is so because the phosphotyrosine-containing

Table 20.1 Functional classes of oncogenes transduced by retroviruses[a]

Transduced oncogene	Viral oncoprotein[b]	Function of cellular homolog
Growth factor		
sis	p28$^{env\text{-}sis}$	Platelet-derived growth factor
Tyrosine kinase growth factor receptors		
erbB	gp65^{erbB}	Epithelial growth factor receptor
fms	gp180$^{gag\text{-}fms}$	Colony-stimulating factor 1 receptor
sea	gpl60$^{env\text{-}sea}$	Receptor; ligand unknown
kit	gp80$^{gag\text{-}kit}$	Hematopoietic receptor; product of the mouse W locus
ros	p68$^{gag\text{-}ros}$	Receptor; ligand unknown
mpl	p31$^{env\text{-}mpl}$	Member of the hematopoietin receptor family
eyk	gp37^{eyk}	Receptor; ligand unknown
Hormone receptor		
erbA	p75$^{gag\text{-}erbA}$	Thyroid hormone receptor
G proteins		
H-*ras*	p21^{ras}	GTPase
K-*ras*	p21^{ras}	GTPase
Adaptor protein		
crk	p47$^{gag\text{-}crk}$	Signal transduction
Nonreceptor tyrosine kinases		
src	pp60^{src}	Signal transduction
abl	p460$^{gag\text{-}abl}$	Signal transduction
fps[c]	p130$^{gag\text{-}fps}$	Signal transduction
	p105$^{gag\text{-}fps}$	
fes[c]	p85$^{gag\text{-}fes}$	Signal transduction
fgr	p70$^{gag\text{-}actin\text{-}fgr}$	Signal transduction
yes	p90$^{gag\text{-}yes}$	Signal transduction
	p80$^{gag\text{-}yes}$	
Serine/threonine kinases		
mos	p37$^{env\text{-}mos}$	Required for germ cell maturation
raf[d]	p75$^{gag\text{-}raf}$	Signal transduction
mil[d]	p100$^{gag\text{-}mil}$	Signal transduction
akt	p86$^{gag\text{-}akt}$	Signal transduction
Nuclear proteins		
jun	p65$^{gag\text{-}jun}$	Transcriptional regulator (Ap-1 complex)
fos	p55^{fos}	Transcriptional regulator (Ap-1 complex)
myc	p100$^{gag\text{-}myc}$	
	p90$^{gag\text{-}myc}$	
	p200$^{gag\text{-}pol\text{-}myc}$	
	p59$^{gag\text{-}myc}$	
myb	p45^{myb}	Transcriptional regulator
	p135$^{gag\text{-}myb\text{-}ets}$	
ets	p135$^{gag\text{-}myb\text{-}ets}$	Transcriptional regulator
rel	p64^{rel}	Transcriptional regulator

(continued)

Table 20.1 Functional classes of oncogenes transduced by retroviruses[a] *(continued)*

Transduced oncogene	Viral oncoprotein[b]	Function of cellular homolog
maf	p100$^{gag\text{-}maf}$	Transcriptional regulator
ski	p110$^{gag\text{-}ski\text{-}pol}$	Transcriptional regulator
qin	p90$^{gag\text{-}qin}$	Transcriptional regulator of the forkhead/Hnk-3 family

[a]Adapted from T. Benjamin and P. Vogt, p. 331, *in* D. M. Knipe and P. M. Howley (ed.), *Fields Virology* (Lippincott-Raven Publishers, Philadelphia, Pa., 1996).
[b]Designations for viral proteins: p, protein; gp, glycoprotein; pp, phosphoprotein. The last is not applied consistently but is used mainly in conjuction with the *src* product. The numbers give the estimated molecular mass in kilodaltons, and the superscript lists the genes from which the coding information is derived in the 5′ → 3′ direction. The listing of more than one protein for an oncogene signifies its inclusion in independent virus isolates.
[c]*fps* and *fes* are the same oncogene derived from the avian and feline genomes, respectively.
[d]*raf* and *mil* are the same oncogene derived from the murine and avian genomes, respectively.

sequence on the target protein might then displace the phosphorylated Src Tyr527 from the Src SH2 domain, in that way releasing the SH3 domain from the polyproline sequence and enabling the SH1 domain to assume its active conformation. Likewise, proteins with polyproline motifs also can activate cellular Src, in that case by binding to the Src SH3 domain, thereby displacing the Src polyproline helix from SH3 and releasing SH1 from its inhibited conformation. Once Src is activated by these means, the activated state can then be stabilized by autophosphorylation of Tyr416 (Box 20.8).

Src is also found in association with **focal adhesions.** These are junctions between a cell and the extracellular matrix at which cellular integrins bind to actin filaments of the extracellular matrix. Interestingly, the initial discovery that the retroviral Src protein associates with focal adhesions (Fig. 20.11) led to the discovery of another nonreceptor protein kinase, the focal adhesion kinase (Fak), which (as its name implies) is also located at focal adhesions. The significance of these facts is as follows. Not only do integrins link cells to the extracellular matrix, but also, when integrins bind to the extracellular matrix, they activate intracellular signaling pathways that link cell adhesion to changes in gene expression. Importantly, some integrin-induced changes in gene expression are analogous to those triggered by growth factor receptors. That is so because the integrin-induced signaling pathway intersects the MAP kinase pathway (Fig. 20.11). Yet unlike the growth factor receptors that trigger the MAP kinase pathway, the integrins have short cytoplasmic domains that do not express any intrinsic tyrosine kinase activity. Instead, the integrins associate with the nonreceptor protein-tyrosine kinases Src and Fak, which, in this instance, actually appear to work in conjunction with each other as follows. When integrins bind to the extracellular matrix, a tyrosine residue on Fak is autophosphorylated. The phosphotyrosine-containing sequence on Fak then displaces the phosphorylated Src Tyr527 from the Src SH2 domain, thereby activating Src. The activated Src then

phosphorylates additional sites on Fak, in that way creating binding sites on Fak for the Grb2/Sos complex. Voila! The MAP kinase pathway is triggered, resulting in the upregulation of genes that promote cell proliferation. Thus, the oncogenic activity of the retroviral Src protein may result in part from its association with focal adhesions and, consequently, its ability to constitutively activate Fak (and, in turn, the MAP kinase pathway). Despite this possibility, bear in mind that most proto-oncogenes encode proteins that play a role in growth factor signaling, not integrin signaling. Still, notice that signals originating from other types of receptors also intersect the MAP kinase pathway, and mediators of these other pathways thus have the potential to become oncogenes (Fig. 20.11).

Why is the retrovirus-encoded Ras protein oncogenic? As discussed above, Ras is a GTP-binding protein. As such, Ras may be present in either an active GTP-bound state or an inactive guanosine diphosphate (GDP)-bound state. GDP-Ras is normally activated by signals that cause SOS to come to the plasma membrane, where it can stimulate GDP/GTP exchange on Ras (Fig. 20.13). Activated GTP-Ras then usually quickly returns to the inactive GDP-bound state because Ras expresses an intrinsic GTPase activity. This ability of Ras to rapidly catalyze its self-inactivation is biologically important because it enables Ras-mediated signal transmission to remain responsive to extracellular signals that cause Grb2/Sos complexes to bind to activated receptors. But in contrast to the cellular Ras protein, activated retroviral Ras proteins hydrolyze GTP relatively slowly. Thus, the retroviral Ras protein remains in the active GTP-bound state for prolonged periods, resulting in constitutive cell growth and proliferation. It is noteworthy that *ras* mutations found in human cancers likewise generally affect the intrinsic Ras GTPase activity. Importantly, mutant *ras* is found in about 15% of all human cancers. But it is found in 25% of lung cancers, 50% of colon cancers, and 90% of prostate cancers. Mutations in cellular *ras* were the first genetic changes linked to human cancer.

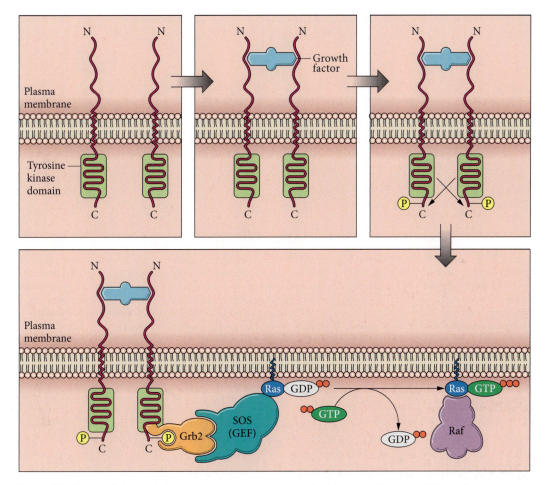

Figure 20.13 Growth factor binding causes receptor protein tyrosine kinases to dimerize, which results in receptor autophosphorylation, as the two polypeptide chains phosphorylate each other. Downstream signaling molecules, with SH2 domains, bind to specific phosphotyrosine-containing sequences on the activated receptors. These proteins serve as adaptor molecules that link the activated receptor to downstream mediators of the signal pathway. In the example shown, a complex of the adaptor protein Grb2 and the guanine nucleotide exchange factor, SOS, binds to a phosphotyrosine-containing sequence in the activated receptor via the Grb2 SH2 domain. This interaction brings SOS to the plasma membrane, where it can stimulate GDP/GTP exchange on Ras, thereby activating Ras. The activated Ras/GTP complex then binds to, and activates, the Raf serine/threonine protein kinase. Activated Raf initiates a protein kinase cascade leading to the activation of MAP kinase (ERK) (Fig. 20.11). Adapted from G. M. Cooper, *The Cell: a Molecular Approach* (ASM Press, Washington, DC, 1997), with permission.

Aberrant *raf* genes (Fig. 20.11) are also found as retrovirus oncogenes. In these instances, a regulatory sequence at the N terminus of the normal Raf protein is deleted, thereby resulting in the constitutive expression of the Raf serine/threonine kinase activity (Fig. 20.11 and 20.13; Table 20.1). Other serine/threonine kinases, which have also become constitutive as a result of mutation, are also found as retroviral oncogenes (Table 20.1).

Still other retroviral oncogenes encode nuclear transcription factors. *myc* is one such notable example. Unlike *ras* and *raf*, which are oncogenic because they contain genetic lesions causing them to be constitutively active, the retroviral *myc* oncogene does not contain such a lesion.

Instead, *myc* is oncogenic because it is overexpressed (Table 20.1). In this regard, the tightly regulated cellular *myc* gene is normally active only during a short period in the G_1 phase of the cell cycle (Box 20.9).

The *erbA* oncogene of AEV encodes a homolog of the cellular thyroid hormone receptor (Fig. 20.12; Table 20.1). Whereas protein growth factors bind to the extracellular domains of receptors at the cell surface, steroid hormones (such as thyroid hormone and retinoic acid) diffuse across the plasma membrane and bind to intracellular receptors located in the cytosol. The ligand-bound steroid receptors then enter the nucleus and directly interact with their target genes. But in contrast to growth factor-triggered

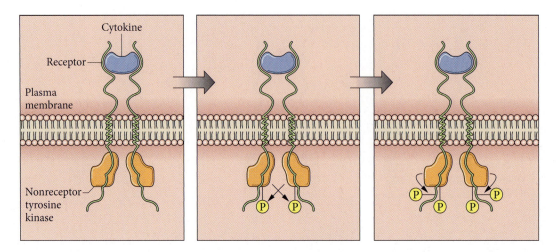

Figure 20.14 Some cellular receptors, including cytokine receptors and the integrins, do not have an intrinsic protein-tyrosine kinase activity. In order for these receptors to transmit signals, they associate with a nonreceptor protein-tyrosine kinase. Ligand binding causes these receptors to dimerize, just as it induces the growth factor receptors to dimerize. However, in the case of these receptors, their associated nonreceptor protein-tyrosine kinases cross-phosphorylate each other, resulting in their activation. The activated nonreceptor protein-tyrosine kinases then phosphorylate the receptor, thereby activating it and triggering subsequent steps in the signaling pathway by creating phosphotyrosine-binding sites for downstream signaling molecules. Modified from G. M. Cooper, *The Cell: a Molecular Approach* (ASM Press, Washington, DC, 1997), with permission.

signals, which stimulate cell proliferation, the host ErbA protein instead induces the differentiation of a variety of cell types. With that in mind, the retroviral ErbA protein was originally thought to act as a dominant-negative inhibitor of the cellular thyroid hormone receptor, in that way blocking the cell differentiation-inducing activity of the host ErbA protein. However, more recent experimental results imply that the oncogenicity of the viral ErbA protein results from its impairment of the closely related retinoic acid receptor. Thus, ErbA is thought to cause erythroleukemia in chickens by preventing the retinoic acid-induced differentiation of erythroblasts into erythrocytes, which instead remain in an actively proliferating predifferentiated state. In humans, mutated forms of the cellular retinoic acid receptor prevent differentiation of promyelocytes to granulocytes, thereby giving rise to acute promyelocytic leukemia. Presumably, the mutated human receptor interferes with the action of the normal homolog, thereby blocking differentiation of promyelocytes (Box 20.10).

Having concluded our survey of selected examples of retroviral oncogenes, we return now to a question alluded to above, i.e., why the oncogene-transducing oncogenic retroviruses derive no apparent benefit from the transduced oncogenes they carry. (The oncogenic Env proteins encoded by Jaagsiekte sheep retrovirus [JSRV] are a special case [Box 20.5].) We might very well have expected that these retroviruses should derive some benefit from

the oncogenes they carry, based upon the fundamental principle that viruses do not usually carry nonbeneficial genes. Moreover, the transforming proteins of the DNA tumor viruses indeed play key roles in the replication of those viruses (although transformation per se does not; Chapters 15, 16, and 17). Thus, we might have expected that retroviral oncogenes would provide some advantage to these viruses, irrespective of the fact that these genes are actually "captured" cellular genes. However, a more perceptive comparison of the retrovirus lifestyle with that of the DNA tumor viruses offers insights into why the retroviruses have little, if any, use for their oncogenes. Recall that the DNA tumor viruses need to express their transforming genes to induce infected cells to enter the S phase of the cell cycle, which enables those cells to efficiently support viral DNA replication. But in contrast to the DNA tumor viruses, retroviruses do not replicate a DNA genome. Instead, reverse transcription is a nonreplicative process that converts the genomic RNA into a double-stranded DNA copy that is integrated into the cellular genome, where it is transcribed by the Pol II RNA polymerase to generate all of the progeny viral RNA genomes. Thus, retroviruses have no critical need to trigger cell proliferation.

Another, perhaps more compelling reason why oncogene-transducing oncogenic retroviruses do not benefit from their oncogenes is that the acquisition of a cellular proto-oncogene generally, but not always, is accompanied by the loss of a viral gene sequence, which leaves the

BOX 20.7

In 1978, Raymond Erikson and coworkers were the first researchers to identify the Src protein. They did so by first preparing lysates from avian and mammalian cells that had been transformed in vitro by Rous sarcoma virus. Their next step was to immunoprecipitate those lysates with antisera from rabbits that bore Rous sarcoma virus-induced tumors, expecting that antisera from the tumor-bearing rabbits might recognize and immunoprecipitate proteins specific to the transformed cells. Shortly afterwards, Erikson's research group and that of Michael Bishop and Harold Varmus independently discovered that if the immunoprecipitates were incubated with $[\gamma$-^{32}P]ATP (adenosine triphosphate), then antibody molecules in the immunoprecipitates would be phosphorylated. Importantly, this finding revealed that there was a protein kinase activity in the immunoprecipitates. (This procedure is now known as the immune complex kinase assay.) Control experiments consisted of using the same rabbit antisera to immunoprecipitate extracts from normal cells or extracts from cells infected with a transformation-defective mutant of Rous sarcoma virus. No signs of protein kinase activity were seen in these control immunoprecipitates.

Significantly, this protein kinase activity was found to be temperature sensitive in immunoprecipitates from cells infected with a Rous sarcoma virus mutant that was temperature sensitive for transformation, thus confirming that the protein kinase activity in the immunoprecipitates was encoded by the virus. Since retroviral *src* expression already was known to be sufficient to induce tumors, taken together, these experimental findings demonstrated that the kinase activity of the retroviral Src protein *plays a key role* in transformation. Moreover, they implied that the cellular Src protein kinase activity plays a role in the control of normal cell proliferation.

At about the time that Erickson's research group was carrying out the above experiments, Walter Eckhart and Tony Hunter were interested in identifying the underlying basis for the transforming activity of the polyomavirus middle T (MT) protein (Chapter 15). Since Src was shown to possess a protein kinase activity, Eckhart and Hunter examined immunoprecipitates of polyomavirus T antigens to see if they too might contain a protein kinase activity and found that indeed they do. (Although not crucial to our story, note that MT per se does not express any intrinsic enzymatic activity of its own, although that fact was not known at the time. Instead, MT interacts with Src, and in so doing it activates the Src protein kinase activity by a mechanism that is still not known. Apropos the interaction of MT with Src, MT is a membrane-associated protein that interacts with several cellular proteins. The end result is a large multiprotein complex comprised of MT, protein phosphatase 2A, and an Src family tyrosine kinase [see Fig. 15.14]. The phosphorylation events carried out by MT-activated Src cause a variety of signal adaptor molecules [e.g., Shc, Grb2, and Sos; see the text] and other signal mediators [e.g., PI3K and PLCγ] to bind to the complex, thereby triggering a variety of mitogenic signaling pathways [Fig. 20.11] [Chapter 15]. Again, these facts were not yet known when Eckhart and Hunter were doing their experiments.)

At the time Src was discovered, serine and threonine were the only amino acids known to be phosphorylated by protein kinases. In fact, Erikson and coworker Marc Collett, as well as Bishop and Varmus, thought that threonine was the amino acid phosphorylated by the Src kinase (see below). Hunter then set out to determine whether the polyomavirus MT protein likewise phosphorylates threonine (again, it was not yet known that MT has no intrinsic kinase activity of its own). Hunter's experimental procedure was relatively straightforward. It involved incubating the immunoprecipitate of MT with $[\gamma$-^{32}P]ATP, hydrolyzing the immunoglobulin, and then separating the amino acids in the hydrolysate by electrophoresis. To Hunter's surprise, the position of the labeled amino acid in the electropherogram did not correspond to that of either threonine or serine.

Knowing that tyrosine is the only remaining amino acid with a free hydroxyl group that might be a target for the putative MT kinase activity, Hunter then considered the possibility that polyomavirus MT might phosphorylate tyrosine. But in point of fact, there was no known precedent for a protein-tyrosine kinase. Regardless, Hunter synthesized a phosphotyrosine molecule to be used as a standard marker against which to compare the labeled amino acid in a repeat of his earlier experiment. Doing so, he found that the amino acid that was phosphorylated by the MT kinase activity ran precisely with the phosphotyrosine standard.

Why then had other researchers not detected tyrosine phosphorylation earlier? In part it is because phosphotyrosine accounts for only about 0.03% of phosphorylated amino acids in normal cells. The remaining 99.97% are phosphoserine and phosphothreonine. But that is not the entire explanation. The rest is truly precious. In Hunter's own words, he was "too lazy to make up fresh buffer" before doing his experiments. Had the buffer been fresh, its pH would have been the usual 1.9, a pH that, unbeknownst to all at the time, does not separate phosphotyrosine from phosphothreonine during the electrophoresis procedure. The pH of the old buffer used by Hunter had inadvertently dropped to 1.7, a pH at which phosphotyrosine is resolved from phosphothreonine.

Box continues

BOX 20.7 *(continued)*

That fact enabled Hunter to resolve phosphotyrosine from phosphothreonine for the first time.

The discovery that tyrosine is the target of the polyomavirus MT protein kinase activity led Hunter and Bart Sefton to then investigate whether Src too might phosphorylate tyrosine, rather than serine or threonine, as previously thought (Hunter and Sefton, 1980). (Bear in mind that it became clear only later that MT actually has no intrinsic enzyme activity of its own and that it acts through Src.) The results of their experiments demonstrated that the normal cellular Src proteins, as well as the retroviral Src, indeed function as protein-tyrosine kinases. Moreover, the levels of phosphotyrosine were 10-fold higher in cells infected with wild-type Rous sarcoma virus than in control cells, consistent with the likelihood that the Src protein-tyrosine kinase activity might account for the altered growth potential of those cells.

Stanley Cohen then discovered that the EGF receptor contains an intrinsic protein-tyrosine kinase activity, further underscoring the importance of protein-tyrosine kinases in the normal control of cell proliferation. Subsequent studies identified additional receptor protein-tyrosine kinases (e.g., the fetal growth factor receptor) and nonreceptor protein-tyrosine kinases (e.g., Abl) that activate mitogenic intracellular signaling pathways.

Tony Hunter and coworkers went on to demonstrate that protein-tyrosine kinases play a key role in other crucial cellular processes, including cellular adhesion, vesicle trafficking, cell communication, the control of gene expression, protein degradation, and immune responses. Moreover, the discoveries regarding the role of protein-tyrosine kinases in cell transformation and cancer gave rise to a promising new rational approach to cancer therapy, i.e., the targeting of protein-tyrosine kinases. For example, the drug Gleevec recently was shown to inhibit the abnormal activation of the Abl and platelet-derived growth factor tyrosine kinases, and in clinical trials it demonstrated potential as a therapeutic agent for the treatment of metastatic melanoma.

Reference
Hunter, T., and B. M. Sefton. 1980. Transforming gene product of Rous sarcoma virus phosphorylates tyrosine. *Proc. Natl. Acad. Sci. USA* 77:1311–1315.

---

resulting oncogene-transducing retrovirus dependent on a helper virus for its replication. Why then do these defective oncogene-transducing retroviruses exist in the first place? They likely arise as by-products of the replication of nondefective, oncogene-deficient retroviruses and are isolated in the laboratory under the selective conditions by which they were detected. As such, they might represent dead ends in Nature.

The defective oncogene-transducing oncogenic retroviruses are propagated in the laboratory in association with their corresponding nondefective, nontransducing counterparts. Consequently, many of the oncogenes referred to above were first identified in such mixed virus stocks. Although it was not known at the time, the original Rous sarcoma virus samples, isolated by Peyton Rous, contained defective oncogene-transducing viruses, in association with nondefective, nontransducing helper viruses. The latter came to be referred to as Rous-associated or leukosis viruses. These nontransducing, Rous-associated viruses were yet able to induce B-cell tumors in chickens after a long incubation period, as described below. (Despite the preceding, nondefective, oncogene-transducing, oncogenic retroviruses do exist. In fact, nondefective AEV carries a complete set of replication genes plus two oncogenes, *erbA* and *erbB*. *erbA* encodes a mutated form of the thyroid hormone receptor, while *erbB* encodes a mutated form of the EGF receptor, as noted above.)

Oncogene-Deficient Oncogenic Retroviruses

So, as just noted, retroviruses that do not carry an oncogene *are yet able to induce tumors*. They can do so by virtue of the fact that all retroviruses generate an integrated DNA provirus, which is flanked by the virus-derived promoter- and enhancer-containing LTR sequence. Consequently, even nontransducing retroviruses can cause neoplasia by integrating in the vicinity of a cellular proto-oncogene, which then might come under the transcriptional control of the highly efficient retroviral promoter and enhancer elements. How close does the provirus actually need to be to the proto-oncogene in order to affect its expression? As a matter of fact, retrovirus insertional events have been found to upregulate cellular genes over distances of more than 10 kb.

This means by which oncogene-deficient oncogenic retroviruses activate cellular proto-oncogenes is referred to as **insertional activation**, or "*cis*-activation." But note that while oncogene-transducing oncogenic retroviruses,

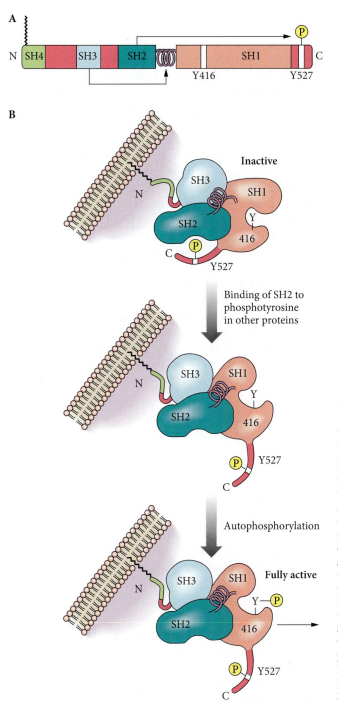

Figure 20.15 The functional domains of the normal cellular Src protein. SH1 is the active kinase domain of Src. Note the tyrosine residues at amino acid positions 416 and 527. Phosphorylation of Tyr527 negatively regulates Src. It does so because phosphorylated Tyr527 binds to the Src SH2 domain, in that way bringing the Src SH3 domain into contact with the polyproline helix located between SH1 and SH2. This interaction between SH3 and the polyproline helix distorts SH1, and in so doing, it blocks the SH1 kinase activity. The viral Src protein is not subject to autoinhibition via the phosphorylated Tyr527 because the region of the viral oncogene encoding the C-terminal regulatory sequence of Src was presumably deleted at the time the virus acquired the cellular *src* gene. Consequently, the viral Src protein is constitutively active. Src can be activated by signals that trigger tyrosine phosphorylation of an unrelated target protein, since the phosphotyrosine on the target protein might then displace the phosphorylated Src Tyr527 from the Src SH2 domain. Proteins with polyproline motifs also can activate cellular Src by binding to the Src SH3 domain, thereby displacing the Src polyproline helix from SH3 and releasing SH1 from its inhibited conformation. Once Src is activated by these means, the activated state can then be stabilized by auto-phosphorylation of Tyr416. Modified from S. J. Flint, L. W. Enquist, R. M. Krug, V. R. Racaniello, and A. M. Skalka, *Principles of Virology: Molecular Biology, Pathogenesis, and Control*, 2nd ed. (ASM Press, Washington, DC, 2004), with permission.

such as Rous sarcoma virus, can induce aggressive neoplasias that can kill an infected host in as little as 2 weeks, oncogene-deficient oncogenic retroviruses induce B-cell, T-cell, or myeloid leukemias and carcinomas only after long incubation periods of weeks to months. Thus, these viruses have been referred to as "weakly oncogenic," "*cis*-activating," or "slowly transforming" retroviruses. (How might you account for the always-long latent period

before the emergence of tumors induced by these viruses? See below.)

The first report to confirm the upregulation of a cellular proto-oncogene by the phenomenon of insertional activation involved the analysis of lymphomas induced by an oncogene-deficient ALV (Hayward et al., 1981). Key experimental findings in that report included the following. First, c-*myc* (i.e., the cellular *myc* gene) was expressed at

BOX 20.8

SH2 and SH3 domains are present on several other components of signal transduction pathways. Their role in these pathways is to mediate the specific binding of these signal mediators to phosphorylated tyrosine residues and polyproline motifs, respectively, on other interacting proteins in the pathway. Some proteins that contain SH2 or SH3 domains may express enzymatic activities, whereas others may serve as linker/adaptor molecules. The adaptor molecule Grb2 is an example of the latter. Grb2 contains an SH2 domain that interacts with phosphotyrosine residues on activated protein-tyrosine kinase receptors. The adaptor Sos has an SH3 domain that interacts with a proline-rich sequence on Grb2. Phospholipase C-γ, which plays a key role at the start of signaling pathways involving signal mediators derived from the membrane phospholipid PIP_2, contains an SH2 domain that enables it to also be stimulated by activated receptor protein-tyrosine kinases. Bearing in mind the critical role of SH2 and SH3 domains in key cell-signaling pathways, note that they were first recognized in the Src protein of Rous sarcoma virus.

BOX 20.9

Unregulated *myc* expression is associated with Burkitt's lymphoma, a malignant B-cell lymphoma of the jaw and face that is linked to Epstein-Barr virus infection (Chapter 18). Cells of Burkitt's tumors contain a chromosomal translocation, in which the cellular *myc* proto-oncogene, normally located on chromosome 8, is translocated to chromosome 14 (or more rarely to chromosome 2 or 22), where it comes under the transcriptional control of immunoglobulin gene enhancers. As a result, the translocated cellular *myc* gene is expressed throughout the cell cycle, resulting in unregulated proliferation of B cells. Epstein-Barr virus is thought to promote Burkitt's lymphoma by stimulating B-cell proliferation during the first stage of B-cell infection, thereby enhancing the probability of chromosomal damage (Chapter 18). Unregulated *c-myc* expression is also associated with several other human cancers, including breast and lung carcinomas. In those instances, the genetic lesion appears to be duplication of the *c-myc* gene.

higher levels in ALV-induced lymphomas than in normal tissue. Second, c-*myc* transcripts from lymphoma cells had the viral R-U_5 sequence at their 5′ ends. Third, the viral LTR sequence and c-*myc* sequence were detected on the same DNA restriction endonuclease fragments of the transformed-cell genomic DNA. Together, these experimental findings showed that the synthesis of the c-*myc* transcripts by the lymphoma cells must have initiated from the LTR sequence at the 3′ end of the provirus, thereby validating the insertional activation hypothesis. The slowly transforming murine and feline leukemia viruses have also been shown to induce lymphomas by insertional activation of the c-*myc* gene.

All cells of tumors induced by oncogene-deficient oncogenic retroviruses contain an integrated provirus, which is integrated at the same chromosomal site in all cells of the particular tumor. However, the exact site of integration varies from one tumor to another. These experimental findings show that tumors induced by oncogene-deficient oncogenic retroviruses are clonal. Moreover, they suggest that the insertion site is, at least to some extent, random. That being so, then perhaps only rare provirus insertional events place a cellular proto-oncogene under the transcriptional control of a provirus, thereby accounting for the inefficiency of transformation by the oncogene-deficient oncogenic retroviruses.

Here is another important point. The proto-oncogenes that are overexpressed in cells transformed by the oncogene-deficient oncogenic retroviruses are the very same proto-oncogenes that are expressed by the oncogene-transducing oncogenic retroviruses. Thus, our understanding of neoplasia as induced by the oncogene-deficient oncogenic retroviruses is consistent with our understanding of neoplasia as induced by the oncogene-transducing oncogenic retroviruses.

Not surprisingly, the efficiency with which an oncogene-deficient oncogenic retrovirus induces tumors depends on the efficiency of the promoter and enhancer elements within its LTR sequence. Moreover, if tumorigenesis via insertional activation results from those rare insertion events that position the provirus near an appropriate cellular proto-oncogene (as is likely), then the ability to replicate and disseminate efficiently may also affect the efficiency with which these viruses induce neoplasia. Consequently, the *gag*, *pol*, and *env* genes of these viruses may also affect their oncogenicity.

As noted just above, it is possible (if not likely) that provirus insertions only rarely result in insertional activation of proto-oncogenes, thereby explaining why oncogene-deficient oncogenic retroviruses rarely induce tumors. But how might we explain the always-long latent period before the emergence of tumors, since one might have

BOX 20.10

In the 1970s, it was hypothesized that it might be possible to rationally treat acute promyelocytic leukemia by using retinoic acid to induce promyelocytes to differentiate into granulocytes. This was a particularly attractive prospect because traditional therapies were based on killing promyelocytes with chemotherapy, an approach that has nasty side effects. Treatment with retinoic acid indeed resulted in the terminal differentiation of the promyelocytes and a 90 to 95% cure rate for acute promyelocytic leukemia patients.

Interestingly, in the 1970s, Chinese researchers demonstrated the efficacy of arsenic trioxide for treating acute promyelocytic leukemia. Moreover, the combination of arsenic trioxide and retinoic acid was found to result in a higher cure rate than either alone. But why was arsenic even tried here, since it is much better known as a poison? Surprisingly, perhaps, arsenic compounds have been used to treat a variety of illnesses. In Traditional Chinese Medicine, arsenic compounds have been used to treat tooth marrow disease, psoriasis, syphilis, and rheumatosis. The rationale in those instances was in a sense "using poison to cure another poison." But arsenic has also been used in Western medicine. For example, arsenic was used to treat syphilis before the advent of penicillin, and it still is used to treat trypanosomiasis involving the central nervous system.

In truth, it is not clear how arsenic trioxide, or even retinoic acid for that matter, effects a cure of promyelocytic leukemia. But one line of thought is as follows. The mutant receptor is thought to block myeloid differentiation by acting as a dominant negative inhibitor of the wild-type receptor response, and there are data showing that both retinoic acid and arsenic trioxide induce the degradation of the mutant retinoic acid receptor, although by different pathways.

Zhu Chen, of the Shanghai Institute of Hematology (Shanghai, China), was the principal figure behind the development of the radical approach that combined arsenic trioxide and retinoic acid for the treatment of acute promyelocytic leukemia; this approach combines the arsenic of Traditional Chinese Medicine with contemporary Western medicine. Chen trained in Western medical science in China and then in France, but his interest in much earlier Chinese studies led to his pioneering fusion of the ancient with the modern. For his contributions he was elected to the U.S. National Academy of Sciences in 2003. (For more on this issue and a biography of Zhu Chen, see Shen et al., 2004.)

Reference
Shen, Z. X., Z. Z. Shi, J. Fang, B. W. Gu, J. M. Li, Y. M. Zhu, J. Y. Shi, P. Z. Zhang, H. Yan, Y. F. Liu, Y. Chen, W. Wu, W. Tang, S. Waxman, H. De Thé, Z. Y. Wang, S. J. Chen, and Z. Chen. 2004. All-trans retinoic acid/As203 combination yields a high quality remission and survival in newly diagnosed acute promyelocytic leukemia. *Proc. Natl. Acad. Sci. USA* **101**:5328–5335.

---

expected at least some proto-oncogene-inducing insertional activation events to occur early? A partial explanation is that full-blown neoplasia requires that several independent events transpire (Box 15.7).

Presuming again that retroviral proviruses insert more or less at random, another puzzling issue is why particular retroviruses appear to induce particular proto-oncogenes. But the observation that particular retroviruses induce particular proto-oncogenes may be more apparent than real. That is, a given retrovirus indeed may activate a variety of proto-oncogenes. However, in a particular cell type, the activation of only a subset of proto-oncogenes might lead to neoplasia, and that subset may vary from one cell type to another. On the other hand, retroviruses might have preferred insertion sites that vary from one cell type to another. This possibility might reflect the experimental findings that insertion of a provirus appears to occur in or near active genes and that different subsets of genes may be active in different cell types.

The human T-lymphotropic virus type 1, which is a complex retrovirus that is associated with adult T-cell lymphocytic leukemia, is thought to induce neoplasia by yet another mechanism. It is believed to encode a regulatory protein that affects the expression of cellular genes in *trans*. Human T-lymphotropic virus type 1 is discussed in more detail in Chapter 21.

RETROVIRUS VECTORS FOR GENE THERAPY

The goal of gene therapy is to transmit a gene to a patient who either is lacking that gene or expresses a defective version of it, with the hope of curing a variety of genetic diseases (see Chapter 17). Apropos the current chapter, a retrovirus-based vector was used in the very first successful human gene therapy trial. The disease that was treated in this instance was X-linked severe combined immunodeficiency, a potentially fatal condition caused by a mutation in the common γc subunit of various **interleukin** receptors. (Interleukins are cytokines that mediate communication between leukocytes [Chapter 4].) (What unique advantage might retroviruses offer as vectors for gene therapy, and what potential hazards as well?)

The vector used in this successful clinical trial was derived from the Moloney murine leukemia virus. It was engineered both to be replication defective and to carry the γc gene. The therapeutic approach was to give infants autologous (i.e., derived from each patient's own body) hematopoietic stem cells that were transduced in vitro with the γc-expressing, replication-defective retroviral vector. The clinical result was that immune function was indeed reconstituted in 10 of 11 treated patients. Moreover, immune function was reconstituted to a level similar to that achieved by more conventional bone marrow transplants.

Although this clinical trial showed much promise for retrovirus-based gene therapy, the use of retrovirus vectors always carried the potential risk of inadvertent insertional activation of a cellular proto-oncogene, as discussed in the preceding subsection. Alternatively, the insertional event might inactivate a tumor suppressor gene. Either of these unintended but plausible events could result in malignancies.

Despite the plausibility of a retroviral insertional event leading to a malignancy, the risk was considered very low, because it had never been observed in earlier clinical trials. However, 3 years into the above trial, three of the gene therapy-treated children developed T-cell leukemias that were attributed to insertional activation of the cellular proto-oncogene *LMO2*. Two of the patients showed integration of the retrovirus vector near *LMO2*, as well as aberrant expression of that proto-oncogene, consistent with the possibility that the viral LTR was causing the upregulated expression of *LMO2* in these instances (Hacein-Bey-Abina et al., 2003). Note that the upregulation of *LMO2* has also been implicated in "spontaneous" human acute T-cell leukemias, which, in those instances, result from chromosomal translocations. The *LMO2* gene encodes a transcription factor that is necessary for normal hematopoiesis, but how its aberrant expression might lead to these cancers is not yet clear (Box 20.11). See Box 17.9 for a recent adverse gene therapy trial using an adenovirus-based vector.

SPUMAVIRUSES

The most intriguing feature of the spumaviruses is their unique replication strategy. Like the other retroviruses, spumaviruses reverse transcribe their RNA genome into a double-stranded DNA copy, which is covalently integrated

---

BOX 20.11

As noted in the text, integrated retroviral proviruses can upregulate cellular genes that are more than 10 kb away. Thus, bearing in mind that there are more than 100 proto-oncogenes in the human genome, it is estimated that about 0.1 to 1% of all retroviral provirus insertion events may cause aberrant expression of a cellular proto-oncogene. This line of argument underscores the potential for unintended adverse events associated with the use of retroviral vectors for gene therapy.

Considering the possible dangers associated with retrovirus-based vectors for gene therapy, might there be any strategies for mitigating those risks? (Give this question some thought before reading on.) Several ways have indeed been proposed to alleviate these dangers. For instance, it should be possible to identify "safe integration sites" in the human genome and to design retrovirus vectors that are more or less specific to those sites. Also, fewer cells might be treated ex vivo with the vector, thereby minimizing the number of integration events that might be needed to correct the cells. Moreover, it should be possible to genetically characterize clones of those treated cells before they are infused back into the patient. For example, the transgene might be used as a tag to identify neighboring cellular sequences in the treated cells, and then those cell clones in which the provirus might neighbor a proto-oncogene that it might potentially upregulate would not be used.

---

into the cellular DNA. However, spumavirus reverse transcription occurs *during final virus assembly* and maturation, within the viral core. Thus, the actual spumaviral genome is double-stranded DNA rather than RNA. (Bear in mind the spumavirus replication strategy when considering the hepadnaviruses [Chapter 22].)

Because of their unique replication strategy, the spumaviruses are considered to be a separate genus within the retrovirus family. Also, their Gag proteins have no sequence similarity to the MA, CA, and NC proteins of other retroviruses.

Spumaviruses are sometimes referred to as foamy viruses because of the characteristic vacuolating cytopathology they induce in cell culture. These viruses are widespread among a variety of species, including nonhuman primates, cats, cows, and horses, in which they establish lifelong, apparently asymptomatic persistent infections.

Spumaviruses do not contain an oncogene, and they are not known to cause disease in any host. The lack of association with any pathology in vivo is somewhat surprising since spumaviruses replicate to high levels and are highly cytopathogenic in cell culture. Perhaps infection in vivo is tightly regulated by host immunity, although there is no evidence that immunosuppression leads to spumavirus-associated pathology.

RETROELEMENTS

When the retroviral reverse transcriptase activity was first discovered, it was thought to be an oddity, unique to retroviruses. However, retroviruses are not the only biological entities that encode or else utilize a reverse transcriptase. As will be seen in Chapter 22, hepadnaviruses, too, have a reverse transcription step in their replication cycle. And, in point of fact, eukaryotic genomes are cluttered with huge numbers (as many as 10^6 in the case of the human *alu* sequence, mentioned above) of so-called **retroelements**, which are vertically transmitted from one generation to the next. An active integrated retroelement can make an RNA copy of itself and then reverse transcribe the RNA into a cDNA, which can then integrate elsewhere into the cellular genome. Moreover, reverse transcription is a normal if not vital cellular process, since the cellular telomerase activity, which maintains the ends of chromosomes, is a reverse transcriptase.

There is a variety of eukaryotic retroelements (Fig. 20.16) and of prokaryote retroelements as well (Box 20.12). Among the eukaryotic retroelements, there are the so-called

endogenous retroviruses. These may have arisen when retroviruses inserted their proviruses into the DNA of germ line cells, thereby creating retroelements that could be vertically transmitted. Many of the established endogenous retroviruses are replication defective, perhaps because they acquired genetic lesions during the insertion event. Endogenous retroviruses encode reverse transcriptase, IN, and Env (Fig. 20.16).

Retrotransposons are another type of eukaryotic retroelement. These are mobile genetic elements that resemble endogenous retroviruses. But unlike endogenous retroviruses, retrotransposons do not contain an *env* gene. Thus, retrotransposons are not able to generate either infectious or noninfectious extracellular particles. A retrotransposon could have been derived from either an exogenous or an endogenous retrovirus. In one scenario, a retrotransposon might have arisen from an improperly spliced Pol II transcript of an integrated provirus. Alternatively, retrotransposons may have been the evolutionary progenitors of retroviruses.

The Ty retrotransposons of yeasts are well-known examples of retrotransposons. They transpose via an RNA intermediate, encode a reverse transcriptase, and produce intracellular, noninfectious, virus-like particles. Their replication cycle, like that of all other retrotransposons, resembles that of a mammalian retrovirus, except that it begins and ends with an integrated form of the retroelement. Interestingly, retrotransposons have been especially successful at establishing themselves in the genomes of higher plants, where they may comprise as much as 80% of the total DNA!

Retroposons, also known as **non-LTR retrotransposons**, differ (not surprisingly) from retrotransposons in not containing LTR sequences. Still, like other transposable genetic elements, retroposons are flanked by direct repeats of host sequences. Also, like SINEs and processed pseudogenes (see below), they contain an A-rich sequence at the end of their coding region.

The so-called **long interspersed nuclear elements** (**LINEs** or **L1s**) are the most abundant class of retroposons in mammalian genomes, in which they are repeated about 850,000 times and comprise about 20% of the genomic DNA. Each LINE is about 6 kb long, and the LINEs are interspersed throughout the genome. Moreover, LINEs have the distinction of being the only active **autonomous retrotransposons** in the human genome. By definition, an autonomous retrotransposon is one able to mediate its own retrotransposition. Thus, autonomous retrotransposons encode both an endonuclease and a reverse transcriptase (Box 20.13). (Of the 850,000 or so human LINES, only about 30 to 60 actually transpose. That is so because the remaining ones have 5′ truncations.)

Figure 20.16 Characteristics of retroelements resident in eukaryotic genomes. Direct repeats of cellular DNA are in purple. See the text for details. Modified from S. J. Flint, L. W. Enquist, R. M. Krug, V. R. Racaniello, and A. M. Skalka, *Principles of Virology: Molecular Biology, Pathogenesis, and Control*, 2nd ed. (ASM Press, Washington, DC, 2004), with permission.

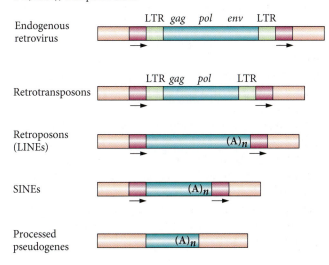

BOX 20.12

Prokaryotes too contain retroelements, three types of which are known. First, there are the **group II introns**, which are mobile genetic elements that encode their own endonuclease and reverse transcriptase activities, which mediate their genetic mobility. Second, there are the **retrons**, which also encode a reverse transcriptase and are able to generate an unusual satellite DNA known as msDNA. Third, there are the remarkable **diversity-generating retroelements (DGRs)**, which combine the basic retroelement life cycle of transcription, reverse transcription, and integration with site-directed, adenine-specific mutagenesis. (The prototype DGR is bacteriophage BPP-1, a temperate phage that infects host *Bordetella* species. By means of a cassette-based, reverse transcriptase-mediated process, BPP-1 can theoretically generate trillions of different amino acid sequences at defined locations in its distal tail fiber protein, thereby producing a vast repertoire of potential virus-receptor interactions. In the case of DGRs, as well as in the case of lymphocytes that are able to create an enormous number of T-cell receptor specificities [Chapter 4], diversity generators have coevolved with protein scaffolds, such that the amino acid variability required to confer distinct binding specificities can be accommodated, while not compromising the fundamental protein structure. BPP-1 is not unique in the prokaryotic world, since homologous DGRs are known to exist in the chromosomes of diverse bacterial species [see Medhekar and Miller, 2007].)

Earlier in the text we noted that the nucleotide sequences that constitute the joint between a retroviral provirus and eukaryotic chromosomal DNA contain features that are strikingly similar to features of the sequences connecting the various prokaryotic transposable genetic elements to the genomes of their bacterial host cells. These similarities include the terminal inverted repeats and conserved dinucleotide sequence at the ends of both eukaryotic and prokaryotic insertion elements and the duplication of a cellular sequence on each side of the insertion (Fig. 20.6). The existence of these similarities is consistent with several fascinating possibilities. First, prokaryotic insertion sequences and retroviruses may have evolved from a common molecular ancestor that existed before the divergence of eukaryotes and prokaryotes. Running with that thought, could the group II, self-splicing introns, which are present in many bacteria (although in small numbers), have moved into evolving eukaryotic genomes, to give rise there to the introns, which are a fundamental characteristic of eukaryotic genomes? Second, these retroelements may be relics of an early pre-DNA, RNA-based biology, or perhaps of an intermediate stage in the evolution of the current DNA-based system.

Reference
Medhekar, B., and J. F. Miller. 2007. Diversity-generating retroelements. *Curr. Opin. Microbiol.* **10:**388–395.

The **short interspersed repeat elements (SINEs)** differ from retroposons in that they are missing *pol.* Thus, SINEs are nonautonomous (i.e., unable to mediate their own retrotransposition), and it has been suggested that they might use the retrotransposition activities of LINEs, working in *trans*, to mediate their transposition. The human *alu* elements are part of the SINE family and are the most abundant mobile elements in the human genome, with somewhat more than 10^6 copies per genome.

Since retroelements insert at random sites within the human genome, each of the classes of retroelements enumerated above might potentially cause chromosomal insertions, deletions, and translocations that might lead to genetic disease. SINEs have been implicated more than either retrotransposons or LINEs as causes of these chromosomal aberrations. The plentiful human *alu* elements in particular have been implicated as mediators of genomic rearrangements that can lead to genetic disease. Since *alu* elements are nonautonomous SINES, it is thought that they might use the LINE machinery to mediate their retrotransposition, as noted above. Thus, LINE elements may cause genetic disease by mobilizing *alu* elements in *trans*, in addition to causing disease by their own retrotransposition (see below) (Box 20.14).

Since retrotransposons, LINEs, and SINEs all pose threats to the integrity of the host genome, might we expect that host organisms have evolved countermeasures to defend against retroelement-induced chromosomal damage? Host mechanisms that silence retroelement transcription and consequently retroelement mobilization indeed have been identified. These mechanisms include RNA interference silencing, DNA methylation, and histone modifications. In addition, A3G and other so-called APOBEC3 proteins can preferentially attack and deaminate deoxycytidines on the minus strand of retroviral DNA during reverse transcription. The deoxyuridines, which are generated by the deamination of deoxycytidines, may then be excised by the cellular uracil-DNA glycosylase, an enzyme that specifically removes the RNA base uracil from DNA. When the deoxyuridines are

BOX 20.13

Since LINEs do not have LTR sequences, how might they initiate reverse transcription and subsequent integration into cellular DNA? In the model depicted here, reverse transcription is primed by a broken end of the target DNA at the integration site. That broken end is generated by the endonuclease activity encoded by the LINE. Reverse transcription then initiates within the poly(A) tail at the 3′ end of the LINE. Synthesis of the opposite strand of the LINE DNA is primed by the other free end of the target DNA at the integration site. As a result of these events, the LINE is simultaneously reverse transcribed and integrated.

Reverse transcription and integration of LINE. From G. M. Cooper and R. E. Hausman, *The Cell, a Molecular Approach* (ASM Press, Washington, DC, 2007), with permission.

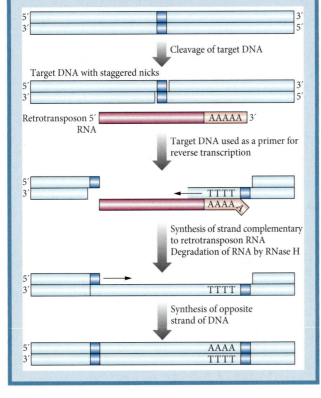

BOX 20.14

The first actual experimental evidence that implicated a retroelement as a cause of human genetic disease involved LINE elements. In 1988, LINE elements were found to give rise to hemophilia, which is an X-linked disorder of blood clotting caused by a deficiency of factor VIII, a serine protease that is one of the central proteins in the coagulation cascade. Specifically, two hemophilia patients were found to have LINES inserted into a particular exon of their factor VIII gene. Each of the parents of these patients was found to have normal factor VIII genes. Thus, the insertions in each of the patients appeared to have arisen de novo.

Reference
Kazazian, H. H., Jr., C. Wong, H. Youssoufian, A. F. Scott, D. G. Phillips, and S. E. Antonarakis. 1988. Haemophilia A resulting from de novo insertion of L1 sequences represents a novel mechanism for mutation in man. *Nature* **332:**164–166.

already undergone posttranscriptional processing, including splicing. Consequently, the processed pseudogene, like the mRNA from which it is derived, does not contain introns. Moreover, since promoters themselves are not transcribed, neither mRNAs nor processed pseudogenes contain promoters. Consequently, processed pseudogenes are nonfunctional when inserted back into the host genome. Since pseudogenes are nonfunctional, they are not under any selective pressure to maintain their integrity. Consequently, they readily accumulate substitutions, insertions, and deletions. Mammalian cells appear to contain thousands of processed pseudogenes.

Plasmids pFOXC2 and pFOXC3 provide examples of an intriguing family of retroelements that are referred to as **retroplasmids**. (Strictly speaking, plasmids are small,

excised, the provirus then becomes a potential target for endonuclease digestion (Box 20.15).

Processed pseudogenes are yet a fifth type of genetic element thought to arise by reverse transcription, but in this instance, by reverse transcription of a cellular mRNA, with reverse transcription likely catalyzed by the activity of one of the numerous autonomous LINEs. Processed pseudogenes are so named because they are generated by reverse transcription of cellular mRNAs, which have

BOX 20.15

APOBEC3 proteins were discovered as antiviral factors against HIV (Chapter 21). They also are effective against other retroviruses and retroelements, as well as against hepatitis B virus (Chapter 22). Interestingly, an enzymatic activity similar to that of APOBEC₃G is responsible for initiating the process of **somatic hypermutation**, which introduces mutations into specific regions of the immunoglobulin heavy- and light-chain genes, thereby greatly increasing the affinity of antibody molecules for their cognate antigens, a process referred to as affinity maturation (Chapter 4).

circular, extrachromosomal DNA molecules that can replicate without having to be inserted into the chromosomal DNA.) They are 1.9-kb linear DNA molecules that *reside in mitochondria* of certain forms of the fungal plant pathogen *Fusarium oxysporum.* They are called retroplasmids, with the prefix "retro," because they each replicate via an RNA intermediate and because each encodes the reverse transcriptase that mediates its replication. Thus far, retroplasmids have been found only in the mitochondria of six species of filamentous fungi. Some, like pFOXC2 and pFOXC3, exist as linear plasmids, whereas others are circular.

Suggested Readings

Baltimore, D. 1970. RNA-dependent DNA polymerase in virions of RNA tumor viruses. *Nature* **226**:1209–1211.

Hacein-Bey-Abina, S., C. Von Kalle, and M. Schmidt. 2003. LMO2-associated clonal T cell proliferation in two patients after gene therapy for SCID-X1. *Science* **302**:415–419.

Hayward, W. S., B. G. Neel, and S. M. Astrin. 1981. Activation of cellular onc gene by promoter insertion in ALV-induced lymphomas. *Nature* **290**:475–480.

Shih, C., and R. A. Weinberg. 1982. Isolation of a transforming sequence from a human bladder carcinoma cell line. *Cell* **29**:161–169.

Stehelin, D., H. E. Varmus, M. Bishop, and P. K. Vogt. 1976. DNA related to the transforming gene(s) of avian sarcoma viruses is present in normal avian DNA. *Nature* **260**:170–173.

Temin, H. M., and S. Mizutani. 1970. RNA-dependent DNA polymerase in virions of Rous sarcoma virus. *Nature* **226**:1211–1213.

21

The Retroviruses: Lentiviruses, Human Immunodeficiency Virus, and AIDS

INTRODUCTION

Historical Background and Social Issues

On 5 June 1981, the *Morbidity and Mortality Weekly Report (MMWR)*, published by the U.S. Centers for Disease Control and Prevention (CDC), contained an account of five sexually active homosexual men who were suffering from a lung disease caused by the protozoan *Pneumocystis carinii*. The CDC report also noted that three of the men had "profoundly depressed numbers of thymus-dependent lymphocyte cells and profoundly depressed . . . responses to mitogens and antigens." That 5 June 1981 article from the CDC marked the start of the acquired immunodeficiency syndrome (AIDS) epidemic in the United States.

In July of the same year, a second report in the CDC's *MMWR* described the presence of a rare cancer, Kaposi's sarcoma, as well as *P. carinii*, in 26 homosexual men, 20 of whom lived in New York City and 6 of whom lived in California. Later that year, similar symptoms were reported in heroin addicts and other intravenous drug users and in hemophiliacs and other blood transfusion recipients. Michael Gottlieb found that the individuals suffering from the *P. carinii* opportunistic infections, as well as those with Kaposi's sarcoma, all were afflicted with an illness that targeted their CD4 T cells for destruction.

By June 1982, there were 355 reported cases of this strange new malady. The clustering of cases and the apparent modes of transmission suggested that the illness was caused by an infectious agent, most likely a virus. Since the new illness appeared to be caused by an infectious agent, in September 1982 it was named *acquired* immunodeficiency disease (AIDS).

In 1983, a "new" virus, later termed the human immunodeficiency virus (HIV), was isolated and identified from a patient with lymphadenopathy. Since then, HIV has been the most intensely studied infectious agent in the world. The reasons are clear. HIV/AIDS is by many criteria the worst outbreak of an infectious disease in history. Approximately 40 million people in the world were infected by the fall of 2007, and the rate of new infections was 2.2 million to 3.2 million per year. (The UNAIDS report notes that these levels and rates are actually down from previous years but that they still remain at an unacceptably high level. Also, while the rate of new infections may have decreased in some regions, these gains are offset by rising rates of infection in other regions [see below and Box 21.1].)

The number of Americans living with HIV/AIDS has been continuously rising and by the fall of 2007 was approaching 1 million individuals. About 56,000 Americans are infected each year, and HIV/AIDS is now the second leading cause of death in African-Americans between the ages of 25 and 44 years. In fact, nearly one-half of new HIV infections in the United States are among African-Americans. However, the impact of HIV/AIDS has been far greater still in the developing world. A 15-year-old in South Africa has a 50% chance of dying from AIDS in his or her lifetime.

Despite the dismal scenario described above, in the developed world, there indeed has been notable progress in the battle against the AIDS epidemic. Many HIV-infected people now live nearly normal lives thanks to new antiretroviral drugs. Moreover, improved diagnosis of pregnant women and the use of new drugs have led to a decline of AIDS in newborns.

Lulled by these and other developments, the media and the public in the developed world may be growing complacent over the AIDS epidemic. Consequently, some readers might view much of what follows as perhaps a bit overstated. But consider that about 40 million people in the world were infected with HIV by the fall of 2007 and that about 56,000 Americans are infected each year, as noted above. Moreover, the risk of HIV infection is actually increasing in some poorer regions of the world. Some 5 million individuals were infected in sub-Saharan Africa and other poorer regions of the world in the year 2000 alone. Moreover, since these regions lack the resources for treatment and prevention, many cases there, including pediatric ones, will go untreated, and these will almost always end in death. Furthermore, even in the United States, AIDS is on the increase in some groups (e.g., African-American and Hispanic-American heterosexuals living in the inner cities).

Also consider the following. On the one hand, effective antiretroviral therapies that can significantly prolong the lives of some HIV-infected individuals have been developed. Nevertheless, while these therapies can impede transmission of the virus, they will not have an immediate large impact on the epidemic worldwide. Moreover, there is no vaccine in sight that might bring the spread of HIV under control and, importantly, there is no cure for HIV/AIDS. Thus, the HIV/AIDS epidemic worsens each day. Also, bear in mind that HIV is one of the most virulent pathogens ever known, inexorably killing 98% or more of all untreated infected individuals. This is a far greater mortality rate than that seen even in the case of the 1918 influenza pandemic (Chapter 12 and Box 21.2). Moreover, the HIV/AIDS epidemic is unique for the profound medical, social, moral, economic, and legal issues that it raises (see below).

BOX 21.1

Shortly before the emergence of HIV/AIDS, the public and many health professionals held the view that the development of "miracle" antibiotics against bacteria and vaccines against poliovirus and other viruses relegated the threat of infectious disease epidemics to history. Indeed, in 1962, Sir MacFarlane Burnet, who won the Nobel Prize in 1960 for his work on immunological tolerance, stated "...one can think of the middle of the 20th century as the end of one of the most important social events in history; the virtual elimination of the infectious disease as a significant factor in social life." However, such thinking would soon prove to be illusory.

HIV/AIDS was not the first new infectious disease to shatter that complacency. Lyme disease (caused by the bacterium *Borrelia burgdorferi*) emerged in 1975, and Legionnaires' disease (caused by the bacterium *Legionella pneumophila*) and Ebola each first emerged in 1976; each of these caught the attention of the public. However, none of these emerging diseases penetrated into the human population or affected it to an extent even remotely similar to HIV/AIDS. On the other hand, rotaviruses are now a major cause of death, especially in children in the developing world, and hepatitis C virus may now affect more people worldwide than HIV. The world has also been beset by the recent emergence of other deadly pathogens, including severe acute respiratory syndrome (SARS) and West Nile virus, among others. Moreover, earlier deadly pathogens, such as *Mycobacterium tuberculosis*, *Corynebacterium diphtheriae*, and the dengue hemorrhagic fever and yellow fever viruses, have reemerged over the past 30 years. During the summer of 2007, the world worriedly eyed the possible emergence of a human "bird flu" pandemic. Currently (summer 2009), the world is anxious over the prospect of a swine flu pandemic.

BOX 21.2

Despite the amount of attention that HIV has attracted since it was first recognized as the cause of AIDS, one might argue that influenza viruses were the most devastating infectious agents of the 20th century. As recounted in Chapter 12, the 1918 influenza pandemic, which killed as many as 50 million people worldwide in a single year, was the severest epidemic in human history. In the United States alone there were an estimated 20 million cases (20% of the population at the time) and 850,000 dead! Moreover, the emergence of another such deadly influenza epidemic remains a distinct possibility. Thus, influenza viruses may remain an even greater threat to human well-being than HIV. Even today, when effective flu vaccines are available, influenza viruses typically kill nearly 40,000 individuals in the United States alone and over half a million worldwide each year.

AIDS was originally known as a disease that affected homosexual men. However, HIV is indeed transmitted during heterosexual encounters, and women are actually much more susceptible to HIV than men. That is so because there is a higher concentration of HIV in semen than in vaginal secretions and the vaginal mucosa is more readily damaged by abrasions than the skin of the penis. Thus, some readers may be surprised to learn that women now face a substantially greater risk of HIV infection than men. In fact, over the past 20 years, HIV/AIDS has affected more women worldwide than any other life-threatening infectious disease.

The risks to women are especially great in developing countries. That is so because of endemic gender inequalities, poverty, cultural opposition to condom usage, lack of education, and in some areas, social instability and violence, all of which contribute to their vulnerability. In actual fact, women constitute about 60% of those infected with HIV in sub-Saharan Africa. What is more, the situation for women in sub-Saharan Africa may in fact be worsening, as shown by the finding that they now constitute 75% of infected individuals between the ages of 15 and 24 in that region. In South Africa, Zambia, and Zimbabwe, young women aged 15 to 24 are three to six times more likely to be infected than young men. This tragedy is compounded further by the vastly increased numbers of infected women who transmit the infection to their newborns during gestation and childbirth or afterwards. But the gender disparity of HIV/AIDS is not seen only in the developing world. In the United States, between 1999 and 2003 the number of estimated AIDS cases increased 15% among women, in comparison to 1% among men.

In addition to being an extraordinarily severe disease, AIDS has also been unique for the amount of fear and stigma engendered against those who were afflicted with this illness. The public does not reject or hate those with cancer, heart disease, diabetes, or other life-threatening diseases. Yet nonhomosexual as well as homosexual AIDS patients often faced rejection, hate, and discrimination, which on occasion turned violent and even deadly.

The stigma associated with AIDS was no doubt due in part to the fact that it is both an infectious and a deadly disease. But that stigma was also due in large measure to the fact that AIDS is associated with human sexuality in all its forms and with intravenous drug use. Largely because of its associations with homosexuality and anal sex, many individuals had difficulty discussing AIDS honestly and openly. It is certain that prejudice against AIDS patients was also due in large part to homophobia and to the fear that HIV might be transmitted by casual contact, despite the fact that there had been convincing evidence to the contrary. Denigration of, or physical violence against, people because of their apparent sexual orientation or gender identity was much more prevalent in the early years of the HIV/AIDS epidemic than it is today (Boxes 21.3 and 21.4).

Many in the gay community became AIDS activists. Importantly, these individuals played a major role in alerting a then-skeptical public that the emergence of the new disease was but the start of an impending crisis. Moreover, they were the first patient group to actually attend scientific conferences en masse, demanding results that might prevent, treat, and cure AIDS. Their efforts were largely responsible for the massive federal funding targeted at AIDS research; such efforts radicalized the relationship between a patient group and the public health system (Box 21.5).

Note that HIV/AIDS has had a major economic impact on health care systems, from the local to the international level. That is so because currently available drugs can cost as much as $20,000 per year per patient. Consider how such a cost would overwhelm a public health system in the developing world.

While HIV/AIDS research has not produced a vaccine or cure for the disease, this is not because of any lack of brilliance or want of trying on the part of researchers. Instead, it is because of the intractable nature of these problems (as discussed in detail below). Still, it is important to note that HIV/AIDS research has led to breakthroughs of singular fundamental and clinical significance. Indeed, in just the first 6 years after AIDS was first recognized as a distinct disease entity, AIDS researchers accomplished the following: (i) HIV was isolated and identified as the cause of AIDS; (ii) the routes of HIV transmission were defined; (iii) the HIV genome was completely sequenced, and individual viral genes were identified; (iv) a blood test for HIV infection was developed, which provided a diagnostic tool and also made the world's blood supply safe from HIV; and (v) the identification of HIV, a retrovirus, as the etiologic agent of AIDS, opened up the therapeutic use of antiretroviral drugs for the treatment of HIV-infected

individuals. Notwithstanding the limitations of these drugs, they have dramatically reduced HIV-related morbidity and death in treated individuals.

HIV and the Lentiviruses

AIDS is now known to be caused by two related retroviruses, HIV types 1 and 2 (HIV-1 and HIV-2, respectively), the latter having been discovered in 1985. HIV-1 is thought to have evolved from a virus of Central African chimpanzees and to have initially entered the human population in the first half of the 20th century. HIV-2 is thought to have originated in West African sooty mangabey monkeys. The HIV-2 genome is only about 40% homologous to that of HIV-1.

Since its emergence, HIV-1 has spread throughout the human population worldwide. In contrast, HIV-2 is found mainly in West Africa and on islands off its coast. Thus, HIV-1 is the most prevalent cause of AIDS worldwide. Moreover, HIV-1 infections progress to clinical AIDS more rapidly and more regularly than HIV-2 infections. Since HIV-1 has had a much greater impact on humans than HIV-2, the remainder of this chapter considers mainly HIV-1, which, for convenience, will be referred to simply as HIV.

HIV is a member of the genus *Lentivirus* of the family *Retroviridae*. The designation "lenti" (Latin for slow) refers to the slow courses of the diseases caused by these viruses. It has no implications regarding their rate of replication. Other lentiviruses include equine infectious anemia virus, visna/maedi virus of sheep, and simian immunodeficiency virus (SIV).

Before we consider the general characteristics of lentiviruses, recall that the RNA tumor viruses (or oncornaviruses), discussed in Chapter 20, are so-called *simple* retroviruses. Their genomes, as typified by that of Rous sarcoma virus, generally consist of only three genes: *gag*, *pol*, and *env*. *gag* encodes the viral structural proteins, *pol* encodes the enzymes (i.e., reverse transcriptase and integrase), and *env* encodes the viral envelope glycoproteins. Simple retroviruses need to generate only two transcripts. These are (i) the full-length, unspliced RNAs, which serve as messenger RNAs (mRNAs) for Gag and Pol and also as progeny viral genomes; and (ii) the singly spliced RNAs, which serve as mRNAs for Env. (Recall that the Rous sarcoma virus also encodes the Src protein, which is translated from a third mRNA, which is singly spliced.) The regulation of simple retrovirus gene expression is mediated entirely by the interplay between *trans*-acting cellular transcription factors and *cis*-acting sequences within the proviral DNA.

The lentiviruses are so-called *complex retroviruses*, reflecting the fact that their genomes and patterns of gene

BOX 21.5

Recall that Harold Varmus shared a 1989 Nobel Prize with Michael Bishop for demonstrating that retrovirus oncogenes have their counterparts (proto-oncogenes) in normal cells (Chapter 20). Varmus served as Director of the U.S. National Institutes of Health from 1993 through 1999 and is credited with having nearly doubled the NIH's research budget during that time period. In his recently published memoir, Varmus recounts some of the political pressures that affected research priorities at the NIH during his tenure as Director (Varmus, 2009). The following short excerpts are from Varmus's memoir.

> ....the priority-setting process can be ugly—for instance, when advocates refuse to recognize, or to care, that funds for their disease must come from funds being spent elsewhere, including funds used for a disease important to another group of advocates.

> For much of my time at the NIH, I was castigated by advocates for research on heart disease because the NIH was spending about as much on AIDS research as on studies of heart disease, even though there were about twenty times more deaths from heart disease than from AIDS in the United States each year. The arguments tended to ignore other important facts: that AIDS was a new and expanding disease, that it is infectious, that it is devastating large parts of the world, or that age-adjusted death rates from heart disease have fallen by two-thirds in the past 50 years.

Varmus notes that the points cited above to justify the expenditures for AIDS research (which might also have included the amount of suffering caused by the disease, the number of susceptible individuals in the population, and the average cost to society), or for any other disease for that matter, are actually "crude tools" for deciding how to apportion research dollars. Nevertheless, such criteria are used (can you think of better ones?), and as Varmus notes, they often led to bad feelings between the directors of the individual institutes at the NIH.

Varmus touches on a number of other interesting issues, including politicians who support the NIH, but whose support is almost always focused on a particular disease. And he makes clear the important point that advocacy for a particular disease is often at odds with how science works best. That is, breakthroughs on particular diseases often stem serendipitously from research that may have had no initial connection to that disease. Indeed, *basic* research, including research aimed at understanding a fundamental principle of biology, which might be funded in the context of a program that is seemingly unrelated to a particular disease, may have more of an effect on that disease than all of the research specifically targeted towards curing it. Yet as Varmus notes, the NIH must have programs that are directed at particular diseases to meet very real public health needs. But he affirms that the NIH (and I would add, politicians as well) also must be sensitive to the fact that scientists are at their creative and productive best when they have the freedom "to exercise their imaginations as fully as possible."

Reference
Varmus, H. 2009. *The Art and Politics of Science.* Norton Books, New York, NY.

---

expression are more complex than those of the oncornaviruses. Like the oncornaviruses, the complex retroviruses also encode Gag, Pol, and Env proteins. But in addition, the complex retroviruses also encode several so-called "regulatory" and "accessory" gene products. For instance, HIV encodes its Tat and Rev proteins, which mainly serve a regulatory function. Moreover, some mRNAs transcribed by complex retroviruses contain *cis*-acting sequences that are recognized by viral gene products. For example, TAR (*trans*-activating response region) and RRE (Rev-responsive element [see below]) are HIV mRNA sequences that are recognized by the HIV Tat and Rev proteins, respectively.

Here are yet other distinctions between the simple retroviruses and the complex retroviruses, the latter as exemplified by HIV. In order for HIV to generate its regulatory and accessory gene products and for those products to carry out their particular functions at appropriate stages of the viral replication cycle, HIV uses a pattern of transcription that differs somewhat from that of the simple retroviruses. Like the simple retroviruses, HIV generates unspliced mRNAs that are translated into its Gag and Pol proteins and singly spliced mRNAs that are translated into its Env proteins. But unlike the simple retroviruses, HIV also transcribes singly spliced mRNAs that are translated into its accessory proteins (i.e., Vif, Vpr, and Vpu). More importantly, perhaps, HIV also generates *multiply spliced* mRNAs that are translated into the regulatory proteins Tat and Rev and accessory protein Nef. (Note that the additional regulatory and accessory proteins can also be differentiated as "early" versus "late," with Tat, Rev, and accessory protein Nef being translated early from multiply spliced mRNAs and the accessory proteins Vif, Vpr, and Vpu being translated later from singly spliced mRNAs.) Simple retroviruses do not generate multiply spliced mRNAs. The significance of the HIV regulatory and accessory gene products and their pattern of transcription are discussed further below.

DISCOVERY OF HIV AS THE CAUSE OF AIDS

The discovery that HIV is the cause of AIDS led to the development of a sensitive blood test for HIV infection, which made it possible to determine who is infected, which in turn made it possible to slow the spread of the infection and to determine the effectiveness of prevention efforts. Moreover, it prevented millions of people from becoming infected through the transfusion of HIV-contaminated blood. In addition, identifying a retrovirus as the cause of AIDS opened up the use of antiretroviral drugs for treating HIV-infected individuals, dramatically reducing morbidity and death.

We reiterate the facts discussed above to underscore the point that the discovery of HIV as the cause of AIDS is one of the major achievements in the history of medical science. Moreover, it was apparent before HIV was discovered that determining the cause of AIDS would be so singularly important that great recognition and awards would surely go to its discoverer. Thus, when Luc Montagnier and his colleagues at the Pasteur Institute in Paris and Robert Gallo and his colleagues at the U.S. National Institutes of Health concurrently laid claim to that discovery, there was much controversy and animosity, particularly between the leaders of these research groups. The claims and counterclaims between the two groups led to sensational accounts in the media over a course of several years. Jay Levy, at the University of California, San Francisco, also published on the discovery of HIV, but he did not get caught up in the dispute involving Gallo and Montagnier, and perhaps for that reason, his contributions are less well known. The following paragraphs contain my attempt to recount the key events involving the discovery of HIV and the basis for the controversy that ensued.

Before HIV was isolated in 1983, very few biomedical scientists thought that AIDS might result from an as yet unknown infectious agent, much less a retrovirus (Box 21.6). Indeed, prior to the discovery of HIV, it was generally thought that there were no human retroviruses, a view based on previous failed attempts to find retroviruses in human cancers. However, in 1980, just before the first patients with AIDS were identified, Robert Gallo and associates discovered the first known human retrovirus, the human T-cell leukemia virus type 1 (also known as the human T-lymphotropic virus [HTLV-1]), which was a

---

BOX 21.6

During the first few years of the AIDS epidemic, before HIV was isolated, very little was known about this new disease, and many hypotheses were put forward to explain its cause. For example, since AIDS first appeared among cohorts of homosexual men, some researchers thought that the disease might be caused by sperm in the male bowel. Others suggested that AIDS might be caused by the excess stress that some individuals placed on their immune systems. Some researchers considered the possibility that cytomegalovirus, Epstein-Barr virus, hepatitis B virus, or herpes simplex viruses might be the cause of AIDS. But few investigators thought that the new disease might be due to a new infectious agent; such an outlook was largely based on the notion that all infectious agents and diseases were already known.

---

virus associated with an adult T-cell leukemia (see below). Two years later, in 1982, Gallo's group reported the discovery of a related virus, HTLV-2, isolated from a cancer described as a "hairy cell T-cell leukemia." (The HTLVs are described in greater detail at the end of this chapter. Unlike HIV, they are not considered to be lentiviruses. Instead, they are classified as deltaretroviruses; *Deltaretrovirus* is a genus of the *Retroviridae* and includes the bovine and feline leukemia viruses.)

Bearing in mind the earlier unsuccessful efforts to isolate a human retrovirus, why was Gallo's research group able to isolate HTLV-1 in 1980? The answer, in part, is as follows. During the previous 15 years, researchers in Gallo's laboratory had been studying other mammalian leukemogenic retroviruses. While engaged in that work, Doris Morgan developed methods for growing T lymphocytes in culture for extended periods; this was a critical breakthrough that depended on her discovery of a T-cell growth factor, now known as interleukin-2 (IL-2). (Kendall Smith at Dartmouth followed up Morgan's initial observations, eventually isolating IL-2 as the factor responsible for the lymphocyte growth that Morgan detected.) Other investigators also made key advances. Important among these was the discovery, only 10 years earlier, of reverse transcriptase by Howard Temin and David Baltimore (Chapter 20). This made it possible to develop highly sensitive PCR-based assays for detecting a retrovirus. Together, these developments enabled Gallo's research group to isolate HTLV-1 in 1980. More importantly, when AIDS emerged, tools and protocols were in place to search for a retrovirus as its causative agent.

But why would Gallo have thought to look for a retrovirus as the cause of AIDS?

Gallo notes several clues, which he thought pointed to a retrovirus that might be similar or identical to HTLV-1, as the etiologic agent of AIDS. First, AIDS is characterized by a loss of CD4 T cells, consistent with the possibility that its cause might be an infectious agent that targets CD4 T cells. In this regard, Gallo noted that HTLV-1 was already known to target CD4 T cells. Second, HTLV-1 was known to be transmitted via blood and sexual activity and from mother to infant, the very modes by which AIDS was proving to be transmitted. Third, a high incidence of AIDS was being reported in Haiti, a region in which HTLV-1 is endemic. Thus, Gallo's initial premise was that AIDS might be caused by a variant of HTLV-1. That premise would prove to be incorrect, but Gallo was indeed correct in hypothesizing that AIDS is caused by a retrovirus. (Arguing against the notion that AIDS might have been caused by a known strain of HTLV-1, this virus does not kill T cells. Instead, it causes them to proliferate. Moreover, there was no detectable AIDS in Japan at that time, despite the fact that HTLV-1 infected at least a million people there.)

Luc Montagnier was, at that time, investigating the possible involvement of retroviruses in human cancers. (In one experimental approach, Montagnier's research group used cloned murine mammary tumor virus sequences to search for related sequences in the DNA of surgically removed human breast tumors.) In 1982, perhaps influenced by Gallo's arguments, Montagnier set out to isolate a retrovirus as the possible etiologic agent of AIDS. But before doing so, a group of French physicians and scientists suggested to Montagnier that the best chance to find and isolate the AIDS agent might be at the beginning of the disease, before the patient's T cells are eliminated. Following that advice, Montagnier and coworkers looked for a retrovirus in a lymph node biopsy specimen from a patient with little indication of immunodeficiency but who displayed persistent lymphadenopathy (swollen lymph glands), which is a clinical sign in patients progressing towards AIDS. Cells from this patient were cultivated in the presence of IL-2 and anti-interferon (anti-IFN) antiserum. The latter was an innovation of Montagnier's research group, based on their earlier finding that retrovirus replication is impeded by endogenous IFN. Two weeks later, in early January 1983, the Montagnier laboratory detected the first evidence of reverse transcriptase activity in the cell culture medium. Next, the virus was seen by electron microscopy, and its morphology indeed resembled that of a lentivirus.

The new retrovirus detected by Montagnier's group was not HTLV-1, as initially shown by the fact that it did not react with anti-HTLV-1 antibodies provided by Gallo and as confirmed afterward by sequence analysis. Importantly, however, antibodies against the virus isolated in Montagnier's laboratory were later found to be present in the sera from most AIDS patients. Moreover, the virus was shown to have a tropism for CD4 T cells. Since this new virus came from an AIDS patient with lymphadenopathy, Montagnier dubbed it "lymphadenopathy-associated virus-1" (LAV-1). This particular isolate of LAV-1 was named "Bru." (Interestingly, Montagnier later obtained a biopsy specimen from another AIDS patient who was infected with HTLV-1, as well as with the virus that Montagnier called LAV-1. If this had been the first patient sample, results might have been very confusing indeed.)

Concurrently, and independently of Montagnier, Gallo, at the U.S. National Institutes of Health, also embarked on isolating a retrovirus from biopsy specimens of AIDS patients. The events which then transpired were truly bizarre. While Gallo was seeking to isolate an AIDS retrovirus, he received a sample of Bru from Montagnier. Shortly afterwards, Gallo announced that he too had isolated a retrovirus from an AIDS patient. The Gallo isolate had properties somewhat different from those of Bru. For example, unlike Bru, which grew only in fresh T-cell cultures, the new isolate also grew in permanent T-cell lines. Thus, Gallo reported that he had isolated a second type of AIDS retrovirus, which he named HTLV-III, based on his belief that it was a member of the HTLV family. But when the nucleotide sequence of HTLV-III was determined soon afterwards, it turned out to be almost identical to that of another LAV-1 sample isolated earlier in Montagnier's laboratory. This was remarkable because HIV has an extraordinarily high mutation rate and, as we will soon see, an untreated HIV-infected individual can produce between 10^8 and 10^{10} new virus particles each day, almost each of which is unique. (Because Montagnier and Gallo gave different names to their AIDS isolates, the AIDS retrovirus was for a time referred to as LAV-1/HTLV-III. Later, an international committee, chaired by Harold Varmus [Chapter 20 and Box 21.5], proposed to rename the virus HIV for "human immunodeficiency virus." From here on, we too refer to the AIDS retrovirus as HIV.)

Priority for the discovery of the AIDS virus was hugely important to both Gallo and Montagnier for the reasons noted above. Not surprisingly, then, accusations flew back and forth between these two scientists, and in the end, it was Gallo's reputation that was tarnished. Interestingly, the governments of the United States and France too weighed in to the fray, since the stakes were high regarding which country might hold patent rights to the HIV blood test that soon would be developed as a result of isolating the virus. (Eventually, U.S. President Ronald Reagan and

French Prime Minister Jacques Chirac signed a declaration that Gallo and Montagnier were codiscoverers.)

But how did it come to pass that Gallo's HIV isolate was so similar to Montagnier's? Here is one likely scenario. After isolating Bru, Montagnier also isolated HIV from biopsy specimens of several other AIDS patients. One such isolate, called "Lai," replicated much more rapidly than Bru in cell culture. Lai replicated more rapidly than Montagnier's other HIV isolates as well. Lai then contaminated stocks of Bru in the Montagnier laboratory. Consequently, and unknowingly, the sample of Bru that Montagnier sent to Gallo (as well as to several other retrovirus researchers, including Levy) was contaminated with Lai, which then inadvertently contaminated the isolate that Gallo's research group thought originated from one of their AIDS patients. (You ask how such accidents can happen in such top-flight laboratories. Unfortunately, accidents of this sort are not as uncommon as you might imagine [Box 21.7].)

So, who discovered the AIDS virus? The answer to this question now appears to be clear.

HIV was discovered by Françoise Barré-Sinoussi in Montagnier's laboratory at the Pasteur Institute, in collaboration with other French clinicians and researchers, including Jean-Claude Chermann, Willy Rozenbaum,

David Klatzmann, and, of course, Montagnier. They published their findings in the journal *Science*, in May 2003, about a year before anyone else (Barré-Sinoussi et al., 1983). Also, note that in 1985 Montagnier's research group, in collaboration with physicians in Lisbon and virologists from Hôpital Claude Bernard in Paris, also discovered HIV-2, which they initially dubbed LAV-II. But also note that concurrent with the above efforts, Jay Levy and colleagues at the University of California, San Francisco, demonstrated that HIV is present in AIDS patients and in healthy carriers as well. As noted above, Levy's contributions are not as well known as those of Montagnier and Gallo, since he did not get involved in the controversy.

While Montagnier's research group is now credited for the initial isolation of HIV, Gallo's research group is generally credited for much of the basic research that made the discovery of HIV possible, particularly the discovery of IL-2, which made it possible to grow T cells in culture and, thus, HIV as well. Indeed, Gallo's group was also the first to grow HIV in an established T-cell line, which was pivotal to the development of blood tests for HIV.

The controversy between Gallo and Montagnier has since subsided, and they appear to be in agreement on all major issues. For his part, Gallo sometimes says that he never claimed to have discovered HIV but rather claims

BOX 21.7

As we see later in this chapter, slow-growing HIV strains, like Bru, are present at early stages of an HIV infection, while rapidly growing strains, like Lai, may become common during later stages of infection in some individuals. Also, fast-growing HIV strains like Lai cause T cells to form syncytia in vitro, and they also can replicate in permanent T-cell lines. In contrast, slow-growing isolates like Bru do not cause syncytia to form and they cannot replicate in permanent T-cell lines. The slow-growing and the fast-growing strains can each grow in primary T-cell cultures.

As described in more detail later, slow-growing, non-syncytium-inducing HIV strains are the most common sexually transmitted form of the virus. These viruses preferentially infect macrophages (which are present in the skin and mucous membrane) rather than T cells, and consequently, they are said to be macrophage-tropic, or M-tropic. Importantly, M-tropic HIV strains typically use the CCR5 β-chemokine receptor for their coreceptor. Sometime later during HIV disease, the virus population within some infected individuals generates variants that use the CXCR4 α-chemokine receptor for their coreceptor. These viruses are

syncytium inducers, and they are able to replicate in permanent T-cell lines. Moreover, they prefer to infect T cells, and consequently, they are referred to as T-tropic strains. T-tropic HIV strains are more aggressive than M-tropic strains, and their prevalence correlates with more-rapid HIV disease progression. Now, recall that LAV-1 was initially isolated from an individual in the early stages of AIDS disease, which usually appears only years after the initial infection.

The discoveries that CCR5 and CXCR4 serve as coreceptors for HIV were prompted by Jay Levy's finding that there are factors in human blood that block HIV infection. Then, in 1995, Gallo's research group identified the β-chemokines RANTES, MIP-1α, and MIP-1β as serum HIV-blocking factors. Importantly, CCR5 is the only receptor on CD4 T cells that binds each of these three ligands. Shortly afterwards, experiments confirmed that CCR5 is the major coreceptor for HIV early in infection and that CXCR4 may serve as the major coreceptor for HIV at late stages of infection in some individuals. Birgitta Asjo and coworkers were the first to show that the M-tropic viruses are prevalent in early-stage infection and that T-tropic viruses can become the major type in late-stage infection.

credit for demonstrating that it is the cause of AIDS. Montagnier concedes that Gallo's evidence in that regard is more convincing than his own. Regardless, the efforts of these two individuals resulted in the identification of a new retrovirus as the cause of AIDS and made it possible to grow large enough amounts of the virus to enable further studies. Moreover, their discovery quickly resulted in a blood test for HIV and opened up the development of antiretroviral drugs as a therapy for AIDS.

During October 2008, while this chapter was undergoing its penultimate revision, the Nobel Committee announced that the winners of the 2008 prize in Physiology or Medicine were Luc Montagnier and Françoise Barré-Sinoussi for the discovery of HIV and Harold Zur Hausen for his work identifying human papillomaviruses as the cause of cervical carcinoma (Chapter 16).

While the decision to exclude Gallo from the Nobel may be somewhat controversial, the Nobel Committee explicitly affirmed who they believe discovered what, and when. They stated that Barré-Sinoussi and Montagnier "made the most important contributions to the discovery." They then refer to Gallo's "detection of a novel ... virus from a vast number of patients with AIDS or pre-AIDS in 1984 ... [which] showed considerable similarities with LAV-1." Montagnier commented that he wished Gallo had been included in the prize. Gallo said that it was a "disappointment" not to be included in the Nobel, but he said that all three recipients deserved the honor. Jay Levy, who was not involved in the controversy, was content. "In the end, what they did was quite, quite fair," he said. "And I congratulate them." But many also regard Jean-Claude Chermann to be equally deserving of the Nobel Award. Chermann supervised Françoise Barré-Sinoussi in the Montagnier group and had the idea of looking at patients with lymphadenopathy, which was a key to isolating HIV. Chermann was the second of the 12 authors of the paper reporting the isolation of HIV (Barré-Sinoussi et al., 1983), implying that he and Barré-Sinoussi did the major portion of the work. Montagnier was last author, the site usually reserved for the director of the group.

One fundamental lesson learned from these experiences was well stated in a report jointly written by Gallo and Montagnier: "Our experience with AIDS underscores the importance of basic research, which gave us the technical and conceptual tools to find the cause less than three years after the disease was first described."

STRUCTURE

Lentivirus particles are organized like those of the simple retroviruses (Chapter 20). However, one noticeable distinction is that in contrast to the isometric cores of the

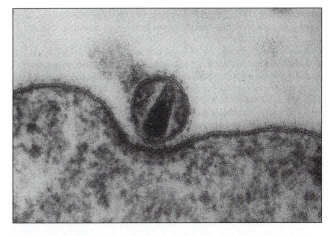

Figure 21.1 Electron micrograph of HIV. From B. S. Weeks and I. E. Alcamo, *AIDS: The Biological Basis*, 4th ed. (Jones and Bartlett, Sudbury, MA, 2006), with permission.

simple retroviruses, lentivirus cores are elongate and predominantly conical (Fig. 21.1).

The structures of most of the proteins that constitute HIV particles have been solved at high resolution (Fig. 21.2). These efforts were motivated in part by the hope that knowledge of the molecular structure might lead to the development of rational therapies to combat HIV/AIDS.

Figure 21.2 High-resolution structures of most of the proteins in an HIV particle and their locations within the particle. From S. J. Flint, L. W. Enquist, R. M. Krug, V. R. Racaniello, and A. M. Skalka, *Principles of Virology: Molecular Biology, Pathogenesis, and Control*, 2nd ed. (ASM Press, Washington, DC, 2004), with permission.

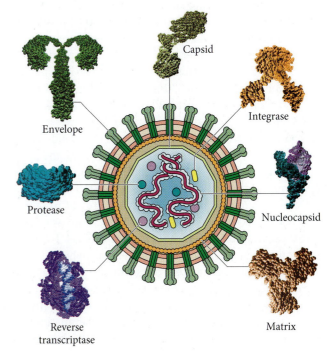

ROUTES OF TRANSMISSION AND MECHANISMS OF INFECTION

Routes of Transmission

HIV/AIDS is transmitted primarily by intimate sexual contact (both heterosexual and homosexual) and by exposure to contaminated blood and blood products. Additionally, HIV may be spread from mother to child, either in utero or postpartum. HIV is transmitted via these routes because it is present mainly in the blood, semen, vaginal secretions, and breast milk of infected individuals.

Sexual transmission accounts for 80% or more of all HIV infections worldwide. Thus, it may come as a surprise that sexual transmission of HIV is actually relatively inefficient. Per coital act, the frequency of HIV transmission is estimated at 1/200 to 1/2,000 from male to female, 1/200 to 1/10,000 from female to male, and 1/10 to 1/1,600 from male to male. Transmission is more efficient from male to female than from female to male because there is a higher concentration of HIV in semen than in vaginal secretions, and the vaginal mucosa is more readily damaged by abrasions than the skin of the penis. Transmission from male to male is higher still because the membranous lining of the rectum is more easily torn than that of the vagina. Moreover, the membranous lining of the rectum is richly vascularized, so that tears can more readily lead to exposure of and to the blood. Together, these factors account for why individuals practicing unprotected anal sex (both males and females) are at greatest risk of sexual transmission. Also, the receptive partner is the one who is most at risk in either vaginal or anal sex.

Not to be lulled into a false sense of security by the above numbers, several factors can dramatically raise the efficiency of sexual transmission, by orders of magnitude. One such factor is other genital infections, particularly those that cause ulceration in either the HIV-infected individual or in the exposed, uninfected individual. That is so for several reasons. First, and most obviously, the ulceration and breaks in the skin caused by some sexually transmitted diseases (STDs) can expose the bloodstream, which contains circulating CD4 T lymphocytes, the principal target cells for HIV. Second, infection is actually transmitted efficiently by HIV-infected CD4 T cells themselves (see below), and CD4 T cells are present in semen and vaginal secretions, as well as in blood. Moreover, some STDs cause an increase in the number of CD4 T cells in cervical secretions. Also, those individuals who have the most sexual partners are the ones who are most likely to have STDs, accounting for the rapid spread of HIV among young, sexually active adults. For these reasons, treatment of STDs to cure genital sores and reduce inflammation can significantly reduce the number of new HIV infections.

Also note that circumcision dramatically lowers transmission (Box 21.8).

Serum HIV levels are highest shortly after an individual is first infected and again in late stages of illness. Between the initial infection and the onset of AIDS-associated symptoms, there is a variably long asymptomatic, chronic stage, which is about 10 years long on average. During the variably long chronic stage of infection, the level of HIV in the blood of an infected individual (the viral "load") stabilizes, but it can vary over a range of several orders of magnitude from one infected individual to another. These points are discussed in detail below. For now, as you might expect, the HIV load in an infected individual is another major determinant of how efficiently that individual might transmit the infection via the sexual route. Also, the long asymptomatic chronic stage, which follows the initial HIV infection, is an extremely important factor in the spread of the virus, since during that time the infected individual may not be aware that he or she is infected and thus may unknowingly be transmitting the infection to others.

Different modes of sexual transmission are preponderant in different parts of the world. In the United States, more than half a million men who have sex with men have been diagnosed with AIDS. These individuals constituted 68% of all men living with HIV in the United States in 2005, even though they account for only about 5 to 7% of the men in the population. In contrast, heterosexual transmission predominates in the developing world. But importantly, the incidence of heterosexual transmission has been increasing worldwide, and consequently, the impact of HIV/AIDS on women has increased dramatically. The percentage of HIV-infected adults who are females was 25% in 1990. It rose to 45% by 1995 and was 50% by 2000. The increase of HIV infections among women has been

BOX 21.8

Apropos the extent to which some STDs might impact the worldwide HIV/AIDS epidemic, consider that there are over 250 million new cases per year of just seven of the most common STDs, which include syphilis, gonorrhea, chlamydia, and genital herpes. In the United States alone, about 60 million people over age 12 have chronic genital herpes. Importantly, among HIV-infected individuals who have genital herpes, HIV is consistently found in their herpetic lesions. Recalling that herpes simplex viruses cause oral as well as genital lesions (Chapter 18), HIV might also be transmitted during oral sex, to either partner. However, there are no published estimates of the risk of transmission via receptive oral exposure.

particularly acute in the developing world, as noted above. Gender inequalities there make it difficult and in some instances impossible for women to protect themselves, even when they have the information and desire to do so. As the frequency of infected women increases, so does the frequency of perinatal transmission, which may occur in utero, during delivery, or via breast-feeding (see below).

The second major route of HIV transmission is via needle sharing by users of intravenous drugs. HIV is transmitted very efficiently via this route (estimated at 0.67% per episode of intravenous syringe or needle exposure), since it delivers the virus directly into the bloodstream of the recipient. In New York City, more than 80% of intravenous drug users are positive for HIV antibodies. Importantly, these people are also major sources for transmission via the sexual and congenital routes.

All blood used for transfusions in the United States comes from unpaid donors and has been screened for HIV. Paid donors may yet sell their blood to be used for making pharmaceuticals such as Rh immunoglobulin, albumin, and intravenous immunoglobulins. (Only whole blood, blood cells, plasma, and blood clotting factors have been implicated in HIV transmission. Fortunately, other blood products, such as pooled immunoglobulin and hepatitis B vaccine, were not a source of HIV transmission, presumably because the virus was destroyed by the production procedures.)

Prior to the introduction of routine antibody screening by blood banks, hemophiliacs were at great risk when receiving transfusions for clotting factor. But note that the risk of receiving HIV-contaminated blood is not zero. The most important reasons are (i) that some infected individuals may fall under the radar because they have not yet begun to produce a detectable level of antibodies and (ii) laboratory error.

The third major route of HIV transmission is from mother to child. In fact, mother-to-child transmission is responsible for nearly all childhood AIDS cases in the United States. Transmission from mother to child can occur (i) in utero, (ii) during delivery, and (iii) postpartum, via the breast milk.

In utero transmission of HIV can be demonstrated by the presence of HIV in fetuses of HIV-infected mothers examined after elective abortion. In this regard, HIV is one of a very few viruses for which infection in utero is a significant mode of transmission. Others are rubella virus, human cytomegalovirus, and parvovirus B19. The transplacental route of HIV infection is discussed in more detail below and in Chapter 3.

Most transmissions from mother to offspring are believed to occur during delivery, at which time the newborn inadvertently ingests the mother's blood or other infected fluids, during passage through the birth canal. For that reason, the issue of whether Caesarean section might decrease the frequency of perinatal transmission has been argued among clinicians. Studies carried out between 1998 and 2002 show that the use of Caesarean section indeed significantly decreases that frequency.

Between 120,000 and 160,000 women of childbearing age in the United States are currently infected with HIV. Nearly one in four of these women does not know that she is infected, and some who do know nevertheless want to become pregnant. What, then, is the overall probability of a fetus being infected by an HIV-infected mother? In the United States, 25% of HIV-infected pregnant women who are not receiving antiretroviral therapy and who are not breast-feeding will transmit the virus to their babies. On the other hand, one of the great successes of antiretroviral therapy has been to decrease the frequency of perinatal transmission of HIV. In one landmark study, reported in 1994, zidovudine therapy (see below), beginning in the second trimester of pregnancy and continuing during delivery and for 6 weeks afterwards, reduced the frequency of HIV transmission to the neonate from 25 to 8%. Regularly testing pregnant women for HIV and providing antiretroviral drugs to those who are infected can now reduce the number of children born with HIV from 25% to less than 2% (see Box 21.52).

For several reasons, including the cost, antiretroviral therapy has not been routinely available in the developing world, where the need for it is greatest. Fortunately, this situation is changing. In 2003, President Bush launched the President's Emergency Plan for AIDS Relief (PEPFAR) to combat global HIV/AIDS, which is the largest commitment by any nation to combat a single disease in human history. New legislation, signed into law on 30 July 2008, authorizes up to $48 billion to combat global HIV/AIDS, tuberculosis, and malaria. Nevertheless, even this commitment does not meet the global need (Box 21.9).

Entry into the Host

While one can examine particular cells and tissues in an individual known to be infected, it is not feasible to examine these cells and tissues at the time of transmission. Thus, it is not possible to critically address the mechanisms by which HIV invades the human host. Nevertheless, several important features of HIV transmission are fairly well established from other lines of evidence, for example, from animal model systems and human explant cultures ex vivo.

HIV can be transmitted either as cell-associated virus or as free virus. That is so because blood, semen, and vaginal secretions contain HIV-infected lymphocytes as well as free HIV particles. In point of fact, seminal fluid contains greater amounts of cell-associated HIV than of free virus.

During sexual transmission, HIV must cross the epithelium of either the genital or anorectal tract in order

BOX 21.9

Casual nonsexual contacts, saliva, coughs, sneezes, urine, sweat, food handling, public toilet seats, swimming pools, and blood-sucking insects are not known to be sources of HIV transmission. But it is interesting that irrespective of assurances from public health experts that HIV is not transmitted by any of these routes, many individuals continued to believe otherwise. Why might that have been? One reason is that journalists (and health officials as well) typically referred to "body fluids," when they should have been specifying semen, blood, and vaginal secretions. Perhaps in those earlier times they were too prudish to be more specific. Another possible explanation is that prior to the emergence of HIV/AIDS in the early 1980s, medical scientists may have led the public to believe that infectious diseases were no longer a threat to human health. The emergence of

HIV/AIDS may then have caused many individuals to doubt the validity of scientific pronouncements. "If they were wrong before about the threat of infectious diseases, why are they not wrong now about the transmission of HIV?" Another reason was the belief among conspiracy theorists that scientists indeed knew that other routes of HIV transmission exist but that they were under political pressure not to say so for fear of instilling panic or for reasons of "political correctness," such as fear of offending HIV-infected individuals or the gay community. Such misimpressions on the part of the public regarding HIV may have largely abated. But does the current public uncertainty regarding global warming, for example, reflect a similar distrust of science? Can you explain to laypeople how scientists come to a consensus with respect to these sorts of issues?

---

to infect underlying CD4 T cells (or macrophages; see below). During transmission via breast-feeding, the virus is believed to penetrate the mucosa of the mouth or gastrointestinal tract. In any case, whether HIV infection is initiated by free virus or by lymphocyte-borne virus, the virus must cross an epithelial barrier to disseminate in the newly infected individual. How then might lymphocyte-borne HIV cross an epithelial barrier to initiate infection of an exposed individual?

The answer to the above question is not clear, but studies in animal models and studies of human explants in culture do not support the premise that infected donor cells per se can cross an epithelium. Thus, infected cells are more likely to provide a local source of free virus particles that are external to the epithelium. How then might free HIV particles cross these epithelia? Before we attempt to answer this question, note that stratified genital epithelial cells are not susceptible to productive HIV infection, nor do they transcytose virus particles. Thus, transport of HIV across these epithelia is thought to be facilitated by tears or lesions in the epithelia, which indeed occur rather frequently.

The anorectal epithelium is more readily damaged than its vaginal counterpart. On the other hand, as part of the gastrointestinal tract, the mucosal surface of the rectum is populated by organized lymphoid tissues referred to as **Peyer's patches**. These are secondary lymphoid organs comprised of B-cell and T-cell areas and also of macrophages and dendritic cells. Peyer's patches are covered by **M cells**, which take up antigens and pathogens from the lumen of the gut and then deliver them to the underlying Peyer's patches by transcytosis (Fig. 21.3). M cells are highly specialized for this function, since they have a

relatively thin cytoplasm and lack the closely packed microvilli and the glycocalyx that covers enterocytes. The human rectum contains a higher density of M cells and Peyer's patches per unit area than any other tissue.

While the M cells and underlying Peyer's patches normally serve a protective function in the host, HIV can subvert them to initiate an infection across the anorectal mucosa. The virus does so by readily making contact with the luminal surface of M cells and then subverting those cells, which deliver it to the CD4 T cells and macrophages of the underlying Peyer's patch. The virus can then infect these migratory lymphoid cells and spread to other sites in the body. (There are reports that intestinal epithelial cells themselves can [in model systems at least] transcytose HIV particles to the underlying lamina propria, when exposed to infected CD4 T cells in the semen. The lamina propria is comprised of a supporting layer of loose connective tissue, which contains large numbers of T cells and macrophages. However, it is not clear whether this transcytosis pathway actually leads to infection in vivo.)

Next, we consider HIV transmission across the vaginal epithelium. Although the vaginal epithelium is less easily damaged than the rectal epithelium, microabrasions have been detected in the vaginal epithelia in 60% of women after consensual vaginal intercourse. Also, as noted above, any ulcerative STDs further compromise the integrity of the vaginal mucosa. In addition, the vaginal mucosa, like other mucosae, contains CD4 T cells and dendritic cells. (The latter may actually be the most important of the antigen-presenting cells because they serve as key sentinel cells in the tissues and, following exposure to microbes, they migrate to lymph nodes and present antigen to virgin

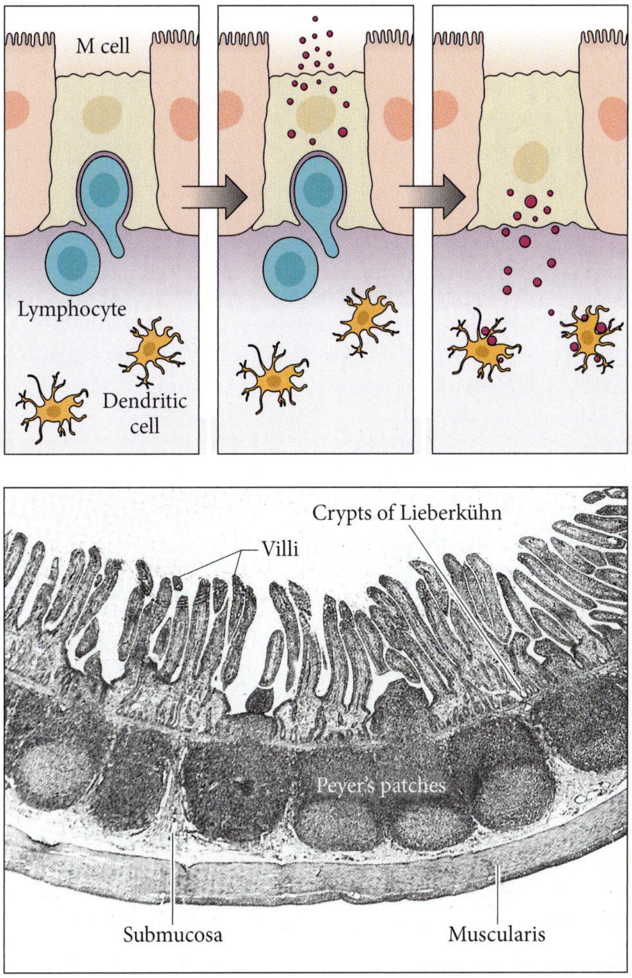

SIV/macaque model enhances infection of these cells at the abraded sites (Box 21.10).

An HIV-infected CD4 T cell may disseminate the infection by directly transferring the virus to an uninfected CD4 T cell. In fact, cell-to-cell transmission of HIV is 1,000 times more efficient than extracellular infection by free virus. The transfer process may be triggered by contacts between viral Env glycoproteins on the surface of the infected CD4 T cell and CD4 and CCR5 molecules on the surface of the target CD4 T cell (as suggested by the ability of blocking antibodies against these proteins to impair cell-to-cell viral transfer).

Interestingly, virus particles in the infected CD4 T cell and CD4 and coreceptor molecules on the target T cell are recruited to the region of cell-to-cell contact (i.e., the immune synapse [Box 21.10]). Virus particles then bud into the "synaptic cleft" between the infected cell and the uninfected cell, followed by fusion of the viral envelope with the plasma membrane of the uninfected cell. (A recent [2009] study followed the simple retrovirus murine leukemia virus in living cells in real time as a model of retroviral cell-to-cell transmission. It found that virus assembly is highly polarized towards sites of cell-to-cell contact. Importantly, the viral Env protein, in particular its cytoplasmic tail, was key to the process. Gag proteins interacted with viral Env proteins, provided that the latter had a cytoplasmic tail, to promote virus assembly at cell-to-cell contacts. If the Env cytoplasmic tail was missing, cell-to-cell contacts still occurred, but there was no virus assembly directed towards those sites.)

The ability of HIV to preferentially produce progeny virus particles at cell-cell interfaces is one reason why cell-to-cell transmission is so efficient. Moreover, this means of transmission through the tight cell-cell interface enables the virus to evade neutralizing antibodies and complement.

Another remarkable means of HIV cell-to-cell transmission has recently come to light, involving dynamic junctions between cells, called tunneling nanotubes. These are membranous channels with a diameter of 50 to 200 nm that transiently connect cells over long distances. They are readily formed between immune cells (as well as some other cell types) and are reported to facilitate HIV transfer from an infected T cell to an uninfected one (Fig. 21.4). Transfer of HIV in this way may be more efficient than diffusion of the virus between cells. Moreover, in contrast to the exposed tethers of the immune synapse, tunneling nanotypes minimize exposure of the virus to extracellular antibodies and complement.

Entry into Cells
The tropism of HIV for CD4 T lymphocytes results from the specific binding of the HIV Env protein to CD4, the

Figure 21.3 M cells. (Top) Drawing of M cells in the intestinal epithelium delivering antigens by endocytosis to underlying lymphoid tissue. Modified from P. Parham, *The Immune System*, 2nd ed. (Garland Science, New York, NY, 2005), with permission. (Bottom) Electron micrograph of a section of a cat's ileum, with its submucosal Peyer's patches. Micrograph from W. Bloom and D. W. Fawcett, *A Textbook of Histology*, 10th ed. (W. B. Saunders Company, Philadelphia, PA, 1975), with permission.

T cells, thereby activating an adaptive immune response [Chapter 4].) Importantly, like CD4 T cells, dendritic cells likewise express CD4 and the coreceptor CCR5, and to a lesser extent coreceptor CXCR4 (see below and Box 21.7). Thus, like CD4 T cells, dendritic cells are target cells for HIV. Moreover, the epithelium might need to be damaged only slightly to expose these cells to HIV. Then, after interacting with the virus, a dendritic cell might migrate to a nearby lymph node, where it could then pass the virus on to CD4 T cells (Box 21.10). Studies of the SIV/macaque animal model and studies of explanted human mucosal tissue ex vivo have indeed identified mucosal macrophages, T cells, and dendritic cells as targets for HIV infection. Moreover, experimental abrasion of the epithelium in the

BOX 21.10

Some researchers believe that dendritic cells that reside in the female genital tract play a key role in the dissemination of HIV. Immature dendritic cells, which are present within and under the epithelium, normally internalize pathogens at the site of an infection. These cells then begin a maturation process, which enables them to process those antigens and migrate to secondary lymphoid organs, where they activate antigen-specific resting T cells. In that way, dendritic cells initiate adaptive immune responses.

Dendritic cells capture HIV early after transmission. Remarkably, however, rather than being processed by dendritic cells, HIV appears to be able to use those cells to mediate its transport to lymphoid tissue, where it might then infect CD4 T cells and, in that way, extensively spread the infection.

How does HIV survive its encounter with dendritic cells and then use those cells as a conduit to infect CD4 T cells? This issue remains unsettled. But recent experimental results show that expression of the HIV Vpr protein impairs dendritic cell maturation and, consequently, the ability of dendritic cells to process and present antigens (see the text). Regardless, dendritic cells do not support HIV replication. Also, note that while immature dendritic cells express the HIV receptor and coreceptors at only very low levels, they, but not macrophages, are able to efficiently transmit HIV to cocultured CD4 T cells.

It may be relevant to this issue that dendritic cells express a novel adhesion receptor, DC-SIGN, which normally binds to ICAM-3, a protein that is expressed constitutively on the surfaces of T lymphocytes. The expression of DC-SIGN may be germane to HIV transmission because this receptor also binds the HIV envelope glycoprotein gp120. In one study, DC-SIGN appeared to be unable to mediate HIV entry into dendritic cells but instead may have functioned as a unique HIV-1 *trans*-receptor, which facilitated HIV infection of CD4 T cells by direct transfer from the dendritic cell surface, followed by fusion of the viral envelope with the plasma membrane of the CD4 T cell (see the text).

In another study, DC-SIGN was seen to mediate HIV entry into dendritic cells by a somewhat more roundabout, exotic route, in which the virus was first taken up into the dendritic cell by an endocytic process that transported it to a low-pH, nonlysosomal compartment, in which the virus remained viable. Then, the virus was passed from the dendritic cell to a CD4 T cell via a so-called immune synapse, which forms between these two cell types. (At an immune synapse, as at a neuronal synapse, specific molecules are released and signals may be transmitted. Dendritic cells present antigens to CD4 T cells at an immune synapse.) The captured virus too was found to be concentrated at the "synapse," ideally positioned for transmission to the T cell.

Proviso: It is not clear whether either of the above two dendritic cell-mediated HIV infection pathways is pertinent in vivo.

---

cell surface protein that characterizes helper T cells (Chapter 4). That tropism is of singular importance since it is the principal determinant of HIV pathogenesis in the human host.

Figure 21.4 Membrane nanotubes readily formed after intercellular contact between HIV-infected (labeled with DiD; red) and uninfected T cells (labeled with DiO; green). From S. Sowinski, C. Jolly, O. Berninghausen, et al., *Nat. Cell Biol.* **10:**211–219, 2008, with permission.

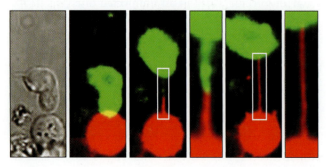

HIV Env is comprised of gp120, which is the surface subunit of Env, and gp41, which is the transmembrane subunit. HIV entry into CD4 T cells begins when gp120 binds to CD4 on the surface of the T cell. (Trimers of Env comprise the spikes seen on native HIV particles in electron micrographs [Fig. 21.1].) See Boxes 21.11 and 21.12.

The interaction of gp120 with CD4 causes gp120 to undergo a conformational transition that exposes binding sites for the coreceptor, either CCR5 or CXCR4 (see below and Box 21.7) (Fig. 21.5). The interaction of gp120 with a coreceptor in turn triggers gp41 to undergo a conformational rearrangement that exposes its fusion peptide, which previously was buried in a hydrophobic region of the protein. The gp41 fusion peptide then inserts into the plasma membrane of the target cell. Next, gp41 undergoes additional conformational transitions that promote fusion. In that regard, gp41 contains two helical heptad repeat regions, HR1 and HR2, which interact with each other, such that a stable six-helix bundle structure is formed

BOX 21.11

The early observation that HIV preferentially infects purified CD4 T cells provided the first suggestion that CD4 might be the receptor for the virus. That premise was then supported by the experimental finding that monoclonal antibodies against CD4 specifically block HIV infection of most susceptible cell types. The critical proof that established CD4 as the receptor for HIV was the demonstration that when CD4-deficient, HIV-resistant cells were genetically engineered to express CD4, they then were susceptible to HIV.

within the Env trimers. This brings the viral envelope into apposition with the plasma membrane, triggering fusion and allowing the viral core to enter the cell.

Interestingly, the six-helix bundle structure formed by gp41 contains a central three-stranded coiled coil that is reminiscent of the rearranged structure of the influenza virus fusion protein (Chapter 12; Fig. 12.5). Thus, the HIV gp41 protein likely promotes fusion via a molecular mechanism similar to that used by influenza viruses and by several other enveloped viruses as well (e.g., simian virus 5 and Ebola virus [Box 12.4]).

When the viral core is released into the cytoplasm, the genomic RNA is then reverse transcribed by the viral reverse transcriptase inside the core. This generates the preintegration complex, comprised of the viral DNA genome and the integrase, which is then transported to the nucleus. Once in the nucleus, the integrase inserts the

BOX 21.12

gp120 contains the principal antigenic determinants on HIV particles. Because of (i) the low fidelity of the reverse transcriptase and (ii) immune selection, gp120 displays very great antigenic variability; this feature of gp120 allows the virus to escape neutralization by antibodies, thereby contributing to its ability to persist in the presence of the host immune response and eventually cause HIV/AIDS (see the text). gp120 variability maps to five distinct regions on the protein, designated V1 to V5. These highly variable regions are separated by the relatively constant regions C1 to C4. Sequences of gp120 that interact with CD4 are in the conserved regions C2 to C4. Why might you expect that the particular sequences of the virus attachment protein which interact with the receptor are the highly conserved ones?

viral DNA into the host genome, where it can carry out subsequent steps in the viral reproductive cycle.

The HIV Coreceptors

The major HIV coreceptors are the β-chemokine receptor, CCR5, and the α-chemokine receptor, CXCR4 (also known as fusin). (Chemokines direct leukocytes to sites of inflammation. RANTES, MIP-1α, and MIP-1β [also known as CCL3, CCL4, and CCL5, respectively] are β-chemokines that bind to CCR5. SDF-1 [also known as CXCL12] is an α-chemokine that binds to CXCR4. In addition to its role in leukocyte chemotaxis, SDF-1 is also involved in hematopoiesis and angiogenesis. These chemokines are characterized by the presence of four conserved cysteines, which form two disulfide bonds in the molecule. In the CC subfamily, the cysteine residues are adjacent to each other. In the CXC subfamily, they are separated by an intervening amino acid.) CCR5 is expressed on macrophages and T cells, while CXCR4 is expressed on many T cells, but usually not on macrophages. Interestingly, only those HIV strains that can use CCR5 for their coreceptor, and thus infect macrophages, can be efficiently transmitted from one individual to another. Those HIV strains are said to be macrophage-tropic, or M-tropic, and they are the non-syncytium-inducing strains referred to above. So-called T-cell-tropic or T-tropic HIV strains prefer to infect CD4 T cells that express CXCR4. These are the syncytium-inducing strains referred to above. CXCR4 does not mediate infection of M-tropic HIV strains.

The premise that HIV might require a coreceptor was the result of the following chain of experimental findings. First, Jay Levy discovered that soluble factors released by CD8 T cells blocked HIV infection of CD4 T cells. Next, Robert Gallo's research group identified these soluble factors as the β-chemokines RANTES, MIP-1α, and MIP-1β. Since these β chemokines blocked HIV infection, and since CCR5 is the only receptor on CD4 T cells that binds these three ligands, it was suggested that the CCR5 β-chemokine receptor might be a coreceptor for HIV. The premise that HIV might require a coreceptor in addition to the CD4 receptor was also supported by the following experimental findings. When a human CD4 gene was expressed in mouse fibroblast cells, HIV was able to attach to those mouse cells but the virus still was not able to enter them; this observation is consistent with the notion that HIV entry might require a second human protein, which might act as a coreceptor. Following on this finding, CCR5 was shown to function as an HIV coreceptor by an experiment in which a human gene for CCR5 was transfected into HIV-resistant, CD4-expressing mouse cells. Expression of the human CCR5 gene in those CD4-expressing mouse cells indeed made them susceptible to non-syncytium-inducing, M-tropic HIV strains but not to syncytium-inducing, T-tropic strains.

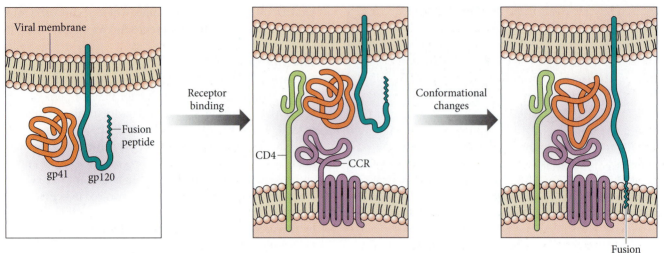

Viral membrane

Fusion peptide

gp41   gp120

Receptor binding

CD4

CCR

Conformational changes

Fusion peptide

Figure 21.5 The interaction of the gp120 subunit of the Env glycoprotein with CD4 induces conformational changes in gp120 that expose coreceptor binding sites on gp120, thereby enabling gp120 to engage the coreceptor, either CCR5 or CXCR4. The interaction of gp120 with the coreceptor in turn triggers a conformational rearrangement in the gp41 transmembrane subunit of Env, which exposes its fusion peptide. The gp41 fusion peptide then inserts into the plasma membrane of the target cell. gp41 then undergoes additional rearrangements that promote fusion. Adapted from S. J. Flint, L. W. Enquist, R. M. Krug, V. R. Racaniello, and A. M. Skalka, *Principles of Virology: Molecular Biology, Pathogenesis, and Control*, 2nd ed. (ASM Press, Washington, DC, 2004), with permission.

Other members of the CC chemokine receptor family, including CCR2b and CCR3, were also shown to serve as coreceptors for M-tropic HIV.

CXCR4 was identified as the coreceptor for T-tropic HIV strains as follows. HIV-resistant mouse cells, which expressed CD4 from a vaccinia virus vector, were transfected with a human HeLa cell plasmid DNA library. The trick was to identify those transfected cells which then expressed the coreceptor for T-tropic HIV. To do this, the transfected cells were cocultured with mouse cells that expressed a T-tropic HIV Env protein from a vaccinia virus vector. CD4-expressing mouse cells, which also expressed the coreceptor for T-tropic HIV from the human DNA library, could fuse with the T-tropic Env-expressing mouse cells. By repeatedly identifying smaller portions of the complementary DNA (cDNA) library that still caused fusion to occur, the researchers were able to isolate a piece of cDNA that encoded the necessary cofactor.

But why might M-tropic HIV strains be the most commonly transmitted form of the virus? Importantly, as described below, HIV can replicate efficiently only in activated CD4 T cells. CCR5 is expressed mainly on memory T cells and activated T cells, while CXCR4 is expressed mainly on naïve T cells. Moreover, memory T cells are more readily activated than naïve ones (Chapter 4). Thus, CCR5-expressing CD4 T cells are more likely to be in an activated state and thus better able to support the HIV lytic cycle. Additionally, dendritic cells, like macrophages, are derived from monocytes and express CCR5, and they may often be the first cells infected by HIV. (See the text below; see also Box 21.14, in particular the second paragraph.)

Bearing in mind that M-tropic HIV is the most commonly transmitted form of the virus, it is interesting that in about one-half of patients HIV may alter its choice of coreceptors during the course of infection. That is, the virus population in some individuals evolves such that some (usually a minority) of the virus particles use CXCR4 rather than CCR5 for their coreceptor.

The factors that influence the emergence of T-tropic HIV strains in some HIV-infected individuals are not well understood. However, T-tropic variants may replicate more aggressively than the original M-tropic virus, and perhaps for that reason their emergence often correlates with the onset of immune system collapse and the opportunistic infections that are the hallmark of AIDS. Nevertheless, since T-tropic virus is not detected in every individual who develops AIDS, the coreceptor switch is not a necessary precondition for disease progression. Also, the greater virulence of the T-tropic virus might be due to its broader target cell range, rather than to more aggressive replication per se.

While much remains to be learned regarding the reasons for the emergence of T-tropic HIV variants and their greater virulence, results from the following experiments illustrate how HIV tropism might affect pathogenesis in vivo. In these experiments, macaques were infected with chimeric simian-human immunodeficiency viruses (SHIV), which have an HIV envelope on an SIV genetic background. T-tropic SHIV spread to lymph nodes, causing marked destruction of T cells and reduction of circulating CD4 T cells. In contrast, M-tropic SHIV infected and depleted gut-associated lymphoid tissue, while not causing a reduction of circulating CD4 T cells. The early loss of gut-associated lymphoid tissue-associated CD4 T cells may be crucially important to the development of the immune deficiency caused by HIV infection (see below).

Despite the emergence of T-tropic variants in about one-half of HIV-infected individuals, the viruses from end-stage disease patients that infect new individuals always employ CCR5 as their coreceptor. Consequently, the viruses, which ultimately devastate the immune system at the end of the disease process in some individuals, may differ in their coreceptor usage from those that are transmitted by those same individuals to others (Boxes 21.13 and 21.14).

Here now is an interesting, unexpected development stemming from the HIV coreceptor story. Prostitutes and sexual partners of HIV-infected individuals, who are repeatedly exposed to HIV, are generally at high risk for infection. Nevertheless, some of these high-risk individuals somehow remained uninfected over time. This phenomenon is rather striking, since 95% or more of HIV-exposed individuals are susceptible to infection. As a matter of human interest, studies to explain these instances of HIV resistance were prompted by two public-spirited homosexual men, who pressed scientists to account for why they themselves were spared infection, despite having had repeated unprotected sexual contacts with companions who had succumbed to HIV/AIDS.

The explanation for this resistance phenomenon was reported in 1996 by Rong Liu and Michael Samson and their respective coworkers, who found that both of these HIV-resistant individuals have the same 32-nucleotide deletion in the gene that encodes CCR5 (the *CCR5-Δ32* allele). The product of this allele is not transported to the cell surface and thus cannot act as a coreceptor for HIV. Thus, individuals homozygous for the 32-nucleotide deletion appeared to be highly resistant to HIV, while heterozygous individuals displayed partial resistance.

There have been conflicting reports regarding the absolute HIV resistance of individuals homozygous for the *CCR5-Δ32* allele. Nevertheless, and not unexpectedly, it appears that these individuals may yet be infected with T-tropic strains that use CXCR4 for their coreceptor. But bear in mind that T-tropic strains are transmitted much less efficiently than M-tropic strains.

What of individuals who are heterozygous for the *CCR5-Δ32* allele? One study, reported in 2001, found that homosexual males heterozygous for the mutant CCR5 allele are 70% less likely to become infected with HIV than those

---

BOX 21.13

The ability of HIV to generate both M-tropic and T-tropic particles is generally attributed to the low fidelity of the retroviral reverse transcriptase, which, unlike other DNA polymerases, lacks a proofreading function. However, a new factor has recently been implicated in HIV mutability: the host cell cytidine deaminase APOBEC3G, which is responsible for G-to-A hypermutation in viral genomes (see the text). Regardless, the ability of HIV to mutate at a high rate also accounts for the antigenic instability of gp120, which plays a key role in the ability of HIV to evade neutralizing antibodies throughout the course of an infection. Infected individuals produce billions of new viral genomes every day (see the text), and virtually every one of those genomes contains at least one mutation. Thus, HIV has ample opportunity to generate the variants that satisfy its purposes.

---

without the mutation. Moreover, if these heterozygous individuals should become infected, then their disease has a slower progression. The partial resistance of heterozygous individuals likely reflects the fact that they express smaller than normal amounts of CCR5 molecules at the cell surface.

As might be expected, children who inherit the defective CCR5 allele are protected from transmission from their mothers via the breast milk. Also as expected, T cells taken from individuals carrying the mutant CCR5 allele are highly resistant to M-tropic HIV ex vivo, but these cells can be infected with T-tropic strains. Thus, the resistance of these children to infection is consistent with the premise that most infections in vivo are initiated by M-tropic virus. (The CCR5 mutation does not necessarily protect against infection transmitted through intravenous drug use. Why might that be so?)

Have you been wondering what sorts of deficits might be suffered by individuals homozygous for the *CCR5-Δ32* allele? Actually, they are otherwise healthy, with no apparent immunodeficiencies (but see Box 21.15). This finding might reflect the redundancy of the chemokines and their receptors. Perhaps even more surprising is the unexpected prevalence in some population groups of individuals who carry the defective CCR5 allele. Among Caucasians of European descent, about 1% are homozygous and about 20% are heterozygous for *CCR5-Δ32*. Yet the defective allele is virtually absent among African, Native American, and East Asian populations (Box 21.15).

Before leaving this subsection, we should note that the *CCR5-Δ32* allele is not the only host factor that affects susceptibility to HIV and the rate of disease progression.

BOX 21.14

Dendritic cells, which are present in the mucosal epithelia of the vagina, exocervix, and anus, express CD4 molecules, as well as CCR5 and CXCR4, and these cells are believed to be the first cells that are infected during sexual transmission of HIV (see the text and Box 21.10). There also are reports that HIV can cross intact monolayers of genital epithelial cells by transcytosis, at least in vitro, and only very inefficiently at that. But despite the alleged ability of HIV to cross genital epithelia by transcytosis, genital epithelial cells do not express CD4 or DC-SIGN. However, these cells do express low levels of CCR5 and CXCR4, as well as certain heparan sulfate proteoglycans that are reported to serve as alternative receptors for HIV. Genital epithelial cells preferentially bind T-tropic HIV. (If you have been following the preceding, this will seem to be a contradiction. Read on.) Still, it is not clear whether the CD4-independent transcytosis pathway across genital epithelial cells contributes to HIV infection in vivo.

Here now is a conundrum. Many T-tropic HIV strains can use CCR5 in addition to CXCR4 for their coreceptor, and some can enter macrophages via CCR5, at least in vitro. Moreover, bearing in mind that sexual intercourse is the major mode of HIV transmission and that genital epithelial cells preferentially bind T-tropic HIV, how might we explain why M-tropic HIV strains are implicated in about 90% of sexual transmissions of the virus? In the text we note that CCR5 is expressed mainly on memory T cells and activated T cells while CXCR4 is expressed mainly on naïve T cells. Thus, CCR5-expressing CD4 T cells are more likely to be in an activated state, able to support HIV replication. Another suggestion is that spermatozoa may selectively carry M-tropic HIV, since those cells express CCR5, but not CXCR4. However, T-tropic as well as M-tropic HIV particles are present in the semen. Another possibility is that the genital epithelial cells do not play a role in transmission per se. Thus, the CXCR4 that they express may act to sequester T-tropic virus, rather than to facilitate its transmission. In that way, M-tropic HIV may be favored to transmit the infection.

Indeed, much information, some of which may have significant clinical value, remains to be learned in these regards.

Entry into Human Placental Trophoblasts: a Special Case?

Nearly all children infected with HIV acquired the virus from their mothers, and in most of those instances, the infection occurred during passage through the birth canal. Neonates may also be infected postpartum, via the breast milk. In addition, a pregnant HIV-infected mother can transmit the virus across the placenta into the fetal bloodstream, by mechanisms that are not yet clear. The frequency of HIV transmission in utero from HIV-infected mothers who are not receiving antiretroviral therapy is estimated to be about 5%. This is in contrast to the overall rate of HIV transmission from infected mothers to their offspring, which is around 25%. Thus, about one-fifth of all transmissions of HIV from mother to offspring occur in utero, and the placenta thus appears to be an effective, if not perfect, barrier against HIV. But how might HIV breach the placenta in those relatively rare instances of in utero transmission?

The maternal and fetal circulations are separated by a polarized epithelium-like monolayer comprised of placental **trophoblast cells** (Fig. 3.8). Thus, if HIV is to enter the fetal circulation and infect the fetus in utero, it must cross the trophoblast cell layer of the placenta. In this regard, there are reports that HIV can infect and replicate in trophoblast cells, in vivo as well as in vitro. In addition, there are reports that HIV can cross the trophoblast cell layer by transcytosis. In either event, HIV would be released into the fetal circulation.

But how might HIV bind to, and enter, a placental trophoblast, since these cells do not express CD4, nor is there evidence that they express CCR5 or CXCR4 at their surface? The answer is not known, and the interaction of HIV with placental trophoblasts may thus represent an alternative, medically relevant means by which HIV interacts with the cells of its human host. What little information there is concerning the interaction of HIV with human placental trophoblasts is summarized as follows.

Whereas HIV enters CD4 T cells by fusion at the plasma membrane, there are reports that HIV enters trophoblasts by endocytosis and that entry into trophoblasts may be followed either by productive infection or by transcytosis across the trophoblast cell layer. Moreover, reports of HIV lytic infection of trophoblasts cite experimental evidence that productive infection of trophoblasts, but not of CD4 T cells, is blocked by endosomal inhibitors. (Endosomal inhibitors used in these experiments included bafilomycin A1, an inhibitor of the vacuolar proton adenosine triphosphatase [ATPase], and the lysosomotropic weak base NH_4Cl. Each of these treatments raises the pH of acidic endosomal compartments.) Treatment of trophoblasts with these agents prevented viral gene expression,

BOX 21.15

Why did the *CCR5-Δ32* allele penetrate into white populations of European descent, and why is it largely restricted to those individuals? A possibility that generated some interest is that the CCR5 allele might have provided some protection against the bubonic plague of 1346 or the smallpox that ravaged Europe during the Middle Ages, thereby explaining why it underwent strong positive selection in Europe but not elsewhere. However, several recent studies based on gene linkage analyses in modern populations and DNA analysis of Bronze Age samples led to the conclusion that the *CCR5-Δ32* allele probably arose more than 5,000 years ago. But these findings do not necessarily argue against more recent selection.

Although the basis for the penetration of *CCR5-Δ32* into some human populations is still speculative, the notion that a mutant host gene might confer resistance to a disease is well established. A familiar example is the mutation that causes sickle cell anemia. Individuals heterozygous for that mutation are resistant to malaria.

Recent reports show that wild-type CCR5 functions as a host defense factor against West Nile virus and other viruses. Might this compromise the utility of CCR5-blocking agents for use in therapies against HIV?

but not HIV entry per se. These experimental results support the premise that HIV productive infection of trophoblasts requires a step in which the virus traffics through a low-pH endocytic compartment.

As noted above, trophoblasts do not express CD4, CCR5, or CXCR4. Moreover, there are reports that HIV infection of trophoblasts might not require the HIV envelope glycoproteins gp120 and gp41. That is, HIV entry and infection of polarized trophoblasts occur with both fusion-incompetent and Env-deficient viruses. Also, heparan sulfate proteoglycans, which may serve as alternative receptors for HIV (Box 21.14), were reported to be required for HIV uptake into trophoblasts. Moreover, HIV infection was increased when placental cells were cocultured with HIV-infected T cells. In addition, infection of placental cells in these cocultures was blocked by antibodies against adhesion molecules (i.e., LFA-1 and β_2 integrins) that mediate the attachment of T cells to placental cells. Together, these experimental findings support the notion that HIV infection of trophoblast cells occurs via an unusual CD4- and gp21-independent pathway that requires contact between HIV-infected T cells and trophoblasts.

It is possible that this atypical, Env-independent pathway of HIV entry into trophoblast cells is rather inefficient, thereby accounting for the low rate of HIV transmission in utero. Regardless, factors such as the immune and hormonal status of the mother, as well as her HIV load, may also affect transmission in utero.

Are there other instances of HIV entry by endocytosis, rather than by fusion? Perhaps yes. There are reports that both free and cell-associated HIV enters epithelial cells of the genital and anorectal tracts by endocytosis. Also, there are reports that HIV may enter dendritic cells by endocytosis (see above) (Box 21.14).

Therapeutic Strategies That Target Entry

When I lecture on HIV entry, a clever student will generally suggest that a rational anti-HIV therapy might be based on using soluble CD4 molecules as a decoy HIV receptor. To the credit of these students, several versions of this strategy have indeed been tried, including the coupling of CD4 to the Fc portion of antibody molecules to engage immunity factors (see Chapter 4). These various strategies work well enough in cell culture but fail in clinical trials. One reason is that each HIV particle has at its surface about 30 copies of the molecular conformation that binds to CD4, and it is not clear how many of these sites need to be blocked on each virus particle to prevent infection. However, a more important negating factor may be that patients with moderate to advanced HIV disease typically produce and release 10^9 or more new infectious particles into the extracellular fluid each day! Moreover, much of this new virus is highly concentrated in lymph nodes, the primary sites in which the virus is generated. Thus, it would be virtually impossible to establish, much less sustain, efficacious doses of soluble CD4 in vivo at these key sites. The difficulty is compounded further by the short half-life of CD4 in the blood.

Another approach to HIV therapy, based on blocking HIV entry, is to impair the receptor or the coreceptor, rather than gp120. A potential advantage of targeting either CD4 or CCR5, rather than gp120, is that both the receptor and the coreceptor are far less variable than the viral gp120 protein (Box 21.16).

Targeting CCR5 may hold an advantage over targeting CD4, since targeting CD4 might impair immune function. In contrast, recall that individuals who lack CCR5 do not show any immune dysfunction. Moreover, they are indeed highly resistant to HIV. With these advantages in mind, a number of molecules that block CCR5 function have been developed. These include chemokines, N-terminally modified chemokine analogs, chemokine-based synthetic peptides, and anti-CCR5 monoclonal antibodies. Also, gene therapy strategies were developed to impair CCR5 expression, including using small interfering RNAs, antisense RNAs, or ribozymes. One recent study demonstrated that a modified form of the CCR5 ligand RANTES can prevent vaginal transmission of SIV in rhesus macaques. Another recent (2008) clinical study showed that a single intravenous injection of PRO 140, a humanized anti-CCR5 monoclonal antibody, has potent and prolonged anti-HIV activity in patients with early-stage HIV infection, while showing only minimal toxicity (Box 21.17).

The gp41-mediated fusion process is another potential target for anti-HIV therapy. In this regard, there are two heptad repeat (HR) regions in gp41, designated HR1 and HR2. These regions interact to form the fusogenic conformation, which contains a thermostable six-helix bundle that is a common fusion intermediate of a number of viral fusion proteins (Box 12.4). **Enfuvirtide** (also known as DP178 or T-20) is a 36-amino-acid peptide that mimics the HR2 region of gp41. It inhibits HIV entry by a mechanism that is thought to involve competitive binding to the HR1 region of gp41 (Fig. 21.6). Enfuvirtide has been approved for treatment of HIV-infected individuals, and it indeed can significantly reduce HIV loads when used in combination with other antiretroviral drugs. Yet, as in the case of other HIV therapies, the emergence of drug-resistant mutants remains a problem (see below).

GENOME ORGANIZATION, EXPRESSION, AND REPLICATION

Genome Organization and Expression

We begin this section by briefly restating some of the key similarities and differences in the organization and expression of the genomes of the simple and complex retroviruses. Then we will take a more detailed look at the HIV gene expression strategy.

The organization and expression of the genomes of the simple and complex retroviruses share the following features (Chapter 20). First, simple and complex retroviruses each contain genes that encode Gag, Pol, and Env proteins,

and these genes are arranged in the same order on the genomes of both simple and complex retroviruses (Fig. 20.1). Second, simple and complex retroviruses each use their virally encoded Pol to reverse transcribe their plus-strand RNA genomes into a double-stranded DNA copy, a process that in each instance generates long terminal repeat (LTR) sequences at each end of the DNA genome. Third, the DNA genomes of both simple and complex retroviruses are then integrated into the cellular DNA by the virally encoded integrase, thereby ensuring the stable association between the DNA provirus and the cellular genome. Fourth, all viral mRNAs (as well as progeny viral genomes) are transcribed using the integrated proviral DNA as the template and using the cellular Pol II RNA polymerase. But the complex retroviruses elaborate on this basic pattern as follows.

Simple retrovirus genomes encode only Gag, Pol, Env, and in some instances a nonessential oncoprotein as well. In contrast, complex retrovirus genomes additionally encode several nonstructural **regulatory and accessory proteins.** For example, HIV encodes regulatory proteins Tat and Rev, which vastly enhance the efficiency of HIV replication. HIV also encodes several other accessory proteins, including Nef, Vif, Vpr, and Vpu, which are not necessary for HIV replication per se, although expression of these proteins does enhance HIV infection in vivo. The

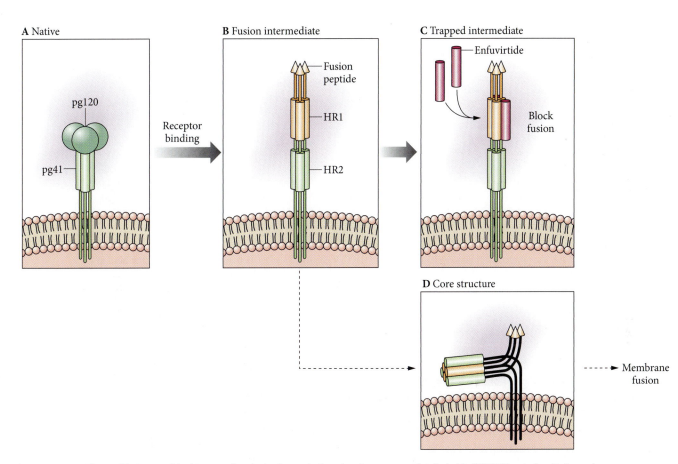

Figure 21.6 A model of Env-mediated membrane fusion, showing a captured enfuvirtide (DP178) complex. In the native form (A), HR2 cannot bind to its target region (HR1). Receptor binding triggers conformational changes (B) that allow HR2 to bind to HR1, creating a trapped fusion intermediate (C). Binding of enfuvirtide to gp41 prevents the formation of the six-helix bundle (D), and fusion is blocked. Modified from R. A. Furuta et al., *Nat. Struct. Biol.* **5**:276–279, 1998, with permission.

functions of the HIV regulatory and accessory proteins are described in detail below.

Here is another key difference between the simple and complex retroviruses. First note that the simple and complex retroviruses each produce a single primary RNA transcript. Also note that simple and complex retroviruses each translate their Gag and Pol proteins from that unspliced primary transcript, and each translates Env from a singly spliced transcript. But the complex retroviruses also generate multiply spliced mRNAs, which are translated into regulatory proteins. For example, HIV regulatory proteins Tat and Rev and the accessory protein Nef are translated from multiply spliced mRNAs (Fig. 20.1). Actually, multiple splicing options lead to over 30 species of HIV RNA, the functions of many of which are not yet known. That is so because the virus is known to generate but nine proteins.

The HIV provirus, like that of all retroviruses, is flanked by LTR sequences that contain viral promoter and enhancer elements (Fig. 21.7). However, while the enhancer elements of the simple retroviruses are activated by transcription fac-

tors widely distributed in a variety of cell types, the HIV enhancer elements are responsive to transcription factors that are preferentially expressed by T cells and macrophages. In both CD4 T cells and macrophages, HIV replication is dependent on transcription factor **NF-κB** (see below) and other transcription factors as well. (The NF-κB/Rel family of transcription factors regulate many genes involved in innate and adaptive immune responses.) Importantly, T cells and macrophages must be activated to express the transcription factors required by HIV. These aspects of HIV transcription, as well as the fact that the virus uses CD4 for its cell surface receptor, restrict the host cell range of HIV to *activated* CD4 T cells and macrophages. This is in contrast to the simple retroviruses, which generally can replicate efficiently regardless of the activation state of their host cells.

Apropos the above, most CD4 T cells are "resting" in vivo. Naïve CD4 T cells have never encountered their cognate antigen, while memory CD4 T cells have undergone clonal expansion following an earlier encounter with their cognate antigen (Chapter 4). In either case, these T cells must

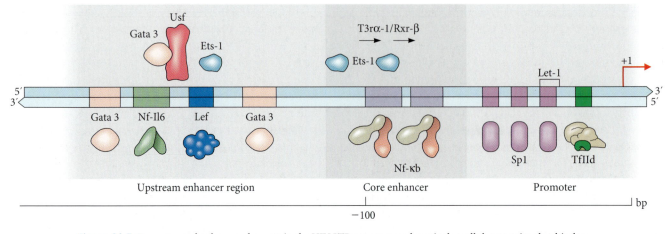

Figure 21.7 Promoter and enhancer elements in the HIV LTR sequence and particular cellular proteins that bind to those elements. Note that Lef is restricted to lymphocytes, and Gata 3 and Ets-1 are restricted to T lymphocytes. Also, notice the binding sites for NF-κB in the HIV core enhancer. Modified from S. J. Flint, L. W. Enquist, R. M. Krug, V. R. Racaniello, and A. M. Skalka, *Principles of Virology: Molecular Biology, Pathogenesis, and Control*, 2nd ed. (ASM Press, Washington, DC, 2004), with permission.

be activated by a professional antigen-presenting cell displaying their specific cognate antigenic peptide, in association with a particular major histocompatibility complex class II (MHC-II) protein (Chapter 4). Again, these T cells express transcription factors needed by HIV only after they have been so activated. As for macrophages, a large number of microbes and their products activate NF-κB in these cells, doing so via their interaction with a family of receptors known as Toll-like receptors, each of which recognizes a particular kind of molecular feature that is common to a broad class of pathogens (Chapter 4).

Because HIV can replicate in CD4 T cells and macrophages only after they have been activated, HIV infection of most CD4 T cells and macrophages initially results in viral latency. A DNA provirus is generated in resting T cells and macrophages, and it is integrated into the cellular genome. However, the provirus remains latent in these cells until they might be activated. (Again, bear in mind that all progeny viral genomes and all viral mRNAs are generated by transcription of the integrated provirus by the Pol II RNA polymerase, and transcription of these viral RNAs requires transcription factors that are present in CD4 T cells only after they have been activated.) Importantly, HIV is immunologically silent in its latent state. Thus, the ability of HIV to persist in a latent state in most infected cells means that neither host immune defenses nor current drug therapies can ever eliminate the virus (see below).

Our next goal is to consider the roles of the HIV regulatory proteins, Tat and Rev, in the HIV replication cycle. But before we do so, note that HIV requires T-cell transcription factor NF-κB to carry out even low basal levels of viral transcription. Indeed, the HIV LTRs are essentially

dormant in the absence of NF-κB. In resting T cells, NF-κB is found in the cytoplasm, bound to its inhibitor protein IκB, which acts by masking the nuclear localization signal on NF-κB (Fig. 18.9). The signals that activate T cells cause IκB to be phosphorylated. As a result, IκB is released from NF-κB, thereby enabling NF-κB to translocate to the nucleus, where it can activate transcription from NF-κB-responsive enhancers, including that of the HIV LTR.

Even when NF-κB enters the nucleus and binds to HIV LTR enhancer elements, viral transcription is still very inefficient. Nevertheless, this relatively low basal level of transcription is sufficient to enable the virus to generate small amounts of its **Tat** protein. Tat contains a nuclear localization signal, and in the nucleus, in conjunction with cellular transcription factors, Tat raises the efficiency of HIV transcription as much as 1,000-fold, thereby setting in motion a positive-feedback loop, which results in vastly enhanced transcription from the viral LTRs. The mechanism of action of Tat and the role played by Rev are described below.

The Regulatory Proteins: Tat and Rev

The HIV regulatory proteins Tat and Rev can be distinguished from the accessory proteins Nef, Vif, Vpr, and Vpu by the fact that the regulatory proteins are essential for viral replication, while the accessory proteins are so named because HIV can replicate in T-cell lines in vitro in the absence of their expression. But why does the virus carry "nonessential" genes? The reason is that each of these four accessory genes in one way or another helps the virus to evade or manipulate innate and adaptive immune responses. Consequently, expression of these genes does affect HIV replication and pathogenesis in vivo. Moreover, since many primary HIV isolates maintain functional, nonessential, accessory genes, selection must act in vivo to preserve their effectiveness. We consider the essential regulatory gene products, Tat and Rev, here and discuss Nef, Vif, Vpu, and Vpr below (see "Immune Evasion and Manipulation").

We begin with Tat, which acts to raise the efficiency of HIV transcription as much as 1,000-fold. Contrary to what you may have expected, Tat does not affect or operate like a typical transcription factor that acts at a promoter/enhancer site. In fact, Tat does not interact with the proviral DNA at all but instead interacts with a sequence near the 5′ end of nascent HIV transcripts. That sequence, which is encoded by a segment within the LTR known as the **transactivation response (TAR) element**, is located just downstream from the initiation site for HIV transcription, and it contains a palindromic sequence that is predicted to give rise to a stem-loop structure in RNA molecules transcribed from it (Fig. 21.8 and 21.9). The importance of the TAR-encoded stem-loop structure is demonstrated by the fact that mutations that disrupt it also impair the Tat-mediated

stimulation of HIV transcription. (That particular experimental finding led to the correct prediction that Tat in fact binds to the TAR-encoded stem-loop structure within nascent HIV transcripts.)

Since Tat binds to the TAR sequence on nascent HIV transcripts, rather than to the TAR element on the DNA provirus, it cannot increase the efficiency of transcription at the level of initiation. How then does Tat enhance the efficiency of HIV transcription? Tat acts by *preventing the premature termination of transcription*. In the absence of Tat, the vast majority of HIV transcripts terminate within 60 base pairs (bp) of the initiation site.

Tat blocks premature termination of HIV transcription by interacting with two different cellular proteins that are present in the transcription complex (Fig. 21.9). One of these cellular proteins is cyclin T, a regulatory subunit of a cyclin-dependent kinase, CDK9. (Recall that the cyclin-dependent kinases are so named because they associate with specific cyclins to promote progression through discrete phases of the cell cycle.) The cyclin-T/CDK9 complex is

Figure 21.8 (A) The HIV site of transcriptional initiation is shown relative to the HIV core enhancer and promoter (Fig. 21.5). Transcription of the proviral DNA generates nascent transcripts that contain the Tar sequence. Note that Tat does not interact with any site on the proviral DNA. Instead, it interacts with the Tar sequence near the 5′ end of nascent HIV transcripts. (B) The Tar sequence is predicted to give rise to a stem-loop structure, which is physiologically relevant, since mutations that would disrupt the stem-loop sequence impair the Tat-mediated stimulation of HIV transcription. Sequences important for recognition of Tar RNA by Tat are indicated in color. Modified from S. J. Flint, L. W. Enquist, R. M. Krug, V. R. Racaniello, and A. M. Skalka, *Principles of Virology: Molecular Biology, Pathogenesis, and Control*, 2nd ed. (ASM Press, Washington, DC, 2004), with permission.

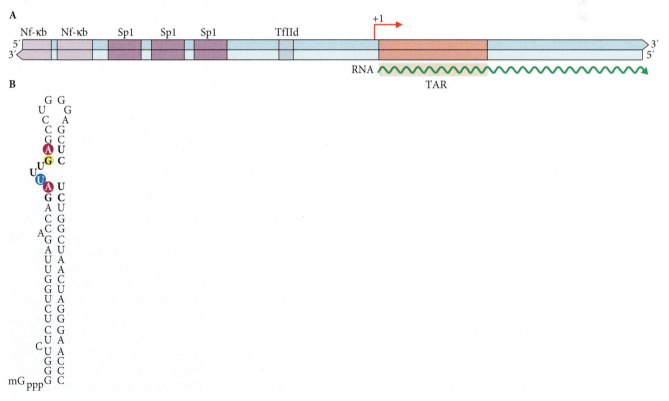

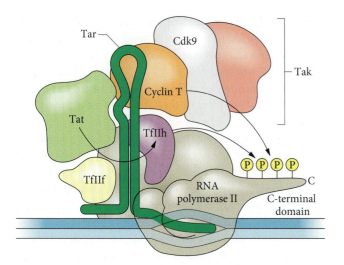

Figure 21.9 Tat blocks premature termination of HIV transcripts by interacting with two different cellular proteins that are present in the transcription complex. One of these cellular proteins is cyclin T, a regulatory subunit of a cyclin-dependent kinase, CDK9. The cyclin-T/CDK9 complex is referred to as the Tat-associated kinase (Tak). The other cellular protein targeted by Tat is the initiation protein TfIIh, which has a kinase activity that is stimulated by association with Tat. These two cellular kinases, Tak and TfIIh, are thought to sequentially phosphorylate sites in the C-terminal domain of the largest subunit of the pol II RNA polymerase, thereby increasing the processivity of the enzyme. Modified from S. J. Flint, L. W. Enquist, R. M. Krug, V. R. Racaniello, and A. M. Skalka, *Principles of Virology: Molecular Biology, Pathogenesis, and Control*, 2nd ed. (ASM Press, Washington, DC, 2004), with permission.

referred to as the Tat-associated kinase (Tak), since it was initially discovered in association with Tat. The other cellular protein targeted by Tat is the initiation protein TfIIh, which has a kinase activity that is stimulated by associating with Tat. These two cellular kinases, Tak and TfIIh, are

thought to sequentially phosphorylate sites in the C-terminal domain of the largest subunit of the Pol II RNA polymerase, thereby increasing the processivity of that enzyme.

Rev, like Tat, plays a key regulatory function, acting to resolve the following conundrum. Since eukaryotic cells produce primary transcripts that generally contain introns, eukaryotic cells have a mechanism in place to ensure that pre-mRNAs do not prematurely exit the nucleus, before all introns are excised and splicing is completed. At first, the export of a pre-mRNA from the nucleus is blocked by the associated splicing machinery. Then, as splicing transpires and a particular mRNA is completely spliced, other proteins bind to the splicing machinery, thereby marking that mRNA for export from the nucleus. In this way, mRNA export is linked to splicing. But since the HIV Gag-Pol and Env proteins are translated only from unspliced and singly spliced mRNAs, respectively, which still contain introns, how might these incompletely spliced HIV mRNAs be exported from the nucleus, before their splicing might otherwise go to completion?

The block to the nuclear export of intron-containing, unspliced, and singly spliced HIV mRNAs is counteracted by the viral Rev protein. Rev binds to a nucleotide sequence on HIV transcripts, which is known as the **Rev-responsive element (RRE)**. The RRE is about 250 nucleotides in length and contains several stem-loops (Fig. 21.10). Importantly, the RRE is located within the Env coding sequence. Thus, the RRE is present on both unspliced and singly spliced HIV mRNAs, but it is not present on the multiply spliced (i.e., completely spliced) viral mRNAs that are translated to yield the accessory proteins (Fig. 20.1).

Rev promotes the nuclear export of incompletely spliced HIV mRNAs by binding both to the RRE and to the cellular

Figure 21.10 (A) Rev binds to a nucleotide sequence on unspliced and singly spliced HIV transcripts, which is known as the Rev-responsive element (RRE). The RRE is about 250 nucleotides in length and is predicted to contain several stem-loops. The high-affinity binding site for Rev is shaded in green. (B) A portion of Rev is shown bound to the RRE, as determined by nuclear magnetic resonance. Interestingly, the major groove of an RNA helix would ordinarily be too narrow to accommodate an α-helical protein domain. However, in the case of the RRE, the groove is widened by two purine-purine base pairs. The purine-purine base pairs are not detected in the solution structure of the RNA alone but are instead induced by the interaction with Rev. Modified from J. L. Battiste et al., *Science* 273:1547–1551, 1996, with permission.

A

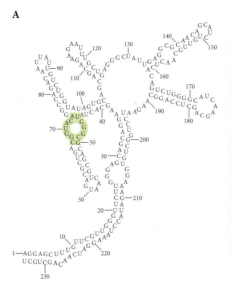

B

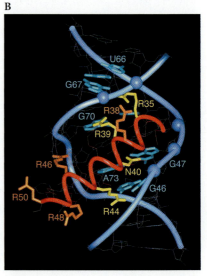

protein exportin-1 (also known as Crm-1), which in turn binds to GTP-Ran. In this way, Rev links the unspliced and singly spliced HIV transcripts to the nuclear export machinery (Fig. 21.10; Box 12.7). In the absence of Rev, these mRNAs either would undergo further splicing or else would be degraded. (The simple retroviruses, which do not encode accessory proteins, also utilize unspliced mRNAs to translate their Gag and Pol proteins and singly spliced mRNAs to translate their Env protein. The mechanism by which simple retroviruses export their unspliced mRNAs from the nucleus is described in Chapter 20.)

Interestingly, Rev appears to commandeer a nuclear export pathway that normally acts to translocate unspliced cellular 5S rRNA to the cytoplasm. This belief is suggested by the finding that the Rev nuclear export signal, which enables it to bind to exportin-1, is similar to, and can be replaced by, the nuclear export signal of the cellular protein TfIIIa, which specifically mediates the export of 5S rRNA from the nucleus. Moreover, peptides that contain the Rev nuclear export signal competitively inhibit the nuclear export of 5S ribosomal RNA (rRNA) but not the export of cellular mRNAs (Box 21.18).

HIV, like other complex retroviruses, must finely balance the relative amounts of its unspliced versus spliced transcripts that are exported from the nucleus. As in the case of the simple retroviruses, the efficiencies of the splice sites on these transcripts have evolved so as to generate the optimal ratio of spliced to unspliced transcripts. Experimentally increasing or decreasing those efficiencies dramatically impairs the replication of the complex retroviruses, as it did in the case of the simple retroviruses.

Next, we review the actions of Tat and Rev, now focusing on how they work in concert to promote HIV replication. HIV transcription begins in the absence of Tat and thus occurs at very low basal levels only. Since Rev too is absent at the start of transcription, only completely spliced transcripts are initially exported from the nucleus. These completely spliced transcripts encode the regulatory proteins Tat and Rev and accessory protein Nef, exclusively. When Tat materializes in the cell and enters the nucleus, HIV transcription then becomes as much as 1,000-fold more efficient. Consequently, vast amounts of mRNA are transcribed. Yet until Rev accumulates in the nucleus to facilitate export of unspliced and singly spliced transcripts, only multiply spliced mRNAs are exported and translated. But once a critical level of Rev is present in the nucleus, a rapid transition from regulatory gene expression to structural gene expression can occur, leading to a burst of virus production. How might this favor the virus? One possibility, although speculative, is that it helps the virus to evade host immunity, as discussed in Box 21.19. If that is so, then Tat and Rev, like the nonessential accessory proteins, are likewise concerned with evading or manipulating host immune responses (Box 21.20).

ASSEMBLY, MATURATION, AND BUDDING

The assembly of HIV and of other complex retroviruses proceeds according to the same principles that govern assembly of the simple retroviruses. That strategy and its considerable advantages are discussed in detail in Chapter 20 and are briefly summarized as follows. First, nascent HIV particles are assembled by interactions between *intact* Gag and Gag-Pol precursor polyproteins. Second, the positioning of the individual HIV structural proteins within the Gag polyprotein is the same as that within the layers of the virus particle (see Fig. 20.10). In this regard, the HIV Gag polyprotein has a membrane-binding myristate in its

BOX 21.18

The orthomyxoviruses, herpesviruses, and hepadnaviruses generate in the cell nucleus mRNAs that contain no introns whatsoever. Since no splicing machinery associates with their intron-less transcripts, how might those transcripts be exported from the nucleus? As you might have expected, the intron-less transcripts encoded by these viruses contain sequences that are recognized by virally encoded proteins, which act like the HIV RRE and Rev, respectively, to promote their export.

BOX 21.19

HIV is a champion among viruses at evading and subverting host immunity. Several of the known HIV immune evasion strategies are discussed in more detail in the text. While the combined action of Tat and Rev is not usually thought of as an immune evasion strategy, consider the following. At first, Tat causes incompletely spliced viral RNAs to accumulate in the nucleus. Some of these RNAs will serve as mRNAs that encode viral proteins, while others will become progeny viral genomes. Next, Rev causes these accumulated RNAs to be rapidly transported to the cytoplasm, resulting in a burst of HIV structural protein production and assembly of progeny virus particles. While admittedly speculative, this short, intense period of viral protein synthesis and assembly might compromise the ability of cell-mediated host defenses to recognize and destroy virus-producing cells before those cells might generate and release progeny virus particles.

Consider the following three points. First, Tat, Rev, and also Nef are translated early in the HIV replicative cycle, before synthesis of the viral structural proteins. Second, epitopes derived from Tat, Rev, and Nef should be recognized on these cells by CD8 cytotoxic T cells. Third, Nef is able to downregulate MHC-I and -II molecules on the surfaces of infected cells (see below). Bearing these points in mind, might it be wise to include these proteins in prospective vaccines, so that the cellular immune response might recognize the infection in a cell before Nef might downregulate MHC-I molecules at the cell surface? As proof of principle, adding Rev, Tat, and Nef to an SIV vaccine, in addition to Gag, Pol, and Env, indeed resulted in enhanced vaccine protection in macaques challenged with SIV. But see the discussion of the Merck STEP vaccine trial in the text.

matrix (MA) segment at the N terminus and a specific RNA-binding domain in its nucleoprotein (NC) segment at the C terminus. The interactions of the MA and NC segments with the plasma membrane and the viral nucleoprotein complex, respectively, also promote the assembly of progeny virus particles. Fourth, the proteolytic processing of Gag, leading to final virus maturation, occurs after budding.

Four of the several advantages of this assembly strategy, all of which are enumerated in Chapter 20, are reiterated here. First, it provides for the efficient incorporation of each of the individual Gag and Gag-Pol cleavage products into progeny virus particles, since the individual cleavage products do not have to come together separately in the nascent virus particle. Second, this strategy efficiently positions each of the structural components of the Gag polyprotein (MA, capsid [CA], and NC) within its proper layer within the nascent virus particle. Third, it enables the virus to solve the problem of how to assemble an inert, stable particle at the end of one replication cycle, which is nevertheless metastable, so that it is poised to undergo the conformational transitions that mediate entry at the start of the next cell. Fourth, it ensures an irreversible, forward direction to the assembly process.

Despite fundamental similarities between the assembly stratagems of the simple and complex retroviruses, there are important differences in the details. First, complex retroviruses also incorporate several of their unique accessory proteins into progeny virus particles. Second, HIV particles acquire their envelopes and bud by a concerted process. That is, HIV Gag polyproteins associate with each other at the same time as they associate with the cytoplasmic face of the plasma membrane, so that assembly and budding occur concurrently (Fig. 20.9). This is in

contrast to some simple retroviruses that employ a sequential process, in which complete cores are first assembled within the interior of the cell before they associate with the plasma membrane and bud.

In the concerted assembly pathway used by HIV, the MA segments of the Gag polyproteins must interact with Env-modified plasma membrane in order for the Gag polyproteins to interact with one another. It seems that the association of Gag with the cytoplasmic tail of Env at the plasma membrane causes Gag to adopt an assembly-competent conformation. This aspect of HIV assembly may be a viral adaptation that ensures against premature, nonproductive interactions between Gag polyproteins, which might otherwise occur in the cell interior.

HIV Assembly in Macrophages: Macrophages in HIV Pathogenesis

The HIV assembly and budding process summarized above is that observed in CD4 T cells, the main target cells of HIV infection. But the course of events is distinctly different in macrophages. In macrophages, HIV assembles on multivesicular bodies (MVBs), which are membrane-bound organelles that serve as intermediates in endocytic and exocytic pathways. After HIV assembly at the surface of the MVB, the virus buds into the MVB interior. The MVB then moves to the plasma membrane and fuses with it, in that way releasing the virus from the cell. (MVBs were discussed in Chapter 10 with respect to rhabdovirus entry. The mechanism that targets an MVB to a nondegradative site [i.e., the plasma membrane in the case of HIV release] versus a degradative site [i.e., a late endosome in the case of rhabdovirus entry] is not yet clear.)

What might be the advantage to HIV of being assembled at, and sequestered within, cytoplasmic vesicles in macrophages? In the case of other enveloped viruses, it is thought that assembly and maturation at an intracellular membrane enable the virus to evade certain host immune defenses. But that may not be the explanation in this instance, since most HIV replication occurs in CD4 T cells, in which maturation and budding occur from the plasma membrane. Also, the immune system is severely impaired at later stages of the infection. Thus, HIV probably does not need to assemble at an intracellular membrane in macrophages for the sole purpose of evading host immunity.

Before considering other possible reasons why HIV might be assembled in cytoplasmic vesicles in macrophages, note the following points. First, macrophages are less susceptible to HIV cytopathic effects and to HIV-induced apoptosis than are CD4 T cells. This enables macrophages to survive for weeks to months after being infected. Second, macrophages that are persistently infected with HIV cross the blood-brain barrier into the central nervous system, where they cause dementia and other neurological disorders in AIDS patients. Persistently infected macrophages are also found in lungs, lymph nodes, spleen, and bone marrow. Together, these macrophages constitute a major reservoir for long-term HIV persistence in infected individuals. Third, the MVBs, into which HIV buds in macrophages, appear to be equivalent to the macrophage compartment in which MHC-II molecules are loaded with peptides. Consequently, when HIV-infected macrophages interact with CD4 T cells, the HIV-containing vesicles may be induced to directly release the virus by exocytosis into the **immune synapse** between the macrophage and the T cell (see above) (Box 21.10). In this way, the virus can be transferred from the macrophage to the T cell, reminiscent of the process described above for the transfer of HIV from dendritic cells to T cells or from one T cell to another. Indeed, this transfer of virus might happen concurrently with the activation of the T cell by the antigen-presenting macrophage. If that is the case, then it is another of HIV's more diabolical adaptations.

Bearing the above arguments in mind, note that most explanations for HIV persistence are concerned with the interaction of the virus with CD4 T cells. Yet these accounts do not explain how viremia is maintained in the late stages of AIDS, when the CD4 T cells are almost completely destroyed. The facts that HIV particles in macrophage cytoplasmic vesicles remain infectious for weeks or longer (even under conditions where further virus replication is blocked) and that those particles can be transmitted from macrophages to T cells suggest that macrophages might contribute to virus persistence throughout the entire course of HIV infection. (Follicular dendritic cells likewise may constitute a major reservoir for HIV, although in this case the virus is actually extracellular [see below].)

THE COURSE OF HIV/AIDS

Untreated HIV infections eventually result in the destruction of the infected individual's CD4 helper T cells, in that way bringing about AIDS, the most severe form of immunodeficiency known. AIDS almost invariably results in death from uncontrolled opportunistic infections.

Early studies that tracked the course of HIV infections used relatively insensitive immunological procedures to monitor virus titers in the blood of infected individuals. Because of the insensitivity of the experimental methods then available, these earlier studies led to the belief that only a very few CD4 T cells in an infected individual harbor or express HIV and that a long period of viral latency precedes the onset of full-blown AIDS. However, with the emergence of more sensitive techniques, in particular quantitative reverse transcription polymerase chain reaction

(RT-PCR) to quantify plasma HIV RNA levels, it became apparent that productive infection is ongoing throughout the entire course of the infection, at levels that were hardly imaginable from the data of the earlier studies. Consequently, results from more recent studies have had a profound impact on the understanding of HIV/AIDS and on clinical practice as well.

The course of an HIV infection can now be thought of as occurring in several stages. The stages of HIV/AIDS, as established by the CDC, are as follows.

The Acute HIV Disease Stage

The first several weeks of an HIV infection are referred to as **the acute** or **the primary phase**. Early in this phase, before an adaptive immune response is activated, large quantities of virus are produced. Typically, as many as 10^7 or more copies of viral genomic RNA (corresponding to 5×10^6 virus particles; i.e., two viral genomes correspond to one virus particle) are detected per ml of plasma (Fig. 21.11). Then, a small fraction of T cells are activated to fight the infection, and as a result, levels of HIV RNA in plasma then drop by 3 or more orders of magnitude. Some of the activated T cells become "resting" memory T cells, and some of these memory cells harbor latent HIV. Thus, the acute phase includes the active proliferation and dissemination of the virus, as well as the establishment of the latent infection.

The steep decline in the viral load seen after the first few weeks is largely due to the action of HIV-specific CD8 cytotoxic T lymphocytes (CTLs). In fact, the CTL population expands well before neutralizing antiviral antibodies can be detected. Indeed, in some cases, neutralizing

Figure 21.11 Schematic view of the course of HIV infection and disease, showing changes in CD4 and CD8 T-cell counts in peripheral blood, antibodies against HIV Env, and the plasma virus load.

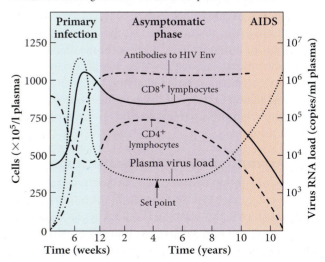

anti-HIV antibodies are not seen for months, and in other cases neutralizing antibodies are not detected at all.

The period of extensive virus replication in the absence of detectable antiviral antibodies is referred to as the **window of infectivity before seroconversion**, or the **window period**. The window period has important clinical consequences, since antibody testing during that time, whether in a clinical laboratory or by use of a home testing kit, will result in a false negative, despite the fact that the patient may be quite infectious. Results from mathematical modeling suggest that 56% or more of all HIV infections may be transmitted during the window period.

During the acute phase, large numbers of virus particles disseminate throughout the body, establishing infection in various organs, particularly lymphoid tissues, such as lymph nodes, spleen, tonsils, and adenoids. Notice that the level of CD4 T cells diminishes during the acute stage but then largely recovers (Fig. 21.11).

Virtually the entire HIV population is M-tropic during the acute phase. Thus, both macrophages and CD4 T cells can be infected. Importantly, infection of the vast majority of CD4 T cells results in virus latency. That is so because viral replication requires immune activation of the cell. This fact is extremely significant because *latently infected cells provide a safe haven for the virus*, in which it can survive for years, even in the face of a potent immune response to the infection or highly active antiretroviral therapy (HAART), which involves the concurrent use of three or more antiretroviral drugs to reduce the viral load in HIV-infected individuals (see below). (Note that HAART cannot prevent the establishment of latent infection of individual cells, even if initiated immediately after exposure to HIV.) Importantly, while HIV latency is established within individual cells of an infected person, a true state of viral latency within that person does not exist at any time during the course of an HIV infection, since there always are cells in that person in which HIV production is occurring, as discussed below.

Clinically, the acute phase is characterized by an influenza- or mononucleosis-like syndrome that includes fever, sore throat, malaise, swollen lymph nodes, headache, arthralgia (joint pain), diarrhea, a maculopapular rash, and thrush (a yeast infection in the mouth). Collectively, these symptoms are referred to as the **acute retroviral syndrome**.

The Asymptomatic HIV Disease Stage

The decline in the plasma HIV RNA levels, which began in the acute phase, continues over the next 6 months or so. Importantly, the plasma HIV RNA levels eventually stabilize around a so-called viral "**set point**," which generally remains fairly constant for a period of years. It also

is significant that the viral set point varies from one individual to another, typically ranging between 10^3 and 10^5 copies of plasma HIV RNA per ml. Some exceptional individuals may have fewer than 200 copies of plasma HIV RNA per ml.

The viral set point is clinically significant since it is one of the best clinical predictors of disease progression over time. Those individuals who display higher viral set points are at greater risk for disease progression. On average, a period of about 10 years elapses between initial infection and the onset of full-blown AIDS. However, at one extreme, about 10% of infected individuals progress to AIDS within 2 to 3 years. At the other extreme, 10% or more will remain disease free for 20 or more years.

The host factors that determine the viral set point and risk of disease progression need to be clarified (Box 21.21). Still, there has been progress towards that end. It is now fairly certain that events occurring during the first 3 weeks or so of infection play a critical role in determining the viral set point and, consequently, the time that elapses before disease progression to full-blown AIDS. This was demonstrated by studies of macaques experimentally infected with SIV, as well as by studies of HIV-infected patients. In each case, the peak plasma viral RNA levels seen during the acute stage correlate directly with the viral set point levels seen during the asymptomatic chronic stage. Moreover, reducing the peak virus levels during the first 3 weeks of infection, by means of antiretroviral therapy, results in long-term reduction of the viral load, even after therapy is terminated. Thus, the factors that determine the extent of early viral replication correlate with, and may largely determine, the long-term course of events (Box 21.22).

While the factors that might determine the degree of virus replication during the acute phase of the infection and, consequently, the viral set point, too, are not yet well known, there are several rather obvious and plausible possibilities, such as the pathogenicity of the HIV strain that initiated the infection. The state of the host's immune system, in particular the CTL response, is another. This premise is supported by new research findings that show that polymorphisms in the MHC may indeed help account for the variability in the viral set point among HIV-infected patients (Box 21.21). Moreover, there is a well-known inverse correlation between the viral set point and the patient's CD4 T-cell count. What is more, HIV-specific CD4 T cells and lymphoproliferative responses are commonly detected in long-term nonprogressors but not in those individuals with progressive HIV infection. One interpretation of these findings is that HIV-specific CD4 T cells might be depleted early in individuals who experience rapidly progressing HIV disease. A corollary is

BOX 21.21

Since the time that elapses before the onset of full-blown AIDS is largely contingent on the viral set point, which differs widely among HIV-infected individuals, several recent studies have sought to identify human genetic differences that might affect this variation. Using a **whole-genome association strategy**, one research group (Fellay et al., 2007) identified four human genetic polymorphisms that explain nearly 15% of the set point variation among HIV-infected individuals. One of these polymorphisms was actually found within an endogenous retroviral element (Chapter 20) and is associated with a major histocompatibility allele of *HLA-B* (Chapter 4). A second polymorphism is located near the *HLA-C* gene. Two other genes, one of which encodes an RNA polymerase I subunit, were also implicated.

The means by which these genetic polymorphisms might affect human susceptibility to HIV are not yet clear. Nevertheless, this study is significant for documenting that host genetic variations affecting the MHC locus can significantly affect the human response to HIV. Moreover, the particular

association of an HLA-C haplotype with slow HIV disease progression may lead to new approaches to anti-HIV therapy. For instance, consider that the HIV accessory protein Nef selectively downregulates the expression of HLA-A and HLA-B, but not HLA-C, on the surfaces of infected cells (see the text). This effect of Nef may benefit the virus because HLA-A and HLA-B present foreign epitopes to CTLs, while HLA-C binds self-peptides and interacts with NK cells to transmit the "don't kill" signal (Chapter 4). Thus, the ability of HIV to selectively downregulate HLA-A and HLA-B may represent another "diabolical" adaptation of this virus. But HLA-C can also present viral peptides to CTLs. Thus, the apparent resistance of HLA-C to downregulation by Nef opens up the prospect for vaccine strategies that might activate HLA-C-mediated host immune responses.

Reference
Fellay, J., K. V. Shiana, D. Ge, S. Colombo, et al. 2007. A whole-genome association study of major determinants for host control of HIV-1. *Science* **317**:944–947.

---

that once HIV-specific CD4 T cells are gone, they may be replenished only slowly, if at all.

Individual differences in chemokines and chemokine receptors among patients may also affect a patient's ability to establish immune control over the infection. The extent to which the patient's immune system is compromised or

BOX 21.22

The point during the course of an HIV infection at which it might be optimal to initiate antiretroviral therapy has been a matter of controversy for some time and is discussed in more detail later in the text. Nevertheless, this is an opportune moment to affirm that initiating antiretroviral therapy during the acute stage of infection may result in a lower viral set point and a better-controlled chronic infection, even if therapy were to be terminated at a later time. Indeed, from a therapeutic standpoint, the notion of a "morning after" pill or, more practically perhaps, the initiation of antiretroviral therapy as early as possible during the acute stage could be very beneficial (Box 21.42). Still, for some time the U.S. National Institutes of Health recommended against early initiation of therapy. What reasons can you think of that might have been the basis for the NIH recommendation?

stimulated by other pathogens may be another determining factor. In that regard, perhaps the fact that HIV can replicate efficiently only in activated CD4 T cells accounts, at least in part, for the more rapid progression of disease in sub-Saharan Africa, where infection by parasites and other pathogens is more common than in the United States and Europe. Age at the time of infection is also a known factor; young age is associated with slower disease progression.

The next portion of this discussion reveals some rather astonishing facts about HIV infection in vivo and the HIV disease process. We begin by noting that before the development of highly sensitive, quantitative, PCR-based procedures, HIV infections were thought to be largely latent during the asymptomatic stage. But newer, PCR-based procedures and the availability of anti-HIV drugs, which upset the balance between virus production and clearance, made it possible to determine the actual rate of virus turnover or, put somewhat differently, to demonstrate just how dynamic the long, asymptomatic stage of an HIV infection can actually be.

Experiments to determine rates of HIV turnover made use of the ability of powerful antiretroviral drugs to quickly and dramatically impede HIV replication in treated individuals. These experiments were carried out by initiating potent antiretroviral therapy in persons who had no prior exposure to therapy and then using the new quantitative

PCR techniques to measure the rate of the initial fall of their viral loads. From the data obtained in this way, it was possible to calculate that the half-life of HIV in the circulation is surprisingly short, only about 6 h. That is, about one-half of the virus in the circulation is replaced by new virus every 6 h. Moreover, to maintain constant or steady-state virus levels in the face of this high rate of virus turnover, 10^{10} to 10^{11} new particles of HIV must be produced, released into the blood circulation, and cleared each day. Also, it was possible to calculate that 30 to 50% of the virus particles present in the circulation of an infected person at any moment were produced by CD4 T cells infected within the previous day. There are no long-lived, chronically infected cells (except latently infected memory cells). Thus, the asymptomatic stage of an HIV infection is by no means a latent infection in the usual sense.

Notably, the high rate of HIV turnover is pretty much the same among all patients. Thus, the rate of viral turnover does not appear to determine the viral set point, nor is it a function of that set point or of the pretreatment CD4 T-cell count.

Here are some additional related points, which are all rather remarkable. Because of the high rate of HIV turnover, together with the low fidelity of the HIV reverse transcriptase, the genetic diversity of HIV in a single infected individual is said to be greater than the genetic diversity of influenza virus in the entire world! Next, consider that 40 million or more individuals are presently infected with HIV worldwide. Consequently, more than 10^{16} distinct HIV genomes are generated in the world each day. The extraordinary genetic diversity of HIV is extremely important clinically (i) because immune escape mutants readily emerge within each infected individual and (ii) because drug-resistant mutants likewise may emerge in each patient (Box 21.23).

By use of the same experimental approach as was used to measure the rate of HIV turnover during the asymptomatic, chronic stage of infection, it was also possible to measure the turnover rate of plasma CD4 T cells. As before, antiretroviral therapy was started in patients who had no prior exposure to therapy. Then, the rate of increase in the numbers of the patients' plasma CD4 T cells was measured. The initiation of antiretroviral therapy indeed caused a rapid increase in the number of circulating CD4 T cells. Based on the rate of increase in circulating CD4 T cells, it was determined that as many as 2×10^9 CD4 T cells, which corresponds to about 5% of the total number of CD4 T cells in the body, are destroyed by HIV and replaced each day, during the *asymptomatic* stage of the infection. Moreover, it was estimated that approximately 0.1% of an infected individual's CD4 T cells are producing HIV during the asymptomatic stage and at least 10% of the total CD4 T-cell population harbors a latent provirus

that potentially can become productive. (Did you notice a discrepancy between the numbers of CD4 T cells that are killed by HIV versus the number that are actually infected? See below.)

These studies, which measured the rates of turnover of HIV and of CD4 T cells in HIV-infected individuals, gave rise to a view of HIV infection that is very different from the latent infection originally envisioned. Moreover, the destruction and replacement of billions of CD4 T cells every day are consistent with the notion that destruction of CD4 T cells over time is the direct cause of the severe immunosuppression that characterizes AIDS.

Astonishingly, perhaps, the effect of HIV infection on CD4 T cells during the asymptomatic stage may be even more severe than that just depicted. That is so because the above experiments measured peripheral blood lymphocytes. Yet peripheral blood lymphocytes constitute only about 2% of the lymphocytes in the body. In fact, the vast majority of CD4 T cells are found in the lymphoid organs, and the percentage of HIV-infected CD4 T cells may be more than 10-fold greater in lymph nodes than in peripheral blood. This reflects the likelihood that an infected cell releases virus that more efficiently infects nearby cells, so that propagation of the infection depends on local cell density and, perhaps, on cell-to-cell interactions as well. At any rate, the damage inflicted by HIV on the immune system occurs primarily in the lymphoid organs, and it may be far greater than what is indicated by analysis of T cells in the peripheral circulation.

Over time, many CD4 T cells in the lymphoid organs are activated by the various opportunistic pathogens that challenge the host, as well as by the ongoing HIV infection. This recurrent activation of CD4 T cells enables latent HIV genomes in activated cells to start up their replication cycles. Moreover, it also increases the numbers of uninfected CD4 T cells that might eventually support productive HIV infection. HIV-producing CD4 T cells (as well as an even greater number of uninfected "bystanders" [see below]) are destroyed as a consequence of HIV replication. Thus, during the relatively asymptomatic period that precedes the onset of AIDS, the CD4 T-cell count slowly but continuously declines, at a loss rate of approximately 60 cells per μl of plasma per year.

Importantly, because of the clonal nature of the antigen-specific immune response (Chapter 4), the CD4 T cells that are directly destroyed by HIV infection are the very ones that are needed to control the HIV infection, as well as the opportunistic pathogens that exploit those individuals with HIV disease. Consequently, the ability of the immune system to control HIV as well as other pathogens declines, and the stage is set for progression to the symptomatic stage of HIV disease.

The Symptomatic HIV Disease Stage

As the number of CD4 T cells falls below 400 per μl of blood (500 per μl is considered normal), the patient enters the **symptomatic HIV disease stage**, characterized by the development of a variety of symptoms that may include weight loss, fatigue, fever, malaise, pain, loss of appetite, diarrhea, headaches, and swollen lymph glands (lymphadenopathy).

In addition to diminished immune function associated with the declining numbers of circulating CD4 T cells during late-stage HIV disease, the lymph nodes themselves undergo a breakdown in anatomical structure, which likewise dramatically affects immune competence. Apropos this point, vast numbers of virus particles have been present in lymph nodes since earlier stages of HIV infection. Moreover, many of these virus particles are associated extracellularly with **follicular dendritic cells**, which play an essential role in trapping pathogens that enter the circulation (Box 21.24). Follicular dendritic cells may also constitute a major extracellular reservoir for HIV. Regardless, in late-stage disease, the follicular dendritic cell network breaks down, concomitant with the breakdown of lymph node structure. Consequently, virus trapping is impaired, allowing large amounts of virus to spill out into the bloodstream. The breakdown of the follicular dendritic cell network illustrates just part of how the breakdown of lymph node structure compromises immune function, beyond that caused by the decline in the number of CD4 T cells (see below).

The symptomatic stage can last for months to years before progression to full-blown AIDS.

BOX 21.24

Follicular dendritic cells are distinct from the tissue dendritic cells that present antigen to naïve T cells and activate them (Chapter 4). For example, follicular dendritic cells do not express MHC-II molecules. In addition, while they play an essential role in trapping pathogens, they do not internalize and process them as tissue dendritic cells do. Instead, follicular dendritic cells display pathogens on their surfaces for long periods of time. It is thought that these follicular dendritic cells may transfer their trapped pathogens to B cells. The B cells then process the pathogen and present it via MHC-II molecules to CD4 T cells, whose T-cell receptors recognize MHC-II/peptide complexes on the antigen-presenting B cell. It also is thought that the ability of follicular dendritic cells to trap and display viral antigens for long periods in lymph nodes plays a role in the maintenance of serological IgG memory.

AIDS: Advanced HIV Disease Stage

The clinical diagnosis of AIDS is based on the patient's clinical condition. However, for epidemiological purposes, HIV-infected individuals are said to have AIDS when their CD4 T-cell count falls below 200 per ml of blood, since clinical AIDS generally occurs when the CD4 T-cell count falls below that level. Clinical AIDS is characterized by the appearance of opportunistic infections and other conditions (Table 21.1). Nevertheless, AIDS patients constitute a heterogeneous group, in which some individuals may feel well for the moment, while others are chronically ill and still others succumb quickly. For many individuals, clinical condition will fluctuate from day to day and from week to week.

This final stage of HIV disease is characterized by the continuing decline in the level of CD4 T cells, coincident with increasing levels of HIV in the blood. Although the basis for the ultimate breakdown in the patient's ability to control the HIV infection is not entirely understood, it is likely that at some point the HIV antigenic diversity, which results from viral mutation, exceeds the capacity of the diminished clonal, HIV-specific CD4 T-cell population to control the infection.

The Viral Set Point and HAART

Before moving on, it is worth reiterating that HIV RNA levels in plasma can drop dramatically and plasma CD4 T cell counts can rise rapidly in patients treated with potent combinations of antiretroviral drugs (HAART). Indeed, when HAART works, it can reduce the viral load to 50

Table 21.1 List of 26 conditions in the AIDS Surveillance Case Definition

---

Candidiasis of bronchi, trachea, or lungs

Candidiasis, esophageal

Cervical cancer, invasive[a]

Coccidioidomycosis, disseminated or extrapulmonary

Cryptococcosis, extrapulmonary

Cryptosporidiosis, chronic intestinal (>1-month duration)

Cytomegalovirus disease (other than liver, spleen, or nodes)

Cytomegalovirus retinitis (with loss of vision)

HIV encephalopathy

Herpes simplex, chronic ulcer(s) (>1-month duration) or bronchitis, pneumonitis, or esophagitis

Histoplasmosis, disseminated or extrapulmonary

Isosporiasis, chronic intestinal (>1-month duration)

Kaposi's sarcoma

Lymphoma, Burkitt's (or equivalent term)

Lymphoma, immunoblastic (or equivalent term)

Lymphoma, primary in brain

Mycobacterium avium complex or *M. kansasii*, disseminated or extrapulmonary

Mycobacterium tuberculosis, disseminated or extrapulmonary

Mycobacterium tuberculosis, any site (pulmonary[a] or extrapulmonary)

Mycobacterium, other species or unidentified species, disseminated or extrapulmonary

Pneumocystis carinii pneumonia

Pneumonia, recurrent[a]

Progressive multifocal leukoencephalopathy

Salmonella septicemia, recurrent

Toxoplasmosis of brain

Wasting syndrome due to HIV

---

[a]Added in the 1993 expansion of the AIDS surveillance case definition. (Adapted from the CDC, Atlanta, GA.)

copies or fewer of viral RNA per ml of blood, within 8 to 24 weeks. Bearing in mind that the viral set point is a strong predictor of disease progression, the early initiation and maintenance of antiretroviral therapy provide hope that advanced AIDS disease might be forestalled indefinitely. Moreover, early initiation of HAART might impair the emergence of drug-resistant HIV mutants. Also, individuals responding to HAART are much less likely to transmit the HIV infection.

Because a patient's viral set point correlates with peak viral levels during the acute stage of infection, many AIDS clinicians recommended that HAART should be initiated during the acute stage. However, in February 2000, the U.S. National Institutes of Health issued guidelines

recommending that HAART be initiated at later times, and only when the CD4 T-cell count falls below 350/µl of blood. What might be the rationale behind such a course? Well, for most patients who have spent months or years on HAART, either the viral load eventually rises again or the toxic side effects of the drugs (e.g., diabetes, bone loss, heart and liver disease, and a variety of cancers) become intolerable. Also, bear in mind that an HIV-infected person may live on average 10 to 11 years without serious discomfort in the absence of treatment. Thus, some clinicians believe that it may be wise to save potent antiretroviral drugs for the time when the patient's immune system begins to fail. If therapy is begun too early, then by the time it otherwise might be needed to actually save the patient's life, its accumulated side effects might force its discontinuation. Also, initiating HAART at earlier stages allows more time for the selection of drug-resistant mutants. These mutants might eventually overwhelm the patient. Moreover, they would be especially difficult to treat if they were transmitted to other individuals.

Apropos the above, many HIV-infected patients indeed have been forced to discontinue treatment because of the side effects of the drugs or because they experienced viral resistance to treatment. In addition, the complexity of the early drug regimens, together with their side effects, often made it difficult for patients to comply with their regimens, further arguing against early initiation of antiretroviral therapy.

On the other hand, others have argued that beginning therapy early might preserve the immune system, so that it too might help to regulate the HIV infection. Indeed, when therapy is delayed, the HAART-induced reversal of profound immune system dysfunction may be a slow and incomplete process in some patients, even when the viral load is reduced by the therapy, supporting the view that therapy should be initiated earlier, rather than later. And of course, waiting too long increases the likelihood of the disease progressing irreversibly to death.

So, the decision of when to initiate antiretroviral therapy depends on weighing the benefits of immediately reducing the viral load against the risks of drug toxicity and the potential emergence of drug-resistant mutants, while bearing in mind that HIV infection is a chronic condition that will likely require continuous therapy, for decades in the case of some individuals. But in 2008, new developments led the International AIDS Society–USA (IAS-USA) to reevaluate their antiretroviral treatment recommendations. First, new treatment regimens had recently been approved, involving newly developed, potent, durable, and less toxic drugs, with long half-lives. Second, new data that enable clinicians to better select the drugs for initial therapy and for the management of treatment failure became available. Third, new experimental findings led to new insights into the role of HIV in the pathogenesis of conditions not previously recognized as HIV related. The following is verbatim from the abstract of the 2008 IAS-USA report (Hammer et al., 2008).

> New data and considerations support initiating therapy before CD4 cell count declines to less than 350/μL. In patients with 350 CD4 cells/μl or more, the decision to begin therapy should be individualized based on the presence of comorbidities, risk factors for progression to AIDS and non-AIDS diseases, and patient readiness for treatment. In addition to the prior recommendation that a high plasma viral load (e.g., >100,000 copies/ml) and rapidly declining CD4 cell count (>100/μl per year) should prompt treatment initiation, active hepatitis B or C virus coinfection, cardiovascular disease risk, and HIV-associated nephropathy increasingly prompt earlier therapy. The initial regimen must be individualized, particularly in the presence of comorbid conditions, but usually will include efavirenz or a ritonavir-boosted protease inhibitor plus 2 nucleoside reverse transcriptase inhibitors (tenofovir/emtricitabine or abacavir/lamivudine). Treatment failure should be identified and managed promptly, with the goal of therapy, even in heavily pretreated patients, being an HIV-1 RNA level below assay detection limits.

Before moving on, it is important to stress that *HAART does not lead to a cure*, nor is it clear yet just how long its suppressive effect might endure. Moreover, not all treated individuals respond, and for those individuals who do not respond, HIV/AIDS remains an unmanageable illness. Finally, even for those who do respond, troublesome side effects often force therapy to be suspended.

IMMUNOPATHOGENESIS

The very earliest reports of AIDS described it as an extraordinary assortment of opportunistic infections in individuals who displayed no previous immune dysfunction. Soon, it became clear that AIDS is a disease characterized by a progressive depletion of CD4 T cells, leading to severe immunodeficiency and then death from unregulated opportunistic infections. No other virus infection of humans is known that causes such a devastating and relentless loss of CD4 T cells and immune system dysfunction. To be sure, *AIDS is the most severe form of immunosuppression known*. It is important to emphasize that, in the end, HIV-infected individuals succumb to opportunistic infections, rather than to HIV per se. These opportunistic pathogens generally pose little threat to individuals with healthy immune systems (Box 21.25).

You may be surprised that, after more than two decades of intense research, there is no consensus among AIDS

BOX 21.25

The crucial roles that CD4 helper T cells play in adaptive immune responses against viruses are described in detail in Chapter 4 and are briefly summarized in this sidebar.

First, CD4 T cells are needed to activate "naïve" or "virgin" CD8 killer cells, also known as CTLs. This is a crucial CD4 T-cell function, because CTLs play an essential role in bringing virus infections under control and ultimately in clearing them. In brief, the steps in the activation of CTLs are as follows. CD4 T cells are themselves initially naïve and must be activated to carry out their activation roles. Activation of CD4 T cells occurs when these cells recognize their cognate antigen, as presented by a professional antigen-presenting cell (e.g., a dendritic cell). Activated CD4 T cells may in turn activate close-by CTLs by releasing cytokines, such as IL-2. The CD4 T cell may do so as a member of a ménage à trois with the dendritic cell and the CTL (Fig. 4.22).

A second major CD4 T-cell function is to activate naïve or virgin B cells, which then become antibody-secreting plasma cells. B cells actually require two signals to be activated. The first signal is transmitted by the B-cell receptor when it recognizes its cognate antigen. Then, the B cell internalizes the receptor/antigen complex and functions as an antigen-presenting cell, displaying antigenic peptide fragments in the context of MHC-II molecules. The second signal that is necessary to activate the B cell is transmitted by an activated CD4 T cell, which has a T-cell receptor able to recognize the antigenic peptide/MHC-II complex on the surface of the antigen-presenting B cell (Fig. 4.21).

Before leaving this sidebar, bear in mind that the antibodies secreted by plasma cells also play a key role in host defenses against bacterial pathogens and thus help to ward off the opportunistic bacterial pathogens that afflict AIDS patients. Bacteria that are decorated (opsonized) by bound antibodies are recognized by the Fc receptors on professional phagocytes, thereby enhancing their destruction. The bound antibodies also activate the classical complement pathway, which also might lead to the destruction of the bacterial pathogen (Chapter 4). Activated CD4 T cells also secrete cytokines that enhance the microbicidal activities of the macrophage and its secretion of inflammatory cytokines. See Box 21.37 for the action of CD4 T cells in the immune response against fungi.

researchers regarding the mechanisms by which HIV causes the depletion of CD4 T cells and the breakdown of the immune system (see Douek et al., 2003). At least some readers may have been assuming, incorrectly, that the characteristic loss of CD4 T cells and the severe immune dysfunction that characterize HIV/AIDS are due solely to lytic HIV infection of CD4 T cells. Moreover, some readers may wrongly have the impression that the decline in CD4 T cells, which leads to AIDS, occurs primarily during the variably long asymptomatic phase of the infection. So, the actual state of affairs, as currently understood, is as follows.

First, and importantly, while HIV indeed lytically infects CD4 T cells, HIV infection is also associated with major losses of *uninfected* CD4 T cells. Indeed, during the chronic phase, more CD4 T cells die as bystanders than as a result of actually being infected by HIV. Second, while the breakdown of host immunity is often thought to result from the long-term chronic infection, important immune deficits are already manifest during the acute stage of the infection. Indeed, they may be seen by the time the primary immune response to HIV is evident. Moreover, among these early immune deficits there appear to be severe and irreversible losses of HIV-specific memory CD4 T cells. That is, an HIV-infected individual may enter the chronic stage of the disease with an already severely depleted population of HIV-specific memory CD4 T cells, all the while contending with a continuing viral onslaught that impairs its renewal capacities.

In view of the above, when attempting to make sense of the eventual depletion of CD4 T cells in HIV-infected patients, it is necessary to consider separately the distinctly different events occurring during the acute and chronic phases of the infection, particularly as they affect the pools of naïve and memory CD4 T cells. With that in mind, we next consider in somewhat more detail the interactions between HIV and the infected individual during the acute phase of the infection and then consider separately those interactions during the chronic phase. One of our purposes is to better appreciate how these early interactions can influence the timing of disease progression later, during the chronic phase of the infection. Also, we want to be able to underscore the ways in which virus-host interactions during the acute phase differ from those during the chronic phase.

First, we must acknowledge that less is known about events during the acute phase of HIV infections than about events during the chronic phase (why is that so?). Also, much of what is known about the acute phase of HIV infection in humans is inferred from the SIV/rhesus macaque model. But it is known that HIV, as well as SIV, initially interacts with host mucosal tissue. Moreover, HIV

initially infects cells that express CCR5. By 2 to 3 days after vaginal inoculation in the SIV/rhesus macaque model, the majority of productively infected cells are CD4 T cells, located in the lamina propria of the endocervix (see above). This lamina propria in fact contains a continuum of CD4 T cells, through which the virus can readily disseminate and multiply. What is more, these CD4 T cells in the periphery are, for the most part, memory T cells and, as such, are more readily activated (e.g., by local inflammatory cytokines) than the naïve T cells in the lymph nodes, thereby enabling them to more readily support HIV replication.

The facts cited in the preceding paragraph begin to explain the sharp drop in the CD4 T-cell count during the acute phase of HIV infection, and in particular, the loss of crucial memory CD4 T cells. Moreover, during the acute stage of SIV infections in rhesus macaques, there is a rapid and nearly total loss of CD4 T cells from the intestinal lamina propria, as well as from other mucosal surfaces, regardless of the route of inoculation. Furthermore, these targeted T cells are, for the most part, memory CD4 T cells. And in contrast to the destruction of CD4 T cells during the chronic phase (see below), these T cells are killed as a consequence of direct lytic SIV infection and CTL-mediated apoptosis.

Remarkably, 60% of mucosal memory CD4 T cells are infected at the peak of the acute SIV infection in rhesus macaques. These findings may be critically significant regarding HIV infections in humans, because the mucosal CD4 T-cell compartment contains most of the CD4 T cells in the body. Moreover, the events noted in SIV-infected rhesus macaques are consistent with the observation that in HIV-infected humans, depleted levels of CD4 T cells are seen in the intestinal mucosa and in other mucosae as well.

Regarding the intestinal mucosa in particular, the gastrointestinal tract may well be the largest lymphoid organ in the body, containing perhaps 60% or more of all T cells in the body. Thus, within only a few weeks of the initial HIV infection, humans may already have severely depleted levels of memory CD4 T cells. Importantly, this is the condition in which humans enter into the chronic phase. Also worthy of note, studies of SIV-infected rhesus macaques show that those monkeys that can partially reconstitute their mucosal T-cell count progress to AIDS at a slower rate than those that are unable to do so. Thus, the integrity of the network of mucosal CD4 T cells may be a major determinant of the rate of progression to AIDS.

Since many readers may have initially thought that no lasting harm occurs to the immune system during the initial acute phase of HIV infection, it is worth reiterating that this is not the case. Indeed, as many as one-half of an individual's memory CD4 T cells can be killed during the acute phase. Moreover, the intestinal network of CD4 T cells is severely compromised, and HIV quickly reduces the ability of the thymus gland to replace the lost CD4 T cells. These facts are crucially important because when an HIV-infected individual progresses to full-blown AIDS, he or she is often afflicted with old infections that earlier were controlled by memory T cells. Additionally, contrary to what you may have presumed, early damage to the immune system cannot be entirely reversed by antiretroviral therapy.

Next, we consider the asymptomatic phase of HIV infection that follows the acute phase. This phase is chronic and generally years-long in duration. Moreover, it is characterized by a slowly falling CD4 T-cell count and by slowly rising plasma HIV RNA levels. Also, in contrast to the state of affairs during the acute phase, in which the death of CD4 T cells results primarily from lytic HIV infection of those cells, the vast majority of CD4 T cells killed during the asymptomatic chronic phase are, in fact, not infected. Thus, the ability of HIV to lytically infect CD4 T cells only partly accounts for the immunodeficiency that manifests itself later as AIDS.

Irrespective of the above, lytic HIV infection of CD4 T cells nevertheless remains crucially important during the asymptomatic phase, because it is the major source of HIV throughout this time. Moreover, HIV infection of CD4 T cells during the asymptomatic phase is particularly insidious for the following reason. Both memory T cells and naïve T cells need to be activated before they might support lytic HIV infection. Since the activation of T cells is antigen specific, the upshot is that HIV replicates in, and destroys, the very CD4 T cells (both naïve and memory) that have been activated against current and ongoing infections, including both the HIV infection and whatever opportunistic pathogens might be infecting the patient. So, we might envision a cycle in which HIV replicates in, and destroys, HIV-specific CD4 T cells, causing a decline in HIV-specific CTL responses, leading to diminished control of the HIV infection, further increases of HIV replication, and so on, eventually terminating in AIDS.

HIV disease progression is also facilitated by the fact that the virus has been chipping away at the CD4 T-cell population under conditions in which the regenerative capacity of the immune system is being severely compromised by the disruptive effects of the infection on the thymus and lymph nodes. The immune system relies on these lymphoid organs to generate new memory CD4 T cells to replenish those that are depleted at peripheral sites. Furthermore, the structural integrity of the lymph nodes is necessary for these organs to support the interactions between T cells, B cells, and antigen-presenting cells that are needed to activate adaptive immune responses (Box 21.25)

and for these cells to respond to cytokine signals within a lymph node. In addition, bear in mind that the continuing production of billions of new HIV particles each day in HIV-infected individuals, many of which are antigenic variants, results in a state of chronic activation of cellular and antibody-mediated immune responses, which stresses the ability of the compromised pool of naïve and memory CD4 T cells to respond to the continually changing antigenic targets, a situation compounded further by the breakdown of the lymph node microenvironment.

Despite the gloomy state of affairs recounted above, note that while viral replication is ongoing and extensive throughout the long asymptomatic stage of the infection, the characteristic loss of CD4 T cells that occurs during this stage, which is thought to underlie the severe immunodeficiency, occurs only slowly, and nearly complete CD4 T-cell depletion is not seen until almost the end.

Since AIDS ultimately results from the depletion of CD4 T cells, we reiterate that more CD4 T cells die *as uninfected bystanders* during the asymptomatic chronic phase of HIV infection than as a result of being lytically infected by HIV. Experimental results from in vitro studies show that HIV may destroy uninfected bystander CD4 T cells by a variety of means. But be forewarned that it is not yet clear whether any of these phenomena seen in vitro actually contribute to the pathology of AIDS in vivo.

Considering that CD4 T cells are the main cell type that HIV kills as uninfected bystanders, it is reasonable to suggest that the HIV envelope glycoproteins gp120 and gp41 themselves might be able to target and kill bystander CD4 T cells. Indeed, under experimental conditions in vitro, these proteins do appear to be able to kill uninfected CD4 T cells by several distinct mechanisms. For example, soluble gp120 can induce apoptosis in primary CD4 T cells in culture. In addition, gp120, expressed on the surfaces of HIV-infected cells, can trigger apoptosis of bystander CD4 T cells through contact with CXCR4 on the bystanders. (Results of genetic studies demonstrate that transmission of the apoptotic signal indeed requires the signal-transmitting cytoplasmic tail of CXCR4, rather than that of CD4.) This apoptotic effect occurs independently of HIV replication, as shown by the fact that cells that are transfected to express Env at their surfaces can induce apoptosis of uninfected CD4 T cells. Note that soluble gp120 has been detected in the body fluids of HIV-infected patients, consistent with the thought that soluble gp120 might actually kill uninfected bystander CD4 T cells in vivo. On the other hand, soluble gp120 is not believed to reach sufficiently high concentrations in vivo to actually induce apoptosis.

gp41 too may be involved in the killing of bystander CD4 T cells, as shown by the finding that bystander cells are killed in vitro when they fuse with HIV-infected cells that are expressing Env. The fused cells die from apoptosis that is perhaps triggered by conflicts in the cell cycle status of the multiple nuclei in those cells, or perhaps when the polyploidy checkpoint is activated. The early timing of apoptosis in some of the syncytia that form in vitro between Env-expressing, infected CD4 T cells and healthy, uninfected CD4 T cells is consistent with a putative apoptotic mechanism referred to as **contagious apoptosis**. This process is thought to happen when, prior to fusion, the HIV-infected cell is in the early throes of apoptosis and thus causes apoptosis to also occur in each of the cells that subsequently fuse with it. Env-mediated apoptosis also may be triggered in bystander cells by the formation of a hemifusion intermediate; this is an early step in the fusion process that can be inhibited by specific peptide inhibitors of gp41-mediated fusion (see below) (Box 21.26).

Still, there is no evidence that any of the above apoptotic outcomes involving cell fusion in vitro plays a role in the depletion of CD4 T cells in vivo. Moreover, the issue of HIV-induced apoptosis in vivo remains controversial in general. Indeed, several observations argue against cell fusion playing a significant role as a cause of CD4 T-cell depletion in vivo. First, there are significant differences in the extent to which various isolates of HIV promote cell fusion and syncytium formation in vitro, and second, syncytia are rarely seen in vivo. On the other hand, in HIV-infected humans, the emergence of syncytium-inducing, T-tropic HIV variants is associated with an accelerated loss of CD4 T cells and more rapid progression to AIDS (Box 21.26). Perhaps more tellingly, the proportion of CD4 T cells that are in the later stages of apoptosis is about twofold higher in HIV-infected individuals than in noninfected control subjects. Thus, one or more of the above processes, and of the several that follow, may well be clinically relevant.

Other HIV gene products in addition to gp120 and gp41, in particular Nef and Tat, may also cause apoptosis of uninfected bystander CD4 T cells, at least in vitro. Before considering how they might do so, first recall that when FasL, on the surface of an immune effector cell (e.g., a CTL or a natural killer [NK] cell), binds to Fas on the surface of a target cell, it transmits an apoptotic signal into the target cell (Chapter 4). Bearing that in mind, note that HIV accessory proteins Nef and Tat upregulate the production of Fas ligand in HIV-infected CD4 T cells. Consequently, when an HIV-infected target cell is engaged either by an HIV-specific CD4 helper T cell or by an HIV-specific CD8 CTL, the target cell can transmit an apoptotic signal into the helper T cell or the CTL, so that the immune effector cell, rather than (or in addition to) the infected

target cell, might undergo Fas-triggered apoptosis. If this mechanism should be relevant in vivo, it would provide yet another rather striking example of just how diabolical HIV can be.

Even if there were no particular HIV gene products per se that triggered apoptosis in vivo, the chronic HIV infection by itself might still trigger apoptosis of uninfected bystander CD4 T cells. That is so because apoptosis actually plays a key role at several points in the normal life of a T cell. One such key role, which is relevant in the current instance, is to control the strength and duration of an immune response. This happens as follows. When a T cell is activated and then reactivated numerous times, it becomes more sensitive to Fas-mediated cell killing, induced either by its own FasL, or by FasL on other T cells (Chapter 4). Thus, uninfected, HIV-specific CD4 T cells might be lost because of the chronic HIV infection, which triggers their activation, proliferation, differentiation, and ultimately their death. This sequence of events is not restricted to HIV-specific T cells but applies to T cells activated in response to other pathogens as well. (Recall that apoptosis is also the means by which autoreactive T cells are eliminated during their maturation in the thymus [Chapter 4]. What might be the clinical consequence if T cells were unable to respond to the fact that an invading pathogen was successfully cleared? One possibility is a disease, known as autoimmune lymphoproliferative syndrome [ALPS], which is characterized by the accumulation of enormous numbers of T cells in the spleen and lymph nodes and an increased risk of lymphoma development. Most [about 70%] ALPS patients are heterozygous for a mutant *fas* allele, while others carry some other mutation. People with ALPS successfully resolve infections. Their problems begin after the infection is cleared.)

In addition to killing uninfected bystander CD4 T cells, in a high percentage of HIV-infected individuals HIV may actually impede the function of uninfected CD4 T cells before there is a measurable decline in the number of T cells. In individuals who are so affected, CD4 T cells do not proliferate in response to antigenic stimulation, and consequently, these individuals cannot mount an effective immune response against HIV or, for that matter, against any other pathogens, despite having sufficient levels of T cells to do so. Immunologically unresponsive T cells are said to be **anergic** (Box 21.27).

Normal CD4 T cells can be made anergic in vitro by exposing them to inactivated HIV particles or to purified gp120. The mechanisms by which the virus or purified virus components might induce T-cell anergy are not yet clear, but here is one possibility. Experiments in vitro show that soluble gp120 impairs expression of the CD4 T-cell gene encoding the cytokine IL-2. In this regard, recall that

BOX 21.26

There are substantial data implicating gp120 in bystander cell death. What then of gp41? While the fusion function of gp41 clearly plays a key role in HIV pathogenesis in vivo, its role in bystander cell death has been less clear.

A recent (2007) study sought to identify regions of the HIV gp41 fusion protein that *might* be necessary for inducing apoptosis in bystander CD4 T cells. In these studies, cells were transfected with different Env constructs and then incubated with target cells. The experimental results unexpectedly added a new twist to the story of bystander cell apoptosis. As predicted by the researchers, fusion-defective gp41 mutants could not induce apoptosis in bystander cells. However, a fusion-defective gp41 mutant that retained the ability to induce **hemifusion** was able to induce apoptosis in bystander cells. Taken together, the experimental findings suggest that gp41-mediated hemifusion is necessary and sufficient to induce apoptosis in bystander cells.

Hemifusion refers to the stage in the membrane fusion process in which the outer leaflets of the two interacting membranes transiently fuse, prior to the formation of a fusion pore. It was considered earlier as a stage in the fusion-mediated entry of influenza virus (Fig. 12.6) and the simple retroviruses (Fig. 20.3). In the case of the fusion-defective gp41 mutant noted here, the membrane fusion process proceeds up to, but not beyond, hemifusion.

In the text we note that the interaction between an HIV-infected cell and an uninfected cell could result in (i) transmission of virus from the infected to the uninfected cell or (ii) syncytium formation that might result in apoptosis by either of several mechanisms. To these possible outcomes between HIV-infected and uninfected cells, we now add a third result, gp41-initiated, hemifusion-mediated apoptosis. But might fusion/hemifusion-mediated apoptosis be relevant to HIV pathogenesis in vivo? The following observations suggest that it might.

Enfuvirtide is a novel anti-HIV drug that acts by binding to gp41 and blocking fusion (Fig. 21.6). Enfuvirtide-resistant HIV mutants commonly arise in patients receiving enfuvirtide therapy, and interestingly, some of these enfuvirtide-resistant mutants are defective at inducing cell-to-cell fusion and hemifusion-mediated apoptosis, while they are yet able to productively infect cells. Apparently, these mutants express sufficient Env fusion activity to mediate some viral entry and replication but not enough to induce cell-to-cell fusion. Other data suggest that the reduction in fusion activity might correlate with reduced bystander cell apoptosis. Importantly, some enfuvirtide-resistant mutants that arise in vivo are associated with a progressive rise in CD4 T cells and improved immune function, consistent with the notion that the reduced fusion-inducing capacity of the mutants might be associated with reduced fusion/hemifusion-dependent bystander apoptosis. Based on these findings, the authors of the study (Garg et al., 2007) suggest that enfuvirtide therapy may have the clinical benefit of selecting for less fusogenic and, consequently, less pathogenic viruses, in contrast to untreated infection, which selects for more pathogenic viruses.

Reference
Garg, H., A. Joshi, E. O. Freed, and R. Blumenthal. 2007. Site-specific mutations in HIV-1 gp41 reveal a correlation between HIV-1-mediated bystander apoptosis and fusion/hemifusion. *J. Biol. Chem.* **282**: 16899–16906.

---

naïve T cells require two signals to be activated (Fig. 4.19). One signal is triggered by the MHC-II/peptide complex on the surface of the professional antigen-presenting cell, which interacts specifically with T-cell receptors on particular T cells. The second signal is triggered by protein B7 on the surface of the professional antigen-presenting cell, which interacts with protein CD28 on the surface of the T cell. Next, note that the first signal leads to the induction of several transcription factors, one of which facilitates transcription of the gene encoding the cytokine IL-2, while the second signal stabilizes the IL-2 mRNA. If the first signal is not accompanied by the second signal, the IL-2 mRNA is rapidly degraded, IL-2 is not made, and the T cell is anergized. (Why might this process be important to the host? Since professional antigen-presenting cells can transmit the second costimulatory signal only if they have been activated at an infected site, the requirement for the costimulatory signal during T-cell activation helps to maintain tolerance to self antigens [Box 21.27; see also Chapter 4].) The preceding points apply to the induction of T-cell anergy by HIV as follows. Since soluble gp120 impairs expression of the CD4 T-cell gene encoding IL-2 (at least in vitro), CD4 T cells from HIV-infected individuals may be anergized, rather than activated, when they interact with professional antigen-presenting cells.

Additional studies show that gp120 can induce a long-lasting anergic state in naïve CD4 T cells by a mechanism

BOX 21.27

A T cell may normally become anergic if its T-cell receptor binds to a peptide/MHC complex on a professional antigen-presenting cell that does not also express the costimulatory signaling molecule B7 (Fig. 4.22). The requirement for the costimulatory signal is part of an elegant, multilayered mechanism that prevents the activation of self-reactive T cells. It works as follows. In non-infected tissues, professional antigen-presenting cells are in a "resting" state, in which they express very low levels of MHC-II molecules and B7. However, when an infection is under way, inflammatory cytokines are produced at the infected site and induce local professional antigen-presenting cells to express elevated levels of MHC-II molecules, and B7 as well. An activated antigen-presenting cell can then deliver its two signals: a general one via B7 in response to the infected state and the specific one via the peptide/MHC-II complex in response to the particular pathogen. The requirement for a nonspecific costimulatory signal, in addition to the pathogen-specific signal, helps to ensure that T cells are activated only in response to an infection, thereby helping to maintain *tolerance* of self antigens (Chapter 4).

that involves the activation of protein kinase A (PKA), doing so through the engagement of CXCR4, not CD4. (Recall that gp120 also transmits an apoptotic signal through CXCR4, rather than through CD4.) PKA is a serine/threonine kinase that regulates a number of cellular processes important for immune activation. It is generally activated by cyclic adenosine monophosphate (cAMP), and most effects of cAMP are mediated via the action of PKA. Now, bearing in mind that soluble gp120 is able to inhibit T-cell IL-2 gene expression, note that stimulation of the cAMP/PKA pathway can likewise impair T-cell responsiveness in vitro by transcriptionally attenuating IL-2 gene expression. So, perhaps HIV modulates T-cell responsiveness by activating the cAMP/PKA pathway. In support of that notion, intracellular cAMP levels are elevated in T cells from HIV-infected individuals and, moreover, PKA inhibitors restore the responsiveness of T cells from HIV-infected patients.

Soluble gp120 was also reported to interfere with T-cell receptor-mediated signaling pathways that activate the Src family tyrosine kinases and the mitogen-activated protein kinase signaling pathway. But, again, the relevance of these phenomena, and those involving IL-2, to HIV disease in vivo remains to be determined.

Thus far, we have been considering the effects of HIV infection on the CD4 T-cell population. But how might HIV infection affect the CD8 CTL population? This question is vitally important, since CD8 T cells play a crucial role in controlling virtually all viral infections, including those involving HIV. In these regards, HIV-infected patients present the following paradox. On the one hand, most untreated HIV-infected patients maintain high levels of HIV-specific CTLs during much of the chronic phase of the infection (Fig. 21.11). On the other hand, these high levels of CTLs provide only partial protection against HIV, and moreover, they do not prevent progression to AIDS.

It is not entirely clear why CTLs are relatively ineffective at controlling HIV infections, but there are several obvious possibilities. For example, CD8 CTLs are activated when they recognize their cognate antigenic peptide, displayed in the context of an MHC-I molecule on the surface of a professional antigen-presenting cell. However, in order to be activated, CD8 T cells also must receive a costimulatory signal, usually from an activated CD4 helper T cell (Fig. 4.22). In fact, antigen binding to the T-cell receptor in the absence of a costimulatory signal not only fails to activate the CTL but also may lead to CTL anergy (see above regarding anergy in CD4 T cells; see also Box 21.27). Thus, the depletion of HIV-specific CD4 helper T cells by the variety of mechanisms enumerated above or by the induction of helper T-cell anergy may cause a parallel loss of function in the very CD8 CTLs that are needed to control the HIV infection. Also, as noted above, HIV infection ultimately compromises the structural and functional integrity of the lymphoid organs, which are necessary for the interactions between antigen-presenting cells and lymphocytes that are necessary to activate adaptive immune responses.

The ineffectiveness of the CTL response against HIV may also reflect several other immune evasion strategies adapted by this most diabolical of viruses. This is the topic that we turn to next.

IMMUNE EVASION AND MANIPULATION

The severe immunodeficiency induced by HIV is only one face of the interaction of this virus with host immunity. The other is the ability of HIV to evade host immunity throughout the entire course of the chronic infection. Studies in vitro suggest that HIV may employ an astonishing repertoire of strategies to evade, subvert, and ultimately destroy host immunity. Some of these strategies were noted above in a somewhat different context, and others are enumerated below, specifically with respect to immune evasion. Once again, be forewarned that the extent to which any of the following phenomena, as seen in vitro, actually contribute to the pathology of AIDS in vivo is not yet clear.

A crucially important and key feature of HIV infection is that during its entire course, the virus is *able to remain latent* in a subset of infected CD4 T cells, a condition in which it is hidden from host immunity as well as from current therapies (Box 21.28). Thus, by irreversibly integrating the DNA provirus into the host cell genome, HIV safeguards its survival for the lifetime of the infected cell. The fact that HIV establishes latency in memory CD4 T cells is especially important in these regards because of (i) their long life span, which may be several years; and (ii) their ability to reactivate when they encounter their cognate antigen, an important fact since lytic virus replication may ensue in a reactivated memory CD4 T cell.

Several reports find that macrophages too harbor latent HIV, which might reactivate when those cells are activated (e.g., by the interaction of a Toll-like receptor with a microbe or a microbial product [Chapter 4]). Macrophages are clinically important in HIV/AIDS, since they are able to invade the immunologically privileged central nervous system, thereby enabling the virus to give rise there to neurological complications.

Importantly, when the virus activates in latently infected cells in response to an appropriate stimulus, it then employs several means for blocking host immune responses. We focus here on those evasion stratagems involving HIV accessory proteins Nef, Vif, Vpu, and Vpr (see Malim and Emerman, 2008, for a detailed discussion of these accessory proteins). We begin with Nef.

The Accessory Proteins: Nef, Vpu, Vif, Vpr, and Vpx

Interestingly, **Nef** derives its name from a laboratory error, which *mistakenly* indicated that the protein inhibits transcription; hence, Nef is an acronym for "negativity factor." But more to the point, Nef is myristoylated and,

consequently, it is anchored to the cytoplasmic face of the plasma membrane and to intracellular membranes as well. This tethering of Nef to cellular membranes is a likely prerequisite for most of that protein's supposed effects, the most often mentioned of which are as follows.

First, Nef promotes the downregulation of several cell surface proteins, in particular MHC-I molecules, and specifically the human leukocyte antigen A (HLA-A) and HLA-B allotypes. Note that the clever Nef has no effect on HLA-C. Thus, infected cells remain protected from NK cell-mediated lysis while also becoming stealthy to CTLs (Chapter 4; Box 21.21). Two mechanisms have been put forward to account for the effect of Nef on HLA-A and HLA-B. In the first, Nef interacts with the cytoplasmic tails of de novo HLA-A and HLA-B class I molecules. By doing so, Nef misdirects their transport out of the *trans*-Golgi apparatus to endosomes rather than to their normal location at the cell surface. In the second, Nef binds to the cytoplasmic tails of cell surface MHC-I molecules to induce their endocytosis from the cell surface. In either case, by impairing cell surface expression of MHC-I molecules, Nef blocks recognition of HIV-infected cells by CTLs. Importantly, like Tat and Rev, Nef is translated from a multiply spliced mRNA. Thus, it is one of the first viral proteins to be expressed following infection, thereby enabling it to effectively carry out this immune evasion stratagem.

A second effect of Nef is to downregulate cell surface MHC-II molecules. Third, the HIV-2 and SIV Nef proteins, but not the HIV-1 Nef, downregulate T-cell receptor (TCR/CD3) complexes (Chapter 4) from the cell surface. What then of the HIV-1 Nef protein? Instead of downregulating cell surface TCR/CD3 complexes from the cell surface, it is thought that the HIV-1 Nef may instead impair the movement of complete TCR/CD3 complexes to immune synapses. This was suggested by the finding that HIV-1-infected T cells inefficiently formed immune synapses with antigen-presenting cells and, when they did so, those complexes were abnormal (Box 21.29). Fourth, as noted above, Nef upregulates Fas ligand expression, so that when an HIV-infected target cell is engaged either by an HIV-specific CD4 helper T cell or by an HIV-specific CD8 CTL, the target cell can transmit an apoptotic signal into the effector cell. Fifth, Nef contains an SH3 domain (Chapter 20), which enables it to bind to cellular protein kinases. Thus, Nef might possibly activate signal transduction cascades that might enhance viral replication. This premise is consistent with the experimental finding that Nef mediates the activation of NF-κB, thereby setting in motion a pattern of cellular gene expression similar to that which results when T cells are activated by professional antigen-presenting cells (Chapter 4). Finally, Nef promotes the degradation of cell surface CD4

Bearing in mind (i) that HIV replication is dependent on activation cues that a latently infected CD4 T cell might receive from an antigen-presenting cell, (ii) that MHC-II molecules on the antigen-presenting cell transmit the specific activation cues to the latently infected CD4 T cell, and (iii) that the MHC-II/antigenic peptide complex on the antigen-presenting cell is recognized by the TCR/CD3 complex on the CD4 T cell, how might HIV benefit from the Nef-mediated downregulation of MHC-II molecules and the TCR/CD3 complex, since downregulating those factors from the cell surface impairs transmission of the activation cues needed by the virus to replicate? Two possibilities come to mind. First, and most obviously, this presentation is also normally essential to initiate the adaptive immune responses that are critical for controlling the HIV infection. Second, and somewhat more speculative, perhaps it benefits the virus by ensuring its prolonged, if less rampant, production, in that way perhaps favoring prolonged host survival and transmission of the virus to new individuals.

and CCR5. This effect of Nef may prevent interactions between gp41 and the receptor and coreceptor at the time of virus budding, thereby facilitating the release of virus particles from the cell surface. (Several other viruses, including measles virus, influenza virus, and hepatitis B virus, also have evolved strategies to remove their entry receptors from the cell surface, a process known as "receptor interference.")

Some of the alleged effects of Nef seen in vitro, and their possible relevance to infection in vivo, are controversial. For example, considering that HIV mutants, which escape CTL recognition, are continually selected during the course of HIV infection, the actual importance of Nef for evading CTL surveillance in vivo may be secondary at best. Also, different *nef* mutants have been studied in different cell lines, under still other different experimental conditions, perhaps accounting for inconsistent experimental results from different researchers. Yet it is almost certain that at least some of the activities of Nef noted above contribute to the pathogenicity of HIV in vivo, as implied by the following. First, SIV mutants, which contain *nef* deletions, did not cause clinical disease in macaques. (This is not a critical argument, since it might be explained in large part by the fact that Nef is required for efficient virus replication, which, in turn, is a requirement for pathogenicity.) Second, Nef's role as a virulence factor was seen in a transgenic mouse model. Finally, HIV strains

containing *nef* deletions were isolated from humans who were experiencing unusually slowly progressing AIDS disease (Box 21.30).

We continue with **Vpu**, which, like Nef, also downregulates cell surface expression of CD4. Moreover, like Nef, Vpu interacts with the cytoplasmic tail of CD4. But while Nef interacts with CD4 at the cell surface or the Golgi apparatus to promote its transport to lysosomes for degradation, Vpu (an 81-amino-acid dimeric integral membrane protein) interacts with CD4 molecules in the endoplasmic reticulum membrane, where it triggers the polyubiquitination and consequent proteasome-mediated degradation of CD4. In doing so, Vpu helps to prevent newly synthesized Env molecules in the endoplasmic reticulum from interacting in that compartment with newly synthesized molecules of CD4. If Env were to interact with CD4 in the endoplasmic reticulum, it might prematurely trigger conformational rearrangements in Env that need to happen later, during HIV entry into a new cell. In addition, an interaction of Env with CD4 in the endoplasmic reticulum might impair the incorporation of Env into progeny virus particles.

Another key function associated with Vpu is to enhance the release of progeny HIV particles from the surfaces of infected cells. Vpu-deficient HIV mutants assemble mature virus particles that bud from the plasma membrane, but those particles then undergo reuptake by endocytosis, rather than release. The release-promoting effect of Vpu is specific to HIV, since Vpu does not inhibit endocytosis in general. Thus, it is all the more surprising that the underlying mechanism of the Vpu release-promoting effect does not involve CD4. Its basis was explained only recently, as follows.

Macaques infected with SIV mutants that contain *nef* deletions did not develop clinical disease. However, infection with those *nef* mutants did generate immunity to wild-type SIV. Bearing these experimental findings in mind, might HIV *nef* deletion mutants have potential as candidate vaccines against HIV? Sadly, the answer is perhaps not. In instances where humans were inadvertently infected with *nef* mutants (from a blood bank donor), they initially did experience a relatively mild disease course, and they also displayed vigorous cellular immunity against HIV. Nevertheless, they eventually developed AIDS. In addition, macaques "immunized" with SIV *nef* mutants have given birth to offspring that developed AIDS-like disease. These outcomes have diminished enthusiasm for basing an HIV vaccine on *nef* mutant strains. Yet Nef remains a potential target for antiretroviral therapy.

In promoting HIV release, Vpu actually counteracts an IFN-α-induced antiviral factor dubbed **tetherin** (previously known as BST-2), which tethers virus particles to the cell surface after they have completely pinched off from the plasma membrane and then redirects them to an endocytic compartment, where they can be degraded (Neil et al., 2008). Relevant experimental findings included the following. First, infecting cells with Vpu-deficient mutants showed that Vpu is required for HIV release only in cells that express tetherin. Second, experimentally blocking tetherin expression in cells that normally produce this protein allowed the release of HIV particles.

How might tetherin entrap mature virus particles at the cell surface, and how might Vpu counteract tetherin? Endogenous tetherin colocalizes with the HIV Gag protein in endosomes, and also at the plasma membrane. Moreover, tetherin is a rather atypical transmembrane protein, in that each of its ends is inserted into the membrane, while its central portion faces out from the cell. These experimental findings are consistent with the premise that tetherin directly snares HIV particles within and on infected cells. Vpu in turn acts by downregulating tetherin from the cell surface. Recalling that Vpu, too, is a transmembrane protein, the Vpu-mediated downregulation of tetherin requires the presence of conserved serine residues in the cytoplasmic domain of Vpu (Van Damme et al., 2008). In the absence of Vpu, the tetherin-mediated reuptake of entrapped virus particles and their subsequent degradation in endocytic compartments are likely explained by the interaction of tetherin with the cellular endocytic machinery.

The fact that cells express tetherin in response to IFN-α, as noted above, implies that tetherin evolved as a component of a broad, type I IFN-induced, innate antiviral mechanism, in this instance, to foil viral dissemination. This notion is supported by the earlier experimental finding that a virus unrelated to HIV, human herpesvirus 8 (HHV-8) (also known as Kaposi's sarcoma-associated herpesvirus; see below), also impairs expression of tetherin. (The HHV-8 K5 protein appears to be responsible for reducing tetherin levels in HHV-8-infected cells. K5 is a ubiquitin ligase and, as such, marks tetherin for degradation by proteasomes.) Thus, tetherin may indeed have broad antiviral activity and, consequently, unrelated viruses may have evolved a variety of countermeasures against this host antiviral factor. (Note that up to this point, we have mainly been considering how HIV subverts and impairs the adaptive immune system. In contrast, this Vpu function provides an example of an HIV countermeasure against an innate antiviral defense mechanism.) Vpu also has been reputed to decrease cell surface expression of MHC-I molecules.

Vif, which stands for viral infectivity factor, is a 192-amino-acid cytoplasmic protein that HIV needs to generate viable proviral DNA. However, HIV requires Vif only in cells that express the cellular (deoxy-)cytidine deaminase APOBEC3G.

To appreciate how Vif acts as a countermeasure against APOBEC3G, first consider the effects of APOBEC3G on HIV replication when Vif is absent. In cells that are infected with *vif*-deficient HIV, APOBEC3G associates with Gag during virus assembly, and it is then packaged into progeny virus particles. During the next round of infection, APOBEC3G deaminates deoxycytidines in the minus-DNA strand that is formed during reverse transcription within the retroviral core. This results in the conversion of minus-DNA strand deoxycytidines to deoxyuridines. These deoxyuridines may then be excised by the cellular uracil-DNA glycosylase, which is an enzyme that specifically removes the RNA base uracil from DNA. If the deoxyuridines are excised, then the reverse-transcribed minus-DNA strand becomes a potential target for endonuclease digestion. If the deoxyuridines are not excised, then deoxyadenosines are incorporated into the plus-DNA strand, instead of deoxyguanosines. Since approximately 10% of G residues can be mutated in this way, the action of APOBEC3G may explain why G-to-A substitutions are the most frequent HIV point mutations. Importantly, in the absence of Vif, the resulting high rate of these nucleotide base substitutions is incompatible with viral viability (Box 21.31). But when Vif is present, it acts in conjunction with a cellular ubiquitin ligase to mediate the polyubiquitination and rapid proteasomal degradation of APOBEC3G (and of Vif itself). In this way, and perhaps by a more direct way as well (see below), Vif prevents the incorporation of APOBEC3G into progeny virus cores.

BOX 21.31

The low fidelity of the HIV reverse transcriptase, due largely to its lack of a proofreading function, enables the virus to readily generate immune escape mutants. This feature of HIV infection can be beneficial to the virus, while being especially problematical to infected individuals, in part because populations of helper T cells and CTLs responding to virus infections typically recognize only a few immunodominant viral epitopes. Yet there are limits to the levels of mutability that are compatible with the viability of the virus. The host exploits that fact by means of its APOBEC3G protein, which is potentially able to increase the viral mutation rate to a level that is incompatible with viability. Yet in some individuals, APOBEC3G may afford the virus an additional means for generating beneficial (to the virus, that is) sequence diversity.

As satisfying as the above story might be, the actual state of affairs may be somewhat less straightforward. Recent studies show that APOBEC3G can restrict HIV, even when the protein is missing its cytidine deaminase activity. Furthermore, fully infectious viruses are produced even under conditions where Vif and APOBEC3G stably coexist. Thus, these experimental findings show that APOBEC3G may impair HIV replication by means other than its cytidine deaminase activity. Moreover, they also imply that the Vif-induced degradation of APOBEC3G may not be the only means by which Vif counteracts APOBEC3G to enhance HIV infectivity. What then might be the other means by which APOBEC3G impairs HIV replication, and how might Vif offset it? In the absence of Vif, APOBEC3G associates with Gag and with HIV genomic RNA and is packaged into virus particles. There is evidence that the particle-associated APOBEC3G may then impair HIV infectivity, independent of its cytidine deaminase activity, by affecting particle morphology and by destabilizing the reverse transcription complex. Thus, by obstructing the inclusion of APOBEC3G into virus cores, Vif may block these anti-HIV activities of APOBEC3G. Other recent experimental results imply that Vif also may counteract APOBEC3G by impeding its translation. Regardless, recent studies show that human genetic polymorphisms in *APOBEC3G* are associated with differences in the rate of HIV/AIDS disease progression, implying that APOBEC3G and Vif play key roles in the HIV/AIDS disease process in vivo.

Interestingly, APOBEC3G was actually discovered as the host restriction factor responsible for limiting the replication of *vif*-deficient HIV (Chapter 4). APOBEC3G (and also APOBEC3F, another member of the APOBEC family of editing enzymes) has since been implicated in restricting other exogenous as well as endogenous retroviruses. What is more, recent studies reported that two very different families of DNA viruses, the parvoviruses and papillomaviruses, are also subject to G-to-A editing by APOBEC family members. Thus, the antiviral effects of these proteins may not be limited to retroviruses, hepadnaviruses, and retroelements (Box 22.6).

Vpr is a 96-amino-acid protein that derives its name from the observation that it affects the rapidity with which HIV replicates in, and destroys, T-cell lines in vitro. While Vpr expression has been associated with a variety of effects in vitro, one agreed-upon Vpr function is that it delays or arrests cells in the G_2 phase of the cell cycle. The Vpr-mediated G_2 arrest effect does not require de novo protein synthesis, which correctly implies that Vpr is a component of virus particles and is consistent with the premise that the protein plays an early role in the infection.

How might the Vpr-mediated G_2 arrest serve the purposes of HIV? While the answer is not known with any certainty, the HIV LTR may be more efficient in the G_2 phase of the cell cycle, and Vpr-induced G_2 arrest does in fact correlate with high levels of viral replication in human CD4 T cells. The molecular mechanism by which Vpr affects G_2 arrest is not yet known.

Another accepted function of Vpr is that it enhances infection of macrophages, perhaps by inducing the activation of transcription factor NF-κB. This effect of Vpr may also promote dissemination of the infection, as follows. Activation of NF-κB results in the secretion by the macrophage of vast amounts of the cytokine IL-8 (also referred to as CXCL8, where the designation CXC indicates that it is a **chemokine** or chemoattractant cytokine). CD4 T cells are very responsive to IL-8, and consequently, they are attracted to the site of HIV infection, where they too might then be infected. These infected CD4 T cells might then traffic to the draining lymph node and then to the circulation, spreading the infection and contributing to CD4 T-cell depletion by the various mechanisms discussed above (Box 21.32).

The following effects have also been attributed to the HIV-1 Vpr. (i) Endogenous Vpr expression in dendritic cells impairs dendritic cell maturation, and consequently, it impairs the ability of the dendritic cell to process and present antigens and to activate an antigen-specific CTL response. (ii) Vpr promotes the nuclear import of the HIV preintegration complex. In this regard, while the preintegration complexes of some simple retroviruses enter the nuclear region when the nuclear membrane breaks down at mitosis (Chapter 20), HIV and other lentiviruses are able to transport their preintegration complexes into the nuclei of resting or nondividing, terminally differentiated cells. Since Vpr was found to bind to nuclear pore proteins, it is thought to facilitate the interaction of the HIV preintegration complex with the nuclear transport machinery, thereby promoting its nuclear entry. (iii) Vpr may upregulate transcription of *nonintegrated* HIV DNA. This effect would appear to be rather odd, since a provirus is usually thought to be required for efficient viral replication and expression. However, recent reports suggest that transcription from nonintegrated HIV DNA may be a normal early step in HIV replication. (It may be relevant in this regard that high levels of nonintegrated HIV DNA are found in the brains of some patients, in association with AIDS dementia [see below].) (iv) Vpr also might act as an adaptor molecule that mediates the efficient recruitment to the HIV LTR promoter of transcriptional coactivators, including NF-κB and Sp1 (Fig. 21.7), thereby enhancing HIV transcription and replication. (v) Vpr might be involved in the activation of host cell signaling pathways, including that which activates NF-κB.

BOX 21.32

HIV-2 and SIV each have an accessory gene, *vpx*, which is not present in HIV-1. The *vpx* gene is closely related to *vpr* and is thought to have arisen as a duplication of *vpr*. So, HIV-2 and SIV contain both *vpr* and *vpx*. The two major unambiguous functions of the HIV-1 Vpr protein are G_2 cell cycle arrest and facilitating infection of macrophages. In HIV-2 and SIV, Vpr is responsible for the first of these functions, and Vpx is responsible for the second. Thus, while these two functions are expressed by the single protein Vpr in HIV-1, they are split between Vpr and Vpx in HIV-2 and SIV.

Recent experimental findings show that the HIV-2 and SIV Vpx proteins promote infection of macrophages by counteracting a macrophage antiviral restriction function (Sharova et al., 2008). For example, when macrophages, in which Vpx is essential for virus infection, are fused with monkey kidney cells, in which Vpx is dispensable, the heterokaryons

support infection by wild-type SIV but not by Vpx deletion mutants. The replication of HIV-1 is likewise strongly impaired in macrophages, and expression of Vpx in macrophages, in *trans*, can counteract the restriction of HIV-1, as well as the restriction of Vpx-deleted SIV.

Interestingly, the HIV-1 Vpr is incapable of relieving the macrophage block to HIV-1 replication to the same extent as the HIV-2 and SIV Vpx proteins can. In fact, studies of the SIV Vpx protein first pointed out the relative inefficiency of HIV-1 replication in macrophages, in comparison to that of HIV-2 and SIV. At any rate, Vpx has evolved in HIV-2 and SIV to effectively counteract the macrophage restriction function.

Reference
Sharova, N., Y. Wu, X. Zhu, R. Stranska, R. Kaushik, M. Sharkey, and M. Stevenson. 2008. Primate lentiviral Vpx commandeers DDB1 to counteract a macrophage restriction. *PLoS Pathog.* 4(5):e1000057.

Taken together, the above effects attributed to Vpr are rather remarkable, particularly when considering that the protein contains only 96 amino acids. Moreover, Vpr may well play a role in the pathogenesis of AIDS in vivo, as implied by the finding that rhesus monkeys experimentally infected with Vpr-deficient SIV mutants showed decreased virus replication and delayed disease progression. Still, the possible relevance of the above Vpr activities to HIV replication and pathogenesis in humans remains to be determined (Box 21.33).

Tat Revisited

Earlier, we noted how Tat and Rev act together to generate a burst of virus production in a relatively brief period, thereby compromising the ability of CTLs to recognize and destroy HIV-producing cells, before they might release progeny virus particles. That well-known effect of Tat and Rev acting together is not generally discussed as an immune evasion strategy. However, Tat, acting alone, is credited with several distinct ways of undermining cellular immune responses against HIV-infected cells. First, in the nucleus, Tat blocks transcription of MHC-I loci. Second, in the cytoplasm, Tat inhibits the proteasome complexes that generate the antigenic peptides presented by MHC-I molecules. Third, and remarkably, Tat also affects the subunit composition of proteasomes, thereby causing them to generate subdominant rather than dominant MHC-I-binding epitopes. (Recall that virus-specific helper T cells and CTLs typically recognize only a few dominant viral epitopes.) Fourth, as noted above, Tat upregulates

production of Fas ligand in infected CD4 T cells. Consequently, when an HIV-infected target cell is engaged by either an HIV-specific CD4 helper T cell or an HIV-specific CD8 CTL, the target cell can transmit an apoptotic signal into the helper T cell or the CTL, so that they, rather than

BOX 21.33

Effects that the HIV accessory proteins might express from the extracellular fluid are controversial. But for the sake of completeness, several putative extracellular effects of Vpr are noted here. (i) Extracellular Vpr induces apoptosis of T cells in culture. Moreover, recent experimental results suggest that Vpr induces apoptosis by directly affecting the permeability of mitochondrial membranes, with consequent release of apoptosis-inducing factors into the cytosol (Chapter 19). But despite these experimental findings, the extent to which Vpr might be responsible for T-cell depletion during the course of HIV infection is not known. (ii) Soluble Vpr is found in the cerebrospinal fluid as well as in the sera of HIV-infected patients and is thought to perhaps play a role in AIDS-associated dementia, possibly as a consequence of its proapoptotic activity. In support of this premise, both intracellular and extracellular recombinant Vpr has neurocytopathic effects on both rat and human neuronal cells. Moreover, these particular cytopathic effects are characteristic of apoptosis. However, most of the direct neurotoxic effects of HIV have been attributed to gp120 and Tat.

the infected target cell, might undergo Fas-triggered apoptosis. (As noted below, Tat also impedes the antiviral effects of IFN-α and IFN-β, the type I IFNs. Thus, Tat also acts as a countermeasure against innate immune defenses.)

Evasion and Subversion of Antibody-Mediated Immunity

Considering HIV's various means for evading and subverting cell-mediated immune defenses, it is no surprise that HIV also has several ways to evade antibody-mediated host defenses. Best known, perhaps, is the ability of HIV to rapidly and continuously generate antibody-resistant (immune escape) variants, an ability which results in part from the low fidelity of the viral reverse transcriptase. In addition, the ability of the virus to rapidly generate these variants also reflects the large population of productively infected cells within the host and the short generation time of the virus. Finally, it reflects the fact that the virus itself can tolerate many amino acid substitutions and still remain viable. In this regard, HIV is said to have **structural plasticity**, a property also of influenza virus and the rhinoviruses, but not of poliovirus or measles virus. (Recall that over 100 different rhinovirus serotypes circulate in the human population, while measles virus and poliovirus are each represented by one prevalent serotype. Why might some viruses display structural plasticity, while others do not?)

HIV also countermands antibody-mediated immune defenses by virtue of the fact that gp120, which contains the major neutralization epitopes of HIV, is one of the most extensively glycosylated proteins known. This has the effect of making gp120 an intrinsically poor antigen for recognition by antibodies. Worse still for the host, gp120 variants that have acquired new serine and asparagine residues, which provide new sites for O-linked and N-linked glycosylation, respectively, may be selected during infection, further enhancing the ability of the virus to evade antibody defenses. Making matters worse yet for the host, during end-stage disease the level of antibody-producing B cells may be diminished, perhaps because of a lack of stimulation by CD4 T cells. NK cell function, which depends on IL-2 produced by CD4 T cells, likewise is diminished during late-stage disease.

HIV also may be able to exploit antibody-mediated defenses to its own advantage by making use of the fact that *nonneutralizing*, anti-HIV antibodies may actually facilitate infection of macrophages and NK cells. That can happen because antibody-decorated (opsonized) virus is efficiently bound by the Fc receptors on those cells (Chapter 4) and then taken up by phagocytosis, independent of the gp120-CD4 interaction. In this way the virus can efficiently infect and destroy those Fc receptor-expressing cells. Antibodies that have that infection-boosting effect are referred to as "enhancing antibodies." Both neutralizing and enhancing antibodies are produced in response to infection. But as in the case of other factors enumerated above, it is not known to what extent enhancing antibodies actually contribute to HIV infection and pathogenesis in vivo. (Enhancing antibodies were found to be generated in individuals receiving a candidate HIV vaccine. That finding may constitute but one of a variety of obstacles impeding the development of an effective HIV vaccine [see below].)

HIV and Complement

When HIV invades a host, it immediately encounters the defenses of the innate immune system, including the complement system. In fact, all the animal retroviruses, including HIV and HTLV, trigger the classical complement activation pathway. Initially, they do so by a direct antibody-independent interaction with complement component C1. This is unusual because the classical pathway of complement activation is generally triggered when C1 binds to the Fc region of an antibody molecule that is bound to the antigen (Chapter 4). But after the patient seroconverts, anti-HIV antibodies may further enhance the activation of the classical complement pathway.

As might be expected, HIV has a means or two for evading the virotoxic effects of the complement cascade. CD55, the complement decay-accelerating factor (a protein expressed on cells exposed to plasma proteins, which normally prevents the activation of the complement cascade on uninfected cells [Chapter 4]), and other cell surface regulators of complement activation are incorporated into HIV envelopes during virus budding. Also, both gp21 and gp41 interact with complement factor H (CFH), a humoral negative regulator of complement activation that also normally acts to inhibit complement activation on uninfected cells. Note that CFH very likely plays a role in HIV infection in vivo, since HIV particles, which are generated in cells cultured in media containing CFH-depleted serum, are particularly sensitive to destruction by anti-HIV antibodies plus complement.

In addition to poking holes in viral envelopes, complement proteins also opsonize viruses to enhance their phagocytosis by macrophages, NK cells, neutrophils, and dendritic cells, all of which express complement receptors at their surfaces (i.e., complement as well as antibody can opsonize a microbe for enhanced phagocytosis [Chapter 4]). However, just as HIV exploits Fc receptors to gain access to Fc receptor-expressing cells, the virus likewise may exploit complement receptors to gain access to, and replicate in, complement receptor-expressing cells. Indeed, a recent study demonstrated that opsonization of HIV with human complement enhances infection of immature dendritic cells and subsequent transmission of the virus to

CD4 T cells. Complement-mediated infection of these dendritic cells is dependent on CR3, the complement receptor that binds complement component C3 (Chapter 4). This was shown by the ability of blocking antibodies against CR3 to inhibit HIV infection of the dendritic cells.

Another example of HIV possibly commandeering complement to bind to and enter cells is as follows. HIV, in infected seminal fluid, might be actively transported across epithelial cells of the genital mucosa, in that way gaining access to dendritic cells and macrophages in the submucosal tissues. But how might HIV bind to these mucosal epithelial cells? Complement receptors on the epithelial cells might play a role in the binding and uptake of the virus, provided that the virus had previously been opsonized by complement.

Since HIV-infected cells express Env at their surfaces, these infected cells, like the free virus, may bind anti-Env antibodies. Thus, like the virus, HIV-infected cells may be potential targets for the lytic activity of the classical complement pathway. But once again, HIV may have a countermeasure. Recall that Env, on free virus, recruits plasma CFH, which inhibits complement activation. Likewise, Env, expressed at the surfaces of HIV-infected cells, also recruits plasma CFH to impede complement.

HIV and the IFNs

HIV also uses several mechanisms to evade IFNs. Recall that infected cells respond to many viruses by producing IFN-α/β. These antiviral cytokines do not protect the cells in which they are produced. Instead, they diffuse to uninfected neighboring cells, in which they induce the upregulation of PKR (double-stranded RNA-dependent protein kinase [Chapter 4]). But PKR remains in a nonactive state in those neighboring cells until they too might be infected. However, when those cells are infected, PKR is activated by the double-stranded RNA molecules that many viruses produce during the course of their replication cycles. Activated PKR then inhibits initiation of protein synthesis in the cell by phosphorylating eukaryotic translational initiation factor, eIF2, and by that means also inhibits virus replication. However, HIV, as a retrovirus, does not generate double-stranded RNA. Why then might HIV need to be concerned with counteracting the antiviral effects of IFN? The answer is that some viruses, like HIV, generate single-stranded RNA molecules that have regions of secondary structure. These RNAs, like double-stranded RNAs, are also able to upregulate and activate PKR. In the case of HIV, the TAR sequence, located at the 5' end of all HIV RNAs, contains stem-loop structures that can activate PKR. How then does HIV countermand the ability of HIV RNA to activate PKR, which then phosphorylates eIF2? Our old acquaintance Tat binds directly to PKR. By that

means, Tat is thought to act as a substrate homologue or decoy, competing with eIF2 and thereby preventing the translational block resulting from phosphorylation of eIF2. In addition, TAR RNA itself has been reported to induce autophosphorylation of PKR, which leads to inactivation of that protein. Remarkably, HIV actually subverts PKR to its own purposes, as follows. When Tat interacts with PKR, it is phosphorylated by PKR, resulting in a tighter binding of Tat to TAR and enhanced levels of viral transcription (see above).

HIV (SIV) in Its Natural Host and Its Origins as a Human Virus

HIV-1 is thought to have crossed into humans from particular subspecies of chimpanzee several times during the 20th century. These transmissions may possibly have resulted from the practice of eating bush meat. The virus in those initial chimpanzee-to-human transmissions would subsequently have mutated to its current form, in which it manifests a tropism for humans (Box 21.34).

BOX 21.34

Before October 2008, the oldest known HIV-1 sequence was that obtained in 1998 by means of RT-PCR amplification of a plasma sample that had been taken in 1959 from a woman in Kinshasa, Democratic Republic of the Congo. The woman was only later found to be seropositive for HIV-1, and the sequence was then derived. Indeed, that HIV-1 sequence (ZR59) was the only known HIV-1 sequence that predated 1976. Then, in October 2008, the sequence of another pre-1976 HIV-1 sample, DRC60, was determined. RT-PCR technology was again used, but this time to amplify RNA sequences in a paraffin-embedded lymph node biopsy specimen obtained in 1960 from another woman in Kinshasa. (See Chapter 12 for a discussion of how this technology was applied to reconstitute the 1918 Spanish influenza pandemic virus.) The researchers then used the two pre-1976 HIV-1 sequences to estimate the date they branched from their common ancestor. Interestingly, they calculated that date to be between 1884 and 1924, with a most likely estimation at about 1908. Intriguingly, the considerable genetic distance between DRC60 and ZR59 implies that diversification of HIV-1 in west-central Africa occurred long before recognition of the AIDS pandemic in the early 1980s.

Reference
Worobey, M., et al. 2008. Direct evidence of extensive diversity of HIV-1 in Kinshasa by 1960. *Nature* 455:661–664.

What then is the origin of HIV-2? Several isolates of HIV-2 are nearly identical in nucleotide sequence to SIV isolated from African sooty mangabeys. Thus, HIV-2 is thought to derive from the sooty mangabey virus. Note that several other monkey species are also known to carry HIV-like viruses.

Ironically, at the time that HIV-1 passed into humans, its primate progenitor virus may have been on its way to extinction. That was so because both the population size and the habitat of its natural chimpanzee host were dwindling to levels below that which might be necessary to sustain the virus. So, by adapting to humans, HIV-1 now has a potential host population of 6 billion individuals worldwide. (See Chapter 12 for a discussion of the minimal number of humans necessary to sustain measles virus in the population.)

Despite the fact that human and chimpanzee genomes display about 98% nucleotide sequence identity, neither chimpanzees nor any other primate suffers from an AIDS-like disease when infected with its own lentivirus. Yet just as humans developed AIDS when infected with the chimpanzee-derived HIV-1 lentivirus, several species of Asian macaques develop AIDS when infected with the lentivirus of African sooty mangabeys, a virus which is nonpathogenic in its sooty mangabey host.

The fact that lentiviruses are not especially pathogenic in their natural hosts should not come as a surprise, since, as noted in earlier chapters, viruses tend *not* to be virulent in their natural hosts. That is so because the virus and its natural host have evolved together over the course of many years. Highly virulent variants of the virus might kill their hosts before they might transmit the infection. Or they might possibly destroy their local host population and, thereby, come to a dead end there. And, as noted, the host too is continually evolving stratagems to keep the virus at bay. (What genes might primates have evolved that enable them to avoid AIDS-like disease when infected by their own lentiviruses?) On the other hand, we have seen several examples of highly pathogenic outcomes when a virus crosses a species barrier to infect a new host.

Among HIV-infected humans, there are rare individuals, termed long-term nonprogressors (LTNPs), who have survived continuing chronic infection with HIV, while also maintaining normal levels of CD4 T cells and without the assistance of antiretroviral therapy. How might these human LTNPs have coped with their HIV infections? They appear to have done so by mounting very strong cellular immune responses against HIV, in that way maintaining very low plasma HIV levels and essentially normal levels of CD4 T cells (Box 21.35).

Considering the example of the human LTNPs, we then might ask if sooty mangabeys, when infected with SIV,

BOX 21.35

A recent study (Migueles et al., 2008) found that LTNPs have 20 times as many CD8 CTLs as progressors. Not surprisingly, then, LTNPs have correspondingly lower viral loads than progressors. What is more, individual CTLs of LTNPs are more efficient killers than their counterparts in progressors, as they produce perforin and granzyme (Chapter 4) at nearly twice the rate as progressor CTLs. Interestingly, CTLs from progressors can yet be activated to destroy HIV-infected cells at close to the rate of CTLs from LTNPs, by treating them with phorbol ester and calcium ionophore, which activate signal transduction pathways. Thus, while CTLs from progressors may be less efficient killers than CTLs from LTNPs, it is not because they are inherently defective for killing. Instead, they are less efficient killers because they are perhaps not being properly activated. Based on these findings, can you suggest any new approaches to anti-HIV therapy?

Reference
Migueles, S. A., C. M. Osborne, C. Royce, A. A. Compton, R. P. Joshi, K. A. Weeks, J. E. Rood, A. M. Berkley, J. B. Sacha, N. A. Cogliano-Shutta, M. Lloyd, G. Roby, R. Kwan, M. McLaughlin, S. Stallings, C. Rehm, M. A. O'Shea, J. Mican, B. Z. Packard, A. Komoriya, S. Palmer, A. P. Wiegand, F. Maldarelli, J. M. Coffin, J. W. Mellors, C. W. Hallahan, D. A. Follman, and M. Connors. 2008. Lytic granule loading of CD8(+) T cells is required for HIV-infected cell elimination associated with immune control. *Immunity* 29: 1009–1021.

likewise mount a strong cellular immune response to maintain nearly normal levels of CD4 T cells. (Recall that the sooty mangabey is the natural host for SIV.) In fact, the sooty mangabey mounts a relatively weak cellular immune response against SIV, one which does not control the high level of viremia. Yet sooty mangabeys still maintain nearly normal levels of CD4 T cells, appearing to do so by means of compensatory mechanisms for replenishing their CD4 T-cell population. This may be a wise adaptation on the part of the sooty mangabey, since in that way it does not have to endure the negative effects of chronic activation of cellular immune responses.

Here is another view of why sooty mangabeys do not undergo depletion of their CD4 T cells, despite supporting SIV replication. Essentially, it is that the rate of CD4 T-cell depletion and disease progression correlates with the level of T-cell activation. As noted above, sooty mangabeys, the natural host of SIV, mount a relatively weak cellular immune response against the virus. In contrast, nonnatural hosts of SIV, such as rhesus macaques, respond to infection with a heightened proliferation of CD4 T cells. Their CD4 T cells are eventually depleted, and they ultimately experience AIDS-like illness. Thus, ironically, immune activation appears to be a precondition for the development of AIDS-like disease. How might this interpretation impact on the prospects for an effective HIV vaccine?

In contrast to the rare human LTNPs, the much more common scenario in HIV-infected humans is for the anti-HIV cellular immune response to only partially control HIV levels and to eventually be overwhelmed by the virus (Fig. 21.11). Nevertheless, while CTLs are not the only factor that controls HIV infections in humans, cellular immune responses are the primary defense in humans against HIV disease progression. Over time, HIV depletes the population of CD4 T cells, which in turn leads to loss of CD8 CTL function. Bearing in mind the crucial role of CD8 CTLs in controlling HIV (Box 21.35), low levels of CD8 CTLs in plasma may be a better predictor of the likelihood of disease progression than high plasma viral load.

CLINICAL MANIFESTATIONS OF HIV/AIDS

Symptoms of the acute stage of HIV infection generally emerge within 2 to 4 weeks of the initial infection and may include fever, fatigue, rash, headache, swollen lymph glands, sore throat, muscle and joint aches, nausea, vomiting, diarrhea, and night sweats. Since these familiar symptoms are caused by several other infections and conditions, such as the more common influenza, many HIV-infected individuals may be inclined to dismiss the possibility that they indeed might be infected with HIV. Also, about one-half of those infected with HIV may not notice anything unusual during the early stages of the infection (Box 21.36).

In contrast to the flu-like symptoms seen during the acute stage of HIV infection, the clinical syndromes associated with later stages of HIV disease consist mainly of opportunistic infections, cancers, and direct effects of the virus on the central nervous system.

Opportunistic infections occur as a direct consequence of the declining immune status of the patient. The manifestation of one or more of these opportunistic infections, including *P. carinii* pneumonia, *Mycobacterium avium-M. intracellulare* complex infection, *Mycobacterium tuberculosis* infection, and human cytomegalovirus infection, among others, are strong indicators of progression of HIV disease to HIV/AIDS (Table 21.2). The same is true for Kaposi's sarcoma, a cancer associated with HHV-8 (Chapter 18).

Notice that most of the opportunistic infections indicative of HIV/AIDS involve pathogens that have an intracellular lifestyle. These include intracellular bacteria and protozoan parasites, as well as viruses. How do you account for that? Next, we look in somewhat more detail

BOX 21.36

Some patients presenting with acute HIV syndrome may not consider the possibility that they have been infected by HIV. Thus, it is important that medical professionals take that prospect into account, particularly when treating patients who are at higher risk of HIV exposure. In those instances, the physician should consider counseling the patient and testing for HIV before a diagnosis of flu or another common illness is made. But even when the physician has no reason to believe that the patient might be at high risk for HIV exposure, if the patient presents with acute HIV syndrome, the physician might still consider carrying out a risk assessment for HIV, rather than take the chance of missing an HIV diagnosis. Failing to make an HIV diagnosis at this point has consequences for the patient and other individuals as well, since this is a time during which the patient is very highly infectious. Then, the only means for diagnosis is an actual AIDS test. Bearing in mind that the HIV antibody blood test will be negative for someone who was infected very recently, tests that measure the viral load (i.e., the RT-PCR test or the branched DNA test [see below]) should be used.

at some of the opportunistic infections associated with HIV/AIDS.

P. carinii pneumonia is caused by the fungus *P. carinii*, an organism that is ubiquitous in the human population. In fact, in the United States, virtually everyone has been exposed to *P. carinii* by age 40. It persists in a dormant state in the lungs of individuals with healthy immune systems, being held in check there by the immune system. Although *P. carinii* is widespread in humans, *P. carinii* pneumonia was very rare before the emergence of the AIDS epidemic in 1981. In fact, fewer than 100 cases were reported per annum in the United States before that time. As might be expected, most of those cases were seen in individuals who were immunosuppressed by conditions such as leukemia and Hodgkin's disease or by chemotherapy. About 80% of AIDS patients develop *P. carinii* pneumonia. In fact, in 1990 it led to the diagnosis of AIDS in more than 65% of all cases (Box 21.37).

Histoplasmosis (caused by *Histoplasma capsulatum*) and **candidiasis** (caused by *Candida albicans*) are other opportunistic fungal infections commonly seen in AIDS patients. Once again, people with healthy immune systems generally have a high resistance to these fungi, which can be life threatening to those whose immunity is compromised. Also, note that fungal infections are difficult to cure in general.

Table 21.2 Some common opportunistic diseases associated with HIV infection and possible therapy[a]

Organism/virus	Clinical manifestation	Possible treatment(s)
Protozoa		
Cryptosporidium muris	Gastroenteritis (inflammation of stomach/intestinal membranes)	Investigational only
Isospora belli	Gastroenteritis	Trimethoprim-sulfamethoxazole (Bactrim)
Toxoplasma gondii	Encephalitis (brain abscess), retinitis, disseminated	Pyrimethamine and leucovorin, plus sulfadiazine, or clindamycin, Bactrim
Fungi		
Candida sp.	Stomatitis (thrush), proctitis, vaginitis, esophagitis	Nystatin, clotrimazole, ketoconazole
Coccidioides immitis	Meningitis, dissemination	Amphotericin B, fluconazole, ketoconazole
Cryptococcus neoformans	Meningitis (membrane inflammation of spinal cord and brain), pneumonia, encephalitis, dissemination (widespread)	Amphotericin B, fluconazole, itraconazole
Histoplasma capsulatum	Pneumonia, dissemination	Amphotericin B, fluconazole, itraconazole
Pneumocystis carinii	Pneumonia	Trimethoprim-sulfamethoxazole (Bactrim, Septra), pentamidine, dapsone, clindamycin, trimetrexate
Bacteria		
Mycobacterium avium complex (MAC)	Dissemination, pneumonia, diarrhea, weight loss, lymphadenopathy, severe gastrointestinal disease	Rifampin + ethambutol + clofazimine + ciprofloxacin +/− amikacin; rifabutin, clarithromycin, and azithromycin
Mycobacterium tuberculosis (tuberculosis)	Pneumonia (tuberculosis), meningitis, dissemination	Isoniazid (INH) + rifampin + ethambutol +/− pyrazinamide
Viruses		
Cytomegalovirus	Fever, hepatitis, encephalitis, retinitis, pneumonia, colitis, esophagitis	Ganciclovir, foscarnet
Epstein-Barr	Oral hairy leukoplakia, B-cell lymphoma	Acyclovir
Herpes simplex	Mucocutaneous (mouth, genital, rectal) blisters and/or ulcers, pneumonia, esophagitis, encephalitis	Acyclovir, famciclovir
Papovavirus J-C	Progressive multifocal leukoencephalopathy	None
Varicella-zoster	Dermatomal skin lesions (shingles), encephalitis	Acyclovir, foscarnet
Cancers		
Kaposi's sarcoma	Disseminated mucocutaneous lesions, often involving skin, lymph nodes, visceral organs (especially lungs and gastrointestinal tract)	Local injection, surgical excision or radiation to small, localized lesions; chemotherapy with vincristine and bleomycin
Primary lymphoma of the brain	Headache, palsies, seizures, hemiparesis, mental status, or personality changes	Radiation and/or chemotherapy
Systemic lymphomas	Fever, night sweats, weight loss, enlarged lymph nodes	Chemotherapy

[a]Patients with compromised immune systems are at increased risk for all known cancers and infections (including bacterial, viral, and protozoal). Most infectious diseases in HIV-infected patients are the result of proliferation of organisms already present in the patient's body. Most of these opportunistic infections are not contagious to others. The notable exception is tuberculosis. (Disclaimer: This table was developed to provide general information only. It is not meant to be diagnostic or to direct treatment.) Adapted from Mountain-Plains Regional Education and Training Center *HIV/AIDS Curriculum*, 4th ed., 1992, updated 1997; and from *Morb. Mortal. Wkly. Rep.* 2002.

Bacterial infections, like fungal infections, are more severe and difficult to treat in AIDS patients than in individuals with healthy immune systems. For example, consider **M. avium-M. intracellulare.** This bacterium is usually confined to the lungs, even in the elderly, provided that their immune systems are otherwise healthy. In contrast, the bacterium gives rise to a systemic infection in AIDS patients, causing severe gastrointestinal and other symptoms. Note that *M. avium-M. intracellulare* normally infects and resides in alveolar macrophages (see below).

Tuberculosis, caused by *Mycobacterium tuberculosis*, is regarded as the classic mycobacterial infection. Some readers may be surprised to learn that this bacterium, which is the leading infectious killer of humans, is actually

Most fungal pathogens cause disease mainly in individuals with impaired immune function. The fact that AIDS patients are more susceptible to most fungal infections (e.g., histoplasmosis, mucosal candidiasis, coccidioidomycosis, and cryptococcosis) than are individuals with normal immune function is evidence for the importance of T-cell function in the control of fungal infections. In contrast, clinical conditions associated with impaired humoral immunity are not generally associated with increased susceptibility to fungal disease.

CD4 T cells play a key role in immune defenses against fungi by producing cytokines such as IFN-γ and IL-2, which activate the phagocytic and fungicidal activities of macrophages (Chapter 4). Moreover, CD4 T cells may be able to function directly as antifungal effector cells, by making direct contact with fungal cells and releasing granules into them. In this regard, perforin-positive, cytotoxic CD4 T cells have been identified in HIV-infected patients, showing that some CD4 T cells can manifest killer activity. Admittedly, the existence of cytotoxic CD4 T cells challenges our general notion of CD4 T-cell function. And, in fact, the normal physiological role of cytotoxic CD4 T cells is still unclear. See Chapter 4 for CTL effector mechanisms and Box 4.12 for a discussion of cytotoxic CD4 T cells in viral infections.

Less than 2% of the CD4 T cells in healthy individuals appear to have killer activity, whereas up to 50% percent of the CD4 T cells in some HIV-infected individuals display recognizable cytotoxic potential. What might you infer from those facts?

widespread in the human population. In fact, about one-third of all individuals worldwide are infected! While 20 million of those infected individuals are active tuberculosis cases, *M. tuberculosis* is generally held in check by *cellular* immunity. Why cellular immunity? That is so because *M. tuberculosis* is a facultative intracellular bacterium. Like *M. avium-M. intracellulare*, it resides within a phagocytic vacuole of alveolar macrophages and is able to survive there by excluding the proton pump from the vacuolar membrane, in that way preventing the vacuole from becoming acidic and thus bactericidal. As expected, the frequency and severity of *M. tuberculosis* illness are greater in HIV/AIDS patients than in otherwise healthy individuals. In HIV-infected individuals, the bacterium is much more likely to spread from the lower respiratory tract to extrapulmonary sites throughout the chest cavity and rapidly lead to death. Extrapulmonary tuberculosis in an HIV-infected individual is also indicative that the HIV infection has progressed to AIDS.

Kaposi's sarcoma was present in about 20% of HIV-infected homosexual men in the 1980s and remains the most frequent cancer seen in AIDS patients. Infection with HHV-8, also known as Kaposi's sarcoma-associated herpesvirus, is necessary but not sufficient to cause this malignancy (Chapter 18). Importantly, AIDS-associated Kaposi's sarcoma is a much more aggressive form of this cancer than the form seen in non-AIDS patients, in which it is rarely lethal. In the so-called "classic," non-AIDS-related, Kaposi's sarcoma, lesions usually appear only on the lower legs, while AIDS-related Kaposi's sarcoma lesions usually appear on the upper body, including the head, neck, and back, and also on the soft palate and gum areas of the mouth (Fig. 18.16). In more advanced AIDS cases, lesions are also found in the stomach and intestines, the lymph nodes, and the lungs. Thus, the immune deficiency associated with HIV infection appears to be responsible for the prevalence as well as the aggressiveness of this cancer in infected patients.

More recently, the frequency of Kaposi's sarcoma cases due to HIV infection has decreased by about 85 to 90%, due to new, more effective AIDS therapies and a greater awareness of how HIV infection is acquired. As noted above in the introduction and in Chapter 18, Kaposi's sarcoma played a key role in the initial recognition of AIDS as a distinct clinical entity (Box 21.38).

B-cell lymphomas and **anogenital carcinomas** are other cancers that occur disproportionately in AIDS patients. Indeed, B-cell lymphomas are 60 to 100 times more prevalent in AIDS patients than in the general population,

Here is an oddity. Kaposi's sarcoma is relatively prevalent in HIV-infected homosexual men (20%) but relatively rare in HIV-infected women (3%) and heterosexual intravenous drug users (3%). How might one account for the preponderance of cases in homosexual men? One suggestion is that it may have resulted in some unknown way from its route of transmission in those individuals, and the small minority of women with Kaposi's sarcoma may have acquired it through sex with homosexual or bisexual men. What other possibilities can you suggest?

while anogenital carcinomas are 2 to 3 times more prevalent in AIDS patients.

Here is an interesting oddity. HIV-associated B-cell lymphomas in the peritoneum and other body cavities are almost always associated with HHV-8, while those in the brain are associated with Epstein-Barr virus. It is not known why these cancers, which are associated with two different herpesviruses, display this particular site specificity.

In contrast to the B-cell lymphomas, which are associated with the herpesviruses HHV-8 and Epstein-Barr virus, the anogenital cancers are associated with human papillomaviruses, which are typically spread via sexual contact (Chapter 16). Still, it is virtually certain that the HIV-associated immunodeficiency in AIDS patients plays no small part in the extraordinary frequency of these cancers in these individuals (Box 21.39).

BOX 21.39

The ability of the immune system to distinguish foreign invaders from self may be its key attribute. Indeed, that ability is crucial since it prevents host tissue from evoking an immune response against itself. However, if tumor cells were to express mutated or unusual products, then, despite being self, they might elicit a cell-mediated immune response. If that were so, then host immunity might be expected to play a role in reducing the incidence of neoplasia. The plausibility of these premises is supported by the observable fact that the incidence of particular cancers (i.e., leukemia, lymphoma, and certain cancers associated with viruses) is higher in humans whose immune systems have been compromised by chemotherapy or by HIV/AIDS.

But why are AIDS patients especially disposed to blood cell cancers and some cancers that are associated with viruses, but not to "spontaneously" occurring **carcinomas** (cancers of epithelial cells) and **sarcomas** (cancers of connective tissues and bone)? The answers may be as follows. CTLs are generally not effective against solid tumors in the tissues because T cells that might recognize epitopes in the tissues do not leave the blood circulation or the lymphatic system unless they have been activated, which generally occurs when they are presented with their cognate antigens in lymphatic organs. Moreover, they also must receive a costimulatory signal from the antigen-presenting cell, which that cell is activated to deliver by cytokines that are generated at a site of infection (Chapter 4). Thus, naïve T cells are not likely to be activated in lymphatic organs to act against solid tumors in the tissues. And even if a naïve T cell were to somehow recognize a tumor antigen in the tissue, it would not receive the costimulatory signal there that it needs to be activated. Thus, the normal trafficking pattern of T cells and the requirement for a costimulatory signal, which are important for the maintenance of self-tolerance (Chapter 4), also may prevent cell-mediated immune responses from being activated against solid tumors in the tissues. In contrast, blood cell cancers do traffic through lymphoid tissues, where they might meet naïve T cells. Furthermore, in contrast to

carcinomas and sarcomas, blood cell cancers (which are derived from hematopoietic stem cells) may express high levels of the costimulatory molecule B7 (Chapter 4). Consequently, these circulating blood cell tumors may be able to activate naïve T cells. Thus, individuals with normal immune function can mount cell-mediated immune responses against blood cell tumors, while those who are immunocompromised by HIV/AIDS cannot. (Nevertheless, the protection provided by CTLs against leukemias and lymphomas in people with healthy immune systems is not 100% effective, since those individuals do develop these cancers.)

Why are AIDS patients also especially disposed to virus-associated cancers, such as Kaposi's sarcoma, which is linked with HHV-8, and anogenital carcinomas, which are caused by human papillomaviruses? First, bear in mind that cell-mediated immune responses are highly specialized to resolve or at least control virus infections (Chapter 4). Second, virus infections that lead to neoplasia generally begin with an acute phase, during which the virus activates cell-mediated immune responses, while also managing to establish a persistent infection, throughout which it is largely hidden from immune surveillance. Over the years, particular viral gene products may promote the growth of latently infected cells, thereby providing enhanced opportunities for cellular mutations to accumulate. The combined effects of viral growth-promoting genes and cellular mutations may eventually result in neoplasia. An immunocompromised state might lead to more viral replication during the acute phase of infection, leading to more latently infected cells and, consequently, a higher rate of neoplasia. Moreover, immunosuppression can lead to the spread of the infection to new cells during episodes of virus reactivation. Importantly, virus-induced tumors will likely express or present viral antigens at their surfaces, which can be recognized by CTLs that were activated by the acute infection, or during episodes of reactivation. Such CTLs will be present in greater numbers in individuals with normal immune function than in those whose immune systems have been compromised by HIV/AIDS.

About two-thirds of all AIDS patients develop **AIDS-related dementia**, a progressive mental deterioration that resembles early stages of Alzheimer's disease. This syndrome is the first sign of clinical AIDS in about 10% of HIV-infected patients. Early symptoms of AIDS-related dementia include forgetfulness, recent memory loss, slowness of thought, social withdrawal, and loss of motor function and balance. Later symptoms include loss of speech, severe fatigue, incontinence, headache, seizures, coma, and finally death.

Interestingly, more HIV-infected individuals show histopathological signs of neural involvement, which might potentially lead to AIDS dementia complex, than actually show symptoms of dementia. In fact, neural pathology is found in 80% of all autopsied AIDS patients, and some clinicians believe that as many as 90% of AIDS patients in terminal stages of the disease actually have AIDS dementia complex.

Does HIV itself have a direct role in causing AIDS-related dementia? Several observations suggest that indeed it does. First, antiretroviral therapy ameliorates AIDS-related dementia in some patients. Second, 95% of all patients with AIDS-related dementia have HIV antibodies in their cerebrospinal fluid. Third, HIV in fact does enter the brain. Moreover, the type of HIV found in the brain is M-tropic, rather than T-tropic. This point is probably clinically significant because the cells targeted by the virus in the brain include microglia (brain macrophages), astrocytes, oligodendrocytes, and capillary endothelial cells. Still, the mechanism by which HIV causes AIDS-related dementia is not clear. Neurons are lost, but they do not appear to be infected by the virus. Perhaps they are affected by cytotoxic viral proteins (e.g., gp120, Nef, Rev, and Tat) or cellular products that might be released from infected microglial cells.

Regardless of the mechanism by which HIV causes dementia, its ability to attack the central nervous system and give rise to neurological disease is fundamentally and clinically important, since it demonstrates that HIV is able to directly impair a second major system of the body. Moreover, since some AIDS patients present with dementia, without the presence of some other opportunistic infection, the initial view of AIDS needed to be modified to include those cases in which there is HIV-associated neurological disease, but no other opportunistic infection.

Neural complications may also be caused by several of the other opportunistic infections associated with AIDS. These include **toxoplasmosis of the brain**, caused by the intracellular protozoan parasite *Toxoplasma gondii*, and **progressive multifocal leukoencephalopathy** (PML), caused by **JC virus**, a polyomavirus. In the United States, 10 to 40% of adults are chronically infected with *T. gondii*, but the vast majority of these individuals are asymptomatic. However, toxoplasmosis develops in over 30% of AIDS patients. If left untreated, this condition can lead to coma and death. The parasite is believed to lie dormant in individuals with healthy immune systems, but it can reactivate in immunocompromised patients.

PML and JC virus are described in Chapters 3 and 15. We reiterate here that JC virus is widespread in the human population but that PML is a very rare disease in patients with healthy immune systems. Yet PML develops in as many as 5% of all AIDS patients, and 1% of HIV-infected individuals present with PML as their first clinical sign of AIDS. PML is almost invariably fatal.

THERAPIES AND VACCINES

Antiretroviral Drugs

When I first began drafting this chapter in 2004, there were about 20 antiretroviral drugs that had been approved by the U.S. Food and Drug Administration (FDA) for treating HIV-infected patients. As of 2008, the U.S. Department of Health and Human Services website lists some 100 approved antiretroviral drugs (see a sampling, in Table 21.3). About three-fourths of these antiretroviral drugs impede the reverse transcription step of HIV replication. Most drugs in this group are **nucleoside analogs**, which work as follows. When these drugs are incorporated at the 3′ end of a growing DNA strand, they prevent further synthesis of that strand. They do so because they are members of the family of 2′,3′-dideoxynucleoside analogs. As such, they lack the free 3′ hydroxyl group that is necessary to form a phosphodiester bond with the would-be next nucleoside 5′-triphosphate (Fig. 21.12; compare with Fig. 18.18). Note that these drugs are not inhibitors of the reverse transcriptase per se but instead work as chain terminators. That is, although they block reverse transcription, they do not block the enzyme.

The nucleoside analog azidothymidine (**AZT**, also known as zidovudine or ZDZ) was the first anti-HIV drug to be licensed. It contains an azido (N3) group in place of the 3′ hydroxyl group of thymidine (Fig. 21.12). Interestingly, AZT was not developed as an antiretroviral drug per se but was instead discovered during a screen for potential anticancer agents. AZT acts against cancer cells by impairing DNA synthesis mediated by the cellular DNA-dependent DNA polymerase. Indeed, all of the antiretroviral, nucleoside analog, reverse transcription inhibitors impair cellular DNA synthesis, which accounts for their side effects when used as either antiretroviral or cancer drugs. Still, reverse transcription is more sensitive to these drugs than cellular DNA synthesis, explaining the partial specificity of the drugs for the virus. Nevertheless, the effect of these drugs on host DNA synthesis can lead to severe side effects. Moreover, the utility of some of these

Table 21.3. Approved HIV medications by class[a]

Multiclass combinations

    Atripla (efavirenz/tenofovir/FTC)

Nucleoside/nucleotide reverse transcriptase inhibitors (NRTIs)

    Combivir (AZT/3TC)

    Emtriva (emtricitabine, FTC)

    Epivir (3TC, lamivudine)

    Epzicom (abacavir/3TC, kivexa)

    Hivid (zalcitabine, ddC)

    Retrovir (zidovudine, AZT)

    Trizivir (AZT/3TC/abacavir)

    Truvada (tenofovir/FTC)

    Videx (didanosine, ddI)

    Viread (tenofovir)

    Zerit (stavudine, d4T)

    Ziagen (abacavir)

Nonnucleoside reverse transcriptase inhibitors (NNRTIs)

    Intelence (etravirine, TMC125)

    Rescriptor (delavirdine)

    Sustiva (efavirenz, stocrin)

    Viramune (nevirapine)

Protease inhibitors (PIs)

    Agenerase (amprenavir)

    Aptivus (tipranavir)

    Crixivan (indinavir)

    Invirase (saquinavir)

    Kaletra (lopinavir/ritonavir)

    Lexiva (fosamprenavir, telzir)

    Norvir (ritonavir)

    Prezista (darunavir, TMC114)

    Reyataz (atazanavir)

    Viracept (nelfinavir)

Entry inhibitors

    Fuzeon (enfuvirtide, T-20)

    Selzentry (maraviroc, celsentri)

Integrase inhibitors

    Isentress (raltegravir, MK-0518)

[a]Drugs are listed by trade name; generic names are given in parentheses.

drugs is further compromised by their short half-lives in patients. For example, the half-life of AZT is only about 1 h, since the drug is efficiently degraded by liver glucuronidation enzymes.

There indeed are other antiretroviral drugs that do target the reverse transcriptase. They are the so-called **non-nucleoside reverse transcriptase inhibitors**, a structurally dissimilar group of agents that are not incorporated into the proviral DNA; instead, they bind noncompetitively to the reverse transcriptase (Fig. 21.12). These drugs were developed with the hope that they might be an important complement to the nucleoside analogs. That was an important goal, since *none* of the drugs that target reverse transcription is effective alone as a long-term therapy. Reasons include (i) the inability of any particular drug to completely block reverse transcription and (ii) the ability of HIV to readily generate mutants that are resistant to any particular drug. Drug resistance is a problem for the individual in whom these mutants arise and for the entire community as well. That is so because drug-resistant mutants are transmitted to new individuals and thus threaten to undermine the entire antiviral effort (Box 21.40).

About 15% of HIV-infected individuals were not able to tolerate the side effects associated with the above drugs, and the remaining individuals were able to tolerate them only for limited periods. For many patients, it was only a matter of a few years before drug side effects and the emergence of drug-resistant mutants began to outweigh the benefits of therapy. These problems underscored the need to identify new approaches to blocking HIV replication.

One such approach was the development of **protease inhibitors**, which bind to the active site of the HIV protease and thus block the ability of the protease to process the Gag and Pol precursor polyproteins. Since the HIV protease is dissimilar to any host protease, it was hoped that specific inhibitors of the viral protease might be developed, which would have only minimal effects on cellular functions. This goal was not realized, since at least 15% of patients experience severe side effects when treated with protease inhibitors. Also, in about 50% of treated individuals, HIV eventually develops resistance to each of the protease inhibitors, just as it does to inhibitors of reverse transcription. What is more, viral mutants that are resistant to one of these inhibitors often display cross-resistance to others of the same class.

The ability of HIV to readily generate mutants that are resistant to each individual antiretroviral drug is the reason that **combination drug therapy** is now the standard anti-HIV treatment. Each drug combination commonly includes two reverse transcription inhibitors (usually a nucleoside analog and a nonnucleoside analog reverse transcriptase inhibitor) and one protease inhibitor. The use of three such drugs in combination is referred to as **highly active antiretroviral therapy** or **HAART**. Before moving on, note that inhibitors that target the integrase and HIV entry have also been developed; these are discussed later (Box 21.41).

HAART is indeed able to reduce the viral load to undetectable levels in about one-half of all treated patients,

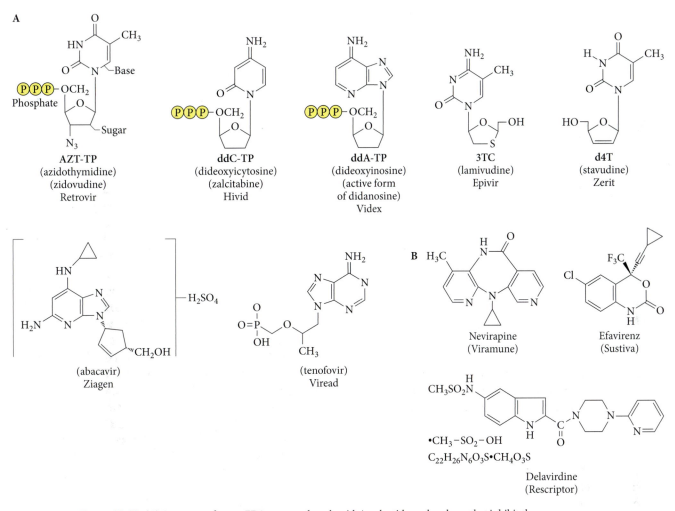

Figure 21.12 (A) Structure of seven FDA-approved nucleoside/nucleotide analog drugs that inhibit the reverse transcription step of HIV replication. (B) Structure of three FDA-approved nonnucleoside reverse transcriptase inhibitors. Modified from G. J. Stine, *AIDS Update 2003, an Annual Overview of Acquired Immune Deficiency Syndrome* (Prentice Hall, Upper Saddle River, NJ, 2003), with permission.

thereby giving some patients a new lease on life. Moreover, for many of these treated patients the immune system recovered sufficiently to allow them to cut back or terminate their prophylactic or suppressive therapies against a variety of opportunistic infections but, and this is crucially important, *not* against their HIV infections (see below).

The development of Atripla, a combination of the nonnucleoside reverse transcriptase inhibitor efavirenz, the nucleoside reverse transcription inhibitor emtricitabine, and the first approved *nucleotide* reverse transcription inhibitor, tenofovir, by Bristol-Myers Squibb and Gilead Sciences, may be a milestone in antiretroviral therapy, since it contains a complete HAART regimen in a single pill that is taken once daily. What is more, the combination of drugs in Atripla is one of the preferred regimens for individuals first beginning their HIV therapy. Importantly, a single dose of Atripla can replace the dozens of pills that some patients may have needed at various times each day, only a decade ago, making compliance very much easier. But Atripla is no more effective than the three drugs taken separately, the side effects are the same, and missing doses can lead to drug resistance. Atripla was approved by the FDA in July 2006.

Despite its successes, HAART is not a panacea. It does not lead to a cure, nor is it clear yet just how long its therapeutic effect might last (see Box 21.28 and below). Moreover, not all treated individuals respond to HAART, and for those individuals who do not respond, HIV/AIDS remains an unmanageable illness. In other individuals, the response to HAART is incomplete. Although the viral load in these individuals is reduced, their CD4 T-cell levels do not recover. Furthermore, as already noted, antiretroviral drugs have serious toxic side effects (these include liver damage, strokes, heart attacks, and kidney

BOX 21.40

HIV resistance to the nucleoside analogs is known to arise in two different ways. First, and perhaps most obviously, a mutation in the reverse transcriptase might reduce the incorporation of a nucleoside analog into the elongating DNA chain. Second, resistance to nucleoside analogs can also arise, in a less obvious way, as follows. The nucleoside analog is incorporated into the elongating DNA, but the drug-resistant mutant is then able to excise the chain-terminating residue. For those who have remained conversant with organic chemistry, the excision occurs by a process in which a molecule of ATP acts as a pyrophosphate donor in an S_N2-like hydrolysis of the 3′-terminal AZT-monophosphate residue, to generate a dinucleoside tetraphosphate and an unblocked elongating chain. Some nucleoside analog-resistant mutants have an enhanced affinity for ATP, which increases the efficiency with which they excise the blocking AZT residue.

failure) that limit their long-term usefulness. In addition, the emergence of drug-resistant mutants poses a major impediment to the long-range effectiveness even of multidrug therapies.

It may appear rather remarkable that HIV mutants that are resistant to all three drugs of a multidrug regimen may yet emerge. But consider the following. First, there is

BOX 21.41

Laypeople who may be infected with HIV or who know someone who is infected with HIV and who know that you are a student of virology may seek your advice regarding treatment. The very best advice you can give, and perhaps the only advice you should give, is that any individual suspecting the possibility of being infected should quickly get to a doctor who is well experienced at treating patients with HIV disease. Experience on the part of the doctor is crucial, since the pace of new drug development is so rapid, there are now so many treatment options, and HIV disease itself is so complicated that only an experienced doctor who follows the latest research and developments may be able to determine the best course for any patient. For the medical practitioner who may be looking at this book to update his or her knowledge of HIV disease and therapies, I emphasize that it is not the purpose of this book to offer advice on either the diagnosis or treatment of HIV, or any other infectious disease.

experimental evidence that about 1 in 10^4 HIV particles in an infected individual will be resistant to any particular antiretroviral drug, even before the initiation of therapy with that drug (Box 21.23). Second, it follows from the first premise that the probability of a virus containing mutations conferring resistance to any three antiretroviral drugs is 1 in $10^4 \times 10^4 \times 10^4$, or 1 in 10^{12}. (In making this argument, I am assuming, incorrectly perhaps, that the multiple mutations in any particular virus do not cumulatively compromise its viability.) This probability is very small indeed. However, HIV-infected individuals generate 10^9 to 10^{10} genetically different progeny virus genomes every day. And 30% of the virus population in newly infected individuals may already be resistant to any one of the standard antiretroviral drugs, before the beginning of therapy. Furthermore, a mutation conferring resistance to a particular inhibitor may confer resistance to one or more inhibitors of the same family of inhibitors. The bottom line is that HAART can fail because of the emergence of multidrug-resistant viral mutants.

The above argument underscores the importance of minimizing virus replication during the entire course of antiretroviral therapy. That is so because mutations can occur only during the replication process. Thus, minimizing virus replication minimizes the chances to generate multidrug-resistant viral mutants. That is a major reason why individuals on antiretroviral therapies *must strictly adhere* to their drug regimen.

Unfortunately, the need for people on HAART to take their drugs strictly on schedule is yet another factor that undermines multidrug therapy for many individuals. Newer regimens, which make use of newer, more potent inhibitors and multiclass combinations (e.g., Atripla), are easier than earlier ones, in which some patients had to take 70 or more pills each day, to control both the HIV infection and the opportunistic infections of HIV disease. While newer anti-HIV medications are easier to take, adhering to newer regimens still means making significant lifestyle adjustments. Medications may still need to be taken several times a day at specific times and may require other changes that may be hard for some to adhere to, such as in diet and meal times. As might be expected, the more difficult the drug regimen, the less likely it is that the patient will adhere to it.

Importantly, even minor noncompliance, such as skipping only a few pills, can have serious consequences both for the individual and for the community. For the individual, multidrug-resistant mutants might arise during the ensuing burst of virus replication. If so, then the particular therapy might be ineffective when reinstated. For the community, the multidrug-resistant mutants might be transmitted to new individuals, making the epidemic even more intractable than it already is. Sadly, research shows that noncompliance is often the norm. And, surprisingly perhaps, that is so irrespective of the socioeconomic status and education of the patient.

In the developed world, more than 1 in 10 patients are now on so-called **salvage therapy**, because they carry HIV that is resistant to standard multidrug HAART regimens. A salvage regimen may include three to four nucleoside reverse transcription inhibitors, one or two nonnucleoside reverse transcriptase inhibitors, three to four protease inhibitors, and hydroxyurea. Remarkably, HIV can generate resistance even against this regimen, and it can do so in only a matter of weeks.

If therapy should fail because of the emergence of multidrug-resistant mutants, it is possible to yet develop effective personalized therapeutic regimens by either a **phenotype analysis** or a **genotype analysis**. The former is based on an assay that measures the rate at which a drug inhibits the replication of the patient's viral sample. However, isolating the virus from the patient sample can be difficult, and the entire procedure may take several weeks to generate results. So, in the alternative procedure, genotype analysis, the viral nucleic acids are purified and converted into cDNA. Then, the portion of the cDNA containing the two most important viral genes, the reverse transcriptase and the protease, is cloned into a standard virus "backbone." The concentration of each drug that inhibits replication of that construct by 50% can then be determined in vitro.

Genotype analysis is potentially enhanced by using sequence data from the isolated virus to identify viral mutations, which are then analyzed by software that predicts the likelihood of resistance to known drugs. Still, phenotyping provides a more direct measure of drug resistance than genotyping, but it is more labor-intensive and expensive and can take as long as 5 weeks as well. In either case, these analyses do not sample all the drug-resistant viruses that may be present in the patient, and they are much better at predicting what will *not* work rather than what will work. And the potential for adverse side effects from any prescribed drug and the eventual emergence of drug-resistant virus still remains.

While multidrug antiretroviral therapies have been dramatically effective at reducing the viral load to virtually undetectable levels in some individuals, current antiretroviral drugs *cannot cure* an HIV infection. This is so largely because reservoirs of latent HIV persist in CD4 T cells, even after HAART may have reduced the serum viral load to undetectable levels. Moreover, the virus can persist in that latent state indefinitely, beyond the reach of current antiretroviral therapies. And the productive infection may fulminate if therapy is interrupted.

It may seem incredible that the HIV infection can persist, even after years of antiretroviral therapy that successfully reduced the viral load to undetectable levels. But consider the results of a 1999 study, which monitored the decay rate of the latent HIV in resting CD4 T cells of 34 adults who were taking effective combination therapy for their HIV infections. For each of these subjects, the drug treatment reduced their plasma HIV RNA to undetectable levels, yet it did not eradicate the virus that was present in a latent form in resting CD4 T cells. The mean half-life of the latent virus was determined to be 43.9 months. Thus, if the reservoir for the latent virus was comprised of only 10^5 cells (which is probably several orders of magnitude less than the actual number), eradication would still take as long as 60 years, even in patients whose antiretroviral therapies were effectively controlling their productive infections. So, for the moment at least, the goal of antiretroviral therapy is to *manage* the infection, *not to cure it*. There is no cure for HIV infection, nor is there a "morning after" pill that might be counted on to prevent infection after an inadvertent or other exposure (Boxes 21.42 and 21.43).

BOX 21.42

Postexposure prophylaxis (PEP) may indeed be administered to health care workers inadvertently exposed to HIV, for example, by a needle prick. It also may be available to the general public in some states (e.g., New York, Massachusetts, New Mexico, and Rhode Island) after high risk of exposure. The rationale for administering early therapy is discussed in the text. Also, see Box 21.22.

Most forms of PEP involve one or several anti-HIV drugs, such as Combivir (AZT and 3TC in one pill) or Combivir plus Crixivan or Kaletra. PEP must be taken within 72 h of high-risk exposure. Moreover, the drugs will need to be taken for a 4- to 6-week period, and they may cause moderate to severe side effects. In addition, PEP must be followed up by testing and counseling. What is more, PEP may be quite costly. And most importantly, PEP by no means guarantees protection. The bottom line is that PEP is *not* a "morning after" pill! You do not just take one pill the morning after and be done with it.

There is currently (2008) no federal funding for PEP, based on the premise that making it available to the public will encourage risky behavior (the same argument that was used against condom distribution). Regardless of whether that position might be motivated more by politics than science, do you think it has merit?

BOX 21.43

Bearing in mind that the reservoir of latent virus is a key factor standing in the way of a therapy that might cure an HIV infection, finding a safe way to flush the latent virus from its reservoir might be a crucial step towards achieving a cure. Such an approach might be used in conjunction with HAART, which might be needed to prevent the flushed virus from restocking the pool of latently infected cells.

While current HAART regimens cannot cure an HIV infection, it is important not to lose sight of the fact that the development of clinically successful HAART regimens has been a major achievement of HIV/AIDS research. Most importantly, HAART has delayed and in many cases prevented clinical progression of HIV disease, while also helping to prevent transmission of the virus. On the other hand, the toxic side effects and the eventual emergence of multidrug-resistant virus mutants have been shortcomings of HAART regimens based on reverse transcription and protease inhibitors. New protease and reverse transcription inhibitors have alleviated these problems somewhat. But the remaining shortcomings of those drugs have led to the development of still newer classes of drugs that target the HIV binding and fusion steps, as well as the integrase, with the hope that these drugs might effectively complement the antiretroviral drugs that target reverse transcription and the protease.

The HIV binding step offers several promising targets for drug intervention. For example, it is possible to rationally design agents to impair the interactions of gp120 with CD4 and the chemokine coreceptor molecules. Indeed, as noted above, a variety of clever approaches to blocking the gp120-CD4 interaction have already been developed and tested in clinical trials. Regrettably, whereas these approaches often work quite well at blocking infection in vitro, they have not been particularly effective in vivo. Obstacles to their effectiveness in vivo may include the high levels of the Env proteins on the surfaces of virus particles, as well as the massive accumulation of virus particles in the lymphoid organs, where most viruses are generated. For these reasons, perhaps, it is not possible to maintain high enough concentrations of these agents to competitively inhibit the gp120-CD4 interaction, particularly at the anatomic sites where they are most needed.

In contrast to agents designed to block the gp120-CD4 interaction, other agents were developed which selectively block the interaction of gp120 with the CCR5 coreceptor.

One such agent, maraviroc (brand-named Selzentry), was shown to effectively reduce the viral load in clinical trials, and it became the first coreceptor antagonist to be approved by the FDA (in 2007). Specifically, maraviroc interacts with, and thereby blocks, CCR5. It was approved for use, in combination with other antiretroviral drugs, in adult patients who have multidrug-resistant virus or whose viral loads otherwise remained detectable while on antiretroviral therapy. It is not yet known whether the drug might be effective in individuals with mixed-tropic or with CXCR4-tropic virus. And as in the case of other antiretroviral drugs, maraviroc can cause side effects such as fever, dizziness, headache, lowered blood pressure, nausea, and bladder irritation.

Other recent efforts have led to drugs that target the HIV fusion step. Recall that when gp120 engages the coreceptor, gp41 then undergoes a conformational rearrangement that exposes its fusion peptide, in that way enabling the fusion peptide to insert into the cell membrane. gp41 then undergoes additional rearrangements that result in membrane fusion (Fig. 21.5 and 21.6). In particular, the heptad repeats HR1 and HR2 undergo an intramolecular interaction, such that a stable six-helix bundle structure is formed within the Env trimers, thereby driving fusion to completion. Enfuvirtide (formerly known as pentafuside, DP178, or T-20) was designed to impair the HR1-HR2 interaction, and it was the first viral fusion inhibitor to be approved for clinical use. It is a 36-amino-acid synthetic peptide whose primary sequence was derived from HR2, and it is believed to act by competitively inhibiting the interactions between HR1 and HR2.

When used as a component of combination therapies, enfuvirtide is indeed an effective antiretroviral drug. In addition, it is effective against multidrug-resistant viruses, and, consequently, it has been used in individuals who were experiencing treatment failure after long-term combination therapy with reverse transcription and protease inhibitors. Moreover, enfuvirtide can help delay the emergence of resistance to other antiretroviral drugs. Nonetheless, enfuvirtide-resistant HIV mutants do emerge, both in vitro and in vivo (see the text above and Box 21.26). (Note that enfuvirtide is administered by injection, using a needle-free injection device. Also, 98% of patients taking enfuvirtide experience skin reactions at the site of injection, and the drug can cause more serious side effects, as well.)

The HIV integrase is a particularly attractive target for the development of antiretroviral drugs because it plays an essential role in the HIV life cycle, and there is no human counterpart to that enzyme. Moreover, structural data are available to facilitate the rational design of inhibitors with a specific affinity for the integrase. Raltegravir, also known as Isentress and MK-0518, is an integrase inhibitor that was approved by the FDA in 2007 for use with other anti-HIV agents. But, like other antiretroviral drugs, raltegravir still can cause side effects, including diarrhea, nausea, headache, and fever (Box 21.44).

For more on the subject of antiretroviral therapy, particularly with regard to its rationale and when it might be best to initiate therapy, see "The viral set point and HAART," above.

Anti-HIV Vaccines

The ideal means for controlling the HIV/AIDS epidemic would be to stop HIV transmission in the first place. If transmission could be prevented, then a time might come when there would be little need for therapies, with their adverse side effects, less than complete efficacy, and very great cost. But at present, no HIV/AIDS vaccines are approved for use, and while several vaccines are in clinical trials, there is currently little promise that an effective anti-AIDS vaccine might be developed. The reasons for this state of affairs are as follows.

First, we note that there is *no vaccine* that is known to be able to prevent *infection* by a virus. But how then do the vaccines that protect us against other viral pathogens, including smallpox virus, rabies virus, and poliovirus, work? Indeed, the development of these vaccines can be counted among the greatest accomplishments in medical history. But before we proceed to answer the question just posed, we might take a moment to consider that these aforementioned vaccines are still in use, although they were developed 50

or more years ago. Then, consider the paradox that despite the advances in knowledge and technology since these vaccines were developed, we still do not have a vaccine against HIV. Returning now to our question, none of these other vaccines is known to block *infection* per se. Instead, attenuated virus vaccines, such as the Sabin polio vaccine, cause antibodies to be in place, while also causing the host to be able to mount a memory response to the pathogen, which includes both memory B cells and memory T cells and which is more rapidly developing and robust than a primary response would be. In that way, the attenuated vaccine enables the host to contain and resolve the infection before the onset of disease, while not blocking the infection per se. Inactivated vaccines, such as the Salk polio vaccine, cause neutralizing antibodies to be in place against the virus, thereby helping to diminish initial dissemination of the virus, and thus enable the infected individual to resolve the infection before disease might develop. But if these vaccine effects suffice to protect us against other viral pathogens while not preventing infection per se, why do they not suffice to protect us against HIV/AIDS? Before we answer this question, first consider the following.

To be completely candid, it is not entirely clear how any vaccine actually confers protection. This fact became glaringly obvious as a variety of approaches to generating an effective AIDS vaccine all proved to be unsuccessful. Indeed, we were fortunate that *empirical* approaches to vaccine development worked well against other deadly viral pathogens, including the aforementioned viruses and others as well. However, once these other vaccines were at hand, there was little incentive to analyze how they actually protected people. Had it been otherwise, virologists might now be in a better position to know how to rationally design a vaccine that might protect against HIV.

Most vaccine trials have been devised based on evidence that cellular immunity likely plays a protective role, even in the absence of neutralizing antibodies. But questions still remain. For example, which particular classes of T cells might be most important? The CD8 CTL response is clearly vital, since the viral load appears to vary inversely with CTL levels in animals and humans (Box 21.35). On the other hand, CD4 T cells are needed to activate CD8 T cells. Moreover, humoral immunity too might be important. If so, then which class of antibodies might be most essential? Is it the immunoglobulin G (IgG) antibodies in the circulation or the IgA antibodies that guard the mucosal surfaces of the body? (Regarding this issue, the vast majority of HIV infections occur via mucosal routes, and it is likely that mucosal immunity might help to contain the virus during early stages of infection.) And once the factors that provide optimal protection have been

BOX 21.44

Antiretroviral drugs, as well as new "cutting edge" drugs that target other human ailments, may seem outrageously expensive. Cost is a particular concern in the case of antiretroviral drugs, since as many as 90% of all new HIV infections are occurring in developing nations, which can hardly afford these medications. And many afflicted individuals in the Western World are indigent. So, we might consider why these drugs are so expensive. In the United States, on average, 12 to 15 years are required to develop a new drug, carry out clinical trials, obtain FDA approval, and then bring the drug to market. The average cost per drug that successfully completes this process is about $500 million! (Merck & Co. claims to have spent $1 billion to bring the protease inhibitor Crixivan to market.) Moreover, after the initial investments, the odds that a new drug will ever make it to market are, by some accounts, about 1 in 10,000.

identified, there still is the matter of devising optimal strata-gems for evoking the expression of those particular factors.

Now, we can return to the key question posed above: why has generating an effective HIV vaccine been so intractable? Even if optimal conditions for generating an HIV vaccine were known, it very well may be that *no vaccine* will be able to prevent HIV/AIDS. First, actually preventing infection by HIV is particularly problematic. That is so because the vaccine would have to be able to block infection by a virus that displays extraordinary antigenic diversity. Moreover, the vaccine would have the unprecedented task of preventing transmission by HIV-infected *cells*, which may be the most important means of HIV transmission.

The above argument, that preventing HIV infection per se is particularly problematical, may seem trivial and perhaps irrelevant, since the fact remains that no antiviral vaccine is known to actually prevent infection. But consider that point in the following context. An HIV-infected individual may harbor the virus for 10 years on the average before AIDS develops. Yet during that entire period, the patient is mounting an adaptive immune response that is far more robust than might be induced by any vaccine but that nevertheless eventually is overwhelmed by the virus. Thus, it is highly unlikely that vaccine-based immunity would facilitate clearance of the early HIV infection, which likely will be necessary to prevent eventual progression to AIDS.

The state of affairs outlined in the preceding paragraph is largely accounted for by the fact that HIV establishes a latent infection in most of the CD4 T cells that it invades, and in that state it is essentially invisible to the immune system. In contrast, most of our available vaccines are targeted against viruses that give rise to acute infections. What is more, these viruses are generally represented by one prevalent serotype (e.g., measles, rubella, and small-pox viruses), and they can be neutralized before they reach the secondary sites of infection, where they might give rise to their characteristic clinical outcomes (e.g., poliovirus, hepatitis B virus, and rabies virus).

Here is yet another potential snag. Recall that immune activation may actually be a precondition for expansion of the infection and the eventual development of AIDS-like disease. Thus, some experts have even suggested that an immune system that has been primed against HIV by a vaccine actually might be more susceptible to HIV infection than a naïve immune system. Also, recall that HIV preferentially infects memory CD4 T cells. So, while one might have thought that the most effective vaccine would produce the most robust memory T-cell response, the expanded memory T-cell population might actually facilitate progression of HIV disease. On the other hand, evidence from the SIV/macaque model suggests that this might not be so. In this model, vaccination prior to infection did reduce the destruction of CD4 T cells, leading to better survival and long-term outcome. The particular vaccine regimen in that study induced antibodies that neutralized the primary isolate of SIV used for the challenge and, importantly, induced extensive CD4 and CD8 T-cell responses in all tissues examined. Still, the vaccine did not prevent either the infection or the eventual AIDS-like disease.

In September 2000, the first of a number of prospective AIDS vaccines entered Phase I clinical trials, but as of 2008, there still was no demonstrably effective AIDS vaccine for use in humans, and the U.S. Department of Health and Human Services AIDS*info* Web site lists 32 completed or terminated vaccine trials (Box 21.45).

The unsuccessful, so-called STEP trial of Merck HIV vaccine V520 is especially noteworthy for several reasons, one being that it induced the best cellular immune responses to HIV proteins seen to date; yet it failed to meet criteria for efficacy. The STEP trial included multiple clinical trials in North and South America, the Caribbean, Australia, and South Africa. The vaccine was based on using a replication-defective adenovirus type-5 vector (Ad5), which delivered the HIV *gag*, *pol*, and *nef* genes for expression in cells (Box 21.20). Unfortunately, about 40% of adults in the United States and Western Europe express significant Ad5 immunity, and as many as 70 to 80% of individuals in countries such as South Africa and Thailand, where HIV prevalence is high, also express high levels of immunity to Ad5. Still, as many as 75% of the volunteers in

---

BOX 21.45

In May 1997, during a commencement address at Morgan State University, President Bill Clinton announced the launching of a national agenda to develop an AIDS vaccine by 2007. Knowledgeable members of the scientific community were taken aback by the President's declaration. Robert Gallo declared that "it is a serious possibility that we may never develop an AIDS vaccine."

The reservations expressed by AIDS experts did not prevent Tommy Thompson, Secretary of Health and Human Services in the George W. Bush administration, from stating later, in June 2001, at a World Health Assembly annual meeting in Geneva, that an HIV vaccine would be in place within 3 to 5 years. Thompson also reiterated a call for a cure. AIDS experts in the audience reacted with a groan. Even today few, if any, AIDS researchers speak of a cure.

the clinical trials displayed a measurable response to the vaccine, although response fell to 50% or less among individuals who had prior immune responses to Ad5. But most importantly, the vaccine did not protect against infection with HIV, nor did it lower the viral load in vaccinated individuals who were subsequently infected. Even worse, the vaccine appeared to increase susceptibility to HIV infection among a subset of trial participants who had preexisting antibody responses to Ad5. (Why might that be so?) The trial was terminated in September 2007.

So, is the cause hopeless? Perhaps not. One reason for optimism is that some people seem to have mounted natural immune responses that have kept them disease free for 25 years or more (i.e., the LTNPs noted above [Box 21.35]). So, perhaps the answer to an effective AIDS vaccine will come from an understanding of how these individuals have remained disease free. Another reason for optimism is the successes achieved, albeit limited, in cases such as the SIV/macaque model noted above. That particular success may have resulted from attempts to specifically stimulate cell-mediated immune defenses, rather than neutralizing antibodies. In that regard, note that the vaccine in that study was a **DNA vaccine**, which carries segments of the SIV Gag gene and carries sequences that encode the immune-system-activating protein IL-2. The DNA vaccine is taken up by cells, in which it is then expressed. Importantly, as in the case of the Merck Ad5-based vaccine, the endogenously produced viral proteins induce the generation of memory B cells and T cells. And while the vaccine does not prevent infection per se, most monkeys in the study demonstrated a measure of protection against disease progression. So then, perhaps a DNA vaccine might hold promise.

To add another optimistic note, a vaccine does not need to protect all recipients to be valuable. Even protection that might spare "only" 20 to 40% of all individuals who otherwise would have been infected and then become transmitters would represent an important step forward in the fight against AIDS. And even if the vaccine did not prevent infection per se, it still might act to extend life and reduce transmission (Box 21.46).

In 2008, the U.S. Department of Health and Human Service website listed three current preventative HIV vaccine trials (for HIV-negative individuals) and four therapeutic AIDS vaccine trials (for HIV-positive individuals). (Readers should check that website for the current state of affairs.)

In September 2009, researchers testing a new HIV vaccine in Thailand generated some excitement when they announced that the vaccine cut the risk of infection by one-third. However, a closer examination of the data showed that the protection was marginal at best.

BOX 21.46

If a *partially* effective HIV vaccine were developed, might its widespread use impede diagnostic testing for actual infection? And if so, should that issue be a consideration before adapting the vaccine? Before dismissing this question out of hand, note that in the United States, the diagnostic value of the tuberculin skin test for existing tuberculosis infection is thought to outweigh the benefits of vaccination against tuberculosis. But that is not a general point of view. For example, in Great Britain, and some other European countries as well, children are routinely vaccinated against tuberculosis. However, before being vaccinated, they are tested for preexisting infection. But then, should there be universal HIV testing, prior to vaccination against HIV? What other solutions might there be? Or is the HIV/AIDS epidemic different in some fundamental way from the tuberculosis problem, so that the value of even a partially effective HIV vaccine clearly outweighs the benefits of testing? Or is the U.S. policy regarding the tuberculosis vaccine misguided?

PREVENTION

There is no cure for AIDS. Drug therapies help to control the infection, but they do not work for all individuals. Moreover, for many patients drug regimens have a limited period of effectiveness, they are difficult to adhere to, and they are very expensive (Box 21.44). What is more, there is no vaccine in sight that might protect uninfected individuals. Thus, it is vitally important to find and employ other effective means to control the spread of HIV/AIDS.

Apropos the above, we reiterate that HIV is known to be transmitted only via an exchange of body fluids during sex, or via HIV-contaminated blood and blood products, or from mother to offspring either in utero, during birth, or via the breast milk. Thus, the great majority of HIV infections can be prevented by avoiding dangerous behaviors such as unprotected sex and the use of shared needles and syringes to inject intravenous drugs. This line of argument leads to the expectation that prevention education ought to be an effective strategy to limit HIV spread. This is demonstrably the case, and prevention education is at least in part the reason why the rate of new infections in the United States has leveled off at about 40,000 per year. Yet many believe we could be doing much better, especially with our young people who are just becoming sexually active. In this regard, it is estimated that 34% of all American heterosexual adults with AIDS were infected with HIV as teenagers (Box 21.47).

A 2005 survey, conducted by the RAND Corporation and Oregon State University, reported that 53% of African-Americans ages 15 to 44 believe that a cure for AIDS exists and is being withheld from the poor. Moreover, 44% believe that people who take antiretroviral drugs are being used by the government as guinea pigs. Nearly 27% agreed with the statement that "AIDS was produced in a government laboratory." About 16% agreed that AIDS was created by the government to control the black population, while about 15% agreed that AIDS is a form of genocide against African-Americans.

Black men holding these beliefs were found to be less likely to use condoms as a precaution against contracting and spreading HIV, but interestingly this was not the case among African-American women. Bearing in mind that African-Americans account for only 12% of the U.S. population but nevertheless accounted for about one-half of all the new AIDS cases in 2002, the authors of this study suggest that distrust of the health care system may be a factor that contributes to the AIDS epidemic among African-Americans. How do you explain the mistrust that many African-Americans have of the biomedical community?

Since HIV/AIDS is transmitted mainly via the sexual route, both abstinence and delaying the onset of sexual activity can diminish the rate of new infections. This was shown most dramatically in Uganda, where abstinence programs had a striking effect in lowering the incidence of new infections.

Abstinence-only programs are advocated by some religious and conservative groups in the United States, but such programs have been derided by other constituencies. So, given that abstinence programs have been effective elsewhere, might abstinence-only programs suffice to prevent the spread of HIV/AIDS in the United States, particularly among the young adult population? This is an important question, since about one-third of all American heterosexual adults with AIDS were infected with HIV as teenagers, as noted above. But as of 2008, the issue of abstinence-only versus other public education programs remained highly politicized in the United States. This is so despite the fact that more than 60% of American teenagers may be sexually active by the time they are high school seniors. Yet fewer than 50% of these young people have received any formal instruction in school regarding AIDS and other STDs. Fewer still were informed in school about condoms, either to prevent STDs or as a means to

prevent pregnancies. The issue of preventing transmission of HIV is made even more compelling by the fact that many infected individuals may be unaware that they are infected and that they are able to transmit the virus (Box 21.48).

Many individuals take our society to task for its disinclination to prepare its teenagers to handle their intimate relationships. These critics generally advocate the view that public school sex education programs that teach explicit facts about STDs dramatically lower HIV transmission among teenagers. But is there evidence to support that seemingly reasonable contention? Unfortunately, studies show that all too many teenagers who are getting the facts about STDs in public school sex education programs are *not* taking that knowledge to heart. Moreover, these teenagers, as well as those who are not in sex education programs, are nevertheless being bombarded with safe-sex messages outside school, to which they also do not respond. Why is that so? For that matter, why doesn't everyone in the industrialized world know how to protect themselves from HIV and other STDs? And how does one account for recently infected homosexual men who did not use condoms and who were surprised to learn that they are in a high-risk group? Perhaps the messages are not frank enough. If so, here is yet another reason for frank sex education. Many teenagers mistakenly believe that oral sex and anal sex are risk-free behaviors. Indeed, many teenagers, as well as a former president of the United States, do not consider oral sex to be sex at all.

If simply getting out the message about safe sex does not appear to change behavior, what then might be done? Are messages promoting abstinence in any way contradictory to those promoting safe sex? If not, should young people receive two messages, one that promotes abstinence and one teaching about safe sex? Or might that be too complex and confusing in the end? And can it be that people are simply tired of hearing about AIDS and are tuning out? See Boxes 21.49 and 21.50.

Intravenous drug users account for one-quarter of all new HIV infections in the United States. They transmit HIV by sharing needles and syringes, and they can also pass their HIV infections on to their sexual partners and, in the case of females, to their children. Needle and syringe exchange programs have been successful at reducing the number of new infections among drug users. However, in the United States, needle exchange programs provoke considerable controversy (reminiscent of the controversy engendered by programs that would distribute condoms in schools). Some states have laws that criminalize the possession of needles and syringes, and the U.S. Government prohibits the use of federal funds to support needle exchange programs. And, reminiscent of the argument over condoms, U.S. Government agencies have regarded efforts to make drug use safer to be the equivalent of

BOX 21.48

The position of the George W. Bush administration was to support abstinence-only education, a position based at least in part on the view that discussing condoms or making them available encourages sexual activity. Yet there is no evidence that abstinence-only programs prevent teen sex, pregnancy, or STDs in the United States. Moreover, the abstinence-until-marriage message has little relevance for homosexual men, for whom marriage is illegal almost everywhere in the United States. But then, do public education programs that promote condom usage to prevent HIV transmission in fact increase the proportion of adolescents who are sexually active? On the contrary, several studies in America and Europe concluded that promoting condom usage can be effective at preventing transmission of HIV, while *not* increasing teen sex activity.

When Colin Powell was serving as Secretary of State in the George W. Bush administration, he expressed a point of view that angered some of the President's closest supporters.

In remarks broadcast worldwide by MTV in 2002, Powell stated, "It is important that the whole international community come together, speak candidly about it, forget taboos, and forget about conservative ideas about what you should tell young people about. It's the lives of young people that are put at risk by unsafe sex."

Powell was speaking from a Washington studio and was connected via satellites to listeners worldwide. A young Italian Catholic woman asked him specifically to state his position regarding condoms. Powell replied "I certainly respect the position of the Holy Father and the Catholic Church. In my own judgment, condoms are a way to prevent infection. Therefore, I not only support their use, I encourage their use among people who are sexually active and need to protect themselves."

If you agree with Colin Powell, how might you feel about programs that would distribute condoms in schools? How might you feel as the parent of a teenager?

---

advocating the use of drugs. Yet just as distributing condoms did not result in an increase in sexually active adolescents, several studies found no evidence that distributing syringes leads to an increase in drug use. Also, many other industrialized nations in fact have such programs in place (Box 21.51).

In truth, the effectiveness of needle and syringe exchange programs is limited by the fact that the next fix may, at times, be more important to a drug user than personal safety. Consequently, some people stress that drug rehabilitation efforts should be a key part of programs to control HIV transmission by drug users. In 2008, only 15% of intravenous drug users in the United States were receiving rehabilitation treatment at any given time. This is troubling because that number is too low to have a significant impact on HIV transmission and because many drug users who want to enter treatment programs are being closed out. Yet other individuals are hard-core users, who will not enter a program even if given the opportunity to do so. Another concern is the shift among drug users to injecting crack cocaine rather than heroine. While heroine users go through only a few needles each day, injection cocaine users may go through 40 to 50 needles daily.

TESTING FOR HIV

Development of the means to test for HIV is a major achievement of HIV/AIDS research. First, knowing who is infected is crucial to preventing transmission of HIV disease. Indeed, prenatal HIV testing provides the best prospect for preventing perinatal HIV transmission (Box 21.52). Second, HIV testing is vital to maintaining the safety of the blood supply. Third, early diagnosis enables infected individuals to initiate treatment options sooner. Fourth, it makes it possible to monitor the effectiveness of the particular drug combination that might be chosen for the patient. Fifth, testing is needed to monitor trends in the pandemic and to evaluate prevention efforts.

Several tests are now available that, when used in combination, can accurately determine if an individual is infected with HIV. These tests include the enzyme-linked immunosorbent assay (ELISA) for detecting HIV antibodies, Western blotting for detecting HIV antigens, and the use of RT-PCR and the branched DNA test for detecting HIV nucleic acids.

The first test for HIV came into use in 1985 and was based on the ELISA procedure. It may be carried out as follows. HIV antigens (derived from HIV grown in cell culture) are preadsorbed to microtiter wells (Fig. 21.13). A sample of the test subject's serum is then added in various dilutions to the microtiter wells. If HIV antibodies are present in the patient's serum, then they bind to the HIV antigens attached to the microtiter wells. The wells are then washed to remove all other components of the serum sample. Any antibodies in the patient's sample, which may have bound to the HIV antigens, are detected using a

BOX 21.49

A personal interjection and other musings: poliovirus has intrigued me for many years, almost certainly because I was a child and young teenager during the annual polio epidemics of the 1940s and 1950s. I remember well the panic that set in at the start of new outbreaks. A cousin (whose infection was fatal), a schoolmate, and a neighbor (who was left crippled) were victims. Also, I recall images of iron lungs (Fig. 6.17), appeals for contributions from the March of Dimes (in the days before the NIH), and the eventual triumphs of Jonas Salk and Albert Sabin (Chapter 6). Yet from the point of view of my students, poliovirus has no more relevance than the plague of the Middle Ages.

In the 1970s, my virology students were especially interested in the herpesviruses, probably because of its relevance to their own lives, and because there were no viruses more compelling to them at the time. But interest in the herpesviruses paled in comparison to the huge interest generated by HIV/AIDS during the 1980s and early 1990s. My guess is that AIDS arrested people's attention at the time because it was new and because, in the pre-HAART era, being diagnosed with AIDS meant a certain death. Also, there were its associations with human sexuality in all its forms. For some of those who witnessed the emergence of the HIV/AIDS epidemic, the disease might still generate a particular interest. But while the emergence of the AIDS epidemic is still vivid in my memory, nearly all of my current students were born well after that event and seem rather indifferent to it. The following anecdote, which may resonate with instructors and older students, may somewhat illustrate the generational difference. During the Fall 2008 semester, when it was announced that Luc Montagnier was the recipient of the Nobel Prize for the discovery of HIV (see the text), I began to recount for my Infectious Disease & Defense class the story of the controversy involving Montagnier and Robert Gallo. Noticing the blank looks on their faces at the mention of Montagnier and Gallo and then inquiring, to my surprise, none of the 60 students, in a class of mostly microbiology majors, had ever heard of either Montagnier or Gallo. Such a response would have been unimaginable in the 1980s.

Bearing in mind that college students in general were hugely interested in HIV/AIDS in the 1980s, how did their knowledge of the epidemic affect their own intimate relationships? A study reported in 1990 traced knowledge and attitudes about AIDS, between the years 1986 and 1988, among undergraduates at a large northeastern public university. Students' knowledge of AIDS increased substantially during this time, and they expressed favorable attitudes regarding "safer-sex" behaviors. But curiously, while they were able to appreciate how others might be vulnerable to the infection, they tended not to view themselves as vulnerable. In fact, during this period, students reported a decrease in the safety of their sexual practices, while their numbers of sexual partners and likelihood of being in intimate relationships had increased. What is more, during the same time, the rate of HIV infection among college students was 10 times higher than the rate in the general heterosexual population. I presume this was not because college students were less informed than the general public.

Influenza virus typically kills 35,000 or more individuals each year in the United States. Yet before recent concern of a possible bird flu outbreak, the public was rather indifferent to influenza virus. At present (late 2009), there is indeed widespread concern over a swine flu outbreak. On the other hand, the public appears to remain indifferent to the annual seasonal flu. Can public complacency towards influenza virus and HIV (at least in the minds of some people) be explained by the fact that for some, these pathogens have always been a presence in their lives? Can that explain the lack of regard for personal safety among many college students in the late 1980s, who were children when HIV first burst onto the scene? Perhaps we tend to become relatively unconcerned with the real threats that we have always lived with. Maybe we simply get tired of hearing about them and lose interest. Yet Americans did not become complacent over polio. I have long thought that was so because polio threatened their children. But then, don't today's parents appreciate the threat that AIDS poses to their teenagers? Despite my invective, the answer may very well be yes, and perhaps I am underestimating their concern. But if the answer to my question is no, at least in the case of some individuals, how might their indifference to AIDS be explained? Might the answer lie in the misperceptions that AIDS is a disease of homosexual men, or that it is an easily treatable disease, or that their children can be trusted not to engage in risky behaviors? Still, despite the seeming indifference of Americans towards AIDS, when pressed, most will say that HIV/AIDS is the number one health problem in the world today, followed closely by cancer.

BOX 21.50

Since syphilis, like HIV/AIDS, is an STD, efforts in the early part of the 20th century to control syphilis transmission largely paralleled those taken today to control HIV/AIDS. Vigorous education programs were established to reduce high-risk behaviors, and the general public was continually exposed to warnings via radio (the television and internet of the day), posters, and pamphlets. Moreover, many states made blood testing for syphilis a prerequisite for obtaining a marriage license. Yet none of these measures had an appreciable effect on the containment of syphilis, which only came with the advent of a cure: penicillin. What might be the implications of these facts for the HIV/AIDS epidemic?

BOX 21.52

As noted in the text, a key strategy for preventing perinatal HIV transmission is prenatal HIV testing of pregnant women, so that infected mothers might be identified and begin antiretroviral therapy. Clinical trial data show that perinatal HIV transmission rates are less than 2% in the cases of HIV-infected women who begin antiretroviral treatment during their pregnancies. This compares with a 12 to 13% transmission rate among HIV-infected women who do not begin preventive treatment until labor and delivery or after birth and 25% among HIV-infected women who receive no preventive treatment.

In the United States there are three different approaches to prenatal HIV testing. First, there is the "opt-in" approach, in which women are given pre-HIV test counseling and then must consent before being given an HIV-antibody test. Second, there is the "opt-out" approach, in which women are notified that an HIV test will be included in a standard battery of prenatal tests, but that they may refuse the HIV test. Third, there is mandatory HIV testing, in which newborns are tested for HIV, with or without the mother's consent, if the mother's HIV status is unknown at delivery. Which approach might be used varies from one state to another. How might you feel about these different approaches to prenatal HIV testing? As you may have expected, opt-out testing results in higher testing rates than the opt-in voluntary testing approach.

secondary antibody that was raised against human IgG and that is conjugated to horseradish peroxidase. A substrate for horseradish peroxidase (*o*-phenylenediamine or azinobenzthiazolium) is then added. Catalysis of the substrate by the enzyme leads to a change in color. The intensity of the color is proportional to the amount of anti-HIV antibody bound to the wells.

A troublesome feature of the ELISA is the uncertainty regarding the proper "cutoff" point between a positive and a negative test result, which can lead to both false-positive and false-negative results. False-positive results can also come about because of possible cross-reactions between patient antibodies against other viruses and HIV antigens. They can also happen as follows. MHC molecules,

expressed at the surface of a cell infected with HIV, may be incorporated into the HIV envelope at budding. Some noninfected people, in particular women who have had multiple pregnancies, may have serum antibodies against the MHC molecules present on the HIV stock virus used as the source of HIV antigen in the ELISA.

The ELISA can give a false-negative result during the window period. (Recall that the window period is the time between the initiation of the infection and when the patient seroconverts.) This is worrisome because an HIV-infected person can transmit the virus to others during the window period and, indeed, may be most infectious during this time. The average window period with antibody tests is 22 days. By 3 months after HIV infection, 99% of infected individuals have detectable antibodies. But again, there are rare individuals who do not produce antibodies for one or more years, while some even rarer infected subjects never generate an antibody response against the HIV infection.

BOX 21.51

In 1998, U.S. Health and Human Services Secretary Donna Shalala advised President Clinton that needle exchange programs were proven to be effective at preventing transmission of HIV, while not increasing drug use. Nevertheless, the President did not lift the then-10-year ban on federal support for needle exchange programs. Clinton did endorse lifting the ban earlier, during the 1992 presidential campaign, but never did so, and (as of 2008) the ban remains in effect. Given the evidence that needle exchange programs save lives and the belief that an important role of government is to save lives, many see the position of the federal government in this instance as one where political concerns trumped science and good public health policy.

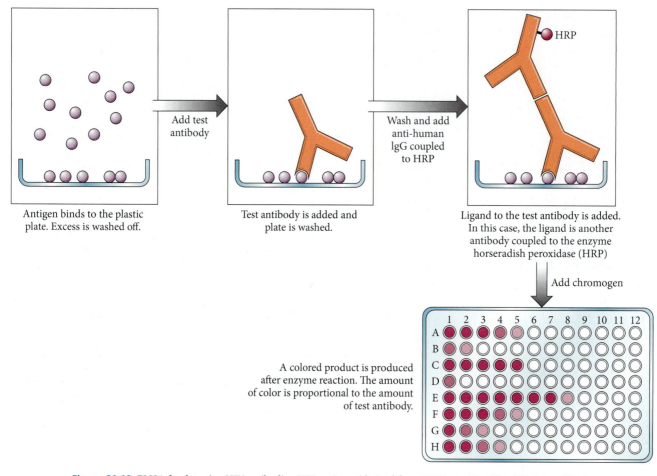

Antigen binds to the plastic plate. Excess is washed off.

Add test antibody

Test antibody is added and plate is washed.

Wash and add anti-human IgG coupled to HRP

HRP

Ligand to the test antibody is added. In this case, the ligand is another antibody coupled to the enzyme horseradish peroxidase (HRP)

Add chromogen

A colored product is produced after enzyme reaction. The amount of color is proportional to the amount of test antibody.

Figure 21.13 ELISA for detecting HIV antibodies. HIV antigens (derived from HIV grown in cell culture) are affixed via adsorption to microtiter wells (A). A sample of the test subject's serum is added in a series of dilutions to the microtiter wells (B). If HIV antibodies are present in the patient's serum, then they may bind to the HIV antigens attached to the microtiter wells. The wells are then washed to remove all other components of the serum sample. Any antibodies in the patient's sample which may have bound to the HIV antigens are detected using a secondary antibody that was raised against human IgG (C). These secondary antibodies are conjugated to horseradish peroxidase (HRP). A substrate for HRP (*o*-phenylenediamine or azinobenzthiazolium) is then added. Catalysis of the substrate by the enzyme leads to a change in color. The intensity of the color is proportional to the amount of anti-HIV antibody bound to the wells (D). Adapted from R. Nairn and M. Helbert, *Immunology for Medical Students* (Mosby International Ltd., Edinburgh, United Kingdom, 2002), with permission.

Because of the possibility of a false-positive ELISA result, a Western blot test has been used in the United States to confirm a positive ELISA result. In the Western blot test, HIV proteins generated in vitro are probed with the test subject's serum. The appearance of bands in the blot, which correspond to HIV proteins, indicates infection. Note that the Western blot test cuts the window period to approximately 16 days. Nevertheless, *all HIV tests are compromised by the window period*. In the case of the highly sensitive nucleic acid-based tests (described below), it is still 12 days.

When ELISA-based screening is followed by a confirmatory Western blot test (or immunofluorescence assay, which uses antibodies conjugated to a fluorescent dye to detect specific antigens in cultures or smears), then false-positive results occur at a frequency of 0.0004 to 0.0007%, and the false-negative rate is 0.003%, for the general population.

Nucleic acid testing is much more sensitive and accurate than either the ELISA or the Western blot test. However, the nucleic acid tests are considerably more expensive. But because of their greater sensitivity and shorter window period, nucleic acid tests are now used routinely to test donated blood in the United States. Because of their expense, they are initially carried out with samples that are pooled from 10 to 20 different donors. If the pool should give a positive result, then individual donor samples are tested separately. (As of 2001, diagnostic tests, combined with careful donor selection, have reduced the risk of transfusion-acquired HIV infection in the United States to approximately 1 per 2.5 million transfusions. But in 2000, the World Health Organization estimated that insufficient blood screening resulted in 1 million new HIV infections worldwide!)

Two approaches to nucleic acid testing are now in use. The first method makes use of RT-PCR, in which RNA extracted from the test subject's plasma is reverse transcribed into DNA and then amplified by multiple cycles of PCR, using a pair of primers that were generated to specifically initiate DNA synthesis on each DNA strand of a highly conserved 142-base-long sequence in the HIV *gag* gene. A positive result is indicated by the presence of an amplified fragment of the expected size in an electropherogram.

The second nucleic acid test is the **branched DNA test**, which is carried out as follows. Virus that may be present in the test subject's plasma is concentrated by centrifugation, and RNA is then extracted from the centrifugation pellet. Oligonucleotides (target probes) are added to the resuspended pellet. These target probes specifically hybridize both to HIV RNA that may be present in the patient sample and to other oligonucleotides (capture probes) that are adsorbed to the surfaces of microwells (Fig. 21.14). In that way, viral RNA in the test sample may be affixed to the microwells. Signal amplification of the HIV RNA is achieved by adding branched oligodeoxyribonucleotides (bDNA), which hybridize to the affixed HIV RNA. Next, alkaline phosphatase-conjugated probes are hybridized to the bDNA amplifier, and the chemiluminescent substrate dioxetane is added. The alkaline phosphatase causes a color reaction which allows quantification of the viral RNA in the original sample. Note that the branching of the bDNA allows for it to be very densely decorated with the enzyme, thereby enabling the assay to detect <100 copies of HIV RNA per ml of blood, all without the need to reverse transcribe and amplify the HIV genome, as in the RT-PCR procedure. Instead, the assay relies entirely on hybridization.

So-called "rapid tests" and "home testing kits" are also in use. The former, which produce results in approximately 20 min, screen for the presence of HIV antibodies in samples taken by a finger prick. They are about as accurate as traditional ELISAs, and for that reason, their results must be confirmed before a final diagnosis is made. All the necessary reagents for the rapid tests are present on a membrane strip, and the reaction begins to develop as soon as the blood sample is added. Rapid tests can be especially useful in the cases of pregnant women who are uncertain of their HIV status at the time of delivery. They are also useful in instances where health care

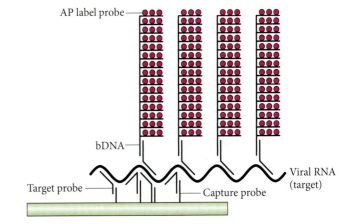

Figure 21.14 Summary of the branched DNA test. In the branched DNA test, virus that may be present in the test subject's plasma is concentrated by centrifugation, and RNA is then extracted from the centrifugation pellet. Oligonucleotides (target probes) are added to the resuspended pellet. These target probes specifically hybridize both to HIV RNA that may be present and to other oligonucleotides (capture probes) that are bound to the surfaces of microwells. In that way the viral RNA becomes affixed to the microwells. Signal amplification of the HIV RNA is achieved by adding bDNAs, which hybridize to the affixed HIV RNA. Next, alkaline phosphatase-conjugated probes are hybridized to the bDNA amplifier, and the chemiluminescent substrate dioxetane is added. The alkaline phosphatase causes a color reaction, which allows quantification of the viral RNA in the original sample. Adapted from M. L. Collins et al., *Nucleic Acids Res.* **25:**2979–2984, 1997, with permission.

workers might have been inadvertently exposed to a patient's blood. Moreover, they are valuable to clinics that serve individuals in a population with a high risk of infection, many of whom might never return for the results of ELISAs. (One study found that about 40% of those tested by ELISA never inquired about their test results.) Considering the speed and utility of rapid tests, they may be the standard by the time this book is in print.

Home testing kits have been available since 1997. Although several are available over the Internet, only the Home Access HIV-1 Test System, which is available in pharmacies, is currently approved by the FDA. The following information regarding this kit is from the CDC Web site.

It is not a true home test, but a home collection kit. The testing procedure involves pricking a finger with a special device, placing drops of blood on a specially treated card, and then mailing the card in to be tested at a licensed laboratory. Customers are given an identification number to use when phoning in for the results. Callers may speak to a counselor before taking the test, while waiting for the test result, and when the results are given. All individuals receiving a positive test result are provided referrals for a follow-up confirmatory test, as well as information and resources on treatment and support services.

The FDA approved the above home testing kit because it provides accurate results while also ensuring patient anonymity and appropriate counseling. Moreover, earlier studies showed that individuals in certain high-risk groups (e.g., intravenous drug users and sexually active homosexual men) would use such a test, while they might be reluctant to come to a clinic or a doctor's office. Yet the use of these test kits has generated some controversy, with some critics questioning whether news of a positive result should be delivered over the telephone. Such news can be both frightening and confusing. Others counter that substandard advice may be given even when standard testing is carried out in a doctor's office. And there are anecdotes of doctors leaving results of standard tests on telephone answering machines (Box 21.53).

HIV/AIDS IN SUB-SAHARAN AFRICA

The General Problem

The HIV/AIDS epidemic is especially severe in sub-Saharan Africa, a region which has just over 10% of the world's population (788 million people), yet which is

---

BOX 21.53

The use of home test kits by adolescents is an especially controversial issue. Some people object to the possibility that they might be available to young teens, in the absence of parental knowledge or consent. In 2003, 27 states had laws stipulating that teenagers must have signed parental consent to be tested for HIV. And there is concern over whether adolescents would be served effectively by telephone counselors. Also, there is the potential that young people might be coerced into using these tests by parents or school officials.

Important questions abound concerning HIV testing in general. For instance, should it ever be mandatory? Before answering, bear in mind that testing is already mandatory for blood donors. This would seem to be fair, considering the risk to the public of contaminated blood. Also, the act of donating blood is voluntary. But then, what of pregnant mothers? Recall that knowing the HIV status of the mother might enable intervention to prevent infection of the newborn (Box 21.52). And what of individuals convicted of prostitution, or sex offenders, particularly before they might be released. And are there circumstances where maintaining confidentiality of test results might not be justified? Is a victim of rape entitled to know the HIV status of the assailant?

---

home to about two-thirds of all people now living with HIV (Fig. 21.15). At the end of 2007, there were an estimated 20 million sub-Saharan African adults and 2 million children infected with HIV. Moreover, the infected children in the region comprise more than 85% of all children in the world now living with HIV. Furthermore, during 2007, about 1.5 million sub-Saharan Africans died as a result of AIDS, approximately 1.9 million others were infected with the virus, and 90% of all instances of mother-to-child transmission worldwide occurred in sub-Saharan Africa. Additionally, the epidemic has left behind an estimated 11.6 million orphaned African children. (See the *UNAIDS 2008 Report on the Global AIDS Epidemic*, available on the Internet.)

The extent of the HIV/AIDS epidemic varies from one sub-Saharan African nation to another (Fig. 21.15). Also, the trend of the epidemic varies as well, such that the percentage of infected individuals is on the rise in some countries while it is stabilizing and even declining in others. Kenya and Zimbabwe provide examples in which declines appear to be under way, probably due to effective prevention campaigns there. On the other hand, the epidemic surged in Cameroon, where the number of HIV-infected pregnant women doubled, to more than 11%, between 1998 and 2000. Such data demonstrate how rapidly the HIV epidemic can expand in the region.

Botswana is one of the most severely affected countries in sub-Saharan Africa. More than one-third of the Botswana population is infected with HIV, and life expectancy in that nation is 39 years, down from 71 years before the HIV/AIDS epidemic. By the year 2010, life expectancy in Botswana is predicted to be 29 years. The HIV/AIDS epidemic is also severe in Swaziland, Namibia, Malawi, and Zimbabwe. In Rwanda, by some estimates, as many as 80% of all women are HIV-positive. This astonishing figure is attributed in large part to widespread instances of rape stemming from the conflict there between the Tutsis and Hutus.

As noted above, Zimbabwe provides an example of a sub-Saharan African nation where the HIV/AIDS epidemic may be on the decline. The HIV prevalence there in pregnant women attending prenatal clinics fell from an astounding 26% in 2002 to a still astounding 18% in 2006, probably due in part to effective prevention campaigns (Fig. 21.15). But while the successes that have been seen in a few countries are encouraging, the overall AIDS death toll in sub-Saharan Africa can be expected to rise for years to come, unless there is much greater commitment of resources and efforts to lower the rates of transmission and to provide treatment for infected individuals.

Beyond the human tragedy of the deaths themselves, the sub-Saharan African HIV/AIDS epidemic has had and

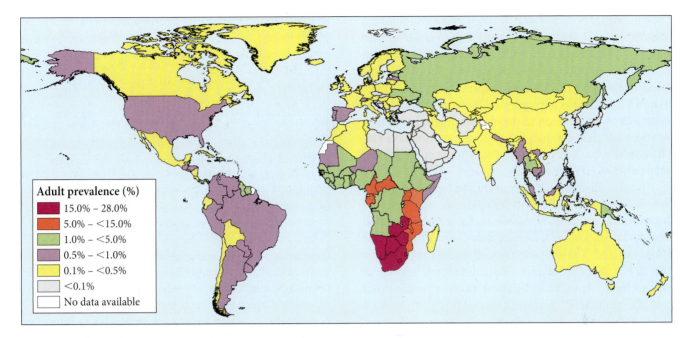

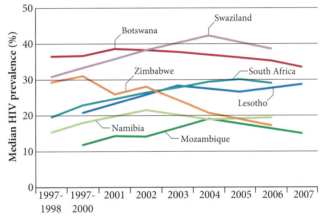

Figure 21.15 (Part 1) A global view of HIV infection, 2007: 33 million people (33 to 36 million) living with HIV, 2007. (Part 2) Median HIV prevalence in selected sub-Saharan African nations. Adapted from the *UNAIDS 2008 Report on the Global AIDS Epidemic.*

will continue to have severe social and economic consequences that will be felt in many areas, including the public health sector, education, industry, agriculture, infrastructure, human resources, and the economy in general.

Many individuals in the Western World have taken individual governments and the international community to task for not providing more resources to fight the sub-Saharan African HIV/AIDS epidemic. For many in the West, the most troubling aspect of the African epidemic has been the extent to which children have been its victims. This disquieting situation persists despite the fact that mother-to-child transmissions can largely be prevented with counseling, testing, and antiretroviral drugs (Box 21.52). Yet antiretroviral drugs are still not available

to about 80% of Africans living with HIV. In addition, treatment is not available for the opportunistic infections that are afflicting many with HIV disease.

The response of the international community to the African AIDS crises has improved somewhat over the last few years. As for the United States, in 2003 George W. Bush instituted the President's Emergency Plan for AIDS Relief (PEPFAR). When the program was announced in 2003, it was estimated that about 50,000 people were receiving treatment for HIV/AIDS in sub-Saharan Africa. The President's program now supports treatment for more than 2 million people; this is hardly enough, but it is 40 times the number receiving treatment only 5 years ago. PEPFAR was reauthorized to continue through fiscal years

2009 to 2013. About $39 billion is targeted towards AIDS, with $4 billion more going towards tuberculosis and $5 billion towards malaria. In addition, the U.S. Government contributes an additional $2 billion yearly to the Global Fund to Fight AIDS, Tuberculosis, and Malaria. In 2003, before PEPFAR was implemented, the United States was contributing $1.5 billion to the global effort against HIV/AIDS. While that sum may seem modest to some, the U.S. Government has been the largest contributor to the global effort against HIV/AIDS.

Next, looking more closely at the epidemiology of the sub-Saharan African HIV/AIDS epidemic, we note that heterosexual transmission accounts for 91% of reported cases in the region, with women now constituting 60% of those infected. These figures, as well as the much greater severity of the sub-Saharan African epidemic relative to that in North America and Europe, underscore the point that the global HIV pandemic is comprised of *multiple distinct epidemics*, each with its own scope and major means of transmission, as influenced by culture and social and political conditions. That being said, what are the cultural and political factors that have made sub-Saharan Africa so vulnerable to HIV/AIDS, and why is the primary mode of transmission in that region distinct from that in the Western World? Most important among the relevant factors may be (i) poverty and (ii) social instability (Boxes 21.54 and 21.55).

The degree of social instability in sub-Saharan Africa varies from one country to another. Rwanda provides an

BOX 21.54

The degree to which sex between men might contribute to the sub-Saharan HIV/AIDS epidemic is not well characterized, but recent studies suggest that it could be a significant factor in several parts of the region. Interestingly, and importantly as well, it appears that the majority of African men who have sex with other men also have sex with women. So, consider how this fact might affect the spread of the infection in the population, while bearing in mind that cultural and religious taboos cause condoms to be used only rarely by men during either anal or vaginal sex.

The use of injecting drugs is relatively new to sub-Saharan Africa. Yet it too may be playing a role in the HIV epidemic there. In fact, in Mauritius, the epidemic is (unique for the region) driven primarily by drug use. As might be expected, drug use is widespread among sex workers, who also do not usually use condoms.

BOX 21.55

Here are yet two more factors that are thought to make the sub-Saharan Africa HIV/AIDS epidemic exceptionally severe. First, bearing in mind that male circumcision dramatically lowers the rate of HIV transmission, the lack of male circumcision among sub-Saharan African men helps to explain the four- to fivefold difference in HIV rates between southern and western African men. Second, there has been much recent discussion about the effect on HIV transmission of having concurrent sexual partners. The comparison is made between being in a concurrent relationship with several partners, versus having "serial" monogamous relationships with an equal number of partners over a longer time frame. This distinction is particularly important if the male, for example, was in concurrent relationships with two or more females during the acute stage of the infection, when his viral load and infectivity are greatest. That recently infected male, on average, will transmit the infection to more partners than an infected male who is in an equal number of "serial" monogamous relationships over a longer period. Indeed, any recently infected individual in the network is a threat to each of the others.

As for African men and women, while they do not necessarily have more sexual partners over a lifetime than men and women elsewhere, epidemiological studies show that they often have more than one (typically two or perhaps three) concurrent relationships that can overlap for months or years. Most African women in a network of concurrent partnerships are not prostitutes. Rather, such relationships are fostered by gender inequality and poverty.

extreme example of social instability, where the key underlying factor has been continuing internecine warfare. More generally, social instability and poverty have been fostered by relatively recent rural-to-urban migrations. And in an urban setting, social instability and poverty lead to a vast increase in prostitution, which results in a greater sharing of sexual partners and a greater prevalence of venereal disease, each a predisposing factor for HIV transmission (see above). As expected, female commercial sex workers are disproportionately affected by HIV, with rates of infection in that group being greater than 80%. Consequently, many men contract the infection from prostitutes. Then, when returning home, most of these men, because of cultural and religious taboos, refuse to use condoms. (Apropos this point, a 2003 study estimated that the total condom supply in sub-Saharan Africa was sufficient to supply each man with 3 condoms per year! Included in this estimation were the 300 million condoms purchased in 2002 for Kenya, at a cost of $7.5 million.) Moreover, because of the unequal status of women, it is difficult or impossible for them to avoid infection, even if they should want to take necessary precautions.

Poverty and social instability also prevent therapeutic intervention on a scale that might appreciably affect the course of the HIV/AIDS epidemic. Considering just poverty, the cost per patient of only the protease inhibitor has ranged between $12,000 and $20,000 per year. In recognition of this state of affairs, special arrangements have been made to make antiretroviral drugs available in parts of sub-Saharan Africa at vastly reduced prices. Yet even with these allowances, costs often remain prohibitive. Thus, as noted above, only a small percentage of individuals in need of HAART actually receive it. Moreover, many of the antiretroviral drugs must be taken with food and clean water, neither of which can be taken for granted in many areas. Furthermore, physicians and nurses are in short supply to diagnose the infection, prescribe the drugs, and monitor patient response, side effects, and compliance with the drug regimen. Bearing in mind the limited effectiveness of public education programs to change high-risk behaviors in the United States, consider how much more difficult that task might be in the developing world, where most individuals have only a vague understanding of the HIV/AIDS epidemic and where other infectious diseases (e.g., malaria), poverty, famine, war, and ethnic hatred pose threats that may seem much more immediate than HIV/AIDS.

The Case of South Africa

South Africa is experiencing one of the world's most devastating AIDS epidemics. In 2007, more than one in five of South African adults were living with HIV, and almost 1,000 AIDS deaths occurred in that country every day. Moreover, the prevalence rate among pregnant South African women rose from 12.2% in 1996 to 30.2% in 2005, and 10% of all new HIV infections in South Africa are of babies. One-fifth of South African nurses and teachers are HIV positive. But then, 70% of its soldiers may be infected as well. What is more, a shocking 71% of all deaths among individuals between 15 and 49 years of age are caused by AIDS. The life expectancy for the population as a whole is now 36 years of age, and a 15-year old South African has a 50:50 chance of dying of AIDS before his or her 30th birthday. With an estimated 5.5 million people living with HIV/AIDS (UNAIDS, 2006), South Africa is the country with the most HIV infections in the world.

South Africa escaped the initial HIV/AIDS epidemic in the 1980s. Why then did the HIV/AIDS epidemic get so out of hand in that nation after it emerged there? One reason is that during the time that HIV prevalence was on the rise in South Africa, the government was preoccupied with major political changes. In particular, South Africa was at the time undergoing its transition from apartheid. But any rational efforts to deal with the South African HIV/AIDS epidemic were severely compromised in 2001, when then South African President Thabo Mbeki embraced the view that HIV is not the cause of AIDS. Mbeki's position can be traced to the convening that year, by his minister of health, of a 36-member international panel of supposed AIDS experts, which included 14 known HIV/AIDS "dissidents," who do not accept that HIV is the cause of AIDS. Mbeki went along with the views of the dissidents. Afterwards, he and his administration were repeatedly accused of failing to respond appropriately to the AIDS epidemic raging in their country (Box 21.56).

South African AIDS patients who used the public health system could get treatment for the opportunistic infections associated with their late-stage HIV disease. However, the Mbeki government issued rulings to prevent those patients from receiving antiretroviral drugs, which might have prevented the breakdown of their immune systems in the first place. Governmental restrictions against the use of antiretroviral drugs stood until August 2003, when the South African Cabinet overrode the President, declared as Cabinet policy that HIV is the cause of AIDS, and promised to formulate a national treatment plan that would include antiretroviral drugs. Yet despite the efforts of the cabinet, the health minister continued to promote nutritional approaches to dealing with AIDS, while also stressing the potential toxicities of antiretroviral drugs. Thus, implementation of the cabinet's enlightened policy has been slow to take effect. By the end of 2007, only about 28% of South Africans in need of antiretroviral drugs had received any. And since many HIV-positive pregnant women are still not getting drugs that might prevent transmission to their babies, HIV remains tragically common among South African children. UNAIDS reported that there were about 280,000 children under 15 years of age living with HIV in South Africa in 2007. Moreover, 1.4 million South African children became orphans in 2007 as a result of the AIDS epidemic, a rise from the 780,000 new orphans reported in 2003. As you well might imagine, the prospects for an orphaned child in South Africa are not good.

President Mbeki resigned in September 2008, after losing the support of his party; this change was welcomed by AIDS activists worldwide. However, given the scale of the South African HIV/AIDS crisis, any change for the better

BOX 21.56

While some AIDS dissidents have legitimate scientific credentials, none, except Peter Duesberg, has had relevant HIV/AIDS expertise. Before Duesberg began expressing his "alternative" views on AIDS, he was a highly regarded retrovirologist, whose numerous contributions included the finding that Rous sarcoma virus contains an oncogene (Chapter 20). For this contribution and others, Duesberg was elected to the prestigious U.S. National Academy of Sciences. However, once Duesberg broke ranks with other scientists over AIDS, he became a scientific outcast, no longer receiving invitations to scientific conferences and no longer able to obtain research grants.

Duesberg was one of the AIDS "experts" with whom Mbeki consulted before taking his own dissident position. Duesberg espoused the belief that AIDS results from drug use, parasitic infections, and malnutrition rather than from a retrovirus and that HIV is just another opportunistic infection. He stated that if he discovered that he was HIV antibody positive, then he would take it as an encouraging sign that his immune system was working. Kary Mullis, who was awarded the Nobel Prize for having invented the PCR, is another AIDS dissident with legitimate scientific credentials.

But, unlike Duesberg, Mullis has no expertise that might be relevant to HIV/AIDS.

So, taking into account Duesberg's very real expertise as a retrovirologist, can he possibly be right about HIV and AIDS, and is his alternative view helpful in the fight against the disease? It would make for a fascinating story if the answers were yes, or just even maybe, but the evidence that HIV causes AIDS is, without exaggeration, overwhelming. Consider just the following. Data from matched groups of homosexual men and hemophiliacs show that only those with HIV develop AIDS. These data have been available for years, and Duesberg should have been well aware of it. Also, consider that individuals who lack the HIV CCR5 coreceptor are highly resistant to HIV. Finally, and most directly, in every known instance where an AIDS patient was examined for HIV infection, there was evidence for presence of the virus. So, as to whether Duesberg's dissident view might have been of any value in the battle against HIV/AIDS, the answer is that to the contrary, it is not known how many people might have been infected or might not have benefited from effective antiretroviral therapy because they heeded Duesberg's arguments. AIDS denialists continue to tout their message, to the detriment of millions of HIV-infected individuals who listen to it.

in that nation's situation is likely to be gradual. Also, considering the prevalence of HIV in South Africa, it may surprise some readers that there is widespread prejudice in the country against those living with AIDS. Thus, Nelson Mandela's action was notable when his son died of AIDS in 2005. Mandela purposely revealed the cause of his son's death to the public, to countermand the stigma associated with being infected with HIV (Box 21.57).

Before moving on, note that there are yet other regions of the world, in addition to sub-Saharan Africa, where the HIV/AIDS epidemic is wreaking havoc (e.g., in areas of India and China). For a thorough account of the worldwide HIV/AIDS problem, I strongly recommend that interested readers have a look at the *UNAIDS 2008 Report on the Global AIDS Epidemic* (or the latest version of this document), available on the Internet.

THE HUMAN T-CELL LEUKEMIA (LYMPHOTROPIC) VIRUSES

During the mid-to-late 1970s, a cluster of CD4 T-cell leukemias was seen in particular prefectures in southwestern Japan. This was the initial recognition of a new

lymphoproliferative disorder, which was named **adult T-cell lymphocytic leukemia (ATLL)**. The clustering of ATLL cases led Japanese researchers to suspect that the new disease might be caused by a virus, and in 1980, M. Yoshida and coworkers isolated a new retrovirus, **HTLV-1**, from neoplastic cells of patients with this T-cell malignancy. The same investigators also demonstrated that an HTLV-1 provirus is integrated into the chromosomal DNA of all ATLL cells, but not normal cells, of ATLL patients.

Concurrent with the discovery of HTLV-1 by the Japanese investigators, Robert Gallo and colleagues at the National Cancer Institute isolated HTLV-1 from a West Indian patient with cutaneous T-cell lymphoma. The discovery of HTLV-1 by the Gallo group was noted above, in the context of the discovery of HIV. And while HIV-1, for reason of its devastating impact, stimulates more interest than HTLV-1, the discovery of the latter virus is especially intriguing in its own right. That is so because HTLV-1 (i) was the first known human retrovirus, (ii) was the first retrovirus directly associated with a human malignancy, and (iii) displays a rather unique lifestyle.

A few years after the discovery of HTLV-1, **HTLV-2** was isolated from a patient suffering from both hairy-cell

BOX 21.57

Why Mbeki held dissident views regarding the cause of the South African AIDS crisis is not known. Some have speculated that Mbeki was reacting to the huge expense of antiretroviral drugs and to the fact that they are sold mainly by powerful Western pharmaceutical companies. Alternatively, Mbeki may have been skeptical of the efficacy of those drugs, an attitude perhaps formed in response to the era of colonial domination and control of South Africa. Indeed, the health policies of the colonial and apartheid governments in South Africa were at times malicious and manipulative. Mbeki also described AIDS as a "disease of poverty," arguing that political attention should be directed towards poverty generally, rather than towards AIDS specifically. Does that argument have merit?

Treating HIV disease has indeed been expensive for South Africa. That is so despite the fact that American and European drug companies have offered antiretroviral drugs to South Africa at greatly reduced prices. Moreover, while South Africa may be among the wealthier countries in Africa, it is still desperately trying to emerge from the poverty of its former oppression. So if you are the President, do you choose to support AIDS therapy, which might exceed the limits of your financial resources for the foreseeable future, or instead, do you invest in water systems, housing, schools, and hospitals? And what of other crucial medical problems that abound in Africa, such as malaria and tuberculosis? You cannot have it all. And even if Mbeki and his government had fully supported antiretroviral therapy, impediments to its effectiveness would still have remained. This is not to defend Mbeki, but rather to briefly illustrate the depth and complexity of the problem.

B-cell leukemia and CD8 large granular lymphocytic leukemia. Yet another human retrovirus, **HTLV-5**, was isolated from a patient with cutaneous T-cell lymphoma/leukemia (mycosis fungoides).

So what is known about the medical significance of these human retroviruses? While HTLV-2 and HTLV-5 have been associated with hairy-cell leukemia and malignant cutaneous leukemia, respectively, HTLV-1 is the only HTLV known with certainty to be a human pathogen. It is clearly the etiologic agent of ATLL, as shown by the facts that (i) ATLL is seen only in individuals infected with HTLV-1 and (ii) all ATLL cells contain an HTLV-1 provirus. HTLV-1 also causes the nonneoplastic neurologic disorder **HTLV-1-associated myelopathy**, also known as

tropical spastic paraparesis (TSP); this disease has important biological similarities to multiple sclerosis.

Regarding their taxonomy, the HTLVs are only distantly related to HIV. HTLV-1 and HTLV-2, together with the bovine leukemia virus, which causes B-cell neoplastic diseases, belong to a genus of the *Retroviridae* referred to as the bovine leukemia virus or genus *Deltaretrovirus*. HTLV-1 and HTLV-2 share cross-reacting internal core proteins, and their proviral genomes are about 60% homologous. In contrast, the HTLV-5 proviral genome hybridizes only weakly with that of HTLV-1 and not at all with that of HTLV-2. (Some sources refer to HTLV-1, HTLV-2, and HTLV-5 as members of a retrovirus subfamily referred to as the "*Oncovirinae*," a term no longer recognized by The International Committee on Taxonomy of Viruses, since members of the subfamily are not necessarily more closely related to each other than they are to members of other subfamilies.)

The remainder of this account of the HTLVs is concerned mainly with HTLV-1, which infects 15 to 25 million individuals worldwide. HTLV-1 is endemic in southwestern Japan (where its prevalence is as high as 35% in some regions), Central Africa, the Caribbean basin, and some regions of South America, Melanesia, and the Middle East. The virus is relatively rare in North America and Europe. However, it is seen among blacks in the southeastern United States (see Poiesz et al., 2003, and Matsuoka and Jeang, 2007, for detailed reviews of HTLV-1).

HTLV-1 is transmitted via the same routes as HIV in regions where it is endemic: by sexual intercourse, via contaminated blood or blood products, and from mother to child via breast-feeding. In the United States, where the virus is not endemic, its primary means of transmission is via needle sharing among intravenous drug users. And in fact, its prevalence in that group is approaching that of HIV.

HTLV-1 also infects the same principal target cell as HIV, that is, the CD4 helper T cell. However, there is a key difference in the principal ways that these two unrelated retroviruses infect a CD4 T cell. Whereas free HIV particles readily infect CD4 T cells, free HTLV-1 particles are extremely inefficient at initiating an infection. Consequently, all three routes of HTLV-1 transmission enumerated above are mediated by the passage of infected cells, rather than free virus particles, from the infected individual. Infection of the recipient then occurs by cell-to-cell contact. In brief, this occurs as follows. When an HTLV-1-infected cell contacts an uninfected cell, a virological synapse forms at the interface between the two cells, and the HTLV-1 Gag/genomic RNA complexes accumulate at the synapse and then invade the uninfected cell.

This direct cell-to-cell mode of infection is important not only with respect to how the virus initiates an infection, but also with respect to the means by which it expands and persists within the host; this key point is considered in more detail below. For now, note that HTLV-1 uses clonal expansion of infected cells as the primary means by which it spreads within an infected individual. In fact, HTLV-1 carriers have almost undetectable levels of the virus particles circulating in their blood. Moreover, that is so despite the fact that as many as 1% of their CD4 T cells carry an HTLV-1 provirus.

Considering that HTLV-1 uses clonal expansion of infected cells as its primary means for replicating and disseminating within an individual, a crucial factor in the HTLV-1 lifestyle is that the HTLV-1 provirus, like the proviruses of other retroviruses, is replicated concurrently with the replication of the cellular genome. And, to enhance the efficiency of its lifestyle, HTLV-1 expresses a factor (the Tax protein [see below]) that stimulates infected cells to proliferate, in that way boosting the clonal expansion of infected cells (see below).

The HTLV-1 infection will persist in the individual for as long as there are infected clones, that is, essentially indefinitely. Reverse transcription inhibitors do not affect the proviral loads in patients suffering from TSP, consistent with the primary mode of HTLV-1 replication via the clonal expansion of infected cells, in which the provirus is replicated as part of, and concurrently with, the host cell genome (Box 21.58).

Cell-free HTLV-1 particles can infect CD4 T cells, at least as demonstrated by experiments in vitro. However, they do so 1,000-fold less efficiently than free HIV particles. Several factors are known that may help to account for the low efficiency of infection of free HTLV-1 particles. First, the HTLV-1 reverse transcriptase is less "robust" than the HIV reverse transcriptase, at least in vitro. The main difference between these polymerases is seen in the efficiency of the "jump" from the "strong stop" DNA to plus-strand DNA synthesis (Chapter 20; Fig. 20.5). Second, HTLV-1 particles contain considerably less envelope protein than HIV particles.

HTLV-1, like HIV, encodes regulatory proteins, one of which, **Tax**, was noted above. As in the case of the HIV regulatory proteins, the HTLV-1 regulatory proteins are required for the efficient expression of the HTLV-1 genome and its replication. Again reminiscent of HIV, the HTLV-1 regulatory proteins are translated from multiply spliced mRNAs.

The aforementioned Tax protein stimulates the HTLV-1 enhancer by activating several cellular transcription factors. CREB is one of the cellular transcription factors activated by Tax. Normally, CREB is activated by intracellular

---

BOX 21.58

HTLV-1 is very stable genetically relative to HIV, probably because the amplification of HTLV-1 within an infected individual occurs by means of clonal expansion of infected cells, rather than by the HIV process of error-prone, reverse transcription-mediated replication of the virus within infected cells.

The genetic stability of HTLV-1 provided the rationale for an investigation into the possible origin of the virus. Key facts and findings are as follows. STLV-1 (the simian counterpart of HTLV-1) is transmitted in the wild between different simian species. In addition, there is a high degree of homology between STLV-1 strains from Central African chimpanzees, as well as from mandrills, and HTLV-1 strains from human inhabitants of the same region. Thus, it is thought that HTLV-1 was transmitted to humans from simians, possibly on several different occasions, over a period of thousands of years. Consistent with that premise, new HTLVs have been isolated from Central Africans who frequently come into contact with the blood of nonhuman primates (e.g., bush meat hunters).

---

signals that increase cAMP levels. cAMP then activates PKA, which enters the nucleus in its activated form, where it activates CREB. CREB then activates transcription of genes that contain a regulatory sequence called CRE (an acronym for "cAMP response element"). (The acronym "CREB" stands for "CRE-binding protein.")

In contrast to the normal mode of CREB activation, as outlined above, CREB can be activated by the direct binding of Tax. HTLV-1 transcription is then activated because the HTLV-1 LTR contains so-called Tax-responsive elements that are homologous to the cellular CRE sequence. As you might then suppose, Tax can likewise activate transcription of particular cellular genes via its interaction with CREB. The Tax-induced upregulation of these cellular genes might further augment virus replication.

Tax can also affect the expression of particular cellular genes by activating NF-κB, doing so directly by binding to NF-κB, as well as indirectly by triggering signaling pathways that result in the degradation of IκB. The degradation of IκB enables NF-κB to translocate from the cytoplasm to the nucleus, where it may promote the expression of NF-κB-responsive genes (see above apropos NF-κB and HIV; see also Fig. 18.13). Since many of the cellular genes activated by Tax are involved in cell growth, differentiation, apoptosis, or cell cycle control, the effects of Tax on these cellular genes are believed to play a role in HTLV-1-induced proliferation of CD4 T cells, thereby promoting virus replication via the clonal expansion of infected cells, and neoplasia as well (see below).

Tax also upregulates expression of the cellular Fas ligand. This helps HTLV-1-infected cells to evade CTL-mediated killing by enabling them to transmit a Fas-mediated, proapoptotic signal into attacking CTLs. Recall that HIV employs a similar mechanism, in which its Nef, Tat, and Env proteins induce expression of Fas ligand (see above).

A second HTLV-1 regulatory protein, **Rex**, is found exclusively in the nucleus, where it is essential for the nuclear export of the two species of incompletely spliced HTLV-1 mRNAs. These are the unspliced mRNA that encodes the HTLV-1 Gag proteins and reverse transcriptase and the singly spliced mRNA that encodes the viral envelope proteins. In the absence of Rex, these incompletely spliced mRNAs would be either completely spliced or else degraded in the nucleus. Rex acts by binding to the so-called Rex responsive element (RxRE) on the incompletely spliced viral mRNAs, thereby facilitating their export from the nucleus. (Recall that Tax and Rex are each translated from completely spliced mRNAs that are readily exported from the nucleus.)

The functions and mechanisms of action of the HTLV-1 Tax and Rex proteins noted above are clearly reminiscent of the HIV Tat and Rev proteins, respectively. What is more, as in the case of the HIV Tat and Rev proteins, the actions of HTLV-1 Tax and Rex give rise to a multiphase virus replication cycle.

In the first phase, before the accumulation of Tax and Rex, there is a low level of overall viral transcription. Moreover, at this time, only completely spliced mRNAs are exported from the nucleus and translated. Thus, Tax and Rex are the only HTLV-1 proteins that are generated at this time. In the second phase, Tax is available to stimulate overall transcription, while Rex enables incompletely spliced mRNAs to be exported to the cytoplasm and translated.

Notice that both Tax and Rex are positive regulators of HTLV-1 gene expression. Also, note that Tax itself is highly immunogenic. Thus, the virus must somehow reduce its expression of Tax in order to reduce the expression of viral genes so as to evade immune defenses and establish latency. More recently, in 2004, the viral p30 protein was described, which binds to and retains the multiply spliced *tax/rex* mRNA in the nucleus, thereby preventing the stimulation of viral transcription by Tax and Rex and leading to the establishment of viral latency.

We might think of p30 as bringing about a third phase in the HTLV-1 replication cycle. p30 downregulates overall viral transcription, while also negatively affecting transcription of certain cellular genes via its effect on Tax. The ongoing interplay between Tax, Rex, p30, and cellular factors may act to determine the pattern of disease development (Box 21.59).

Most HTLV-1 infections are asymptomatic. However, about 5% of infected individuals develop ATLL after an extremely long incubation period of 30 or more years. ATLL is a neoplasm involving CD4 helper T cells, and it is

BOX 21.59

Since the HTLV-1 and HIV regulatory proteins display key functional similarities, and since they each are translated from multiply spliced mRNAs, it is fitting to reiterate that these viruses are not closely related and, indeed, occupy separate genera of the *Retroviridae*. One prominent difference in the biology of these viruses is that HTLV-1 produces mRNAs and proteins at much lower levels than HIV does. This may be explained, at least in part, by the fact that the HIV LTR contains a much more robust enhancer than that of HTLV-1. That difference should not be understated, since differences in the efficiencies of gene expression between HTLV-1 and HIV may underlie the key difference in their "lifestyles." That is, HTLV-1 is amplified within an infected individual primarily by the clonal expansion of infected cells, while lytic HIV infection produces free virus particles that disseminate the infection to new cells.

Here is another difference in the biology of these viruses. HIV makes use of numerous stratagems to evade host immunity. And while not usually regarded as an immune evasion strategy per se, the effects of the HIV Tat and Rev proteins enable HIV to produce a burst of free progeny virus particles before the infected cell might be recognized by immune effector cells. In addition, the dependence of each HIV replication cycle on the activity of the error-prone viral reverse transcriptase enables HIV to readily generate antigenic variants that might evade host immunity. In contrast, the relatively inefficient expression of HTLV-1 genes helps that virus to evade immune detection, thereby helping to make possible the clonal expansion of infected cells. Yet HTLV-1 does express an accessory protein, p12, that suppresses expression of MHC-I molecules. Nevertheless, the above differences between HTLV-1 and HIV may account for the difference in the clinical outcomes of their infections: a lymphocytic leukemia in the case of HTLV-1 and depletion of CD4 T cells and immunodeficiency in the case of HIV.

usually fatal within 1 year of diagnosis. As noted above, antiretroviral therapy does not prevent the clinical outcome, which is not surprising, considering the means by which the infection is sustained. Also, since ATLL occurs only after decades of infection, the outcome may be irreversibly determined before the onset of symptoms. Thus, even if antiretroviral therapy were able to block virus replication at the first sign of symptoms, it still might not affect disease progression.

The mechanism by which HTLV-1 causes ATLL is not yet understood, but it must be different from the transforming mechanisms of the simple retroviruses (oncornaviruses [Chapter 20]). That much is clear, since the HTLV-1 genome does not contain an oncogene, nor does the virus transform cells by insertionally upregulating cellular proto-oncogenes. Also, it is not clear why most HTLV-1-infected individuals become asymptomatic carriers while others develop ATLL. In this regard the individual's particular MHC-I haplotype appears to be a factor. Perhaps the MHC-I alleles expressed by ATLL patients do not mediate a sufficiently strong CTL response against virus-carrying cells.

HTLV-1 tumors are clonal, as indicated by the fact that the provirus has no preferred chromosomal location yet is found at the same chromosomal site in all leukemic cells from a given case of ATLL. That point and the facts that (i)

most HTLV-1-infected individuals are asymptomatic and (ii) there is an incubation period of decades before the emergence of ATLL in HTLV-1-infected individuals are together consistent with the notion that HTLV-1-induced leukemogenesis must involve multiple events. Perhaps the numerous cycles of T-cell proliferation induced by the virus eventually lead to sufficient chromosomal alterations to result in neoplasia in some individuals, thereby explaining the remarkably long incubation period that precedes the onset of ATLL. In this regard, it may be useful to compare ATLL to hepatitis B virus-induced hepatocellular carcinoma (Chapter 22).

Several different lines of experimental evidence support a role for Tax in bringing about ATLL. For example, genetic studies using mutant viruses demonstrated that the Tax proteins of HTLV-1 (and of HTLV-2 as well) are necessary for inducing transformation in vitro. Another experimental approach was to generate transgenic mice that expressed Tax under the control of the HTLV-1 LTR. These mice synthesized Tax in muscle tissue, where its expression gave rise to sarcomas.

The Tax mutants in the genetic studies noted in the preceding paragraph were unable to activate cellular genes that are upregulated by NF-κB and CREB, in agreement with the plausible premise that Tax promotes T-cell leukemia by activating cellular genes involved in the regulation of

BOX 21.60

While the HTLV-1 Tax protein is required by the virus to transform cells, *tax* transcripts are seen in the leukemic cells of only 40% of all ATLL cases. Mutations in the *tax* gene per se or in the proviral 5′ LTR and DNA methylation of the proviral DNA have all been found to silence *tax* expression in ATLL cells in vivo. Importantly, these findings show that while Tax is necessary to initiate transformation, the continued expression of *tax* is not necessary to maintain the transformed cell phenotype.

Considering the high frequency of ATLL cases in which *tax* is silenced, one might propose that the silencing of *tax* may actually provide some advantage to ATLL cells in vivo. For example, Tax is the main target of cell-mediated immune responses against ATLL cells. Thus, cells that shut down *tax* expression may be selected for clonal expansion during disease progression. (Apropos the issue of the dependence of the transformed cell phenotype on the continuous expression of a viral gene product, compare transformation by HTLV-1 to transformation by the polyomaviruses [Chapter 15] and papillomaviruses [Chapter 16].)

Recently, and remarkably, the *minus* strand of the HTLV-1 provirus was found to encode a protein, HBZ, which heterodimerizes with CREB, in that way preventing the direct binding of Tax to CREB, thereby impairing Tax-mediated transactivation of viral transcription from the 5′ LTR (see the text). Intriguingly, and consistent with the notion that silencing of *tax* promotes virus proliferation by the clonal expansion of infected cells, when HBZ expression was experimentally reduced using short hairpin RNAs, the proliferation of ATLL cells was slowed. Conversely, when HBZ was experimentally overexpressed, cellular proliferation was enhanced. Also, in contrast to *tax* mRNA, which is detected in only 40% of all ATLL cases, HBZ mRNA has been detected in all ATLL cases examined to date. These results imply that the HTLV-1 HBZ protein may enhance the proliferation of ATLL cells in vivo. Other recent intriguing results imply that HBV mRNA per se, in addition to the HBV protein, may have its own unique role in promoting cell proliferation.

T-cell proliferation. Consistent with that premise, Tax has also been shown to upregulate the cellular genes encoding IL-2 and the IL-2 receptor, resulting in the autocrine stimulation of cell proliferation. Tax also upregulates IL-3, the proliferating cell nuclear antigen, and the proto-oncogenes c-*fos* and c-*sis*, among others. In addition, Tax activates cyclin-dependent kinases (e.g., CDK4 and CDK6), doing so by direct protein binding. Activation of these cyclins leads to hyperphosphorylation of the Rb tumor-suppressor protein (Chapter 15). Tax also upregulates expression of antiapoptotic factors. (Interestingly, both Tax and the E6 proteins of highly oncogenic papillomaviruses [Chapter 16] are able to target proteins with so-called PDZ protein-protein interaction domains, which are found in nearly 200 human signaling proteins, in which they help to stabilize signaling complexes. The interactions of Tax and E6 with PDZ-containing proteins enable the respective viruses to inactivate additional host tumor-suppressor proteins. Thus, these unrelated and distinctly different cancer viruses appear to share mechanisms for bringing about cell transformation.) (See Box 21.60).

Lastly, we consider TSP, a demyelinating disease associated with HTLV-1 that involves the long tracts of motor neurons in the spinal cord. Initial symptoms of TSP include lumbar back pain that radiates down the legs, followed by weakness and spastic paralysis of both legs and sometimes vision problems. Similar symptoms may be seen in multiple sclerosis. However, in contrast to multiple sclerosis, there are no remissions in TSP.

The mechanism by which HTLV-1 causes TSP is not yet known. But HTLV-1 has been isolated from the cerebrospinal fluid of TSP patients, and the viral genome can be detected in TSP lesions by PCR and in situ hybridization. Moreover, chronic inflammation of the spinal cord and the presence of lymphocytes within TSP spinal cord lesions are characteristic pathological features of the disease. This infiltration of lymphocytes into the central nervous system may be induced by cytokines that are generated in response to the HTLV-1 infection there. Also, there is evidence that the HTLV-1 provirus is expressed by lymphocytes that infiltrate into TSP spinal cord lesions. Expression of the HTLV-1 provirus by these lymphocytes

might also be a factor in the immunopathology seen in TSP patients.

There also is evidence that TSP might be associated with immune responses to the Tax protein. Molecular mimicry (Chapter 3) has been suggested as well. That premise is supported by the observations that TSP patients make antibodies to Tax that cross-react with heterogeneous nuclear ribonuclear protein A1 (hnRNP A1), a neuron-specific autoantigen.

Suggested Readings

Barre-Sinoussi, F., J. C. Chermann, F. Rey, M. T. Nugeyre, S. Chamaret, J. Gruest, C. Dauguet, C. Axler-Blin, F. Vezinet-Brun, C. Rouzioux, W. Rozenbaum, and L. Montagnier. 1983. Isolation of a T-lymphotropic retrovirus from a patient at risk for acquired immune deficiency syndrome (AIDS). *Science* **220:** 868–871.

Douek, D. C., L. J. Picker, and R. A. Koup. 2003. T cell dynamics in HIV-1 infection. *Annu. Rev. Immunol.* **21:**265–304.

Hammer, S. M., J. J. Eron, Jr., P. Reiss, R. T. Schooley, M. A. Thompson, S. Walmsley, P. Cahn, M. A. Fischl, J. M. Gatell, M. S. Hirsch, D. M. Jacobsen, J. S. Montaner, D. D. Richman, P. G. Yeni, and P. A. Volberding. 2008. Antiretroviral treatment of adult HIV infection: recommendations of the International AIDS Society–USA Panel. *JAMA* **300:**555–570.

Malim, M. H., and M. Emerman. 2008. HIV-1 accessory proteins—ensuring viral survival in a hostile environment. *Cell Host Microbe* **3:**388–398.

Matsuoka, M., and K. T. Jeang. 2007. Human T-cell leukaemia virus type 1 (HTLV-1) infectivity and cellular transformation. *Nat. Rev. Cancer* **7:**270–280.

Neil, S. J. D., T. Zang, and P. D. Bieniasz. 2008. Tetherin inhibits retrovirus release and is antagonized by HIV-1 Vpu. *Nature* **451:**425–430.

Poiesz, B. J., M. J. Poiesz, and D. Choi. 2003. Human T-cell lymphoma/leukemia virus-associated T-cell lymphoma and leukemia, p. 141–163. *In* P. H. Wiernik, J. M. Goldman, J. P. Dutcher, and R. A. Kyle (ed.), *Neoplastic Diseases of the Blood.* Cambridge University Press, Cambridge, England.

UNAIDS. 2008. *UNAIDS 2008 Report on the Global AIDS Epidemic.* http://www.unaids.org/en/KnowledgeCentre/HIVData/GlobalReport/2008/2008_Global_report.asp.

Van Damme, N., D. Goff, C. Katsura, R. L. Jorgenson, R. Mitchell, M. C. Johnson, E. B. Stephens, and J. Guatelli. 2008. The interferon-induced protein BST-2 restricts HIV-1 release and is downregulated from the cell surface by the viral Vpu protein. *Cell Host Microbe* **3:**254–252.

Hepadnaviruses and Hepatitis Delta Virus

THE HEPADNAVIRUSES: DNA RETROVIRUSES

INTRODUCTION

The *Hepadnaviridae* are distinguished from all other virus families by what well may be the most remarkable, and perhaps bizarre, viral molecular biology and replication strategy known. Moreover, the **human hepatitis B virus (HBV)** (previously known as serum hepatitis virus), the prototype for the family, is one of the most important of all human pathogens.

As to the hepadnavirus replication strategy, despite having double-stranded DNA genomes, hepadnaviruses replicate *via RNA intermediates*. In brief, hepadnavirus circular, double-stranded DNA genomes are transcribed in the nucleus by the cellular RNA polymerase II (RNA pol II) enzyme. Several distinct species of viral RNA transcripts are thus generated, all of which are exported to the cytoplasm. The largest of these RNA species, the pregenomic RNA, is a complete transcript of the circular DNA genome that is longer than the genome itself, as it contains terminal redundant sequences. Remarkably, the pregenomic RNA is packaged within nascent virus capsids, in which it is reverse transcribed by a virus-encoded reverse transcriptase activity. In that way, progeny hepadnavirus particles that contain double-stranded DNA genomes are generated. Thus, the hepadnaviruses appear to turn the already extraordinary retroviral replication strategy (Chapter 20) on its head. That is, whereas the retroviruses are RNA viruses that replicate via a DNA intermediate, the hepadnaviruses are DNA viruses that replicate via an RNA intermediate. (The hepadnavirus replication cycle outlined above was determined after the Baltimore classification scheme was first put forward. Since the hepadnavirus replication cycle differs fundamentally from that of all of the other double-stranded DNA viruses that constitute the class I viruses, should the hepadnavirus family constitute a Baltimore class of its own? See Box 22.1.)

As noted above, HBV is the prototype of the hepadnavirus family. The family also includes the woodchuck hepatitis virus and the ground squirrel hepatitis virus, as well as the more distantly related duck hepatitis virus. As the family name implies, all of the known hepadnaviruses have a special tropism for hepatocytes. These viruses may infect other cell types, too, but only inefficiently.

Even if the hepadnaviruses did not have an out of the ordinary replication strategy to rouse our interest, HBV would still command our attention, since it gives rise to one of the world's most widespread and medically important chronic viral infections, with about *500 million chronic HBV carriers worldwide*. In the United States more than 300,000 individuals are infected with HBV each year, resulting in about 4,000 fatalities. However, in some areas of the world, including Southeast Asia and southern Africa, as many as 50% of all individuals are infected. Moreover, each year HBV infections lead to between 1 and 2 million cases of fatal liver disease, despite the existence of effective vaccines against the agent. What is more, chronic HBV infection may also lead to fatal **hepatocellular carcinoma** in a significant number of individuals. Indeed, this virus ranks high after tobacco among known environmental carcinogens. In fact, hepatocellular carcinoma is one of the three most common causes of cancer fatalities. (The World Health Organization estimates that 80% of all cases of hepatocellular carcinoma can be attributed to chronic HBV infection.) And, where HBV is prevalent, chronic liver disease and hepatocellular carcinoma are among the

region's major health problems. All things considered, HBV is one of the most important of all human pathogens.

The discovery of the remarkable hepadnavirus replication strategy by Summers and Mason (1982) came as a great surprise because, until then, reverse transcription was considered to be a peculiarity of the retroviruses. It is also remarkable that, despite the fact that the hepadnaviruses contain double-stranded DNA genomes while the retroviruses contain plus-strand RNA genomes, these seemingly disparate virus families actually appear to be phylogenetically related. For example, the genes encoding the reverse transcriptase of each of these virus families share similar sequences. In fact, the enzymatic domains of the hepadnaviral reverse transcriptase were first identified by amino acid sequence alignment with the retroviral reverse transcriptases. Moreover, the structure of the HBV capsid protein (see below) has a notable resemblance to the structure of the human immunodeficiency virus (HIV) capsid protein (Zlotnick et al., 1998). Furthermore, the motifs that are common to the HBV and HIV capsid proteins are not seen in the capsid proteins of most other virus families. In addition, the genomes of these two virus families contain a similar number of genes, which are similarly arranged according to their respective common functions. As then might be expected, the hepadnaviruses also appear to be related to the foamy viruses and to some transposable retroelements (Chapter 20). Indeed, the foamy viruses, which also replicate via reverse transcription, contain a similarly nicked, double-stranded DNA genome (Chapter 20; see also below).

The hepadnaviruses are among the smallest known animal viruses. Their double-stranded DNA genomes contain only between 3.0 and 3.4 kilobases (kb), and that of HBV contains 3.2 kb (Box 22.2). Moreover, the hepadnavirus genomes present a most unusual structure, as described below.

HBV was discovered by Baruch S. Blumberg in 1966. Blumberg's actual goal at the time was to identify a host gene that might predispose some individuals to hepatitis. Towards that end, Blumberg's group reported on the familial clustering and segregation of a serum antigen present at high frequency in certain tropical human populations that were predisposed to hepatitis. Because the antigen was first identified in an Australian aborigine, it was termed **Australia antigen**. At first, Blumberg's finding led to the thought that the Australia antigen is a host gene product that predisposes one to an infectious agent that causes hepatitis. However, it soon became clear that rather than being a host gene product, the Australia antigen is, in fact, associated with an infectious agent that occurs in viral hepatitis. The antigen is actually the surface protein of the 42-nm HBV particles, and it is now known as the

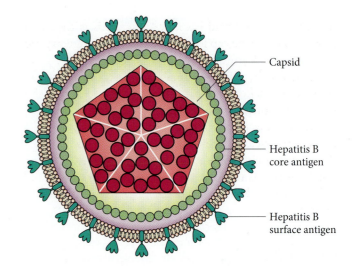

Figure 22.1 HBV particles are spherical, enveloped, 42-nm-diameter structures that enclose an inner 27-nm icosahedral nucleocapsid. The envelope consists of hepatocyte membrane lipids and the viral surface glycoprotein HBsAg. Between the envelope and the nucleocapsid there is the core, comprised of the hepatitis B core antigen HBcAg.

hepatitis B surface antigen (HBsAg). HBV particles per se were first described in 1970 by the British pathologist David Dane, and thus, they are sometimes referred to as **Dane particles**.

In the sections that follow, wherever possible, HBV will be used to exemplify the hepadnaviruses.

STRUCTURE

HBV particles are spherical, enveloped, 42-nm-diameter structures, which enclose an inner 27-nm-diameter nucleocapsid (Fig. 22.1). The envelope consists of hepatocyte membrane lipids and the viral surface glycoproteins, the HBsAg noted above. (The surface glycoproteins actually are comprised of three different glycoproteins. As described below in more detail, each of these glycoproteins is encoded by the same open reading frame but initiated at a different start codon within that reading frame. Thus, that reading frame generates the large [L], medium [M], and small [S] surface proteins.) The inner core of the virus particle contains the single core or capsid (C) protein (also referred to as the HBcAg), the viral DNA genome, and the reverse transcriptase (P protein).

The HBV genome, at only 3.2 kb, is one of the smallest viral genomes to be found in a replication-competent virus (as noted above and in Box 22.1). Also, the HBV genome has a rather atypical structure (Fig. 22.2). While it is circular, the plus strand is incomplete, such that 15 to 50% of the genome is actually single stranded. In contrast, the minus strand is complete but contains a gap. The plus strand always overlaps the discontinuity in the minus strand, thereby maintaining the circularity of the genome.

That overlap region is comprised of the 5′ ends of the two strands. The minus strand is actually slightly longer than unit length, and both strands contain the 10- to 12-base pair (bp) direct repeats, DR1 and DR2, which span the gap in the minus strand. In the capsid, a single molecule of the P (polymerase) protein is covalently linked to the 5′ end of the gapped minus strand, and the 5′ end of the

Figure 22.2 Depiction of a hepadnavirus genome, showing the full-length but nicked DNA minus strand and the gapped DNA plus strand. The circularity is maintained by the overlapping 3′ ends. Notice that the minus strand is actually somewhat longer than full length and that the polymerase is attached at its 5′ end. The plus strand terminates with a cap at its 5′ end. Also notice the direct repeat sequences (DR1 and DR2) located near the 5′ ends of each strand.

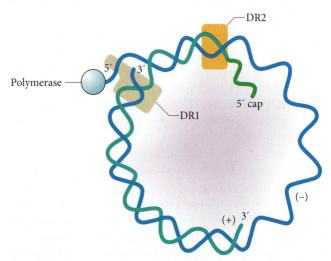

incomplete plus strand is capped. (Yes, this is the same 7-methylguanosine cap found at the 5′ ends of eukaryotic messenger RNA [mRNA] molecules. See below.)

Interestingly, HBV icosahedral capsids come in two sizes. The smaller of these contains 180 subunits; the larger one contains 240 subunits. The arrangement of the surface glycoproteins, which project from the lipid envelope, appears to be nonicosahedral, suggesting that the glycoproteins do not interact with the underlying capsid. If that is the case, then this is a way in which the HBV structure differs from that of most other enveloped viruses.

REPLICATION

Details of the HBV replication cycle are not known with certainty, since there has been no suitable cell culture system in which productive infection by this virus might be studied. Consequently, much of what is believed to be true for HBV is inferred from studies of the woodchuck, squirrel, and duck hepatitis viruses, as well as from studies of cells transfected with various recombinant HBV constructs.

Entry

Before proceeding with hepadnavirus entry, we need to pause for a moment to reiterate that the HBV envelope gene encodes three envelope proteins from the same S open reading frame. These three proteins are the S, M, and L proteins, which are translated from different in-frame start codons within the single S open reading frame. The smallest of these proteins, the S protein, contains only the surface or S domain, while the M protein has an N-terminal 55-amino-acid extension called the pre-S2 domain and the L protein contains an additional N-terminal 108-amino-acid extension called the pre-S1 domain. (These facts are discussed in detail below under the subheadings "Genome Organization" and "Gene Expression." Also, see Fig. 22.4.) In contrast to HBV, the duck hepatitis virus encodes only an L protein and an S protein from its corresponding open reading frame. The duck hepatitis virus L protein, which contains all 167 amino acids of its S protein, also contains 160 additional amino acids at its N terminus; this is the so-called pre-S domain.

The first steps of HBV entry may be conventional, but at some point the HBV entry pathway may become rather unconventional. As for the conventional part, the virus is thought to bind to heparan sulfate proteoglycans on the hepatocyte surface, doing so via the pre-S1 domain of its L protein. Unfortunately, for reasons noted above, details of subsequent steps in HBV entry are not known

with certainty. Thus, we continue our discussion of hepadnavirus entry, now using duck hepatitis virus as our case in point.

Duck hepatitis virus attaches to hepatocytes via the pre-S domain of its L surface protein. The plasma membrane-associated enzyme, carboxypeptidase D, is currently considered to be a putative receptor for duck hepatitis virus, since it is essential for infection by that virus. After attaching to the cell surface, duck hepatitis virus is taken up into the cell by receptor-mediated endocytosis.

The duck hepatitis virus entry pathway (and presumably that of HBV as well) appears to become nonconventional at the point where internalized virus particles need to escape from the endocytic pathway. That is so for the following reason. Although hepadnaviruses are enveloped, duck hepatitis virus does not use membrane fusion to escape from endosomes. Consistent with that fact, the duck hepatitis virus L protein does not contain a fusion peptide. Instead, the pre-S domain of that protein (as well as the pre-S1 domain of HBV) appears to include a cell-permeable domain, which contains a so-called **translocation motif** (TLM) (Stoeckl et al., 2006). The TLM per se contains a 12-amino-acid amphipathic α-helix that mediates the transfer across membranes of attached peptides, nucleic acids, and proteins. Surprisingly, the TLM appears to be able to do so by an energy- and receptor-independent mechanism, and it does not affect the integrity of the cargo fused to it. (In point of fact, the duck hepatitis virus entry pathway may become nonconventional even prior to the events recounted above, since, as in the case of HBV, essentially nothing is known of the preceding steps of duck hepatitis virus entry.)

The TLM motifs are masked on native hepadnavirus particles, so as not to act prematurely. However, during the time that the virus is in the endosomal compartment, the TLM motifs are unmasked by endosomal proteases (Fig. 22.3). This unmasking of the TLMs enables the translocation of the virus particles across the endosomal membrane and into the cytosol, so that infection might then proceed. As might be expected, TLM-deficient mutant viruses bind to cells and are taken up into the endosomal compartment, but they cannot escape from endosomes.

Once again, and importantly, the TLM functions as a *cell permeabilizer*, not as a fusogenic peptide, and the escape of duck hepatitis virus from the endosome is not a fusogenic process. This is supported by the following experimental findings. (i) The role of the TLM in duck hepatitis virus infection as a cell permeabilizer was initially suggested by the finding that experimental exposure of the TLM on the surface of viral particles may actually

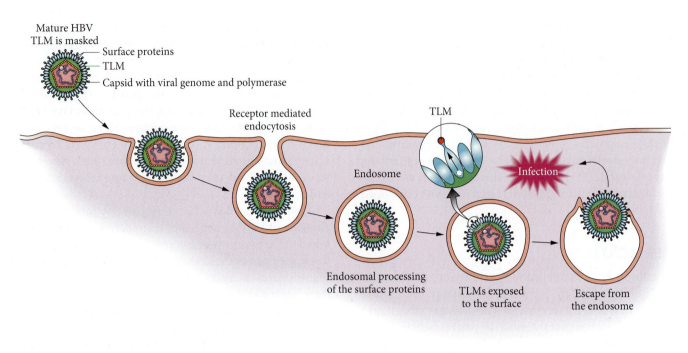

Mature HBV
TLM is masked

— Surface proteins
— TLM
— Capsid with viral genome and polymerase

Receptor mediated
endocytosis

TLM

Endosome

Infection

Endosomal processing
of the surface proteins

TLMs exposed
to the surface

Escape from
the endosome

Figure 22.3 Model of the endosomal processing of hepadnaviral particles. Hepadnaviruses are internalized by receptor-mediated endocytosis. In the endosomal compartment, proteolytic cleavage of the surface protein occurs, resulting in a conformational change that exposes the TLMs (shown as red circles) on the surface of the viral particle. The high density of TLMs exposed on the surface of the particle allows endosomal escape into the cytosol to initiate infection. Note that the TLM-mediated escape model may not be valid for HBV. Adapted from L. Stoeckl, A. Funk, A. Kopitzki, et al., *Proc. Natl. Acad. Sci. USA* **103:**6730–6734, 2006, with permission.

lead to entry. This is in contrast to the results seen in the cases of enveloped viruses that enter by a conventional fusogenic process. In those instances, premature exposure of the fusion peptide aborts fusion and entry. That is so because fusion proteins must undergo their conformational rearrangements concurrently with fusion (see Chapter 12 for a detailed discussion of this issue). (ii) In instances where escape from the endosomal compartment is mediated by viral fusogenic proteins, those proteins remain in the endosomal membranes. In contrast, the TLM-containing duck hepatitis virus envelope proteins appear to remain associated with the nucleocapsids during their escape from the endocytotic pathway. The mechanism by which the capsid is released from the enveloped particle is not yet known.

May we presume that HBV likewise escapes from the endosomal compartment by a similar nonconventional mechanism? The answer is not clear. On the one hand, the surface proteins of all members of the hepadnavirus family appear to contain the TLM motif. On the other hand, while infectivity assays seem to prove that the duck hepatitis virus TLM motif is necessary for infection by that virus, results of a recent genetic analysis show that the HBV TLM motif is not necessary for the infectivity of HBV. Moreover, other more recent data imply that the HBV pre-S1 domain is able to insert into

the hydrophobic core of a membrane bilayer (as is usual for a fusion protein) and to undergo a conformational rearrangement that could be involved in a fusion mechanism. So, here is another story to follow for future developments.

While the viral surface proteins are responsible for the entry of the virus into the host cell, the capsid is responsible for transporting the viral genome to the nucleus. The nuclear transport of the capsid occurs via an active, microtubule-dependent process. Release of the genome occurs exclusively in the nucleus. The mechanisms for nuclear entry and release of the genome are not yet known.

Genome Organization

The HBV genetic map, although short in length, is extraordinarily complex. That is so because it contains four overlapping reading frames within its length of only 3.2 kb (Fig. 22.4). Moreover, *every* nucleotide of the genome is part of at least one open reading frame, and every *cis*-acting promoter and enhancer sequence is contained within one or more open reading frames. As might be expected, such genetic complexity made for challenging genetic analysis.

All four HBV open reading frames are encoded on the minus strand. Reading in the clockwise direction from the

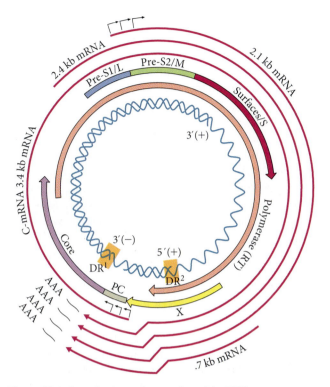

Figure 22.4 Organization and expression of the HBV genome. Salient features are as follows. The genome contains four overlapping reading frames within its 3.2-kb length. All four open reading frames are on the minus strand. These are the C (core) reading frame, the S (envelope protein) reading frame, the P (polymerase) reading frame, and the X protein reading frame. No open reading frame crosses the gap in the minus strand. The C gene has two initiation sites that divide it into the pre-C and C regions. As a consequence of these multiple initiation sites, the *C* gene produces two distinct proteins. These are the pre-C protein, also known as the e antigen or HBeAg, and the C or core protein, also known as the HBcAg. The single S open reading frame encodes the three viral glycoproteins, which are contained in the HBV envelope. These three proteins are the small (S), middle (M), and large (L) proteins, which are translated from different in-frame start codons within the S open reading frame. The S protein contains only the surface or S domain, while the M protein has an N-terminal 55-amino-acid extension called the pre-S2 domain and the L protein contains an additional N-terminal 108-amino-acid extension called the pre-S1 domain. HBV transcribes four classes of RNAs, which are under the control of four different RNA pol II promoters. These four RNAs are the pregenomic RNA and three classes of subgenomic RNAs, i.e., the pre-S1, S, and X RNAs, which are 3.4, 2.4, 2.1, and 0.7 kb in length, respectively. Since the genome is 3.2 kb in length, the pregenomic RNA actually is longer than the genome itself, containing terminal redundant sequences. Since each of the hepadnavirus RNAs is synthesized by the cellular RNA pol II, then like all other transcripts made by this enzyme, they are capped at their respective 5′ ends by cellular capping enzymes. None of the hepadnavirus RNA species is spliced. All HBV transcripts use a common polyadenylation signal located within the region encoding the C protein (indicated by AAA). Modified from E. K. Wagner, M. J. Hewlett, D. C. Bloom, and D. Camerini, *Basic Virology*, 3rd ed. (Blackwell Publishing, Malden, MA, 2008), with permission.

gap in the minus strand, these are the C (core) reading frame, the P (polymerase) reading frame, the S (envelope protein) reading frame, and the X protein reading frame. No open reading frame crosses the gap in the minus strand.

Note that open reading frames for the hepadnavirus C, P, and S proteins lie in the same relative positions as the open reading frames of the functionally equivalent Gag, Pol, and Env proteins of the retroviruses (Chapter 20), in keeping with the premise that the hepadnaviruses are biologically related to the retroviruses. However, the genetic organizations of the hepadnaviruses and the retroviruses differ in the following respects. The hepadnavirus S open reading frame overlaps that of the P protein and is read in a different reading frame. This contrasts with the retroviruses, in which the reading frame for Env is entirely downstream from that for Pol. Importantly, like the retroviral *pol* gene product, the hepadnavirus P protein has reverse transcriptase, DNA polymerase, and ribonuclease (RNase) H activities. (However, it does not contain an integrase activity.)

The reading frame for the X protein, like the S reading frame, also overlaps that of the P protein, but from the other end, and spans the cohesive ends of the genome but not the gap in the minus strand. The X protein has transactivating functions that activate a wide range of viral and cellular promoters, including those for proto-oncogenes. Moreover, it activates the transcription regulator nuclear factor kappa B (NF-κB). These facts led to the belief that the viral X protein may be important in the etiology of hepatocellular carcinoma (see below).

The *C* gene has two initiation sites that divide it into the pre-C and C regions. Because of these multiple initiation sites, the *C* gene produces two distinct proteins. These are the pre-C protein, also known as the e antigen or HBeAg, and the C or core protein, also known as the HBcAg. The pre-C protein contains the entire C protein, extended at its N terminus by the pre-C-specific amino acids. The biological function of the pre-C protein is not known with certainty, but it is secreted from cells and may act to modulate the immune response against the virus (see below).

The pre-C protein and the X protein are the only hepadnavirus gene products that are not incorporated into virus particles.

The single S open reading frame encodes the three glycoproteins, which are all contained in the hepatitis B virus envelope. These three proteins are the S, M, and L proteins. As described below, they are translated from different in-frame start codons within the S open reading frame. Thus, the S protein contains only the surface or S domain, while the M protein has an N-terminal 55-amino-acid extension called the pre-S2 domain, and the L protein contains an additional N-terminal 108-amino-acid extension called the pre-S1 domain.

Gene Expression

After the partially double-stranded viral DNA genome is released into the nucleus, the DNA plus strand is synthesized to completion and the gaps are repaired. In that way, a covalently closed, circular, double-stranded DNA molecule is generated. The mechanisms by which these steps are carried out are not yet clear (see below). The small, double-stranded viral DNA molecule then associates with cellular histones, thereby becoming a minichromosome, which can serve as a template for transcription. It does so as a nonreplicating, nonintegrated, plasmid-like template, which is transcribed by the cellular RNA pol II (Box 22.3).

Hepadnaviruses transcribe four classes of RNAs, which are under the control of four different RNA pol II promoters (Fig. 22.4). These four RNAs are the 3.4-kb pregenomic RNA and the three classes of subgenomic RNAs, i.e., the pre-S1, S, and X RNAs, which are 2.4, 2.1, and 0.7 kb long, respectively. Since the genome is 3.2 kb in length, the pregenomic RNA actually is longer than the genome itself, as it contains terminal redundant sequences.

The genome also contains two enhancer elements, enhancer I and enhancer II, which regulate the viral promoters. Enhancer I is adjacent to the *X* gene promoter and stimulates transcription of all viral promoters. Enhancer II is adjacent to the pregenomic RNA promoter and stimulates transcription from the pregenomic and pre-S promoters. Each of these enhancers is activated by hepatocyte-specific enhancer-binding proteins, and the promoters too are regulated by cell-type-specific transcription factors. These facts largely account for why the hepadnaviruses have a tropism for hepatocytes.

Since each of the hepadnavirus RNAs is transcribed by the cellular RNA pol II, then like all other transcripts made by this enzyme, they are capped at their respective 5′ ends by cellular capping enzymes. Moreover, all HBV transcripts are polyadenylated at their 3′ ends, and they use a common polyadenylation signal located within the region encoding the C protein.

However, in contrast to cellular mRNAs, none of the hepadnavirus RNA species is spliced, which raises the following issue. Bearing in mind that the export of mRNAs from the nucleus is associated with the splicing process (Chapter 20), how might hepadnaviral RNAs, all of which are unspliced, be exported? The answer is that each of the hepadnaviral RNAs contains a 433-nucleotide sequence, referred to as the **posttranscriptional regulatory element** (**PRE**), which allows for the export of unspliced HBV mRNAs. It is thought that the PRE enables hepadnaviral RNAs to use a cellular pathway for the export of rare, unspliced cellular RNAs. (Interestingly, the hepadnaviral PRE can substitute for the HIV Rev-responsive element to promote export of unspliced HIV mRNAs [Chapter 21].)

The pregenomic RNA, which is transcribed from the pregenomic promoter, plays multiple roles in virus replication, serving as the template for viral DNA synthesis via reverse transcription (see below), and also as the mRNA that encodes the pre-C, C, and P proteins. As noted above, the pregenomic RNA actually is somewhat longer than the circular viral genome from which it is transcribed. That is possible because of the location of the polyadenylation signal (Fig. 22.4). Because the pregenomic RNA is longer than the circular template from which it is transcribed, it contains sequences repeated at each end, reminiscent of the retrovirus long terminal repeat sequences (Chapter 20). As then might be expected, these repeated sequences enable the pregenomic RNA to fulfill its role as the template from which the viral DNA is reverse transcribed (see below).

The production of progeny hepadnavirus genomes by reverse transcription occurs in the cytoplasm. Yet early in infection, some of the progeny viral DNA molecules, which are generated in the cytoplasm by reverse transcription, are transported back to the nucleus, where they augment the number of viral DNA molecules that serve as templates for transcription of hepadnaviral RNAs. (The mechanism that enables progeny genomes to be transported back to the nucleus at early times is described below.) Since transcription of the hepadnavirus DNA genome to generate the pregenomic RNA is the first step in the generation of progeny hepadnavirus DNA genomes, the steps in the making of the pregenome are described in more detail in the following subsection.

Looking at the HBV genetic map, and bearing in mind that eukaryotic translation generally initiates at the particular AUG codon that is closest to the 5′ end of the

BOX 22.3

Retroviral transcription too is mediated by the cellular RNA pol II. However, whereas all retroviral RNAs are transcribed from the integrated DNA provirus (Chapter 20), hepadnavirus RNAs are transcribed from a nonintegrated, plasmid-like template. Consistent with that difference, hepadnaviruses, unlike retroviruses, do not encode an integrase activity. However, neither the integrated retroviral DNA provirus nor the free hepadnavirus minichromosome directly replicates to generate progeny viral genomes. Instead, each is transcribed to generate plus-strand RNA molecules, which are the progeny genomes in the case of the retroviruses and are reverse transcribed to yield progeny genomes in the case of the hepadnaviruses.

mRNA, you might be asking how the pre-C, C, and P proteins are all translated from the pregenomic RNA. The answer for the pre-C and C proteins is that the mRNAs that initiate from the genomic promoter have heterogeneous 5′ ends. On one subclass of these mRNAs, the most 5′ AUG is the translation initiation codon for the pre-C protein, and on another mRNA, the most 5′ AUG is the initiation codon for the C protein. These two proteins have very different fates. Whereas the C protein is the principal structural component of the viral core, the pre-C-specific region of the pre-C protein contains a signal sequence that targets the protein to the secretory pathway, leading to its release from the cell as the HBV e antigen, also referred to as HBeAg. Along the way, it is cleaved by a cellular signal peptidase at its 5′ end, and it is also cleaved at its 3′ end to remove a basic sequence. The role of the secreted HBeAg is described below.

We still need to account for how the P protein is translated from the C protein mRNA, particularly since the *P* gene open reading frame overlaps the 3′ end of the upstream *C* gene and is in a +1 reading frame relative to the C reading frame. Note that this aspect of the hepadnavirus gene organization is reminiscent of the retroviruses, in which the *pol* gene overlaps the upstream *gag* gene, but in which it is in a −1 reading frame relative to the *gag* reading frame. In the case of the retroviruses, a ribosomal frameshift occurs within the *gag/pol* overlap region to produce a Gag-Pol polyprotein, from which mature Pol is generated by viral protease-mediated processing (Chapter 20). Experimental evidence was sought, and none was found, which might support a ribosomal frameshift as the means by which the hepadnavirus P protein might be generated. For example, a putative C/P precursor polypeptide was looked for but could not be detected. Other known mechanisms for the translation of polycistronic mRNAs were considered as well. These included a simple leaky scanning mechanism. However, this possibility appeared to be unlikely, since there are too many AUG codons upstream from the P protein initiation site (4 in the case of HBV and 13 in the case of duck hepatitis virus). Another possibility is that translation of the P protein might initiate via an internal ribosome entry site, as in the case of the picornaviruses (Chapter 6). However, no experimental evidence was seen that might support this premise.

Although the mechanism for generating the P protein is not known with certainty for all of the hepadnaviruses, the available data for HBV are consistent with the following modified leaky scanning mechanism. (Recall that there are four AUG codons upstream from the P protein initiation site on the C protein mRNA [i.e., the pregenome RNA].) Most ribosomes scanning from the 5′ end of the pregenome RNA initiate translation at the AUG codon for the C protein. However, a *few* ribosomes bypass this start site and the next AUG codon as well, which is in a weak initiation context. These few ribosomes then initiate translation at the next AUG codon, which is in a different reading frame. From this AUG, they translate a 7-codon-long open reading frame, which enables them to bypass the next AUG, which is in a strong translation initiation context. After ribosomes complete translation of this minicistron, they are thought to resume scanning and to reinitiate translation at the AUG codon for the P protein. Having thus been translated, the P protein binds to a specific site at the 3′ end of its own mRNA (i.e., the pregenome), where it serves to prime reverse transcription (see below). Note that this mechanism results in a lower level of P expression relative to C, in keeping with the different amounts of these two proteins that are required by HBV. (Recalling that HBV has 4 AUG codons upstream from the P protein initiation site whereas the duck hepatitis virus has 13 such AUG codons, could this modified ribosomal scanning model, in which leaky scanning is combined with translation of a minicistron, work for the duck hepatitis virus?)

There are three HBV envelope proteins, the L, M, and S proteins, as described above. The L protein is generated from the dedicated pre-S1 promoter. The M and the S proteins are each generated from the S promoter. The S promoter is able to generate both the M and S proteins, since, like the pregenomic promoter, it is able to initiate transcription at multiple sites. The envelope proteins contain signal sequences, and these proteins are thus translated by ribosomes bound to the endoplasmic reticulum, and they then enter the secretory pathway. During transit of the L protein through the secretory pathway, it is modified at the amino-terminal glycine of the pre-S1 domain with a myristate residue, a step which is necessary for viral infectivity. Recall that the pre-S1 domain of the L protein is thought to be involved in binding to the HBV receptor.

Reverse Transcription

Recall that the HBV genome contained within the parental virus particle is comprised of an incomplete plus strand and a complete minus strand that is not closed. In order for the parental genome to transcribe the viral RNAs, including the pregenomic RNA, it must be repaired to generate a circular, covalently closed, double-stranded DNA molecule. Thus, this repair process is a precondition for reverse transcription. Yet the mechanisms that produce the circular, covalently closed molecule are not yet known. While it is clear that the process requires several steps, including removal of the covalently linked P protein from the minus strand of the genome, the filling of the gaps, and the ligation of the ends of both strands, it is not clear which steps might be performed by the viral P protein and which are carried out by cellular enzymes.

After the covalently closed, circular DNA genome is generated, it associates with cellular histones to become a minichromosome. Next, the pregenomic RNA is transcribed from the minus strand of the genomic DNA by the cellular RNA pol II. This transcription initiates on the genomic minus strand, approximately 6 bp upstream of DR1 (Fig. 22.4). Transcription then continues around the circular genome, terminating at the common polyadenylation signal, just downstream of DR1. Thus, the pregenomic RNA is actually about 200 nucleotides longer than the template DNA and contains a copy of DR1 at each of its ends. (Just as the retroviral terminal repeat [R] sequences play a key role in the retroviral reverse transcription process, the presence of the 11- to 12-nucleotide-long DR sequences in the hepadnaviral pregenomic RNA likewise plays a key role in the hepadnaviral reverse transcription process, as described below.)

Once the unspliced 3.4-kb pregenomic RNA has been transcribed, then, like all hepadnavirus RNAs, it is recognized via its PRE (see above) for export to the cytoplasm. In the cytoplasm, the pregenomic RNA is translated to yield the C and P proteins, also as described above.

Next, bearing in mind that the process of reverse transcription per se takes place entirely within nascent core particles, we note the following events that explain how the pregenome is encapsidated together with the reverse transcriptase (i.e., the P protein). The de novo P protein binds to the pregenomic RNA molecule from which it was translated. It does so specifically at a 100-nucleotide-long stem-loop structure called ε (epsilon), located at the 5′ end of the pregenomic RNA (Fig. 22.5 and 22.6). In order for the P protein to bind to ε, P must first interact with cellular heat shock proteins, which induce a conformational change in P that enables it to recognize and bind to the ε structure (Fig. 22.5).

Note that the ε sequence is also present at the 3′ end of all hepadnavirus transcripts, including that of the pregenomic RNA. However, the pregenomic RNA is the only hepadnavirus RNA species in which ε also is present at the 5′ end. Importantly, and for reasons not yet clear, P promotes encapsidation only when it is bound to the ε sequence at the 5′ end of the pregenomic RNA. Moreover, the binding of P to the 5′ ε sequence also initiates reverse transcription. In that regard, the P protein serves as the primer, as well as the polymerase (see below). (ε was originally identified as the RNA encapsidation signal but later was also found to be the origin of reverse transcription.)

Noting that encapsidation of the pregenome is a precondition for reverse transcription, consider the following points regarding the C protein, which encapsidates the pregenome. The C protein contains within its 183 amino

Figure 22.5 Model for the assembly of hepadnavirus nucleocapsids. (A) TP denotes the N-terminal domain of the P protein, which harbors the primer tyrosine residue for the initiation of reverse transcription (see the text). RT denotes the polymerase and RNase H domains of the P protein. The heat shock protein 90 (Hsp90) complex is shown together with the chaperone partner, p23, and possible additional proteins (X) that may be components of the complex (as described by Hu et al., 1997). The function of Hsp90 also depends on the ATPase Hsp70 and its partner Hsp40. Interaction of the polymerase with the chaperone complex induces a conformational change (B) that triggers the binding of the P protein to ε RNA (C), which in turn provides the signal for nucleocapsid assembly and initiation of viral DNA synthesis (D). A detailed description of this pathway is provided by Hu et al., 1997. Modified from Hu et al., *EMBO J.* **16**:59–68, 1997, with permission.

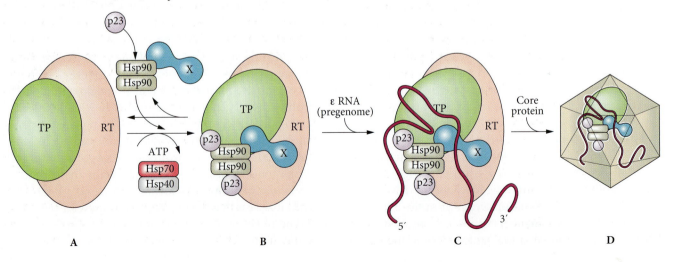

acids two domains that facilitate core formation. These domains are (i) the 140-amino-acid, N-terminal domain that is responsible for the formation of core protein dimers and their assembly into capsid structures; and (ii) the C-terminal region that is rich in arginine residues that have nucleic acid-binding activity. The capsid is assembled from either 90 or 120 dimers of the core protein, and assembly is thought to occur as the C protein dimers recognize the pregenomic RNA/polymerase complex. Also note that the core contains only one copy of the pregenomic RNA and one copy of the polymerase protein. Encapsidation in turn triggers the initiation of reverse transcription, which takes place entirely within the icosahedral capsid structures. The nascent particle-associated P protein contains all the activities required for reverse transcription.

As noted previously, it has been difficult to study HBV replication directly, because in vitro systems that might support HBV replication have not been available. Thus, the description of reverse transcription that follows is that for duck hepatitis virus. The essential features of this process are presumably very similar among all members of the hepadnavirus family. As you soon will see, the process is not what one might call straightforward. If you are wondering why that is so, bear in mind the common problem that exists whenever a DNA molecule is generated: how to produce its ends (see Box 14.4).

From its position at the ε sequence, at the 5′ end of the pregenomic RNA, the P protein initiates reverse transcription by incorporating the first four nucleotides of the DNA minus strand (Fig. 22.6, step 1). Note that the P protein itself primes the reverse transcription of the DNA minus strand. (Adenoviral DNA synthesis likewise is primed by a protein. In that instance, the primer is the precursor to the adenoviral terminal protein [Chapter 17; Box 22.4].) In order for the P protein to carry out its primer function, the N-terminal domain of the P protein (sometimes misleadingly referred to as the **terminal protein**) contains a tyrosine residue that covalently binds the first nucleotide of the DNA minus strand. Consequently, the short four-nucleotide-long DNA segment that is transcribed from the ε sequence at initiation is likewise covalently bound to the P protein. Indeed, the P protein remains covalently attached to the 5′ end of the nascent DNA minus strand throughout the entire course of the reverse transcription process, and it remains so attached in the mature virus particle.

Notice that by initiating reverse transcription from a position near the 5′ end of the pregenomic RNA, the polymerase soon would run out of template if it were to continue reverse transcribing to the 5′ end of the pregenomic RNA. This potential dilemma is avoided by the translocation of the first four nucleotides of the nascent DNA minus strand and the covalently attached P protein to the DR1 sequence near the 3′ end of the pregenomic RNA (Fig. 22.6, step 2), a step referred to as the **first template exchange**. This template exchange is possible because the DR1 at the 3′ end of the pregenomic RNA contains the same tetranucleotide sequence as that in the ε sequence from which the first four nucleotides of the DNA minus strand were reverse transcribed. The mechanism that underlies this template exchange is not known, but it is thought to depend on the organization of the template within the core particles. One then might ask why reverse transcription does not simply initiate at the ε sequence at the 3′ end of the pregenome. The answer is not clear since, for unknown reasons, only the 5′ ε sequence of the pregenomic RNA is able to serve as the initiation site for reverse transcription (Box 22.4).

Following the first template exchange to the 3′ DR1 sequence, reverse transcription of the DNA minus strand then continues all the way to the 5′ end of the pregenomic RNA molecule. As reverse transcription continues, the pregenomic RNA template is degraded by the RNase H activity of the P protein (Fig. 22.6, steps 3 and 4). (Recall that RNase H is a special RNase that degrades the RNA component of RNA/DNA heteroduplexes.) In this regard, as in the case of the RNase H activity of retroviral reverse transcriptases, the RNase H activity of the hepadnaviral P protein likewise operates 15 to 18 nucleotides behind the DNA polymerase activity of the protein. Consequently, the 5′ end of the capped pregenomic RNA template is spared from RNase H digestion (Fig. 22.6, step 4).

The spared pregenomic RNA sequence will play a crucial role in DNA plus-strand synthesis. But first, notice that the nucleotides at the 5′ end of the pregenomic RNA, which were spared from RNase H digestion, contain the DR1 sequence. Second, notice that a sequence that is complementary to the 5′ end of the pregenomic RNA sequence also is present near the 5′ end of the new DNA minus strand. That is so because sequences near the 5′ end of the new DNA minus strand were reverse transcribed from the DR2 sequence (which is identical to DR1) at the 3′ end of the pregenomic RNA. Consequently, the spared pregenomic RNA is able to undergo the second template exchange, in which it base pairs to the DR2 sequence near the 5′ end of the DNA minus strand (Fig. 22.6, step 5). From that position, the spared pregenomic RNA sequence carries out its crucial function as the primer of DNA plus-strand synthesis (Figure 22.6, step 6). The mechanism promoting this template exchange also is not yet known. However, it may be facilitated in part by the presumptive ability of a portion of the complementary sequence at the 3′ end of the DNA minus strand to form a hairpin loop, the result of which would be to displace the 3′ end of the

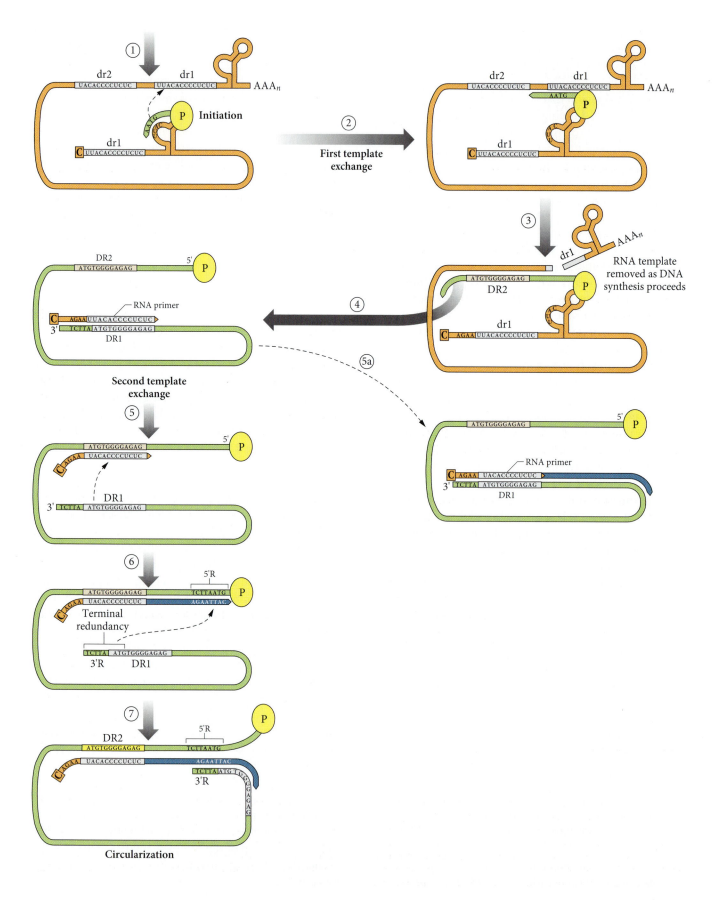

①

dr2　　　　　dr1
UACACCCCUCUC　　UUACACCCCUCUC　　AAAₙ

P　Initiation

dr1
C UUACACCCCUCUC

②

First template
exchange

dr2　　　　　dr1
UACACCCCUCUC　　UUACACCCCUCUC　　AAAₙ
AATG　P

dr1
C UUACACCCCUCUC

③

dr1　AAAₙ
RNA template
removed as DNA
synthesis proceeds

ATGTGGGGAGAG
DR2　　P

dr1
C AGAA UUACACCCCUCUC

DR2
ATGTGGGGAGAG　　5' P

④

RNA primer
C AGAA UUACACCCCUCUC
3' TCTTA ATGTGGGGAGAG
DR1

Second template
exchange

⑤a

ATGTGGGGAGAG　　5' P

RNA primer
C AGAA UACACCCCUCUC
3' TCTTA ATGTGGGGAGAG
DR1

⑤

ATGTGGGGAGAG　　5' P
C AGAA UACACCCCUCUC

3' TCTTA ATGTGGGGAGAG
DR1

⑥

ATGTGGGGAGAG　　5'R
TCTTAATG　P
C AGAA UACACCCCUCUC　AGAATTAC

Terminal
redundancy

TCTTA ATGTGGGGAGAG
3'R　DR1

⑦

DR2　　　　5'R
ATGTGGGGAGAG　　TCTTAATG　P
C AGAA UACACCCCUCUC　AGAATTAC
TCTTAATG TGGGGAGAG
3'R

Circularization

682

BOX 22.4

The initiation of hepadnaviral reverse transcription near the 5′ terminus of the RNA template is reminiscent of the initiation of retroviral reverse transcription, which likewise occurs near the 5′ terminus of the RNA template, and which similarly is followed by a template exchange reaction (Chapter 20). On the other hand, whereas retrovirus reverse transcription is primed by a cellular transfer RNA (tRNA), hepadnaviral reverse transcription is primed by the polymerase itself.

Initiating DNA synthesis by means of a protein primer (or, for that matter, from a tRNA primer in the case of the retroviruses) has the advantage of simplicity, in comparison to priming via RNA synthesis. Nevertheless, hepadnaviruses and adenoviruses (Chapter 17) are the only DNA animal viruses that are known to use proteins to prime DNA synthesis. Still, the hepadnaviruses remain unique, since only in their case are the protein primer and the polymerase one and the same protein.

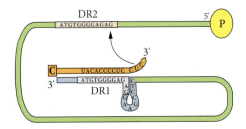

Figure 22.7 Why does the second template exchange occur (Fig. 22.6, step 5) more often than the generation of a linear duplex molecule from a fixed primer (Fig. 22.6, step 5a)? The formation of a small DNA hairpin, which overlaps the 5′ end of DR1 in the minus strand, contributes to the regulation of primer translocation, by making the 3′ end of the minus-strand DNA a poor template for initiation. Modified from J. J. Habig and D. D. Loeb, *J. Virol.* **76:**980–989, 2002, with permission.

As synthesis of the DNA plus strand proceeds from the translocated RNA primer, it soon reaches the 5′ end of the DNA minus-strand template (Fig. 22.6, step 6). Thus, another template exchange is needed to enable DNA plus-strand synthesis to carry on. The template exchange in this instance is facilitated by complementarity between the AGAAT sequence near the 3′ end of the nascent DNA plus strand (i.e., at the point where it runs out of template) and the TCTTA nucleotide sequence at the 3′ end of the DNA minus strand (Fig. 22.6, steps 6 and 7). That complementarity exists because the pregenomic RNA is longer than the circular DNA template from which it is transcribed and, thus, contains sequences that are repeated at each end.

The translocation of the 3′ end of the nascent DNA plus strand to the 3′ end of the DNA minus strand (Fig. 22.6,

spared pregenomic RNA segment (Fig. 22.7). Indeed, directed mutations, which disrupt base pairing in this hairpin, lead to increased levels of linear duplexes (Fig. 22.7). Thus, as in the case of the first template exchange, the conformational organization of the nucleotide chains within the core particles may also facilitate the second template exchange. In any event, the stage is now set for the synthesis of the DNA plus strand.

Figure 22.6 Key steps in the reverse transcription of the hepadnavirus pregenomic RNA. The entire process of reverse transcription takes place within the nascent core particle. From its position at the ε sequence, at the 5′ end of the capped and polyadenylated pregenomic RNA, the polymerase initiates reverse transcription by incorporating the first four nucleotides of the DNA minus strand (step 1). Note that the hepadnaviral polymerase itself primes the reverse transcription of the DNA minus strand. For this purpose, the N-terminal domain of the P protein contains a tyrosine residue that covalently binds the first nucleotide of the DNA minus strand, and consequently, the P protein remains covalently attached to the 5′ end of the nascent DNA minus strand during the entire remainder of the reverse transcription process, and it remains so attached in the mature virus particle. The first four nucleotides of the nascent DNA minus strand and the covalently attached P protein translocate to the DR1 sequence near the 3′ end of the pregenomic RNA (step 2), a step referred to as the first template exchange. Reverse transcription of the DNA minus strand then continues all the way to the 5′ end of the pregenomic RNA. As reverse transcription continues, the copied pregenomic RNA template is degraded by the RNase H activity of the P protein (steps 3 and 4). However, since the RNase H activity of the hepadnaviral P protein operates 15 to

18 nucleotides behind the DNA polymerase activity of the protein, the 5′ end of the capped pregenomic RNA template is spared from RNase H digestion (step 4). The spared pregenomic RNA then undergoes a second template exchange, in which it base pairs to the DR2 sequence near the 5′ end of the DNA minus strand (step 5). From that position, the spared pregenomic RNA sequence serves as the primer of DNA plus strand synthesis (step 6). When synthesis of the DNA plus strand reaches the 5′ end of the DNA minus-strand template (step 6), another template exchange occurs to enable DNA plus-strand synthesis to carry on (step 7). This translocation of the 3′ end of the nascent DNA plus strand to the 3′ end of the DNA minus strand gives rise to a reverse transcription intermediate that now has a circular molecular structure. However, synthesis of the DNA plus strand terminates before it is completed. Step 5a depicts an alternative pathway for generating the DNA plus strand, in which the RNA primer remains in place and DNA plus-strand synthesis ensues to produce a linear duplex molecule. A circular molecule is then generated by means of nonhomologous recombination. Linear duplex molecules are generated only a small fraction of the time (Fig. 22.7). Modified from J. J. Habig and D. D. Loeb, *J. Virol.* **76:**980–989, 2002, with permission.

step 7) gives rise to a reverse-transcription intermediate that now has a circular molecular structure. Consequently, reverse transcription may resume and continue around the circular molecule. Still, the molecular mechanisms underlying these events indeed are mysterious, particularly so since the polymerase remains covalently bound to the 5′ end of the DNA minus strand, which needs to be displaced to permit the template exchange. Once again, specific conformational adjustments of the nucleic acids within the core would seem to be needed.

Note that synthesis of the DNA plus strand terminates before its completion, again for reasons that are not yet known with certainty. Perhaps maturation of the nascent virus particle might somehow be responsible for the premature termination of DNA plus-strand synthesis. One possibility is that DNA plus-strand synthesis induces a conformational change in the core proteins, which in turn triggers envelopment. Once the viral core is enveloped, deoxyribonucleoside triphosphate precursors no longer would be able to enter the particle, so that further DNA synthesis would not be possible.

Despite the fact that several important details of the hepadnavirus reverse-transcription process remain obscure, the scheme outlined above accounts for the following key distinctive features of the hepadnaviral genome: (i) the DNA minus strand contains a gap; (ii) its 5′ terminus overlaps the circular genome; (iii) it is covalently bound to the P protein; (iv) the incomplete DNA plus strand is covalently linked to a short, capped RNA sequence at its 5′ end; and importantly, (v) the conservation of the ends of the genome and of the DR sequences is accounted for.

An alternative pathway, which also leads to production of the DNA plus strand, is depicted in Fig. 22.6, step 5A. In this instance, the RNA primer remains in place, and DNA plus-strand synthesis ensues to produce a linear duplex molecule. A circular molecule can be generated from this structure. However, circularization in this instance is accomplished by means of nonhomologous recombination. This pathway is not likely to be an efficient means for generating infectious circular genomes, since nonhomologous recombination is inherently inefficient and, moreover, is characterized by a high level of mutations at the recombinational joints. Thus, this pathway will yield mostly defective genomes. Perhaps for that reason, 90% or more of the hepadnavirus reverse-transcription reactions involve the template exchanges depicted in Fig. 22.6, steps 6 and 7. In fact, a mechanism may be in place to ensure that the template exchange pathway is favored. That is, the formation of a small DNA hairpin that overlaps the 5′ end of DR1 in the DNA minus strand makes the 3′ end of the minus-strand DNA a poor template for initiation (Boxes 22.5 and 22.6).

The Foamy Viruses: Reprise

The foamy viruses, also known as spumaviruses, are described in Chapter 20. These viruses have a unique replication pathway, with key features that appear to have been taken from both the retroviruses and hepadnaviruses. Foamy viruses resemble retroviruses in that they carry out reverse transcription to generate a DNA provirus, which is stably integrated into the host genome. Also like the retroviruses, all foamy virus RNAs are then transcribed from the integrated provirus by the cellular RNA pol II. (Recall that hepadnavirus genomes remain nonintegrated, although they, too, are transcribed by pol II.) However, in contrast to the retroviruses, but like the hepadnaviruses, foamy viruses package a pregenomic RNA, and reverse transcription then occurs in nascent virus particles, so that like in the case of the hepadnaviruses, the actual foamy virus genome is DNA rather than RNA.

Might foamy viruses be an evolutionary link between retroviruses and hepadnaviruses? What other evolutionary trees might you envision relating these viruses?

Maturation and Release

The events enumerated above result in hepadnavirus DNA genomes encapsidated within cores that are comprised of the C protein. Next, the hepadnaviruses acquire their envelopes by budding into the endoplasmic reticulum. The means by which this envelopment step is triggered is not known with certainty. One possibility, noted above, is that a late stage of reverse transcription may elicit a conformational change in the associated core proteins, which in turn prompts envelopment of the core. In any case, as a general principle, viruses that acquire envelopes from internal cellular compartments generally do so via the interaction between viral structural components associated with the viral genomes and virus envelope proteins that have been targeted to the particular intracellular membrane. Thus, it is thought that when sufficient levels of the hepadnavirus L glycoprotein have accumulated in the endoplasmic reticulum membrane, then the activated viral cores can interact with those L proteins, thereby bringing about budding into the endoplasmic reticulum. Next, the now-enveloped virus particles might be packaged within cellular transport vesicles for transport via the secretory pathway for release from the cell.

The premise that envelopment might be triggered by a late stage of reverse transcription is attractive because it neatly accounts for why only those cores that have sufficiently completed the reverse transcription of the pregenome are able to bud into the endoplasmic reticulum and enter the secretory pathway. A possible advantage to the virus of budding into an internal cellular compartment rather than budding from the plasma membrane is that it

BOX 22.5

The discovery that hepadnaviruses replicate their DNA via the bizarre reverse transcription of an RNA intermediate was prompted by the rather strange finding of Jesse Summers and William Mason, in 1982, that the hepadnaviral DNA genome is not totally double stranded but instead contains a large single-stranded gap of variable size. This finding caught the attention of Summers and Mason, since it is not consistent with semiconservative DNA synthesis. Moreover, hepadnavirus particles were found to contain an endogenous DNA polymerase activity. This too seemed strange, since virus particle-associated polymerases generally are needed only to carry out an early step in infection that cannot be carried out by an existing cellular enzyme. For example, reoviruses contain an RNA transcriptase to transcribe mRNAs from their double-stranded RNA genomes, a process that cannot be mediated by any host enzyme. In contrast, DNA viruses usually are able to use the cellular DNA polymerase to replicate their genomes and the cellular RNA pol II to transcribe them. This line of thought suggested that the hepadnaviral polymerase might have unique roles in replication, in addition to serving as a DNA-dependent DNA polymerase. What is more, the replication intermediates isolated from hepadnavirus-infected cells also were unusual. These intermediates included mainly DNA minus strands that were less than unit length. Moreover, only a few of these minus strands were associated with plus strands.

Summers and Mason recognized that these peculiarities might be accounted for by replication via reverse transcription of an RNA intermediate. Indeed, the hepadnavirus genome structure itself resembled one of the postulated intermediates in the synthesis of a retroviral DNA provirus, which is reverse transcribed from an RNA template (Chapter 20). Drug sensitivity studies also yielded experimental results consistent with reverse transcription. For example, synthesis of the hepadnavirus minus DNA strand was *insensitive* to the drug actinomycin D. Since this drug acts by intercalating between the stacked bases of double-stranded DNA molecules, thereby blocking transcription of double-stranded DNA, the experimental result implied that the DNA minus strand was not transcribed from a double-stranded DNA template. In contrast, synthesis of the plus strand was inhibited by actinomycin D, implying that the plus strand is synthesized using a double-stranded nucleic acid template. Synthesis of the minus and plus strands of the retroviral DNA provirus was earlier shown to have the same respective sensitivities to actinomycin D as the hepadnaviral minus and plus DNA strands.

---

does not expose viral envelope proteins at the cell surface, where they might be recognized by host immune defenses.

Early in infection, some of the progeny nucleocapsids are transported back to the nucleus. Their encapsidated genomes then are released and converted to the covalently closed configuration. In that way, a pool of viral DNA molecules is generated in the nucleus, which amplifies viral transcription. This is clearly beneficial to the virus. It happens because at early times during infection, the concentration of the L protein in the endoplasmic reticulum membrane is too low to facilitate efficient envelopment of the cores. Rather than being inefficiently enveloped at those times, the cores instead deliver their mature genomes to the nucleus, where they might enhance the efficiency of the replication cycle (Box 22.7).

ROUTES OF TRANSMISSION AND PATHOGENESIS

The routes of HBV transmission and the mechanisms of pathogenesis are described in some detail in Chapter 3. In brief, HBV is spread mainly via blood, but the virus also may be present in sweat, tears, saliva, semen, vaginal secretions, menstrual blood, and breast milk. As expected, transmission is efficient between intravenous drug users who share needles and syringes, and the virus indeed is prevalent in this group. Prior to 1975, and the introduction of blood screening, HBV was transmitted by transfusions of contaminated blood. The virus can also be transmitted by acupuncture, body piercing, and tattooing. Medical personnel are at risk from inadvertent transmission via a needle stick with contaminated blood. Individuals with multiple sex partners also are at high risk. Indeed, the only natural routes of infection are through very close personal contact involving exposure to blood, semen, vaginal secretions, and saliva. A most import means of transmission is from a chronically infected mother to her child at birth or via breast milk (see below).

All of the hepadnaviruses have a restricted host range and display a marked tropism for the liver. HBV infection may be either acute or chronic and either symptomatic or asymptomatic. Chronic HBV infection is characterized by persistent viremia, in which total HBV production in an infected individual can be as high as 10^{11} virus particles per day, and this rate of virus production can continue for

BOX 22.6

We noted earlier that the cellular APOBEC3 cytidine deaminase proteins are potent inhibitors of simple retroviruses and retrotransposons (Chapter 20) and of HIV in the absence of its Vif protein (Chapter 21). In those instances, APOBEC3 proteins preferentially attack and deaminate deoxycytidines on the minus strand of reverse-transcribed retroviral DNA. The resulting deoxuridines then may be excised by the cellular uracil-DNA glycosylase, an enzyme that specifically removes the RNA base uracil from DNA. The double-stranded DNA product of reverse transcription then becomes a potential target for endonuclease digestion.

Recent experimental results show that APOBEC3 proteins also, to some extent, control HBV infection (Jost et al., 2007). However, the mechanisms by which they do so are no clearer than they are in the cases of the retroviruses and retroelements. Whereas APOBEC3B inhibits viral core-associated HBV DNA synthesis, it also inhibits HBV gene expression, impairing synthesis of both HBsAg and HBeAg. APOBEC3B appears to express this effect indirectly, via its interaction with the heterogeneous nuclear ribonucleoprotein K

(hnRNP K), which HBV uses as a positive regulator of *S* gene transcription. The interaction of APOBEC3B with hnRNP K may thus inhibit *S* gene transcription. In addition, APOBEC3B directly suppresses the activity of the HBV *S* gene promoter.

Interestingly, other recent experimental results show that APOBEC3 proteins also inhibit cytomegalovirus and simian virus 40 promoter-mediated gene expression, suggesting that these proteins may have even more widespread antiviral activity. Consistent with that notion, yet other results show that type I and type II IFN treatment induces the production of APOBEC3 proteins in human hepatocytes (Zhang et al., 2008).

References

Jost, S., P. Turelli, B. Mangeat, U. Protzer, and D. Trono. 2007. Induction of antiviral cytidine deaminases does not explain the inhibition of hepatitis B virus replication by interferons. *J. Virol.* **81:** 10588–10596.

Zhang, W., X. Zhang, C. Tian, T. Wang, P. T. Sarkis, Y. Fang, S. Zheng, X. F. Yu, and R. Xu. 2008. Cytidine deaminase APOBEC3B interacts with heterogeneous nuclear ribonucleoprotein K and suppresses hepatitis B virus expression. *Cell. Microbiol.* **10:**112–121.

---

years. The outcome of the infection (chronic versus self-limited and symptomatic versus clinically unapparent) may be determined by several factors, as discussed below (Box 22.8).

Such high levels of virus production, over indefinite periods, in a crucial organ like the liver, would not be possible if HBV were cytopathic. Indeed, the noncytopathogenicity of the virus is thought to be a factor enabling it to establish and maintain a chronic infection (Chapter 3).

BOX 22.7

For the sake of comparison, the "clever" amplification of the pool of DNA templates on the part of the hepadnaviruses does not occur in the case of the retroviruses, since the retroviruses generate but a single copy of the DNA provirus, which then is integrated into the cellular DNA. All progeny retroviral genomes are then generated by reverse transcription of the integrated provirus, while the provirus is replicated exclusively in step with the host DNA as an integrated nuclear form.

Since HBV is noncytopathic in hepatocytes, what underlies the liver damage and cirrhosis that may result from infection by this virus? HBV provides an important example wherein *host immunity is largely responsible* for viral pathogenesis. During the many years of chronic HBV infection, virus-specific cytotoxic T lymphocytes (CTLs) infiltrate the liver and kill virus-infected cells. In addition, the vast amounts of virus that are produced and released into the bloodstream may lead to the depositing of immune complexes in the kidneys, causing immune complex disease in that organ.

Despite the fact that CTLs are the major effectors of HBV-induced liver disease, bear in mind that these cells normally play a key role in the control of the HBV infection in many individuals. Moreover, and importantly, CTLs also have noncytolytic means by which to control the infection, such as via the antiviral effects of cytokines like gamma interferon (IFN-γ) and tumor necrosis factor alpha, which CTLs generate and release. See Chapter 4. ·

The ability of CTLs to express noncytolytic defenses against virus infections is thought to be especially important in vital organs such as the liver, where it is imperative to avoid widespread tissue destruction. Now, consider the likelihood that the number of HBV-infected hepatocytes in the liver is vastly greater than the number of virus-specific

BOX 22.8

The long-term persistent viremia that may characterize some HBV infections was once thought to be exceptional, since viremias were thought to be of generally limited duration. However, other viruses are now known to also give rise to persistent viremias. HIV (Chapter 21) is a prime example. Other examples include hepatitis C virus and hepatitis G virus (a flavivirus like hepatitis C virus, but only loosely related to it). The rate of HBV production in a chronically infected individual (as high as 10^{11} virus particles per day) is rather astonishing. But recall that similar rates of HIV production are seen in HIV-infected individuals.

CTLs in that organ. Thus, the host could probably not control the infection if its only means of doing so was via the direct killing of infected cells by CTLs. However, the ability of CTLs to produce effective antiviral cytokines may enable them to control the infection in a much larger number of infected cells. Still, the CTL-mediated destruction of hepatocytes over time does result in pathogenesis. Moreover, tissue damage may also result from the inflammation induced by the cytokines released by CTLs and by other immune effector cells recruited to the infected sites by those cytokines.

Note that recovery from an acute HBV infection also depends on a strong B-cell response, in which antibodies are produced against the viral envelope glycoproteins. In contrast, a strong T-cell response needs to be directed against multiple human leukocyte antigen (HLA)-restricted epitopes (Chapter 4) in the viral C and P proteins as well. Interestingly, the HBV-specific CTL response is maintained for decades after patients have apparently recovered from an acute infection, suggesting that complete elimination of the infection is rarely achieved. In contrast to those patients who appear to recover from an acute infection, those who are chronically infected appear to mount only a weak CTL response, directed against one or a few epitopes.

Considering that HBV is able to establish long-term persistent infections, you may well have correctly expected that it employs several strategies for evading host immune defenses. One notable evasion strategy used by the virus is to generate large numbers of noninfectious "decoy" particles, which are comprised of HBsAg and bind to, and block, the action of neutralizing antibodies. Indeed, the serum of an infected person may contain 1,000 to 1,000,000 empty decoy particles for every particle that contains the viral nucleocapsid.

Another immune evasion strategy used by HBV is to impair the expression of IFNs, which, as noted above, play an important role in the noncytolytic, CTL-mediated host defense against the virus. The exact mechanism by which HBV impairs IFN expression is not known. One possibility is that the viral C protein inhibits IFN expression by means of its ability to alter cytokine expression in general. The means by which the C protein affects cytokine expression is not known, but there is experimental evidence that it may occur at the transcriptional level. Regardless of the underlying mechanism, IFN levels are suppressed in the serum of chronic HBV carriers. Other experimental results suggest that the viral P protein inhibits the activities of the STAT proteins of the JAK/STAT signaling pathway (Chapter 4; Fig. 4.2), which could lead to the inhibition of IFN-α and IFN-β signaling. Furthermore, the HBV X protein inhibits expression of the p53 tumor suppressor gene, thereby impairing p53-mediated apoptosis and possibly accounting in part for the development of HBV-associated hepatocellular carcinoma (see below).

Notwithstanding HBV immune evasion strategies, host immune defenses appear to eventually resolve the infection in more than 90% of all adult cases. Many of these individuals experience few, if any, symptoms, and all have lifelong protective immunity. In the cases of the remaining infected adults, there is sufficient liver damage to cause the characteristic nausea, jaundice, and liver pain associated with hepatitis. Nevertheless, in some of these individuals, eventual resolution of the infection allows the parenchyma to regenerate. But severe infections, activation of chronic infections, or coinfection with the delta agent (see below) can lead to permanent liver damage and cirrhosis. Still, it is not entirely clear why so many individuals escape disease when infected with this potentially virulent virus. Consider the following example. During the Second World War, approximately 45,000 American soldiers were accidentally infected with HBV when they were injected with a contaminated yellow fever vaccine. Yet only about 900 of these soldiers, or 2%, developed clinical hepatitis, and only 36 of these hepatitis cases were deemed serious.

Age at the time of infection is one factor that affects the clinical outcome. For instance, only about 3% of individuals infected as adults become chronically infected, while nearly all infected newborns go on to become chronic carriers themselves. The difference in outcome between adults and newborns might be due to differences in the maturity of their immune systems. Consistent with that premise (and also the notion that host immunity is largely responsible for viral pathogenesis), newborns suffer less tissue damage and have milder symptoms than individuals contracting the infection as adults. Indeed, the status of the host's immune system at the time the individual is exposed to the virus likely affects the clinical outcome in general.

Females who are chronically infected as newborns will one day be able to efficiently pass their chronic infections on to their offspring. Importantly, transmission from a chronically infected mother to her child at birth is the major *natural* route of HBV transmission. Thus, since nearly all infected newborns go on to become chronic carriers themselves, HBV has evolved a very efficient means to ensure its maintenance in the human population. (Mother-to-child transmission of HBV can be largely prevented by vaccinating the newborn, which is an important reason for screening expectant mothers for HBV infection [see below].)

Clinical manifestations of chronic HBV infections may range from an asymptomatic carrier state to chronic hepatitis. Importantly, chronic hepatitis may progress to potentially fatal cirrhosis and usually fatal hepatocellular carcinoma. (As noted above, the clinical outcome is affected by the age at infection and the immune status of the individual.) It also is noteworthy that the latency period for the development of hepatocellular carcinoma is between 9 and 35 years. The possible molecular mechanisms that might underlie this neoplasm are considered below.

Hepatocellular Carcinoma

One million or more long-term HBV carriers die each year from hepatocellular carcinoma. Yet for several reasons, the underlying mechanism by which HBV induces this neoplasm in humans is not understood with any certainty. First, HBV-induced cancer develops only after a very long latent period, measured in decades. Consequently, there is no particular "smoking gun" that might be identified. Second, HBV replicates only in humans, so that there is no animal model in which to study hepatocellular carcinoma induced by HBV per se. Woodchucks that have been infected with the woodchuck hepatitis virus have served as a model system for human hepatocellular carcinoma. However, the woodchuck system is far from ideal, since it is generally characterized by acute infection. (It is possible to experimentally induce a chronic infection in woodchucks, which might lead to neoplasia, by treating the animals with immunosuppressive drugs.) Despite such experimental obstacles, likely mechanisms that might underlie HBV-induced neoplasia have been considered. Current understanding of this malignancy is summarized below.

No HBV gene has unequivocally been shown to have transforming activity. Instead, the development of hepatocellular carcinoma is believed to result from the extensive cell proliferation occurring in the liver, in that organ's attempt to replenish the cells that are continually being destroyed by the immune response to the infection. This extensive cell proliferation is thought to increase the risk that a cell will acquire cancer-causing mutations. The chronic inflammation or cirrhosis of the liver also may be a factor, possibly because it is associated with potentially mutagenic, high, local concentrations of superoxides and free radicals. Indeed, hepatocellular carcinoma in the absence of cirrhosis is a relatively rare event.

On the other hand, all HBV-associated liver cancers contain at least fragments of integrated viral DNA; this finding is consistent with the possibility that a viral gene product in some way promotes cancer development. The HBV X protein, in particular, has been implicated in that regard, since it is expressed in many human hepatocellular carcinomas and because it stimulates expression of many cellular genes, including proto-oncogenes. Moreover, the viral X protein appears to inhibit the p53 tumor suppressor activity. (There are reports that the X protein may inhibit the p53 tumor suppressor gene at the level of transcription. Also, there are reports that the X protein may directly interact with p53 to inhibit its function, and recent reports from laboratories in China state that the X protein blocks p53-induced apoptosis by activating cyclooxygenase-2. Cyclooxygenase-2 is known to inhibit apoptosis. One way in which it may do so is by increasing the level of Bcl-2 [Chapter 4]. Moreover, it is angiogenic [i.e., it promotes the vascularization of tumors, thereby enhancing tumorigenesis] and it is activated in a range of neoplasms, including lung, breast, colon, and pancreatic cancers. Cyclooxygenase-2 inhibitors may have therapeutic potential against a variety of cancers, including hepatocellular carcinomas.)

The HBV X protein does not bind to DNA. Therefore, it is thought to affect cellular gene expression via protein-protein interactions involving transcription factors such as AP-1 and also by activation of signaling via NF-κB and other signal transduction pathways. For instance, the X protein activates Src and downstream mitogen-activated protein kinase cascades. In addition, the X protein binds to the ultraviolet (UV)-damaged DNA-binding protein, thereby inhibiting its DNA repair function and perhaps accelerating the accumulation of mutations in cellular genes that regulate cell proliferation. Regardless, the exact function, if any, of the X protein in the development of human hepatocellular carcinoma remains controversial. Furthermore, studies with transgenic mice show that X gene expression alone is not sufficient for carcinogenesis in that system. However, transgenic mice that express the HBV X protein are more susceptible to carcinogen-induced liver malignancies.

As noted above, all HBV-associated liver cancers contain at least fragments of integrated HBV DNA. The integrated HBV DNA may encode two envelope glycoproteins, which, like the X protein, are also transcriptional activators of cellular proto-oncogenes. One of these proteins is the L protein, and the other is a C-terminally truncated version of the M protein. Each of these proteins contains the pre-S2 domain, which is the factor responsible for their transcriptional activator function. These truncated M proteins are found only in liver cells that carry integrated HBV DNA.

The presence of integrated HBV genome fragments in neoplastic cells raises the possibility that oncogenesis might result from the insertional activation of cellular proto-oncogenes. Indeed, in the woodchuck model, over 90% of hepatocellular carcinomas contain integrated woodchuck hepatitis virus DNA, positioned so as to activate the *myc* proto-oncogene (Chapter 20). Nevertheless, in humans, HBV DNA does not integrate at specific sites in the host genome, thereby diminishing the likelihood that integrated virus DNA might induce neoplasia via a *cis* effect on a particular flanking cellular gene. Moreover, there is no evidence for the insertional activation by HBV of any of the common proto-oncogenes. Additionally, some hepatocellular carcinomas from patients may not carry any HBV sequences, raising the possibility that HBV is necessary to initiate, but not maintain, the malignancy.

Regardless of which, if any, of the above propositions might be correct, the long period of chronic HBV infection that precedes the development of human hepatocellular carcinoma is consistent with the likelihood that this neoplasm ultimately results from the accumulated effects of several low-probability events.

HBV Vaccine

The HBV surface antigen self-associates into 22-nm particles, which may be either circular or filamentous. These particles are released from cells and accumulate in the blood of chronically infected individuals, reaching levels as high as 10^{13} per ml. As noted above, these particles are noninfectious and act as immune decoys. These points are reiterated here because these S-protein-containing particles, prepared from the plasma of chronically infected individuals, were used in the original HBV vaccine.

Although the S-protein-containing particles are noninfectious, there was the danger that samples of the original vaccine might inadvertently contain infectious virus, since they were prepared from the plasmas of chronically infected individuals. The vaccine was prepared that way because the virus cannot be grown in vitro, nor does it have a nonhuman host in which it might have been grown.

The current vaccine was genetically engineered. It is produced by a plasmid that expresses the HBV *S* gene in the yeast *Saccharomyces cerevisiae*. Such a recombinant vaccine has the advantage that there is no danger of it containing any infectious virus.

Since the current vaccine contains a single virus component rather than the entire virus, it is referred to as a **subunit vaccine**. The *S* gene product is the virus component of choice, since it is expressed on the surfaces of virus particles and contains the virus's neutralization epitopes. Moreover, since the *S* gene product spontaneously assembles into virus-like particles, it induces a stronger protective response than might be induced by monomeric virus components. (Why is that so?) The recombinant vaccine has been refined over the years. More recently, the *pre-S1* and *pre-S2* gene sequences are also expressed, since they enhance the immunogenicity of the virus-like particles that form.

The current vaccine is given in a series of three injections. It is highly effective, such that more than 95% of those receiving all three injections develop protective antibodies. It is recommended for all infants, particularly in countries of the developing world where the carrier rate is high. Indeed, routine immunization of newborns against HBV is standard practice in over 80 countries. Vaccination is recommended for children, and especially for individuals in high-risk groups. Immunized mothers are less likely to transmit the infection to their newborns and to their older children as well. This is significant from a public health standpoint, since it reduces the number of chronically infected individuals (see above), who are a major source of the infection. But an important proviso is that the duration of the protection afforded by the vaccine is still unknown.

The efficacy of the vaccine may be explained in part by the facts that there is only one HBV serotype and that the virus has no animal reservoir.

More recently, an HBV recombinant DNA vaccine has been under development. The potential of this vaccine was demonstrated by its ability to protect chimpanzees against viral challenge. A would-be advantage of the DNA vaccine is that it generates virus gene products within cells of the vaccinated host, thereby inducing cellular as well as humoral immunity.

HBV Therapies

The primary aim of HBV therapy is to suppress HBV replication in chronically infected patients before there is irreversible liver damage. Antiviral drugs capable of blocking HBV replication are now available and are widely used to treat individuals living with chronic HBV infection. In the United States (as of 2008), seven drugs have been approved for adults, two have been approved for children, and other promising new drugs are in development. Still, none of these drugs can bring about a complete cure (except perhaps in very rare cases). Treatment seems to be most effective in individuals showing signs of active liver disease. These drugs are not used to treat acute infections.

The drugs that have been approved for use in the United States (as of 2008) are IFN-α (for adults and children), pegylated interferon (under the trade name Pegasys; only for adults), lamivudine (under the trade names Epivir-HBV, Zeffix, or Heptodin; approved for adults; pediatric clinical trials are in progress), adefovir dipivoxil (under the trade name Hepsera; approved for adults; pediatric clinical trials are in progress), entecavir (trade name, Baraclude; approved for adults; pediatric clinical trials are in progress), telbivudine (trade names, Tyzeka and Sebivo; for adults), and tenofovir (trade name, Viread; for adults).

Lamivudine and tenofovir are reverse transcriptase inhibitors, and each is used to treat patients infected with HIV, as well as those chronically infected with HBV. Adefovir, entecavir, and telbivudine are nucleoside analogs, used primarily to treat chronic HBV infections. These drugs usually must be taken for a year, and sometimes longer. The two IFN products can cause flu-like side effects, while the others may cause serious liver or other organ damage.

While these drugs generally do not bring about a complete cure, they do significantly decrease the risk of liver damage from HBV by impairing virus replication. On the other hand, patients must be continually monitored for potentially serious side effects. As in the case of antiretroviral therapy against HIV (Chapter 21), combination therapy may be the most effective means of treating chronic HBV infections. For an update on anti-HBV therapies and clinical trials now in progress, visit the Hepatitis B Foundation website (http://www.hepb.org/).

HEPATITIS DELTA VIRUS: A SUBVIRAL PATHOGEN

INTRODUCTION

In 1977, a new antigen was detected in the nuclei of hepatocytes from individuals who were experiencing especially severe cases of HBV infection. This antigen, which was dubbed the delta antigen, was soon found to be present inside 36-nm particles, which, significantly, also contained the HBsAg on their surfaces. It then became apparent that these 36-nm particles are, in fact, a defective **satellite virus** of HBV, now known as the **hepatitis delta virus** or **hepatitis D virus** (HDV), which is found only in individuals who also are infected with HBV. By 1980, it was evident that HDV is a factor that might increase the severity of liver disease in individuals who are coinfected with HBV. HDV infects many millions of people worldwide, but it is not as widespread as HBV. In the United States, it is estimated that 4% of acute HBV infections are coinfections with HDV.

HDV is not a hepadnavirus, and it has no sequence similarity to HBV or to any other hepadnavirus, nor does it show any sequence similarity to cellular DNA. Instead, it is a minus-strand RNA virus, classified as the only known representative of the free-standing genus *Deltavirus*. Because HDV is a satellite virus of HBV, and because it enhances the severity of HBV infections, we consider it in our chapter on the hepadnaviruses. Moreover, it is somewhat fitting that we should end our perusal of the animal virus families and genera by considering HDV, since, as we soon will see, this remarkable agent displays features that may well be as extraordinary as those of its HBV helper.

HDV is referred to as a **subviral pathogen** for several reasons. First, in order for HDV to form a mature virus particle that can reenter hepatocytes for a second round of infection, it must replicate in a cell that is coinfected with HBV, which serves as the HDV helper virus (see below and Box 22.9). Second, the HDV genome contains only 1.7 kb and thus is considerably smaller than the genomes of nearly all other animal viruses (Box 22.2).

REPLICATION

HBV, in its role as the HDV helper, provides HDV with the three HBV envelope glycoproteins. In all other respects, the replication of HDV is independent of HBV. In this regard, the HDV genome encodes only the HDV delta antigen, a relatively small protein of only 195 amino acids. Moreover, the HDV delta antigen is unique in that it is the only viral structural protein that is known to be isoprenylated, a feature that is necessary for the assembly of HDV particles (Box 22.9).

HDV is especially intriguing regarding its similarities to the remarkable plant **viroids**, which also are categorized

---

BOX 22.9

The dependency of one virus on another, unrelated, virus was seen before in the case of the dependovirus genus of the parvoviruses (Chapter 14). Also see Box 19.4 for an account of Sputnik, a satellite virus of mamavirus and mimivirus. The plant viroids too (see the text) are subviral pathogens. Also, in Chapter 3, we briefly discussed the unique transmissible spongiform encephalopathies, which are caused by the proteinaceous infectious agents or prions. We recall the prions here only because they are smaller and simpler than viruses, and hence, they too are referred to as subviral pathogens. Importantly, HDV contains a nucleic acid genome and an envelope, whereas prions contain neither.

as subviral pathogens. Viroids are infectious agents of plants that do not contain a capsid. Instead they are comprised solely of a closed, circular, single-stranded RNA genome that forms base pairs within itself to generate a rod-like RNA structure. The HDV genome resembles that of the viroids with respect to each of these physical attributes (Fig. 22.8). But the similarities between HDV and the plant viroids do not end there. Remarkably, the HDV RNA genome is replicated by the cellular RNA pol II, and viroids are similarly replicated by the plant RNA pol II. These are the only known examples in which a cellular enzyme transcribes RNA from an RNA template (Box 22.10).

But before the HDV genome might be expressed, the HDV particle first needs to enter the cell and the HDV genome needs to be uncoated. Then, the delta antigen, which is stably complexed to the genome and which contains a nuclear localization signal, mediates the transport of the genome to the nucleus. In the nucleus, the HDV minus-strand RNA genome is then transcribed into full-length, closed-circular, plus-strand RNA molecules, which in turn serve as templates for transcribing minus-strand, progeny HDV RNA genomes (Box 22.11).

Here now is another notable feature of the HDV RNA genome and replication cycle. The HDV minus-strand RNA genome and the plus-strand RNA molecules transcribed from it express ribozyme activities. (A **ribozyme** is an enzyme in which an RNA molecule, rather than a protein, expresses catalytic activity.) Each of the HDV RNA strands catalyzes self-cleavage in the absence of any protein and may catalyze self-ligation as well. The roles of these ribozyme activities in HDV replication are as follows.

Replication of the circular, single-stranded, internally hydrogen-bonded HDV RNA genome occurs by means of a rolling-circle mechanism (Box 14.5), in which full-length, plus-strand RNA monomers are released from the replicative intermediate by a unique ribozyme-mediated self-cleavage process. The termini of the released plus-strand monomers are then self-ligated by the same ribozyme (Fig. 22.8). As might be expected, the full-length, ligated plus strands then serve as templates for the synthesis of minus-strand genomes, via the same rolling-circle process. (Interestingly, viroid RNAs likewise express ribozyme activity, in keeping with the possibility that the disparate HDV and viroids are biologically related.)

Before moving on, we underscore several rather extraordinary features of the replication of the HDV closed-circle, minus-strand RNA genome. First, HDV is the only known animal virus that uses the cellular DNA-dependent RNA pol II to replicate its genome via a process that does not include a DNA intermediate (i.e., compare it to the retroviruses, which likewise use RNA pol II to replicate

their RNA genomes). Second, HDV uses a rolling-circle mechanism to replicate its *RNA* (italics for emphasis) genome. Third, the HDV RNA genome is itself a ribozyme that catalyzes steps in its own replication.

Here is a tricky situation for HDV. Before HDV can generate full-length plus strands, which might then serve as templates for the transcription of progeny minus-strand genomes, the virus must first generate the delta antigen, since a form of the delta antigen is a transactivator of genome replication (see below). (Note that the production of the delta antigen by HDV is a feature that distinguishes this agent from plant viroids, which do not encode any protein.) This appears to be somewhat of a snag for HDV because HDV full-length, plus-strand RNA molecules themselves do not have mRNA activity. Thus, HDV must first generate delta antigen-encoding mRNAs, which are distinct from full-length plus strands and, presumably, generate them by a mechanism that is distinct from that which subsequently generates full-length plus strands. These events play out as follows.

The initiation site for transcription of the 0.8-kb delta antigen-encoding mRNA molecules is the same as that for the 1.7-kb full-length plus strands. It is located just upstream from the coding region for the delta antigen. Once RNA pol II has transcribed the polyadenylation signal on the nascent delta antigen mRNA, host cell enzymes carry out 5′-capping and polyadenylation (Fig. 22.8). But how then does HDV switch from mRNA production to synthesis of full-length plus strands? The answer is that the de novo delta antigen is thought to suppress the polyadenylation signal present on the nascent plus strands.

Here is another unique feature of the HDV replication cycle, in this instance involving generation of the delta antigen. The virus actually translates two forms of the delta antigen, the small (SDAg) and large (LDAg) forms, from the same single open reading frame on the 0.8-kb delta antigen-encoding mRNA. The SDAg contains 195 amino acids, and the LDAg contains an additional 19 amino acids at its C terminus (Fig. 22.8).

SDAg promotes the switch from mRNA synthesis to the generation of antigenomes, as just noted. In contrast, LDAg is thought to suppress genome replication and promote HDV assembly. The longer form, by virtue of its ability to limit replication and cell destruction, may also promote persistent infection. It is the longer form that is isoprenylated (see above). To facilitate its replication activities, the delta antigen contains RNA-binding domains and a *leucine zipper* dimerization domain, as well as its nuclear localization signal. The leucine zipper dimerization domain is characteristic of some proteins that regulate transcription.

But we have seen in this book other examples of viruses encoding multiple proteins from a single open reading

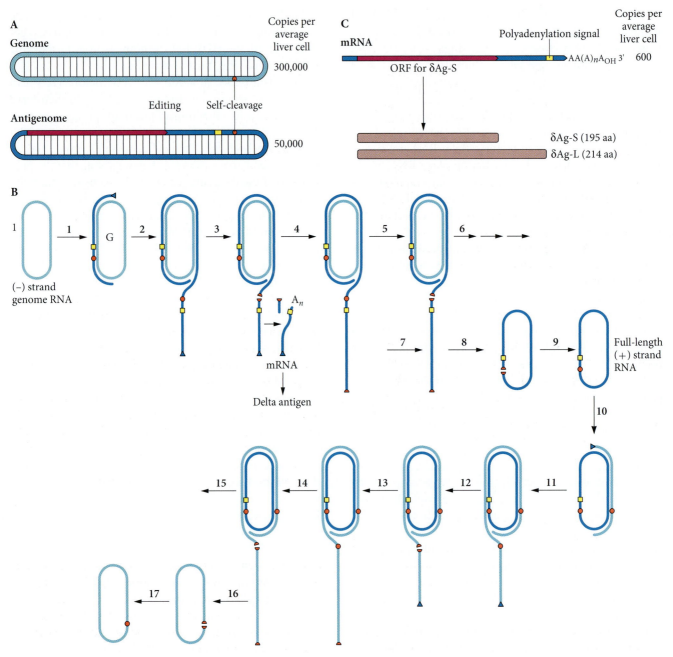

Figure 22.8 (A) Schematic showing the covalently closed, hydrogen-bonded, minus-strand HDV RNA genome and the similarly configured plus-strand antigenome. Note the self-cleavage (orange) and polyadenylation (yellow) signals and the ORF for δAg-S (red). (B) Depiction of HDV mRNA synthesis and the rolling-circle process by which the HDV genome is replicated. (steps 1 through 6) Replication of the HDV genome requires de novo translation of the delta antigen. Since full-length HDV plus strands do not have mRNA activity, HDV mRNAs must first be produced by a mechanism distinct from that which generates full-length plus strands. The initiation site for transcription of HDV mRNA is the same as that for the full-length plus strands. It is located just upstream from the coding region for the delta antigen. Once the polymerase has transcribed the polyadenylation signal on the nascent delta antigen mRNA, host cell enzymes carry out 5'-capping and polyadenylation. HDV is thought to switch from mRNA production to synthesis of full-length plus strands by suppressing the polyadenylation signal present on the nascent plus strands. (Steps 5 through 9) Full-length, plus-strand RNA monomers are generated by the rolling-circle mechanism and are released from the replicative intermediate by the unique ribozyme-mediated self-cleavage process. The termini of the released plus-strand monomers are then self-ligated by the same ribozyme, and these molecules then serve as templates for the synthesis of genomic minus strands, via the same rolling-circle process (steps 10 through 17). (C) Some of the delta antigen mRNAs undergo RNA editing that changes a UAG (amber) termination codon into a UGG tryptophan codon. The result is that two forms of the delta antigen are generated: one containing 195 amino acids and one containing 214 amino acids. Modified from J. M. Taylor, *Curr. Top. Microbiol. Immunol.* **239:**107–122, 1999, with permission.

The fundamental similarities between HDV and the viroids raise the possibility that HDV, a subviral pathogen of humans, which is dependent for its propagation on coinfection with HBV, somehow arose from a plant viroid! Alternatively, perhaps plant viroids arose from a delta-like agent of animals, or perhaps each derives from a common ancestor. Regardless, since the cellular RNA pol II is otherwise known to use only DNA for a template, HDV and viroids provide striking exceptions to the generalization that every instance in which a cellular activity mediates viral transcription or replication will have a normal cellular counterpart. While HDV and viroids are categorized as subviral entities rather than true viruses, in any case, they remain a striking exception to the generalization.

What inferences might you draw from the fact that the one exception we might note to the above generalization involves an RNA-dependent RNA polymerase activity of the cellular RNA pol II, as utilized by HDV and viroids? Might the following paragraph lead to a suggestion?

Potato spindle tuber viroid is the prototype of the viroid class of agents. All are disseminated in nature via insects. They also may be transmitted via cultivation practices, such as via cuttings from infected plants or by mechanical harvesters. The mechanisms by which viroids induce disease are still not clear, although specific sequences in the viroid RNA are known to be necessary. That is a most interesting point because viroids, unlike satellite viruses, do not encode any protein product.

The fact that the HDV RNA genome is transcribed and replicated in the nucleus is somewhat unique, since influenza virus (Chapter 12) and Borna disease virus (a paramyxovirus [Chapter 11]) provide the only other known instances of RNA viruses that carry out transcription and replication in the nucleus. The HDV RNA genome must be transcribed and replicated in the nucleus, since those processes are catalyzed for HDV by the cellular RNA pol II, which functions solely in the nucleus. Influenza virus and Borna disease virus replicate in the nucleus for other reasons, as recounted in Chapter 12.

approximately 30-fold higher than that of antigenomic, plus-strand RNA synthesis. Still another, and perhaps related, issue is why synthesis of genomic RNA is sensitive to α-amanitin (a fungal toxin that inhibits the cellular RNA pol II), while synthesis of the antigenome is completely resistant to that drug. One possibility is that genomic and antigenomic HDV RNA syntheses are catalyzed by two different host cell enzymes. Based on the differing sensitivities of genomic and antigenomic RNA syntheses to α-amanitin, genomic RNA is likely generated by RNA pol II, while the enzyme responsible for synthesizing antigenomic RNA remains to be determined with certainty.

ROUTES OF TRANSMISSION AND PATHOGENESIS

Approximately 20 million people worldwide are infected with HDV. For reasons noted above, HDV can only be transmitted as a coinfection with HBV or as a superinfection of an individual already infected with HBV. Thus, like HBV, HDV is spread via blood, semen, and vaginal secretions. (The seroprevalence of HDV among intravenous drug users is estimated to be between 20 and 53%.) Moreover, HDV selectively binds to, and is internalized by, hepatocytes, as expected since the HDV envelope contains the three HBV envelope glycoproteins.

HDV is medically important because it may exacerbate acute and chronic liver diseases during coinfection with HBV or following superinfection of an HBV-infected individual. For instance, HDV superinfection of a patient manifesting HBV cirrhosis can lead to extensive necrosis and liver failure. Yet the great majority (97 to 99%) of adults who are infected with HBV and HDV resolve the infections, as do individuals who are singly infected with HBV. (Note that at least one study did not find an increase

frame. However, the mechanism used by HDV for this purpose is thus far unique. It happens as follows. Some of the rod-structured, plus-strand, antigenomic RNA molecules undergo RNA editing that changes a UAG (amber) termination codon into a UGG tryptophan codon, resulting in the translation of the delta antigen containing 214 amino acids. This RNA editing is carried out by host enzymes called ADARs, for "adenosine deaminases that act on double-stranded RNA."

To be completely candid, key details of the process by which HDV RNA is replicated are not known with certainty and remain to be clarified. Thus, the above depiction of events is almost certainly an oversimplification.

Another issue that remains to be explained is why genomic, minus-strand RNA synthesis occurs at a rate in HBV pathogenesis associated with HDV coinfection. Nevertheless, most HBV researchers agree that the extent of HBV liver degeneration is greater in individuals coinfected with HDV.)

Interestingly, people who have been chronically infected with HBV and who then are superinfected with HDV appear to experience more severe symptoms than those who are simultaneously infected with both viruses. A possible explanation is that the satellite virus can begin to replicate immediately and more extensively in individuals who are already infected with HBV, since many hepatocytes in those individuals are already producing HBV envelope glycoproteins. In either case, persistent HDV infection is established in HBV-infected persons, thereby enabling HDV to potentially amplify the cytotoxicity and liver damage caused by HBV.

Since HBV damages the liver by an immunopathological process, the more severe disease that might be seen in individuals who are also infected with HDV may result from simultaneous immune responses against the two viruses, rather than from a direct cytotoxic effect of HDV, such as, for example, might result from a form of the delta antigen. This premise is supported by results of studies carried out with transgenic mice that expressed the small and large delta antigens. These mice displayed no biological or histopathological evidence of liver disease during 18 months of observation. (By what other means might HDV exacerbate HBV infections?)

Because HDV can replicate only in HBV-infected individuals, immunization against HBV affords protection against HDV as well. There is no treatment for HDV. However, HDV infection and disease are coming under control, thanks to effective vaccines against HBV. Indeed, most cases of chronic HDV infection are largely holdovers from prevaccination times, and the prevalence of HDV has been decreasing worldwide in recent years.

Suggested Readings

Stoeckl, L., A. Funk, A. Kopitzki, B. Brandenburg, S. Oess, H. Will, H. Sirma, and E. Hildt. 2006. Identification of a structural motif crucial for infectivity of hepatitis B viruses. *Proc. Natl. Acad. Sci. USA* **103:**6730–6734.

Summers, J., and W. S. Mason. 1982. Replication of a hepatitis B-like virus by reverse transcription of an RNA intermediate. *Cell* **9:**403–415.

Zlotnick, A., S. J. Stahl, P. T. Wingfield, J. F. Conway, N. Cheng, and A. C. Steven. 1998. Shared motifs of the capsid proteins of hepadnaviruses and retroviruses suggest a common evolutionary origin. *FEBS Lett.* **431:**301–304.

Index